COLD SPRING HARBOR SYMPOSIA
ON QUANTITATIVE BIOLOGY

VOLUME LXXXII

COLD SPRING HARBOR SYMPOSIA ON QUANTITATIVE BIOLOGY

VOLUME LXXXII

Chromosome Segregation and Structure

symposium.cshlp.org

Symposium Organizers: Terri Grodzicker, David Stewart, and
Bruce Stillman (*Cold Spring Harbor Laboratory*)

Editors: David Stewart and Bruce Stillman (*Cold Spring Harbor Laboratory*)

COLD SPRING HARBOR LABORATORY PRESS
2017

COLD SPRING HARBOR SYMPOSIA ON QUANTITATIVE BIOLOGY VOLUME LXXXII

© 2017 by Cold Spring Harbor Laboratory Press
International Standard Book Number 978-1-621822-86-8 (cloth)
International Standard Book Number 978-1-621822-87-5 (paper)
International Standard Serial Number 0091-7451
Library of Congress Catalog Card Number 34-8174

COLD SPRING HARBOR SYMPOSIA ON QUANTITATIVE BIOLOGY
Founded in 1933 by
REGINALD G. HARRIS
Director of the Biological Laboratory 1924 to 1936
Previous Symposia Volumes

I (1933) Surface Phenomena
II (1934) Aspects of Growth
III (1935) Photochemical Reactions
IV (1936) Excitation Phenomena
V (1937) Internal Secretions
VI (1938) Protein Chemistry
VII (1939) Biological Oxidations
VIII (1940) Permeability and the Nature of Cell Membranes
IX (1941) Genes and Chromosomes: Structure and Organization
X (1942) The Relation of Hormones to Development
XI (1946) Heredity and Variation in Microorganisms
XII (1947) Nucleic Acids and Nucleoproteins
XIII (1948) Biological Applications of Tracer Elements
XIV (1949) Amino Acids and Proteins
XV (1950) Origin and Evolution of Man
XVI (1951) Genes and Mutations
XVII (1952) The Neuron
XVIII (1953) Viruses
XIX (1954) The Mammalian Fetus: Physiological Aspects of Development
XX (1955) Population Genetics: The Nature and Causes of Genetic Variability in Population
XXI (1956) Genetic Mechanisms: Structure and Function
XXII (1957) Population Studies: Animal Ecology and Demography
XXIII (1958) Exchange of Genetic Material: Mechanism and Consequences
XXIV (1959) Genetics and Twentieth Century Darwinism
XXV (1960) Biological Clocks
XXVI (1961) Cellular Regulatory Mechanisms
XXVII (1962) Basic Mechanisms in Animal Virus Biology
XXVIII (1963) Synthesis and Structure of Macromolecules
XXIX (1964) Human Genetics
XXX (1965) Sensory Receptors
XXXI (1966) The Genetic Code
XXXII (1967) Antibodies
XXXIII (1968) Replication of DNA in Microorganisms
XXXIV (1969) The Mechanism of Protein Synthesis
XXXV (1970) Transcription of Genetic Material
XXXVI (1971) Structure and Function of Proteins at the Three-dimensional Level
XXXVII (1972) The Mechanism of Muscle Contraction
XXXVIII (1973) Chromosome Structure and Function
XXXIX (1974) Tumor Viruses
XL (1975) The Synapse

XLI (1976) Origins of Lymphocyte Diversity
XLII (1977) Chromatin
XLIII (1978) DNA: Replication and Recombination
XLIV (1979) Viral Oncogenes
XLV (1980) Movable Genetic Elements
XLVI (1981) Organization of the Cytoplasm
XLVII (1982) Structures of DNA
XLVIII (1983) Molecular Neurobiology
XLIX (1984) Recombination at the DNA Level
L (1985) Molecular Biology of Development
LI (1986) Molecular Biology of Homo sapiens
LII (1987) Evolution of Catalytic Function
LIII (1988) Molecular Biology of Signal Transduction
LIV (1989) Immunological Recognition
LV (1990) The Brain
LVI (1991) The Cell Cycle
LVII (1992) The Cell Surface
LVIII (1993) DNA and Chromosomes
LVIX (1994) The Molecular Genetics of Cancer
LX (1995) Protein Kinesis: The Dynamics of Protein Trafficking and Stability
LXI (1996) Function & Dysfunction in the Nervous System
LXII (1997) Pattern Formation during Development
LXIII (1998) Mechanisms of Transcription
LXIV (1999) Signaling and Gene Expression in the Immune System
LXV (2000) Biological Responses to DNA Damage
LXVI (2001) The Ribosome
LXVII (2002) The Cardiovascular System
LXVIII (2003) The Genome of Homo sapiens
LXIX (2004) Epigenetics
LXX (2005) Molecular Approaches to Controlling Cancer
LXXI (2006) Regulatory RNAs
LXXII (2007) Clocks and Rhythms
LXXIII (2008) Control and Regulation of Stem Cells
LXXIV (2009) Evolution: The Molecular Landscape
LXXV (2010) Nuclear Organization and Function
LXXVI (2011) Metabolism and Disease
LXXVII (2012) The Biology of Plants
LXXVIII (2013) Immunity and Tolerance
LXXIX (2014) Cognition
LXXX (2015) 21st Century Genetics: Genes at Work
LXXXI (2016) Targeting Cancer

Front cover (*paperback*): Original artwork by Beata Edyta Mierzwa. www.BeataScienceArt.com.

For a complete catalog of all Cold Spring Harbor Laboratory Press publications, visit our website www.cshlpress.org.

Online access: Please visit our companion website at symposium.cshlp.org. For access issues, please contact Cold Spring Harbor Laboratory Press at subscriptions@cshl.edu.

Symposium Participants

ABBATEMARCO, SIMONA, University of Geneva, Geneva, Switzerland

ABDENNUR, NEZAR, Massachusetts Institute of Technology, Cambridge, Massachusetts

AHMADI-PIRSHAHID, DADAR, Children's Medical Research Institute, Westmead, New South Wales, Australia

AKERA, TAKASHI, University of Pennsylvania, Philadelphia, Pennsylvania

ALFIERI, CLAUDIO, Laboratory of Molecular Biology, Cambridge, United Kingdom

ALHAJ ABED, JUMANA, Harvard Medical School, Boston, Massachusetts

ALMOUZNI, GENEVIEVE, Institut Curie/Research Centre, Paris, France

AMON, ANGELIKA, Massachusetts Institute of Technology, Cambridge, Massachusetts

ANDRIANI, GRASIELLA, Albert Einstein College of Medicine, Bronx, New York

AWAL, SUSHIL, IGBMC Gie CERBM, Illkirch, France

AZUMA, YOSHIAKI, University of Kansas, Lawrence, Kansas

BAKHOUM, SAMUEL, Memorial Sloan Kettering Cancer Center, New York, New York

BALLEW, OLIVIA, Indiana University, Bloomington, Indiana

BARRAL, YVES, ETH Zürich, Zürich, Switzerland

BARTOLOMEI, MARISA, University of Pennsylvania, Perelman School of Medicine, Philadelphia, Pennsylvania

BATEMAN, JACK, Bowdoin College, Brunswick, Maine

BAUMANN, KIM, Springer Nature, London, United Kingdom

BELL, JASON, Stanford University, Stanford, California

BERNARD, PASCAL, CNRS, Lyon, France

BERNARDI, GIORGIO, Roma Tre University, Rome, Italy

BETTENCOURT-DIAS, MÓNICA, Instituto Gulbenkian de Ciencia, Oeiras, Portugal

BIAN, QIAN, University of California, Berkeley, Berkeley, California

BLACK, BEN, University of Pennsylvania, Philadelphia, Pennsylvania

BLOBEL, GERD, Children's Hospital of Philadelphia, Philadelphia, Pennsylvania

BLOOM, KERRY, University of North Carolina at Chapel Hill, Chapel Hill, North Carolina

BLUNT, ENDIA, Cornell University, Ithaca, New York

BOEKHOUT, MICHIEL, Memorial Sloan Kettering Cancer Center, New York, New York

BONORA, GIANCARLO, University of Washington, Seattle, Washington

BORTVIN, ALEX, Carnegie Institution for Science, Baltimore, Maryland

BRAHMACHARI, SUMITABHA, Northwestern University, Evanston, Illinois

BRANDAO, HUGO, Harvard University, Cambridge, Massachusetts

BRAVO NÚNEZ, MÁRIA, Stowers Institute for Medical Research, Kansas City, Missouri

BROOKS, KELSEY, Oregon Health and Science University, Beaverton, Oregon

CANTOR, SHARON, University of Massachusetts Medical School, Worcester, Massachusetts

CAO, SHENGYA, Stanford University, Stanford, California

CARBONE, LUCIA, Oregon Health and Science University, Portland, Oregon

CARONE, BENJAMIN, Rowan University, Glassboro, New Jersey

CARONE, DAWN, Swarthmore College, Swarthmore, Pennsylvania

CHAMPION, LYSIE, ETH Zürich, Zürich, Switzerland

CHAPARD, CHRISTOPHE, University of Oxford, Oxford, United Kingdom

CHAVEZ, SHAWN, Oregon Health and Science University, Beaverton, Oregon

CHEESEMAN, IAIN, Whitehead Institute for Biomedical Research, Cambridge, Massachusetts

CHEN, JINGXUN, University of California, Berkeley, Berkeley, California

CHEN, XIN, Johns Hopkins University, Baltimore, Maryland

CHEN, YUPING, Stony Brook University, Stony Brook, New York

CHENG, NINGYAN, University of Texas Southwestern Medical Center, Dallas, Texas

CHI, MAOYEN, Cold Spring Harbor Asia, Suzhou, China

CHOU, HSIANG CHEN, Cold Spring Harbor Laboratory, Cold Spring Harbor, New York

CHOY, JOHN, The Catholic University of America, Washington, D.C.

CLEVELAND, DON, University of California, San Diego, La Jolla, California

COLAIACOVO, MONICA, Harvard Medical School, Boston, Massachusetts

CORBETT, KEVIN, University of California, San Diego, La Jolla, California

DASSO, MARY, National Institutes of Health, National Institute of Child Health and Human Development, Bethesda, Maryland

DAUGHTRY, BRITTANY, Oregon Health and Science University, Portland, Oregon

DAVIS, TRISHA, University of Washington, Seattle, Washington

DE MASSY, BERNARD, CNRS, Montpellier, France

DE REGT, ANNA, Fred Hutchinson Cancer Research Center, Seattle, Washington

DE VOER, RICHARDA, Radboudumc, Nijmegen, Netherlands

DELUCA, JENNIFER, Colorado State University, Fort Collins, Colorado

DENG, XIEXIONG, Michigan State University, East Lansing, Michigan

DESAI, ARSHAD, University of California, San Diego, La Jolla, California

DHATCHINAMOORTHY, KARTHIK, Stowers Institute for Medical Research, Kansas City, Missouri

DICK, FRED, Western University, London, Ontario, Canada

DONG, QIANHUA, New York University, New York, New York

DUROCHER, DANIEL, The Lunenfeld–Tanenbaum Research Institute, Toronto, Ontario, Canada

EARNSHAW, WILLIAM, University of Edinburgh, Edinburgh, United Kingdom

EDGERTON, HEATHER, University of Minnesota, Minneapolis, Minnesota

EDWARDS, GARRETT, University of Colorado Boulder, Boulder, Colorado

ELEWA, AHMED, Karolinska Institute, Stockholm, Sweden

FALK, MARTIN, Massachusetts Institute of Technology, Cambridge, Massachusetts

FENG, HAIYANG, University of Bergen, Bergen, Norway

FINGERHUT, JACLYN, University of Michigan, Ann Arbor, Michigan

FRENCH, BRADLEY, Stanford University, Stanford, California

FROBERG, JOHN, Harvard Medical School, Boston, Massachusetts

FUJIOKA, MIKI, Thomas Jefferson University, Philadelphia, Pennsylvania

FUNABIKI, HIRONORI, Rockefeller University, New York, New York

FUTCHER, BRUCE, Stony Brook University, Stony Brook, New York

FUTIA, RAYMOND, Stanford University, Stanford, California

GAILLARD, MARIE-CÉCILE, Johns Hopkins University School of Medicine, Baltimore, Maryland

GALASSO, JENNIFER, Amityville Memorial High School, Amityville, New York

GALLI, MARTINA, University of Milan, Milan, Italy

GAN, LU, National University of Singapore, Singapore, Singapore

GARTENBERG, MARC, Robert Wood Johnson Medical School, Piscataway, New Jersey

GASSER, SUSAN, Friedrich Miescher Institute for Biomedical Research, Basel, Switzerland

GASSLER, JOHANNA, IMBA—Institute of Molecular Biotechnology GmbH, Vienna, Austria

GEORGE, ANUJA, Rutgers Robert Wood Johnson Medical School, Piscataway, New Jersey

GERBI, SUSAN, Brown University BioMed Division, Providence, Rhode Island

GERLICH, DANIEL, Institute of Molecular Biotechnology, Vienna, Austria

GLEASON, RYAN, Johns Hopkins University, Baltimore, Maryland

GOLOBORODKO, ANTON, Massachusetts Institute of Technology, Cambridge, Massachusetts

GRISHCHUK, EKATERINA, University of Pennsylvania, Philadelphia, Pennsylvania

GRODZICKER, TERRI, Cold Spring Harbor Laboratory, Cold Spring Harbor, New York

GUERIN, THOMAS, CEA/FAR, Fontenay-aux-Roses, France

GUPTA, DEEPESH KUMAR, Uppsala University, Uppsala, Sweden

GUTBROD, MICHAEL, Cold Spring Harbor Laboratory, Cold Spring Harbor, New York

HAMMELL, CHRISTOPHER, Cold Spring Harbor Laboratory, Cold Spring Harbor, New York

HARASYMIW, LAUREN, University of Minnesota, Minneapolis, Minnesota

HARRISON, STEPHEN, Harvard Medical School, Boston, Massachusetts

HATCH, EMILY, Fred Hutchinson Cancer Research Center, Seattle, Washington

HAVALDA, ROBERT, Institute of Molecular Genetics of the ASCR, BIOCEV, Vestec, Czech Republic

HEARD, EDITH, Curie Institute, CNRS, Paris, France

HEASLEY, LYDIA, Colorado State University, Fort Collins, Colorado

HEGARAT, NADIA, Genome Damage and Stability Centre, Brighton, United Kingdom

HENIKOFF, STEVEN, Fred Hutchinson Cancer Research Center, Seattle, Washington

HETZER, MARTIN, The Salk Institute for Biological Studies, La Jolla, California

HEWAWASAM, GEETHA, Stowers Institute for Medical Research, Kansas City, Kansas

HICKSON, IAN, University of Copenhagen, Copenhagen, Denmark

HOLLAND, ANDREW, Johns Hopkins University School of Medicine, Baltimore, Maryland

HOLLERER, INA, University of California, Berkeley, Berkeley, California

HONG, JINGJUN, National Institutes of Health, Bethesda, Maryland

HOPKINS, JESSICA, Johns Hopkins Bloomberg School of Public Health, Baltimore, Maryland

HU, YIXIN, Cold Spring Harbor Laboratory, Cold Spring Harbor, New York

HUNTER, NEIL, University of California, Davis, Davis, California

HYMAN, ANTHONY, Max-Planck Institute of Molecular Cell Biology and Genetics, Dresden, Germany

IKUI, AMY, Brooklyn College, Brooklyn, New York

IMAKAEV, MAKSIM, Massachusetts Institute of Technology, Cambridge, Massachusetts

INOUE, AZUSA, Harvard Medical School, Boston, Massachusetts

INOUE, TOMOKO, Hiroshima University Hospital, Hiroshima, Japan

ISHIDA, SEIKO, Hiroshima University Hospital, Hiroshima, Japan

JACOB, ETAI, Dana-Farber Cancer Institute, Boston, Massachusetts

JAGANNATHAN, MADHAV, Howard Hughes Medical Institute/University of Michigan, Ann Arbor, Michigan

JAKIMO, ALAN, Hofstra University, Hempstead, New York

JANG, CHANG-YOUNG, Sookmyung Women's University, Seoul, South Korea

JARAMILLO-LAMBERT, AIMEE, National Institute of Diabetes and Digestive and Kidney Diseases/National Institutes of Health, Bethesda, Maryland

JASIN, MARIA, Memorial Sloan Kettering Cancer Center, New York, New York

JAYNES, JAMES, Thomas Jefferson University, Philadelphia, Pennsylvania

JERABKOVA, KATERINA, IGBMC GIE CERBM, Illkirch, France

JIMENEZ, DAVID, New York University, New York, New York

JOHNSON, RON, National Institutes of Health/National Cancer Institute, Bethesda, Maryland

JOHNSON, WHITNEY, Stanford University, Stanford, California

KAPOOR, TARUN, The Rockefeller University, New York, New York

KARASU, MEHMET ERMAN, Memorial Sloan Kettering Cancer Center, New York, New York

KAUR, HARDEEP, Center for Cancer Research/National Cancer Institute/National Institutes of Health, Bethesda, Maryland

KEENEY, SCOTT, Memorial Sloan Kettering Cancer Center, New York, New York

KHONDKER, SHOILY, CUNY Graduate Center, New York, New York

KIM, DONG HYUN, Ludwig Institute for Cancer Research, La Jolla, California

KIM, JAE OOK, University of Washington, Seattle, Washington

KIM, KYUNGTAE, National Cancer Center, Goyang, South Korea

KIM, SEUL, Sookmyung Women's University, Seoul, South Korea

KIM, YUMI, Johns Hopkins University, Baltimore, Maryland

KROSCHEWSKI, RUTH, ETH Zürich, Zürich, Switzerland

KULKARNI, DHANANJAYA, University of California, Davis, Davis, California

KURBIDAEVA, AMINA, Princeton University, Princeton, New Jersey

KWON, MIJUNG, Dana-Farber Cancer Institute, Boston, Massachusetts

LAMPSON, MICHAEL, University of Pennsylvania, Philadelphia, Pennsylvania

LARSCHAN, ERICA, Brown University, Providence, Rhode Island

LASCARIDES, KIMBERLIE, Sanford H. Calhoun High School, Merrick, New York

LEE, JAE-HO, Ajou University School of Medicine, Suwon, South Korea

LERA, ROBERT, University of Wisconsin–Madison, Madison, Wisconsin

LI, GUOHONG, Chinese Academy of Sciences, Beijing, China

LI, SHUANGSHUANG, University of Saskatchewan, Saskatoon, Saskatchewan, Canada

LIHM, JAYON, Cold Spring Harbor Laboratory, Woodbury, New York

LIMZERWALA, JAZEEL, Mayo Clinic Graduate School of Biomedical Science, Rochester, Minnesota

LIU, HONG, Tulane University SOM, New Orleans, Louisiana

LIU, JINGHUI, Massachusetts Institute of Technology, Cambridge, Massachusetts

LIU, SHIWEI, Dana-Faber Cancer Institute/Harvard Medical School, Boston, Massachusetts

LIU, YU, University of Massachusetts Medical School, Worcester, Massachusetts

LOGARINHO, ELSA, Instituto de Biologia Molecular e Celular (IBMC), Porto, Portugal

LONG, HAIZHEN, Institute of Biophysics, CAS, Beijing, China

LUO, JINGCHUAN, NYU Langone Medical Center, New York, New York

LY, PETER, Ludwig Institute for Cancer Research, University of California, San Diego, La Jolla, California

MARSHALL, AREN, University of Western Ontario, London, Ontario, Canada

MARTIN-SERRANO, JUAN, King's College London, School of Medicine, London, United Kingdom

MASAMSETTI, V. PRAGATHI, Children's Medical Research Institute, Sydney, New South Wales, Australia

MERKEL, FABIAN, European Molecular Biology Laboratory (EMBL), Heidelberg, Germany

MEYER, BARBARA, Howard Hughes Medical Institute/University of California, Berkeley, Berkeley, California

MILLER, RICHARD, Princeton University, Princeton, New Jersey

MIRNY, LEONID, Massachusetts Institute of Technology, Cambridge, Massachusetts

MISTELI, TOM, National Cancer Institute, National Institutes of Health, Bethesda, Maryland

MITCHISON, TIMOTHY, Harvard Medical School, Boston, Massachusetts

MONTAGNA, CRISTINA, Albert Einstein College of Medicine, Bronx, New York

MOOREFIELD, BETH, *Nature Structural & Molecular Biology*, New York, New York

MUNOZ, ADRIANA, Cold Spring Harbor Laboratory, Cold Spring Harbor, New York

NAMBIAR, MRIDULA, Fred Hutchinson Cancer Research Center, Seattle, Washington

NASMYTH, KIM, University of Oxford, Oxford, United Kingdom

NAVARRO, ALEXANDRA, Massachusetts Institute of Technology, Cambridge, Massachusetts

NECHEMIA-ARBELY, YAEL, Ludwig Institute for Cancer Research, University of California, San Diego, La Jolla, California

NEUROHR, GABRIEL, Massachusetts Institute of Technology, Cambridge, Massachusetts

NGO, BRYAN, Weill Cornell/Memorial Sloan Kettering, New York, New York

NIGG, ERICH, University of Basel, Basel, Switzerland

NIR, GUY, Harvard Medical School, Boston, Massachusetts

NITISS, JOHN, University of Illinois College of Pharmacy, Rockford, Illinois

NUEBLER, JOHANNES, Massachusetts Institute of Technology, Cambridge, Massachusetts

OLAFSSON, GUDJON, The Francis Crick Institute, London, United Kingdom

ON, KIN, Cold Spring Harbor Laboratory, Cold Spring Harbor, New York

OU, HORNG, Salk Institute for Biological Studies, La Jolla, California

PAGE, DAVID, Whitehead Institute; Massachusetts Institute of Technology/Howard Hughes Medical Institute, Cambridge, Massachusetts

PAL, DEBJANI, Cold Spring Harbor Laboratory, Cold Spring Harbor, New York

PALOU MARÍN, ROGER, Université de Montréal, Montréal, Canada

PARK, JI EUN, Sookmyung Women's University, Seoul, South Korea

PELLMAN, DAVID, Dana-Farber Cancer Institute, Boston, Massachusetts

PENTAKOTA, SATYAKRISHNA, Max Planck Institute of Molecular Physiology, Dortmund, Germany

PEREA-RESA, CARLOS, Massachusetts General Hospital-HMS, Boston, Massachusetts

PETERS, ANTOINE, Friedrich Miescher Institute for Biomedical Research, Basel, Switzerland

PETERS, JAN-MICHAEL, Research Institute of Molecular Pathology, Vienna, Austria

PHAN, SEBASTIEN, University of California, San Diego, La Jolla, California

PISKADLO, EWA, Fundacao Gulbenkian-Instituto Gulbenkian Ciencia, Oeiras, Portugal

PODSYPANINA, KATRINA, Institut Curie, Paris, France

POLAK, BRUNO, Ruder Boskovic Institute, Zagreb, Croatia

POLLOCK, MILA, Cold Spring Harbor Laboratory, Cold Spring Harbor, New York

PRAJAPATI, HEMANT, Indian Institute of Technology Bombay, Mumbai, India

RANDO, OLIVER, University of Massachusetts Medical School, Worcester, Massachusetts

ROCA, AMANDA, Northwestern University, Evanston, Illinois

ROELENS, BAPTISTE, Stanford University, Stanford, California

ROHRBERG, JULIA, University of California, San Francisco, San Francisco, California

RYNDITCH, ALLA, Institute of Molecular Biology and Genetics, Kiev, Ukraine

SADLER, JESSICA, King's College London, London, United Kingdom

SAITOU, MITINORI, Graduate School of Medicine, Kyoto University, Kyoto, Japan

SALIM, DEVIKA, Stowers Institute for Medical Research, Kansas City, Missouri

SCHIKLENK, CHRISTOPH, European Molecular Biology Laboratory (EMBL), Heidelberg, Germany

SCHINDLER, KAREN, Rutgers University, Piscataway, New Jersey

SCHLISSEL, GAVIN, University of California, Berkeley, Berkeley, California

SCHNEIDER, JAY, UT Southwestern Medical Center, Dallas, Texas

SCHVARZSTEIN, MARA, City University of New York, Brooklyn College, Brooklyn, New York

SEPANIAC, LESLIE, University of Vermont, Burlington, Vermont

SEVER, RICHARD, Cold Spring Harbor Laboratory Press, Woodbury, New York

SHEINWALD, LORIE, Farmingdale High School, Farmingdale, New York

SHELTZER, JASON, Cold Spring Harbor Laboratory, Cold Spring Harbor, New York

SHINTOMI, KEISHI, RIKEN, Wako, Saitama, Japan

SHIROLE, NITIN, Cold Spring Harbor Laboratory, Cold Spring Harbor, New York

SHOSHANI, OFER, University of California, San Diego, La Jolla, California

SILIO, VIRGINIA, University of Warwick, Coventry, United Kingdom

SING, TINA, University of Toronto, Toronto, Ontario, Canada

SITBON, DAVID, Institut Curie, Paris, France

SMITH, OWEN, Stanford University, Stanford, California

SNEDEKER, JONATHAN, Johns Hopkins University, Baltimore, Maryland

SONG, HAIYU, Sookmyung Women's University, Seoul, South Korea

SONNETT, MATTHEW, Harvard Medical School, Boston, Massachusetts

SPRADLING, ALLAN, Howard Hughes Medical Institute, Carnegie Institution, Baltimore, Maryland

STEARNS, TIMOTHY, Stanford University, Stanford, California

STEWART, DAVID, Cold Spring Harbor Laboratory, Cold Spring Harbor, New York

STILLMAN, BRUCE, Cold Spring Harbor Laboratory, Cold Spring Harbor, New York

STOJIC, LOVORKA, Cancer Research UK Cambridge Institute, Cambridge, United Kingdom

STRAIGHT, AARON, Stanford University, Stanford, California

STUKENBERG, TODD, University of Virginia School of Medicine, Charlottesville, Virginia

STUMPFF, JASON, University of Vermont, Burlington, Vermont

SUSSMAN, HILLARY, *Genome Research*, Executive Editor, Cold Spring Harbor Laboratory Press, Woodbury, New York

SZEWCZAK, LARA, *Cell*, Cambridge, Massachusetts

TAILEB, MOUFIDA, University of California, San Francisco, San Francisco, California

TAYLOR, ALISON, Dana-Farber Cancer Institute, Boston, Massachusetts

TERAN, NIKKI, Stanford Medical School, Stanford, California

TEWELDE, BLOSSOM, Johns Hopkins School of Medicine, Baltimore, Maryland

THAKUR, JITENDRA, Fred Hutchinson Cancer Research Center, Seattle, Washington

TRIVEDI, PRASAD, University of Virginia, Charlottesville, Virginia

TROMER, EELCO, Hubrecht Institute, Utrecht, Netherlands

UHLMANN, FRANK, The Francis Crick Institute, London, United Kingdom

URBAN, JENNIFER, Brown University, Providence, Rhode Island

VACCARIELLO, MICHAEL, Sachem High School East, Farmingville, New York

VAN DEN BERG, AAFKE, Massachusetts Institute of Technology, Cambridge, Massachusetts

VAN DEURSEN, JAN, Mayo Clinic Rochester, Rochester, Minnesota

VAN RUITEN, MARJON, Netherlands Cancer Institute, Amsterdam, Netherlands

VIETS, KAYLA, Johns Hopkins University, Baltimore, Maryland

WALTHER, NIKE, European Molecular Biology Laboratory (EMBL), Heidelberg, Germany

WANG, FANGWEI, Zhejiang University, Hangzhou, China

WANG, JIN, Stony Brook University, Stony Brook, New York

WANG, LIN-ING, Rutgers University, New Brunswick, Piscataway, New Jersey

WARSINGER-PEPE, NATALIE, University of Michigan/Life Sciences Institute, Ann Arbor, Michigan

WATANABE, YOSHINORI, The University of Tokyo, Tokyo, Japan

WATASE, GEORGE, University of Michigan/Howard Hughes Medical Institute, Ann Arbor, Michigan

WEAVER, BETH, University of Wisconsin–Madison, Madison, Wisconsin

WELLARD, STEPHEN, Johns Hopkins Bloomberg School of Public Health, Baltimore, Maryland

WEN, ZENGQI, Institute of Biophysics, CAS, Beijing, China

WEST, STEPHEN, The Francis Crick Institute, London, United Kingdom

WHEELAN, SARAH, The Johns Hopkins University School of Medicine, Baltimore, Maryland

WIELENGA, BAS

WITKOWSKI, JAN, Cold Spring Harbor Laboratory, Cold Spring Harbor, New York

WOOTEN, MATTHEW, Johns Hopkins University, Baltimore, Maryland

XIE, KATHLEEN, Stanford University, Stanford, California

XU, RUI-MING, Chinese Academy of Sciences, Beijing, China

XU, X. MIKE, University of Miami, Miami, Florida

YAMASHITA, YUKIKO, University of Michigan, Ann Arbor, Ann Arbor, Michigan

YANG, JINPU, New York University, New York, New York

YOO, TAE YEON, Harvard University, Cambridge, Massachusetts

ZARET, KENNETH, University of Pennsylvania, Perelman School of Medicine, Philadelphia, Pennsylvania

ZHANG, BI NING, The Chinese University of Hong Kong, Kowloon City, Hong Kong

ZHANG, XIANGHUA, Sookmyung Women's University, Seoul, South Korea

ZHANG, ZHIGUO, Columbia University, New York, New York

ZHAO, HANG, University of Oklahoma, Norman, Oklahoma

ZIERHUT, CHRISTIAN, The Rockefeller University, New York, New York

ZUK, DORIT, Health and Human Services/National Institutes of Health/National General Medical Science, Bethesda, Maryland

Row 1: J. Watson, S. Harrison; M. Lampson, S. Bakhoum; S. Phan, H. Ou
Row 2: Y. Watanabe; N. Walther; M. Bettencourt-Dias, M. Kwon, D. Pellman
Row 3: R. Miller, E. Blunt; K. Corbett, A. Holland; A. George, D. Salim, A. Goloborodko
Row 4: M. Gartenberg, G. Schlissel; M.E. Karasu, R. Havalda; G. Andriani, R. de Voer

Row 1: M. Hetzer, D. Cleveland; A. Bortvin, B. Black; R. Johnson, S. Ishida

Row 2: A. Spradling; J. van Deursen, A. Holland, A. Amon; T. Stukenberg, K. Bloom; E. Tromer, C. Alfieri

Row 3: G. Edwards, R. Miller; H. Funabiki, E. Nigg; B. Moorefield, C. Chapard

Row 4: B. French, Y. Nechemia-Arbely; W. Earnshaw, F. Uhlmann

Row 1: G. Almouzni, Y. Barral, Y. Neurohr; O. Rando, G. Li; K. Nasmyth, J.-M. Peters
Row 2: C. Perea-Resa, C. Alfieri; D. Gerlich, L. Mirny, B. Moorefield
Row 3: C. Chapard, B.N. Zhang; B. de Massy, G. Nir; E. Heard, K. Zaret
Row 4: Symposium picnic

Row 1: S. West, J. Bateman; C. Montagna, G. Andriani, J. Schneider; J. Chen, M. Taileb
Row 2: In session; T. Mitchison, A. Desai, A. Amon
Row 3: S. Keeney, D. Page; L. Szewczak, I. Cheeseman; J. Jaynes, M. Fujioka
Row 4: H. Liu, X. Deng; S. Henikoff, R. Sever; T.Y. Yoo

Row 1: M. Gartenberg, P. Bernard; L. Heasley; G. Blobel, M. Bartolomei
Row 2: S. Cantor; J. DeLuca; Y. Yamashita, K. Baumann
Row 3: Y. Kim, B. Black; K. Nasmyth, B. Stillman; B. Moorefield, I. Hickson
Row 4: C. Montagna, G. Andriani; S. Gerbi, J. Bateman; S. Gasser, T. Misteli

Foreword

The decision to focus the 82nd Symposium on Chromosome Segregation and Structure reflects the enormous progress in our understanding of segregation, dynamics, and stability of chromosomes, the essential packaging for the organism's hereditary material. Many previous Cold Spring Harbor Symposia have been devoted to chromosome biology, notably Genes and Chromosomes: Structure and Organization (1941); Genetic Mechanisms: Structure and Function (1956); Chromosome Structure and Function (1973); Chromatin (1977); Structures of DNA (1982); DNA and Chromosomes (1993); Epigenetics (2004); and Nuclear Organization and Function (2010). Every decade has seen important advances in the field, so it is unsurprising that past Symposia have shed light from many angles on the structure and function of chromosomes. Current research in chromosome biology and dynamics includes many different fields, models, and approaches, and this Symposium offered the timely opportunity to bring together researchers using a multitude of advanced tools and techniques to investigate key chromosomal mechanisms and their aberrant function in disease. The scope of the meeting included the structure, division, and segregation of chromosomes in germ cell and somatic lineages in a variety of organisms.

Topics addressed at the 2017 Symposium included meiosis; mitosis; chromosome segregation; centrosomes and centrioles, ploidy, chromosome segregation errors, and disease; asymmetric cell division; nuclear architecture; chromosome structure and condensation; sister chromatid cohesion; genome stability; and germ cells. The Symposium attracted more than 280 participants and provided an extraordinary five-day synthesis of current understanding in the field. Opening night talks setting the scene for later sessions included Steven Henikoff on the inner kinetochore complex, Angelika Amon on cell nonautonomous regulation of chromosome segregation, Scott Keeney on DNA dynamics during meiosis, and Mónica Bettencourt-Dias on centrosome biogenesis. David Page delivered a fascinating Dorcas Cummings lecture on "Sex and Disease—Do Males and Females Read Their Genomes Differently?" for the Laboratory's friends and neighbors. Rising to the challenging task of condensing more than 50 talks over the prior five days, David Pellman provided a masterly summary of the state of the field at the conclusion of the Symposium. Interviews with leading scientists by participating editors, including Kim Baumann, Beth Moorefield, Richard Sever, Lara Szewczak, and Jan Witkowski, were conducted throughout the Symposium to provide a snapshot of the state of current research and are available on the CSHL Leading Strand YouTube channel (https://www.youtube.com/user/LeadingStrand). Transcripts of these Symposium conversations are provided here.

We thank Val Pakaluk, Mary Smith, and Ed Campodonico and his staff in the Meetings & Courses Program for their assistance in organizing and running the Symposium, and John Inglis and his staff at Cold Spring Harbor Laboratory Press, particularly Inez Sialiano, Jan Argentine, Kathleen Bubbeo, and Denise Weiss, for publishing the printed and online versions of the Symposium proceedings. Photographer Connie Brukin captured candid snapshots throughout the meeting.

Symposium Organizers
Terri Grodzicker, David Stewart, and Bruce Stillman
Cold Spring Harbor Laboratory

Symposium Editors
David Stewart and Bruce Stillman
Cold Spring Harbor Laboratory

Financial support from the corporate sponsors of our meetings program is essential for these Symposia to remain a success and we are most grateful for their continued support:

Corporate Benefactor
Regeneron

Corporate Sponsors
Agilent Technologies
Bristol-Myers Squibb Company
Calico Labs
Celgene
Genentech, Inc.
Merck
Monsanto Company
New England BioLabs
Pfizer
Thermo Fisher Scientific

Corporate Partners
Alexandria Real Estate
Gilead Sciences
Novartis Institutes for Biomedical Research
Sanofi

Contents

Meiosis

Germ Cells, Imprinting, Gene Dosage, and Regulation

Heterochromatin, Errors, And Damage

Summary

Dorcas Cummings Lecture

Conversations at the Symposium

Shaping Chromatin in the Nucleus: The Bricks and the Architects

DAVID SITBON,[1,2] KATRINA PODSYPANINA,[1,2] TEJAS YADAV,[1,2] AND GENEVIÈVE ALMOUZNI[1,2]

[1]*Institut Curie, PSL Research University, CNRS, UMR3664, Equipe Labellisée Ligue contre le Cancer, Paris, France*

[2]*Sorbonne Universités, UPMC Univ Paris 06, CNRS, UMR3664, Paris, France*

Correspondence: genevieve.almouzni@curie.fr

Chromatin organization in the nucleus provides a vast repertoire of information in addition to that encoded genetically. Understanding how this organization impacts genome stability and influences cell fate and tumorigenesis is an area of rapid progress. Considering the nucleosome, the fundamental unit of chromatin structure, the study of histone variants (the bricks) and their selective loading by histone chaperones (the architects) is particularly informative. Here, we report recent advances in understanding how relationships between histone variants and their chaperones contribute to tumorigenesis using cell lines and *Xenopus* development as model systems. In addition to their role in histone deposition, we also document interactions between histone chaperones and other chromatin factors that govern higher-order structure and control DNA metabolism. We highlight how a fine-tuned assembly line of bricks (H3.3 and CENP-A) and architects (HIRA, HJURP, and DAXX) is key in adaptation to developmental and pathological changes. An example of this conceptual advance is the exquisite sensitivity displayed by p53-null tumor cells to modulation of HJURP, the histone chaperone for CENP-A (CenH3 variant). We discuss how these findings open avenues for novel therapeutic paradigms in cancer care.

Over the past decades, the study of chromatin structure and regulation has tremendously improved our understanding of overall nuclear organization, ranging from the basic repeating unit—the nucleosome core particle comprised of DNA wrapped around histones—up to beads-on-a-string and higher-order structures (Fig. 1A; Kornberg 1974; Olins and Olins 1974; Oudet et al. 1975; Laskey et al. 1978; McGhee and Felsenfeld 1980; Luger et al. 1997; Nora et al. 2012; Naumova et al. 2013). Together, these structures define the "epigenome" or specific chromatin landscapes that inform cell fate decisions and govern cell identity (Fig. 1B). Here, we refer to epigenetic features as the "structural adaptation of chromosomal regions to register, signal or perpetuate altered activity states" (Bird 2007). Unlike the embedded genetic code, epigenetic features of chromatin are plastic and can be reversed enabling genomic reprogramming (Gurdon 1962; Takahashi and Yamanaka 2006). Establishment, propagation, and maintenance of different layers in the hierarchical chromatin structure directly impinge upon gene function. At the most fundamental level, the nucleosome, histone variants provide a choice of distinct bricks in the construction of chromatin architecture. Accordingly, selection of histone variants and their handling by distinct histone chaperones (architects) contribute to the marking of specific genomic locations during various cell cycle phases (Fig. 2). A network of assembly lines involving chaperone–variant interactions regulates chromatin organization in various developmental stages and disease states (Gurard-Levin et al. 2014). Importantly, his-

tone variants (Buschbeck and Hake 2017; Talbert and Henikoff 2017), histone-modifying enzymes (that catalyze distinct PTMs [posttranslational modifications] on histone variants) (Gurard-Levin and Almouzni 2014), and chromatin remodeling factors (Bowman and Poirier 2014; Swygert and Peterson 2014) are constituents of defined complexes that organize chromatin. Furthermore, these large, multicomponent complexes bring together transcription factors, chromatin-associated proteins, and DNA-modifying enzymes to interface with bricks and architects (De Koning et al. 2007). Hence, an ensemble is orchestrated in order to control gene regulation and cellular fate depending on temporal or spatial cues and a delicate balance of chromatin and transcription factor availability. Consequently, the stoichiometry of specific histone bricks, architects, and accessory factors is of paramount importance for chromatin regulation.

Notably, changes in histone variant and chaperone dosage are well known in the context of development. For example, in *Xenopus*, unusually high levels of Npm2 (nucleoplasmin 2), the first described histone chaperone (Laskey et al. 1978), are present in the oocyte and during early embryogenesis (Burglin et al. 1987; Litvin and King 1988). This elevated dosage of the architect Npm2 ensures storage of soluble histones H2A–H2B contributing to chromatin assembly after fertilization during the rapid cell divisions in early development (Finn et al. 2012). In adult tissues, concentration of Npm2 decreases in parallel with the lower rate of cell division. During early development, a switch from storage to rapid deposition of histones

Published by Cold Spring Harbor Laboratory Press; doi: 10.1101/sqb.2017.82.033753

Cold Spring Harbor Symposia on Quantitative Biology, Volume LXXXII

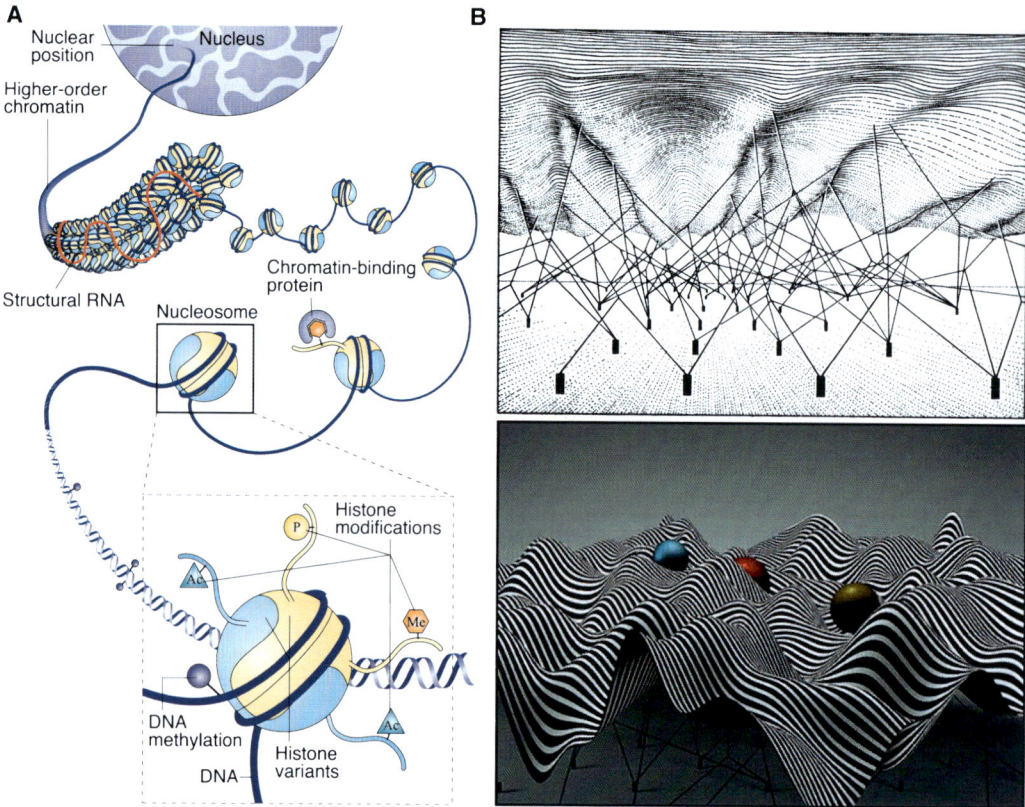

Figure 1. Epigenetic components of chromatin architecture and a complex network of interactions underlying cell fate determination. (*A*) Numerous epigenetic factors help organize chromatin in the 4D nucleus: along with the linker histone H1, DNA wrapped around one (H3–H4)$_2$ tetramer capped by two H2A–H2B dimers forms the nucleosome—the fundamental repeating unit of chromatin. DNA can be methylated and histones can be posttranslationally modified (e.g., by methylation [Me], acetylation [Ac], and phosphorylation [P]). Chromatin-binding proteins such as methyl- or histone-recognizing factors read the information encoded by these covalent marks. The presence of histone variants adds further complexity. Arrays of nucleosomes fold into higher-order chromatin structures, potentially guided by noncoding RNA. The nuclear localization of a given chromosomal domain represents an additional level of regulatory information. (*B*) View from below of the historical Waddington landscape (*top*). Pegs represent genes and the strings represent a complex system of genetic interaction that determines cell fate. The epigenetic landscape is shaped by the strings, which are ultimately anchored to the genes. (*Bottom*) A modern re-interpretation of the Waddington landscape. The valleys represent ongoing decision events in cells (represented by colored beads) on their path to a final cell fate. (*A*, Adapted, with permission, from Probst et al. 2009; *B*, *top*, reprinted, with permission, from Waddington 1957; *bottom*, reproduced, with permission, from Paul L. Harrison and the Epigenesys Network.)

is controlled by both the level of chaperones and the amount of soluble histones. Early studies showed that pre-MBT (midblastula transition) transcriptional repression could be alleviated by titration with exogenous DNA (Newport and Kirschner 1982). This is in line with the importance of histone concentration as depletion of H3 protein results in premature cell cycle elongation in the embryo, whereas excess H3–H4 maintains shorter cell cycles after MBT (Amodeo et al. 2015). Notably, interfering with chromatin assembly pathways leads to aberrant development in *Xenopus* embryos (Quivy et al. 2001; Ray-Gallet et al. 2002; Szenker et al. 2012). In somatic mammalian cells, the soluble pool of histones is regulated by histone chaperones NASP (nuclear autoantigenic sperm protein) (Cook et al. 2011) and Asf1 (antisilencing factor 1) (Groth et al. 2005) that act as a reservoir for excess H3–H4. These chaperones mitigate fluctuations in supply and demand of histones upon replication stress, when the replication fork arrests and newly synthesized histones cannot be loaded onto new DNA. Moreover, his-

tone management is particularly critical at telomeres where down-regulation of Asf1 activates the ALT (alternative lengthening of telomeres) pathway, thereby compromising genome integrity (O'Sullivan and Almouzni 2014). The importance of histone variant handling is further underscored in mice defective for replication-coupled histone chaperone CAF-1 (chromatin assembly factor 1), composed of p150, p60, p48 subunits in mammals (Houlard et al. 2006). CAF-1 mutant animals display heterochromatin disorganization and arrest during early embryonic development. Interestingly, fine-tuning levels of CAF-1 influence somatic cell identity, allowing efficient production of induced pluripotent stem cells (Cheloufi et al. 2015; Ishiuchi et al. 2015; Yang et al. 2015). The above examples highlight the inherently plastic nature of the chaperone–variant network during growth and development, an aspect that needs to be further explored in various contexts and cell lineages.

The study of diseases provides another case where the role of variants and chaperones is at stake. Indeed, identi-

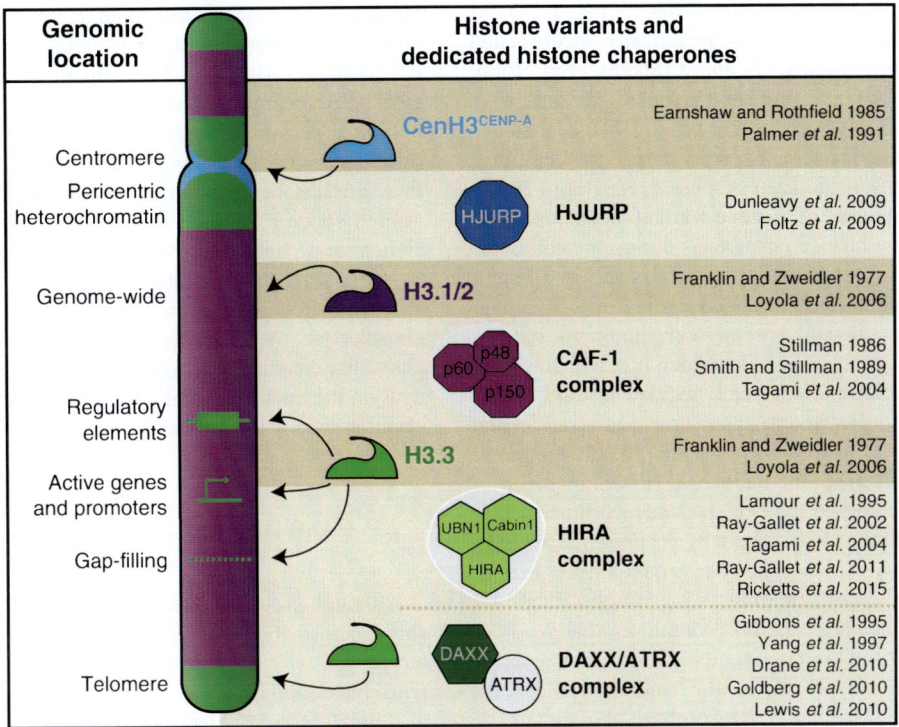

Figure 2. Histone variants (bricks) and their associated histone chaperones (architects) mark specific chromosomal regions. Among H3 variants in humans, at least four are handled by specific histone chaperones during particular cell cycle stages: new CENP-A, the centromeric variant, is incorporated at centromeres by HJURP during late mitosis–early G_1; H3.1/2, the replicative variants, are incorporated genome-wide during DNA synthesis by CAF-1 complex; H3.3, the replacement form, is placed throughout the cell cycle at regulatory elements, gene bodies by the HIRA complex, and at telomeres and pericentric heterochromatin by the DAXX/ATRX complex.

fication of H3 variant mutations in very aggressive pediatric tumors exemplifies a major role for histone variants in cancer (Schwartzentruber et al. 2012; Wu et al. 2012; Lewis et al. 2013). Furthermore, a variety of tumors carry mutations in the H3.3 histone chaperone DAXX (death-domain associated protein 6)/ATRX (α-thalassemia mental retardation syndrome X-linked) complex (Heaphy et al. 2011; Jiao et al. 2011; Schwartzentruber et al. 2012). How distinct mutations alter the chaperone–variant balance and contribute to tumorigenesis is an active area of research. Interestingly, at the transcriptional level, cancers also often display alterations in the expression of histone regulators including key histone variants and chaperones (Montes de Oca et al. 2015; Gurard-Levin et al. 2016). For example, a series of cancer types show up-regulation of Asf1b, one of the two human paralogs of the Asf1 chaperone (Corpet et al. 2011). Notably, overexpression of CENP-A, the centromere-specific histone variant (Tomonaga et al. 2003; Amato et al. 2009), or its cognate histone chaperone HJURP (Holliday junction recognition protein) (Kato et al. 2007; Montes de Oca et al. 2015) is a hallmark of tumor progression and poor prognosis. Here, we summarize recent work suggesting that therapies targeted against specific histone chaperones and their "client" variants represent a promising opportunity for improved cancer care. In addition, we highlight how dynamics of CENP-A in cancer (Lacoste et al. 2014; Filipescu et al. 2017) and H3.3 in development (Szenker et al. 2012) can open new avenues for fundamental and translational research in the future.

H3 VARIANT DYNAMICS: VARIATIONS ON A THEME IN THE 4D NUCLEUS

Histone variants are forms of the four core histones (H3, H4, H2A, and H2B), originally characterized based on migration properties using Triton acid urea gel electrophoresis (Franklin and Zweidler 1977). For H3 histones, this technique enabled the distinction of two major categories —namely, the replicative and replacement variants. In humans, the two closely related replicative variants H3.1 and H3.2—later referred to as H3.1/2—show a peak of expression in S phase and are incorporated into chromatin via the DSC (DNA synthesis–coupled) pathway during replication (Stillman 1986; Smith and Stillman 1989; Groth et al. 2007) or DNA repair (Gaillard et al. 1996; Polo et al. 2006). The replacement variant H3.3, incorporated via the DSI (DNA synthesis–independent) pathway (Ray-Gallet et al. 2002), is expressed throughout all cell cycle phases, including quiescence, and is associated with transcription (Ahmad and Henikoff 2002) and major rearrangements of the sperm chromatin during fertilization (Katagiri and Ohsumi 1994; Ahmad and Henikoff 2002; Loppin et al. 2005; Torres-Padilla et al. 2006; Bonnefoy et al. 2007; Santenard et al. 2010; Orsi et al. 2013). Finally, the most divergent H3 variant CenH3 (centromeric histone 3), called CENP-A (histone H3-like centromeric protein A) in humans (Earnshaw and Rothfield 1985; Palmer et al. 1991), is specifically associated with centromeric regions (Muller and Almouzni 2017). In mammals,

CENP-A shows a peak of expression during the G_2/M phases, and new CENP-A is specifically incorporated during late mitosis–early G_1 phase (Shelby et al. 2000; Jansen et al. 2007). In humans, eight H3 variants have been identified to date (Maze et al. 2014); the existence of corresponding homologs in other organisms highlights remarkable conservation across species (Waterborg 2012). A detailed analysis of histone variant deposition and maintenance by histone chaperones during the cell cycle (both at specific chromosomal territories and in different cell types) is central to our understanding of local nucleosome organization and genome-wide chromatin states.

A particular group of histone chaperones, which we call "dedicated" architects, selectively regulate histone variant availability and placement across time and space in the nucleus (Fig. 2; Gurard-Levin et al. 2014; Hammond et al. 2017). As mentioned above, CAF-1 is involved in chromatin assembly in the DSC pathway (Stillman 1986; Smith and Stillman 1989) shown by the presence of CAF-1 in complex with the replicative variant (Tagami et al. 2004). Similarly, isolation of the H3.3 complex identified two distinct histone chaperone complexes that work in combination to orchestrate incorporation of the replacement histone variant H3.3. On one hand, the HIRA (histone cell cycle regulator A) complex—composed of HIRA and at least two other subunits, Cabin1 (calcineurin-binding protein 1) and UBN1 (ubinuclein 1) (Ricketts and Marmorstein 2017), and possibly also UBN2 (ubinuclein 2) (Banumathy et al. 2009)—deposits H3.3 at gene bodies, promoters, and regulatory elements (Lamour et al. 1995; Ray-Gallet et al. 2002; Tagami et al. 2004). On the other hand, the DAXX/ATRX complex contributes to the H3.3 presence at peri-centromeric regions and telomeres (Gibbons et al. 1995; Yang et al. 1997; Drane et al. 2010; Goldberg et al. 2010; Lewis et al. 2010). Finally, we used a similar strategy for isolation of the most divergent variant, CENP-A, in a soluble predeposition complex. This led to the identification of the chaperone HJURP (Dunleavy et al. 2009; Foltz et al. 2009; Shuaib et al. 2010), as being responsible for the specific loading of CENP-A at centromeres in late mitosis–early G_1 phase.

While dedicated histone chaperones can ensure the delivery of specific histones at the right time and place, other chaperones do not show such selectivity. Yet, such "casual" chaperones play critical functions in histone supply and demand during DNA replication and genotoxic events. For example, Asf1 acts upstream of dedicated histone chaperones, handling old and newly synthesized soluble core histones (Tyler et al. 1999; Munakata et al. 2000; Sharp et al. 2001; Mello et al. 2002). Asf1 is also involved in nucleosome assembly during temporally constrained processes such as DNA replication (Groth et al. 2005, 2007). Additionally, Asf1 influences chromosome stability at distinct landmarks, such as chromosome ends, by regulating telomere length maintenance (O'Sullivan and Almouzni 2014). NASP (nuclear autoantigenic sperm protein) acts as an emergency reservoir of soluble histones (Welch et al. 1990; Richardson et al. 2000), and MCM2 (DNA replication licensing factor, a component of the

DNA helicase), newly identified as a histone chaperone, coordinates parental histone recycling at the replisome (Clement and Almouzni 2015; Huang et al. 2015; Richet et al. 2015). Importantly, casual chaperones display remarkable versatility as they can potentially interact with different histone variants. In summary, we propose that there are histone chaperones that care more about "who" (which specific histone variant), others that care about "when and why" (DNA replication, transcription, or developmental fate changes), and yet others that remain "multifunctional," in some cases even indiscriminatingly nonselective. Within such a multipartner network, it appears that the choice between these alternatives is dependent on the nuclear availability and relative abundance of specific histone variants and chaperone molecules.

A CASE OF CELLULAR ADAPTABILITY TO CHANGES IN DOSAGE

Although previous studies emphasize selective partnerships, perturbations to histone variant and chaperone interactions exist in physiological and disease contexts (Buschbeck and Hake 2017). Recent studies on centromeric chromatin assembly provide a striking example of polyvalent partner switching as a result of altered histone variant dosage (Lacoste et al. 2014; Shrestha et al. 2017). In healthy human, mouse, and most other eukaryotic cells, CENP-A marks the centromeres (Fig. 3A; Fukagawa and Earnshaw 2014; McKinley and Cheeseman 2016). Indeed, centromeric histone variant CENP-A is an integral epigenetic determinant of centromere identity and genome maintenance (Allshire and Karpen 2008). Loss of CENP-A gives rise to defects in mitosis and leads to senescence (Maehara et al. 2010). Yet, the proportion of CENP-A-containing nucleosomes at the centromere is low, with current estimates of only 2%–4% of total α-satellite DNA wrapped around CENP-A-containing nucleosomes in cultured diploid human cells (Bodor et al. 2013; Nechemia-Arbely et al. 2017). Thus, even at a steady state, total CENP-A levels in proliferating RPE (retinal pigment epithelium) cells are almost 100 times higher than the relative amount associated with centromeric DNA (Bodor et al. 2014). One hypothesis proposes that surplus CENP-A safeguards against chromosome loss during random redistribution of old and new CENP-A following cell division. Altogether, these findings point toward the central role of HJURP in de novo CENP-A deposition and, consequently, in centromere establishment, maintenance, and propagation.

Recent experimentation has helped to uncover interesting changes associated with a pathological excess of either histone variants or chaperones or both in cases of co-regulated expression. Increased levels of CENP-A are commonly reported in a range of aggressive cancers (Ma et al. 2003; Tomonaga et al. 2003; Li et al. 2011; Qiu et al. 2013; Gu et al. 2014; Benito et al. 2015; Sun et al. 2016). The excess CENP-A can be incorporated both at the centromere and also in the chromosome arms (Fig. 3B). We have generated a model of oncogene-transformed MEFs (mouse embryonic fibroblasts) that recapitulates the

A

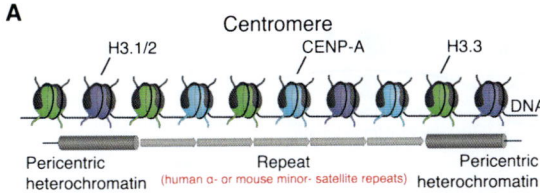

B

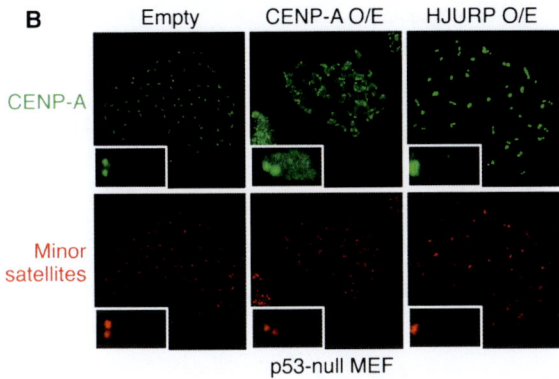

C

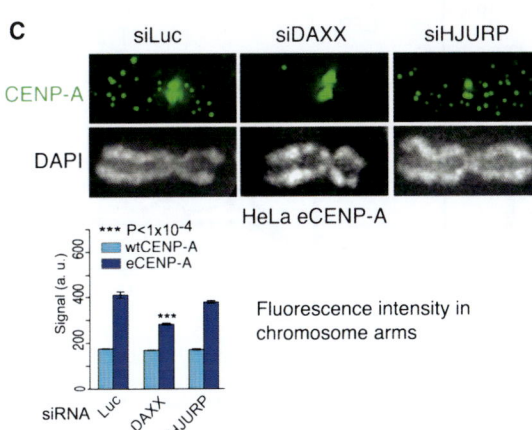

Figure 3. DAXX mediates ectopic CENP-A deposition in chromosome arms. (*A*) Schematic representation of centromeric chromatin. Centromeres comprise CENP-A nucleosomes interspersed with H3.1/2 and H3.3 nucleosomes. Centromeric chromatin contains minor- (in mice) and α- (in humans) satellite repeats flanked by pericentric heterochromatin. (*B*) DNA FISH on metaphase spreads of p53-null mouse embryonic fibroblasts (MEFs) transduced with the indicated retroviral construct. Cells were stained with an antibody against CENP-A and LNA FISH (fluorescence in situ hybridization) probes for minor satellites (site of centromeric CENP-A deposition). *Insets* show individual magnified chromosomes. Overexpression (O/E) of CENP-A, but not HJURP, leads to ectopic deposition of CENP-A on chromosome arms. (*C*) Fluorescent microscopy visualization of CENP-A after treatment with the indicated siRNA in a human cell line overexpressing CENP-A (HeLa eCENP-A). Depletion of histone chaperone DAXX, but not HJURP, rescues the ectopic CENP-A deposition phenotype. Histogram representing the fluorescence of CENP-A present in chromosome arms of metaphase spreads from two cell lines (wild-type, wtCENP-A and overexpression, eCENP-A) treated with the indicated siRNA. *P* values represent pairwise comparison of siLuc (control), and all other siRNAs tested with the same cell line with a Mann–Whitney *U*-test. (*A*, Adapted, with permission, from Muller and Almouzni 2017; *B*, adapted, with permission from Filipescu et al. 2017; *C*, adapted, with permission, from Lacoste et al. 2014.)

increase in CENP-A and HJURP levels (Filipescu et al. 2017). Notably, in MEFs, altering the dose of the brick or its architect does not necessarily lead to the same outcome. Indeed, overexpressing HJURP in cultured cells leads to increased CENP-A deposition at the centromere, but not in the arms, in line with its specific role in targeting CENP-A to the centromere. Conversely, experimentally induced or cancer-associated surplus CENP-A enables ectopic deposition of additional CENP-A molecules at the chromosome arms. Presumably, this effect is observed because the cognate chaperone HJURP, required for the specific targeting, is limiting. Our data show that another multifaceted histone chaperone DAXX (part of the DAXX/ATRX complex) is involved in the ectopic deposition in the arms (Fig. 3C; Lacoste et al. 2014). Arguably, chaperone–variant polyvalency in centromere assembly indicates flexibility in the interaction modules of various architects and bricks. Partner switching, wherein histone chaperones promiscuously interface with additional histone variants, contributes to a new dimension of adaptation to shifting physiological needs and nucleic acid transactions. Disease scenarios represent one such context wherein inherently plastic histone variant–chaperone relationships become critical in the face of pathological challenges to growth and survival.

ALTERED DOSAGE OF ARCHITECTS AND BRICKS POTENTIATES TUMORIGENESIS: FROM MOUSE MODELS TO HUMAN PATHOLOGY

Substantial changes in the levels of molecular components (including chromatin regulators and transcription factors) have been characterized in several genetic diseases including cancer (Montes de Oca et al. 2015; Gurard-Levin et al. 2016). Epigenetic factors may be altered as a consequence of cancer-causing mutations, without themselves being mutated (Fig. 4A). We analyzed TCGA data to show that a genetic defect in the tumor suppressor p53 selectively up-regulates centromeric histone variant CENP-A and its chaperone HJURP in cancers of different origins (Fig. 4B). In a reporter system, p53 can associate at the promoters of CENP-A and HJURP, as part of the E2F-DREAM repressive complex (Filipescu et al. 2017). This finding indicates that increased CENP-A and HJURP levels are linked to loss of the transcriptional function of p53 in cancer cells. Interestingly, CENP-A and HJURP are overexpressed not only in cancers with p53-inactivating genetic lesions, but also in a subset of tumors with gain-of-function mutations in p53 (Filipescu et al. 2017). To determine whether this aspect of p53-mediated up-regulation involves other transcription factors, such as FoxM1 (a p53-partner), it will be necessary to extend experiments to new mutant mouse models (Barsotti and Prives 2009; Bieging et al. 2014). Importantly, although CENP-A and HJURP levels are high in p53-deficient transformed cells, this elevation cannot be merely attributed to hyperproliferation (Filipescu et al. 2017). To determine if an increased dose of a histone variant or its chaperone could

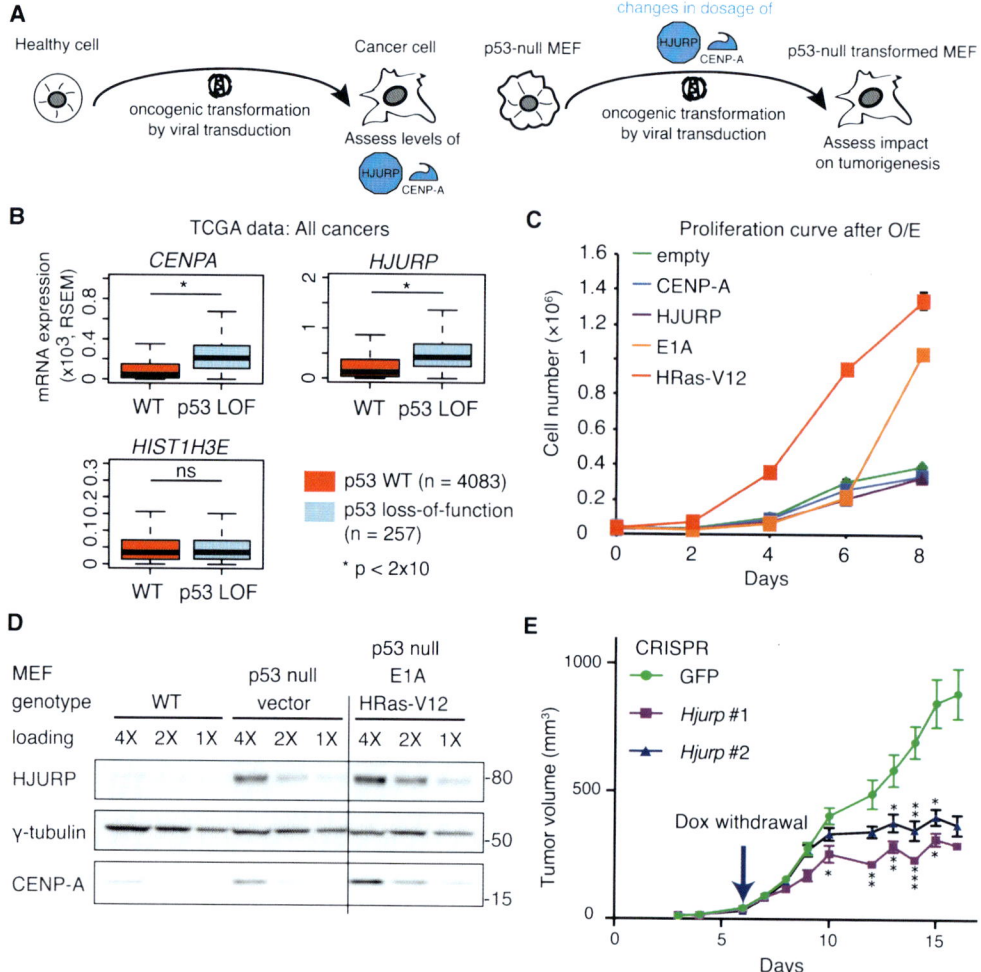

Figure 4. Cancer cells display addiction to chromatin architect HJURP. (*A*) Key chromatin factors, including important bricks and architects, display variations in expression during tumorigenesis. HJURP and CENP-A levels were assessed in experimentally transformed cells and in human cancer cells. Then, their levels were modified in cultured cells and in tumors to assess effects on tumorigenesis. (*B*) Gene expression levels of CENP-A and HJURP are higher in cancers with defective p53. Box plot comparisons of relative expression (mRNA) of genes coding for CENP-A (CENPA), HJURP (HJURP), and H3.1 (HIST1H3E) from all cancers (28 cancer types), classified according to p53 status (TCGA provisional data). Tumors are either wild type for TP53 ($n = 4083$) or p53 loss of function (LOF) ($n = 257$). Significance was computed using Wilcoxon rank sum tests. (*C*) HJURP or CENP-A overexpression alone does not transform p53-null MEFs. Proliferation curve of p53-null MEFs transduced with the indicated retroviral construct. The graph displays the quantified cell number ± SEM of triplicates. (*D*) HJURP and CENP-A are overexpressed in p53-null wild-type and transformed MEFs. Western blot of HJURP and CENP-A levels in RIPA-soluble extracts from MEFs transduced with empty vector or sequentially with E1A and HRas-V12. (*E*) Conditional depletion of exogenous HJURP stops the growth of established tumors lacking endogenous HJURP. Allograft assay measuring tumor growth following HJURP knockout over time. p53-null HRas-V12 MEFs were generated with an inducible CRISPR-resistant HJURP transgene maintained in the presence of doxycycline. Doxycycline was withdrawn on day 6 after injection (*Hjurp* switched off) and allograft tumor volume was measured over time. Data represent mean tumor volume ± SEM. *, $P < 0.05$; **, $P < 0.005$; ***, $P < 0.0005$, *t*-test. (Adapted, with permission, from Filipescu et al. 2017.)

drive transformation, we overproduced the two factors individually in mouse cells. These experiments proved that the inducible overexpression of CENP-A and HJURP do not act as "drivers" of transformation in a p53-null background, unlike a second hit with a viral oncogene E1A or a mutant form of HRas (Fig. 4C).

Until recently, the specific role of centromeric variant–chaperone partnerships has remained largely unexplored in cells lacking a proficient p53 checkpoint pathway. In human and mouse cells, CENP-A and HJURP levels are co-regulated through a dual control mechanism involving both transcriptional and posttranslational modes. Early

studies had found that loss of either CENP-A or HJURP leads to growth arrest with features of senescence and p53 pathway activation (Maehara et al. 2010; Heo et al. 2013; Valente et al. 2013). As mentioned above, in p53-defective cells, there is a burst of expression of both centromeric factors CENP-A and HJURP (Fig. 4D). Instead of undergoing growth arrest, both mouse and human p53-deficient cells undergo rapid apoptosis upon removal of HJURP (Filipescu et al. 2017). Remarkably, in a p53-deficient background, severely reduced growth of established tumors occurs upon conditional depletion of HJURP (using a doxycycline-dependent system) (Fig. 4E). Thus, in the absence of p53-dependent checkpoint activation, cells develop an "epigenetic" addiction to HJURP, likely because of its important role in CENP-A loading at centromere. This is the first reported example of a histone chaperone being necessary to support tumor growth and kinetics of tumor progression. However, epigenetic addiction is also a part of other cancer phenotypes, including evidence for acquired dependence on histone-modifying enzyme, MLL, in tumor cells with gain-of-function mutations in p53 (Zhu et al. 2015). Thus, the case of cancers addicted to high levels of epigenetic factors reflects a wider phenomenon of context-specific requirement for chromatin organization. Realizing the full promise of targeting the epigenome (Gurard-Levin et al. 2016; Morel et al. 2017) and its components in cancer necessitates in-depth exploration of chaperone–variant imbalance and downstream consequences. Patient-derived samples and animal models in conjunction with the latest genome editing tools (such as CRISPR–Cas9) and rapid protein-degradation methods (like auxin-inducible or photo-activated degrons) will be instrumental in the determination of relevant drug targets that serve as the Achilles' heel for epigenetically addicted tumors.

A CENTRAL ROLE FOR HISTONE VARIANTS IN DEVELOPMENT

Given the high degree of sequence similarity between histone variants, contextual differences raise the issue of variant interchangeability and redundancy (Waterborg 2012). Indeed, mammalian replicative histone variants H3.1 and H3.2 only differ from the replacement form H3.3 by five and four residues, respectively. Despite these seemingly minor variations in amino acid composition, replicative and replacement histone variants cannot directly substitute for each other without causing deleterious effects in mammals (Couldrey et al. 1999; Bush et al. 2013; Tang et al. 2015). Both H3.1 and H3.2 variants are used during DNA synthesis to provide a full complement of histones necessary to assemble newly replicated chromatin (MacAlpine and Almouzni 2013). Besides contributing to the bulk of new H3.1 and H3.2 histones in S phase, the unique roles of replicative H3 variants have remained largely uncharacterized. Intriguingly, in *Arabidopsis thaliana*, the H3.1 variant mediates the restoration of the transcriptional repressive mark H3K27me3 during DNA replication (Jacob et al. 2014; Jiang and Berger 2017). This specificity observed in plants merits careful examination in mammals to uncover similar dedicated functions.

The replacement variant H3.3, which can be incorporated at any time during the cell cycle, has been linked to many critical cell processes ranging from transcription and DNA repair to reprogramming (Ahmad and Henikoff 2002; Schwartz and Ahmad 2005; Ng and Gurdon 2008; Jullien et al. 2012; Adam et al. 2013). However, it has been difficult to evaluate whether the identity of H3.3 variant or its specific mode of incorporation is crucial for these processes. Evolutionary analyses can help shed light

on distinct and universal roles of H3.3 variants. Interestingly, yeast possesses a single form of noncentromeric histone H3, closely related to the mammalian H3.3 variant and capable of performing both replicative and replacement functions (Dion et al. 2007; Jamai et al. 2007; Rufiange et al. 2007). In contrast, the need for specific histone variants H3.1 and H3.3 is observed in *A. thaliana* (Jacob et al. 2014; Jiang and Berger 2017). Here, we have compiled phenotypes linked to alteration of H3.3 and its associated chaperones observed across different species (Table 1). In most multicellular eukaryotes, different developmental stages are sensitive to removal of key chromatin players. Thus, each of these developmental time periods represents critical windows for interrogating context-dependent cellular functions of histone variants. In *A. thaliana*, complete loss of H3.3 leads to embryonic lethality and partial sterility due to defective male gametogenesis (Wollmann et al. 2017). Albeit replicative histone forms in *Drosophila melanogaster* can compensate for the absence of H3.3 variant in somatic tissues, both male and female flies are sterile (Hodl and Basler 2009; Sakai et al. 2009). A possible explanation for H3.3-related sterility is the lack of a sufficient pool of maternal H3.3 to help in replacing protamines in the sperm after fertilization. Unlike flies, worms remain viable and fertile despite a complete loss of the H3.3 variant (Piazzesi et al. 2016). Indeed, *Caenorhabditis elegans* lacking H3.3 display only mild phenotypes such as "bagging" (eggs hatching inside the adult body) and heightened sensitivity to stress. In vertebrate models, such as *Mus musculus,* deletion of one of the two copies of H3.3 leads not only to sterility but also to developmental defects (Couldrey et al. 1999; Bush et al. 2013; Tang et al. 2015). Notably, point mutations in human H3 variants, such as K27M and G34R/V (Schwartzentruber et al. 2012; Wu et al. 2012), or their associated chaperones (Heaphy et al. 2011; Jiao et al. 2011) are linked to aggressive pediatric glioblastoma. The question remains whether individual histone H3 residues are functional during development (loss of function) or provide adaptive advantages following (gain-of-function) mutation in cancer cells. Taken together, these findings underline the role of specific bricks and architects during cell fate determination and metazoan development. Finally, the striking case of brain tumors occurring in children highlights the important link between developmental contexts and cancer.

In vertebrates, research on chromatin assembly was pioneered in *Xenopus laevis* (Laskey et al. 1977) enabling further characterization of distinct nucleosome deposition pathways (Almouzni and Mechali 1988) and the key roles of histones chaperones (Quivy et al. 2001; Ray-Gallet et al. 2002). The high yield of eggs supplies a large quantity of material available for experimentation both in vitro and in vivo (Gurdon 1976). This widely used model provides several salient advantages for the study of chromatin (Almouzni et al. 1990). Given the recurrent sterility phenotype in many species, *Xenopus* offers a convenient setting to investigate other functions of H3.3 as sperm chromatin retains H3.3–H4 dimers instead of protamines (Katagiri and Ohsumi 1994). In addition, the presence of a

Table 1. Alteration in histone variant H3.3 and associated-chaperones across species

Species	Strategy of H3.3 alteration	Associated phenotype(s)	Strategy of chaperone alteration	Associated phenotype(s)
Tetrahymena thermophila	Deletion of the only H3.3-encoding gene (*HHT3*) — Cui et al. 2006	H3.3 KO is viable but leads to sterility	TTHERM_00046490 annotated as HirA homolog — Miao et al. 2009	No mutant alleles tested
Arabidopsis thaliana	Deletion of two histone H3.3 genes (*HTR4*, *HTR8*), artificial mRNAs target third copy (*HTR5*) — Wollmann et al. 2017	H3.3 KD impairs male gametogenesis; H3.3 KO is lethal	No reported homolog of DAXX — Lewis et al. 2010; T-DNA insertional mutant alleles (*HIRA, UBN1, UBN2, CABIN1*) and artificial mRNA against *HIRA* — Duc et al. 2015	No mutant alleles tested; Individual mutants are viable; HIRA KO and KD show reduced fertility
Saccharomyces cerevisiae	Deletion of both histone H3 genes (*HHT1, HHT2*) — Jamai et al. 2007	H3 (ancestral H3.3-like histone) KO is lethal	No reported homolog of DAXX — Wollmann et al. 2012; Deletion of individual HIR complex genes (*HIR1, HIR2, HIR3, HPC2*) — Osley 1991; Kaufman et al. 1998; Qian et al. 1998	No mutant alleles tested; Individual and double KOs are viable but result in derepression of histone genes outside of late G1/S phase
	Galactose-inducible *HHT2* allele; remaining H3 copy (*HHT1*) is deleted — Gossett and Lieb 2012		No reported homolog of DAXX — Lewis et al. 2010	No mutant alleles tested
Caenorhabditis elegans	Loss-of-function alleles of H3.3-encoding genes (*HIS72, HIS71*) — Piazzesi et al. 2016	H3.3 KO is viable but leads to bagging phenotype (eggs hatching within animal) and stress sensitivity	K10D2.1 or ceHILP annotated as HIRA homolog — Wilming et al. 1997; No reported homolog of DAXX — Lewis et al. 2010	No mutant alleles tested; No mutant alleles tested
Drosophila melanogaster	Deletion of the two histone H3.3-encoding genes (*His3.3A, His3.3B*) — Sakai et al. 2009; Hodl and Basler 2009	H3.3 KO is viable but leads to sterility	Loss of function *Hira* or *UBN1* allele(s) — Loppin et al. 2005; Bonnefoy et al. 2007; Loss-of-function allele of DAXX-encoding *dlp* gene — Fromental-Ramain et al. 2017	HIRA KO is viable but females are sterile; HIRA deposits H3.3 in the decondensed sperm pronucleus; DAXX KO is viable
Xenopus laevis	Morpholino targeting H3.3-encoding *h3f3a* mRNA — Szenker et al. 2012	H3.3 KD leads to gastrulation defects	Morpholino targeting *hira* mRNA — Szenker et al. 2012	HIRA KD is viable but leads to gastrulation defects
Mus musculus	Insertional mutagenesis by gene trapping of H3f3a in 129Sv/Ev background — Couldrey et al. 1999	H3f3a KO results in significant neonatal lethality	*daxx* annotated as DAXX homolog — Lewis et al. 2010; Targeted deletion of *Hira* gene in 129Sv, CD1, C57BL6/129Sv backgrounds — Roberts et al. 2002	No mutant alleles tested; HIRA KO is embryonic lethal
	Conditional KO allele of *H3f3b* (floxed exons 2–4 using germline deleter Zp3-Cre) in C57BL/6N background — Bush et al. 2013	H3f3b KO results in significant neonatal lethality, females are subfertile, males are infertile	Targeted deletion of *Daxx* gene in 129Sv/Ev, 129Sv/Ev/BLKSW backgrounds — Michaelson et al. 1999	DAXX KO is embryonic lethal
	Conditional KO allele of H3f3a or H3f3b (floxed exons 2 using germline deleter Hprt-Cre) in 129S1/Sv background — Tang et al. 2015	H3f3a KO is viable, males are subfertile; H3f3b KO is lethal		
Homo sapiens	Mutations H3.3 K27M or H3.3 G34R/V in pediatric cancers — Schwartzentruber et al. 2012; Wu et al. 2012	Glioblastoma multiforme (GBM)	Point mutation (L786M) in *UBN1* in consanguineous families — Shamseldin et al. 2015	HIRA complex subunit recessive mutant is embryonic lethal
		Diffuse intrinsic pontine glioma (DIPG)	Mutations in pancreatic neuroendocrine tumors (PanNETs) and GBM — Jiao et al. 2011; Heaphy et al. 2011; Schwartzentruber et al. 2012	Mutant DAXX/ATRX complex is linked to cancer

KO, knockout; KD, knockdown.

single replicative H3 variant—namely H3.2—enables a simple comparison of the replicative and replacement variant contributions to a given phenotype. Also, both H3.2 and H3.3 protein sequences are conserved between *Xenopus* and human. Finally, the extensive characterization of *Xenopus* development is a major advantage to analyze phenotypes at distinct developmental transitions (Nieuwkoop and Faber 1994). After fertilization, the embryo starts to divide quickly without any gap phases, whereas no zygotic transcription is detected (Fig. 5A; Dumont 1972; Keller 1991). This time window provides a unique setting to explore the importance of histone variant dynamics independent of any role in transcription. Following this period of rapid embryonic cell cycles, MBT marks

the activation of zygotic transcription. At this transition, cell division time slows down, intervening G_1 and G_2 phases are incorporated into the cell cycle to match the duration observed in somatic cells (Newport and Kirschner 1982) and cell migration begins.

With this model system, we could explore specific roles for H3 variants in development and cell fate determination. Following knockdown of endogenous H3.3 or its chaperone HIRA using a morpholino strategy, we found that *Xenopus* embryos developed up to the early gastrula stage but showed severe defects at late gastrulation (Szenker et al. 2012). Consequently, mutant embryos underwent massive apoptosis and did not develop to term. Furthermore, depletion of either H3.3 or HIRA led to chromatin disorganization as attested by MNase digestion (Fig. 5B), and we observed aberrant transcription of gastrulation-specific genes (Szenker et al. 2012). Interestingly, rescue experiments using exogenous H3.2 variant failed to overcome the defects due to H3.3 loss (Fig. 5C). Thus, our findings reveal a critical developmental context wherein H3.3 exerts an essential, nonredundant function. Similar observations in mouse models corroborate the inability of replicative variants to substitute for the loss of the replacement variant during early development (Couldrey et al. 1999; Bush et al. 2013; Tang et al. 2015). This is in contrast with findings from *Tetrahymena thermophila* (Cui et al. 2006) and *D. melanogaster* (Sakai et al. 2009), in which the increased expression of the replicative histone variant can overcome the lack of H3.3 from the histone pool. Considering these discrepancies, we hypothesized that our results may be explained by the restricted ability of certain cognate chaperones to interact with only one histone variant, rather than any intrinsic properties of the incorporated variants. In order to test this hypothesis, rescue experiments conducted in *Xenopus* offer attractive possibilities to explore combinatorial mutations of distinct chaperone–variant pairs. Using this developmental system, we hope to decipher how the fine-tuned balance between brick quality and quantity and the choice made by a given architect underpin the homeostasis of functional chromatin.

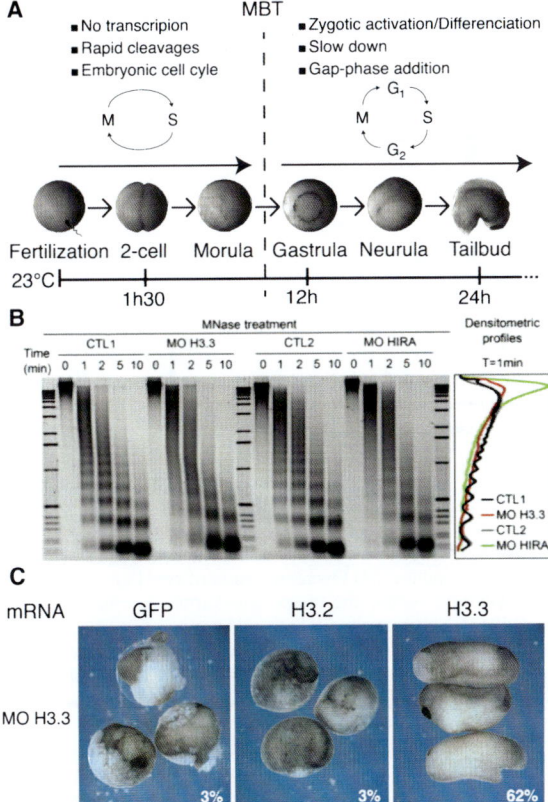

Figure 5. H3.3 plays a nonredundant role in *Xenopus* early development. (*A*) Well-characterized *Xenopus laevis* embryonic development provides a crucial window for studying chromatin. The embryo undergoes rapid synchronous cell divisions without any zygotic transcription until the midblastula transition (MBT). After the MBT, gap phases are introduced in the cell cycle along with the onset of zygotic transcription. (*B*) Chromatin disorganization upon depletion of a crucial brick (H3.3) or architect (HIRA). MNase digestion profile of H3.3 and HIRA morpholino (MO) injected embryos. Stage 14 embryos were used to prepare nuclei. For each time point of MNase digestion, purified DNA fragments were analyzed by agarose gel electrophoresis. Densitometric profiles of the 1-min digestion products are shown on the *right* (Szenker et al. 2012). (*C*) H3.2 expression does not rescue loss of H3.3 during *Xenopus* early development. H3.3 MO and indicated mRNAs were injected into two-cell stage embryos. Stage 25 embryos were used for image acquisition. Percentage viability of 30 embryos is shown. (Adapted from Szenker et al. 2012.)

CONCLUSION

In this review, we highlight two principal mechanisms that influence chromatin architecture and function: (1) dosage fluctuations of individual bricks and architects and (2) context-specific regulation of these factors. The nuclear abundance of most histone variants exceeds the concentration of their cognate histone chaperones. Underlying this stoichiometric disparity, cells maintain a basal equilibrium in the dosage of architects and bricks. We envisage that this principle could apply to wide-ranging events involved in numerous developmental changes and diseases. Nonetheless, future experimental perturbation of this delicate balance between variants and architects, at different loci and times, will help characterize the link to human pathology. In the face of potential disequilibrium, the chaperone–variant circuitry employs adaptive buffering mechanisms during specific cellular contexts. Our work on CENP-A and H3.3 modulation in cancer (Lacoste et al. 2014; Filipescu et al. 2017) and development (Szenker et al. 2012), respectively, illustrates a remarkable, in-built contingency plan within the system (Fig. 6).

Experimental overexpression of CENP-A can lead to ectopic deposition in the chromosome arms by the histone chaperone DAXX, instead of HJURP (Fig. 6A). The most critical situation is encountered when tumor cells lack p53 (Fig. 6B). Our model integrates CENP-A deposition by HJURP with the activity of the p53 checkpoint pathway whereby tumors become addicted to the high level of HJURP in the absence of p53. In this scheme, HJURP and CENP-A should be considered as major anticancer targets. A similar scenario likely plays out in several, albeit not all, cancers. Moving toward precision medicine, a perspective on interactions between histones chaperones and the canonical cancer drivers (oncogenes and tumor suppressors) could help identify cases of synthetic lethality and circumvent emergence of drug resistance.

The critical role of H3.3 variant during gastrulation in the tractable *Xenopus* development model (Fig. 6C) raises several fundamental and clinical issues. In light of the importance of variant–chaperone balance, it will be interesting to dissect H3.3 functions in the context of its chaperone interactions. It will also be important to address whether intrinsic properties of this specific variant determine its behavior in the chromatin. In the future, characterization of disease-associated mutants will provide insight into pathology of developmental diseases, such as pediatric cancers.

Here, we have shown the importance of chaperone–variant interactions in common cellular and developmental models. Further research will benefit from altering the chaperone–variant balance in other organismal systems, including additional human and mouse models. Extending these findings to other critical players such as Polycomb repressive complexes, regulators of cellular differentiation during development, will play a role in delineating a link with development and disease. Indeed, stem cell differentiation, neurogenesis and immune cell lineage commitment represent promising new directions to harness the insights from the current work on chaperone–variant interactions.

In conclusion, dosage regulation, specific histone variant characteristics, and chaperone choice prove to be integral elements of chromatin regulation. Combined together, these findings raise several unresolved questions: "When" (during cell cycle and development process), "where" (at distinct genomic landmarks and in specific cell types), and "how" (in a dose-dependent manner or under the influence of genotoxic stress such as DNA damage) do vital partner switches occur? Future studies will have to elaborate upon a unifying consideration that now surrounds the critical driving forces for shaping chromatin in the nucleus: is it the means of nucleosome assembly or the choice of histone variant that matters, or both? Ultimately, these properties will inform our understanding of the complex aspects underlying development and disease in different cellular contexts.

"All things are poison and nothing is without poison; only the dose makes a thing not a poison" Paracelsus

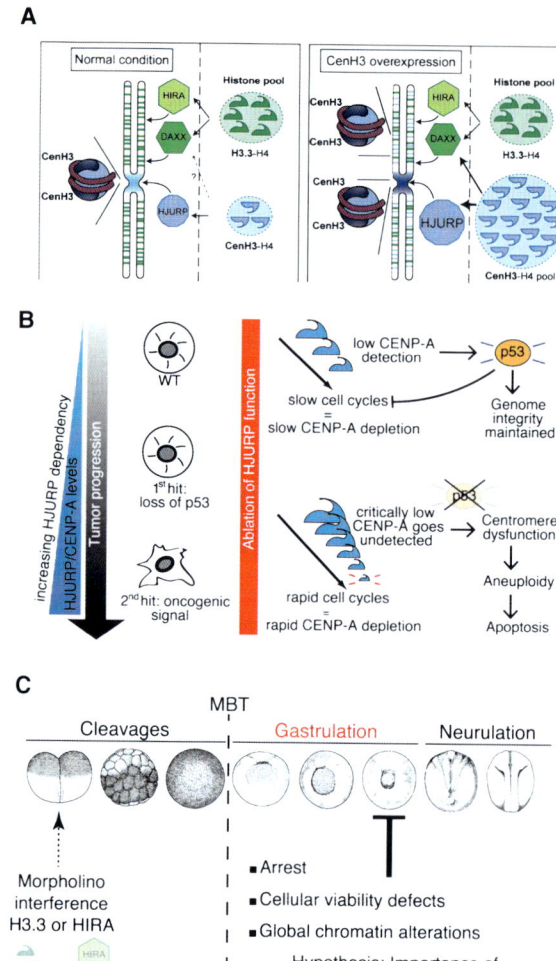

Figure 6. Dosage and context matter: important tenets of chromatin maintenance. (*A*) Overexpression of cenH3 results in partner switching within the chaperone–variant circuitry (Lacoste et al. 2014). Schematic model of cenH3 overexpression in human cells. In physiological conditions, cenH3-H4 is loaded at the centromere by HJURP. The H3.3–H4 pool is taken care of by DAXX or HIRA for loading in chromosome arms. When the pool of cenH3-H4 increases, HJURP is overexpressed and greater concentration of cenH3 is found at the centromere. The excess of cenH3 is handled by DAXX that loads or maintains it in chromosome arms. (*B*) Increasing levels of HJURP maintain tumorigenesis in cells lacking functional p53. The model integrates CENP-A deposition by HJURP with the activity of the p53 checkpoint pathway. There are different outcomes, following CENP-A or HJURP knockout, dependent on the p53 status. When cell cycle is slow and CENP-A levels are low in p53-proficient cells, genome integrity remains intact and is protected by the function of p53. Tumor growth arrest, due to loss of CENP-A at centromeres, in established p53-deficient tumors is striking and underlines a case of epigenetic-addiction to high levels of HJURP. If HJURP is depleted in this context, it leads to massive aneuploidy and apoptosis. Thus, tumors become addicted to the high level of HJURP in the absence of p53. (*C*) In *Xenopus laevis,* selective loss of H3.3 variant causes global chromatin dysregulation during early development (Szenker et al. 2012). Depletion of endogenous H3.3 impairs cellular viability leading to gastrulation defects. Histone chaperone HIRA knockdown gives rise to similar phenotypes, implicating the DNA synthesis-independent H3.3 incorporation pathway during development. (*A*, Adapted, with permission, from Lacoste et al. 2014; *B*, reprinted, with permission, from Filipescu et al. 2017.)

ACKNOWLEDGMENTS

We apologize to all our colleagues whose valuable work in this field was not cited because of space constraints. This work was supported by la Ligue Nationale contre le Cancer (Equipe labellisée Ligue), ANR-11-LABX-0044_DEEP and ANR-10-IDEX-0001-02 PSL, ANR-12-BSV5-0022-02 "CHAPINHIB," ANR-14-CE16-0009 "Epicure," ANR-14-CE10-0013 "CELLECTCHIP," EU project 678563 "EPOCH28," ERC-2015-ADG-694694 "ChromADICT," ANR-16-CE15-0018 "CHRODYT," ANR-16-CE12-0024 "CHIFT," "Parisian Alliance of Cancer Research Institutes." D.S. was supported by PSL.

REFERENCES

Adam S, Polo SE, Almouzni G. 2013. Transcription recovery after DNA damage requires chromatin priming by the H3.3 histone chaperone HIRA. *Cell* **155:** 94–106.

Ahmad K, Henikoff S. 2002. The histone variant H3.3 marks active chromatin by replication-independent nucleosome assembly. *Mol Cell* **9:** 1191–1200.

Allshire RC, Karpen GH. 2008. Epigenetic regulation of centromeric chromatin: Old dogs, new tricks? *Nat Rev Genet* **9:** 923–937.

Almouzni G, Mechali M. 1988. *Xenopus* egg extracts: A model system for chromatin replication. *Biochim Biophys Acta* **951:** 443–450.

Almouzni G, Clark DJ, Mechali M, Wolffe AP. 1990. Chromatin assembly on replicating DNA in vitro. *Nucleic Acids Res* **18:** 5767–5774.

Amato A, Schillaci T, Lentini L, Di Leonardo A. 2009. CENPA overexpression promotes genome instability in pRb-depleted human cells. *Mol Cancer* **8:** 119.

Amodeo AA, Jukam D, Straight AF, Skotheim JM. 2015. Histone titration against the genome sets the DNA-to-cytoplasm threshold for the *Xenopus* midblastula transition. *Proc Natl Acad Sci* **112:** E1086–E1095.

Banumathy G, Somaiah N, Zhang R, Tang Y, Hoffmann J, Andrake M, Ceulemans H, Schultz D, Marmorstein R, Adams PD. 2009. Human UBN1 is an ortholog of yeast Hpc2p and has an essential role in the HIRA/ASF1a chromatin-remodeling pathway in senescent cells. *Mol Cell Biol* **29:** 758–770.

Barsotti AM, Prives C. 2009. Pro-proliferative FoxM1 is a target of p53-mediated repression. *Oncogene* **28:** 4295–4305.

Benito JM, Godfrey L, Kojima K, Hogdal L, Wunderlich M, Geng H, Marzo I, Harutyunyan KG, Golfman L, North P, et al. 2015. MLL-rearranged acute lymphoblastic leukemias activate BCL-2 through H3K79 methylation and are sensitive to the BCL-2-specific antagonist ABT-199. *Cell Rep* **13:** 2715–2727.

Bieging KT, Mello SS, Attardi LD. 2014. Unravelling mechanisms of p53-mediated tumour suppression. *Nat Rev Cancer* **14:** 359–370.

Bird A. 2007. Perceptions of epigenetics. *Nature* **447:** 396–398.

Bodor DL, Valente LP, Mata JF, Black BE, Jansen LE. 2013. Assembly in G1 phase and long-term stability are unique intrinsic features of CENP-A nucleosomes. *Mol Biol Cell* **24:** 923–932.

Bodor DL, Mata JF, Sergeev M, David AF, Salimian KJ, Panchenko T, Cleveland DW, Black BE, Shah JV, Jansen LE. 2014. The quantitative architecture of centromeric chromatin. *Elife* **3:** e02137.

Bonnefoy E, Orsi GA, Couble P, Loppin B. 2007. The essential role of *Drosophila* HIRA for de novo assembly of paternal chromatin at fertilization. *PLoS Genet* **3:** 1991–2006.

Bowman GD, Poirier MG. 2014. Post-translational modifications of histones that influence nucleosome dynamics. *Chem Rev* **115:** 2274–2295.

Burglin TR, Mattaj IW, Newmeyer DD, Zeller R, De Robertis EM. 1987. Cloning of nucleoplasmin from *Xenopus laevis* oocytes and analysis of its developmental expression. *Genes Dev* **1:** 97–107.

Buschbeck M, Hake SB. 2017. Variants of core histones and their roles in cell fate decisions, development and cancer. *Nat Rev Mol Cell Biol* **18:** 299–314.

Bush KM, Yuen BT, Barrilleaux BL, Riggs JW, O'Geen H, Cotterman RF, Knoepfler PS. 2013. Endogenous mammalian histone H3.3 exhibits chromatin-related functions during development. *Epigenetics Chromatin* **6:** 7.

Cheloufi S, Elling U, Hopfgartner B, Jung YL, Murn J, Ninova M, Hubmann M, Badeaux AI, Euong Ang C, Tenen D, et al. 2015. The histone chaperone CAF-1 safeguards somatic cell identity. *Nature* **528:** 218–224.

Clement C, Almouzni G. 2015. MCM2 binding to histones H3-H4 and ASF1 supports a tetramer-to-dimer model for histone inheritance at the replication fork. *Nat Struct Mol Biol* **22:** 587–589.

Cook AJ, Gurard-Levin ZA, Vassias I, Almouzni G. 2011. A specific function for the histone chaperone NASP to fine-tune a reservoir of soluble H3-H4 in the histone supply chain. *Mol Cell* **44:** 918–927.

Corpet A, De Koning L, Toedling J, Savignoni A, Berger F, Lemaitre C, O'Sullivan RJ, Karlseder J, Barillot E, Asselain B, et al. 2011. Asf1b, the necessary Asf1 isoform for proliferation, is predictive of outcome in breast cancer. *EMBO J* **30:** 480–493.

Couldrey C, Carlton MBL, Nolan PM, Colledge WH, Evans MJ. 1999. A retroviral gene trap insertion into the histone 3.3A gene causes partial neonatal lethality, stunted growth, neuromuscular deficits and male sub-fertility in transgenic mice. *Hum Mol Genet* **8:** 2489–2495.

Cui B, Liu Y, Gorovsky MA. 2006. Deposition and function of histone H3 variants in *Tetrahymena thermophila*. *Mol Cell Biol* **26:** 7719–7730.

De Koning L, Corpet A, Haber JE, Almouzni G. 2007. Histone chaperones: An escort network regulating histone traffic. *Nat Struct Mol Biol* **14:** 997–1007.

Dion MF, Kaplan T, Kim M, Buratowski S, Friedman N, Rando OJ. 2007. Dynamics of replication-independent histone turnover in budding yeast. *Science* **315:** 1405–1408.

Drane P, Ouararhni K, Depaux A, Shuaib M, Hamiche A. 2010. The death-associated protein DAXX is a novel histone chaperone involved in the replication-independent deposition of H3.3. *Genes Dev* **24:** 1253–1265.

Duc C, Benoit M, Le Goff S, Simon L, Poulet A, Cotterell S, Tatout C, Probst AV. 2015. The histone chaperone complex HIR maintains nucleosome occupancy and counterbalances impaired histone deposition in CAF-1 complex mutants. *Plant J* **81:** 707–722.

Dumont JN. 1972. Oogenesis in *Xenopus laevis* (Daudin). I. Stages of oocyte development in laboratory maintained animals. *J Morphol* **136:** 153–179.

Dunleavy EM, Roche D, Tagami H, Lacoste N, Ray-Gallet D, Nakamura Y, Daigo Y, Nakatani Y, Almouzni-Pettinotti G. 2009. HJURP is a cell-cycle-dependent maintenance and deposition factor of CENP-A at centromeres. *Cell* **137:** 485–497.

Earnshaw WC, Rothfield N. 1985. Identification of a family of human centromere proteins using autoimmune sera from patients with scleroderma. *Chromosoma* **91:** 313–321.

Filipescu D, Naughtin M, Podsypanina K, Lejour V, Wilson L, Gurard-Levin ZA, Orsi GA, Simeonova I, Toufektchan E, Attardi LD, et al. 2017. Essential role for centromeric factors following p53 loss and oncogenic transformation. *Genes Dev* **31:** 463–480.

Finn RM, Ellard K, Eirin-Lopez JM, Ausio J. 2012. Vertebrate nucleoplasmin and NASP: Egg histone storage proteins with multiple chaperone activities. *FASEB J* **26:** 4788–4804.

Foltz DR, Jansen LE, Bailey AO, Yates JR III, Bassett EA, Wood S, Black BE, Cleveland DW. 2009. Centromere-specific assembly of CENP-A nucleosomes is mediated by HJURP. *Cell* **137:** 472–484.

Franklin SG, Zweidler A. 1977. Non-allelic variant of histone 2a, 2b and 3 in mammals. *Nature* **266**: 273–275.

Fromental-Ramain C, Ramain P, Hamiche A. 2017. The *Drosophila* DAXX-like protein (DLP) cooperates with ASF1 for H3.3 deposition and heterochromatin formation. *Mol Cell Biol* **37**: e00597-16.

Fukagawa T, Earnshaw WC. 2014. The centromere: Chromatin foundation for the kinetochore machinery. *Dev Cell* **30**: 496–508.

Gaillard PH, Martini EM, Kaufman PD, Stillman B, Moustacchi E, Almouzni G. 1996. Chromatin assembly coupled to DNA repair: A new role for chromatin assembly factor I. *Cell* **86**: 887–896.

Gibbons RJ, Picketts DJ, Villard L, Higgs DR. 1995. Mutations in a putative global transcriptional regulator cause X-linked mental retardation with α-thalassemia (ATR-X syndrome). *Cell* **80**: 837–845.

Goldberg AD, Banaszynski LA, Noh KM, Lewis PW, Elsaesser SJ, Stadler S, Dewell S, Law M, Guo X, Li X, et al. 2010. Distinct factors control histone variant H3.3 localization at specific genomic regions. *Cell* **140**: 678–691.

Gossett AJ, Lieb JD. 2012. In vivo effects of histone H3 depletion on nucleosome occupancy and position in *Saccharomyces cerevisiae*. *PLoS Genet* **8**: e1002771.

Groth A, Ray-Gallet D, Quivy JP, Lukas J, Bartek J, Almouzni G. 2005. Human Asf1 regulates the flow of S phase histones during replicational stress. *Mol Cell* **17**: 301–311.

Groth A, Corpet A, Cook AJ, Roche D, Bartek J, Lukas J, Almouzni G. 2007. Regulation of replication fork progression through histone supply and demand. *Science* **318**: 1928–1931.

Gu XM, Fu J, Feng XJ, Huang X, Wang SM, Chen XF, Zhu MH, Zhang SH. 2014. Expression and prognostic relevance of centromere protein A in primary osteosarcoma. *Pathol Res Pract* **210**: 228–233.

Gurard-Levin ZA, Almouzni G. 2014. Histone modifications and a choice of variant: A language that helps the genome express itself. *F1000Prime Rep* **6**: 76.

Gurard-Levin ZA, Quivy JP, Almouzni G. 2014. Histone chaperones: Assisting histone traffic and nucleosome dynamics. *Annu Rev Biochem* **83**: 487–517.

Gurard-Levin ZA, Wilson LO, Pancaldi V, Postel-Vinay S, Sousa FG, Reyes C, Marangoni E, Gentien D, Valencia A, Pommier Y, et al. 2016. Chromatin regulators as a guide for cancer treatment choice. *Mol Cancer Ther* **15**: 1768–1777.

Gurdon JB. 1962. The developmental capacity of nuclei taken from intestinal epithelium cells of feeding tadpoles. *J Embryol Exp Morphol* **10**: 622–640.

Gurdon JB. 1976. Injected nuclei in frog oocytes—Fate, enlargement, and chromatin dispersal. *J Embryol Exp Morphol* **36**: 523–540.

Hammond CM, Stromme CB, Huang H, Patel DJ, Groth A. 2017. Histone chaperone networks shaping chromatin function. *Nat Rev Mol Cell Biol* **18**: 141–158.

Heaphy CM, de Wilde RF, Jiao Y, Klein AP, Edil BH, Shi C, Bettegowda C, Rodriguez FJ, Eberhart CG, Hebbar S, et al. 2011. Altered telomeres in tumors with ATRX and DAXX mutations. *Science* **333**: 425.

Heo JI, Cho JH, Kim JR. 2013. HJURP regulates cellular senescence in human fibroblasts and endothelial cells via a p53-dependent pathway. *J Gerontol A Biol Sci Med Sci* **68**: 914–925.

Hodl M, Basler K. 2009. Transcription in the absence of histone H3.3. *Curr Biol* **19**: 1221–1226.

Houlard M, Berlivet S, Probst AV, Quivy JP, Hery P, Almouzni G, Gerard M. 2006. CAF-1 is essential for heterochromatin organization in pluripotent embryonic cells. *PLoS Genet* **2**: e181.

Huang H, Stromme CB, Saredi G, Hodl M, Strandsby A, Gonzalez-Aguilera C, Chen S, Groth A, Patel DJ. 2015. A unique binding mode enables MCM2 to chaperone histones H3–H4 at replication forks. *Nat Struct Mol Biol* **22**: 618–626.

Ishiuchi T, Enriquez-Gasca R, Mizutani E, Boskovic A, Ziegler-Birling C, Rodriguez-Terrones D, Wakayama T, Vaquerizas JM, Torres-Padilla ME. 2015. Early embryonic-like cells are induced by downregulating replication-dependent chromatin assembly. *Nat Struct Mol Biol* **22**: 662–671.

Jacob Y, Bergamin E, Donoghue MT, Mongeon V, LeBlanc C, Voigt P, Underwood CJ, Brunzelle JS, Michaels SD, Reinberg D, et al. 2014. Selective methylation of histone H3 variant H3.1 regulates heterochromatin replication. *Science* **343**: 1249–1253.

Jamai A, Imoberdorf RM, Strubin M. 2007. Continuous histone H2B and transcription-dependent histone H3 exchange in yeast cells outside of replication. *Mol Cell* **25**: 345–355.

Jansen LE, Black BE, Foltz DR, Cleveland DW. 2007. Propagation of centromeric chromatin requires exit from mitosis. *J Cell Biol* **176**: 795–805.

Jiang D, Berger F. 2017. DNA replication-coupled histone modification maintains *Polycomb* gene silencing in plants. *Science* **357**: 1146–1149.

Jiao Y, Shi C, Edil BH, de Wilde RF, Klimstra DS, Maitra A, Schulick RD, Tang LH, Wolfgang CL, Choti MA, et al. 2011. DAXX/ATRX, MEN1, and mTOR pathway genes are frequently altered in pancreatic neuroendocrine tumors. *Science* **331**: 1199–1203.

Jullien J, Astrand C, Szenker E, Garrett N, Almouzni G, Gurdon JB. 2012. HIRA dependent H3.3 deposition is required for transcriptional reprogramming following nuclear transfer to *Xenopus* oocytes. *Epigenetics Chromatin* **5**: 17.

Katagiri C, Ohsumi K. 1994. Remodeling of sperm chromatin induced in egg extracts of amphibians. *Int J Dev Biol* **38**: 209–216.

Kato T, Sato N, Hayama S, Yamabuki T, Ito T, Miyamoto M, Kondo S, Nakamura Y, Daigo Y. 2007. Activation of Holliday junction recognizing protein involved in the chromosomal stability and immortality of cancer cells. *Cancer Res* **67**: 8544–8553.

Kaufman PD, Cohen JL, Osley MA. 1998. Hir proteins are required for position-dependent gene silencing in *Saccharomyces cerevisiae* in the absence of chromatin assembly factor I. *Mol Cell Biol* **18**:8 4793–4806.

Keller R. 1991. Early embryonic development of *Xenopus laevis*. *Methods Cell Biol* **36**: 61–113.

Kornberg RD. 1974. Chromatin structure: A repeating unit of histones and DNA. *Science* **184**: 868–871.

Lacoste N, Woolfe A, Tachiwana H, Villar Garea A, Barth T, Cantaloube S, Kurumizaka H, Imhof A, Almouzni G. 2014. Mislocalization of the centromeric histone variant CenH3/CENP-A in human cells depends on the chaperone DAXX. *Mol Cell* **53**: 631–644.

Lamour V, Lécluse Y, Desmaze C, Spector M, Bodescot M, Aurias A, Osley MA, Lipinski M. 1995. A human homolog of the *S. cerevisiae* HIR1 and HIR2 transcriptional repressors cloned from the DiGeorge syndrome critical region. *Hum Mol Genet* **4**: 791–799.

Laskey RA, Mills AD, Morris NR. 1977. Assembly of SV40 chromatin in a cell-free system from *Xenopus* eggs. *Cell* **10**: 237–243.

Laskey RA, Honda BM, Mills AD, Finch JT. 1978. Nucleosomes are assembled by an acidic protein which binds histones and transfers them to DNA. *Nature* **275**: 416–420.

Lewis PW, Elsaesser SJ, Noh KM, Stadler SC, Allis CD. 2010. Daxx is an H3.3-specific histone chaperone and cooperates with ATRX in replication-independent chromatin assembly at telomeres. *Proc Natl Acad Sci* **107**: 14075–14080.

Lewis PW, Muller MM, Koletsky MS, Cordero F, Lin S, Banaszynski LA, Garcia BA, Muir TW, Becher OJ, Allis CD. 2013. Inhibition of PRC2 activity by a gain-of-function H3 mutation found in pediatric glioblastoma. *Science* **340**: 857–861.

Li Y, Zhu Z, Zhang S, Yu D, Yu H, Liu L, Cao X, Wang L, Gao H, Zhu M. 2011. ShRNA-targeted centromere protein A inhibits hepatocellular carcinoma growth. *PLoS One* **6**: e17794.

Litvin J, King ML. 1988. Expression and segregation of nucleoplasmin during development in *Xenopus*. *Development* **102**: 9–21.

Loppin B, Bonnefoy E, Anselme C, Laurencon A, Karr TL, Couble P. 2005. The histone H3.3 chaperone HIRA is essential for chromatin assembly in the male pronucleus. *Nature* **437:** 1386–1390.

Loyola A, Bonaldi T, Roche D, Imhof A, Almouzni G. 2006. PTMs on H3 variants before chromatin assembly potentiate their final epigenetic state. *Mol Cell* **24:** 309–316.

Luger K, Mäder AW, Richmond RK, Sargent DF, Richmond TJ. 1997. Crystal structure of the nucleosome core particle at 2.8 Å resolution. *Nature* **389:** 251–560.

Ma XJ, Salunga R, Tuggle JT, Gaudet J, Enright E, McQuary P, Payette T, Pistone M, Stecker K, Zhang BM, et al. 2003. Gene expression profiles of human breast cancer progression. *Proc Natl Acad Sci* **100:** 5974–5979.

MacAlpine DM, Almouzni G. 2013. Chromatin and DNA replication. *Cold Spring Harb Perspect Biol* **5:** a010207.

Maehara K, Takahashi K, Saitoh S. 2010. CENP-A reduction induces a p53-dependent cellular senescence response to protect cells from executing defective mitoses. *Mol Cell Biol* **30:** 2090–2104.

Maze I, Noh KM, Soshnev AA, Allis CD. 2014. Every amino acid matters: Essential contributions of histone variants to mammalian development and disease. *Nat Rev Genet* **15:** 259–271.

McGhee JD, Felsenfeld G. 1980. Nucleosome structure. *Annu Rev Biochem* **49:** 1115–1156.

McKinley KL, Cheeseman IM. 2016. The molecular basis for centromere identity and function. *Nat Rev Mol Cell Biol* **17:** 16–29.

Mello JA, Silljé HHW, Roche D, Kirschner DB, Nigg EA, Almouzni G. 2002. Human Asf1 and CAF-1 interact and synergize in a repair-coupled nucleosome assembly pathway. *EMBO Rep* **3:** 329–334.

Miao W, Xiong J, Bowen J, Wang W, Liu Y, Braguinets O, Grigull J, Pearlman RE, Orias E, Gorovsky MA. 2009. Microarray analyses of gene expression during the *Tetrahymena thermophila* life cycle. *PLoS One* **4:** e4429.

Michaelson JS, Bader D, Kuo F, Kozak C, Leder P. 1999. Loss of Daxx, a promiscuously interacting protein, results in extensive apoptosis in early mouse development. *Genes Dev* **13:** 1918–1923.

Montes de Oca R, Gurard-Levin ZA, Berger F, Rehman H, Martel E, Corpet A, de Koning L, Vassias I, Wilson LO, Meseure D, et al. 2015. The histone chaperone HJURP is a new independent prognostic marker for luminal A breast carcinoma. *Mol Oncol* **9:** 657–674.

Morel D, Almouzni G, Soria JC, Postel-Vinay S. 2017. Targeting chromatin defects in selected solid tumors based on oncogene addiction, synthetic lethality and epigenetic antagonism. *Ann Oncol* **28:** 254–269.

Muller S, Almouzni G. 2017. Chromatin dynamics during the cell cycle at centromeres. *Nat Rev Genet* **18:** 192–208.

Munakata T, Adachi N, Yokoyama N, Kuzuhara T, Horikoshi M. 2000. A human homologue of yeast anti-silencing factor has histone chaperone activity. *Genes Cells* **5:** 221–233.

Naumova N, Imakaev M, Fudenberg G, Zhan Y, Lajoie BR, Mirny LA, Dekker J. 2013. Organization of the mitotic chromosome. *Science* **342:** 948–953.

Nechemia-Arbely Y, Fachinetti D, Miga KH, Sekulic N, Soni GV, Kim DH, Wong AK, Lee AY, Nguyen K, Dekker C, et al. 2017. Human centromeric CENP-A chromatin is a homotypic, octameric nucleosome at all cell cycle points. *J Cell Biol* **216:** 607–621.

Newport J, Kirschner M. 1982. A major developmental transition in early *Xenopus* embryos: I. Characterization and timing of cellular changes at the midblastula stage. *Cell* **30:** 675–686.

Ng RK, Gurdon JB. 2008. Epigenetic memory of an active gene state depends on histone H3.3 incorporation into chromatin in the absence of transcription. *Nat Cell Biol* **10:** 102–109.

Nieuwkoop PD, Faber J, eds. 1994. *Normal table of Xenopus laevis (Daudin): A systematic and chronological survey of the development from the fertilized egg till the end of metamorphosis.* Garland, New York.

Nora EP, Lajoie BR, Schulz EG, Giorgetti L, Okamoto I, Servant N, Piolot T, van Berkum NL, Meisig J, Sedat J, et al. 2012. Spatial partitioning of the regulatory landscape of the X-inactivation centre. *Nature* **485:** 381–385.

Olins AD, Olins DE. 1974. Spheroid chromatin units (v bodies). *Science* **183:** 330–333.

Orsi GA, Algazeery A, Meyer RE, Capri M, Sapey-Triomphe LM, Horard B, Gruffat H, Couble P, Ait-Ahmed O, Loppin B. 2013. *Drosophila* yemanuclein and HIRA cooperate for de novo assembly of H3.3-containing nucleosomes in the male pronucleus. *PLoS Genet* **9:** e1003285.

Osley MA. 1991. The regulation of histone synthesis in the cell cycle. *Annu Rev Biochem* **60:** 827–861.

O'Sullivan RJ, Almouzni G. 2014. Assembly of telomeric chromatin to create ALTernative endings. *Trends Cell Biol* **24:** 675–685.

Oudet P, Gross-Bellard M, Chambon P. 1975. Electron microscopic and biochemical evidence that chromatin structure is a repeating unit. *Cell* **4:** 281–300.

Palmer DK, O'Day K, Trong HL, Charbonneau H, Margolis RL. 1991. Purification of the centromere-specific protein CENP-A and demonstration that it is a distinctive histone. *Proc Natl Acad Sci* **88:** 3734–3738.

Piazzesi A, Papić D, Bertan F, Salomoni P, Nicotera P, Bano D. 2016. Replication-independent histone variant H3.3 controls animal lifespan through the regulation of pro-longevity transcriptional programs. *Cell Rep* **17:** 987–996.

Polo SE, Roche D, Almouzni G. 2006. New histone incorporation marks sites of UV repair in human cells. *Cell* **127:** 481–493.

Probst AV, Dunleavy E, Almouzni G. 2009. Epigenetic inheritance during the cell cycle. *Nat Rev Mol Cell Biol* **10:** 192–206.

Qian Z, Huang H, Hong JY, Burck CL, Johnston SD, Berman J, Carol A, Liebman SW. 1998. Yeast Ty1 retrotransposition is stimulated by a synergistic interaction between mutations in chromatin assembly factor I and histone regulatory proteins. *Mol Cell Biol* **18:** 4783–4792.

Qiu JJ, Guo JJ, Lv TJ, Jin HY, Ding JX, Feng WW, Zhang Y, Hua KQ. 2013. Prognostic value of centromere protein-A expression in patients with epithelial ovarian cancer. *Tumour Biol* **34:** 2971–2975.

Quivy JP, Grandi P, Almouzni G. 2001. Dimerization of the largest subunit of chromatin assembly factor 1—Importance in vitro and during *Xenopus* early development. *EMBO J* **20:** 2015–2027.

Ray-Gallet D, Quivy JP, Scamps C, Martini EM, Lipinski M, Almouzni G. 2002. HIRA is critical for a nucleosome assembly pathway independent of DNA synthesis. *Mol Cell* **9:** 1091–1100.

Ray-Gallet D, Woolfe A, Vassias I, Pellentz C, Lacoste N, Puri A, Schultz DC, Pchelintsev NA, Adams PD, Jansen LE, et al. 2011. Dynamics of histone H3 deposition in vivo reveal a nucleosome gap-filling mechanism for H3.3 to maintain chromatin integrity. *Mol Cell* **44:** 928–941.

Richardson RT, Batova IN, Widgren EE, Zheng LX, Whitfield M, Marzluff WF, O'Rand MG. 2000. Characterization of the histone H1-binding protein, NASP, as a cell cycle–regulated somatic protein. *J Biol Chem* **275:** 30378–30386.

Richet N, Liu D, Legrand P, Velours C, Corpet A, Gaubert A, Bakail M, Moal-Raisin G, Guerois R, Compper C, et al. 2015. Structural insight into how the human helicase subunit MCM2 may act as a histone chaperone together with ASF1 at the replication fork. *Nucleic Acids Res* **43:** 1905–1917.

Ricketts MD, Marmorstein R. 2017. A molecular prospective for HIRA complex assembly and H3.3-specific histone chaperone function. *J Mol Biol* **429:** 1924–1933.

Ricketts DM, Frederick B, Hoff H, Tang Y, Schultz DC, Singh Rai T, Grazia Vizioli M, Adams PD, Marmorstein R. 2015. Ubinuclein-1 confers histone H3.3-specific-binding by the HIRA histone chaperone complex. *Nat Commun* **6:** 7711.

Roberts C, Sutherland HF, Farmer H, Kimber W, Halford S, Carey A, Brickman JM, Wynshaw-Boris A, Scambler PJ. 2002. Targeted mutagenesis of the *Hira* gene results in gastrulation defects and patterning abnormalities of mesoendodermal derivatives prior to early embryonic lethality. *Mol Cell Biol* 22: 2318–2328.

Rufiange A, Jacques PE, Bhat W, Robert F, Nourani A. 2007. Genome-wide replication-independent histone H3 exchange occurs predominantly at promoters and implicates H3 K56 acetylation and Asf1. *Mol Cell* 27: 393–405.

Sakai A, Schwartz BE, Goldstein S, Ahmad K. 2009. Transcriptional and developmental functions of the H3.3 histone variant in *Drosophila*. *Curr Biol* 19: 1816–1820.

Santenard A, Ziegler-Birling C, Koch M, Tora L, Bannister AJ, Torres-Padilla ME. 2010. Heterochromatin formation in the mouse embryo requires critical residues of the histone variant H3.3. *Nat Cell Biol* 12: 853–862.

Schwartz BE, Ahmad K. 2005. Transcriptional activation triggers deposition and removal of the histone variant H3.3. *Genes Dev* 19: 804–814.

Schwartzentruber J, Korshunov A, Liu XY, Jones DT, Pfaff E, Jacob K, Sturm D, Fontebasso AM, Quang DA, Tonjes M, et al. 2012. Driver mutations in histone H3.3 and chromatin remodelling genes in paediatric glioblastoma. *Nature* 482: 226–231.

Shamseldin HE, Tulbah M, Kurdi W, Nemer M, Alsahan N, Al Mardawi E, Khalifa O, Hashem A, Kurdi A, Babay Z, et al. 2015. Identification of embryonic lethal genes in humans by autozygosity mapping and exome sequencing in consanguineous families. *Genome Biol* 16: 116.

Sharp JA, Fouts ET, Krawitz DC, Kaufman PD. 2001. Yeast histone deposition protein Asf1p requires Hir proteins and PCNA for heterochromatic silencing. *Curr Biol* 11: 463–473.

Shelby RD, Monier K, Sullivan KF. 2000. Chromatin assembly at kinetochores is uncoupled from DNA replication. *J Cell Biol* 151: 1113–1118.

Shrestha RL, Ahn GS, Staples MI, Sathyan KM, Karpova TS, Foltz DR, Basrai MA. 2017. Mislocalization of centromeric histone H3 variant CENP-A contributes to chromosomal instability (CIN) in human cells. *Oncotarget* 8: 46781–46800.

Shuaib M, Ouararhni K, Dimitrov S, Hamiche A. 2010. HJURP binds CENP-A via a highly conserved N-terminal domain and mediates its deposition at centromeres. *Proc Natl Acad Sci* 107: 1349–1354.

Smith S, Stillman B. 1989. Purification and characterization of CAF-I, a human cell factor required for chromatin assembly during DNA replication in vitro. *Cell* 58: 15–25.

Stillman B. 1986. Chromatin assembly during SV40 DNA replication in vitro. *Cell* 45: 555–565.

Sun X, Clermont PL, Jiao W, Helgason CD, Gout PW, Wang Y, Qu S. 2016. Elevated expression of the centromere protein-A (CENP-A)-encoding gene as a prognostic and predictive biomarker in human cancers. *Int J Cancer* 139: 899–907.

Swygert SG, Peterson CL. 2014. Chromatin dynamics: Interplay between remodeling enzymes and histone modifications. *Biochim Biophys Acta* 1839: 728–736.

Szenker E, Lacoste N, Almouzni G. 2012. A developmental requirement for HIRA-dependent H3.3 deposition revealed at gastrulation in *Xenopus*. *Cell Rep* 1: 730–740.

Tagami H, Ray-Gallet D, Almouzni G, Nakatani Y. 2004. Histone H3.1 and H3.3 complexes mediate nucleosome assembly pathways dependent or independent of DNA synthesis. *Cell* 116: 51–61.

Takahashi K, Yamanaka S. 2006. Induction of pluripotent stem cells from mouse embryonic and adult fibroblast cultures by defined factors. *Cell* 126: 663–676.

Talbert PB, Henikoff S. 2017. Histone variants on the move: Substrates for chromatin dynamics. *Nat Rev Mol Cell Biol* 18: 115–126.

Tang MC, Jacobs SA, Mattiske DM, Soh YM, Graham AN, Tran A, Lim SL, Hudson DF, Kalitsis P, O'Bryan MK, et al. 2015. Contribution of the two genes encoding histone variant H3.3 to viability and fertility in mice. *PLoS Genet* 11: e1004964.

Tomonaga T, Matsushita K, Yamaguchi S, Oohashi T, Shimada H, Ochiai T, Yoda K, Nomura F. 2003. Overexpression and mistargeting of centromere protein-A in human primary colorectal cancer. *Cancer Res* 63: 3511–3516.

Torres-Padilla ME, Bannister AJ, Hurd PJ, Kouzarides T, Zernicka-Goetz M. 2006. Dynamic distribution of the replacement histone variant H3.3 in the mouse oocyte and preimplantation embryos. *Int J Dev Biol* 50: 455–461.

Tyler JK, Adams CR, Chen SR, Kobayashi R, Kamakaka RT, Kadonaga JT. 1999. The RCAF complex mediates chromatin assembly during DNA replication and repair. *Nature* 402: 555–560.

Valente V, Serafim RB, de Oliveira LC, Adorni FS, Torrieri R, Tirapelli DP, Espreafico EM, Oba-Shinjo SM, Marie SK, Paco-Larson ML, et al. 2013. Modulation of HJURP (Holliday Junction-Recognizing Protein) levels is correlated with glioblastoma cells survival. *PLoS One* 8: e62200.

Waddington CH. 1957. *The strategy of the genes: A discussion of some aspects of theoretical biology.* Routledge, Taylor & Francis Group, ©Allen and Unwin, New York.

Waterborg JH. 2012. Evolution of histone H3: Emergence of variants and conservation of post-translational modification sites. *Biochem Cell Biol* 90: 79–95.

Welch JE, Zimmerman LJ, Joseph DR, O'Rand MG. 1990. Characterization of a sperm-specific nuclear autoantigenic protein. I. Complete sequence and homology with the *Xenopus* protein, N1/N2. *Biol Reprod* 43: 559–568.

Wilming LG, Snoeren CAS, van Rijswijk A, Grosveld F, Meijers C. 1997. The murine homologue of *HIRA*, a DiGeorge syndrome candidate gene, is expressed in embryonic structures affected in CATCH22 patients. *Hum Mol Genet* 6: 247–258.

Wollmann H, Holec S, Alden K, Clarke ND, Jacques PE, Berger F. 2012. Dynamic deposition of histone variant H3.3 accompanies developmental remodeling of the *Arabidopsis* transcriptome. *PLoS Genet* 8: e1002658.

Wollmann H, Stroud H, Yelagandula R, Tarutani Y, Jiang D, Jing L, Jamge B, Takeuchi H, Holec S, Nie X, et al. 2017. The histone H3 variant H3.3 regulates gene body DNA methylation in *Arabidopsis thaliana*. *Genome Biol* 18: 94.

Wu G, Broniscer A, McEachron TA, Lu C, Paugh BS, Becksfort J, Qu C, Ding L, Huether R, Parker M, et al. 2012. Somatic histone H3 alterations in pediatric diffuse intrinsic pontine gliomas and non-brainstem glioblastomas. *Nat Genet* 44: 251–253.

Yang X, Khosravi-Far R, Chang HY, Baltimore D. 1997. Daxx, a novel Fas-binding protein that activates JNK and apoptosis. *Cell* 89: 1067–1076.

Yang BX, El Farran CA, Guo HC, Yu T, Fang HT, Wang HF, Schlesinger S, Seah YF, Goh GY, Neo SP, et al. 2015. Systematic identification of factors for provirus silencing in embryonic stem cells. *Cell* 163: 230–245.

Zhu J, Sammons MA, Donahue G, Dou Z, Vedadi M, Getlik M, Barsyte-Lovejoy D, Al-awar R, Katona BW, Shilatifard A, et al. 2015. Gain-of-function p53 mutants co-opt chromatin pathways to drive cancer growth. *Nature* 525: 206–211.

The Importance of Satellite Sequence Repression for Genome Stability

PETER ZELLER[1,2] AND SUSAN M. GASSER[1,2]

[1]*Friedrich Miescher Institute for Biomedical Research, CH-4058 Basel, Switzerland*
[2]*Faculty of Natural Sciences, University of Basel, CH-4056 Basel, Switzerland*
Correspondence: susan.gasser@fmi.ch

Up to two-thirds of eukaryotic genomes consist of repetitive sequences, which include both transposable elements and tandemly arranged simple or satellite repeats. Whereas extensive progress has been made toward understanding the danger of and control over transposon expression, only recently has it been recognized that DNA damage can arise from satellite sequence transcription. Although the structural role of satellite repeats in kinetochore function and end protection has long been appreciated, it has now become clear that it is not only these functions that are compromised by elevated levels of transcription. RNA from simple repeat sequences can compromise replication fork stability and genome integrity, thus compromising germline viability. Here we summarize recent discoveries on how cells control the transcription of repeat sequence and the dangers that arise from their expression. We propose that the link between the DNA damage response and the transcriptional silencing machinery may help a cell or organism recognize foreign DNA insertions into an evolving genome.

In eukaryotic organisms, DNA is packaged by proteins into a structure called chromatin that influences genomic interactions with the transcription machinery. Chromatin structure is modulated by the accumulation of covalent modifications to the DNA itself, to histones and nonhistone chromatin factors, and by protein composition. Essentially it is the accessibility of a DNA sequence to the transcription machinery that regulates gene expression, although additional regulation can also be imposed co- and posttranscriptionally (Tippmann et al. 2012).

The main DNA modification in vertebrate genomes is methylation on the relatively underrepresented dinucleotide, CpG, producing ^{5me}CpG or 5-methylcytosine. Although much of the mammalian genome carries this modification constitutively, it is regulated at promoters that carry CpG clusters (CpG islands), as well as over the gene bodies (Weber et al. 2005). At both sites, ^{5me}CpG is usually associated with transcriptional inhibition (Razin and Riggs 1980). It silences either by recruiting ^{5me}C-binding factors (Jones et al. 1998; Bird and Wolffe 1999) or by disabling transcription factor binding sites (e.g., at CpG islands [Watt and Molloy 1988; Bell and Felsenfeld 2000]).

In addition to DNA modification, posttranslational histone tail modifications are especially well-studied. Depending on the nature and position of the modification on the histone, they can either work on nucleosome–nucleosome or nucleosome–DNA interactions by changing the charge of the highly basic histone tail or by generating specific binding sites for proteins that recognize modified lysine or arginine residues. The range of characterized chemical modifications on histones has been expanding steadily, with the most common being lysine methylation, acetylation, ubiquitination, SUMOylation, and ribosylation or serine or threonine phosphorylation. Different chromatin states are defined by the combination of histone modifications that some considered to be an instructive "histone code" (Strahl and Allis 2000). The proteins that recognize covalent histone modifications often act in *trans* by promoting or inhibiting the recruitment of additional regulators either of transcription or chromatin compaction. Here as well, there is a growing list of motifs that characterize the readers of specific histone modifications (Taverna et al. 2007).

HETEROCHROMATIN

Early microscopy experiments in moss by Heitz in 1928 distinguished two chromatin "states" in the interphase nucleus. Heterochromatic regions at the nuclear envelope and around the nucleolus stained strongly during the whole cell cycle, which was interpreted as a constant high level of compaction, whereas euchromatic regions stained strongly only during mitosis, suggesting that they "unfold" in interphase. This observation was the foundation for the model that the higher-order packaging in heterochromatic regions could be refractory for the binding of the transcription machinery. Later, heterochromatic regions were further separated into constitutive and facultative heterochromatin, often correlated with either methyl-K9 (Noma et al. 2001; Schotta et al. 2002) or methyl-K27 on histone H3, respectively (Bernstein et al. 2006; Kalantry et al. 2006). Detailed chromatin immunoprecipitation (ChIP) experiments mapped H3K9 methylation to both the centromere-flanking satellite repeats and interspersed repetitive elements (REs) throughout the genome in all cells, thus defining constitutive heterochromatin (Pimpinelli et al. 1995; Gerstein et al. 2010; Liu et al.

Published by Cold Spring Harbor Laboratory Press; doi: 10.1101/sqb.2017.82.033662

Cold Spring Harbor Symposia on Quantitative Biology, Volume LXXXII

2011). Genome-wide comparisons between transcriptional activity and the presence of histone modifications showed a clear correlation of gene expression with euchromatin, whereas repressed cell type–specific genes were often marked with H3K27me3.

Two main approaches have been used to examine if chromatin composition can indeed regulate access to the underlying DNA sequence. One method was to expose chromatin to exogenously added DNase1. In early experiments using isolated nuclei, DNase1 showed a preferential digestion of the actively transcribed albumin gene in liver tissue (Weintraub and Groudine 1976). Combining this approach with whole genome sequencing, it became clear that this method primarily identifies nucleosome-free regions, particularly at the enhancers and transcription start sites of active genes (Boyle et al. 2008). Outside of these two classes of elements, the sensitivity of this approach turned out to be rather limited, as it did not clearly distinguish euchromatin and heterochromatin. The second approach was based on the expression of an *E. coli* DNA methyl transferase (DAM) in intact cells, whose modification on adenine can be quantified as the degree of protection against the methyl-sensitive restriction enzyme *Dpn*I. Once again, in a genome-wide study, Bell et al. (2010) were able to show a small reduction in DAM methylation over H3K27me3 regions, but no difference between H3K9-methylated regions and unexpressed euchromatic loci could be observed. One caveat of these techniques may be that the chromatin compaction induced by heterochromatin does not interfere strongly with the temporal interaction of a single protein but rather hinders the assembly of multiprotein complexes, such as the general transcription machinery. Indeed, these methods primarily mapped nucleosome density, which may not represent the organizational level affected by heterochromatin. Recent work suggests that nucleosome turnover rates may be the more important criterion that distinguishes euchromatin from heterochromatin (Taneja et al. 2017), given that local histone turnover is greatly reduced in heterochromatic domains (Aygün et al. 2013; Toyama et al. 2013). This could either reflect a reduced action of nucleosome remodelers or of demethylating enzymes that remove the repressive H3K9me mark, such as KDM4b (Tsurumi et al. 2013).

THE TWO-FACED ROLE OF HETEROCHROMATIN IN DNA DAMAGE AND REPAIR

Not only transcription, but DNA damage as well, occurs in the context of chromatin. Intriguingly, during the DNA damage response, heterochromatin seems to play both positive and negative roles. On one hand, nucleosomes are thought to be an obstacle for the DNA damage repair machinery and thus must be removed. Indeed, following UV damage ubiquitination-mediated histone mobilization has been reported (Wang et al. 2006; Lan et al. 2012; Adam et al. 2013) and in the case of DNA double-strand breaks (DSBs) in yeast, histone ChIP experiments showed a local depletion around an induced DSB (van Attikum et al.

2004, 2007), as well as a global loss of histones at high levels of Zeocin- or γIR-induced DNA damage (Hauer et al. 2017). Local histone release was shown to depend on the activity of nucleosome remodelers BRG1/RSC and/or INO80 (van Attikum et al. 2004, 2007; Zhao et al. 2009; Jiang et al. 2010). On the other hand, multiple repressive factors, including HP1 (Luijsterburg et al. 2009), Polycomb (Hong et al. 2008), and HDAC1/2 (Miller et al. 2010), were reported to be recruited to sites of DNA damage. Consistently, animals lacking HP1 (Luijsterburg et al. 2009) or Polycomb components (Hong et al. 2008) are hypersensitive to genotoxic stress. Besides a potential role of these factors in the recruitment of the repair machinery, they are also thought to promote silencing around sites of damage to prevent conflicts or interference between the repair and transcription machineries (Ui et al. 2015; Vissers et al. 2012) reviewed in Polo (2017).

An interesting twist to this role was nicely shown in *Drosophila*. In the absence of the fly heterochromatin components Su(var)3-9 and HP1α, an increase of spontaneous RAD51 foci was scored specifically in DAPI-dense regions of the nucleus, suggesting that heterochromatin itself might prevent spontaneous DNA DSBs (Peng and Karpen 2009; Chiolo et al. 2011). This effect is thought to stem from the loss of transcriptional repression coincident with loss of H3K9me and its reader. One of the main sources of DSBs in eukaryotic cells in the absence of exogenous insult is the stalling of replication forks. Moreover, it has been clear for many years that transcription itself can be a major impediment to replication fork progression (Brewer 1988; French 1992; Liu and Alberts 1995). Thus, to avoid collisions of the replication fork with the transcription machinery, cells coordinate these two nuclear events. This is achieved in part by delaying the firing of certain origins of replication; transcriptionally active sites were shown to replicate early in S phase, whereas transcriptional silent sites were replicated late (Schübeler et al. 2002; Rivera et al. 2014).

The importance of this coordination of transcription and replication was elegantly shown in a study that identified DNA breaks by ligating sequencing adapters onto the open ends of unfragmented DNA (Break-seq). The authors could show that fragile sites occurred most frequently at sites where replication and transcription coincide. The shift of collision points by perturbing replication timing or by inducing unscheduled transcription resulted in a corresponding change in position or intensity of the breaks (Hoffman et al. 2015). At the longest human genes, such conflicts seem to be impossible to avoid, given that a full-length transcript takes more than one cell cycle in some cases, and this in turn leads to the formation of fragile sites that break in a transcription-dependent manner (Helmrich et al. 2011).

RNA:DNA HYBRIDS: TRANSCRIPTION INTERMEDIATES IMPAIR REPLICATION FORK PROGRESSION

One characteristic feature that identifies sites of conflict between the replication and transcription machineries is the formation of RNA:DNA hybrids, also called R-loops. These hybrid nucleic acid structures result from the displacement of the second DNA strand by a transcribed RNA. Studies in yeast and mammalian cells have mapped RNA:DNA hybrid accumulation to highly transcribed genes (Wahba et al. 2016). Studies performed to elucidate the mechanisms that protect cells from RNA:DNA hybrid accumulation led to the identification of factors involved in transcriptional processivity, as well as factors known to be involved in the restart of collapsed forks (Santos-Pereira and Aguilera 2015).

The dangers posed by RNA:DNA hybrid formation was first shown in cells depleted for certain RNA biogenesis and processing factors, such as the THO complex in yeast (Huertas and Aguilera 2003) and *Caenorhabditis elegans* (Castellano-Pozo et al. 2012), or the serine/arginine-rich splicing factor 1 (SRSF1; previously known as ASF and SF2) in vertebrates (Li and Manley 2005). Additionally, there is evidence for a replication independent role of R-loops in generating DNA breaks, as the nucleotide excision repair (NER) nucleases XPG and XPF were shown to be able to process R-loops into DSBs in some situations (Sollier et al. 2014). Recently, our laboratory found that the absence of H3K9me leads to an accumulation of RNA:DNA hybrids in *C. elegans* on repetitive elements (Zeller et al. 2016), and that R-loop occurrence correlated strikingly with the mapping of genomic mutations in REs.

The presence of DSBs in repetitive regions of the genome constitutes a particular challenge for repair mechanisms. Repair by homologous recombination (HR) can result in translocations between chromosomes that bear common repeats, rather than restoring intact chromatids, and single-strand annealing can lead to small inserts and deletions or copy-number variation. It was proposed that the methylation of H3K9 in these repetitive regions might sequester damage away from the HR machinery, thus reducing the risk of inappropriate repair. Consistently, the groups of Karpen (Chiolo et al. 2011; Janssen et al. 2016), Chiolo (Ryu et al. 2015), and Soutoglou (Tsouroula et al. 2016) found that breaks within heterochromatic sequences were repaired differently than breaks in euchromatin, at least when it came to HR. They described an HP1a-dependent pathway for repair in heterochromatin (Chiolo et al. 2011) that allows early steps of the DNA damage response to occur, but then ensures that later steps (i.e., Rad51 binding) occurred only after the break site had been relocated outside of the heterochromatic domain. Nonhomologous end joining, on the other hand, occurred at normal rates within heterochromatin, given that end joining requires no template (Janssen et al. 2016). The suppression of HR within heterochromatin depends on both the H3K9 methyltransferase Su(var)3-9 and its reader HP1a.

Although it is still unclear exactly how heterochromatin factors impact the repair pathway (Janssen et al. 2016), in the absence of H3K9me there is an increase in repeat-specific DNA damage. Indeed, H3K9 methylation can both reduce damage by suppressing the frequency of replication fork collisions with the transcription machinery and repress homology-driven recombination in a repeat-rich domain.

REPETITIVE ELEMENTS AND THE DANGER OF THEIR EXPRESSION

In constitutive heterochromatin, the main sequence classes are repetitive elements, which make up a large fraction of the human genome (de Koning et al. 2011). Repetitive elements can be subdivided into tandem repeats and transposable elements, based on their sequence characteristics. Tandem repeats are head-to-tail repetitions of 2–200 bp long sequence elements, present either as micro- and minisatellites of 5–150 bp in length, dispersed around the genome, or as megabase long stretches of major satellite sequence around centromeres. There is no protein encoded by tandem repeats and they do not contain canonical RNA-PII promoters, even though transcription factors occasionally bind in repeats. In contrast to tandem repeats, complete transposable elements do harbor promoter regions and encode for proteins that facilitate their transposition in the genome. Based on the intermediates formed during transposition, these elements can be further subdivided into RNA (copy–paste) and DNA (cut–paste) transposons. Among both RNA and DNA transposon classes are some that are autonomous (encoding all the proteins necessary for transposition) and others that are nonautonomous. These depend on autonomous family members for transposition, as they contain coding regions that are incomplete or mutated.

RNA transposons make up the largest fraction of repetitive elements in the human genome and are the most active transposon class. They either do or do not contain long terminal repeats (LTRs; retrovirus-like transposons). Of special importance is the non-LTR-transposon LINE-1, as it makes up ~17% of our genomes and is one of the most transpositionally active transposons in the human genome (Lander et al. 2001; Boissinot and Furano 2005).

A variety of heterochromatin-generating pathways mediate transcriptional or posttranscriptional silencing of these sequences, with the implementation and choice depending on the sequence composition of the repeat, the organism, the cell type concerned, stage of development, and other biological features of the repeat (Grewal and Rice 2004; Martens et al. 2005; Padeken et al. 2015; Nishibuchi and Dejardin 2017; Papin et al. 2017). Several of the repression mechanisms are mentioned below, although here we focus primarily on the repercussions of repeat expression and not on the range of mechanisms that lead to their repression.

The existence and transcriptional activity of these repetitive elements were shown to pose a serious threat to the genome's integrity (summarized in Fig. 1). First of all, the de novo integration of transposable elements into coding regions was found at the origin of multiple inheritable diseases. To date 124 genetic diseases have been identified that are caused by transposon insertions (Hancks and Kazazian 2016). The first of these identified was a form of hemophilia A characterized in 1988 (Kazazian et al. 1988). With respect to cancer, most studies have focused on the most abundant repeat elements in man, the autonomous LINE-1 and the nonautonomous Alu RNA transposons. A large genome sequencing study of 244 cancer

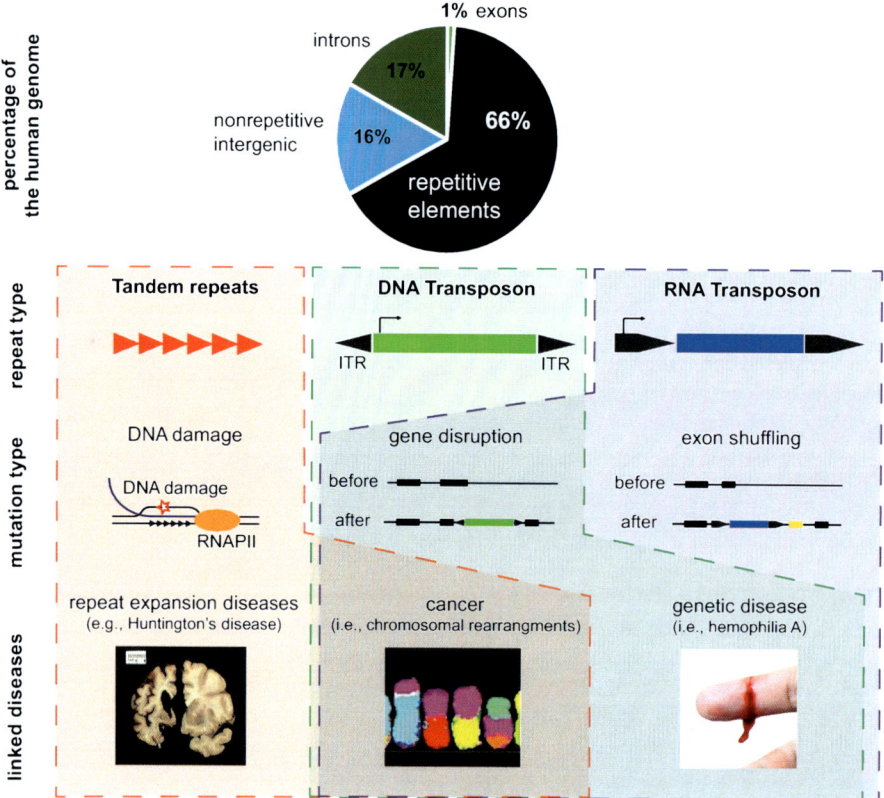

Figure 1. Repeat classes and related human diseases. Summary of the contribution of the three major repeat classes to the human genome: tandem repeats, DNA transposons, and RNA transposons based on de Koning et al. (2011). Dashed lines indicate the mutation types and diseases associated with the expression of the repeat type. Transcription of tandem repeats was shown to cause RNA:DNA hybrid (R-loop) accumulation and DNA damage, which in turn leads to neurodegenerative diseases and cancer. DNA transposition in principle can disrupt gene function, thereby causing cancer and genetic diseases. Potentially because of the low activity of DNA transposons in humans, they have not been implicated in diseases as yet. In contrast, in *Caenorhabditis elegans*, where DNA transposons are still active, their transposition was implicated in the occurrence of spontaneous genetic disorders. RNA transposons, especially of the LINE-1 family, are the most active transposon class in humans and their transposition is frequently observed in cancer. As a consequence of 3′ readthrough during the first step of transposition, new coding sequences from their original position can move with them. Black bars indicate wild-type exons; the yellow bar indicates the new exon carried over by the adjacent transposon. Because of the frequent reoccurrence of all three repeat types in the genome, they can serve as a basis for homology-driven chromosome rearrangements.

patients bearing 12 different cancer types came to the surprising discovery that in 50% of these cases somatic transposition had occurred (Tubio et al. 2014). In a few cases of colorectal cancer the transposition was considered causative for tumorigenesis, as it disrupted the APC tumor-suppressor gene (Miki et al. 1992). An interesting observation arising from these studies was that nearly one-fourth of the observed transposition events also copied additional nonrepetitive sequence pieces, because of transposon transcription continuing after the 3′ end of the transposable element (Tubio et al. 2014). When the original transposon was situated upstream of a coding exon, this led to a phenomenon called exon shuffling.

In most cases, however, a causal link between transposon hopping and oncogenic transformation is unlikely, and transposon expression is more likely a reflection of the general loss of heterochromatic silencing that accompanies oncogenic transformation (Burns 2017). Indeed, studies have linked LINE-1 promoter DNA hypomethylation, the expression and transposition of the transposon and a loss of genome integrity, with poor overall prognosis in non–small

cell lung cancer (Daskalos et al. 2009; Saito et al. 2010). This can be extended to generally poor outcomes in many other cancer types as well (Burns 2017). Although it is rarely the initial trigger for oncogenesis, the increased mutation rate caused by the transposon activity may enhance tumor progression and contribute to drug resistance. It is important to note that all repetitive elements can serve as a basis for homology-mediated chromosome rearrangements because of their frequent occurrence in the genome.

Recently, we found that the unscheduled transcription of repetitive elements alone can compromise genome integrity (Zeller et al. 2016). Using a mutant of the nematode *C. elegans* that lacks all histone H3K9 methylation, we could specifically detect the accumulation of insertions and deletions in derepressed RE, which correlates with a repeat-specific increase in RNA:DNA hybrids (Fig. 2A). Although some genes are also derepressed (234 genes in embryos lacking H3K9me at 20°C are derepressed more than twofold), there was a striking increase in transcripts from all three RE classes (Zeller et al. 2016). Moreover, a DRIP-seq analysis to map RNA:DNA hybrids genome-

A Heterochromatic factors prevent DNA damage and R-loop accumulation

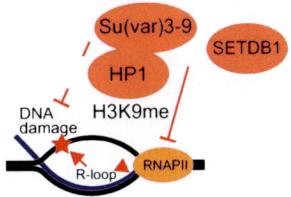

B Heterochromatic factors are recruited by DNA damage and/or are involved in repair

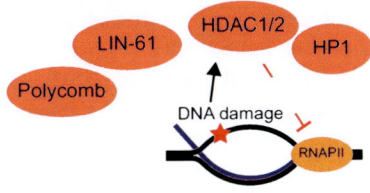

C DNA damage factors require H3K9me for recruitment/activity

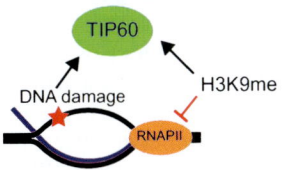

D DNA damage response factors act on R-loops

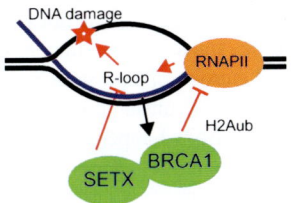

E DNA damage response factors in senesence-induced heterochromatization

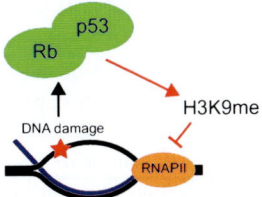

Figure 2. The interplay of DNA damage and heterochromatin machinery. Transcription of tandem repeat sequences was shown to provoke an accumulation of DNA damage causing RNA:DNA hybrids. This figure summarizes pathways (*A–E*, described in the text) that play a role in both the DNA damage and heterochromatin machinery. Black connections indicate the recruitment; red connections indicate the effector function of the proteins. The dashed line indicates that a direct dependency has not yet been proven.

wide showed their accumulation on DNA transposons and tandem repeats, specifically in the H3K9me-deficient worms. We also scored a detectable level of R-loops in wild-type worms on a subset of highly transcribed genes,

as earlier reported (Wahba et al. 2016), yet these do not increase in the absence of H3K9me. It is not clear if replication is necessary for RNA:DNA hybrid formation or whether this stems from the specific character of the repeat RNA.

Different sequences may be differentially prone to form RNA:DNA hybrids when transcribed, as previously suggested from studies in yeast and mammalian cells. Whereas the accumulation of RNA:DNA hybrids generally correlates with the strength of gene expression (Wahba et al. 2016), sequence features that facilitate hybrid formation, such as GC content and poly(A) tracks, were also found (Ginno et al. 2012; Wahba et al. 2016). Of particular note is the accumulation of RNA:DNA hybrids on telomeres and Thy1 transposons (Chan et al. 2014), despite the relatively low expression level of those sequences.

In recent years several studies have tried to identify the mechanisms that protect cells from RNA:DNA hybrid accumulation. These efforts identified a role for many factors involved in transcriptional processivity and RNA degradation (Santos-Pereira and Aguilera 2015). A screen for synthetic sterility with the loss of H3K9me in *C. elegans,* identified many of the same factors, as well as factors involved in replication fork stability (J Padeken, P Zeller, and SM Gasser, pers. comm.). Based on such screens we postulate that four parameters regulate RNA:DNA hybrid formation: first is the affinity of the transcribed RNA for the homologous DNA, a feature dictated by sequence; second, the nature of the transcript itself with respect to introns, secondary structures and the nuclear RNA degradation machinery; third, the factors that bind, process and protect the RNA; and finally, the processivity and/or abundance of the RNA polymerase and its coordination with the replication fork.

It is intriguing that under the conditions that generate the strongest increase in the level of RNA:DNA hybrids, that is, higher temperatures and the loss of histone H3K9 methylation, the major transcriptional difference is a derepression of tandem repeats (Zeller et al. 2016) (J Padeken, P Zeller, and SM Gasser, pers. comm.). In *C. elegans*, this class of repetitive elements is composed of linear repeats of 2–5-bp sequence units, which in theory allows for RNA:DNA pairing in multiple positions. As mentioned above, they neither possess promoter sequences, open reading frames, nor the sequence elements that promote the recruitment of RNA processing factors. These characteristics therefore facilitate the formation of RNA:DNA hybrids upon transcription. A pyrimidine versus purine preference in RNA:DNA hybrid formation could not be detected in REs.

Interestingly, tandem repeats have previously been shown to be very fragile if transcribed, particularly if transcription is bidirectional (Wierdl et al. 1996; Lin et al. 2006; Nakamori et al. 2011). This was mostly attributed to their potential to form higher-order structures such as stem-loop structures (Pearson et al. 2005), given that the torsional stress produced by bidirectional transcription can favor stem loops and G-quadruplex formation. These factors indeed favor RNA:DNA hybrid accumulation, as well (Roy and Lieber 2009; El Hage et al. 2010; Hamperl and

Cimprich 2014). Because RNA:DNA pairing is more stable thermodynamically than DNA:DNA pairing, RNA:DNA hybrids are an obstacle for DNA replication fork progression (Thomas et al. 1976). It is still unclear, however, exactly what the harmful structures that form at transcribed tandem repeats are. Experiments that combine systems for induced tandem repeat expression with a depletion of RNA:DNA hybrids through RNAseH overexpression, or specific targeting of a RNAseH fusion protein, may be able to answer this question.

The only other repeat elements that are enriched for H3K9me2 over me3 and enriched for RNA:DNA hybrids are DNA transposons. The most prominent group of DNA transposons in *C. elegans* works by a cut-and-paste mechanism leaving a double-strand break behind upon relocation (Vos et al. 1996; Bessereau 2006). Thus, in *C. elegans*, the two classes of repeat elements that are preferentially modified by H3K9me2 rather than H3K9me3 (i.e., DNA transposons and tandem repeats) can potentially generate a DSB if expressed. The former may activate a transposition pathway, whereas the latter provokes replication fork collapse.

Whereas RNA transposons are also derepressed in the absence of H3K9 methylation, they do not accumulate RNA:DNA hybrids based on DRIP-seq in *C. elegans* embryos (Zeller et al. 2016). Moreover, although *C. elegans* contains full-length RNA transposons of the LTR class (Ganko et al. 2001), their transposition has not been reported (Bessereau 2006). There may be a link between the absence of RNA:DNA hybrid formation on RNA transposons and their lack of transposition. Indeed, cotranscriptional RNA:DNA hybrid formation (Huertas and Aguilera 2003) may be prevented by cotranscriptional RNA degradation. A likely player in this is the 5′-3′ exonuclease XRN-2, which was shown to be involved in the cotranscriptional degradation of nascent RNA as part of the transcriptional termination process (West et al. 2004). More recent studies additionally showed that XRN-2 is involved in the degradation of many endogenous transcripts when transcriptional processivity is impaired (Davidson et al. 2012). Given that the RNA transposons in *C. elegans* are considered to be evolutionary very young (Ganko et al. 2001), their transcripts may not be efficiently processed for export and thus be preferred targets for XRN-2 mediated degradation. Given that retrotransposons are particularly enriched for H3K9me3, a mark that frequently co-occurs with H3K27me3 in worms, it is conceivable that the transcripts from RNA transposons are specifically targeted for degradation by one or the other mark. This hypothesis could be tested by analyzing sequence-specific RNA:DNA hybrid accumulation in worms lacking both H3K27me and H3K9me.

DOES INSTABILITY INITIATE TRANSCRIPTIONAL SILENCING?

For both transposable elements and genes, the mechanisms that are known to recruit H3K9 histone methyltransferases (HMTs) include small RNA pathways like the PIWI pathway in the germline (Haynes et al. 2006; Sienski

et al. 2012), dsRNA transcripts at satellite repeats in *Schizosaccharomyces pombe* (Keller et al. 2012), and transcription factors, including zinc-finger proteins or the orphan nuclear receptor SHP (Fang et al. 2007; Garcia-Bassets et al. 2007; Bulut-Karslioglu et al. 2012). Considering the risk of genome instability that arises from the transcription of tandem repeats, it is relevant to ask how cells ensure that this class of RE is efficiently modified with H3K9me2. Indeed, beyond centromeric satellite repeats, very little is known about the control of dispersed tandem repeats.

One intriguing possibility is the existence of a feedback loop from the DNA damage caused by the transcription of tandem repeats, which could recruit HMTs to deposit H3K9me. Several arguments support such an idea. First, repressive factors, including HP1 (Luijsterburg et al. 2009), Polycomb (Hong et al. 2008) and HDAC1/2 (Miller et al. 2010), are recruited to sites of DNA damage and the MBT-domain containing protein LIN-61 was shown to be crucial for repair of DSBs by homologous recombination (Fig. 2B; Johnson et al. 2013). Besides a role for heterochromatin factors in repair enzyme recruitment, they might also be implicated in the local transcriptional silencing around damage sites to prevent conflicts between the repair and the transcription machineries (Vissers et al. 2012; Ui et al. 2015). In this context, the Almouzni group showed that a transient transcriptional silencing occurs during DNA damage repair (Adam et al. 2013). In their study, they identified the histone chaperone HIRA as a crucial component that ensures transcriptional reactivation after UVC-induced DNA damage repair. It remains to be seen what would happen in a region that does not contain a strong and specific transcription factor that recruits HIRA. Would the heterochromatic state persist, marking the region as a potentially unstable domain? Would this prevent future breaks from happening or target it for a specific pathway of repair? UV damage and replication-fork-associated damage are repaired by distinct mechanisms, but it is possible that fork-associated damage also recruits heterochromatin components to silence and protect regions that have a bias for fork collapse.

The histone acetyltransferase TIP-60 truly is another example how heterochromatin and the DNA repair machinery are linked. A crucial step during the DNA damage response is the activation of the DNA damage checkpoint protein ATM through acetylation by TIP-60 (Fig. 2C). Sun et al. (2009) could show that the enzymatic activity of TIP-60 depends on its binding to H3K9me. They further showed that under damaging conditions histone H3K9me levels are not increasing globally, but rather the mark is made more accessible by the release of HP1β. Tying the activity of TIP-60-mediated ATM activation to the presence of ligand-free H3K9me could in principle allow the chromatin state to stimulate selectively one repair pathway over another.

REPAIR FACTORS WITH A ROLE IN TRANSCRIPTIONAL SILENCING

The cross talk between DNA damage and heterochromatin formation, would predict that loss of repair factors might cause defects in de novo silencing, particularly with respect to tandem repeats. Recent studies on the human breast-cancer susceptibility gene 1 (*BRCA1*) support such a functional connection. The *BRCA1* protein is a tumor suppressor with E3 ubiquitin ligase activity that contributes to DNA repair by homologous recombination (Scully et al. 1997; Moynahan et al. 1999), and by regulating the DNA damage transcriptional response (Chapman and Verma 1996; Anderson et al. 1998). Mutations in *BRCA1* give a strong predisposition toward breast and ovarian cancers. *C. elegans* BRC-1, as well as mammalian BRCA-1, forms a heterodimer with BRD-1/BARD1, respectively (Boulton et al. 2004), and its dimerization is essential both for preserving genome integrity and for depositing histone H2A ubiquitination (Polanowska et al. 2006).

New evidence suggests that the repressive function of BRCA1 is of central importance for the preservation of genome integrity (Fig. 2D). Zhu et al. (2011) could show that loss of BRCA1 in mice leads to the derepression of tandemly repeated satellite DNA but not of other heterochromatic sequences, such as transposable elements. The function of BRCA1 in repeat silencing is largely mediated by its ability to monoubiquitinate H2A, because the expression of a constitutively ubiquitinated histone H2A could restore satellite expression to wild-type levels. Strikingly, artificial expression of these tandem repeats from a transgene can phenocopy the defects arising from *BRCA1* mutation, including centrosome amplification, cell cycle checkpoint defects, DNA damage, and genomic instability (Zhu et al. 2011). This argues that transcriptional control of tandem repeats is one of the main functions of this central and well-studied tumor suppressor and DNA damage response factor. In addition to its role in the DNA damage response and in transcriptional silencing, BRCA1 was also found to directly initiate RNA:DNA hybrid removal by the recruitment of the RNA helicase Senataxin (Hatchi et al. 2015).

If a cell accumulates too much DNA damage or if certain oncogenic pathways become hyperactive, the last resort a cell has to protect the organism from the onset of cancer is to enter the nonproliferative state of senescence. This process depends on the DNA damage checkpoint proteins Rb and p53, which silence the target genes controlled by E2F, as their expression drives the cell division cycle. These genes accumulate H3K9 methylation and HP1 binding upon repression (Beausejour et al. 2003; Narita et al. 2003). On a cellular level the accumulation of heterochromatin can also be seen as the appearance of senescence-associated heterochromatic foci (SAHFs), although it is still a matter of debate how crucial SAHF formation is for the state of cellular senescence (Kosar et al. 2011). On the other hand, this would be an example of how the DNA damage checkpoint can use heterochromatin to prevent the proliferation of transformed cells.

OPEN QUESTIONS: SILENCING AND SEQUESTRATION

Many questions remain to be resolved concerning the role of repeat elements and their repression in genome stability and in disease. The first unanswered question is how either H3K9me2 or H3K9me3 is specifically targeted to repeats. In *C. elegans*, it is clear that the H3K9me2 HMT, MET-2, can methylate sites independently of the second HMT, SET-25 (J Padeken, P Zeller, and SM Gasser, pers. comm.). A small but specific subset of genomic loci depends on the H3K9me3 mark deposited by SET-25 for repression, and among these are genes whose silent state is stably inherited across generations, following repression by the addition of exogenous RNAi (Ashe et al. 2012; Buckley et al. 2012). Although the authors did not prove that the MET-2 HMT has no role in this process, they showed the ability of the H3K9me3 machinery to target sequences de novo for methylation, independent of their position in the genome. This pathway may be crucial for the survival of a species in an environment where new retrotransposons can infect at any time, and require de novo repression.

This interplay of two parallel methylation pathways is not unlike that of DNA methylation in mammalian cells, where one CpG methyltransferase, DNMT1, is dedicated to maintenance of ^{me}CpG at the replication fork (Gruenbaum et al. 1982; Bestor and Ingram 1983), whereas DNMT3a and DNMT3b exist to methylate sequences de novo (Okano et al. 1998; Lyko et al. 1999). Links between DNA methylation and histone H3K9 methylation exist, but it is poorly understood how they cooperate and how this depends on chromosomal context.

A final open question concerns the role of spatial sequestration of heterochromatin at the nuclear envelope for facilitating transcriptional repression. Peripheral localization of genomic regions in general correlates with low-level transcription (Pickersgill et al. 2006) and silencing factors such as HDAC3 (Somech et al. 2005) and HP1 (Ye and Worman 1996) have been shown to interact with components of the nuclear envelope. In worms the level of derepression of satellite or simple repeat sequences provoked by loss of H3K9me2 correlates with the proximity of the sequence to the nuclear periphery, which is not true for sequences bearing H3K9me3 (J Padeken, P Zeller, and SM Gasser, pers. comm.). Indeed, not every sequence tethered to the nuclear periphery is silenced (Finlan et al. 2008; Reddy et al. 2008; Ruault et al. 2008). Consistently, the loss of heterochromatin anchoring in *C. elegans* embryos did not alter the transcription of genes or repeats dramatically (Gonzalez-Sandoval et al. 2015). Thus the role of heterochromatin sequestration in genome stability remains an open question. It may be that the proper control of replication timing and origin usage requires association with the nuclear envelope (Guelen et al. 2008; Hansen et al. 2010). Moreover, depletion of the origin of replication-associated protein (ORCA) was shown to interfere with H3K9me propagation (Wang et al. 2017). Thus, the link between repeat element repression, replication timing, and the spatial organization of chromatin in the nucleus remains an active area of research.

ACKNOWLEDGMENTS

We thank Jan Padeken, Robin van Schendel, Marcel Tijsterman, and the FMI Genomics and Microscopy facilities for advice and discussion and Jan Padeken for communicating unpublished results. S.M.G. thanks the Swiss National Science Foundation, the European Research Council, and the Novartis Research Foundation for support.

REFERENCES

Adam S, Polo SE, Almouzni G. 2013. Transcription recovery after DNA damage requires chromatin priming by the H3.3 histone chaperone HIRA. *Cell* **155:** 94–106.

Anderson SF, Schlegel BP, Nakajima T, Wolpin ES, Parvin JD. 1998. BRCA1 protein is linked to the RNA polymerase II holoenzyme complex via RNA helicase A. *Nat Genet* **19:** 254–256.

Ashe A, Sapetschnig A, Weick E-M, Mitchell J, Bagijn MP, Cording AC, Doebley A-L, Goldstein LD, Lehrbach NJ, Le Pen J. 2012. piRNAs can trigger a multigenerational epigenetic memory in the germline of *C. elegans*. *Cell* **150:** 88–99.

Aygün O, Mehta S, Grewal SIS. 2013. HDAC-mediated suppression of histone turnover promotes epigenetic stability of heterochromatin. *Nat Struct Mol Biol* **20:** 547–554.

Beausejour CM, Krtolica A, Galimi F, Narita M, Lowe SW, Yaswen P, Campisi J. 2003. Reversal of human cellular senescence: Roles of the p53 and p16 pathways. *EMBO J* **22:** 4212–4222.

Bell AC, Felsenfeld G. 2000. Methylation of a CTCF-dependent boundary controls imprinted expression of the Igf2 gene. *Nature* **405:** 482–485.

Bell O, Schwaiger M, Oakeley EJ, Lienert F, Beisel C, Stadler MB, Schubeler D. 2010. Accessibility of the *Drosophila* genome discriminates PcG repression, H4K16 acetylation and replication timing. *Nat Struct Mol Biol* **17:** 894–900.

Bernstein E, Duncan EM, Masui O, Gil J, Heard E, Allis CD. 2006. Mouse polycomb proteins bind differentially to methylated histone H3 and RNA and are enriched in facultative heterochromatin. *Mol Cell Biol* **26:** 2560–2569.

Bessereau JL. 2006. Transposons in *C. elegans*. *WormBook*: 1–13.

Bestor TH, Ingram VM. 1983. Two DNA methyltransferases from murine erythroleukemia cells: Purification, sequence specificity, and mode of interaction with DNA. *Proc Natl Acad Sci* **80:** 5559–5563.

Bird AP, Wolffe AP. 1999. Methylation-induced repression—Belts, braces, and chromatin. *Cell* **99:** 451–454.

Boissinot S, Furano AV. 2005. The recent evolution of human L1 retrotransposons. *Cytogenet Genome Res* **110:** 402–406.

Boulton SJ, Martin JS, Polanowska J, Hill DE, Gartner A, Vidal M. 2004. BRCA1/BARD1 orthologs required for DNA repair in *Caenorhabditis elegans*. *Curr Biol* **14:** 33–39.

Boyle AP, Davis S, Shulha HP, Meltzer P, Margulies EH, Weng Z, Furey TS, Crawford GE. 2008. High-resolution mapping and characterization of open chromatin across the genome. *Cell* **132:** 311–322.

Brewer BJ. 1988. When polymerases collide: Replication and the transcriptional. *Cell* **53:** 679–686.

Buckley BA, Burkhart KB, Gu SG, Spracklin G, Kershner A, Fritz H, Kimble J, Fire A, Kennedy S. 2012. A nuclear Argonaute promotes multigenerational epigenetic inheritance and germline immortality. *Nature* **489:** 447–451.

Bulut-Karslioglu A, Perrera V, Scaranaro M, de la Rosa-Velazquez IA, van de Nobelen S, Shukeir N, Popow J, Gerle B, Opravil S, Pagani M, et al. 2012. A transcription factor-based mechanism for mouse heterochromatin formation. *Nat Struct Mol Biol* **19:** 1023–1030.

Burns KH. 2017. Transposable elements in cancer. *Nat Rev Cancer* **17:** 415–424.

Castellano-Pozo M, García-Muse T, Aguilera A. 2012. R-loops cause replication impairment and genome instability during meiosis. *EMBO Rep* **13:** 923–929.

Chan YA, Aristizabal MJ, Lu PY, Luo Z, Hamza A, Kobor MS, Stirling PC, Hieter P. 2014. Genome-wide profiling of yeast DNA:RNA hybrid prone sites with DRIP-chip. *PLoS Genet* **10:** e1004288.

Chapman MS, Verma IM. 1996. Transcriptional activation by BRCA1. *Nature* **382:** 678–679.

Chiolo I, Minoda A, Colmenares SU, Polyzos A, Costes SV, Karpen GH. 2011. Double-strand breaks in heterochromatin move outside of a dynamic HP1a domain to complete recombinational repair. *Cell* **144:** 732–744.

Daskalos A, Nikolaidis G, Xinarianos G, Savvari P, Cassidy A, Zakopoulou R, Kotsinas A, Gorgoulis V, Field JK, Liloglou T. 2009. Hypomethylation of retrotransposable elements correlates with genomic instability in non-small cell lung cancer. *Int J Cancer* **124:** 81–87.

Davidson L, Kerr A, West S. 2012. Co-transcriptional degradation of aberrant pre-mRNA by Xrn2. *EMBO J* **31:** 2566–2578.

de Koning AJ, Gu W, Castoe TA, Batzer MA, Pollock DD. 2011. Repetitive elements may comprise over two-thirds of the human genome. *PLoS Genet* **7:** e1002384.

El Hage A, French SL, Beyer AL, Tollervey D. 2010. Loss of topoisomerase I leads to R-loop-mediated transcriptional blocks during ribosomal RNA synthesis. *Genes Dev* **24:** 1546–1558.

Fang S, Miao J, Xiang L, Ponugoti B, Treuter E, Kemper JK. 2007. Coordinated recruitment of histone methyltransferase G9a and other chromatin-modifying enzymes in SHP-mediated regulation of hepatic bile acid metabolism. *Mol Cell Biol* **27:** 1407–1424.

Finlan LE, Sproul D, Thomson I, Boyle S, Kerr E, Perry P, Ylstra B, Chubb JR, Bickmore WA. 2008. Recruitment to the nuclear periphery can alter expression of genes in human cells. *PLoS Genet* **4:** e1000039.

French S. 1992. Consequences of replication fork movement through transcription units in vivo. *Science* **258:** 1362–1365.

Ganko EW, Fielman KT, McDonald JF. 2001. Evolutionary history of Cer elements and their impact on the *C. elegans* genome. *Genome Res* **11:** 2066–2074.

Garcia-Bassets I, Kwon Y-S., Telese F, Prefontaine GG, Hutt KR, Cheng CS, Ju B-G., Ohgi KA, Wang J, Escoubet-Lozach L. 2007. Histone methylation-dependent mechanisms impose ligand dependency for gene activation by nuclear receptors. *Cell* **128:** 505–518.

Gerstein MB, Lu ZJ, Van Nostrand EL, Cheng C, Arshinoff BI, Liu T, Yip KY, Robilotto R, Rechtsteiner A, Ikegami K, et al. 2010. Integrative analysis of the *Caenorhabditis elegans* genome by the modENCODE project. *Science* **330:** 1775–1787.

Ginno PA, Lott PL, Christensen HC, Korf I, Chédin F. 2012. R-loop formation is a distinctive characteristic of unmethylated human CpG island promoters. *Mol Cell* **45:** 814–825.

Gonzalez-Sandoval A, Towbin BD, Kalck V, Cabianca DS, Gaidatzis D, Hauer MH, Geng L, Wang L, Yang T, Wang X, et al. 2015. Perinuclear anchoring of H3K9-methylated chromatin stabilizes induced cell fate in *C. elegans* embryos. *Cell* **163:** 1333–1347.

Grewal SI, Rice JC. 2004. Regulation of heterochromatin by histone methylation and small RNAs. *Curr Opin Cell Biol* **16:** 230–238.

Gruenbaum Y, Cedar H, Razin A. 1982. Substrate and sequence specificity of a eukaryotic DNA methylase. *Nature* **295:** 620–622.

Guelen L, Pagie L, Brasset E, Meuleman W, Faza MB, Talhout W, Eussen BH, de Klein A, Wessels L, de Laat W, et al. 2008. Domain organization of human chromosomes revealed by mapping of nuclear lamina interactions. *Nature* **453:** 948–951.

Hamperl S, Cimprich KA. 2014. The contribution of co-transcriptional RNA:DNA hybrid structures to DNA damage and genome instability. *DNA Repair (Amst)* **19:** 84–94.

Hancks DC, Kazazian HH Jr. 2016. Roles for retrotransposon insertions in human disease. *Mob DNA* **7:** 9.

Hansen RS, Thomas S, Sandstrom R, Canfield TK, Thurman RE, Weaver M, Dorschner MO, Gartler SM, Stamatoyannopoulos JA. 2010. Sequencing newly replicated DNA reveals widespread plasticity in replication timing. *Proc Natl Acad Sci* **107:** 139–144.

Hatchi E, Skourti-Stathaki K, Ventz S, Pinello L, Yen A, Kamie-niarz-Gdula K, Dimitrov S, Pathania S, McKinney KM, Eaton ML. 2015. BRCA1 recruitment to transcriptional pause sites is required for R-loop-driven DNA damage repair. *Mol Cell* **57:** 636–647.

Hauer MH, Seeber A, Singh V, Thierry R, Sack R, Amitai A, Kryzhanovska M, Eglinger J, Holcman D, Owen-Hughes T, et al. 2017. Histone degradation in response to DNA damage enhances chromatin dynamics and recombination rates. *Nat Struct Mol Biol* **24:** 99–107.

Haynes KA, Caudy AA, Collins L, Elgin SC. 2006. Element 1360 and RNAi components contribute to HP1-dependent silencing of a pericentric reporter. *Curr Biol* **16:** 2222–2227.

Helmrich A, Ballarino M, Tora L. 2011. Collisions between replication and transcription complexes cause common fragile site instability at the longest human genes. *Mol Cell* **44:** 966–977.

Hoffman EA, McCulley A, Haarer B, Arnak R, Feng W. 2015. Break-seq reveals hydroxyurea-induced chromosome fragility as a result of unscheduled conflict between DNA replication and transcription. *Genome Res* **25:** 402–412.

Hong Z, Jiang J, Lan L, Nakajima S, Kanno S-i, Koseki H, Yasui A. 2008. A polycomb group protein, PHF1, is involved in the response to DNA double-strand breaks in human cell. *Nucleic Acids Res* **36:** 2939–2947.

Huertas P, Aguilera A. 2003. Cotranscriptionally formed DNA: RNA hybrids mediate transcription elongation impairment and transcription-associated recombination. *Mol Cell* **12:** 711–721.

Janssen A, Breuer GA, Brinkman EK, van der Meulen AI, Borden SV, van Steensel B, Bindra RS, LaRocque JR, Karpen GH. 2016. A single double-strand break system reveals repair dynamics and mechanisms in heterochromatin and euchromatin. *Genes Dev* **30:** 1645–1657.

Jiang Y, Wang X, Bao S, Guo R, Johnson DG, Shen X, Li L. 2010. INO80 chromatin remodeling complex promotes the removal of UV lesions by the nucleotide excision repair pathway. *Proc Natl Acad Sci* **107:** 17274–17279.

Johnson NM, Lemmens BB, Tijsterman M. 2013. A role for the malignant brain tumour (MBT) domain protein LIN-61 in DNA double-strand break repair by homologous recombination. *PLoS Genet* **9:** e1003339.

Jones PL, Veenstra GCJ., Wade PA, Vermaak D, Kass SU, Landsberger N, Strouboulis J, Wolffe AP. 1998. Methylated DNA and MeCP2 recruit histone deacetylase to repress transcription. *Nat Genet* **19:** 187–191.

Kalantry S, Mills KC, Yee D, Otte AP, Panning B, Magnuson T. 2006. The Polycomb group protein Eed protects the inactive X-chromosome from differentiation-induced reactivation. *Nat Cell Biol* **8:** 195–202.

Kazazian HH Jr., Wong C, Youssoufian H, Scott AF, Phillips DG, Antonarakis SE. 1988. Haemophilia A resulting from de novo insertion of L1 sequences represents a novel mechanism for mutation in man. *Nature* **332:** 164–166.

Keller C, Adaixo R, Stunnenberg R, Woolcock KJ, Hiller S, Buhler M. 2012. HP1(Swi6) mediates the recognition and destruction of heterochromatic RNA transcripts. *Mol Cell* **47:** 215–227.

Kosar M, Bartkova J, Hubackova S, Hodny Z, Lukas J, Bartek J. 2011. Senescence-associated heterochromatin foci are dispensable for cellular senescence, occur in a cell type- and insult-dependent manner and follow expression of p16(ink4a). *Cell Cycle* **10:** 457–468.

Lan L, Nakajima S, Kapetanaki MG, Hsieh CL, Fagerburg M, Thickman K, Rodriguez-Collazo P, Leuba SH, Levine AS, Rapic-Otrin V. 2012. Monoubiquitinated histone H2A destabilizes photolesion-containing nucleosomes with concomitant release of UV-damaged DNA-binding protein E3 ligase. *J Biol Chem* **287:** 12036–12049.

Lander ES, Linton LM, Birren B, Nusbaum C, Zody MC, Baldwin J, Devon K, Dewar K, Doyle M, FitzHugh W, et al. 2001. Initial sequencing and analysis of the human genome. *Nature* **409:** 860–921.

Li X, Manley JL. 2005. Inactivation of the SR protein splicing factor ASF/SF2 results in genomic instability. *Cell* **122:** 365–378.

Lin Y, Dion V, Wilson JH. 2006. Transcription promotes contraction of CAG repeat tracts in human cells. *Nat Struct Mol Biol* **13:** 179–180.

Liu B, Alberts BM. 1995. Head-on collision between a DNA replication apparatus and RNA polymerase transcription complex. *Science* **267:** 1131–1137.

Liu T, Rechtsteiner A, Egelhofer TA, Vielle A, Latorre I, Cheung MS, Ercan S, Ikegami K, Jensen M, Kolasinska-Zwierz P, et al. 2011. Broad chromosomal domains of histone modification patterns in *C. elegans. Genome Res* **21:** 227–236.

Luijsterburg MS, Dinant C, Lans H, Stap J, Wiernasz E, Lagerwerf S, Warmerdam DO, Lindh M, Brink MC, Dobrucki JW. 2009. Heterochromatin protein 1 is recruited to various types of DNA damage. *J Cell Biol* **185:** 577–586.

Lyko F, Ramsahoye BH, Kashevsky H, Tudor M, Mastrangelo MA, Orr-Weaver TL, Jaenisch R. 1999. Mammalian (cytosine-5) methyltransferases cause genomic DNA methylation and lethality in Drosophila. *Nat Genet* **23:** 363–366.

Martens JH, O'Sullivan RJ, Braunschweig U, Opravil S, Radolf M, Steinlein P, Jenuwein T. 2005. The profile of repeat-associated histone lysine methylation states in the mouse epigenome. *EMBO J* **24:** 800–812.

Miki Y, Nishisho I, Horii A, Miyoshi Y, Utsunomiya J, Kinzler KW, Vogelstein B, Nakamura Y. 1992. Disruption of the APC gene by a retrotransposal insertion of L1 sequence in a colon cancer. *Cancer Res* **52:** 643–645.

Miller KM, Tjeertes JV, Coates J, Legube G, Polo SE, Britton S, Jackson SP. 2010. Human HDAC1 and HDAC2 function in the DNA-damage response to promote DNA nonhomologous end-joining. *Nat Struct Mol Biol* **17:** 1144–1151.

Moynahan ME, Chiu JW, Koller BH, Jasin M. 1999. Brca1 controls homology-directed DNA repair. *Mol Cell* **4:** 511–518.

Nakamori M, Pearson CE, Thornton CA. 2011. Bidirectional transcription stimulates expansion and contraction of expanded (CTG)*(CAG) repeats. *Hum Mol Genet* **20:** 580–588.

Narita M, Nunez S, Heard E, Narita M, Lin AW, Hearn SA, Spector DL, Hannon GJ, Lowe SW. 2003. Rb-mediated heterochromatin formation and silencing of E2F target genes during cellular senescence. *Cell* **113:** 703–716.

Nishibuchi G, Dejardin J. 2017. The molecular basis of the organization of repetitive DNA-containing constitutive heterochromatin in mammals. *Chromosome Res* **25:** 77–87.

Noma K, Allis CD, Grewal SI. 2001. Transitions in distinct histone H3 methylation patterns at the heterochromatin domain boundaries. *Science* **293:** 1150–1155.

Okano M, Xie S, Li E. 1998. Cloning and characterization of a family of novel mammalian DNA (cytosine-5) methyltransferases. *Nat Genet* **19:** 219–220.

Padeken J, Zeller P, Gasser SM. 2015. Repeat DNA in genome organization and stability. *Curr Opin Genet Dev* **31:** 12–19.

Papin C, Ibrahim A, Gras SL, Velt A, Stoll I, Jost B, Menoni H, Bronner C, Dimitrov S, Hamiche A. 2017. Combinatorial DNA methylation codes at repetitive elements. *Genome Res* **27:** 934–946.

Pearson CE, Nichol Edamura K, Cleary JD. 2005. Repeat instability: Mechanisms of dynamic mutations. *Nat Rev Genet* **6:** 729–742.

Peng JC, Karpen GH. 2009. Heterochromatic genome stability requires regulators of histone H3 K9 methylation. *PLoS Genet* **5:** e1000435.

Pickersgill H, Kalverda B, de Wit E, Talhout W, Fornerod M, van Steensel B. 2006. Characterization of the *Drosophila melanogaster* genome at the nuclear lamina. *Nat Genet* **38:** 1005–1014.

Pimpinelli S, Berloco M, Fanti L, Dimitri P, Bonaccorsi S, Marchetti E, Caizzi R, Caggese C, Gatti M. 1995. Transposable elements are stable structural components of *Drosophila melanogaster* heterochromatin. *Proc Natl Acad Sci* **92:** 3804–3808.

Polanowska J, Martin JS, Garcia-Muse T, Petalcorin MI, Boulton SJ. 2006. A conserved pathway to activate BRCA1-dependent ubiquitylation at DNA damage sites. *EMBO J* **25:** 2178–2188.

Polo SE. 2017. Switching genes to silent mode near DNA double-strand breaks. *EMBO Rep* **18:** 659–660.

Razin A, Riggs AD. 1980. DNA methylation and gene function. *Science* **210:** 604–610.

Reddy K, Zullo J, Bertolino E, Singh H. 2008. Transcriptional repression mediated by repositioning of genes to the nuclear lamina. *Nature* **452:** 243–247.

Rivera C, Gurard-Levin ZA, Almouzni G, Loyola A. 2014. Histone lysine methylation and chromatin replication. *Biochim Biophys Acta* **1839:** 1433–1439.

Roy D, Lieber MR. 2009. G clustering is important for the initiation of transcription-induced R-loops in vitro, whereas high G density without clustering is sufficient thereafter. *Mol Cell Biol* **29:** 3124–3133.

Ruault M, Dubarry M, Taddei A. 2008. Re-positioning genes to the nuclear envelope in mammalian cells: Impact on transcription. *Trends Genet* **24:** 574–581.

Ryu T, Spatola B, Delabaere L, Bowlin K, Hopp H, Kunitake R, Karpen GH, Chiolo I. 2015. Heterochromatic breaks move to the nuclear periphery to continue recombinational repair. *Nat Cell Biol* **17:** 1401–1411.

Saito K, Kawakami K, Matsumoto I, Oda M, Watanabe G, Minamoto T. 2010. Long interspersed nuclear element 1 hypomethylation is a marker of poor prognosis in stage IA non–small cell lung cancer. *Clin Cancer Res* **16:** 2418–2426.

Santos-Pereira JM, Aguilera A. 2015. R loops: New modulators of genome dynamics and function. *Nat Rev Genet* **16:** 583–597.

Schotta G, Ebert A, Krauss V, Fischer A, Hoffmann J, Rea S, Jenuwein T, Dorn R, Reuter G. 2002. Central role of *Drosophila* SU (VAR) 3–9 in histone H3-K9 methylation and heterochromatic gene silencing. *EMBO J* **21:** 1121–1131.

Schübeler D, Scalzo D, Kooperberg C, van Steensel B, Delrow J, Groudine M. 2002. Genome-wide DNA replication profile for *Drosophila melanogaster*: A link between transcription and replication timing. *Nat Genet* **32:** 438–442.

Scully R, Chen J, Plug A, Xiao Y, Weaver D, Feunteun J, Ashley T, Livingston DM. 1997. Association of BRCA1 with Rad51 in mitotic and meiotic cells. *Cell* **88:** 265–275.

Sienski G, Dönertas D, Brennecke J. 2012. Transcriptional silencing of transposons by Piwi and maelstrom and its impact on chromatin state and gene expression. *Cell* **151:** 964–980.

Sollier J, Stork CT, García-Rubio ML, Paulsen RD, Aguilera A, Cimprich KA. 2014. Transcription-coupled nucleotide excision repair factors promote R-loop-induced genome instability. *Mol Cell* **56:** 777–785.

Somech R, Shaklai S, Geller O, Amariglio N, Simon AJ, Rechavi G, Gal-Yam EN. 2005. The nuclear-envelope protein and transcriptional repressor LAP2β interacts with HDAC3 at the nuclear periphery, induces histone H4 deacetylation. *J Cell Sci* **118:** 4017–4025.

Strahl BD, Allis CD. 2000. The language of covalent histone modifications. *Nature* **403:** 41–45.

Sun Y, Jiang X, Xu Y, Ayrapetov MK, Moreau LA, Whetstine JR, Price BD. 2009. Histone H3 methylation links DNA damage detection to activation of the tumour suppressor Tip60. *Nat Cell Biol* **11:** 1376–1382.

Taneja N, Zofall M, Balachandran V, Thillainadesan G, Sugiyama T, Wheeler D, Zhou M, Grewal SI. 2017. SNF2 Family protein Fft3 suppresses nucleosome turnover to promote epigenetic inheritance and proper replication. *Mol Cell* **66:** 50–62 e56.

Taverna SD, Li H, Ruthenburg AJ, Allis CD, Patel DJ. 2007. How chromatin-binding modules interpret histone modifications: Lessons from professional pocket pickers. *Nat Struct Mol Biol* **14:** 1025–1040.

Thomas M, White RL, Davis RW. 1976. Hybridization of RNA to double-stranded DNA: Formation of R-loops. *Proc Natl Acad Sci* **73:** 2294–2298.

Tippmann SC, Ivanek R, Gaidatzis D, Scholer A, Hoerner L, van Nimwegen E, Stadler PF, Stadler MB, Schubeler D. 2012. Chromatin measurements reveal contributions of synthesis and decay to steady-state mRNA levels. *Mol Syst Biol* **8:** 593.

Toyama BH, Savas JN, Park SK, Harris MS, Ingolia NT, Yates JR 3rd, Hetzer MW. 2013. Identification of long-lived proteins reveals exceptional stability of essential cellular structures. *Cell* **154:** 971–982.

Tsouroula K, Furst A, Rogier M, Heyer V, Maglott-Roth A, Ferrand A, Reina-San-Martin B, Soutoglou E. 2016. Temporal and spatial uncoupling of DNA double strand break repair pathways within mammalian heterochromatin. *Mol Cell* **63:** 293–305.

Tsurumi A, Dutta P, Yan S-J., Shang R, Li WX. 2013. *Drosophila* Kdm4 demethylases in histone H3 lysine 9 demethylation and ecdysteroid signaling. *Sci Rep* **3:** 2894.

Tubio JMC., Li Y, Ju YS, Martincorena I, Cooke SL, Tojo M, Gundem G, Pipinikas CP, Zamora J, Raine K, et al. 2014. Mobile DNA in cancer. Extensive transduction of nonrepetitive DNA mediated by L1 retrotransposition in cancer genomes. *Science* **345:** 1251343.

Ui A, Nagaura Y, Yasui A. 2015. Transcriptional elongation factor ENL phosphorylated by ATM recruits polycomb and switches off transcription for DSB repair. *Mol Cell* **58:** 468–482.

van Attikum H, Fritsch O, Hohn B, Gasser SM. 2004. Recruitment of the INO80 complex by H2A phosphorylation links ATP-dependent chromatin remodeling with DNA double-strand break repair. *Cell* **119:** 777–788.

van Attikum H, Fritsch O, Gasser SM. 2007. Distinct roles for SWR1 and INO80 chromatin remodeling complexes at chromosomal double-strand breaks. *EMBO J* **26:** 4113–4125.

Vissers JH, van Lohuizen M, Citterio E. 2012. The emerging role of Polycomb repressors in the response to DNA damage. *J Cell Sci* **125:** 3939–3948.

Vos JC, De Baere I, Plasterk RH. 1996. Transposase is the only nematode protein required for in vitro transposition of Tc1. *Genes Dev* **10:** 755–761.

Wahba L, Costantino L, Tan FJ, Zimmer A, Koshland D. 2016. S1-DRIP-seq identifies high expression and polyA tracts as major contributors to R-loop formation. *Genes Dev* **30:** 1327–1338.

Wang H, Zhai L, Xu J, Joo HY, Jackson S, Erdjument-Bromage H, Tempst P, Xiong Y, Zhang Y. 2006. Histone H3 and H4 ubiquitylation by the CUL4-DDB-ROC1 ubiquitin ligase facilitates cellular response to DNA damage. *Mol Cell* **22:** 383–394.

Wang Y, Khan A, Marks AB, Smith OK, Giri S, Lin YC, Creager R, MacAlpine DM, Prasanth KV, Aladjem MI, et al. 2017. Temporal association of ORCA/LRWD1 to late-firing origins during G1 dictates heterochromatin replication and organization. *Nucleic Acids Res* **45:** 2490–2502.

Watt F, Molloy PL. 1988. Cytosine methylation prevents binding to DNA of a HeLa cell transcription factor required for optimal expression of the adenovirus major late promoter. *Genes Dev* **2:** 1136–1143.

Weber M, Davies JJ, Wittig D, Oakeley EJ, Haase M, Lam WL, Schubeler D. 2005. Chromosome-wide and promoter-specific analyses identify sites of differential DNA methylation in normal and transformed human cells. *Nat Genet* **37:** 853–862.

Weintraub H, Groudine M. 1976. Chromosomal subunits in active genes have an altered conformation. *Science* **193:** 848–856.

West S, Gromak N, Proudfoot NJ. 2004. Human 5′→3′exonuclease Xrn2 promotes transcription termination at co-transcriptional cleavage sites. *Nature* **432:** 522–525.

Wierdl M, Greene CN, Datta A, Jinks-Robertson S, Petes TD. 1996. Destabilization of simple repetitive DNA sequences by transcription in yeast. *Genetics* **143:** 713–721.

Ye Q, Worman HJ. 1996. Interaction between an integral protein of the nuclear envelope inner membrane and human chromodomain proteins homologous to *Drosophila* HP1. *J Biol Chem* **271:** 14653–14656.

Zeller P, Padeken J, van Schendel R, Kalck V, Tijsterman M, Gasser SM. 2016. Histone H3K9 methylation is dispensable for *Caenorhabditis elegans* development but suppresses RNA: DNA hybrid-associated repeat instability. *Nat Genet* **48:** 1385–1395.

Zhao Q, Wang QE, Ray A, Wani G, Han C, Milum K, Wani AA. 2009. Modulation of nucleotide excision repair by mammalian SWI/SNF chromatin-remodeling complex. *J Biol Chem* **284:** 30424–30432.

Zhu Q, Pao GM, Huynh AM, Suh H, Tonnu N, Nederlof PM, Gage FH, Verma IM. 2011. BRCA1 tumour suppression occurs via heterochromatin-mediated silencing. *Nature* **477:** 179–184.

Structure and Epigenetic Regulation of Chromatin Fibers

PING CHEN AND GUOHONG LI

*National Laboratory of Biomacromolecules, CAS Center for Excellence in Biomacromolecules,
Institute of Biophysics, Chinese Academy of Sciences, Beijing 100101, China*

Correspondence: liguohong@ibp.ac.cn

In eukaryotes, genomic DNA is hierarchically packaged by histones into chromatin on several levels to fit inside the nucleus. As a central-level structure between nucleosomal arrays and higher-order chromatin organizations, the 30-nm chromatin fiber and its dynamics play a crucial role in gene regulation. However, despite considerable efforts over the past three decades, the fundamental structure and its dynamic regulation of chromatin fibers still remain as a big challenge in molecular biology. Here, we mainly summarize the most recent progress in elucidating the structure of the 30-nm chromatin fiber in vitro and epigenetic regulation of chromatin fibers by chromatin factors, particularly histone variants. In addition, we also discuss recent studies in unraveling the three-dimensional organization of chromatin fibers in situ by genomic approaches and electron microscopy.

In eukaryotic cells, the accessibility of DNA is dependent on the packing density of chromatin fibers. Genomic DNA firstly wraps around a histone octamer to form a nucleosome, which is connected by linker DNA to form the primary chromatin structure, the "beads-on-a-string" nucleosomal array. Nucleosomes in the array are further compacted by linker histone H1 to form a 30-nm chromatin fiber—typically regarded as the secondary structure of chromatin. Chromatin fibers are further organized into other higher-levels of chromatin structures, but so far details of these structural levels still remain obscure. The three-dimensional (3D) organization of genomic DNA plays a critical role in regulating DNA-related biological processes, such as gene transcription and DNA replication, repair, and recombination. Elucidating the structure and dynamics of chromatin fibers in molecular details is the key to understanding the epigenetic regulation of gene expression by different chromatin factors.

HIGH-RESOLUTION STRUCTURE OF 30-NM CHROMATIN FIBERS

It is still a puzzle how genomic DNA is hierarchically organized in eukaryotic cells. From a structural point of view, the DNA double-helix structure discovered by Watson and Crick is surely the most important milestone in molecular biology (Watson and Crick 1953). After more than 40 years, the high-resolution structure of the nucleosome core particle (NCP) has been defined by crystal X-ray studies (Luger et al. 1997), which undoubtedly reveals the structural details of histone–histone and histone–DNA interactions within nucleosomes (Fig. 1). Other high-resolution structures of NCPs containing core histones from different species or with a different DNA sequence, different histone variants, or different nucleosome-binding proteins/peptides have been resolved subsequently to in-

vestigate the regulation of nucleosome structure as reviewed previously (Cutter and Hayes 2015; McGinty and Tan 2015; Zhu and Li 2016).

Afterward, the manner of how a "beads-on-a-string" nucleosomal array folds into a condensed 30-nm chromatin fiber remains to be determined. Two basic classes of structural models had been proposed previously based on the studies of native 30-nm fibers in nuclei or isolated from nuclei (Finch and Klug 1976; Langmore and Paulson 1983; Widom and Klug 1985; Gerchman and Ramakrishnan 1987; Ghirlando and Felsenfeld 2008). One is the solenoid model, in which nucleosomes are arranged linearly in a one-start solenoid-type helix with bend linker DNA, and the other is the zigzag model, in which nucleosomes zigzag back and forth in a two-start stack of nucleosomes connected by a relatively straight DNA linker (Woodcock et al. 1984; Widom and Klug 1985; Williams et al. 1986). To discriminate between these two structural models, the detailed structure of chromatin fibers needs to be resolved. However, the heterogeneous properties of nucleosomes in native chromatin with different DNA sequences/linker lengths and different histone compositions/modifications make this difficult. The reconstitution of chromatin fibers in vitro using regular tandem repeats of unique nucleosome-positioning DNA sequences and purified histone proteins greatly improves the reproducibility and uniformity for structural analysis (Dorigo et al. 2004). Using this system, Richmond and colleagues first resolved the crystal X-ray structure of chromatin fibers reconstituted by tetranucleosomal arrays with a 20-bp linker DNA at a high concentration of Mg^{2+} (120 mM). The structure resolved at ~9 Å resolution reveals two stacks of nucleosomes connected by straight linker DNA, which agrees with the zigzag model (Schalch et al. 2005). Within each nucleosome stack, strong interactions between the H2B–helix $\alpha1/\alpha C$ and the adjacent H2A–helix $\alpha2$ stabi-

Published by Cold Spring Harbor Laboratory Press; doi: 10.1101/sqb.2017.82.033795

Cold Spring Harbor Symposia on Quantitative Biology, Volume LXXXII

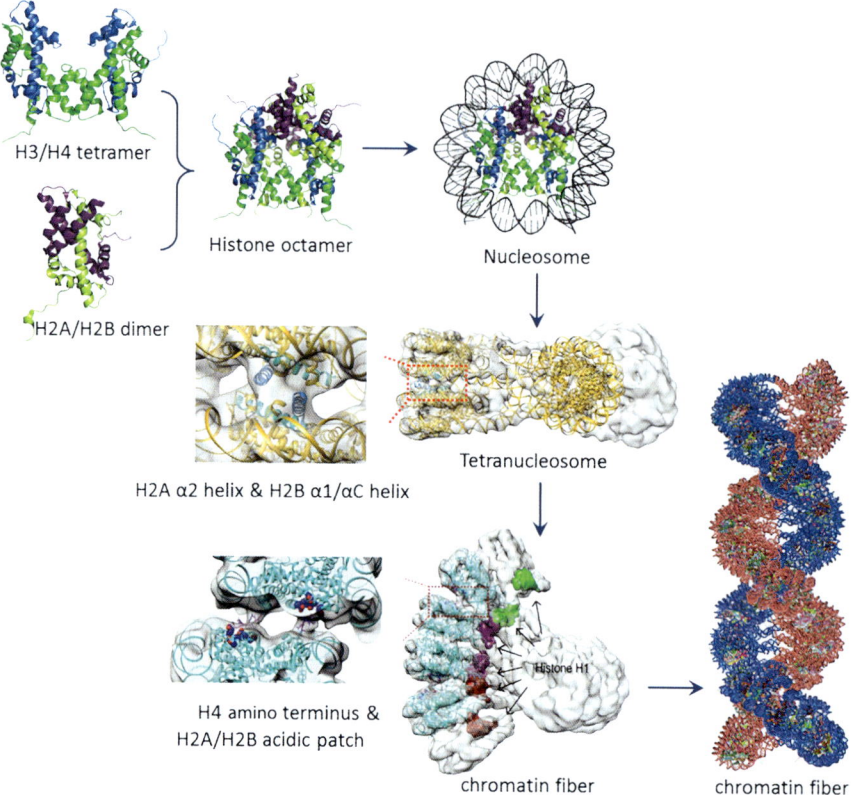

H3/H4 tetramer

H2A/H2B dimer

Histone octamer

Nucleosome

H2A α2 helix & H2B α1/αC helix

Tetranucleosome

H4 amino terminus &
H2A/H2B acidic patch

Histone H1

chromatin fiber   chromatin fiber

Figure 1. The hierarchical organization of chromatin fibers: 147 base pairs (bp) of DNA wrapped around a histone octamer (with one histone H3/H4 tetramer and two H2A/H2B dimers) ∼1.7 times in a left-handed manner to form nucleosome, the basic repeating unit of chromatin (Luger et al. 1997). Nucleosomes interact with each other to form the tetranucleosomal structural units, within which the four nucleosomes zigzag back and forth to form two stacks of two nucleosome cores connected by straight linker DNA (Schalch et al. 2005; Song et al. 2014). Strong interactions between the H2B–helix α1/αC and the adjacent H2A–helix α2 stabilize each nucleosome stack of tetranucleosomal units. The tetranucleosomal units are twisted against each other in a left-handed manner to form the final double-helical structure of chromatin fiber (Song et al. 2014). Between the structural units, the H1–H1 interaction and the interactions of H4 amino-terminal tail with the H2A/H2B acidic patch between the nucleosomal interfaces play central roles.

lize the chromatin fibers (Fig. 1). However, the crystal structure was determined for a tetranucleosomal array, which is too short to form a solenoid structure, and with a very short nucleosome repeat length (NRL; 167 bp) in the absence of H1, which is uncommon in nature.

Recently, we had determined the 3D cryo-electron microscopy (EM) structures at ∼11 Å resolution of longer chromatin fibers reconstituted in vitro from 12-nucleosomal arrays with two different NRLs (177 bp and 187 bp), which reveal a left-handed double helix twisted with the repeating tetranucleosomal structural units (Song et al. 2014). The structures constitute the largest fragments of chromatin resolved at this high resolution so far and provide new insights into the nucleosome arrangement in helical structure of 30-nm chromatin fibers. The four nucleosomes within the structural unit zigzag back and forth to form two stacks of two nucleosome cores connected by straight linker DNA, which appear very similar to the resolved X-ray structure of a tetranucleosome (Schalch et al. 2005). The tetranucleosomal units are twisted against each other in a left-hand manner to form the final helical structure of 30-nm fibers. The H1–H1 interaction and the interactions of the H4 amino-terminal tail with the H2A/H2B acidic patch between the neighboring nucleosomal

interfaces play central roles in interaction and twisting between the structural units (Fig. 1). In comparison to the closely stacked nucleosomes within the tetranucleosomal structural unit, the apparently formed gaps between the structural units may provide a platform for histone modifications or other architectural proteins to modulate the internucleosomal surface interactions in the regulation of higher-order chromatin structures.

The linker histone has been considered to play an essential role in the formation of compact chromatin fiber (Thoma et al. 1979; Allan et al. 1980; Bates and Thomas 1981; Thomas 1999). But the molecular details of how linker histones bind to nucleosome and compact chromatin fibers remain to be determined. Our cryo-EM structures, for the first time, clearly address the location and the role of H1 in the formation of the 30-nm chromatin fiber (Song et al. 2014). H1 directly interacts with both the dyad and the entry/exit nucleosomal DNA in a three-contact mode. The binding of H1, which locks the outer DNA wraps of nucleosomes at the enter/exit point as previously discussed (van Holde and Zlatanova 1996; Syed et al. 2010), enhances the stability of the outer nucleosomal wrap with the increase of ∼20 kJ/mol free energy and accelerates the folding/unfolding rate of the outer wrap

(Li et al. 2016). In addition, the H1 locates asymmetrically in each nucleosome core with its bulk of globular domain pointed outside the structural units, which allows the self-association of H1 by its globular domain between tetra-nucleosomal units and imparts an additional twist between each structural unit. However, the 11 Å resolution of the resolved cryo-EM structures cannot give the structural details of H1 at the atomic levels. Higher resolution structures of chromatin fibers still need to be determined to solve these problems.

SINGLE-MOLECULE STUDY ON THE DYNAMICS OF CHROMATIN FIBERS

The 30-nm chromatin fiber has been shown to be the first level of the transcriptionally dormant platform (Li et al. 2010). Thus, the structural transition between the 30-nm chromatin fiber and the nucleosomal array plays a critical role in regulating the accessibility of the DNA template (Li et al. 2010; Li and Reinberg 2011). Static conformations obtained by our cryo-EM structures of 30-nm chromatin fibers in vitro cannot provide much detailed information for such a dynamic process. In addition, because of the highly dynamic and heterogeneous properties of the chromatin fiber, it is technically challenging to study the dynamics of the chromatin fibers. Recently, several single-molecule techniques have been applied to investigate the

dynamics of nucleosome/chromatin structures (Cui and Bustamante 2000; Pope et al. 2005; Kruithof et al. 2009). Mechanical manipulations of single mononucleosome show that a nucleosome in the absence of linker histone unravels in two major stages: the outer nucleosomal DNA unwrapped at ~3 pN and the inner turn unwrapped at ~10 pN (Brower-Toland et al. 2002; Hall et al. 2009; Bintu et al. 2012). For the mechanical decompaction of higher-order chromatin fibers, Bustamante and colleagues observed a distinct structural transition between "compacted" and "extended" states for the isolated native chicken erythrocyte chromatin fibers at medium ionic strengths (>40 mM NaCl) at 5–6 pN, which was attributed to the disruption of internucleosomal interactions for stabilizing the higher-order structures of chromatin fibers (Cui and Bustamante 2000). Van Noort and colleagues used magnetic tweezers to investigate the mechanical stretching of single reconstituted chromatin fibers and also observed the dynamic unfolding behavior of chromatin fibers at low forces (<6 pN) (Kruithof et al. 2009; Meng et al. 2015). They found that the chromatin fibers with 197-bp NRL stretch like a Hookian spring, which supports a solenoid topology with a high nucleosome–nucleosome stacking energy. Using magnetic tweezers (Fig. 2A), we investigated the hierarchical organization and dynamics of 30-nm chromatin fibers reconstituted in vitro in the presence of H1, whose high-resolution 3D cryo-EM structures

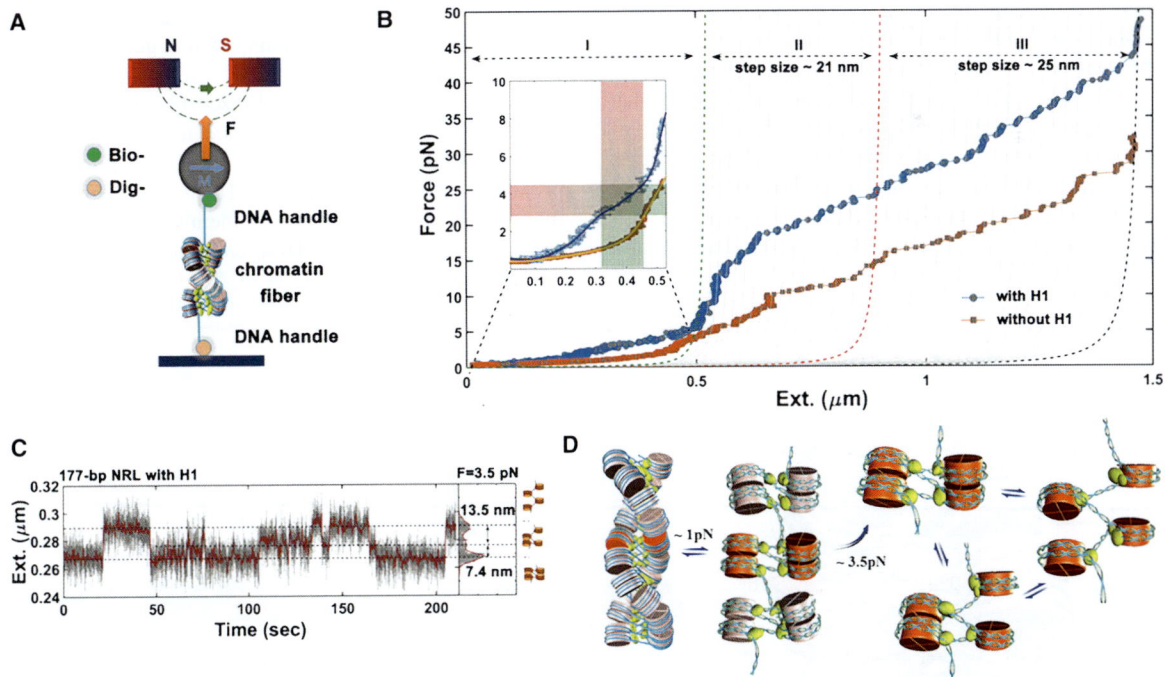

Figure 2. Mechanical unfolding of 30-nm chromatin fibers by single-molecular magnetic tweezers. (*A*) Schematic setup of the magnetic tweezers used in chromatin fiber studies (not to scale). (*B*) Comparison of two typical force–extension curves of a chromatin fiber with H1 (blue curve) and without H1 (orange curve) in HE buffer. Three major distinct stages as labeled can be recognized in the force–extension curve of the chromatin fiber with H1. The *inset* shows the details of Stage I at low forces (<8 pN). (*C*) Stepwise folding/unfolding dynamics of tetranucleosomal units with two alternative pathways at 3.5 pN for the 177-bp nucleosome repeat length (NRL)'s chromatin fiber with H1. (*D*) Model for the dynamic organization of chromatin fibers. The left-handed double-helical chromatin fiber unfolds to a "tetranucleosomes-on-a-string" extended structure; then the tetranucleosomal unit unfolds to a complete open nucleosomal array in one or two steps. (Adapted, with permission, from Li et al. 2016.)

have been resolved recently (Song et al. 2014; Li et al. 2016). We showed that mechanical unfolding of a 30-nm fiber is a multistep process (Fig. 2B). Under increasing tensile force applied, the compacted chromatin fiber first unfolds to a "tetranucleosomes-on-a-string" extended structure by disrupting the internucleosome interactions between tetranucleosomal units. The tetranucleosomal unit exists as a stable structural intermediate of the 30-nm chromatin fiber and further unfolds to a more extended "beads-on-a-string" conformation by disrupting the nucleosome–nucleosome interactions within the tetranucleosomal unit in a two-step, three-transition-state process at force of ~3.5 pN, which is characteristic of three transitions with the step sizes of $1L$, $2L$, and $3L$ (in which L is the linker length between the adjacent nucleosomes), respectively (Fig. 2C,D). Further increasing force causes nucleosome unwrapping in two major stages, including the one-step disruption of the outer DNA turn and the one-step unwrapping of the inner turn, consistent with previous investigations (Brower-Toland et al. 2002; Hall et al. 2009; Bintu et al. 2012; Li et al. 2016). Importantly, similar dynamic processes can be observed in the chromatin fibers assembled not only on regular tandem repeat of the 601 DNA sequence but also on the scrambled (nonrepetitive) DNA sequence, suggesting that the existence of tetranucleosomal units is not dependent on DNA sequence. Moreover, the low energy required to disrupt the interaction between two tetranucleosomal units (~1.8 k_BT), which is comparable to the thermal fluctuations, suggests that chromatin fibers may undergo spontaneously rapid folding/unfolding dynamics between a compact regular 30-nm chromatin fiber and an extended "tetranucleosomes-on-a-string" at physiological conditions in vivo. Thus, it is likely that the chromatin fibers in vivo probably adopt an irregular "tetranucleosomes-on-a-string" structure, which combines fully folded regular zigzag tetranucleosomal clusters/stacks with partially unfolded "nucleosomal array" or "nucleosomal clutches" regions (Collepardo-Guevara and Schlick 2011).

EPIGENETIC REGULATION OF CHROMATIN FIBERS BY CHROMATIN FACTORS

In the past 20 years, many chromatin factors, including chromatin modifications, histone chaperones, histone variants, chromatin remodelers, and chromatin architectural proteins, have been shown to be involved in regulating chromatin dynamics (Luger and Hansen 2005; Hake and Allis 2006; Greaves et al. 2007). Deciphering the structure and dynamics of the 30-nm chromatin fibers in molecular details is essential for understanding such regulations. Our cryo-EM structures clearly imply the presence of three important interaction interfaces in the 30-nm fiber—namely, DNA–histone interaction interfaces within the nucleosome, internucleosome interaction interfaces within the tetranucleosomal unit, and interaction interfaces between tetranucleosomal units, which can be potentially regulated by different chromatin factors (Fig. 3). The newly developed in vitro single-molecule techniques and in vivo genomic approaches enable us to investigate the epigenetic regulations of 30-nm fibers by different chromatin factors (Hsieh et al. 2015; Li et al. 2016; Risca et al. 2017).

Previously, FACT, a conserved histone chaperone for H2A–H2B dimers, has been shown to destabilize nucleosomes for RNA polymerase progression on chromatin templates (Orphanides et al. 1998, 1999; Belotserkovskaya et al. 2003) and maintain chromatin structure in vivo during DNA transcription, replication, and repair (Saunders et al. 2003; Fujimoto et al. 2012; Formosa 2013; McCullough et al. 2015). However, the molecular mechanisms of how FACT remodels chromatin remain largely unclear. As discussed above, the nucleosomal stacks within tetranucleosomal unit are mainly stabilized by the interactions between the H2B–helix $\alpha1/\alpha C$ and the adjacent H2A–helix $\alpha2$ (Song et al. 2014). A recent structural study revealed that the recognition of H2A–H2B heterodimer by FACT is mainly mediated by the interactions between the U-turn motif of Spt16M and the amino-terminal $\alpha1$ helix of H2B (Hondele et al. 2013), suggesting that

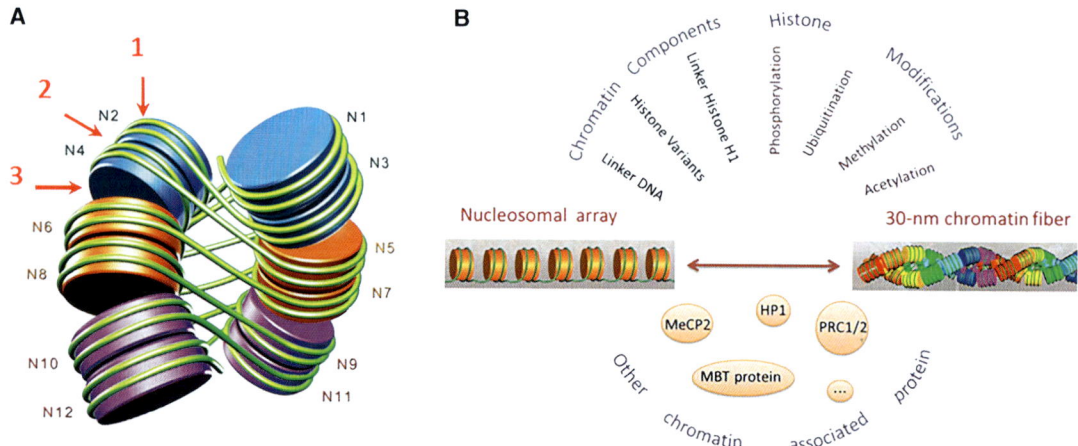

Figure 3. Epigenetic regulation of chromatin fibers by chromatin factors. (*A*) Schematic model of 30-nm chromatin fibers with the three important interaction interfaces as indicated by arrows, including the DNA–histone interaction interfaces within nucleosome (1), internucleosome interaction interfaces within tetranucleosomal unit (2), and interaction interfaces between tetranucleosomal units (3). (*B*) The different chromatin factors potentially regulate the interaction interfaces of 30-nm chromatin fibers.

FACT may remodel chromatin fibers via interfering these interfacial interactions. Indeed, our single-molecule measurements showed that FACT can remodel higher-order chromatin structure by destabilizing the tetranucleosomal unit (Li et al. 2016). More importantly, genomic analyses showed that FACT destabilizes the N/N+2 Micro-C interactions (tetranucleosomal motifs) to facilitate gene transcription in yeast. Distinct domains of FACT have been shown to interact with H2A/H2B or H3/H4 separately (Stuwe et al. 2008; Zhang et al. 2015; Tsunaka et al. 2016), suggesting that these distinct domains of FACT might be responsible for their different activities in remodeling the nucleosome and/or the tetranucleosomal unit. It will be of great interest to generate a series of FACT mutants to resolve the distinctive activity in remodeling the nucleosome and/or the tetranucleosome using single-molecule measurements. Beside FACT, several other chromatin remodelers (such as INO80, which recognizes the H2B αC helix) (Tosi et al. 2013), or certain repressive chromatin factors (such as PRC1, which has been shown to compact three to four nucleosomes) (Francis et al. 2004) may also fulfill their biological functions via modulating the dynamics or stability of tetranucleosomal units. The tetranucleosomal unit may provide an additional level of gene regulation beyond the nucleosome. It will be of great interest to identify these factors and decipher their regulatory interactions with the tetranucleosomal units.

The incorporation of histone variants has been shown to regulate the structure and dynamics of chromatin fibers and create architecturally distinct chromatin states that play diverse functions in genome-associated biological processes. Among them, histone variant H3.3 has been largely considered as a mark of transcriptionally activated genes and has been deposited into transcribed genes, promoters, and gene-regulatory elements (Schwartz and Ahmad 2005; Jin et al. 2009; Goldberg et al. 2010). However, a few recent studies showed that H3.3 was also incorporated at regions of the genome that are typically thought to be relatively transcriptionally inactive, such as the telomere and pericentric heterochromatin in mouse embryonic stem (mES) cells (van der Heijden et al. 2007; Goldberg et al. 2010; Wong et al. 2010). H3.3 differs from canonical H3 at only four amino acid residues, three are hidden inside the NCP in region 87–90 and one residue, Ser31, is exposed outside of the NCP. We and others have showed that Ala87 and Gly90 are the principal determinants of human H3.3 specificity in DAXX recognition (Elsässer et al. 2012; Liu et al. 2012). We also investigated how H3.3 regulates the nucleosome/chromatin dynamics and gene transcription (Fig. 4; Chen et al. 2013). Our FRET and magnetic tweezers experiments showed that H3.3 has little effect on the stability of nucleosomes, which was consistent with previous findings (Flaus et al. 2004; Thakar et al. 2009), but it greatly impairs the compaction of chromatin by analytical ultracentrifugation (AUC) and EM analyses (Chen et al. 2013). Interestingly, we also showed that the histone variant H2A.Z not only stabilizes the nucleosome but also facilitates compaction of nucleosomal arrays into a featured "ladder-liker" chromatin fiber. Although the incorporation of H3.3 does not

affect the stabilization effect of H2A.Z on nucleosomes, it counteracts H2A.Z-mediated chromatin compaction (Fig. 4A,B). Residues 89 and 90 of H3.3 were mainly responsible for the counteractivity of H2A.Z-mediated chromatin compaction. Moreover, we found that H3.3 could antagonize the inhibitory effects of H2A.Z on chromatin transcription in vitro by RNA Pol II, which partially resulted from the counteractivity of compaction by H2A.Z. Our results suggest that H3.3 may play a dominant role in regulation of the dynamics of higher-ordered chromatin and transcriptional activity in the chromatin context. To this end, we further analyzed the dynamic depositions and/or replacement of H2A.Z and H3.3 and the corresponding structural changes of chromatin at the enhancer and promoter regions of RAR/RXR targeted genes during gene activation by all-*trans*-retinoid acid (tRA) induction. Our results (Fig. 4C) revealed that deposition of H2A.Z results in compaction of chromatin at promoter regions, which inhibits the transcription of the Cyp26a1 and Hoxa1 genes before tRA induction, whereas the incorporation of H3.3 at enhancer regions decorates the chromatin architecture to a relatively open conformation, allowing the recognition and binding of transcriptional activators, which subsequently recruits the ATP-dependent chromatin remodeling complexes and/or histone-modifying enzymes to remodel the nucleosome architecture at promoter regions upon gene induction, as previously reported (Li et al. 2010). Our results provide new insights into the molecular mechanism of how histone variants function cooperatively to establish featured chromatin structures at enhancer and promoter regions to prime inducible genes for rapidly activation in response to environmental stimulation.

The centromere-specific histone H3 variant (CenH3 [CENP-A in humans]), is an epigenetic factor essential for the centromere identity and function (Allshire and Karpen 2008). In human cells, CENP-A is specifically recognized and deposited into centromeres by its chaperone and assembly factor HJURP (Dunleavy et al. 2009; Foltz et al. 2009). Our structural and biochemical analysis showed that other than the CATD of CENP-A, previously identified as the exclusive region responsible for HJURP binding (Black et al. 2004; Bassett et al. 2012), the residue Ser68 also plays essential roles for HJURP binding (Hu et al. 2011). We further showed that the dynamic phosphorylation/dephosphorylation of Ser68 in CENP-A, which is mediated by the Cdk1/Cyclin B and PP1α complex, temporally controls the HJURP-mediated assembly of CENP-A into centromeric regions (Yu et al. 2015). CENP-N, an indispensable member of CCAN (the constitutive centromere-associated network), was identified as the first "reader" of epigenetic marks present in the CENP-A-containing chromatin (Carroll et al. 2009). CENP-N is dynamically recruited to centromeres/kinetochores during the cell cycle (Hellwig et al. 2011) and plays an essential role in the assembly of kinetochore complex at centromeres and faithful segregation of sister chromatids during cell division (Foltz et al. 2006; Carroll et al. 2009). By using biochemical, biophysical, and cell-based assays, we showed that the RG loop, the two CENP-A-specific residues Arg80 and Gly81 in loop1 located at the lateral surface of CENP-A

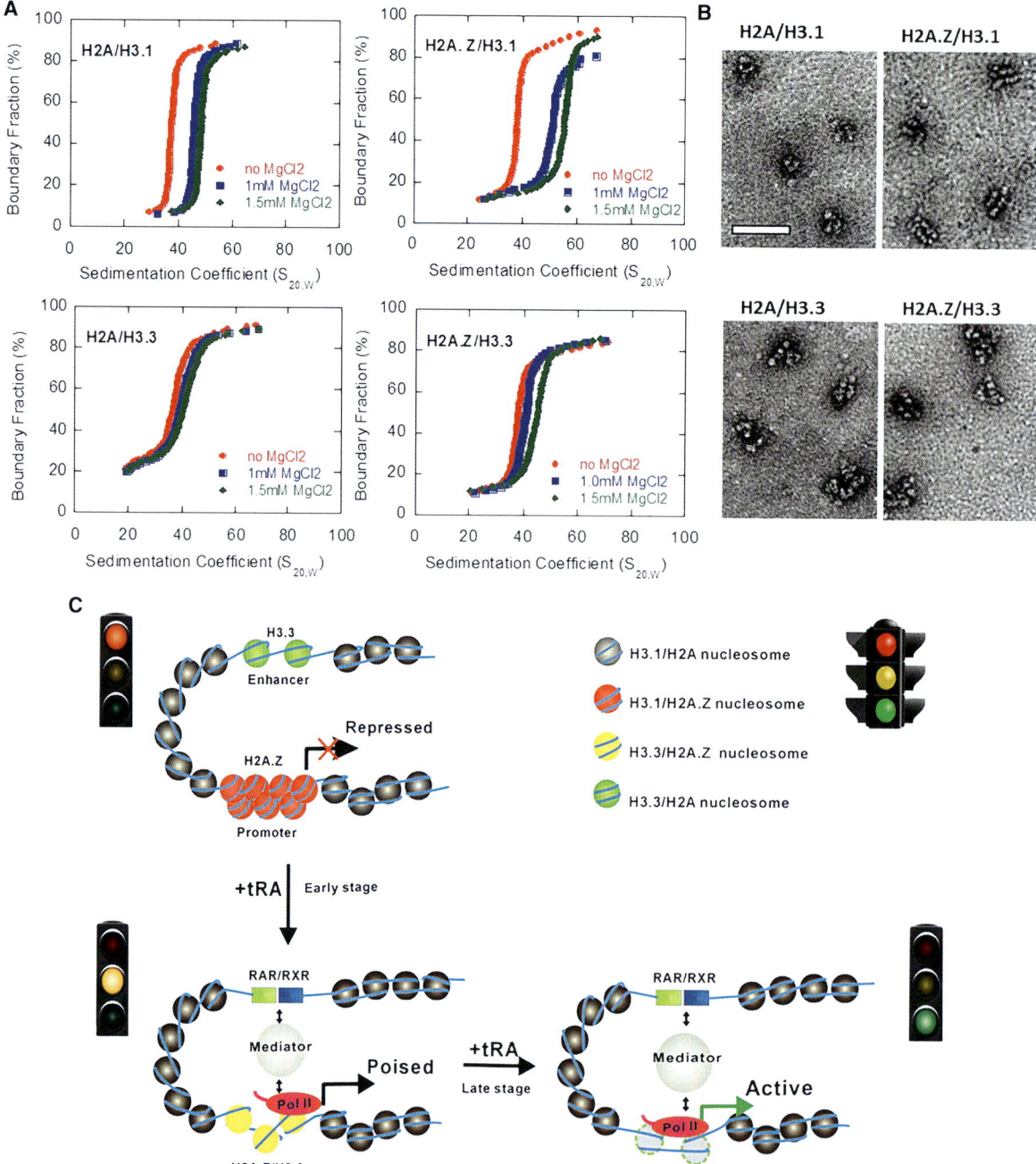

Figure 4. Histone variants H3.3 and H2A.Z regulate chromatin structures cooperatively for gene activation. (*A*) Sedimentation coefficient distribution plots for the canonical, H2A.Z-, H3.3-, and double-variant H2A.Z/H3.3-containing nucleosomal arrays at 0, 1.0, and 1.5 mM $MgCl_2$. (*B*) Negatively stained electron microscopic (EM) images of the canonical, H2A.Z-, H3.3-, and double-variant H2A.Z/H3.3-containing chromatin fibers in 1.0 mM $MgCl_2$. Scale bar, 100 nm. (*C*) The model for the dynamic regulation of H2A.Z and H3.3 on chromatin structures at the enhancer and promoter regions during the gene activation process by the addition of *trans*-retinoic acid (tRA). (Adapted, with permission, from Chen et al. 2013.)

nucleosome (Tachiwana et al. 2011), not only provides the recognition site for the binding of CENP-N to the CENP-A nucleosome, but also facilitates the folding of CENP-A arrays into a compact "ladder-like" chromatin structure in the presence of $MgCl_2$, as revealed by AUC and EM analyses (Fig. 5A–C; Fang et al. 2015). Interestingly, we found that the formation of a compact "ladder-like" structure of CENP-A chromatin inhibits the binding and recruitment of

CENP-N through concealing/hiding the RG loop within chromatin fiber. More importantly, our fluorescence resonance energy transfer (FRET) analysis and SNAP imaging showed that upon G_1/S phase transition, centromeric chromatin switches from the compact to an open state because of the dilution of CENP-A nucleosome during DNA replication, which enables the now-exposed RG loop to recruit CENP-N at the S phase (Fig. 5D).

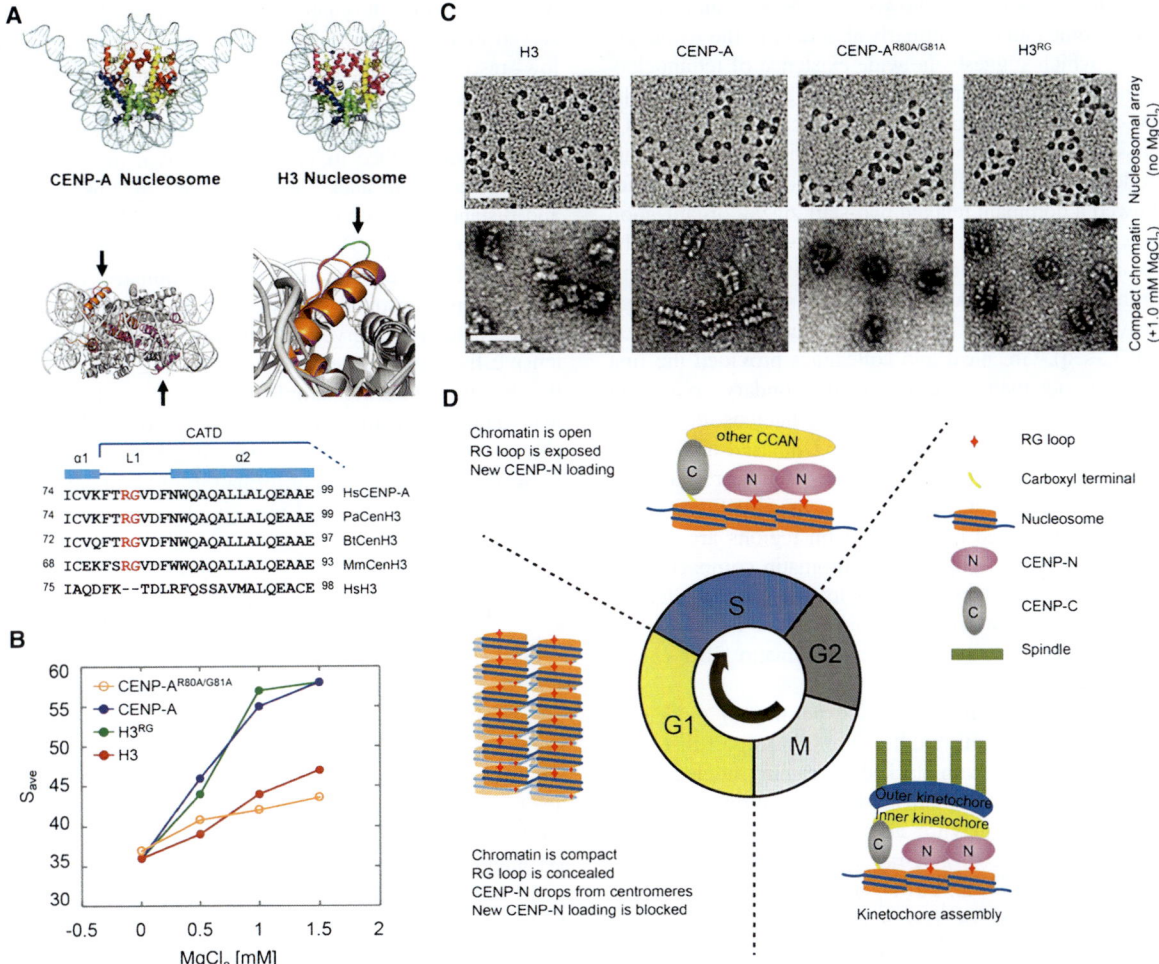

Figure 5. Histone variant CENP-A regulates centromeric chromatin structures for cell-dependent recruitment of CENP-N. (*A*) The comparison between the X-ray structures of nucleosomes containing either CENP-A or canonical H3 (*top*) and the side view of CENP-A nucleosome (*bottom*) with the CENP-A molecule shown in magenta and superimposition of H3 loop1 regions (orange). Arrows indicate the tip of the CENP-A loop1 containing Arg80 and Gly81 residues (green). ClustalX2 alignment of the CATD domain is shown with residues R80G81 highlighted (red) in CENP-A orthologs from *Homo sapiens* (Hs), *Pongo abelii* (Pa), *Bos taurus* (Bt), and *Mus musculus* (Mm), and the corresponding region in human H3. (*B*) Distribution plots of S_{ave} values of chromatins containing canonical H3, CENP-A, CENP-A$^{R80A/G81A}$, and H3^{RG} at different concentration of MgCl$_2$. (*C*) EM images of the canonical H3, CENP-A, CENP-A$^{R80A/G81A}$, and H3^{RG} containing nucleosomal arrays (by the metal-shadowing method; scale bar, 50 nm) and their corresponding compact states in 1.0 mM MgCl$_2$ (by the negatively stained method; scale bar, 100 nm). (*D*) The model of structural transitions of centromeric chromatin regulate the cell cycle–dependent recruitment of CENP-N via modulating the accessibility of the RG loop of CENP-A. (Adapted, with permission, from Fang et al. 2015.)

ORGANIZATION OF CHROMATIN FIBERS IN NUCLEUS

The 30-nm chromatin fiber has long been thought to be the first level of the hierarchical chromatin compaction, but the existence of the 30-nm fiber in vivo still remains controversial. Studies of purified or reconstituted chromatin have provided evidence supporting the existence of longitudinally compacted 30-nm chromatin fibers; however, until now high-resolution imaging of chromatin in living cells had not been possible (Eltsov et al. 2008; Fussner et al. 2011a). Early cryo-EM studies have shown that a 30-nm fiber is indeed the most predominant form of starfish sperm and nucleated chicken erythrocyte chromatin (Horowitz et al. 1994; Woodcock 1994; Scheffer et al.

2011). Using cryo-EM tomography of vitreous sections, Frangakis and colleagues showed that the most predominant form of chromatin in chicken erythrocyte nuclei is indeed a 30-nm fiber arranged in a two-start helix formation with ~6.7 nucleosomes per turn, in which the nucleosomes are juxtaposed face to face (Scheffer et al. 2011). In addition, the stacked nucleosomes were shown to be shifted off their superhelical axes, with an axial translation of ~3.4 nm and an azimuthal rotation of ~54°, very consistent with the high-resolution structure of in vitro reconstituted chromatin fibers (Schalch et al. 2005; Song et al. 2014). In addition, Rando and colleagues recently developed a novel Hi-C-based method named "Micro-C" to probe nucleosome organization at nucleosome resolution in yeast (Hsieh et al. 2015). Despite the lack of periodicity in their data set, they showed that N/N+2 and N/N+1 nucleosome pairs are similarly abundant in the whole genome, which suggests the wide existence of tetranucleosomal folding motifs in yeast genome. Similarly, using the EM-assisted nucleosome interaction capture (EMANIC) cross-linking experiments in combination with mesoscale modeling of chromatin fibers, Schlick and colleagues showed a dominant relaxed two-start zigzag organization rather than longitudinal compaction associated with the 30-nm fiber (Grigoryev et al. 2009). Most recently, combining the ionizing radiation-induced spatially correlated cleavage of DNA with deep-sequencing technique (RICC-seq), Greenleaf and colleagues provided the first genome-wide map of the chromatin secondary structure in living human cells at the one- to three-nucleosome (50- to 500-bp) scale (Risca et al. 2017). Unbiased analysis of RICC-seq signals in intact interphase nuclei reveals that RICC-seq fragmentation patterns in H3K9me3- and H3K27me3-marked heterochromatin regions are consistent with variable longitudinal chromatin compaction of two-start helical fibers with face-to-face stacked alternating nucleosomes in tri- or tetranucleosome units, as seen in the structure of reconstituted chromatin fiber (Song et al. 2014).

This overwhelming body of experimental evidences from EM and genomic studies strongly support that two-start helical fibers with stacked alternating nucleosomes are an important mechanism for generating chromatin compaction both in vitro and in vivo. However, cryo-EM (Gan et al. 2013), X-ray scattering (Nishino et al. 2012; Maeshima et al. 2014), and electron spectroscopy imaging (ESI) studies (Bazett-Jones et al. 2008; Fussner et al. 2011b, 2012) of the nucleus do not support the existence of the regular 30-nm fiber within intact cells (Eltsov et al. 2008; Fussner et al. 2011a,b; Nishino et al. 2012). Because of the poor contrast of DNA in vitreous ice and the limited 3D sampling volume of ESI, it is technically difficult to identify chromatin unambiguously and to reconstruct 3D organization of chromatin through large nuclear volume in cryo-EM experiments. Most recently, O'Shea and colleagues developed a novel ChromEMT technique by combining electron microscopy tomography (EMT) with a labeling method, which explores a fluorescent dye (DRAQ5) that binds to DNA and enhances the contrast of DNA, enabling chromatin to be visualized with OsO_4 in EM (Ou et al. 2017). Using this technique, they only observed a flexible and disordered granular chromatin chain with diameters between 5 and 24 nm, but no regular higher-order chromatin fibers in human interphase and mitotic cells in situ. However, a few important issues still need to be clarified in the technique. The first is whether DRAQ5 also binds to other type nucleic acids in situ, such as RNA (particularly double-stranded RNA [dsRNA]). It is unclear how to distinguish chromatin fibers from other nucleic acid–protein particles, which are abundant and distributed widely in nuclei. Second, no evidence in vitro or in vivo showed that DRAQ5 can bind equally to free DNA and chromatin DNA. It is likely that the DNA in compacted chromatin fragments may not be accessible for staining by the dye. In addition, the path

of DNA in the chromatin fiber cannot be directly observed in vivo by the ChromEMT technique under such low resolution. As discussed above, we and others showed that chromatin fibers mainly adopt a two-started helical conformation with alternating nucleosomes stacked face to face in vitro and in vivo, in which the two nucleosomal stacks are connected by linker DNA and distanced from each by ~7.5–10 nm (Bednar et al. 1998; Song et al. 2014; Li et al. 2016). In this context, when the linker DNA cannot be discriminated, only two separated nucleosomal stacks with a diameter of ~10 nm can be actually observed in ChromEMT. Moreover, extensive fixation by chemical cross-linking and dehydration by ethanol during sample preparation in ChromEMT may damage the ultrastructure and 3D organization of chromatin fiber. Therefore, it is still a big challenge to preserve the natural 3D conformation of chromatin fiber in ChromEMT or cryo-EM tomography.

PERSPECTIVES AND CONCLUSION

Our 3D cryo-EM structures at 11 Å resolution have provided the fundamental structural features of the elusive 30-nm chromatin fiber and a solid foundation for understanding the basic principle of chromatin compaction, whereas higher-resolution structures of chromatin fiber are needed to uncover much more structural details for the nucleosome–nucleosome, nucleosome–H1, and H1–H1 interactions in chromatin fibers. In addition, our cryo-EM structures clearly imply the presence of three important interaction interfaces in the 30-nm fiber, whereas it still remains unclear how these interfaces are regulated by different chromatin factors. Regarding the variation of NRLs in vivo, reconstituted chromatin fibers with a combination of different NRLs will also be good candidates for single-molecule and cryo-EM studies in the future. These further studies will not only enhance our understanding of the diversity of chromatin structures in vivo but also provide structural basis for how different combinations of DNA sequences, NRLs, histone variants, chromatin modifications, and chromatin architectural proteins can be coordinated to precisely regulate the biological function of genomic DNA in the nucleus.

It is still a puzzle as to whether the structural results from the in vitro studies can represent the actual structure of chromatin fibers in situ. Therefore, the 3D organization of chromatin fibers in the intact nuclei needs to be further studied by using newly developed techniques. Well-characterized chromatin fibers reconstituted in vitro, including compacted 30-nm fibers and open nucleosomal arrays, would be perfect structural references for analyzing the 3D organization of chromatin fibers in situ in these studies. The combination of cryo-EM with super-resolution fluorescence imaging techniques has been recently developed to visualize and quantify the ultrastructure of cryo-preserved cells (Chang et al. 2014; Liu et al. 2015). The combination of genomic approaches (such as micro-C and RICC-seq) and CRISPR (clustered regularly interspaced short palindromic repeat)-based imaging techniques may enable us to probe the ultrastructure and 3D organization of chromatin fiber at defined genomic regions in the intact nuclei. Undoubtedly, more structural detail for the 3D organization of chromatin fibers in situ can be obtained by the application of these advanced imaging techniques in the future.

ACKNOWLEDGMENTS

This work was supported by grants to G.L. from the National Natural Science Foundation of China (31630041, 31525013, and 31521002), the Ministry of Science and Technology of China (2017YFA0504200, 2015CB856200), and the Chinese Academy of Sciences (CAS) Strategic Priority Research Program (XDB1904 0202); and to P.C. from the National Natural Science Foundation of China (31471218) and the Youth Innovation Promotion Association CAS (2015071). The work was also supported by the CAS Key Research Program on Frontier Science (QYZDY-SSW-SMC020) and HHMI international research scholar grant (55008737) for G.L. The authors declare that there are no conflicts of interest.

REFERENCES

Allan J, Hartman PG, Crane-Robinson C, Aviles FX. 1980. The structure of histone H1 and its location in chromatin. *Nature* **288:** 675–679.

Allshire RC, Karpen GH. 2008. Epigenetic regulation of centromeric chromatin: Old dogs, new tricks? *Nat Rev Genet* **9:** 923–937.

Bassett EA, DeNizio J, Barnhart-Dailey MC, Panchenko T, Sekulic N, Rogers DJ, Foltz DR, Black BE. 2012. HJURP uses distinct CENP-A surfaces to recognize and to stabilize CENP-A/histone H4 for centromere assembly. *Dev Cell* **22:** 749–762.

Bates DL, Thomas JO. 1981. Histones H1 and H5: One or two molecules per nucleosome? *Nucleic Acids Res* **9:** 5883–5894.

Bazett-Jones DP, Li R, Fussner E, Nisman R, Dehghani H. 2008. Elucidating chromatin and nuclear domain architecture with electron spectroscopic imaging. *Chromosome Res* **16:** 397–412.

Bednar J, Horowitz RA, Grigoryev SA, Carruthers LM, Hansen JC, Koster AJ, Woodcock CL. 1998. Nucleosomes, linker DNA, and linker histone form a unique structural motif that directs the higher-order folding and compaction of chromatin. *Proc Natl Acad Sci* **95:** 14173–14178.

Belotserkovskaya R, Oh S, Bondarenko VA, Orphanides G, Studitsky VM, Reinberg D. 2003. FACT facilitates transcription-dependent nucleosome alteration. *Science* **301:** 1090–1093.

Bintu L, Ishibashi T, Dangkulwanich M, Wu YY, Lubkowska L, Kashlev M, Bustamante C. 2012. Nucleosomal elements that control the topography of the barrier to transcription. *Cell* **151:** 738–749.

Black BE, Foltz DR, Chakravarthy S, Luger K, Woods VL Jr, Cleveland DW. 2004. Structural determinants for generating centromeric chromatin. *Nature* **430:** 578–582.

Brower-Toland BD, Smith CL, Yeh RC, Lis JT, Peterson CL, Wang MD. 2002. Mechanical disruption of individual nucleosomes reveals a reversible multistage release of DNA. *Proc Natl Acad Sci* **99:** 1960–1965.

Carroll CW, Silva MC, Godek KM, Jansen LE, Straight AF. 2009. Centromere assembly requires the direct recognition of CENP-A nucleosomes by CENP-N. *Nat Cell Biol* **11:** 896–902.

Chang YW, Chen S, Tocheva EI, Treuner-Lange A, Löbach S, Sogaard-Andersen L, Jensen GJ. 2014. Correlated cryogenic photoactivated localization microscopy and cryo-electron tomography. *Nat Methods* **11:** 737–739.

Chen P, Zhao J, Wang Y, Wang M, Long H, Liang D, Huang L, Wen Z, Li W, Li X, et al. 2013. H3.3 actively marks enhancers and primes gene transcription via opening higher-ordered chromatin. *Genes Dev* **27:** 2109–2124.

Collepardo-Guevara R, Schlick T. 2011. The effect of linker histone's nucleosome binding affinity on chromatin unfolding mechanisms. *Biophys J* **101:** 1670–1680.

Cui Y, Bustamante C. 2000. Pulling a single chromatin fiber reveals the forces that maintain its higher-order structure. *Proc Natl Acad Sci* **97:** 127–132.

Cutter AR, Hayes JJ. 2015. A brief review of nucleosome structure. *FEBS Lett* **589:** 2914–2922.

Dorigo B, Schalch T, Kulangara A, Duda S, Schroeder RR, Richmond TJ. 2004. Nucleosome arrays reveal the two-start organization of the chromatin fiber. *Science* **306:** 1571–1573.

Dunleavy EM, Roche D, Tagami H, Lacoste N, Ray-Gallet D, Nakamura Y, Daigo Y, Nakatani Y, Almouzni-Pettinotti G. 2009. HJURP is a cell-cycle-dependent maintenance and deposition factor of CENP-A at centromeres. *Cell* **137:** 485–497.

Elsässer SJ, Huang H, Lewis PW, Chin JW, Allis CD, Patel DJ. 2012. DAXX envelops a histone H3.3-H4 dimer for H3.3-specific recognition. *Nature* **491:** 560–565.

Eltsov M, MacLellan KM, Maeshima K, Frangakis AS, Dubochet J. 2008. Analysis of cryo-electron microscopy images does not support the existence of 30-nm chromatin fibers in mitotic chromosomes in situ. *Proc Natl Acad Sci* **105:** 19732–19737.

Fang J, Liu Y, Wei Y, Deng W, Yu Z, Huang L, Teng Y, Yao T, You Q, Ruan H, et al. 2015. Structural transitions of centromeric chromatin regulate the cell cycle–dependent recruitment of CENP-N. *Genes Dev* **29:** 1058–1073.

Finch JT, Klug A. 1976. Solenoidal model for superstructure in chromatin. *Proc Natl Acad Sci* **73:** 1897–1901.

Flaus A, Rencurel C, Ferreira H, Wiechens N, Owen-Hughes T. 2004. *Sin* mutations alter inherent nucleosome mobility. *EMBO J* **23:** 343–353.

Foltz DR, Jansen LE, Black BE, Bailey AO, Yates JR III, Cleveland DW. 2006. The human CENP-A centromeric nucleosome-associated complex. *Nat Cell Biol* **8:** 458–469.

Foltz DR, Jansen LE, Bailey AO, Yates JR III, Bassett EA, Wood S, Black BE, Cleveland DW. 2009. Centromere-specific assembly of CENP-a nucleosomes is mediated by HJURP. *Cell* **137:** 472–484.

Formosa T. 2013. The role of FACT in making and breaking nucleosomes. *Biochim Biophys Acta* **1819:** 247–255.

Francis NJ, Kingston RE, Woodcock CL. 2004. Chromatin compaction by a polycomb group protein complex. *Science* **306:** 1574–1577.

Fujimoto M, Takaki E, Takii R, Tan K, Prakasam R, Hayashida N, Iemura S, Natsume T, Nakai A. 2012. RPA assists HSF1 access to nucleosomal DNA by recruiting histone chaperone FACT. *Mol Cell* **48:** 182–194.

Fussner E, Ching RW, Bazett-Jones DP. 2011a. Living without 30 nm chromatin fibers. *Trends Biochem Sci* **36:** 1–6.

Fussner E, Djuric U, Strauss M, Hotta A, Perez-Iratxeta C, Lanner F, Dilworth FJ, Ellis J, Bazett-Jones DP. 2011b. Constitutive heterochromatin reorganization during somatic cell reprogramming. *EMBO J* **30:** 1778–1789.

Fussner E, Strauss M, Djuric U, Li R, Ahmed K, Hart M, Ellis J, Bazett-Jones DP. 2012. Open and closed domains in the mouse genome are configured as 10-nm chromatin fibres. *EMBO Rep* **13:** 992–996.

Gan L, Ladinsky MS, Jensen GJ. 2013. Chromatin in a marine picoeukaryote is a disordered assemblage of nucleosomes. *Chromosoma* **122:** 377–386.

Gerchman SE, Ramakrishnan V. 1987. Chromatin higher-order structure studied by neutron scattering and scanning transmission electron microscopy. *Proc Natl Acad Sci* **84:** 7802–7806.

Ghirlando R, Felsenfeld G. 2008. Hydrodynamic studies on defined heterochromatin fragments support a 30-nm fiber having six nucleosomes per turn. *J Mol Biol* **376:** 1417–1425.

Goldberg AD, Banaszynski LA, Noh KM, Lewis PW, Elsaesser SJ, Stadler S, Dewell S, Law M, Guo X, Li X, et al. 2010. Distinct factors control histone variant H3.3 localization at specific genomic regions. *Cell* **140**: 678–691.

Greaves IK, Rangasamy D, Ridgway P, Tremethick DJ. 2007. H2A.Z contributes to the unique 3D structure of the centromere. *Proc Natl Acad Sci* **104**: 525–530.

Grigoryev SA, Arya G, Correll S, Woodcock CL, Schlick T. 2009. Evidence for heteromorphic chromatin fibers from analysis of nucleosome interactions. *Proc Natl Acad Sci* **106**: 13317–13322.

Hake SB, Allis CD. 2006. Histone H3 variants and their potential role in indexing mammalian genomes: The "H3 barcode hypothesis". *Proc Natl Acad Sci* **103**: 6428–6435.

Hall MA, Shundrovsky A, Bai L, Fulbright RM, Lis JT, Wang MD. 2009. High-resolution dynamic mapping of histone–DNA interactions in a nucleosome. *Nat Struct Mol Biol* **16**: 124–129.

Hellwig D, Emmerth S, Ulbricht T, Döring V, Hoischen C, Martin R, Samora CP, McAinsh AD, Carroll CW, Straight AF, et al. 2011. Dynamics of CENP-N kinetochore binding during the cell cycle. *J Cell Sci* **124**: 3871–3883.

Hondele M, Stuwe T, Hassler M, Halbach F, Bowman A, Zhang ET, Nijmeijer B, Kotthoff C, Rybin V, Amlacher S, et al. 2013. Structural basis of histone H2A–H2B recognition by the essential chaperone FACT. *Nature* **499**: 111–114.

Horowitz RA, Agard DA, Sedat JW, Woodcock CL. 1994. The three-dimensional architecture of chromatin in situ: Electron tomography reveals fibers composed of a continuously variable zig-zag nucleosomal ribbon. *J Cell Biol* **125**: 1–10.

Hsieh TH, Weiner A, Lajoie B, Dekker J, Friedman N, Rando OJ. 2015. Mapping nucleosome resolution chromosome folding in yeast by Micro-C. *Cell* **162**: 108–119.

Hu H, Liu Y, Wang M, Fang J, Huang H, Yang N, Li Y, Wang J, Yao X, Shi Y, et al. 2011. Structure of a CENP-A-histone H4 heterodimer in complex with chaperone HJURP. *Genes Dev* **25**: 901–906.

Jin C, Zang C, Wei G, Cui K, Peng W, Zhao K, Felsenfeld G. 2009. H3.3/H2A.Z double variant-containing nucleosomes mark "nucleosome-free regions" of active promoters and other regulatory regions. *Nat Genet* **41**: 941–945.

Kruithof M, Chien FT, Routh A, Logie C, Rhodes D, van Noort J. 2009. Single-molecule force spectroscopy reveals a highly compliant helical folding for the 30-nm chromatin fiber. *Nat Struct Mol Biol* **16**: 534–540.

Langmore JP, Paulson JR. 1983. Low angle x-ray diffraction studies of chromatin structure in vivo and in isolated nuclei and metaphase chromosomes. *J Cell Biol* **96**: 1120–1131.

Li G, Reinberg D. 2011. Chromatin higher-order structures and gene regulation. *Curr Opin Genet Dev* **21**: 175–186.

Li G, Margueron R, Hu G, Stokes D, Wang YH, Reinberg D. 2010. Highly compacted chromatin formed in vitro reflects the dynamics of transcription activation in vivo. *Mol Cell* **38**: 41–53.

Li W, Chen P, Yu J, Dong L, Liang D, Feng J, Yan J, Wang PY, Li Q, Zhang Z, et al. 2016. FACT remodels the tetranucleosomal unit of chromatin fibers for gene transcription. *Mol Cell* **64**: 120–133.

Liu CP, Xiong C, Wang M, Yu Z, Yang N, Chen P, Zhang Z, Li G, Xu RM. 2012. Structure of the variant histone H3.3-H4 heterodimer in complex with its chaperone DAXX. *Nat Struct Mol Biol* **19**: 1287–1292.

Liu B, Xue Y, Zhao W, Chen Y, Fan C, Gu L, Zhang Y, Zhang X, Sun L, Huang X, et al. 2015. Three-dimensional super-resolution protein localization correlated with vitrified cellular context. *Sci Rep* **5**: 13017.

Luger K, Hansen JC. 2005. Nucleosome and chromatin fiber dynamics. *Curr Opin Struct Biol* **15**: 188–196.

Luger K, Mäder AW, Richmond RK, Sargent DF, Richmond TJ. 1997. Crystal structure of the nucleosome core particle at 2.8 A resolution. *Nature* **389**: 251–260.

Maeshima K, Imai R, Hikima T, Joti Y. 2014. Chromatin structure revealed by X-ray scattering analysis and computational modeling. *Methods* **70**: 154–161.

McCullough L, Connell Z, Petersen C, Formosa T. 2015. The Abundant histone chaperones Spt6 and FACT collaborate to assemble, inspect, and maintain chromatin structure in *Saccharomyces cerevisiae*. *Genetics* **201**: 1031–1045.

McGinty RK, Tan S. 2015. Nucleosome structure and function. *Chem Rev* **115**: 2255–2273.

Meng H, Andresen K, van Noort J. 2015. Quantitative analysis of single-molecule force spectroscopy on folded chromatin fibers. *Nucleic Acids Res* **43**: 3578–3590.

Nishino Y, Eltsov M, Joti Y, Ito K, Takata H, Takahashi Y, Hihara S, Frangakis AS, Imamoto N, Ishikawa T, et al. 2012. Human mitotic chromosomes consist predominantly of irregularly folded nucleosome fibres without a 30-nm chromatin structure. *EMBO J* **31**: 1644–1653.

Orphanides G, LeRoy G, Chang C-H, Luse DS, Reinberg D. 1998. FACT, a factor that facilitates transcript elongation through nucleosomes. *Cell* **92**: 105–116.

Orphanides G, Wu WH, Lane WS, Hampsey M, Reinberg D. 1999. The chromatin-specific transcription elongation factor FACT comprises human SPT16 and SSRP1 proteins. *Nature* **400**: 284–288.

Ou HD, Phan S, Deerinck TJ, Thor A, Ellisman MH, O'Shea CC. 2017. ChromEMT: Visualizing 3D chromatin structure and compaction in interphase and mitotic cells. *Science* **357**: eaag0025.

Pope LH, Bennink ML, van Leijenhorst-Groener KA, Nikova D, Greve J, Marko JF. 2005. Single chromatin fiber stretching reveals physically distinct populations of disassembly events. *Biophys J* **88**: 3572–3583.

Risca VI, Denny SK, Straight AF, Greenleaf WJ. 2017. Variable chromatin structure revealed by in situ spatially correlated DNA cleavage mapping. *Nature* **541**: 237–241.

Saunders A, Werner J, Andrulis ED, Nakayama T, Hirose S, Reinberg D, Lis JT. 2003. Tracking FACT and the RNA polymerase II elongation complex through chromatin in vivo. *Science* **301**: 1094–1096.

Schalch T, Duda S, Sargent DF, Richmond TJ. 2005. X-ray structure of a tetranucleosome and its implications for the chromatin fibre. *Nature* **436**: 138–141.

Scheffer MP, Eltsov M, Frangakis AS. 2011. Evidence for short-range helical order in the 30-nm chromatin fibers of erythrocyte nuclei. *Proc Natl Acad Sci* **108**: 16992–16997.

Schwartz BE, Ahmad K. 2005. Transcriptional activation triggers deposition and removal of the histone variant H3.3. *Genes Dev* **19**: 804–814.

Song F, Chen P, Sun D, Wang M, Dong L, Liang D, Xu RM, Zhu P, Li G. 2014. Cryo-EM study of the chromatin fiber reveals a double helix twisted by tetranucleosomal units. *Science* **344**: 376–380.

Stuwe T, Hothorn M, Lejeune E, Rybin V, Bortfeld M, Scheffzek K, Ladurner AG. 2008. The FACT Spt16 "peptidase" domain is a histone H3-H4 binding module. *Proc Natl Acad Sci* **105**: 8884–8889.

Syed SH, Goutte-Gattat D, Becker N, Meyer S, Shukla MS, Hayes JJ, Everaers R, Angelov D, Bednar J, Dimitrov S. 2010. Single-base resolution mapping of H1-nucleosome interactions and 3D organization of the nucleosome. *Proc Natl Acad Sci* **107**: 9620–9625.

Tachiwana H, Kagawa W, Shiga T, Osakabe A, Miya Y, Saito K, Hayashi-Takanaka Y, Oda T, Sato M, Park SY, et al. 2011. Crystal structure of the human centromeric nucleosome containing CENP-A. *Nature* **476**: 232–235.

Thakar A, Gupta P, Ishibashi T, Finn R, Silva-Moreno B, Uchiyama S, Fukui K, Tomschik M, Ausio J, Zlatanova J. 2009. H2A.Z and H3.3 histone variants affect nucleosome structure: Biochemical and biophysical studies. *Biochemistry* **48**: 10852–10857.

Thoma F, Koller T, Klug A. 1979. Involvement of histone H1 in the organization of the nucleosome and of the salt-dependent superstructures of chromatin. *J Cell Biol* **83**: 403–427.

Thomas JO. 1999. Histone H1: Location and role. *Curr Opin Cell Biol* **11**: 312–317.

Tosi A, Haas C, Herzog F, Gilmozzi A, Berninghausen O, Unge-wickell C, Gerhold CB, Lakomek K, Aebersold R, Beckmann R, et al. 2013. Structure and subunit topology of the INO80 chromatin remodeler and its nucleosome complex. *Cell* **154:** 1207–1219.

Tsunaka Y, Fujiwara Y, Oyama T, Hirose S, Morikawa K. 2016. Integrated molecular mechanism directing nucleosome reorga-nization by human FACT. *Genes Dev* **30:** 673–686.

van der Heijden GW, Derijck AA, Pósfai E, Giele M, Pelczar P, Ramos L, Wansink DG, van der Vlag J, Peters AH, de Boer P. 2007. Chromosome-wide nucleosome replacement and H3.3 incorporation during mammalian meiotic sex chromosome in-activation. *Nat Genet* **39:** 251–258.

van Holde K, Zlatanova J. 1996. Chromatin architectural proteins and transcription factors: A structural connection. *Bioessays* **18:** 697–700.

Watson JD, Crick FH. 1953. Molecular structure of nucleic acids; a structure for deoxyribose nucleic acid. *Nature* **171:** 737–738.

Widom J, Klug A. 1985. Structure of the 300A chromatin filament: X-ray diffraction from oriented samples. *Cell* **43:** 207–213.

Williams SP, Athey BD, Muglia LJ, Schappe RS, Gough AH, Langmore JP. 1986. Chromatin fibers are left-handed double helices with diameter and mass per unit length that depend on linker length. *Biophys J* **49:** 233–248.

Wong LH, McGhie JD, Sim M, Anderson MA, Ahn S, Hannan RD, George AJ, Morgan KA, Mann JR, Choo KH. 2010. ATRX interacts with H3.3 in maintaining telomere structural integrity in pluripotent embryonic stem cells. *Genome Res* **20:** 351–360.

Woodcock CL. 1994. Chromatin fibers observed in situ in frozen hydrated sections. Native fiber diameter is not correlated with nucleosome repeat length. *J Cell Biol* **125:** 11–19.

Woodcock CL, Frado LL, Rattner JB. 1984. The higher-order structure of chromatin: Evidence for a helical ribbon arrange-ment. *J Cell Biol* **99:** 42–52.

Yu Z, Zhou X, Wang W, Deng W, Fang J, Hu H, Wang Z, Li S, Cui L, Shen J, et al. 2015. Dynamic phosphorylation of CENP-A at Ser68 orchestrates its cell-cycle-dependent deposition at centromeres. *Dev Cell* **32:** 68–81.

Zhang W, Zeng F, Liu Y, Shao C, Li S, Lv H, Shi Y, Niu L, Teng M, Li X. 2015. Crystal structure of human SSRP1 middle domain reveals a role in DNA binding. *Sci Rep* **5:** 18688.

Zhu P, Li G. 2016. Structural insights of nucleosome and the 30-nm chromatin fiber. *Curr Opin Struct Biol* **36:** 106–115.

The Role of Bromodomain and Extraterminal Motif (BET) Proteins in Chromatin Structure

SARAH C. HSU[1] AND GERD A. BLOBEL[1,2]

[1]Perelman School of Medicine, University of Pennsylvania, Philadelphia, Pennsylvania 19104
[2]Division of Hematology, Children's Hospital of Philadelphia, Philadelphia, Pennsylvania 19104
Correspondence: blobel@email.chop.edu

Bromodomain and extraterminal motif (BET) proteins have been widely investigated for their roles in gene regulation and their potential as therapeutic targets in cancer. Pharmacologic BET inhibitors target the conserved bromodomain–acetyllysine interaction and do not distinguish between BRD2, BRD3, and BRD4. Thus, comparatively little is known regarding the distinct roles played by individual family members, as well as the underlying mechanisms that drive the transcriptional effects of BET inhibitors. Here we review studies regarding the contributions of BET proteins to genome structure and function, including recent work identifying a role for BRD2 as a component of functional and physical chromatin domain boundaries. We also discuss directions of future studies aimed at providing insights into broader architectural functions of BET proteins and their roles in chromatin domain boundary formation.

Bromodomain and extraterminal motif (BET) proteins bind chromatin and play critical roles in transcription. The BET family consists of BRD2, BRD3, and BRD4, which are broadly expressed across tissues, as well as BRDT, which is predominantly present in the testes (Shang et al. 2004). BET proteins are characterized by their conserved domain structure, which consists of two amino-terminal bromodomains that bind acetylated lysine residues on histones and other proteins, as well as an extraterminal domain that mediates further protein–protein interactions (for review, see Belkina and Denis 2012). BET proteins are generally considered to be chromatin "readers" that translate the chromatin state into gene activity by recruiting transcriptional regulatory complexes to their binding sites. BRD4 in particular has been widely investigated as a transcriptional activator. BRD4 is present in protein complexes such as PTEF-b (Jang et al. 2005; Yang et al. 2005; Bisgrove et al. 2007), which phosphorylates control elements of the elongation machinery and RNA polymerase II (Pol II) to stimulate productive transcription elongation. The functions of BRD2 and BRD3 are less well-characterized, although BRD2 has also been associated with gene activation as a member of E2F-containing complexes (Denis et al. 2000; Sinha et al. 2005).

The advent of pharmacologic inhibitors of the BET bromodomain–acetyllysine interaction has revealed that targeting BET proteins has therapeutic potential in a wide variety of disorders including cancer (Filippakopoulos et al. 2010; Dawson et al. 2011; Delmore et al. 2011; Zuber et al. 2011; Shi and Vakoc 2014), heart failure (Anand et al. 2013), and inflammatory conditions (Nicodeme et al. 2011). Interestingly, despite the ubiquitous expression of BET proteins and the presence of one or more at essentially all active genes (Anders et al. 2013),

BET inhibition does not result in global transcriptional down-regulation but rather has gene-specific effects that seem to depend largely on cellular context (for review, see Shi and Vakoc 2014). BRD2, BRD3, and BRD4 can bind to overlapping as well as distinct regions of the genome (Anders et al. 2013; Asangani et al. 2015; Stonestrom et al. 2015), yet whether they regulate different genes or perform separate functions has yet to be clearly delineated. This point is also relevant when interpreting the biological functions of BET inhibitors, most of which target all members of the BET family.

Many questions remain regarding the mechanisms by which BET proteins regulate transcription. In particular, what explains the gene selectivity of BET inhibitors? What distinguishes the roles of BRD2, BRD3, and BRD4? Do they function at distinct stages of transcription? Which BET proteins account for the phenotype of BET inhibition in a particular context? Here we discuss a role for BET proteins in the maintenance of chromatin structure, from the local chromatin environment to large-scale 3D genome architecture, and how architectural functions of BET proteins may influence gene regulation.

BET PROTEINS ARE LINKED TO THE MAINTENANCE OF LARGE-SCALE NUCLEAR STRUCTURE

BET proteins are most widely known for their ability to "read" or "interpret" acetylated chromatin. Most work regarding BET proteins has focused on their roles as transcriptional activators; however, a number of studies have also highlighted functions of BET proteins important for large-scale chromatin structure. Specifically, either deple-

Published by Cold Spring Harbor Laboratory Press; doi: 10.1101/sqb.2017.82.033829
Cold Spring Harbor Symposia on Quantitative Biology, Volume LXXXII

tion of BRD4 or disruption of BRD4 binding using a dominant-negative form consisting of the tandem bromodomains leads to chromatin decondensation and fragmentation as measured by MNase and imaging techniques (Wang et al. 2012). Of note, given the conserved structure of the bromodomains between family members, it is possible that such a method of inhibition could also interfere with BRD2 or BRD3 function. Regions in the carboxyl terminus of BRD4 are important for maintaining chromatin structure (Wang et al. 2012), suggesting that BRD4 may recruit additional proteins to perform this function. Other studies have reported a role for BRD4 in chromatin compaction, yet have described conflicting results. Specifically, recruitment (rather than disruption) of BRD4 to an artificial locus using a LacI fusion system resulted in expansion of the region in imaging studies (Zhao et al. 2011), suggesting that BRD4 is linked to chromatin decompaction. Overexpression of BRD4 in HeLa cells resulted in increased nuclear volume, and BRD4 levels were also associated with increased sensitivity of chromatin to MNase digestion in vitro (Devaiah et al. 2016), suggesting that BRD4's regulation of global chromatin structure is likely complex and context-specific.

BRDT, another member of the BET family that is expressed primarily in the testes, plays a role in chromatin structure during spermatogenesis. Spermatogenesis is a dynamic process during which chromatin organization and cellular morphology undergo dramatic changes—namely, nuclear compaction and histone hyperacetylation and replacement with transition proteins and protamines (Rajender et al. 2011). Ectopic expression of BRDT in non–germ cells resulted in global changes in chromatin organization, including inducing the formation of condensed nuclear foci in a manner requiring the bromodomains (Pivot-Pajot et al. 2003). Disruption of *Brdt* in mice results in disrupted nuclear morphology in sperm cells as revealed by electron microscopy, as well as fragmentation of the chromocenter, a structure formed by pericentromeric heterochromatin (Shang et al. 2007; Berkovits and Wolgemuth 2011). These observations suggest that BRDT has critical roles in the maintenance of chromatin structure. Notably, in spermatids BRD4 is enriched in a unique ring-like structure at the base of the acrosome, and it has been speculated that this ring might contain a slightly smaller spermatid-specific BRD4 isoform that binds to acetylated proteins to promote chromatin compaction and reorganization during spermatid maturation (Bryant et al. 2015).

In addition, a number of studies have detected persistent BET binding to chromatin during mitosis, a phase of the cell cycle associated with extreme changes in 3D chromatin structure (Naumova et al. 2013). BRD4 (Dey et al. 2000, 2009; Zhao et al. 2011), BRD2 (Kanno et al. 2004), and BRD3 (Garcia-Gutierrez et al. 2012) have been shown to remain bound to mitotic chromosomes by various methods including chromatin immunoprecipitation and imaging studies. Whether or how BET proteins influence mitotic chromosome organization has yet to be tested.

Although manipulation of BET proteins results in clear defects in chromatin organization in the nucleus, the mechanism by which they contribute to chromatin architecture remains to be elucidated. In particular, do BET proteins themselves play a structural or architectural role? Alternatively, are the structural changes observed secondary to transcriptional disruption? How do effects on chromatin structure relate to gene regulation by BET proteins? Observations regarding these questions are discussed in the sections that follow.

BET PROTEINS MODULATE THE LOCAL CHROMATIN ENVIRONMENT

One mechanism by which BET proteins may control chromatin organization is through effects on the surrounding chromatin environment. BET proteins bind to acetylated residues on histones and other proteins, and a number of lines of evidence suggest that BETs can either directly or indirectly influence histone modifications and DNA accessibility. Proteomics studies have shown that BETs reside in nuclear complexes with chromatin-modifying enzymes (Dawson et al. 2011; Rahman et al. 2011) such as histone methyltransferases and chromatin remodeling machinery. BRD4 associates with the H3K36 methyltransferase NSD3 (Rahman et al. 2011), and BRD4 depletion results in a reduction in H3K36me3 at specific loci. In addition, a short version of NSD3 lacking catalytic activity connects BRD4 to CHD8, an ATP-dependent chromatin-remodeling enzyme (Shen et al. 2015), indicating that BET proteins can indirectly impact the local structure of the genes they regulate. BRD2 and BRD3 associate with hyperacetylated nucleosomes and assist Pol II transcription elongation in vitro, suggesting they have histone chaperone activity (LeRoy et al. 2008).

Recent reports have suggested that BRD4 also has inherent activity as a protein kinase (Devaiah et al. 2012) and histone acetyltransferase (Devaiah et al. 2016). The latter has been proposed to deposit H3K122ac, a modification on the surface of the globular histone domain. This study showed that BRD4 was associated with chromatin decompaction, nucleosome eviction in vitro, and decreased nucleosome occupancy in vivo. It should be noted, however, that although H3K122ac stimulates transcription, it did not significantly impact chromatin compaction in in vitro chromatin assembly assays in other reports (Tropberger et al. 2013), suggesting that BRD4's effects on chromatin compaction may be propagated through multiple mechanisms.

BRD4, and likely BRD3, also plays a role in defective chromatin regulation in the setting of malignancy. NUT midline carcinoma, a rare squamous cell cancer, is driven by translocations generating oncogenic fusion proteins that bring BRD3 or BRD4 into close association with the protein NUT. BRD4-NUT fusions form foci in the nuclei of NUT midline carcinoma cells and can generate large, hyperacetylated "megadomains" that are associated with broad transcriptional activation of the genes they contain (Alekseyenko et al. 2015), likely through a mechanism involving the p300 acetyltransferase (Alekseyenko et al. 2017). Interestingly these megadomains appear to initiate from active enhancers and are largely cell type–specific, indicating that the impact of BET proteins on chromatin structure may be strongly influenced by the existing cellular transcriptional program.

Thus, one mechanism by which BETs alter chromatin structure may be through modulation of histone modifications that govern the local chromatin environment. However, whether such changes are directly linked to gene expression awaits further study. In particular, can the transcriptional effects of BET inhibitors be rescued by restoring the binding of BET-associated histone modifiers? Targeted tethering of BET proteins or their downstream cofactors to specific genomic loci may provide insight into establishing a causal relationship. In addition, the effects of BET disruption on surrounding histone modifications may be secondary to changes in the transcriptional state of a given gene. Using systems such as the auxin-inducible or FKBP-derived degrons to rapidly and specifically degrade BRD2, BRD3, or BRD4 has the potential to provide insight into the direct effects of BET proteins on local chromatin structure.

BETs AS INSULATOR/ARCHITECTURAL PROTEINS

Early work in yeast provided evidence that BETs may play a role in chromatin boundary integrity or barrier function. Yeast have two BET homologs, Bdf1 and Bdf2, that share domain structure similar to mammalian BETs (Wu and Chiang 2007). Microarray analysis of gene expression in yeast strains with either deletion of *BDF1* or mutations in Bdf1 that abrogate binding to acetylated residues revealed that nearly one-third of down-regulated genes are within 100 kb of telomeres (Ladurner et al. 2003). The authors showed that this occurs through a mechanism involving competitive binding between Bdf1 and the SIR silencing complex, with loss of acetyl-histone binding by Bdf1 allowing the expansion of telomere-proximal binding of Sir3 with concomitant reduction of H4 acetylation. This mechanism is further supported by the observation that yeast lacking SIR3 have the opposite phenotype, with increased subtelomeric binding of Bdf1. Thus in yeast, Bdf1 performs a type of barrier function to prevent the inappropriate spread of heterochromatin. Whether BET inhibition similarly allows the spread of heterochromatin or the blurring of heterochromatin–euchromatin boundaries in mammalian cells has not been explored in detail. Interestingly, this model implies that the overall levels of BET proteins within the cell must be carefully titrated to ensure an appropriate balance between the active and repressed chromatin state.

A barrier function for BRDT was also suggested by studies of spermatogenesis, where loss of the first bromodomain of BRDT in mice led to increased levels of chromatin-associated HP1α in the testes, suggesting an overall increase in heterochromatin (Berkovits and Wolgemuth 2011). By immunofluorescence imaging, BRDT and the mammalian sirtuin SIRT1 were localized adjacent to the heterochromatic chromocenter, suggesting a similar mechanism for BRDT in preventing the spread of heterochromatin as described above for yeast.

In addition to barrier functions, BET proteins associate with insulator proteins in multiple organisms. *Drosophila* have a single BET homolog, F(s)1h, which is expressed in two isoforms—F(s)1h-S (short) and F(s)1h-L (long) that contains an additional extended carboxy-terminal domain. Genome-wide analysis of differential binding by these two isoforms revealed that F(s)1h-S is present at active enhancers and promoters, while F(s)1-h-L additionally binds to sites occupied by *Drosophila* insulator proteins (Kellner et al. 2013). Fs(1)h-L interacts with the insulator proteins GAF, Su(Hw), Mod(mdg4), and CP190 in co-immunoprecipitation experiments. The ability of insulator proteins to form homo- and heterotypic interactions is thought to enable chromosome folding in a manner that constrains enhancer–promoter contacts to shield genes from inappropriate regulation (Schwartz and Cavalli 2017).

In mammalian cells we recently observed that the BET protein BRD2 occupies sites bound by the architectural/insulator protein CTCF in erythroid cells (Hsu et al. 2017). This is in line with findings from another recent study showing significant overlap between BRD2 and CTCF binding in Th17 cells (Cheung et al. 2017) and in the Kaposi's sarcoma–associated herpesvirus genome (Chen et al. 2017), indicating that the relationship between BRD2 and CTCF is a widespread phenomenon. We observed that BRD3 also binds to CTCF sites, consistent with previous work from our laboratory demonstrating partial functional redundancy between BRD2 and BRD3 in erythroid maturation (Stonestrom et al. 2015). BRD4 did not similarly localize to CTCF sites in either erythroid (Hsu et al. 2017) or Th17 (Cheung et al. 2017) cells, providing evidence for distinct roles of BET family members. We showed that mutation of a specific CTCF site using genome editing led to loss of both CTCF and BRD2 binding, whereas depletion of BRD2 appeared to have little effect on CTCF occupancy, indicating that CTCF recruits BRD2 to co-bound sites.

CTCF has known roles as an insulator protein by preventing the inappropriate spreading of chromatin states and as an enhancer blocker that inhibits errant pairing of enhancers and promoters (for review, see Phillips and Corces 2009). To test whether BRD2 and CTCF are functionally linked, we mutated a CTCF/BRD2 site that sits between two unrelated genes. One gene, *Slc25a37*, is driven by an erythroid-specific enhancer, whereas the gene on the other side, *Entpd4*, is independently regulated (Huang et al. 2016). Deletion of the CTCF/BRD2 site between these two genes led to inappropriate activation of *Entpd4*, but only under conditions in which the *Slc25a37* enhancer was active. Accordingly, the enhancer formed ectopic contacts with the *Entpd4* promoter, indicating that CTCF performs a boundary or enhancer-blocking function at this locus. Single-molecule mRNA FISH confirmed these findings by demonstrating that, at the single-cell level, the expression of these two genes becomes increasingly correlated with either deletion of the intervening CTCF/BRD2 co-occupied site or with BRD2 depletion. Conceptually analogous findings were recently reported at another erythroid gene locus, in which mutation of CTCF sites at the α-globin gene cluster in mice resulted in the formation of new contacts between the α-globin enhancer and the promoters of adjacent genes normally protected from its influence (Hanssen et al. 2017).

More recently CTCF has been shown to frequently mark the borders of topologically associating domains (TADs), which separate the genome into distinct regions that appear to limit the space within which chromatin interactions are permitted to form (Dixon et al. 2012; Nora et al. 2012). Prolonged or acute depletion of CTCF leads to increased cross-TAD boundary contacts and varying degrees of reduction in TAD boundary insulation (Zuin et al. 2014; Kubo et al. 2017; Nora et al. 2017). This is consistent with CTCF-dependent and -independent mechanisms of boundary formation. Notably, sustained BRD2 depletion in erythroid cells also led to increased contacts across chromatin domain boundaries, specifically at those boundaries occupied by BRD2 (Hsu et al. 2017), suggesting that BRD2 may potentiate CTCF's ability to maintain domain border integrity. The effects of BRD2 depletion may be less pronounced compared to those observed upon CTCF depletion in other contexts, which might be due to compensatory mechanisms, perhaps mediated by BRD3, or reflect that CTCF likely functions with multiple cofactors to form or reinforce boundaries. It is worth noting that BRD2 depletion did not reduce CTCF or cohesin binding to chromatin (Hsu et al. 2017), suggesting a parallel or downstream function of BRD2 in establishing chromatin contacts.

Our observations indicate that BRD2 participates in CTCF's ability to form chromatin domains—both in a locus-specific fashion by ensuring an appropriate range of enhancer activity and on a genome-wide scale by maintaining the integrity of architectural boundaries. BRD3 also associates with CTCF sites and thus may have similar functions. BRD4 likely regulates chromatin structure through a distinct mechanism, as it does not appear to bind CTCF sites to the same degree. Many questions remain regarding how BRD2 modulates the activity of CTCF, and the mechanisms by which it contributes to boundary formation. In particular, understanding how and under what contexts BRD2 is recruited to CTCF sites may provide valuable insight. Further study of how BRD2 and CTCF interact —and whether the interaction is direct—could allow dissection of BRD2's function through the construction of mutated forms that are unable to bind CTCF sites.

BET proteins can be found at enhancers and promoters where they can be recruited by acetylated transcription factors. For example, the transcription factor GATA1, which binds to numerous enhancers and promoters in erythroid cells, is involved in chromatin looping (Vakoc et al. 2005), is acetylated (Hung et al. 1999), and is associated with BET proteins (Lamonica et al. 2011). However, preliminary experiments showed that treatment with BET inhibitors did not disrupt GATA1-dependent enhancer-promoter contacts at the β-globin locus (unpublished observation). Whether BET proteins directly contribute to the formation of specific loops at other genes remains to be explored. Finally, similar to CTCF, BET proteins are located at diverse genomic sites including those unrelated to boundaries, raising the question as to what contextual features specify architectural from nonarchitectural functions of BET proteins. Further study of such questions has the potential to shed light on the transcriptional effects of BRD2 depletion, and by extension may lead to better predictions of the transcriptional and biological responses to BET inhibitors.

POTENTIAL MECHANISMS OF BET PROTEINS IN CHROMATIN DOMAIN BOUNDARY FORMATION

Our recent work identified a functional association between BRD2 and CTCF in the maintenance of chromatin domain boundaries, but how BRD2 exerts boundary function remains unclear. One potential mechanism may be through the formation of loops that physically isolate regions of chromatin from one another. Crystal structures of the first bromodomain of BRD2 showed that it has the potential to form homodimers (Nakamura et al. 2007), suggesting that BRD2 molecules bound at different genomic locations may be able to associate with one another. BET proteins, including BRD2, have an additional conserved domain termed "motif b," which can mediate the formation of heterodimers with BRD3 and BRD4 (Garcia-Gutierrez et al. 2012). BET proteins' dual bromodomains may also be able to bind to and bring distant regions of the genome into closer proximity. Recent work in Kaposi's sarcoma–associated herpesvirus (KSHV) supports a role, albeit likely indirect, for BET proteins in sustaining looping interactions. Specifically, BRD2 and BRD4 associate with CTCF/cohesin binding sites in the latent and lytic control regions of the viral genome, and treatment with the BET inhibitor JQ1 activated the lytic cycle transcriptional program and reduced looped viral genomic contacts (Chen et al. 2017). Loop disruption was associated with diminished binding of cohesin, likely explaining the loss of contacts and contrasting with the above-mentioned findings in erythroid cells.

TAD boundaries have been proposed to form through a number of mechanisms. In one model, CTCF complexes at either end of a TAD form a looping interaction that separates the intervening chromatin from surrounding regions (Rao et al. 2014; Rudan and Hadjur 2015; Sanborn et al. 2015; Dekker and Mirny 2016). Others have suggested that intra-TAD interactions also contribute to the boundaries between domains (Giorgetti et al. 2014). As described above, BRD2 may facilitate or stabilize loops involved in either scenario. Additional models of TAD boundary formation hypothesize that such boundaries represent regions of the genome with specific structural requirements (Dixon et al. 2016a). TAD boundaries are enriched for transcription start sites (TSSs) and housekeeping genes, suggesting that they are associated with highly active transcription (Dixon et al. 2012). Such models propose that factors such as nucleosome spacing around TSSs or CTCF sites, which are known to be associated with specific nucleosome spacing parameters

(Fu et al. 2008), may impose limits on chromatin flexibility that subsequently generate a boundary (Dixon et al. 2016a). It is possible that BRD2 may promote high levels of transcription at certain boundaries. In addition, BRD2 was recently reported to bind the histone variant H2A.Z (Vardabasso et al. 2015; Surface et al. 2016), which is enriched in the nucleosomes arrayed around CTCF sites (Fu et al. 2008), suggesting that it might contribute to structural features of boundary regions.

CONCLUSION

Members of the BET protein family clearly exert both overlapping and distinct functions that are partly reflected in their genomic localization patterns (Stonestrom et al. 2015). Roles of BET proteins at enhancers and promoters and in gene bodies are well described, but recent reports summarized here further suggest that BRD2, and perhaps BRD3, carry out separate functions at chromatin domain boundaries. BRD2 appears to strengthen boundaries alone or in the context of CTCF by helping to confine long-range chromatin contacts to within domains (Hsu et al. 2017). Although CTCF is required for BRD2 recruitment to their colocalized sites, what tethers BRD2 to boundaries devoid of CTCF is currently unknown. Future studies will need to address whether BRD2 assists looped contacts between boundaries, and whether such contacts are required for boundary activity. Also, assessing whether BET proteins may be sufficient for boundary formation is an important next step. Genome editing to perturb existing boundaries combined with the use of gain-of-function tethering experiments might shed light on these questions. If indeed BET proteins contribute to boundary function in *Drosophila* as discussed above, whole organismal studies may be within reach. Another unresolved question is why protein association studies of CTCF have not detected BET proteins and, conversely, why assessments of BET-associated protein complexes do not seem to involve CTCF. However, contacts might be indirect or may be stabilized in the context of a permissive chromatin environment. Sub-TADs, or subdomains, are structures that display more variation between cell types (Phillips-Cremins et al. 2013; Dixon et al. 2016b). Given the association of BET proteins with tissue-specific or inducible gene expression programs, one possibility may be that they form similarly "inducible" boundaries that constrain tissue-specific enhancers upon their activation.

Most current BET inhibitors show little specificity for individual BET family members. Although experiments have been performed to attribute the phenotypic consequences of BET inhibition to the impairment of particular BET proteins, interpretation of the results has been limited in part because of the overlapping functions of BET proteins defined in cellular assays and may also be complicated by BRD2's role as a boundary factor. This underscores the need for further studies to understand and predict the consequences of BET inhibition in a given tissue or disease context.

ACKNOWLEDGMENTS

This work was supported by National Institutes of Health (NIH) grants RO1DK54937, R24DK106766, and U01HL129998A to G.A.B. The Patel Family Scholar Award supported S.C.H.

REFERENCES

Alekseyenko AA, Walsh EM, Wang X, Grayson AR, Hsi PT, Kharchenko PV, Kuroda MI, French CA. 2015. The oncogenic BRD4-NUT chromatin regulator drives aberrant transcription within large topological domains. *Genes Dev* **29:** 1507–1523.

Alekseyenko AA, Walsh EM, Zee BM, Pakozdi T, Hsi P, Lemieux ME, Dal Cin P, Ince TA, Kharchenko PV, Kuroda MI, et al. 2017. Ectopic protein interactions within BRD4–chromatin complexes drive oncogenic megadomain formation in NUT midline carcinoma. *Proc Natl Acad Sci* **114:** E4184–E4192.

Anand P, Brown JD, Lin CY, Qi J, Zhang R, Artero PC, Alaiti MA, Bullard J, Alazem K, Margulies KB, et al. 2013. BET bromodomains mediate transcriptional pause release in heart failure. *Cell* **154:** 569–582.

Anders L, Guenther MG, Qi J, Fan ZP, Marineau JJ, Rahl PB, Lovén J, Sigova AA, Smith WB, Lee TI, et al. 2013. Genome-wide localization of small molecules. *Nat Biotechnol* **32:** 92–96.

Asangani IA, Dommeti VL, Wang X, Malik R, Cieslik M, Yang R, Escara-Wilke J, Wilder-Romans K, Dhanireddy S, Engelke C, et al. 2015. Therapeutic targeting of BET bromodomain proteins in castration-resistant prostate cancer. *Nature* **510:** 278–282.

Belkina AC, Denis GV. 2012. BET domain co-regulators in obesity, inflammation and cancer. *Nat Rev Cancer* **12:** 465–477.

Berkovits BD, Wolgemuth DJ. 2011. The first bromodomain of the testis-specific double bromodomain protein Brdt is required for chromocenter organization that is modulated by genetic background. *Dev Biol* **360:** 358–368.

Bisgrove DA, Mahmoudi T, Henklein P, Verdin E. 2007. Conserved P-TEFb-interacting domain of BRD4 inhibits HIV transcription. *Proc Natl Acad Sci* **104:** 13690–13695.

Bryant JM, Donahue G, Wang X, Meyer-Ficca M, Luense LJ, Weller AH, Bartolomei MS, Blobel GA, Meyer RG, Garcia BA, et al. 2015. Characterization of BRD4 during mammalian postmeiotic sperm development. *Mol Cell Biol* **35:** 1433–1448.

Chen H-S, De Leo A, Wang Z, Kerekovic A, Hills R, Lieberman PM. 2017. BET-inhibitors disrupt Rad21-dependent conformational control of KSHV latency. *PLoS Pathog* **13:** e1006100.

Cheung KL, Zhang F, Jaganathan A, Sharma R, Zhang Q, Konuma T, Shen T, Lee J-Y, Ren C, Chen C-H, et al. 2017. Distinct roles of Brd2 and Brd4 in potentiating the transcriptional program for Th17 cell differentiation. *Mol Cell* **65:** 1068–1080.e5.

Dawson MA, Prinjha RK, Dittmann A, Giotopoulos G, Bantscheff M, Chan W-I, Robson SC, Chung C-W, Hopf C, Savitski MM, et al. 2011. Inhibition of BET recruitment to chromatin as an effective treatment for MLL-fusion leukaemia. *Nature* **478:** 529–533.

Dekker J, Mirny L. 2016. The 3D genome as moderator of chromosomal communication. *Cell* **164:** 1110–1121.

Delmore JE, Issa GC, Lemieux ME, Rahl PB, Shi J, Jacobs HM, Kastritis E, Gilpatrick T, Paranal RM, Qi J, et al. 2011. BET bromodomain inhibition as a therapeutic strategy to target c-Myc. *Cell* **146:** 904–917.

Denis GV, Vaziri C, Guo N, Faller DV. 2000. RING3 kinase transactivates promoters of cell cycle regulatory genes through E2F. *Cell Growth Differ* **11:** 417–424.

Devaiah BN, Lewis BA, Cherman N, Hewitt MC, Albrecht BK, Robey PG, Ozato K, Sims RJ, Singer DS. 2012. BRD4 is an atypical kinase that phosphorylates serine2 of the RNA polymerase II carboxy-terminal domain. *Proc Natl Acad Sci* **109:** 6927–6932.

Devaiah BN, Case-Borden C, Gegonne A, Hsu CH, Chen Q, Meerzaman D, Dey A, Ozato K, Singer DS. 2016. BRD4 is a histone acetyltransferase that evicts nucleosomes from chromatin. *Nat Struct Mol Biol* **23:** 540–548.

Dey A, Ellenberg J, Farina A, Coleman AE, Maruyama T, Sciortino S, Lippincott-Schwartz J, Ozato K. 2000. A bromodomain protein, MCAP, associates with mitotic chromosomes and affects G$_2$-to-M transition. *Mol Cell Biol* **20:** 6537–6549.

Dey A, Nishiyama A, Karpova T, McNally J, Ozato K. 2009. Brd4 marks select genes on mitotic chromatin and directs postmitotic transcription. *Mol Biol Cell* **20:** 4899–4909.

Dixon JR, Selvaraj S, Yue F, Kim A, Li Y, Shen Y, Hu M, Liu JS, Ren B. 2012. Topological domains in mammalian genomes identified by analysis of chromatin interactions. *Nature* **485:** 376–380.

Dixon JR, Gorkin DU, Ren B. 2016a. Chromatin domains: The unit of chromosome organization. *Mol Cell* **62:** 668–680.

Dixon JR, Jung I, Selvaraj S, Shen Y, Antosiewicz-Bourget JE, Lee AY, Ye Z, Kim A, Rajagopal N, Xie W, et al. 2016b. Chromatin architecture reorganization during stem cell differentiation. *Nature* **518:** 331–336.

Filippakopoulos P, Qi J, Picaud S, Shen Y, Smith WB, Fedorov O, Morse EM, Keates T, Hickman TT, Felletar I, et al. 2010. Selective inhibition of BET bromodomains. *Nature* **468:** 1067–1073.

Fu Y, Sinha M, Peterson CL, Weng Z. 2008. The insulator binding protein CTCF positions 20 nucleosomes around its binding sites across the human genome. *PLoS Genet* **4:** e1000138.

Garcia-Gutierrez P, Mundi M, Garcia-Dominguez M. 2012. Association of bromodomain BET proteins with chromatin requires dimerization through the conserved motif B. *J Cell Sci* **125:** 3671–3680.

Giorgetti L, Galupa R, Nora EP, Piolot T, Lam F, Dekker J, Tiana G, Heard E. 2014. Predictive polymer modeling reveals coupled fluctuations in chromosome conformation and transcription. *Cell* **157:** 950–963.

Hanssen LLP, Kassouf MT, Oudelaar AM, Biggs D, Preece C, Downes DJ, Gosden M, Sharpe JA, Sloane-Stanley JA, Hughes JR, et al. 2017. Tissue-specific CTCF–cohesin-mediated chromatin architecture delimits enhancer interactions and function in vivo. *Nat Cell Biol* **19:** 952–961.

Hsu SC, Gilgenast TG, Bartman CR, Edwards CR, Stonestrom AJ, Huang P, Emerson DJ, Evans P, Werner MT, Keller CA, et al. 2017. The BET protein BRD2 cooperates with CTCF to enforce transcriptional and architectural boundaries. *Mol Cell* **66:** 102–116.e7.

Huang J, Liu X, Li D, Shao Z, Cao H, Zhang Y, Trompouki E, Bowman TV, Zon LI, Yuan G-C, et al. 2016. Dynamic control of enhancer repertoires drives lineage and stage-specific transcription during hematopoiesis. *Dev Cell* **36:** 9–23.

Hung HL, Lau J, Kim AY, Weiss MJ, Blobel GA. 1999. CREB-binding protein acetylates hematopoietic transcription factor GATA-1 at functionally important sites. *Mol Cell Biol* **19:** 3496–3505.

Jang MK, Mochizuki K, Zhou M, Jeong H-S, Brady JN, Ozato K. 2005. The bromodomain protein Brd4 is a positive regulatory component of P-TEFb and stimulates RNA polymerase II–dependent transcription. *Mol Cell* **19:** 523–534.

Kanno T, Kanno Y, Siegel RM, Jang MK, Lenardo MJ, Ozato K. 2004. Selective recognition of acetylated histones by bromodomain proteins visualized in living cells. *Mol Cell* **13:** 33–43.

Kellner WA, Van Bortle K, Li L, Ramos E, Takenaka N, Corces VG. 2013. Distinct isoforms of the *Drosophila* Brd4 homologue are present at enhancers, promoters and insulator sites. *Nucleic Acids Res* **41:** 9274–9283.

Kubo N, Ishii H, Gorkin D, Meitinger F, Xiong X, Fang R, Liu T, Ye Z, Li B, Dixon J, et al. 2017. Preservation of chromatin organization after acute loss of CTCF in mouse embryonic stem cells. *bioRxiv* doi: 10.1101.118737.

Ladurner AG, Inouye C, Jain R, Tjian R. 2003. Bromodomains mediate an acetyl-histone encoded antisilencing function at heterochromatin boundaries. *Mol Cell* **11:** 365–376.

Lamonica JM, Deng W, Kadauke S, Campbell AE, Gamsjaeger R, Wang H, Cheng Y, Billin AN, Hardison RC, Mackay JP, et al. 2011. Bromodomain protein Brd3 associates with acetylated GATA1 to promote its chromatin occupancy at erythroid target genes. *Proc Natl Acad Sci* **108:** E159–E168.

LeRoy G, Rickards B, Flint SJ. 2008. The double bromodomain proteins Brd2 and Brd3 couple histone acetylation to transcription. *Mol Cell* **30:** 51–60.

Nakamura Y, Umehara T, Nakano K, Jang MK, Shirouzu M, Morita S, Uda-Tochio H, Hamana H, Terada T, Adachi N, et al. 2007. Crystal structure of the human BRD2 bromodomain: Insights into dimerization and recognition of acetylated histone H4. *J Biol Chem* **282:** 4193–4201.

Naumova N, Imakaev M, Fudenberg G, Zhan Y, Lajoie BR, Mirny LA, Dekker J. 2013. Organization of the mitotic chromosome. *Science* **342:** 948–953.

Nicodeme E, Jeffrey KL, Schaefer U, Beinke S, Dewell S, Chung C-W, Chandwani R, Marazzi I, Wilson P, Coste H, et al. 2011. Suppression of inflammation by a synthetic histone mimic. *Nature* **468:** 1119–1123.

Nora EP, Lajoie BR, Schulz EG, Giorgetti L, Okamoto I, Servant N, Piolot T, van Berkum NL, Meisig J, Sedat J, et al. 2012. Spatial partitioning of the regulatory landscape of the X-inactivation centre. *Nature* **485:** 381–385.

Nora EP, Goloborodko A, Valton A-L, Gibcus JH, Uebersohn A, Abdennur N, Dekker J, Mirny LA, Bruneau BG. 2017. Targeted degradation of CTCF decouples local insulation of chromosome domains from genomic compartmentalization. *Cell* **169:** 930–933.e22.

Phillips JE, Corces VG. 2009. CTCF: Master weaver of the genome. *Cell* **137:** 1194–1211.

Phillips-Cremins JE, Sauria MEG, Sanyal A, Gerasimova TI, Lajoie BR, Bell JSK, Ong C-T, Hookway TA, Guo C, Sun Y, et al. 2013. Architectural protein subclasses shape 3D organization of genomes during lineage commitment. *Cell* **153:** 1281–1295.

Pivot-Pajot C, Caron C, Govin J, Vion A, Rousseaux S, Khochbin S. 2003. Acetylation-dependent chromatin reorganization by BRDT, a testis-specific bromodomain-containing protein. *Mol Cell Biol* **23:** 5354–5365.

Rahman S, Sowa ME, Ottinger M, Smith JA, Shi Y, Harper JW, Howley PM. 2011. The Brd4 extraterminal domain confers transcription activation independent of pTEFb by recruiting multiple proteins, including NSD3. *Mol Cell Biol* **31:** 2641–2652.

Rajender S, Avery K, Agarwal A. 2011. Epigenetics, spermatogenesis and male infertility. *Mutat Res* **727:** 62–71.

Rao SSP, Huntley MH, Durand NC, Stamenova EK, Bochkov ID, Robinson JT, Sanborn AL, Machol I, Omer AD, Lander ES, et al. 2014. A 3D map of the human genome at kilobase resolution reveals principles of chromatin looping. *Cell* **159:** 1665–1680.

Rudan MV, Hadjur S. 2015. Genetic tailors: CTCF and cohesin shape the genome during evolution. *Trends Genet* **31:** 651–660.

Sanborn AL, Rao SSP, Huang S-C, Durand NC, Huntley MH, Jewett AI, Bochkov ID, Chinnappan D, Cutkosky A, Li J, et al. 2015. Chromatin extrusion explains key features of loop and domain formation in wild-type and engineered genomes. *Proc Natl Acad Sci* **112:** E6456–E6465.

Schwartz YB, Cavalli G. 2017. Three-dimensional genome organization and function in *Drosophila*. *Genetics* **205:** 5–24.

Shang E, Salazar G, Crowley TE, Wang X, Lopez RA, Wang X, Wolgemuth DJ. 2004. Identification of unique, differentiation stage-specific patterns of expression of the bromodomain-containing genes Brd2, Brd3, Brd4, and Brdt in the mouse testis. *Gene Expr Patterns* **4:** 513–519.

Shang E, Nickerson HD, Wen D, Wang X, Wolgemuth DJ. 2007. The first bromodomain of Brdt, a testis-specific member of the BET sub-family of double-bromodomain-containing proteins, is essential for male germ cell differentiation. *Development* **134:** 3507–3515.

Shen C, Ipsaro JJ, Shi J, Milazzo JP, Wang E, Roe J-S, Suzuki Y, Pappin DJ, Joshua-Tor L, Vakoc CR. 2015. NSD3-short is an adaptor protein that couples BRD4 to the CHD8 chromatin remodeler. *Mol Cell* **60:** 847–859.

Shi J, Vakoc CR. 2014. The mechanisms behind the therapeutic activity of BET bromodomain inhibition. *Mol Cell* **54:** 728–736.

Sinha A, Faller DV, Denis GV. 2005. Bromodomain analysis of Brd2-dependent transcriptional activation of cyclin A. *Biochem J* **387:** 257–269.

Stonestrom AJ, Hsu SC, Jahn KS, Huang P, Keller CA, Giardine BM, Kadauke S, Campbell AE, Evans P, Hardison RC, et al. 2015. Functions of BET proteins in erythroid gene expression. *Blood* **125:** 2825–2834.

Surface LE, Fields PA, Subramanian V, Behmer R, Udeshi N, Peach SE, Carr SA, Jaffe JD, Boyer LA. 2016. H2A.Z.1 mono-ubiquitylation antagonizes BRD2 to maintain poised chromatin in ESCs. *Cell Rep* **14:** 1142–1155.

Tropberger P, Pott S, Keller C, Kamieniarz-Gdula K, Caron M, Richter F, Li G, Mittler G, Liu ET, Bühler M, et al. 2013. Regulation of transcription through acetylation of H3K122 on the lateral surface of the histone octamer. *Cell* **152:** 859–872.

Vakoc CR, Letting DL, Gheldof N, Sawado T, Bender MA, Groudine M, Weiss MJ, Dekker J, Blobel GA. 2005. Proximity among distant regulatory elements at the β-globin locus requires GATA-1 and FOG-1. *Mol Cell* **17:** 453–462.

Vardabasso C, Gaspar-Maia A, Hasson D, Pünzeler S, Valle-Garcia D, Straub T, Keilhauer EC, Strub T, Dong J, Panda T, et al. 2015. Histone variant H2A.Z.2 mediates proliferation and drug sensitivity of malignant melanoma. *Mol Cell* **59:** 75–88.

Wang R, Li Q, Helfer CM, Jiao J, You J. 2012. Bromodomain protein Brd4 associated with acetylated chromatin is important for maintenance of higher-order chromatin structure. *J Biol Chem* **287:** 10738–10752.

Wu SY, Chiang CM. 2007. The double bromodomain-containing chromatin adaptor Brd4 and transcriptional regulation. *J Biol Chem* **282:** 13141–13145.

Yang Z, Yik JHN, Chen R, He N, Jang MK, Ozato K, Zhou Q. 2005. Recruitment of P-TEFb for stimulation of transcriptional elongation by the bromodomain protein Brd4. *Mol Cell* **19:** 535–545.

Zhao R, Nakamura T, Fu Y, Lazar Z, Spector DL. 2011. Gene bookmarking accelerates the kinetics of post-mitotic transcriptional re-activation. *Nat Cell Biol* **13:** 1295–1304.

Zuber J, Shi J, Wang E, Rappaport AR, Herrmann H, Sison EA, Magoon D, Qi J, Blatt K, Wunderlich M, et al. 2011. RNAi screen identifies Brd4 as a therapeutic target in acute myeloid leukaemia. *Nature* **478:** 524–528.

Zuin J, Dixon JR, van der Reijden MIJA, Ye Z, Kolovos P, Brouwer RWW, van de Corput MPC, van de Werken HJG, Knoch TA, van IJcken WFJ, et al. 2014. Cohesin and CTCF differentially affect chromatin architecture and gene expression in human cells. *Proc Natl Acad Sci* **111:** 996–1001.

Emerging Evidence of Chromosome Folding by Loop Extrusion

Geoffrey Fudenberg,[1,5] Nezar Abdennur,[2,3,5] Maxim Imakaev,[3] Anton Goloborodko,[3,4] and Leonid A. Mirny[3,4]

[1]Gladstone Institute of Data Science and Technology, University of California, San Francisco, California 94158

[2]Computational and Systems Biology Program, Massachusetts Institute of Technology, Cambridge, Massachusetts 02139

[3]Institute for Medical Engineering and Science (IMES), Massachusetts Institute of Technology, Cambridge, Massachusetts 02139

[4]Department of Physics, Massachusetts Institute of Technology, Cambridge, Massachusetts 02139

Correspondence: geoff.fudenberg@gmail.com; leonid@mit.edu

Chromosome organization poses a remarkable physical problem with many biological consequences: How can molecular interactions between proteins at the nanometer scale organize micron-long chromatinized DNA molecules, insulating or facilitating interactions between specific genomic elements? The mechanism of active loop extrusion holds great promise for explaining interphase and mitotic chromosome folding, yet remains difficult to assay directly. We discuss predictions from our polymer models of loop extrusion with barrier elements and review recent experimental studies that provide strong support for loop extrusion, focusing on perturbations to CTCF and cohesin assayed via Hi-C in interphase. Finally, we discuss a likely molecular mechanism of loop extrusion by structural maintenance of chromosomes complexes.

Mammalian interphase chromosomes exhibit both cell type- and locus-specific organizations that manifest characteristic patterns on Hi-C maps. These include square areas of enriched contact frequency along the diagonal, termed topologically associating domains (TADs) (Dixon et al. 2012; Nora et al. 2012), often elaborated with *peaks* at their corners (Rao et al. 2014), *grids* of peaks within and between TADs, and enriched lines or *tracks* of contact frequency emanating from a boundary (Fudenberg et al. 2016) (Fig. 1C; for reviews, see Bonev and Cavalli 2016; Merkenschlager and Nora 2016). We distinguish TADs from compartmental segments of the genome, which also appear as squares along the diagonal of Hi-C maps but differ in that they associate to form a checkered pattern in *cis* and in *trans*. Indeed, TADs, peaks, and tracks have an independent mechanistic origin from the patterns associated with the compartmental segregation of active and inactive chromatin (Schwarzer et al. 2017), and we discuss the interplay of these two mechanisms elsewhere (Nuebler et al. 2017). TAD boundaries are frequently demarcated by binding sites of the transcription factor *CTCF*, and are enriched for the structural maintenance of chromosomes (SMC) complex *cohesin*. Functionally, TADs are believed to demarcate coherent *cis* neighborhoods of gene-regulatory activity and hence are crucial for development (Spielmann and Mundlos 2016). To explain how such neighborhoods can be formed, we put forward a mechanism based on a still-hypothetical process of *loop extrusion*.

Here we present emerging evidence that interphase chromosomes are organized by loop extrusion, an active ATP-dependent process that allows nanometer-size molecular machines to organize chromosomes at much larger scales. We review how loop extrusion by cohesins can explain the formation of TADs, peaks, and tracks visible in interphase Hi-C maps. We then detail specific predictions made by the polymer model of loop extrusion, and discuss recent experimental perturbations to CTCF and cohesin that test these predictions and provide strong support for the loop extrusion mechanism. Although we focus on comparisons to mammalian interphase Hi-C experiments, loop extrusion likely plays important roles in other organisms and parts of the cell cycle. We also discuss imaging experiments, single-molecule experiments, and a possible molecular mechanism of loop extrusion.

POLYMER MODEL OF LOOP EXTRUSION WITH BARRIER ELEMENTS

We frame our discussion around how we originally implemented the mechanism of loop extrusion limited by directional barriers as a polymer model (Fudenberg et al. 2016). In the process of loop extrusion, loop extruding factors (LEFs) translocate along the chromosomes, holding together progressively more genomically distant loci along a chromosome, thus producing dynamically expanding chromatin loops (see Supplemental Movie 1).

---

[5]These authors contributed equally to this work.

Supplemental material is available for this article at symposium.cshlp.org.

Published by Cold Spring Harbor Laboratory Press; doi: 10.1101/sqb.2017.82.034710

Cold Spring Harbor Symposia on Quantitative Biology, Volume LXXXII

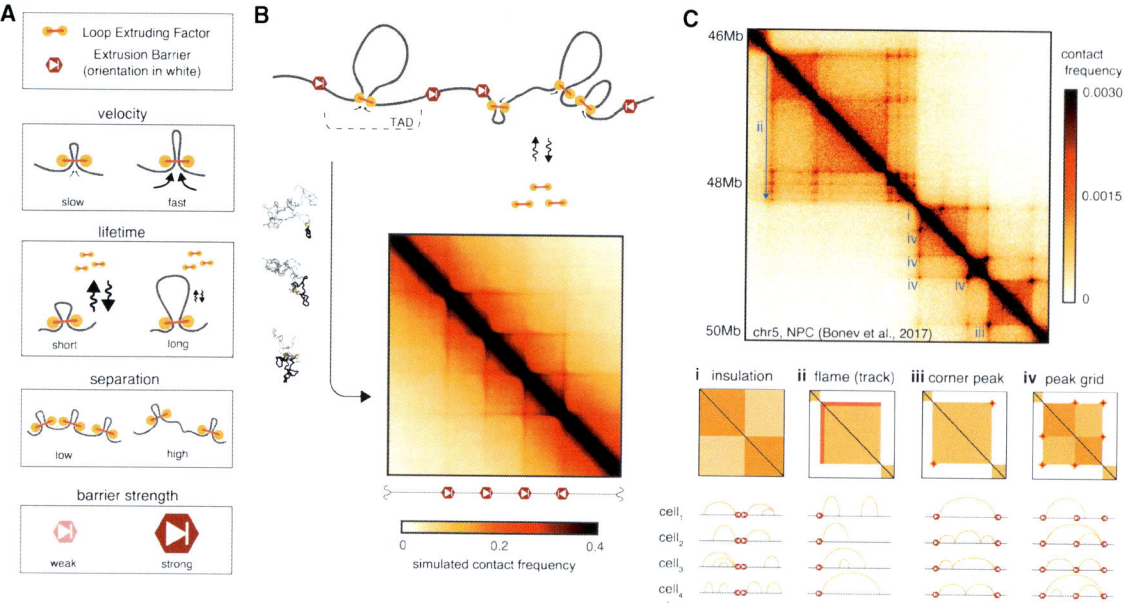

Figure 1. Polymer model of loop extrusion with barrier elements recapitulates features of interphase chromosome folding (see also Supplemental Movie 1). (*A*) Illustrations of the four key parameters governing the dynamics of interphase loop extrusion: LEF velocity, LEF lifetime, LEF separation, and barrier strength. Characterizing how changes to these parameters affect Hi-C maps in silico allows us to make experimental predictions for perturbations. (*B*) To compare our models with Hi-C experiments, we generate ensembles of conformations for each set of parameters, and then compute average contact maps. To compare with imaging experiments, we can calculate other observables (e.g., pairwise distance between loci). (*C*) Interphase Hi-C data from mouse neural progenitor cells (Bonev et al. 2017), plotted with HiGlass (Kerpedjiev et al. 2017), annotated with features that can emerge via loop extrusion in blue (*i–iv*). Arc diagrams depict how stochastic configurations of LEF-mediated loops in distinct nuclei can lead to the population-averaged features. Chromatin loops directly held by LEFs are depicted with yellow arcs, whereas dashed gray arcs depict "transitive loops" from sets of adjacent LEFs. (*i*) Insulation, observed as squares along the diagonal of Hi-C maps (i.e., TADs), arises when extrusion barriers halt LEF translocation. LEFs then facilitate additional contacts within TADs, but not between TADs. (*ii*) Flames (or tracks), observed as straight lines often emerging from the borders of TADs, arise when LEFs become halted at a barrier while continuing to extrude from the other side (referred to as "lines" in Fudenberg et al. 2016). (*iii*) Peaks of enriched contact frequency often appear at the corners of TADs and also often coincide with intersection points of flames. These peaks emerge as a result of LEFs being halted on both sides by extrusion barriers. (*iv*) Peak grids can emerge either when internal boundaries are skipped or via transitive sets of LEF-mediated loops.

LEF translocation is either halted by encounters with other LEFs or probabilistically halted at specific genomic loci that contain *extrusion barriers*. We assume that if halted only on one side, a LEF may continue to extrude chromatin from its other side. LEFs continue to extrude until they dissociate from the chromatin fiber, releasing the extruded loop, as they dynamically exchange with the nucleoplasm.

The minimal system of LEFs limited by extrusion barriers that we implement is defined by four parameters (Fig. 1A):

- *lifetime* on chromatin (sec)
- *velocity* along the chromatin fiber (kb/sec)
- *separation* between LEFs (kb)
- *permeability* of the extrusion barriers (probability)

For comparison to ensemble-averaged Hi-C experiments, that capture a snapshot of contacts occurring at a particular point in time, it is also useful to define the product of lifetime and velocity, *processivity* (kb), which indicates the average size of a loop that a LEF would extrude if left unobstructed. Motivated by observations of CTCF motif orientations at TAD boundaries and at

peaks (Rao et al. 2014; Vietri Rudan et al. 2015), we implement barriers as being *directional*, (i.e., halting LEFs approaching it from only one side). Barriers can be modeled as either halting LEFs as long as the blocking factor is present, or stalling them until LEF dissociation (see Supplemental Movie 1). In our models, the permeability can be thought to represent the probability that a barrier locus is occupied by a blocking factor.

To compare predictions from our simulations with experiments, we generate a simulated ensemble of chromatin conformations for a given set of parameters (Fig. 1B). To accurately capture features of chromatin folding at high resolutions we typically use monomers representing several nucleosomes to simulate 10–50 Mb of chromatin. From these conformations we can extract experimentally relevant observables (Imakaev et al. 2015). These include maps of contact frequency that can be compared to Hi-C contact maps, as well as distributions of spatial distances between pairs of loci, that can be compared with FISH experiments (Fudenberg and Imakaev 2017). From the simulated contact maps, we can then quantify features such as TADs, peaks, and contact frequency decay, as done for experimental Hi-C maps. By comparing simulated and experimental features, we can then define a set of wild-type parameters, from which perturbations, and hence predictions, can be made.

The mechanism of loop extrusion limited by directional barriers recapitulates many features of interphase chromosome folding visible in Hi-C maps (Fig. 1C), including:

- TADs: regions of enriched contact frequency between neighboring barriers
- Tracks: lines emerging from one side of a barrier
- Peaks and grids of peaks, occurring between proximal barriers in *cis* but not between chromosomes
- Presence of inward-oriented CTCF motifs at TAD boundaries and at peak bases

Further support comes from site-specific disruptions of TAD boundaries and peak bases, which respectively result in merging of adjacent TADs (Nora et al. 2012; Narendra et al. 2015; Rodríguez-Carballo et al. 2017) and orientation-dependent losses of peaks (de Wit et al. 2015; Guo et al. 2015; Sanborn et al. 2015). To our knowledge, no alternative mechanism of interphase chromosome organization currently agrees with all the above.

Although we focus here on interphase loop extrusion, we note that loop extrusion by SMCs appears to have important consequences in mitosis (Naumova et al. 2013; Goloborodko et al. 2016a; Gibcus et al. 2018), where the term was coined and first mathematically modeled (Alipour and Marko 2012). The closely related concepts of reeling (Riggs 1990), facilitated tracking (Blackwood and Kadonaga 1998), loop expansion (Kimura et al. 1999) and progressive loop enlargement (Nasmyth 2001) have a rich history. Loop extrusion also appears relevant in bacteria (Gruber 2014; Wang et al. 2015, 2017). There are also related proposals for interphase loop extrusion (Nichols and Corces 2015; Sanborn et al. 2015; Yamamoto and Schiessel 2017; Brackley et al. 2018), which we discuss briefly below.

We note that although the terms "contact," "loop," and "interaction" are often used interchangeably in the chromosome organization literature, they are often used to describe very different features of Hi-C contact maps (Forcato et al. 2017). In the context of loop extrusion, we reserve the term "loop" in the very narrow sense of two regions of a continuous chromatin fiber brought together by a LEF at a given point in time. Moreover, simulations (Benedetti et al. 2014; Doyle et al. 2014; Hofmann and Heermann 2015; Fudenberg et al. 2016) and data analyses (Giorgetti et al. 2014; Cattoni et al. 2017; Finn et al. 2017; Fudenberg and Imakaev 2017) show that peaks of contact frequency in interphase Hi-C maps are not consistent with stable chromatin loops. Therefore, we refrain from using "loop" to describe any feature of Hi-C contact maps.

Challenges for Testing Models of Loop Extrusion

The stochastic nature of loop extrusion poses an experimental challenge for testing predictions from the model. Extruded loops are not directly visible via population-average Hi-C approaches because they are located at different genomic positions in different cells at any given time. Even with single-cell Hi-C methods an individual pair of loci linked by an extruding loop would not appear particularly different from any other captured contact. Visualization of extruded loops by microscopy is similarly challenging due to their continually changing locations both along the genome and in 3D space. Direct confirmation that a particular chromatin loop has been extruded in vivo will require methods that can simultaneously track multiple DNA loci as well as the loop extruders themselves. Nevertheless, much of the strongest evidence to date supporting the role of loop extrusion in interphase comes from changes in Hi-C maps upon perturbations that affect specific components of the loop extrusion machinery.

PREDICTIONS FROM THE MODEL OF INTERPHASE LOOP EXTRUSION

To make experimental predictions, we must first identify components of the interphase loop extrusion machinery with their biological candidates. Several lines of evidence make us hypothesize that cohesin complexes play the role of LEFs, and CTCF plays the role of an extrusion barrier (Fudenberg et al. 2016). Cohesin is enriched at TAD boundaries in interphase and is highly homologous to condensins, the main complexes responsible for compacting mitotic chromosomes. CTCF is enriched at TAD boundaries at preferentially oriented motifs, and, compared with other transcriptional regulators, binds relatively stably to its cognate sites (for review, see Hansen et al. 2018). With these identities, we discuss how our model of loop extrusion predicts different outcomes for three perturbations: depletion of CTCF, depletion of cohesins, and increased processivity of cohesins (Fig. 2).

LEF Depletion

For the depletion of the LEF, cohesin, our simulations display two phenomena (Fig. 2B) (i) the loss of TADs and associated Hi-C peaks; and (ii) decompaction of chromatin at the scales of individual extruded loops (<200 kb). Changes in local compaction, in turn, can be studied by observing changes in the contact probability, $P(s)$, as a function of genomic separation, s. Local compaction is seen as a region of $P(s)$ with a shallow slope (~100–500 kb), which we refer to as the *shoulder* (Fig. 2A); decompaction leads to reduction or loss of the shoulder region. We note that our models predict that a sharp decrease in LEF processivity would similarly lead to a loss of TADs, peaks, and compaction.

Extrusion Barrier Depletion

For the depletion of site-specific extrusion barriers, as imposed by CTCF, our simulations also predict the loss of TADs and associated Hi-C peaks (Fig. 2C). However, our simulations predict that other consequences of this perturbation should be very different from depletion of LEFs. This is because in our model, extrusion barriers only impose an instructive function (i.e., their major effect is on

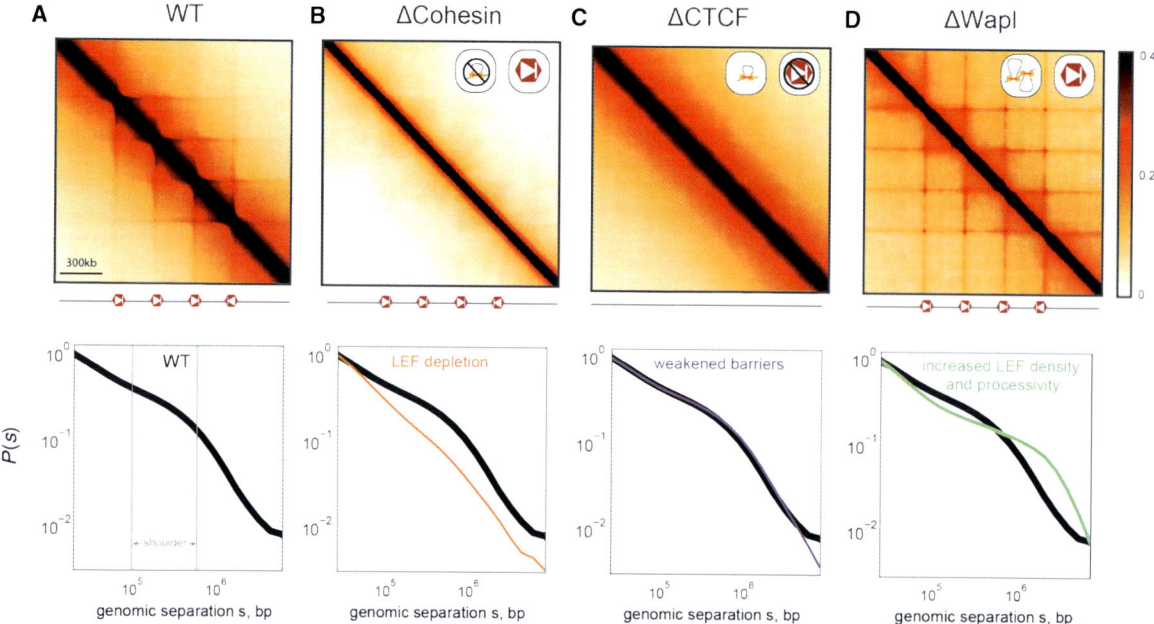

Figure 2. Loop extrusion polymer simulations predict the consequences of cohesin and CTCF perturbations. (*Top* row) Simulated Hi-C maps for indicated perturbations. (*Bottom* row) P(s) for indicated perturbation compared to WT P(s). All simulations considered a 36-Mb chain (3600 monomers) with the same positions and orientations of CTCF barriers (separated by 300 kb) and the same LEF velocity (250 3D-per-1D steps). (*A*) WT simulations used processivity 200 kb, separation 200 kb, and barrier strength 0.995. The shoulder in P(s), indicative of compaction via loop extrusion, is indicated in gray. (*B*) For ΔCohesin, our simulations predict the loss of TADs, peaks, flames, and the shoulder of P(s). ΔCohesin was simulated using processivity 200 kb, separation 2 Mb, and boundary strength 0.995. This can represent the loss of actively extruding cohesins via ΔNipbl, ΔRad21, or other cohesin subunits. (*C*) For ΔCTCF, our simulations predict the loss of TADs, peaks, flames, yet no discernible change to P(s). This arises because CTCF plays an instructive role for the activity of extrusion. ΔCTCF was simulated using processivity 200 kb, separation 200 kb, and boundary strength 0.9. (*D*) For ΔWapl, our simulations predict the emergence of additional peaks, including at further genomic separations, as well as an extension of the shoulder in P(s). ΔWapl was simulated using processivity 1 Mb, separation 150 kb, and boundary strength 0.995.

the localization of extruded loops rather than on their sizes or abundance). We therefore predict little effect on overall compaction, and hence little change in the P(s) curve. This differentiates our predictions for CTCF depletion from those for cohesin depletion.

Increased LEF Density and Processivity

For the depletion of a cohesin unloading factor, like Wapl, our model predicts that the consequent increased processivity and number of LEFs would lead to several phenotypes (Fig. 2D): (i) peaks at corners of TADs become stronger and appear between more distal barrier loci, creating extended grids of peaks; (ii) the orientational preference of barrier loci will become weaker, as LEFs halted at a directional barrier for long durations can stop traffic from the opposing direction as well. Finally, (iii), our model predicts that sufficiently increased coverage by extruded loops will overcompact chromosomes. In Hi-C this would be detected as an extension of the shoulder in P(s), as opposed to how it recedes in the case of cohesin depletion. Macroscopically, sufficient compaction would cause chromosomes to condense into a prophase-like state with a cohesin-rich central scaffold.

Crucially, our model predicts that the loss of cohesin loop extruders and the loss of CTCF extrusion barriers should both lead to the loss of TADs and Hi-C peaks, yet in completely distinct fashions. Furthermore, in-creased processivity of cohesin extruders is predicted to manifest distinct phenotypes on Hi-C maps and macroscopic chromosome organization.

EXPERIMENTAL PERTURBATIONS CONSISTENT WITH INTERPHASE LOOP EXTRUSION

Whereas perturbing CTCF and cohesin dynamics is crucial for testing predictions of loop extrusion, depletion of such essential complexes poses many experimental challenges. For CTCF, cells begin dying after ~4 days of stringent depletion (Nora et al. 2017). For cohesin, there are additional challenges related to its role in sister chromatid cohesion and chromosome segregation during mitosis (Peters and Nishiyama 2012), and its multiple dynamically exchanging subunits and regulators (Peters and Nishiyama 2012; Rhodes et al. 2017) that can be present in different abundances and likely have unique impacts on loop extrusion dynamics. Despite these challenges, recent studies have achieved modulation of cohesin and CTCF that result in dramatic changes, consistent with predictions from polymer models of loop extrusion (Table 1).

Cohesin Depletion

Consistent with our predictions for decreasing the number of active LEFs, depletion of the cohesin loader Nipbl

Table 1. List of recent experimental perturbations, prediction from loop extrusion, effects in recent Hi-C experiments, and effect on overall chromatin density

Perturbation[a]	Prediction from loop extrusion	Effect on Hi-C	Effect on compaction
ΔCTCF	Barriers become more permeable	Loss of TADs and peaks, same $P(s)$ (Nora et al. 2017; Wutz et al. 2017)	Little change overall (Nozaki et al. 2017)
ΔNipbl	Increase separation, possibly decrease velocity	Loss of TADs, peaks and $P(s)$ shoulder (Schwarzer et al. 2017)	Decompaction (Nozaki et al. 2017)
ΔRad21	Increase separation	Loss of TADs, peaks and $P(s)$ shoulder (Rao et al. 2017; Wutz et al. 2017; Gassler et al. 2017)	Decompaction (Nozaki et al. 2017)
ΔWapl	Increase processivity, possibly decrease separation	New peaks, extend $P(s)$ shoulder (Haarhuis et al. 2017; Wutz et al. 2017)	Vermicelli (Tedeschi et al. 2013; Haarhuis et al. 2017; Wutz et al. 2017)

[a]See Supplemental Table S1 for additional experimental perturbations and details.

(Scc2) (Schwarzer et al. 2017) and acute degradation of the cohesin kleisin Rad21 (Scc1) (Rao et al. 2017; Wutz et al. 2017) during interphase led to both: (i) complete erasure of TADs and Hi-C peaks (ii) and decompaction, as evidenced by loss of the $P(s)$ shoulder (Fig. 3A). Decompaction is further supported by imaging, showing loss of H2B clustering by PALM following both RNAi

knockdown of NIPBL and AID-mediated degradation of Rad21 (Nozaki et al. 2017). We note that earlier Hi-C studies (Seitan et al. 2013; Sofueva et al. 2013; Zuin et al. 2014) saw limited impact following the depletion of Rad21, potentially due to incomplete depletion.

A corollary of the Nipbl depletion result is that cohesin must be constantly loaded on chromatin to maintain TADs

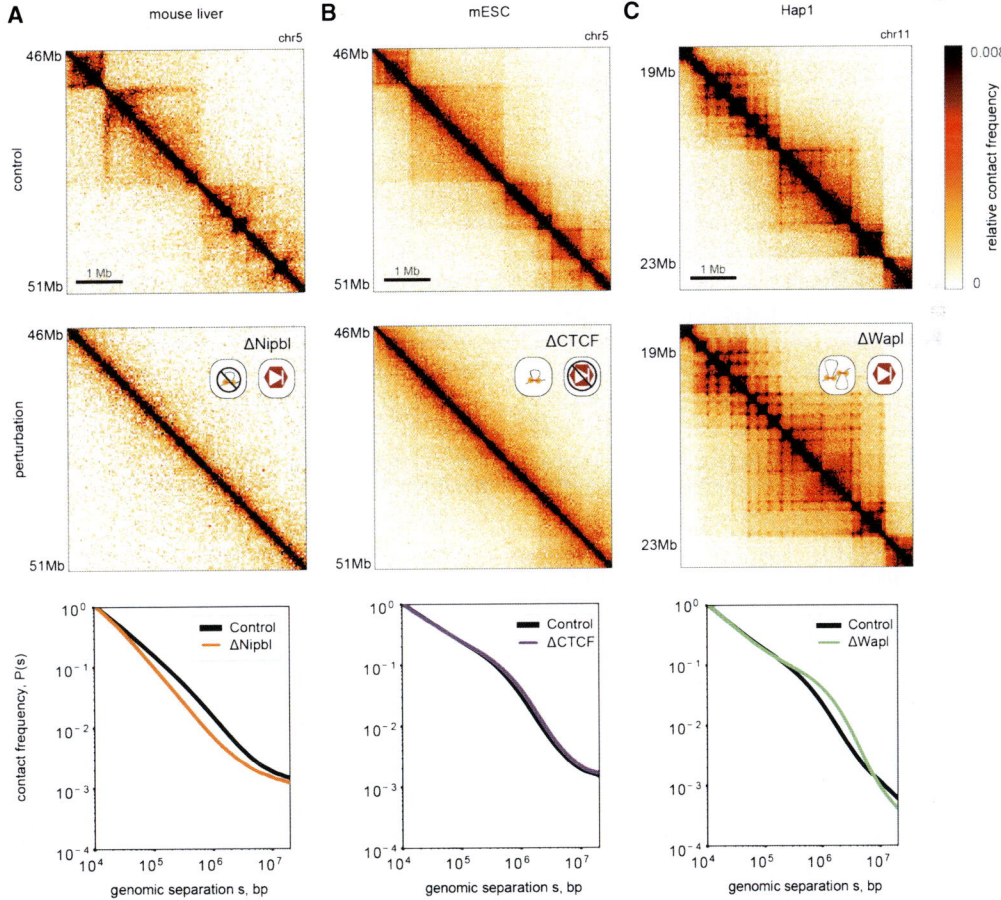

Figure 3. Experimental phenotypes are consistent with predictions from loop extrusion simulations. (*Top* row) Unperturbed experimental Hi-C maps, replotted from indicated studies (see Supplemental Methods; also see interactive HiGlass displays, http://mirnylab .mit.edu/projects/emerging-evidence-for-loop-extrusion). (*Middle* row) Hi-C maps for indicated perturbations. (*Bottom* row) $P(s)$ for indicated perturbation compared to unperturbed $P(s)$ normalized to contact frequency at 10 kb. (*A*) Schwarzer et al. (2017) used tissue-specific CRE-inducible gene deletion in mouse liver cells to deplete Nipbl. (*B*) Nora et al. (2017) used an auxin-inducible degron system to deplete CTCF in mESCs. (*C*) Haarhuis et al. (2017) deleted Wapl in the Hap1 haploid human cell line, via CRISPR.

and associated corner peaks. Consistently, TADs and Hi-C peaks are both rapidly lost upon AID-mediated degradation of Rad21 (<3 h (Wutz et al. 2017)) and reestablished after auxin wash-off (40–60 min (Rao et al. 2017)). These consequences follow directly from our loop extrusion models, and the turnover time of cohesin in G1 (~5–30 min [Gerlich et al. 2006; Hansen et al. 2017; Wutz et al. 2017]).

Future studies will be useful to dissect the dynamics of the processes and the potential role of Nipbl beyond that of a loader (Petela et al. 2017; Rhodes et al. 2017). In particular, although Nipbl depletion appears to have a dramatic effect on extrusion, knockout of its cofactor Mau2 (Scc4) appears to have a much weaker effect on loading yet a fairly strong effect on processivity (Haarhuis et al. 2017). Moreover, we note that different components of the interphase extrusion machinery could be limiting at different concentrations and in different contexts. We hypothesize that, via its consequences on loop extrusion, modulation of the levels of various cohesin subunits and interactors can serve to fine-tune overall gene regulation across cell-types and tissues.

CTCF Depletion

Consistent with our predictions for the loss of site-specific barriers to extrusion, acute auxin-induced degradation of CTCF in mESCs (Nora et al. 2017) and HeLa cells (Wutz et al. 2017) led to a dramatic loss of TADs and Hi-C peaks (Fig. 3B). However, the $P(s)$ curve did not change, implying that although demarcation of contact-insulating boundaries in Hi-C maps was lost, the same degree of chromatin compaction was maintained. In support of the dynamic exchange of LEFs in our model, the effect of CTCF depletion was fully reversible following a 2-day wash-off period (Nora et al. 2017). We note that stringent dosage depletion was necessary to observe dramatic insulation defects: even a 15% preservation of CTCF showed a relatively mild phenotype (Nora et al. 2017). Similar loss of TADs and peaks were reported in vivo for an inducible CTCF knockout in cardiomyocytes (Lee et al. 2017). Weaker effects have also been reported recently (Kubo et al. 2017; Rosa-Garrido et al. 2017) and earlier (Zuin et al. 2014), but this may have been due to relatively inefficient depletion or lower starting levels of CTCF.

The predicted lack of decondensation following CTCF depletion is further supported by imaging. PALM shows little difference in H2B clustering (Nozaki et al. 2017). Imaging of FISH probes at selected loci upon CTCF degradation show that inter-TAD distances increased, whereas intra-TAD distances remained the same (Nora et al. 2017). Together these results are consistent with global compaction levels being unchanged but with diminished insulation across CTCF sites. Importantly, the lack of chromatin decompaction in CTCF depletion rules out models in which CTCF is strictly required for the loading (Nichols and Corces 2015) of chromatin-bound cohesin and any ensuing cohesin-mediated loops. Instead, the differences in imaging and Hi-C maps upon CTCF versus cohesin depletion are consistent with the loop extrusion model we describe (Fudenberg et al. 2016), in which CTCF barriers serve an instructive function (Wendt and Peters 2009) and cohesin is loaded onto chromatin and can compact chromosomes through extrusion even in the absence of CTCF.

Wapl Depletion

Consistent with our predictions for increasing the processivity and density of active LEFs, depletion of the cohesin unloader Wapl led to multiple phenotypes observed in Hi-C maps (Gassler et al. 2017; Haarhuis et al. 2017; Wutz et al. 2017) and by imaging (Tedeschi et al. 2013). For Hi-C (Fig. 3C) this includes (i) strengthened peaks at TAD corners, (ii) emergence of new peaks between boundaries at greater separations, creating extended grids of corner peaks; (iii) a weakened correspondence between these features and CTCF motif orientation. Increased local compaction upon Wapl depletion is reflected by (iv) extension of the shoulder in the $P(s)$ curve and, (v) the emergence of prophase-like vermicelli chromatids via imaging (Tedeschi et al. 2013). This remarkable observation provides further evidence for a *universal molecular mechanism*—loop extrusion—underlying both metaphase and interphase chromosome organization (Imakaev et al. 2015; Dekker and Mirny 2016).

Depletion of another component of the cohesin unloading machinery, Pds5A and Pds5B (Pds5A/B), led to many of the same phenotypes (Wutz et al. 2017). However, there were also intriguing differences that may prove instructive for determining exactly how CTCF halts the progression of cohesin along the chromosome—for example, Pds5 may instruct directional cohesin stalling (Petela et al. 2017; Wutz et al. 2017), and competition between the two HAWK family proteins, Nipbl and Pds5, may regulate cohesin translocation velocity (Petela et al. 2017). The observation that Wapl depletion appears to largely rescue the Hi-C phenotype of Mau2 depletion provides further support to the proposal that the Nipbl/Mau2 "loading complex" also has roles in promoting cohesin processivity for loop extrusion (Haarhuis et al. 2017). Finally, consistent with loop extrusion simulations with increased processivity, the joint depletion of Wapl and Pds5A/B showed even stronger effects in terms of shifting the shoulder in $P(s)$ and in the emergence of vermicelli.

Collectively, the congruence of both Hi-C and imaging experiments following the perturbation of CTCF and cohesin dynamics strongly supports the role of loop extrusion in interphase. Future simulations and experiments will be valuable for probing the consequences of multiple simultaneous perturbations (Busslinger et al. 2017; Wutz et al. 2017).

SINGLE-MOLECULE EXPERIMENTS SUPPORT ACTIVE LOOP EXTRUSION

Although providing strong support for chromosome folding by loop extension in vivo, the studies discussed above do not directly probe the molecular details of loop extrusion. Molecularly realizing the process of loop extrusion presents a considerable challenge, namely, that the protein complexes performing loop extrusion need to *track consistently* in *cis* along chromatin, over large distances (up to tens-of-thousands of nucleosomes) without falling off. Moreover, the substrate, chromatin, is highly disordered due to nucleosomes and other DNA-bound proteins, likely posing a greater challenge than tracking along microtubules performed by cytoplasmic motors. Here we discuss how recent single molecule experiments argue that loop extrusion likely occurs via an active process, driven by molecular motors. Although many of these observations were made with condensin and bacterial SMCs, they illustrate that loop-extrusion is a plausible mechanism of action for the whole family of SMC proteins, including cohesin.

ATP-Dependent Translocation

Recently, (Terakawa et al. 2017) showed that a single yeast condensin has motor activity and is able to translocate processively along naked DNA in vitro. Using a DNA curtain assay, they found individual condensin complexes travel unidirectionally, rapidly (~4 kb/min) and processively (~10 kb) in an ATP-dependent manner with 10 nm steps (30 bp on naked DNA). As previous single-molecule studies only reported sliding dynamics of SMCs (Davidson et al. 2016; Kanke et al. 2016; Kim and Loparo 2016; Stigler et al. 2016; for review, see Eeftens and Dekker 2017), the directional translocation observed by Terakawa et al. (2017) is incredibly important.

The high structural homology of cohesin to condensin makes it likely that the same physical mechanism would govern its processive motion, in addition to its established role of mediating sister chromatin cohesion (Peters and Nishiyama 2012). Indeed, the ability of these SMCs to compact chromosomes appears to be remarkably coherent over evolutionary timescales and cellular contexts (Schalbetter et al. 2017). Due to its dual roles, and more elaborate set of subunits, however, reconstituting this activity for cohesin may be more difficult in vitro. Nevertheless, we believe that the in vitro observations of ATP-driven processive condensin translocation argue against the likelihood of motor-free mechanisms (Yamamoto and Schiessel 2017; Brackley et al. 2018) of SMC processivity in general, including for cohesin.

Although strongly supporting the loop extrusion mechanism, the single-molecule experiments leave open several questions of how loop extrusion can work in vivo:

- How can SMCs translocate on chromatinized rather than naked DNA?
- How can translocation result in loop extrusion?
- Is the measured speed of translocation sufficient to generate TADs and peaks?
- Do cells have sufficient ATP budgets to support extrusion during interphase?

Walking Hypothesis

In particular, it remains to be understood how SMC complexes can translocate on chromatin fibers rather than naked DNA. Translocations while maintaining constant contact with DNA may not always be possible due to the complexity of chromatin fiber and abundance of DNA-bound proteins. Although the size of an SMC complex (~50 nm) exceeds that of a single nucleosome (~10 nm), nucleosomes would constitute challenging obstacles for SMC translocation if maintaining constant contact with DNA is required for translocation.

A possible solution comes from the structural similarity of SMC domain organization to that of kinesin and myosin motors (Guacci et al. 1993; Peterson 1994) that walk on microtubules and actin, which suggests that SMCs can similarly walk on chromatinized DNA. Importantly, a *walking mechanism* would allow translocation where obstacles such as nucleosomes and other DNA-bound proteins can be passed over, avoiding disruptions of the underlying nucleosomal array. During each step of the walking process, one SMC head can remain DNA-bound, whereas the other hops forward and rebinds nearby DNA (Fig. 4A). SMC walking is consistent with the rapid and flexible dynamics of their arms (Eeftens et al. 2017), and the 10 nm step size (Terakawa et al. 2017) would allow passing over nucleosomes (e.g., by hopping from linker to linker) and other DNA-bound complexes, avoiding the need for unwinding nucleosomal DNA or nucleosome eviction (Fig. 4B).

A walking mechanism would be greatly aided by the known ability of SMCs to *topologically entrap* DNA (Peters and Nishiyama 2012), which can ensure that the walker tracks in *cis*, along the same chromatin fiber (Fig. 4C). Pseudo-topological (Srinivasan et al. 2017) entrapment can similarly help maintain extrusive cohesins on the same DNA molecule over long genomic distances (Kschonsak et al. 2017). In other words, SMC complexes may translocate along the chromatin fiber and accomplish loop extrusion as *shackled walkers* (Fig. 4A).

An important open question is how CTCF, and possibly other chromatin-bound proteins, can halt cohesin translocation whereas nucleosomes do not, when they are fairly similar in size. Although they probed diffusive sliding rather than processive tracking dynamics, Davidson et al. (2016) report that cohesin can rapidly slide over some DNA-bound proteins and nucleosomes, but becomes obstructed by DNA-bound CTCF and transcriptional machinery; a similar, yet more restrictive, dependence of sliding on the size of DNA-bound factors has been reported in other single-molecule studies probing sliding dynamics (Kanke et al. 2016; Stigler et al. 2016). This suggests that CTCF blocks translocation of cohesin by a specific mechanism rather than by steric exclusion—for example, by inhibiting the ATPase action of the cohesin machinery (Petela et al. 2017; Wutz et al. 2017) directly or via other cohesin interactors (e.g., via Pds5) and potentially in concert with cofactors (Hsu et al. 2017). Alternatively, CTCF may recruit additional cofactors to increase its physical size or pose a greater challenge for walking due to its DNA binding geometry (Hashimoto et al. 2017).

From Translocation to Loop Extrusion

Multiple possibilities exist as to how the translocation of motor complexes along a chromosome can realize the

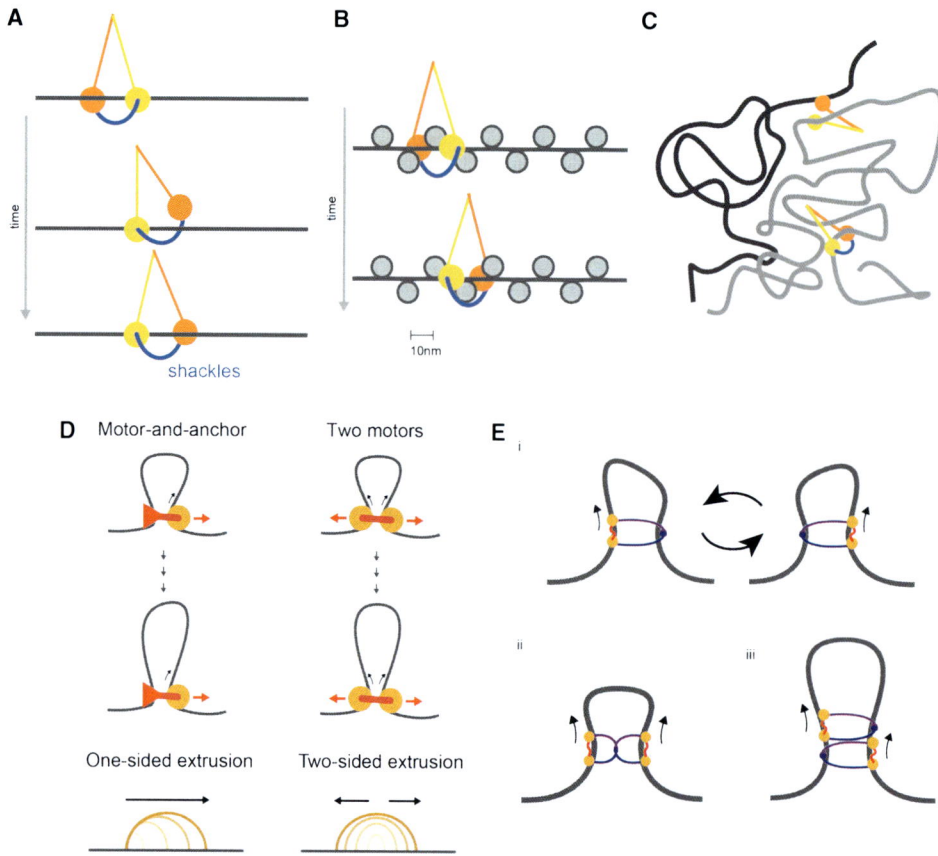

Figure 4. (*A*) Walking as a possible mechanism of SMC translocation, with SMC arms in yellow and orange and kleisin in blue, creating a *shackled walker*. (*B*) Walking along a chromatin fiber, by hopping from linker to linker without disrupting nucleosomal DNA. (*C*) Benefit of topological entrapment: An SMC walker without a kleisin can step from one chromatin strand (gray) to another in its vicinity (black), whereas a shackled SMC walker with a kleisin is able to track in *cis* over long distances. (*D*) Two possible mechanisms for converting translocation to extrusion: The first involves a single translocating motor attached to an anchor, leading to single-sided extrusion; the second involves two motors translocating in opposite directions, leading to two-sided extrusion. (*E*) Possible realizations of motor activity by SMCs (*i–iii*). (*i*) A single SMC acting as single motor that switches between entrapped chromatin strands, effectively performing two-sided extrusion; (*ii*) dimerized SMCs performing two-sided extrusion; (*iii*) alternatively dimerized SMCs performing two-sided extrusion.

process of loop extrusion. These include (1) a single trans-locating motor attached to a chromatin anchor; (2) two connected motors translocating in opposite directions; (3) a single motor that switches between two chromatin fiber substrates (Fig. 4D). These architectures for the action of SMC motors can lead to different consequences for the processive dynamics of extrusion. Unidirectional extrusion could result from a single motor-and-anchor architecture. Bidirectional extrusion would emerge from the latter two possibilities. We note there are multiple possibilities for how many SMC complexes are required to realize motor activity (Fig. 4E), either as monomers or, potentially, oligomers (Keenholtz et al. 2017). One advantage of two-motor extrusion is that it naturally allows one motor to continue extruding if the other becomes blocked. Because models discussed here and elsewhere (in Sanborn et al. 2015; Fudenberg et al. 2016; Goloborodko et al. 2016a,b) assumed independent bidirectional extrusion, it remains unclear if one-sided loop extrusion is sufficient to form TADs, peaks, and tracks, as well as to compact mitotic chromatids.

Velocity of Loop Extruders

The measured rates of stepping and step sizes for condensin (Terakawa et al. 2017) agree well with the expectations of the loop extrusion theory in interphase for cohesin. Using ~2 steps/sec and ~10-nm step size measured in vitro, this gives ~18 kb/min if cohesin moves one nucleosome per step (~150 bp). This is further doubled if cohesin extrusion occurs via a two-motor mechanism, yielding ~36 kb/min. These values are compatible with the ~10–30 kb/min predicted by polymer models as sufficient to generate TADs and corner peaks in vivo. There are several ways to arrive at this estimate. The first involves dividing the size of the largest TADs (~1 Mb [Bonev and Cavalli 2016]) by time to reestablish TADs following exit from mitosis (~0.5–2 h [Naumova et al. 2013], ~30 min [Nagano et al. 2017]) or following auxin wash-off (~30 min [Rao et al. 2017]). Alternatively, one can use the processivity of cohesin estimated from fitting Hi-C data with loop extrusion models (~200–400 kb [Fudenberg et al. 2016]), and divide this by the cohesin turnover time (~5–30 min [Gerlich et al. 2006; Hansen et al. 2017; Wutz et al. 2017]).

We note that pushing by RNA Pol II alone, at its reported velocities, would be too slow (~1.5–3 kb/min [Danko et al. 2013; Jonkers et al. 2014; Veloso et al. 2014]). The observation of cohesin-dependent features in both active and inactive chromatin (Haarhuis et al. 2017; Schwarzer et al. 2017), as well as the transcriptionally inactive maternal zygotic pronucleus (Gassler et al. 2017), further argues against Pol II providing the primary motive force for loop extrusion.

Energy Budget

A simple estimate shows that the energy burden of ATP consumption by loop-extruding cohesins in interphase is negligible as compared to ATP production in a mammalian cell. Again using 2 ATP per sec per SMC complex (Terakawa et al. 2017), and the total number of actively extruding cohesin molecules, either measured (~100,000 per cell [Hansen AS, pers commun]) or estimated from fitting simulations to Hi-C data (~1 loop-extruder/200 kb, i.e., ~60,000/diploid G2 cell), one obtains a very low rate of ATP consumption ($<2 \times 10^5$ ATP/sec). This constitutes <0.02% of the 10^9 ATP/sec production rate by a fibroblast (Flamholz et al. 2014). Thus the energy burden of chromosome organization by cohesin is marginal.

Direct Observation of Loop Extrusion

While in proofs, a paper (Ganji et al. 2018) appeared that reported a direct observation of loop extrusion in vitro by single purified yeast condensin complexes on DNA. In their experiments, condensins extruded loops of up to tens of kilobases at a speed of up to 1.5 kb/sec in an ATP-dependent fashion. Surprisingly, the extrusion observed was strictly one-sided, which prompts further investigation. Overall, this exciting new study provides the first direct evidence of active loop extrusion by SMC complexes.

CONCLUSION

Although the key role of molecular motors in the cytoplasm is broadly appreciated (Phillips et al. 2012), there is now a growing appreciation for loop extrusion by SMC complexes as an active processes organizing and compacting chromatin in the nucleus (Haarhuis and Rowland 2017; Nasmyth 2017). Analogous to the myriad uses for the contractile dynamics of active actin and tubulin networks, we hypothesize that interphase loop extrusion has been repurposed for a variety of biological ends (Dekker and Mirny 2016; Fudenberg et al. 2016), including targeting VDJ recombination, and regulation of enhancer–promoter interactions.

Hi-C maps and videos are available at http://mirnylab .mit.edu/projects/emerging-evidence-for-loop-extrusion.

ACKNOWLEDGMENTS

We thank Elphege Nora for detailed feedback and Anders Sejr Hansen for insight into cohesin dynamics. This work was supported by the National Institutes of Health (NIH) (GM114190), the National Science Foundation (NSF), Physics of Living Systems (15049420) grants, and the Center for 3D Structure and Physics of the Genome of NIH 4DN Consortium (DK107980). G.F. was supported by the San Simeon Fund (PI: K. Pollard).

REFERENCES

Alipour E, Marko JF. 2012. Self-organization of domain structures by DNA-loop-extruding enzymes. *Nucleic Acids Res* **40:** 11202–11212.

Benedetti F, Dorier J, Burnier Y, Stasiak A. 2014. Models that include supercoiling of topological domains reproduce several known features of interphase chromosomes. *Nucleic Acids Res* **42:** 2848–2855.

Blackwood EM, Kadonaga JT. 1998. Going the distance: A current view of enhancer action. *Science* **281:** 60–63.

Bonev B, Cavalli G. 2016. Organization and function of the 3D genome. *Nat Rev Genet* **17:** 661–678.

Bonev B, Mendelson Cohen N, Szabo Q, Fritsch L, Papadopoulos GL, Lubling Y, Xu X, Lv X, Hugnot J-P, Tanay A, et al. 2017. Multiscale 3D genome rewiring during mouse neural development. *Cell* **171:** 557–572.e24.

Brackley CA, Johnson J, Michieletto D, Morozov AN, Nicodemi M, Cook PR, Marenduzzo D. 2018. Extrusion without a motor: A new take on the loop extrusion model of genome organization. *Nucleus* **9:** 95–103.

Busslinger GA, Stocsits RR, van der Lelij P, Axelsson E, Tedeschi A, Galjart N, Peters J-M. 2017. Cohesin is positioned in mammalian genomes by transcription, CTCF and Wapl. *Nature* **544:** 503–507.

Cattoni DI, Cardozo Gizzi AM, Georgieva M, Di Stefano M, Valeri A, Chamousset D, Houbron C, Déjardin S, Fiche J-B, González I, et al. 2017. Single-cell absolute contact probability detection reveals chromosomes are organized by multiple low-frequency yet specific interactions. *Nat Commun* **8:** 1753.

Danko CG, Hah N, Luo X, Martins AL, Core L, Lis JT, Siepel A, Kraus WL. 2013. Signaling pathways differentially affect RNA polymerase II initiation, pausing, and elongation rate in cells. *Mol Cell* **50:** 212–222.

Davidson IF, Goetz D, Zaczek MP, Molodtsov MI, Huis In 't Veld PJ, Weissmann F, Litos G, Cisneros DA, Ocampo-Hafalla M, Ladurner R, et al. 2016. Rapid movement and transcriptional re-localization of human cohesin on DNA. *EMBO J* **35:** 2671–2685.

Dekker J, Mirny L. 2016. The 3D genome as moderator of chromosomal communication. *Cell* **164:** 1110–1121.

de Wit E, Vos ESM, Holwerda SJB, Valdes-Quezada C, Verstegen MJAM, Teunissen H, Splinter E, Wijchers PJ, Krijger PHL, de Laat W. 2015. CTCF binding polarity determines chromatin looping. *Mol Cell* **60:** 676–684.

Dixon JR, Selvaraj S, Yue F, Kim A, Li Y, Shen Y, Hu M, Liu JS, Ren B. 2012. Topological domains in mammalian genomes identified by analysis of chromatin interactions. *Nature* **485:** 376–380.

Doyle B, Fudenberg G, Imakaev M, Mirny LA. 2014. Chromatin loops as allosteric modulators of enhancer–promoter interactions. *PLoS Comput Biol* **10:** e1003867.

Eeftens J, Dekker C. 2017. Catching DNA with hoops-biophysical approaches to clarify the mechanism of SMC proteins. *Nat Struct Mol Biol* **24:** 1012–1020.

Eeftens JM, Bisht S, Kerssemakers J, Kschonsak M, Haering CH, Dekker C. 2017. Real-time detection of condensin-driven DNA compaction reveals a multistep binding mechanism. *EMBO J* **36:** 3448–3457.

Finn E, Pegoraro G, Brandao HB, Valton A-L, Oomen ME, Dekker J, Mirny L, Misteli T. 2017. Heterogeneity and Intrinsic Variation in Spatial Genome Organization. *bioRxiv* 171801. https://www.biorxiv.org/content/early/2017/08/03/171801.

Flamholz A, Phillips R, Milo R. 2014. The quantified cell. *Mol Biol Cell* 25: 3497–3500.

Forcato M, Nicoletti C, Pal K, Livi CM, Ferrari F, Bicciato S. 2017. Comparison of computational methods for Hi-C data analysis. *Nat Methods* 14: 679–685.

Fudenberg G, Imakaev M. 2017. FISH-ing for captured contacts: Towards reconciling FISH and 3C. *Nat Methods* 14: 673–678.

Fudenberg G, Imakaev M, Lu C, Goloborodko A, Abdennur N, Mirny LA. 2016. Formation of chromosomal domains by loop extrusion. *Cell Rep* 15: 2038–2049.

Ganji M, Shaltiel IA, Bisht S, Kim E, Kalichava A, Haering CH, Dekker C. 2018. Real-time imaging of DNA loop extrusion by condensin. *Science* 360: 102–105.

Gassler J, Brandão HB, Imakaev M, Flyamer IM, Ladstätter S, Bickmore WA, Peters J-M, Mirny LA, Tachibana K. 2017. A mechanism of cohesin-dependent loop extrusion organizes zygotic genome architecture. *EMBO J* 36: 3600–3618.

Gerlich D, Koch B, Dupeux F, Peters J-M, Ellenberg J. 2006. Live-cell imaging reveals a stable cohesin–chromatin interaction after but not before DNA replication. *Curr Biol* 16: 1571–1578.

Gibcus JH, Samejima K, Goloborodko A, Samejima I, Naumova N, Nuebler J, Kanemaki MT, Xie L, Paulson JR, Earnshaw WC, et al. 2018. A pathway for mitotic chromosome formation. *Science* 359: eaao6135.

Giorgetti L, Galupa R, Nora EP, Piolot T, Lam F, Dekker J, Tiana G, Heard E. 2014. Predictive polymer modeling reveals coupled fluctuations in chromosome conformation and transcription. *Cell* 157: 950–963.

Goloborodko A, Imakaev MV, Marko JF, Mirny L. 2016a. Compaction and segregation of sister chromatids via active loop extrusion. *Elife* 5: e14864.

Goloborodko A, Marko JF, Mirny LA. 2016b. Chromosome compaction by active loop extrusion. *Biophys J* 110: 2162–2168.

Gruber S. 2014. Multilayer chromosome organization through DNA bending, bridging and extrusion. *Curr Opin Microbiol* 22: 102–110.

Guacci V, Yamamoto A, Strunnikov A, Kingsbury J, Hogan E, Meluh P, Koshland D. 1993. Structure and function of chromosomes in mitosis of budding yeast. *Cold Spring Harb Symp Quant Biol* 58: 677–685.

Guo Y, Xu Q, Canzio D, Shou J, Li J, Gorkin DU, Jung I, Wu H, Zhai Y, Tang Y, et al. 2015. CRISPR inversion of CTCF sites alters genome topology and enhancer/promoter function. *Cell* 162: 900–910.

Haarhuis JHI, van der Weide RH, Blomen VA, Yáñez-Cuna JO, Amendola M, van Ruiten MS, Krijger PHL, Teunissen H, Medema RH, van Steensel B, et al. 2017. The cohesin release factor WAPL restricts chromatin loop extension. *Cell* 169: 693–707.e14.

Haarhuis J, Rowland BD. 2017. Cohesin: Building loops, but not compartments. *EMBO J* 36: 3549–3551.

Hansen AS, Pustova I, Cattoglio Claudia, Tjian Robert, Darzacq X. 2017. CTCF and cohesin regulate chromatin loop stability with distinct dynamics. *Elife* 6: e25776.

Hansen AS, Cattoglio C, Darzacq X, Tjian R. 2018. Recent evidence that TADs and chromatin loops are dynamic structures. *Nucleus* 9: 20–32.

Hashimoto H, Wang D, Horton JR, Zhang X, Corces VG, Cheng X. 2017. Structural basis for the versatile and methylation-dependent binding of CTCF to DNA. *Mol Cell* 66: 711–720.e3.

Hofmann A, Heermann DW. 2015. The role of loops on the order of eukaryotes and prokaryotes. *FEBS Lett* 589: 2958–2965.

Hsu SC, Gilgenast TG, Bartman CR, Edwards CR, Stonestrom AJ, Huang P, Emerson DJ, Evans P, Werner MT, Keller CA, et al. 2017. The BET protein BRD2 cooperates with CTCF to enforce transcriptional and architectural boundaries. *Mol Cell* 66: 102–116.e7.

Imakaev MV, Fudenberg G, Mirny LA. 2015. Modeling chromosomes: Beyond pretty pictures. *FEBS Lett* 589: 3031–3036.

Jonkers I, Kwak H, Lis JT. 2014. Genome-wide dynamics of Pol II elongation and its interplay with promoter proximal pausing, chromatin, and exons. *Elife* 3: e02407.

Kanke M, Tahara E, Huis In't P, Nishiyama T. 2016. Cohesin acetylation and Wapl-Pds5 oppositely regulate translocation of cohesin along DNA. *EMBO J* 35: 2686–2698.

Keenholtz RA, Dhanaraman T, Palou R, Yu J, D'Amours D, Marko JF. 2017. Oligomerization and ATP stimulate condensin-mediated DNA compaction. *Sci Rep* 7: 14279.

Kerpedjiev P, Abdennur N, Lekschas F, McCallum C, Dinkla K, Strobelt H, Luber JM, Ouellette SB, Ahzir A, Kumar N, et al. 2017. HiGlass: Web-based visual comparison and exploration of genome interaction maps. *bioRxiv* 121889. http://biorxiv.org/content/early/2017/03/31/121889.

Kim H, Loparo JJ. 2016. Multistep assembly of DNA condensation clusters by SMC. *Nat Commun* 7: 10200.

Kimura K, Rybenkov VV, Crisona NJ, Hirano T, Cozzarelli NR. 1999. 13S condensin actively reconfigures DNA by introducing global positive writhe: Implications for chromosome condensation. *Cell* 98: 239–248.

Kschonsak M, Merkel F, Bisht S, Metz J, Rybin V, Hassler M, Haering CH. 2017. Structural basis for a safety-belt mechanism that anchors condensin to chromosomes. *Cell* 171: 588–600.e24.

Kubo N, Ishii H, Gorkin D, Meitinger F, Xiong X, Fang R, Liu T, Ye Z, Li B, Dixon J, et al. 2017. Preservation of chromatin organization after acute loss of CTCF in mouse embryonic stem cells. *bioRxiv* 118737. https://www.biorxiv.org/content/early/2017/03/20/118737.

Lee D, Tan W, Anene G, Li P, Danh T, Tiang Z, Ng SL, Efthymios M, Autio M, Jiang J, et al. 2017. Gene neighbourhood integrity disrupted by CTCF loss in vivo. *bioRxiv* 187393. https://www.biorxiv.org/content/early/2017/09/12/187393.

Merkenschlager M, Nora EP. 2016. CTCF and cohesin in genome folding and transcriptional gene regulation. *Annu Rev Genomics Hum Genet* 17: 17–43.

Nagano T, Lubling Y, Várnai C, Dudley C, Leung W, Baran Y, Mendelson Cohen N, Wingett S, Fraser P, Tanay A. 2017. Cell-cycle dynamics of chromosomal organization at single-cell resolution. *Nature* 547: 61–67.

Narendra V, Rocha PP, An D, Raviram R, Skok JA, Mazzoni EO, Reinberg D. 2015. CTCF establishes discrete functional chromatin domains at the Hox clusters during differentiation. *Science* 347: 1017–1021.

Nasmyth K. 2001. Disseminating the genome: Joining, resolving, and separating sister chromatids during mitosis and meiosis. *Annu Rev Genet* 35: 673.

Nasmyth K. 2017. How are DNAs woven into chromosomes? *Science* 358: 589–590.

Naumova N, Imakaev M, Fudenberg G, Zhan Y, Lajoie BR, Mirny LA, Dekker J. 2013. Organization of the mitotic chromosome. *Science* 342: 948–953.

Nichols MH, Corces VG. 2015. A CTCF code for 3D genome architecture. *Cell* 162: 703–705.

Nora EP, Lajoie BR, Schulz EG, Giorgetti L, Okamoto I, Servant N, Piolot T, van Berkum NL, Meisig J, Sedat J, et al. 2012. Spatial partitioning of the regulatory landscape of the X-inactivation centre. *Nature* 485: 381–385.

Nora EP, Goloborodko A, Valton A-L, Gibcus JH, Uebersohn A, Abdennur N, Dekker J, Mirny LA, Bruneau BG. 2017. Targeted degradation of CTCF decouples local insulation of chromosome domains from genomic compartmentalization. *Cell* 169: 930–944.e22.

Nozaki T, Imai R, Tanbo M, Nagashima R, Tamura S, Tani T, Joti Y, Tomita M, Hibino K, Kanemaki MT, et al. 2017. Dynamic organization of chromatin domains revealed by super-resolution live-cell imaging. *Mol Cell* 67: 282–293.e7.

Nuebler J, Fudenberg G, Imakaev M, Abdennur N, Mirny L. 2017. Chromatin organization by an interplay of loop extrusion and compartmental segregation. *bioRxiv* 2017. https://doi.org/101101/196261.

Petela N, Gligoris TG, Metson JS, Lee B-G, Voulgaris M, Hu B, Kikuchi S, Chapard C, Chen W, Rajendra E, et al. 2017. Multiple interactions between Scc1 and Scc2 activate cohesin's DNA dependent ATPase and replace Pds5 during loading. *bioRxiv* 205914.

Peters J-M, Nishiyama T. 2012. Sister chromatid cohesion. *Cold Spring Harb Perspect Biol* **4:** a011130.

Peterson CL. 1994. The SMC family: Novel motor proteins for chromosome condensation? *Cell* **79:** 389–392.

Phillips R, Kondev J, Theriot J, Garcia H. 2012. *Physical biology of the cell*, 2nd edn. Garland Science.

Rao SSP, Huntley MH, Durand NC, Stamenova EK, Bochkov ID, Robinson JT, Sanborn AL, Machol I, Omer AD, Lander ES, et al. 2014. A 3D map of the human genome at kilobase resolution reveals principles of chromatin looping. *Cell* **159:** 1665–1680.

Rao SSP, Huang S-C, Glenn St Hilaire B, Engreitz JM, Perez EM, Kieffer-Kwon K-R, Sanborn AL, Johnstone SE, Bascom GD, Bochkov ID, et al. 2017. Cohesin loss eliminates all loop domains. *Cell* **171:** 305–320.e24.

Rhodes J, Mazza D, Nasmyth K, Uphoff S. 2017. Scc2/Nipbl hops between chromosomal cohesin rings after loading. *Elife* **6:** e30000.

Riggs AD. 1990. DNA methylation and late replication probably aid cell memory, and type 1 DNA reeling could aid chromosome folding and enhancer function. *Philos Trans R Soc Lond B Biol Sci* **326:** 285–297.

Rodríguez-Carballo E, Lopez-Delisle L, Zhan Y, Fabre PJ, Beccari L, El-Idrissi I, Huynh THN, Ozadam H, Dekker J, Duboule D. 2017. The HoxD cluster is a dynamic and resilient TAD boundary controlling the segregation of antagonistic regulatory landscapes. *Genes Dev* **31:** 2264–2281.

Rosa-Garrido M, Chapski DJ, Schmitt AD, Kimball TH, Karbassi E, Monte E, Balderas E, Pellegrini M, Shih T-T, Soehalim E, et al. 2017. High-resolution mapping of chromatin conformation in cardiac myocytes reveals structural remodeling of the epigenome in heart failure. *Circulation* **136:** 1613–1625.

Sanborn AL, Rao SSP, Huang S-C, Durand NC, Huntley MH, Jewett AI, Bochkov ID, Chinnappan D, Cutkosky A, Li J, et al. 2015. Chromatin extrusion explains key features of loop and domain formation in wild-type and engineered genomes. *Proc Natl Acad Sci* **112:** E6456–E6465.

Schalbetter SA, Goloborodko A, Fudenberg G, Belton J-M, Miles C, Yu M, Dekker J, Mirny L, Baxter J. 2017. SMC complexes differentially compact mitotic chromosomes according to genomic context. *Nat Cell Biol* **19:** 1071–1080.

Schwarzer W, Abdennur N, Goloborodko A, Pekowska A, Fudenberg G, Loe-Mie Y, Fonseca NA, Huber W, Haering CH, Mirny L, et al. 2017. Two independent modes of chromatin organization revealed by cohesin removal. *Nature* **551:** 51–56.

Seitan VC, Faure AJ, Zhan Y, McCord RP, Lajoie BR, Ing-Simmons E, Lenhard B, Giorgetti L, Heard E, Fisher AG, et al. 2013. Cohesin-based chromatin interactions enable regulated gene expression within preexisting architectural compartments. *Genome Res* **23:** 2066–2077.

Sofueva S, Yaffe E, Chan W-C, Georgopoulou D, Rudan MV, Mira-Bontenbal H, Pollard SM, Schroth GP, Tanay A, Hadjur S. 2013. Cohesin-mediated interactions organize chromosomal domain architecture. *EMBO J* **32:** 3119–3129.

Spielmann M, Mundlos S. 2016. Looking beyond the genes: The role of non-coding variants in human disease. *Hum Mol Genet* **25:** R157–R165.

Srinivasan M, Scheinost J, Petela N, Gligoris T, Wissler M, Ogushi S, Collier J, Voulgaris M, Kurze A, Chan K-L, et al. 2017. The cohesin ring uses its hinge to organize DNA using non-topological as well as topological mechanisms. *bioRxiv* 197848.

Stigler J, Çamdere GÖ, Koshland DE, Greene EC. 2016. Single-molecule imaging reveals a collapsed conformational state for DNA-bound cohesin. *Cell Rep* **15:** 988–998.

Tedeschi A, Wutz G, Huet S, Jaritz M, Wuensche A, Schirghuber E, Davidson IF, Tang W, Cisneros DA, Bhaskara V, et al. 2013. Wapl is an essential regulator of chromatin structure and chromosome segregation. *Nature* **501:** 564–568.

Terakawa T, Bisht S, Eeftens JM, Dekker C, Haering CH, Greene EC. 2017. The condensin complex is a mechanochemical motor that translocates along DNA. *Science* **358:** 672–676.

Veloso A, Kirkconnell KS, Magnuson B, Biewen B, Paulsen MT, Wilson TE, Ljungman M. 2014. Rate of elongation by RNA polymerase II is associated with specific gene features and epigenetic modifications. *Genome Res* **24:** 896–905.

Vietri Rudan M, Barrington C, Henderson S, Ernst C, Odom DT, Tanay A, Hadjur S. 2015. Comparative Hi-C reveals that CTCF underlies evolution of chromosomal domain architecture. *Cell Rep* **10:** 1297–1309.

Wang X, Le TBK, Lajoie BR, Dekker J, Laub MT, Rudner DZ. 2015. Condensin promotes the juxtaposition of DNA flanking its loading site in *Bacillus subtilis*. *Genes Dev* **29:** 1661–1675.

Wang X, Brandao HB, Le TBK, Laub MT, Rudner DZ. 2017. *Bacillus subtilis* SMC complexes juxtapose chromosome arms as they travel from origin to terminus. *Science* **355:** 524.

Wendt KS, Peters J-M. 2009. How cohesin and CTCF cooperate in regulating gene expression. *Chromosome Res* **17:** 201–214.

Wutz G, Várnai C, Nagasaka K, Cisneros DA, Stocsits RR, Tang W, Schoenfelder S, Jessberger G, Muhar M, Hossain MJ, et al. 2017. Topologically associating domains and chromatin loops depend on cohesin and are regulated by CTCF, WAPL, and PDS5 proteins. *EMBO J* **36:** 3573–3599.

Yamamoto T, Schiessel H. 2017. Osmotic mechanism of the loop extrusion process. *Phys Rev E* **96:** 030402.

Zuin J, Dixon JR, van der Reijden MIJA, Ye Z, Kolovos P, Brouwer RWW, van de Corput MPC, van de Werken HJG, Knoch TA, van IJcken WFJ, et al. 2014. Cohesin and CTCF differentially affect chromatin architecture and gene expression in human cells. *Proc Natl Acad Sci* **111:** 996–1001.

SpotLearn: Convolutional Neural Network for Detection of Fluorescence In Situ Hybridization (FISH) Signals in High-Throughput Imaging Approaches

Prabhakar R. Gudla,[1,2] Koh Nakayama,[2,3] Gianluca Pegoraro,[1,2] and Tom Misteli[2]

[1]High-Throughput Imaging Facility, National Cancer Institute, National Institutes of Health, Bethesda, Maryland 20892

[2]Cell Biology of Genomes Group, National Cancer Institute, National Institutes of Health, Bethesda, Maryland 20892

[3]Oxygen Biology Laboratory, Medical Research Institute, Tokyo Medical and Dental University, Tokyo, Japan 1138510

Correspondence: mistelit@mail.nih.gov

DNA fluorescence in situ hybridization (FISH) is the technique of choice to map the position of genomic loci in three-dimensional (3D) space at the single allele level in the cell nucleus. High-throughput DNA FISH methods have recently been developed using complex libraries of fluorescently labeled synthetic oligonucleotides and automated fluorescence microscopy, enabling large-scale interrogation of genomic organization. Although the FISH signals generated by high-throughput methods can, in principle, be analyzed by traditional spot-detection algorithms, these approaches require user intervention to optimize each interrogated genomic locus, making analysis of tens or hundreds of genomic loci in a single experiment prohibitive. We report here the design and testing of two separate machine learning–based workflows for FISH signal detection in a high-throughput format. The two methods rely on random forest (RF) classification or convolutional neural networks (CNNs), respectively. Both workflows detect DNA FISH signals with high accuracy in three separate fluorescence microscopy channels for tens of independent genomic loci, without the need for manual parameter value setting on a per locus basis. In particular, the CNN workflow, which we named SpotLearn, is highly efficient and accurate in the detection of DNA FISH signals with low signal-to-noise ratio (SNR). We suggest that SpotLearn will be useful to accurately and robustly detect diverse DNA FISH signals in a high-throughput fashion, enabling the visualization and positioning of hundreds of genomic loci in a single experiment.

The genome is nonrandomly organized in the cell nucleus (Bonev and Cavalli 2016). Spatial genome organization occurs in a hierarchical fashion: Genomic loci with similar transcriptional activity and epigenetic profiles preferentially fold into domains, known as topologically associated domains (TADs); these, in turn, form larger domains, which are then further organized into chromosome territories. The three-dimensional (3D) organization of the genome allows the compaction of ~2 m of linear DNA in human cell nuclei with an ~10 μm diameter, and it provides a regulatory layer for key cellular pathways such as transcription, replication, and DNA damage and repair (Cavalli and Misteli 2013). Alterations in genome folding and organization have been linked to cancer (Flavahan et al. 2016) and developmental syndromes (Lupiáñez et al. 2015; Franke et al. 2016), highlighting the importance of 3D genome architecture in physiological and diseased states.

DNA fluorescence in situ hybridization (FISH) is one of the widely used tools of choice to study genome organization, because it directly visualizes the position of genomic loci in 3D space in the nucleus (Solovei et al. 2002). Traditional DNA FISH uses enzymatically labeled fluorescent probes, which hybridize in a sequence-specific manner to the genomic region of interest. In contrast to other biochemical techniques used to study genome organization, such as chromosome conformation capture (3C), DNA FISH allows visualization and measurement of actual physical distances between multiple genomic loci at the single-allele level. Despite this advantage, DNA FISH has mostly been used as a semiquantitative technique to validate a select few genomic interactions, mostly because of the need for laborious generation of fluorescent probes and the limited throughput of traditional fluorescence microscopy.

Two recent technical developments have helped overcome these limitations and enable large-scale FISH detection. The first is the substitution of enzymatically labeled DNA FISH probes with large libraries of chemically synthesized DNA oligos, a technique named Oligopaint (Beliveau et al. 2012, 2015; Joyce et al. 2012). Oligopaint allows the precise and flexible selection of computationally designed primary oligonucleotides binding to nonrepetitive genomic regions, increases the resolution of DNA FISH to as little as 5 kb, and, because of the use of combinatorial labeling schemes involving secondary fluorescent oligo DNA barcodes (Beliveau et al. 2015; Chen et al. 2015), increases the potential number of genomic loci that

Supplemental material is available for this article at symposium.cshlp.org.

This is a work of the US Government.

Published by Cold Spring Harbor Laboratory Press; doi: 10.1101/sqb.2017.82.033761

can be visualized to a few hundred in a single experiment (Wang et al. 2016). The second technical innovation is high-throughput imaging (HTI), which uses multiwell imaging plates, automated liquid handling, and high-throughput 3D confocal fluorescence image acquisition to generate hundreds of thousands of images relative to thousands of cells for each of hundreds of experimental conditions (Pegoraro and Misteli 2017).

Despite these experimental advances, challenges remain to the reliable and automated detection and quantification of DNA FISH signals, which appear as diffraction-limited fluorescent spots in the nucleus. Several image-processing algorithms, such as difference of Gaussians (Bright and Steel 1987), multiscale wavelet-based (Olivo 1996), and radial symmetry (Parthasarathy 2012), detect spot-like objects in 2D fluorescence microscopy images. However, for efficient spot detection performance, the investigator needs to empirically determine appropriate sets of values for the algorithm parameters. The optimal parameter values vary with the signal-to-noise ratio (SNR) of the fluorescent spot signal, which itself varies between different DNA FISH probe sets and fluorophores in a single experiment. Although manual value optimization of spot detection parameters is feasible for one or a few DNA FISH probe sets, it is extremely laborious and subject to bias when used in the analysis of image data sets of hundreds of probes. These drawbacks, thus, negate the gains of using Oligopaint with HTI to visualize genomic organization on a large scale. In an effort to overcome this limitation, we have implemented two supervised machine learning–based analysis workflows for the high-throughput segmentation and classification of large and diverse sets of FISH signals generated by Oligopaint DNA FISH and high-throughput confocal imaging. The first method uses a manually optimized spot detection algorithm coupled with a supervised random forest (RF) classifier (Breiman 2001), whereas the second method is based on a different class of supervised machine learning (ML) algorithms, deep convolutional neural networks (CNNs) (Szegedy et al. 2016; Krizhevsky et al. 2017; Shelhamer et al. 2017). Here we describe and report on the performance of the RF- and CNN-based algorithms in DNA FISH spot detection and classification tasks. Our results indicate that SpotLearn will be readily adaptable to high-throughput FISH data sets, thus allowing fully automated, single-allele analysis of genome organization using HTI.

MATERIALS AND METHODS

Cell Culture

MDA-MB-231 cells (ATCC, Cat. HTB-26) were maintained in DMEM medium, 10% FBS, penicillin 100 U/mL, streptomycin 100 µg/mL in a humidified incubator at 37°C and 5% CO_2. Cells were seeded in CellCarrier-Ultra 384-well plates (PerkinElmer, Cat. 6057500) at a seeding density of 5000 cells per well. Cells were cultured for 72 h before direct fixation in the medium with 4% paraformaldehyde (PFA) in PBS.

Oligopaint DNA FISH

After fixation, cells were washed in PBS three times for 5 min, permeabilized with 0.5% saponin, 0.5% Triton X-100 for 20 min, washed in PBS three times for 3 min, treated with 0.1 N HCl for 15 min, washed in 2× SSC buffer once for 5 min, and then preincubated in 2× SSC/ 50% formamide for at least 30 min. The DNA oligo library including encoding probes were synthesized by Twist Bioscience and amplified according to a previously published protocol (Chen et al. 2015). The 5′-labeled (Alexa488, ATTO565, or Cy5) decoding probes were synthesized by Eurofin Genomics. Both the encoding oligo library and the fluorescent decoding oligos were added to cells in a 15 µL volume of hybridization buffer (50% formamide, 20% dextran sulfate, 1× Denhardt's solution, 2× SSC) per well. The encoding oligo library was used at a final concentration of 330 nM in every well. Different three-way combinations of fluorescently labeled readout oligos were used in each well at a final concentration of 6.6 nM each. Cells and oligo DNA probes were denatured for 7 min at 85°C on a heating block, and then immediately transferred for a 16 h incubation at 37°C. After oligo DNA FISH probe hybridization, cells were washed with 2× SSC three times for 5 min at 42°C and three times for 5 min at 60°C. Finally, nuclei were stained with DAPI (4′6-diamidino-2-phenylindole), washed in PBS three times for 3 mi, and stored in PBS at 4°C until imaging.

Experimental Layout

Cells were grown and stained with Oligopaint DNA FISH probes in three colors in 28 wells on a single 384-well plate as described above (training plate, Train-P1). The images from Train-P1 were used for training a supervised RF classifier (Ho 1998; Breiman 2001) for filtering mis-segmented and/or overlapping nuclei, optimizing parameters of the FISH spot detection algorithm, training a supervised RF classifier for filtering false-positive FISH spots from spot detection, and training and validation of supervised fully CNN-based spot segmentation algorithm. For testing, cells were grown and stained with Oligopaint DNA FISH in three colors in 48 wells (12 unique three-color probe sets, four wells as technical controls per probe set) on two separate 384-well plates on different days as described above (Testing Plates plate, Test-P1 and Test-P2). Test-P1 and Test-P2 were biological replicates. Three-color Oligopaint DNA FISH sets were designated with an "i-j-k" scheme, where "i" is an identifier for the gene locus labeled with Alexa488, "j" is an identifier for the gene locus labeled with ATTO565, and "k" is an identifier for the gene locus labeled with Cy5. The Oligopaint DNA FISH probe sets used for Test-P1 and Test-P2 were different from those used for Train-P1.

High-Throughput Image Acquisition

Images were acquired using an automated high-throughput spinning disk microscope (Yokogawa Cell Voyager 7000) to acquire four spectral channels: DAPI, Alexa-488,

ATTO565, and Cy5. We used a 40× dry objective (0.95 NA), four excitation lasers (405, 488, 561, and 640 nm), a quad-band dichroic mirror for excitation, a fixed 568-nm dichroic mirror for detection, two 16-bit Andor Neo 5.5 sCMOS cameras (5.5 Mp; pixel binning, 2; field-of-view covering 1276 × 1076 pixels), and switchable matched bandpass filters for each channel in front of the cameras (DAPI, BP445/45; Alexa488, BP525/50; ATTO565, BP600/37; and Cy5, BP676/29). In addition, Z stacks of four images at every 1.0 μm were acquired for each channel in each field. In these imaging conditions, the pixel size was 323 nm. These channels were imaged in a sequential mode to minimize spectral potential bleed-through. We imaged six locations (fields of view) for each well. Identical acquisition settings were used for every well on the same plate—namely, laser intensity and exposure time on the CMOS cameras.

Nucleus Segmentation and Filtering

The nuclei from the maximum intensity projected DAPI channel were segmented using a seeded watershed algorithm (Vincent and Soille 1991). The preliminary segmentation boundaries from the seeded watershed were further refined using ultrametric contour maps (UCMs) to minimize oversegmentation (Arbelaez 2006). Briefly, UCMs achieve this by combining several types of low-level image information (e.g., gradients and intensity) to construct hierarchical representation of the image boundaries. Under this representation, boundary pixels along the nucleus periphery typically receive a higher score than other pixels associated with internal structures of the nucleus. Global thresholding (Otsu 1979; Sezgin 2004) of UCMs eliminates weaker, internal boundaries, and it minimizes oversegmentation. UCMs, however, cannot resolve boundaries between overlapping nuclei. To filter out overlapping nuclei from subsequent analysis, as well as any remaining oversegmented nuclei, we used a binary RF classifier (*Good* and *Bad*). To generate the training data for the RF classifier, we used an interactive KNIME (Berthold et al. 2008) workflow to annotate 441 segmented objects (class-*Good*, 304; class-*Bad*, 137) selected randomly from different wells of the training plate, Train-P1. Next, we extracted 14 morphometric features (e.g., circularity, solidity, area, perimeter, and major elongation) for each of the labeled object using the 2D geometric feature set from the KNIME Image Processing (KNIP) Feature Calculator Node (Dietz and Berthold 2016). The extracted features along with the class labels were used to train a RF classifier using KNIME's Tree Ensemble Learner node. The goal of this supervised classifier was to filter out overlapping and mis-segmented objects from the nucleus segmentation.

Spot Detection and Filtering

DNA FISH signals in each spectral channel were segmented using the undecimated multiscale wavelet transform algorithm (UMSWT) (Olivo 1996; Olivo-Marin 2002). We used two wavelet scales for segmenting DNA

FISH signals. The per-scale threshold parameters of the spot detection algorithm were manually adjusted so that DNA FISH signals in all three channels (Alexa488, ATTO565, and Cy5), which could be reliably detected with the same set of parameters. We found that the values of 2.0 and 1.0 for scale 0 and 1, respectively, gave the best results across all three FISH channels. These per-scale threshold parameters were intentionally set to lower values for detecting FISH signals with low SNR. Therefore, the spot detection algorithm also segmented background regions as potential FISH signals (false positives). To filter out these background regions we used a binary supervised RF classifier. To filter out false-positive FISH signals, we incorporated normalized spot intensity features in addition to the spot morphometric features. For generating the training data for random classifier per FISH channel, objects detected by the UMSWT spot detection algorithm were manually annotated as either correctly (class-*GoodFISH*) or incorrectly (class-*BadFISH*) segmented FISH spots. FISH images from training plate Train-P1 were used for generating this training data.

Fully CNN Model for FISH Spot Detection

The fully CNN model is based on the U-Net autoencoder architecture (Ronneberger et al. 2015). All convolutional layers in our CNN model used a 3×3 kernel size and *ReLU* (Rectified Linear Unit) activation, except for the last layer, which used a *sigmoid* activation function to generate an output image containing pixel-level probability values. Convolutional layers were followed by a max-pooling (2×2 window size) layer for downsampling the input. The max-pooling layers were used to generate discriminating features at multiple scales/resolutions. We also introduced a dropout layer (dropout rate = 0.2) in between convolution layers within each CNN block to minimize overfitting (Srivastava et al. 2014). These minor modifications resulted in a model with a total of 402,625 training parameters. We used the modified Dice coefficient as the loss function (see Eq. 1) for optimizing the model.

For training (and validation) of the CNN we used 222 DNA FISH images and their corresponding binary masks of the DNA FISH signals collected from all three DNA FISH channels. These images were a subset of annotated spots data used for training the RF classifiers for spot filtering: Only those nuclei where all DNA FISH signals were annotated as *GoodFISH* were retained for CNN training. The CNN model was optimized using the Adam optimizer (Kingma and Ba 2014) with a learning rate of 10^{-5}, a batch size of two images, and 5000 epochs with early stopping criteria (validation loss change of $<10^{-7}$ across 500 epochs). The collection of 222 images and their corresponding binary masks were randomly split (90%–10%) to generate training (198) and validation (23) data sets. Note that the FISH images and the ground truth binary masks from the validation set are never used by the CNN model for optimizing the training parameters. The U-Net-2L model was implemented as a stand-alone Python script (https://github.com/jocicmarko/ultrasound-nerve-segmentation) and run on a HPC compute node with Nvidia Tesla K80 GPU.

Quantitative Assessment Methodology

We used the out-of-bag training accuracy calculations to assess the performance of the RF classifiers. For these classifiers, out-of-bag accuracy on the training sets is a good approximation of testing accuracy for similar sets of the same size. The training set was generated using the data (nuclei and spots) from training plate, Train-P1. We used the modified Dice coefficient (Dice 1945; Sørensen 1948; Zijdenbos et al. 1994) as the loss function for the CNN model:

Modified Dice coefficient

$$
\begin{aligned}
&= -\mathrm{loss(CNN)} \\
&= \frac{2*(G \cap P) + 1}{(G \cup P) + 1} \in [0, 1],
\end{aligned}
\tag{1}
$$

where G is the ground truth binary image of DNA FISH signals in the input grayscale image, and P is the predicted probability image of each pixel belonging to a DNA FISH signal. The symbols $\cap$ and $\cup$ denote intersection and union operations, respectively. The smoothing factor 1 in both the numerator and denominator helps the CNN model to handle the special case of input grayscale DNA FISH image not having any segmented objects.

To quantitatively compare the two FISH spot detection methods (RF and CNN), we generated ground truth images (binary masks) corresponding to FISH signals from a randomly sampled FISH images in Test-P2. The ground truth images were generated for all three FISH channels (Alexa488, ATTO565, and Cy5) and are used to enumerate the total number of true positives (TP, FISH signal correctly detected as a spot), false positives (FP, background signal detected as FISH spot), and false negatives/missing spot(s) (FN, FISH signal not detected as spot) for both ML-based spot detection methods. We then calculated the following performance metrics for both methods:

$$
\mathrm{Accuracy} = \frac{\mathrm{TP}}{\mathrm{TP} + \mathrm{FP} + \mathrm{FN}},
$$

$$
\mathrm{Precision} = \frac{\mathrm{TP}}{\mathrm{TP} + \mathrm{FP}},
$$

$$
\mathrm{Recall} = \frac{\mathrm{TP}}{\mathrm{TP} + \mathrm{FN}},
$$

$$
F\text{-score} = \frac{2 \times \mathrm{Precision} \times \mathrm{Recall}}{\mathrm{Precision} + \mathrm{Recall}}.
$$

We validated the robustness of both spot detection approaches by comparing the number of FISH signals identified in each cell (copy number) per probe set in separate technical and biological replicates.

Implementation

We implemented all image analysis and ML classification workflows in the Konstanz Information Miner, KNIME (Berthold et al. 2008), Analytics Platform (version 3.2.1, 64-bit) using compatible KNIME Image Processing Nodes (KNIP, version 1.5.2.201610061254) (Berthold et al. 2008; Dietz and Berthold 2016). All the KNIME image analysis workflows were run on dedicated computing nodes from a high-performance batch cluster (Biowulf, National Institutes of Health) either interactively or in a batch mode. The computing nodes had the following specifications: 64-Bit RedHat Enterprise Linux 6.9 (Santiago), 28 cores (56 threads) Intel X2680 processor, 256 GB RAM, 4 Nvidia Tesla K80 GPUs (12 GB VRAM), and 800 GB of Solid State Storage. The CNN-based spot segmentation algorithm (U-Net-2L) was implemented as a Python script (Python version, 2.7.13, 64-bit) using Keras (version 2.0.5) with the Tensorflow backend (GPU-enabled, version 1.1.0, 64-bit). The trained CNN model was used in the KNIME workflow for spot segmentation using the Python Scripting KNIME node.

Code Availability

All the KNIME workflows, the training and validating data, and the models are available via GitHub repository at https://github.com/CBIIT/Misteli-Lab-CCR-NCI/tree/master/Gudla_CSH_2017.

RESULTS

Complex Oligopaint Libraries of Probes Generate Diverse FISH Signals

Our laboratory has extensively used enzymatically labeled fluorescent DNA FISH probes and traditional spot detection algorithms for HTI applications (Burman et al. 2015; Shachar et al. 2015; Finn et al. 2017). In these cases, although the number of experimental conditions reached up to approximately 700 wells per experiment (Shachar et al. 2015), the number of different genomic loci labeled in any single experiment never exceeded 21 (Finn et al. 2017), making it feasible to empirically determine the most appropriate sets of spot detection parameters on a per genomic locus basis. Oligopaint DNA FISH (Beliveau et al. 2012, 2015; Joyce et al. 2012), on the other hand, allows one to label and detect hundreds of different genomic loci in a single experiment on one or more 384-well plates by using combinatorial hybridization of pools of chemically synthesized oligo probes to genomic targets, together with fluorescently labeled decoding oligo probes (Fig. 1A,B; Chen et al. 2015; Moffitt et al. 2016; Wang et al. 2016). While developing HTI assays for the detection of tens of genomic loci in the same experiment, we observed that FISH signals generated by Oligopaint probe sets targeting different genomic regions can have considerably different SNRs, ranging from ~2 to 9 (Fig. 1C). Consequently, manual parameter setting of traditional spot detection algorithms for potentially hundreds of different genomic targets did not seem to be either a practical or a robust solution to this computational problem. To tackle this issue, we tested and compared the performance of two independent supervised ML approaches for the

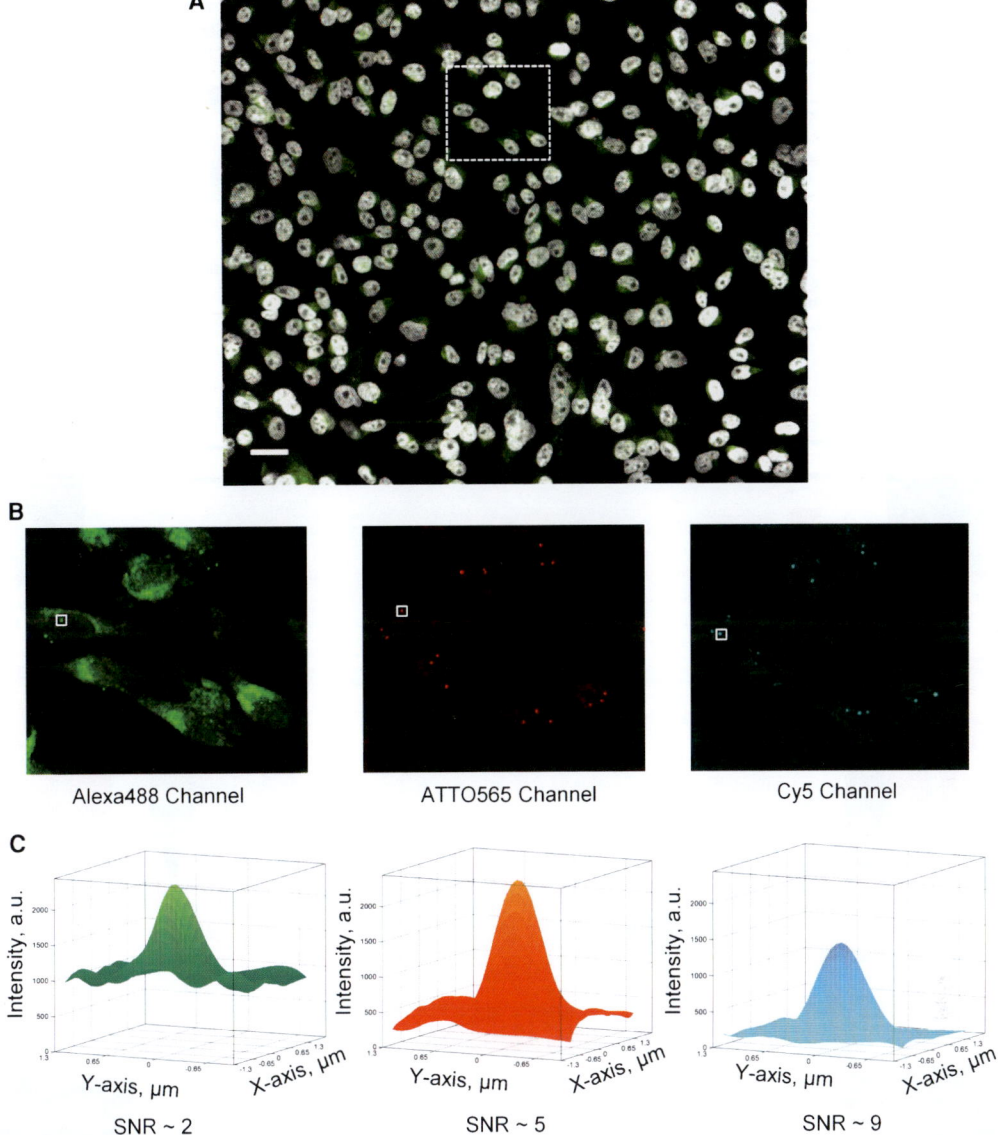

Figure 1. (*A*) Representative maximum intensity projection images of MDA-MB-231 cells from one of the 48 wells with a unique Oligopaint DNA FISH set from test plate Test-P2. The DAPI-stained nuclei (blue channel) are displayed in grayscale, whereas the Alexa488 (green channel), ATTO565 (red channel), and Cy5 (far-red channel) channels are displayed using green, red, and cyan look-up tables, respectively. Scale bar, 10 μm. (*B*) Three Oligopaint DNA FISH channels corresponding to the region enclosed by the dashed box in *A*. (*C*) Representative surface rendering of the DNA FISH signals intensity identified by the solid-white colored boxes in *B*. The surface plots were generated by interpolating the raw DNA FISH signal intensities (*Z*-axis, arbitrary units) of 8 × 8 pixels region centered around the brightest pixel of the DNA FISH spot (1 pixel = 0.325 μm). The DNA FISH signal intensity for the Alexa488-labeled genomic locus had higher background around 1000 arbitrary units (a.u.) with peak signal of ~2000 a.u., thus leading to a signal-to-(background/)noise ratio (SNR) of ~2. The ATTO565- and Cy5-labeled DNA had lower background, 400 and 160 a.u., respectively, and SNR of 5 and 9, respectively.

rapid and robust detection of DNA FISH spots from high-throughput fluorescence microscopy images (Fig. 2A,B).

Random Forest Filtering for Nuclear Segmentation

As a first step in the image analysis workflow, both spot detection methods rely on supervised RF (for details, see Materials and Methods) for nuclear segmentation using the DAPI channel images as input (Fig. 2A). We trained the RF binary classifier to distinguish bona fide segment-

ed nuclei from debris and segmentation errors by using 441 user-annotated images of segmented nuclei from the training plate Train-P1 and achieved an out-of-bag classification accuracy of 97.3% (Fig. 3A,B; see Materials and Methods for details). The trained RF classifier was then tested on a larger data set of images originating from two biological replicate plates, Test-P1 and Test-P2, containing 48 wells each stained with DAPI and Oligopaint DNA FISH probe sets in three colors targeting 12 genomic loci. The workflow detected 23,235 and 21,208 nuclei

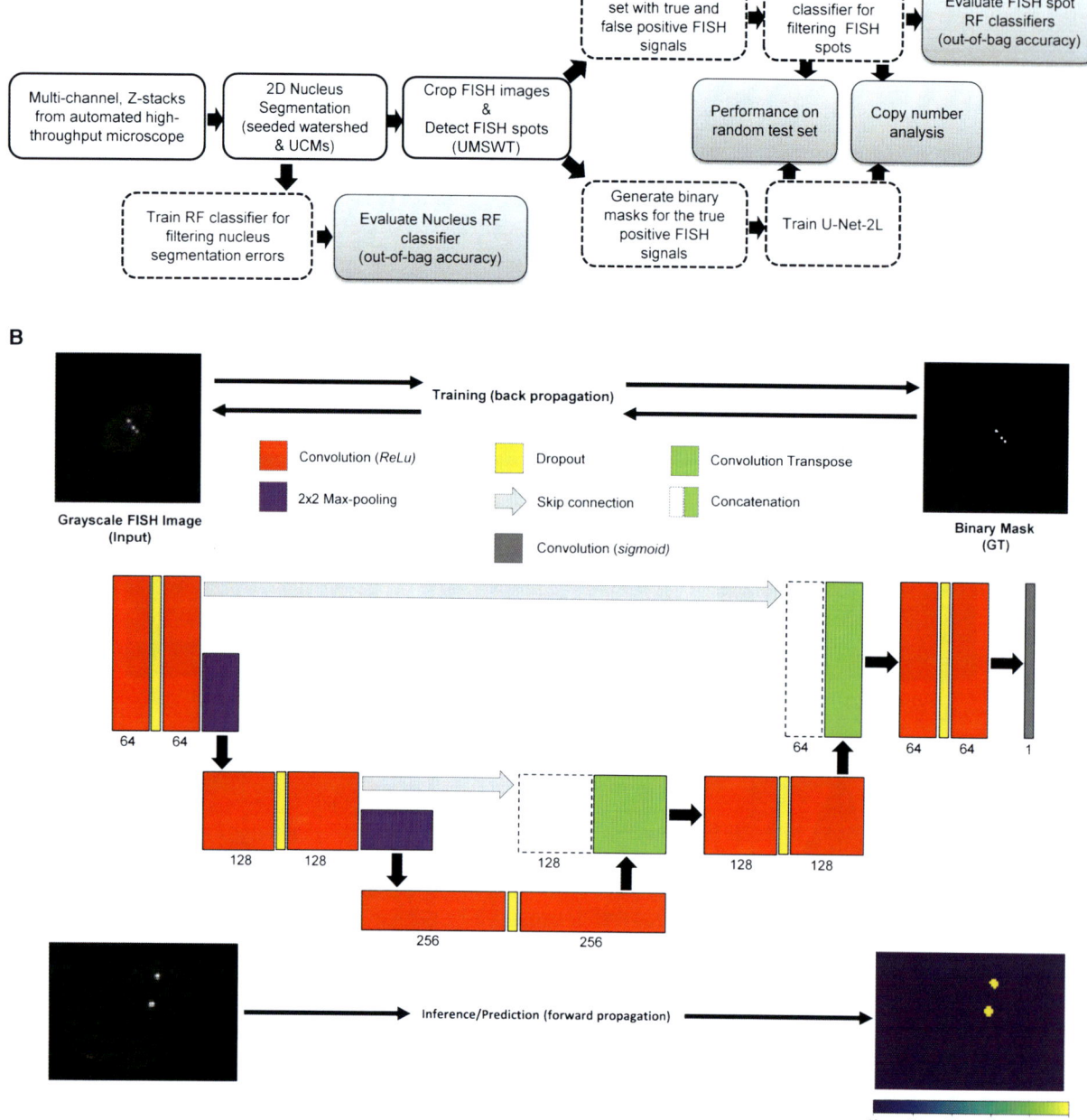

Figure 2. (*A*) Schematic of the image analysis workflow for detecting DNA FISH signals from high-throughput microscope images (blue-shaded block). The *top* branch after the spot detection block corresponds to a first approach for detecting and filtering DNA FISH signal using a random forest (RF) classifier, whereas the bottom branch corresponds to a second approach using fully CNN (U-Net-2L). UCMs, ultrametric contour maps; RF, random forest(s); UMSWT, undecimated multiscale wavelet transform. (*B*) The U-Net-2L architecture used for DNA FISH signal segmentation. During the training phase, the CNN model uses a grayscale DNA FISH image (*top left*) along with binary mask (GT; *top right*) corresponding to DNA FISH signals for optimizing the model parameters using error back-propagation. The numbers below each layer in the CNN correspond to the number of convolution filters and the height of each layer is proportional to the height of the input image(s). During the prediction (inference) phase, for a given grayscale DNA FISH image (*bottom left*) the CNN model predicts the probability of each pixel belonging to a DNA FISH signal (*bottom right*). Red boxes represent the 2D convolutional layer with 3 × 3 filter size and *ReLU* activation; purple boxes represent the maximum-pooling layers using 2 × 2 window size; green boxes represent the up-convolutional (convolutional transpose) layer; yellow boxes correspond to dropout layers; the gray box corresponds to a 1 × 1 convolution layer with sigmoid activation function for generating probability for each pixel belonging to a DNA FISH signal; green boxes along with dashed boxes represent concatenation layer to merge information from different resolutions/ scales.

A

Nucleus RF Classifier		Ground Truth	
		Good	Bad
Predicted	Good	TP = 299	FP = 5
	Bad	FN = 6	TN = 131

$$\text{Accuracy} = \frac{100 * (TP + TN)}{(TP + FP + FN + TN)} = 97.3\%$$

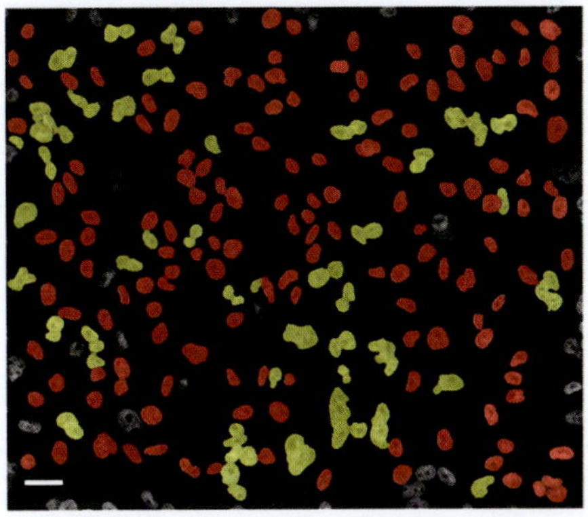

Figure 3. (*A*) Confusion matrix for calculating the out-of-bag accuracy for the random forest (RF) classifier for filtering out mis-segmented nuclei from the DAPI channel. TP, true positives; FP, false positives; FN, false negatives; TN, true negatives. (*B*) A representative example of classes (class-Good and class-Bad) assigned by RF classifier to the segmented objects in the DAPI channel using seeded watershed with ultrametric contour maps. Red and yellow colors represent class-Good and class-Bad nuclei, respectively. Nuclei along the edge of the field of view are filtered out before applying the RF classifier. Scale bar, 10 μm.

from Test-P1 and Test-P2, respectively. The run time of the complete workflow for each plate, including reading the multichannel 3D stacks of images, was ~2400 sec (60 sec per well) on a 56-core compute node (see Materials and Methods for details). The workflow detected 493 ± 52 (mean ± SD) and 442 ± 39 (mean ± SD) nuclei per well in testing plates Test-P1 and Test-P2, respectively (Supplemental Fig. S1). We conclude that the RF classifier for nuclear segmentation can classify nuclei with high accuracy.

Training Strategy and Initial Testing for the Performance of Two Different ML-Aided Spot Detection Algorithms

We then used the nuclear masks generated in the previous step as the search region for a traditional wavelet-based spot detection algorithm (UMSWT; see Materials and Methods for details) in images for all three DNA FISH channels (Fig. 4A). The parameters for spot detection were set for low object detection stringency to provide both positive (real FISH signals) and negative (background) examples for subsequent classifier training (Fig. 4B). In these conditions, the spot detection algorithm generated 619, 110, and 459 spot segmentation masks for the Alexa488, ATTO565, Cy5 channels, respectively. These sets were then user-annotated in two classes depending on whether they contained a real FISH signal (GoodFISH) or not (BadFISH), to generate a ground truth set for the ML algorithms (Fig. 4C).

In the first of the two spot detection workflows (Fig. 2A), three separate supervised RF binary classifiers were trained using the spot masks images from the Alexa488, ATTO565, and Cy5 channels to filter out false positives (i.e., background regions detected as FISH signals). These

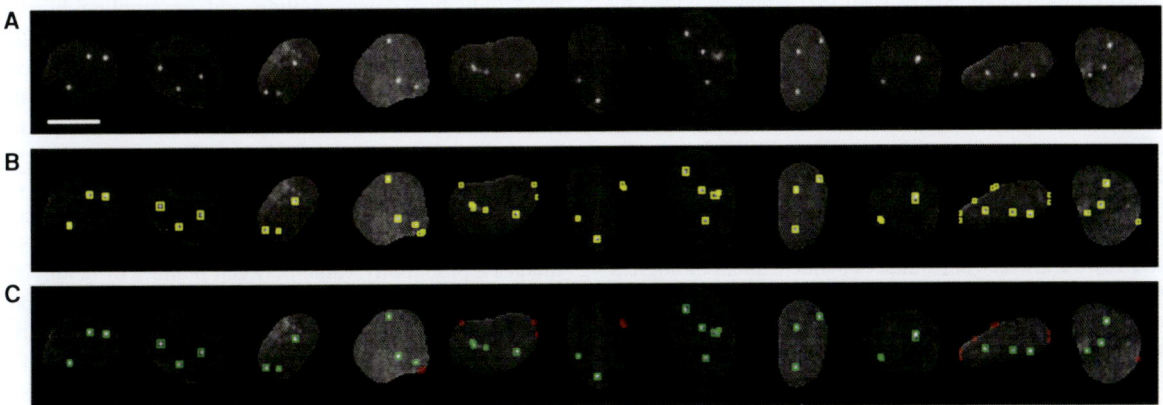

Figure 4. (*A*) Representative examples of DNA FISH images used for generating the training set(s) for (a) spot filtering using random forest classifier(s) and (b) binary masks for CNN spot segmentation. Scale bar, 10 μm. (*B*) Spots detected by the wavelet-based method with low stringency detection parameters. Each detected object in the nucleus is labeled with a yellow color bounding box. (*C*) Detected objects after user annotation using an interactive KNIME workflow. Spots labeled class-goodFISH and class-badFISH by the user are shown in green and red colors, respectively. Nuclei in which all the DNA FISH signals were labeled as class-goodFISH were used for generating the binary masks for CNN.

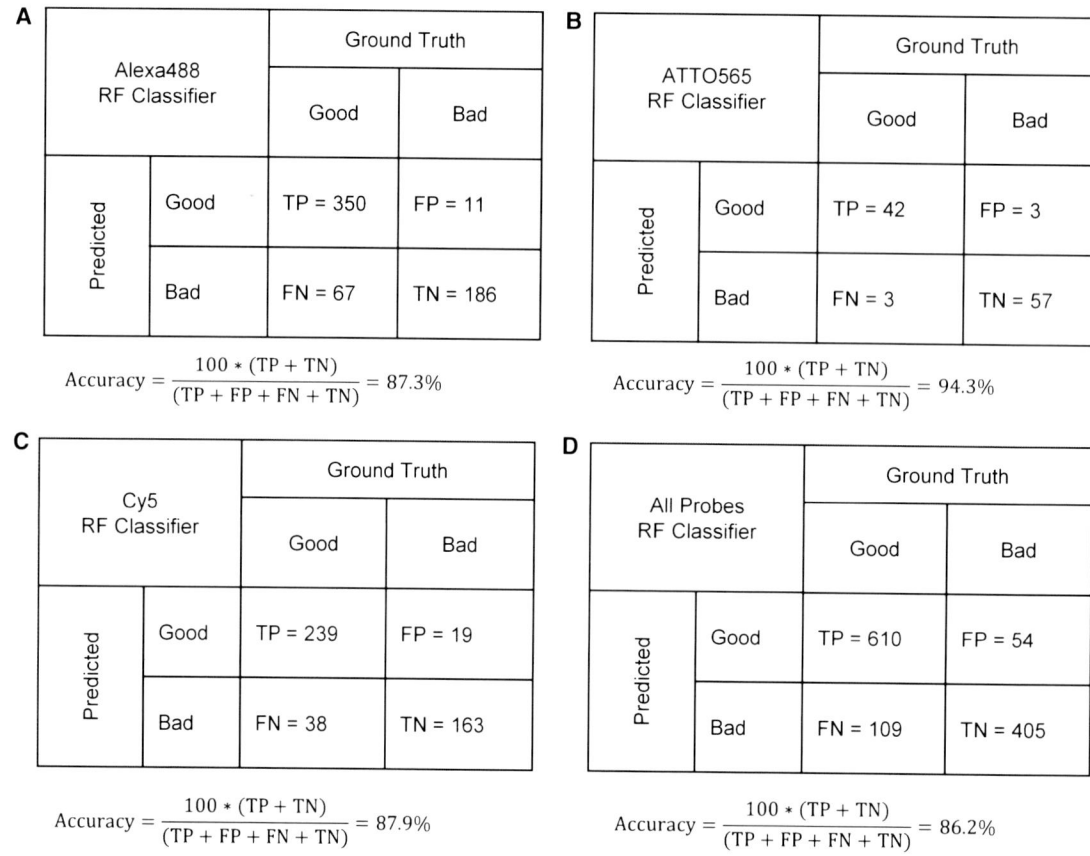

Figure 5. Confusion matrices for calculating the out-of-bag accuracy for random forest (RF) classifier(s) for filtering out background signals in the (*A*) Alexa488 channel, (*B*) ATTO565 channel, and (*C*) Cy5 channel. (*D*) Confusion matrix for the RF classifier trained with images from all the DNA FISH channels. TP, true positives; FP, false positives; FN, false negatives; TN, true negatives.

RF classifiers used a predetermined set of spot morphology and intensity features, which were user-specified and extracted by a traditional image processing algorithm that was a part of the upstream image analysis workflow. The extracted features for the annotated objects were used to learn a binary classification scheme for FISH signals based on the GoodFISH or BadFISH labels provided by user annotation. The out-of-bag accuracy for the spot detection using the RF classifiers in the Alexa488, ATTO565, and Cy5 was 87.3%, 94.3%, and 87.5%, respectively (Fig. 5A–C). Training a single RF classifier using an ensemble of spots from the three different DNA FISH fluorescent channels did not improve the overall out-of-bag accuracy (86.2%; Fig. 5D). We then tested the trained RF classifiers for spot classification using the two test plates (Test-P1 and Test-P2), which constitute two independent biological replicates and were previously unseen by the RF models. As a result, the three RF spot classifiers detected a total of 38,796 (59,854), 61,1121 (57,341), and 56,488 (51,750) DNA FISH spots in the Alexa488, ATTO565, and Cy5 channels from the Test-P1 (Test-P2) plates, respectively, which contain four replicate wells and 12 separate three-color probe sets each. The total run times were 2.0 and 2.5 h for each plate on a 56-core compute node (approximately 10 nuclei per second) and included FISH spot detection, spot feature extraction, and RF filtering. The

spot detection, feature calculation, and filtering out false positives steps using the RF classifiers took ∼18–22 min per 10,000 nuclei (i.e., 8–10 nuclei per second).

The second approach for FISH spot detection was based on the CNN architecture originally named U-Net (Ronneberger et al. 2015), with the important distinction that it used only two downsampling and upsampling blocks, resulting in a shallow convolution network (U-Net-2L) (Fig. 2B). This allowed the use of smaller training data sets and to optimize only 402,625 parameters when compared with state-of-the-art deep learning architectures for semantic segmentation of objects from digital images (Jegou et al. 2017). The main difference for this CNN model, as opposed to the RF classifiers for spot filtering described above, is that it assigns each pixel in DNA FISH images a probability value between 0 and 1 of belonging to a FISH signal. The CNN achieves this result by autonomously extracting an optimized set of image features directly from the FISH images of the training set for pixel-level classification. We trained the CNN for approximately 3000 epochs using an ensemble data set of DNA FISH images in three channels and their corresponding candidate spot masks previously generated by the wavelet-based spot detection algorithm. These were the same spot candidate regions previously used to train the RF classifiers (Plate Train-P1). We assessed the accuracy of the CNN in segmenting DNA FISH spots by calculating the Dice coefficient (a measure of pixel classification accuracy; see Materials and Methods for details). The calculated Dice coefficient values for the trained CNN model were 0.993 for the training data (198 FISH images) and 0.913 for the validation data (23 FISH images), indicating that it can classify pixels as spots with high classification accuracy (Supplemental Fig. S2). Furthermore, and similar to previous observations on RF classifiers, we tested the CNN method to detect FISH signals from 48 wells in the Test-P1 and Test-P2 plates. For this task, we implemented an image analysis workflow that, when compared with the RF filtering approach, used the CNN method to replace (i) the wavelet-based spot detection, (ii) the feature calculation of detected spots, and (iii) the RF predictor for filtering out false-positive spots. Using this workflow, we identified a total of 63,100 (53,698), 69,244 (64,025), and 72,549 (66,260) Alexa488, ATTO565, and Cy5-labeled FISH spots from Test-P1 (Test-P2) images, respectively. The total runtime for CNN FISH detection was ~15 min for each plate on a 56-core, HPC compute node with 4 Tesla K80 GPUs (approximately 80 nuclei per second) of which the CNN spot detection was only ~42 sec for 10,000 nuclei (i.e., 230 nuclei per second). These results indicate that the CNN classifier is 25–30-fold faster than the RF at detecting spots when compared with the RF models.

Comparison between RF and CNN for FISH Spot Detection Accuracy

Having determined the feasibility and computational speed of applying either the RF or the CNN methods for detection of FISH signals in three spectral channels on two independent test plates, we conducted a side-by-side comparison of their accuracy in the identification and classification of FISH spots not present in the training data set (Train-P1). To this end, we first randomly selected 184, 186, and 153 cropped nuclear Alexa488, ATTO565, and Cy5 images, respectively, from Test-P2 plate (Fig. 6A–C, top rows), and spot regions generated by the wavelet-based spot detection algorithm were manually annotated as belonging to the GoodFISH or BadFISH classes as previously described (Fig. 6A–6C, second row). Finally, spots were detected using either the RF filter method (Fig. 6A–C, third row) or the CNN method (Fig. 6A–C, fourth row).

A quantitative assessment of the results of this comparison showed that for the Alexa488 images, in which the FISH signals were dimmer (SNR ~ 2), the RF method detected 579 FISH spots. Of these, 518 were true positives, whereas the RF method failed to detect nine bona fide DNA FISH spots. Altogether, testing the RF model on this subset of annotated spots from the Test-P2 plate resulted in an accuracy of 88.1%, precision of 89.5%, recall of 98.3%, and F-score of 93.7% (Fig. 7A; see Materials and Methods for details). On the same set of images, the CNN method detected a total of 530 spots, of which 524 (out of 527) were true positives and failed to detect only three DNA FISH signals, thus resulting in an

accuracy of 98.3%, precision of 98.9%, recall of 99.4%, and F-score of 99.1% (Fig. 7A). For the random set of ATTO565 images, in which the DNA FISH signals had a higher SNR (~5) than did the Alexa488 signals, the RF classifier detected 501 spots out of 523 (ground truth) without any false positives, but failed to detect 22 bona fide DNA FISH spots (false negatives) (Fig. 7B). The CNN classifier detected all the 523 spots (true positives) from the ground truth along with 12 false positives. Thus, the performance metrics for the RF versus the CNN spot detection methods on the ATTO565 images were, respectively: accuracy 95.80% versus 97.76%, precision 100% versus 97.78%, recall 95.79% versus 100%, and F-score 97.85% versus 98.87% (Fig. 7B). The performance of the two methods on Cy5-labeled DNA FISH signals images followed a similar trend as for ATTO565. In fact, out of 153 Cy5 images (SNR ~ 9) containing 440 FISH spots, the RF method detected 393 FISH spots of which only three were false positives and failed to detect 50 true FISH spots (false negatives). Most of the false negatives were due to corresponding RF classifier filtering out detected spots by the wavelet-based spot detection algorithm (data not show). This resulted in accuracy of 88.03%, precision of 99.24%, recall of 88.64%, and F-score of 93.64% (Fig. 7C). In contrast, the CNN method (Fig. 7C) detected all the 440 true positives, but also detected 27 false positives, thus, leading to performance metrics of 94.22% accuracy, 94.21% precision, 100% recall, and 97.02% F-score.

Altogether, these results suggest that both the RF and the CNN methods can detect FISH signals originating from diverse genomic loci and in three different spectral channels with high accuracy and precision, especially when the SNR of the FISH signals is high (≥5). The CNN method, however, outperformed the RF-based spot classifier based on quantitative metrics on images with weaker FISH signal (SNR ~ 2).

Distribution of the Oligopaint DNA FISH Signals Sets as Detected by RF and CNN Models

To further compare the performance of the RF and CNN spot detection algorithms on different sets of Oligopaint DNA FISH probes, we measured the number of DNA FISH signals per nucleus in Alexa488, ATTO565, and Cy5 FISH images obtained from the two biological replicate plates Test-P1 and Test-P2 (Fig. 8). The cells used in this study, MDA-MB-231, are hypotriploid and contain a total of 59 to 66 highly rearranged chromosomes, with an autosomal chromosome copy number ranging from 2 to 4, depending on the chromosome (American Tissue Culture Collection 2012). The copy number profile generated by the RF and CNN spot detection algorithms for each OliGoPaint DNA FISH probe set directed against a particular genomic target was concordant for all probes in all channels (Fig. 8). As expected from MDA-MB-231 karyotype analysis, most genomic loci probed here with Oligopaint and detected either with RF or with CNN in the ATTO565 or Cy5 channels, and to a lesser extent in the Alexa488 channel, showed a sharp peak, in at least 50% of the nuclei,

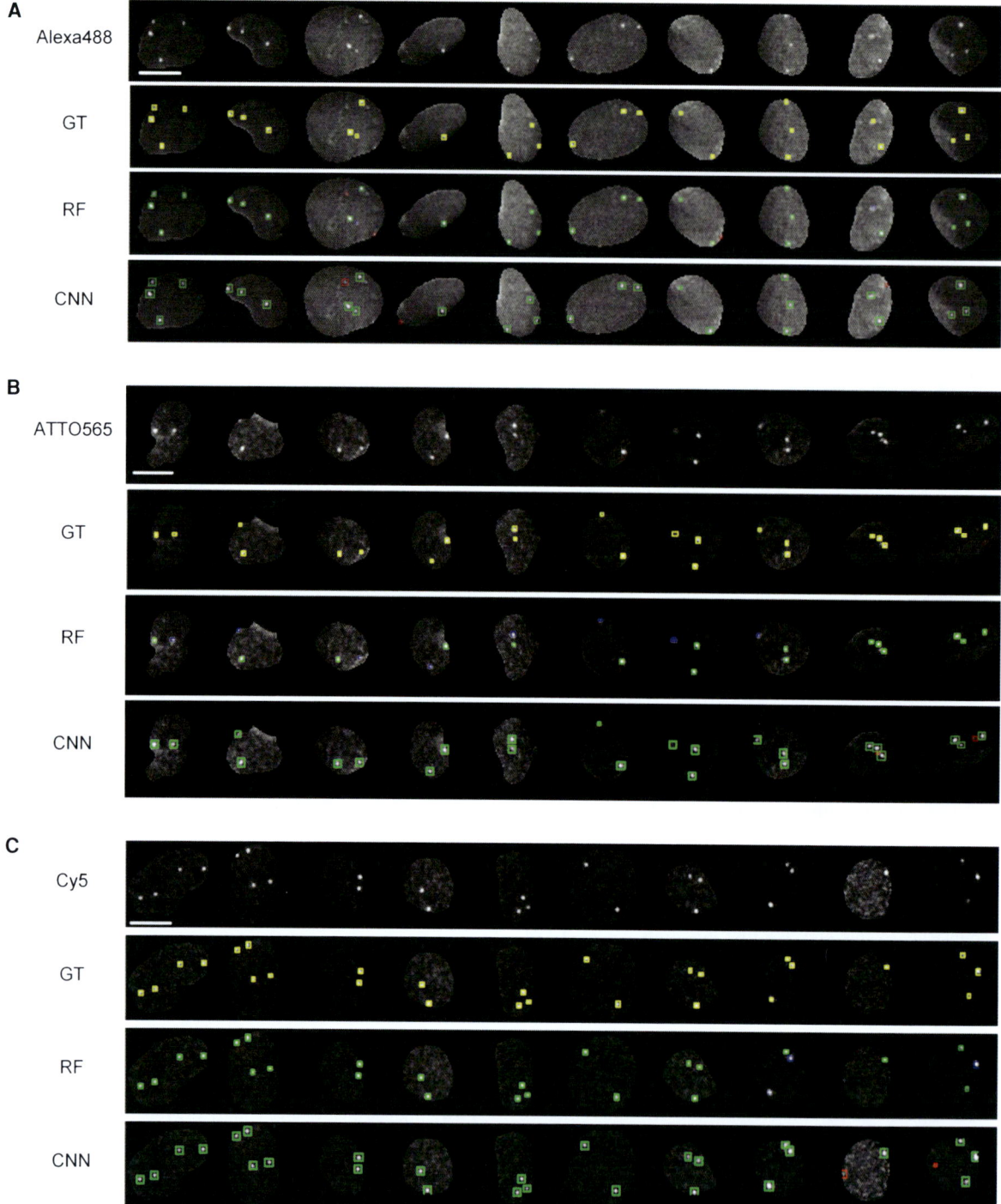

Figure 6. Illustrative examples of DNA FISH spot detection by random forest (RF) and CNN methods on a random set of nuclei from test plate Test-P2. (*A–C*) Representative images with Alexa488, ATTO565, and Cy5-labeled DNA FISH (rows: Alexa488, ATTO565, Cy5). "GT" rows correspond to ground truth FISH regions (yellow color) identified by a user. "RF" rows correspond to DNA FISH spot regions after applying RF filtering to remove background regions, green colored spot regions represent true DNA FISH signals (true positives), red colored spot regions represent background signals even after RF filtering (false positives), and blue colored spot regions represent true DNA FISH signals that did not get detected (false negatives). CNN rows correspond to DNA FISH spots detected by the U-NET-2L spot detection method. The color scheme of the spot regions in CNN rows is identical to in the RF rows. Scale bar, 10 μm.

A

Alexa488 Ground Truth Spots = 527	Detected	True Positives	Missing (False Negatives)	False Positives
RF	579	518	9	61
CNN	530	524	3	6

B

ATTO565 Ground Truth Spots = 523	Detected	True Positives	Missing (False Negatives)	False Positives
RF	501	501	22	0
CNN	535	523	0	12

C

Cy5 Ground Truth Spots = 440	Detected	True Positives	Missing (False Negatives)	False Positives
RF	393	390	50	3
CNN	467	440	0	27

Figure 7. Summary of DNA FISH signal detection, using (a) random forest (RF) filtering on spots identified by the wavelet-based method and (b) CNN, on randomly selected set of DNA FISH images from test plate Test-P2. (*A*) Results for 184 Alexa488-labeled DNA FISH images with 527 spots. (*B*) Results for 186 ATTO565-labeled DNA FISH images with 523 spots. (*C*) Results for 153 Cy5-labeled DNA FISH images with 440 spots.

at either 2 (e.g., locus 49 in the Alexa488 channel, locus 31 in the ATTO565 channel, and locus 1 in the Cy5 channel; Fig. 8) or 3 FISH spots per nucleus (e.g., locus 49 in the Alexa488 channel, locus 39 in the ATTO565 channel, or locus 19 in the Cy5 channel; Fig. 8). Importantly, the standard deviation among technical replicates was low (Fig. 8), and the measurements on the two biological replicates (Test-P1 and Test-P2) were highly concordant (Fig. 8). We conclude that both RF and CNN are robust spot detection algorithms that can be trained once on an ensemble of different Oligopaint DNA FISH probe sets and then reused in multiple independent biological replicates to visualize multiple genomic loci with high accuracy without the need for a manual search of the best detection parameters.

DISCUSSION

We present here two ML-based image analysis workflows that can be used to detect multiple FISH signals from three fluorescence microscopy channels at high throughput with diverse SNR. More importantly, the workflows do not require manual parameter tweaking for each individual genomic loci tested (Figs. 4–7). Because the RF and CNN spot classifiers presented here are supervised ML methods, the main effort for the user is in the manual annotation of sets of DNA FISH channels images to create a ground truth data set of spots regions of

interest to train the spot ML classifiers. Given that spot-like features generally are limited to small image regions (e.g., 10×10 pixels) we reduced this burden by selecting an RF model for spot classification task and a shallow CNN autoencoder model (U-Net-2L) for spot segmentation. This resulted in a smaller number of model parameters to train and in smaller training data sets for efficient performance. In addition, we greatly facilitated manual annotation of the FISH images by using a wavelet-based spot detection algorithm with nonstringent detection parameters, to automatically generate a diverse set of spot segmentation masks that can be directly used by the user for training ML methods. Overall this resulted in a manual annotation time of the order of 30–60 min. This initial investment dramatically reduces the time needed for downstream image analysis setup when compared with manual tweaking of traditional spot detection parameters, and we expect that ML methods for spot detection will greatly reduce this bottleneck in the analysis of image data sets including tens or hundreds of different genomic loci tagged with fluorescently labeled Oligopaint DNA FISH probes. Furthermore, it is likely that these ML spot detection methods explore a much larger parameter optimization space when compared with manual parameter tweaking of traditional spot detection methods and will likely result in higher spot detection accuracy. Finally, given the robustness of these ML approaches to reproducibly detect spots from images generated in different tech-

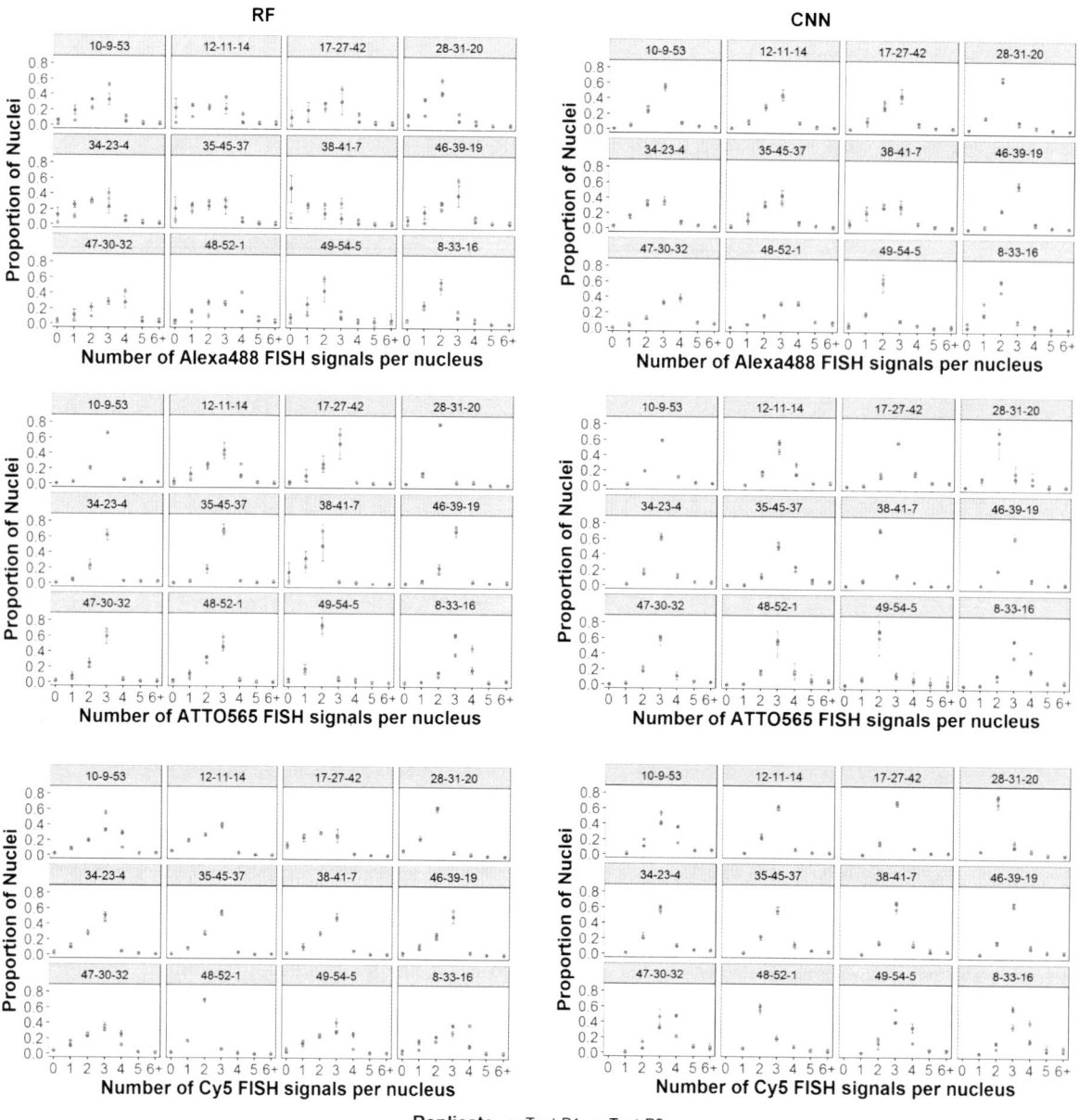

Figure 8. Plots comparing the number of DNA FISH signals (copy number) detected per nucleus in MDA-MB-231 cells between biological replicates. The results for 36 genomic loci labeled using 12 unique combinations of Alexa488-, ATTO565-, and Cy5-labeled DNA Oligopaint probes are displayed in the first, second, and third row, respectively. DNA FISH probe sets were designated with an "i-j-k" scheme, where "i" is an identifier for the gene locus labeled with Alexa488, "j" is an identifier for the gene locus labeled with ATTO565, and "k" is an identifier for the gene locus labeled with Cy5. The *left* column shows the copy number histograms using the RF classifier for spot filtering, and the corresponding results from CNN-based spot-detection method (U-Net-2L) are in the *right* column. Each dot in the histogram represents the mean proportion of nuclei with the detected copy number calculated over the four technical replicates. The bars represent the standard deviation (SD) of the mean across the technical replicates.

nical and biological replicates (Fig. 8), we expect that, provided that the experimental conditions remain the same, it will be possible to apply these models to different experiments performed on different days without the need for retraining, thus making it a "train once and reuse multiple times" approach for DNA FISH spot detection.

Although both the more traditional RF classifier and the CNN model show comparable high performance in terms of spot segmentation accuracy (Figs. 4–7), we expect that

the CNN model will be the approach of choice because it performs better than RF in both high- and low-SNR conditions (Fig. 7). More importantly, the CNN approach eliminates the need for computationally intensive image feature calculations, takes advantage of highly efficient GPU hardware acceleration for spot segmentation, and leads to a 25–30-fold increase in the spot detection speed. This last feature is essential to process large data sets of images generated by high-throughput microscopes (up to

10^5 images per 24 h). For these reasons, the CNN model, which we named SpotLearn, is an ideal choice for ML-mediated spot detection of Oligopaint DNA FISH signals. We expect that SpotLearn could be further improved by tuning the hyperparameters of the model—namely, the number of filters in the convolutional layers, the filter activation functions, the number of downsampling and upsampling levels—and by including batch normalization (Ioffe and Szegedy 2015). In addition, it might be possible to improve the SpotLearn model accuracy and to further strengthen it against overfitting by performing K-fold cross-validation by training K SpotLearn models on training-validation data sets (He et al. 2016). Finally, it might also be worthwhile to explore data augmentation strategies (vertical and horizontal flips and shifting) for increasing the size of the data training sets (Huang et al. 2017). More generally, it should also be possible to use SpotLearn for a range of cellular and biomolecular imaging applications requiring spot detection as an intermediate step. For instance, SpotLearn could be used in lieu of traditional spot detection algorithms for detecting single molecules in super-resolution and localization microscopy (Betzig et al. 2006; Smith et al. 2010; Liu et al. 2017) and for detecting nascent RNA transcripts in live cell images (Larson et al. 2011). Altogether, we anticipate that SpotLearn will become an important tool for the detection of genomic loci in a high-throughput fashion and in the study of the 3D genome organization.

ACKNOWLEDGMENTS

We thank the High-Performance Computing Group, CIT, National Institutes of Health for their computational hardware and support. This research was supported by funding from the Intramural Research Program of the National Institutes of Health (NIH), National Cancer Institute, and Center for Cancer Research.

REFERENCES

American Tissue Culture Collection. 2012. *SOP: Thawing, propagation and cryopreservation of NCI-PBCF-HTB26 (MDA-MB-231)*. Version 1.5. https://physics.cancer.gov/docs/bioresource/breast/NCI-PBCF-HTB26_MDA-MB-231_SOP-508.pdf.

Arbelaez P. 2006. Boundary extraction in natural images using ultrametric contour maps. In *Conference on Computer Vision and Pattern Recognition Workshop*, pp. 182–189.

Beliveau BJ, Joyce EF, Apostolopoulos N, Yilmaz F, Fonseka CY, McCole RB, Chang Y, Li JB, Senaratne TN, Williams BR, et al. 2012. Versatile design and synthesis platform for visualizing genomes with Oligopaint FISH probes. *Proc Natl Acad Sci* **109:** 21301–21306.

Beliveau BJ, Boettiger AN, Avendaño MS, Jungmann R, McCole RB, Joyce EF, Kim-Kiselak C, Bantignies F, Fonseka CY, Erceg J, et al. 2015. Single-molecule super-resolution imaging of chromosomes and in situ haplotype visualization using Oligopaint FISH probes. *Nat Commun* **6:** 7147.

Berthold MR, Cebron N, Dill F, Gabriel TR, Kötter T, Meinl T, Ohl P, Sieb C, Thiel K, Wiswedel B. 2008. KNIME: The Konstanz information miner. In *Data analysis, machine learning and applications, studies in classification, data analysis, and knowledge organization*, pp. 319–326. Springer, Berlin.

Betzig E, Patterson GH, Sougrat R, Lindwasser OW, Olenych S, Bonifacino JS, Davidson MW, Lippincott-Schwartz J, Hess HF. 2006. Imaging intracellular fluorescent proteins at nanometer resolution. *Science* **313:** 1642–1645.

Bonev B, Cavalli G. 2016. Organization and function of the 3D genome. *Nat Rev Genet* **17:** 661–678.

Breiman L. 2001. Random forests. *Mach Learn* **45:** 5–32.

Bright DS, Steel EB. 1987. Two-dimensional top hat filter for extracting spots and spheres from digital images. *J Microsc* **146:** 191–200.

Burman B, Zhang ZZ, Pegoraro G, Lieb JD, Misteli T. 2015. Histone modifications predispose genome regions to breakage and translocation. *Genes Dev* **29:** 1393–1402.

Cavalli G, Misteli T. 2013. Functional implications of genome topology. *Nat Struct Mol Biol* **20:** 290–299.

Chen KH, Boettiger AN, Moffitt JR, Wang S, Zhuang X. 2015. RNA imaging. Spatially resolved, highly multiplexed RNA profiling in single cells. *Science* **348:** aaa6090.

Dice LR. 1945. Measures of the amount of ecologic association between species. *Ecology* **26:** 297–302.

Dietz C, Berthold MR. 2016. KNIME for open-source bioimage analysis: A tutorial. *Adv Anat Embryol Cell Biol* **219:** 179–197.

Finn E, Pegoraro G, Brandao HB, Valton A-L, Oomen ME, Dekker J, Mirny L, Misteli T. 2017. Heterogeneity and intrinsic variation in spatial genome organization. *bioRxiv* doi: 10.1101/171801 (accessed August 28, 2017).

Flavahan WA, Drier Y, Liau BB, Gillespie SM, Venteicher AS, Stemmer-Rachamimov AO, Suvà ML, Bernstein BE. 2016. Insulator dysfunction and oncogene activation in IDH mutant gliomas. *Nature* **529:** 110–114.

Franke M, Ibrahim DM, Andrey G, Schwarzer W, Heinrich V, Schöpflin R, Kraft K, Kempfer R, Jerković I, Chan W-L, et al. 2016. Formation of new chromatin domains determines pathogenicity of genomic duplications. *Nature* **538:** 265–269.

He K, Zhang X, Ren S, Sun J. 2016. Deep residual learning for image recognition. In *Proceedings of the IEEE conference on computer vision and pattern recognition*, pp. 770–778. IEEE, Piscataway, NJ.

Ho TK. 1998. The random subspace method for constructing decision forests. *IEEE Trans Pattern Anal Mach Intell* **20:** 832–844.

Huang Z, Pan Z, Lei B. 2017. Transfer learning with deep convolutional neural network for SAR target classification with limited labeled data. *Remot Sens* **9:** 907.

Ioffe S, Szegedy C. 2015. Batch normalization: Accelerating deep network training by reducing internal covariate shift. In *International Conference on Machine Learning*, pp. 448–456.

Jegou S, Drozdzal M, Vazquez D, Romero A, Bengio Y. 2017. The one hundred layers tiramisu: Fully convolutional DenseNets for semantic segmentation. In *2017 IEEE Conference on Computer Vision and Pattern Recognition Workshops (CVPRW)* http://dx.doi.org/10.1109/cvprw.2017.156.

Joyce EF, Williams BR, Xie T, Wu C-T. 2012. Identification of genes that promote or antagonize somatic homolog pairing using a high-throughput FISH–based screen. *PLoS Genet* **8:** e1002667.

Kingma DP, Ba J. 2014. Adam: A method for stochastic optimization. *arXiv [csLG]*. http://arxiv.org/abs/1412.6980.

Krizhevsky A, Sutskever I, Hinton GE. 2017. ImageNet classification with deep convolutional neural networks. *Commun ACM* **60:** 84–90.

Larson DR, Zenklusen D, Wu B, Chao JA, Singer RH. 2011. Real-time observation of transcription initiation and elongation on an endogenous yeast gene. *Science* **332:** 475–478.

Liu S, Mlodzianoski MJ, Hu Z, Ren Y, McElmurry K, Suter DM, Huang F. 2017. sCMOS noise-correction algorithm for microscopy images. *Nat Methods* **14:** 760–761.

Lupiáñez DG, Kraft K, Heinrich V, Krawitz P, Brancati F, Klopocki E, Horn D, Kayserili H, Opitz JM, Laxova R, et al. 2015. Disruptions of topological chromatin domains cause pathogenic rewiring of gene-enhancer interactions. *Cell* **161:** 1012–1025.

Moffitt JR, Hao J, Wang G, Chen KH, Babcock HP, Zhuang X. 2016. High-throughput single-cell gene-expression profiling with multiplexed error-robust fluorescence in situ hybridization. *Proc Natl Acad Sci* **113:** 11046–11051.

Olivo J-C. 1996. Automatic detection of spots in biological images by a wavelet-based selective filtering technique. In *Proceedings of 3rd IEEE International Conference on Image Processing*, pp. I:311–I:314. IEEE, Piscataway, NJ.

Olivo-Marin J-C. 2002. Extraction of spots in biological images using multiscale products. *Pattern Recognit* **35:** 1989–1996.

Otsu N. 1979. A threshold selection method from gray-level histograms. *IEEE Trans Syst Man Cybern* **9:** 62–66.

Parthasarathy R. 2012. Rapid, accurate particle tracking by calculation of radial symmetry centers. *Nat Methods* **9:** 724–726.

Pegoraro G, Misteli T. 2017. High-throughput imaging for the discovery of cellular mechanisms of disease. *Trends Genet* **33:** 604–615.

Ronneberger O, Fischer P, Brox T. 2015. U-Net: Convolutional networks for biomedical image segmentation. In *Lecture notes in computer science*, pp. 234–241.

Sezgin M. 2004. Survey over image thresholding techniques and quantitative performance evaluation. *J Electron Imaging*. http://electronicimaging.spiedigitallibrary.org/article.aspx?articleid=1098183.

Shachar S, Voss TC, Pegoraro G, Sciascia N, Misteli T. 2015. Identification of gene positioning factors using high-throughput imaging mapping. *Cell* **162:** 911–923.

Shelhamer E, Long J, Darrell T. 2017. Fully convolutional networks for semantic segmentation. *IEEE Trans Pattern Anal Mach Intell* **39:** 640–651.

Smith CS, Joseph N, Rieger B, Lidke KA. 2010. Fast, single-molecule localization that achieves theoretically minimum uncertainty. *Nat Methods* **7:** 373–375.

Solovei I, Cavallo A, Schermelleh L, Jaunin F, Scasselati C, Cmarko D, Cremer C, Fakan S, Cremer T. 2002. Spatial preservation of nuclear chromatin architecture during three-dimensional fluorescence in situ hybridization (3D-FISH). *Exp Cell Res* **276:** 10–23.

Sørensen T. 1948. [A method of establishing groups of equal amplitude in plant sociology based on similarity of species and its application to analyses of the vegetation on Danish commons]. *Biol Skr* **5:** 1–34.

Srivastava N, Hinton G, Krizhevsky A, Sutskever I, Salakhutdinov R. 2014. Dropout: A simple way to prevent neural networks from overfitting. *J Mach Learn Res* **15:** 1929–1958.

Szegedy C, Vanhoucke V, Ioffe S, Shlens J, Wojna Z. 2016. Rethinking the inception architecture for computer vision. In *2016 IEEE Conference on Computer Vision and Pattern Recognition (CVPR)* http://dx.doi.org/10.1109/cvpr.2016.308.

Vincent L, Soille P. 1991. Watersheds in digital spaces: An efficient algorithm based on immersion simulations. *IEEE Trans Pattern Anal Mach Intell* **13:** 583–598.

Wang S, Su J-H, Beliveau BJ, Bintu B, Moffitt JR, Wu C-T, Zhuang X. 2016. Spatial organization of chromatin domains and compartments in single chromosomes. *Science* **353:** 598–602.

Zijdenbos AP, Dawant BM, Margolin RA, Palmer AC. 1994. Morphometric analysis of white matter lesions in MR images: Method and validation. *IEEE Trans Med Imaging* **13:** 716–724.

Remarkable Evolutionary Plasticity of Centromeric Chromatin

Steven Henikoff,[1,2] Jitendra Thakur,[1,2] Sivakanthan Kasinathan,[2,3] and Paul B. Talbert[1,2]

[1]Howard Hughes Medical Institute, Fred Hutchinson Cancer Research Center, Seattle, Washington 98109

[2]Basic Sciences Division, Fred Hutchinson Cancer Research Center, Seattle, Washington 98109

[3]Medical Scientist Training Program, University of Washington School of Medicine, Seattle, Washington 98195

Correspondence: steveh@fhcrc.org

Centromeres were familiar to cell biologists in the late 19th century, but for most eukaryotes the basis for centromere specification has remained enigmatic. Much attention has been focused on the cenH3 (CENP-A) histone variant, which forms the foundation of the centromere. To investigate the DNA sequence requirements for centromere specification, we applied a variety of epigenomic approaches, which have revealed surprising diversity in centromeric chromatin properties. Whereas each point centromere of budding yeast is occupied by a single precisely positioned tetrameric nucleosome with one cenH3 molecule, the "regional" centromeres of fission yeast contain unphased presumably octameric nucleosomes with two cenH3s. In *Caenorhabditis elegans*, kinetochores assemble all along the chromosome at sites of cenH3 nucleosomes that resemble budding yeast point centromeres, whereas holocentric insects lack cenH3 entirely. The "satellite" centromeres of most animals and plants consist of cenH3-containing particles that are precisely positioned over homogeneous tandem repeats, but in humans, different α-satellite subfamilies are occupied by CENP-A nucleosomes with very different conformations. We suggest that this extraordinary evolutionary diversity of centromeric chromatin architectures can be understood in terms of the simplicity of the task of equal chromosome segregation that is continually subverted by selfish DNA sequences.

Centromeres have been familiar to cell biologists even before the rediscovery of Mendel's laws. In 1882, Walther Flemming used cytological stains to observe mitotic figures, which he called "chromatin" (Flemming 1882). In his drawings of lily chromosomes at metaphase and anaphase, Flemming darkly shaded the chromatin corresponding to each site of spindle fiber attachment, which later was referred to as the centromere or kinetochore. By current usage, the centromere refers to the genetic locus, whereas the kinetochore is the large complex of proteins specifically bound to centromeric DNA (Talbert et al. 2008). A further distinction can be made between the inner kinetochore proteins that remain associated with the DNA throughout the cell cycle, referred to as the centromeric chromatin-associated network (CCAN), and protein complexes of the outer kinetochore that assemble at mitosis.

The classical work of Morgan and his students genetically defined the centromere as a locus (Sturtevant 1913). But the molecular study of the centromere began in earnest in 1980, when Clarke and Carbon genetically mapped the centromere of *Saccharomyces cerevisiae* (budding yeast) Chromosome III to a 1.6-kb DNA segment (Clarke and Carbon 1980). When they inserted this DNA segment into a circular plasmid vector with a replication origin, they observed faithful segregation through mitosis and meiosis.

The molecular description of proteins of the CCAN began with the discovery by Earnshaw and Rothfield of what they termed Centromere Proteins A, B, and C (CENP-A, CENP-B, and CENP-C) using human autoimmune antibodies (Earnshaw and Rothfield 1985). Meanwhile, Palmer and Margolis (1985) showed that bovine CENP-A is a histone H3 variant (Palmer et al. 1987), one that remains with the DNA in mature sperm (Palmer et al. 1990). This was the first suggestion that centromeres might be "epigenetic"—that is, defined by the presence of the centromere-specific histone variant, which we refer to generically as "cenH3" to conform to the well-established histone nomenclature (Bradbury 1977; Talbert et al. 2012). Understanding the basis for this dual definition of centromeres as genetic and/or epigenetic has been a driving force for work in the laboratory for almost two decades.

Centromeres are generally conserved in position along the chromosome over evolutionary time; however, surprisingly this is not the case for centromeric DNA sequence. The realization that centromeric DNA has been rapidly evolving came from studies in the 1960s of satellite DNA, tandem repeats that could be separated from main-band DNA in buoyant density gradients and that differed between even closely related species (Yunis and Yasmineh 1971). The discovery that the sequence of the *S. cerevisiae* cenH3 (Cse4) is as divergent from H3 as it is from human CENP-A (Meluh et al. 1998) was another clue suggesting that centromeres are not under the same evolutionary constraints as are genes and their DNA-bind-

Published by Cold Spring Harbor Laboratory Press; doi: 10.1101/sqb.2017.82.033605

Cold Spring Harbor Symposia on Quantitative Biology, Volume LXXXII

ing proteins. These considerations led us to identify and characterize cenH3s from diverse model organisms, including *Caenorhabditis elegans* (Buchwitz et al. 1999), *Drosophila melanogaster* (Henikoff et al. 2000), *Arabidopsis thaliana* (Talbert et al. 2002), rice (Nagaki et al. 2004), and maize (Jin et al. 2004). Indeed, we found that evolutionary divergence is a general feature of cenH3s, despite the fact that H3 and its general replacement variant, H3.3, are among the most highly conserved proteins known. We referred to the remarkable divergence of both centromeric satellite DNA and its dedicated histone as the centromere paradox: rapid evolution with a conserved function that is essential through every cell division (Henikoff et al. 2001).

Harmit Malik in the laboratory made the key observation that *Drosophila* cenH3 (Cid) is evolving adaptively using standard population genetic measures, suggesting an arms race (Malik and Henikoff 2001). Reasoning that the arms race would be between centromeric satellites and cenH3, we posited that centromeres compete for inclusion into the egg at female meiosis I (Henikoff et al. 2001). Our "centromere drive" hypothesis combined the concept of female meiotic drive described by Rhoades for the selfish segregation behavior of distal heterochromatic knobs in maize (Rhoades 1952) with that of centromere "strength" for the inferred female meiotic orientation of heterochromatin-rich centromeres by Novitski (1955). Centromere drive as an explanation for the centromere paradox has since become the standard explanation for centromere divergence based on evidence from humans (Daniel 2002), monkey flowers (Fishman and Willis 2005), and mice (Chmátal et al. 2014) and an attractive general mechanism for postzygotic reproductive isolation as species diverge (Henikoff and Malik 2002; Burt and Trivers 2006).

The realizations that satellite centromeres are selfish and that cenH3s and other CCAN proteins may act to suppress drive (Talbert et al. 2004) raise the question as to how these CCAN proteins interact with their DNA substrates. We started by studying *S. cerevisiae* centromeres, with simple point centromeres that provide a basis for comparison to the more complex regional centromeres of *Schizosaccharomyces pombe*, the holocentromeres of *C. elegans* and holocentric insects, and the satellite centromeres of humans. In each case we have applied high-resolution mapping technologies to address the basic problem of centromere specificity. We find that the evolutionary divergence of centromeric chromatin is reflected in the plasticity of CCAN complex composition between different organisms and sometimes of different centromeres of the same individual.

BUDDING YEAST CENTROMERES ARE OCCUPIED BY HEMISOMES

Based on a comparison of centromeres from different budding yeast chromosomes, Clarke and Carbon (1985) identified consensus motifs that they termed centromere-determining element (CDE) I, II, and III. Subsequent work identified the ~120-bp segment spanning CDEI–II–III as the sequence-specific core of each of the 16

S. cerevisiae centromeres. The 8-bp CDEI consensus is the binding site for the Cbf1 sequence-specific transcription factor, and the 26-bp CDEIII consensus is the binding site for the centromere-specific multisubunit CBF3 complex. These elements flank the 82 ± 4-bp CDEII, which lacks a specific consensus, but consists of $\geq 90\%$ AT-rich DNA. Micrococcal nuclease (MNase) mapped a single Cse4 nucleosome to the functional centromere (Furuyama and Biggins 2007). This observation is difficult to reconcile with the availability of only ~80 bp of DNA for wrapping, suggesting the presence of a particle with fewer than the eight subunits in an H3 nucleosome, which wraps 147 bp of DNA.

To address this conundrum, Takehito Furuyama determined the chirality of the DNA superhelical writhe around the Cse4-containing core, using superhelical density mapping of circular chromosomes. This analysis revealed that the budding yeast centromere induces positive DNA supercoils, indicative of a right-handed writhe around the particle, opposite to that of left-handed H3 nucleosomes (Furuyama and Henikoff 2009). Our supercoiling measurements were later confirmed and extended in a study suggesting that the right-handed configuration is enforced by the formation of a DNA loop held together by the CBF3 complex, excluding CDEI (Díaz-Ingelmo et al. 2015). A possible rationale for the right-handed superhelical wrap is that DNA overwinds when stretched (Gore et al. 2006), so that pulling on the centromere at anaphase would be expected to cause the (right-handed) DNA to tighten around the particle.

Using a new high-resolution native chromatin immunoprecipitation (ChIP) method with V-plot analysis (Henikoff et al. 2011), Krassovsky et al. (2012) succeeded in mapping the Cse4-containing particle directly over CDEII. Meanwhile, some cytological studies detected more fluorescent Cse4-GFP particles over centromeres than could plausibly fit within CDEII (Coffman et al. 2011; Lawrimore et al. 2011), challenging the "point" interpretation of yeast centromeres (Furuyama and Biggins 2007). However, our quantitative ChIP-seq data could exclude the possibility that additional molecules are incorporated outside of the genetically defined centromere (Henikoff and Henikoff 2012), rather suggesting that unincorporated Cse4 molecules are closely associated with functional centromeres. Thus, only two plausible models for the budding yeast centromere remained: a $(Cse4/H4)_2$ "tetrasome" or a (Cse4/H4/H2A/H2B) "hemisome," either of which could wrap the ~80-bp CDEII DNA with right-handed chirality.

To determine whether hemisomes can be assembled on short segments of DNA, Takehito Furuyama adopted a classical salt-assembly method (Tatchell and Van Holde 1979), demonstrating that Cse4 but not H3 hemisomes that are stable in 4 M urea can be readily produced on CDEII DNA (Furuyama et al. 2013). Our finding suggested that the extreme AT richness of CDEII is an adaptation for exclusion of conventional nucleosomes from the functional centromere, as poly(dA:dT) tracts in the yeast genome are known to exclude nucleosomes (Struhl and Segal 2013).

To definitively determine the composition and conformation of the budding yeast centromeric nucleosome in vivo, we applied the Widom chemical cleavage mapping method, in which histone H4 is derivatized such that it cleaves DNA next to the dyad (Brogaard et al. 2012). By mapping the sites of cleavage, we could determine the precise location of H4 with base-pair resolution and also determine whether there is one H4 or two H4s. Results were unequivocal: Mapping of H4 cleavages at all 16 yeast centromeres showed only a single H4 with cleavage sites within CDEII consistent with a hemisome and only with a hemisome among all proposed structures (Henikoff et al. 2014) (Fig. 1, middle panel). More recently, we have used our novel CUT&RUN chromatin profiling method (Skene and Henikoff 2017) to show the presence of histone H2A in the CDEII particle (Fig. 2), definitively excluding the $(Cse4/H4)_2$ tetrasome model (Mizuguchi et al. 2007; Wisniewski et al. 2014).

FISSION YEAST CENTROMERES LACK POSITIONING AND ARE POPULATED BY OCTAMERIC NUCLEOSOMES

Examination of the fungal phylogeny argues that point centromeres, which are exclusive to the Saccharomycetes, have evolved from "regional" centromeres that lack obvious sequence specificity (Malik and Henikoff 2009). For example, centromeres of the parasitic budding yeast *Candida albicans* show no sequence similarity to those of its close relative *Candida dubliniensis* (Padmanabhan et al. 2008), and deletion or replacement results in efficient formation of cenH3-enriched neocentromeres nearby (Baum et al. 2006; Ketel et al. 2009; Thakur and Sanyal 2013). Centromeres of the fission yeast *S. pombe* consist

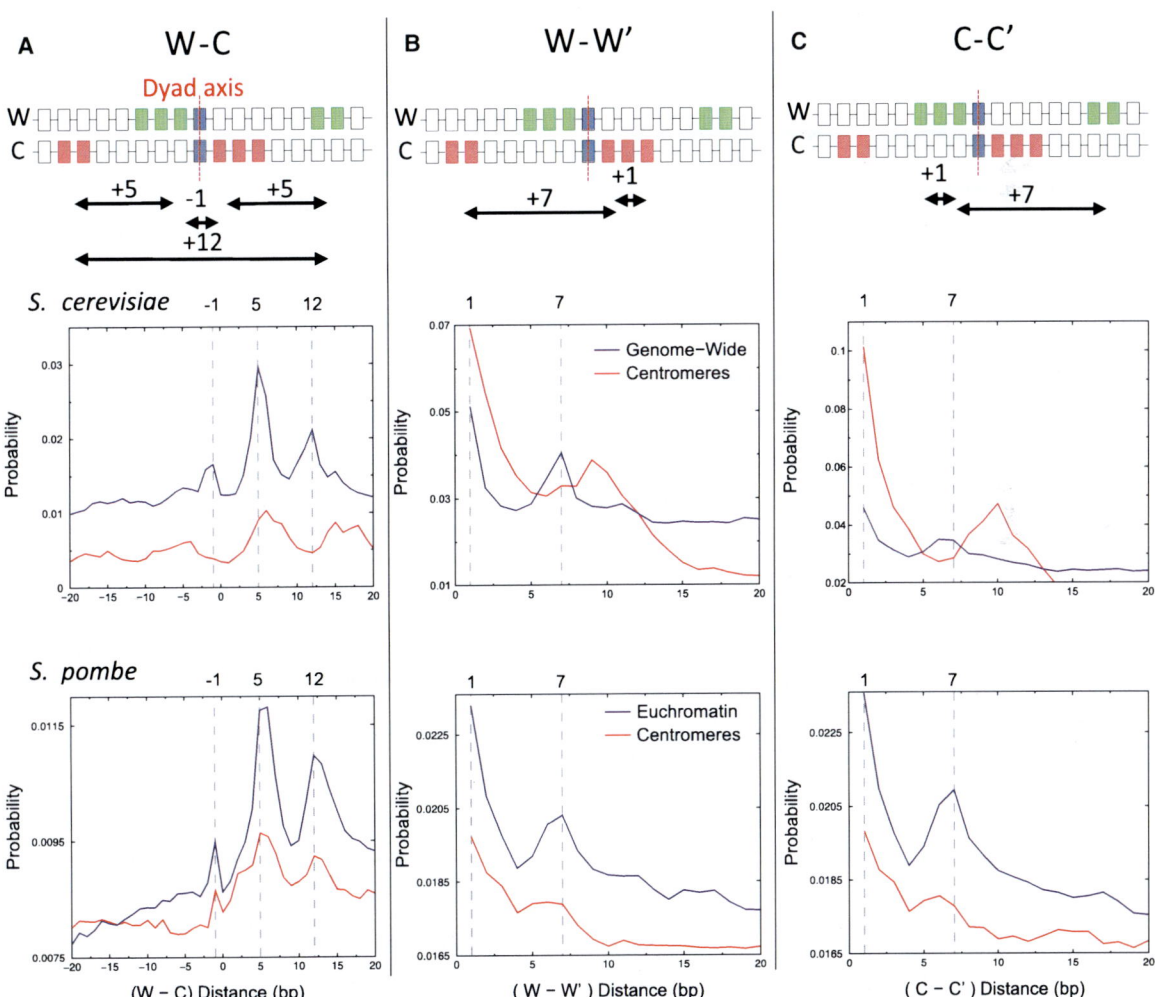

Figure 1. Budding yeast centromeric nucleosomes have one histone H4, but those in fission yeast have two. In vivo chemical cleavage mapping shows one CENP-A H4 molecule for *Saccharomyces cerevisiae* but two for *Schizosaccharomyces pombe*. (*A–C*) (*Top*) A structure-based model of where H4S47C-anchored cleavages are predicted to occur on the Watson (W, 5′→3′) and Crick (C, 3′→5′) strands around the dyad axis of an H3 nucleosome (Henikoff et al. 2014). Each box represents a single nucleotide position, where the filled boxes are predicted cleavage sites on the W strand (green) and the C strand (pink). The predicted distances between cleavage sites are indicated below. (*Middle*) Distribution of distances between cleavage fragment ends genome-wide (blue lines) and for budding yeast centromeres (red lines) (Henikoff et al. 2014). (*Bottom*) Comparable distributions for the central core of fission yeast centromeres (red lines) and euchromatin (blue lines) (Thakur et al. 2015). Distributions were normalized by the total number of combinations within the 40-bp distance range. (*A*) W–C distances. (*B*) W–W′ distances. (*C*) C–C′ distances. (Adapted from Thakur et al. 2015, with permission from the Genetics Society of America.)

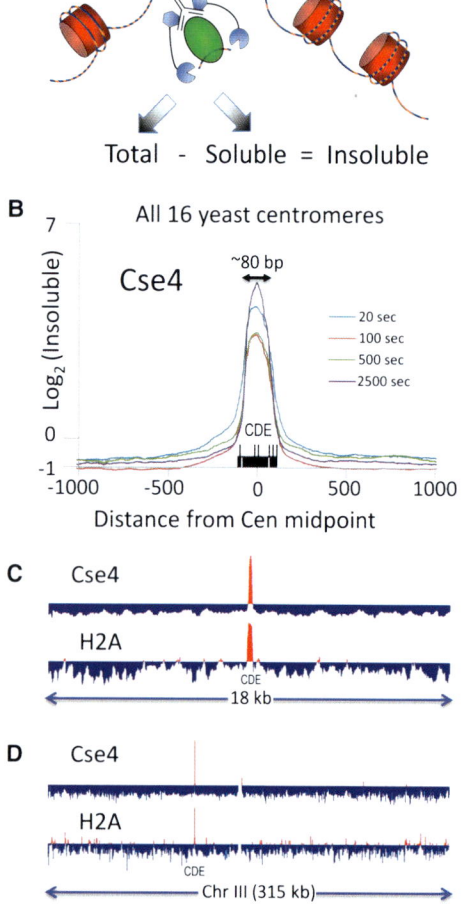

Figure 2. Budding yeast centromeres are occupied by insoluble kinetochore-bound 80-bp particles containing Cse4 and H2A. (*A*) CUT&RUN releases chromatin particles by Ca^{2+}-dependent cleavage on either side when ProteinA/MNase (blue) is tethered to an antibody (Skene and Henikoff 2017). Insoluble chromatin is measured by splitting the sample after MNase digestion and extracting the soluble and total DNA, where the difference represents insoluble chromatin. (*B*) Cse4 (cenH3) CUT&RUN log-ratio profiles of insoluble chromatin over a >2 order of magnitude time course shows the presence of a stable ~80-bp Cse4-containing particle over CDEII. (*C*) Similar CUT&RUN profiles are seen for Cse4 and H2A over Centromere III (2500-sec time point), ruling out the (Cse4/H4)$_2$ tetrasome model (Mizuguchi et al. 2007). (*D*) The Cse4/H2A-containing centromeric nucleosome is the most insoluble nucleosome on the entire chromosome.

of a series of heterochromatic outer repeats flanking a cenH3-rich core that lacks obvious sequence-specific features. Thus it would appear that budding yeast point centromeres have emerged from a fungal ancestor that lacks sequence-specific centromeres, perhaps the result of colonization by the 2-μm parasitic element that segregates autonomously within the Saccharomycetes (Malik and Henikoff 2009).

Consistent with an independent origin of point centromeres from regional centromeres, we have found that the cenH3 nucleosomes at the centromeric core of *S. pombe* regional centromeres are completely different from those of *S. cerevisiae*. Fission yeast cenH3 nucleosomes show no

detectable positioning within the central core, and H4 chemical cleavage mapping reveals that there are two H4s per particle (Fig. 1, bottom panel), indicative of octameric nucleosomes (Thakur et al. 2015). Thus, the extreme diversity of centromere architecture is reflected in the diversity of centromeric chromatin within the fungal lineage, where point centromeres are bound by well-phased hemisomes and regional centromeres are bound by unphased octasomes.

C. ELEGANS HOLOCENTROMERES ARE POLYCENTRIC

The most fundamental distinction between centromeres is the distinction between familiar monocentromeres and holocentromeres, in which microtubule attachments occur throughout the length of the chromosome. Our cloning and cytological characterization of cenH3 from the nematode worm *C. elegans* revealed that it occupies the leading edge of each sister chromosome as it segregates to the pole at mitosis, and absence of cenH3 within the bulk of chromatin implied that holocentromeres are discontinuous (Buchwitz et al. 1999). Standard ChIP-seq uncovered a large domain structure of low-density cenH3 (Gassmann et al. 2012), and Florian Steiner's high-resolution native ChIP revealed the presence of approximately 100 well-positioned cenH3 loci dispersed throughout each chromosome (Steiner and Henikoff 2014). Remarkably, these centromeric loci resembled yeast point centromeres with particle sizes of ~80 bp flanked by well-phased H3 nucleosomes (Fig. 3). Motif analysis showed that holocentromeres correspond to previously described GA-rich transcription factor hotspots (Gerstein et al. 2010), leading us to speculate that low-affinity binding of transcription factors at these sites during interphase prevents encroachment by flanking nucleosomes, thus maintaining holocentromere sites accessible at mitosis throughout development (Steiner and Henikoff 2014).

The mapping of dispersed point centromeres at transcription factor hotspots throughout *C. elegans* chromosomes addressed a long-standing question in chromosome biology by showing that holocentromeres are polycentric as opposed to being diffuse (Schrader 1935). More recently, native ChIP-seq of centromeres of the parasitic nematode *Ascaris suum* suggested that these centromeres are also polycentric, but they do not appear to be point centromeres in that cenH3-rich regions consist of high-density arrays of 1–15 kb, with particle sizes of ~140 bp, consistent with conventional octasomes (Kang et al. 2016).

INSECT HOLOCENTROMERES LACK cenH3

Holocentricity has evolved in multiple lineages of animals and plants (Melters et al. 2012). For example, holocentromeres have evolved independently in insects at least four times (Drinnenberg et al. 2014). When Ines Drinnenberg was a postdoctoral fellow jointly with the Malik laboratory, she discovered that cenH3 and other

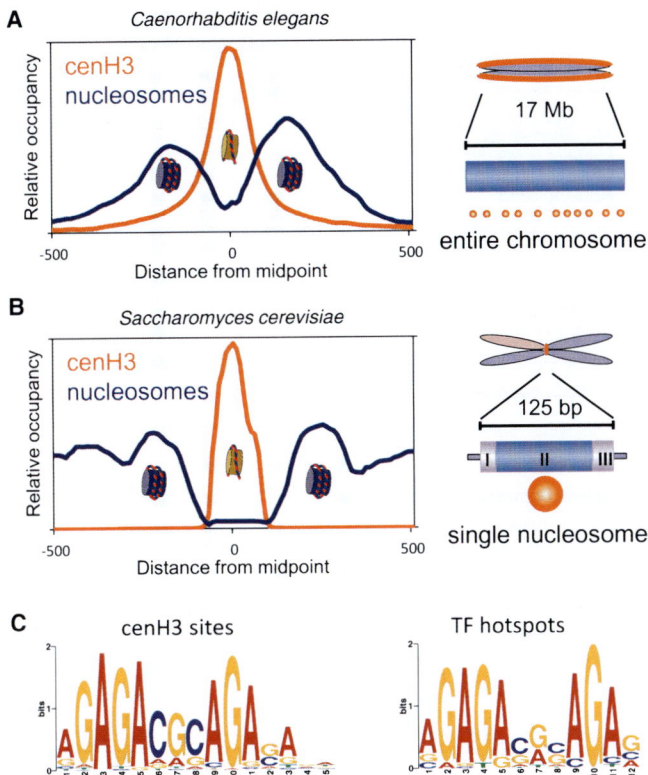

Figure 3. *Caenorhabditis elegans* holocentromeres are polycentric with dispersed point centromeres at transcription factor hotspots. (*A*) cenH3 native MNase ChIP-seq peaks (*n* = 707) were called and aligned at their midpoints. The orange curve represents the average cenH3/input profile, and the blue curve represents the input profile for paired-end reads in the 141–500-bp range. (*B*) For comparison, the equivalent profile is shown for alignment of MNase ChIP-seq and input data from *Saccharomyces cerevisiae*. (*C*) Sequence logo representations of the consensus motif at cenH3 peaks (*left*) and transcription factor hotspots (*right*) called using MEME (Bailey and Elkan 1994). (Adapted from Steiner and Henikoff 2014.)

proteins of the CCAN are absent from holocentric Lepidopteran lineages. Remarkably, loss of these CCAN proteins coincided with all four known transitions from monocentricity to holocentricity in insects (Fig. 4). In contrast, proteins of the outer kinetochore are nearly universally conserved, including in the kinetoplastid *Trypanosoma*, which also lacks canonical CCAN proteins (Akiyoshi and Gull 2014; Senaratne and Drinnenberg 2017). Identifying the chromatin counterparts of the CCAN in these lineages and determining whether insect holocentromeres are polycentric or diffuse are issues that remain to be addressed (Drinnenberg et al. 2016).

PRECISE POSITIONING OF cenH3 NUCLEOSOMES AT SATELLITE CENTROMERES

The identification of CENP-A nucleosomes as marking functional mammalian centromeres (Palmer and Margolis 1985) was an important advance in centromere biology. However, it was not until 1997 that the equivalent of the Clarke and Carbon construction of a functional centromere was accomplished for a satellite centromere (Harrington et al. 1997). Most human artificial centromeres require hundreds of kilobases of an ~170-bp α-satellite repeat array organized as higher-order repeats (HORs), including ~17-bp consensus binding sites for CENP-B protein (Hayden et al. 2013). Several studies have used ChIP-seq to show that CENP-A nucleosomes occupy α-

satellite sequences, although the exact composition and conformation of the particles continue to be debated (Bui et al. 2012; Hasson et al. 2013; Lacoste et al. 2014; Athwal et al. 2015; Henikoff et al. 2015; Thakur and Henikoff 2016; Nechemia-Arbely et al. 2017). We have found that much of the disagreement stems from the difficulty of mapping ChIP-seq reads to tandemly repeated DNA sequences, which have proven to be intractable using standard tools for sequence assembly. To circumvent this problem, we clustered CENP-A ChIP-seq reads de novo, identifying two families of homogeneous dimeric repeats with CENP-B boxes that dominate human centromeres (Henikoff et al. 2015). This extended earlier work of Alexandrov et al. (2001), who originally identified these two "suprachromosomal" α-satellite families, SF1 and SF2, as homogeneous dimeric arrays present on 20 of the 24 different human centromeres. Indeed, we identified a preponderance of precisely positioned 100-bp MNase protected particles for these two dominant families, but we also identified larger particles for other HORs, such as the DXZ1 SF3 subfamily (Fig. 5), that are less homogeneous in CENP-A ChIP-seq data (Henikoff et al. 2015). Our findings suggested that particle conformation detected by native MNase ChIP-seq can differ greatly between different human α-satellite families, with precise positioning a characteristic of the most homogeneous repeats. Precise rotational positioning of cenH3 nucleosomes is a feature of homogeneous satellite centromeres in rice, as determined in a collaborative ChIP-seq study from the Jiming Jiang laboratory (Zhang et al. 2013).

Figure 4. Insect holocentromeres lack cenH3. Four separate transitions from monocentric to holocentric kinetochores have occurred within insect evolution (red H). In each case, the transition involves loss of the inner kinetochore proteins cenH3 and CENP-C, except in Odonata, where CENP-C is retained but no longer has the CENPC motif that recognizes cenH3. Outer kinetochore components, notably Ndc80, which attaches directly to microtubules, are generally preserved. (Adapted from Drinnenberg et al. 2014.)

A COHERENT INNER KINETOCHORE COMPLEX OCCUPIES YOUNG HUMAN CENTROMERIC REPEATS

Whereas we and others had observed 100-bp α-satellite particles using native MNase-based ChIP-seq (Hasson et al. 2013; Henikoff et al. 2015), we also found that formaldehyde cross-linking resulted in protection of particles that were larger than nucleosome size (Thakur and Henikoff 2016). This observation suggested that MNase cleavage under the conditions we used for native ChIP-seq was disrupting particle integrity, but that cross-linking held together a particle consisting of additional components. In our ChIP-seq study of fission yeast centromeres, we had identified CENP-T as an integral component of CENP-A and CENP-C enriched chromatin (Thakur et al. 2015). However, we initially failed to identify human CENP-T by native MNase-based ChIP-seq, consistent with the prevailing view that in mammals CENP-T makes connections with H3 but not CENP-A nucleosomes (McKinley and Cheeseman 2016). As connections to the outer kinetochore are made independently by both CENP-C and CENP-T, the concept of separate anchors for the kinetochore on the DNA was a central issue in understanding centromere biology. The possibility of independent connections between centromeric DNA and the outer kinetochore seemed plausible, insofar as CENP-T and its three partner proteins in the CENP-TWSX complex should be sufficient to directly anchor the outer kinetochore: All four are histone-fold proteins that can be assembled in vitro into nucleosome-like particles that stably wrap DNA in a right-handed orientation. But we wondered: Could it be that the CENP-TWSX particle accounts for the difference in protection by CENP-A/CENP-C particles using native versus formaldehyde cross-linking MNase-ChIP?

To directly address this possibility, we reasoned that the conditions that are used for MNase ChIP-seq might leave behind more condensed particles. Differential salt solubility is a feature of conventional nucleosomes, which can be fractionated into classical "active" (low-NaCl), histone H1–rich (high-NaCl), and "nuclear matrix" (insoluble) components (Sanders 1978; Henikoff et al. 2009). Indeed, using cross-linking and light sonication after MNase digestion, we found that CENP-T containing α-satellite particles could be recovered in robust amounts, and CENP-T particles precisely co-mapped with CENP-A and CENP-C (Thakur and Henikoff 2016). It is possible that estimated particle size differences reported in studies using native MNase ChIP-seq (Hasson et al. 2013; Lacoste et al. 2014; Henikoff et al. 2015; Nechemia-Arbely et al. 2017) can be explained by the lability of CENP-TWSX-containing particles when subjected to MNase treatment.

To confirm that a coherent CCAN particle containing CENP-A, CENP-C, and CENP-T occupies functional human centromeres, we performed tandem ChIP-seq on FLAG-CENP-A particles subjected to cross-linking and heavy digestion with MNase to produce single coherent particles. After first immunoprecipitating with an anti-FLAG antibody and eluting with FLAG peptide, we immunoprecipitated with CENP-A, CENP-B, CENP-C, and CENP-T antibodies, obtaining strong enrichment for each component (Fig. 6). Enrichment was seen for consensus sequences representing SF1, SF2, and SF3 subfamilies and for the Y centromere, which indicates that at all human centromeres a single CCAN complex contains four of these centromere-specific DNA-binding components.

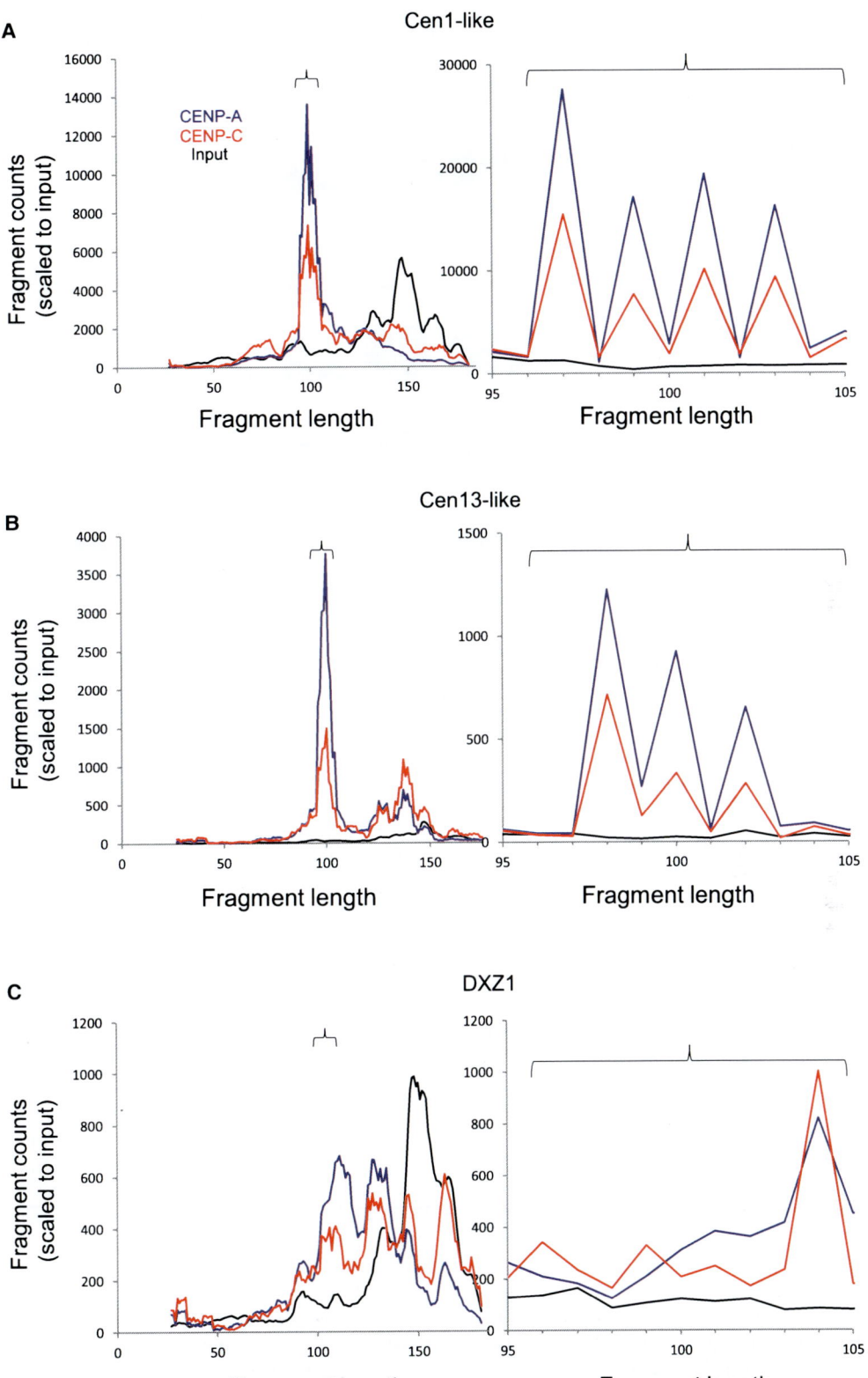

Figure 5. Precise positioning of cenH3 nucleosomes at human α-satellite centromeres. Young α-satellite dimers precisely position ~100-bp CENP-A nucleosome particles. (*A–C*) Size distributions of fragments mapping to an SF1 consensus dimer (*A*) and an SF2 consensus dimer (*B*) and to the most proximal 6-kb region of DXZ1 (*C*), which belongs to SF3. Graphs on the *right* are expansions of graphs on the *left* (indicated by brackets). The *y*-axis scale is for input normalized counts, and the areas under the other curves were equalized to that for input. (Reprinted from Henikoff et al. 2015.)

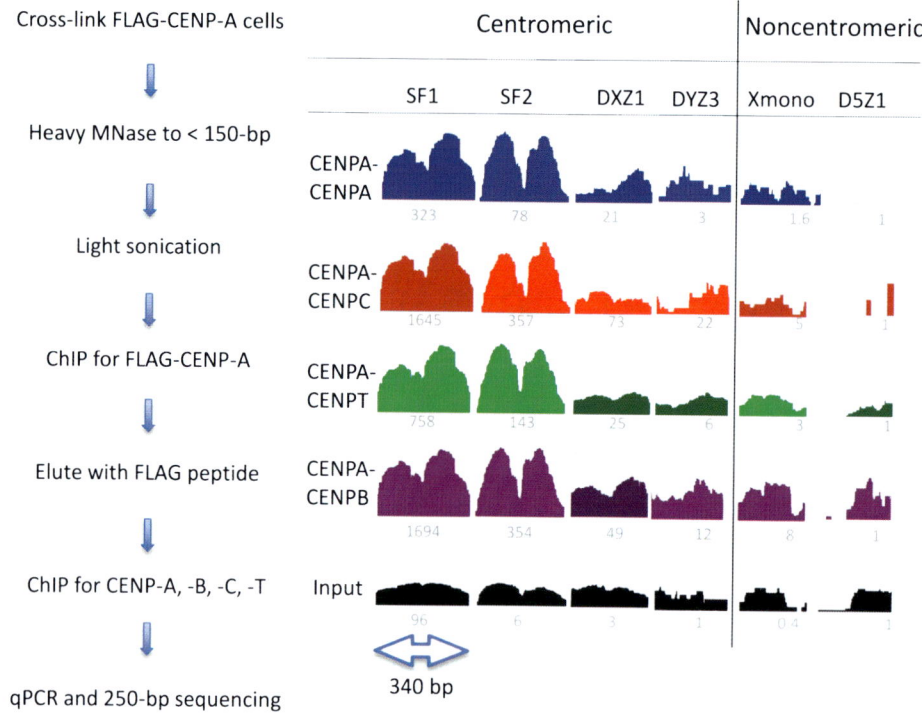

Figure 6. A coherent inner kinetochore complex occupies young human centromeric repeats. Sequential ChIP-seq profiles of CCAN components are nearly identical. A single α-satellite dimer from each array is shown, and the relative scale is the area of the indicated profile divided by the area of the D5Z1 profile, in which the numbers reflect the product of the total sequence abundance and enrichment. Because DXZ1, DYZ3, D19Z1, Xmono, and D5Z1 are not dimeric units, we chose pairs of tandem monomers as representatives. (Based on data from Thakur and Henikoff 2016.)

UNDERSTANDING THE EVOLUTIONARY PLASTICITY OF CENTROMERIC CHROMATIN

These studies of centromeric chromatin in a wide variety of organisms have uncovered an astonishing diversity that is difficult to reconcile with the conserved function of centromeric chromatin, which is to anchor the outer kinetochore complex (Steiner and Henikoff 2015). What accounts for such remarkable centromeric chromatin plasticity?

Evolutionary plasticity of centromeric chromatin might arise from selfish processes. Centromere drive is a cyclical process, in which satellite sequences expand and compete for the egg pole at meiosis I but are suppressed by mutations in host CCAN and other proteins (Malik and Henikoff 2001). Successful drive of a satellite centromere may eventually result in fixation of that satellite sequence in the species evolving into a genetically defined centromere, whereas successful suppression of drive over evolutionary time tends to reduce centromere sequence specificity evolving into an epigenetically defined centromere (Dawe and Henikoff 2006). Interestingly, the cenH3s of the holocentric plant genus *Luzula* show no evidence of centromere drive, in contrast to the cenH3s of monocentric plants (Zedek and Bureš 2016), which suggests that elimination of a fixed position on the chromosome that can be colonized by selfish DNA may be an effective defense

against centromere drive (Fig. 7; Talbert et al. 2008). In contrast, the holocentromeres in nematode oocyte meiosis are axially oriented so that the chromosome end functionally resembles a monocentromere (Albertson and Thomson 1993). This telokinetic form of chromosome disjunction may be less effective at avoiding centromere drive than the holokinetic disjunction of equatorially orientated *Luzula* chromosomes (Heckmann et al. 2014). Although the role of cenH3 in the cup-like kinetochore of *Caenorhabditis* meiosis is disputed (Chan et al. 2004; Monen et al. 2005), it differs between oocyte and spermatocyte meiosis (Shakes et al. 2009), and cenH3 shows evidence of positive selection (Zedek and Bureš 2012), suggesting that centromere drive may affect telokinetic holocentromeres. This may explain the massive accumulation of heterochromatic sequences at the ends of the single chromosome in *Parascaris univalens* (Talbert et al. 2008).

Whereas native human centromeres are genetically defined in that they are dominated by specific α-satellite arrays on each centromere, rare cases of epigenetically defined human neocentromeres suggests that centromeric chromatin plasticity is an inevitable consequence of an ever-changing DNA sequence substrate for CCAN assembly. Other selfish processes may be responsible for transitions between different fungal centromere types, such as the proposed colonization of Saccharomycetes centromeres by its 2-μm plasmid (Malik and Henikoff 2009).

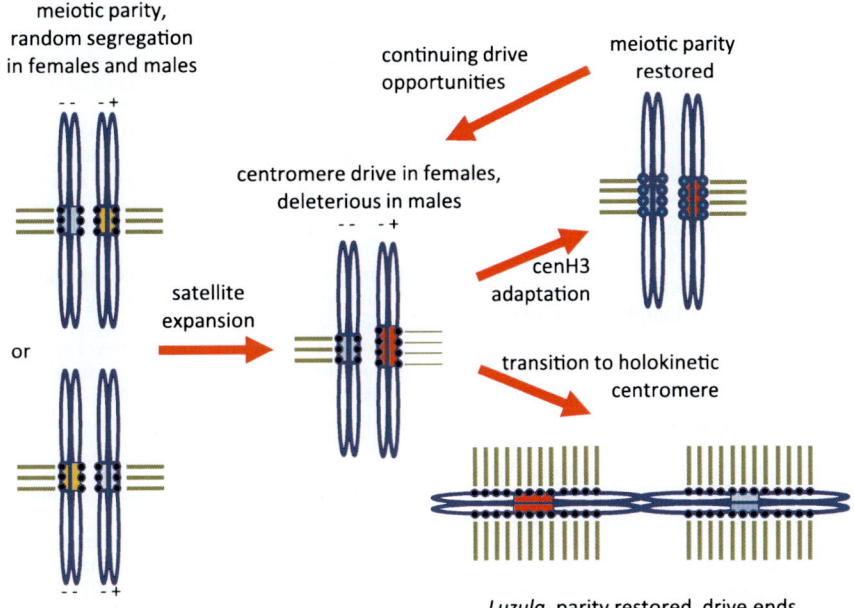

Figure 7. Holokinetic meiosis suppresses centromere drive. In female meiosis, three meiotic products (−) degenerate and one (+) survives. With random segregation each chromatid in the meiotic tetrad has an equal opportunity to be oriented toward the egg pole. However, selfish DNA that influences centromere "strength" through greater assembly of kinetochore components may be preferentially transmitted (centromere drive), while simultaneously creating deleterious centromere imbalance in males. Adaptation of kinetochore proteins to restore centromere parity may suppress drive, but cycles of drive and suppression will repeat. In the holokinetic chromosomes of *Luzula*, the equatorial orientation of many broadly distributed kinetochore sites makes it difficult for any localized selfish element to gain a transmission advantage, effectively ending cycles of drive and suppression (Zedek and Bureš 2016).

One common feature of genetically defined centromeres, whether the point centromeres of Saccharomycetes or the satellite centromeres of animals and plants, is that they are well-positioned. In contrast, the central core of fission yeast regional centromeres lacks obvious sequence specificity and shows random positioning of nucleosomes (Thakur et al. 2015). Precise positioning of nucleosomes, which is especially evident in the satellite centromeres of rice (Zhang et al. 2013), is a sequence-specific adaptation of satellite DNA that has been exploited for the first successful crystallization of H3 nucleosomes, which were produced using a human α-satellite DNA derivative (Harp et al. 1996; Luger et al. 1997). Precise phasing, together with the action of satellite-specific DNA-binding proteins, such as CENP-B, may represent the means whereby selfish centromeres perpetuate themselves. CENP-B is derived from the transposase encoded by the pogo DNA transposon, and it is attractive to imagine that its original domestication as a centromere protein was via recruitment by a driving centromere.

The domestication of selfish elements appears to be responsible for the most remarkable example of centromere evolution, the partial or complete replacement of the satellite centromeres of the wild ancestor of maize by a selfish element in maize inbred lines (Schneider et al. 2016). During domestication of maize from teosinte over the past ~9000 years, selection for agronomic traits was accompanied by dense insertion of CR2 centromere-specific retrotransposons and frequent loss by small inversions and deletions of the ancestral 155-bp CentC satellite in all 10 maize centromeres, 57 times independently in 26 inbred lines. Domestication and inbreeding would thus appear to be a powerful driver of centromere plasticity and could account for the near-complete loss of satellite arrays at several rice (Nagaki et al. 2004) and potato (Gong et al. 2012) centromeres and the complete loss of repetitive sequences from horse Centromere 11 (Purgato et al. 2015). Thus, whether the direct result of centromere competition, the colonization by selfish elements, or the indirect result of domestication, strong selective forces appear to be drivers of the remarkable plasticity of centromeric chromatin.

CONCLUSION

Our investigations into the molecular basis for centromere function have uncovered an unexpected plasticity in chromatin architecture, with evidence for histone-containing particles that are unlike any nucleosomes observed on chromosome arms. These studies have led to evidence for centromeres with right-handed writhes, hemisomes, and coherent CCAN particles containing multiple centromere-specific histone-fold and nonhistone proteins. Such structural diversity might seem surprising considering the simplicity of the conserved function of centromeric chromatin to connect to the outer kinetochore. However, the centromere's control of its chromosome's fate at every cell division has unleashed powerful selfish and selective forces over evolutionary time and even during human domestication that have resulted in such interesting molecular complexity.

ACKNOWLEDGMENTS

We thank Mitchell Smith, who approximately 20 years ago pointed out that yeast Cse4 and human CENP-A are no more similar to one another than to H3, a curious fact that led to our subsequent focus on histone variants and centromere evolution. We also thank former members of our laboratory whose work and insights over the years have contributed to this narrative.

REFERENCES

Akiyoshi B, Gull K. 2014. Discovery of unconventional kinetochores in kinetoplastids. *Cell* **156:** 1247–1258.

Albertson DG, Thomson JN. 1993. Segregation of holocentric chromosomes at meiosis in the nematode, *Caenorhabditis elegans*. *Chromosome Res* **1:** 15–26.

Alexandrov I, Kazakov A, Tumeneva I, Shepelev V, Yurov Y. 2001. α-satellite DNA of primates: Old and new families. *Chromosoma* **110:** 253–266.

Athwal RK, Walkiewicz MP, Baek S, Fu S, Bui M, Camps J, Ried T, Sung MH, Dalal Y. 2015. CENP-A nucleosomes localize to transcription factor hotspots and subtelomeric sites in human cancer cells. *Epigenetics Chromatin* **8:** 2.

Bailey T, Elkan C. 1994. Fitting a mixture model by expectation maximization to discover motifs in biopolymers. In *Proceedings of the Second International Conference on Intelligent Systems for Molecular Biology*, pp. 28–36. AAAI Press, Menlo Park, CA.

Baum M, Sanyal K, Mishra PK, Thaler N, Carbon J. 2006. Formation of functional centromeric chromatin is specified epigenetically in *Candida albicans*. *Proc Natl Acad Sci* **103:** 14877–14882.

Bradbury EM. 1977. Histone nomenclature. *Methods Cell Biol* **16:** 179–181.

Brogaard K, Xi L, Wang JP, Widom J. 2012. A map of nucleosome positions in yeast at base-pair resolution. *Nature* **486:** 496–501.

Buchwitz BJ, Ahmad K, Moore LL, Roth MB, Henikoff S. 1999. A histone-H3-like protein in *C. elegans*. *Nature* **401:** 547–548.

Bui M, Dimitriadis EK, Hoischen C, An E, Quénet D, Giebe S, Nita-Lazar A, Diekmann S, Dalal Y. 2012. Cell-cycle-dependent structural transitions in the human CENP-A nucleosome in vivo. *Cell* **150:** 317–326.

Burt A, Trivers R. 2006. *Genes in conflict: The biology of selfish genetic elements*. Harvard University Press, Cambridge, MA.

Chan RC, Severson AF, Meyer BJ. 2004. Condensin restructures chromosomes in preparation for meiotic divisions. *J Cell Biol* **167:** 613–625.

Chmátal L, Gabriel SI, Mitsainas GP, Martínez-Vargas J, Ventura J, Searle JB, Schultz RM, Lampson MA. 2014. Centromere strength provides the cell biological basis for meiotic drive and karyotype evolution in mice. *Curr Biol* **24:** 2295–2300.

Clarke L, Carbon J. 1980. Isolation of a yeast centromere and construction of functional small circular chromosomes. *Nature* **287:** 504–509.

Clarke L, Carbon J. 1985. The structure and function of yeast centromeres. *Annu Rev Genet* **19:** 29–55.

Coffman VC, Wu P, Parthun MR, Wu JQ. 2011. CENP-A exceeds microtubule attachment sites in centromere clusters of both budding and fission yeast. *J Cell Biol* **195:** 563–572.

Daniel A. 2002. Distortion of female meiotic segregation and reduced male fertility in human Robertsonian translocations: Consistent with the centromere model of co-evolving centromere DNA/centromeric histone (CENP-A). *Am J Med Genet* **111:** 450–452.

Dawe RK, Henikoff S. 2006. Centromeres put epigenetics in the driver's seat. *Trends Biochem Sci* **31:** 662–669.

Díaz-Ingelmo O, Martínez-Garcia B, Segura J, Valdés A, Roca J. 2015. DNA topology and global architecture of point centromeres. *Cell Rep* **13:** 667–677.

Drinnenberg IA, deYoung D, Henikoff S, Malik HS. 2014. Recurrent loss of CenH3 is associated with independent transitions to holocentricity in insects. *Elife* **3:** e03676.

Drinnenberg IA, Henikoff S, Malik HS. 2016. Evolutionary turnover of kinetochore proteins: A ship of Theseus? *Trends Cell Biol* **26:** 498–510.

Earnshaw WC, Rothfield N. 1985. Identification of a family of human centromere proteins using autoimmune sera from patients with scleroderma. *Chromosoma* **91:** 313–321.

Fishman L, Willis JH. 2005. A novel meiotic drive locus almost completely distorts segregation in *Mimulus* (monkeyflower) hybrids. *Genetics* **169:** 347–353.

Flemming W. 1882. *Zellsubstanz, Kern und Zelltheilung*. F.C.W. Vogel, Leipzig.

Furuyama S, Biggins S. 2007. Centromere identity is specified by a single centromeric nucleosome in budding yeast. *Proc Natl Acad Sci* **104:** 14706–14711.

Furuyama T, Henikoff S. 2009. Centromeric nucleosomes induce positive DNA supercoils. *Cell* **138:** 104–113.

Furuyama T, Codomo CA, Henikoff S. 2013. Reconstitution of hemisomes on budding yeast centromeric DNA. *Nucleic Acids Res* **41:** 5769–5783.

Gassmann R, Rechtsteiner A, Yuen KW, Muroyama A, Egelhofer T, Gaydos L, Barron F, Maddox P, Essex A, Monen J, et al. 2012. An inverse relationship to germline transcription defines centromeric chromatin in *C. elegans*. *Nature* **484:** 534–537.

Gerstein MB, Lu ZJ, Van Nostrand EL, Cheng C, Arshinoff BI, Liu T, Yip KY, Robilotto R, Rechtsteiner A, Ikegami K, et al. 2010. Integrative analysis of the *Caenorhabditis elegans* genome by the modENCODE project. *Science* **330:** 1775–1787.

Gong Z, Wu Y, Koblízková A, Torres GA, Wang K, Iovene M, Neumann P, Zhang W, Novák P, Buell CR, et al. 2012. Repeatless and repeat-based centromeres in potato: Implications for centromere evolution. *Plant Cell* **24:** 3559–3574.

Gore J, Bryant Z, Nöllmann M, Le MU, Cozzarelli NR, Bustamante C. 2006. DNA overwinds when stretched. *Nature* **442:** 836–839.

Harp JM, Uberbacher EC, Roberson AE, Palmer EL, Gewiess A, Bunick GJ. 1996. X-ray diffraction analysis of crystals containing twofold symmetric nucleosome core particles. *Acta Crystallogr D Biol Crystallogr* **52:** 283–288.

Harrington JJ, Van Bokkelen G, Mays RW, Gustashaw K, Willard HF. 1997. Formation of de novo centromeres and construction of first-generation human artificial microchromosomes. *Nat Genet* **15:** 345–355.

Hasson D, Panchenko T, Salimian KJ, Salman MU, Sekulic N, Alonso A, Warburton PE, Black BE. 2013. The octamer is the major form of CENP-A nucleosomes at human centromeres. *Nat Struct Mol Biol* **20:** 687–695.

Hayden KE, Strome ED, Merrett SL, Lee HR, Rudd MK, Willard HF. 2013. Sequences associated with centromere competency in the human genome. *Mol Cell Biol* **33:** 763–772.

Heckmann S, Schubert V, Houben A. 2014. Holocentric plant meiosis: First sisters, then homologues. *Cell Cycle* **13:** 3623–3624.

Henikoff S, Henikoff JG. 2012. "Point" centromeres of *Saccharomyces* harbor single centromere-specific nucleosomes. *Genetics* **190:** 1575–1577.

Henikoff S, Malik HS. 2002. Centromeres: Selfish drivers. *Nature* **417:** 227.

Henikoff S, Ahmad K, Platero JS, van Steensel B. 2000. Heterochromatic deposition of centromeric histone H3-like proteins. *Proc Natl Acad Sci* **97:** 716–721.

Henikoff S, Ahmad K, Malik HS. 2001. The centromere paradox: Stable inheritance with rapidly evolving DNA. *Science* **293:** 1098–1102.

Henikoff S, Henikoff JG, Sakai A, Loeb GB, Ahmad K. 2009. Genome-wide profiling of salt fractions maps physical properties of chromatin. *Genome Res* **19:** 460–469.

Henikoff JG, Belsky JA, Krassovsky K, Macalpine DM, Henikoff S. 2011. Epigenome characterization at single base-pair resolution. *Proc Natl Acad Sci* **108:** 18318–18323.

Henikoff S, Ramachandran S, Krassovsky K, Bryson TD, Codomo CA, Brogaard K, Widom J, Wang JP, Henikoff JG. 2014. The budding yeast Centromere DNA Element II wraps a stable Cse4 hemisome in either orientation in vivo. *Elife* **3:** e01861.

Henikoff JG, Thakur J, Kasinathan S, Henikoff S. 2015. A unique chromatin complex occupies young α-satellite arrays of human centromeres. *Sci Adv* **1:** e14000234.

Jin W, Melo JR, Nagaki K, Talbert PB, Henikoff S, Dawe RK, Jiang J. 2004. Maize centromeres: Organization and functional adaptation in the genetic background of oat. *Plant Cell* **16:** 571–581.

Kang Y, Wang J, Neff A, Kratzer S, Kimura H, Davis RE. 2016. Differential chromosomal localization of centromeric histone CENP-A contributes to nematode programmed DNA elimination. *Cell Rep* **16:** 2308–2316.

Ketel C, Wang HS, McClellan M, Bouchonville K, Selmecki A, Lahav T, Gerami-Nejad M, Berman J. 2009. Neocentromeres form efficiently at multiple possible loci in *Candida albicans*. *PLoS Genet* **5:** e1000400.

Krassovsky K, Henikoff JG, Henikoff S. 2012. Tripartite organization of centromeric chromatin in budding yeast. *Proc Natl Acad Sci* **109:** 243–248.

Lacoste N, Woolfe A, Tachiwana H, Garea AV, Barth T, Cantaloube S, Kurumizaka H, Imhof A, Almouzni G. 2014. Mislocalization of the centromeric histone variant CenH3/CENP-A in human cells depends on the chaperone DAXX. *Mol Cell* **53:** 631–644.

Lawrimore J, Bloom KS, Salmon ED. 2011. Point centromeres contain more than a single centromere-specific Cse4 (CENP-A) nucleosome. *J Cell Biol* **195:** 573–582.

Luger K, Mader AW, Richmond RK, Sargent DF, Richmond TJ. 1997. Crystal structure of the nucleosome core particle at 2.8 A resolution. *Nature* **389:** 251–260.

Malik HS, Henikoff S. 2001. Adaptive evolution of Cid, a centromere-specific histone in *Drosophila*. *Genetics* **157:** 1293–1298.

Malik HS, Henikoff S. 2009. Major evolutionary transitions in centromere complexity. *Cell* **138:** 1067–1082.

McKinley KL, Cheeseman IM. 2016. The molecular basis for centromere identity and function. *Nat Rev Mol Cell Biol* **17:** 16–29.

Melters DP, Paliulis LV, Korf IF, Chan SW. 2012. Holocentric chromosomes: Convergent evolution, meiotic adaptations, and genomic analysis. *Chromosome Res* **20:** 579–593.

Meluh PB, Yang P, Glowczewski L, Koshland D, Smith MM. 1998. Cse4p is a component of the core centromere of *Saccharomyces cerevisiae*. *Cell* **94:** 607–613.

Mizuguchi G, Xiao H, Wisniewski J, Smith MM, Wu C. 2007. Nonhistone Scm3 and histones CenH3-H4 assemble the core of centromere-specific nucleosomes. *Cell* **129:** 1153–1164.

Monen J, Maddox PS, Hyndman F, Oegema K, Desai A. 2005. Differential role of CENP-A in the segregation of holocentric *C. elegans* chromosomes during meiosis and mitosis. *Nat Cell Biol* **7:** 1148–1155.

Nagaki K, Cheng Z, Ouyang S, Talbert PB, Kim M, Jones KM, Henikoff S, Buell CR, Jiang J. 2004. Sequencing of a rice centromere uncovers active genes. *Nat Genet* **36:** 138–145.

Nechemia-Arbely Y, Fachinetti D, Miga KH, Sekulic N, Soni GV, Kim DH, Wong AK, Lee AY, Nguyen K, Dekker C, et al. 2017. Human centromeric CENP-A chromatin is a homotypic, octameric nucleosome at all cell cycle points. *J Cell Biol* **216:** 607–621.

Novitski E. 1955. Genetic measures of centromere activity in Drosophila melanogaster. *J Cell Comp Physiol* **45:** 151–169.

Padmanabhan S, Thakur J, Siddharthan R, Sanyal K. 2008. Rapid evolution of Cse4p-rich centromeric DNA sequences in closely related pathogenic yeasts, *Candida albicans* and *Candida dubliniensis*. *Proc Natl Acad Sci* **105:** 19797–19802.

Palmer DK, Margolis RL. 1985. Kinetochore components recognized by human autoantibodies are present on mononucleosomes. *Mol Cell Biol* **5:** 173–186.

Palmer DK, O'Day K, Wener MH, Andrews BS, Margolis RL. 1987. A 17-kD centromere protein (CENP-A) copurifies with nucleosome core particles and with histones. *J Cell Biol* **104:** 805–815.

Palmer DK, O'Day K, Margolis RL. 1990. The centromere specific histone CENP-A is selectively retained in discrete foci in mammalian sperm nuclei. *Chromosoma* **100:** 32–36.

Purgato S, Belloni E, Piras FM, Zoli M, Badiale C, Cerutti F, Mazzagatti A, Perini G, Della Valle G, Nergadze SG, et al. 2015. Centromere sliding on a mammalian chromosome. *Chromosoma* **124:** 277–287.

Rhoades MM. 1952. Preferential segregation in maize. In *Heterosis* (ed. Gowen JW), pp. 66–80. Iowa State College Press, Ames.

Sanders MM. 1978. Fractionation of nucleosomes by salt elution from micrococcal nuclease-digested nuclei. *J Cell Biol* **79:** 97–109.

Schneider KL, Xie Z, Wolfgruber TK, Presting GG. 2016. Inbreeding drives maize centromere evolution. *Proc Natl Acad Sci* **113:** E987–E996.

Schrader F. 1935. Notes on the mitotic behavior of long chromosomes. *Cytologia (Tokyo)* **6:** 422–430.

Senaratne AP, Drinnenberg IA. 2017. All that is old does not wither: Conservation of outer kinetochore proteins across all eukaryotes? *J Cell Biol* **216:** 291–293.

Shakes DC, Wu JC, Sadler PL, Laprade K, Moore LL, Noritake A, Chu DS. 2009. Spermatogenesis-specific features of the meiotic program in *Caenorhabditis elegans*. *PLoS Genet* **5:** e1000611.

Skene PJ, Henikoff S. 2017. An efficient targeted nuclease strategy for high-resolution mapping of DNA binding sites. *Elife* **6:** e21856.

Steiner FA, Henikoff S. 2014. Holocentromeres are dispersed point centromeres localized at transcription factor hotspots. *Elife* **3:** e02025.

Steiner FA, Henikoff S. 2015. Diversity in the organization of centromeric chromatin. *Curr Opin Genet Dev* **31:** 28–35.

Struhl K, Segal E. 2013. Determinants of nucleosome positioning. *Nat Struct Mol Biol* **20:** 267–273.

Sturtevant AH. 1913. The linear arrangement of six sex-linked factors in *Drosophila*, as shown by their mode of association. *J Exp Zool* **14:** 43–59.

Talbert PB, Masuelli R, Tyagi AP, Comai L, Henikoff S. 2002. Centromeric localization and adaptive evolution of an Arabidopsis histone H3 variant. *Plant Cell* **14:** 1053–1066.

Talbert PB, Bryson TD, Henikoff S. 2004. Adaptive evolution of centromere proteins in plants and animals. *J Biol* **3:** 18.

Talbert PB, Bayes JJ, Henikoff S. 2008. Evolution of centromeres and kinetochores: A two-part fugue. In *The kinetochore* (ed. De Wulf P, Earnshaw WC). Springer, Berlin.

Talbert PB, Ahmad K, Almouzni G, Ausio J, Berger F, Bhalla PL, Bonner WM, Cande WZ, Chadwick BP, Chan SW, et al. 2012. A unified phylogeny-based nomenclature for histone variants. *Epigenetics Chromatin* **5:** 7.

Tatchell K, Van Holde KE. 1979. Nucleosome reconstitution: Effect of DNA length on nucleosome structure. *Biochemistry (Mosc)* **18:** 2871–2880.

Thakur J, Henikoff S. 2016. CENPT bridges adjacent CENPA nucleosomes on young human α-satellite dimers. *Genome Res* **26:** 1178–1187.

Thakur J, Sanyal K. 2013. Efficient neocentromere formation is suppressed by gene conversion to maintain centromere function at native physical chromosomal loci in *Candida albicans*. *Genome Res* **23:** 638–652.

Thakur J, Talbert PB, Henikoff S. 2015. Interactions of inner kinetochore proteins with fission yeast regional centromeres. *Genetics* **201**: 543–561.

Wisniewski J, Hajj B, Chen J, Mizuguchi G, Xiao H, Wei D, Dahan M, Wu C. 2014. Imaging the fate of histone Cse4 reveals de novo replacement in S phase and subsequent stable residence at centromeres. *Elife* **3**: e02203.

Yunis JJ, Yasmineh WG. 1971. Heterochromatin, satellite DNA, and cell function. Structural DNA of eucaryotes may support and protect genes and aid in speciation. *Science* **174**: 1200–1209.

Zedek F, Bureš P. 2012. Evidence for centromere drive in the holocentric chromosomes of *Caenorhabditis*. *PLoS One* **7**: e30496.

Zedek F, Bureš P. 2016. Absence of positive selection on CenH3 in Luzula suggests that holokinetic chromosomes may suppress centromere drive. *Ann Bot* **118**: 1347–1352.

Zhang T, Talbert PB, Zhang W, Wu Y, Yang Z, Henikoff JG, Henikoff S, Jiang J. 2013. The CentO satellite confers translational and rotational phasing on cenH3 nucleosomes in rice centromeres. *Proc Natl Acad Sci* **110**: E4875–E4883.

Molecular Structures of Yeast Kinetochore Subcomplexes and Their Roles in Chromosome Segregation

Simon Jenni,[1] Yoana N. Dimitrova,[1,2] Roberto Valverde,[1,3] Stephen M. Hinshaw,[1,2] and Stephen C. Harrison[1,2]

[1]*Department of Biological Chemistry and Molecular Pharmacology, Harvard Medical School, Boston, Massachusetts 02115*

[2]*Howard Hughes Medical Institute, Boston, Massachusetts 02115*

Correspondence: harrison@crystal.harvard.edu

Kinetochore molecular architecture exemplifies "form follows function." The simplifications that generated the one-chromosome:one-microtubule linkage in point-centromere yeast have enabled strategies for systematic structural analysis and high-resolution visualization of many kinetochore components, leading to specific proposals for molecular mechanisms. We describe here some structural features that allow a kinetochore to remain attached to the end of a depolymerizing microtubule (MT) and some characteristics of the connections between substructures that permit very sensitive regulation by differential kinase activities. We emphasize in particular the importance of flexible connections between rod-like structural members and the integration of these members into a compliant cage-like assembly anchored on the MT by a sliding molecular ring.

Kinetochores attach chromosomes to microtubules (MTs) of the mitotic spindle, thereby coupling chromosome movement and MT dynamics. Studies of the molecular organization of kinetochores became possible when Carbon and coworkers showed that budding-yeast centromeres are relatively short, ~125-bp, regions with defined DNA sequences (Clarke and Carbon 1980) recognized by specific proteins, some of which they proceeded to isolate (Lechner and Carbon 1991). These proteins in turn direct deposition of a single, centromere-specific nucleosome (Meluh et al. 1998). The "parts list" for the kinetochores that then assemble on the so-called "point centromeres" of budding yeast now includes more than 50 distinct gene products (depending on how one chooses to define a "part"), most of them associated into distinct, multiprotein complexes (De Wulf et al. 2003; Musacchio and Desai 2017). Conservation of many of these complexes among eukaryotes has reinforced the view that a budding-yeast kinetochore represents a simplified module of the kinetochores distributed over much longer centromeres in other organisms (Fig. 1).

MT ATTACHMENT

Kinetochore protein complexes fall into three groups: MT-proximal components, chromatin-proximal components, and intermediate adaptors. We concentrate in this report on some molecular–structural characteristics of MT attachment and adaptor interactions. Two of us have reviewed elsewhere the molecular activities of chromatin-proximal components (Hinshaw and Harrison 2017). Because point-centromere kinetochores capture a single spindle MT (Winey et al. 1995), they create a one-to-one centromere-to-MT bridge. The molecular architecture of the kinetochore–MT connection is thus more easily described than it is for kinetochores on the larger, multinucleosomal, "regional" centromeres of most other eukaryotes. Ndc80c and DASH/Dam1c are the principal MT attachment complexes (Fig. 1). Ndc80c is a long (~620-Å) rod with small, globular ends (Ciferri et al. 2005; Wei et al. 2005). The shaft of the rod is largely a parallel, two-chain α-helical coiled-coil. DASH/Dam1c encircles the MT by assembling into a sliding ring (Miranda et al. 2005; Westermann et al. 2005). It is likely that the ring is both a processivity factor and an organizer for the approximately eight Ndc80c rods (Joglekar et al. 2006) that extend from each kinetochore to capture the plus end of a single MT.

Ndc80c

The complex is a heterotetramer (Fig. 2A; Ciferri et al. 2005; Wei et al. 2005). Ndc80 and Nuf2 pair at one end and contribute ~70% of the coiled-coil shaft. An extended segment of about 100 amino acid residues at the amino terminus of Ndc80 and the calponin homology (CH) domain that follows it together create the MT contact (Cheeseman et al. 2006; Wei et al. 2007); a CH domain at the amino terminus of Nuf2 supports the MT contacting module but does not reach the MT surface (Ciferri et al. 2008). Spc24 and Spc25 pair at the other end of the rod, with the same N-to-C polarity as Ndc80:Nuf2, so that the globular tip contains the tightly associated, carboxy-terminal, RWD domains of the two subunits (Wei et al. 2006).

[3]Present address: Relay Therapeutics, Cambridge, Massachusetts 02142

Published by Cold Spring Harbor Laboratory Press; doi: 10.1101/sqb.2017.82.033738

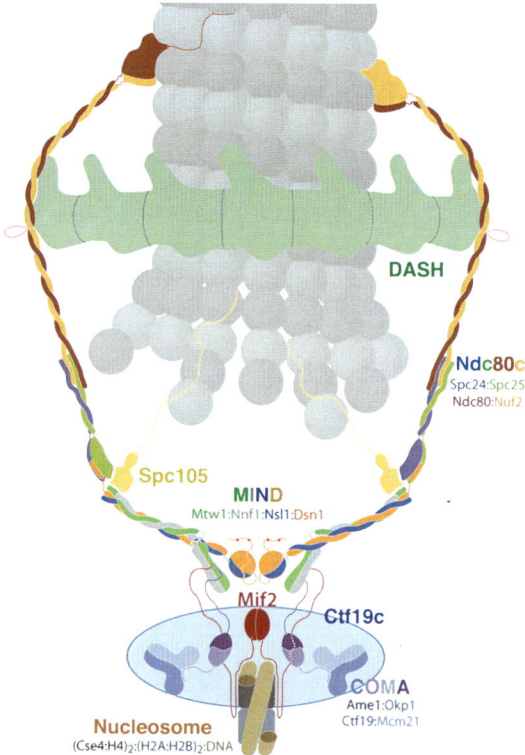

Figure 1. Schematic diagram of the principal molecular components of a yeast kinetochore. This two-dimensional representation does not show relative stoichiometries: Although there is just one centromeric (Cse4-containing) nucleosome and one microtubule, there are approximately four to six MIND complexes and approximately eight Ndc80 complexes, distributed, like the two shown here, around the DASH/Dam1c ring (Joglekar et al. 2006); some of the latter may be recruited by Cnn1, part of the Ctf19 complex, which includes a total of 13 components (Pekgöz Altunkaya et al. 2016). The contacts from Mif2 and Ame1 may be to distinct MIND complexes (accounting for up to four copies of MIND), rather than to the same one, as drawn for simplicity here.

part because of likely interactions with DASH/Dam1c (Maure et al. 2011).

Images of negatively stained Ndc80c suggest a tendency to bend at about 160 Å from the Ndc80:Nuf2 globular tip (Wang et al. 2008). In previous work, this "kink" has been assigned to the loop position, but the analysis here suggests that the mean position of the loop should be at ~270 Å from the tip instead (Fig. 2C). Preferential proteolytic cleavages at residues 380 and 410 in Ndc80 indicate some variation in the strength of coiled-coil interactions along the Ndc80:Nuf2 shaft; the positions of cleavage would be at about 160 Å and 200 Å from the tip, respectively (Wei et al. 2005). Measurements performed on recently published images of rotary-shadowed Ndc80c are more consistent with a somewhat distributed tendency to bend, rather than a uniquely positioned kink (Huis In 't Veld et al. 2016). An additional point of local flexibility is at the head–shaft junction, as indicated by comparison of crystal structures of two different dwarf Ndc80c constructs (Valverde et al. 2016). These properties may allow Ndc80c to bend around the DASH/Dam1c ring, as suggested schematically in Figure 1.

The MT interface of the Ndc80 head includes both its CH domain and amino-terminal extension; the K_d for the human ortholog (Hec1) is ~2 μM. Phosphorylation by Aurora B/Ipl1 or deletion of the extension weakens the interaction, raising the K_d by about an order of magnitude (DeLuca et al. 2006; Wei et al. 2007). Bead and single-molecule tracking experiments have shown that Ndc80c in clusters of two or more diffuses on the surface of a MT lattice (Powers et al. 2009). A biased diffusion mechanism can then account for in vitro measurements of assembly- and disassembly-coupled movement, perhaps reflecting the properties of initial, side-on MT capture by kinetochores, which depends on Ndc80c but not on DASH/Dam1c (Tanaka et al. 2005, 2007).

DASH/Dam1c

The DASH/Dam1c protomer is an assembly of 10 distinct polypeptide chains, all of which are needed to reconstitute the complex in vitro (Li et al. 2002; Miranda et al. 2005). In the presence of MTs, it encircles them by polymerizing into rings of about twice the outer diameter of the MT (Miranda et al. 2005; Westermann et al. 2005). The MT contacts are from extended carboxy-terminal "arms" of two of the subunits, Dam1 and Duo1 (Miranda et al. 2007). Redundancy and flexibility of these contacts can allow the ring to track the end of the MT, by rapidly detaching and reattaching.

Moderate-resolution cryo-EM reconstructions of DASH/Dam1c rings show 15- and 16-protomer assemblies when reconstituted free in solution (at relatively high protein concentration) or around MTs, respectively (Ramey et al. 2011). The protomer has the shape of a thick rod, with a short protrusion at its center. Analysis of the amino acid sequences of the ten subunits and of their conservation among fungal species suggests that the body of the protomer contains largely α-helical regions of the com-

The crystal structure of a "dwarf" Ndc80c, with a large deletion in the coiled-coil shaft of Ndc80:Nuf2 and a somewhat shorter deletion in the shaft of Spc24:Spc25, shows the molecular details of the end-to-end joint between the two coiled-coils (Fig. 2B; Valverde et al. 2016). The transition between a two-chain coiled-coil of Ndc80 and Nuf2 to a four-helix bundle and back to a two-chain coiled-coil of Spc24 and Scp25 includes some irregularities for which the contributing amino acid residues are conserved among point-centromere fungi (Fig. 2B, inset). These idiosyncratic features suggest interaction with another, yet-to-be-determined kinetochore component.

The heptad registers established on both sides of the joint by the dwarf Ndc80c structure have allowed us to predict the structures of the deleted coiled-coiled regions. Direct coupling analysis (DCA), based on ~1200 fungal Ndc80c sequences, yields correlations consistent with heptad counting (Morcos et al. 2011; Ekeberg et al. 2013, 2014). Opposite Nuf2 residue 309 is an insertion in Ndc80, called the "Ndc80 loop" (Fig. 2C; Maiolica et al. 2007). Deletion of loop residues impairs end-on MT attachment but does not affect MT side binding, at least in

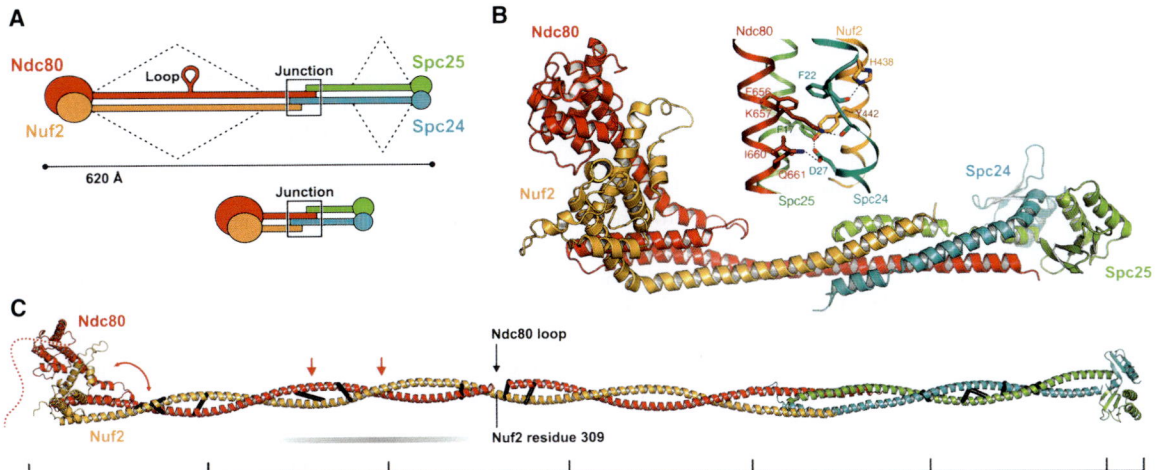

Figure 2. Structure of Ndc80c. (*A*) Diagram of Ndc80c subunit organization. Dashed lines show deletions used to create a dwarf Ndc80c that retains the tetramer junction. (*B*) Structure of dwarf Ndc80c (*Saccharomyces cerevisiae*) (Valverde et al. 2016). *Inset* shows detail of the junction region, with residues conserved in point-centromere yeast shown as sticks on the ribbon backbone. (*C*) Model of complete Ndc80c structure, based on the structure of dwarf Ndc80c, heptad counting across the deleted coiled-coil segments, and inferences from published cross-linking data for the human ortholog (Maiolica et al. 2007). Results of direct-coupling analysis (see text) are shown as heavy black lines. The scale bar below corresponds to 620 Å, with 100-Å intervals marked. Dotted red curve, amino-terminal extension of Ndc80; red arrowheads, preferential cleavage points in Ndc80 (Wei et al. 2005); curved, double-headed arrow, region of hinge-like joint indicated by comparison of two different dwarf constructs; and gray bar, approximate distribution of bends in published rotary-shadowed images (Huis In't Veld et al. 2016: Fig. 6, figure supplement 2) of human Ndc80c (140 particles; 72 with visually identifiable kink or bend; 38 unbent; 30 poorly contrasted or not measurable; mean position of bend ~200 Å from one end). (*A,B*, Modified, with permission, from Valverde et al. 2016.)

ponents, with long, potentially flexible extensions at the carboxy-terminal ends of Dam1, Duo1, and Ask1, consistent with their protease sensitivity, and shorter extensions at the amino termini of many of the subunits (Fig. 3).

Chemical cross-linking of Ndc80 with DASH/Dam1c, analyzed by mass spectrometry, has defined three sets of cross-linked contact points (Kim et al. 2017). Residues near the carboxyl terminus of Dam1 cross-link with residues in the helical hairpin connecting the CH domain of Nc80 with the coiled-coil shaft; residues in the carboxy-terminal half of Ask1 cross-link with residues in Ndc80 between the helical hairpin and the loop; and residues in the carboxy-terminal half of Spc34 cross-link with residues in Ndc80 near the junction with Spc24:Spc25. These positions on Ndc80 span ~300 Å along the coiled-coil shaft. In vivo fluorescence resonance energy transfer (FRET) analysis also places the carboxyl terminus of Dam1 close to the globular end of Ndc80:Nuf2 (Arava-

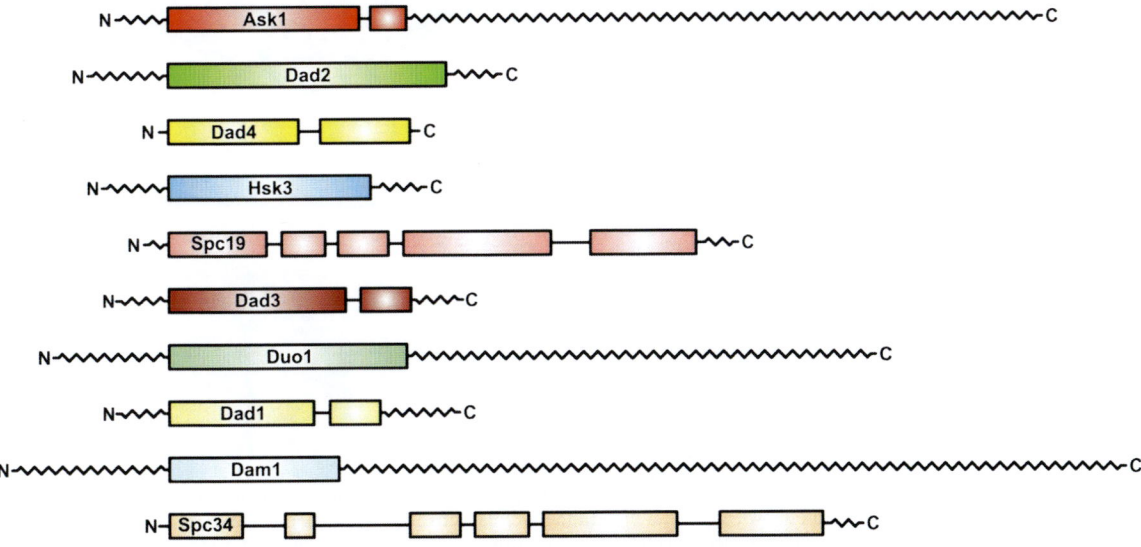

Figure 3. DASH/Dam1c subunits. Boxes show regions of predicted fold, largely α-helical; zig-zags show prediction of flexible amino- and carboxy-terminal extensions.

mudhan et al. 2014). Moreover, the extensions of Dam1, Duo1, and Ask1 are long enough (Fig. 3) that flexible projections from even a single, relatively compact ring could span the required distance, so that all the cross-links could originate from concurrent contacts by the same ring and even by the same protomers in that ring. Indeed, the reported amounts per cell of several of the 10 DASH/Dam1c components are enough for just one 15- or 16-protomer ring per kinetochore MT (Ghaemmaghami et al. 2003).

A stable transition from side-on to end-on MT attachment requires DASH/Dam1c (Tanaka et al. 2007). The geometry of end-on attachment couples the curled ends of disassembling protofilaments with the position of the attached kinetochore (Asbury et al. 2006), and MT depolymerization, rather than motor activity, can indeed drive poleward motion of chromosomes in anaphase (Koshland et al. 1988; Grishchuk and McIntosh 2006; Tytell and Sorger 2006). The diameter of the DASH ring will prevent it from pulling off the MT, and recent laser-trap experiments have shown that the curled protofilaments have enough elastic strength to exert the required force (Driver et al. 2017). Flexible connections from the carboxy-terminal extensions of Dam1 and Duo1 will facilitate biased motion in response to MT depolymerization, as protofilament curling will experience relatively little resistance. Moreover, because an individual arm can reach multiple tubulin binding sites, dissociation and reassociation of just one or a few of these contacts will allow the ring (and the attached chromosome) to translocate (Miranda et al. 2007).

DASH/Dam1c is present primarily in fungi, but also in representative species of other eukaryotic taxa (van Hooff et al. 2017). Most of the species, including nearly all metazoans, that lack DASH/Dam1c have the three-subunit Ska complex instead, and very few show evidence of both. Tracking the two complexes back along the evolutionary tree appears to favor a model in which Ska was present in the last eukaryotic common ancestor and that DASH/Dam1c appeared in an ancestral fungal species, displacing Ska and spreading to certain other eukaryotes by lateral transfer (van Hooff et al. 2017). The authors of the evolutionary analysis have suggested that the two structurally unrelated complexes are functional analogs and that both coordinate Ndc80c attachment to spindle MTs.

ADAPTORS

Two adaptor complexes—MIND[MIS12c] and Cnn1: Wip1[CENP-T:CENP-W]—associate with the Spc24:Spc25 tip of Ndc80c and connect the MT attachment assembly with chromatin-associated complexes (Hori et al. 2008; Malvezzi et al. 2013). The former is essential in yeast; the latter is not. Thus, Cnn1:Wip1, presumably with associated Ndc80 complexes, may enhance kinetochore strength or connectivity but probably does not modify its fundamental molecular organization. The two adaptors bind Spc24:Spc25 through related motifs—at the carboxy-terminal end of the MIND subunit Dsn1 and near the amino-terminal end of Cnn1 (Bock et al. 2012; Malvezzi et al. 2013; Dimitrova et al. 2016; Petrovic et al. 2016). Mps1

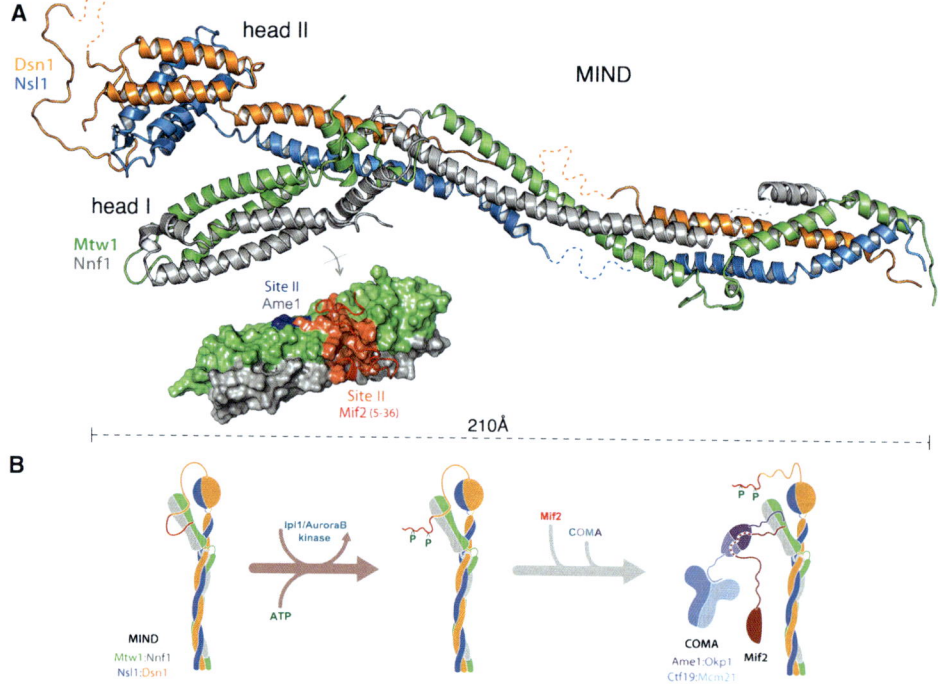

Figure 4. Structure and regulation of MIND. (*A*) Ribbon diagram of the MIND heterotetramer with surface representation of head I. Positions of Mif2 and Ame1 binding are shown as patches of red and blue, respectively; the backbone of Mif2 (residues 5–36) is shown in its bound conformation. (*B*) Scheme showing mechanism by which phosphorylation of Dsn1 relieves MIND autoinhibition, allowing Mif2 and Ame1 to interact with head I. (Modified, with permission, from Dimitrova et al. 2016; original structure generated with PyMOL [http://www.pymol.org].)

phosphorylation of a serine residue in the binding segment of Cnn1 blocks the interaction (Thapa et al. 2015); serine residues in the related Dsn1 segment are potential sites of phosphoregulation, but a role for their modification, if any, has yet to be identified (Dimitrova et al. 2016).

MIND is a 210-Å-long, heterotetrameric rod, with two globular, four-helix bundles ("heads") at one end (Dimitrova et al. 2016; Petrovic et al. 2016). Mtw1[MIS12] and Nnf1[PMF1] contribute to "head 1"; Dsn1 and Nsl1, to "head 2." Head 1 binds projecting segments from two different, chromosome-proximal kinetochore components—Mif2[CENP-C] and Ame1[CENP-U] (Fig. 4). Both these interactions are sensitive to phosphorylation of residues in a Dsn1 amino-terminal extension that projects from head 2 (Akiyoshi et al. 2013). When not phosphorylated, this segment of Dsn1 binds head 1 and blocks the sites for Mif2 and Ame1 (Dimitrova et al. 2016; Petrovic et al. 2016). Phosphorylation at two positions (serines 240 and 250 in *Saccharomyces cerevisiae*) by Ipl1[AuroraB] reverses the interaction and allows Mif2 and Ame1 to bind instead (Fig. 4B).

Ipl1 also phosphorylates the amino-terminal regions of Ndc80 and Dam1, presumably to promote their dissociation from the MT surface when sister kinetochores are not under tension from bi-orientation (Biggins et al. 1999; Cheeseman et al. 2002; Akiyoshi et al. 2009). At the level of molecular mechanism, this activity resembles the action of Ipl1 on Dsn1 by enhancing dissociation of the modified segment from a target site. At the level of the larger assembly, the outcome is different: Ipl1 promotes MIND attachment to chromosome-proximal components by reversing an autoinhibition, but it promotes Ndc80c and DASH/Dam1c detachment from MTs in the absence of tension (Tanaka et al. 2002; Kalantzaki et al. 2015). The specific properties of the Ndc80c–DASH/Dam1c assembly that allow it to respond to tension and protect the amino-terminal segments of Ndc80 and Dam1 from phosphorylation, either by increasing the lifetimes of their bound states or by shielding them from Ipl1, remain to be determined. The same or related structural properties might also account for the observation that tension stabilizes attachment independently of Ipl1 (Akiyoshi et al. 2010).

INTERCOMPLEX CONNECTIONS

The various complexes just described all have a conformationally well-defined core with flexibly projecting extensions, often from one or both termini of a subunit. The extensions link the complexes into larger substructures by docking into specific binding sites. These "docked-peptide" interactions (see Table 1) have functionally important properties: (1) adaptability of the contact to alternative conformational contexts, (2) mechanical compliance of the larger assembly, (3) ease of regulation by phosphorylation or other modification, and (4) potential for intervention by a disassembly-inducing modification.

1. Adaptability is the most general of these properties. In an assembly composed of multiple copies of a structure unit, but without strict symmetry, the relative orientations of connected components are necessarily variable.

Table 1. Docked-peptide interactions between yeast kinetochore subcomplexes discussed in this article

Donor (docked peptide)	Acceptor (docking site)
Mif2 (N-term[a])	MIND (head 1)
COMA: Ame1	MIND (head 1)
MIND: Dsn1 (N-term[a])	MIND (head 1)
MIND: Dsn1 (C-term)	Ndc80c (Spc24:Spc25)
Cnn1 (N-term[a])	Ndc80c (Spc24:Spc25)
DASH: Dam1 (C-term[a])	MT
DASH: Duo1 (C-term)	MT
Ndc80c: Ndc80 (N-term[a])	MT

[a]Shown to be regulated by phosphorylation of peptide.
N-term, amino-terminal; C-term, carboxy-terminal; MT, microtubule.

Even the stoichiometry of some of the kinetochore components may vary. Cnn1 recruitment seems to vary with cell cycle timing (Bock et al. 2012), and therefore the number of Ndc80 complexes probably does also. Moreover, redundancy of contacts implies that not all opportunities for recruitment need be fulfilled.

2. Mechanical compliance could in principle account for some of the kinetochore's apparent tension-sensing properties. Tension can be detected by the extent to which it displaces one part of an assembly with respect to another. The yeast kinetochore appears to be an open, cage-like structure (Akiyoshi et al. 2010); flexible connections between its rod-like members (especially Ndc80c and MIND) imply that the cage can distort (e.g., elongate or contract) without major changes within the individual subcomplexes (Fig. 5). It remains to be determined whether this property is indeed part of kinetochore-signaling mechanisms.

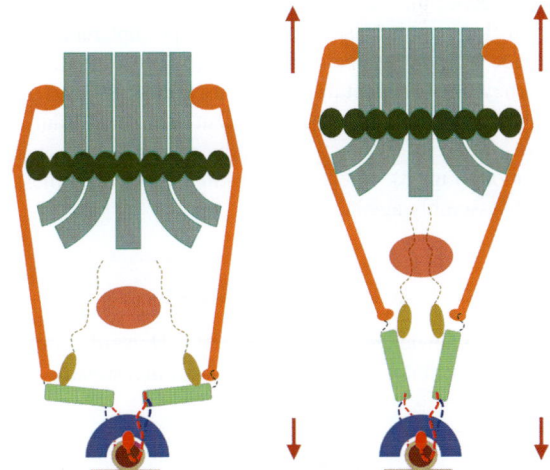

Figure 5. Example of how a kinetochore might respond to tension. (*Left*) Relaxed structure in the absence of tension. (*Right*) Extended structure in the presence of tension. Key hinge points are at docked-peptide interfaces between subcomplexes and (perhaps) at positions of preferential bending within Ndc80c (see Fig. 2). Colors of components correspond approximately to those in Figure 1; the red oval in the center symbolizes that tension can in principle affect the spindle-assembly checkpoint proteins, which associate (directly or indirectly) with Mps1-phosphorylated repeats on Spc105 (Aravamudhan et al. 2016).

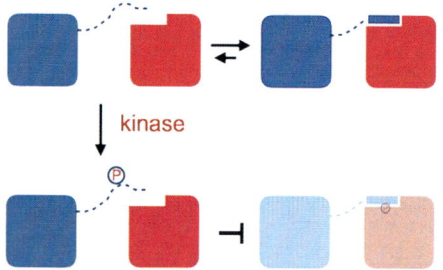

Figure 6. Docked-peptide association and reversal by phosphorylation. A flexible peptide (e.g., at the amino- or carboxyl terminus) extended from a stably folded core docks into a target site on another protein (*top*). When transiently dissociated, capture by a kinase phosphorylates a target residue, blocking reassociation (*bottom*).

3. Regulation by modification requires a segment of unfolded and extended polypeptide chain, because of the active site properties of the modifying enzymes. Protein kinases specify, by the characteristics of their substrate docking site, the sequence context of the residue to be phosphorylated (e.g., Moore et al. 2003). Switching of docked-peptide contacts by phosphorylation is therefore a familiar mode of regulation, either within a multidomain protein (e.g., the carboxy-terminal tail of c-Src) or between members of a multicomponent assembly (e.g., the amino-terminal region of Dsn1).

4. A disassembly inducing modification requires a transient dissociation ("breathing"), so that the modifier can access a residue that participates in the contact. The buried surface in a docked peptide contact is relatively modest, usually resulting in equilibrium dissociation constants of between 1 and 10 μM and a residence time of ~1 sec. In a kinetochore, redundancy in the organized structure ensures prompt rebinding, but a suitably localized kinase (or other modifying enzyme) can capture and modify the peptide (e.g., the amino-terminal region of Dsn1 or Ndc80) during its transient dissociation, thereby preventing reassociation (Fig. 6). The localized kinase then becomes a disassembly agent.

ACKNOWLEDGMENTS

We acknowledge support from the Howard Hughes Medical Institute, in which S.C.H. is an Investigator.

REFERENCES

Akiyoshi B, Nelson CR, Ranish JA, Biggins S. 2009. Analysis of Ipl1-mediated phosphorylation of the Ndc80 kinetochore protein in *Saccharomyces cerevisiae*. *Genetics* **183:** 1591–1595.

Akiyoshi B, Sarangapani KK, Powers AF, Nelson CR, Reichow SL, Arellano-Santoyo H, Gonen T, Ranish JA, Asbury CL, Biggins S. 2010. Tension directly stabilizes reconstituted kinetochore-microtubule attachments. *Nature* **468:** 576–579.

Akiyoshi B, Nelson CR, Biggins S. 2013. The aurora B kinase promotes inner and outer kinetochore interactions in budding yeast. *Genetics* **194:** 785–789.

Aravamudhan P, Felzer-Kim I, Gurunathan K, Joglekar AP. 2014. Assembling the protein architecture of the budding yeast kinetochore-microtubule attachment using FRET. *Curr Biol* **24:** 1437–1446.

Aravamudhan P, Chen R, Roy B, Sim J, Joglekar AP. 2016. Dual mechanisms regulate the recruitment of spindle assembly checkpoint proteins to the budding yeast kinetochore. *Mol Biol Cell* **27:** 3405–3417.

Asbury CL, Gestaut DR, Powers AF, Franck AD, Davis TN. 2006. The Dam1 kinetochore complex harnesses microtubule dynamics to produce force and movement. *Proc Natl Acad Sci* **103:** 9873–9878.

Biggins S, Severin FF, Bhalla N, Sassoon I, Hyman AA, Murray AW. 1999. The conserved protein kinase Ipl1 regulates microtubule binding to kinetochores in budding yeast. *Genes Dev* **13:** 532–544.

Bock LJ, Pagliuca C, Kobayashi N, Grove RA, Oku Y, Shrestha K, Alfieri C, Golfieri C, Oldani A, Dal Maschio M, et al. 2012. Cnn1 inhibits the interactions between the KMN complexes of the yeast kinetochore. *Nat Cell Biol* **14:** 614–624.

Cheeseman IM, Anderson S, Jwa M, Green EM, Kang J, Yates JR III, Chan CS, Drubin DG, Barnes G. 2002. Phospho-regulation of kinetochore-microtubule attachments by the Aurora kinase Ipl1p. *Cell* **111:** 163–172.

Cheeseman IM, Chappie JS, Wilson-Kubalek EM, Desai A. 2006. The conserved KMN network constitutes the core microtubule-binding site of the kinetochore. *Cell* **127:** 983–997.

Ciferri C, De Luca J, Monzani S, Ferrari KJ, Ristic D, Wyman C, Stark H, Kilmartin J, Salmon ED, Musacchio A. 2005. Architecture of the human ndc80-hec1 complex, a critical constituent of the outer kinetochore. *J Biol Chem* **280:** 29088–29095.

Ciferri C, Pasqualato S, Screpanti E, Varetti G, Santaguida S, Dos Reis G, Maiolica A, Polka J, De Luca JG, De Wulf P, et al. 2008. Implications for kinetochore-microtubule attachment from the structure of an engineered Ndc80 complex. *Cell* **133:** 427–439.

Clarke L, Carbon J. 1980. Isolation of a yeast centromere and construction of functional small circular chromosomes. *Nature* **287:** 504–509.

DeLuca JG, Gall WE, Ciferri C, Cimini D, Musacchio A, Salmon ED. 2006. Kinetochore microtubule dynamics and attachment stability are regulated by Hec1. *Cell* **127:** 969–982.

De Wulf P, McAinsh AD, Sorger PK. 2003. Hierarchical assembly of the budding yeast kinetochore from multiple subcomplexes. *Genes Dev* **17:** 2902–2921.

Dimitrova YN, Jenni S, Valverde R, Khin Y, Harrison SC. 2016. Structure of the MIND complex defines a regulatory focus for yeast kinetochore assembly. *Cell* **167:** 1014–1027 e1012.

Driver JW, Geyer EA, Bailey ME, Rice LM, Asbury CL. 2017. Direct measurement of conformational strain energy in protofilaments curling outward from disassembling microtubule tips. *Elife* **6:** e28433.

Ekeberg M, Lovkvist C, Lan Y, Weigt M, Aurell E. 2013. Improved contact prediction in proteins: Using pseudolikelihoods to infer Potts models. *Phys Rev E Stat Nonlin Soft Matter Phys* **87:** 012707.

Ekeberg M, Hartonen T, Aurell E. 2014. Fast pseudolikelihood maximization for direct-coupling analysis of protein structure from many homologous amino-acid sequences. *J Comput Phys* **276:** 341–356.

Ghaemmaghami S, Huh WK, Bower K, Howson RW, Belle A, Dephoure N, O'Shea EK, Weissman JS. 2003. Global analysis of protein expression in yeast. *Nature* **425:** 737–741.

Grishchuk EL, McIntosh JR. 2006. Microtubule depolymerization can drive poleward chromosome motion in fission yeast. *EMBO J* **25:** 4888–4896.

Hinshaw SM, Harrison SC. 2017. Kinetochore function from the bottom up. *Trends Cell Biol.* doi: 10.1016/j.tcb.2017.09.002.

Hori T, Amano M, Suzuki A, Backer CB, Welburn JP, Dong Y, McEwen BF, Shang WH, Suzuki E, Okawa K, et al. 2008. CCAN makes multiple contacts with centromeric DNA to provide distinct pathways to the outer kinetochore. *Cell* **135:** 1039–1052.

Huis In 't Veld PJ, Jeganathan S, Petrovic A, Singh P, John J, Krenn V, Weissmann F, Bange T, Musacchio A. 2016. Molecular basis of outer kinetochore assembly on CENP-T. *Elife* **5:** e21007.

Joglekar AP, Bouck DC, Molk JN, Bloom KS, Salmon ED. 2006. Molecular architecture of a kinetochore-microtubule attachment site. *Nat Cell Biol* **8:** 581–585.

Kalantzaki M, Kitamura E, Zhang T, Mino A, Novak B, Tanaka TU. 2015. Kinetochore-microtubule error correction is driven by differentially regulated interaction modes. *Nat Cell Biol* **17:** 530.

Kim JO, Zelter A, Umbreit NT, Bollozos A, Riffle M, Johnson R, MacCoss MJ, Asbury CL, Davis TN. 2017. The Ndc80 complex bridges two Dam1 complex rings. *Elife* **6:** e21069.

Koshland DE, Mitchison TJ, Kirschner MW. 1988. Polewards chromosome movement driven by microtubule depolymerization in vitro. *Nature* **331:** 499–504.

Lechner J, Carbon J. 1991. A 240 kd multisubunit protein complex, CBF3, is a major component of the budding yeast centromere. *Cell* **64:** 717–725.

Li Y, Bachant J, Alcasabas AA, Wang Y, Qin J, Elledge SJ. 2002. The mitotic spindle is required for loading of the DASH complex onto the kinetochore. *Genes Dev* **16:** 183–197.

Maiolica A, Cittaro D, Borsotti D, Sennels L, Ciferri C, Tarricone C, Musacchio A, Rappsilber J. 2007. Structural analysis of multiprotein complexes by cross-linking, mass spectrometry, and database searching. *Mol Cell Proteomics* **6:** 2200–2211.

Malvezzi F, Litos G, Schleiffer A, Heuck A, Mechtler K, Clausen T, Westermann S. 2013. A structural basis for kinetochore recruitment of the Ndc80 complex via two distinct centromere receptors. *EMBO J* **32:** 409–423.

Maure JF, Komoto S, Oku Y, Mino A, Pasqualato S, Natsume K, Clayton L, Musacchio A, Tanaka TU. 2011. The Ndc80 loop region facilitates formation of kinetochore attachment to the dynamic microtubule plus end. *Curr Biol* **21:** 207–213.

Meluh PB, Yang P, Glowczewski L, Koshland D, Smith MM. 1998. Cse4p is a component of the core centromere of *Saccharomyces cerevisiae*. *Cell* **94:** 607–613.

Miranda JJ, De Wulf P, Sorger PK, Harrison SC. 2005. The yeast DASH complex forms closed rings on microtubules. *Nat Struct Mol Biol* **12:** 138–143.

Miranda JJ, King DS, Harrison SC. 2007. Protein arms in the kinetochore-microtubule interface of the yeast DASH complex. *Mol Biol Cell* **18:** 2503–2510.

Moore MJ, Adams JA, Taylor SS. 2003. Structural basis for peptide binding in protein kinase A. Role of glutamic acid 203 and tyrosine 204 in the peptide-positioning loop. *J Biol Chem* **278:** 10613–10618.

Morcos F, Pagnani A, Lunt B, Bertolino A, Marks DS, Sander C, Zecchina R, Onuchic JN, Hwa T, Weigt M. 2011. Direct-coupling analysis of residue coevolution captures native contacts across many protein families. *Proc Natl Acad Sci* **108:** E1293–E1301.

Musacchio A, Desai A. 2017. A molecular view of kinetochore assembly and function. *Biology (Basel)* **6:** E5.

Pekgöz Altunkaya G, Malvezzi F, Demianova Z, Zimniak T, Litos G, Weissmann F, Mechtler K, Herzog F, Westermann S. 2016. CCAN assembly configures composite binding interfaces to promote cross-linking of Ndc80 complexes at the kinetochore. *Curr Biol* **26:** 2370–2378.

Petrovic A, Keller J, Liu Y, Overlack K, John J, Dimitrova YN, Jenni S, van Gerwen S, Stege P, Wohlgemuth S, et al. 2016. Structure of the MIS12 complex and molecular basis of its interaction with CENP-C at human kinetochores. *Cell* **167:** 1028–1040.

Powers AF, Franck AD, Gestaut DR, Cooper J, Gracyzk B, Wei RR, Wordeman L, Davis TN, Asbury CL. 2009. The Ndc80 kinetochore complex forms load-bearing attachments to dynamic microtubule tips via biased diffusion. *Cell* **136:** 865–875.

Ramey VH, Wong A, Fang J, Howes S, Barnes G, Nogales E. 2011. Subunit organization in the Dam1 kinetochore complex and its ring around microtubules. *Mol Biol Cell* **22:** 4335–4342.

Tanaka TU, Rachidi N, Janke C, Pereira G, Galova M, Schiebel E, Stark MJ, Nasmyth K. 2002. Evidence that the Ipl1-Sli15 (Aurora kinase-INCENP) complex promotes chromosome bi-orientation by altering kinetochore-spindle pole connections. *Cell* **108:** 317–329.

Tanaka TU, Stark MJ, Tanaka K. 2005. Kinetochore capture and bi-orientation on the mitotic spindle. *Nat Rev Mol Cell Biol* **6:** 929–942.

Tanaka K, Kitamura E, Kitamura Y, Tanaka TU. 2007. Molecular mechanisms of microtubule-dependent kinetochore transport toward spindle poles. *J Cell Biol* **178:** 269–281.

Thapa KS, Oldani A, Pagliuca C, De Wulf P, Hazbun TR. 2015. The Mps1 kinase modulates the recruitment and activity of Cnn1(CENP-T) at *Saccharomyces cerevisiae* kinetochores. *Genetics* **200:** 79–90.

Tytell JD, Sorger PK. 2006. Analysis of kinesin motor function at budding yeast kinetochores. *J Cell Biol* **172:** 861–874.

Valverde R, Ingram J, Harrison SC. 2016. Conserved tetramer junction in the kinetochore Ndc80 complex. *Cell Rep* **17:** 1915–1922.

van Hooff JJE, Snel B, Kops G. 2017. Unique phylogenetic distributions of the Ska and Dam1 complexes support functional analogy and suggest multiple parallel displacements of Ska by Dam1. *Genome Biol Evol* **9:** 1295–1303.

Wang HW, Long S, Ciferri C, Westermann S, Drubin D, Barnes G, Nogales E. 2008. Architecture and flexibility of the yeast Ndc80 kinetochore complex. *J Mol Biol* **383:** 894–903.

Wei RR, Sorger PK, Harrison SC. 2005. Molecular organization of the Ndc80 complex, an essential kinetochore component. *Proc Natl Acad Sci* **102:** 5363–5367.

Wei RR, Schnell JR, Larsen NA, Sorger PK, Chou JJ, Harrison SC. 2006. Structure of a central component of the yeast kinetochore: The Spc24p/Spc25p globular domain. *Structure* **14:** 1003–1009.

Wei RR, Al-Bassam J, Harrison SC. 2007. The Ndc80/HEC1 complex is a contact point for kinetochore-microtubule attachment. *Nat Struct Mol Biol* **14:** 54–59.

Westermann S, Avila-Sakar A, Wang HW, Niederstrasser H, Wong J, Drubin DG, Nogales E, Barnes G. 2005. Formation of a dynamic kinetochore-microtubule interface through assembly of the Dam1 ring complex. *Mol Cell* **17:** 277–290.

Winey M, Mamay CL, O'Toole ET, Mastronarde DN, Giddings TH Jr, McDonald KL, McIntosh JR. 1995. Three-dimensional ultrastructural analysis of the *Saccharomyces cerevisiae* mitotic spindle. *J Cell Biol* **129:** 1601–1615.

Aurora A Kinase Function at Kinetochores

JENNIFER G. DELUCA

*Department of Biochemistry and Molecular Biology, Colorado State University, Fort Collins,
Colorado 80523-1870*

Correspondence: jdeluca@colostate.edu

One of the most important regulatory aspects of chromosome segregation is the ability of kinetochores to precisely control their attachment strength to spindle microtubules. Central to this regulation is Aurora B, a mitotic kinase that phosphorylates kinetochore substrates to promote microtubule turnover. A critical target of Aurora B is the kinetochore protein Ndc80/Hec1, which is a component of the NDC80 complex, the primary force-transducing link between kinetochores and microtubules. Although Aurora B is regarded as the "master regulator" of kinetochore–microtubule attachment, it is becoming clear that this kinase is not solely responsible for phosphorylating Hec1 and other kinetochore substrates to facilitate microtubule turnover. In particular, there is growing evidence that Aurora A kinase, whose activities at spindle poles have been extensively described, has additional roles at kinetochores in regulating the kinetochore–microtubule interface.

REGULATION OF KINETOCHORE–MICROTUBULE ATTACHMENT STABILITY

Kinetochores are large protein structures assembled on centromeric chromatin that power and regulate chromosome segregation during mitosis. These orchestrators of mitosis physically connect chromosomes to spindle microtubules and transduce forces through the connections to align and segregate chromosomes. Successful mitosis requires that kinetochores precisely regulate their attachment strength to microtubules. In early mitosis, kinetochores ensure that attachments are labile and short-lived so that improper attachments are corrected, whereas in late mitosis, kinetochores form stable, persistent attachments to microtubules so that forces can be generated to drive chromosome movements and to silence the spindle assembly checkpoint (Funabiki and Wynne 2013; Godek et al. 2015; Lampson and Grishchuk 2017; Musacchio and Desai 2017). Central to this regulation is Aurora B, a mitotic kinase that phosphorylates kinetochore substrates to promote microtubule turnover (Carmena et al. 2012; Krenn and Musacchio 2015; Hindriksen et al. 2017). A key Aurora B target involved in this regulation is the Hec1 protein of the kinetochore-associated NDC80 complex, a primary contributor to the generation of stable, end-on attachments to spindle microtubules. Hec1 is phosphorylated by Aurora B kinase on as many as nine target sites within its unstructured "tail" domain to tune the affinity of kinetochores for microtubules throughout mitosis (Cheeseman et al. 2006; DeLuca et al. 2006, 2011; Alushin et al. 2012; Zaytsev et al. 2014, 2015). High levels of phosphorylation promote kinetochore–microtubule turnover, which facilitates attachment error correction in early mitosis, whereas low levels of Hec1 phosphorylation result in stabilized attachments in late mitosis. Hec1 is arguably a main effector of Aurora B kinase-mediated regulation of attach-ment stability, however additional kinetochore proteins (e.g., Dsn1, Knl1, Ska1, Ska3, and CENP-E) are phosphorylated by Aurora B, which likely contributes to kinetochore–microtubule attachment regulation (Kim et al. 2010; Welburn et al. 2010; Chan et al. 2012).

Although Aurora B is well-established as the "master regulator" of kinetochore–microtubule attachment stability, there is growing evidence that it does not work alone to phosphorylate kinetochore substrates to promote microtubule turnover. A number of recent studies suggest that Aurora A kinase, which has well-known functions at spindle poles in mitosis, additionally plays a role at the kinetochore–microtubule interface to ensure proper regulation of attachments during mitotic progression. How the division of labor between Aurora A and Aurora B is coordinated to ensure proper spatial and temporal control of kinetochore–microtubule attachment stability is an important question that has gained attention in recent years.

THE AURORA KINASE FAMILY

The Aurora kinases are a highly conserved family of serine/threonine kinases that have essential functions in organisms across evolution. Budding and fission yeast have just one Aurora kinase (Ipl1 and Ark1, respectively), whereas mammals typically have three family members: Aurora A, B, and C (Chan and Botstein 1993; Bischoff and Plowman 1999; Giet and Prigent 1999; Adams et al. 2001; Petersen et al. 2001). In humans, Aurora A and Aurora B have essential functions in mitotic cell division (Bolanos-Garcia 2005; Barr and Gergely 2007; Vader and Lens 2008; Krenn and Musacchio 2015). The third human family member, Aurora C, plays a role in meiotic cell division, as well as the first mitotic divisions of the mammalian zygote (Fernandez-Miranda et al. 2011; Yang et al.

Published by Cold Spring Harbor Laboratory Press; doi: 10.1101/sqb.2017.82.034991

2015; Nguyen and Schindler 2017). The Aurora kinases share ~70% homology in their catalytic domains, although their localization patterns and substrates are distinct, resulting in unique roles during cell division (Carmena et al. 2009; Nguyen and Schindler 2017).

Aurora A Kinase

Aurora A localizes primarily to spindle poles throughout mitosis and has roles in centrosome maturation, centrosome separation, and bipolar spindle assembly (Katayama et al. 2003; Crane et al. 2004; Ducat and Zheng 2004; Barr and Gergely 2007). Activation of Aurora A is promoted by interaction with its well-characterized cofactor TPX2, a microtubule binding protein that localizes to spindle poles and microtubules (Kufer et al. 2002; Bayliss et al. 2003; Eyers and Maller 2003, 2004; Ozlu et al. 2005). Although TPX2 is considered the predominant cofactor for Aurora A, other proteins also participate in its activation including Ajuba, Bora, Pak2, and Inhibitor-2 (Hirota et al. 2003; Satinover et al. 2004; Zhao et al. 2005; Hutterer et al. 2006; Seki et al. 2008; Carmena et al. 2009). The ability of Aurora A to carry out its functions in centrosome separation and spindle assembly is facilitated through its targeting to these locations by binding partners (e.g., to spindle microtubules by TPX2; to centrosomes by Cep192) (Kufer et al. 2002; Glover 2003; Joukov et al. 2010; Bertolin et al. 2016). Although the majority of studies regarding Aurora A have focused on its functions in centrosome separation and spindle assembly, there is growing evidence that it has additional roles in facilitating successful chromosome alignment and segregation, which will be the focus of this review.

Aurora B Kinase

Aurora B localizes to the centromere/kinetochore region of mitotic chromosomes from late prophase to metaphase and then relocates to the spindle midzone at anaphase onset (Earnshaw and Bernat 1991; Bischoff and Plowman 1999; Giet and Prigent 1999). All functions of Aurora B studied to date require its incorporation into the chromosome passenger complex (CPC), which additionally contains INCENP, Survivin, and Borealin (Adams et al. 2001; Wheatley et al. 2001; Bolton et al. 2002; Honda et al. 2003; Gassmann et al. 2004; Vader et al. 2006). The carboxyl terminus of INCENP binds to and increases the catalytic activity of Aurora B, whereas Survivin, Borealin, and the amino terminus of INCENP form a module that facilitates localization of the entire CPC to centromeres in early mitosis and to the spindle midzone in late mitosis (Bishop and Schumacher 2002; Honda et al. 2003; Yasui et al. 2004; Sessa et al. 2005; Klein et al. 2006; Vader et al. 2006; Jeyaprakash et al. 2007). In early mitosis, Aurora B has roles in kinetochore assembly, activation of the spindle assembly checkpoint, and kinetochore–microtubule error correction (Biggins et al. 1999; Kallio et al. 2002; Murata-Hori et al. 2002; Murata-Hori and Wang 2002; Tanaka

et al. 2002; Carvalho et al. 2003; Hauf et al. 2003; Lens et al. 2003; Liu et al. 2006; Santaguida et al. 2011). To execute its error correction function, Aurora B phosphorylates kinetochore substrates, including the microtubule attachment factor Hec1, to promote kinetochore–microtubule turnover (Cheeseman et al. 2006; DeLuca et al. 2006; Welburn et al. 2010). Rapid turnover of kinetochore microtubules in early mitosis, facilitated through high levels of Hec1 phosphorylation, prevents the accumulation of erroneous kinetochore–microtubule attachments during mitotic progression. Although Aurora B kinase-mediated phosphorylation of Hec1 markedly decreases by metaphase, some level of phosphorylation of Hec1 is required in this later stage of mitosis for kinetochores to fluidly track the growing and shortening ends of attached microtubules (DeLuca et al. 2011; Zaytsev et al. 2014). At anaphase onset, Aurora B relocates to the spindle midzone where it functions in stabilization of the central spindle and regulation of cytokinesis (Cooke et al. 1987; Eckley et al. 1997; Schumacher et al. 1998; Tatsuka et al. 1998; Terada et al. 1998; Bischoff and Plowman 1999; Adams et al. 2000; Kaitna et al. 2000; Giet and Glover 2001). For further discussion of Aurora B kinase and its mitotic functions, the reader is directed to recent reviews (Carmena et al. 2012; van der Horst and Lens 2014; Krenn and Musacchio 2015; Afonso et al. 2017).

Aurora C Kinase

Aurora C is primarily expressed in germ cells and participates in the regulation of chromosome–microtubule interactions during meiosis and mitotic divisions of the early embryo (Tang et al. 2006; Sharif et al. 2010; Yang et al. 2010; Fernandez-Miranda et al. 2011; Schindler et al. 2012; Balboula and Schindler 2014). Meiotic oocytes contain two forms of the CPC: one containing Aurora B and one containing Aurora C. Based on early studies demonstrating that Aurora C could compensate for mitotic Aurora B functions, Aurora C was presumed to be the meiotic homolog of Aurora B (Sasai et al. 2004). However, recent studies have suggested nonredundant functions of the kinases and also showed that the two kinases localize to distinct structures in meiotic cells. Aurora B predominantly resides on kinetochores, where it contributes to chromosome alignment, and Aurora C localizes to spindle poles, centromeres, and the interchromatid axis, where it is reported to have roles in spindle assembly and in correction of kinetochore–microtubule attachment errors (Shuda et al. 2009; Balboula and Schindler 2014; Nguyen et al. 2014; Fellmeth et al. 2015; Balboula et al. 2016; Sasai et al. 2016; Quartuccio et al. 2017). It is also noteworthy that although Aurora C is primarily expressed in germ line cells, it is also expressed in human tumor cells, and its overexpression in noncancerous cells can result in cell transformation (Ehara et al. 2003; Dutertre et al. 2005; Ulisse et al. 2006; Khan et al. 2011; Tsou et al. 2011). Discussions of Aurora C kinase and its functions can be found in recent reviews (Yang et al. 2015; Nguyen and Schindler 2017).

KINETOCHORE FUNCTIONS OF AURORA A KINASE

A little over 15 years ago, two studies reported a role for Aurora A kinase in chromosome alignment, independent of its role in spindle pole separation. Kunitoku et al. (2003) found that a small population of human cells depleted of Aurora A by RNAi managed to enter mitosis and form bipolar spindles. These cells experienced defects in chromosome alignment, with most mitotic cells exhibiting pole-localized chromosomes. Similarly, Marumoto et al. (2003) reported chromosome alignment defects after microinjection of Aurora A antibodies into prometaphase human cells with separated spindle poles. In the former study, the authors identified the kinetochore protein CENP-A as an Aurora A binding partner through a yeast two-hybrid approach and mapped Ser7 as a bona fide Aurora A substrate whose phosphorylation in prophase is dependent on Aurora A activity. The authors went on to show that expression of a nonphosphorylatable CENP-A (Ser7 mutated to Alanine) also resulted in chromosome misalignment. Because CENP-A Ser7 is also a target of Aurora B (Zeitlin et al. 2001), it remains unclear whether the observed defects in cells expressing a nonphosphorylatable CENP-A Ser7 mutant are related to Aurora A or to Aurora B–mediated phosphorylation. Still, these studies uncovered a potential new role for Aurora A in mitotic chromosome alignment and segregation beyond mitotic entry and spindle pole separation. More recently, additional Aurora A kinase substrates have been identified at the kinetochore, and their phosphorylation specifically by Aurora A has been shown to be essential for regulation of kinetochore–microtubule attachments. Work from multiple laboratories has demonstrated that there are at least two mechanisms for Aurora A–mediated phosphorylation of kinetochore substrates. In the first, spindle pole-localized Aurora A phosphorylates kinetochores on chromosomes that occupy a position, even transiently, near one of the two spindle poles (Kim et al. 2010; Chmátal et al. 2015; Ye et al. 2015). In the second, Aurora A is recruited to the kinetochore region to phosphorylate substrates there, independent of chromosome proximity to a spindle pole (DeLuca et al. 2018). Both mechanisms likely play an important role in temporally regulating kinetochore–microtubules during mitosis, and are discussed in detail below.

Kinetochore Substrate Phosphorylation by Pole-Associated Aurora A

The kinetochore-associated, plus end-directed microtubule motor protein CENP-E transports chromosomes along spindle microtubules from spindle poles to the metaphase plate in mitosis (Kapoor et al. 2006). In 2010, Kim et al. demonstrated that CENP-E Thr422 is targeted by both Aurora A and Aurora B kinases, and phosphorylation at this site decreases CENP-E-mediated recruitment of PP1 (a phosphatase that promotes attachment stability by counteracting Aurora B kinase activity) and reduces affinity of CENP-E for microtubules. The authors found that centrosome-proximal chromosomes exhibited high levels of Thr422 phosphorylation, presumably due to active Aurora A at spindle poles and active Aurora B at the centromere/kinetochore region. In this scenario, the authors reasoned that incorrect attachments near spindle poles (i.e., syntelic attachments, in which both sister kinetochores are attached to microtubules from a single spindle pole) become destabilized, allowing for subsequent formation of correct attachments and productive CENP-E-powered movements of chromosomes toward the spindle equator (Kim et al. 2010). These findings suggest a role for spindle pole-associated Aurora A kinase in phosphorylating kinetochore substrates to promote correction of attachment errors specifically on chromosomes that become "trapped" near the centrosome region.

Hec1 was also recently identified as a kinetochore substrate of spindle-pole associated Aurora A kinase both in mitotic and meiotic cells. In a study by Ye et al., the authors found that in *Drosophila* S2 cells, overexpression of Aurora A resulted in enrichment of the kinase at spindle poles and a reduced incidence of experimentally induced syntelic attachments (Ye et al. 2015). In a reciprocal experiment, RNAi-mediated depletion of Aurora A resulted in increased pole-associated, syntelically attached kinetochores. The authors expanded their study into PtK1 cells, and reported a significant increase in pole-oriented chromosomes in cells treated with an Aurora A–specific kinase inhibitor (Ye et al. 2015). They went on to confirm that Hec1 Ser55 is not only an Aurora B target site, but also a substrate of Aurora A whose phosphorylation decreased in cells treated with Aurora A inhibitors (DeLuca et al. 2011; Kettenbach et al. 2011; Ye et al. 2015). In a companion study, Chmátal et al. (2015) demonstrated a similar role for spindle pole-associated Aurora A kinase in dividing meiotic cells. The authors found that during meiosis I in mouse oocytes, microtubule attachments to kinetochores on pole-proximal chromosomes were specifically destabilized, and this effect was reduced upon inhibition of Aurora A kinase (Chmátal et al. 2015). Combined with the observation that Aurora B kinase activity is high on pole-proximal chromosomes and contributes to the destabilization of kinetochore–microtubule attachments on these chromosomes (Maia et al. 2010), these studies suggest that Hec1 phosphorylation by both Aurora B kinase and pole-associated Aurora A kinase plays an important role in the correction of syntelic attachments in early mitotic cells.

It is well-established that Aurora B phosphorylates substrates at the centromere/kinetochore region. From recent work it has become clear that spindle pole-associated Aurora A also phosphorylates kinetochore substrates, and in fact, the two kinases may act together on the same substrates. This leads to a model whereby the degree of Aurora-mediated phosphorylation on a particular substrate is dictated by the landscape of chromosome distribution within a mitotic cell. This model reinforces the dogma that Aurora A and Aurora B are specifically compartmentalized within mitotic cells and that phosphorylation depends on substrate localization to these compartments.

Kinetochore Substrate Phosphorylation by Centromere/Kinetochore-Associated Aurora A

In a recent study, we found that Hec1 Ser69, which was previously identified as an in vitro substrate of Aurora B kinase (DeLuca et al. 2006), is primarily phosphorylated by Aurora A kinase in both human and PtK1 cells. Inhibition of Aurora A, but not Aurora B kinase, resulted in markedly reduced levels of phosphorylation at this site on kinetochores. Unlike the other characterized Hec1 tail domain target sites, phosphorylation of Ser69 persists at high levels throughout the entirety of mitosis (DeLuca et al. 2018). Furthermore, we found that Aurora A–mediated phosphorylation of Ser69 is required for normal kinetochore oscillatory movements at metaphase. Given the well-established localization of Aurora A at spindle poles, persistent phosphorylation of an Aurora A–specific kinetochore substrate in late mitosis was surprising, because kinetochores are maximally distant from poles in metaphase. We first hypothesized, consistent with the compartmentalization model described above, that Aurora A phosphorylates Ser69 in early mitosis when chromosomes are likely to encounter a spindle pole, and this phosphorylation persists throughout mitosis. However, when we allowed mitotic cells to progress to metaphase and added Aurora A inhibitors only after chromosomes had completely aligned, immunofluorescence experiments revealed a near-complete loss of both Ser69 phosphorylation and kinetochore oscillations (DeLuca et al. 2018). These studies demonstrate that Aurora A kinase is able to phosphorylate kinetochores on chromosomes distal from the spindle poles, and that this activity is required for regulation of kinetochore–microtubule attachments. As such, they compel a re-evaluation of the Aurora kinase compartmentalization model and beg the question of how Aurora A is recruited to kinetochores in metaphase.

AURORA A RECRUITMENT TO THE CENTROMERE/KINETOCHORE REGION

It is relatively easy to envision a scenario in which Aurora A kinase is spatially restricted in a mitotic cell, and its phosphorylation of kinetochore substrates is dictated by chromosome location. In this case, the kinase resides at spindle poles and phosphorylates kinetochores only on chromosomes that migrate within range of the localized kinase. Aurora A would be ideally positioned to reduce the incidence of chromosomes that are trapped at spindle poles through both prevention and correction of erroneous kinetochore–microtubule attachments, a phenomenon reported in previous studies (Chmátal et al. 2015; Ye et al. 2015). However, it has become clear that Aurora A also phosphorylates kinetochore substrates on chromosomes that are aligned at the spindle equator, distant from spindle poles (DeLuca et al. 2018). It is more difficult to envision a scenario in which a spindle pole-restricted kinase is able to phosphorylate distal kinetochore substrates. Several possibilities for how this might be achieved are discussed below.

In the first, Aurora A is directly recruited to the centromere/kinetochore region via INCENP (in complex with the other CPC components). In this model, a population of Aurora A is bound to INCENP rather than TPX2. In support of this model, the Sen laboratory demonstrated through immunoprecipitation experiments that Aurora A forms distinct complexes with both TPX2 and INCENP during mitosis and that Aurora A directly interacts with INCENP in vitro (Katayama et al. 2008). In contrast, Aurora B kinase interacted only with INCENP, and not TPX2 (Katayama et al. 2008). In this study, the authors carried out a time-course immunoprecipitation experiment to determine the timing of Aurora A and Aurora B complex formation with INCENP. Interestingly, Aurora B/INCENP was detectable over a broader timescale during mitosis than Aurora A/INCENP, which was evident during only a short window during mitosis, and whose abundance peaked after Aurora B/INCENP (Katayama et al. 2008). This suggests that Aurora A may only bind INCENP in late mitosis, lending support to the idea that the kinase might switch from binding TPX2 to INCENP, allowing for recruitment to the centromere region specifically in metaphase. Our recent results confirm that Aurora A associates with INCENP during mitosis, and specifically that the interaction of INCENP with Aurora A is mediated through INCENP's IN-box domain, the domain required for Aurora B binding (Fu et al. 2009; DeLuca et al. 2018). Furthermore, depletion of INCENP by RNAi in human cells significantly reduced phosphorylation of Hec1 Ser69. Because this site is primarily phosphorylated by Aurora A in cells, this result suggests that an INCENP-Aurora A kinase complex may indeed phosphorylate Hec1 (DeLuca et al. 2018). It will be important in future studies to determine how a potential switch between TPX2- and INCENP-Aurora A association is regulated in cells to ensure a timely hand-off between the two regulators.

A second possibility is that Aurora A is directly recruited to outer kinetochores, independent of INCENP, to phosphorylate kinetochore proteins. Chmátal et al. found that Aurora A kinase localizes to kinetochores in mouse meiotic cell division, suggesting kinetochores have the inherent ability to bind Aurora A (Chmátal et al. 2015). In addition, Hans et al. showed that a truncated version of Aurora A kinase missing its amino-terminal 120 amino acids localizes readily to kinetochores in human mitotic cells depleted of endogenous Aurora B (Hans et al. 2009). Strikingly, this localization was not dependent on the ability of Aurora A to bind INCENP, and furthermore, the kinetochore-associated, truncated Aurora A kinase did not exhibit "passenger"-like activity at anaphase onset, but instead remained bound to kinetochores (Hans et al. 2009). These studies raise the interesting possibility that Aurora A may be temporally regulated, such that in early mitosis, it is largely spindle pole-associated, and in late mitosis a population is modified such that it directly binds kinetochores. However, this may be unlikely, because Aurora A is not readily detectable at kinetochores in unperturbed mitotic cells.

A third possibility is that kinetochore substrates are phosphorylated by a population of Aurora A kinase that is delivered to the vicinity of kinetochores by spindle microtubules. Both Aurora A and its activator TPX2 associate with spindle microtubules, and in fact, an additional cofactor, RHAMM (a TPX2 and microtubule binding protein) has been identified both on spindle microtubules and kinetochores (Chen et al. 2014). Finally, in a fourth possibility, it is feasible that Aurora A does not require recruitment, per se, to kinetochores, but instead a population of soluble, cytoplasmic kinase is sufficient to phosphorylate kinetochores during late mitosis. A recent study demonstrated such a phenomenon for Aurora B. In this case, an experimentally activated version of the kinase was mutated such that it could not bind the centromere/kinetochore region; however, it maintained the ability to phosphorylate outer kinetochore substrates during mitosis (Haase et al. 2017).

Phosphorylation of kinetochore substrates by Aurora A kinase occurs when chromosomes are near spindle poles (in prometaphase) and distal from spindle poles (in metaphase), and both activities contribute to accurate, error-free chromosome segregation. The mechanism by which Aurora A phosphorylates kinetochores in early mitosis on pole-associated chromosomes is likely straightforward, given the high levels of Aurora A at spindle poles. In contrast, it has been difficult to detect endogenous Aurora A at kinetochores or centromeres during mitosis, therefore the mechanism by which the kinase phosphorylates kinetochore substrates in metaphase remains unresolved. Future studies using approaches to detect transient protein–protein interactions at kinetochores should help further our understanding of these mechanisms.

AURORA KINASE SUBSTRATE SPECIFICITY

Consistent with the ability of Aurora A and B to phosphorylate the same substrates in vitro (Vader and Lens 2008; Carmena et al. 2009), the consensus sequences recognized by the two kinases are quite similar, with only subtle distinctions (Meraldi et al. 2004; Ohashi et al. 2006; Kim et al. 2010; Kettenbach et al. 2011; Koch et al. 2011). Given comparable consensus sequences and no compelling evidence indicating substrate preference in vitro, how is specificity achieved in cells? As discussed above, the prevailing model is that Aurora kinases are recruited to specific subcellular localizations to phosphorylate targets at these sites. This compartmentalization model predicts that targeting either kinase to the other kinase's endogenous location should result in functional replacement—meaning that each kinase should be able to phosphorylate the other kinase's substrates, as long as it is targeted to the appropriate location in the cell. This was tested by two groups in 2009. In both studies, the authors demonstrated that mutation of a single amino acid in Aurora A, Gly198, to the analogous residue in Aurora B, an Asparagine, resulted in loss of Aurora A localization to spindle poles, and subsequent relocalization to centromeres prior to anaphase and to the spindle midzone after anaphase onset. Strikingly, in Aurora B–depleted cells, this mutant version of Aurora A was able to functionally replace Aurora B, and cells progressed through mitosis without detectable chromosome segregation defects (Fu et al. 2009; Hans et al.

2009). Furthermore, in vitro studies revealed that this single mutation increased Aurora A's ability to bind INCENP and Survivin (Fu et al. 2009; Hans et al. 2009). The authors concluded that the activity of Aurora A and Aurora B toward spindle pole and centromere/kinetochore substrates, respectively, in unperturbed mitotic cells is a consequence of spatial compartmentalization. In further support of this idea, a recent study showed that ectopically targeting Aurora A to either mitotic chromatin (fusion to Histone H2B), centromeres (fusion to CENP-B), or kinetochores (fusion to Hec1) resulted in rescue of Aurora B function at each of those locations in cells depleted of endogenous Aurora B (Li et al. 2015). Conversely, ectopically targeting Aurora B to spindle poles (fusion to a fragment of centrosome component PLK4), rescued the function of Aurora A kinase at spindle poles in cells depleted of Aurora A (Li et al. 2015). Together, these studies support a model in which the specificity of the Aurora kinases for their substrates is dictated in large part by location within the cell.

Our recent work, however, suggests that this model may not entirely explain why certain substrates are phosphorylated preferentially by one Aurora kinase over the other during mitosis. We found that Aurora A and Aurora B contribute to phosphorylation of Hec1 tail domain target sites (of the three tested) to varying degrees in mammalian cells. In early mitosis, Aurora A kinase robustly phosphorylates Hec1 Ser69 and Ser55, but only minimally contributes to phosphorylation of closely neighboring Ser44 (DeLuca et al. 2018). Thus, in early mitosis, when Hec1 tail domain substrates are similarly exposed to both Aurora A and Aurora B, the kinases exhibit preference for certain sites, indicating that spatial compartmentalization is not the only determinant of kinase specificity in cells. Inherent bias for one site over another likely does not explain this phenomenon, because we found that in vitro, all sites are phosphorylated by both kinases, with little preference for any individual site(s) (DeLuca et al. 2018). We cannot rule out the possibility, however, that our in vitro assay could not resolve such preferences. Perhaps even more intriguing, during metaphase, we observed a marked reduction in Ser55 phosphorylation, whereas Ser69 phosphorylation remained high, and this sustained phosphorylation was dependent on Aurora A activity. At this point in mitosis, Aurora B activity is low at kinetochores, and our data demonstrating that Ser44 phosphorylation levels are maximally low at metaphase support this premise (DeLuca et al. 2011). These results suggest that the specificity of Aurora A kinase toward distinct kinetochore target sites may be differentially regulated. We tested to see if the position of a particular target site within the Hec1 tail may give rise to this specificity (i.e., Ser69 may be "accessible" to Aurora A, whereas Ser55 may not). For this experiment, we generated a tail "domain swap" mutant in which regions of the Hec1 tail were transposed, resulting in Ser55 occupying a position near the Hec1 globular domain (similar to Ser69's position in the wild-type tail), and Ser69 occupying a position toward the amino-terminal, distal region of the tail. The temporal phosphorylation patterns of Ser55 and Ser69 during mitotic progression were identical in wild-type and mutant Hec1, and the sensitivity of phosphorylation at each site to Aurora A and Aurora B inhibitors was also unchanged (DeLuca et al. 2018), leading us to conclude that site specificity is not dictated by spatial positioning within the tail. How then is specificity achieved? Although this remains an important area for future investigation, an interesting possibility is that phosphorylation on individual Hec1 sites varies as a consequence of differential dephosphorylation. The kinetochore phosphatases responsible for dephosphorylation of the Hec1 target sites remain unknown, but their identification and functional characterization will no doubt aid in understanding how different phosphorylation states of individual Aurora substrates are achieved in mitosis.

WHY DO CELLS REQUIRE AURORA A–MEDIATED KINETOCHORE REGULATION?

These studies raise the question of why mammalian cells require Aurora A, in addition to Aurora B, to facilitate kinetochore–microtubule attachment regulation? In early mitosis, it is important that global kinetochore–microtubule attachment turnover is high to prevent premature stabilization of kinetochore–microtubules. It is well-established that Aurora B kinase, which is highly concentrated at the centromere/kinetochore region in early mitosis, contributes to this error–correction (and error–prevention) mechanism. During prometaphase, mono-oriented chromosomes are typically transported to spindle poles by the minus end-directed microtubule motor protein dynein to promote chromosome alignment and biorientation. However, while localized near a spindle pole, chromosomes risk forming syntelic attachments. Thus, it is important that cells have a robust mechanism to ensure that these erroneous attachments are corrected (or avoided altogether) on pole-localized chromosomes. Aurora A kinase at spindle poles helps solve this problem and works together with Aurora B (Maia et al. 2010; Krenn and Musacchio 2015) to ensure that kinetochore substrates such as Hec1 are highly phosphorylated to promote kinetochore–microtubule turnover.

The role for Aurora A kinase in metaphase appears to be distinct from its role in early mitosis. As discussed above, mutation of the Aurora A substrate Hec1 Ser69 to Alanine or cell-wide inhibition of Aurora A kinase activity results in markedly decreased kinetochore oscillations. These perturbations result in a small but significant increase in chromosome segregation errors—about a twofold increase in lagging chromosomes in anaphase in human cells (DeLuca et al. 2018). A likely cause for these errors is a failure to correct merotelic attachments (i.e., one kinetochore of a sister pair is bound to microtubules emanating from both poles), which is known to increase the incidence of lagging chromosomes in anaphase (Cimini et al. 2001). In support of this notion, Ye et al. (2015) found in PtK1 cells that inhibition of Aurora A kinase activity resulted in a 3.6-fold increase in merotelic attachments at metaphase, and a corresponding 3.2-fold increase in lagging chromosomes in anaphase. Together, these studies suggest a role for Aurora A kinase in regu-lating kinetochore–microtubule dynamics in late mitosis. These results also support the idea that kinetochore oscillations in late mitosis contribute to mitotic fidelity. This is noteworthy, because the role for oscillations in mitosis has yet to be resolved. One possibility is that oscillations provide increased kinetochore–microtubule turnover to detach any merotelic attachments that may remain in metaphase (Cimini et al. 2003). However, this mechanism may not be universally conserved, because there are cell types that do not exhibit metaphase oscillations (e.g., embryonic cells of *Drosophila* and *Xenopus*) (Desai et al. 1998; Maddox et al. 2002) but have no obvious increased incidence of lagging chromosomes. Interestingly, spindles in these cells have high rates of flux (i.e., the continual addition of tubulin dimers at microtubule plus ends and a corresponding loss from microtubule minus ends at spindle poles), which may provide the high turnover of tubulin subunits at the kinetochore–microtubule interface to promote release of erroneous attachments (Desai et al. 1998; Brust-Mascher and Scholey 2002; Maddox et al. 2002, 2003). Although the physiological reason for kinetochore oscillations remains an important topic to be explored, it is clear that Aurora A kinase is required for late mitotic kinetochore–microtubule attachment regulation to ensure error-free chromosome segregation.

CONCLUSION

Precise regulation of kinetochore-microtubule attachment stability is essential for ensuring the fidelity of chromosome segregation during mitosis. Research over the last several decades has established that Aurora B kinase is the master controller of this process, however, recent studies have revealed that its sister kinase, Aurora A, likely plays a supporting role in this regulation. From these studies, a model emerges in which Aurora A, concentrated at mitotic spindle poles, phosphorylates kinetochore substrates on pole-proximal chromosomes (together with centromere/kinetochore-localized Aurora B) to ensure both the prevention and correction of attachment errors in early mitosis. As chromosomes align and Aurora B kinase activity decreases at kinetochores, Aurora A continues to phosphorylate outer kinetochore substrates which permits the fluid tracking of dynamic microtubule plus-ends and chromosome oscillatory movements. Many questions regarding this duel regulation still remain: How is Aurora A recruited to the centromere/kinetochore region? What determines site specificity for Aurora A versus Aurora B in cells? Are unique phosphatase complexes used to dephosphorylate different Aurora-mediated phosphorylation events? Addressing these questions will be critical in unraveling how the Aurora kinases ensure correct and timely chromosome segregation during mitotic cell division.

ACKNOWLEDGMENTS

This work is supported by grant R01GM088371 from the National Institutes of Health. I thank Keith DeLuca and Dr. Susanne Lens for critical comments on the manuscript.

REFERENCES

Adams RR, Wheatley SP, Gouldsworthy AM, Kandels-Lewis SE, Carmena M, Smythe C, Gerloff DL, Earnshaw WC. 2000. INCENP binds the Aurora-related kinase AIRK2 and is required to target it to chromosomes, the central spindle and cleavage furrow. *Curr Biol* **10:** 1075–1078.

Adams RR, Carmena M, Earnshaw WC. 2001. Chromosomal passengers and the (aurora) ABCs of mitosis. *Trends Cell Biol* **11:** 49–54.

Afonso O, Figueiredo AC, Maiato H. 2017. Late mitotic functions of Aurora kinases. *Chromosoma* **126:** 93–103.

Alushin GM, Musinipally V, Matson D, Tooley J, Stukenberg PT, Nogales E. 2012. Multimodal microtubule binding by the Ndc80 kinetochore complex. *Nat Struct Mol Biol* **19:** 1161–1167.

Balboula AZ, Schindler K. 2014. Selective disruption of aurora C kinase reveals distinct functions from aurora B kinase during meiosis in mouse oocytes. *PLoS Genet* **10:** e1004194.

Balboula AZ, Nguyen AL, Gentilello AS, Quartuccio SM, Drutovic D, Solc P, Schindler K. 2016. Haspin kinase regulates microtubule-organizing center clustering and stability through Aurora kinase C in mouse oocytes. *J Cell Sci* **129:** 3648–3660.

Barr AR, Gergely F. 2007. Aurora-A: The maker and breaker of spindle poles. *J Cell Sci* **120:** 2987–2996.

Bayliss R, Sardon T, Vernos I, Conti E. 2003. Structural basis of Aurora-A activation by TPX2 at the mitotic spindle. *Mol Cell* **12:** 851–862.

Bertolin G, Sizaire F, Herbomel G, Reboutier D, Prigent C, Tramier M. 2016. A FRET biosensor reveals spatiotemporal activation and functions of aurora kinase A in living cells. *Nat Commun* **7:** 12674.

Biggins S, Severin FF, Bhalla N, Sassoon I, Hyman AA, Murray AW. 1999. The conserved protein kinase Ipl1 regulates microtubule binding to kinetochores in budding yeast. *Genes Dev* **13:** 532–544.

Bischoff JR, Plowman GD. 1999. The Aurora/Ipl1p kinase family: Regulators of chromosome segregation and cytokinesis. *Trends Cell Biol* **9:** 454–459.

Bishop JD, Schumacher JM. 2002. Phosphorylation of the carboxyl terminus of inner centromere protein (INCENP) by the Aurora B Kinase stimulates Aurora B kinase activity. *J Biol Chem* **277:** 27577–27580.

Bolanos-Garcia VM. 2005. Aurora kinases. *Int J Biochem Cell Biol* **37:** 1572–1577.

Bolton MA, Lan W, Powers SE, McCleland ML, Kuang J, Stukenberg PT. 2002. Aurora B kinase exists in a complex with survivin and INCENP and its kinase activity is stimulated by survivin binding and phosphorylation. *Mol Biol Cell* **13:** 3064–3077.

Brust-Mascher I, Scholey JM. 2002. Microtubule flux and sliding in mitotic spindles of *Drosophila* embryos. *Mol Biol Cell* **13:** 3967–3975.

Carmena M, Ruchaud S, Earnshaw WC. 2009. Making the Auroras glow: Regulation of Aurora A and B kinase function by interacting proteins. *Curr Opin Cell Biol* **21:** 796–805.

Carmena M, Wheelock M, Funabiki H, Earnshaw WC. 2012. The chromosomal passenger complex (CPC): From easy rider to the godfather of mitosis. *Nat Rev Mol Cell Biol* **13:** 789–803.

Carvalho A, Carmena M, Sambade C, Earnshaw WC, Wheatley SP. 2003. Survivin is required for stable checkpoint activation in taxol-treated HeLa cells. *J Cell Sci* **116:** 2987–2998.

Chan CS, Botstein D. 1993. Isolation and characterization of chromosome-gain and increase-in-ploidy mutants in yeast. *Genetics* **135:** 677–691.

Chan YW, Jeyaprakash AA, Nigg EA, Santamaria A. 2012. Aurora B controls kinetochore-microtubule attachments by inhibiting Ska complex-KMN network interaction. *J Cell Biol* **196:** 563–571.

Cheeseman IM, Chappie JS, Wilson-Kubalek EM, Desai A. 2006. The conserved KMN network constitutes the core microtubule-binding site of the kinetochore. *Cell* **127:** 983–997.

Chen H, Mohan P, Jiang J, Nemirovsky O, He D, Fleisch MC, Niederacher D, Pilarski LM, Lim CJ, Maxwell CA. 2014. Spatial regulation of Aurora A activity during mitotic spindle assembly requires RHAMM to correctly localize TPX2. *Cell Cycle* **13:** 2248–2261.

Chmátal L, Yang K, Schultz RM, Lampson MA. 2015. Spatial regulation of kinetochore microtubule attachments by destabilization at spindle poles in meiosis I. *Curr Biol* **25:** 1835–1841.

Cimini D, Howell B, Maddox P, Khodjakov A, Degrassi F, Salmon ED. 2001. Merotelic kinetochore orientation is a major mechanism of aneuploidy in mitotic mammalian tissue cells. *J Cell Biol* **153:** 517–527.

Cimini D, Moree B, Canman JC, Salmon ED. 2003. Merotelic kinetochore orientation occurs frequently during early mitosis in mammalian tissue cells and error correction is achieved by two different mechanisms. *J Cell Sci* **116:** 4213–4225.

Cooke CA, Heck MM, Earnshaw WC. 1987. The inner centromere protein (INCENP) antigens: Movement from inner centromere to midbody during mitosis. *J Cell Biol* **105:** 2053–2067.

Crane R, Gadea B, Littlepage L, Wu H, Ruderman JV. 2004. Aurora A, meiosis and mitosis. *Biol Cell* **96:** 215–229.

DeLuca JG, Gall WE, Ciferri C, Cimini D, Musacchio A, Salmon ED. 2006. Kinetochore microtubule dynamics and attachment stability are regulated by Hec1. *Cell* **127:** 969–982.

DeLuca KF, Lens SM, DeLuca JG. 2011. Temporal changes in Hec1 phosphorylation control kinetochore-microtubule attachment stability during mitosis. *J Cell Sci* **124:** 622–634.

DeLuca KF, Meppelink A, Broad AJ, Mick JE, Peersen OB, Pektas S, Lens SMA, DeLuca JG. 2018. Aurora A kinase phosphorylates Hec1 to regulate metaphase kinetochore-microtubule dynamics. *J Cell Biol* **217:** 163–177.

Desai A, Maddox PS, Mitchison TJ, Salmon ED. 1998. Anaphase A chromosome movement and poleward spindle microtubule flux occur at similar rates in *Xenopus* extract spindles. *J Cell Biol* **141:** 703–713.

Ducat D, Zheng Y. 2004. Aurora kinases in spindle assembly and chromosome segregation. *Exp Cell Res* **301:** 60–67.

Dutertre S, Hamard-Peron E, Cremet JY, Thomas Y, Prigent C. 2005. The absence of p53 aggravates polyploidy and centrosome number abnormality induced by Aurora-C overexpression. *Cell Cycle* **4:** 1783–1787.

Earnshaw WC, Bernat RL. 1991. Chromosomal passengers: Toward an integrated view of mitosis. *Chromosoma* **100:** 139–146.

Eckley DM, Ainsztein AM, Mackay AM, Goldberg IG, Earnshaw WC. 1997. Chromosomal proteins and cytokinesis: Patterns of cleavage furrow formation and inner centromere protein positioning in mitotic heterokaryons and mid-anaphase cells. *J Cell Biol* **136:** 1169–1183.

Ehara H, Yokoi S, Tamaki M, Nishino Y, Takahashi Y, Deguchi T, Kimura M, Yoshioka T, Okano Y. 2003. Expression of mitotic Aurora/Ipl1p-related kinases in renal cell carcinomas: An immunohistochemical study. *Urol Res* **31:** 382–386.

Eyers PA, Maller JL. 2003. Regulating the regulators: Aurora A activation and mitosis. *Cell Cycle* **2:** 287–289.

Eyers PA, Maller JL. 2004. Regulation of *Xenopus* Aurora A activation by TPX2. *J Biol Chem* **279:** 9008–9015.

Fellmeth JE, Gordon D, Robins CE, Scott RT Jr, Treff NR, Schindler K. 2015. Expression and characterization of three Aurora kinase C splice variants found in human oocytes. *Mol Hum Reprod* **21:** 633–644.

Fernandez-Miranda G, Trakala M, Martin J, Escobar B, Gonzalez A, Ghyselinck NB, Ortega S, Canamero M, Perez de Castro I, Malumbres M. 2011. Genetic disruption of aurora B uncovers an essential role for aurora C during early mammalian development. *Development* **138:** 2661–2672.

Fu J, Bian M, Liu J, Jiang Q, Zhang C. 2009. A single amino acid change converts Aurora-A into Aurora-B-like kinase in terms of partner specificity and cellular function. *Proc Natl Acad Sci* **106:** 6939–6944.

Funabiki H, Wynne DJ. 2013. Making an effective switch at the kinetochore by phosphorylation and dephosphorylation. *Chromosoma* **122:** 135–158.

Gassmann R, Carvalho A, Henzing AJ, Ruchaud S, Hudson DF, Honda R, Nigg EA, Gerloff DL, Earnshaw WC. 2004. Borealin: A novel chromosomal passenger required for stability of the bipolar mitotic spindle. *J Cell Biol* **166:** 179–191.

Giet R, Glover DM. 2001. *Drosophila* aurora B kinase is required for histone H3 phosphorylation and condensin recruitment during chromosome condensation and to organize the central spindle during cytokinesis. *J Cell Biol* **152:** 669–682.

Giet R, Prigent C. 1999. Aurora/Ipl1p-related kinases, a new oncogenic family of mitotic serine-threonine kinases. *J Cell Sci* **112:** 3591–3601.

Glover DM. 2003. Aurora A on the mitotic spindle is activated by the way it holds its partner. *Mol Cell* **12:** 797–799.

Godek KM, Kabeche L, Compton DA. 2015. Regulation of kinetochore-microtubule attachments through homeostatic control during mitosis. *Nat Rev Mol Cell Biol* **16:** 57–64.

Haase J, Bonner MK, Halas H, Kelly AE. 2017. Distinct roles of the chromosomal passenger complex in the detection of and response to errors in kinetochore-microtubule attachment. *Dev Cell* **42:** 640–654.e645.

Hans F, Skoufias DA, Dimitrov S, Margolis RL. 2009. Molecular distinctions between Aurora A and B: A single residue change transforms Aurora A into correctly localized and functional Aurora B. *Mol Biol Cell* **20:** 3491–3502.

Hauf S, Cole RW, LaTerra S, Zimmer C, Schnapp G, Walter R, Heckel A, van Meel J, Rieder CL, Peters JM. 2003. The small molecule Hesperadin reveals a role for Aurora B in correcting kinetochore-microtubule attachment and in maintaining the spindle assembly checkpoint. *J Cell Biol* **161:** 281–294.

Hindriksen S, Lens SMA, Hadders MA. 2017. The ins and outs of Aurora B inner centromere localization. *Front Cell Dev Biol* **5:** 112.

Hirota T, Kunitoku N, Sasayama T, Marumoto T, Zhang D, Nitta M, Hatakeyama K, Saya H. 2003. Aurora-A and an interacting activator, the LIM protein Ajuba, are required for mitotic commitment in human cells. *Cell* **114:** 585–598.

Honda R, Korner R, Nigg EA. 2003. Exploring the functional interactions between Aurora B, INCENP, and survivin in mitosis. *Mol Biol Cell* **14:** 3325–3341.

Hutterer A, Berdnik D, Wirtz-Peitz F, Zigman M, Schleiffer A, Knoblich JA. 2006. Mitotic activation of the kinase Aurora-A requires its binding partner Bora. *Dev Cell* **11:** 147–157.

Jeyaprakash AA, Klein UR, Lindner D, Ebert J, Nigg EA, Conti E. 2007. Structure of a Survivin-Borealin-INCENP core complex reveals how chromosomal passengers travel together. *Cell* **131:** 271–285.

Joukov V, De Nicolo A, Rodriguez A, Walter JC, Livingston DM. 2010. Centrosomal protein of 192 kDa (Cep192) promotes centrosome-driven spindle assembly by engaging in organelle-specific Aurora A activation. *Proc Natl Acad Sci* **107:** 21022–21027.

Kaitna S, Mendoza M, Jantsch-Plunger V, Glotzer M. 2000. Incenp and an aurora-like kinase form a complex essential for chromosome segregation and efficient completion of cytokinesis. *Curr Biol* **10:** 1172–1181.

Kallio MJ, McCleland ML, Stukenberg PT, Gorbsky GJ. 2002. Inhibition of aurora B kinase blocks chromosome segregation, overrides the spindle checkpoint, and perturbs microtubule dynamics in mitosis. *Curr Biol* **12:** 900–905.

Kapoor TM, Lampson MA, Hergert P, Cameron L, Cimini D, Salmon ED, McEwen BF, Khodjakov A. 2006. Chromosomes can congress to the metaphase plate before biorientation. *Science* **311:** 388–391.

Katayama H, Brinkley WR, Sen S. 2003. The Aurora kinases: Role in cell transformation and tumorigenesis. *Cancer Metastasis Rev* **22:** 451–464.

Katayama H, Sasai K, Kloc M, Brinkley BR, Sen S. 2008. Aurora kinase-A regulates kinetochore/chromatin associated microtubule assembly in human cells. *Cell Cycle* **7:** 2691–2704.

Kettenbach AN, Schweppe DK, Faherty BK, Pechenick D, Pletnev AA, Gerber SA. 2011. Quantitative phosphoproteomics identifies substrates and functional modules of Aurora and Polo-like kinase activities in mitotic cells. *Sci Signal* **4:** rs5.

Khan J, Ezan F, Cremet JY, Fautrel A, Gilot D, Lambert M, Benaud C, Troadec MB, Prigent C. 2011. Overexpression of active Aurora-C kinase results in cell transformation and tumour formation. *PLoS One* **6:** e26512.

Kim Y, Holland AJ, Lan W, Cleveland DW. 2010. Aurora kinases and protein phosphatase 1 mediate chromosome congression through regulation of CENP-E. *Cell* **142:** 444–455.

Klein UR, Nigg EA, Gruneberg U. 2006. Centromere targeting of the chromosomal passenger complex requires a ternary subcomplex of Borealin, Survivin, and the N-terminal domain of INCENP. *Mol Biol Cell* **17:** 2547–2558.

Koch A, Krug K, Pengelley S, Macek B, Hauf S. 2011. Mitotic substrates of the kinase aurora with roles in chromatin regulation identified through quantitative phosphoproteomics of fission yeast. *Sci Signal* **4:** rs6.

Krenn V, Musacchio A. 2015. The Aurora B kinase in chromosome bi-orientation and spindle checkpoint signaling. *Front Oncol* **5:** 225.

Kufer TA, Sillje HH, Korner R, Gruss OJ, Meraldi P, Nigg EA. 2002. Human TPX2 is required for targeting Aurora-A kinase to the spindle. *J Cell Biol* **158:** 617–623.

Kunitoku N, Sasayama T, Marumoto T, Zhang D, Honda S, Kobayashi O, Hatakeyama K, Ushio Y, Saya H, Hirota T. 2003. CENP-A phosphorylation by Aurora-A in prophase is required for enrichment of Aurora-B at inner centromeres and for kinetochore function. *Dev Cell* **5:** 853–864.

Lampson MA, Grishchuk EL. 2017. Mechanisms to avoid and correct erroneous kinetochore-microtubule attachments. *Biology (Basel)* **6:** pii: E1. doi: 10.3390/biology6010001.

Lens SM, Wolthuis RM, Klompmaker R, Kauw J, Agami R, Brummelkamp T, Kops G, Medema RH. 2003. Survivin is required for a sustained spindle checkpoint arrest in response to lack of tension. *EMBO J* **22:** 2934–2947.

Li S, Deng Z, Fu J, Xu C, Xin G, Wu Z, Luo J, Wang G, Zhang S, Zhang B, et al. 2015. Spatial compartmentalization specializes the function of Aurora A and Aurora B. *J Biol Chem* **290:** 17546–17558.

Liu ST, Rattner JB, Jablonski SA, Yen TJ. 2006. Mapping the assembly pathways that specify formation of the trilaminar kinetochore plates in human cells. *J Cell Biol* **175:** 41–53.

Maddox P, Desai A, Oegema K, Mitchison TJ, Salmon ED. 2002. Poleward microtubule flux is a major component of spindle dynamics and anaphase A in mitotic *Drosophila* embryos. *Curr Biol* **12:** 1670–1674.

Maddox P, Straight A, Coughlin P, Mitchison TJ, Salmon ED. 2003. Direct observation of microtubule dynamics at kinetochores in *Xenopus* extract spindles: Implications for spindle mechanics. *J Cell Biol* **162:** 377–382.

Maia AF, Feijao T, Vromans MJ, Sunkel CE, Lens SM. 2010. Aurora B kinase cooperates with CENP-E to promote timely anaphase onset. *Chromosoma* **119:** 405–413.

Marumoto T, Honda S, Hara T, Nitta M, Hirota T, Kohmura E, Saya H. 2003. Aurora-A kinase maintains the fidelity of early and late mitotic events in HeLa cells. *J Biol Chem* **278:** 51786–51795.

Meraldi P, Honda R, Nigg EA. 2004. Aurora kinases link chromosome segregation and cell division to cancer susceptibility. *Curr Opin Genet Dev* **14:** 29–36.

Murata-Hori M, Wang YL. 2002. The kinase activity of aurora B is required for kinetochore-microtubule interactions during mitosis. *Curr Biol* **12:** 894–899.

Murata-Hori M, Tatsuka M, Wang YL. 2002. Probing the dynamics and functions of aurora B kinase in living cells during mitosis and cytokinesis. *Mol Biol Cell* **13:** 1099–1108.

Musacchio A, Desai A. 2017. A molecular view of kinetochore assembly and function. *Biology (Basel)* **6:** pii: E5. doi: 10.3390/biology6010005.

Nguyen AL, Schindler K. 2017. Specialize and divide (twice): Functions of three Aurora kinase homologs in mammalian oocyte meiotic maturation. *Trends Genet* **33:** 349–363.

Nguyen AL, Gentilello AS, Balboula AZ, Shrivastava V, Ohring J, Schindler K. 2014. Phosphorylation of threonine 3 on histone H3 by haspin kinase is required for meiosis I in mouse oocytes. *J Cell Sci* **127:** 5066–5078.

Ohashi S, Sakashita G, Ban R, Nagasawa M, Matsuzaki H, Murata Y, Taniguchi H, Shima H, Furukawa K, Urano T. 2006. Phospho-regulation of human protein kinase Aurora-A: Analysis using anti-phospho-Thr288 monoclonal antibodies. *Oncogene* **25:** 7691–7702.

Ozlu N, Srayko M, Kinoshita K, Habermann B, O'Toole ET, Muller-Reichert T, Schmalz N, Desai A, Hyman AA. 2005. An essential function of the *C. elegans* ortholog of TPX2 is to localize activated aurora A kinase to mitotic spindles. *Dev Cell* **9:** 237–248.

Petersen J, Paris J, Willer M, Philippe M, Hagan IM. 2001. The *S. pombe* aurora-related kinase Ark1 associates with mitotic structures in a stage dependent manner and is required for chromosome segregation. *J Cell Sci* **114:** 4371–4384.

Quartuccio SM, Dipali SS, Schindler K. 2017. Haspin inhibition reveals functional differences of interchromatid axis-localized AURKB and AURKC. *Mol Biol Cell* **28:** 2233–2240.

Santaguida S, Vernieri C, Villa F, Ciliberto A, Musacchio A. 2011. Evidence that Aurora B is implicated in spindle checkpoint signalling independently of error correction. *EMBO J* **30:** 1508–1519.

Sasai K, Katayama H, Stenoien DL, Fujii S, Honda R, Kimura M, Okano Y, Tatsuka M, Suzuki F, Nigg EA, et al. 2004. Aurora-C kinase is a novel chromosomal passenger protein that can complement Aurora-B kinase function in mitotic cells. *Cell Motil Cytoskeleton* **59:** 249–263.

Sasai K, Katayama H, Hawke DH, Sen S. 2016. Aurora-C interactions with Survivin and INCENP reveal shared and distinct features compared with Aurora-B chromosome passenger protein complex. *PLoS One* **11:** e0157305.

Satinover DL, Leach CA, Stukenberg PT, Brautigan DL. 2004. Activation of Aurora-A kinase by protein phosphatase inhibitor-2, a bifunctional signaling protein. *Proc Natl Acad Sci* **101:** 8625–8630.

Schindler K, Davydenko O, Fram B, Lampson MA, Schultz RM. 2012. Maternally recruited Aurora C kinase is more stable than Aurora B to support mouse oocyte maturation and early development. *Proc Natl Acad Sci* **109:** E2215–E2222.

Schumacher JM, Golden A, Donovan PJ. 1998. AIR-2: An Aurora/Ipl1-related protein kinase associated with chromosomes and midbody microtubules is required for polar body extrusion and cytokinesis in *Caenorhabditis elegans* embryos. *J Cell Biol* **143:** 1635–1646.

Seki A, Coppinger JA, Jang CY, Yates JR, Fang G. 2008. Bora and the kinase Aurora a cooperatively activate the kinase Plk1 and control mitotic entry. *Science* **320:** 1655–1658.

Sessa F, Mapelli M, Ciferri C, Tarricone C, Areces LB, Schneider TR, Stukenberg PT, Musacchio A. 2005. Mechanism of Aurora B activation by INCENP and inhibition by hesperadin. *Mol Cell* **18:** 379–391.

Sharif B, Na J, Lykke-Hartmann K, McLaughlin SH, Laue E, Glover DM, Zernicka-Goetz M. 2010. The chromosome passenger complex is required for fidelity of chromosome transmission and cytokinesis in meiosis of mouse oocytes. *J Cell Sci* **123:** 4292–4300.

Shuda K, Schindler K, Ma J, Schultz RM, Donovan PJ. 2009. Aurora kinase B modulates chromosome alignment in mouse oocytes. *Mol Reprod Dev* **76:** 1094–1105.

Tanaka TU, Rachidi N, Janke C, Pereira G, Galova M, Schiebel E, Stark MJ, Nasmyth K. 2002. Evidence that the Ipl1-Sli15 (Aurora kinase-INCENP) complex promotes chromosome biorientation by altering kinetochore-spindle pole connections. *Cell* **108:** 317–329.

Tang CJ, Lin CY, Tang TK. 2006. Dynamic localization and functional implications of Aurora-C kinase during male mouse meiosis. *Dev Biol* **290:** 398–410.

Tatsuka M, Katayama H, Ota T, Tanaka T, Odashima S, Suzuki F, Terada Y. 1998. Multinuclearity and increased ploidy caused by overexpression of the aurora- and Ipl1-like midbody-associated protein mitotic kinase in human cancer cells. *Cancer Res* **58:** 4811–4816.

Terada Y, Tatsuka M, Suzuki F, Yasuda Y, Fujita S, Otsu M. 1998. AIM-1: A mammalian midbody-associated protein required for cytokinesis. *EMBO J* **17:** 667–676.

Tsou JH, Chang KC, Chang-Liao PY, Yang ST, Lee CT, Chen YP, Lee YC, Lin BW, Lee JC, Shen MR, et al. 2011. Aberrantly expressed AURKC enhances the transformation and tumourigenicity of epithelial cells. *J Pathol* **225:** 243–254.

Ulisse S, Delcros JG, Baldini E, Toller M, Curcio F, Giacomelli L, Prigent C, Ambesi-Impiombato FS, D'Armiento M, Arlot-Bonnemains Y. 2006. Expression of Aurora kinases in human thyroid carcinoma cell lines and tissues. *Int J Cancer* **119:** 275–282.

Vader G, Kauw JJ, Medema RH, Lens SM. 2006. Survivin mediates targeting of the chromosomal passenger complex to the centromere and midbody. *EMBO Rep* **7:** 85–92.

Vader G, Lens SM. 2008. The Aurora kinase family in cell division and cancer. *Biochim Biophys Acta* **1786:** 60–72.

van der Horst A, Lens SM. 2014. Cell division: Control of the chromosomal passenger complex in time and space. *Chromosoma* **123:** 25–42.

Welburn JP, Vleugel M, Liu D, Yates JR III, Lampson MA, Fukagawa T, Cheeseman IM. 2010. Aurora B phosphorylates spatially distinct targets to differentially regulate the kinetochore-microtubule interface. *Mol Cell* **38:** 383–392.

Wheatley SP, Carvalho A, Vagnarelli P, Earnshaw WC. 2001. INCENP is required for proper targeting of Survivin to the centromeres and the anaphase spindle during mitosis. *Curr Biol* **11:** 886–890.

Yang KT, Li SK, Chang CC, Tang CJ, Lin YN, Lee SC, Tang TK. 2010. Aurora-C kinase deficiency causes cytokinesis failure in meiosis I and production of large polyploid oocytes in mice. *Mol Biol Cell* **21:** 2371–2383.

Yang KT, Tang CJ, Tang TK. 2015. Possible role of Aurora-C in meiosis. *Front Oncol* **5:** 178.

Yasui Y, Urano T, Kawajiri A, Nagata K, Tatsuka M, Saya H, Furukawa K, Takahashi T, Izawa I, Inagaki M. 2004. Autophosphorylation of a newly identified site of Aurora-B is indispensable for cytokinesis. *J Biol Chem* **279:** 12997–13003.

Ye AA, Deretic J, Hoel CM, Hinman AW, Cimini D, Welburn JP, Maresca TJ. 2015. Aurora A kinase contributes to a pole-based error correction pathway. *Curr Biol* **25:** 1842–1851.

Zaytsev AV, Sundin LJ, DeLuca KF, Grishchuk EL, DeLuca JG. 2014. Accurate phosphoregulation of kinetochore-microtubule affinity requires unconstrained molecular interactions. *J Cell Biol* **206:** 45–59.

Zaytsev AV, Mick JE, Maslennikov E, Nikashin B, DeLuca JG, Grishchuk EL. 2015. Multisite phosphorylation of the NDC80 complex gradually tunes its microtubule-binding affinity. *Mol Biol Cell* **26:** 1829–1844.

Zeitlin SG, Shelby RD, Sullivan KF. 2001. CENP-A is phosphorylated by Aurora B kinase and plays an unexpected role in completion of cytokinesis. *J Cell Biol* **155:** 1147–1157.

Zhao ZS, Lim JP, Ng YW, Lim L, Manser E. 2005. The GIT-associated kinase PAK targets to the centrosome and regulates Aurora-A. *Mol Cell* **20:** 237–249.

RotoStep: A Chromosome Dynamics Simulator Reveals Mechanisms of Loop Extrusion

Josh Lawrimore, Brandon Friedman, Ayush Doshi, and Kerry Bloom

Department of Biology, University of North Carolina at Chapel Hill, North Carolina 27599-3280

Correspondence: kerry_bloom@unc.edu

ChromoShake is a three-dimensional simulator designed to explore the range of configurational states a chromosome can adopt based on thermodynamic fluctuations of the polymer chain. Here, we refine ChromoShake to generate dynamic simulations of a DNA-based motor protein such as condensin walking along the chromatin substrate. We model walking as a rotation of DNA-binding heat-repeat proteins around one another. The simulation is applied to several configurations of DNA to reveal the consequences of mechanical stepping on taut chromatin under tension versus loop extrusion on single-tethered, floppy chromatin substrates. These simulations provide testable hypotheses for condensin and other DNA-based motors functioning along interphase chromosomes. Our model reveals a novel mechanism for condensin enrichment in the pericentromeric region of mitotic chromosomes. Increased condensin dwell time at centromeres results in a high density of pericentric loops that in turn provide substrate for additional condensin.

There has been a revolution in understanding the higher-order structure and organization of chromosome in the past decade. Several major approaches (3C, ChromEMT, and super-resolution microscopy) are indicative of a disordered array of loopy fibers that emanate from an axial core (Dostie and Bickmore 2012; Dekker et al. 2013; Ou et al. 2017). The hierarchical models of structural intermediates building from 11 to 30 nm and larger fibers are not borne out in these recent 3D and live-cell studies (Ou et al. 2017). DNA looping was first observed in squash preparations of salamander eggs under the light microscope by the embryologist Oskar Hertwig in the early 1900s (Hertwig 1906). Paulson and Laemmli (1977) observed DNA loops when examining chromosome spreads in isolated mammalian cells. In metaphase, the loops emanate from a protein-rich chromosome scaffold. The chromosome scaffold is enriched in topology-adjusting proteins, such as topoisomerase II and the SMC (structural maintenance of chromosomes) proteins, known as condensin (Earnshaw et al. 1985; Hirano 2006).

Loops are a natural consequence of the entropic fluctuations and excluded volume interactions of tethered polymer chains in a confined space, such as the nucleus (Vasquez et al. 2016). If we consider the genome as a ball of yarn, the formation of loops can be appreciated as chains that randomly collide and wiggle around one another. Energy-requiring processes are also involved in loop formation. The earliest suggestion of loop extrusion came from the trombone model of DNA replication (Sinha et al. 1980; Alberts et al. 1983) and direct visualization of DNA looping at the replication fork (Park et al. 1998). More recently, SMC proteins (e.g., cohesin and condensin), which bind and hydrolyze ATP, have been cited as having loop extrusion potential (Alipour and Marko 2012). Condensin has garnered attention based on recent studies showing it to be a DNA translocase (Terekawa et al. 2017).

Condensin is composed of five subunits, two coiled-coils SMC2 and 4, a kleisin (Brn1), and two heat-repeat-containing proteins (Ycs4 and Ycg1). The heat-repeat proteins are likely to be sites of DNA-binding within the condensin complex. Terekawa et al. (2017) showed the ability of condensin to move processively along linear DNA sheets (Fazio et al. 2008). The challenge ahead is to understand, first, how local motion and topological constraints of the DNA exerted by condensin will be dissipated along the length of a long-chain polymer and, second, how the fluctuating, loopy genome feeds back to the ability of condensin to translocate processively.

ChromoShake is a statistical mechanics model depicting the motions of arrays of bead-springs (DNA) that show Brownian dynamics in a viscous environment (Lawrimore et al. 2016). The simulation has been applied to the centromere in budding yeast to reveal how the density of pericentromeric loops stiffen centromeric chromatin, imparting an active function to the centromere in mitosis (Lawrimore et al. 2016). In previous studies condensin was implemented as static springs (Lawrimore et al. 2016). Here, we introduce a new model, RotoStep, as a first-principles statistical mechanics approach to simulate condensin dynamics. Terekawa et al. (2017) have recently examined the behavior of single condensin molecules on DNA sheets. Terekawa et al. (2017) provides critical experimental metrics for evaluating results from simulation. Using RotoStep to simulate hand-over-hand motion (e.g., microtubule-based kinesin motor [Kull et al. 1996]), our simulations indicate that condensin can translocate along taut linear DNA and compact singly tethered DNA chains.

Published by Cold Spring Harbor Laboratory Press; doi: 10.1101/sqb.2017.82.033696

Cold Spring Harbor Symposia on Quantitative Biology, Volume LXXXII

The dynamics of condensin stepping along single-tethered DNA result in extrusion of DNA loops. The same parameters for motor stepping result in drastically different geometries that are dictated by chromatin substrate dynamics. These simulations provide the first glimpse of how loop extrusion might work in living organisms.

RESULTS

Model Assumptions

Condensin is a DNA-based motor protein with the ability to translocate along double-stranded DNA at relatively high velocities (60 bp/sec) (Terekawa et al. 2017). To simulate condensin tracking DNA we used the array of bead-springs to depict both the DNA (long chain linear bead-spring configuration) and condensin (bead-spring chain). There are 11 beads in the condensin holocomplex; each bead is ~10 nm in diameter. Condensin is a very flexible molecule, showing a persistence length of ~4 nm (Eeftens et al. 2016), about one-tenth that of DNA (50 nm). The chains of the antiparallel coiled-coils fold back on one another to length of 45 nm. The complex adopts a number of configurations, including circular, V-shaped, and globular. The SMC protein coiled-coils are represented by beads 2–8. To achieve DNA binding, we assert that the heat-repeat proteins (Ycs4, Ycg1) bind DNA (Piazza et al. 2014). These are beads 1 (representing Ycg1, red in Fig. 1) and beads 9 and 10 (representing Ycs4, white and red in Fig. 1) in simulations. The kleisin (Brn1) is represented as bead 11 (pink, Fig. 1) between the heat-repeat proteins connected via springs (not shown). This feature provides mechanical linkage between the leading and trailing points of contact.

The RotoStep program parses the coordinates, spring attachments, and indices of all beads in a ChromoShake simulation model. The binding mechanics of condensin changes when the condensin molecule becomes extended. We consider condensin–DNA binding unstable if the distance between beads 2 and 8 is >30 nm. If the distance between beads 2 and 8 is <30 nm (Fig. 1A), the program determines the distances of the Ycs4 beads, 9 and 10 (rightmost white and red beads, Fig. 1A), to the Ycg1 bead, 1 (leftmost red bead, Fig. 1A). The closest Ycs4

bead to the Ycg1 bead is labeled proximal, and the other distal. A vector is drawn from the center of the proximal Ycs4 bead to the center of the distal Ycs4 bead and extended 10 nm. The proximal Ycs4 bead is bound to the closest DNA bead to this point in space as the previous proximal Ycs4 bead DNA attachment is removed. This results in condensin stepping along the substrate resulting in loop extrusion (Fig. 1B).

If the distance between beads 2 and 8 is >30 nm (Fig. 1C), the spring constant of the spring connecting the Ycg1 bead to the DNA is weakened 1000-fold. ChromoShake is then run to introduce thermal noise, allowing the springs between kleisin and beads 2 and 8, to pinch the condensin molecule together (Fig. 1D). On the next iteration of Roto-Step, the Ycg1 bead is joined to the nearest DNA bead with a spring as the previous weak spring is simultaneously deleted. This results in the destabilization of the previous condensin-mediated loop and causes the Ycg1 end of condensin to bind DNA at a new location. After every iteration of RotoStep, ChromoShake is run on simulations to input thermal noise. The springs and kleisin-based threshold (30 nm) are set to yield processive motion on a doubly tethered substrate (Fig. 2A), as described in Terekawa et al. (2017).

Consequences of Processive Motion on Tethered Substrate with *cis*- and *trans*-Binding

We set our condensin parameters (see Box 1) such that a single condensin could processively walk on an extended and pinned DNA substrate with a translocation speed of 60 bp/sec as reported in Terekawa et al. (2017). As shown in Figure 2A and Fig. 3A–C, upon loading condensin at one end, it processively migrates to the other end of the polymer chain.

Terekawa et al. (2017) reported a singly pinned DNA substrate incubated with condensin resulted in compacted DNA. We confirmed that our RotoStep simulation caused a singly tethered strand to compact (Fig. 2B,C and Fig. 3D, E). We added a single condensin to the tethered end of a singly tethered strand (Fig. 3D,E). Given that the kleisin spring never becomes taut, the strand is extruded as a loop. After several steps, a loop is spontaneously extruded as the strand adopts a random coil. Despite the random orientation

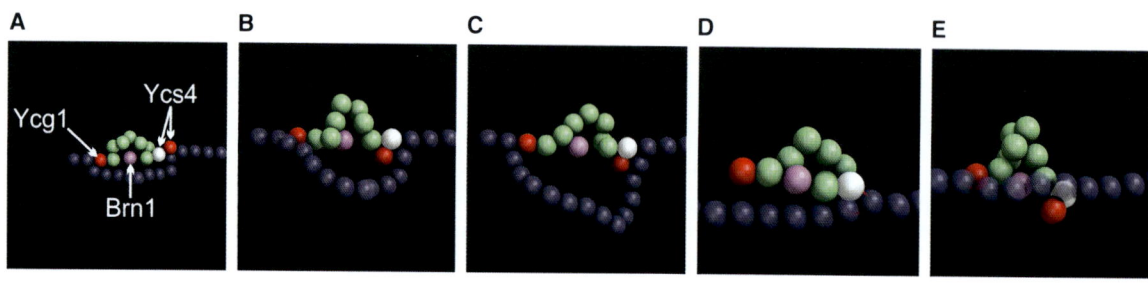

Figure 1. Simulated step and release algorithm. (*A*) Coiled-coil α-helices of SMC proteins (SMC2,4) are in green. Heat-repeat proteins Ycg1 (*left* red bead) and Ycs4 (white and right red beads) are bound to DNA (purple beads). The kleisin, Brn1 (pink bead), bridges the SMC containing subunits. (*B*) One heat-repeat in Ycs4 (white bead) detaches and rebinds DNA depending on the projection of vector (10 nm) between the two heat repeats in Ycs4 relative to Ycg1 (see text). (*C*) When springs linking kleisin (pink bead) to beads 2 and 8 in the SMC coiled-coils are extended ~30 nm, Ycg1 (trailing red bead) is released (shown in *D*). (*E*) Rebinding of Ycg1 to DNA following kleisin spring recoil. This results in directed motion on a taut chain.

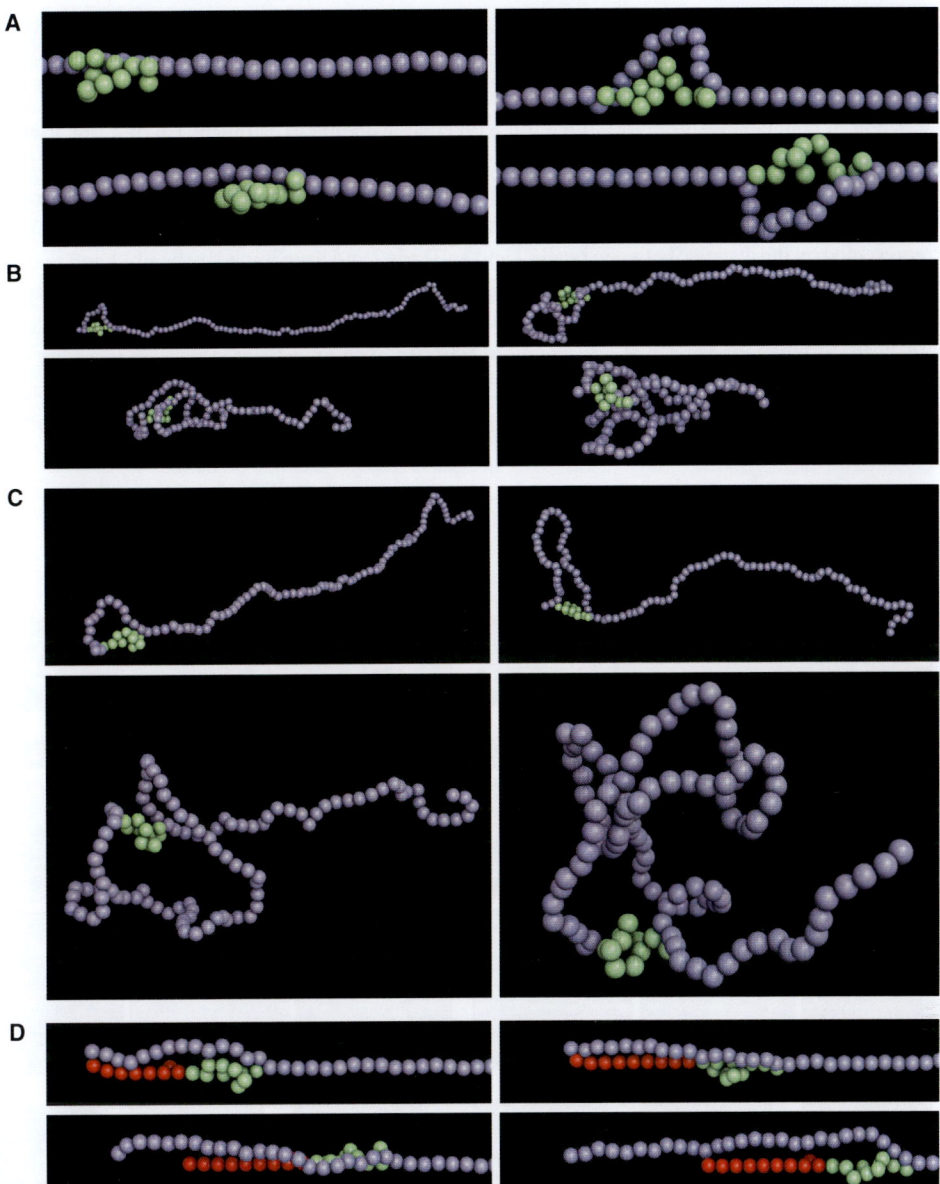

Figure 2. Simulation of condensin walking along a chromatin substrate. (*A*) Stepping: The DNA (purple beads) with a single condensin (green beads). The DNA ends are tethered to watch how a single condensin molecule steps along the DNA. (*B*) Looping: The DNA is only tethered at one end, so that its behavior is more like a fluctuating chain in vivo. Note that when condensin steps, loops are extruded as a consequence of the floppiness of the chain. (*C*) Looping (different camera perspective from *B*). (*D*) Translocation: Condensin can bind in *cis* (the same molecule) or in *trans* (different molecules). When it binds in *trans* and takes steps, it will move one DNA strand relative to the other. This is a cross-linking and mobilization function. The two DNA strands are depicted in purple and red. The red strand is fluctuating but is equivalent to the mass of lambda DNA (Terekawa et al. 2017), ~10× greater than the mass of the purple beads.

of the floppy strand, which we hypothesized would cause condensin to flip its stepping orientation, a single condensin is able to extrude a loop in a processive manner with simple, proximity-based binding mechanics. This simulation demonstrates loop extrusion is a consequence of processive motion on a floppy substrate. This is an important insight and provides a mechanism for how condensin alters the topology of DNA in vivo to form chromosomes.

Last, Terekawa et al. (2017) reported that condensin is able to proceed along an extended DNA substrate while bound to a separate DNA strand. We replicated this pro-

cessive motion of a *trans*-substrate by placing the Ycs4 beads on a doubly tethered strand and the Ycg1 bead on the other strand. In this situation (Fig. 2D), as condensin walks, the *trans*-associated molecule is transported along the substrate molecule, recapitulating the phenomena discovered in Terekawa et al. (2017).

Model Convolution

To translate simulations into experimental images, we implement a method known as model convolution (Gard-

BOX 1: PARAMETER LIST

Persistence length of DNA = 50 nm, set by hinge force.

Persistence length of condensin = unset, no hinge force on condensin due to estimates Lp of 4 nm. Two heat-repeat subunit = beads 9 and 10, condensin to DNA spring strength is same as DNA (2 GPa), rest length 10 nm.

One heat-repeat subunit = bead 1, two binding states, strong state is 2 GPa, weak state is 2 MPa, rest length is 10 nm, weak state activated when kleisin spring extension is >30 nm (3× mass separations [thresh in code]).

Kleisin is represented by the distance between beads 2 and 8. Beads 2 and 8 are joined by bead 11 and two springs (i.e., 2-11-8). Spring strength is 200 MPa; rest length of springs are 10 nm. Kleisin threshold is 30 nm.

Step size/rate: Beads 9 and 10 will rotate about each other every 35 μsec of simulation time, for one bead to bind to a new DNA bead. Simulations are run at a viscosity of 0.01 P. Given an estimated nuclear viscosity of 141 Poise (Fisher et al. 2009), 35 μsec of simulation time is equivalent to 0.5 sec. Given that the most common step size is one bead per rotation, this results in a step rate of 2 beads per second. Each bead represents ~30 bp, so condensin has a step rate of 60 bp/sec, as described in Terekawa et al. (2017).

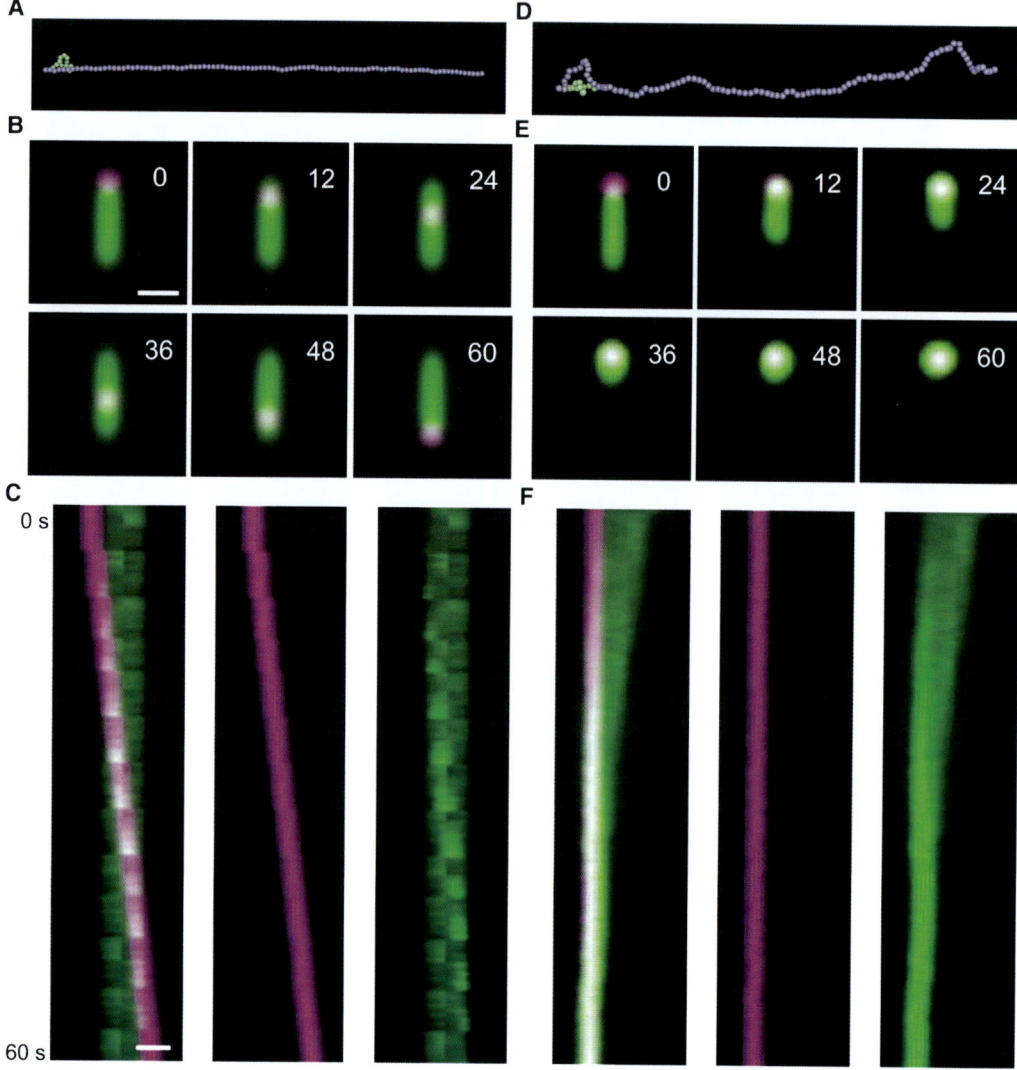

Figure 3. Kymographs of simulated motion convolved through objective point spread function. Rendering of double-tethered DNA (*A*) and single-tethered DNA (*D*), simulations containing a single condensin. Condensin is green; DNA is blue. Time series of simulated fluorescence images of double-tethered (*B*) and single-tethered (*E*) simulations. Condensin is magenta; DNA is green. Kymographs of double-tethered DNA (*C*) and single-tethered DNA (*F*) simulations. Time in seconds.

ner et al. 2010). We use the microscope objective point spread function to render the simulation into a format that can be directly compared with microscope images of live cells. Figure 3 shows the behavior of condensin walking on a pinned DNA substrate. Note the correspondence between this image and those obtained by Terekawa et al. (2017) for condensin translocating on DNA. The extrusion of DNA loops when condensin translocates along an unpinned substrate readily shows that the DNA compacts into a diffraction-limited spot (Fig. 3E,F). These simulations (Fig. 2) and model convolution (Fig. 3) provide proof of principle that RotoStep can recapitulate the behavior of condensin in vitro.

Strand Compaction via Multiple Condensins

In vivo, there are multiple condensin molecules on each chromosome. In budding yeast, the density of condensin is about 1 molecule/10 kb in the bulk of the chromosome and about 1 molecule/3 kb in the pericentromere. We placed 10 molecules of condensin on a single-tethered chain (5 μm, 15 kb). The rules for condensin steps are as previously described, and the step orientation for each molecule is random (Fig. 4). As individual condensin's take steps, loops are extruded along the strand (as shown for a single molecule, Fig. 2C,D), and the strand is rapidly compacted into a chromosome structure. Loop extrusion is independent of the direction that condensin steps. With multiple condensin molecules, as multiple loops form, overall chain length shortens. The backbone is fed into the loops via the step function of each molecule. The surprise is that no new features were implemented beyond the simplest of step functions. Compaction is simply the transfer of the backbone chain into extruded loops (Fig. 4).

Tension along the DNA Polymer in a Metaphase Configuration Results in Transient, Pericentric Loops

Centromere DNA is tethered to microtubules via their kinetochore attachment sites. The enrichment of condensin can be visualized in vivo as a bar or one or two foci along the spindle axis (Bachellier-Bassi et al. 2008; Ste-

phens et al. 2011, 2013). To try to gain insight into the thermodynamics governing condensin localization within the pericentromere we provided a dicentric plasmid substrate (two centromeres) for simulation. The actual plasmid is 11.4 kb (~3.86 micron, 386 beads in simulation). The position of the two centromeres (tethered ends) are indicated as pink beads (Fig. 5) and lie 800 nm from one another, roughly the distance between separated sister kinetochores in budding yeast metaphase (Pearson et al. 2001). The color of beads on the polymer reflect yeast DNA (blue) and a repeat array of tetracycline operator (white) as described in Lawrimore et al. (2015).

In our simulations, condensin walks along the chain as described above, randomly oriented with respect to one another (Fig. 5). It is difficult for condensin to step through the centromere (simulated as tether sites in pink) because of the geometry of the beads, which increased the dwell time of condensin near the centromeres. Thus, there is kinetic delay for a single condensin molecule at the centromere. This reflects the situation in cells, in which the centromere is at the apex of a stereotypic loop where the kinetochore attaches to microtubules (Yeh et al. 2008). Over time, additional condensin molecules arrive, also extruding loops. The concentration of loops increase and now each condensin can "jump" from one loop to another, amplifying the kinetic delay. The emergent phenomenon is that condensin tends to accumulate near the centromere tether sites. This reflects the density of loops at tether sites. We predict that the accumulation of loops at the centromere reflects the in vivo situation. Because of stochastics of motion in the loops and condensin stepping, a single condensin molecule will escape one position and start toward the other centromere. At this juncture, the loop density decreases, which biases the remaining condensin molecules to "follow" the initial escapee. In silico, this results in condensin molecules "chasing" one another until they accumulate at the other centromere and the cycle continues.

There are several in vivo behaviors that now can be accounted for. Condensin appears as foci that could have been interpreted as oligomerization. The alternative explanation based on simple thermodynamics is that the concentration of condensin reflects the local density of loops. Where loop density is high, condensin will accumulate

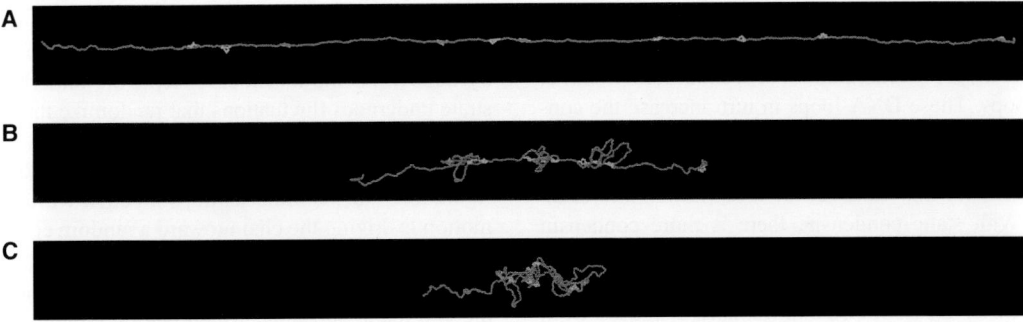

Figure 4. Chromosome compaction. Several condensin molecules were distributed in random position and direction. The DNA molecule is unpinned. As condensin steps (*A–C*, over time), the DNA gets reeled in and the entire ensemble rapidly condenses into a dense aggregate of DNA and condensin.

A   **B**   **C**

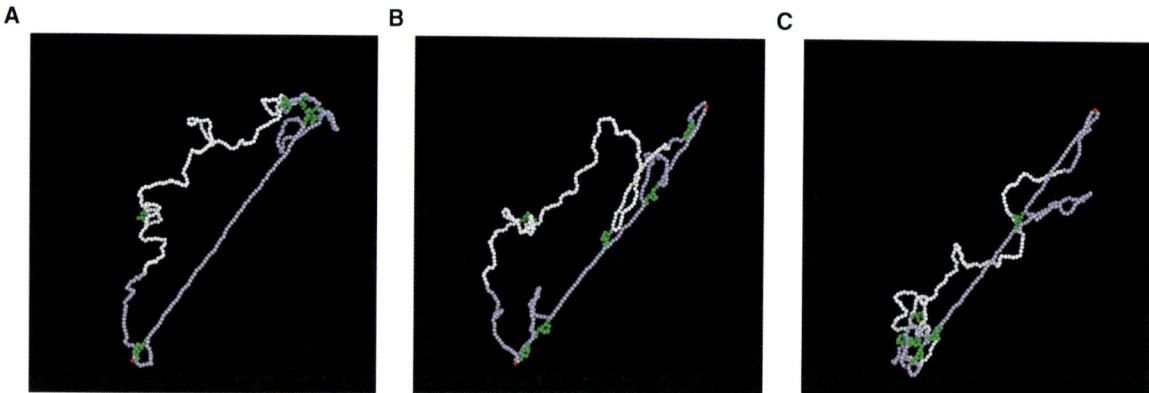

Figure 5. Motion on a dicentric plasmid. A dicentric plasmid in metaphase is simulated by tethering two DNA beads at the position of centromeres within the plasmid (DNA/bead coordinates) and in space at the position of separated kinetochores in yeast metaphase (Lawrimore et al. 2015). The purple beads represent plasmid DNA sequences, the white beads the position of repeat arrays of Tet operator used to visualize with plasmid with TetR-GFP. The number of condensin molecules (green) represents the average density of condensin within the pericentric region of the centromere. Condensin is randomly positioned with respect to DNA sequence and direction. Upon simulated motion (A–C, over time), condensin travels from one tether to the other. Condensin aggregates at the tethers, as a function of the number of loops (i.e., beads). Random fluctuations result in individual molecules migrating from one tether to the other. Once one condensin escapes the aggregate, the DNA density decreases (fewer loops, i.e., fewer beads), biasing additional molecules to migrate away. The appearance is that condensins chase one another. This phenocopies what is observed in live cells, although we cannot observe the behavior of single molecules.

because of the increase in substrate concentration. The second behavior is the directional instability observed in vivo. This can be accounted for through stochastics of a single molecule that escapes the region of loops. As one molecule leaves, the loop density is reduced triggering a cascade that will allow it to escape and accumulate on the other side.

Condensin Stepping Creates Condensin-Rich DNA Loops near Tethered Centromeres

We performed statistical analysis on two dicentric plasmids with the same initial placement of condensin, but with mobile (Fig. 6A,B) and immobile condensin (Fig. 6C,D). The probability distribution of condensin (Fig. 6A,C) and DNA beads (Fig. 6B,D) over all time points of the simulations are shown. In simulations with dynamic condensin there is a greater correlation between condensin number and DNA-strand density than immobile condensin (Fig. 6E). Mobile condensin colocalizes with centromeres more often than immobile condensin (Fig. 6F). The enrichment of condensin observed at centromeres may be due to the increased condensin dwell time at tether sites, resulting in the accumulation of centromere-proximal DNA loops. These DNA loops in turn increase the concentration of condensin near centromeres, providing a positive-feedback loop. This pattern of condensin is consistent with that observed in vivo (Lawrimore et al. 2015). With static condensin, there is more condensin along the axis (Fig. 6A,C). Consequently, the DNA has more freedom of movement, which is reflected in the DNA density distribution being radially displaced from the axis (Fig. 6B,D), recapitulating in vivo experiments of the temperature-sensitive condensin mutation brn1-9 (Stephens et al. 2011).

Figure 7 demonstrates that the motion of the substrate confounds the analysis of condensin turnover because of motion of the substrate. Paradoxically, even a highly processive motor does not appear to recover after photobleaching because of a dense, floppy DNA substrate.

CONCLUSION

We have used a statistical mechanics model of chromatin to examine the consequences of translocation of a motor protein such as condensin on a floppy chromatin substrate. Terekawa et al. (2017) recently reported that condensin is a mechanochemical motor protein. Through simulations that recapitulate in vitro findings we are able to query the consequences of motor stepping on a thermodynamically mobile substrate. Unlike the behavior of motor proteins on DNA sheets under flow, or kinesin-like microtubule-based motor proteins walking on a rigid microtubule, the behavior of the chromatin substrate has a disproportionate contribution to the behavior of a generic DNA-based translocase. Translocation along a floppy substrate is more akin to the random walk of a segment of a fluctuating polymer chain (i.e., subdiffusive Rouse chain [Rubinstein and Colby 2003]). As the translocase steps, the underlying substrate undergoes fluctuations that randomize the direction of the next step (compare Figs. 2 and 3).

At each step, the chromatin is constantly exploring configurations that maximize entropy. In other words, thermal motion is driving the chain toward a random coil. The net consequence is the extrusion of DNA loops (Fig. 2B,C). A key parameter in terms of cell physiology is the stiffness of the chromatin substrate, which can be altered by tethers and the presence of other DNA motors/loop extruders generating regions of taut substrate. If the local substrate is floppy, there will be little if any directed motion. In

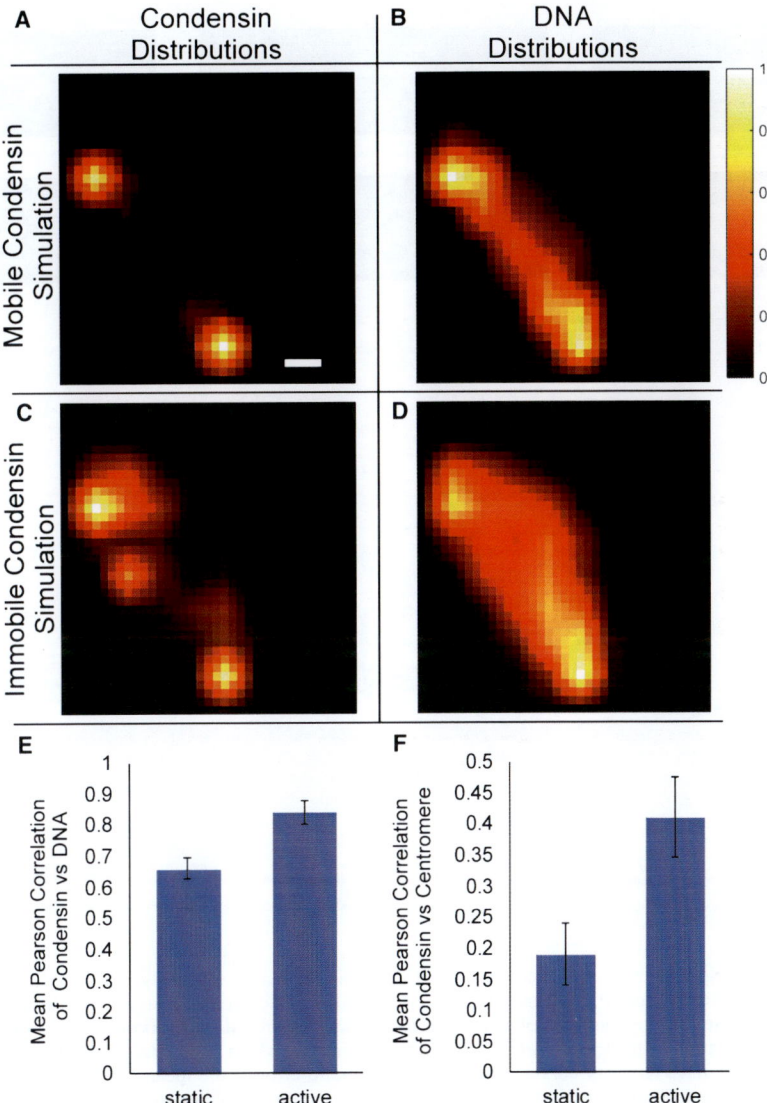

Figure 6. Condensin enrichment at centromere-proximal DNA loops in simulation with mobile vs. immobile condensin. The DNA substrate is the dicentric plasmid described in Figure 5 (both ends pinned at centromeres). Condensin is randomly positioned and oriented along the dicentric molecule. Condensin can step along the DNA in *A,B*, but is static in *C,D*. The starting positions of condensin are the same in both simulations. A probability density map of condensin is shown in *A,C* and the plasmid DNA in *B,D*. (*E,F*) Statistical analysis of the correlation of condensin with DNA (*E*) and centromere (tethered beads) (*F*). Mobile condensin tends to accumulate at the sites of tethering near the centromeres, resulting in condensin-rich, centromere-proximal DNA loops. DNA tends to show greater radial freedom of motion in simulation with mobile condensin (width of *B* vs. *D*). Scale bar, 100 nm.

contrast, if the substrate is stiffer because of extension, the motor will be more prone to display directed motion. The behavior is also dependent on whether the substrate ends are tethered. If not, even a flexible translocase will enhance the ability of the substrate to adopt a random coil (Fig. 3E). It is reported that condensin is very flexible (Eeftens et al. 2016) (short persistence length, 4 nm), and thus may show bursts of directed motion in vivo, depending on the distribution of tethers along the chromosome.

A second feature of translocation along a floppy substrate is compaction. Because there is no vectorial motion, the motor will decrease the time required for the chain to

adopt a random coil. This may be perceived as active condensation, but it is a natural consequence of random stepping on a thermally fluctuating chain. Likewise, condensin can appear to concentrate in particular locations, such as within the pericentromere and the nucleolus (Bachellier-Bassi et al. 2008; Stephens et al. 2013; Snider et al. 2014). This has been interpreted as oligomerization between condensin holoenzymes or other cross-linking proteins. The alternative interpretation from the statistical mechanics model is that the concentration of DNA loops increases because of condensin's inability to quickly traverse the centromere region. As condensin steps and extrudes loops, in an environment where there are a plethora

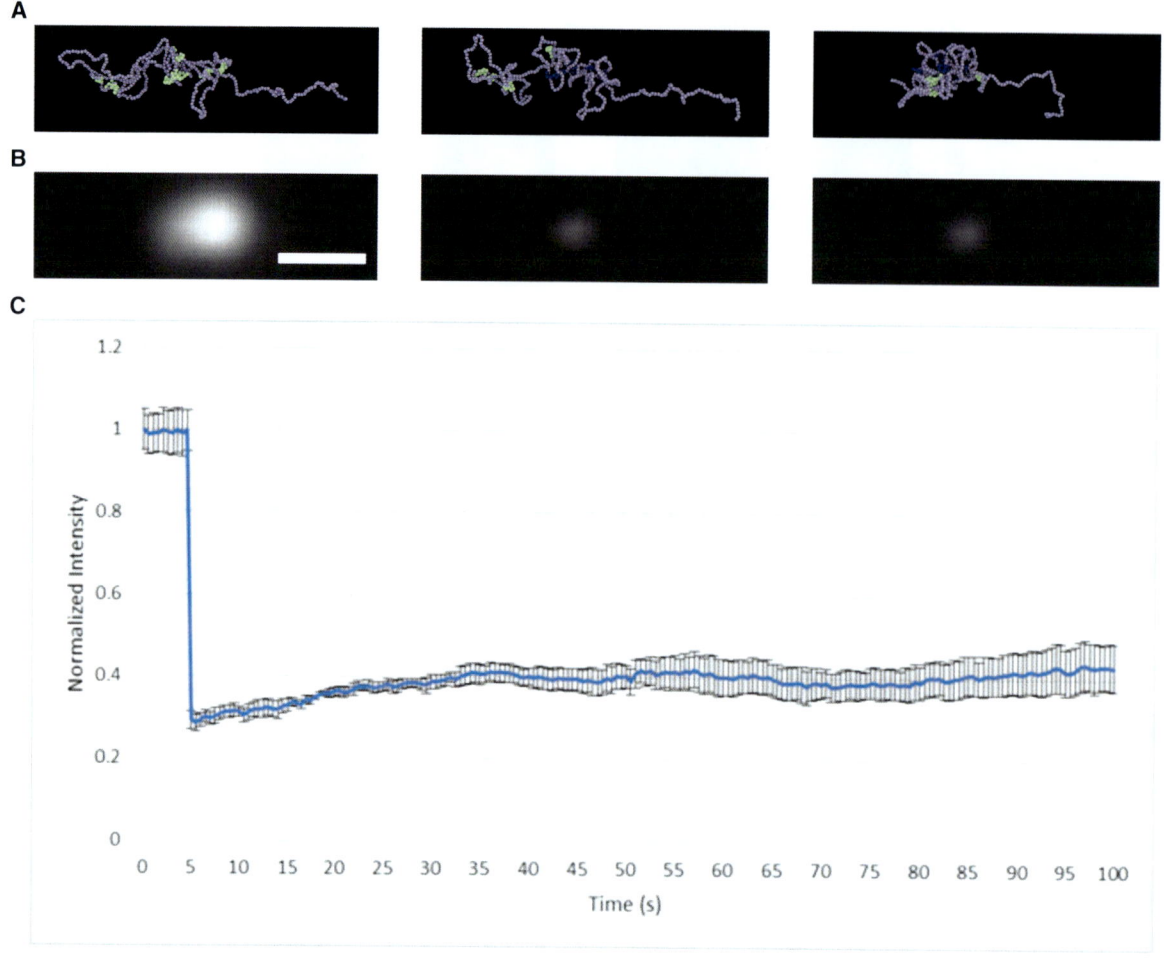

Figure 7. Simulated FRAP experiment by model convolution. Condensin molecules randomly positioned on a 5-kb DNA chain and convolved with the point spread function (PSF) (described above) to simulate a microscopic image. (*A*) A diffraction limited spot (~250 micron) was drawn around several condensin molecules. Bleaching was simulated by marking these molecules dark (blue). (*B*) Fluorescence pre- and postbleach (from *left* to *right*). Scale bar, 0.5 μm. (*C*) Mean normalized quantitative fluorescence recovery. Curve is mean of three normalized recovery curves. Error bars are standard error of the mean. The finding in simulation is that any recovery can be attributed to fluctuation of the chain (and associated condensin) into the bleach zone. Thus FRAP is confounded by the thermodynamics of the substrate. (Compare *B* to Fig. 5 in Lawrimore et al. [2015].)

of molecules, condensin on one loop can readily step to a physically adjacent loop. This is evident in Figure 6, showing accumulation of condensin at sites of tethers, corresponding with increased concentration of DNA loops. Thus, perceived condensin accumulation may reflect the propensity for loops to congregate condensin, which produces more loops. Condensin's ability to step from one loop to another provides a positive-feedback mechanism to increase the duration of these transient events and the local concentration of DNA loops. However, the stochastics of the system will result in occasional molecules escaping sites of loop accumulation by traversing the taut regions of the substrate. As shown in Figure 5, when one molecule escapes, the loop density decreases, thus biasing additional molecules to exit sites of loop accumulation. This unstable positive-feedback loop may contribute to condensin's enrichment at the pericentromere and nucleolus in vivo. Such a transient, positive feedback may explain the existence of TADs in population

studies, but not in single cells (Flyamer et al. 2017). In essence, the organization of the DNA biases the occasional enrichment of condensin based on the presence of a tether causing a loop.

The statistical mechanics of polymer behavior and protein translocation provides a powerful tool to build intuition for understanding experimental observation and making physically accurate hypotheses. Using the simple assumptions of proximity-based substrate binding and extensional-dependent substrate unbinding, we provide alternative explanations for several experimental observations and make new predictions about the rules for loop persistence and extrusion. The model highlights potential new functions for tethers along the chromosome and how the distribution of tethers will be intimately involved in loop formation at a distance (Fig. 3B,E). Finally, the mechanical feedback between loops in the same proximity provides new perspectives on interchromosomal communication.

ACKNOWLEDGMENTS

We thank Dr. Paula Vasquez (University of South Carolina) and Dr. Elaine Yeh (University of North Carolina, Chapel Hill) for discussion and critical reading of the manuscript. This work was supported by the National Institutes of Health General Medicine, R37GM32238 to K.B. and T32CA201159-01 and T32GM007092-39 to J.L.

REFERENCES

Alberts BM, Barry J, Bedinger P, Formosa T, Jongeneel CV, Kreuzer KN. 1983. Studies on DNA replication in the bacteriophage T4 in vitro system. *Cold Spring Harb Symp Quant Biol* **47 Pt 2:** 655–668.

Alipour E, Marko JF. 2012. Self-organization of domain structures by DNA-loop-extruding enzymes. *Nucleic Acids Res* **40:** 11202–11212.

Bachellier-Bassi S, Gadal O, Bourout G, Nehrbass U. 2008. Cell cycle-dependent kinetochore localization of condensin complex in *Saccharomyces cerevisiae*. *J Struct Biol* **162:** 248–259.

Dekker J, Marti-Renom MA, Mirny LA. 2013. Exploring the three-dimensional organization of genomes: Interpreting chromatin interaction data. *Nat Rev Genet* **14:** 390–403.

Dostie J, Bickmore WA. 2012. Chromosome organization in the nucleus—Charting new territory across the Hi-Cs. *Curr Opin Genet Dev* **22:** 125–131.

Earnshaw WC, Halligan B, Cooke CA, Heck MM, Liu LF. 1985. Topoisomerase II is a structural component of mitotic chromosome scaffolds. *J Cell Biol* **100:** 1706–1715.

Eeftens JM, Katan AJ, Kschonsak M, Hassler M, de Wilde L, Dief EM, Haering CH, Dekker C. 2016. Condensin Smc2-Smc4 dimers are flexible and dynamic. *Cell Rep* **14:** 1813–1818.

Fazio T, Visnapuu ML, Wind S, Greene EC. 2008. DNA curtains and nanoscale curtain rods: High-throughput tools for single molecule imaging. *Langmuir* **24:** 10524–10531.

Fisher JK, Ballenger M, O'Brien ET, Haase J, Superfine R, Bloom K. 2009. DNA relaxation dynamics as a probe for the intracellular environment. *Proc Natl Acad Sci* **106:** 9250–9255.

Flyamer IM, Gassler J, Imakaev M, Brandao HB, Ulianov SV, Abdennur N, Razin SV, Mirny LA, Tachibana-Konwalski K. 2017. Single-nucleus Hi-C reveals unique chromatin reorganization at oocyte-to-zygote transition. *Nature* **544:** 110–114.

Gardner MK, Sprague BL, Pearson CG, Cosgrove BD, Bicek AD, Bloom K, Salmon ED, Odde DJ. 2010. Model convolution: A computational approach to digital image interpretation. *Cell Mol Bioeng* **3:** 163–170.

Hertwig O. 1906. *Lehrbuch der Entwicklungsgeschichte des Menschen und der Wirbeltiere* [Textbook of developmental history of humans and vertebrates]. Reissue 2016. Wentworth Press, Sydney.

Hirano T. 2006. At the heart of the chromosome: SMC proteins in action. *Nat Rev Mol Cell Biol* **7:** 311–322.

Kull FJ, Sablin EP, Lau R, Fletterick RJ, Vale RD. 1996. Crystal structure of the kinesin motor domain reveals a structural similarity to myosin. *Nature* **380:** 550–555.

Lawrimore J, Vasquez PA, Falvo MR, Taylor RM II, Vicci L, Yeh E, Forest MG, Bloom K. 2015. DNA loops generate intracentromere tension in mitosis. *J Cell Biol* **210:** 553–564.

Lawrimore J, Aicher JK, Hahn P, Fulp A, Kompa B, Vicci L, Falvo M, Taylor RM II, Bloom K. 2016. ChromoShake: A chromosome dynamics simulator reveals that chromatin loops stiffen centromeric chromatin. *Mol Biol Cell* **27:** 153–166.

Ou HD, Phan S, Deerinck TJ, Thor A, Ellisman MH, O'Shea CC. 2017. ChromEMT: Visualizing 3D chromatin structure and compaction in interphase and mitotic cells. *Science* **357:** 370–382.

Park K, Debyser Z, Tabor S, Richardson CC, Griffith JD. 1998. Formation of a DNA loop at the replication fork generated by bacteriophage T7 replication proteins. *J Biol Chem* **273:** 5260–5270.

Paulson JR, Laemmli UK. 1977. The structure of histone-depleted metaphase chromosomes. *Cell* **12:** 817–828.

Pearson CG, Maddox PS, Salmon ED, Bloom K. 2001. Budding yeast chromosome structure and dynamics during mitosis. *J Cell Biol* **152:** 1255–1266.

Piazza I, Rutkowska A, Ori A, Walczak M, Metz J, Pelechano V, Beck M, Haering CH. 2014. Association of condensin with chromosomes depends on DNA binding by its HEAT-repeat subunits. *Nat Struct Mol Biol* **21:** 560–568.

Rubinstein M, Colby RH. 2003. *Polymer physics*. Oxford University Press, Oxford.

Sinha NK, Morris CF, Alberts BM. 1980. Efficient in vitro replication of double-stranded DNA templates by a purified T4 bacteriophage replication system. *J Biol Chem* **255:** 4290–4293.

Snider CE, Stephens AD, Kirkland JG, Hamdani O, Kamakaka RT, Bloom K. 2014. Dyskerin, tRNA genes, and condensin tether pericentric chromatin to the spindle axis in mitosis. *J Cell Biol* **207:** 189–199.

Stephens AD, Haase J, Vicci L, Taylor RM II, Bloom K. 2011. Cohesin, condensin, and the intramolecular centromere loop together generate the mitotic chromatin spring. *J Cell Biol* **193:** 1167–1180.

Stephens AD, Quammen CW, Chang B, Haase J, Taylor RM II, Bloom K. 2013. The spatial segregation of pericentric cohesin and condensin in the mitotic spindle. *Mol Biol Cell* **24:** 3909–3919.

Terekawa T, Bisht S, Eeftens JM, Dekker C, Haering CH, Greene EC. 2017. The condensin complex is a mechanochemical motor that translocates along DNA. *Science* **358:** 672–676.

Vasquez PA, Hult C, Adalsteinsson D, Lawrimore J, Forest MG, Bloom K. 2016. Entropy gives rise to topologically associating domains. *Nucleic Acids Res* **44:** 5540–5549.

Yeh E, Haase J, Paliulis LV, Joglekar A, Bond L, Bouck D, Salmon ED, Bloom KS. 2008. Pericentric chromatin is organized into an intramolecular loop in mitosis. *Curr Biol* **18:** 81–90.

Taming the Beast: Control of APC/C^{Cdc20}-Dependent Destruction

Pablo Lara-Gonzalez,[1,2] Taekyung Kim,[1,2] and Arshad Desai[1,2]

[1]*Ludwig Institute for Cancer Research, La Jolla, California 92093*

[2]*Department of Cellular and Molecular Medicine, University of California San Diego, La Jolla, California 92093*

Correspondence: abdesai@ucsd.edu

The anaphase-promoting complex/cyclosome (APC/C) is a large multisubunit ubiquitin ligase that triggers the metaphase-to-anaphase transition in the cell cycle by targeting the substrates cyclin B and securin for destruction. APC/C activity toward these two key substrates requires the coactivator Cdc20. To ensure that cells enter mitosis and partition their duplicated genome with high accuracy, APC/C^{Cdc20} activity must be tightly controlled. Here, we discuss the mechanisms that regulate APC/C^{Cdc20} activity both before and during mitosis. We focus our discussion primarily on the chromosomal pathways that both accelerate and delay APC/C activation by targeting Cdc20 to opposing fates. The findings discussed provide an overview of how cells control the activation of this major cell cycle regulator to ensure both accurate and timely cell division.

During cell division, genome stability depends on tight regulation of anaphase, the mitotic stage in which sister chromatids are separated. Anaphase should only occur after sister chromatids of all replicated chromosomes have correctly attached to opposite poles of the mitotic spindle (Fig. 1A). Progression into anaphase before achieving this fully attached state can lead to errors in chromosome segregation and aneuploidy, a hallmark of birth defects and cancer (Holland and Cleveland 2012; Santaguida and Amon 2015; Funk et al. 2016).

In eukaryotes, anaphase onset is triggered by the anaphase-promoting complex/cyclosome (APC/C), a large E3 ubiquitin ligase (Fig. 1A,B; Peters 2006; Pines 2011; Primorac and Musacchio 2013; Barford 2015). When the APC/C is active, it promotes the polyubiquitination of its substrates, which leads to their proteasome-mediated degradation. The essential APC/C substrates for anaphase onset are securin and cyclin B. Securin is the inhibitor of separase, the cysteine protease that cleaves a subunit of the cohesin complex that holds sister chromatids together. Cyclin B is the activator of Cdk1, the essential kinase that drives mitotic entry. Therefore, degradation of securin and cyclin B simultaneously results in chromosome segregation and exit from mitosis.

APC/C activity requires binding to a class of proteins known as coactivators that all harbor a carboxy-terminal WD40 domain (Fig. 1C). Although there are species-specific APC/C coactivators that participate in meiosis, such as Ama1 in *Saccharomyces cerevisiae* (Cooper et al. 2000) and Cortex in *Drosophila melanogaster* (Chu et al. 2001), the two widely conserved APC/C coactivators are Cdc20 and Cdh1. Cdc20 is essential for mitotic pro-

gression and Cdc20 depletion or mutation results in highly penetrant metaphase arrest and lethality (Dawson et al. 1995; Sigrist et al. 1995; Lim et al. 1998; Kitagawa et al. 2002; Li et al. 2007). In contrast, depletion or mutation of Cdh1 results in milder cell cycle defects (Schwab et al. 1997; Sigrist and Lehner 1997; Fay et al. 2002; Garcia-Higuera et al. 2008). The current view in the field is that Cdh1 has important roles in postmitotic contexts, including during cell differentiation and in the formation of the nervous system (Eguren et al. 2011). In this perspective, we discuss Cdc20 and the control of Cdc20-activated APC/C during the cell cycle, with a focus on new findings on the control of this key activity by chromosomes during mitosis.

APC/C^{Cdc20} activity is linked to cell cycle progression (Fig. 2). During the majority of the cell cycle, APC/C^{Cdc20} activity is inhibited (Peters 2006; Pines 2011). At mitotic entry, APC/C^{Cdc20} is activated by cyclin B-Cdk1, the essential mitotic kinase complex (see below). This activation itself could explain the cell cycle oscillator: the increase in Cdk1-cyclin B gradually activates APC/C^{Cdc20} and, once its activity reached a critical threshold, APC/C^{Cdc20} degrades cyclin B to inactivate Cdk1 leading to mitotic exit and reverting APC/C^{Cdc20} back to its inhibited state. However, this view is too simplistic, as APC/C^{Cdc20} activity is tightly regulated, most importantly by the chromosomal cargo of cell division. Here, we briefly review the structure and activity of the APC/C and then discuss mechanisms that control APC/C^{Cdc20}, with a focus on the chromosomal mechanisms that balance the need for accurate segregation with timely mitotic progression.

Published by Cold Spring Harbor Laboratory Press; doi: 10.1101/sqb.2017.82.033712

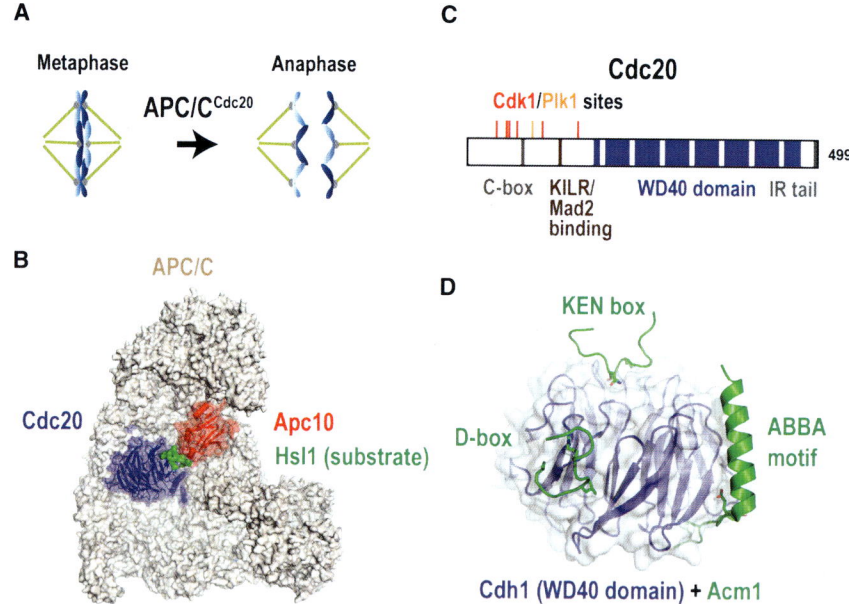

Figure 1. Anaphase-promoting complex/cyclosome (APC/C) structure and mechanism of substrate recognition. (*A*) Cartoon illustrating the metaphase-to-anaphase transition, which is promoted by APC/C^{Cdc20} activity. Microtubules are in yellow, chromosomes in blue, and kinetochores in gray. (*B*) Structure of APC/C^{Cdc20} bound to a D-box-containing substrate, Hsl1 (Zhang et al. 2016). The substrate binds to the interphase between Cdc20 and the APC/C subunit Apc10 (adapted from Corbett 2017). (*C*) Schematic illustrating the domains in human Cdc20. The C-box, KILR, and IR-tail motifs contribute to APC/C binding, whereas the WD40 domain is involved in substrate recognition. Inhibitory Cdk1 phosphorylation sites are shown in red, whereas S92, which is phosphorylated by Plk1, is in orange. Note that the KILR motif is also the Mad2 interacting motif. (*D*) Structure of the WD40 domain of *Saccharomyces cerevisiae* Cdh1 bound to an inhibitor, Acm1 (He et al. 2013). The structure shows the interaction sites for the three APC/C degrons: D-box, KEN box, and ABBA motif. (Adapted from Corbett 2017.)

A BRIEF OVERVIEW OF APC/C STRUCTURE AND MECHANISM OF PROTEIN UBIQUITINATION

The APC/C is a large complex composed of 14–16 subunits, depending on the species (Peters 2006; Pines 2011). Coactivator binding is essential for APC/C activity and for the recruitment of substrates. All APC/C coactivators possess a carboxy-terminal WD40 domain that is required for substrate recognition. In addition, the C-box and IR-tail motifs participate in APC/C binding (Schwab et al. 2001; Passmore et al. 2003; Vodermaier et al. 2003). Other coactivator-specific motifs correspond to the KLLR motif in Cdh1 (Chang et al. 2015) and KILR motif in Cdc20 (Izawa and Pines 2012), which also contribute to APC/C binding (Fig. 1C). Understanding of APC/C regulation has been greatly advanced in recent years by high-resolution cryo-EM studies that have revealed how each

subunit is assembled into the complex, how coactivators promote APC/C activity, and how different regulators control APC/C activity (Fig. 1B; Buschhorn et al. 2011; da Fonseca et al. 2011; Frye et al. 2013; Barford 2015; Chang et al. 2015; Alfieri et al. 2016; Yamaguchi et al. 2016; Zhang et al. 2016).

The APC/C recognizes substrates that possess degrons (Fig. 1D) known as the D-box [RXXL] and KEN box (Glotzer et al. 1991; Pfleger and Kirschner 2000). Other substrate-specific degrons such as the A-box in Aurora A (Littlepage and Ruderman 2002) and the O-box in Orc1 (Araki et al. 2005) have been described, although the latter was subsequently found to function as a D-box (He et al. 2013). In addition, Cdc20 and yeast Cdh1 interact with proteins containing a motif known as the Phe box or ABBA motif (for Acm1, Bub1, BubR1, and cyclin A) (Lu et al. 2014; Di Fiore et al. 2015; Diaz-Martinez et al. 2015) that can serve as a degron in some cases,

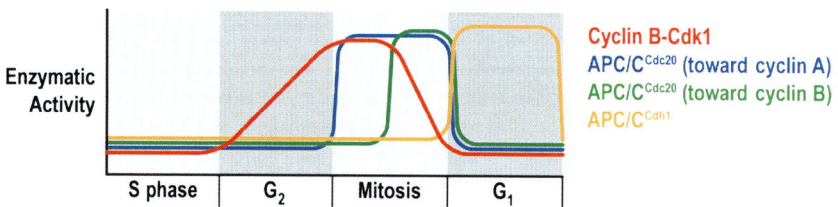

Figure 2. Regulation of APC/C activity during the cell cycle. Schematic illustrates the current model for the temporal regulation of the activities of Cdk1–Cyclin B (red), APC/C^{Cdh1} (orange), and APC/C^{Cdc20} toward cyclin A (blue) or cyclin B (green).

such as for cyclin A (Di Fiore et al. 2015). When bound by coactivators, the APC/C forms a bipartite receptor for D-box substrates that comprises the side of the WD40 barrel and the subunit Apc10/Doc1 (Fig. 1B; Passmore et al. 2003; Carroll et al. 2005; Kraft et al. 2005; Matyskiela and Morgan 2009; Buschhorn et al. 2011; da Fonseca et al. 2011). In addition, the WD40 barrel serves as the receptor for KEN box and ABBA substrates (Fig. 1D; Chao et al. 2012; He et al. 2013). In vertebrates, the APC/C functions with two E2 ubiquitin conjugating enzymes: UbcH10/Ube2C, and UbcH5/Ube2D (King et al. 1995; Aristarkhov et al. 1996); in addition, Ube2S participates in ubiquitin chain extension (Garnett et al. 2009; Williamson et al. 2009; Wu et al. 2010). On the other hand, budding yeast uses Ubc4 for mono-ubiquitination and Ubc1 for ubiquitin chain extension (Rodrigo-Brenni and Morgan 2007). Inhibitors of the APC/C, such as Emi1 or the mitotic checkpoint complex (discussed below), are known to regulate multiple aspects of APC/C function, including coactivator binding, substrate recognition, and activity/binding of the E2 enzymes that act in conjunction with the APC/C to catalyze substrate ubiquitination.

APC/C^{Cdc20} INHIBITION DURING G$_2$

A major target of the APC/C activated by Cdc20 is cyclin B, the activator of the essential mitosis-promoting kinase Cdk1 (Peters 2006; Pines 2011). Thus, APC/C^{Cdc20} activity must be kept in check during interphase to allow sufficient accumulation of cyclin B for mitotic entry.

Cdc20 itself is only synthesized in late S phase and its levels reach a maximum in mitosis (Weinstein 1997; Prinz et al. 1998; Shirayama et al. 1998), which may partially contribute to limiting APC/C^{Cdc20} activity in interphase. In addition, interphase APC/C is precluded from binding to Cdc20 by an auto-inhibitory mechanism that is released upon mitotic phosphorylation (see below). However in vitro, Cdc20 can efficiently activate interphase, non-phosphorylated APC/C (Fang et al. 1998), suggesting that other mechanisms also contribute to inhibiting Cdc20 before mitotic entry.

An initial candidate was the APC/C inhibitor Emi1 (Rca1 in *Drosophila*). (Dong et al. 1997; Reimann et al. 2001a, b; Grosskortenhaus and Sprenger 2002). However, although in vitro Emi1 can inhibit both APC/C^{Cdc20} and APC/C^{Cdh1}, its physiological target appears to be Cdh1 (Di Fiore and Pines 2007; Machida and Dutta 2007). An ortholog of Emi, called Emi2 or XErp1, inhibits APC/C^{Cdc20} to maintain the metaphase II arrest of mature *Xenopus* eggs (Schmidt et al. 2005; Tung et al. 2005). A role for Emi2 beyond meiotic arrest has been reported in developing *Xenopus* embryos, where it inhibits APC/C^{Cdc20} to promote cyclin B accumulation (Tischer et al. 2012). However, a mouse knockout of Emi2 is sterile and shows defects in meiotic progression but develops normally (Gopinathan et al. 2017), suggesting that it does not make a major contribution to somatic divisions in other systems. Thus, at present, Cdc20 synthesis and auto-inhibition of the APC/C that is relieved by mitotic phosphorylation are the major mechanisms implicated in keeping APC/C^{Cdc20} in check to allow sufficient building of cyclin B and mitotic entry.

CYTOPLASMIC AND CHROMOSOMAL REGULATION OF APC/C^{Cdc20} ACTIVITY IN MITOSIS

Upon nuclear envelope breakdown, the APC/C binds to Cdc20 and immediately becomes active toward substrates such as cyclin A and Nek2A (van Zon and Wolthuis 2010). However, cyclin B and securin are only degraded once all chromosomes have attached to spindle microtubules via their kinetochores, the protein assemblies build on their centromere regions to connect to spindle microtubules (Cheeseman 2014; Musacchio and Desai 2017). In addition to forming a dynamic microtubule interface, kinetochores function as signaling hubs where kinase and phosphatase activities are integrated to correct attachment errors and both promote as well as inhibit APC/C^{Cdc20} activation. A tight connection exists between microtubule attachment at kinetochores and APC/C^{Cdc20}-mediated degradation of securin and cyclin B, which ensures coordinated segregation of all chromosomes and prevents chromosome loss. Below we discuss both cytoplasmic and kinetochore-based mechanisms that control APC/C^{Cdc20} activity.

Cytosolic APC/C^{Cdc20} Activation by Phosphorylation

Studies in the late 1990s and early 2000s showed that APC/C phosphorylation during mitosis was a prerequisite for its activation by Cdc20 (Lahav-Baratz et al. 1995; Peters et al. 1996; Patra and Dunphy 1998; Shteinberg et al. 1999; Golan et al. 2002; Kraft et al. 2003). The mitotic kinases Cdk1 and Plk1 phosphorylate multiple APC/C subunits and this phosphorylation increases binding affinity for Cdc20 (Kraft et al. 2003). However, the biochemical and structural mechanism of this phospho-dependent regulation has only recently been elucidated (Fig. 3; Fujimitsu et al. 2016; Qiao et al. 2016; Zhang et al. 2016). In brief, the APC/C subunit Apc1 possesses an internal loop that blocks the binding of the C-box of Cdc20 to the APC/C subunit Apc8. Thus, apo-APC/C is normally in an auto-inhibited state (Fig. 3A). Phosphorylation of the Apc1 loop by Cdk1 and Plk1 releases it from Apc8 and thereby promotes Cdc20 binding (Fig. 3B). In agreement with this model, mutation or deletion of the Apc1 loop permits Cdc20 binding regardless of APC/C phosphorylation status (Fujimitsu et al. 2016; Qiao et al. 2016; Zhang et al. 2016).

Interestingly, Apc1 phosphorylation is facilitated by an initial priming phosphorylation of the Apc3 subunit by Cdk1, which then recruits Cdk1-Cks complexes to further phosphorylate Apc3 and then Apc1. Moreover, the APC/C has been shown to be a weak substrate for Cdk1 in vitro and in vivo (i.e., it is only phosphorylated once high Cdk1 activity is achieved right before mitotic entry) (Lindqvist et al. 2007; Deibler and Kirschner 2010). These mechanisms may enforce a dependence on high Cdk1 activity

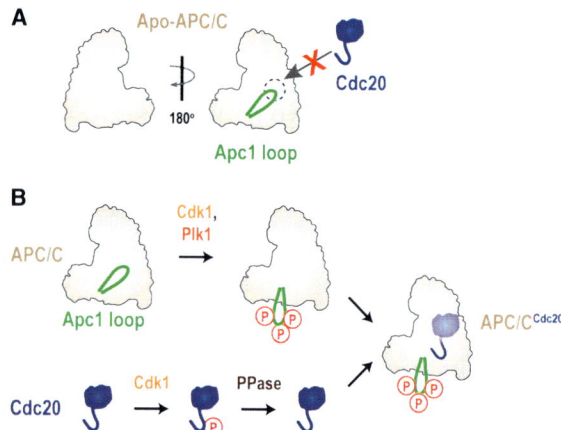

Figure 3. Control of APC/C^{Cdc20} activity by phosphorylation. (*A*) Schematic of apo-APC/C, showing auto-inhibition by the Apc1 loop. (*B*) The Cdk1 and Plk1 kinases phosphorylate the Apc1 loop, which releases the APC/C auto-inhibition mechanism. At the same time, Cdk1 phosphorylates the amino-terminal tail of Cdc20, which prevents its interaction with the APC/C. Dephosphorylation of Cdc20 by phosphatases (PPase) would cause its activation and binding to the APC/C.

and make APC/C^{Cdc20} activation kinetically lag behind initial Cdk1 activation, which may explain why APC/C^{Cdc20} only starts degrading substrates upon nuclear envelope breakdown. Although this model has good support from biochemical experiments in *Xenopus* egg extracts (Fujimitsu et al. 2016; Qiao et al. 2016), it will be important to assess whether phosphorylation of the Apc1 loop represents a conserved mechanism restraining activation of APC/C^{Cdc20} to mitosis in a cellular context. Interestingly, the interaction between APC/C and Cdh1 does not appear to be significantly affected by APC/C phosphorylation. This may be due to the fact that Cdh1 binds to the APC/C with higher affinity (Zhang et al. 2016), enabling it to efficiently displace the Apc1 loop from Apc8. This feature may explain the switch from APC/C^{Cdc20} to APC/C^{Cdh1} in late mitosis (see below).

Kinetochore-Mediated Cdc20 Activation through Dephosphorylation

As discussed above, Cdk1 activity promotes the interaction between APC/C and Cdc20. However, paradoxically Cdk1/2 proteins also block the binding of both Cdc20 and Cdh1 to the APC/C (Fig. 3B; Kramer et al. 2000; Yudkovsky et al. 2000). Phosphorylation sites near the amino-terminal C-box prevent the interaction of coactivators with the APC/C (Fig. 1C; Labit et al. 2012; Chang et al. 2015). Phosphorylated Cdc20 is found already in G$_2$ and, in human tissue culture cells, its phosphorylation may be important for the accumulation of cyclins and mitotic entry (Hein and Nilsson 2016). Interestingly, in *C. elegans* embryos, preventing Cdc20 phosphorylation significantly accelerates anaphase onset (Kim et al. 2017), indicating that Cdc20 phosphorylation is an important mechanism restraining APC/C^{Cdc20} activity in mitosis. These observations suggest that Cdc20 must be dephosphorylated to allow full APC/C activation (Fig. 3B).

In recent work, we showed that Cdc20 dephosphorylation, which contributes to APC/C activation, is promoted by kinetochores in *C. elegans* embryos (Fig. 4A; Kim et al. 2017). During mitosis, Cdc20 is recruited to kinetochores through its interaction with Bub1 (Di Fiore et al. 2015; Vleugel et al. 2015; Kim et al. 2017), a conserved component implicated in both the spindle assembly checkpoint and chromosome segregation (Bolanos-Garcia and Blundell 2011; Elowe 2011). Bub1, along with its binding partner Bub3, is recruited to kinetochores through the kinetochore scaffold Knl1, which is phosphorylated on "MELT" repeats in its amino terminus by the kinases Mps1 and Plk1 (London et al. 2012; Shepperd et al. 2012; Yamagishi et al. 2012; Espeut et al. 2015; von Schubert et al. 2015). At its extreme amino terminus, Knl1 possesses "SILK" and "RVxF" motifs that recruit the catalytic subunit of protein phosphatase 1 (PP1c) (Liu et al. 2010; Meadows et al. 2011; Rosenberg et al. 2011; Espeut et al. 2012). Our findings suggest that by bringing Cdc20 to the vicinity of Knl1-bound PP1, kinetochores catalyze Cdc20 activation by removing the inhibitory phosphorylation in its amino terminus (Fig. 4A). In support of this model, blocking Cdc20 or PP1 recruitment to kinetochores delays anaphase onset, an effect that can be bypassed by mutating the Cdk phosphorylation sites on Cdc20 (Kim et al. 2015, 2017). Notably, a role for kinetochore-localized Bub1-Bub3 in promoting APC/C^{Cdc20} activation has also been reported in budding yeast (Yang et al. 2015). As preventing Cdc20 recruitment to kinetochores does not result in a mitotic arrest, cytosolic phosphatases likely also promote Cdc20 dephosphorylation independently of kinetochores; alternatively, Cdc20 amino-terminal phosphorylation may not be sufficient to fully block its binding and activation of the APC/C.

In addition to its regulation by Cdk1/2, human Cdc20 is also inhibited by phosphorylation on Ser92 by Plk1 (Fig. 1C; Craney et al. 2016; Jia et al. 2016; Lee et al. 2017). This phosphorylation is facilitated by Bub1 and is suggested to inhibit the recruitment of the E2 ubiquitin conjugating enzyme Ube2S to the APC/C. In late mitosis, Ser92 phosphorylation is reversed by PP2A-B56 docked onto either BubR1 or the APC/C itself. Although this mechanism has biochemical support (Craney et al. 2016; Jia et al. 2016), its importance in an in vivo context is unclear, given that deletion of Ube2S (Wild et al. 2016) or mutation of Ser92 in Cdc20 (Lee et al. 2017) result in relatively mild effects on mitotic exit.

Kinetochore-Dependent APC/C^{Cdc20} Inhibition by the Spindle Assembly Checkpoint

Phosphorylation of the APC/C at mitotic entry that relieves inhibition of Cdc20 binding might explain why degradation of APC/C^{Cdc20} substrates such as cyclin A and Nek2A begins right at nuclear envelope breakdown, when mitotic kinases are active (van Zon and Wolthuis 2010). However, degradation of cyclin B and securin only occurs after microtubule binding to all kinetochores to

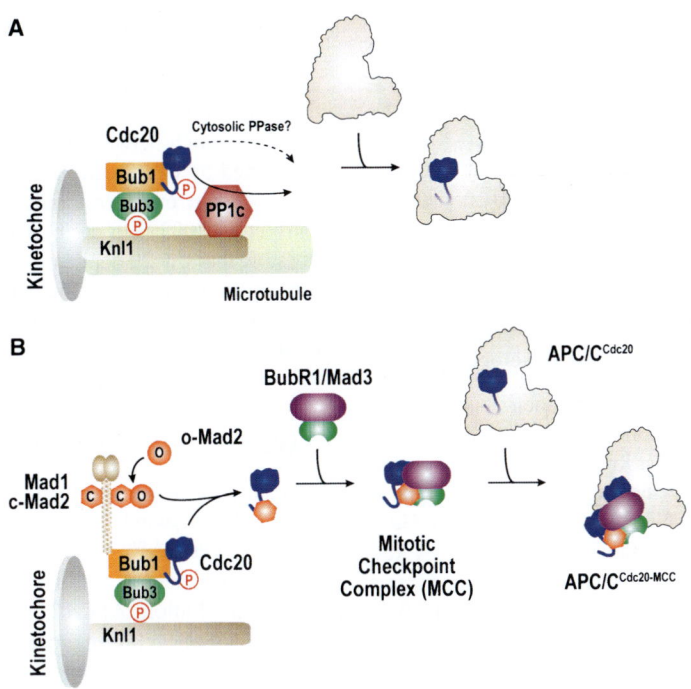

Figure 4. The two fates of Cdc20 at kinetochores. During mitosis, Cdc20 is recruited to kinetochores by Bub1/Bub3, which is bound to phospho-Knl1. (*A*) When kinetochores are attached by microtubules, kinetochores promote Cdc20 dephosphorylation by kinetochore-localized PP1c, which allows its activation. Cdc20 may also be dephosphorylated at the cytosol, likely through PP2A-B56. (*B*) When microtubules are unattached, signal from the spindle assembly checkpoint catalyzes the incorporation of Cdc20 into the mitotic checkpoint complex (MCC), which binds and inhibits APC/C^{Cdc20} activity. See text for more details.

prevent errors in chromosome segregation (Fig. 2). A large body of work has focused on how chromosomes regulate APC/C^{Cdc20} to prevent premature cyclin B and securin degradation, which is discussed below.

The spindle checkpoint is the mechanism that inhibits degradation of cyclin B and securin by APC/C^{Cdc20} in the presence of chromosomes with unattached kinetochores (Fig. 4B). When unattached, kinetochores catalyze the formation of an APC/C^{Cdc20} inhibitor known as the mitotic checkpoint complex or MCC, composed of BubR1 (Mad3 in yeast and nematodes), Bub3, Mad2, and Cdc20 (Sudakin et al. 2001). The spindle checkpoint has been subjected to extensive mechanistic analysis (for detailed reviews, see Lara-Gonzalez et al. 2012; Jia et al. 2013; Musacchio 2015; Etemad and Kops 2016; Corbett 2017). Here, we briefly summarize current understanding of how kinetochores control formation of the MCC and an interesting intertwining with the kinetochore-based APC/C activation mechanism that acts on Cdc20.

The initiation step in spindle checkpoint signaling is recruitment of the Mad1–Mad2 complex to kinetochores (Fig. 4B; Chen et al. 1996, 1998; Li and Benezra 1996; Maldonado and Kapoor 2011). Once localized, the Mad1–Mad2 complex recruits free, cytosolic Mad2, which is in an "open" conformation, and converts it into a "closed" form that captures Cdc20 (Mapelli and Musacchio 2007; Luo and Yu 2008; Rosenberg and Corbett 2015). The Mad2–Cdc20 dimer interacts with the BubR1–Bub3 complex to assemble the full MCC (Sudakin et al. 2001). The MCC then binds APC/C^{Cdc20} and inhibits the recruitment

of substrates; this is achieved by a conserved D-box–ABBA–KEN–ABBA cassette in BubR1 that prevents the formation of the bipartite Cdc20–Apc10 D-box substrate receptor, directly blocks the binding of KEN-box substrates, and partially prevents the recruitment of E2 enzymes (Burton and Solomon 2007; Sczaniecka et al. 2008; Malureanu et al. 2009; Elowe et al. 2010; Lara-Gonzalez et al. 2011; Chao et al. 2012; Izawa and Pines 2015; Alfieri et al. 2016; Di Fiore et al. 2016; Yamaguchi et al. 2016; Sewart and Hauf 2017). Mps1, an essential spindle checkpoint kinase, is required for many steps in spindle checkpoint signaling, including the recruitment of the Bub1–Bub3 and Mad1–Mad2 complexes to kinetochores (Hewitt et al. 2010; Maciejowski et al. 2010; Santaguida et al. 2010; London et al. 2012; Shepperd et al. 2012; Yamagishi et al. 2012).

Microtubule attachment silences spindle checkpoint signaling employing at least three different mechanisms. First, microtubules promote the dynein motor-dependent "stripping" of spindle checkpoint proteins from the kinetochore (Howell et al. 2001; Wojcik et al. 2001). Second, microtubule attachment promotes PP1c recruitment to the kinetochore, which dephosphorylates the MELT repeats on Knl1 and therefore removes Bub1–Bub3 from kinetochores (Liu et al. 2010; Lesage et al. 2011; London et al. 2012; Espert et al. 2014; Nijenhuis et al. 2014). Third, microtubule binding by Ndc80 displaces Mps1 from kinetochores (Hiruma et al. 2015; Ji et al. 2015) (although this is not the case in budding yeast, where Mps1 persists at kinetochores even after microtubule attachment; Aravamudhan et al. 2015). In addition, cytosolic mechanisms such as p31-TRIP13 and Apc15-mediated Cdc20 autoubiquitination contribute to APC/C^{Cdc20} activation by catalyzing MCC disassembly (Habu et al. 2002; Xia et al. 2004; Reddy et al. 2007; Yang et al. 2007; Hagan et al. 2011; Jia et al. 2011; Mansfeld et al. 2011; Teichner et al. 2011; Westhorpe et al. 2011; Foster and Morgan 2012; Uzunova et al. 2012; Eytan et al. 2014; Ye et al. 2015; Yamaguchi et al. 2016; Zhang et al. 2016).

Integration of Mechanisms Activating and Inhibiting APC/C^{Cdc20} at the Kinetochore

As mentioned above, unattached kinetochores signal through the spindle checkpoint to inhibit APC/C^{Cdc20}. However, we have found that kinetochores also promote APC/C^{Cdc20} activation by removing inhibitory phosphates on the amino terminus of Cdc20 (Kim et al. 2017). How then can these opposing functions be reconciled? A key observation is that both mechanisms depend on the recruitment of Cdc20 to kinetochores (Fig. 4). Cdc20 is recruited through Bub1, which possesses a Cdc20-binding "ABBA" motif (Di Fiore et al. 2015; Vleugel et al. 2015; Kim et al. 2017). Notably, this recruitment is highly dynamic with kinetochore-bound Cdc20 showing a half-life of 0.5–2 sec (Kallio et al. 2002; Kim et al. 2017). Thus, Cdc20 is rapidly fluxing through kinetochores via interaction with Bub1's ABBA motif. Mutation of the ABBA motif on Bub1 not only prevents the kinetochore-dependent anaphase promoting function but also abolishes spindle checkpoint signaling (Di Fiore et al. 2015; Vleugel et al. 2015; Kim et al. 2017). Bub1 is critical to recruit the Mad1–Mad2 complex to unattached kinetochores (Klebig et al. 2009; London and Biggins 2014; Moyle et al. 2014; Zhang et al. 2017), although this function is independent of the ABBA motif (Vleugel et al. 2015; Kim et al. 2017). Therefore, recruitment of Cdc20 to the ABBA motif of Bub1 likely promotes formation of the MCC by bringing it in close proximity to active Mad1–Mad2 that is also bound to Bub1 (Fig. 4B). Interestingly, Mps1 phosphorylation of the carboxyl terminus of Mad1, which is essential for Mad1–Mad2 activation (Faesen et al. 2017), may also create a binding site for Cdc20 (Ji et al. 2017). Thus, Bub1's ABBA motif may help generate a locally high concentration of Cdc20 at kinetochores that, if Mad1–Mad2 is present and phosphorylated, places Cdc20 on the Mad1 carboxyl terminus in close proximity to the conformationally converting Mad2 and promotes formation of the Mad2–Cdc20 complex that matures into the MCC (Fig. 4B).

The above-mentioned data suggest that Cdc20 recruited to kinetochores on a single site has two opposite fates: APC/C activation through Cdc20 dephosphorylation and APC/C inhibition through its incorporation on the MCC (Kim et al. 2017). Given that the spindle assembly checkpoint is only active at unattached kinetochores, the choice between these two fates is dependent on the status of kinetochore-microtubule interactions (Fig. 4A,B). At unattached kinetochores, spindle checkpoint signaling would cause Cdc20 to be primarily incorporated onto

the MCC to prevent premature APC/C^{Cdc20} activation, whereas following microtubule attachment, when the spindle checkpoint is silenced, Cdc20 would be primarily dephosphorylated and activated to promote anaphase onset. The switch between these two fates could be further sharpened by PP1c recruitment, which may be promoted or dependent on microtubule attachment (Trinkle-Mulcahy et al. 2003; Liu et al. 2010; Kim et al. 2017). It is possible that Cdc20 dephosphorylation occurs throughout mitosis, regardless of kinetochore–microtubule interactions. Regardless, the responsiveness of checkpoint signaling to microtubule attachment would still shift the balance between the opposing Cdc20 fates.

APC/C^{Cdc20} INACTIVATION IN LATE MITOSIS

Once securin and cyclin B are degraded, the APC/C is thought to switch coactivators from Cdc20 to Cdh1 (Fig. 2). APC/C^{Cdh1} activity in late mitosis is essential for the degradation of Aurora kinases (Floyd et al. 2008). In addition, APC/C^{Cdh1} is required in G_1 for the degradation of cyclins in order to allow the loading of pre-replication complexes onto chromatin for the subsequent S phase (for review, see Sivaprasad et al. 2007).

The Cdc20–Cdh1 switch is likely explained by the decline in Cyclin B-Cdk1 activity, enabling phosphatases to dephosphorylate the APC/C and reduce its affinity for Cdc20. At the same time, Cdh1, which is kept inactivated by Cdk-dependent phosphorylation throughout most of the cell cycle, would become dephosphorylated and bind to and activate the APC/C (Peters 2006; Pines 2011). However, some APC/C^{Cdc20} activity persists in late mitosis and indeed, many late APC/C substrates, such as Plk1, survivin and Cenp-F are reliant on Cdc20 for their degradation (Floyd et al. 2008; Gurden et al. 2010). Regardless, at anaphase onset, Cdc20 itself becomes an APC/C substrate and therefore, by G_1, the APC/C is mostly Cdh1-bound.

CONCLUSION

Since its discovery in the early 1990s as the machine that drives mitotic exit (King et al. 1995; Sudakin et al. 1995), the APC/C and its coactivator Cdc20 have been extensively studied. In the last 5 years, advances in high-resolution cryo-EM combined with biochemical and cell-based assays have led to an explosive increase in our understanding of APC/C^{Cdc20} enzymology and mechanisms of its regulation.

Interestingly, the APC/C is not only required in dividing cells but also plays important roles in differentiated tissues, such as the nervous system (Huang and Bonni 2016). Although most of these functions depend on Cdh1, Cdc20 is expressed in some neuronal types and is required for their differentiation (Kim et al. 2009; Yang et al. 2009; Kowalski et al. 2014; Watanabe et al. 2014; Mao et al. 2015). These findings highlight the potential for new studies focused on understanding how postmitotic APC/C functions are regulated. For example, a cyclin-dependent kinase called Cdk5 is present in sensory neurons, where it regulates multiple signaling events (Kawauchi 2014);

therefore, Cdk5 may substitute for Cdk1 in neurons to regulate the interaction between APC/C and its coactivators in a manner similar to what has been observed during cell cycle progression (Maestre et al. 2008; Veas-Perez de Tudela et al. 2015). Given that Cdk5 has garnered a significant amount of interest for its role in Alzheimer's disease progression (Fuchsberger et al. 2017), its mechanistic connection with the APC/C in the nervous system is likely to be the focus of future work.

Finally, understanding of APC/C^{Cdc20} mechanism and regulation has opened the possibility for new therapies targeting the APC/C in cancer (Wang et al. 2015; Zhou et al. 2016). Current treatments use spindle poisons to activate the spindle assembly checkpoint and induce apoptosis but are limited by cells slipping out of mitosis because of residual APC/C activity (Brito and Rieder 2006; Gascoigne and Taylor 2008). A number of studies have shown that directly inhibiting mitotic exit is a more efficient approach to killing cancer cells (Huang et al. 2009; Manchado et al. 2010). Two small-molecule APC/C inhibitors have been developed, proTAME and Apcin (Zeng et al. 2010; Sackton et al. 2014), which block the interaction between coactivators and the APC/C. When added to cells in combination, proTAME and Apcin efficiently block mitotic exit (Sackton et al. 2014). Once optimized to act in a clinical context, these drugs have the potential to synergize with commonly used microtubule poisons that activate the spindle checkpoint (Giovinazzi et al. 2013; de Lange et al. 2015) and contribute to improving this widely used chemotherapeutic strategy.

ACKNOWLEDGMENTS

Work in the Desai laboratory is supported by a National Institutes of Health (NIH) grant (GM074215) and the Ludwig Institute for Cancer Research. P.L.-G. was supported by a Pew Latin American fellowship.

REFERENCES

Alfieri C, Chang L, Zhang Z, Yang J, Maslen S, Skehel M, Barford D. 2016. Molecular basis of APC/C regulation by the spindle assembly checkpoint. *Nature* **536:** 431–436.

Araki M, Yu H, Asano M. 2005. A novel motif governs APC-dependent degradation of *Drosophila* ORC1 in vivo. *Genes Dev* **19:** 2458–2465.

Aravamudhan P, Goldfarb AA, Joglekar AP. 2015. The kinetochore encodes a mechanical switch to disrupt spindle assembly checkpoint signalling. *Nat Cell Biol* **17:** 868–879.

Aristarkhov A, Eytan E, Moghe A, Admon A, Hershko A, Ruderman JV. 1996. E2-C, a cyclin-selective ubiquitin carrier protein required for the destruction of mitotic cyclins. *Proc Natl Acad Sci* **93:** 4294–4299.

Barford D. 2015. Understanding the structural basis for controlling chromosome division. *Philos Trans A Math Phys Eng Sci* **373:** 20130392.

Bolanos-Garcia VM, Blundell TL. 2011. BUB1 and BUBR1: Multifaceted kinases of the cell cycle. *Trends Biochem Sci* **36:** 141–150.

Brito DA, Rieder CL. 2006. Mitotic checkpoint slippage in humans occurs via cyclin B destruction in the presence of an active checkpoint. *Curr Biol* **16:** 1194–1200.

Burton JL, Solomon MJ. 2007. Mad3p, a pseudosubstrate inhibitor of APCCdc20 in the spindle assembly checkpoint. *Genes Dev* **21:** 655–667.

Buschhorn BA, Petzold G, Galova M, Dube P, Kraft C, Herzog F, Stark H, Peters JM. 2011. Substrate binding on the APC/C occurs between the coactivator Cdh1 and the processivity factor Doc1. *Nat Struct Mol Biol* **18:** 6–13.

Carroll CW, Enquist-Newman M, Morgan DO. 2005. The APC subunit Doc1 promotes recognition of the substrate destruction box. *Curr Biol* **15:** 11–18.

Chang L, Zhang Z, Yang J, McLaughlin SH, Barford D. 2015. Atomic structure of the APC/C and its mechanism of protein ubiquitination. *Nature* **522:** 450–454.

Chao WC, Kulkarni K, Zhang Z, Kong EH, Barford D. 2012. Structure of the mitotic checkpoint complex. *Nature* **484:** 208–213.

Cheeseman IM. 2014. The kinetochore. *Cold Spring Harb Perspect Biol* **6:** a015826.

Chen RH, Waters JC, Salmon ED, Murray AW. 1996. Association of spindle assembly checkpoint component XMAD2 with unattached kinetochores. *Science* **274:** 242–246.

Chen RH, Shevchenko A, Mann M, Murray AW. 1998. Spindle checkpoint protein Xmad1 recruits Xmad2 to unattached kinetochores. *J Cell Biol* **143:** 283–295.

Chu T, Henrion G, Haegeli V, Strickland S. 2001. Cortex, a *Drosophila* gene required to complete oocyte meiosis, is a member of the Cdc20/fizzy protein family. *Genesis* **29:** 141–152.

Cooper KF, Mallory MJ, Egeland DB, Jarnik M, Strich R. 2000. Ama1p is a meiosis-specific regulator of the anaphase promoting complex/cyclosome in yeast. *Proc Natl Acad Sci* **97:** 14548–14553.

Corbett KD. 2017. Molecular mechanisms of spindle assembly checkpoint activation and silencing. *Prog Mol Subcell Biol* **56:** 429–455.

Craney A, Kelly A, Jia L, Fedrigo I, Yu H, Rape M. 2016. Control of APC/C-dependent ubiquitin chain elongation by reversible phosphorylation. *Proc Natl Acad Sci* **113:** 1540–1545.

da Fonseca PC, Kong EH, Zhang Z, Schreiber A, Williams MA, Morris EP, Barford D. 2011. Structures of APC/C^{Cdh1} with substrates identify Cdh1 and Apc10 as the D-box co-receptor. *Nature* **470:** 274–278.

Dawson IA, Roth S, Artavanis-Tsakonas S. 1995. The *Drosophila* cell cycle gene fizzy is required for normal degradation of cyclins A and B during mitosis and has homology to the *CDC20* gene of *Saccharomyces cerevisiae*. *J Cell Biol* **129:** 725–737.

Deibler RW, Kirschner MW. 2010. Quantitative reconstitution of mitotic CDK1 activation in somatic cell extracts. *Mol Cell* **37:** 753–767.

de Lange J, Faramarz A, Oostra AB, de Menezes RX, van der Meulen IH, Rooimans MA, Rockx DA, Brakenhoff RH, van Beusechem VW, King RW, et al. 2015. Defective sister chromatid cohesion is synthetically lethal with impaired APC/C function. *Nat Commun* **6:** 8399.

Diaz-Martinez LA, Tian W, Li B, Warrington R, Jia L, Brautigam CA, Luo X, Yu H. 2015. The Cdc20-binding Phe box of the spindle checkpoint protein BubR1 maintains the mitotic checkpoint complex during mitosis. *J Biol Chem* **290:** 2431–2443.

Di Fiore B, Pines J. 2007. Emi1 is needed to couple DNA replication with mitosis but does not regulate activation of the mitotic APC/C. *J Cell Biol* **177:** 425–437.

Di Fiore B, Davey NE, Hagting A, Izawa D, Mansfeld J, Gibson TJ, Pines J. 2015. The ABBA motif binds APC/C activators and is shared by APC/C substrates and regulators. *Dev Cell* **32:** 358–372.

Di Fiore B, Wurzenberger C, Davey NE, Pines J. 2016. The mitotic checkpoint complex requires an evolutionary conserved cassette to bind and inhibit active APC/C. *Mol Cell* **64:** 1144–1153.

Dong X, Zavitz KH, Thomas BJ, Lin M, Campbell S, Zipursky SL. 1997. Control of G$_1$ in the developing *Drosophila* eye: rca1 regulates Cyclin A. *Genes Dev* **11:** 94–105.

Eguren M, Manchado E, Malumbres M. 2011. Non-mitotic functions of the anaphase-promoting complex. *Semin Cell Dev Biol* **22:** 572–578.

Elowe S. 2011. Bub1 and BubR1: At the interface between chromosome attachment and the spindle checkpoint. *Mol Cell Biol* **31:** 3085–3093.

Elowe S, Dulla K, Uldschmid A, Li X, Dou Z, Nigg EA. 2010. Uncoupling of the spindle-checkpoint and chromosome-congression functions of BubR1. *J Cell Sci* **123:** 84–94.

Espert A, Uluocak P, Bastos RN, Mangat D, Graab P, Gruneberg U. 2014. PP2A-B56 opposes Mps1 phosphorylation of Knl1 and thereby promotes spindle assembly checkpoint silencing. *J Cell Biol* **206:** 833–842.

Espeut J, Cheerambathur DK, Krenning L, Oegema K, Desai A. 2012. Microtubule binding by KNL-1 contributes to spindle checkpoint silencing at the kinetochore. *J Cell Biol* **196:** 469–482.

Espeut J, Lara-Gonzalez P, Sassine M, Shiau AK, Desai A, Abrieu A. 2015. Natural loss of Mps1 kinase in nematodes uncovers a role for polo-like kinase 1 in spindle checkpoint initiation. *Cell Rep* **12:** 58–65.

Etemad B, Kops GJ. 2016. Attachment issues: Kinetochore transformations and spindle checkpoint silencing. *Curr Opin Cell Biol* **39:** 101–108.

Eytan E, Wang K, Miniowitz-Shemtov S, Sitry-Shevah D, Kaisari S, Yen TJ, Liu ST, Hershko A. 2014. Disassembly of mitotic checkpoint complexes by the joint action of the AAA-ATPase TRIP13 and p31^{comet}. *Proc Natl Acad Sci* **111:** 12019–12024.

Faesen AC, Thanasoula M, Maffini S, Breit C, Müller F, van Gerwen S, Bange T, Musacchio A. 2017. Basis of catalytic assembly of the mitotic checkpoint complex. *Nature* **542:** 498–502.

Fang G, Yu H, Kirschner MW. 1998. Direct binding of CDC20 protein family members activates the anaphase-promoting complex in mitosis and G$_1$. *Mol Cell* **2:** 163–171.

Fay DS, Keenan S, Han M. 2002. fzr-1 and lin-35/Rb function redundantly to control cell proliferation in *C. elegans* as revealed by a nonbiased synthetic screen. *Genes Dev* **16:** 503–517.

Floyd S, Pines J, Lindon C. 2008. APC/C^{Cdh1} targets aurora kinase to control reorganization of the mitotic spindle at anaphase. *Curr Biol* **18:** 1649–1658.

Foster SA, Morgan DO. 2012. The APC/C subunit Mnd2/Apc15 promotes Cdc20 autoubiquitination and spindle assembly checkpoint inactivation. *Mol Cell* **47:** 921–932.

Frye JJ, Brown NG, Petzold G, Watson ER, Grace CR, Nourse A, Jarvis MA, Kriwacki RW, Peters JM, Stark H, et al. 2013. Electron microscopy structure of human APC/C^{CDH1}-EMI1 reveals multimodal mechanism of E3 ligase shutdown. *Nat Struct Mol Biol* **20:** 827–835.

Fuchsberger T, Lloret A, Vina J. 2017. New functions of APC/C ubiquitin ligase in the nervous system and its role in Alzheimer's disease. *Int J Mol Sci* **18:** 1057.

Fujimitsu K, Grimaldi M, Yamano H. 2016. Cyclin-dependent kinase 1-dependent activation of APC/C ubiquitin ligase. *Science* **352:** 1121–1124.

Funk LC, Zasadil LM, Weaver BA. 2016. Living in CIN: Mitotic infidelity and its consequences for tumor promotion and suppression. *Dev Cell* **39:** 638–652.

Garcia-Higuera I, Manchado E, Dubus P, Cañamero M, Méndez J, Moreno S, Malumbres M. 2008. Genomic stability and tumour suppression by the APC/C cofactor Cdh1. *Nat Cell Biol* **10:** 802–811.

Garnett MJ, Mansfeld J, Godwin C, Matsusaka T, Wu J, Russell P, Pines J, Venkitaraman AR. 2009. UBE2S elongates ubiquitin chains on APC/C substrates to promote mitotic exit. *Nat Cell Biol* **11:** 1363–1369.

Gascoigne KE, Taylor SS. 2008. Cancer cells display profound intra- and interline variation following prolonged exposure to antimitotic drugs. *Cancer Cell* **14:** 111–122.

Giovinazzi S, Bellapu D, Morozov VM, Ishov AM. 2013. Targeting mitotic exit with hyperthermia or APC/C inhibition to increase paclitaxel efficacy. *Cell Cycle* **12:** 2598–2607.

Glotzer M, Murray AW, Kirschner MW. 1991. Cyclin is degraded by the ubiquitin pathway. *Nature* **349:** 132–138.

Golan A, Yudkovsky Y, Hershko A. 2002. The cyclin-ubiquitin ligase activity of cyclosome/APC is jointly activated by protein kinases Cdk1-cyclin B and Plk. *J Biol Chem* **277:** 15552–15557.

Gopinathan L, Szmyd R, Low D, Diril MK, Chang HY, Coppola V, Liu K, Tessarollo L, Guccione E, van Pelt AMM, et al. 2017. Emi2 is essential for mouse spermatogenesis. *Cell Rep* **20:** 697–708.

Grosskortenhaus R, Sprenger F. 2002. Rca1 inhibits APC-Cdh1^{Fzr} and is required to prevent cyclin degradation in G$_2$. *Dev Cell* **2:** 29–40.

Gurden MD, Holland AJ, van Zon W, Tighe A, Vergnolle MA, Andres DA, Spielmann HP, Malumbres M, Wolthuis RM, Cleveland DW, et al. 2010. Cdc20 is required for the post-anaphase, KEN-dependent degradation of centromere protein F. *J Cell Sci* **123:** 321–330.

Habu T, Kim SH, Weinstein J, Matsumoto T. 2002. Identification of a MAD2-binding protein, CMT2, and its role in mitosis. *EMBO J* **21:** 6419–6428.

Hagan RS, Manak MS, Buch HK, Meier MG, Meraldi P, Shah JV, Sorger PK. 2011. p31^{comet} acts to ensure timely spindle checkpoint silencing subsequent to kinetochore attachment. *Mol Biol Cell* **22:** 4236–4246.

He J, Chao WC, Zhang Z, Yang J, Cronin N, Barford D. 2013. Insights into degron recognition by APC/C coactivators from the structure of an Acm1-Cdh1 complex. *Mol Cell* **50:** 649–660.

Hein JB, Nilsson J. 2016. Interphase APC/C-Cdc20 inhibition by cyclin A2-Cdk2 ensures efficient mitotic entry. *Nat Commun* **7:** 10975.

Hewitt L, Tighe A, Santaguida S, White AM, Jones CD, Musacchio A, Green S, Taylor SS. 2010. Sustained Mps1 activity is required in mitosis to recruit O-Mad2 to the Mad1-C-Mad2 core complex. *J Cell Biol* **190:** 25–34.

Hiruma Y, Sacristan C, Pachis ST, Adamopoulos A, Kuijt T, Ubbink M, von Castelmur E, Perrakis A, Kops GJ. 2015. CELL DIVISION CYCLE. Competition between MPS1 and microtubules at kinetochores regulates spindle checkpoint signaling. *Science* **348:** 1264–1267.

Holland AJ, Cleveland DW. 2012. Losing balance: The origin and impact of aneuploidy in cancer. *EMBO Rep* **13:** 501–514.

Howell BJ, McEwen BF, Canman JC, Hoffman DB, Farrar EM, Rieder CL, Salmon ED. 2001. Cytoplasmic dynein/dynactin drives kinetochore protein transport to the spindle poles and has a role in mitotic spindle checkpoint inactivation. *J Cell Biol* **155:** 1159–1172.

Huang J, Bonni A. 2016. A decade of the anaphase-promoting complex in the nervous system. *Genes Dev* **30:** 622–638.

Huang HC, Shi J, Orth JD, Mitchison TJ. 2009. Evidence that mitotic exit is a better cancer therapeutic target than spindle assembly. *Cancer Cell* **16:** 347–358.

Izawa D, Pines J. 2012. Mad2 and the APC/C compete for the same site on Cdc20 to ensure proper chromosome segregation. *J Cell Biol* **199:** 27–37.

Izawa D, Pines J. 2015. The mitotic checkpoint complex binds a second CDC20 to inhibit active APC/C. *Nature* **517:** 631–634.

Ji Z, Gao H, Yu H. 2015. Cell division cycle. Kinetochore attachment sensed by competitive Mps1 and microtubule binding to Ndc80C. *Science* **348:** 1260–1264.

Ji Z, Gao H, Jia L, Li B, Yu H. 2017. A sequential multi-target Mps1 phosphorylation cascade promotes spindle checkpoint signaling. *Elife* **6:** e22513.

Jia L, Li B, Warrington RT, Hao X, Wang S, Yu H. 2011. Defining pathways of spindle checkpoint silencing: Functional redundancy between Cdc20 ubiquitination and p31^{comet}. *Mol Biol Cell* **22:** 4227–4235.

Jia L, Kim S, Yu H. 2013. Tracking spindle checkpoint signals from kinetochores to APC/C. *Trends Biochem Sci* **38:** 302–311.

Jia L, Li B, Yu H. 2016. The Bub1-Plk1 kinase complex promotes spindle checkpoint signalling through Cdc20 phosphorylation. *Nat Commun* **7:** 10818.

Kallio MJ, Beardmore VA, Weinstein J, Gorbsky GJ. 2002. Rapid microtubule-independent dynamics of Cdc20 at kinetochores and centrosomes in mammalian cells. *J Cell Biol* **158:** 841–847.

Kawauchi T. 2014. Cdk5 regulates multiple cellular events in neural development, function and disease. *Dev Growth Differ* **56:** 335–348.

Kim AH, Puram SV, Bilimoria PM, Ikeuchi Y, Keough S, Wong M, Rowitch D, Bonni A. 2009. A centrosomal Cdc20-APC pathway controls dendrite morphogenesis in postmitotic neurons. *Cell* **136:** 322–336.

Kim T, Moyle MW, Lara-Gonzalez P, De Groot C, Oegema K, Desai A. 2015. Kinetochore-localized BUB-1/BUB-3 complex promotes anaphase onset in *C. elegans*. *J Cell Biol* **209:** 507–517.

Kim T, Lara-Gonzalez P, Prevo B, Meitinger F, Cheerambathur DK, Oegema K, Desai A. 2017. Kinetochores accelerate or delay APC/C activation by directing Cdc20 to opposing fates. *Genes Dev* **31:** 1089–1094.

King RW, Peters JM, Tugendreich S, Rolfe M, Hieter P, Kirschner MW. 1995. A 20S complex containing CDC27 and CDC16 catalyzes the mitosis-specific conjugation of ubiquitin to cyclin B. *Cell* **81:** 279–288.

Kitagawa R, Law E, Tang L, Rose AM. 2002. The Cdc20 homolog, FZY-1, and its interacting protein, IFY-1, are required for proper chromosome segregation in *Caenorhabditis elegans*. *Curr Biol* **12:** 2118–2123.

Klebig C, Korinth D, Meraldi P. 2009. Bub1 regulates chromosome segregation in a kinetochore-independent manner. *J Cell Biol* **185:** 841–858.

Kowalski JR, Dube H, Touroutine D, Rush KM, Goodwin PR, Carozza M, Didier Z, Francis MM, Juo P. 2014. The Anaphase-Promoting Complex (APC) ubiquitin ligase regulates GABA transmission at the *C. elegans* neuromuscular junction. *Mol Cell Neurosci* **58:** 62–75.

Kraft C, Herzog F, Gieffers C, Mechtler K, Hagting A, Pines J, Peters JM. 2003. Mitotic regulation of the human anaphase-promoting complex by phosphorylation. *EMBO J* **22:** 6598–6609.

Kraft C, Vodermaier HC, Maurer-Stroh S, Eisenhaber F, Peters JM. 2005. The WD40 propeller domain of Cdh1 functions as a destruction box receptor for APC/C substrates. *Mol Cell* **18:** 543–553.

Kramer ER, Scheuringer N, Podtelejnikov AV, Mann M, Peters JM. 2000. Mitotic regulation of the APC activator proteins CDC20 and CDH1. *Mol Biol Cell* **11:** 1555–1569.

Labit H, Fujimitsu K, Bayin NS, Takaki T, Gannon J, Yamano H. 2012. Dephosphorylation of Cdc20 is required for its C-box-dependent activation of the APC/C. *EMBO J* **31:** 3351–3362.

Lahav-Baratz S, Sudakin V, Ruderman JV, Hershko A. 1995. Reversible phosphorylation controls the activity of cyclosome-associated cyclin-ubiquitin ligase. *Proc Natl Acad Sci* **92:** 9303–9307.

Lara-Gonzalez P, Scott MI, Diez M, Sen O, Taylor SS. 2011. BubR1 blocks substrate recruitment to the APC/C in a KEN-box-dependent manner. *J Cell Sci* **124:** 4332–4345.

Lara-Gonzalez P, Westhorpe FG, Taylor SS. 2012. The spindle assembly checkpoint. *Curr Biol* **22:** R966–R980.

Lee SJ, Rodriguez-Bravo V, Kim H, Datta S, Foley EA. 2017. The PP2A^{B56} phosphatase promotes the association of Cdc20 with APC/C in mitosis. *J Cell Sci* **130:** 1760–1771.

Lesage B, Qian J, Bollen M. 2011. Spindle checkpoint silencing: PP1 tips the balance. *Curr Biol* **21:** R898–R903.

Li Y, Benezra R. 1996. Identification of a human mitotic checkpoint gene: hsMAD2. *Science* **274:** 246–248.

Li M, York JP, Zhang P. 2007. Loss of Cdc20 causes a securin-dependent metaphase arrest in two-cell mouse embryos. *Mol Cell Biol* **27:** 3481–3488.

Lim HH, Goh PY, Surana U. 1998. Cdc20 is essential for the cyclosome-mediated proteolysis of both Pds1 and Clb2 during M phase in budding yeast. *Curr Biol* **8:** 231–234.

Lindqvist A, van Zon W, Karlsson Rosenthal C, Wolthuis RM. 2007. Cyclin B1-Cdk1 activation continues after centrosome separation to control mitotic progression. *PLoS Biol* **5:** e123.

Littlepage LE, Ruderman JV. 2002. Identification of a new APC/C recognition domain, the A box, which is required for the Cdh1-dependent destruction of the kinase Aurora-A during mitotic exit. *Genes Dev* **16:** 2274–2285.

Liu D, Vleugel M, Backer CB, Hori T, Fukagawa T, Cheeseman IM, Lampson MA. 2010. Regulated targeting of protein phosphatase 1 to the outer kinetochore by KNL1 opposes Aurora B kinase. *J Cell Biol* **188:** 809–820.

London N, Biggins S. 2014. Mad1 kinetochore recruitment by Mps1-mediated phosphorylation of Bub1 signals the spindle checkpoint. *Genes Dev* **28:** 140–152.

London N, Ceto S, Ranish JA, Biggins S. 2012. Phosphoregulation of Spc105 by Mps1 and PP1 regulates Bub1 localization to kinetochores. *Curr Biol* **22:** 900–906.

Lu D, Hsiao JY, Davey NE, Van Voorhis VA, Foster SA, Tang C, Morgan DO. 2014. Multiple mechanisms determine the order of APC/C substrate degradation in mitosis. *J Cell Biol* **207:** 23–39.

Luo X, Yu H. 2008. Protein metamorphosis: The two-state behavior of Mad2. *Structure* **16:** 1616–1625.

Machida YJ, Dutta A. 2007. The APC/C inhibitor, Emi1, is essential for prevention of rereplication. *Genes Dev* **21:** 184–194.

Maciejowski J, George KA, Terret ME, Zhang C, Shokat KM, Jallepalli PV. 2010. Mps1 directs the assembly of Cdc20 inhibitory complexes during interphase and mitosis to control M phase timing and spindle checkpoint signaling. *J Cell Biol* **190:** 89–100.

Maestre C, Delgado-Esteban M, Gomez-Sanchez JC, Bolaños JP, Almeida A. 2008. Cdk5 phosphorylates Cdh1 and modulates cyclin B1 stability in excitotoxicity. *EMBO J* **27:** 2736–2745.

Maldonado M, Kapoor TM. 2011. Constitutive Mad1 targeting to kinetochores uncouples checkpoint signalling from chromosome biorientation. *Nat Cell Biol* **13:** 475–482.

Malureanu LA, Jeganathan KB, Hamada M, Wasilewski L, Davenport J, van Deursen JM. 2009. BubR1 N terminus acts as a soluble inhibitor of cyclin B degradation by APC/C^{Cdc20} in interphase. *Dev Cell* **16:** 118–131.

Manchado E, Guillamot M, de Cárcer G, Eguren M, Trickey M, García-Higuera I, Moreno S, Yamano H, Cañamero M, Malumbres M. 2010. Targeting mitotic exit leads to tumor regression in vivo: Modulation by Cdk1, Mastl, and the PP2A/B55α,δ phosphatase. *Cancer Cell* **18:** 641–654.

Mansfeld J, Collin P, Collins MO, Choudhary JS, Pines J. 2011. APC15 drives the turnover of MCC-CDC20 to make the spindle assembly checkpoint responsive to kinetochore attachment. *Nat Cell Biol* **13:** 1234–1243.

Mao DD, Gujar AD, Mahlokozera T, Chen I, Pan Y, Luo J, Brost T, Thompson EA, Turski A, Leuthardt EC, et al. 2015. A CDC20-APC/SOX2 signaling axis regulates human glioblastoma stem-like cells. *Cell Rep* **11:** 1809–1821.

Mapelli M, Musacchio A. 2007. MAD contortions: Conformational dimerization boosts spindle checkpoint signaling. *Curr Opin Struct Biol* **17:** 716–725.

Matyskiela ME, Morgan DO. 2009. Analysis of activator-binding sites on the APC/C supports a cooperative substrate-binding mechanism. *Mol Cell* **34:** 68–80.

Meadows JC, Shepperd LA, Vanoosthuyse V, Lancaster TC, Sochaj AM, Buttrick GJ, Hardwick KG, Millar JB. 2011. Spindle checkpoint silencing requires association of PP1 to both Spc7 and kinesin-8 motors. *Dev Cell* **20:** 739–750.

Moyle MW, Kim T, Hattersley N, Espeut J, Cheerambathur DK, Oegema K, Desai A. 2014. A Bub1-Mad1 interaction targets the Mad1-Mad2 complex to unattached kinetochores to initiate the spindle checkpoint. *J Cell Biol* **204:** 647–657.

Musacchio A. 2015. The molecular biology of spindle assembly checkpoint signaling dynamics. *Curr Biol* **25:** R1002–R1018.

Musacchio A, Desai A. 2017. A molecular view of kinetochore assembly and function. *Biology (Basel)* **6:** 5.

Nijenhuis W, Vallardi G, Teixeira A, Kops GJ, Saurin AT. 2014. Negative feedback at kinetochores underlies a responsive spindle checkpoint signal. *Nat Cell Biol* **16:** 1257–1264.

Passmore LA, McCormack EA, Au SW, Paul A, Willison KR, Harper JW, Barford D. 2003. Doc1 mediates the activity of the anaphase-promoting complex by contributing to substrate recognition. *EMBO J* **22**: 786–796.

Patra D, Dunphy WG. 1998. Xe-p9, a *Xenopus* Suc1/Cks protein, is essential for the Cdc2-dependent phosphorylation of the anaphase-promoting complex at mitosis. *Genes Dev* **12**: 2549–2559.

Peters JM. 2006. The anaphase promoting complex/cyclosome: A machine designed to destroy. *Nat Rev Mol Cell Biol* **7**: 644–656.

Peters JM, King RW, Hoog C, Kirschner MW. 1996. Identification of BIME as a subunit of the anaphase-promoting complex. *Science* **274**: 1199–1201.

Pfleger CM, Kirschner MW. 2000. The KEN box: An APC recognition signal distinct from the D box targeted by Cdh1. *Genes Dev* **14**: 655–665.

Pines J. 2011. Cubism and the cell cycle: The many faces of the APC/C. *Nat Rev Mol Cell Biol* **12**: 427–438.

Primorac I, Musacchio A. 2013. Panta rhei: The APC/C at steady state. *J Cell Biol* **201**: 177–189.

Prinz S, Hwang ES, Visintin R, Amon A. 1998. The regulation of Cdc20 proteolysis reveals a role for the APC components Cdc23 and Cdc27 during S phase and early mitosis. *Curr Biol* **8**: 750–760.

Qiao R, Weissmann F, Yamaguchi M, Brown NG, VanderLinden R, Imre R, Jarvis MA, Brunner MR, Davidson IF, Litos G, et al. 2016. Mechanism of APC/C^{CDC20} activation by mitotic phosphorylation. *Proc Natl Acad Sci* **113**: E2570–E2578.

Reddy SK, Rape M, Margansky WA, Kirschner MW. 2007. Ubiquitination by the anaphase-promoting complex drives spindle checkpoint inactivation. *Nature* **446**: 921–925.

Reimann JD, Freed E, Hsu JY, Kramer ER, Peters JM, Jackson PK. 2001a. Emi1 is a mitotic regulator that interacts with Cdc20 and inhibits the anaphase promoting complex. *Cell* **105**: 645–655.

Reimann JD, Gardner BE, Margottin-Goguet F, Jackson PK. 2001b. Emi1 regulates the anaphase-promoting complex by a different mechanism than Mad2 proteins. *Genes Dev* **15**: 3278–3285.

Rodrigo-Brenni MC, Morgan DO. 2007. Sequential E2s drive polyubiquitin chain assembly on APC targets. *Cell* **130**: 127–139.

Rosenberg SC, Corbett KD. 2015. The multifaceted roles of the HORMA domain in cellular signaling. *J Cell Biol* **211**: 745–755.

Rosenberg JS, Cross FR, Funabiki H. 2011. KNL1/Spc105 recruits PP1 to silence the spindle assembly checkpoint. *Curr Biol* **21**: 942–947.

Sackton KL, Dimova N, Zeng X, Tian W, Zhang M, Sackton TB, Meaders J, Pfaff KL, Sigoillot F, Yu H, et al. 2014. Synergistic blockade of mitotic exit by two chemical inhibitors of the APC/C. *Nature* **514**: 646–649.

Santaguida S, Amon A. 2015. Short- and long-term effects of chromosome mis-segregation and aneuploidy. *Nat Rev Mol Cell Biol* **16**: 473–485.

Santaguida S, Tighe A, D'Alise AM, Taylor SS, Musacchio A. 2010. Dissecting the role of MPS1 in chromosome biorientation and the spindle checkpoint through the small molecule inhibitor reversine. *J Cell Biol* **190**: 73–87.

Schmidt A, Duncan PI, Rauh NR, Sauer G, Fry AM, Nigg EA, Mayer TU. 2005. *Xenopus* polo-like kinase Plx1 regulates XErp1, a novel inhibitor of APC/C activity. *Genes Dev* **19**: 502–513.

Schwab M, Lutum AS, Seufert W. 1997. Yeast Hct1 is a regulator of Clb2 cyclin proteolysis. *Cell* **90**: 683–693.

Schwab M, Neutzner M, Möcker D, Seufert W. 2001. Yeast Hct1 recognizes the mitotic cyclin Clb2 and other substrates of the ubiquitin ligase APC. *EMBO J* **20**: 5165–5175.

Sczaniecka M, Feoktistova A, May KM, Chen JS, Blyth J, Gould KL, Hardwick KG. 2008. The spindle checkpoint functions of Mad3 and Mad2 depend on a Mad3 KEN box–mediated interaction with Cdc20-anaphase-promoting complex (APC/C). *J Biol Chem* **283**: 23039–23047.

Sewart K, Hauf S. 2017. Different functionality of Cdc20 binding sites within the mitotic checkpoint complex. *Curr Biol* **27**: 1213–1220.

Shepperd LA, Meadows JC, Sochaj AM, Lancaster TC, Zou J, Buttrick GJ, Rappsilber J, Hardwick KG, Millar JB. 2012. Phosphodependent recruitment of Bub1 and Bub3 to Spc7/KNL1 by Mph1 kinase maintains the spindle checkpoint. *Curr Biol* **22**: 891–899.

Shirayama M, Zachariae W, Ciosk R, Nasmyth K. 1998. The Polo-like kinase Cdc5p and the WD-repeat protein Cdc20p/fizzy are regulators and substrates of the anaphase promoting complex in *Saccharomyces cerevisiae*. *EMBO J* **17**: 1336–1349.

Shteinberg M, Protopopov Y, Listovsky T, Brandeis M, Hershko A. 1999. Phosphorylation of the cyclosome is required for its stimulation by Fizzy/cdc20. *Biochem Biophys Res Commun* **260**: 193–198.

Sigrist SJ, Lehner CF. 1997. *Drosophila* fizzy-related down-regulates mitotic cyclins and is required for cell proliferation arrest and entry into endocycles. *Cell* **90**: 671–681.

Sigrist S, Jacobs H, Stratmann R, Lehner CF. 1995. Exit from mitosis is regulated by *Drosophila* fizzy and the sequential destruction of cyclins A, B and B3. *EMBO J* **14**: 4827–4838.

Sivaprasad U, Machida YJ, Dutta A. 2007. APC/C–the master controller of origin licensing? *Cell Div* **2**: 8.

Sudakin V, Ganoth D, Dahan A, Heller H, Hershko J, Luca FC, Ruderman JV, Hershko A. 1995. The cyclosome, a large complex containing cyclin-selective ubiquitin ligase activity, targets cyclins for destruction at the end of mitosis. *Mol Biol Cell* **6**: 185–197.

Sudakin V, Chan GK, Yen TJ. 2001. Checkpoint inhibition of the APC/C in HeLa cells is mediated by a complex of BUBR1, BUB3, CDC20, and MAD2. *J Cell Biol* **154**: 925–936.

Teichner A, Eytan E, Sitry-Shevah D, Miniowitz-Shemtov S, Dumin E, Gromis J, Hershko A. 2011. p31^{comet} Promotes disassembly of the mitotic checkpoint complex in an ATP-dependent process. *Proc Natl Acad Sci* **108**: 3187–3192.

Tischer T, Hormanseder E, Mayer TU. 2012. The APC/C inhibitor XErp1/Emi2 is essential for *Xenopus* early embryonic divisions. *Science* **338**: 520–524.

Trinkle-Mulcahy L, Andrews PD, Wickramasinghe S, Sleeman J, Prescott A, Lam YW, Lyon C, Swedlow JR, Lamond AI. 2003. Time-lapse imaging reveals dynamic relocalization of PP1γ throughout the mammalian cell cycle. *Mol Biol Cell* **14**: 107–117.

Tung JJ, Hansen DV, Ban KH, Loktev AV, Summers MK, Adler JR III, Jackson PK. 2005. A role for the anaphase-promoting complex inhibitor Emi2/XErp1, a homolog of early mitotic inhibitor 1, in cytostatic factor arrest of *Xenopus* eggs. *Proc Natl Acad Sci* **102**: 4318–4323.

Uzunova K, Dye BT, Schutz H, Ladurner R, Petzold G, Toyoda Y, Jarvis MA, Brown NG, Poser I, Novatchkova M, et al. 2012. APC15 mediates CDC20 autoubiquitylation by APC/C^{MCC} and disassembly of the mitotic checkpoint complex. *Nat Struct Mol Biol* **19**: 1116–1123.

van Zon W, Wolthuis RM. 2010. Cyclin A and Nek2A: APC/C-Cdc20 substrates invisible to the mitotic spindle checkpoint. *Biochem Soc Trans* **38**: 72–77.

Veas-Perez de Tudela M, Maestre C, Delgado-Esteban M, Bolaños JP, Almeida A. 2015. Cdk5-mediated inhibition of APC/C-Cdh1 switches on the cyclin D1-Cdk4-pRb pathway causing aberrant S-phase entry of postmitotic neurons. *Sci Rep* **5**: 18180.

Vleugel M, Hoek TA, Tromer E, Sliedrecht T, Groenewold V, Omerzu M, Kops GJ. 2015. Dissecting the roles of human BUB1 in the spindle assembly checkpoint. *J Cell Sci* **128**: 2975–2982.

Vodermaier HC, Gieffers C, Maurer-Stroh S, Eisenhaber F, Peters JM. 2003. TPR subunits of the anaphase-promoting complex mediate binding to the activator protein CDH1. *Curr Biol* **13**: 1459–1468.

von Schubert C, Cubizolles F, Bracher JM, Sliedrecht T, Kops GJ, Nigg EA. 2015. Plk1 and Mps1 cooperatively regulate the spindle assembly checkpoint in human cells. *Cell Rep* **12**: 66–78.

Wang L, Zhang J, Wan L, Zhou X, Wang Z, Wei W. 2015. Targeting Cdc20 as a novel cancer therapeutic strategy. *Pharmacol Ther* **151**: 141–151.

Watanabe Y, Khodosevich K, Monyer H. 2014. Dendrite development regulated by the schizophrenia-associated gene FEZ1 involves the ubiquitin proteasome system. *Cell Rep* **7:** 552–564.

Weinstein J. 1997. Cell cycle-regulated expression, phosphorylation, and degradation of p55Cdc. A mammalian homolog of CDC20/Fizzy/slp1. *J Biol Chem* **272:** 28501–28511.

Westhorpe FG, Tighe A, Lara-Gonzalez P, Taylor SS. 2011. p31comet-mediated extraction of Mad2 from the MCC promotes efficient mitotic exit. *J Cell Sci* **124:** 3905–3916.

Wild T, Larsen MS, Narita T, Schou J, Nilsson J, Choudhary C. 2016. The spindle assembly checkpoint is not essential for viability of human cells with genetically lowered APC/C activity. *Cell Rep* **14:** 1829–1840.

Williamson A, Wickliffe KE, Mellone BG, Song L, Karpen GH, Rape M. 2009. Identification of a physiological E2 module for the human anaphase-promoting complex. *Proc Natl Acad Sci* **106:** 18213–18218.

Wojcik E, Basto R, Serr M, Scaërou F, Karess R, Hays T. 2001. Kinetochore dynein: Its dynamics and role in the transport of the Rough deal checkpoint protein. *Nat Cell Biol* **3:** 1001–1007.

Wu T, Merbl Y, Huo Y, Gallop JL, Tzur A, Kirschner MW. 2010. UBE2S drives elongation of K11-linked ubiquitin chains by the anaphase-promoting complex. *Proc Natl Acad Sci* **107:** 1355–1360.

Xia G, Luo X, Habu T, Rizo J, Matsumoto T, Yu H. 2004. Conformation-specific binding of p31^{comet} antagonizes the function of Mad2 in the spindle checkpoint. *EMBO J* **23:** 3133–3143.

Yamagishi Y, Yang CH, Tanno Y, Watanabe Y. 2012. MPS1/Mph1 phosphorylates the kinetochore protein KNL1/Spc7 to recruit SAC components. *Nat Cell Biol* **14:** 746–752.

Yamaguchi M, VanderLinden R, Weissmann F, Qiao R, Dube P, Brown NG, Haselbach D, Zhang W, Sidhu SS, Peters JM, et al. 2016. Cryo-EM of mitotic checkpoint complex-bound APC/C reveals reciprocal and conformational regulation of ubiquitin ligation. *Mol Cell* **63:** 593–607.

Yang M, Li B, Tomchick DR, Machius M, Rizo J, Yu H, Luo X. 2007. p31comet blocks Mad2 activation through structural mimicry. *Cell* **131:** 744–755.

Yang Y, Kim AH, Yamada T, Wu B, Bilimoria PM, Ikeuchi Y, de la Iglesia N, Shen J, Bonni A. 2009. A Cdc20-APC ubiquitin signaling pathway regulates presynaptic differentiation. *Science* **326:** 575–578.

Yang Y, Tsuchiya D, Lacefield S. 2015. Bub3 promotes Cdc20-dependent activation of the APC/C in S. cerevisiae. *J Cell Biol* **209:** 519–527.

Ye Q, Rosenberg SC, Moeller A, Speir JA, Su TY, Corbett KD. 2015. TRIP13 is a protein-remodeling AAA+ ATPase that catalyzes MAD2 conformation switching. *Elife* **4:** e07367.

Yudkovsky Y, Shteinberg M, Listovsky T, Brandeis M, Hershko A. 2000. Phosphorylation of Cdc20/fizzy negatively regulates the mammalian cyclosome/APC in the mitotic checkpoint. *Biochem Biophys Res Commun* **271:** 299–304.

Zeng X, Sigoillot F, Gaur S, Choi S, Pfaff KL, Oh DC, Hathaway N, Dimova N, Cuny GD, King RW. 2010. Pharmacologic inhibition of the anaphase-promoting complex induces a spindle checkpoint-dependent mitotic arrest in the absence of spindle damage. *Cancer Cell* **18:** 382–395.

Zhang S, Chang L, Alfieri C, Zhang Z, Yang J, Maslen S, Skehel M, Barford D. 2016. Molecular mechanism of APC/C activation by mitotic phosphorylation. *Nature* **533:** 260–264.

Zhang G, Kruse T, López-Méndez B, Sylvestersen KB, Garvanska DH, Schopper S, Nielsen ML, Nilsson J. 2017. Bub1 positions Mad1 close to KNL1 MELT repeats to promote checkpoint signalling. *Nat Commun* **8:** 15822.

Zhou Z, He M, Shah AA, Wan Y. 2016. Insights into APC/C: From cellular function to diseases and therapeutics. *Cell Div* **11:** 9.

Noncanonical Biogenesis of Centrioles and Basal Bodies

Catarina Nabais, Sónia Gomes Pereira, and Mónica Bettencourt-Dias

Cell Cycle Regulation Lab, Instituto Gulbenkian de Ciência (IGC), 2780-156 Oeiras, Portugal

Correspondence: cnabais@igc.gulbenkian.pt; sgpereira@igc.gulbenkian.pt; mdias@igc.gulbenkian.pt

Centrioles and basal bodies (CBBs) organize centrosomes and cilia within eukaryotic cells. These organelles are composed of microtubules and hundreds of proteins performing multiple functions such as signaling, cytoskeleton remodeling, and cell motility. The CBB is present in all branches of the eukaryotic tree of life and, despite its ultrastructural and protein conservation, there is diversity in its function, occurrence (i.e., presence/absence), and modes of biogenesis across species. In this review, we provide an overview of the multiple pathways through which CBBs are formed in nature, with a special focus on the less studied, noncanonical ways. Despite the differences among each mechanism herein presented, we highlighted some of their common principles. These principles, governing different steps of biogenesis, ensure that CBBs may perform a multitude of functions in a huge diversity of organisms but yet retained their robustness in structure throughout evolution.

Centrioles and basal bodies (CBBs) are microtubule-based structures that assemble centrosomes and cilia. The centrosome is the dominant microtubule organizing center (MTOC) in most animal cells, thereby regulating intracellular transport, spindle pole formation, and establishing cellular polarity and migration. Each centrosome is composed of two cylindrical centrioles, often ninefold symmetric, surrounded by dynamic pericentriolar material (PCM). The PCM is responsible for anchoring and nucleating microtubules. Centrioles, then called basal bodies, can also anchor to the cell membrane and template the growth of motile and immotile cilia. In animals, most cell types form only one cilium (the primary cilium), but others can form hundreds (multiciliogenesis). These organelles are required for both cell and flow motility and sensing environmental cues.

It is essential that a cell regulates CBBs biogenesis to ensure they assemble at the right place, time, and number. Failure to regulate this process can lead to cellular defects and diseases. If cells possess more than two centrosomes at mitotic onset, they may assemble multipolar spindles and segregate the genome unevenly. This leads to aneuploidy, genomic instability, and cancer (Peel et al. 2007; Ganem et al. 2009; Silkworth et al. 2009; Godinho and Pellman 2014; Levine et al. 2017). Similarly, problems in cilia assembly cause a plethora of ciliopathies (Badano et al. 2006), which, in some cases, may arise from structural defects in the basal bodies (e.g., some mutations causing Bardet–Biedl syndrome [Ansley et al. 2003]).

CBBs are well-conserved structures present across the eukaryotic tree of life and probably derived from a basal body–like organelle already present in the last eukaryotic ancestor (LECA) (Cavalier-Smith 2002; Hodges et al. 2010). They have been lost within plant, fungi, and amoebae lineages or reduced to particular tissues or life-cycle stages in other groups, acquiring new morphologies and modes of biogenesis.

CBBs can assemble by several pathways; the best-characterized one is centriole duplication (Loncarek and Bettencourt-Dias 2018). This, hereafter called canonical pathway, occurs through the formation of two daughter centrioles close to two preexisting ones. In mitosis, one centrosome is segregated to each daughter cell, ensuring that cells maintain a correct centriole number when they proliferate. Canonical biogenesis is always coupled to the cell cycle, ensuring that CBBs only form once. Centrioles can also assemble through noncaninocal pathways, but less is know in terms of their regulation and origin, though they are widespread in nature.

In this review, we describe the diverse pathways through which CBBs are formed. We focus mostly on the noncanonical strategies, which have been less explored in the literature. We differentiate these strategies into two categories: deuterosome-mediated biogenesis, when centrioles form in bulk in the presence of preexisting centrioles, and de novo, strictly referring to biogenesis without any previously existing centrioles in the cell/organism. We highlight the similarities and differences between these pathways and discuss both their evolution and underlying molecular and cellular mechanisms.

PATHWAYS OF BIOGENESIS

The Canonical Pathway (Centriole Duplication)

In cycling cells, centrioles assemble in G_1 to S transition, forming one daughter centriole orthogonally to each mother. The daughter centrioles elongate and, in late G_2, undergo centriole-to-centrosome conversion losing the cartwheel (in vertebrate cells) and recruiting PCM (Fu et al. 2016). Then, the two centrosomes migrate toward opposite poles of the cell organizing the mitotic spindle. After mitosis, each daughter cell inherits exactly one pair of centrioles (Fig. 1).

Supplemental material is available for this article at symposium.cshlp.org.

Published by Cold Spring Harbor Laboratory Press; doi: 10.1101/sqb.2017.82.034694

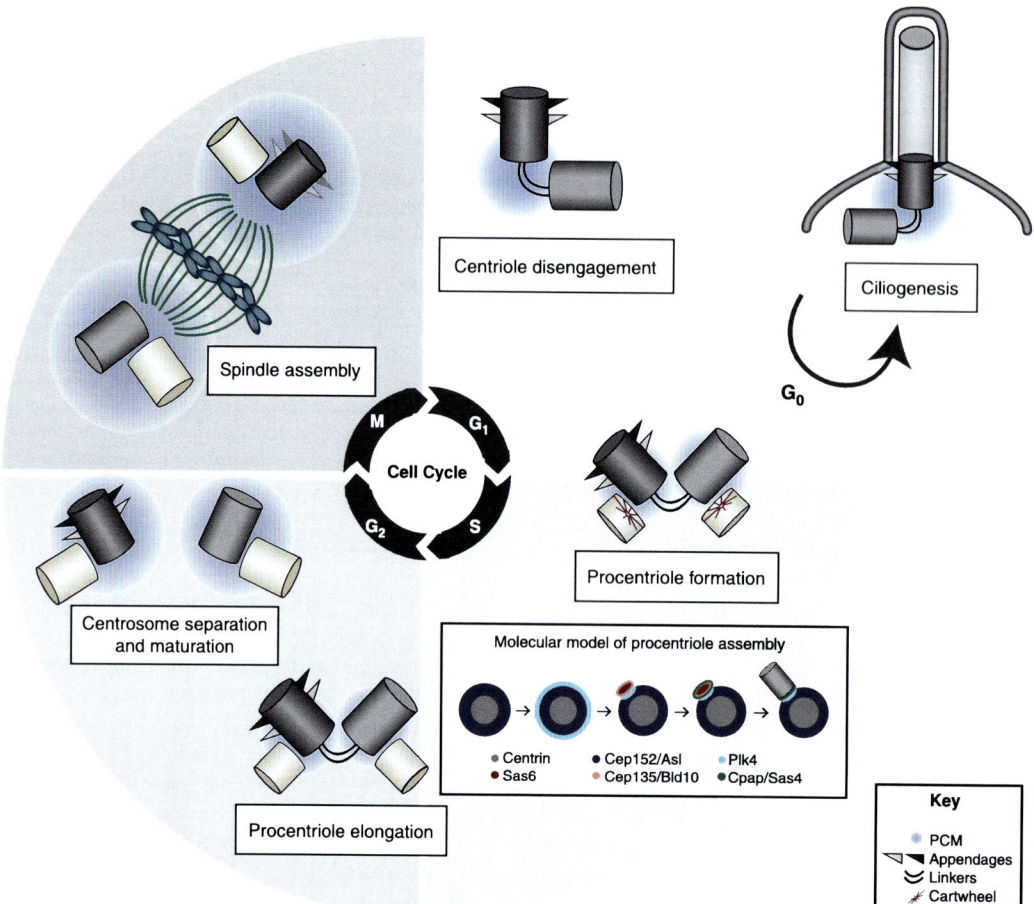

Figure 1. Canonical biogenesis in cycling cells. In early G_1, cells have one centrosome with two centrioles (mother and daughter) orthogonally oriented. Before duplication, the two centrioles disengage (G_1), losing their orthogonal configuration, and both become mother centrioles (Robbins et al. 1968). From G_1 to S transition, one procentriole forms orthogonally to each mother. The procentrioles elongate during the G_2 phase and each centrosome starts recruiting PCM components (Robbins et al. 1968; Kuriyama and Borisy 1981). From G_2 to mitosis, the two centrosomes separate and migrate toward opposite poles of the cell. Mitotic centrosomes recruit more PCM allowing them to organize the mitotic spindle. Upon mitotic completion, each daughter inherits exactly one pair of centrioles. At the beginning of each G_1 phase, the oldest centriole acquires both distal and subdistal appendages (Kong et al. 2014). Procentriole assembly relies on the stepwise incorporation of conserved molecules (depicted in the *inset*). Cep152/Asl recruits Plk4, which phosphorylates downstream substrates, allowing the formation of the Sas6 and Cep135/Bld10 cartwheel, thus building the centriole core. Cep152/Asl also interacts with Cpap/Sas4, promoting the incorporation of PCM components (Cizmecioglu et al. 2010; Dzhindzhev et al. 2010; Gopalakrishnan et al. 2011; Sonnen et al. 2013).

Although we are not yet aware of all the details governing this process and preventing reduplication, the molecular pathways involved in triggering and coupling centriole duplication to the cell cycle have been extensively studied in recent years (Matsumoto et al. 1999; Meraldi et al. 1999; Harrison et al. 2011; Zitouni et al. 2016). Such mechanisms are not detailed here, but they have been covered by numerous reviews (Loncarek and Bettencourt-Dias 2018; Nigg and Holland 2018).

Deuterosome-Mediated Biogenesis

Postmitotic cells containing two resident centrioles can differentiate into multicilated cells (MCCs), assembling CBBs in large scale through the deuterosome-mediated pathway (Fig. 2; Meunier and Azimzadeh 2016). Many multiciliated vertebrate tissues—the respiratory tract, the

oviduct, skin, efferent ducts, and the brain ependymal—are composed of MCCs. These cells produce fluid flow and particle movement, through the coordinated beating of their motile cilia. We hereby describe multiciliogenesis in vertebrate MCCs, whose molecular aspects have been characterized in recent years, showing that deuterosome-mediated and canonical biogenesis share part of their molecular cascade (Vladar and Stearns 2007; Azimzadeh et al. 2012; Klos Dehring et al. 2013; Zhao et al. 2013; Mori et al. 2017). We also speculate that a similar mechanism might contribute to the formation of multiciliated sperm in some invertebrates, such as in mollusks (*Cipangopaludina malleata* [Gall 1961] and *Pyrazus ebeninus* [Healy and Jamieson 1981]) and the insect *Mastotermes darwiniensis* (Baccetyi and Dallai 1978; Riparbelli et al. 2009).

In primary ciliogenesis, a single cilium derives directly from a CBB formed canonically, whereas in multicilio-

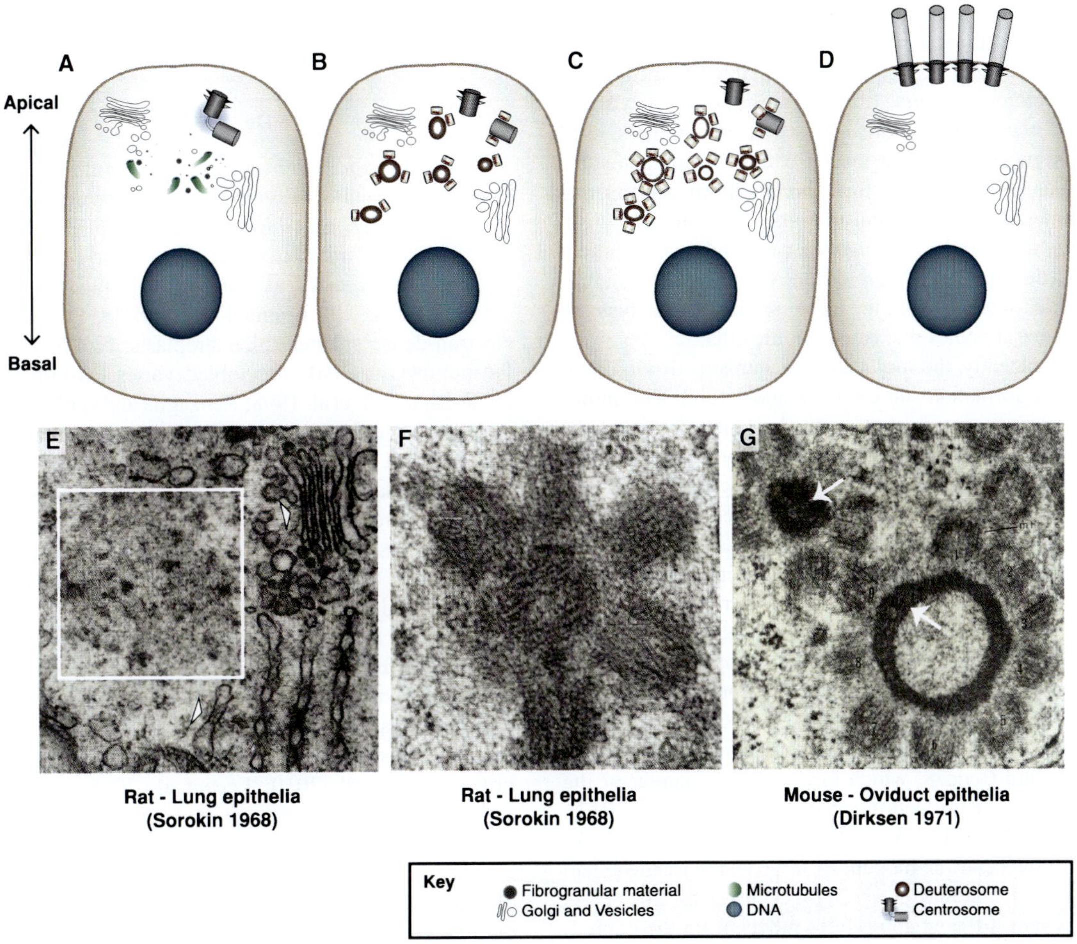

Rat - Lung epithelia
(Sorokin 1968)

Rat - Lung epithelia
(Sorokin 1968)

Mouse - Oviduct epithelia
(Dirksen 1971)

Key
● Fibrogranular material   ◗ Microtubules   ◎ Deuterosome
▥ Golgi and Vesicles   ● DNA   ⏚ Centrosome

Figure 2. Deuterosome-mediated biogenesis in vertebrate multicilated cells (MCCs). Multiciliogenesis starts with the formation of electron-dense "fibrogranular material" (in *A* and depicted within the white square in the EM micrograph, *E*) in the cytosol, close to preexisting centrioles. This dense material is usually enriched with microtubules (MTs), Golgi cisternae, and vesicles (*A,E,* arrowheads). The "fibrogranular material" condenses and deuterosomes—electron-dense hollow spheres—are formed (*B,G,* arrows). A recent study in ependymal cells demonstrated that the resident daughter centriole is capable of generating multiple deuterosomes, which detach from its wall and give rise to many procentrioles (*B,C,G*) (Al Jord et al. 2014). Additionally, procentrioles assemble directly around the resident centrioles (*C*), as shown in the EM micrograph (*F*). Hundreds of CBBs are formed in the cytosol, which then migrate and dock to the cell membrane assembling hundreds of cilia (*D*). (*E* [×37,000] and *F* [×50,000]: Adapted, with permission, from Sorokin 1968, *Journal of Cell Science*, 3: 207–230; *G* [×96,000]: adapted, with permission, from Dirksen 1971, *Journal of Cell Biology*, J51(1): 286–302 DOI: 10.1083/jcb.51.1.286.)

genesis, hundreds of basal bodies are generated, which nucleate hundreds of cilia. Centriole biogenesis in MCCs does not rely only on the association with preexisting centrioles but instead depends on additional specialized structures (deuterosomes) to efficiently assemble a large number of CBBs. Electron microscopy (EM) studies described the formation of electron-dense granules ("fibrogranular material") in the cytosol—usually in the vicinity of resident centrioles, in the apical region of the cell—as the first morphological evidence of ciliogenesis (Fig. 2A,E; Sorokin 1968; Steinman 1968; Kalnins and Porter 1969; Dirksen 1971; Hagiwara et al. 2004; Vladar and Stearns 2007). Progressively, these granules increase in size and condense into large spherical bodies, the deuterosomes, which show no discernible structure but are extremely electron-dense (Fig. 2B,C,G); suggesting they consist of concentrated proteins. Frequently, numerous

Golgi cisternae, small vesicles, and microtubules were seen in the vicinity of deuterosomes (Fig. 2A,E; Sorokin 1968; Kalnins and Porter 1969; Dirksen 1971; Vladar and Stearns 2007), suggesting these organelles might contribute to deuterosome formation and procentriole biogenesis. Although Golgi and vesicles, together with microtubule activity, can supply the deuterosome with precursors, preexisting centrioles might contribute with activating enzymes catalyzing biogenesis from the centriolar precursors. One such case, can be mediated by the activity of the Polo-like kinase 4 (Plk4), a master regulator and upstream player in centriole assembly (Bettencourt-Dias et al. 2005; Habedanck et al. 2005).

Several evenly spaced procentrioles assemble simultaneously from each deuterosome (Fig. 2B,C,G). In most tissues, procentrioles form both around the amorphous deuterosome (acentriolar-mediated) (Fig. 2G) and the preexisting centrioles (centriolar-mediated) (Fig. 2F; Sorokin 1968; Anderson and Brenner 1971; Hagiwara et al. 2004; Al Jord et al. 2014). During ependymal MCC differentiation, deuterosomes arise from the wall of the (preexisting) daughter centriole (Al Jord et al. 2014). Nonetheless, in all tissues, most of the centrioles (70%–90%) are generated via deuterosomes rather than directly from centrosomal centrioles. The specific centriole amplification mechanism used by different MCCs might then depend on the number of cilia they produce (Meunier and Azimzadeh 2016). Procentrioles separate from the clusters, mature, and become typical basal bodies nucleating motile cilia.

Only recently, the molecular mechanisms driving deuterosome formation started to be understood. The multiciliogenesis program starts with down-regulation of the Notch signaling pathway in MCCs precursors. Then, MCCs activate a cascade, mediated by the GemC1–Multicilin–E2f4/5 complex, triggering cell cycle exit, cytoskeleton remodeling, and up-regulation of several centriole biogenesis components, including Cep152/Asl, Plk4, Cpap/Sas4, Sas6, Stil/Sas5, and centrin (Vladar and Stearns 2007; Hoh et al. 2012; Zhao et al. 2013; Mori et al. 2017; Arbi et al. 2017). These proteins are usually at very low abundance in cycling cells, hence limiting the number of centrioles that are formed. MCCs also express deuterosome-specific components, Deup1 (a paralog of Cep63) and Ccdc78, which localize to the center of the deuterosome (Klos Dehring et al. 2013; Zhao et al. 2013). Deup1 binds Cep152/Asl, which then recruits Plk4, kick-starting the centriole biogenesis molecular cascade (Zhao et al. 2013; Al Jord et al. 2014; Mori et al. 2017). As MCCs start differentiating, E2f4 moves from the nucleus to the cytosol, where it interacts with Deup1 (Mori et al. 2017). Cep152/Asl, Plk4, and centrin are subsequently enriched at the deuterosome and at the preexisting centrioles, seeding the biogenesis of multiple CBBs. E2f4 has a dual role in the cell; first driving the transcription of centrosomal components and later participating in their assembly in the cytoplasm.

Nevertheless, it is still left to determine how centriole amplification stops. Is there a feedback mechanism that terminates centriole amplification? Or does it simply result from exhaustion of centrosomal components?

De Novo

Centrioles can assemble de novo (i.e., without centriolar structures present in the cell) in several species. However, in most naturally occurring cases (see Fig. 6; Supplemental Table S1), the mechanisms remain poorly understood. Centrioles may arise as single units (Fig. 3), as two centrioles coaxially oriented (bicentriole; Fig. 4), or in electron-dense spheres (blepharoplasts; Fig. 5) in which the number of CBBs assembled varies (Miki-Noumura 1977; Riparbelli et al. 1998; Renzaglia and Garbary 2001).

Amoebae to flagellate transition in *Naegleria gruberi* is accompanied by the biogenesis of two centrioles. Because amoebae lack centrioles and microtubules, and so far no basal body precursor has been found, it was proposed that centrioles assemble de novo (Dingle and Fulton 1966; Fulton and Dingle 1971). By studying the localization of centrin and γ-tubulin during this transition, Fritz-Laylin et al. (2016) have shown that only the first centriole assembles de novo, whereas the second one appears to duplicate from the first. There is no EM support for the underlying pathway and, despite some molecular insights from recent studies (Suh et al. 2002; Kim et al. 2005; Fritz-Laylin et al. 2010; Lee et al. 2015; Fritz-Laylin and Fulton 2016), the exact molecular cascade is still unknown.

Other examples of de novo biogenesis of single centrioles take place in parthenogenetic insect eggs (in *Muscidifurax uniraptor* [Fig. 3; Riparbelli et al. 1998], and *Drosophila mercatorum* [Riparbelli and Callaini 2003]) and artificially activated eggs of sea urchin (Dirksen 1961; Miki-Noumura 1977) and in the surf clam *Spisula solidissima* (Fig. 6; Supplemental Table S1; Kuriyama et al. 1986; Palazzo et al. 1992). As in most animals, centrioles are lost during oogenesis (Fig. 3A) and are delivered to the egg by the sperm upon fertilization. In activated hemynopteran eggs, multiple microtubule asters containing single centrioles are formed along the cortex

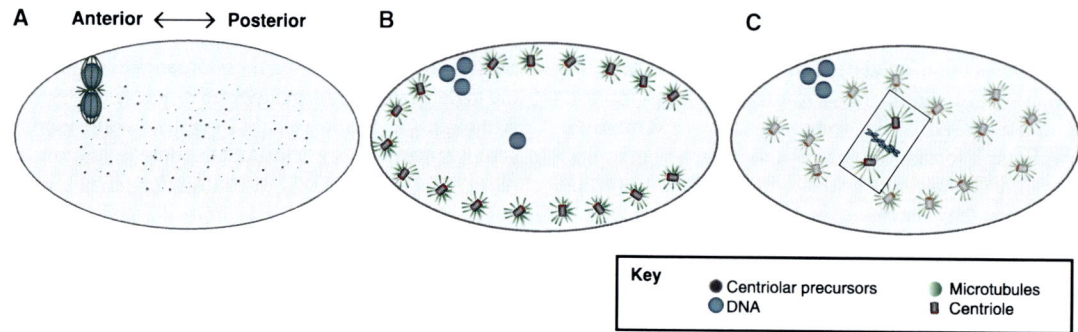

Figure 3. De novo centriole biogenesis in parthenogenic insect eggs. Unfertilized eggs do not have centrioles but contain high levels of centriolar precursors (*A*). Upon egg activation and meiotic resumption, centrioles are formed de novo along the cell cortex (*B*). These single centrioles nucleate MT asters. Meiosis is completed and the free centrosomes migrate toward the egg center (*C*). Two asters interact with the female pronucleus, assembling the first mitotic division and triggering embryonic development (*C*, black rectangle). The remaining centrosomes degenerate (Riparbelli et al. 1998).

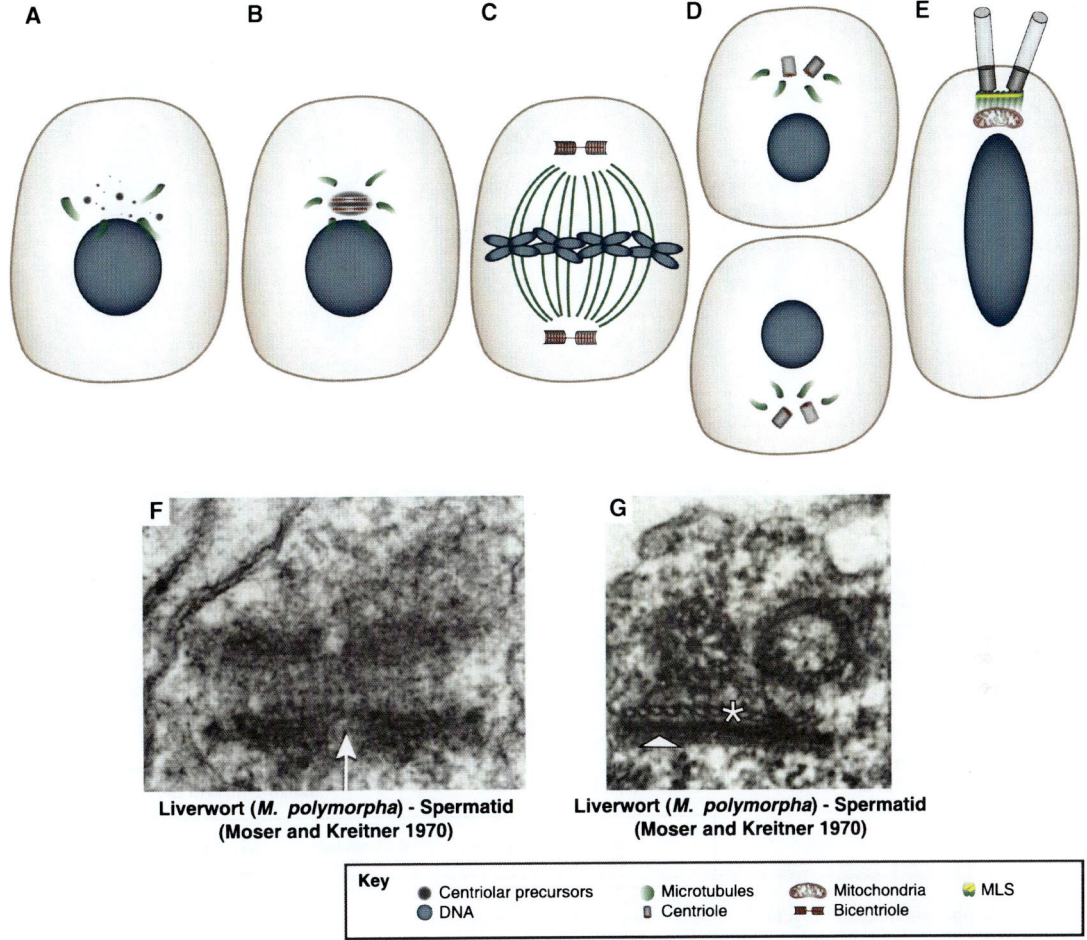

Key
● Centriolar precursors ● Microtubules ⬭ Mitochondria ▪ MLS
● DNA ▯ Centriole ▪▬▪ Bicentriole

F Liverwort (*M. polymorpha*) - Spermatid
(Moser and Kreitner 1970)

G Liverwort (*M. polymorpha*) - Spermatid
(Moser and Kreitner 1970)

Figure 4. Bicentriole-mediated biogenesis in land plants with biciliated sperm. During spermatogenesis, electron-dense material enriched in microtubules (MTs) is found near the nuclear envelope (*A*). This material assembles into two light lobes, surrounded by a darker matrix (*B*). As mitosis begins, the two lobes separate, migrate toward the spindle poles and mature into bicentrioles (*C*). Bicentrioles are composed of two coaxial centrioles connected by their central hub and with discontinuous MT triplets (*F*, white arrow). Each daughter cell (spermatid) inherits one bicentriole that breaks in half and separates into two centrioles (*D*) that will migrate to the edge of the cell and anchor to the multilayered structure (MLS), serving as basal bodies during ciliogenesis (*E,G*). The MLS is composed of a bundle of parallel MTs—the spline (*G*, asterisk)—and layers of electron-dense material—the lamellar strip (*G*, arrowhead). (*F* [×50,000] and *G* [×50,000]: Adapted, with permission, from Moser and Kreitner 1970, *Journal of Cell Biology*, 44(2): 454-458 DOI: 10.1083/jcb.44.2.454.)

(Fig. 3B). These migrate toward the center of egg. Parthenogenetic development is initiated when two asters are captured by the female pronuclei forming the first mitotic spindle (Fig. 3C; Riparbelli et al. 1998; Tram and Sullivan 2000).

The centriole in the mouse sperm is unable to nucleate microtubules after fertilization (Schatten et al. 1985; Gueth-Hallonet et al. 1993), so the first embryonic divisions are acentrosomal (Gueth-Hallonet et al. 1993; Courtois et al. 2012) and centrioles are only detected by EM from 64-cell stage onward (Gueth-Hallonet et al. 1993). Throughout the first mitotic divisions, the spindles become progressively more focused and are enriched with PCM and centriolar components, such as centrin, pericentrin, and CP110. Nevertheless, the trigger underlying centriole assembly is still unclear. A gradual concentration of PCM and centriolar components throughout the mitotic cycles could allow crossing a molecular threshold that

finally enables the formation of centrioles (Courtois et al. 2012).

Oocytes represent a very particular cell type that is loaded with centriolar components; therefore, mechanisms blocking spontaneous centriole assembly could be present. Although in most eggs, centrioles do not assemble spontaneously, overexpression of Plk4 is enough to drive de novo formation of multiple centrioles (Peel et al. 2007; Rodrigues-Martins et al. 2007).

In most cases, centrioles assembled de novo seem to be able to replicate through the canonical pathway (Palazzo et al. 1992; Rodrigues-Martins et al. 2007; Fritz-Laylin et al. 2016). Therefore, in cases where several centrioles are observed, we cannot exclude that some could result from duplication following de novo biogenesis. Moreover, in *Naegleria*, both CBBs form cilia, indicating that centrioles formed de novo and canonically are equally capable of nucleating cilia without the need of a full cell cycle to mature.

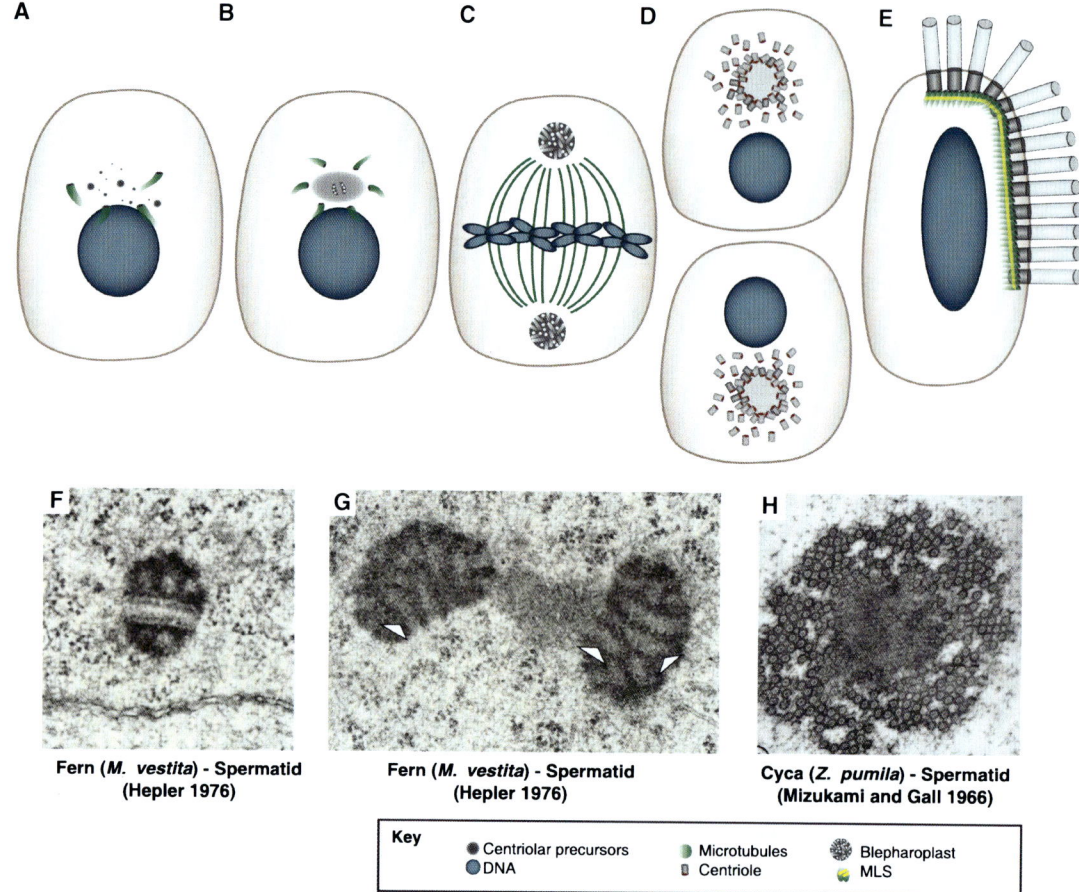

Figure 5. Blepharoplast-mediated biogenesis in land plants with multiciliated sperm. In plants with multiciliated sperm, an electron-dense agglomerate of material and microtubules (MTs) is first detected near the nuclear envelope of the sperm mother cell (*A*). This material develops into two darker hemispherical lobes, intercalated by lighter cylinders (*B,F,G*, arrowheads). As the cell approaches mitosis, the lobes enlarge and separate (*G*). Each lobe migrates to a pole of the mitotic spindle and assembles a blepharoplast (*C*). Each spermatid inherits one blepharoplast, where many centrioles are assembled. The blepharoplast eventually collapses releasing the individual centrioles (*D,H*) that will migrate and anchor to the MLS, giving rise to the basal bodies of the several cilia (*E*). (*F* [×37,000] and *G* [×37,000]: Adapted, with permission, from Hepler 1976, *Journal of Cell Science*, 21: 361–390; H [×21,000]: adapted, with permission, from Mizukami and Gall 1966, *Journal of Cell Biology*, 29(1): 97-111 DOI: 10.1083/jcb.29.1.97.)

Bicentriole. De novo centriole biogenesis through bicentrioles is known to occur in plants with biflagellated sperm, such as bryophytes, as well as in the protist *Labyrinthula* spp. (Fig. 6; Supplemental Table S1; Perkins 1970). A bicentriole is composed of two centrioles oriented end-to-end, aligned along the same axis, and connected by a continuous cartwheel hub and discontinuous triplet microtubules (Fig. 4C,F; Moser and Kreitner 1970; Robbins 1984).

In land plants, two bicentrioles appear simultaneously in the sperm mother cell. First, an electron-dense body without any recognizable structure is detected in the outer surface of the nucleus. Microtubules emanate from this structure, suggesting that it has MTOC activity (Fig. 4A). Next, it separates into two different lobes (pro-bicentrioles) with a lighter stained central core surrounded by a darker matrix (Fig. 4B; Robbins 1984). Before mitosis, the two pro-bicentrioles separate, migrate toward the poles of the cell, and mature into bicentrioles, assembling MT triplets (Robbins 1984; Renzaglia and Duckett 1987). Each bicentriole at the spindle pole contains two

coaxial centrioles (Fig. 4C,F; Moser and Kreitner 1970; Robbins 1984).

Each spermatid inherits one bicentriole. The central hub breaks at its midpoint and the two resulting centrioles undergo planar rotation becoming almost parallel to each other, with their proximal ends facing the same direction (Fig. 4D; Moser and Kreitner 1970; Kreitner and Carothers 1976; Robbins 1984). Centriole reorientation is accompanied by the development of the multilayered structure (MLS), immediately below the centrioles (Fig. 4E,G). The MLS is composed of a bundle of parallel microtubule singlets—the spline (Fig. 4G, asterisk)—and by the lamellar strip (layers of electron-dense material) (Fig. 4G, arrowhead). The centrioles anchor to the MLS and become basal bodies for ciliogenesis (Fig. 4E; Moser et al. 1977; Renzaglia and Duckett 1987).

There is no available molecular data on centriole assembly through bicentrioles, except that these structures appear to contain γ-tubulin (Shimamura et al. 2004). The only study reporting the early stages of de novo bicentriole assembly is from Robbins (1984) on spermatogenesis in

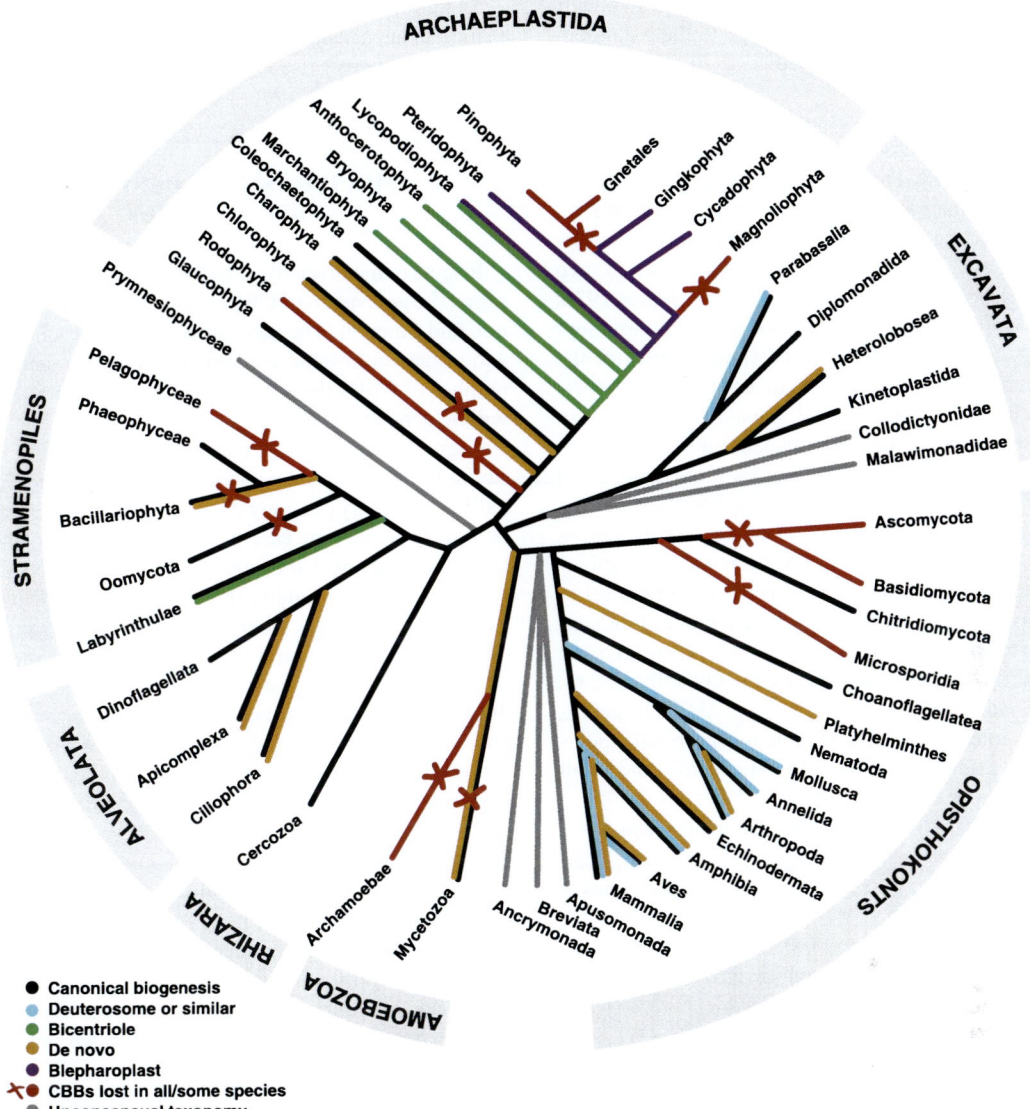

Figure 6. Consensus eukaryotic tree of life (selected groups; following Burki 2014 and Worden et al. 2015). The distinct centriole biogenesis pathways are represented in different colors. Canonical biogenesis (in black) is the most prevalent pathway and probably, the ancestral one. Deuterosomes (blue), the bicentriole (green), and blepharoplast (purple) are all evolutionary innovations, arising relatively recently in the eukaryotic history. Some pathways are more restricted to some groups—for example, the canonical and deuterosome pathways are predominant in vertebrates, whereas most plants assemble CBBs through a bicentriole or a blepharoplast. There are some striking exceptions, like the presence of a deuterosome-like mechanism in the sperm of some invertebrates. Whereas in gastropods (Mollusca) the noncanonical pathway seems to be centriolar, where up to 20 CBBs assemble only around the existing centrioles, the sperm from annelids and *Mastotermes darwiniensis* (Arthropoda) possesses a very high number of CBBs, likely formed via both centriolar and acentriolar ways. Similarly, within the class Parabasalia (Excavata) some protists undergo massive centriole amplification. It is proposed that biogenesis is driven by resident centrioles along a "ladder"-like configuration (Tamm and Tamm 1980). In all these studies, no typical deuterosomes were detected, only occasional clouds of electron-dense material containing microtubules. There are other examples of convergent evolution among pathways, such as the presence of a bicentriole in Labyrinthulae (Stramenopila). Future studies should be expanded to more species in less known groups to clarify the mechanism involved in de novo biogenesis (orange) and understanding if they are all a result of lineage-specific evolution (convergent evolution). CBBs were lost in multiple lineages (red lines and crosses, absent in all species within the groups; red crosses, lost in only some species within the lineage).

the bryophyte *Riella americana*. Early land plants, such as *Marchantia polymorpha, Physcomitrella patens*, and *Selaginella moellendorffii* are model organisms that assemble CBBs through the bicentriole pathway and therefore, could be used to better describe this pathway and understand its regulatory mechanisms.

Blepharoplast. In land plants with multiciliated sperm such as ferns, cycads, and *Ginkgo* (Fig. 6; Supplemental Table S1), CBBs are formed through blepharoplasts. The blepharoplast arises de novo as a spherical electron-dense organelle that is initially amorphous (Fig. 5A), and during maturation it becomes intercalated by lighter cylinders embedded in an electron-opaque matrix. These cylinders mature into centrioles that later give rise to the basal bodies of multiple cilia (Fig. 5; Hepler 1976; Gifford and Larson 1980).

Blepharoplast biogenesis starts with the appearance of two hemispherical densely stained structures near the cell nucleus (Fig. 5B,F). Then, cylinders organize within the electron-dense matrix (Fig. 5G, arrowheads), with microtubules emanating from the blepharoplast. These structures grow and become spherical, giving rise to two blepharoplasts (Mizukami and Gall 1966; Hepler 1976; Hoffman and Vaughn 1995). The two blepharoplasts separate (Fig. 5G) and migrate to the spindle poles of the mitotic cell, where they appear to act as MTOC (Fig. 5C; Hepler 1976; Gifford and Larson 1980; Doonan et al. 1986). In the metaphase–anaphase transition of the last mitosis, the blepharoplast becomes more diffuse and loses its MT-nucleating ability. The cylinders acquire a ninefold symmetry and a hub-and-spokes configuration, therefore resembling procentrioles. Each daughter cell inherits one blepharoplast (Norstog 1967; Gifford and Lin 1975; Hepler 1976). Sperm development proceeds as centrioles are formed (Fig. 5D,H; Hepler 1976; Renzaglia and Maden 2000). The blepharoplast eventually collapses, resulting in individualized centrioles (Fig. 5H). The centrioles dock into the MLS and function as basal bodies nucleating axonemes (Fig. 5E; Mizukami and Gall 1966; Doonan et al. 1986; Norstog 1986).

Molecular characterization of blepharoplast assembly is still scarce. However, a few studies have reported the localization of centrin, acetylated, tyrosinated, and β-tubulins at the blepharoplast (Doonan et al. 1986; Klink and Wolniak 2001; Vaughn and Renzaglia 2006). Centrin's function was studied in *Marsilea vestita*, in which RNAi experiments highlighted its requirement for proper blepharoplast and centriole biogenesis (Klink and Wolniak 2001).

To this date, there is no evidence for centriole duplication in multiciliated plant cells. It appears that each CBB formed de novo only gives rise to one cilium (Mizukami and Gall 1966; Norstog 1967, 1986; Gifford and Lin 1975).

MECHANISMS UNDERLYING CBBs ASSEMBLY

In spite of the diversity of pathways, their outcome is the same: the generation of CBBs with a conserved ultrastructure and function. The mechanism used by each cell type and organism to build it seems highly dependent on the number of CBBs they have to begin with and how many will be generated.

Regulation of centriole number is still not fully understood. In the canonical pathway number regulation is partially achieved by coupling of the centriole and cell cycles, but this cannot be the case in the noncanonical pathways. One possibility, is that centriole number only depends on the amount of its building blocks, and as centrioles are assembled, these are depleted. Under this hypothesis, number regulation would take place mostly at the levels of transcription and translation. Another strategy would be the activation of a negative feedback mechanism wherein, once the right amounts of centrioles are assembled, any further biogenesis is inhibited. Studies indicate that even noncanonical pathways show some centriole number regulation because each multicilated cell type assembles a consistent CBB number.

Nevertheless, canonical and noncanonical pathways share many striking similarities. Two centriolar proteins—Sas6 and centrin—and pericentriolar components γ-tubulin and pericentrin have been shown to be present in both canonical and noncanonical pathways in multiple species (Fig. 7). Sas6 is the most conserved centriolar protein and

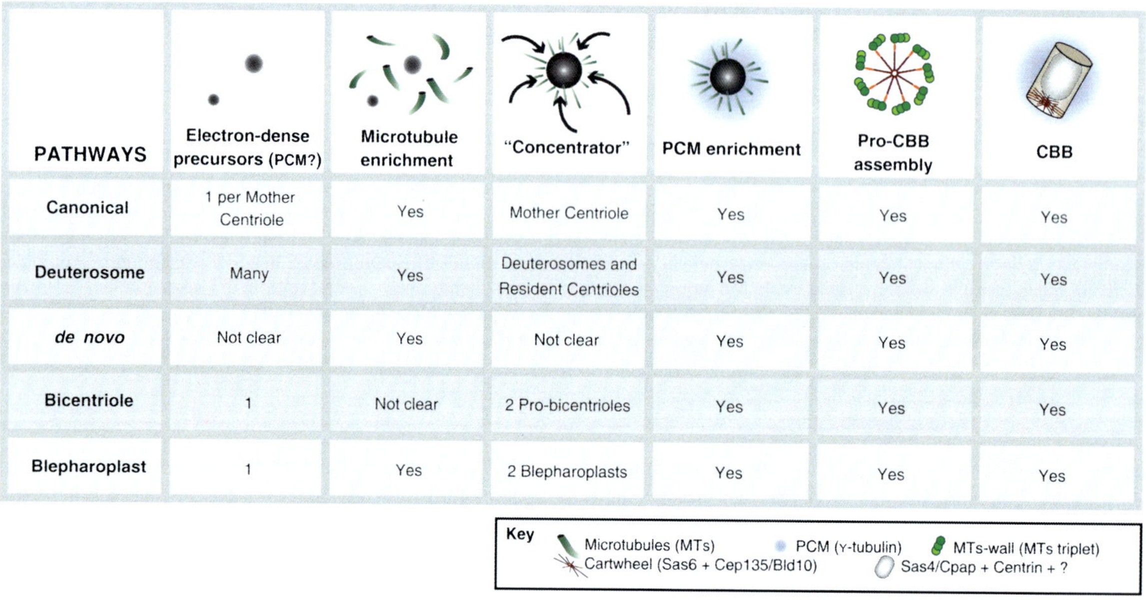

PATHWAYS	Electron-dense precursors (PCM?)	Microtubule enrichment	"Concentrator"	PCM enrichment	Pro-CBB assembly	CBB
Canonical	1 per Mother Centriole	Yes	Mother Centriole	Yes	Yes	Yes
Deuterosome	Many	Yes	Deuterosomes and Resident Centrioles	Yes	Yes	Yes
de novo	Not clear	Yes	Not clear	Yes	Yes	Yes
Bicentriole	1	Not clear	2 Pro-bicentrioles	Yes	Yes	Yes
Blepharoplast	1	Yes	2 Blepharoplasts	Yes	Yes	Yes

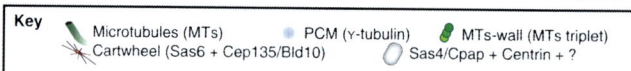

Key
Microtubules (MTs)   PCM (γ-tubulin)   MTs-wall (MTs triplet)
Cartwheel (Sas6 + Cep135/Bld10)   Sas4/Cpap + Centrin + ?

Figure 7. Common principles underlying centriole biogenesis among known pathways.

the major molecular component of the cartwheel, forming ninefold symmetrical stacks at the core of the centriolar barrel (Nakazawa et al. 2007; van Breugel et al. 2011; Kitagawa et al. 2011). In plants, centrin and γ-tubulin are enriched in the blepharoplast of *Ceratopteris richardii* (Hoffman et al. 1994), and functional studies demonstrated that centrin is needed to form the blepharoplast and therefore the ciliary apparatus in *M. vestita* sperm (Klink and Wolniak 2001). De novo CBB formation in *N. gruberi* is preceded by the formation of a γ-tubulin, pericentrin, and myosin II complex, at the site where Sas6 and centrin-positive centrioles assemble (Fritz-Laylin et al. 2010; Lee et al. 2015; Fritz-Laylin and Fulton 2016). In vertebrates, all of these previously mentioned components along with others localize to centrioles generated de novo in mammalian culture cells (Khodjakov et al. 2002; La Terra et al. 2005; Uetake et al. 2007) and are up-regulated in multiciliogenesis (Vladar and Stearns 2007; Klos Dehring et al. 2013; Zhao et al. 2013; Mori et al. 2017). Though the molecules are the same, differential regulation of their levels allows overcoming the canonical biogenesis regulation and assembling multiple CBBs.

The location where procentrioles assemble is determined by the site where its precursors concentrate, herein called "concentrator." Even though the "concentrator" might be morphologically distinct in each centriolar or acentriolar pathways, components must first accumulate in a defined location in the cytosol, and subsequently seed the growth of CBBs. In the canonical pathway the mother centriole acts as a concentrator, whereas in the noncanonical pathways organisms evolved multiple structures where centriolar components are specifically enriched—the blepharoplast, the deuterosome, and other electron-dense structures. This way, the concentrator regulates the location and number of CBBs assembled (Fig. 7).

The microtubule cytoskeleton helps transporting components to the concentrator (Fig. 7). CHO cells, upon centriolar removal and if treated with nocodazole, no longer form centrioles de novo (Khodjakov et al. 2002). Multiciliogenesis is accompanied by cytoskeleton remodeling that promotes assembly of stable cytoplasmic microtubules (more resistant to depolymerization) (Vladar and Stearns 2007). Microtubule enrichment is also detected close to the fibrogranular material preceding deuterosome formation (Steinman 1968; Dirksen 1971) and microtubules regrow from the blepharoplast after depolymerization (Vaughn and Bowling 2008). Overall, multiple observations hint that microtubules are important for CBBs assembly, however it is still left to determine when exactly they are critical. Are they needed in the very early stages of precursor concentration? Or do they only facilitate recruitment once there is already a centriolar primordium? Some components might have evolved affinity for the MTs, naturally concentrating at the MTOCs and facilitating the process. Among those components, PCM proteins are known to be required to stabilize centrioles and allow efficient centriole duplication (Dammermann et al. 2004; Pimenta-Marques et al. 2016). Proteins like chTOG/XMAP215, members of the Tacc family, Cpap/Sas4, and γ-tubulin are important for PCM assembly and

microtubule organization and are widely present in eukaryotes (Dammermann et al. 2004; Peset and Vernos 2008; Hodges et al. 2010). PCM might help concentrating centriolar proteins; therefore stable PCM aggregates in the cytosol might create a suitable environment for CBBs biogenesis (Fig. 7; Varmark et al. 2007; Dzhindzhev et al. 2010).

Finally, self-assembly and catalytic activity of centrosomal components are important in driving CBBs biogenesis. In several animals, Plk4 is the main kinase triggering centriole biogenesis. Plk4 controls its own activation by *trans*-autophosphorylation, which results in a positive feedback loop dependent on Plk4 concentration (Lopes et al. 2015). Self-assembling properties facilitate Sas6 oligomerization in vitro (Kitagawa et al. 2011). Together with Cep135/Bld10, these two *Chlamydomonas* proteins are able to assemble a cartwheel, the first step in building the centriolar core (Guichard et al. 2017). Recent studies have also shown that some centrosomal components spontaneously form condensates in vitro. Above a critical concentration, *C. elegans* Spd5 (a master PCM recruiter), forms a supramolecular scaffold where other PCM proteins can bind (Woodruff et al. 2017). Spd5 condensates enriched with chTOG and TPX2 are capable of concentrating α- and β-tubulin and organizing microtubule asters. Future work should dissect the role of self-assembling in vivo.

EVOLUTIONARY HISTORY OF CBBs AND THEIR PATHWAYS

Several lines of evidence support that CBBs are the same identity that was co-opted throughout evolution to perform different functions within the eukaryotic cell. Not only CBBs are ultrastructurally similar and co-occur across distinct "taxa," but the same gene network, the core centriolar assembly, is conserved in the genome of ciliated species (Woodland and Fry 2008; Carvalho-Santos et al. 2010; Hodges et al. 2010). Indeed, CBBs are found in all seven major eukaryotic lineages (Fig. 6; Supplemental Table S1), suggesting they were already present in the LECA but apparently not before (Carvalho-Santos et al. 2010). The ancestral CBB was most likely a basal body-like organelle composed of nine microtubule triplets arranged in a radially symmetrical cylinder (Beisson and Wright 2003) involved in the nucleation of motile cilia (Carvalho-Santos et al. 2011; Azimzadeh 2014). CBBs (and their gene repertoire) have been independently lost in several lineages and are frequently absent in some plants (Archaeplastida), fungi (Opisthokonts), and amoebae (Amoebozoa) (Fig. 6; Renzaglia and Garbary 2001; Woodland and Fry 2008; Carvalho-Santos et al. 2011; Judelson et al. 2012; Yubuki and Leander 2013).

Throughout evolution, the requirement for ciliary motility imposed a functional constrain on basal body architecture, as absence of cilia allowed for complete centriole loss and the generation of MTOCs with very distinct morphology like the spindle pole body (SPB) of fungi and the nuclear-associated body (NAB) of amoebae

(Supplemental Table S1; Hodges et al. 2010; Azimzadeh 2014).

Although cilia are seemingly ancestral structures, centrosomes most probably are not. A good example is the animal centrosome, which is mostly composed by Holozan-specific components (Holozoa is an Opisthokont subdivision including animals and closely related organisms except fungi) (Hodges et al. 2010). Recently, Gouw et al. (M Gouw, unpubl.) used maximum parsimony landscapes to assess the probability of the cilium and the centriole-based centrosome being ancestral in specific eukaryotic lineages. This analysis favored a convergent evolution hypothesis for the origin of centriole-based centrosomes, suggesting that centrioles were co-opted as part of the centrosomes independently in different eukaryotic lineages. The acquisition of centrosomal functions might have occurred in a stepwise manner. First, by becoming part of the spindle poles, CBBs could segregate equally to daughter cells upon cell division. This could favor an enrichment in PCM, potentiating MTOC activity. Finally, the acquisition of cell cycle components (Lange 2002) would link centrosome biogenesis and segregation to cell cycle progression, allowing a much tighter regulation of its activity and copy number in cells (Nigg and Holland 2018).

All pathways share components; a specific set of centriolar proteins—Sas6, Cpap/Sas4, Cep135/Bld10, Poc1, centrin—and α-, β-, and γ-tubulin are found in the genome of most eukaryotic species that assemble CBBs (Fig. 7; Carvalho-Santos et al. 2010; Hodges et al. 2010). Functional studies and expression data are still scarce outside Opisthokonts, but are needed to validate the function of these components in each pathway.

Canonical duplication is the most prevalent pathway and probably, the ancestral one. It is present in every main branch of the eukaryotic tree, though the mechanism is somewhat different in specific taxa. In some oomycetes such as *Saprolegnia ferax* and *Phytophthora infestans* (Stramenopiles) and in *Plasmodiophora* spp. (Rhizaria) (Fig. 6; Supplemental Table S1), daughter centrioles assemble in a 180° angle from their mother (coaxial orientation), rather than the usual 90°, forming a bicentriole, similar to the one found in some plants (Heath and Greenwood 1970; Heath 1974a,b; Garber and Aist 1979).

Not only the centriole-based centrosomes, but also deuterosomes, bicentrioles, and blepharoplasts are all evolutionary innovations, arising relatively recently in eukaryotic history (Fig. 6). A recent study argued that the deuterosome-mediated pathway is vertebrate-specific, arising just before tetrapode divergence. That is based on evidence that Deup1, a specific component of the deuterosome resulting from Cep63 duplication, is only found in the genomes of lobe-finned fish and tetrapods (Zhao et al. 2013). Some gastropodes (*C. malleatus* and *P. ebeninus*), annellides (*Tubifex* spp.), and the termite *Mastotermes darwiniensis* produce multiciliated sperm (Fig. 6; Supplemental Table S1; Gall 1961; Baccetyi and Dallai 1978; Healy and Jamieson 1981; Ferraguti et al. 2002; Riparbelli et al. 2009). In these naturally occurring cases, the sperm basal bodies might derive from a mechanism similar to the deuterosome. In all these studies, no typical

deuterosomes were detected, only occasional clouds of electron-dense material containing microtubules.

Archaeplastida, the group including plants and some algae, suffered multiple events of centriole loss, both in basal groups (in some green algae and in red algae altogether) and in gymnosperms after the split of conifers and gnetales from cycads and ginkgophytes and once again before angiosperm evolution (Magnoliophyta) (Bremer et al. 1987; Finet et al. 2010). Within this vast group, de novo mechanisms are the most prevalent, based either on the bicentriole or the blepharoplast, as most plants lack CBBs throughout their life cycle except in sperm. The bicentriole appeared in land plants; it is present in most Marchantiophyta and Bryophyta, and in some species of Anthocerotophyta and Lycopodiophyta, but it is absent in the basal Archaeplastida species (Fig. 6; Supplemental Table S1; for review, see Renzaglia and Garbary 2001). Interestingly, a bicentriole is also formed de novo in *Labyrinthula* spp., a Stramenopila (Fig. 6; Supplemental Table S1). It is possible that the blepharoplast from the Pteridophyta and some gymnosperms derived from the bicentriole. Interestingly, the blepharoplast is mechanistically very similar to the deuterosome, suggesting a scenario of convergent evolution. CBBs are required for species that form motile cilia and somehow depend on a moist environment for fertilization. Gymnosperms (Pinaceae and Gnetales) and all angiosperms (Magnoliophyta) no longer use motile cilia, because fertilization takes place by means of a pollen tube with immotile sperm cells.

It also remains to be understood if, in all the species of Amoebozoa assembling CBBs de novo upon ameboid to flagellate transition (for e.g., *Physarum* spp.) the mechanisms resemble those found in animals (e.g., in female eggs) or if these have evolved their own specific precursor and uncharacterized pathway. Fungi with CBBs seem to conserve the ancestral canonical pathway of biogenesis, but likely suffered more than one event of centriole loss (Fig. 6).

Throughout the eukaryotic tree, there are several examples of convergent evolution where unrelated groups appear to share similar strategies to assemble CBBs. This suggests that the possibilities for how to make CBBs are somewhat limited, indicating some sort of morphological (perhaps even molecular) constraint inherent to the process.

CONCLUSION

In this review, we have highlighted that noncanonical modes of CBBs assembly are widespread in the eukaryotic tree. Although some pathways are more lineage-specific, there are several examples of convergent evolution, suggesting that when it comes to making centrioles, the options are limited and mostly governed by numbers.

Most descriptions of noncanonical assembly were done by EM in chemically fixed samples. However, new techniques are now available, such as high-pressure freezing followed by freeze substitution (HPF + FS) and Cryo-EM, which can improve the quality of the data and help to unravel the true representation of each step of these processes. Super-resolution microscopy, in particular 3D-structured illumination microscopy, allows correlating different proteins within the organelles at much better resolution and, potentially, following CBBs biogenesis live.

Molecular studies on noncanonical centriole biogenesis are scarce and focused on a few species (such as *N. gruberi* and *Drosophila* spp.) and biased toward the deuterosome-mediated pathway in vertebrate multiciliated cells. One reason is the absence of tools to study other systems, which can now be overcome with CRISPR–Cas9 technology and the increasing availability of genomic data. More gene expression data and functional studies should expand our molecular knowledge outside the Opisthokonts, in order to understand what are the universal principles underlying centriole assembly as well as the specific properties inherent to each pathway.

Many of the core centriolar components and some regulators (Polo-like kinases, PCM components, and MT regulators) appear to be conserved across evolution (Hodges et al. 2010; Carvalho-Santos et al. 2010, 2011), suggesting an ancestral molecular cascade, common to most centriole assembly pathways. However, noncanonical centriole biogenesis seems more confined to specific cell types during differentiation (multiciliated cells in vertebrates—deuterosome-mediated pathway) or life-cycle stages (*N. gruberi* and spermatogenesis in plants—de novo pathways), suggesting that centriole assembly must be under developmental regulation. In the future, it will be important to unravel how the multiple pathways operate in different organisms; how the PCM components, the MT cytoskeleton, and centriolar precursors create a suitable environment that forms a scaffold for centriole assembly. Only then we will fully understand CBBs function and its upstream and downstream molecular machinery.

ACKNOWLEDGMENTS

The laboratory is funded by the European Research Council Consolidator Grant (CoG683528_Centriole-BirthDeath). C.N. is funded by the Boehringer Ingelheim Fonds and S.G.P. by a Fundação para a Ciência e Tecnologia Grant (PD/BD/114350/2016). We thank Eduardo Marabuto for confirming all the taxonomic information and Maria Francia for clarifying centriole biology in unicellular protozoa and Marc Gouw and collaborators for sharing unpublished information.

REFERENCES

Al Jord A, Lemaître AI, Delgehyr N, Faucourt M, Spassky N, Meunier A. 2014. Centriole amplification by mother and daughter centrioles differs in multiciliated cells. *Nature* **516:** 104–107.

Anderson RGW, Brenner RM. 1971. The formation of basal bodies (Centrioles) in the Rhesus Monkey oviduct. *J Cell Biol* **50:** 10–34.

Ansley SJ, Badano JL, Blacque OE, Hill J, Hoskins BE, Leitch CC, Kim JC, Ross AJ, Eichers ER, Teslovich TM, et al. 2003. Basal body dysfunction is a likely cause of pleiotropic Bardet–Biedl syndrome. *Nature* **425:** 628–633.

Arbi M, Pefani DE, Taraviras S, Lygerou Z. 2017. Controlling centriole numbers: Geminin family members as master regulators of centriole amplification and multiciliogenesis. *Chromosoma* doi:10.1007/s00412-017-0652-7.

Azimzadeh J. 2014. Exploring the evolutionary history of centrosomes. *Philos Trans R Soc Lond B Biol Sci* **369:** 20130453.

Azimzadeh J, Wong ML, Downhour DM, Alvarado AS, Marshall WF. 2012. Centrosome loss in the evolution of planarians. *Science* **335:** 461–463.

Baccetyi B, Dallai R. 1978. The spermatozoon of arthropoda. XXX. The multiflagellate spermatozoon in the termite *Mastotermes darwiniensis*. *J Cell Biol* **76:** 569–576.

Badano JL, Mitsuma N, Beales PL, Katsanis N. 2006. The ciliopathies: An emerging class of human genetic disorders. *Annu Rev Genomics Hum Genet* **7:** 125–148.

Beisson J, Wright M. 2003. Basal body/centriole assembly and continuity. *Curr Opin Cell Biol* **15:** 96–104.

Bettencourt-Dias M, Rodrigues-Martins A, Carpenter L, Riparbelli M, Lehmann L, Gatt MK, Carmo N, Balloux F, Callaini G, Glover DM. 2005. SAK/PLK4 is required for centriole duplication and flagella development. *Curr Biol* **15:** 2199–2207.

Bremer K, Humphries CJ, Mishler BD, Churchill SP. 1987. On cladistic relationships in green plants. *Taxon* **36:** 339–349.

Burki F. 2014. The eukaryotic tree of life from a global phylogenomic perspective. *Cold Spring Harb Perspect Biol* **6:** a016147.

Carvalho-Santos Z, Machado P, Branco P, Tavares-Cadete F, Rodrigues-Martins A, Pereira-Leal JB, Bettencourt-Dias M. 2010. Stepwise evolution of the centriole-assembly pathway. *J Cell Sci* **123:** 1414–1426.

Carvalho-Santos Z, Azimzadeh J, Pereira-Leal JB, Bettencourt-Dias M. 2011. Tracing the origins of centrioles, cilia, and flagella. *J Cell Biol* **194:** 165–175.

Cavalier-Smith T. 2002. The phagotrophic origin of eukaryotes and phylogenetic classification on protozoa. *Int J Syst Evol Microbiol* **52:** 297–354.

Cizmecioglu O, Arnold M, Bahtz R, Settele F, Ehret L, Haselmann-Weiß U, Antony C, Hoffmann I. 2010. Cep152 acts as a scaffold for recruitment of Plk4 and CPAP to the centrosome. *J Cell Biol* **191:** 731–739.

Courtois A, Schuh M, Ellenberg J, Hiiragi T. 2012. The transition from meiotic to mitotic spindle assembly is gradual during early mammalian development. *J Cell Biol* **198:** 357–370.

Dammermann A, Müller-Reichert T, Pelletier L, Habermann B, Desai A, Oegema K. 2004. Centriole assembly requires both centriolar and pericentriolar material proteins. *Dev Cell* **7:** 815–829.

Dingle AD, Fulton C. 1966. Development of the flagellar apparatus of *Naegleria*. *J Cell Biol* **31:** 43–54.

Dirksen ER. 1961. The presence of centrioles in artificially activated sea urchin eggs. *J Cell Biol* **11:** 244–247.

Dirksen ER. 1971. Centriole morphogenesis in developing ciliated epithelium of the mouse oviduct. *J Cell Biol* **51:** 286–302.

Doonan JH, Lloyd CW, Duckett JG. 1986. Anti-tubulin antibodies locate the blepharoplast during spermatogenesis in the fern *Platyzoma microphyllum* R.Br.: A correlated immunofluorescence and electron-microscopic study. *J Cell Sci* **81:** 243–265.

Dzhindzhev NS, Yu QD, Weiskopf K, Tzolovsky G, Cunha-Ferreira I, Riparbelli M, Rodrigues-Martins A, Bettencourt-Dias M, Callaini G, Glover DM. 2010. Asterless is a scaffold for the onset of centriole assembly. *Nature* **467:** 714–718.

Ferraguti M, Fascio U, Boi S. 2002. Mass production of basal bodies in paraspermiogenesis of Tubificinae (Annelida, Oligochaeta). *Biol Cell* **94:** 109–115.

Finet C, Fourquin C, Vinauger M, Berne-Dedieu A, Chambrier P, Paindavoine S, Scutt CP. 2010. Parallel structural evolution of auxin response factors in the angiosperms. *Plant J* **63:** 952–959.

Fritz-Laylin LK, Fulton C. 2016. *Naegleria*: A classic model for de novo basal body assembly. *Cilia* **5:** 10.

Fritz-Laylin LK, Assaf ZJ, Chen S, Cande WZ. 2010. *Naegleria gruberi* de novo basal body assembly occurs via stepwise incorporation of conserved proteins. *Eukaryot Cell* **9:** 860–865.

Fritz-Laylin LK, Levy YY, Levitan E, Chen S, Cande WZ, Lai EY, Fulton C. 2016. Rapid centriole assembly in *Naegleria* reveals conserved roles for both de novo and mentored assembly. *Cytoskeleton* **73**: 109–116.

Fu J, Lipinszki Z, Rangone H, Min M, Mykura C, Chao-Chu J, Schneider S, Dzhindzhev NS, Gottardo M, Riparbelli MG, et al. 2016. Conserved molecular interactions in centriole-to-centrosome conversion. *Nat Cell Biol* **18**: 87–99.

Fulton C, Dingle AD. 1971. Basal bodies, but not centrioles, in *naegleria. J Cell Biol* **51**: 826–835.

Gall JG. 1961. Centriole replication. A study of spermatogenesis in the snail *Viviparus. J Biophys Biochem Cytol* **10**: 163–193.

Ganem NJ, Godinho SA, Pellman D. 2009. A mechanism linking extra centrosomes to chromosomal instability. *Nature* **460**: 278–282.

Garber RC, Aist JR. 1979. The ultrastructure of mitosis in *Plasmodiophora brassicae* (Plasmodiophorales). *J Cell Sci* **40**: 89–110.

Gifford EM, Larson S. 1980. Developmental features of the spermatogenous cell in *Ginkgo biloba. Am J Bot* **67**: 119–124.

Gifford EM, Lin J. 1975. Light microscope and ultrastructural studies of the male gametophyte in *Ginkgo biloba*: The spermatogenous cell. *Am J Bot* **62**: 974–981.

Godinho SA, Pellman D. 2014. Causes and consequences of centrosome abnormalities in cancer. *Philos Trans R Soc B Biol Sci* **369**: 20130467.

Gopalakrishnan J, Mennella V, Blachon S, Zhai B, Smith AH, Megraw TL, Nicastro D, Gygi SP, Agard DA, Avidor-Reiss T. 2011. Sas-4 provides a scaffold for cytoplasmic complexes and tethers them in a centrosome. *Nat Commun* **2**: 359.

Gueth-Hallonet C, Antony C, Aghion J, Santa-Maria A, Lajoie-Mazenc I, Wright M, Maro B. 1993. γ-Tubulin is present in acentriolar MTOCs during early mouse development. *J Cell Sci* **105**: 157–166.

Guichard P, Hamel V, Le Guennec M, Banterle N, Iacovache I, Nemcíková V, Flückiger I, Goldie KN, Stahlberg H, Lévy D, et al. 2017. Cell-free reconstitution reveals centriole cartwheel assembly mechanisms. *Nat Commun* **8**: 14813.

Habedanck R, Stierhof YD, Wilkinson CJ, Nigg EA. 2005. The Polo kinase Plk4 functions in centriole duplication. *Nat Cell Biol* **7**: 1140–1146.

Hagiwara H, Ohwada N, Takata K. 2004. Cell biology of normal and abnormal ciliogenesis in the ciliated epithelium. *Int Rev Cytol* **234**: 101–141.

Harrison MK, Adon AM, Saavedra HI. 2011. The G_1 phase Cdks regulate the centrosome cycle and mediate oncogene-dependent centrosome amplification. *Cell Div* **6**: 2.

Healy JM, Jamieson BGM. 1981. An ultrastructural examination of developing and mature paraspermatozoa in *Pyrazus ebeninus* (Mollusca, Gastropoda, Potamididae). *Zoomorphology* **98**: 101–119.

Heath IB. 1974a. Centrioles and mitosis in some oömycetes. *Mycologia* **66**: 354–359.

Heath IB. 1974b. Mitosis in the fungus *Thraustotheca clavata. J Cell Biol* **60**: 204–220.

Heath IB, Greenwood AD. 1970. Centriole replication and nuclear division in *Saprolegnia. J Gen Microbiol* **62**: 139–148.

Hepler PK. 1976. The blepharoplast of Marsilea: its de novo formation and spindle association. *J Cell Sci* **21**: 361–90.

Hodges ME, Scheumann N, Wickstead B, Langdale JA, Gull K. 2010. Reconstructing the evolutionary history of the centriole from protein components. *J Cell Sci* **123**: 1407–1413.

Hoffman JC, Vaughn KC. 1995. Using the developing spermatogenous cells of ceratopteris to unlock the mysteries of the plant cytoskeleton. *Int J Plant Sci* **156**: 346–358.

Hoffman JC, Vaughn KC, Joshi HC. 1994. Structural and immunocytochemical characterization of microtubule organizing centers in pteridophyte spermatogenous cells. *Protoplasma* **179**: 46–60.

Hoh RA, Stowe TR, Turk E, Stearns T. 2012. Transcriptional program of ciliated epithelial cells reveals new cilium and centrosome components and links to human disease. *PLoS One* **7**: e52166.

Judelson HS, Shrivastava J, Manson J. 2012. Decay of genes encoding the oomycete flagellar proteome in the downy mildew *Hyaloperonospora arabidopsidis. PLoS One* **7**: e47624.

Kalnins VI, Porter KR. 1969. Centriole replication during ciliogenesis in the chick tracheal epithelium. *Z Zellforsch* **100**: 1–30.

Khodjakov A, Rieder CL, Sluder G, Cassels G, Sibon O, Wang C-L. 2002. De novo formation of centrosomes in vertebrate cells arrested during S phase. *J Cell Biol* **158**: 1171–1181.

Kim HK, Kang JG, Yumura S, Walsh CJ, Jin WC, Lee J. 2005. De novo formation of basal bodies in *Naegleria gruberi*: Regulation by phosphorylation. *J Cell Biol* **169**: 719–724.

Kitagawa D, Vakonakis I, Olieric N, Hilbert M, Keller D, Olieric V, Bortfeld M, Erat MC, Flückiger I, Gönczy P, et al. 2011. Structural basis of the 9-fold symmetry of centrioles. *Cell* **144**: 364–375.

Klink VP, Wolniak SM. 2001. Centrin is necessary for the formation of the motile apparatus in spermatids of *Marsilea. Mol Biol Cell* **12**: 761–776.

Klos Dehring DA, Vladar EK, Werner ME, Mitchell JW, Hwang P, Mitchell BJ. 2013. Deuterosome mediated centriole biogenesis. *Dev Cell* **27**: 103–112.

Kong D, Farmer V, Shukla A, James J, Gruskin R, Kiriyama S, Loncarek J. 2014. Centriole maturation requires regulated Plk1 activity during two consecutive cell cycles. *J Cell Biol* **206**: 855–865.

Kreitner GL, Carothers ZB. 1976. Studies of spermatogenesis in the Hepaticae V. Blepharoplast development in *Marchantia polymorpha. Am J Bot* **63**: 545–557.

Kuriyama R, Borisy GG. 1981. Microtubule-nucleating activity of centrosomes in Chinese hamster ovary cells is independent of the centriole cycle but coupled to the mitotic cycle. *J Cell Biol* **91**: 822–826.

Kuriyama R, Borisy GG, Masui Y. 1986. Microtubule cycles in oocytes of the surf clam, *Spisula solidissima*: An immunofluorescence study. *Dev Biol* **114**: 151–160.

La Terra S, English CN, Hergert P, McEwen BF, Sluder G, Khodjakov A. 2005. The de novo centriole assembly pathway in HeLa cells: Cell cycle progression and centriole assembly/maturation. *J Cell Biol* **168**: 713–722.

Lange BMH. 2002. Integration of the centrosome in cell cycle control, stress response and signal transduction pathways. *Curr Opin Cell Biol* **14**: 35–43.

Lee J, Kang S, Choi YS, Kim HK, Yeo CY, Lee Y, Roth J, Lee J. 2015. Identification of a cell cycle-dependent duplicating complex that assembles basal bodies de novo in *Naegleria. Protist* **166**: 1–13.

Levine MS, Bakker B, Boeckx B, Moyett J, Lu J, Vitre B, Spierings DC, Lansdorp PM, Cleveland DW, Lambrechts D, et al. 2017. Centrosome amplification is sufficient to promote spontaneous tumorigenesis in mammals. *Dev Cell* **40**: 313–322.

Loncarek J, Bettencourt-Dias M. 2018. Building the right centriole for each cell type. *J Cell Biol.* doi: jcb.201704093.

Lopes CAM, Jana SC, Cunha-Ferreira I, Zitouni S, Bento I, Duarte P, Gilberto S, Freixo F, Guerrero A, Francia M, et al. 2015. PLK4 trans-autoactivation controls centriole biogenesis in space. *Dev Cell* **35**: 222–235.

Matsumoto Y, Hayashi K, Nishida E. 1999. Cyclin-dependent kinase 2 (Cdk2) is required for centrosome duplication in mammalian cells. *Curr Biol* **9**: 429–432.

Meraldi P, Lukas J, Fry AM, Bartek J, Nigg EA. 1999. Centrosome duplication in mammalian somatic cells requires E2F and Cdk2-cyclin A. *Nat Cell Biol* **1**: 88–93.

Meunier A, Azimzadeh J. 2016. Multiciliated cells in animals. *Cold Spring Harb Perspect Biol* **8**: a028233.

Miki-Noumura T. 1977. Studies on the de novo formation of centrioles: Aster formation in the activated eggs of sea urchin. *J Cell Sci* **24**: 203–216.

Mizukami I, Gall J. 1966. Centriole replication. II. Sperm formation in the fern, *Marsilea*, and the cycad, *Zamia. J Cell Biol* **29**: 97–111.

Mori M, Hazan R, Danielian PS, Mahoney JE, Li H, Lu J, Miller ES, Zhu X, Lees JA, Cardoso W V. 2017. Cytoplasmic E2f4 forms organizing centres for initiation of centriole amplification during multiciliogenesis. *Nat Commun* **8:** 15857.

Moser JW, Kreitner GL. 1970. Centrosome structure in *Anthoceros laevis* and *Marchantia polymorpha*. *J Cell Biol* **44:** 454–458.

Moser JW, Duckett JG, Carothers ZB. 1977. Ultrastructural studies of spermatogenesis in the anthocerotales. I. The blepharoplast and anterior mitochondrion in *Phaeoceros laevis*: Early development. *Am J Bot* **64:** 1097–1106.

Nakazawa Y, Hiraki M, Kamiya R, Hirono M. 2007. SAS-6 is a cartwheel protein that establishes the 9-fold symmetry of the centriole. *Curr Biol* **17:** 2169–2174.

Nigg EA, Holland AJ. 2018. Once and only once: Mechanisms of centriole duplication and their deregulation in disease. *Nat Rev Mol Cell Biol* doi:10.1038/nrm.2017.127.

Norstog K. 1967. Fine structure of the spermatozoid of *Zamia* with special reference to the flagellar apparatus. *Am J Bot* **54:** 831–840.

Norstog KJ. 1986. The blepharoplast of *Zamia pumila* L. *Bot Cazette* **147:** 40–46.

Palazzo RE, Vaisberg E, Cole RW, Rieder CL. 1992. Centriole duplication in lysates of *Spisula solidissima* oocytes. *Science* **256:** 219–221.

Peel N, Stevens NR, Basto R, Raff JW. 2007. Overexpressing centriole-replication proteins in vivo induces centriole overduplication and de novo formation. *Curr Biol* **17:** 834–843.

Perkins FO. 1970. Formation of centriole and centriole-like structures during meiosis and mitosis in *Labyrinthula* Sp. (Rhizopodea, Labyrinthulida). An electron-microscope study. *J Cell Sci* **6:** 629–653.

Peset I, Vernos I. 2008. The TACC proteins: TACC-ling microtubule dynamics and centrosome function. *Trends Cell Biol* **18:** 379–388.

Pimenta-Marques A, Bento I, Lopes CAM, Duarte P, Jana SC, Bettencourt-Dias M. 2016. A mechanism for the elimination of the female gamete centrosome in *Drosophila melanogaster*. *Science* **353:** aaf4866.

Renzaglia KS, Duckett JG. 1987. Spermatogenesis in *Blasia pusilla*: From young antheridium through mature spermatozoid. *Bryologist* **90:** 419–449.

Renzaglia KS, Garbary DJ. 2001. Motile gametes of land plants: Diversity, development, and evolution. *CRC Crit Rev Plant Sci* **20:** 107–213.

Renzaglia KS, Maden AR. 2000. Microtubule organizing centers and the origin of centrioles during spermatogenesis in the pteridophyte *Phylloglossum*. *Microsc Res Tech* **49:** 496–505.

Riparbelli MG, Callaini G. 2003. *Drosophila* parthenogenesis: A model for de novo centrosome assembly. *Dev Biol* **260:** 298–313.

Riparbelli MG, Stouthamer R, Dallai R, Callaini G. 1998. Microtubule organization during the early development of the parthenogenetic egg of the hymenopteran *Muscidifurax uniraptor*. *Dev Biol* **195:** 89–99.

Riparbelli MG, Callaini G, Mercati D, Hertel H, Dallai R. 2009. Centrioles to basal bodies in the spermiogenesis of *Mastotermes darwiniensis* (Insecta, Isoptera). *Cell Motil Cytoskeleton* **66:** 248–259.

Robbins RR. 1984. Origin and behavior of bicentriolar centrosomes in the bryophyte *Riella americana*. *Protoplasma* **121:** 114–119.

Robbins E, Jentzsch G, Micali A. 1968. The centriole cycle in synchronized HeLa cells. *J Cell Biol* **36:** 329–339.

Rodrigues-Martins A, Riparbelli M, Callaini G, Glover DM, Bettencourt-Dias M. 2007. Revisiting the role of the mother centriole in centriole biogenesis. *Science* **316:** 1046–1050.

Schatten G, Simerly C, Schatten H. 1985. Microtubule configurations during fertilization, mitosis, and early development in the mouse and the requirement for egg microtubule-mediated motility during mammalian fertilization. *Proc Natl Acad Sci* **82:** 4152–4156.

Shimamura M, Brown RC, Lemmon BE, Akashi T, Mizuno K, Nishihara N, Tomizawa K-I, Yoshimoto K, Deguchi H, Hosoya H, et al. 2004. γ-Tubulin in basal land plants: Characterization, localization, and implication in the evolution of acentriolar microtubule organizing centers. *Plant Cell* **16:** 45–59.

Silkworth WT, Nardi IK, Scholl LM, Cimini D. 2009. Multipolar spindle pole coalescence is a major source of kinetochore misattachment and chromosome mis-segregation in cancer cells. *PLoS One* **4:** e6564.

Sonnen KF, Gabryjonczyk A-M, Anselm E, Stierhof Y-D, Nigg EA. 2013. Human Cep192 and Cep152 cooperate in Plk4 recruitment and centriole duplication. *J Cell Sci* **126:** 3223–3233.

Sorokin SP. 1968. Reconstructions of centriole formation and ciliogenesis in mammalian lungs. *J Cell Sci* **3:** 207–230.

Steinman RM. 1968. An electron microscopic study of ciliogenesis in developing epidermis and trachea in the embryo of *Xenopus laevis*. *Am J Anat* **122:** 19–55.

Suh MR, Han JW, No YR, Lee J. 2002. Transient concentration of a γ-tubulin-related protein with a pericentrin-related protein in the formation of basal bodies and flagella during the differentiation of *Naegleria gruberi*. *Cell Motil Cytoskeleton* **52:** 66–81.

Tamm S, Tamm SL. 1980. Origin and development of free kinetosomes in the flagellates *Deltotrichonympha* and *Koruga*. *J Cell Sci* **42:** 189–205.

Tram U, Sullivan W. 2000. Reciprocal inheritance of centrosomes in the parthenogenetic hymenopteran *Nasonia vitripennis*. *Curr Biol* **10:** 1413–1419.

Uetake Y, Lončarek J, Nordberg JJ, English CN, La Terra S, Khodjakov A, Sluder G. 2007. Cell cycle progression and de novo centriole assembly after centrosomal removal in untransformed human cells. *J Cell Biol* **176:** 173–182.

van Breugel M, Hirono M, Andreeva A, Yanagisawa H, Yamaguchi S, Nakazawa Y, Morgner N, Petrovich M, Ebong I, Robinson C V, et al. 2011. Structures of SAS-6 suggest its organization in centrioles. *Science* **331:** 1196–1199.

Varmark H, Llamazares S, Rebollo E, Lange B, Reina J, Schwarz H, Gonzalez C. 2007. Asterless is a centriolar protein required for centrosome function and embryo development in *Drosophila*. *Curr Biol* **17:** 1735–1745.

Vaughn KC, Bowling AJ. 2008. Recovery of microtubules on the blepharoplast of *Ceratopteris* spermatogenous cells after oryzalin treatment. *Protoplasma* **233:** 231–240.

Vaughn KC, Renzaglia KS. 2006. Structural and immunocytochemical characterization of the *Ginkgo biloba* L. sperm motility apparatus. *Protoplasma* **227:** 165–173.

Vladar EK, Stearns T. 2007. Molecular characterization of centriole assembly in ciliated epithelial cells. *J Cell Biol* **178:** 31–42.

Woodland HR, Fry AM. 2008. Pix proteins and the evolution of centrioles. *PLoS One* **3:** e3778.

Woodruff JB, Ferreira Gomes B, Widlund PO, Mahamid J, Honigmann A, Hyman AA. 2017. The centrosome is a selective condensate that nucleates microtubules by concentrating tubulin. *Cell* **169:** 1066–1077.

Worden AZ, Follows MJ, Giovannoni SJ, Wilken S, Zimmerman AE, Keeling PJ. 2015. Rethinking the marine carbon cycle: Factoring in the multifarious lifestyles of microbes. *Science* **347.**

Yubuki N, Leander BS. 2013. Evolution of microtubule organizing centers across the tree of eukaryotes. *Plant J* **75:** 230–244.

Zhao H, Zhu L, Zhu Y, Cao J, Li S, Huang Q, Xu T, Huang X, Yan X, Zhu X. 2013. The cep63 paralogue deup1 enables massive de novo centriole biogenesis for vertebrate multiciliogenesis. *Nat Cell Biol* **15:** 1434–1444.

Zitouni S, Francia ME, Leal F, Gouveia SM, Nabais C, Duarte P, Gilberto S, Brito D, Moyer T, Ohta M, et al. 2016. CDK1 prevents unscheduled PLK4-STIL complex assembly in centriole biogenesis. *Curr Biol* **26:** 1127–1137.

Impact of Centrosome Aberrations on Chromosome Segregation and Tissue Architecture in Cancer

Erich A. Nigg,[1] Dominik Schnerch,[1,2] and Olivier Ganier[1]

[1]Biozentrum, University of Basel, Basel CH-4056, Switzerland

Correspondence: erich.nigg@unibas.ch

Centrosomes determine the disposition of microtubule networks and thereby contribute to regulate cell shape, polarity, and motility, as well as chromosome segregation during cell division. Additionally, centrioles, the core components of centrosomes, are required for the formation of cilia and flagella. Mutations in genes coding for centrosomal and centriolar proteins are responsible for several human diseases, foremost ciliopathies and developmental disorders resulting in small brains (primary microcephaly) or small body size (dwarfism). Moreover, a long-standing postulate implicates numerical and/or structural centrosome aberrations in the etiology of cancer. In this review, we will discuss recent work on the role of centrosome aberrations in the promotion of genome instability and the disruption of tissue architecture, two hallmarks of human cancers. We will emphasize recent studies on the impact of centrosome aberrations on the polarity of epithelial cells cultured in three-dimensional spheroid models. Collectively, the results from these in vitro systems suggest that different types of centrosome aberrations can promote invasive behavior through different pathways. Particularly exciting is recent evidence indicating that centrosome aberrations may trigger the dissemination of potentially metastatic cells through a non-cell-autonomous mechanism.

Centrosomes function as major organizers of intracellular microtubule networks in animal cells (Bornens 2012). Hence, they influence cell shape, polarity, and motility. Furthermore, centrosomes contribute to the assembly of the bipolar spindle apparatus during mitosis, with important consequences for chromosome segregation and the positioning of the cleavage furrow. Aberrant centrosome numbers cause the formation of mono- or multipolar spindles, often with dire consequences for the fidelity of chromosome segregation (Nigg 2002; Godinho and Pellman 2014; Gönczy 2015). A typical centrosome comprises two centrioles, embedded in a proteinaceous matrix known as the pericentriolar material (PCM) (Woodruff et al. 2014; Conduit et al. 2015). Centrioles are cylindrical organelles built of exceptionally stable microtubules, and they display an evolutionarily conserved ninefold rotational symmetry (Azimzadeh and Marshall 2010; Gönczy 2012; Guichard et al. 2013). The PCM is a dynamic structure harboring more than 200 distinct proteins, including proteins important for microtubule nucleation (Paz and Luders 2017) and components of signaling pathways (Arquint et al. 2014). Importantly, centrioles are required not only for centrosome assembly but also for the formation of cilia and flagella (Sánchez and Dynlacht 2016). In these latter roles, centrioles are often referred to as basal bodies (Garcia and Reiter 2016). In recent years, mutations in several genes coding for centrosomal proteins or proteins of the basal body ciliary apparatus were shown to cause human diseases, notably ciliopathies and developmental disorders affecting either the brain (primary microcephaly) or the entire body (dwarfism). Accordingly, much research effort is currently devoted to elucidating the pathways that link specific alterations in centrosomal or centriolar proteins to the above disease phenotypes (Nigg and Raff 2009; Bettencourt-Dias et al. 2011; Braun and Hildebrandt 2017).

Researchers also continue to explore purported causal relationships between centrosome aberrations and cancer. In a seminal monograph, published at the beginning of the 20th century, Boveri (1914) had proposed that tumorigenesis might be a consequence of chromosome missegregation, which in turn he attributed to multipolar spindles caused by extra centrosomes. Many studies have since followed up on Boveri's hypothesis, and, to date, the prevalence of centrosome aberrations in both solid and hematologic cancers is firmly established. Although centrosome aberrations generally affect only subpopulations of tumor cells, they can be observed already at early stages of tumorigenesis (Lingle et al. 2002; Pihan et al. 2003; Guo et al. 2007; Chan 2011). Traditionally, centrosome aberrations are classified as either numerical or structural, depending on whether the abnormalities concern centrosome numbers or structures, but, importantly, these aberrations often occur together (Lingle et al. 1998; Pihan et al. 2003; Guo et al. 2007). Although the origins and consequences of numerical aberrations have been studied extensively, research on structural aberrations has barely begun (Nigg 2002; Godinho and Pellman 2014; Gönczy 2015). Genetic evidence implicating centrosomal proteins in the etiology of human cancer remains scarce, but experimentally induced centrosome aberrations are able to trigger tumorigenesis in both *Drosophila* and mouse models

---

[2]Present address: Department of Medicine I, Medical Center—University of Freiburg, Faculty of Medicine, University of Freiburg, Freiburg 79106, Germany

Published by Cold Spring Harbor Laboratory Press; doi: 10.1101/sqb.2017.82.034421

Cold Spring Harbor Symposia on Quantitative Biology, Volume LXXXII

(Basto et al. 2008; Coelho et al. 2015; Serçin et al. 2016; Levine et al. 2017). What remain to be clarified are to what extent centrosome aberrations contribute to tumor formation and/or progression in human patients and through what mechanisms.

THE CENTROSOME DUPLICATION CYCLE AND ITS ABERRATIONS

Cell cycle research traditionally focuses on DNA replication during S phase and chromosome segregation during M phase. However, it is important to bear in mind that this chromosome duplication–segregation cycle is accompanied by a centrosome duplication–segregation cycle, and that preservation of genome integrity during cell proliferation requires coordination between these two cycles (Nigg and Stearns 2011; Firat-Karalar and Stearns 2014; Fu et al. 2015). This coordination is achieved through a number of key regulatory proteins that control both processes (Nigg and Holland 2018). In brief, a typical G_1 phase cell harbors a single centrosome, comprising two centrioles. These two centrioles are then duplicated during S phase, so that by G_2, the cell harbors two centrosomes, each comprising a pair of centrioles. By virtue of their association with the poles of the mitotic spindle apparatus, the two centrosomes then segregate during M phase. To keep centriole and centrosome numbers constant in each cell cycle, regulatory mechanisms need to ensure that centriole duplication occurs once and only once, and that only one new centriole is built per preexisting centriole (Nigg 2007). In recent years much has been learned about the regulation of centriole duplication, as well as structural aspects of centriole biogenesis (Fu et al. 2015; Nigg and Holland 2018). For the sake of simplicity, we summarize that centriole biogenesis relies on a core module composed of three crucial centriole duplication factors, Polo-like kinase 4 (PLK4), SAS-6, and STIL (Arquint and Nigg 2016). PLK4 functions as the master regulator of centriole biogenesis (Bettencourt-Dias et al. 2005; Habedanck et al. 2005; Kleylein-Sohn et al. 2007; Rodrigues-Martins et al. 2007), SAS-6 represents a core component of a scaffolding structure, termed cartwheel, that imparts the evolutionarily conserved ninefold symmetry to centrioles (Kitagawa et al. 2011; van Breugel et al. 2011; Hirono 2014; Guichard et al. 2017), and STIL plays crucial roles in centriole assembly through interactions with both PLK4 and SAS-6 (Arquint and Nigg 2014; Ohta et al. 2014; Arquint et al. 2015; Kratz et al. 2015; Moyer et al. 2015). Importantly, depletion of PLK4, SAS-6, or STIL stops centriole duplication, and, conversely, overexpression of either protein triggers centrosome amplification (Nigg and Holland 2018).

Because centrosomes are important for continued proliferation of normal (untransformed) mammalian cells, centrosome aberrations are generally deleterious. Thus, experimentally induced loss of centrosomes triggers robust cell cycle arrest in both cultured cells and mouse embryos (Hinchcliffe et al. 2001; Khodjakov and Rieder 2001; Bazzi and Anderson 2014). This arrest can be over-come by inactivation of the p53 pathway, explaining why p53-deficient cancer cells fail to arrest in response to centrosome loss (Bazzi and Anderson 2014; Lambrus et al. 2015; Wong et al. 2015). Recent genome-wide screens have identified a p53-dependent pathway, including the p53-binding protein 53BP1, the deubiquitinase USP38, and the CDK inhibitor p21, that imposes a proliferation arrest on centrosome-deficient cells (Fong et al. 2016; Lambrus et al. 2016; Meitinger et al. 2016). Most likely, this newly discovered pathway detects the extended mitotic duration that is caused by the absence of centrosomes, leading to its designation as "mitotic surveillance pathway" (Lambrus and Holland 2017). In future, it will be interesting to elucidate the exact contributions of this pathway to normal physiology and disease.

Cells respond not only to loss of centrosomes but also to extra centrosomes. Accordingly, centrosome amplification also suppresses cell proliferation and, again, the arrest depends on p53 (Holland et al. 2012; Levine et al. 2017). Importantly, however, the response to centrosome amplification does not require USP28 or 53BP1, demonstrating that centrosome loss and centrosome amplification activate distinct pathways. How cells detect extra centrosomes and how they respond to this anomaly remains to be fully understood, but a recent study identifies the PIDDosome, a multiprotein complex responsible for caspase-2 activation, as a key component of the pathway (Fava et al. 2017). PIDDosome components were found to colocalize with mature centrioles, suggesting that proximity-induced activation stimulates caspase-2, which then results in caspase-2-mediated MDM2 cleavage, p53 stabilization, and p21-dependent cell cycle arrest (Fava et al. 2017). It has also been reported that extra centrosomes trigger p53 stabilization through the kinase LATS2 (Ganem et al. 2014), and it appears safe to predict that additional components linking centrosome amplification to p53 activation await discovery.

IMPACT OF CENTROSOME ABERRATIONS ON FIDELITY OF CHROMOSOME SEGREGATION

Although centrosomes play a major role in bipolar spindle formation and chromosome segregation in somatic human cells, centrosomes are not required for spindle formation in all organisms or cell types (Szollosi et al. 1972; Basto et al. 2006; Bornens 2012). This is illustrated best by higher land plants or oocytes of many animal species, which form bipolar spindles even though they lack centrosomes. Similarly, cell division occurs in the absence of centrosomes in planarians, where centrioles are only assembled in terminally differentiated multiciliated cells (Azimzadeh et al. 2012). Research on the mechanisms underlying centrosome-independent spindle formation has revealed that these rely on microtubule nucleation in the vicinity of chromatin and involve the small GTPase Ran as well as several microtubule-binding proteins and microtubule-dependent motor proteins (Prosser and Pelletier 2017). Importantly, centrosome-dependent and centrosome-independent mechanisms for spindle formation coexist in somatic human cells (Karsenti and Vernos 2001; Khodjakov and Rieder 2001), and this has important implications for the clustering of extra centrosomes in tumor cells (see below).

Although structural centrosome aberrations are thought to reflect the unbalanced expression of PCM components in tumors, most numerical centrosome aberrations are likely to result from centriole overduplication or division failure (Nigg 2002; Godinho and Pellman 2014; Gönczy 2015). A priori, both types of aberrations can lead to the formation of multipolar spindles. Spindle multipolarity often reflects numerical aberrations, but occasionally can arise also through PCM fragmentation (Fig. 1). The two etiologies can readily be distinguished by monitoring of centriolar markers: whereas numerical aberrations will produce multipolar spindles with each pole containing at least one centriole, PCM fragmentation will result in extra spindle poles devoid of centrioles. When multipolarity can be attributed to numerical aberrations, a question of potential clinical importance concerns the origin of these aberrations. If centriole overduplication represents the root cause, any subsequent multipolar division will lead to the segregation of a diploid genome into more than two cells, and the resulting hypodiploid progeny is not generally expected to be viable (Ganem et al. 2009; Silkworth et al. 2009). In contrast, when centrosome amplification resulted from division failure, a subsequent multipolar division will concern a tetraploid cell, implying that two complete diploid sets of chromosomes can be distributed among three or more cells. A priori, this would seem to set the stage for the occasional production of viable, aneuploid cells. Considering that tetraploid intermediates are commonly observed in carcinogenesis, it seems plausible that abnormal divisions of tetraploid cells, induced by

centrosome aberrations, may constitute a prominent path to aneuploid progeny (Meraldi et al. 2002; Nigg 2002; Ganem et al. 2007).

Spindle multipolarity does not always lead to multipolar divisions. Instead, as a consequence of centrosome clustering, multipolar spindles often coalesce to bipolar spindles prior to cell division (Nigg 2002; Quintyne et al. 2005). This prominent phenomenon of centrosome clustering in tumor cells almost certainly reflects up-regulation of centrosome-independent spindle assembly pathways (see above). Therefore, inhibition of proteins or mechanisms required for centrosome clustering might constitute an attractive strategy for targeting tumor cells harboring extra centrosomes (Nigg 2002; Ganem et al. 2007; Kwon et al. 2008). Importantly, the ability of tumor cells to cluster extra centrosomes increases not only their viability, but also the frequency of chromosome mis-segregation (Ganem et al. 2009; Silkworth et al. 2009). This is because centrosome clustering enhances the occurrence of so-called merotelic microtubule–kinetochore interactions (i.e., interactions of a single chromosomal kinetochore with microtubules emanating from opposite spindle poles), which then increases the frequency of lagging chromosomes, resulting in aneuploidy (Salmon et al. 2005). Thus, centrosome aberrations are now recognized to constitute a prominent cause of whole chromosome instability in cancer.

IMPACT OF CENTROSOME ABERRATIONS ON TISSUE ARCHITECTURE

Centrosomes function not only in mitotic spindle formation and chromosome segregation but also in the organization of microtubule networks in interphase cells. This

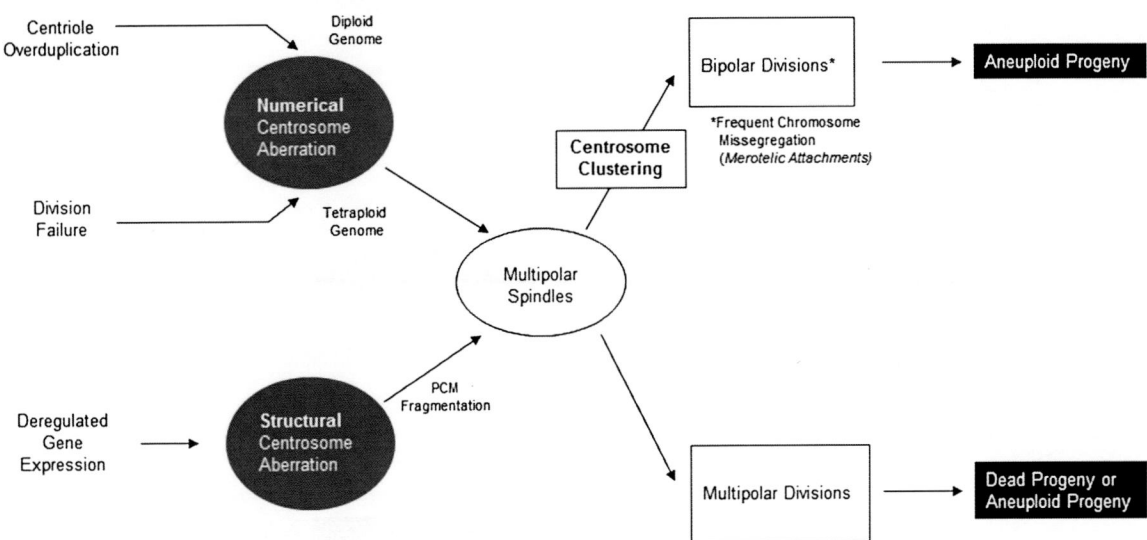

Figure 1. Impact of centrosome aberrations on chromosome segregation. Both numerical and structural aberrations can give rise to multipolar spindles. Multipolar divisions will generally produce dead cells, but centrosome clustering will allow bipolar divisions, albeit at the cost of increased chromosome missegregation. A priori, multipolar divisions are more likely to produce viable progeny if the cells harboring centrosome aberrations carry a tetraploid genome. Hence, the root cause of spindle multipolarity is not irrelevant (Meraldi et al. 2002; Nigg 2002). PCM, pericentriolar material.

has important consequences for cell shape, polarity, and motility. Thus, centrosome aberrations may influence not only the fidelity of chromosome segregation but also tissue architecture and, potentially, the propensity of tumor cells to metastasize. To explore this possibility, several recent studies used three-dimensional (3D) culture systems to study the impact of centrosome aberrations on the organization of epithelial structures (Godinho et al. 2014; Schnerch and Nigg 2016; Ganier et al. 2017). Collectively, these studies establish that centrosome aberrations profoundly disrupt epithelial architecture. Moreover, they lend support to the hypothesis that centrosome aberrations may contribute to tissue invasion and the dissemination of metastatic cells.

Overexpression of PLK4, the master regulator of centriole duplication, is widely used as the method of choice for inducing numerical centrosome aberrations. Likewise, overexpression of the PCM component Ninein-like protein (NLP) (Casenghi et al. 2003) can be used to explore the impact of structural centrosome aberrations on cell fate (Schnerch and Nigg 2016; Ganier et al. 2017). NLP is frequently overexpressed in human tumors and NLP overexpression in transgenic mice was reported to trigger tumorigenesis (Shao et al. 2010). In the studies described below, effects of centrosome aberrations were studied in 3D spheroids grown in hydrogel matrices (Matrigel) or in 2D monolayer cultures. As summarized in Figure 2, the

available evidence suggests that centrosome aberrations can confer invasive properties to epithelial cells through at least three distinct mechanisms, (i) invadopodia formation, (ii) the non-cell-autonomous budding of mitotic cells, and (iii) basal extrusion. In the following, we will briefly describe each of these three mechanisms, and we will highlight methodological and functional differences.

Analysis of human mammary epithelial cells grown in 3D cultures revealed that centrosome amplification, triggered by overexpression of PLK4, causes the formation of striking invasive protrusions known as invadopodia (Godinho et al. 2014). The observed invasive phenotypes were remarkably similar to those seen upon overexpression of ERBB2, a prominent breast cancer oncogene. Moreover, dissection of the underlying pathway revealed that PLK4-induced centrosome amplification resulted in increased centrosomal microtubule nucleation, which then enhanced the activity of the small GTPase Rac1, resulting in the disruption of cell–cell adhesion (Godinho et al. 2014). In full agreement with the original findings, we have subsequently reproduced invadopodia formation in response to numerical centrosome aberrations induced by PLK4 overexpression (Ganier et al. 2017). Remarkably, however, we found that an indistinguishable invasive phenotype could also be triggered by structural centrosome aberrations, as induced by NLP overexpression (Ganier et al. 2017). This suggests that different types of centro-

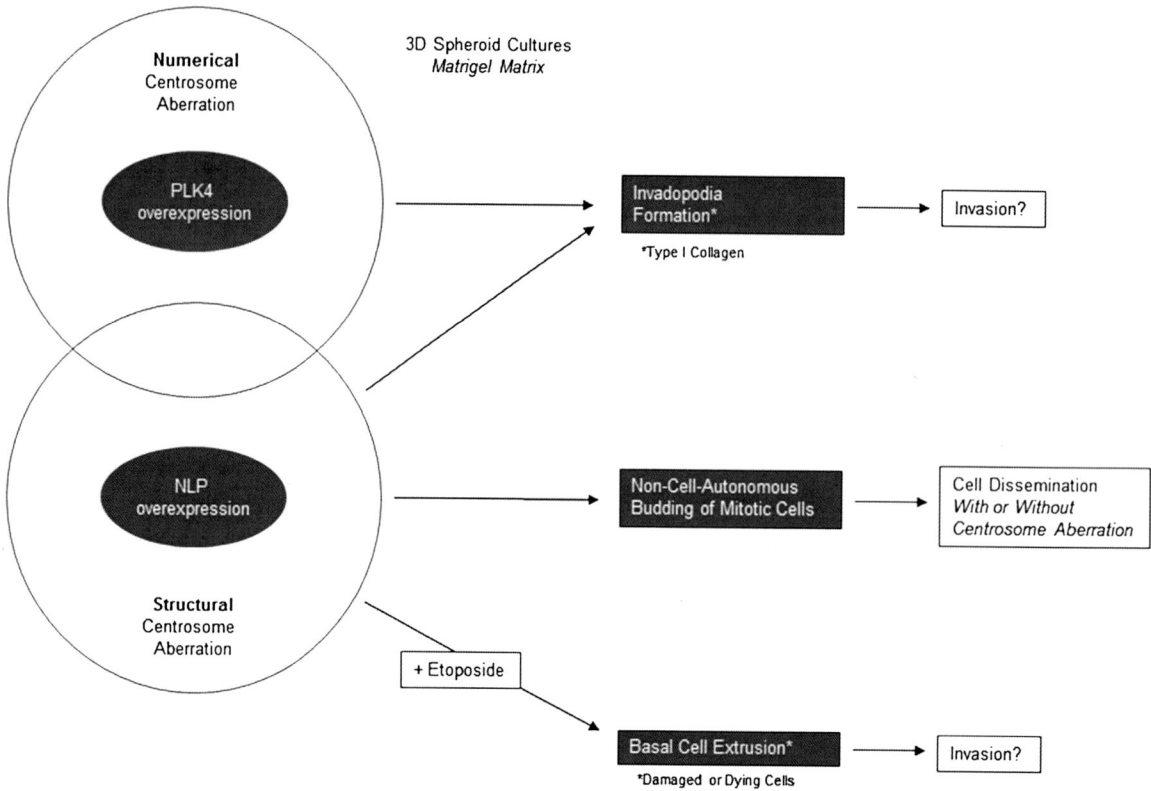

Figure 2. Impact of centrosome aberrations on tissue architecture and invasion. In vitro studies using three-dimensional (3D) spheroid models suggest that centrosome aberrations can confer invasive properties. The schematic distinguishes three distinct mechanisms, invadopodia formation, the non-cell-autonomous budding of mitotic cells, and basal extrusion. PLK4, Polo-like kinase 4; NLP, Ninein-like protein.

some aberrations can cause alterations in the cytoskeleton that then lead to invadopodia formation. One intriguing methodological aspect of the original study was the use of type I collagen to sensitize invasion assays (Godinho et al. 2014). Type I collagen is known to increase the stiffness of Matrigel matrices, thereby favoring invadopodia formation (Di Martino et al. 2015; Artym 2016). In our laboratory, we found that PLK4- or NLP-induced invadopodia formation was strictly dependent on addition of type I collagen to Matrigel preparations (Schnerch and Nigg 2016; Ganier et al. 2017). This dependence on type I collagen could be interpreted to suggest that invadopodia formation represents a rather subtle phenotype, but we emphasize that type I collagen frequently accumulates within tumor stroma (Artym 2016). Thus, addition of this extracellular matrix component to Matrigel may well mimic a pathophysiologically relevant milieu.

In our laboratory we have recently discovered a novel process, termed "budding," that also offers the potential to contribute to the spreading of metastatic cells (Ganier et al. 2017). This process is mechanistically distinct from invadopodia formation and does not require addition of type I collagen to Matrigel. Specifically, we found that structural centrosome aberrations, induced by overexpression of NLP, promote the dissemination of living cells from epithelial spheroid cultures. Most remarkably, live cell imaging showed that all budding cells were invariably undergoing mitosis. Induction of NLP transgene expression in spheroid cultures caused the formation of structural centrosome aberrations that closely resembled those seen in malignant tumors. Furthermore, just like in tumors, these aberrations were not seen in all cells of a given spheroid. In fact, because the NLP transgene was not expressed in all cells, this resulted in epithelia harboring cells with centrosome aberrations interspersed with cells lacking such aberrations. A high proportion of cells (>50%) with structural centrosome aberrations was a prerequisite for cell budding from spheroids, but, remarkably, only half of the disseminating mitotic cells were found to carry such aberrations. This strongly suggests that budding results from multicellular cooperation and represents a non-cell-autonomous phenomenon (Fig. 3). Regarding the mechanisms leading to cell budding, we found that NLP-induced structural centrosome aberrations triggered cytoskeletal reorganizations, resulting in weakened E-cadherin junctions and randomization of spindle orientation. Moreover, cells budding from NLP-overexpressing spheroids showed marked delays in mitotic progression and extensive membrane blebbing. As both mitotic delays and membrane blebbing were previously recognized to reflect pressure induced by confinement (Cattin et al. 2015), we used atomic force microscopy to measure the stiffness of epithelial cells with or without centrosome aberrations. The analysis of epithelial cells at different cell cycle stages revealed that NLP-induced structural centrosome aberrations markedly increased cellular stiffness. In contrast, no major difference in these mechanobiological properties could be detected in response to numerical centrosome aberrations induced by PLK4 (Ganier et al. 2017). These data led us to propose that NLP-induced structural centrosome aberrations trigger the

dissemination of mitotic cells from mosaic epithelia, but that the escaping cells do not necessarily have to carry any centrosome aberration (Fig. 3). One attraction of this newly discovered mechanism for cell dissemination is that it may explain a long-standing conundrum in the centrosome field. Considering that centrosome aberrations are expected to impair cell viability, their prominence in aggressive tumors is puzzling. However, it is well recognized that centrosome aberrations are hardly ever present in all cells of a tumor cell population. Thus, a non-cell-autonomous mechanism for cell dissemination offers an explanation for how centrosome aberrations might conceivably promote metastasis, without the invading cells themselves necessarily carrying deleterious centrosome aberrations (Fig. 3).

Removal of damaged or dying cells is critical for the preservation of the barrier function of epithelia. Research into the mechanism underlying this cell extrusion has revealed that cells destined to undergo apoptosis signal to their living neighbors, which then causes actomyosin ring contractility and the delamination of dying cells from the epithelial sheet (Rosenblatt et al. 2001; Gibson and Perrimon 2005; Rodriguez-Boulan and Macara 2014). In wild-type epithelia, most apoptotic cells are squeezed apically into the glandular lumen, and this requires the repositioning of the actomyosin ring from an apical to a basal location. However, a conspicuous change in extrusion directionality toward the basal side has been observed in epithelia carrying oncogenic mutations, and in this case, the repositioning of the actomyosin ring is inhibited (Slattum et al. 2009, 2014; Gu et al. 2015). Considering that apoptosis is blocked in some tumors, this raises the exciting possibility that "basal cell extrusion" could enable disseminating cells to initiate metastasis (Slattum et al. 2009, 2014; Marshall et al. 2011; Slattum and Rosenblatt 2014; Gudipaty et al. 2017). Having observed that about one-third of cells disseminating from spheroids in response to NLP-induced centrosome aberrations stained positively for activated capase-3, a marker of apoptosis, we asked whether the dissemination of some of the dying cells might represent "basal extrusion." To facilitate mechanistic studies, we adopted an assay commonly used to study basal extrusion and treated spheroids harboring NLP-induced centrosome aberrations with the DNA-damaging drug etoposide (Slattum et al. 2014). Analysis of the apoptotic cells disseminating in response to a combination of NLP overexpression and etoposide treatment revealed all the hallmarks of basal extrusion, notably the absence of basal repositioning of the constricting actomyosin ring and sensitivity to agonists of the sphingosine 1 receptor 2 pathway (D Schnerch, O Ganier, and EA Nigg, unpubl.). From these results we conclude that structural centrosome aberrations can sensitize cells to the activation of the basal extrusion pathway. The observed microtubule and actomyosin reorganizations are highly reminiscent of the cytoskeletal rearrangements observed in cells that lack the adenomatous polyposis coli (APC) gene product or express an oncogenic version of APC (Marshall et al. 2011). We conclude, therefore, that structural centrosome aberrations can also contribute to promote basal extrusion of tumor cells (Fig. 2).

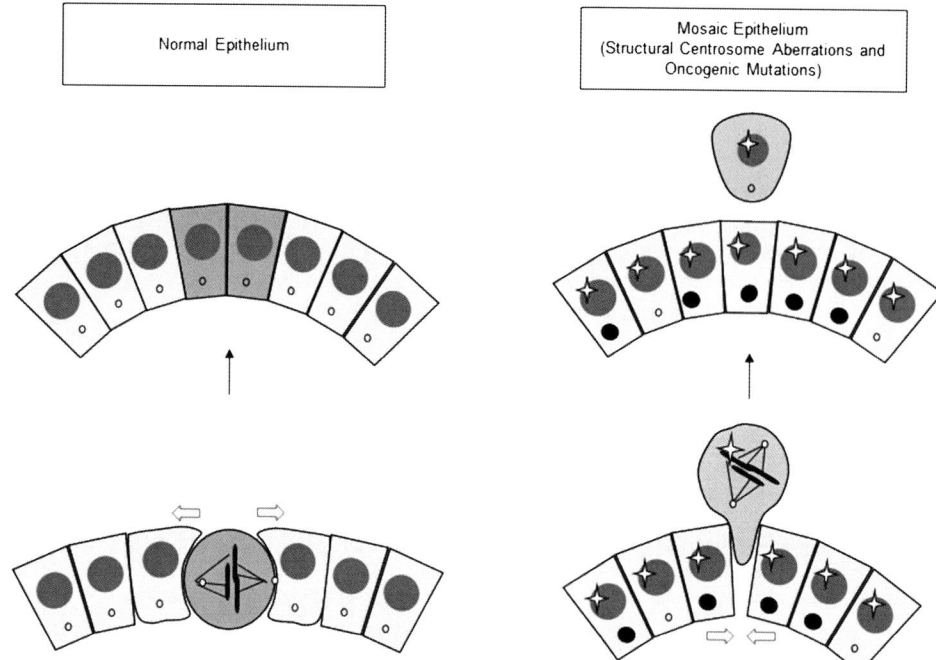

Figure 3. Structural centrosome aberrations cause cell "budding" through non-cell-autonomous mechanism. (*Left*) When cells divide in a wild-type epithelium, mitotic spindles are positioned in the plane of the sheet and surrounding cells are sufficiently deformable to accommodate the products of cell division. (*Right*) In a mosaic epithelium comprising cells with and without structural centrosome aberrations, spindle orientation is randomized and surrounding cells show reduced deformability when they harbor structural centrosome aberrations. As a consequence, mitotic cells are occasionally squeezed out of the epithelium through budding, regardless of whether they carry centrosome aberrations. If these cells additionally harbor oncogenic mutations, this proposed non-cell-autonomous process for cell dissemination may promote metastasis. Small open circles, normal centrosomes; large black circles, structurally aberrant centrosomes (due to NLP overexpression); stars, oncogenic mutations. Horizontal arrows refer to forces.

The studies discussed above show that numerical and structural centrosome aberrations cause profound reorganizations of the cytoskeleton and, moreover, exert drastic effects on the polarity of epithelial layers. As a result, cells disseminating from 3D spheroid structures acquire potentially invasive properties. The available evidence suggests that different types of centrosome aberrations can induce fundamentally different cellular behaviors. In particular, we emphasize that NLP-induced structural centrosome aberrations trigger cell dissemination through a non-cell-autonomous mechanism (Ganier et al. 2017). According to our model (Fig. 3), cell dissemination can be triggered in response to centrosome aberrations without the evading cell itself harboring any centrosomal anomalies. Instead, it suffices that neighboring cells carry NLP-induced centrosome aberrations and thereby change the mechanobiological properties of epithelia. This hypothesis has the potential to explain how centrosome aberrations can promote carcinogenesis, even though centrosome aberrations are expected to jeopardize cellular fitness and never affect all cells within a given tumor.

CONCLUSION

Centrosome aberrations have been implicated in tumorigenesis for more than 100 years. To date, it is firmly established that numerical and structural centrosome aberrations are prominent features of many tumor cells.

Furthermore, experimentally induced centrosome amplification was convincingly shown to trigger tumorigenesis in animals. What remains to be fully understood is the impact of centrosome aberrations on the development and progression of human tumors. Work performed over the past decades supports the notion that centrosome aberrations constitute one prominent root cause of chromosome missegregation. It is now widely accepted that centrosome anomalies likely contribute to explain the aneuploidy and chromosomal instability phenotypes commonly associated with human cancers, vindicating Boveri's early hypothesis. In more recent years, increasing attention is focused on the possibility that centrosome aberrations may disrupt tissue architecture and potentially contribute to confer metastatic properties to cancer cells. One important insight emerging from studies on 2D and 3D culture models is that different types of centrosome aberrations can trigger mechanistically distinct types of invasive behaviors, including invadopodia formation, the dissemination ("budding") of mitotic cells, and basal extrusion. Although this should caution against premature generalizations, it also offers the exciting prospect that careful classification of centrosome aberrations in human tumors might ultimately have diagnostic or prognostic value. Clearly, further work will be required before data obtained in cell culture models can be extrapolated to clinically relevant settings. Continued use of in vitro models for studying centrosome aberrations will be indispensable for elucidating the molecular mechanisms underlying chromosome missegregation and/or the acquisition of invasive properties, but animal models and intravital imaging will increasingly become important for testing emerging models in vivo.

ACKNOWLEDGMENTS

We thank our laboratory members and colleagues in the centrosome field for helpful discussions. Work in the author's laboratory was supported by the University of Basel and a grant from the Swiss National Science Foundation (310030B-149641) to E.A.N.

REFERENCES

Arquint C, Nigg EA. 2014. STIL microcephaly mutations interfere with APC/C-mediated degradation and cause centriole amplification. *Curr Biol* 24: 351–360.

Arquint C, Nigg EA. 2016. The PLK4-STIL-SAS-6 module at the core of centriole duplication. *Biochem Soc Trans* 44: 1253–1263.

Arquint C, Gabryjonczyk AM, Nigg EA. 2014. Centrosomes as signalling centres. *Philos Trans R Soc Lond B Biol Sci* 369: 20130464.

Arquint C, Gabryjonczyk AM, Imseng S, Böhm R, Sauer E, Hiller S, Nigg EA, Maier T. 2015. STIL binding to Polo-box 3 of PLK4 regulates centriole duplication. *ELife* 4: e07888.

Artym VV. 2016. Dense fibrillar collagen is a master activator of invadopodia. *Mol Cell Oncol* 3: e1035476.

Azimzadeh J, Marshall WF. 2010. Building the centriole. *Curr Biol* 20: R816–R825.

Azimzadeh J, Wong ML, Downhour DM, Sanchez Alvarado A, Marshall WF. 2012. Centrosome loss in the evolution of planarians. *Science* 335: 461–463.

Basto R, Lau J, Vinogradova T, Gardiol A, Woods CG, Khodjakov A, Raff JW. 2006. Flies without centrioles. *Cell* 125: 1375–1386.

Basto R, Brunk K, Vinadogrova T, Peel N, Franz A, Khodjakov A, Raff JW. 2008. Centrosome amplification can initiate tumorigenesis in flies. *Cell* 133: 1032–1042.

Bazzi H, Anderson KV. 2014. Acentriolar mitosis activates a p53-dependent apoptosis pathway in the mouse embryo. *Proc Natl Acad Sci* 111: E1491–E1500.

Bettencourt-Dias M, Rodrigues-Martins A, Carpenter L, Riparbelli M, Lehmann L, Gatt MK, Carmo N, Balloux F, Callaini G, Glover DM. 2005. SAK/PLK4 is required for centriole duplication and flagella development. *Curr Biol* 15: 2199–2207.

Bettencourt-Dias M, Hildebrandt F, Pellman D, Woods G, Godinho SA. 2011. Centrosomes and cilia in human disease. *Trends Genet* 27: 307–315.

Bornens M. 2012. The centrosome in cells and organisms. *Science* 335: 422–426.

Boveri T. 1914. *Zur Frage der Entstehung maligner Tumoren.* Gustav Fischer (Herausgeber), Jena.

Braun DA, Hildebrandt F. 2017. Ciliopathies. *Cold Spring Harb Perspect Biol* 9: a028191.

Casenghi M, Meraldi P, Weinhart U, Duncan PI, Korner R, Nigg EA. 2003. Polo-like kinase 1 regulates Nlp, a centrosome protein involved in microtubule nucleation. *Dev Cell* 5: 113–125.

Cattin CJ, Düggelin M, Martinez-Martin D, Gerber C, Muller DJ, Stewart MP. 2015. Mechanical control of mitotic progression in single animal cells. *Proc Natl Acad Sci* 112: 11258–11263.

Chan JY. 2011. A clinical overview of centrosome amplification in human cancers. *Int J Biol Sci* 7: 1122–1144.

Coelho PA, Bury L, Shahbazi MN, Liakath-Ali K, Tate PH, Wormald S, Hindley CJ, Huch M, Archer J, Skarnes WC, et al. 2015. Over-expression of Plk4 induces centrosome am-

plification, loss of primary cilia and associated tissue hyperplasia in the mouse. *Open Biol* 5: 150209.

Conduit PT, Wainman A, Raff JW. 2015. Centrosome function and assembly in animal cells. *Nat Rev Mol Cell Biol* 16: 611–624.

Di Martino J, Moreau V, Saltel F. 2015. Type I collagen fibrils: An inducer of invadosomes. *Oncotarget* 6: 28519–28520.

Fava LL, Schuler F, Sladky V, Haschka MD, Soratroi C, Eiterer L, Demetz E, Weiss G, Geley S, Nigg EA, et al. 2017. The PIDDosome activates p53 in response to supernumerary centrosomes. *Genes Dev* 31: 34–45.

Firat-Karalar EN, Stearns T. 2014. The centriole duplication cycle. *Philos Trans R Soc Lon B Biol Sci* 369: 20130460.

Fong CS, Mazo G, Das T, Goodman J, Kim M, O'Rourke BP, Izquierdo D, Tsou MF. 2016. 53BP1 and USP28 mediate p53-dependent cell cycle arrest in response to centrosome loss and prolonged mitosis. *ELife* 5: e16270.

Fu J, Hagan IM, Glover DM. 2015. The centrosome and its duplication cycle. *Cold Spring Harb Perspect Biol* 7: a015800.

Ganem NJ, Storchova Z, Pellman D. 2007. Tetraploidy, aneuploidy and cancer. *Curr Opin Genet Dev* 17: 157–162.

Ganem NJ, Godinho SA, Pellman D. 2009. A mechanism linking extra centrosomes to chromosomal instability. *Nature* 460: 278–282.

Ganem NJ, Cornils H, Chiu SY, O'Rourke KP, Arnaud J, Yimlamai D, Thery M, Camargo FD, Pellman D. 2014. Cytokinesis failure triggers hippo tumor suppressor pathway activation. *Cell* 158: 833–848.

Ganier O, Schnerch D, Oertle P, Lim RYH, Plodinec M, Nigg EA. 2017. Structural centrosome aberrations promote non-cell-autonomous invasiveness. *bioRxiv* 216804; doi: 10.1101/216804.

Garcia G, Reiter JF. 2016. A primer on the mouse basal body. *Cilia* 5: 17.

Gibson MC, Perrimon N. 2005. Extrusion and death of DPP/BMP-compromised epithelial cells in the developing *Drosophila* wing. *Science* 307: 1785–1789.

Godinho SA, Pellman D. 2014. Causes and consequences of centrosome abnormalities in cancer. *Philos Trans R Soc Lond B Biol Sci* 369: 20130467.

Godinho SA, Picone R, Burute M, Dagher R, Su Y, Leung CT, Polyak K, Brugge JS, Théry M, Pellman D. 2014. Oncogene-like induction of cellular invasion from centrosome amplification. *Nature* 510: 167–171.

Gönczy P. 2012. Towards a molecular architecture of centriole assembly. *Nat Rev Mol Cell Biol* 13: 425–435.

Gönczy P. 2015. Centrosomes and cancer: Revisiting a long-standing relationship. *Nat Rev Cancer* 15: 639–652.

Gu Y, Shea J, Slattum G, Firpo MA, Alexander M, Mulvihill SJ, Golubovskaya VM, Rosenblatt J. 2015. Defective apical extrusion signaling contributes to aggressive tumor hallmarks. *ELife* 4: e04069.

Gudipaty SA, Lindblom J, Loftus PD, Redd MJ, Edes K, Davey CF, Krishnegowda V, Rosenblatt J. 2017. Mechanical stretch triggers rapid epithelial cell division through Piezo1. *Nature* 543: 118–121.

Guichard P, Hachet V, Majubu N, Neves A, Demurtas D, Olieric N, Fluckiger I, Yamada A, Kihara K, Nishida Y, et al. 2013. Native architecture of the centriole proximal region reveals features underlying its 9-fold radial symmetry. *Curr Biol* 23: 1620–1628.

Guichard P, Hamel V, Le Guennec M, Banterle N, Iacovache I, Nemčiková V, Fl I, Goldie KN, Stahlberg H, Lévy D, et al. 2017. Cell-free reconstitution reveals centriole cartwheel assembly mechanisms. *Nat Commun* 8: 14813.

Guo HQ, Gao M, Ma J, Xiao T, Zhao LL, Gao Y, Pan QJ. 2007. Analysis of the cellular centrosome in fine-needle aspirations of the breast. *Breast Cancer Res* 9: R48.

Habedanck R, Stierhof YD, Wilkinson CJ, Nigg EA. 2005. The Polo kinase Plk4 functions in centriole duplication. *Nat Cell Biol* 7: 1140–1146.

Hinchcliffe EH, Miller FJ, Cham M, Khodjakov A, Sluder G. 2001. Requirement of a centrosomal activity for cell cycle progression through G_1 into S phase. *Science* 291: 1547–1550.

Hirono M. 2014. Cartwheel assembly. *Philos Trans R Soc Lond B Biol Sci* **369:** 20130458.

Holland AJ, Fachinetti D, Zhu Q, Bauer M, Verma IM, Nigg EA, Cleveland DW. 2012. The autoregulated instability of Polo-like kinase 4 limits centrosome duplication to once per cell cycle. *Genes Dev* **26:** 2684–2689.

Karsenti E, Vernos I. 2001. The mitotic spindle: A self-made machine. *Science* **294:** 543–547.

Khodjakov A, Rieder CL. 2001. Centrosomes enhance the fidelity of cytokinesis in vertebrates and are required for cell cycle progression. *J Cell Biol* **153:** 237–242.

Kitagawa D, Vakonakis I, Olieric N, Hilbert M, Keller D, Olieric V, Bortfeld M, Erat MC, Flückiger I, Gönczy P, et al. 2011. Structural basis of the 9-fold symmetry of centrioles. *Cell* **144:** 364–375.

Kleylein-Sohn J, Westendorf J, Le Clech M, Habedanck R, Stierhof YD, Nigg EA. 2007. Plk4-induced centriole biogenesis in human cells. *Dev Cell* **13:** 190–202.

Kratz AS, Bärenz F, Richter KT, Hoffmann I. 2015. Plk4-dependent phosphorylation of STIL is required for centriole duplication. *Biol Open* **4:** 370–377.

Kwon M, Godinho SA, Chandhok NS, Ganem NJ, Azioune A, Thery M, Pellman D. 2008. Mechanisms to suppress multipolar divisions in cancer cells with extra centrosomes. *Genes Dev* **22:** 2189–2203.

Lambrus BG, Holland AJ. 2017. A new mode of mitotic surveillance. *Trends Cell Biol* **27:** 314–321.

Lambrus BG, Uetake Y, Clutario KM, Daggubati V, Snyder M, Sluder G, Holland AJ. 2015. p53 protects against genome instability following centriole duplication failure. *J Cell Biol* **210:** 63–77.

Lambrus BG, Daggubati V, Uetake Y, Scott PM, Clutario KM, Sluder G, Holland AJ. 2016. A USP28-53BP1-p53-p21 signaling axis arrests growth after centrosome loss or prolonged mitosis. *J Cell Biol* **214:** 143–153.

Levine MS, Bakker B, Boeckx B, Moyett J, Lu J, Vitre B, Spierings DC, Lansdorp PM, Cleveland DW, Lambrechts D, et al. 2017. Centrosome amplification is sufficient to promote spontaneous tumorigenesis in mammals. *Dev Cell* **40:** 313–322.e315.

Lingle WL, Lutz WH, Ingle JN, Maihle NJ, Salisbury JL. 1998. Centrosome hypertrophy in human breast tumors: Implications for genomic stability and cell polarity. *Proc Natl Acad Sci* **95:** 2950–2955.

Lingle WL, Barrett SL, Negron VC, D'Assoro AB, Boeneman K, Liu W, Whitehead CM, Reynolds C, Salisbury JL. 2002. Centrosome amplification drives chromosomal instability in breast tumor development. *Proc Natl Acad Sci* **99:** 1978–1983.

Marshall TW, Lloyd IE, Delalande JM, Näthke I, Rosenblatt J. 2011. The tumor suppressor adenomatous polyposis coli controls the direction in which a cell extrudes from an epithelium. *Mol Biol Cell* **22:** 3962–3970.

Meitinger F, Anzola JV, Kaulich M, Richardson A, Stender JD, Benner C, Glass CK, Dowdy SF, Desai A, Shiau AK, et al. 2016. 53BP1 and USP28 mediate p53 activation and G_1 arrest after centrosome loss or extended mitotic duration. *J Cell Biol* **214:** 155–166.

Meraldi P, Honda R, Nigg EA. 2002. Aurora-A overexpression reveals tetraploidization as a major route to centrosome amplification in $p53^{-/-}$ cells. *EMBO J* **21:** 483–492.

Moyer TC, Clutario KM, Lambrus BG, Daggubati V, Holland AJ. 2015. Binding of STIL to Plk4 activates kinase activity to promote centriole assembly. *J Cell Biol* **209:** 863–878.

Nigg EA. 2002. Centrosome aberrations: Cause or consequence of cancer progression? *Nat Rev Cancer* **2:** 815–825.

Nigg EA. 2007. Centrosome duplication: Of rules and licenses. *Trends Cell Biol* **17:** 215–221.

Nigg EA, Holland AJ. 2018. Once and only once: Mechanisms of centriole duplication and their deregulation in disease. *Nat Rev Mol Cell Biol* (in press).

Nigg EA, Raff JW. 2009. Centrioles, centrosomes, and cilia in health and disease. *Cell* **139:** 663–678.

Nigg EA, Stearns T. 2011. The centrosome cycle: Centriole biogenesis, duplication and inherent asymmetries. *Nat Cell Biol* **13:** 1154–1160.

Ohta M, Ashikawa T, Nozaki Y, Kozuka-Hata H, Goto H, Inagaki M, Oyama M, Kitagawa D. 2014. Direct interaction of Plk4 with STIL ensures formation of a single procentriole per parental centriole. *Nat Commun* **5:** 5267.

Paz J, Luders J. 2017. Microtubule-organizing centers: Towards a minimal parts list. *Trends Cell Biol.* doi: 10.1016/j.tcb.2017.10.005.

Pihan GA, Wallace J, Zhou Y, Doxsey SJ. 2003. Centrosome abnormalities and chromosome instability occur together in pre-invasive carcinomas. *Cancer Res* **63:** 1398–1404.

Prosser SL, Pelletier L. 2017. Mitotic spindle assembly in animal cells: A fine balancing act. *Nat Rev Mol Cell Biol* **18:** 187–201.

Quintyne NJ, Reing JE, Hoffelder DR, Gollin SM, Saunders WS. 2005. Spindle multipolarity is prevented by centrosomal clustering. *Science* **307:** 127–129.

Rodriguez-Boulan E, Macara IG. 2014. Organization and execution of the epithelial polarity programme. *Nat Rev Mol Cell Biol* **15:** 225–242.

Rodrigues-Martins A, Riparbelli M, Callaini G, Glover DM, Bettencourt-Dias M. 2007. Revisiting the role of the mother centriole in centriole biogenesis. *Science* **316:** 1046–1050.

Rosenblatt J, Raff MC, Cramer LP. 2001. An epithelial cell destined for apoptosis signals its neighbors to extrude it by an actin- and myosin-dependent mechanism. *Curr Biol* **11:** 1847–1857.

Salmon ED, Cimini D, Cameron LA, DeLuca JG. 2005. Merotelic kinetochores in mammalian tissue cells. *Philos Trans R Soc Lond B Biol Sci* **360:** 553–568.

Sánchez I, Dynlacht BD. 2016. Cilium assembly and disassembly. *Nat Cell Biol* **18:** 711–717.

Schnerch D, Nigg EA. 2016. Structural centrosome aberrations favor proliferation by abrogating microtubule-dependent tissue integrity of breast epithelial mammospheres. *Oncogene* **35:** 2711–2722.

Serçin Ö, Larsimont JC, Karambelas AE, Marthiens V, Moers V, Boeckx B, Le Mercier M, Lambrechts D, Basto R, Blanpain C. 2016. Transient PLK4 overexpression accelerates tumorigenesis in p53-deficient epidermis. *Nat Cell Biol* **18:** 100–110.

Shao S, Liu R, Wang Y, Song Y, Zuo L, Xue L, Lu N, Hou N, Wang M, Yang X, et al. 2010. Centrosomal Nlp is an oncogenic protein that is gene-amplified in human tumors and causes spontaneous tumorigenesis in transgenic mice. *J Clin Invest* **120:** 498–507.

Silkworth WT, Nardi IK, Scholl LM, Cimini D. 2009. Multipolar spindle pole coalescence is a major source of kinetochore mis-attachment and chromosome mis-segregation in cancer cells. *PLoS One* **4:** e6564.

Slattum GM, Rosenblatt J. 2014. Tumour cell invasion: An emerging role for basal epithelial cell extrusion. *Nat Rev Cancer* **14:** 495–501.

Slattum G, McGee KM, Rosenblatt J. 2009. P115 RhoGEF and microtubules decide the direction apoptotic cells extrude from an epithelium. *J Cell Biol* **186:** 693–702.

Slattum G, Gu Y, Sabbadini R, Rosenblatt J. 2014. Autophagy in oncogenic K-Ras promotes basal extrusion of epithelial cells by degrading S1P. *Curr Biol* **24:** 19–28.

Szollosi D, Calarco P, Donahue RP. 1972. Absence of centrioles in the first and second meiotic spindles of mouse oocytes. *J Cell Sci* **11:** 521–541.

van Breugel M, Hirono M, Andreeva A, Yanagisawa HA, Yamaguchi S, Nakazawa Y, Morgner N, Petrovich M, Ebong IO, Robinson CV, et al. 2011. Structures of SAS-6 suggest its organization in centrioles. *Science* **331:** 1196–1199.

Wong YL, Anzola JV, Davis RL, Yoon M, Motamedi A, Kroll A, Seo CP, Hsia JE, Kim SK, Mitchell JW, et al. 2015. Cell biology. Reversible centriole depletion with an inhibitor of Polo-like kinase 4. *Science* **348:** 1155–1160.

Woodruff JB, Wueseke O, Hyman AA. 2014. Pericentriolar material structure and dynamics. *Philos Trans R Soc Lond B Biol Sci* **369:** 20130459.

The ABCs of Centriole Architecture: The Form and Function of Triplet Microtubules

JENNIFER T. WANG[1] AND TIM STEARNS[1,2]

[1]Department of Biology, Stanford University, Stanford, California 94305-5020
[2]Department of Genetics, Stanford School of Medicine, Stanford, California 94305

Correspondence: stearns@stanford.edu

The centriole is a defining feature of many eukaryotic cells. It nucleates a cilium, organizes microtubules as part of the centrosome, and is duplicated in coordination with the cell cycle. Centrioles have a remarkable structure, consisting of microtubules arranged in a barrel with ninefold radial symmetry. At their base, or proximal end, centrioles have unique triplet microtubules, formed from three microtubules linked to each other. This microtubule organization is not found anywhere else in the cell, is conserved in all major branches of the eukaryotic tree, and likely was present in the last eukaryotic common ancestor. At their tip, or distal end, centrioles have doublet microtubules, which template the cilium. Here, we consider the structures of the compound microtubules in centrioles and discuss potential mechanisms for their formation and their function. We propose that triplet microtubules are required for the structural integrity of centrioles, allowing the centriole to serve as the essential nucleator of the cilium.

In most animal cells, centrioles come in pairs, and in cells that are traversing the cell cycle, this pair of centrioles is precisely duplicated and segregated to ensure that each daughter cell receives exactly one pair of centrioles. Many of the molecular players in this cycle of centriole duplication have been identified and are conserved through evolution (Carvalho-Santos et al. 2011). Birth of a new centriole occurs in S phase, when a procentriole forms off the side of each existing mother centriole, near the proximal end. The triplet microtubules are added to a structural hub, known as the cartwheel, early in the process, resulting in a structure $\sim$0.2 μm long and 0.2 μm wide (Kuriyama and Borisy 1981; Vorobjev and Chentsov 1982; Chrétien et al. 1997). In G_2, the triplet microtubules elongate, reaching a maximum length of $\sim$0.35 μm. At the G_2/M transition, the inner two microtubules of the triplet elongate to their final length of 0.5 μm, forming a distal end with only doublet microtubules. In the following cell cycle, this new centriole recruits the pericentriolar material components of the centrosome and acquires appendages, allowing it to serve as a nucleator of the cilium in the subsequent G_0/G_1. The doublet microtubules of the centriole extend to form the axoneme of the cilium. Here, we refer to the doublet and triplet microtubules of the centriole and axoneme as compound microtubules. The centriole structure at the base of the cilium is often termed a basal body; we will use the term centriole to refer to all such structures. Although conserved in structure, centrioles, and the associated cilium and centrosome, have been lost from several key eukaryotic groups, including higher fungi and higher plants (Carvalho-Santos et al. 2011).

Centrioles can also be made without an existing centriole. The mechanisms underlying this de novo centriole formation are less understood, but have been observed in multiple species: the flatworm *Planaria*, the amoeba *Naegleria*, and the plants *Ginkgo biloba* and *Zamia* (Norstog 1967; Fulton and Dingle 1971; Gifford and Lin 1975; Azimzadeh et al. 2012). Centrioles formed de novo in these organisms have the typical ninefold structure with triplet microtubules, indicating that it is not necessary to have an existing centriole to create a new one. Human cells usually do not form centrioles de novo but can be induced to do so by experimental manipulation (Khodjakov et al. 2002; La Terra et al. 2005; Uetake et al. 2007). Centrioles formed under these conditions often have structural defects (Khodjakov et al. 2002; Wang et al. 2015). Last, specialized cells in animals can generate many centrioles by centriole amplification. In differentiating multiciliated cells, centriole amplification occurs from unique structures known as the deuterosome (Sorokin 1968; Steinman 1968; Kalnins and Porter 1969), which is formed from an existing centriole (Al Jord et al. 2014).

Regardless of the mechanism of formation, centrioles share two defining structural features: ninefold symmetry and the presence of compound microtubules (Fig. 1). Ninefold symmetry is derived from the intrinsic assembly properties of proteins making up a ninefold symmetric hub and spoke structure known as the cartwheel. The cartwheel is the first structure to form in the nascent centriole, and its major component, SASS6, is capable of self-assembling in vitro into ninefold symmetric, cartwheel-like structures (Kitagawa et al. 2011; van Breugel et al. 2011). During the normal centriole duplication cycle, the mother centriole templates cartwheel formation (Fong et al. 2014). Although the cartwheel is the basis of centriolar ninefold symmetry, there are likely other interactions that reinforce

Published by Cold Spring Harbor Laboratory Press; doi: 10.1101/sqb.2017.82.034496

Cold Spring Harbor Symposia on Quantitative Biology, Volume LXXXII

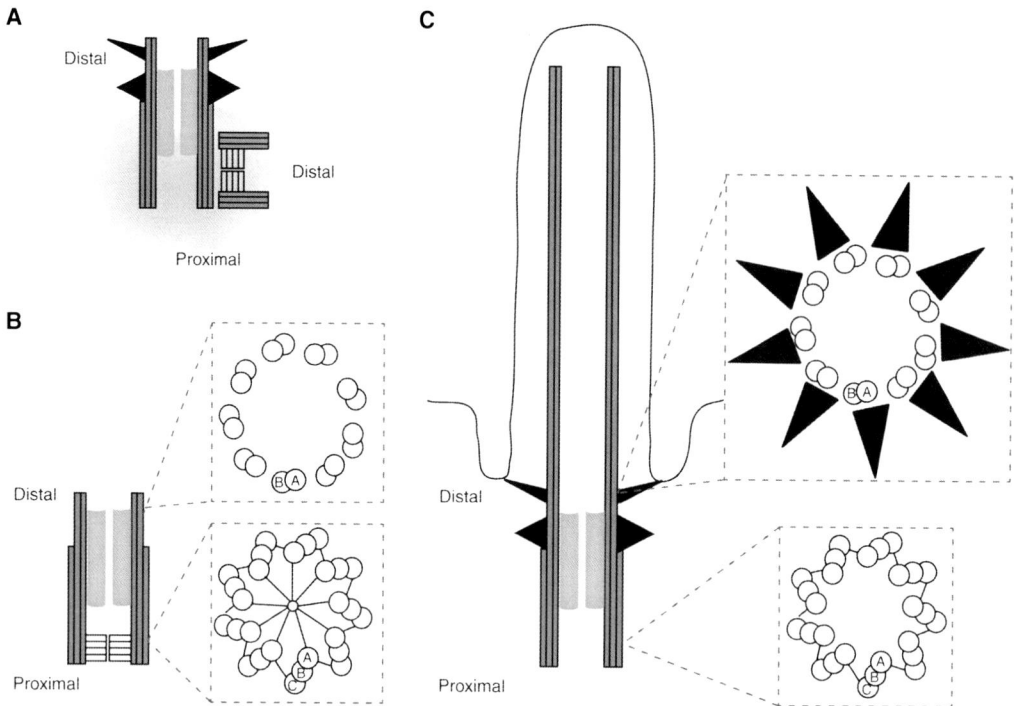

Figure 1. Centriole structures at key stages of the cell cycle. (*A*) Centrosome structure in S-phase cells. Each centrosome is composed of two centrioles, surrounded by pericentriolar material (gray cloud). The mother centriole is vertical and has appendages (black triangles) on its distal end. The procentriole is at a right angle and has a cartwheel structure within its proximal end. (*B*) Centriole structure in procentriole at G_2–M. The centriole has elongated to its full length. Cross sections through the centriole are shown at *right*. A-, B-, and C-tubules are labeled. (*C*) Mother centriole at the base of a cilium. The centriole has appendages and lacks a cartwheel. Cross sections through the centriole are shown at *right*. A-, B-, and C-tubules are labeled.

this symmetry, such as those between the cartwheel and centriolar microtubules (Hilbert et al. 2016). Interestingly, ~15% of *Chlamydomonas* mutants lacking SASS6 have ninefold symmetric centrioles (Nakazawa et al. 2007), and human cells expressing mutant forms of SASS6 deficient in oligomerization can form centrioles with a normal structure (Wang et al. 2015). Other mechanisms that impart ninefold symmetry may include structural constraints imposed by the A–C linker that bridges adjacent centriolar microtubule triplets (Hirono 2014; Meehl et al. 2016).

In contrast with the radial symmetry of centrioles, the mechanisms by which centrioles form compound microtubule are not understood. The triplet microtubules, in particular, define centrioles in most organisms, differentiating them from the ciliary axoneme, which has doublet microtubules that extend directly from the doublets at the distal end of the centriole. Here, we discuss the structure of compound microtubules, potential mechanisms for their formation and elongation, and their function. Because centriole structure is largely conserved in eukaryotes, we have incorporated information from many organisms, unless otherwise noted.

COMPOUND MICROTUBULE STRUCTURE: TRIPLETS AND DOUBLETS

The proximal end of the centriole in most organisms has a unique triplet microtubule structure not found anywhere

else in the cell. These triplets consist of three linked microtubules, named the A-, B-, and C-tubules (Fig. 1). The A-tubule is connected to the cartwheel through pinhead structures (Anderson and Brenner 1971; Cavalier-Smith 1974; Paintrand et al. 1992; Guichard et al. 2013). In human centrioles, the triplets are ~350 nm long (Paintrand et al. 1992), and only the A-tubule and B-tubule extend through the remaining one-third of the centriole. When the centriole forms a cilium, the A- and B-tubules extend to form the axonemal doublet microtubules (Fig. 1C).

The compound microtubules of the centriole share some similarities with standard microtubules. Each is composed of α-tubulin and β-tubulin subunits (Kuriyama 1976; Nogales et al. 1998), which form heterodimers that associate head-to-tail in a protofilament of a microtubule. Standard microtubules are a tube formed from 13 protofilaments (Tilney et al. 1973), each associating with its neighboring protofilaments through lateral contacts between the M-loop and H1–S2/H2–S3 lateral surfaces of tubulin (Nogales et al. 1999; Sui and Downing 2010). Most of these lateral contacts are formed from connections between like tubulins: between α-tubulin and α-tubulin or between β-tubulin and β-tubulin, with the exception of the seam that closes the tube, where α-tubulin of one protofilament associates laterally with β-tubulin of its neighbor (Song and Mandelkow 1993; Kikkawa et al. 1994; McIntosh et al. 2009).

In the centriolar compound microtubules, the A-tubule has a very similar structure to standard microtubules: 13-

protofilaments form a ring (Fig. 2A; Linck and Stephens 2007; Li et al. 2012; Maheshwari et al. 2015). Unlike standard microtubules, which are circular in cross section, the A-tubule is a deformed oval (Li et al. 2012; Ichikawa et al. 2017). The lateral contacts between protofilaments have been proposed to be similar to those of standard microtubules, consisting of connections between like tubulin subunits (Song and Mandelkow 1995; Maheshwari et al. 2015). The seam has been proposed to lie between protofilaments A1 and A2, between A9 and A10, or between A10 and A11 (Song and Mandelkow 1995; Maheshwari et al. 2015; Ichikawa et al. 2017). Unlike the A-tubule, the B-, and C-tubules are incomplete microtubules, each sharing walls with the adjoining tubules: the B-tubule is made of 10 protofilaments (B1–B10) and shares four additional protofilaments (A10–A13) with the A-tubule, and the C-tubule is also made of 10 protofilaments (C1–C10) and shares four additional protofilaments with the B-tubule (B5–B8). Within each tubule, the protofilaments interact through their M-loop and H1–S2/H2–S3 lateral surfaces, as in standard microtubules, and lateral interactions between protofilaments consist of connections between like tubulin subunits (Song and Mandelkow 1995; Maheshwari et al. 2015).

The defining features of the triplet microtubules are the junctions between the three tubules. The inner junction, nearest the cartwheel, differs from the outer junction, located toward the cytoplasm. The outer junction between the A- and B-tubules consists of interactions between protofilament B1 of the B-tubule, and protofilaments A10 and A11 of the A-tubule (Fig. 2A). The interactions between these protofilaments differ from those between standard microtubule protofilaments (Li et al. 2012; Ichikawa et al. 2017) in that here protofilament B1 interacts with the cytoplasm-facing sidewalls of protofilaments A10

and A11. Though the amino acid residues involved in these side wall interactions have not been identified in centrioles, recent high-resolution cryo-EM of ciliary axonemes has identified several conserved residues that may be involved (Ichikawa et al. 2017). At the inner junction between the A-tubule and B-tubule, nontubulin proteins form additional structure, known as the Y-shaped linker (Li et al. 2012). The identities of these proteins are unknown, though a recent report suggests that Poc16/WDR90 may form part of this structure (Hamel et al. 2017).

The outer junction between the B- and C-tubules varies between organisms. In the proximal region of *Chlamydomonas* centrioles, protofilament C1 is a microtubule protofilament like B1 (Li et al. 2012), and the interface between protofilaments C1 and B4/B5 is likely similar to that between B1 and A10/A11. Toward the distal end, however, the C1 microtubule protofilament is absent and in its place is an elongated linker, which has ~8-nm periodicity along the centriole, suggesting that it may bind to only one tubulin molecule in the heterodimer. In *Trichonympha* centrioles, an extended linker with 8.5-nm periodicity spans the B4 and C1 protofilaments in the proximal region of the centriole (Guichard et al. 2013). The identities of the proteins that form these structures are unknown. It is also unknown whether similar structures are present in human centrioles. The inner junction between the B- and C-tubules consists of structures, unique to the C-tubule, bridging protofilaments B8 and C10 (Li et al. 2012; Guichard et al. 2013).

In addition to the junctional structures in the centriolar triplets, multiple nonmicrotubule densities are associated with the triplet microtubule structure (Li et al. 2012). These include densities both inside and outside the microtubule protofilaments, some spanning the protofilaments longitudinally, and others bridging protofilaments. Some of these

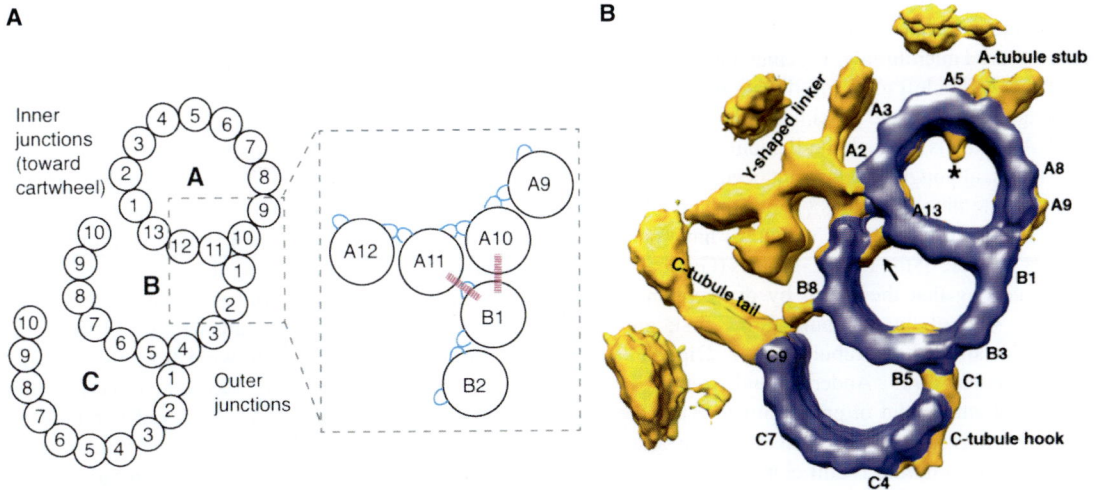

Figure 2. Triplet microtubule structure. (*A*) (*Left*) Protofilaments of the triplet microtubules, numbered according to convention by Linck and Stephens (2007). Inner junctions and outer junctions are labeled. (*Right*) Enlarged region of the outer junction between the A-tubule and B-tubule. Blue, M-loop and H1–S2/H2–S3 loop connections in canonical lateral interactions; pink lines, noncanonical external side wall interactions between protofilament B1 and protofilaments A10 and A11. (*B*) Centriole triplet microtubule cryo-EM structure. Blue, microtubule protofilaments; yellow, nonmicrotubule densities. (*B*, Reprinted from Li et al. 2012, with permission from John Wiley and Sons, © 2012 European Molecular Biology Organization.)

structures are also present in ciliary axonemes (Sui and Downing 2006; Nicastro et al. 2011; Pigino et al. 2012; Maheshwari et al. 2015; Ichikawa et al. 2017). Although the proteins that form these densities have not been definitively assigned, they may include ribbon proteins and tektins, intermediate-filament-like proteins that localize to both the centriole and cilium (Amos 2008; Linck et al. 2014). Altogether, we suggest three nonexclusive possibilities for the function of these proteins: formation of the B- and C-tubules, stabilization of the triplet structure, or recruitment of other proteins for centriole function. Further work will be needed to distinguish between these possibilities.

FORMING THE TRIPLETS

The triplet microtubules of the centriole form in the context of the centriole duplication cycle (Firat-Karalar and Stearns 2014). Centriole assembly occurs at the G_1–S transition and begins with formation of the cartwheel structure, followed by attachment of microtubules (Kalnins and Porter 1969; Vorobjev and Chentsov 1982). The pinhead structure, which attaches the A-tubule to the cartwheel (Anderson and Brenner 1971; Cavalier-Smith 1974; Guichard et al. 2013), is likely composed of the carboxyl terminus of SASS6 and CEP135 (Matsuura et al. 2004; Hiraki et al. 2007; Guichard et al. 2017). CEP135 is able to bind and bundle microtubules (Kraatz et al. 2016) and interacts with CPAP/CenpJ/Sas-4, another microtubule-binding protein that affects centriolar microtubule dynamics (Lin et al. 2013a; Sharma et al. 2016; Zheng et al. 2016). Thus, it is very likely that these proteins are involved in recruitment and stabilization of microtubules at the proximal end of the centriole.

In human cells, a conical cap, proposed to be the γ-tubulin ring complex, is present at the base of the A-tubule (Guichard et al. 2010). The distal end of the A-tubule during elongation appears similar to the plus-end of a growing standard microtubule. Together, these results suggest that the A-tubule grows uni-directionally from a γ-tubulin ring complex at the proximal end of the centriole. Once the A-tubule is formed, the B-tubule and C-tubules are then added, although this does not occur synchronously at each of the nine A-tubules (Vorobjev and Chentsov 1982). γ-Tubulin ring complex structures have not been observed at the base of the B- or C-tubules (Guichard et al. 2010), suggesting that they form by different methods. The outside wall of the B-tubule forms first, growing from the sidewall of the A-tubule (Fig. 2; Dippell 1968; Kalnins and Porter 1969; Anderson and Brenner 1971), indicating that attachment of protofilament B1 is the initiating event in B-tubule formation. C-tubule assembly begins while the B-tubule is still elongating, likely from the outer junction (Kalnins and Porter 1969; Anderson and Brenner 1971; Guichard et al. 2010). The B- and C-tubules of human centrioles are initiated about 50 to 100 nm from the proximal base of the A-tubule, then grow bidirectionally, elongating toward both the proximal and distal ends (Guichard et al. 2010).

Any mechanism for triplet microtubule assembly must incorporate the observations that these structures are only found at centrioles, and that only triplets form, not higher-order microtubule structures. We speculate that the following factors may be involved: (1) centriole-specific tubulin proteins, and (2) the cartwheel.

Centriolar Tubulins

A possible explanation for the unique structure of the compound microtubules is that they are assembled from centriole-specific tubulin superfamily members, or from their isoforms. α-Tubulin and β-tubulin form heterodimers that are the basis of all microtubules, including the compound microtubules of axonemes and centrioles (Kuriyama 1976; Nogales et al. 1998). Most organisms have multiple α-tubulin and β-tubulin isoforms, which mainly differ in the amino acid composition of their disordered, highly charged carboxy-terminal tails that project from the cytoplasmic surface of the microtubule (Redeker 2010). Different α-tubulin and β-tubulin isoforms impart different dynamics to standard microtubules (Vemu et al. 2017), and it is possible that specific isoforms are required for the formation of compound microtubules. Indeed, early work on axonemes from multiple organisms suggests that the axonemal microtubules may contain specific isoforms relative to standard microtubules (Stephens 1970; Bibring et al. 1976; Lai et al. 1979). In *Drosophila*, a sperm-specific β-tubulin isoform forms the axonemal compound microtubules, as well as other microtubules in sperm (Kemphues et al. 1982; Raff et al. 2000; Nielsen et al. 2001). A more divergent β-tubulin isoform, found in a minority of tissues, cannot compensate for loss of the sperm-specific β-tubulin (Hoyle and Raff 1990). It is not clear whether any specific isoforms are strictly necessary for the formation of compound microtubules in other organisms.

Regulation of the carboxy-terminal tails of tubulin is likely critical to compound microtubule formation. This region is highly charged, and is also the site of several post-translational modifications (Wloga and Gaertig 2010; Yu et al. 2015; Gadadhar et al. 2017). Remarkably, sperm of a *Drosophila* mutant that expresses β-tubulin lacking its carboxy-terminal tail were shown to have compound microtubules with many additional tubules (Fackenthal et al. 1993), suggesting that the carboxyl terminus is normally inhibitory to formation. In vitro experiments showing that removal of the tails with subtilisin facilitated the formation of "hooks" growing from the sidewalls of microtubules (Serrano et al. 1984), a structure similar to that occurring during the early stages of B-tubule assembly. The formation of hooks on microtubules has been used as a means of determining microtubule polarity in cells, as the handedness of the hook reveals the polarity of the microtubule on which it has formed (Euteneuer and McIntosh 1980; Heidemann and McIntosh 1980; McIntosh and Euteneuer 1984; Baas and Lin 2011). Interestingly, the standard protocol for hook formation in these experiments involved high ionic-strength buffer conditions, which we propose might facilitate hook formation by neutralizing the inhibitory effect of the carboxy-terminal tails on the unique sidewall interaction that defines compound microtubules. Indeed, the first observation of higher-order microtubule assembly in vitro was by Burton and Himes (1978), who showed that pH regulates their formation. These results suggest that regulation of electrostatic interactions that involve the carboxy-terminal tails of tubulin is important for compound microtubule formation.

In addition to α-tubulin and β-tubulin, the tubulin superfamily contains four additional members: γ-tubulin, delta (δ)-tubulin, epsilon (ε)-tubulin, and zeta (ζ)-tubulin. γ-Tubulin is absolutely conserved in all eukaryotes, including those that have lost the ability to make centrioles and cilia in evolution. However, the other three, termed the ZED-tubulins, are found only in a subset of eukaryotes, all of which form centrioles and cilia. The ZED-tubulins form an evolutionarily coconserved module, suggesting that they may function together (Turk et al. 2015). δ-Tubulin was identified in *Chlamydomonas* as the protein encoded by *UNI3*, (Dutcher and Trabuco 1998), ε-tubulin was first found in the human genome by sequence-alignment (Chang and Stearns 2000), and ζ-tubulin was found by evolutionary comparison of sequenced genomes (Findeisen et al. 2014; Turk et al. 2015).

Experiments probing the function of δ-tubulin and ε-tubulin reveal that they are required for triplet microtubule formation and/or stability in a range of organisms from protists to humans (Goodenough and St Clair 1975; Dutcher and Trabuco 1998; Garreau de Loubresse et al. 2001; Chang et al. 2002; Dupuis-Williams et al. 2002; Dutcher et al. 2002; Gadelha et al. 2006; Ross et al. 2013; Wang et al. 2017). The limited information on localization of these tubulins indicates that they are limited to centrioles, suggesting that they might be involved in interactions that constrain the occurrence of triplet microtubules to the centriole. The functions of δ-tubulin and ε-tubulin have been tested by a variety of methods, from depletion of the protein to mutations likely to be strong loss-of-function alleles. Although some aspects of the phenotypes observed differ, there are commonalities in most of the experiments that suggest a conserved function. In most cases, cells in which either δ-tubulin or ε-tubulin has been lost make centrioles that lack triplet microtubules and are unstable over time.

We recently tested the roles of δ-tubulin and ε-tubulin in human cells by using CRISPR/Cas9 to make null mutations (Wang et al. 2017). Null mutations in δ-tubulin or ε-tubulin could only be recovered in a *p53*-null background, indicating that cells had defects in creating functional centrioles, which is known to trigger a *p53*-dependent cell cycle arrest in mammalian cells (Lambrus et al. 2015; Wong et al. 2015). Cells with mutations in either δ-tubulin or ε-tubulin have in a similar phenotype, with centrioles that have only have singlet microtubules, and that disintegrate each time the cell transitions from mitosis to interphase (Wang et al. 2017). Less is known about the function of the more recently described ζ-tubulin, but experiments in *Xenopus* multiciliated cells suggest that, in

that cell type at least, it is involved in elaboration of centriolar appendages rather than centriole formation (Turk et al. 2015).

Mutations in δ-tubulin and ε-tubulin are among the only mutations known to specifically affect the formation of the triplet microtubules. How might they function in forming compound microtubules? A defining feature of all characterized tubulins is that they interact with other tubulins in the context of structures of the microtubule cytoskeleton. δ-Tubulin and ε-tubulin have been shown to interact with each other in directed experiments (Breslow et al. 2017; Wang et al. 2017) and high-throughput screens (Hein et al. 2015) and have been proposed to share particular interaction interfaces with α-tubulin and β-tubulin (Inclán and Nogales 2001). A simple hypothesis for the function of δ-tubulin and ε-tubulin is that they replace α-tubulin and/or β-tubulin in specific sites critical to the interactions that uniquely occur in compound microtubules. Another model would be that they form a nonprotofilament structure, such as the extended linker found between the B- and C-tubules (Li et al. 2012). Breslow et al. (2017) identified two previously uncharacterized proteins as stoichiometric interactors with δ-tubulin and ε-tubulin, which might also take part in such structures.

As attractive as these models are for the function of δ-tubulin and ε-tubulin, there are indications from evolutionary analysis that they cannot fully account for all instances of compound microtubules. First, nematodes and dipteran insects lack any of the ZED tubulins but make doublet microtubules as part of ciliary axonemes. Second, *Drosophila* and the primitive plant *Ginkgo biloba* lack the ZED tubulins, but form centrioles with triplet microtubules in their sperm (Gifford and Lin 1975; Gottardo et al. 2015; Wang et al. 2017). Since the mechanism of action of δ-tubulin and ε-tubulin at centrioles is not known, we cannot yet say whether these are examples in which the requirement for δ-tubulin and ε-tubulin has been bypassed evolutionarily, perhaps by gain-of-function in another tubulin, or in which the compound microtubules differ from those in other organisms that have δ-tubulin and ε-tubulin. We note that a mutation in α-tubulin can partially suppress the centriole defects of the *uni3* δ-tubulin mutation in *Chlamydomonas*, consistent with the possibility of evolutionary adaptation (Fromherz et al. 2004).

The Role of the Cartwheel

One possible mechanism to limit triplet microtubule assembly to centrioles would be to restrict formation to preexisting structures already formed at the nascent procentriole, such as the cartwheel. In theory, the cartwheel may help to promote triplet microtubule assembly in several ways, including: modification of the nascent A-tubule, alteration of microtubule growth kinetics, creation of a template for the triplet microtubules, or stabilization of the triplets against depolymerization.

The A-tubule differs subtly in structure from a standard microtubule—it is a deformed oval in cross section, rather than round (Li et al. 2012). If this deformation were the result of attachment of the cartwheel or its associated proteins, it might limit compound microtubule formation to that site by altering the lattice of the A-tubule, and making it a suitable substrate for addition of the B1 protofilament. However, it is not known whether A-tubule deformation is due to cartwheel interaction, or whether such deformation aids in B-tubule formation.

The cartwheel, or associated proteins, may alter the growth kinetics of centriolar microtubules, aiding in compound microtubule formation. Compared to standard dynamic microtubules, the centriolar microtubules appear to grow very slowly (Kuriyama and Borisy 1981; Chrétien et al. 1997; Kinoshita et al. 2001), and it is possible that rapid dynamic growth would be incompatible with protofilament–sidewall interactions. Recently, it has been proposed that the centriole assembly protein CPAP enforces slow, processive growth on microtubules (Sharma et al. 2016; Zheng et al. 2016). CPAP interacts with the pinhead protein CEP135 (Lin et al. 2013a), and thus the cartwheel may spatially constrain CPAP activity by localization. It is not clear whether CPAP acts on all three microtubules of the triplet, or whether other mechanisms are involved in restricting the growth of the B- and C-tubules.

Other, yet more speculative possibilities might also involve the cartwheel. For example, the cartwheel, or associated proteins, could provide a direct template for the triplet microtubules, controlling both their structure and their spatial distribution. Or it might stabilize triplet microtubules that form constantly in the cytoplasm, but disassemble before being observed without that stabilization. Finally, the localization of the cartwheel adjacent to the mother centriole during centriole duplication places it within the pericentriolar material of the centrosome, which might be a permissive environment for triplet microtubule assembly. The pericentriolar material has been proposed to act as a tubulin concentrator (Woodruff et al. 2017), and high concentrations of tubulin are used to form microtubule hook structures in vitro (Euteneuer and McIntosh 1980; Heidemann and McIntosh 1980; McIntosh and Euteneuer 1984; Baas and Lin 2011). We note that the pericentriolar material is unlikely to be instructive for triplet microtubule assembly: centrioles with triplet microtubules can form de novo without a mother centriole in many contexts. However, in cycling cells, de novo formation of centrioles often leads to centriole structural aberrations, including loss of whole triplets (Khodjakov et al. 2002; Wang et al. 2015).

REGULATING THE MICROTUBULES: ELONGATION

The compound microtubules of the centriole are initiated in G_1/S, soon after the initiation of the procentrioles, but they continue to elongate slowly until the completion of cell division. Given the slow rate of elongation, compared to the growth rate of standard microtubules, this suggests that centriolar triplet growth is regulated, perhaps coordinately with formation of other centriole structures. In human cells, the C-tubule is only a partial tubule near

the end of the triplets (Paintrand et al. 1992), and in many contexts, including human cells, the C-tubule terminates before the doublets end, forming a distal end extension consisting only of doublets. The doublets extend during ciliogenesis, forming the doublet axoneme of the cilium. It seems likely that there is a structural distinction between the centriole and axoneme, because the centriolar microtubules are stable over time spans as long as the life of the organism (Kochanski and Borisy 1990; Balestra et al. 2015), whereas the axoneme is capable of disassembly and assembly within a single cell cycle (Sánchez and Dynlacht 2016).

The picture of growth of the centriolar compound microtubules is a complex one, with several points of regulation likely required to limit growth of one or more tubules but also to relieve that limitation for axoneme formation. Several proteins have been implicated in control of centriole length, perhaps directly though controlling centriole microtubule growth. Overexpression or depletion of several proteins (CPAP, CEP120, or CEP295, CP110, OFD1, RTTN, POC5) results in centrioles with a variety of length defects (Azimzadeh et al. 2009; Kohlmaier et al. 2009; Schmidt et al. 2009; Tang et al. 2009; Singla et al. 2010; Comartin et al. 2013; Lin et al. 2013b; Chang et al. 2016; Chen et al. 2017).These proteins have been placed in a recruitment pathway (Chen et al. 2017). Among these, CPAP, OFD1, CEP295, and CEP120 have been reported to directly interact with standard microtubules, and therefore might be candidates for direct action on centriolar compound microtubules (Hsu et al. 2008; Cormier et al. 2009; Singla et al. 2010; Lin et al. 2013b; Chang et al. 2016). CPAP may be a particularly critical protein for the regulation of centriolar microtubules, with recent work showing that it acts both to promote and to limit microtubule growth during growth of the centriole (Sharma et al. 2016; Zheng et al. 2016). Members of the kinesin-13 family—KIF24 in humans and Klp10A in flies—have also been shown to be involved in centriolar microtubule structure (Kobayashi et al. 2011; Delgehyr et al. 2012). Kinesins of this family depolymerize microtubules, and such an activity might be important for limiting the extent of centriole microtubules relative to other centriole structures, or in enforcing equal length across all nine of the centriolar compound microtubules.

Finally, axoneme growth is restricted to mother centrioles that are primed to form a cilium (Fig. 1C). Initiation of axonemal microtubule growth from the centriolar microtubules is a multistep process, initiated by vesicle docking and the recruitment of a kinase, TTBK2, to distal appendages (Sorokin 1962; Goetz et al. 2012; Schmidt et al. 2012; Joo et al. 2013; Sillibourne et al. 2013; Tanos et al. 2013; Čajánek and Nigg 2014; Kobayashi et al. 2014; Lu et al. 2015). These processes result in the removal of CP110, which is thought to form a cap at the distal end of the centriole, limiting elongation of the microtubules (Kleylein-Sohn et al. 2007; Spektor et al. 2007; Kobayashi and Dynlacht 2011). CP110 interacts with other proteins that are important for its function, including KIF24 (Tsang et al. 2006, 2008, 2009; Spektor et al. 2007;

Kobayashi et al. 2011, 2014; Franz et al. 2013; Al-Jassar et al. 2017).

FUNCTIONS OF COMPOUND MICROTUBULES

What properties do compound microtubules impart on the structures that have them? We propose several possibilities: centriole structure stabilization, recruitment of distinct proteins to these structures, elaboration of tracks for protein movement in the axoneme, and differentiation of the basal body from the axoneme.

First, it is possible that compound microtubules are intrinsically more stable than standard microtubules, and that the triplet centriolar microtubules are required for long-term stability of the centriole structure. Axonemal microtubules are relatively stable to depolymerization in vitro, but these have many associated proteins that might affect their stability, and remarkably little is known about the dynamic properties of unadorned doublet or triplet microtubules. In cells, mutations in δ-tubulin or ε-tubulin, which lack triplet microtubules, result in centriole instability in *Chlamydomonas*, *Tetrahymena*, and human cells (Goodenough and St Clair 1975; Ross et al. 2013; Wang et al. 2017). In human cells lacking δ-tubulin or ε-tubulin, centrioles with singlet microtubules formed in S phase, and elongated in G_2–M, forming centrioles of approximately normal length (Wang et al. 2017). However, these centrioles disintegrated during the transition from mitosis to interphase of the next cell cycle. Treating these cells with the microtubule-stabilizing drug paclitaxel suppressed centriole disintegration, suggesting that centriolar microtubules are essential to the integrity of the centriole. This is consistent with the previous work of Bobinnec et al. (1998), indicating that the triplet microtubules are required for normal centriole stability. We note that *Caenorhabditis elegans* has singlet microtubule centrioles in most of its cells (Pelletier et al. 2006), yet has long-lived centrioles (Balestra et al. 2015), challenging the idea that the triplet microtubules are universally required for centriole stability.

Second, it is possible that the compound microtubules are required to recruit proteins that help stabilize the centriole, or contribute to centriolar function. These might include many of the proteins and structures known to localize to centrioles, including the A–C linker that connects triplet microtubules to each other, the centriolar appendages, and the pericentriolar material. Each of these structures fail to form in centrioles with only singlet microtubules from δ-tubulin or ε-tubulin mutant cells (Wang et al. 2017). These mutants also failed to recruit POC5, a component of the distal extension, and *Chlamydomonas* ε-tubulin mutants failed to localize katanin (Esparza et al. 2013). It is possible that these missing structures help stabilize the centriole, and such a role has been identified for the pericentriolar material (for review, see Werner et al. 2017). Presumably, each of these structures involves a protein directly interacting with the compound microtubules. In the pericentriolar material, a protein that interacts

with both polyglutamylated tubulin and pericentrin, ATF5, has been proposed to link it to the centriole (Madarampalli et al. 2015). The mechanisms by which specificity for doublet or triplet microtubules is achieved are unknown.

Third, the centriole is the essential nucleator of the cilium and determinant of axoneme structure. Every organism that makes cilia has compound microtubules as part of the cilium, even *C. elegans* (Nechipurenko et al. 2017; Serwas et al. 2017). Thus, compound microtubules are likely intrinsic to the function of the cilium. This is especially apparent in motile cilia, in which many proteins complexes associated with motility are attached to the doublet microtubules (Nicastro et al. 2011; Pigino et al. 2012). In addition, the doublets of the axoneme in motile cilia have recently been shown to be double-track "railways" for movement of cargoes (Stepanek and Pigino 2016), and kinesin-2 motors move differently upon standard microtubules and doublet microtubules from motile cilia (Stepp et al. 2017). Axonemes have doublets, never triplets, and it is possible that the triplet microtubules of the centriole are required to distinguish the centriole from the cilium. The cilium assembles and disassembles each cell cycle, but the centriole is stable and is segregated in cell division. Thus, the triplet microtubules may serve as a point of recognition to limit depolymerization, allowing the axoneme to be dynamic and its nucleator stable.

CONCLUSION

Much work has been done on the architecture of the centriole, the basis for the ninefold symmetry of the structure, and the means by which it is duplicated each cell cycle. Less is understood about the compound microtubules found at the centriole, which are also defining properties of the organelle. Here we have considered possible mechanisms for the formation of compound microtubules, and for limiting such microtubule structures to centrioles and axonemes, as well as possible functions imparted by compound microtubules on the structures that have them. Much of this review is necessarily speculative, because, although there is much morphological information, there is little information bearing on mechanism. Future work will focus on the molecular basis for compound microtubule structure, function and regulation. This information will be critical to understanding the centrosome and cilium, key organelles with strong ties to human disease.

ACKNOWLEDGMENTS

We thank members of the Stearns laboratory for comments on the manuscript and helpful discussions. This work was supported by National Research Service Award grant 5 F32 GM117678 to J.T.W. and National Institutes of Health (NIH) grant R01GM052022 to T.S.

REFERENCES

Al-Jassar C, Andreeva A, Barnabas DD, McLaughlin SH, Johnson CM, Yu M, van Breugel M. 2017. The ciliopathy-associated Cep104 protein interacts with tubulin and Nek1 kinase. *Structure* **25:** 146–156.

Al Jord A, Lemaître A-I, Delgehyr N, Faucourt M, Spassky N, Meunier A. 2014. Centriole amplification by mother and daughter centrioles differs in multiciliated cells. *Nature* **516:** 104–107.

Amos LA. 2008. The tektin family of microtubule-stabilizing proteins. *Genome Biol* **9:** 229.

Anderson RG, Brenner RM. 1971. The formation of basal bodies (centrioles) in the Rhesus monkey oviduct. *J Cell Biol* **50:** 10–34.

Azimzadeh J, Hergert P, Delouvée A, Euteneuer U, Formstecher E, Khodjakov A, Bornens M. 2009. hPOC5 is a centrin-binding protein required for assembly of full-length centrioles. *J Cell Biol* **185:** 101–114.

Azimzadeh J, Wong ML, Downhour DM, Sánchez Alvarado A, Marshall WF. 2012. Centrosome loss in the evolution of planarians. *Science* **335:** 461–463.

Baas PW, Lin S. 2011. Hooks and comets: The story of microtubule polarity orientation in the neuron. *Dev Neurobiol* **71:** 403–418.

Balestra FR, von Tobel L, Gönczy P. 2015. Paternally contributed centrioles exhibit exceptional persistence in *C. elegans* embryos. *Cell Res* **25:** 642–644.

Bibring T, Baxandall J, Denslow S, Walker B. 1976. Heterogeneity of the α subunit of tubulin and the variability of tubulin within a single organism. *J Cell Biol* **69:** 301–312.

Bobinnec Y, Khodjakov A, Mir LM, Rieder CL, Eddé B, Bornens M. 1998. Centriole disassembly in vivo and its effect on centrosome structure and function in vertebrate cells. *J Cell Biol* **143:** 1575–1589.

Breslow DK, Hoogendoorn S, Kopp AR, Morgens DW, Vu BK, Han K, Li A, Hess GT, Bassik MC, Chen JK, et al. 2017. A comprehensive portrait of cilia and ciliopathies from a CRISPR-based screen for Hedgehog signaling. *bioRxiv* doi: 10.1101/156059.

Burton PR, Himes RH. 1978. Electron microscope studies of pH effects on assembly of tubulin free of associated proteins. Delineation of substructure by tannic acid staining. *J Cell Biol* **77:** 120–133.

Čajánek L, Nigg EA. 2014. Cep164 triggers ciliogenesis by recruiting Tau tubulin kinase 2 to the mother centriole. *Proc Natl Acad Sci* **111:** E2841–E2850.

Carvalho-Santos Z, Azimzadeh J, Pereira-Leal JB, Bettencourt-Dias M. 2011. Evolution: Tracing the origins of centrioles, cilia, and flagella. *J Cell Biol* **194:** 165–175.

Cavalier-Smith T. 1974. Basal body and flagellar development during the vegetative cell cycle and the sexual cycle of *Chlamydomonas reinhardtii*. *J Cell Sci* **16:** 529–556.

Chang P, Stearns T. 2000. δ-tubulin and ε-tubulin: Two new human centrosomal tubulins reveal new aspects of centrosome structure and function. *Nat Cell Biol* **2:** 30–35.

Chang P, Giddings TH Jr, Winey M, Stearns T. 2002. ε-Tubulin is required for centriole duplication and microtubule organization. *Nat Cell Biol* **5:** 71–76.

Chang C-W, Hsu W-B, Tsai J-J, Tang C-JC, Tang TK. 2016. CEP295 interacts with microtubules and is required for centriole elongation. *J Cell Sci* **129:** 2501–2513.

Chen H-Y, Wu C-T, Tang C-JC, Lin Y-N, Wang W-J, Tang TK. 2017. Human microcephaly protein RTTN interacts with STIL and is required to build full-length centrioles. *Nat Commun* **8:** 247.

Chrétien D, Buendia B, Fuller SD, Karsenti E. 1997. Reconstruction of the centrosome cycle from cryoelectron micrographs. *J Struct Biol* **120:** 117–133.

Comartin D, Gupta GD, Fussner E, Coyaud É, Hasegan M, Archinti M, Cheung SWT, Pinchev D, Lawo S, Raught B, et al. 2013. CEP120 and SPICE1 cooperate with CPAP in centriole elongation. *Curr Biol* **23:** 1360–1366.

Cormier A, Clément M-J, Knossow M, Lachkar S, Savarin P, Toma F, Sobel A, Gigant B, Curmi PA. 2009. The PN2-3 domain of centrosomal P4.1-associated protein implements a novel mechanism for tubulin sequestration. *J Biol Chem* **284:** 6909–6917.

Delgehyr N, Rangone H, Fu J, Mao G, Tom B, Riparbelli MG, Callaini G, Glover DM. 2012. Klp10A, a microtubule-depolymerizing kinesin-13, cooperates with CP110 to control *Drosophila* centriole length. *Curr Biol* **22:** 502–509.

Dippell RV. 1968. The development of basal bodies in paramecium. *Proc Natl Acad Sci* **61:** 461–468.

Dupuis-Williams P, Fleury-Aubusson A, De Loubresse NG, Geoffroy H, Vayssié L, Galvani A, Espigat A, Rossier J. 2002. Functional role of ε-tubulin in the assembly of the centriolar microtubule scaffold. *J Cell Biol* **158:** 1183–1193.

Dutcher SK, Trabuco EC. 1998. The UNI3 gene is required for assembly of basal bodies of *Chlamydomonas* and encodes δ-tubulin, a new member of the tubulin superfamily. *Mol Biol Cell* **9:** 1293–1308.

Dutcher SK, Morrissette NS, Preble AM, Rackley C, Stanga J. 2002. ε-Tubulin is an essential component of the centriole. *Mol Biol Cell* **13:** 3859–3869.

Esparza JM, O'Toole E, Li L, Giddings TH, Kozak B, Albee AJ, Dutcher SK. 2013. Katanin localization requires triplet microtubules in *Chlamydomonas reinhardtii*. *PLoS One* **8:** e53940.

Euteneuer U, McIntosh JR. 1980. Polarity of midbody and phragmoplast microtubules. *J Cell Biol* **87:** 509–515.

Fackenthal JD, Turner FR, Raff EC. 1993. Tissue-specific microtubule functions in *Drosophila* spermatogenesis require the β 2-tubulin isotype-specific carboxy terminus. *Dev Biol* **158:** 213–227.

Findeisen P, Mühlhausen S, Dempewolf S, Hertzog J, Zietlow A, Carlomagno T, Kollmar M. 2014. Six subgroups and extensive recent duplications characterize the evolution of the eukaryotic tubulin protein family. *Genome Biol Evol* **6:** 2274–2288.

Firat-Karalar EN, Stearns T. 2014. The centriole duplication cycle. *Philos Trans R Soc Lond B Biol Sci* **369:** 20130460.

Fong CS, Kim M, Yang TT, Liao JC, Tsou MFB. 2014. SAS-6 assembly templated by the lumen of cartwheel-less centrioles precedes centriole duplication. *Dev Cell* **30:** 238–245.

Franz A, Roque H, Saurya S, Dobbelaere J, Raff JW. 2013. CP110 exhibits novel regulatory activities during centriole assembly in *Drosophila*. *J Cell Biol* **203:** 785–799.

Fromherz S, Giddings TH Jr, Gomez-Ospina N, Dutcher SK. 2004. Mutations in α-tubulin promote basal body maturation and flagellar assembly in the absence of δ-tubulin. *J Cell Sci* **117**(Pt 2)**:** 303–314.

Fulton C, Dingle AD. 1971. Basal bodies, but not centrioles, in *Naegleria*. *J Cell Biol* **51:** 826–836.

Gadadhar S, Bodakuntla S, Natarajan K, Janke C. 2017. The tubulin code at a glance. *J Cell Sci* **130:** 1347–1353.

Gadelha C, Wickstead B, McKean PG, Gull K. 2006. Basal body and flagellum mutants reveal a rotational constraint of the central pair microtubules in the axonemes of trypanosomes. *J Cell Sci* **119**(Pt 12)**:** 2405–2413.

Garreau de Loubresse N, Ruiz F, Beisson J, Klotz C. 2001. Role of δ-tubulin and the C-tubule in assembly of *Paramecium* basal bodies. *BMC Cell Biol* **2:** 4.

Gifford EM Jr, Lin J. 1975. Light microscope and ultrastructural studies of the male gametophyte in Ginkgo biloba: The spermatogenous cell. *Am J Bot* **62:** 974–981.

Goetz SC, Liem KF Jr, Anderson KV. 2012. The spinocerebellar ataxia-associated gene Tau tubulin kinase 2 controls the initiation of ciliogenesis. *Cell* **151:** 847–858.

Goodenough UW, St Clair HS. 1975. BALD-2: A mutation affecting the formation of doublet and triplet sets of microtubules in *Chlamydomonas reinhardtii*. *J Cell Biol* **66:** 480–491.

Gottardo M, Callaini G, Riparbelli MG. 2015. The *Drosophila* centriole—Conversion of doublets into triplets within the stem cell niche. *J Cell Sci* **128:** 2437–2442.

Guichard P, Chrétien D, Marco S, Tassin A-M. 2010. Procentriole assembly revealed by cryo-electron tomography. *EMBO J* **29:** 1565–1572.

Guichard P, Hachet V, Majubu N, Neves A, Demurtas D, Olieric N, Fluckiger I, Yamada A, Kihara K, Nishida Y, et al. 2013. Native architecture of the centriole proximal region reveals features underlying its 9-fold radial symmetry. *Curr Biol* **23:** 1620–1628.

Guichard P, Hamel V, Le Guennec M, Banterle N, Iacovache I, Nemčíková V, Flückiger I, Goldie KN, Stahlberg H, Lévy D, et al. 2017. Cell-free reconstitution reveals centriole cartwheel assembly mechanisms. *Nat Commun* **8**: 14813.

Hamel V, Steib E, Hamelin R, Armand F, Borgers S, Flückiger I, Busso C, Olieric N, Sorzano COS, Steinmetz MO, et al. 2017. Identification of *Chlamydomonas* central core centriolar proteins reveals a role for human WDR90 in ciliogenesis. *Curr Biol* **27**: 2486–2498.e6.

Heidemann SR, McIntosh JR. 1980. Visualization of the structural polarity of microtubules. *Nature* **286**: 517–519.

Hein MY, Hubner NC, Poser I, Cox J, Nagaraj N, Toyoda Y, Gak IA, Weisswange I, Mansfeld J, Buchholz F, et al. 2015. A human interactome in three quantitative dimensions organized by stoichiometries and abundances. *Cell* **163**: 712–723.

Hilbert M, Noga A, Frey D, Hamel V, Guichard P, Kraatz SHW, Pfreundschuh M, Hosner S, Flückiger I, Jaussi R, et al. 2016. SAS-6 engineering reveals interdependence between cartwheel and microtubules in determining centriole architecture. *Nat Cell Biol* **18**: 393–403.

Hiraki M, Nakazawa Y, Kamiya R, Hirono M. 2007. Bld10p constitutes the cartwheel-spoke tip and stabilizes the 9-fold symmetry of the centriole. *Curr Biol* **17**: 1778–1783.

Hirono M. 2014. Cartwheel assembly. *Philos Trans R Soc Lond B Biol Sci* **369**: 20130458.

Hoyle HD, Raff EC. 1990. Two *Drosophila* β tubulin isoforms are not functionally equivalent. *J Cell Biol* **111**: 1009–1026.

Hsu W-B, Hung L-Y, Tang C-JC, Su C-L, Chang Y, Tang TK. 2008. Functional characterization of the microtubule-binding and -destabilizing domains of CPAP and d-SAS-4. *Exp Cell Res* **314**: 2591–2602.

Ichikawa M, Liu D, Kastritis PL, Basu K, Hsu TC, Yang S, Bui KH. 2017. Subnanometre-resolution structure of the doublet microtubule reveals new classes of microtubule-associated proteins. *Nat Commun* **8**: 15035.

Inclán YF, Nogales E. 2001. Structural models for the self-assembly and microtubule interactions of γ-, δ- and ε-tubulin. *J Cell Sci* **114**(Pt 2): 413–422.

Joo K, Kim CG, Lee M-S, Moon H-Y, Lee S-H, Kim MJ, Kweon H-S, Park W-Y, Kim C-H, Gleeson JG, et al. 2013. CCDC41 is required for ciliary vesicle docking to the mother centriole. *Proc Natl Acad Sci* **110**: 5987–5992.

Kalnins VI, Porter KR. 1969. Centriole replication during ciliogenesis in the chick tracheal epithelium. *Z Zellforsch Mikrosk Anat* **100**: 1–30.

Kemphues KJ, Kaufman TC, Raff RA, Raff EC. 1982. The testis-specific β-tubulin subunit in *Drosophila melanogaster* has multiple functions in spermatogenesis. *Cell* **31**(3 Pt 2): 655–670.

Khodjakov A, Rieder CL, Sluder G, Cassels G, Sibon O, Wang CL. 2002. De novo formation of centrosomes in vertebrate cells arrested during S phase. *J Cell Biol* **158**: 1171–1181.

Kikkawa M, Ishikawa T, Nakata T, Wakabayashi T, Hirokawa N. 1994. Direct visualization of the microtubule lattice seam both in vitro and in vivo. *J Cell Biol* **127**(6 Pt 2): 1965–1971.

Kinoshita K, Arnal I, Desai A, Drechsel DN, Hyman AA. 2001. Reconstitution of physiological microtubule dynamics using purified components. *Science* **294**: 1340–1343.

Kitagawa D, Vakonakis I, Olieric N, Hilbert M, Keller D, Olieric V, Bortfeld M, Erat MC, Flückiger I, et al. 2011. Structural basis of the 9-fold symmetry of centrioles. *Cell* **144**: 364–375.

Kleylein-Sohn J, Westendorf J, Le Clech M, Habedanck R, Stierhof YD, Nigg EA. 2007. Plk4-induced centriole biogenesis in human cells. *Dev Cell* **13**: 190–202.

Kobayashi T, Dynlacht BD. 2011. Regulating the transition from centriole to basal body. *J Cell Biol* **193**: 435–444.

Kobayashi T, Tsang WY, Li J, Lane W, Dynlacht BD. 2011. Centriolar kinesin Kif24 interacts with CP110 to remodel microtubules and regulate ciliogenesis. *Cell* **145**: 914–925.

Kobayashi T, Kim S, Lin Y-C, Inoue T, Dynlacht BD. 2014. The CP110-interacting proteins Talpid3 and Cep290 play overlapping and distinct roles in cilia assembly. *J Cell Biol* **204**: 215–229.

Kochanski RS, Borisy GG. 1990. Mode of centriole duplication and distribution. *J Cell Biol* **110**: 1599–1605.

Kohlmaier G, Lončarek J, Meng X, McEwen BF, Mogensen MM, Spektor A, Dynlacht BD, Khodjakov A, Gönczy P. 2009. Overly long centrioles and defective cell division upon excess of the SAS-4-related protein CPAP. *Curr Biol* **19**: 1012–1018.

Kraatz S, Guichard P, Obbineni JM, Olieric N, Hatzopoulos GN, Hilbert M, Sen I, Missimer J, Gönczy P, Steinmetz MO. 2016. The human centriolar protein CEP135 contains a two-stranded coiled-coil domain critical for microtubule binding. *Structure* **24**: 1358–1371.

Kuriyama R. 1976. In vitro polymerization of flagellar and ciliary outer fiber tubulin into microtubules. *J Biochem* **80**: 153–165.

Kuriyama R, Borisy GG. 1981. Centriole cycle in Chinese hamster ovary cells as determined by whole-mount electron microscopy. *J Cell Biol* **91**(3 Pt 1): 814–821.

Lai EY, Walsh C, Wardell D, Fulton C. 1979. Programmed appearance of translatable flagellar tubulin mRNA during cell differentiation in *Naegleria*. *Cell* **17**: 867–878.

Lambrus BG, Uetake Y, Clutario KM, Daggubati V, Snyder M, Sluder G, Holland AJ. 2015. P53 protects against genome instability following centriole duplication failure. *J Cell Biol* **210**: 63–77.

La Terra S, English CN, Hergert P, McEwen BF, Sluder G, Khodjakov A. 2005. The de novo centriole assembly pathway in HeLa cells: Cell cycle progression and centriole assembly/maturation. *J Cell Biol* **168**: 713–722.

Li S, Fernandez J-J, Marshall WF, Agard DA. 2012. Three-dimensional structure of basal body triplet revealed by electron cryo-tomography. *EMBO J* **31**: 552–562.

Lin Y-C, Chang C-W, Hsu W-B, Tang C-JC, Lin Y-N, Chou E-J, Wu C-T, Tang TK. 2013a. Human microcephaly protein CEP135 binds to hSAS-6 and CPAP, and is required for centriole assembly. *EMBO J* **32**: 1141–1154.

Lin YN, Wu CT, Lin YC, Hsu WB, Tang CJC, Chang CW, Tang TK. 2013b. CEP120 interacts with CPAP and positively regulates centriole elongation. *J Cell Biol* **202**: 211–219.

Linck RW, Stephens RE. 2007. Functional protofilament numbering of ciliary, flagellar, and centriolar microtubules. *Cell Motil Cytoskeleton* **64**: 489–495.

Linck R, Fu X, Lin J, Ouch C, Schefter A, Steffen W, Warren P, Nicastro D. 2014. Insights into the structure and function of ciliary and flagellar doublet microtubules: Tektins, Ca^{2+}-binding proteins, and stable protofilaments. *J Biol Chem* **289**: 17427–17444.

Lu Q, Insinna C, Ott C, Stauffer J, Pintado PA, Rahajeng J, Baxa U, Walia V, Cuenca A, Hwang Y-S, et al. 2015. Early steps in primary cilium assembly require EHD1/EHD3-dependent ciliary vesicle formation. *Nat Cell Biol* **17**: 531.

Madarampalli B, Yuan Y, Liu D, Lengel K, Xu Y, Li G, Yang J, Liu X, Lu Z, Liu DX. 2015. ATF5 connects the pericentriolar materials to the proximal end of the mother centriole. *Cell* **162**: 580–592.

Maheshwari A, Obbineni JM, Bui KH, Shibata K, Toyoshima YY, Ishikawa T. 2015. α- and β-tubulin lattice of the axonemal microtubule doublet and binding proteins revealed by single particle cryo-electron microscopy and tomography. *Structure* **23**: 1584–1595.

Matsuura K, Lefebvre PA, Kamiya R, Hirono M. 2004. Bld10p, a novel protein essential for basal body assembly in *Chlamydomonas*: Localization to the cartwheel, the first ninefold symmetrical structure appearing during assembly. *J Cell Biol* **165**: 663–671.

McIntosh JR, Euteneuer U. 1984. Tubulin hooks as probes for microtubule polarity: An analysis of the method and an evaluation of data on microtubule polarity in the mitotic spindle. *J Cell Biol* **98**: 525–533.

McIntosh JR, Morphew MK, Grissom PM, Gilbert SP, Hoenger A. 2009. Lattice structure of cytoplasmic microtubules in a cultured Mammalian cell. *J Mol Biol* **394**: 177–182.

Meehl JB, Bayless BA, Giddings TH Jr, Pearson CG, Winey M. 2016. Tetrahymena Poc1 ensures proper intertriplet microtubule linkages to maintain basal body integrity. *Mol Biol Cell* **27:** 2394–2403.

Nakazawa Y, Hiraki M, Kamiya R, Hirono M. 2007. SAS-6 is a cartwheel protein that establishes the 9-fold symmetry of the centriole. *Curr Biol* **17:** 2169–2174.

Nechipurenko IV, Berciu C, Sengupta P, Nicastro D. 2017. Centriolar remodeling underlies basal body maturation during ciliogenesis in *Caenorhabditis elegans*. *Elife* **6:** e25686.

Nicastro D, Fu X, Heuser T, Tso A, Porter ME, Linck RW. 2011. Cryo-electron tomography reveals conserved features of doublet microtubules in flagella. *Proc Natl Acad Sci* **108:** E845–E853.

Nielsen MG, Turner FR, Hutchens JA, Raff EC. 2001. Axoneme-specific β-tubulin specialization: A conserved C-terminal motif specifies the central pair. *Curr Biol* **11:** 529–533.

Nogales E, Wolf SG, Downing KH. 1998. Structure of the α β tubulin dimer by electron crystallography. *Nature* **391:** 199–203.

Nogales E, Whittaker M, Milligan RA, Downing KH. 1999. High-resolution model of the microtubule. *Cell* **96:** 79–88.

Norstog K. 1967. Fine structure of the spermatozoid of Zamia with special reference to the flagellar apparatus. *Am J Bot* **54:** 831–840.

Paintrand M, Moudjou M, Delacroix H, Bornens M. 1992. Centrosome organization and centriole architecture: Their sensitivity to divalent cations. *J Struct Biol* **108:** 107–128.

Pelletier L, O'Toole E, Schwager A, Hyman AA, Müller-Reichert T. 2006. Centriole assembly in *Caenorhabditis elegans*. *Nature* **444:** 619–623.

Pigino G, Maheshwari A, Bui KH, Shingyoji C, Kamimura S, Ishikawa T. 2012. Comparative structural analysis of eukaryotic flagella and cilia from *Chlamydomonas, Tetrahymena*, and sea urchins. *J Struct Biol* **178:** 199–206.

Raff EC, Hutchens JA, Hoyle HD, Nielsen MG, Turner FR. 2000. Conserved axoneme symmetry altered by a component β-tubulin. *Curr Biol* **10:** 1391–1394.

Redeker V. 2010. Mass spectrometry analysis of C-terminal posttranslational modifications of tubulins. *Methods Cell Biol* **95:** 77–103.

Ross I, Clarissa C, Giddings TH Jr, Winey M. 2013. ε-tubulin is essential in *Tetrahymena thermophila* for the assembly and stability of basal bodies. *J Cell Sci* **126:** 3441–3451.

Sánchez I, Dynlacht BD. 2016. Cilium assembly and disassembly. *Nat Cell Biol* **18:** 711–717.

Schmidt TI, Kleylein-Sohn J, Westendorf J, Le Clech M, Lavoie SB, Stierhof YD, Nigg EA. 2009. Control of centriole length by CPAP and CP110. *Curr Biol* **19:** 1005–1011.

Schmidt KN, Kuhns S, Neuner A, Hub B, Zentgraf H, Pereira G. 2012. Cep164 mediates vesicular docking to the mother centriole during early steps of ciliogenesis. *J Cell Biol* **199:** 1083–1101.

Serrano L, de la Torre J, Maccioni RB, Avila J. 1984. Involvement of the carboxyl-terminal domain of tubulin in the regulation of its assembly. *Proc Natl Acad Sci* **81:** 5989–5993.

Serwas D, Su TY, Roessler M, Wang S, Dammermann A. 2017. Centrioles initiate cilia assembly but are dispensable for maturation and maintenance in *C. elegans*. *J Cell Biol* **216:** 1659–1671.

Sharma A, Aher A, Dynes NJ, Frey D, Katrukha EA, Jaussi R, Grigoriev I, Croisier M, Kammerer RA, Akhmanova A. 2016. Centriolar CPAP/SAS-4 imparts slow processive microtubule growth. *Dev Cell* **37:** 362–376.

Sillibourne JE, Hurbain I, Grand-Perret T, Goud B, Tran P, Bornens M. 2013. Primary ciliogenesis requires the distal appendage component Cep123. *Biol Open* **2:** 535–545.

Singla V, Romaguera-Ros M, Garcia-Verdugo JM, Reiter JF. 2010. Ofd1, a human disease gene, regulates the length and distal structure of centrioles. *Dev Cell* **18:** 410–424.

Song YH, Mandelkow E. 1993. Recombinant kinesin motor domain binds to β-tubulin and decorates microtubules with a B surface lattice. *Proc Natl Acad Sci* **90:** 1671–1675.

Song YH, Mandelkow E. 1995. The anatomy of flagellar microtubules: Polarity, seam, junctions, and lattice. *J Cell Biol* **128:** 81–94.

Sorokin S. 1962. Centrioles and the formation of rudimentary cilia by fibroblasts and smooth muscle cells. *J Cell Biol* **15:** 363–377.

Sorokin SP. 1968. Reconstructions of centriole formation and ciliogenesis in mammalian lungs. *J Cell Sci* **3:** 207–230.

Spektor A, Tsang WY, Khoo D, Dynlacht BD. 2007. Cep97 and CP110 suppress a cilia assembly program. *Cell* **130:** 678–690.

Steinman RM. 1968. An electron microscopic study of ciliogenesis in developing epidermis and trachea in the embryo of *Xenopus laevis*. *Am J Anat* **122:** 19–55.

Stepanek L, Pigino G. 2016. Microtubule doublets are double-track railways for intraflagellar transport trains. *Science* **352:** 721–724.

Stephens RE. 1970. Thermal fractionation of outer fiber doublet microtubules into A- and B-subfiber components. A- and B-tubulin. *J Mol Biol* **47:** 353–363.

Stepp WL, Merck G, Mueller-Planitz F, Ökten Z. 2017. Kinesin-2 motors adapt their stepping behavior for processive transport on axonemes and microtubules. *EMBO Rep* **18:** 1947–1956.

Sui H, Downing KH. 2006. Molecular architecture of axonemal microtubule doublets revealed by cryo-electron tomography. *Nature* **442:** 475–478.

Sui H, Downing KH. 2010. Structural basis of interprotofilament interaction and lateral deformation of microtubules. *Structure* **18:** 1022–1031.

Tang C-JC, Fu R-H, Wu K-S, Hsu W-B, Tang TK. 2009. CPAP is a cell-cycle regulated protein that controls centriole length. *Nat Cell Biol* **11:** 825–831.

Tanos BE, Yang H-J, Soni R, Wang W-J, Macaluso FP, Asara JM, Tsou M-FB. 2013. Centriole distal appendages promote membrane docking, leading to cilia initiation. *Genes Dev* **27:** 163–168.

Tilney LG, Bryan J, Bush DJ, Fujiwara K, Mooseker MS, Murphy DB, Snyder DH. 1973. Microtubules: Evidence for 13 protofilaments. *J Cell Biol* **59**(2 Pt 1)**:** 267–275.

Tsang WY, Spektor A, Luciano DJ, Indjeian VB, Chen Z, Salisbury JL, Sánchez I, Dynlacht BD. 2006. CP110 cooperates with two calcium-binding proteins to regulate cytokinesis and genome stability. *Mol Biol Cell* **17:** 3423–3434.

Tsang WY, Bossard C, Khanna H, Peränen J, Swaroop A, Malhotra V, Dynlacht BD. 2008. CP110 suppresses primary cilia formation through its interaction with CEP290, a protein deficient in human ciliary disease. *Dev Cell* **15:** 187–197.

Tsang WY, Spektor A, Vijayakumar S, Bista BR, Li J, Sanchez I, Duensing S, Dynlacht BD. 2009. Cep76, a centrosomal protein that specifically restrains centriole reduplication. *Dev Cell* **16:** 649–660.

Turk E, Wills AA, Kwon T, Sedzinski J, Wallingford JB, Stearns T. 2015. ζ-tubulin is a member of a conserved tubulin module and is a component of the centriolar basal foot in multiciliated cells. *Curr Biol* **25:** 2177–2183.

Uetake Y, Lončarek J, Nordberg JJ, English CN, La Terra S, Khodjakov A, Sluder G. 2007. Cell cycle progression and de novo centriole assembly after centrosomal removal in untransformed human cells. *J Cell Biol* **176:** 173–182.

van Breugel M, Hirono M, Andreeva A, Yanagisawa H-A, Yamaguchi S, Nakazawa Y, Morgner N, Petrovich M, Ebong I-O, Robinson CV, et al. 2011. Structures of SAS-6 suggest its organization in centrioles. *Science* **331:** 1196–1199.

Vemu A, Atherton J, Spector JO, Moores CA, Roll-Mecak A. 2017. Tubulin isoform composition tunes microtubule dynamics. *Mol Biol Cell* **28:** 3564–3572.

Vorobjev IA, Chentsov Y. 1982. Centrioles in the cell cycle. I. Epithelial cells. *J Cell Biol* **93:** 938–949.

Wang W-J, Acehan D, Kao C-H, Jane W-N, Uryu K, Tsou M-FB. 2015. De novo centriole formation in human cells is error-prone and does not require SAS-6 self-assembly. *Elife* **4:** e10586.

Wang JT, Kong D, Hoerner CR, Loncarek J, Stearns T. 2017. Centriole triplet microtubules are required for stable centriole formation and inheritance in human cells. *Elife* **6:** e29061.

Werner S, Pimenta-Marques A, Bettencourt-Dias M. 2017. Maintaining centrosomes and cilia. *J Cell Sci* **130:** 3789–3800.

Wloga D, Gaertig J. 2010. Post-translational modifications of microtubules. *J Cell Sci* **123:** 3447–3455.

Wong YL, Anzola JV, Davis RL, Yoon M, Motamedi A, Kroll A, Seo CP, Hsia JE, Kim SK, Mitchell JW, et al. 2015. Cell biology. Reversible centriole depletion with an inhibitor of Polo-like kinase 4. *Science* **348:** 1155–1160.

Woodruff JB, Ferreira Gomes B, Widlund PO, Mahamid J, Honigmann A, Hyman AA. 2017. The centrosome is a selective condensate that nucleates microtubules by concentrating tubulin. *Cell* **169:** 1066–1077.e10.

Yu I, Garnham CP, Roll-Mecak A. 2015. Writing and reading the tubulin code. *J Biol Chem* **290:** 17163–17172.

Zheng X, Ramani A, Soni K, Gottardo M, Zheng S, Gooi LM, Li W, Feng S, Mariappan A, Wason A, et al. 2016. Molecular basis for CPAP-tubulin interaction in controlling centriolar and ciliary length. *Nat Commun* **7:** 11874.

Mitotic Chromosome Assembly In Vitro: Functional Cross Talk between Nucleosomes and Condensins

KEISHI SHINTOMI AND TATSUYA HIRANO

Chromosome Dynamics Laboratory, RIKEN, Wako, Saitama 351-0198, Japan

Correspondence: hiranot@riken.jp

The mitotic chromosome is a macromolecular assembly that ensures error-free transmission of the genome during cell division. It has long been a big mystery how long stretches of DNA might be folded into rod-shaped chromosomes or how such an elaborate process might be accomplished at a mechanistic level. Cell-free extracts made from frog eggs offer a unique opportunity to address these questions by enabling mitotic chromosomes to be assembled in a test tube. Moreover, the core part of the chromosome assembly reaction can now be reconstituted with a limited number of purified factors. A combination of these in vitro assays makes it possible not only to prepare a complete list of proteins required for chromosome assembly but also to dissect functions of individual proteins and their cooperation with unparalleled clarity. Emerging lines of evidence underscore the paramount importance of condensins in building mitotic chromosomes and shed new light on the functional cross talk between nucleosomes and condensins in this process.

In eukaryotic cells, chromatin undergoes a series of dynamic structural changes throughout the cell cycle, which culminate in the assembly of chromosomes during mitosis. The dramatic transformation of an amorphous mass of chromatin into a discrete set of rod-shaped chromosomes is thought to be an essential process for faithful segregation of the genome during cell division. Among various experimental systems for studying mitotic chromosome organization, the cell-free system derived from *Xenopus* egg extracts is unique in the sense that the whole process of mitotic chromosome assembly is recapitulated in a test tube (e.g., Hirano and Mitchison 1993). Remarkably, at least the core part of the chromosome assembly reaction achieved in the cell-free system can now be reconstituted using purified factors in vitro (Shintomi et al. 2015). We will summarize what we have learned from these in vitro assays and what we will need to learn by extending these efforts.

IDENTIFICATION OF CONDENSINS USING *XENOPUS* EGG EXTRACTS

The cell cycle of unfertilized eggs of the frog *Xenopus laevis* is naturally arrested at metaphase of meiosis II, where the activity of cyclin B–Cdk1 is kept at a high level. Upon fertilization, a transient increase of cytosolic calcium ions (Ca^{2+}) triggers the degradation of cyclin B, resulting in mitotic exit. Therefore, when the eggs are crushed by centrifugation in a buffer containing the Ca^{2+}-chelator EGTA, a metaphase egg extract can be obtained (Lohka and Maller 1985; Murray 1991). When demembranated *Xenopus* sperm nuclei are incubated with this metaphase extract, they quickly swell and then turn into a cluster of fibrous structures. These chromatin fibers progressively get thickened and individualized and are eventually converted into rod-shaped chromatids (Fig. 1A; Hirano and Mitchison 1993). Because the chromatin does not undergo DNA replication in this experimental setup, the resulting structures are composed of "single" chromatids. Isolation of these chromatids from the reaction mixture by single-step centrifugation led to the identification of their major proteinaceous components, collectively referred to as *Xenopus* chromosome-associated polypeptides (XCAPs). The composition of XCAPs turned out to be surprisingly simple, being composed of the core and linker histones, topoisomerase II (topo II), and five subunits of condensin I (Hirano and Mitchison 1994; Hirano et al. 1997). Following these pioneering studies, a second condensin complex, termed condensin II, was identified in *Xenopus* egg extracts although it was less abundant than condensin I (Ono et al. 2003). It is now known that the two condensin complexes distribute widely in many if not all eukaryotic cells (Hirano 2016; Uhlmann 2016; Kalitsis et al. 2017).

One of the most powerful applications of this cell-free system is the so-called immunodepletion assay. In this assay, a protein of interest is depleted from the extracts using its specific antibody, and then substrate nuclei are added to test how loss of that particular protein affects the chromatid assembly reaction. This in vitro immunodepletion approach contrasts various in vivo depletion methods (e.g., siRNA-mediated and drug-induced transcriptional repression), often producing "clean" defective phenotypes that enable researchers to make their straightforward interpretations and to draw solid conclusions. For example, when sperm nuclei were incubated with egg extracts depleted of both condensin I and condensin II, they completely failed to be transformed into mitotic chromatids, resulting in amorphous cloud-like masses of chroma-

Published by Cold Spring Harbor Laboratory Press; doi: 10.1101/sqb.2017.82.033639

Cold Spring Harbor Symposia on Quantitative Biology, Volume LXXXII

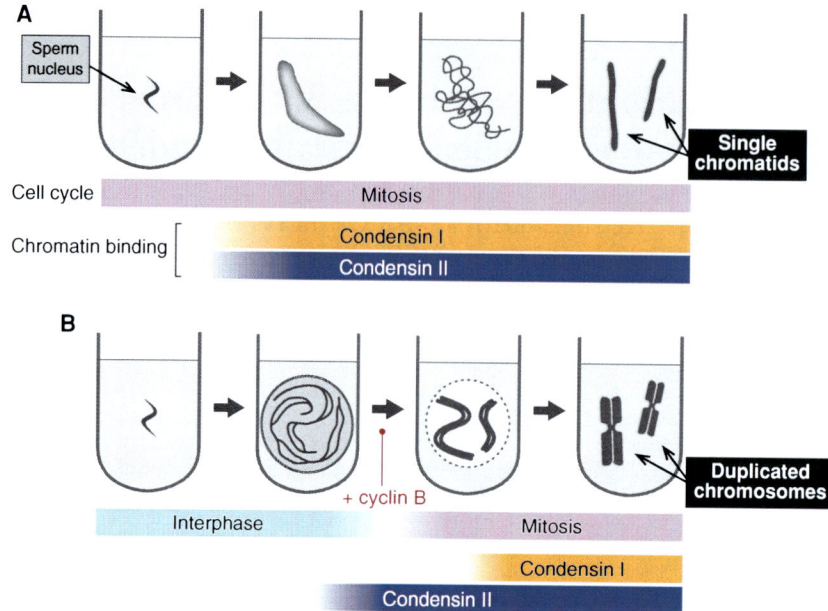

Figure 1. Two different cell-free assays for chromosome assembly in *Xenopus* egg extracts. (*A*) When demembranated *Xenopus* sperm nuclei are directly incubated with a mitotic extract, they are transformed into single chromatids without the process of DNA replication. (*B*) Sperm nuclei are first incubated with an interphase extract to allow nuclear envelope formation and DNA replication. Cyclin B is then added to convert the cell cycle state of the extract into mitosis, resulting in the assembly of duplicated chromosomes. Temporal orders of chromatin binding of condensins I and II in each assay are shown at the *bottom* of each panel.

tin (Hirano et al. 1997; Wignall et al. 2003; Takemoto et al. 2006; Shintomi and Hirano 2011). This result convincingly demonstrated that condensins I and II are indispensable for chromatid assembly during mitosis. Furthermore, individual depletion experiments showed that condensins I and II play nonoverlapping functions although the contribution of condensin I is dominant over that of condensin II in this cell-free system (Ono et al. 2003; Shintomi and Hirano 2011).

BALANCING ACTS OF CONDENSINS I AND II

Xenopus egg cell-free extracts can also be used to assemble "duplicated" chromosomes, in which a pair of sister chromatids are juxtaposed with each other. To this end, unfertilized eggs are treated with a calcium ionophore to mimic fertilization, and they are crushed by centrifugation to prepare an "interphase" extract (Shintomi and Hirano 2017). When sperm nuclei are incubated with this interphase extract, the nuclear envelope is assembled around them, in which a single round of DNA replication takes place and the physical linkage between replicated chromatids (i.e., cohesion) is created by the action of cohesin (Fig. 1B). Addition of cyclin B into the reaction mixture then triggers entry into mitosis, where bulk cohesin is dissociated from chromosome arms and condensin II accumulates on each chromatid axis. Upon nuclear envelope breakdown, condensin I gains access to chromatin and promotes the assembly of fully compacted chromosomes (Shintomi and Hirano 2011). This spatiotemporal regulation of condensins is reminiscent of that observed in mammalian tissue culture cells (Hirota et al. 2004; Ono et al. 2004; Gerlich et al. 2006). Thus, the duplicated chromo-

some assembly assay represents a more physiologically relevant reaction than the single chromatid assembly assay where the ordered action of condensins I and II cannot be assured (Fig. 1A).

Immunodepletion experiments clearly demonstrated that condensin I makes a greater contribution to chromosome assembly compared with condensin II in the duplicated chromosome assembly assay, as had been shown in the single chromatid assembly assay (Shintomi and Hirano 2011). To gain further insight into the functional differences between the two complexes, a protocol was devised that made it possible to precisely manipulate the levels of condensins I and II in the egg extracts. For example, when the original ratio of condensin I to II (5:1) in the extracts was changed to 1:1, shorter and thicker chromatids were assembled. This series of manipulation experiments indicated that condensin II primarily acts to promote axial shortening of chromatids, whereas condensin I acts to compact them laterally. Thus, the shape of chromosomes is determined by an exquisite balance between condensins I and II. This conclusion has been supported by a subsequent in vivo depletion study using chicken DT40 cells (Green et al. 2012).

CONDENSIN-MEDIATED CHROMATID AXIS FORMATION

How do condensins contribute to the assembly of rod-shaped chromatids? Condensins I and II were enriched along the central axes of chromatids (Ono et al. 2003), and perturbation of condensin functions by antibody addition converted rod-shaped chromatids into random-coiled chromatin masses (Hirano and Mitchison 1994). These

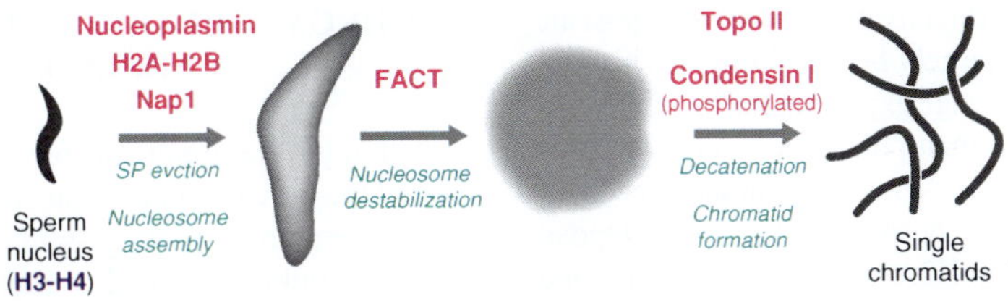

Figure 2. Reconstitution of mitotic chromatids with six purified factors. Deduced roles for each factor in the whole reaction and the resulting (intermediate) structures of chromatin are indicated. In short, nucleoplasmin evicts sperm-specific proteins (SPs) from sperm chromatin, and Nap1 loads histone H2A–H2B onto preexisting histone H3–H4. FACT destabilizes nucleosomes to support subsequent actions of topo II and condensin I, the latter of which needs to be phosphorylated by cyclin B–Cdk1.

observations suggested that the formation of condensin-positive axes is a key step for linear organization of mitotic chromatids. This notion was addressed by a recent study using a panel of recombinant condensin I complexes in the single chromatid assembly assay (Kinoshita et al. 2015). Although the wild-type holocomplex efficiently rescued defective phenotypes observed in an extract depleted of endogenous condensins, ATPase mutant complexes failed to do so, demonstrating that the ATP-binding and hydrolysis cycle is essential for proper actions of condensin I. Interestingly, a mutant "subcomplex" lacking CAP-G, one of the two HEAT-repeat subunits, produced highly characteristic shape of chromatids, in which a discrete axis positive for this subcomplex was surrounded by hazy chromatin loops. In contrast, another subcomplex lacking CAP-D2 produced a poorly organized chromatin structure with no axes. Together with other results, it was proposed that the CAP-G and CAP-D2 subunits of condensin I have seemingly antagonistic impacts on chromatid axis formation, and that balancing actions of these two subunits support proper assembly of mitotic chromatids. More recent experiments under a different setup have demonstrated that condensin II also participates in chromatid axis assembly by collaborating with condensin I (Shintomi et al. 2017), details of which will be discussed later.

Topo II was also enriched along chromatid axes. Although depletion of topo II impaired the early step of the single chromatid assembly process, it was possible to remove this protein from chromatids without disrupting their overall morphology once their assembly was complete (Hirano and Mitchison 1993). Moreover, although budding yeast topo II could functionally rescue a *Xenopus* egg extract depleted of endogenous topo II, it failed to display continuous, axial localization along chromatids (Shintomi et al. 2015). Thus, although topo II plays a vital role in ongoing chromatid assembly process (most likely by catalyzing removal of entanglements between chromatids), its contribution to the structural maintenance of chromatid axes remains not fully understood.

RECONSTITUTION OF MITOTIC CHROMATIDS WITH PURIFIED FACTORS

As mentioned above, the single chromatids assembled in mitotic egg extracts had exhibited a very simple protein composition (Hirano and Mitchison 1994). Importantly, no factor other than condensin I or topo II had been shown to be required for chromatid assembly in this cell-free

system. It should be also noted that the linker histone B4 was found to be dispensable for such a reaction (Ohsumi et al. 1993; Maresca et al. 2005). These pieces of information prompted us to recapitulate the assembly reaction using a limited number of purified factors in vitro. *Xenopus* sperm nuclei contained core histone H3-H4 and highly basic, sperm-specific proteins (SPs) (Shechter et al. 2009b). Previous studies had shown that nucleoplasmin removes SPs from sperm nuclei (Ohsumi and Katagiri 1991; Philpott and Leno 1992) and that the histone chaperone Nap1 loads core histone H2A–H2B onto preexisting histone H3–H4 to assemble octameric nucleosomes (Shintomi et al. 2005). We therefore initiated our attempt by mixing *Xenopus* sperm nuclei with a cocktail of five factors (H2A–H2B, nucleoplasmin, Nap1, topo II, and condensin I), which turned out to be unsuccessful. We then biochemically fractionated egg extracts and identified the missing factor as another histone chaperone FACT (Shintomi et al. 2015), thereby establishing a protocol in which mitotic chromatids could be reconstituted in vitro with only six purified factors (Fig. 2).

The reconstitution system was instrumental in further addressing cell-cycle regulation of mitotic chromatid assembly. Five out of the six factors used were recombinant proteins, and no considerations had been given for their cell cycle–dependent modifications. The only exception was condensin I, which had been purified as a phosphorylated form from mitotic egg extracts. When a nonphosphorylated form of condensin I purified from interphase egg extracts was used instead, chromatids failed to be reconstituted. Remarkably, addition of the mitotic kinase cyclin B–Cdk1 fully restored the reconstitution reaction (Shintomi et al. 2015), demonstrating that the Cdk1 phosphorylation of condensin I is the sole posttranslational modification important for chromatid assembly, at least in the current setup. Cdk1 phosphorylation had been shown to stimulate the positive supercoiling activity of condensin I in vitro (Kimura et al. 1998; St-Pierre et al. 2009), implicating that condensin I–mediated manipulation of DNA topology could underlie large-scale mitotic chromatid assembly.

NUCLEOSOME DYNAMICS DURING CHROMATID ASSEMBLY

Although the core histones account for approximately half the weight of the whole protein components of mitotic chromatids (Hirano and Mitchison 1994; Ohta et al. 2010), to what extent they might directly contribute to large-scale chromatid assembly had largely been unknown. This was mainly because conventional experimental systems had not allowed researchers to manipulate the level of histones or nucleosomes at will. By using the chromatid reconstitution assays, however, it became possible to deposit any pairs of recombinant H2A and H2B (variants and mutants) along the entire genome. It was found that, among the histone H2A–H2B dimers tested, only the combination of amino-terminally truncated versions of H2A.X-F (an embryo-specific variant of H2A; Shechter et al. 2009a) and canonical H2B supported successful chromatid reconstitution (Shintomi et al. 2015). This result provided us with two implications. First, deleting the amino termini of H2A and H2B might bypass potential requirements for posttranslational modifications of the corresponding regions in the current setup. Second, unique characteristics of H2A.X-F (e.g., its extended, acidic carboxyl terminus) could weaken or destabilize interactions between histones and DNA, consequently facilitating the productive action of topo II and/or condensin I on nucleosome arrays.

A similar argument may also be applied to the action of the histone chaperone FACT, as recent structural studies suggested that FACT facilitates nucleosome reorganization by partially disrupting histone–DNA interactions (Hondele et al. 2013; Kemble et al. 2015; Tsunaka et al. 2016; Valieva et al. 2016). It is therefore reasonable to speculate that FACT confers structural flexibility on nucleosomes in the chromatid reconstitution reaction. Collectively, large-scale assembly of mitotic chromatids is likely to be dependent on the dynamic nature of nucleosomes.

CHROMATID AXIS ASSEMBLY WITHOUT NUCLEOSOMES

Although the chromatid reconstitution system had revealed the hitherto underappreciated importance of nucleosome dynamics in large-scale chromatid assembly, it remained unknown whether nucleosome assembly per se is an essential prerequisite for this process. To address this question, we made a simple modification to the single chromatid assembly assay using the *Xenopus* egg cell-free extracts; mouse sperm nuclei, which barely contain all core histones (Brykczynska et al. 2010), were used as a substrate instead of *Xenopus* sperm nuclei. It was first confirmed that mitotic extracts have the ability to support H3–H4 deposition on mouse sperm DNAs, followed by full nucleosome assembly, further converting them into a cluster of rod-shaped single chromatids (Shintomi et al. 2017).

The use of mouse sperm nuclei allowed us to block the whole process of nucleosome assembly by depleting the histone chaperone Asf1 from the egg extracts (Ray-Gallet et al. 2007). Although virtually no nucleosome was assembled on mouse sperm DNA as expected, we were surprised to find that mitotic chromatid-like structures were built under this condition (Shintomi et al. 2017). The resultant "nucleosome-depleted" chromatids were composed of DAPI-dense central axes and fuzzy "loop" regions surrounding them (Fig. 3, DAPI). The axes were positive for topo II and condensins, and the overall structures were sparser and more fragile than normal nucleosome-containing chromatids. No protamine was detectable in the nucleosome-depleted chromatids as well as in the control chromatids (Fig. 3, Protamine-1), excluding the

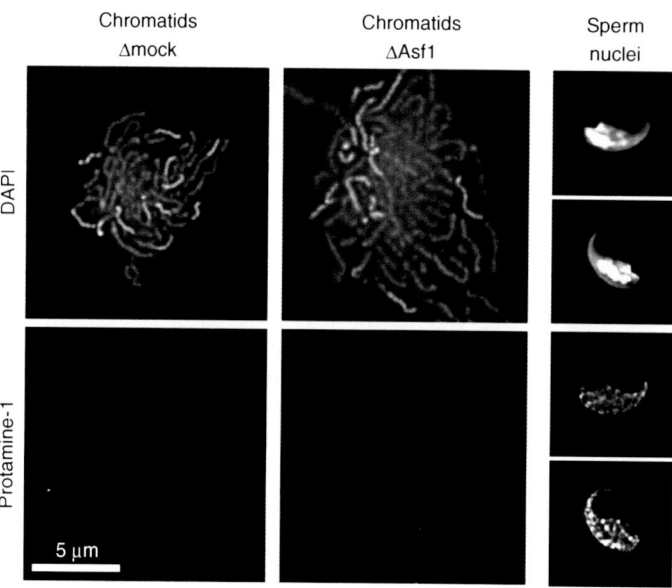

Figure 3. Immunofluorescence analysis with an antibody against protamine. Demembranated mouse sperm nuclei were incubated with mock-depleted and Asf1-depleted egg extracts for 3 h at 22°C to assemble nucleosome-containing chromatids (Δmock) and nucleosome-depleted chromatids (ΔAsf1), respectively. The resultant chromatids, along with the original sperm nuclei (*right* column), were labeled with anti-protamine-1 (HAL Technologies, Mab-001; *bottom* row). DNA was counterstained with DAPI (*upper* row). The results clearly show that protamine present in sperm nuclei is completely displaced from DNA during the chromatid assembly reactions, regardless of the presence or absence of nucleosomes.

possibility that residual protamine contributed to the assembly of this unique structure.

It is intriguing to note that the nucleosome-depleted chromatids assembled in the *Xenopus* egg cell-free extracts were reminiscent, in terms of both their morphology and protein compositions, of the "chromosome scaffold," a substructure obtained after chemical treatments of mitotic chromosomes (Laemmli et al. 1978; Earnshaw et al. 1985; Saitoh et al. 1994; Shintomi et al. 2017). Close comparison of the two structures prepared under completely different conditions will be of great help to get deep insights into the roles of condensins and topo II in chromatid axis formation. The peculiar morphology of the nucleosome-depleted chromosomes was also in good accordance with Kruitwagen et al. (2015), who used yeast genetics to address functional cross talk between nucleosomes and condensin. Taken together, these results converge on the very simple view that nucleosomes compact chromatin, whereas condensins shape chromosomes.

FUNCTIONAL CROSS TALK BETWEEN NUCLEOSOMES AND CONDENSINS

To gain deep insight into functional cross talk between nucleosomes and the two different condensin complexes, Asf1 depletion was combined with condensin I depletion or condensin II depletion (Shintomi et al. 2017). Each reaction produced characteristic morphology of chromatids and condensins' anomalous localization, as summarized in Fig. 4. In short, double depletion of Asf1 and condensin I did not significantly affect the axial distribution of condensin II, whereas depletion of Asf1 and condensin II severely impaired the action of condensin I. These observations clearly showed that nucleosome

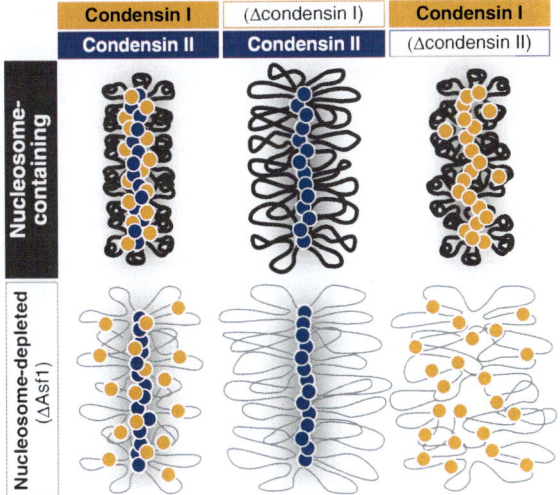

Figure 4. Chromatids assembled from mouse sperm nuclei with egg extracts depleted of condensins and/or Asf1 in different combinations. Their presumed structures, along with the localization of condensin I (yellow ovals) and condensin II (blue ovals), are depicted. The thick and thin lines indicate nucleosome and non-nucleosome fibers, respectively.

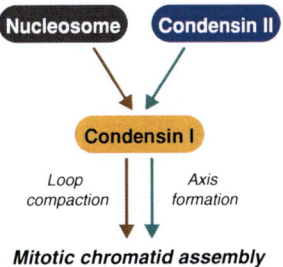

Figure 5. Schematic presentation of the proposed functional cross talk between nucleosomes and condensins. Details are described in the text.

assembly is an essential prerequisite for condensin I–mediated compaction of chromatin loops, whereas condensin II's ability to form chromatid axes is largely independent on nucleosomes (Fig. 5). This scheme is also consistent with the results from the reconstitution assay demonstrating that the productive action of condensin I relies on proper assembly of nucleosomes (Shintomi et al. 2015).

The functional cross talk described above raises a pair of new questions. First, why does condensin I appear to function more efficiently on nucleosomal DNA than on nonnucleosomal DNA? Although early studies proposed the model that the nucleosome itself serves as a "receptor" for targeting condensins to chromatin (Liu et al. 2010; Tada et al. 2011), accumulating lines of recent evidence from diverse experimental approaches support the idea that condensins' binding to chromosomal DNA is nucleosome-independent (Piazza et al. 2014; Zierhut et al. 2014; Shintomi et al. 2015, 2017; Toselli-Mollereau et al. 2016). One of the plausible scenarios is that condensin I first gets targeted to nucleosome-free regions of chromosomal DNA regions and initiates looping of their surrounding nucleosomal regions, possibly by imposing positive superhelical torsion (Kimura and Hirano 1997; Hirano 2014) or through an extrusion mechanism (Goloborodko et al. 2016). In either case, mechanical responses to the proposed action of condensin I would be substantially different between nucleosomal and nonnucleosomal DNAs.

Second, does condensin II indeed have an ability to form chromatid axes independently of nucleosomes and, if so, how? It has been proposed that formation of consecutive loops could be sufficient to bring distantly located condensins together and to accumulate them in axial regions (Goloborodko et al. 2016). An alternative, but not mutually exclusive, possibility is that intermolecular attractions between multiple condensin II complexes, possibly through their HEAT-repeat subunits, help to generate crowded environment at the central part of chromosomes, leading to the formation and stabilization of chromatid axes (Yoshimura and Hirano 2016). Given the amphiphilic and highly flexible nature of the HEAT-repeat subunits, it will be of great interest to consider potential mechanistic parallels between chromatid axes and phase-separated organelles made by intrinsically disordered proteins and other factors (Hyman et al. 2014).

CONCLUSION AND FUTURE CHALLENGES

In the current manuscript, we have summarized the history and development of the in vitro assays for mitotic chromosome assembly. Recent major breakthroughs in this approach include the successful reconstitution of mitotic chromatids using purified factors in vitro (Shintomi et al. 2015) and the demonstration that nucleosomes can be dispensable for building chromosome-like structures under certain conditions (Shintomi et al. 2017). These studies strongly suggest that the mechanism of mitotic chromosome assembly may be much simpler than previously thought and further emphasize the central importance of condensins in this process. Despite the exciting progress, a number of outstanding questions remain ahead of us.

It has become increasingly clear that the condensin complexes are very elaborate molecular machines. As discussed in the previous section, understanding how condensins might work at a mechanistic level remains one of the most important challenges in the years to come. Could condensins act as a putative loop extruder (Goloborodko et al. 2016)? How could gigantic protein complexes such as condensins function in a crowded environment of the interior of chromosomes (Hihara et al. 2012; Yoshimura and Hirano 2016)? To address these questions, it is worth testing the functions of recombinant condensin II and its mutants in the cell-free add-back assay and in the chromatid reconstitution assay. Also, there is much to be learned about the exact function of topo II in the process of building mitotic chromosomes. For instance, we need to keep in mind that we still know very little about how topo II acts on nucleosomal DNA (Salceda et al. 2006) or how it collaborates with condensins (Baxter et al. 2011).

There remains room for further modification and refinement of the first-generation chromatid reconstitution system (Shintomi et al. 2015). It will be possible to further advance the reaction by supplementing it with additional factors such as linker histones, chromatin remodeling factors, and condensin II. Equally important, some agents that mimic macromolecular crowding in crude extracts could further improve the reconstitution reaction (Hancock 2012). Moreover, we must admit that we are still largely ignorant of ion atmospheres (Mg^{2+}, Ca^{2+}, Na^+, and K^+) that contribute to large-scale chromosome assembly in vivo (Mathieson and Olayemi 1975; Strick et al. 2001; Hudson et al. 2003; Phengchat et al. 2016; Ono et al. 2017). The reconstitution system will be instrumental in sharply addressing this issue with an unprecedented precision. In addition, the use of mouse sperm nuclei as a substrate will allow us to engineer H3–H4 as well as H2A–H2B in the reconstitution assays and provide us with an opportunity to learn much more about nucleosome-mediated regulation of chromatid assembly.

Finally, it should be noted that the current in vitro assays rely solely on the morphology of chromosomes as judged by light microscopy. Only limited efforts have been made to analyze the in vitro assembled chromosomes by electron microscopy (Konig et al. 2007). It will be of great interest to analyze them, for instance, by recently developed electron microscopy (EM) tomography, known as ChromEMT (Ou et al. 2017). There is no doubt that genome-based approaches such as Hi-C techniques (Naumova et al. 2013; Kakui et al. 2017; Lazar-Stefanita et al. 2017; Schalbetter et al. 2017) and mechanical stretching approaches (Almagro et al. 2004; Yan et al. 2007; Xiao et al. 2012) will also provide valuable information regarding the architecture and physical properties of the in vitro assembled chromosomes. Powerful combinations of the sophisticated in vitro assays and the emerging analytical technologies will provide us with a key to unlock one of the most important questions in biology: How is a centimeters-long genomic DNA folded into a micrometers-long chromosome?

ACKNOWLEDGMENTS

We are grateful to members of the Hirano laboratory for critically reading the manuscript. Work in the authors' laboratory was funded by Grant-in-Aid for Scientific Research, KAKENHI (grant numbers 16H01323 and 26440012 to K.S. and 15H05971 and 15K14455 to T.H.).

REFERENCES

Almagro S, Riveline D, Hirano T, Houchmandzadeh B, Dimitrov S. 2004. The mitotic chromosome is an assembly of rigid elastic axes organized by structural maintenance of chromosomes (SMC) proteins and surrounded by a soft chromatin envelope. *J Biol Chem* **279:** 5118–5126.

Baxter J, Sen N, Martínez VL, De Carandini ME, Schvartzman JB, Diffley JF, Aragón L. 2011. Positive supercoiling of mitotic DNA drives decatenation by topoisomerase II in eukaryotes. *Science* **331:** 1328–1332.

Brykczynska U, Hisano M, Erkek S, Ramos L, Oakeley EJ, Roloff TC, Beisel C, Schübeler D, Stadler MB, Peters AH. 2010. Repressive and active histone methylation mark distinct promoters in human and mouse spermatozoa. *Nat Struct Mol Biol* **17:** 679–687.

Earnshaw WC, Halligan B, Cooke CA, Heck MM, Liu LF. 1985. Topoisomerase II is a structural component of mitotic chromosome scaffolds. *J Cell Biol* **100:** 1706–1715.

Gerlich D, Hirota T, Koch B, Peters JM, Ellenberg J. 2006. Condensin I stabilizes chromosomes mechanically through a dynamic interaction in live cells. *Curr Biol* **16:** 333–344.

Goloborodko A, Imakaev MV, Marko JF, Mirny L. 2016. Compaction and segregation of sister chromatids via active loop extrusion. *Elife* **5:** e14864.

Green LC, Kalitsis P, Chang TM, Cipetic M, Kim JH, Marshall O, Turnbull L, Whitchurch CB, Vagnarelli P, Samejima K, et al. 2012. Contrasting roles of condensin I and condensin II in mitotic chromosome formation. *J Cell Sci* **125:** 1591–1604.

Hancock R. 2012. Structure of metaphase chromosomes: A role for effects of macromolecular crowding. *PLoS One* **7:** e36045.

Hihara S, Pack CG, Kaizu K, Tani T, Hanafusa T, Nozaki T, Takemoto S, Yoshimi T, Yokota H, Imamoto N, et al. 2012. Local nucleosome dynamics facilitate chromatin accessibility in living mammalian cells. *Cell Rep* **2:** 1645–1656.

Hirano T. 2014. Condensins and the evolution of torsion-mediated genome organization. *Trends Cell Biol* **24:** 727–733.

Hirano T. 2016. Condensin-based chromosome organization from bacteria to vertebrates. *Cell* **164:** 847–857.

Hirano T, Mitchison TJ. 1993. Topoisomerase II does not play a scaffolding role in the organization of mitotic chromosomes assembled in *Xenopus* egg extracts. *J Cell Biol* **120:** 601–612.

Hirano T, Mitchison TJ. 1994. A heterodimeric coiled-coil protein required for mitotic chromosome condensation in vitro. *Cell* **79:** 449–458.

Hirano T, Kobayashi R, Hirano M. 1997. Condensins, chromosome condensation protein complexes containing XCAP-C, XCAP-E and a *Xenopus* homolog of the *Drosophila* Barren protein. *Cell* **89:** 511–521.

Hirota T, Gerlich D, Koch B, Ellenberg J, Peters JM. 2004. Distinct functions of condensin I and II in mitotic chromosome assembly. *J Cell Sci* **117:** 6435–6445.

Hondele M, Stuwe T, Hassler M, Halbach F, Bowman A, Zhang ET, Nijmeijer B, Kotthoff C, Rybin V, Amlacher S, et al. 2013. Structural basis of histone H2A-H2B recognition by the essential chaperone FACT. *Nature* **499:** 111–114.

Hudson DF, Vagnarelli P, Gassmann R, Earnshaw WC. 2003. Condensin is required for nonhistone protein assembly and structural integrity of vertebrate mitotic chromosomes. *Dev Cell* **5:** 323–336.

Hyman AA, Weber CA, Jülicher F. 2014. Liquid-liquid phase separation in biology. *Annu Rev Cell Dev Biol* **30:** 39–58.

Kakui Y, Rabinowitz A, Barry DJ, Uhlmann F. 2017. Condensin-mediated remodeling of the mitotic chromatin landscape in fission yeast. *Nat Genet* **49:** 1553–1557.

Kalitsis P, Zhang T, Marshall KM, Nielsen CF, Hudson DF. 2017. Condensin, master organizer of the genome. *Chromosome Res* **25:** 61–76.

Kemble DJ, McCullough LL, Whitby FG, Formosa T, Hill CP. 2015. FACT disrupts nucleosome structure by binding H2A-H2B with conserved peptide motifs. *Mol Cell* **60:** 294–306.

Kimura K, Hirano T. 1997. ATP-dependent positive supercoiling of DNA by 13S condensin: A biochemical implication for chromosome condensation. *Cell* **90:** 625–634.

Kimura K, Hirano M, Kobayashi R, Hirano T. 1998. Phosphorylation and activation of 13S condensin by Cdc2 in vitro. *Science* **282:** 487–490.

Kinoshita K, Kobayashi TJ, Hirano T. 2015. Balancing acts of two HEAT subunits of condensin I support dynamic assembly of chromosome axes. *Dev Cell* **33:** 94–106.

Konig P, Braunfeld MB, Sedat JW, Agard DA. 2007. The three-dimensional structure of in vitro reconstituted *Xenopus laevis* chromosomes by EM tomography. *Chromosoma* **116:** 349–372.

Kruitwagen T, Denoth-Lippuner A, Wilkins BJ, Neumann H, Barral Y. 2015. Axial contraction and short-range compaction of chromatin synergistically promote mitotic chromosome condensation. *Elife* **4:** e1039.

Laemmli UK, Cheng SM, Adolph KW, Paulson JR, Brown JA, Baumbach WR. 1978. Metaphase chromosome structure: The role of nonhistone proteins. *Cold Spring Harb Symp Quant Biol* **42**(Pt 1): 351–360.

Lazar-Stefanita L, Scolari VF, Mercy G, Muller H, Guérin TM, Thierry A, Mozziconacci J, Koszul R. 2017. Cohesins and condensins orchestrate the 4D dynamics of yeast chromosomes during the cell cycle. *EMBO J* **36:** 2684–2697.

Liu W, Tanasa B, Tyurina OV, Zhou TY, Gassmann R, Liu WT, Ohgi KA, Benner C, Garcia-Bassets I, Aggarwal AK, et al. 2010. PHF8 mediates histone H4 lysine 20 demethylation events involved in cell cycle progression. *Nature* **466:** 508–512.

Lohka MJ, Maller JL. 1985. Induction of nuclear envelope breakdown, chromosome condensation, and spindle formation in cell-free extracts. *J Cell Biol* **101:** 518–523.

Maresca TJ, Freedman BS, Heald R. 2005. Histone H1 is essential for mitotic chromosome architecture and segregation in *Xenopus laevis* egg extracts. *J Cell Biol* **169:** 859–869.

Mathieson AR, Olayemi JY. 1975. The interaction of calcium and magnesium ions with deoxyribonucleic acid. *Arch Biochem Biophys* **169:** 237–243.

Murray AW. 1991. Cell cycle extracts. *Methods Cell Biol* **36:** 581–605.

Naumova N, Imakaev M, Fudenberg G, Zhan Y, Lajoie BR, Mirny LA, Dekker J. 2013. Organization of the mitotic chromosome. *Science* **342:** 948–953.

Ohsumi K, Katagiri C. 1991. Characterization of the ooplasmic factor inducing decondensation of and protamine removal from toad sperm nuclei: Involvement of nucleoplasmin. *Dev Biol* **148:** 295–305.

Ohsumi K, Katagiri C, Kishimoto T. 1993. Chromosome condensation in *Xenopus* mitotic extracts without histone H1. *Science* **262:** 2033–2035.

Ohta S, Bukowski-Wills JC, Sanchez-Pulido L, Alves Fde L, Wood L, Chen ZA, Platani M, Fischer L, Hudson DF, Ponting CP, et al. 2010. The protein composition of mitotic chromosomes determined using multiclassifier combinatorial proteomics. *Cell* **142:** 810–821.

Ono T, Losada A, Hirano M, Myers MP, Neuwald AF, Hirano T. 2003. Differential contributions of condensin I and condensin II to mitotic chromosome architecture in vertebrate cells. *Cell* **115:** 109–121.

Ono T, Fang Y, Spector DL, Hirano T. 2004. Spatial and temporal regulation of Condensins I and II in mitotic chromosome assembly in human cells. *Mol Biol Cell* **15:** 3296–3308.

Ono T, Sakamoto C, Nakao M, Saitoh N, Hirano T. 2017. Condensin II plays an essential role in reversible assembly of mitotic chromosomes in situ. *Mol Biol Cell* (in press). doi: 10.1091/mbc.E17-04-0252.

Ou HD, Phan S, Deerinck TJ, Thor A, Ellisman MH, O'Shea CC. 2017. ChromEMT: Visualizing 3D chromatin structure and compaction in interphase and mitotic cells. *Science* **357:** eaag0025.

Phengchat R, Takata H, Morii K, Inada N, Murakoshi H, Uchiyama S, Fukui K. 2016. Calcium ions function as a booster of chromosome condensation. *Sci Rep* **6:** 38281.

Philpott A, Leno GH. 1992. Nucleoplasmin remodels sperm chromatin in *Xenopus* egg extracts. *Cell* **69:** 759–767.

Piazza I, Rutkowska A, Ori A, Walczak M, Metz J, Pelechano V, Beck M, Haering CH. 2014. Association of condensin with chromosomes depends on DNA binding by its HEAT-repeat subunits. *Nat Struct Mol Biol* **21:** 560–568.

Ray-Gallet D, Quivy JP, Silljé HW, Nigg EA, Almouzni G. 2007. The histone chaperone Asf1 is dispensable for direct de novo histone deposition in *Xenopus* egg extracts. *Chromosoma* **116:** 487–496.

Saitoh N, Goldberg IG, Wood ER, Earnshaw WC. 1994. ScII: An abundant chromosome scaffold protein is a member of a family of putative ATPases with an unusual predicted tertiary structure. *J Cell Biol* **127:** 303–318.

Salceda J, Fernández X, Roca J. 2006. Topoisomerase II, not topoisomerase I, is the proficient relaxase of nucleosomal DNA. *EMBO J* **25:** 2575–2583.

Schalbetter SA, Goloborodko A, Fudenberg G, Belton JM, Miles C, Yu M, Dekker J, Mirny L, Baxter J. 2017. SMC complexes differentially compact mitotic chromosomes according to genomic context. *Nat Cell Biol* **19:** 1071–1080.

Shechter D, Chitta RK, Xiao A, Shabanowitz J, Hunt DF, Allis CD. 2009a. A distinct H2A.X isoform is enriched in *Xenopus laevis* eggs and early embryos and is phosphorylated in the absence of a checkpoint. *Proc Natl Acad Sci* **106:** 749–754.

Shechter D, Nicklay JJ, Chitta RK, Shabanowitz J, Hunt DF, Allis CD. 2009b. Analysis of histones in *Xenopus laevis*. I. A distinct index of enriched variants and modifications exists in each cell type and is remodeled during developmental transitions. *J Biol Chem* **284:** 1064–1074.

Shintomi K, Hirano T. 2011. The relative ratio of condensin I to II determines chromosome shapes. *Genes Dev* **25:** 1464–1469.

Shintomi K, Hirano T. 2017. A sister chromatid cohesion assay using *Xenopus* egg extracts. *Methods Mol Biol* **1515:** 3–21.

Shintomi K, Iwabuchi M, Saeki H, Ura K, Kishimoto T, Ohsumi K. 2005. Nucleosome assembly protein-1 is a linker histone chaperone in *Xenopus* eggs. *Proc Natl Acad Sci* **102:** 8210–8215.

Shintomi K, Takahashi TS, Hirano T. 2015. Reconstitution of mitotic chromatids with a minimum set of purified factors. *Nat Cell Biol* **17:** 1014–1023.

Shintomi K, Inoue F, Watanabe H, Ohsumi K, Ohsugi M, Hirano T. 2017. Mitotic chromosome assembly despite nucleosome depletion in *Xenopus* egg extracts. *Science* **356:** 1284–1287.

St-Pierre J, Douziech M, Bazile F, Pascariu M, Bonneil E, Sauve V, Ratsima H, D'Amours D. 2009. Polo kinase regulates mitotic chromosome condensation by hyperactivation of condensin DNA supercoiling activity. *Mol Cell* **34:** 416–426.

Strick R, Strissel PL, Gavrilov K, Levi-Setti R. 2001. Cation-chromatin binding as shown by ion microscopy is essential for the structural integrity of chromosomes. *J Cell Biol* **155:** 899–910.

Tada K, Susumu H, Sakuno T, Watanabe Y. 2011. Condensin association with histone H2A shapes mitotic chromosomes. *Nature* **474:** 477–483.

Takemoto A, Kimura K, Yanagisawa J, Yokoyama S, Hanaoka F. 2006. Negative regulation of condensin I by CK2-mediated phosphorylation. *EMBO J* **25:** 5339–5348.

Toselli-Mollereau E, Robellet X, Fauque L, Lemaire S, Schiklenk C, Klein C, Hocquet C, Legros P, N'Guyen L, Mouillard L, et al. 2016. Nucleosome eviction in mitosis assists condensin loading and chromosome condensation. *EMBO J* **35:** 1565–1581.

Tsunaka Y, Fujiwara Y, Oyama T, Hirose S, Morikawa K. 2016. Integrated molecular mechanism directing nucleosome reorganization by human FACT. *Genes Dev* **30:** 673–686.

Uhlmann F. 2016. SMC complexes: From DNA to chromosomes. *Nat Rev Mol Cell Biol* **17:** 399–412.

Valieva ME, Armeev GA, Kudryashova KS, Gerasimova NS, Shaytan AK, Kulaeva OI, McCullough LL, Formosa T, Geor-giev PG, Kirpichnikov MP, et al. 2016. Large-scale ATP-independent nucleosome unfolding by a histone chaperone. *Nat Struct Mol Biol* **23:** 1111–1116.

Wignall SM, Deehan R, Maresca TJ, Heald R. 2003. The condensin complex is required for proper spindle assembly and chromosome segregation in *Xenopus* egg extracts. *J Cell Biol* **161:** 1041–1051.

Xiao B, Freedman BS, Miller KE, Heald R, Marko JF. 2012. Histone H1 compacts DNA under force and during chromatin assembly. *Mol Biol Cell* **23:** 4864–4871.

Yan J, Maresca TJ, Skoko D, Adams CD, Xiao B, Christensen MO, Heald R, Marko JF. 2007. Micromanipulation studies of chromatin fibers in *Xenopus* egg extracts reveal ATP-dependent chromatin assembly dynamics. *Mol Biol Cell* **18:** 464–474.

Yoshimura SH, Hirano T. 2016. HEAT repeats—Versatile arrays of amphiphilic helices working in crowded environments? *J Cell Sci* **129:** 3963–3970.

Zierhut C, Jenness C, Kimura H, Funabiki H. 2014. Nucleosomal regulation of chromatin composition and nuclear assembly revealed by histone depletion. *Nat Struct Mol Biol* **21:** 617–625.

Spindle-to-Cortex Communication in Cleaving Frog Eggs

Timothy J. Mitchison[1,2] and Christine M. Field[1,2]

[1]*Department of Systems Biology, Harvard Medical School, Boston, Massachusetts 02115*
[2]*Marine Biological Laboratory, Woods Hole, Massachusetts 02543*
Correspondence: timothy_mitchison@hms.harvard.edu

During cytokinesis, the mitotic spindle communicates with the cell cortex to position a cleavage furrow that will cut through the cell in the plane defined by the metaphase plate. We investigated the molecular basis of this communication in *Xenopus laevis* eggs, where the signal has to travel ~400 μm in ~30 min to reach the cortex from the first anaphase spindle. At anaphase onset, huge microtubule asters grow out from the poles of the spindle and meet at the plane previously defined by the metaphase plate. This disc-shaped boundary plane recruits the chromosome passenger complex (CPC) and centralspindlin to antiparallel microtubule bundles. It grows out to the cell cortex as the asters expand, where it induces the furrow. CPC and centralspindlin were not recruited to boundaries between asters from different spindles, suggesting a role of chromatin in triggering the CPC-positive state. Recruitment of CPC to aster boundaries was reconstituted in an extract system, and we observed that recruitment was stimulated by proximity to chromatin. Finally, we discuss models for molecular processes involved in initiation and growth of the CPC-positive disc that communicates the position of the metaphase plate to the cortex over hundreds of micrometers in frog eggs.

During cytokinesis in animal cells an actomyosin-powered cleavage furrow cleaves the cell into two daughters. The furrow is precisely positioned in space and time to ensure that each daughter is approximately equal in volume and contains an intact, single genome. It has long been recognized that the cleavage furrow initiates after anaphase onset and tends to cut through the plane defined by the metaphase plate. Our focus is on the positioning of the furrow. Rappaport (1971, 1986) used elegant micromanipulation experiments in echinoderm eggs to show that the anaphase spindle transmits a spatially restricted signal to the egg cortex, which is uniformly receptive to furrow induction during a defined cell cycle window (Fig. 1A). Similar signaling occurs in somatic cells, and spindle-to-cortex communication during cytokinesis is perhaps the most famous example of spatial communication between two parts of the same cell. Rappaport showed that the signal propagates outward from the anaphase spindle at a rate of 6–7 μm/min and involves the concerted action of the pair of microtubule asters that emanate from the poles of the spindle, which we will call sister asters. He and others considered models where the aster pair promotes contractility at the equator ("equatorial stimulation") or inhibits it at the poles ("polar relaxation"). Part of Rappaport's evidence supporting that the signal is transmitted by a pair of asters came from a famous micromanipulation experiment where asters from two different spindles were brought into proximity during anaphase, whereupon they induced an ectopic furrow (Fig. 1B). This experiment also showed that chromatin is not required for spindle-to-cortex communication in some systems.

Here, we will consider spindle-to-cortex communication in frog eggs and focus entirely on equatorial stimulation, because there is currently little evidence for polar relaxation in frog eggs. Given the molecular conservation of cell division proteins in general, there is every reason to expect that frog eggs use the same molecules and mechanisms to signal from the spindle to the cortex as echinoderm eggs and somatic cells. However, their emergent behavior may differ in interesting ways. For example, classic observations in polyspermic frog eggs showed that furrows are only induced between asters that emanate from the same spindle and are not induced between spindles (Fig. 1C; Brachet 1910; Herlant 1911). This suggests a role for chromatin in spindle-to-cortex communication, which we explore below.

Frog eggs are also huge. The *Xenopus* eggs we study are ~1.2 mm in diameter, more than 10 times the size of the Rappaport's echinoderm eggs. The distance between the first metaphase spindle and the nearest cortex is ~500 μm in *Xenopus laevis*, and the time interval between anaphase onset and ingression of the first furrow is ~30 min. These large spatial and temporal scales likely require special adaptation of the conserved mechanisms that mediate spindle-to-cortex communication, and they might make it easier to observe the relevant molecules in action. *Xenopus* eggs are opaque because of distributed yolk platelets and lipid droplets, which makes live imaging of the intact egg difficult. To get around this problem, we either image-fixed and stained eggs in a high refractive index clearing solvent or used a cell-free extract system that reconstitutes the microtubule and actin organization characteristic of cytokinesis (Nguyen et al. 2014; Field et al. 2017).

Published by Cold Spring Harbor Laboratory Press; doi: 10.1101/sqb.2017.82.033654

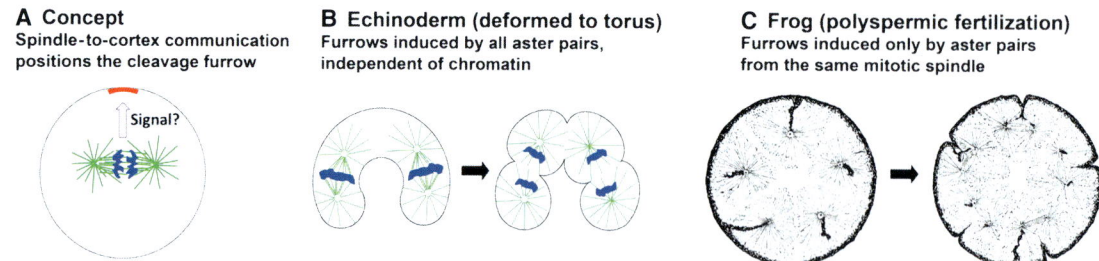

Figure 1. Spindle-to-cortex communication in different systems. (*A*) In most animal cells the spindle sends a signal to localize the cleavage furrow, which then cuts through the cell in the plane defined by the earlier metaphase plate. (*B*) In mechanically deformed echinoderm eggs, aster pairs trigger furrows whether they grow from the same or different spindles. This shows that chromatin is not essential to general the furrow signal. (*C*) In polyspermic frog eggs, only aster pairs from the same spindle trigger furrows. These images are drawn from sections of fixed frog eggs after polyspermic fertilization. The *left* egg was fixed before first mitosis illustrating five sperm asters. Black trails mark the path of the sperm as it moves into the cytoplasm after entering the egg. The *right* egg was fixed during first cleavage. It illustrates cleavage with five spindles, where each furrow cuts through the sites previously occupied by a metaphase plate. Note the lack of furrows between spindles, unlike the Echinoderm system. (*B*, Redrawn from Rappaport and Conrad 1963, with estimated chromatin localization added in blue; *C*, modified from Brachet 1910.)

ASTER GROWTH AND RECRUITMENT OF FURROW-STIMULATING COMPLEXES

The first metaphase spindle is much smaller than the egg in *X. laevis* (Fig. 2A,A′). Its position and orientation are determined by dynein-dependent forces acting on the sperm aster before nuclear envelope breakdown (Wühr et al. 2010). At anaphase onset, a pair of huge sister asters grow progressively outward from the poles of the spindle, reaching the cortex ~30 min later (Fig. 2B,C). The microtubules in these asters appear "bushy," and their density at the aster periphery remains approximately constant as the aster grows (Fig. 2B–D). Egg extract experiments suggest this morphology is due to the aster being built of a network of short microtubules mostly nucleated away from the centrosome (Ishihara et al. 2016). Furrowing initiates when the

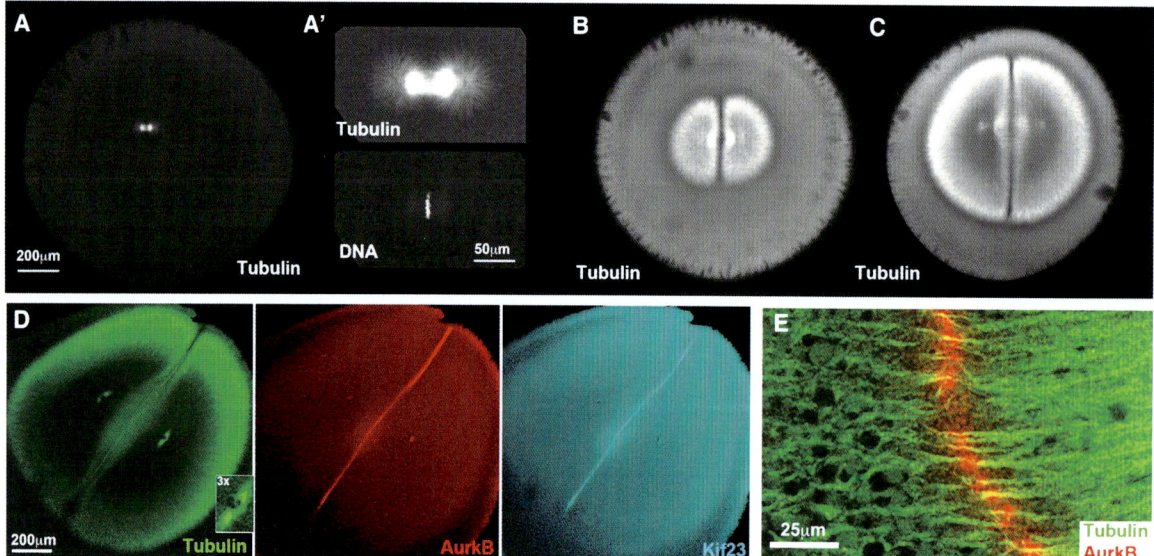

Figure 2. Aster growth and recruitment of furrow-stimulating complexes in *Xenopus laevis* eggs. Images are confocal sections viewed with the animal–vegetal axis in the microscope *z*-axis. The first cleavage furrow cuts through the animal pole, parallel to the *z*-axis in these images. (*A*,*A′*) Egg fixed ~60 min after fertilization, at metaphase of first mitosis. Note the spindle is small compared to the egg. (*B*, *C*) Eggs fixed ~70 and ~80 min after fertilization, between first anaphase and first cleavage. Note aster pairs grow out from the poles of the spindle and meet at the midplane. Each aster is dome-shaped, with a sharp plane of lower microtubule density where they meet. In cross section this geometry appears as two D-shaped asters arranged back-to-back with a line between them. (*D*) Egg fixed at cleavage initiation, ~90 min after fertilization. The asters are just touching the cortex at the animal pole. AurkB is a subunit of the CPC; Kif23 is a subunit of centralspindlin. Both furrow-stimulating complexes are enriched at a disc between the asters that marks the future path of the cleavage furrow. Note the centrosomes are already positioned for the second round of mitosis and cleavage that will occur orthogonal to the first (3× *inset* in tubulin panel). (*E*) Higher-magnification view of the boundary region between asters. Note the CPC localizes to the center of presumably antiparallel microtubule bundles. (*A–C*, Modified, with permission, from Mitchison et al. 2012; *D*, modified, with permission, from Field et al. 2015; *E*, TJ Mitchison and CM Field, unpubl. data.)

asters touch the cortex, consistent with spindle-to-cortex communication occurring via aster growth. Treatment of eggs with nocodazole, which inhibits microtubule polymerization, completely blocks furrowing (Hara et al. 1980).

A characteristic feature of sister aster pairs in fixed *Xenopus* eggs is the presence of a sharp boundary between the asters where the microtubule density is lower (Fig. 2B–D). The boundary region contains microtubule bundles that we interpret as antiparallel overlaps (Fig. 2E). It defines a plane at the center of the egg that extends to the cortex, and it predicts the path that the furrow will later follow as it ingresses. Because sister asters grow symmetrically from the poles of the spindle, the boundary plane between them coincides with the plane previously occupied by the metaphase plate. Therefore, radial growth of the boundary plane between asters provides a structural mechanism for propagating information on the localization of the metaphase plate to the cortex. As the asters grow, the boundary between them remains at the midplane of the egg while the centrosomes and nuclei at their centers move apart, so they are approximately centered in between the midplane and the cortex by the time the furrow ingresses (Wühr et al. 2010; Mitchison et al. 2012). This centering movement ensures that the furrow will cut between the separating nuclei and also positions the centrosomes for the next round of mitosis and cleavage. Prepositioning of centrosomes for the next round of cleavage is evident in the tubulin image, Figure 2D (3× inset), where the two centrosomes within each aster have moved to the center of the half-cell and split apart parallel to the boundary between the asters. After first cleavage, each centrosome pair will establish a second mitotic spindle on this axis, which will cause the second furrows to cleave orthogonal to the first. We hypothesized that this prepositioning of centrosomes was accomplished by dynein pulling on astral microtubules (Wühr et al. 2010).

To understand how the boundary between sister asters instructs cleavage furrow assembly, we localized conserved cytokinesis proteins that are known to have this function in other systems. Two conserved microtubule binding complexes, chromosome passenger complex (CPC) and centralspindlin, promote furrow assembly in somatic cells. CPC is a 1:1:1:1 complex of AURKB, INCENP, borealin/dasra/CDCA8, and survivin/BIRC5 (Ruchaud et al. 2007). In *Xenopus* eggs, CDCA8 is replaced by the egg-specific ortholog Dasra2/CDCA9 (Sampath et al. 2004). The AURKB subunit of the CPC is a multifunctional kinase that promotes autophosphorylation and autoactivation of the CPC when it clusters on chromatin or microtubules. CPC is transported to microtubule plus ends by MKLP2/Kif20A, where it is thought to activate other cytokinesis proteins including centralspindlin (Gruneberg et al. 2004). Centralspindlin is a 2:2 complex of RACGAP1 and MKLP1/KIF23. It is transported to plus ends by its intrinsic kinesin activity and functions to activate RhoA and contractility (White and Glotzer 2012; Mishima 2016). Both CPC and centralspindlin accumulate on antiparallel microtubule bundles at the boundary between sister asters in frog eggs, defining a disc between the asters that will trigger the furrow when it touches the cortex (Fig. 2D,E). We believe that this CPC-centralspindlin-positive disc between the asters, which initiates at anaphase and then grows with the asters, mediates spindle-to-cortex communication in frog eggs.

ONLY ASTER PAIRS FROM THE SAME SPINDLE RECRUIT CPC AND CENTRALSPINDLIN

In polyspermic frog eggs, only aster pairs from the same spindle triggered furrows (Fig. 1C). We repeated this classic experiment and stained for CPC and centralspindlin. Figure 3 shows polyspermic eggs fixed at different times between first anaphase and first cleavage. CPC and centralspindlin were recruited to the boundary plane between asters from the same spindle but not to planes where asters from different spindle met. Recruitment of CPC is better visualized, in part because it is more abundant in eggs (Field et al. 2015). These images also provide a clear view of the growth of the furrow-simulating plane between sister asters, starting as a small disc making the plane previously occupied by the metaphase plate, then growing outward normal to the axis of chromosome separation. These observations provide a molecular explanation for the classic observation that only aster pairs from the same spindle induce furrows in frog eggs, but they prompt the question of why only these asters pairs recruit furrow-simulating complexes to their shared boundary. This could be due to the presence of chromatin at anaphase, timing differences in when the asters meet, or other factors. This selectivity is not absolute and can be overridden by experimental perturbation. Artificially boosting CPC activation with an injected antibody, reducing the distance between asters by highly polyspermic fertilization, or stabilizing microtubules with paclitaxel all triggered indiscriminant recruitment of CPC to aster boundaries independent of their origin (Field et al. 2015). Evidently, the requirement that the two spindles originate from the same spindle to induce furrows only holds under normal conditions.

The CPC was discovered by Earnshaw and Cooke (1991), and named to reflect its localization to chromatin in metaphase and midzone microtubule bundles in anaphase–telophase. They proposed that the CPC was transported to the metaphase plate by chromosomes, where it later instructs the furrow. This model was questioned by experiments showing that the CPC can be recruited to microtubule bundles between spindles in somatic cells (Savoian et al. 1999) in a somatic cell version of Figure 1B. The apparent chromatin requirement for CPC localization in polyspermic *Xenopus* eggs (Fig. 3) triggered our interest in testing if proximity to chromatin could stimulate formation of aster boundaries with furrow-inducing potential in an egg extract system.

RECRUITMENT OF CPC AND CENTRALSPINDLIN TO ASTER BOUNDARIES IN EGG EXTRACT

We turned to an actin-intact egg extract system to probe how the boundary region between asters forms, grows,

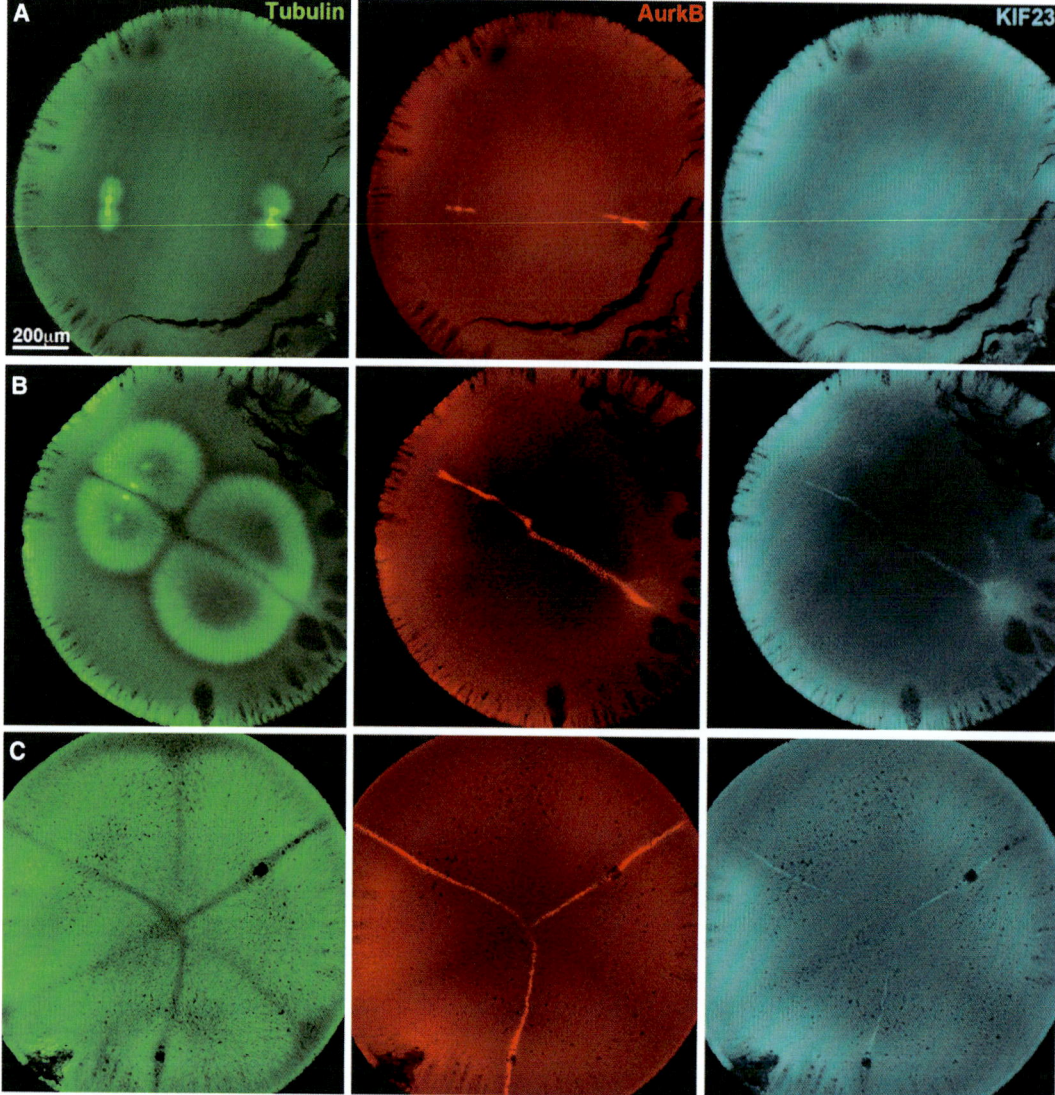

Figure 3. Furrow-stimulating complexes in polyspermic *Xenopus laevis* eggs. Eggs fixed at successive times between first anaphase and first cleavage, with confocal sections oriented as in Figure 2. Cleavage has just initiated at the animal pole in *C*. The eggs in *A* and *B* were fertilized with two sperm and in *C* with four, but asters are only visible for three. Note recruitment of CPC and centralspindlin to a subset of boundaries between asters in *B* and *C*. In *B*, the aster morphology shows that the CPC-positive boundaries are between asters from the same spindle. In *C*, we infer the same is true given classic evidence that furrows cut through spindles in frog eggs and not between them (Fig. 1C). (Modified, with permission, from Field et al. 2015.)

accumulates CPC and centralspindlin, and signals to the cortex. Figure 4A shows a typical experiment where the extract was squashed to a ~20-µm-deep layer between two passivated coverslips, illustrating aster growth and recruitment of CPC to a boundary between asters. Centralspindlin localized in a similar manner (Nguyen et al. 2014). CPC recruitment to aster boundaries was stimulated by low concentrations of paclitaxel or inhibition of the catastrophe factor MCAK, suggesting that microtubule stabilization plays a role in CPC recruitment (Field et al. 2015). To probe how CPC is recruited we used total internal reflection fluorescence (TIRF) microscopy on reactions similar to Figure 4A. We observed individual CPC aggregates moving toward microtubule plus ends and accumulating at the aster boundary. Figure 4B shows kymograph

analysis of the movement, in which the red line in the first panel traces the center of the boundary region. Note multiple CPC aggregates moving inward. This movement required two plus end–directed kinesins, Kif20A and Kif4. When Kif20A was depleted, we observed no CPC recruitment to microtubules, consistent with a central role of this kinesin, which is also called MKLP2, in CPC recruitment (Nguyen et al. 2014). This observation is consistent with a Kif20A/MKLP2 requirement for CPC localization to midzones in somatic cells (Gruneberg et al. 2004). Removal of Kif4A, which is involved in microtubule length regulation in midzones (Hu et al. 2011), still allowed CPC accumulation as boundaries between asters but appeared to block the active transport we could visualize (Fig. 4B). In more complex experiments designed to reconstitute signaling to

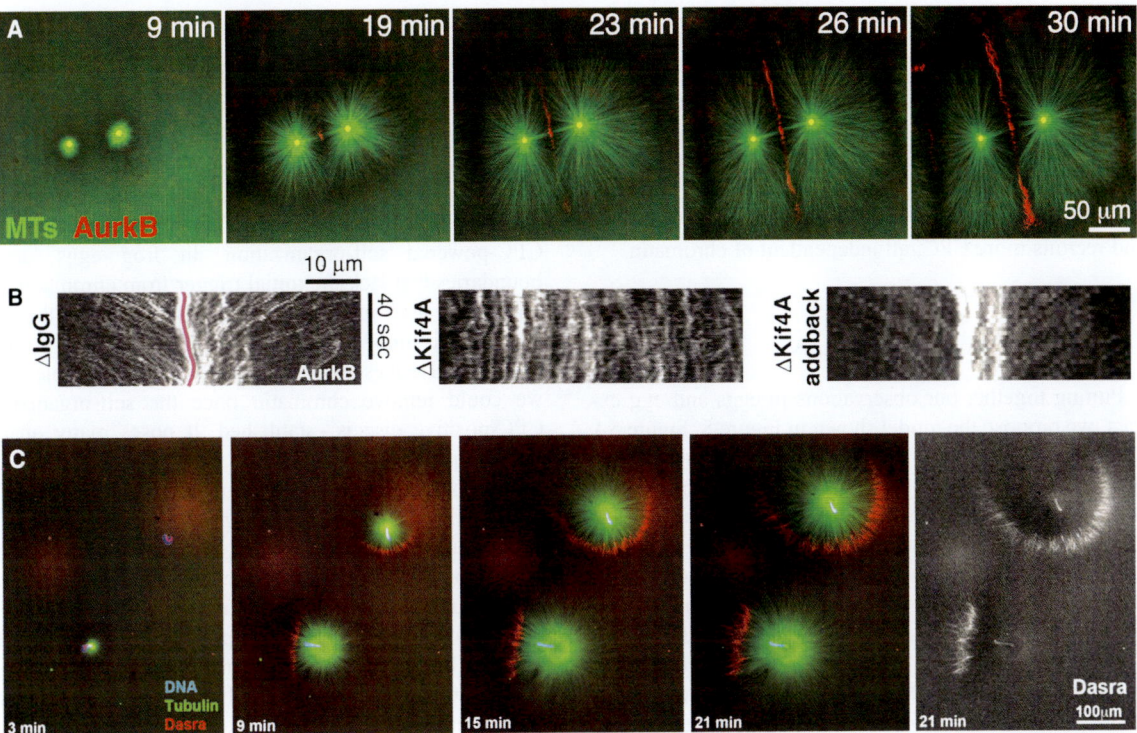

Figure 4. Recruitment of CPC to aster boundaries in egg extract. Actin-intact extract from unfertilized eggs was supplemented with imaging probes, converted to interphase with a calcium transient, and squashed between passivated coverslips. (*A*) Widefield imaging of asters nucleated by anti-AurkA coated beads. CPC was visualized with anti-AurkB. Note recruitment of CPC to antiparallel bundles at the boundary between asters. (*B*) Total internal reflection fluorescence (TIRF) imaging of CPC movement in reactions similar to *A*. Movies were recorded at CPC-positive boundaries between asters, and kymograph analysis was used to visualize movement of CPC aggregates. The pink line denotes the center of the boundary region. Diagonal streaks represent movement of CPC aggregates toward plus ends at ~15 µm/min. CPC recruitment to microtubules depended on an egg ortholog of Kif20A we called Kif20AE, and movement depended on Kif4A as shown by this depletion-add back experiment. (*C*) Widefield imaging of asters nucleated by sperm centrosomes. CPC was imaged with GFP-DasraB. Note formation of polarized monopolar asters with CPC recruited to a crescent at the periphery. The CPC-positive crescent always formed on the chromatin-proximal side of asters and usually expanded radially as the asters grew. (*A,B*, Modified, with permission, from Nguyen et al. 2014; *C*, modified, with permission, from Field et al. 2015.)

the cortex, we layered extract over a supported lipid bilayer and observed recruitment of RhoA.GTP to the bilayer in proximity to CPC recruited to boundaries between asters, showing that the extract system can fully reconstitute spindle-to-cortex signaling (Nguyen et al. 2014).

Chromatin was not required for assembly of CPC- and centralspindlin-positive zones between asters in the extract system (Fig. 4A). However, in the absence of chromatin there was a delay between aster–aster contact and CPC recruitment, and only a subset of aster boundaries recruited CPC. CPC recruitment to zones between asters was greatly enhanced by artificially stabilizing microtubules or activating the CPC (Field et al. 2015). These observations suggested chromatin might promote CPC recruitment to aster boundaries if it was present.

To test if chromosomes could stimulate CPC recruitment, as we suspect occurs in eggs, we nucleated asters from permeabilized sperm nuclei with attached centrosomes. We followed the growth of isolated sperm and noted that, as they grew, they usually accumulated CPC on a crescent-shaped region of the aster periphery (Fig. 4C). When this CPC-positive crescent formed it always did so on the side of the aster closest to the sperm chromatin. The chromatin is visible as a blue dot in the color panels and is shown more clearly by the Dasra probe alone in the last panel. CPC was recruited, in a chromatin-polarized manner, to unipolar microtubules bundles at the aster periphery in this experiment. Polarized recruitment of CPC to the periphery of monopolar asters was observed previously during monopolar cytokinesis in tissue culture cells (Canman et al. 2003; Hu et al. 2008). In that system, CPC and centralspindlin are recruited during anaphase to the side of a monopolar microtubule array that is closest to chromatin, and a partial furrow is later induced on the CPC-positive side.

In the extract monopolar system, CPC was first recruited to the aster periphery around the time when the aster periphery was growing past chromatin (Fig. 4C, 9 min). As asters grew out further, CPC-positive crescents persisted and tended to expand radially (see the upper aster in Fig. 4C). By the end of the experiment, the CPC-positive crescent at the aster periphery was >100 µm from the nearest chromatin, spanned more than half that aster circumference, and contained much more CPC that was present when it initially formed close to chromatin. These observations do not support a simple chromosome passenger model, because the amount of CPC present at the aster periphery at late times is much larger than that which could have been initially recruited from chromatin. CPC recruitment to aster boundaries far from chromatin in frog eggs also seems inconsistent with a chromosome passenger model. Rather, we prefer a model where proximity to chromatin triggers initial formation of a CPC-positive crescent, and, once formed, this crescent persists, grows, and recruits more CPC, all independent of chromatin.

CONCLUSION

Putting together our observations in eggs and egg extract, we propose the model shown in Figure 5. Figure 5A illustrates an early stage in propagation of a spatial signal from the anaphase spindle toward the cortex, where directional arrows and estimated location of chromatin are superimposed on a late-anaphase spindle taken from the earliest time point in Figure 3. The arrows indicate chromatin movement (blue), aster growth (green), and growth of the CPC-positive disc of microtubule bundles (red). This disc expands normal to the axis of chromosome movement and thus transmits information on the position of the metaphase plate outward toward the cortex. Figure 5B illustrates working models for the molecular subprocesses involved in generation, growth, and action of the CPC-positive disc. It initiates in the anaphase spindles at antiparallel microtubule close to chromatin, in an update of Earnshaw's classic chromosome passenger model. CPC is recruited to metaphase chromatin in *Xenopus* eggs by binding of its BIRC5/survivin subunit to histone H3 T3 phosphosites that are generated by haspin kinase (Kelly et al. 2010). We hypothesize that haspin-dependent recruitment of CPC to chromatin is important for the chromatin-to-microtubule handoff during anaphase. Once initiated, the CPC-positive disc grows laterally as the asters grow radially. Reactions involved in growth of the disc likely include lateral recruitment of microtubules from the growing asters by bundle formation, CPC recruitment from solution both directly and via KIF20A-mediated microtubule transport, CPC autoactivation by proximity-driven autophosphorylation, and perhaps direct CPC aggregation. When the disc of CPC-coated microtubule bundles reaches the plasma membrane, it triggers furrow

assembly by some combination of CPC and centralspindlin stimulation of conserved furrow assembly pathways.

The model in Figure 5 provides a preliminary structural and molecular basis for Rappaport's spindle-to-cortex communication in frog eggs. The signal consists of a disc-shaped array of CPC and centralspindlin-coated antiparallel microtubule bundles that grow out from the spindle to the cortex by a combination of aster growth and CPC-powered self-organization. In frog eggs, aster boundaries that lack the initial trigger from chromatin do not acquire aster-inducing potential (Figs. 1C, 3), though they do in echinoderm eggs (Fig. 1B). The model in Figure 5 model makes testable predictions—for example, that we could remove chromatin once the self-organizing CPC-positive disc is established. It poses many unanswered questions at the molecular level—for example, how chromatin triggers the initial assembly of the CPC-positive disc, and how the disc grows by recruiting more microtubules and CPC.

Our focus in investigating the Rappaport signal from spindle to cortex has been on the role of the CPC, whereas others have focused on centralspindlin (Mishima 2016). The CPC is more abundant than centralspindlin in frog eggs (~100 nM vs. ~25 nM) (Wühr et al. 2014), and in our experiments appears to enrich more on microtubules. The CPC plays a major role in organizing microtubule bundles between asters in egg extracts, whereas depleting the Kif23 subunit of centralspindlin had little effect (Nguyen et al. 2014). However, in other systems centralspindlin plays a central role in stimulating RhoA activity at the cortex and inducing furrows. Both complexes are clearly important for spindle-to-cortex communication, although their precise functions differ, as may their relative importance in different systems. Spindle-to-cortex communication has to propagate over an unusually large distance in frog eggs, which might explain their greater reliance on CPC. In particular, the propensity of CPC to auto-activate and aggregate may help the signal travel hundreds of microns. An interesting future question is how the CPC auto-recruitment system is tuned in frog eggs (but not echinoderm eggs) such that the CPC-positive state spreads robustly away from chromatin along antiparallel microtubule bundles, yet does not emerge spontaneously among similar bundles when they are formed between asters that were not part of the same spindle.

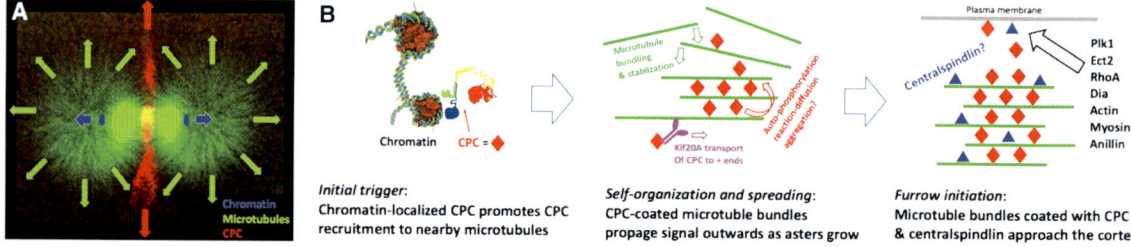

Figure 5. Working model for spindle-to-cortex communication in frog eggs. (*A*) Late-anaphase spindle morphology, shortly after initiating spindle-to-cortex communication. The spindle image is taken from the earliest time point in Figure 3. Arrows indicate directions of movement/growth. Note that the furrow-stimulating CPC-positive disc (red arrows) grows outward on a plane orthogonal to the axis of chromosome separation (blue arrows). (*B*) Models for the molecular events involved in triggering formation of the CPC-positive disc near chromatin (*left*), growing it outward (*middle*), and signaling to the cortex (*right*).

ACKNOWLEDGMENTS

This work was supported by National Institutes of Health (NIH) grant no. GM39565. Microscopy was supported at Harvard Medical School by the Nikon Imaging Center and at the Marine Biological Laboratory by Nikon Inc. We thank the National Xenopus Resource (NXR) for *Xenopus* animals and care.

REFERENCES

Brachet A. 1910. La polyspermie experimental comme moyen d'analyse de la fecondacion. *Arch Entwicklungsmech Org* **30:** 261–303.

Canman JC, Cameron LA, Maddox PS, Straight A, Tirnauer JS, Mitchison TJ, Fang G, Kapoor TM, Salmon ED. 2003. Determining the position of the cell division plane. *Nature* **424:** 1074–1078.

Earnshaw WC, Cooke CA. 1991. Analysis of the distribution of the INCENPs throughout mitosis reveals the existence of a pathway of structural changes in the chromosomes during metaphase and early events in cleavage furrow formation. *J Cell Sci* **98:** 443–461.

Field CM, Groen AC, Nguyen PA, Mitchison TJ. 2015. Spindle-to-cortex communication in cleaving, polyspermic *Xenopus* eggs. *Mol Biol Cell* **26:** 3628–3640.

Field CM, Pelletier JF, Mitchison TJ. 2017. *Xenopus* extract approaches to studying microtubule organization and signaling in cytokinesis. *Methods Cell Biol* **137:** 395–435.

Gruneberg U, Neef R, Honda R., Nigg EA, Barr FA. 2004. Relocation of Aurora B from centromeres to the central spindle at the metaphase to anaphase transition requires MKlp2. *J Cell Biol* **166:** 167–172.

Hara K, Tydeman P, Kirschner M. 1980. A cytoplasmic clock with the same period as the division cycle in *Xenopus* eggs. *Proc Natl Acad Sci* **77:** 462–466.

Herlant M. 1911. Recherches sur les oeufs di-et-trispermiques de grenouille. *Archs Biol* **26:** 103–328.

Hu C-K, Coughlin M, Field CM, Mitchison TJ. 2008. Cell polarization during monopolar cytokinesis. *J Cell Biol* **181:** 195–202.

Hu C-K, Coughlin M, Field CM, Mitchison TJ. 2011. KIF4 regulates midzone length during cytokinesis. *Curr Biol* **21:** 815–824.

Ishihara K, Korolev KS, Mitchison TJ. 2016. Physical basis of large microtubule aster growth. *Elife* **5:** e19145.

Kelly AE, Ghenoiu C, Xue JZ, Zierhut C, Kimura H, Funabiki H. 2010. Survivin reads phosphorylated histone H3 threonine 3 to activate the mitotic kinase Aurora B. *Science* **330:** 235–239.

Mishima M. 2016. Centralspindlin in Rappaport's cleavage signaling. *Semin Cell Dev Biol* **53:** 45–56.

Mitchison T, Wühr M, Nguyen P, Ishihara K, Groen A, Field CM. 2012. Growth, interaction, and positioning of microtubule asters in extremely large vertebrate embryo cells. *Cytoskeleton* **69:** 738–750.

Nguyen PA, Groen AC, Loose M, Ishihara K, Wühr M, Field CM, Mitchison TJ. 2014. Spatial organization of cytokinesis signaling reconstituted in a cell-free system. *Science* **346:** 244–247.

Rappaport R. 1971. Cytokinesis in animal cells. *Int Rev Cytol* **31:** 169–213.

Rappaport R. 1986. Establishment of the mechanism of cytokinesis in animal cells. *Int Rev Cytol* **105:** 245–281.

Rappaport R, Conrad GW. 1963. An experimental analysis of unilateral cleavage in invertebrate eggs. *J Exp Zool* **153:** 99–112.

Ruchaud S, Carmena M, Earnshaw WC. 2007. Chromosomal passengers: Conducting cell division. *Nat Rev Mol Cell Biol* **8:** 798–812.

Sampath SC, Ohi R, Leismann O, Salic A, Pozniakovski A, Funabiki H. 2004. The chromosomal passenger complex is required for chromatin-induced microtubule stabilization and spindle assembly. *Cell* **118:** 187–202.

Savoian MS, Earnshaw WC, Khodjakov A, Rieder CL. 1999. Cleavage furrows formed between centrosomes lacking an intervening spindle and chromosomes contain microtubule bundles, INCENP, and CHO1 but not CENP-E. *Mol Biol Cell* **10:** 297–311.

White EA, Glotzer M. 2012. Centralspindlin: At the heart of cytokinesis. *Cytoskeleton* **69:** 882–892.

Wühr M, Tan ES, Parker SK, Detrich HW III, Mitchison TJ. 2010. A model for cleavage plane determination in early amphibian and fish embryos. *Curr Biol* **20:** 2040–2045.

Wühr M, Freeman RM, Presler M, Horb ME, Peshkin L, Gygi SP, Kirschner MW. 2014. Deep proteomics of the *Xenopus laevis* egg using an mRNA-derived reference database. *Curr Biol* **24:** 1467–1475.

Nucleosome-Dependent Pathways That Control Mitotic Progression

Hironori Funabiki, Christopher Jenness, and Christian Zierhut

Laboratory of Chromosome and Cell Biology, The Rockefeller University, New York, New York 10065

Correspondence: funabih@rockefeller.edu

The majority of eukaryotic chromosomal DNA exists in the form of nucleosomes, where ~147 bp DNA wraps around histone hetero-octamers, composed of histone H3, H4, H2A, and H2B. Despite their obvious importance in DNA compaction and accessibility, studying their specific roles, such as regulation of mitotic progression, in a physiological environment is associated with critical caveats because of their major contributions in transcriptional control. Through establishing a method to deplete endogenous histones H3 and H4 from frog egg extracts and complementing their functions using recombinant nucleosome arrays, we are now able to analyze their roles in mitotic progression without affecting overall transcriptomic profiles. Here we summarize advancements learned from this system, illustrating that microtubule and nuclear envelope assembly can be regulated by two major nucleosome-bound protein complexes, RCC1–Ran and the chromosomal passenger complex (CPC) containing the mitotic protein kinase Aurora B. We also discuss roles of the CPC on the proteomic composition of mitotic chromatin. The CPC promotes dissociation of a variety of nucleosome remodelers and DNA repair pathway proteins, suggesting its role in suppressing DNA processing activities on mitotic chromosomes. We speculate that this suppression particularly on chromosomes under microtubule tension may be important to preserve genome integrity.

Major characteristics that define eukaryotes are the intracellular membrane system, which forms the nucleus to encapsulate genomic DNA, and cell division through mitosis, where replicated genomic DNA is organized into topologically distinct multiple chromosomal threads, which are distributed equally to two daughter cells. DNA replication and chromosome segregation in eukaryotes rely on formation of microscopic-scale architectures around chromosomes, the nuclear envelope, and spindle microtubules, respectively. Pioneering studies in *Xenopus* eggs and their extracts have shown that exposure of DNA to cytoplasm can trigger formation of the nuclear envelope in interphase and spindle microtubule assembly in M phase, in a DNA sequence-independent manner (Forbes et al. 1983; Karsenti et al. 1984; Heald et al. 1996). Here we will review our current understanding of how these DNA-induced processes are controlled by nucleosomes, the fundamental unit that folds genomic DNAs in eukaryotes.

A SYSTEM TO DIRECTLY MANIPULATE HISTONES H3 AND H4 IN *XENOPUS* EGG EXTRACTS

DNA added to *Xenopus* egg extracts is rapidly chromatinized with the large excess of maternally stored histones in egg cytoplasm. To study the roles of nucleosomes, we therefore established a method to deplete histones H3 and H4 from egg extracts and complement them with recombinant proteins (Fig. 1; Zierhut et al. 2014). We used monoclonal antibodies against histone H4 acetylated at Lys12 (H4K12ac) to deplete the H3–H4 complex, and

depleted extracts were complemented with synthetic DNA preassembled with nucleosomes (Fig. 1A,B; Zierhut et al. 2014). As the great majority of histone H4 in eggs is acetylated at Lys5 and Lys12 and forms a complex with H3 (Nicklay et al. 2009), monoclonal antibodies against H4K12ac were able to deplete >90% of histones H3 and H4. Under this experimental condition, reconstituted nucleosome arrays, coupled to magnetic beads, supported spindle assembly in M phase extracts and nuclear envelope formation with nuclear import activity in interphase extracts (Fig. 1C; Zierhut et al. 2014). Naked DNA beads in interphase ΔH3–H4 extracts were able to recruit membranes but were defective in recruiting the nuclear pore complex (NPC). Naked DNA-beads also failed to induce spindle microtubule assembly in M phase ΔH3–H4 extracts. These experiments directly showed the importance of nucleosomes in spindle formation and NPC. An independent study in mouse oocyte, where de novo formation of nucleosomes on sperm nuclei can be inhibited by depleting H3.3 or HIRA, also showed the importance of nucleosome formation for NPC assembly (Inoue and Zhang 2014).

MASS SPECTROMETRY ANALYSIS OF NUCLEOSOME-DEPENDENT BINDING PROTEINS

ΔH3–H4 egg extracts offered a unique opportunity to compare protein constituents that assemble on nucleosomes and nucleosome-free DNA under physiological conditions using quantitative mass spectrometry (MS)

Published by Cold Spring Harbor Laboratory Press; doi: 10.1101/sqb.2017.82.034512

Cold Spring Harbor Symposia on Quantitative Biology, Volume LXXXII

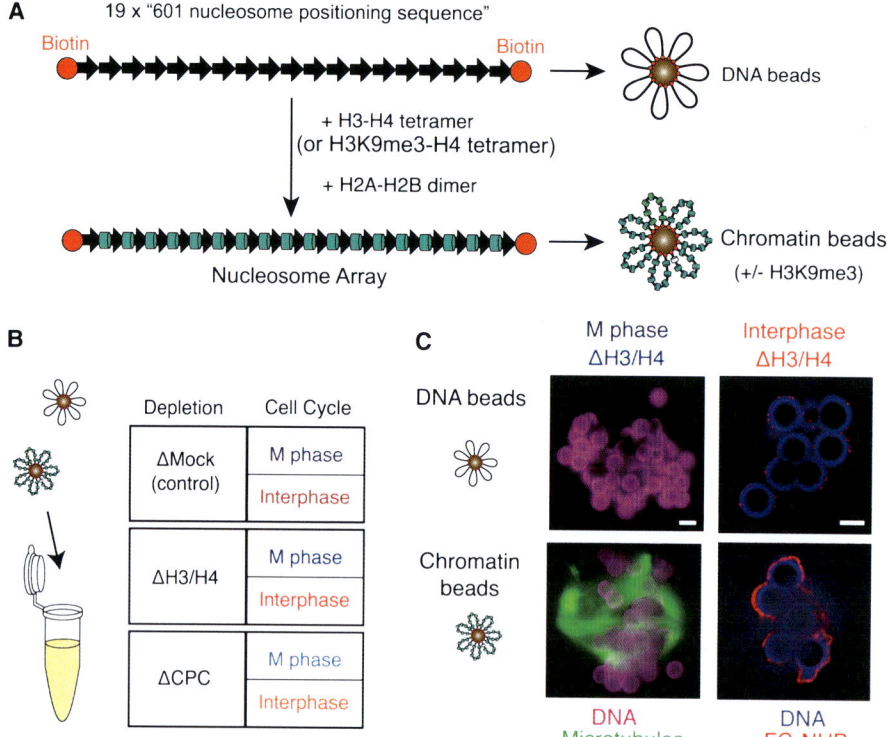

Figure 1. *Xenopus* egg extract system to determine functions of the nucleosome and specific histone modifications. (*A*) End-biotinylated tandem 19-mer arrays of Widom's 601 sequences were coupled to streptavidin-coated magnetic beads, with or without nucleosome assembly by salt dialysis using recombinant core histones. (*B*) *Xenopus* egg extract conditions. (*C*) Spindle formation in M phase and nuclear pore complex (NPC) assembly in interphase on DNA/chromatin beads in the absence of histones H3/H4 were visualized by rhodamine-tubulin and antibodies against the NPC component ELYS. Scale bars, 3 μM. (Modified from Zierhut et al. 2014; Zierhut and Funabiki 2015.)

analysis. In both M phase and interphase extracts, the list of most abundant nucleosome-dependent binders are similar; the linker histone H1M, the FACT complex, Ran–RCC1, DDB1, and the chromosomal passenger complex (CPC) (Zierhut et al. 2014; C Jenness and H Funabiki, unpubl.). Unlike the linker histone and potential nucleosome regulators, FACT and DDB1 (Belotserkovskaya and Reinberg 2004; He et al. 2006; Winkler and Luger 2011), RCC1–Ran and the CPC play critical roles beyond structural organization of mitotic chromatin, as discussed below.

Intriguingly, SMC-family protein complexes, condensin and cohesin, show a mild preference for nucleosome-free DNA over nucleosomal DNA (Zierhut et al. 2014; Hirano 2016; C Jenness and H Funabiki, unpubl.). This may reflect the evolutionary conservation of SMC proteins in prokaryotes, which lack nucleosomes. Similarly, the MCM complex, the AAA family ATPase required for DNA replication initiation, effectively binds to nucleosomal and nucleosome-free DNA (Zierhut et al. 2014; C Jenness and H Funabiki, unpubl.), although it is not clear at present if this reflects functional, topological, binding. It seems that these protein complexes evolved to acquire additional modules/factors to interact and deal with nucleosomal DNA, such as FACT (Kinoshita et al. 2015; Shintomi et al. 2015; Hirano 2016; Kurat et al. 2017). In contrast, major critical roles in eukaryote-specific events, nuclear envelope formation, and spindle microtubule as-

sembly are performed by nucleosome-dependent chromatin proteins, RCC1–Ran and the CPC.

THE RCC1–Ran PATHWAY

The GTP-bound form of the small GTPase Ran controls a number of processes by modulating karyopherin family proteins, such as importins and exportins (Cavazza and Vernos 2015). RanGTP disrupts the interaction between importin β and importin α, which recognize a variety of nuclear proteins that contain classical nuclear localization signals (Fig. 2, steps a and b). In the context of nuclear envelope formation, RanGTP liberates components of NPCs from importins (Walther et al. 2003), whereas in M phase, it releases proteins that promote spindle assembly (Cavazza and Vernos 2015). One of many mechanisms involve TPX2, which promotes microtubule nucleation by tethering tubulin dimers and by activating Aurora A and the γ-tubulin ring complex (γTuRC) (Fig. 2, step c; Groen et al. 2004; Tsai and Zheng 2005; Pinyol et al. 2013; Roostalu et al. 2015; Scrofani et al. 2015; Zhang et al. 2017).

GDP bound to Ran is exchanged with GTP with the help of RCC1 associated with nucleosomes (Fig. 2, step a; Nemergut et al. 2001; Makde et al. 2010). Although RCC1 can directly interact with DNA, an additional interaction with the acidic patch of H2A–H2B in the nucleosome is critical

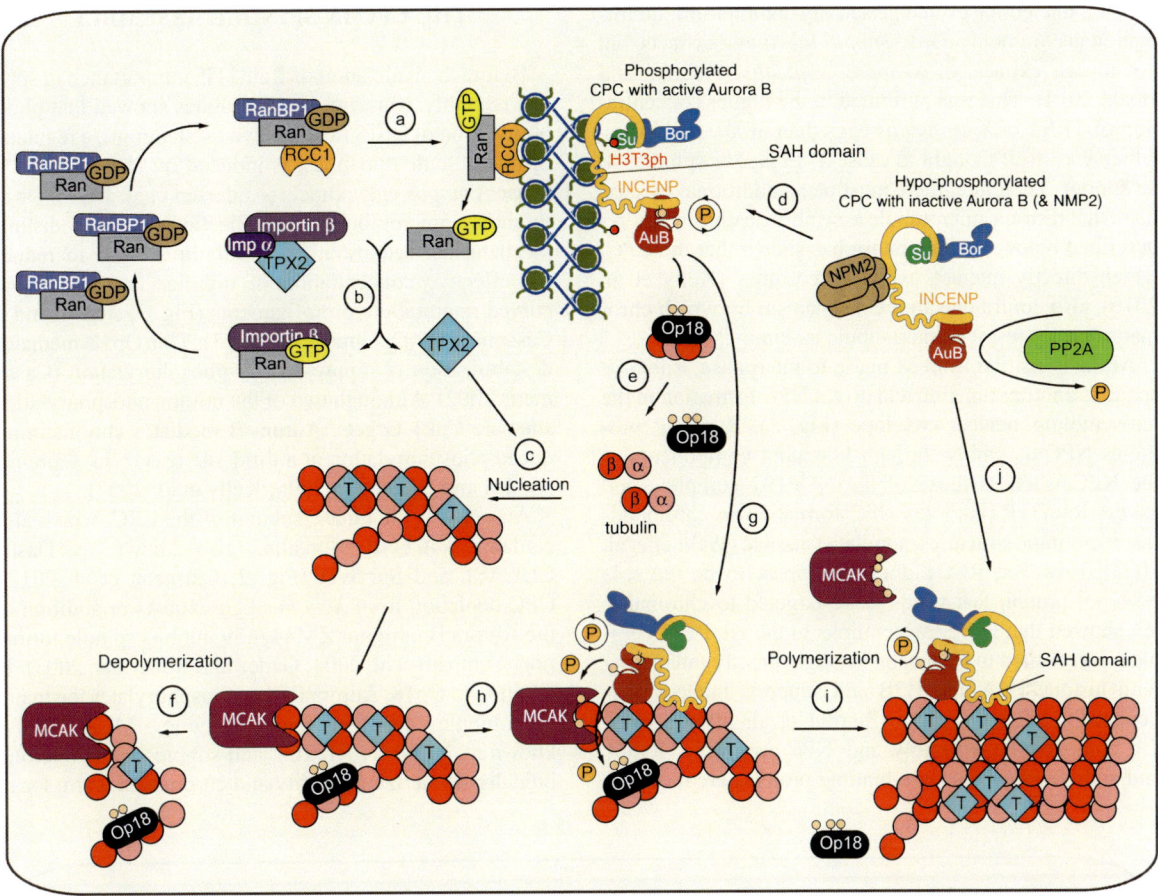

Figure 2. Schematics of chromatin-induced microtubule nucleation. In the cytoplasm distal to chromatin, RanGDP forms a trimeric complex with RCC1 and RanBP1 and is targeted to chromatin via RCC1-nucleosome interaction (step a). On chromatin, GDP is replaced with GTP by RCC1, whose activity is stimulated by nucleosome binding. RanGTP releases a number of proteins, such as TPX2 and other proteins promoting microtubule assembly, from Importins (step b). TPX2 binds and stabilizes interfaces between adjacent α/β tubulin dimers to promote microtubule nucleation (step c). In the cytoplasm distal to chromosomes, the chromosomal passenger complex (CPC; composed of Aurora B, INCENP, Borealin, and Survivin) is kept in hypophosphorylated status by PP2A and is bound to NPM2 oligomers. Upon binding to chromatin through Survivin–H3T3ph interaction and the SAH–chromatin interaction, Aurora B and INCENP become phosphorylated, leading to Aurora B activation (step d). Activated Aurora B phosphorylates Op18, which sequesters tubulin dimers. Aurora B-dependent Op18 phosphorylation releases tubulins to promote microtubule assembly (step e). Hypophosphorylated MCAK and Op18 interact with microtubule ends and stimulate depolymerization (step f). The CPC interacts with microtubules through the SAH domain of INCENP. Upon interaction with microtubules, Aurora B is activated (step g). MCAK and Op18 are phosphorylated by Aurora B (step h). Phosphorylated MCAK and Op18 are inactivated, leading to microtubule polymerization (step i). The CPC can directly bind to microtubules and become activated (step j). (Modified from Zierhut and Funabiki 2015; Wheelock et al. 2017.)

(Makde et al. 2010). Ran has been shown to directly interact with histones H3–H4, but this interaction is very weak, and accordingly, our quantitative MS analysis shows that equivalent amount of Ran and RCC1 exist on mitotic chromatin (Bilbao-Cortés et al. 2002; Zierhut et al. 2014; Jenness et al. 2018), even though the concentration of Ran (5 μM) in egg cytoplasm is more than 30-fold higher than that of RCC1 (150 nM) (Wuhr et al. 2015)

A bead coupled directly with RCC1 proteins can promote bipolar spindle assembly, suggesting that local enrichment of RCC1 can act as an effective trigger for microtubule nucleation and subsequent spindle assembly (Halpin et al. 2011). However, spindles are shorter and spindle microtubule density is lower on an RCC1 bead than on a chromatin bead. This may reflect the involvement of RCC1-independent regulation (e.g., through activation of the CPC, see below) or the presence of additional

mechanisms by which chromatin activates RCC1. Indeed, the catalytic activity of RCC1 is stimulated by nucleosome interaction (Nemergut et al. 2001). Furthermore, RanBP1, whose concentration in eggs is 2 μM, forms a trimeric complex with Ran and RCC1 to inhibit the catalytic activity of RCC1 (Zhang et al. 2014). Despite its abundance, no RanBP1 was detected on purified chromatin (Zierhut et al. 2014; Jenness et al. 2018), indicating that binding of RCC1–RanGDP–RanBP1 to chromatin releases RanBP1, and thus licenses RanGDP for conversion to RanGTP. It would be interesting to know if RanBP1 is defective in binding to RCC1 that is coupled to beads.

Adding a Ran mutant defective in GTP hydrolysis to egg extracts is sufficient to drive microtubule nucleation and assembly, highlighting the importance of local enrichment of RanGTP (Cavazza and Vernos 2015). However, the dominant negative RanT24N mutant, which inhibits

RCC1's nucleotide exchange activity, inhibits spindle formation in commonly used *Xenopus laevis* egg extracts, but not in egg extracts of *Xenopus tropicalis* (Helmke and Heald 2014). This was attributed to the higher concentration of TPX2 in *X. tropicalis* eggs than in *X. laevis* eggs, which was itself thought to cause *X. tropicalis* spindles to be shorter. Therefore, there must be an additional mechanism that restricts microtubule assembly on chromatin. As described below, our laboratory has shown that the CPC, which directly interacts with nucleosomes (Kelly et al. 2010), also contributes to the mechanism by which chromatin locally restricts microtubule assembly.

At the transition from M phase to interphase, RanGTP acquires another function and drives NPC formation in the reassembling nuclear envelope (Fig. 3). RanGTP promotes NPC assembly through liberating components of the NPC, such as those of the NUP107 complex, and excess RanGTP promotes NPC formation in chromatin-free membrane structures, annulate lamellae (Walther et al. 2003). However, RCC1 does not appear to be the sole essential protein that needs to be targeted to chromatin. We showed that ELYS, which links to the NUP107 complex, is recruited to chromatin through directly interacting with histone H2A and H2B and supports nucleosome-dependent NPC formation (Zierhut et al. 2014). Thus, for both spindle assembly and NPC assembly, RCC1 and additional nucleosome-binding proteins are required.

THE CPC IN SPINDLE ASSEMBLY

Before the realization of RanGTP's importance in spindle assembly, Karsenti and colleagues showed that phosphorylation of Op18 (also known as Stathmin), a regulator of microtubule dynamics, is induced by chromatin in M phase *Xenopus* egg extracts (Andersen et al. 1997). Op18 promotes microtubule destabilization by two distinct mechanisms: sequestration of tubulin dimers to reduce the effective concentration of tubulins and binding to curved microtubule protofilaments (Fig. 2, steps e and f; Cassimeris 2002; Gupta et al. 2013). This Op18-mediated destabilization is suppressed by phosphorylation (Cassimeris 2002). Although two of the mitotic phosphorylation sites are Cdk1 targets, Aurora B mediates chromatin-induced phosphorylation at a third site (Ser16 in *Xenopus*) (Gadea and Ruderman 2006; Kelly et al. 2007).

Aurora B is the kinase subunit of the CPC, which also contains INCENP, Borealin (also known as Dasra, CDCA8), and Survivin (Fig. 2; Carmena et al. 2012). CPC depletion from *Xenopus* egg extracts or addition of the Aurora B inhibitor ZM447439 inhibits spindle formation (Sampath et al. 2004; Gadea and Ruderman 2005). In addition to Op18, Aurora B also phosphorylates the major microtubule depolymerizing enzyme, MCAK (also known as XKCM1, KIF2C), and suppresses its microtubule depolymerizing activity and chromosome arm local-

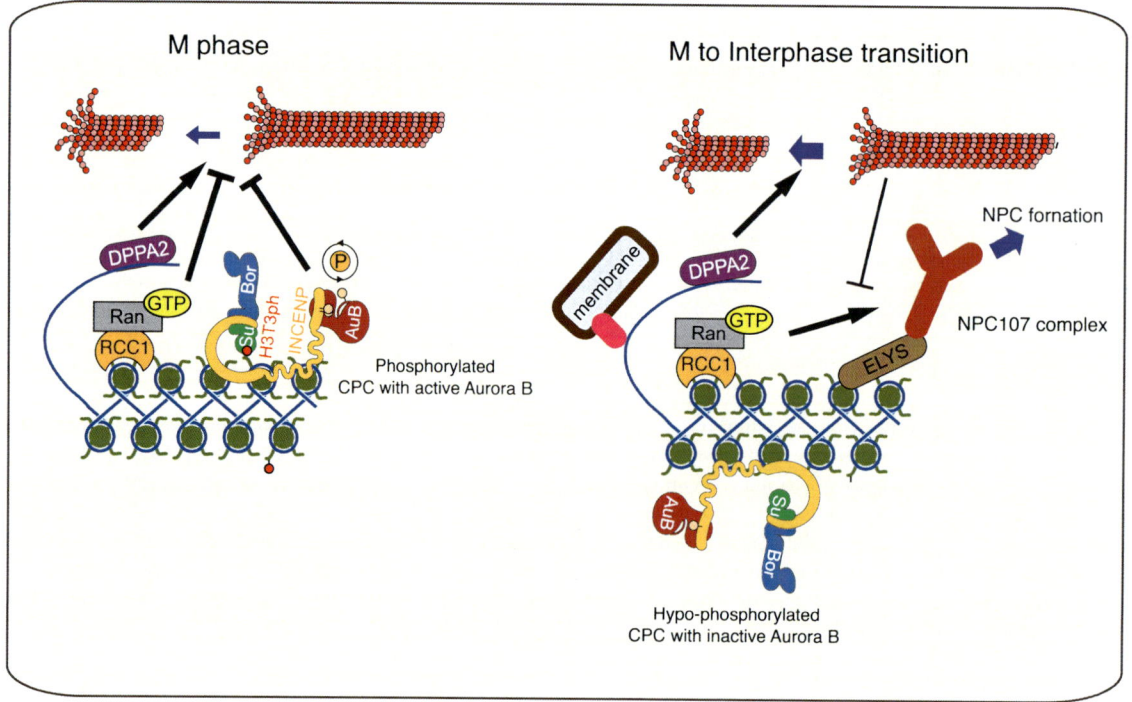

Figure 3. Mechanism of nucleosome-dependent nuclear envelope formation and interference by microtubules. (*Left*) In M phase, the RanGTP pathway and the chromosomal passenmger complex (CPC) pathways suppress microtubule depolymerization, whereas DNA-binding protein DPPA2 promotes it. (*Right*) At the transition into interphase, H3T3 is dephosphorylated, resulting in suppression of Aurora B on chromatin. Microtubules inhibit functional nuclear envelope assembly, but DPPA2-mediated microtubule disassembly facilitates nuclear envelope formation. Nuclear membrane can associate nucleosome-free DNA, whereas NPC formation is supported by at least two nucleosome-dependent mechanisms. First, RanGTP liberates building blocks of nuclear pore complex (NPC) microtubule assembly, such as subunits of the NPC107 complex. Second, specifically during interphase, ELYS, which interacts with the NPC107 complex, is recruited to nucleosomes. (Modified from Zierhut and Funabiki 2015; Wheelock et al. 2017.)

ization (Andrews et al. 2004; Lan et al. 2004; Ohi et al. 2004; Zhang et al. 2007) (Fig. 2, step i; also see Table 2).

The kinase activity of Aurora B is regulated in multiple steps. By itself, Aurora B is a poor protein kinase, but binding of the carboxy-terminal IN-Box module of INCENP allosterically stimulates Aurora B (Bishop and Schumacher 2002; Honda et al. 2003; Sessa et al. 2005). However, binding of INCENP to Aurora B is insufficient to promote effective phosphorylation in egg extracts, where the majority of Aurora B exists in the complex with INCENP along with Dasra A (egg form of Borealin family protein) and Survivin (Bolton et al. 2002; Sampath et al. 2004; Kelly et al. 2007). For full activation of Aurora B, phosphorylation at its catalytic loop (T-loop) and IN-Box must be phosphorylated by Aurora B, but those sites are generally not phosphorylated in egg extracts because of active type 2A (and also likely type 1) phosphatases (Kelly et al. 2007). Nucleoplasmin/nucleophosmin (NPM2) proteins also interact with unphosphorylated, cytoplasmic CPC, although its functional significance remains to be tested (Hanley et al. 2017). However, Aurora B autophosphorylation (a hallmark of kinase activation) can be induced by chromatin or taxol (a microtubule stabilizing drug) in M phase egg extracts (Fig. 2, steps d and j; Kelly et al. 2010; Tseng et al. 2010). Because adding antibodies that cluster the CPC can also promote Aurora B activation, we have proposed that local enrichment of the CPC can activate Aurora B (Kelly et al. 2007). Thus, Aurora B activity can be coupled to intracellular localization.

How can the CPC be recruited to chromatin? We and others showed that Survivin, a CPC subunit important for chromatin targeting (Carvalho et al. 2003; Lens et al. 2003; Yue et al. 2008), directly interacts with the H3 tail when phosphorylated at threonine 3 (H3T3ph) (Fig. 2, step d; Kelly et al. 2010; Wang et al. 2010; Yamagishi et al. 2010). The H3 amino-terminal tail binds to a cleft of the BIR domain of Survivin, in a coordination similar to how the BIR domain of XIAP interacts with SMAC (DIABLO) and caspase-9 (Kelly et al. 2010; Jeyaprakash et al. 2011; Du et al. 2012; Niedzialkowska et al. 2012). In egg extracts, we showed that H3T3ph is critical for chromatin-induced activation of Aurora B, as nucleosomes with a phosphorylation-defective H3T3A mutant fail to activate Aurora B, whereas nucleosomes with a phosphomimetic H3T3E mutant bypass the requirement for the H3T3 kinase Haspin (Kelly et al. 2010; Zierhut et al. 2014). Surprisingly, although Aurora B, INCENP, and Dasra A bind to interphase chromatin at a level comparable to mitotic chromatin (Jenness et al. 2018), activated Aurora B can be seen only on mitotic chromatin and not on interphase chromatin (C Jenness and H Funabiki, unpubl.). This interphase chromatin association may be supported by the single α helix (SAH) domain of INCENP, which also contributes to chromatin binding (Wheelock et al. 2017), but this is not sufficient to activate Aurora B. Thus, beyond the chromatin enrichment, the M phase–specific H3T3ph-Survivin interaction activates Aurora B by an additional mechanism, perhaps involving structural reorganization of the CPC on chromatin related to dimerization capacity of Borealin (Bourhis et al. 2009; Bekier et al. 2015).

The requirement for chromatin-induced mechanisms to activate Aurora B for spindle assembly in *Xenopus* egg extracts can be bypassed by artificial activation of Aurora B through antibody-mediated clustering (Kelly et al. 2007). Adding the CPC clustering anti-INCENP antibody also facilitates assembly of microtubules that are not attached to chromatin. However, Aurora B activation is only sufficient to promote spindle assembly if INCENP interacts with microtubules through the SAH domain (Fig. 2, steps g and j; Tseng et al. 2010). Therefore, the SAH–microtubule interaction may be required for optimal substrate phosphorylation, whereby the microtubule-binding capacity of the CPC facilitates phosphorylation of substrates on microtubules (Noujaim et al. 2014). It may be worth noting that Aurora B–dependent phosphorylation often weakens microtubule-binding activity of substrates (Cheeseman et al. 2006; Wang et al. 2007; Gestaut et al. 2008; Alushin et al. 2010), and this is related to the fact that Aurora B is a basophilic kinase (Alexander et al. 2011), whereas basic amino acids are often used to recognize negatively charged E hooks of tubulins. Thus, it is possible that the Aurora B substrate–microtubule interaction may limit the substrate accessibility by Aurora B, necessitating the CPC–microtubule interaction for effective Aurora B substrate recognition.

A microtubule-targeted fluorescence resonance energy transfer (FRET)-based sensor revealed that Aurora B–dependent phosphorylation on microtubule-bound substrates can be broadly observed across the mitotic spindle in human tissue culture cells (Tseng et al. 2010; Tan and Kapoor 2011). Before the metaphase to anaphase transition, higher phosphorylation levels occur near bulk chromosomes than at regions close to poles. However, on metaphase spindles, the gradient of Aurora B–dependent phosphorylation is not obvious unless Aurora B activity is partially inhibited (Tan and Kapoor 2011; Wang et al. 2011). In contrast, during anaphase, a clear gradient of Aurora B–dependent phosphorylation was seen, centering on the spindle midzone, where the CPC is relocalized from chromosomes at the metaphase to anaphase transition (Fuller et al. 2008; Tan and Kapoor 2011). This difference likely reflects the more stable Aurora B enrichment at the anaphase spindle midzone than in preanaphase mitotic stages when the CPC preferentially enriches on the inner centromere over microtubules.

The weakly tuned microtubule-binding property of INCENP during preanaphase is functionally important. Robust microtubule binding by the INCENP SAH domain requires the adjacent phospho-regulatory domain (PRD) (Wheelock et al. 2017). CDK-dependent phosphorylation of the PRD suppresses, but not completely inhibits, microtubule binding. Strikingly, although deleting the SAH domain prevents the CPC from supporting spindle assembly, replacing the SAH domain with alternative microtubule-binding domain from PRC1 or Tau promotes spontaneous microtubule assembly in the absence of chromatin (Tseng et al. 2010). We therefore speculate that feedback activation of Aurora B by assembled microtubules causes chromatin-independent microtubule assembly, explaining why the SAH–microtubule interaction must be tuned by Cdk1-dependent phosphorylation.

In HeLa cells, where spindle assembly does not require Aurora B activity, this tuned microtubule-binding capacity of INCENP is critical for activation and suppression of the spindle assembly checkpoint (SAC) (Wheelock et al. 2017). We have shown that SAH-microtubule binding is important to support SAC activation in HeLa cells upon taxol treatment, likely through promoting phosphorylation of kinetochore proteins, such as Hec1, and through destabilizing kinetochore–microtubule interactions. When the microtubule–SAH interaction is enhanced by mutating CDK-dependent phosphorylation sites in the PRD to alanines, cells show a difficulty in silencing the SAC even after metaphase plate formation in otherwise untreated cells. Similarly, the microtubule-binding capacity of INCENP, which is under the negative control of CDK1-dependent phosphorylation but not centromere targeting of the CPC, is critical for essential functions of the CPC in budding yeast (Campbell and Desai 2013; Fink et al. 2017). In addition, it was proposed that the interaction between the microtubule plus-end tracking protein EB1 and Aurora B mediates microtubule-dependent Aurora B activation and CPC recruitment to the centromere (Banerjee et al. 2014), and this interaction may also contribute to the function of Aurora B in the SAC.

Despite the capacity of chromatin to stimulate RanGTP production and Aurora B activation, which are both thought to stabilize microtubules, no spatial difference in microtubule stability (and instability) can be found across the metaphase spindle in *Xenopus* egg extracts (Brugués et al. 2012). Based on a series of experiments and mathematical simulations, Needleman and colleagues have proposed that the spindle size and microtubule length distribution can be explained by spatial regulation of microtubule nucleation but not by that of microtubule stabilization. Incorporation of new microtubules is preferred at the center of the spindle, and microtubules are shorter at the poles. This can be explained by enhanced microtubule nucleation at the spindle equator and the transport of the microtubules by motor proteins (Brugués et al. 2012). A chromatin-centered gradient of microtubule nucleation activity can be generated by RanGTP-mediated activation of spindle assembly factors (SAFs) that interact with microtubules (Carazo-Salas et al. 2001; Groen et al. 2004; Cavazza and Vernos 2015; Oh et al. 2016). The Aurora B pathway may also contribute to local microtubule nucleation around chromatin by inhibiting Op18 and MCAK, which can prevent the formation of elongation-competent microtubule plus ends (Wieczorek et al. 2015). Consistent with this idea, it was reported that the Op18–tubulin interaction is suppressed near mitotic chromatin (Niethammer et al. 2004), and that depletion of Haspin, required for chromatin-induced Aurora B activity, decreases spindle size (Kelly et al. 2010). However, unhydrolyzable RanGTP can promote microtubule nucleation in CPC-depleted extracts (Sampath et al. 2004), suggesting that the CPC is not required in the presence of excess RanGTP. It is likely that RanGTP and Aurora B both act in a cooperative manner to recognize chromatin and promote microtubule

assembly, although their importance in providing spatial information of chromatin can be redundant depending on the system. This may explain why depletion of the CPC subunits or Aurora B inactivation in somatic cells and mammalian oocytes may cause chromosome misalignment and spindle morphology defect but usually does not inhibit spindle assembly (Zierhut and Funabiki 2015), whereas the CPC supports kinetochore-induced microtubule nucleation in somatic cells (Tulu et al. 2006) and is absolutely essential for spindle assembly in *Xenopus* egg extracts (Sampath et al. 2004).

FUNCTIONAL COORDINATION OF MICROTUBULE DYNAMICS AND NUCLEAR ENVELOPE FORMATION BY CHROMATIN FACTORS

Although the RCC1-Ran pathway promotes both spindle assembly and nuclear envelope formation (Cavazza and Vernos 2015), the CPC acts to promote spindle assembly but inhibits nuclear envelope formation (Fig. 3). The CPC, which is enriched on centromeres during the preanaphase stages of mitosis, is relocalized to the spindle midzone in anaphase. Preventing this process by either inhibition of the p97–Cdc48 pathway or excessive activity of Haspin, which retains H3T3ph and thus Aurora B activity on chromosomes, delays chromosome decompaction and nuclear envelope formation (Ramadan et al. 2007; Kelly et al. 2010). One of the mechanisms by which Aurora B prevents proper nuclear envelope formation is through microtubules. In *Xenopus* egg extracts, we have shown that a DNA-bound microtubule-destabilizing protein, Dppa2, is required for formation of a sperm pronucleus of proper size and shape (Xue et al. 2013). This function is opposed by the CPC, suggesting that suppression of microtubule–chromosome interaction by chromatin recruitment of Dppa2, as well as concomitant CPC inhibition, is important for proper nuclear assembly. Although artificial microtubule stabilization with taxol recapitulated this effect, suppression of microtubule assembly by nocodazole also delayed kinetics of nuclear expansion, indicating that microtubules play both positive and negative roles in reformation of the nucleus after mitosis, highlighting the importance of spatiotemporal control of microtubule assembly (Xue et al. 2013). The carboxy-terminal microtubule-destabilizing domain of Dppa2 is not conserved in mammalian orthologs, and Dppa2 family proteins are not found in fish and birds, indicating that the role of xDppa2 in microtubule destabilization may not be evolutionarily conserved. However, the inhibitory function of microtubules in nuclear formation appears to be universal for eukaryotes undergoing open mitosis (Lu et al. 2011; Xue and Funabiki 2014).

ROLE OF THE CPC ON THE PROTEOMIC COMPOSITION OF CHROMATIN

Major chromatin substrates of Aurora B are the Ser10 and Ser28 residues of histone H3 (Hsu et al. 2000; Goto et al. 2002), but their molecular functions are largely mysterious. In budding yeast, it was shown that H3S10ph promotes mitotic chromosome compaction through recruiting the histone H4 deacetylase Hst2 (Wilkins et al. 2014), but it is unclear if this process is conserved in vertebrates. In human cells, it was suggested that the serine/arginine-rich splicing factors (SRSFs), SRp20 (SFRS3), and ASF/SF2 (SRFS1), dissociate from mitotic chromosomes by Aurora B–dependent H310 phosphorylation (Loomis et al. 2009). At heterochoromatin, H3S10ph, inhibits bind-

ing of chromodomain of HP1 to its adjacent heterochromatin associated modification, trimethylated Lys9 (H3K9me3) (Fischle et al. 2005; Hirota et al. 2005).

To verify these reported phenomena and to seek novel factors whose binding is influenced by Aurora B–dependent phosphorylation, we examined the impact of CPC depletion on the proteomic profile of chromatin-binding proteins in the presence or absence of H3K9me3 (Fig. 4; Tables 1–3; Jenness et al. 2018). Consistent with our previous findings, we found that HP1 exclusively interacts

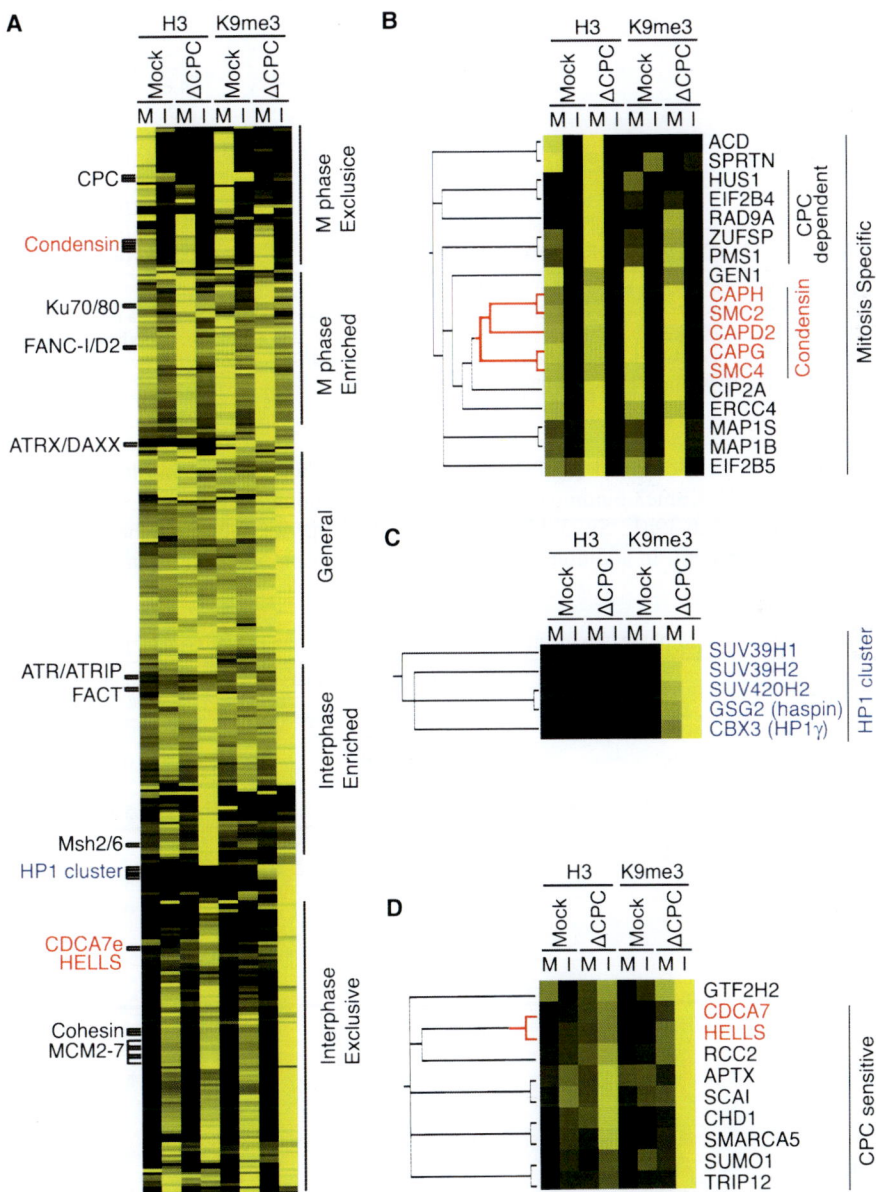

Figure 4. Effect of the chromosomal passenger complex (CPC) depletion, H3K9me3, and cell cycle on the proteomic profile of chromatin. Chromatin beads with or without H3K9me3 were incubated with four different egg extract conditions (control Δmock extracts or ΔCPC extracts in M phase or in interphase) and isolated and then identity and quantity of bound proteins were determined by liquid chromatography–tandem mass chromatography (LC-MS/MS). (*A*) Based on the similarity of relative abundance across the varying conditions, proteins were clustered using hierarchical clustering algorism. Relative abundance was expressed as a heatmap (low/black to high/yellow). (*B*) Clades of proteins exclusively enriched on chromatin in M phase, including a few showing CPC dependency. (*C*) Clades of proteins showing exclusive binding to H3K9me3-nucleosomes in ΔCPC extracts. (*D*) Clade of proteins showing reduced binding to M phase chromatin in a manner dependent on the CPC. (Modified from Jenness et al. 2018.)

Table 1. Most abundant M phase chromatin-associated proteins in a manner dependent on the chromosomal passenger complex (CPC)

Gene	Alt name	Complex	ΔCPC chromatin abundance (A.U.)	Δmock chromatin abundance (A.U.)	Δmock/ΔCPC
CCNB2	Cyclin B2		0	1.8×10^8	—
CDCA8	Dasra A, Borealin	CPC	0	1.7×10^8	—
AURKB	Aurora B	CPC	0	1.3×10^8	—
ZFP161	ZBTB14, ZNF478		0	1.1×10^8	—
TPX2		HURP	3.0×10^7	1.1×10^8	3.8
ZMYM4	ZNF262		0	1.1×10^8	—
DLGAP5	HURP	HURP	1.7×10^7	1.0×10^8	—
PSMB8		Proteasome	0	8.7×10^7	—
INCENP		CPC	0	8.3×10^7	—
EML4	EMAP-4		0	6.2×10^7	—
EVC			0	6.1×10^7	—
TRIM2			0	6.0×10^7	—
RDBP	NELFE		0	5.9×10^7	—
FAM98B			0	5.7×10^7	—
KIFC1	HSET		0	4.6×10^7	—
NME2	NDP-kinase 2		0	3.8×10^7	—
BIRC5	Survivin	CPC	0	3.8×10^7	—
RPSA	Laminin Receptor 1		1.0×10^7	3.7×10^7	3.5
CCNB1	Cyclin B1		0	3.1×10^7	—
MAP4			0	3.0×10^7	—
EEF2	EF2		0	3.0×10^7	—
HMMR	RHAMM	RHAMM-γTuRC-TPX2	0	3.0×10^7	—

Data are generated from Supplemental Table S1 in Jenness et al. 2018.

Abundance (arbitrary units) of proteins that associate with chromatin beads in M phase *Xenopus* egg extracts in a manner dependent on the CPC are shown. The top 22 most abundant proteins (except for tubulins, actins, mitochondrial proteins, and highly abundant glycolytic enzymes) showing at least threefold enrichment on chromatin in control Δmock egg extracts over ΔCPC extracts are listed. Subunits of the CPC and the HURP complex are colored in red and blue, respectively. Other known microtubule-binding proteins are shown in purple.

with H3K9me3 particular in CPC-depleted (ΔCPC) extracts (Fig. 4C). Other known H3K9me3-binding proteins containing a chromodomain, Suv39h1, Suv39h2, and Suv420h2, also showed this pattern. In addition, Haspin (GSG2), whose fission yeast homolog is known to interact with HP1 homolog Swi6 (Yamagishi et al. 2010), showed similar behavior, indicating the evolutionary conservation of HP1-Haspin interaction. In contrast, despite their high

Table 2. Proteins that dissociate from M phase chromatin in a manner dependent on the chromosomal passenger complex (CPC) (I)

Gene	Alt name	Complex	ΔCPC chromatin abundance (A.U.)	Δmock chromatin abundance (A.U.)	ΔCPC/Δmock
CDCA7		CHIRRC	4.6×10^8	0	—
HELLS	LSH, SMARCA6	CHIRRC	4.6×10^8	2.0×10^7	22
CHD1			1.9×10^8	0	—
SMARCA5	SNF2H, ISWI	ASF/WICH/CHRAC/RSF	1.9×10^8	5.5×10^7	3.3
EIF2B3			1.6×10^8	0	—
BAZ1B	WSTF	WICH	1.5×10^8	4.5×10^7	3.3
CCDC39			1.4×10^8	0	—
UBE3C			1.4×10^8	0	—
CUL9			1.3×10^8	0	—
PLCH2			1.1×10^8	0	—
CDH23	Cadherin-23		9.5×10^7	0	—
BAZ1A	ACF1	ASF/CHRAC	7.4×10^7	0	—
TRIP12			6.5×10^7	1.8×10^7	3.6
MDC1			6.3×10^7	1.8×10^7	3.5
ATAD2B			6.3×10^7	0	—
DAZAP2			6.0×10^7	0	—
TTN			5.6×10^7	0	—
AIM1	CRYBG1		5.3×10^7	0	—
PSMD8	RPN12	26S proteasome	5.2×10^7	1.1×10^7	4.8
KIF2C	MCAK, XKCM1		5.0×10^7	8.4×10^6	5.9
RCC2	TD-60		5.0×10^7	1.4×10^7	3.4
IGF2BP3			4.8×10^7	4.2×10^6	11
ATAD2			4.1×10^7	0	—

Data are generated from Supplemental Table S1 in Jenness et al. 2018.

Abundance (arbitrary units) of proteins that associate with chromatin-beads in M phase *Xenopus* egg extracts preferentially in ΔCPC extracts are shown. The top 23 most abundant proteins showing at least threefold enrichment on chromatin in ΔCPC egg extracts over control Δmock extracts are listed. Subunits of the CHIRRC are colored in red. Proteins forming a complex with SMARCA5 are in blue.

Table 3. Proteins that dissociate from M phase chromatin in a manner dependent on the chromosomal passenger complex (CPC) (II)

	Alt name	Complex	ΔCPC chromatin abundance (A.U.)	Δmock chromatin abundance (A.U.)
RAD1		9-1-1	3.2×10^7	0
SGOL1	Sgo1	Shugoshin	3.1×10^7	0
POLE3	CHRAC17	CHRAC	2.9×10^7	0
HUS1		9-1-1	2.9×10^7	0
FEN1			2.1×10^7	0
PRPF19		PRP19	2.0×10^7	0
KPNA2	Importin α	Importin	2.0×10^7	0
MCM7		MCM	2.0×10^7	0
MMS22L		TONSL-MMS22L	1.9×10^7	0
APEX1		SET	1.9×10^7	0
CSNK2A1	Casein kinase 2		1.9×10^7	0
RAD9A		9-1-1	1.8×10^7	0
HIST1H1D	Histone H1.3		1.8×10^7	0
THOC2		THO	1.8×10^7	0
MCM2		MCM	1.8×10^7	0
UHRF1			1.6×10^7	0
TONSL		TONSL-MMS22L	1.6×10^7	0
RDM1	RAD52B		1.6×10^7	0
THOC1		THO	1.5×10^7	4.5×10^6
PPP2R5D	PP2A B56Delta	Shugoshin	1.5×10^7	0
RSF1		RSF	1.4×10^7	0
MCM3		MCM	1.4×10^7	0
PMS2		MutL α	1.1×10^7	0
KPNB1	Importin β	Importin	1.0×10^7	0

Data are generated from Supplemental Table S1 in Jenness et al. 2018.

Abundance (arbitrary units) of notable proteins that associate with chromatin-beads in M phase *Xenopus* egg extracts preferentially in ΔCPC extracts are shown. Proteins forming a complex with SMARCA5 are in blue.

abundance in egg extracts (100–1000 nM) (Wuhr et al. 2015), none of the SRSF proteins could be copurified with chromatin at any conditions, even in ΔCPC extracts (Jenness et al. 2018). SIRT2 is one of the most abundant known histone deacetylases in *Xenopus* eggs (178 nM) (Wuhr et al. 2015), but SIRT2 could also not be detected on purified chromatin at any condition including M phase (Jenness et al. 2018). The only histone deacetylase that could be detected on chromatin was HDAC1, but its chromatin association was restricted to interphase (Jenness et al. 2018). Thus, the proposed interaction between Hst2 and H3S10ph may be too dynamic to detect by our method, limited to anaphase, not conserved in *Xenopus* egg extracts, or require additional conditions that may not be met on the DNA beads that we used.

Several proteins showed CPC-dependent binding to or dissociation from chromatin beads (Tables 1–3). Many of the CDC-dependent binders were microtubule-binding proteins, likely reflecting the role of the CPC in microtubule assembly (Table 1; Jenness et al. 2018). It remains to be clarified if these interactions depend on microtubules that survived our bead-washing procedures. Although it has been suggested that Aurora B contributes to condensin association with mitotic chromosomes (Tada et al. 2011), CPC depletion did not show any impact on chromatin enrichment of condensin and DNA topoisomerase II in *Xenopus* egg extracts (Fig. 4B; MacCallum et al. 2002; Jenness et al. 2018). Proteins whose chromatin association is negatively regulated by the CPC include several nucleosome-remodeling complexes such as CHIRRC (HELLS-CDCA7), WICH (ISWI-WSTF), CHRAC (ISWI-ACF1-POLE3), and RSF (ISWI-RSF1) (Fig. 4D; Tables 2,3; MacCallum et al. 2002; Jenness et al. 2018), but neither H3S10ph nor H3S28ph contributed to dissociation of HELLS and ISWI from chromatin in M phase (Jenness et al. 2018), indicating that these effects are not mediated through phosphorylation of H3S10 or H3S28. In addition to nucleosome remodelers, the CPC also dissociates a variety of proteins involved in DNA repair and replication (Tables 2 and 3). Although future investigations are required to establish the role of H3S10ph and H3S28ph outside the context of H3K9me3, these data suggest that the CPC (and perhaps H3S10ph/H3S28ph) contribute to suppressing DNA processing during mitosis.

An intriguing possibility is that local suppression of the CPC pathway enables targeting some of these factors at specific chromosomes or loci (Fig. 5). For example, chromosome arm binding of Sgo1 and MCAK are suppressed by the CPC (Tables 2,3; Zhang et al. 2007; Rivera et al. 2012), but their centromeric enrichment can be positively regulated by Aurora B (Tanno et al. 2010). A similar mechanism may be relevant when chromosomes undergo missegregation and are lagged behind during anaphase. Aurora B–dependent phosphorylation on chromosome substrates (e.g., H3S10) is maintained on lagging chromosomes even when segregated chromosomes are dephosphorylated (Su et al. 1999; Fuller et al. 2008). As proposed for regulation of nuclear envelope reassembly (Afonso et al. 2014), preventing loading of nucleosome remodelers and other DNA processing proteins to lagging chromosomes may be important to preserve genome integrity, particularly, when these chromosomes are interacting with microtubules (Fig. 5). The pushing and pulling forces generated by a single microtubule fiber are estimated to be ~50 pN (Nicklas 1983; Jannink et al. 1996; Grishchuk et al. 2005; Bloom 2008). This is comparable to the

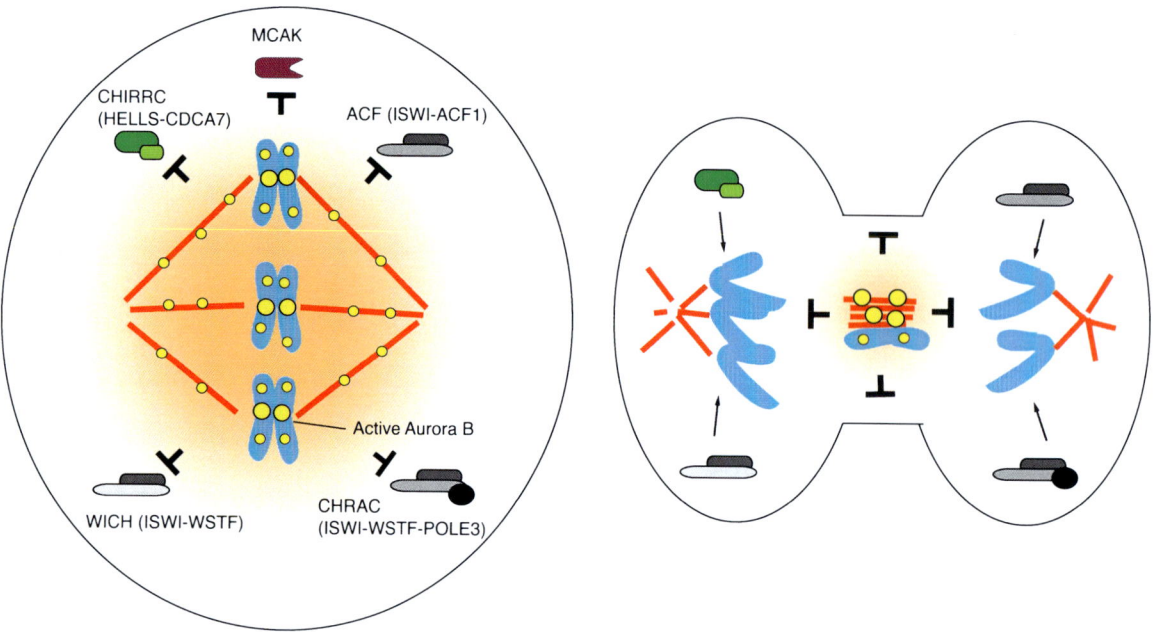

Figure 5. Hypothetical role of Aurora B–dependent removal of mitotic chromatin proteins. During preanaphase stages of mitosis (*left*), Aurora B activated by chromosomes and/or microtubules promote dissociation of the microtubule-destabilizing protein MCAK and chromatin-remodeling complexes, CHIRRC, ACF, CHRAC, and WICH. In an anaphase/telophase cell with a lagging chromosome associated with spindle midzone (*right*), Aurora B–dependent phosphorylation is restricted to regions adjacent to the midzone, where reloading of the chromatin remodeling complexes is inhibited.

measured force (20 pN) required to evict nucleosomes from DNA in vitro (Cui and Bustamante 2000; Bloom 2008). In *Xenopus* interphase egg extracts, as little as 4 pN is sufficient to unwrap nucleosomes in the absence of ATP, and the presence of ATP further destabilizes nucleosomes, making them vulnerable to a force as low as 1 pN (Yan et al. 2007). It would be interesting to explore the possibility that displacement of chromatin remodeling or processing proteins is important to suppress microtubule-dependent alternation of chromatin structure and the formation of DNA damage in lagging chromosomes.

CONCLUSION

Nucleosomes are critical for microtubule formation during M phase and NPC assembly in interphase. RCC1–Ran and the CPC are key nucleosome-binding components that regulate both of these processes. Although RCC1–Ran acts positively for both events, the CPC promotes spindle assembly but suppresses nuclear formation. Many nucleosome-binding factors, such as nucleosome-remodeling proteins and histone chaperones, are used to control DNA accessibility. Unlike these regulators, RCC1–RanGTP and the CPC help to form eukaryote-specific macrostructure assembly through regulating proteins that are not necessarily directly interacting with chromatin. In addition, it has been suggested that RanGTP converts the chromatin-remodeling factor ISWI into a microtubule-binding protein to control anaphase spindle stability (Yokoyama et al. 2009). Therefore, it is plausible that

other nucleosome regulators may also have other distinct functions.

Our efforts to understand nucleosome-dependent and -independent processes raise the question of why nuclear envelope formation is broken into steps that can be mediated by nucleosome-free DNA, which can recruit membranes, and steps that require nucleosomes, which are necessary for NPC assembly. We speculate that this mechanism may be actively used to avoid spontaneous formation of functional nuclei from exogenous nucleosome-free DNA, such as that of viruses and other pathogens. An exception is the DNA provided by sperm, where sperm-specific protamines tightly pack DNA in a manner that prevents DNA replication and transcription. Thus, fertilization is a unique developmental event where external DNAs that penetrate into the egg cytoplasm rapidly assemble into nucleosomes. This is made possible by pre-stored soluble histone pools and histone chaperones that support de novo nucleosome assembly. In contrast, invasion of foreign DNA into somatic cells, whose cytoplasm harbors little soluble histones, may trigger innate immune response (Chen et al. 2016). We hypothesized that nucleosome-dependency of spindle assembly and NPC formation allows eukaryotic cells to distinguish between foreign nucleosome-free DNAs, such as those from virus and bacteria, and genomic nucleosomal DNAs, so that foreign DNAs cannot easily hijack the replication and segregation system (Zierhut et al. 2014; Zierhut and Funabiki 2015). We are currently exploring the possibility that nucleosomes could suppress a response by a cytoplasmic DNA sensor involved in innate immune response (Zierhut and Funabiki 2017). It is tempting to speculate that suppression of nucleosome remodeling and unwrapping on lagging chromosomes inhibits such sensors that recognize pathogenic or foreign DNA. In this sense, nucleosome loss in mitosis can be sensed as an aberration, and if difficult to be repaired, it may trigger a mechanism to purge these problematic cells.

ACKNOWLEDGMENTS

The research of H.F. is supported by grants from the National Institutes of Health (R01GM075249 and R01GM125302).

REFERENCES

Afonso O, Matos I, Pereira AJ, Aguiar P, Lampson MA, Maiato H. 2014. Feedback control of chromosome separation by a midzone Aurora B gradient. *Science* **345:** 332–336.

Alexander J, Lim D, Joughin BA, Hegemann B, Hutchins JR, Ehrenberger T, Ivins F, Sessa F, Hudecz O, Nigg EA, et al. 2011. Spatial exclusivity combined with positive and negative selection of phosphorylation motifs is the basis for context-dependent mitotic signaling. *Sci Signal* **4:** ra42.

Alushin GM, Ramey VH, Pasqualato S, Ball DA, Grigorieff N, Musacchio A, Nogales E. 2010. The Ndc80 kinetochore complex forms oligomeric arrays along microtubules. *Nature* **467:** 805–810.

Andersen SS, Ashford AJ, Tournebize R, Gavet O, Sobel A, Hyman AA, Karsenti E. 1997. Mitotic chromatin regulates phosphorylation of Stathmin/Op18. *Nature* **389:** 640–643.

Andrews PD, Ovechkina Y, Morrice N, Wagenbach M, Duncan K, Wordeman L, Swedlow JR. 2004. Aurora B regulates MCAK at the mitotic centromere. *Dev Cell* **6:** 253–268.

Banerjee B, Kestner CA, Stukenberg PT. 2014. EB1 enables spindle microtubules to regulate centromeric recruitment of Aurora B. *J Cell Biol* **204:** 947–963.

Bekier ME, Mazur T, Rashid MS, Taylor WR. 2015. Borealin dimerization mediates optimal CPC checkpoint function by enhancing localization to centromeres and kinetochores. *Nat Commun* **6:** 6775.

Belotserkovskaya R, Reinberg D. 2004. Facts about FACT and transcript elongation through chromatin. *Curr Opin Genet Dev* **14:** 139–146.

Bilbao-Cortés D, Hetzer M, Längst G, Becker PB, Mattaj IW. 2002. Ran binds to chromatin by two distinct mechanisms. *Curr Biol* **12:** 1151–1156.

Bishop JD, Schumacher JM. 2002. Phosphorylation of the carboxyl terminus of inner centromere protein (INCENP) by the Aurora B Kinase stimulates Aurora B kinase activity. *J Biol Chem* **277:** 27577–27580.

Bloom KS. 2008. Beyond the code: The mechanical properties of DNA as they relate to mitosis. *Chromosoma* **117:** 103–110.

Bolton MA, Lan W, Powers SE, McCleland ML, Kuang J, Stukenberg PT. 2002. Aurora B kinase exists in a complex with survivin and INCENP and its kinase activity is stimulated by survivin binding and phosphorylation. *Mol Biol Cell* **13:** 3064–3077.

Bourhis E, Lingel A, Phung Q, Fairbrother WJ, Cochran AG. 2009. Phosphorylation of a borealin dimerization domain is required for proper chromosome segregation. *Biochemistry* **48:** 6783–6793.

Brugués J, Nuzzo V, Mazur E, Needleman DJ. 2012. Nucleation and transport organize microtubules in metaphase spindles. *Cell* **149:** 554–564.

Campbell CS, Desai A. 2013. Tension sensing by Aurora B kinase is independent of survivin-based centromere localization. *Nature* **497:** 118–121.

Carazo-Salas RE, Gruss OJ, Mattaj IW, Karsenti E. 2001. RanGTP coordinates regulation of microtubule nucleation and dynamics during mitotic-spindle assembly. *Nat Cell Biol* **3:** 228–234.

Carmena M, Wheelock M, Funabiki H, Earnshaw WC. 2012. The chromosomal passenger complex (CPC): From easy rider to the godfather of mitosis. *Nat Rev Mol Cell Biol* **13:** 789–803.

Carvalho A, Carmena M, Sambade C, Earnshaw WC, Wheatley SP. 2003. Survivin is required for stable checkpoint activation in taxol-treated HeLa cells. *J Cell Sci* **116**(Pt 14): 2987–2998.

Cassimeris L. 2002. The oncoprotein 18/stathmin family of microtubule destabilizers. *Curr Opin Cell Biol* **14:** 18–24.

Cavazza T, Vernos I. 2015. The RanGTP pathway: From nucleocytoplasmic transport to spindle assembly and beyond. *Front Cell Dev Biol* **3:** 82.

Cheeseman IM, Chappie JS, Wilson-Kubalek EM, Desai A. 2006. The conserved KMN network constitutes the core microtubule-binding site of the kinetochore. *Cell* **127:** 983–997.

Chen Q, Sun L, Chen ZJ. 2016. Regulation and function of the cGAS-STING pathway of cytosolic DNA sensing. *Nat Immunol* **17:** 1142–1149.

Cui Y, Bustamante C. 2000. Pulling a single chromatin fiber reveals the forces that maintain its higher-order structure. *Proc Natl Acad Sci* **97:** 127–132.

Du J, Kelly AE, Funabiki H, Patel DJ. 2012. Structural basis for recognition of H3T3ph and Smac/DIABLO N-terminal peptides by human Survivin. *Structure* **20:** 185–195.

Fink S, Turnbull K, Desai A, Campbell CS. 2017. An engineered minimal chromosomal passenger complex reveals a role for INCENP/Sli15 spindle association in chromosome biorientation. *J Cell Biol* **216:** 911–923.

Fischle W, Tseng BS, Dormann HL, Ueberheide BM, Garcia BA, Shabanowitz J, Hunt DF, Funabiki H, Allis CD. 2005. Regulation of HP1-chromatin binding by histone H3 methylation and phosphorylation. *Nature* **438:** 1116–1122.

Forbes DJ, Kirschner MW, Newport JW. 1983. Spontaneous formation of nucleus-like structures around bacteriophage DNA microinjected into *Xenopus* eggs. *Cell* **34:** 13–23.

Fuller BG, Lampson MA, Foley EA, Rosasco-Nitcher S, Le KV, Tobelmann P, Brautigan DL, Stukenberg PT, Kapoor TM. 2008. Midzone activation of aurora B in anaphase produces an intracellular phosphorylation gradient. *Nature* **453:** 1132–1136.

Gadea BB, Ruderman JV. 2005. Aurora kinase inhibitor ZM447439 blocks chromosome-induced spindle assembly, the completion of chromosome condensation, and the establishment of the spindle integrity checkpoint in *Xenopus* egg extracts. *Mol Biol Cell* **16:** 1305–1318.

Gadea BB, Ruderman JV. 2006. Aurora B is required for mitotic chromatin-induced phosphorylation of Op18/Stathmin. *Proc Natl Acad Sci* **103:** 4493–4498.

Gestaut DR, Graczyk B, Cooper J, Widlund PO, Zelter A, Wordeman L, Asbury CL, Davis TN. 2008. Phosphoregulation and depolymerization-driven movement of the Dam1 complex do not require ring formation. *Nat Cell Biol* **10:** 407–414.

Goto H, Yasui Y, Nigg EA, Inagaki M. 2002. Aurora-B phosphorylates Histone H3 at serine28 with regard to the mitotic chromosome condensation. *Genes Cells* **7:** 11–17.

Grishchuk EL, Molodtsov MI, Ataullakhanov FI, McIntosh JR. 2005. Force production by disassembling microtubules. *Nature* **438:** 384–388.

Groen AC, Cameron LA, Coughlin M, Miyamoto DT, Mitchison TJ, Ohi R. 2004. XRHAMM functions in ran-dependent microtubule nucleation and pole formation during anastral spindle assembly. *Curr Biol* **14:** 1801–1811.

Gupta KK, Li C, Duan A, Alberico EO, Kim OV, Alber MS, Goodson HV. 2013. Mechanism for the catastrophe-promoting activity of the microtubule destabilizer Op18/stathmin. *Proc Natl Acad Sci* **110:** 20449–20454.

Halpin D, Kalab P, Wang J, Weis K, Heald R. 2011. Mitotic spindle assembly around RCC1-coated beads in *Xenopus* egg extracts. *PLoS Biol* **9:** e1001225.

Hanley ML, Yoo TY, Sonnett M, Needleman DJ, Mitchison TJ. 2017. Chromosomal passenger complex hydrodynamics suggests chaperoning of the inactive state by nucleoplasmin/nucleophosmin. *Mol Biol Cell* **28:** 1444–1456.

He YJ, McCall CM, Hu J, Zeng Y, Xiong Y. 2006. DDB1 functions as a linker to recruit receptor WD40 proteins to CUL4-ROC1 ubiquitin ligases. *Genes Dev* **20:** 2949–2954.

Heald R, Tournebize R, Blank T, Sandaltzopoulos R, Becker P, Hyman A, Karsenti E. 1996. Self-organization of microtubules into bipolar spindles around artificial chromosomes in *Xenopus* egg extracts. *Nature* **382:** 420–425.

Helmke KJ, Heald R. 2014. TPX2 levels modulate meiotic spindle size and architecture in *Xenopus* egg extracts. *J Cell Biol* **206:** 385–393.

Hirano T. 2016. Condensin-based chromosome organization from bacteria to vertebrates. *Cell* **164:** 847–857.

Hirota T, Lipp JJ, Toh BH, Peters JM. 2005. Histone H3 serine 10 phosphorylation by Aurora B causes HP1 dissociation from heterochromatin. *Nature* **438:** 1176–1180.

Honda R, Korner R, Nigg EA. 2003. Exploring the functional interactions between Aurora B, INCENP, and survivin in mitosis. *Mol Biol Cell* **14:** 3325–3341.

Hsu JY, Sun ZW, Li X, Reuben M, Tatchell K, Bishop DK, Grushcow JM, Brame CJ, Caldwell JA, Hunt DF, et al. 2000. Mitotic phosphorylation of histone H3 is governed by Ipl1/aurora kinase and Glc7/PP1 phosphatase in budding yeast and nematodes. *Cell* **102:** 279–291.

Inoue A, Zhang Y. 2014. Nucleosome assembly is required for nuclear pore complex assembly in mouse zygotes. *Nat Struct Mol Biol* **21:** 609–616.

Jannink G, Duplantier B, Sikorav JL. 1996. Forces on chromosomal DNA during anaphase. *Biophys J* **71:** 451–465.

Jenness C, Giunta S, Müller MM, Kimura H, Muir TW, Funabiki H. 2018. HELLS and CDCA7 comprise a bipartite nucleosome remodeling complex defective in ICF syndrome. *Proc Natl Acad Sci* doi: 10.1073/pnas.1717509115.

Jeyaprakash AA, Basquin C, Jayachandran U, Conti E. 2011. Structural basis for the recognition of phosphorylated histone h3 by the survivin subunit of the chromosomal passenger complex. *Structure* **19:** 1625–1634.

Karsenti E, Newport J, Kirschner M. 1984. Respective roles of centrosomes and chromatin in the conversion of microtubule arrays from interphase to metaphase. *J Cell Biol* **99**(1 Pt 2): 47s–54s.

Kelly AE, Sampath SC, Maniar TA, Woo EM, Chait BT, Funabiki H. 2007. Chromosomal enrichment and activation of the aurora B pathway are coupled to spatially regulate spindle assembly. *Dev Cell* **12:** 31–43.

Kelly AE, Ghenoiu C, Xue JZ, Zierhut C, Kimura H, Funabiki H. 2010. Survivin reads phosphorylated histone H3 threonine 3 to activate the mitotic kinase Aurora B. *Science* **330:** 235–239.

Kinoshita K, Kobayashi TJ, Hirano T. 2015. Balancing acts of two HEAT subunits of condensin I support dynamic assembly of chromosome axes. *Dev Cell* **33:** 94–106.

Kurat CF, Yeeles JT, Patel H, Early A, Diffley JF. 2017. Chromatin controls DNA replication origin selection, lagging-strand synthesis, and replication fork rates. *Mol Cell* **65:** 117–130.

Lan W, Zhang X, Kline-Smith SL, Rosasco SE, Barrett-Wilt GA, Shabanowitz J, Hunt DF, Walczak CE, Stukenberg PT. 2004. Aurora B phosphorylates centromeric MCAK and regulates its localization and microtubule depolymerization activity. *Curr Biol* **14:** 273–286.

Lens SM, Wolthuis RM, Klompmaker R, Kauw J, Agami R, Brummelkamp T, Kops G, Medema RH. 2003. Survivin is required for a sustained spindle checkpoint arrest in response to lack of tension. *Embo J* **22:** 2934–2947.

Loomis RJ, Naoe Y, Parker JB, Savic V, Bozovsky MR, Macfarlan T, Manley JL, Chakravarti D. 2009. Chromatin binding of SRp20 and ASF/SF2 and dissociation from mitotic chromosomes is modulated by histone H3 serine 10 phosphorylation. *Mol Cell* **33:** 450–461.

Lu L, Ladinsky MS, Kirchhausen T. 2011. Formation of the postmitotic nuclear envelope from extended ER cisternae precedes nuclear pore assembly. *J Cell Biol* **194:** 425–440.

MacCallum DE, Losada A, Kobayashi R, Hirano T. 2002. ISWI remodeling complexes in *Xenopus* egg extracts: Identification as major chromosomal components that are regulated by INCENP-aurora B. *Mol Biol Cell* **13:** 25–39.

Makde RD, England JR, Yennawar HP, Tan S. 2010. Structure of RCC1 chromatin factor bound to the nucleosome core particle. *Nature* **467:** 562–566.

Nemergut ME, Mizzen CA, Stukenberg T, Allis CD, Macara IG. 2001. Chromatin docking and exchange activity enhancement of RCC1 by histones H2A and H2B. *Science* **292:** 1540–1543.

Nicklas RB. 1983. Measurements of the force produced by the mitotic spindle in anaphase. *J Cell Biol* **97:** 542–548.

Nicklay JJ, Shechter D, Chitta RK, Garcia BA, Shabanowitz J, Allis CD, Hunt DF. 2009. Analysis of histones in *Xenopus laevis*. II. Mass spectrometry reveals an index of cell type-specific modifications on H3 and H4. *J Biol Chem* **284:** 1075–1085.

Niedzialkowska E, Wang F, Porebski PJ, Minor W, Higgins JM, Stukenberg PT. 2012. Molecular basis for phosphospecific recognition of histone H3 tails by Survivin paralogues at inner centromeres. *Mol Biol Cell* **23:** 1457–1466.

Niethammer P, Bastiaens P, Karsenti E. 2004. Stathmin-tubulin interaction gradients in motile and mitotic cells. *Science* **303:** 1862–1866.

Noujaim M, Bechstedt S, Wieczorek M, Brouhard GJ. 2014. Microtubules accelerate the kinase activity of Aurora-B by a reduction in dimensionality. *PloS One* **9:** e86786.

Oh D, Yu CH, Needleman DJ. 2016. Spatial organization of the Ran pathway by microtubules in mitosis. *Proc Natl Acad Sci* **113:** 8729–8734.

Ohi R, Sapra T, Howard J, Mitchison TJ. 2004. Differentiation of cytoplasmic and meiotic spindle assembly MCAK functions by Aurora B-dependent phosphorylation. *Mol Biol Cell* **15:** 2895–2906.

Pinyol R, Scrofani J, Vernos I. 2013. The role of NEDD1 phosphorylation by Aurora A in chromosomal microtubule nucleation and spindle function. *Curr Biol* **23:** 143–149.

Ramadan K, Bruderer R, Spiga FM, Popp O, Baur T, Gotta M, Meyer HH. 2007. Cdc48/p97 promotes reformation of the nucleus by extracting the kinase Aurora B from chromatin. *Nature* **450:** 1258–1262.

Rivera T, Ghenoiu C, Rodriguez-Corsino M, Mochida S, Funabiki H, Losada A. 2012. *Xenopus* Shugoshin 2 regulates the spindle assembly pathway mediated by the chromosomal passenger complex. *EMBO J* **31:** 1467–1479.

Roostalu J, Cade NI, Surrey T. 2015. Complementary activities of TPX2 and chTOG constitute an efficient importin-regulated microtubule nucleation module. *Nat Cell Biol* **17:** 1422–1434.

Sampath SC, Ohi R, Leismann O, Salic A, Pozniakovski A, Funabiki H. 2004. The chromosomal passenger complex is required for chromatin-induced microtubule stabilization and spindle assembly. *Cell* **118:** 187–202.

Scrofani J, Sardon T, Meunier S, Vernos I. 2015. Microtubule nucleation in mitosis by a RanGTP-dependent protein complex. *Curr Biol* **25:** 131–140.

Sessa F, Mapelli M, Ciferri C, Tarricone C, Areces LB, Schneider TR, Stukenberg PT, Musacchio A. 2005. Mechanism of Aurora B activation by INCENP and inhibition by hesperadin. *Mol Cell* **18:** 379–391.

Shintomi K, Takahashi TS, Hirano T. 2015. Reconstitution of mitotic chromatids with a minimum set of purified factors. *Nat Cell Biol* **17:** 1014–1023.

Su TT, Campbell SD, O'Farrell PH. 1999. *Drosophila* grapes/CHK1 mutants are defective in cyclin proteolysis and coordination of mitotic events. *Curr Biol* **9:** 919–922.

Tada K, Susumu H, Sakuno T, Watanabe Y. 2011. Condensin association with histone H2A shapes mitotic chromosomes. *Nature* **474:** 477–483.

Tan L, Kapoor TM. 2011. Examining the dynamics of chromosomal passenger complex (CPC)-dependent phosphorylation during cell division. *Proc Natl Acad Sci* **108:** 16675–16680.

Tanno Y, Kitajima TS, Honda T, Ando Y, Ishiguro K, Watanabe Y. 2010. Phosphorylation of mammalian Sgo2 by Aurora B recruits PP2A and MCAK to centromeres. *Genes Dev* **24:** 2169–2179.

Tsai MY, Zheng Y. 2005. Aurora A kinase-coated beads function as microtubule-organizing centers and enhance RanGTP-induced spindle assembly. *Curr Biol* **15:** 2156–2163.

Tseng BS, Tan L, Kapoor TM, Funabiki H. 2010. Dual detection of chromosomes and microtubules by the chromosomal passenger complex drives spindle assembly. *Dev Cell* **18:** 903–912.

Tulu US, Fagerstrom C, Ferenz NP, Wadsworth P. 2006. Molecular requirements for kinetochore-associated microtubule formation in mammalian cells. *Curr Biol* **16:** 536–541.

Walther TC, Askjaer P, Gentzel M, Habermann A, Griffiths G, Wilm M, Mattaj IW, Hetzer M. 2003. RanGTP mediates nuclear pore complex assembly. *Nature* **424:** 689–694.

Wang HW, Ramey VH, Westermann S, Leschziner AE, Welburn JP, Nakajima Y, Drubin DG, Barnes G, Nogales E. 2007. Architecture of the Dam1 kinetochore ring complex and implications for microtubule-driven assembly and force-coupling mechanisms. *Nat Struct Mol Biol* **14:** 721–726.

Wang F, Dai J, Daum JR, Niedzialkowska E, Banerjee B, Stukenberg PT, Gorbsky GJ, Higgins JM. 2010. Histone H3 Thr-3 phosphorylation by Haspin positions Aurora B at centromeres in mitosis. *Science* **330:** 231–235.

Wang E, Ballister ER, Lampson MA. 2011. Aurora B dynamics at centromeres create a diffusion-based phosphorylation gradient. *J Cell Biol* **194:** 539–549.

Wheelock MS, Wynne DJ, Tseng BS, Funabiki H. 2017. Dual recognition of chromatin and microtubules by INCENP is important for mitotic progression. *J Cell Biol* **216:** 925–941.

Wieczorek M, Bechstedt S, Chaaban S, Brouhard GJ. 2015. Microtubule-associated proteins control the kinetics of microtubule nucleation. *Nat Cell Biol* **17:** 907–916.

Wilkins BJ, Rall NA, Ostwal Y, Kruitwagen T, Hiragami-Hamada K, Winkler M, Barral Y, Fischle W, Neumann H. 2014. A cascade of histone modifications induces chromatin condensation in mitosis. *Science* **343:** 77–80.

Winkler DD, Luger K. 2011. The histone chaperone FACT: Structural insights and mechanisms for nucleosome reorganization. *J Biol Chem* **286:** 18369–18374.

Wuhr M, Güttler T, Peshkin L, McAlister GC, Sonnett M, Ishihara K, Groen AC, Presler M, Erickson BK, Mitchison TJ,

et al. 2015. The nuclear proteome of a vertebrate. *Curr Biol* **25:** 2663–2671.

Xue JZ, Funabiki H. 2014. Nuclear assembly shaped by microtubule dynamics. *Nucleus* **5:** 40–46.

Xue JZ, Woo EM, Postow L, Chait BT, Funabiki H. 2013. Chromatin-bound *Xenopus* Dppa2 shapes the nucleus by locally inhibiting microtubule assembly. *Dev Cell* **27:** 47–59.

Yamagishi Y, Honda T, Tanno Y, Watanabe Y. 2010. Two histone marks establish the inner centromere and chromosome bi-orientation. *Science* **330:** 239–243.

Yan J, Maresca TJ, Skoko D, Adams CD, Xiao B, Christensen MO, Heald R, Marko JF. 2007. Micromanipulation studies of chromatin fibers in *Xenopus* egg extracts reveal ATP-dependent chromatin assembly dynamics. *Mol Biol Cell* **18:** 464–474.

Yokoyama H, Rybina S, Santarella-Mellwig R, Mattaj IW, Karsenti E. 2009. ISWI is a RanGTP-dependent MAP required for chromosome segregation. *J Cell Biol* **187:** 813–829.

Yue Z, Carvalho A, Xu Z, Yuan X, Cardinale S, Ribeiro S, Lai F, Ogawa H, Gudmundsdottir E, Gassmann R, et al. 2008. Deconstructing Survivin: Comprehensive genetic analysis of Survivin function by conditional knockout in a vertebrate cell line. *J Cell Biol* **183:** 279–296.

Zhang X, Lan W, Ems-McClung SC, Stukenberg PT, Walczak CE. 2007. Aurora B phosphorylates multiple sites on mitotic centromere-associated kinesin to spatially and temporally regulate its function. *Mol Biol Cell* **18:** 3264–3276.

Zhang MS, Arnaoutov A, Dasso M. 2014. RanBP1 governs spindle assembly by defining mitotic Ran-GTP production. *Dev Cell* **31:** 393–404.

Zhang R, Roostalu J, Surrey T, Nogales E. 2017. Structural insight into TPX2-stimulated microtubule assembly. *Elife* **6:** e30959.

Zierhut C, Funabiki H. 2015. Nucleosome functions in spindle assembly and nuclear envelope formation. *Bioessays* **37:** 1074–1085.

Zierhut C, Funabiki H. 2017. The cytoplasmic DNA sensor cGAS promotes mitotic cell death. *bioRxiv.* doi: https://doi .org/10.1101/168070.

Zierhut C, Jenness C, Kimura H, Funabiki H. 2014. Nucleosomal regulation of chromatin composition and nuclear assembly revealed by histone depletion. *Nat Struct Mol Biol* **21:** 617–625.

Knotty Problems during Mitosis: Mechanistic Insight into the Processing of Ultrafine DNA Bridges in Anaphase

Kata Sarlós,[1] Andreas Biebricher,[2] Erwin J.G. Petermann,[2] Gijs J.L. Wuite,[2] and Ian D. Hickson[1]

[1]Center for Chromosome Stability and Center for Healthy Aging, Department of Cellular and Molecular Medicine, University of Copenhagen, 2200 Copenhagen N, Denmark

[2]Department of Physics and Astronomy and LaserLab, Vrije Universiteit Amsterdam, 1081 HV Amsterdam, The Netherlands

Correspondence: iandh@sund.ku.dk

To survive and proliferate, cells have to faithfully segregate their newly replicated genomic DNA to the two daughter cells. However, the sister chromatids of mitotic chromosomes are frequently interlinked by so-called ultrafine DNA bridges (UFBs) that are visible in the anaphase of mitosis. UFBs can only be detected by the proteins bound to them and not by staining with conventional DNA dyes. These DNA bridges are presumed to represent entangled sister chromatids and hence pose a threat to faithful segregation. A failure to accurately unlink UFB DNA results in chromosome segregation errors and binucleation. This, in turn, compromises genome integrity, which is a hallmark of cancer. UFBs are actively removed during anaphase, and most known UFB-associated proteins are enzymes involved in DNA repair in interphase. However, little is known about the mitotic activities of these enzymes or the exact DNA structures present on UFBs. We focus on the biology of UFBs, with special emphasis on their underlying DNA structure and the decatenation machineries that process UFBs.

Visible evidence of mitotic chromosome segregation problems, such as lagging chromatin or bulky (chromatinized) DNA bridges, has long been used as a marker of genomic instability (McClintock 1938, 1942; Gisselsson et al. 2000, 2002; Hoffelder et al. 2004; Thompson and Compton 2011). These structures are generally revealed by staining with DNA dyes such as DAPI. This explains why ultrafine bridges (UFBs) had escaped detection for decades because they cannot be visualized using any of the commonly used dyes (Fig. 1A). Furthermore, because they are dechromatinized, they also cannot be detected by staining for histones. Instead, they were originally revealed through studies of the mitotic localization of DNA processing enzymes, such as the BLM helicase defective in Bloom's syndrome (Chan et al. 2007) or the Polo-like kinase 1 interacting checkpoint helicase (PICH) (Fig. 1A; Baumann et al. 2007). One curious feature of UFBs is the fact that they are generally coated along their length with PICH/BLM even when they are several microns in length in late anaphase.

A number of studies have investigated the mechanisms by which UFBs are generated and resolved (Wang et al. 2008, 2010b; Chan and Hickson 2009; Naim and Rosselli 2009; Nielsen et al. 2015). It is known that UFBs can be induced by exposure to a range of stressors, and that they often arise from defined genomic loci (centromeres, common fragile sites [CFSs], telomeres, and ribosomal DNA [rDNA]). Moreover, interfering with the functions of UFB-binding proteins has serious consequences for mitosis and genome integrity, such as the generation of aneu-

ploidy, binucleation, and micronucleus formation (Lukas et al. 2011; Nielsen et al. 2015). In our laboratory, we are developing tools to reconstitute anaphase chromosome segregation in vitro. For this, we are investigating the action of recombinant enzymes present at UFBs by combining ensemble biochemical methods with single-molecule optical tweezers combined with fluorescence microscopy. Here, we summarize our current knowledge on UFBs based on cellular observations and introduce our in vitro approaches to build a mechanistic model of sister chromatid disjunction.

THE ORIGINS OF UFBs

Replication causes the newly replicated strands to be interlinked/catenated (Schvartzman and Stasiak 2004; Vos et al. 2011). In parallel to this, the cohesin complex is deposited along the chromosomes to encircle and hold the sister chromatids together (Tanaka et al. 2001; Nasmyth 2011). Most of the DNA catenanes are removed by topoisomerase IIα (TopIIα); either during S phase or during early mitosis when DNA condensation occurs (Hirano 2015). In the prophase of mitosis, most of the cohesin located on chromosome arms (but not at centromeres) is released in a condensation-dependent manner (Hirano 2015). The activities of TopIIα, condensin I and II, and cohesin are tightly coordinated and give rise to the classical, X-shaped, chromosome structure, where the arms are devoid of both DNA cohesion and DNA cate-

Published by Cold Spring Harbor Laboratory Press; doi: 10.1101/sqb.2017.82.033647

Cold Spring Harbor Symposia on Quantitative Biology, Volume LXXXII

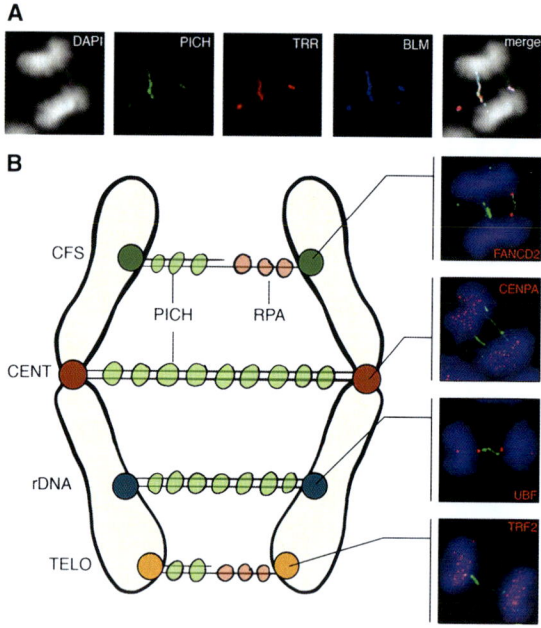

Figure 1. Ultrafine bridges (UFBs) arise in anaphase from various chromosomal loci. (*A*) Immunofluorescence image of a UFB coated by PICH (green), BLM (blue), and TRR (red). The bulk DNA was stained using DAPI. (*B, left*) Schematic representation of the chromosomal origins of UFBs, marked by surrogate markers at their termini. Examples are shown on the *right*. From *top* to *bottom*: a common fragile state (CFS)-UFB marked by FANCD2, a C-UFB marked by CENPA, an R-UFB marked by UBF, and a T-UFB marked by TRF2. UFBs were visualized using antibodies to PICH (green).

nanes (Hirano 2015). Interfering with either cohesion or condensation gives rise to UFBs connecting the chromosome arms, suggesting that these processes give directionality to decatenation by TopIIα (Baxter and Aragón 2012; Minocherhomji et al. 2015; Piskadlo et al. 2017).

UFBs at Centromeres

UFBs arising from the centromeres (C-UFBs) can be identified by the presence of centromeric markers such as CENP-A at their termini (Fig. 1B). C-UFBs are by far the most prevalent of all UFBs (Chan et al. 2007, 2009) and exist in every mitosis. Their number is minimized by an active removal process that occurs at anaphase onset (Wang et al. 2008). Importantly, inhibition of TopIIα by specific drugs, such as ICRF-193, induces the persistence and number of centromeric UFBs (Baumann et al. 2007; Chan et al. 2007; Wang et al. 2008). Moreover, TopIIα colocalizes with PICH on ICRF-193-induced UFBs (Nielsen et al. 2015), indicating that TopIIα is required for their decatenation. Because centromeric cohesin is protected from release by Sgo1-PP2A (Kitajima et al. 2006), and is only cleaved at anaphase onset by Separase (Uhlmann et al. 2000), it is thought that TopIIα only has a brief period in which to decatenate any C-UFBs after cohesin cleavage (Wang et al. 2010b). However, based on their frequency, it is also conceivable that the persistence of centromeric UFBs until anaphase is not simply an unwant-

ed side effect of the masking of DNA catenation by cohesin, but rather has a physiological role in maintaining DNA-based sister chromatid "cohesion" until the metaphase-to-anaphase transition.

UFBs at Common Fragile Sites

CFSs are viewed as an Achilles' heel of the genome. They are frequently deleted or rearranged in cancer cells and can appear as gaps or breaks on mitotic chromosomes following replication perturbation (termed CFS expression) (Glover et al. 1984; Durkin and Glover 2007). CFSs are regions where replication is problematic and delayed (Debatisse et al. 2012). According to recent modeling studies, cells with large genomes enter mitosis with, on average, three underreplicated sites per cell, even in unperturbed growth conditions. This problem is exacerbated by conditions that induce replication stress or by reducing the number of origins (Al Mamun et al. 2016; Moreno et al. 2016).

The Fanconi anemia (FA) DNA repair proteins, FANCD2 and FANCI (Sims et al. 2007; Smogorzewska et al. 2007), associate with CFSs after replication stress and serve as surrogate markers for these loci (Chan et al. 2009; Naim and Rosselli 2009). In contrast to centromeric UFBs, CFS–UFBs (Fig. 1B) rarely appear spontaneously and cannot be induced by inhibiting TopIIα (Chan et al. 2009). Rather, CFS–UFBs accumulate after perturbation of DNA replication by the DNA polymerase inhibitor aphidicolin (Chan and Hickson 2009; Naim and Rosselli 2009). This suggests that CFS–UFBs are composed of underreplicated DNA.

UFBs at Telomeres

The ends of linear eukaryotic chromosomes are organized into well-defined structures called telomeres (Doksani and de Lange 2014; Arnoult and Karlseder 2015). The telomeric DNA is looped back in a DNA structure called a T-loop that prevents the DNA end from being exposed (Griffith et al. 1999; Doksani et al. 2013). T-loops are stabilized by the shelterin complex, which comprises several proteins including telomeric repeat-binding factors 1 and 2 (TRF1 and TRF2) (Palm and de Lange 2008). Telomeres show similarities to CFSs in that replication stress induces the so-called "fragile-telomere" phenotype, where the chromosomes appear to be broken at the very end (Sfeir et al. 2009), indicating that these loci are also inherently difficult to replicate (Martinez and Blasco 2015; Higa et al. 2017). Telomeres give rise to BLM-coated UFBs (T-UFBs) (Fig. 1B) following exposure to aphidicolin (Chan and Hickson 2009; Barefield and Karlseder 2012; d'Alcontres et al. 2014). Interfering with the integrity of the shelterin complex via changing the levels of TRF1 or TRF2 also induces telomere fragility (Martinez et al. 2009; Sfeir et al. 2009) and gives rise to T-UFBs (d'Alcontres et al. 2014; Nera et al. 2015). TRF1 was shown to protect against fragility by recruiting BLM to the telomeres (Sfeir et al. 2009), suggesting that BLM facilitates replication (Drosopoulos et al. 2015) or disentangles late-replicating structures at these regions (Chan et al. 2009; Barefield and Karlseder 2012). In contrast to CFS-UFBs, inhibition of TopIIα by ICRF-193 induces T-UFBs, and TRF1 has been shown to recruit TopIIα to telomeres. These findings suggest that at least a subset of T-UFBs are likely be completely replicated, double-stranded DNA (dsDNA) catenanes (d'Alcontres et al. 2014).

UFBs at the rDNA

PICH is present at rDNA loci in chicken and human cells in early mitosis (Nielsen et al. 2015). The rDNA loci also occasionally give rise to PICH- and TopIIα-decorated R-UFBs (Fig. 1B; Nielsen and Hickson 2016). The number of R-UFBs increases following inhibition of TopIIα by ICRF-193. This suggests that the structure and decatenation mechanism of R-UFBs are similar to those of C-UFBs. The rDNA locus is known to be segregated late during mitosis in yeast (Sullivan et al. 2004; Wang et al. 2004; Clemente-Blanco et al. 2009) and has been shown to be transcriptionally active even in early mitosis in humans (Gebrane-Younes et al. 1997; Sirri et al. 1999; Voit et al. 2015). Because active transcription interferes with condensation, which, in turn, is required for decatenation by TopIIα (Lukas et al. 2011; Baxter and Aragón 2012), this would leave cells only a short time window in which to decatenate the rDNA during mitosis, thus potentially explaining the appearance of UFBs from these loci.

UFB RECOGNITION AND PROCESSING MACHINERIES

PICH

PICH was first identified as an interacting factor of the mitotic kinase Plk1 (Baumann et al. 2007). PICH is excluded from the nucleus during interphase and is only recruited to chromatin after nuclear envelope breakdown, whereupon it accumulates at centromeric loci. PICH seems to be the main recognition and recruitment factor for UFBs, as several other UFB-processing factors fail to localize to UFBs in the absence of PICH, such as members of the Bloom syndrome complex (Chan et al. 2007) and RIF1 (Hengeveld et al. 2015). This makes it difficult to detect or analyze UFBs in the absence of PICH. PICH was reported to influence chromosome condensation, as chromosome structure is abnormal in the absence of PICH (Leng et al. 2008; Kurasawa and Yu-Lee 2010; Rouzeau et al. 2012; Nielsen et al. 2015). PICH belongs to the SNF2 family of translocases (Singleton et al. 2007) and contains a motor domain typical in this enzyme family (Fig. 2A). Consistent with this, PICH possesses ATP-dependent dsDNA translocase activity (Biebricher et al. 2013). In addition to the SNF2 region, PICH has accessory domains, including the PICH-family domain, and two TPR motifs reported to be involved in protein interactions (Hengeveld et al. 2015; Pitchai et al. 2017). PICH appears to have a high affinity for stretched dsDNA, which is consistent with the idea that UFBs must be under considerable tension created by the mitotic spindle (Baumann

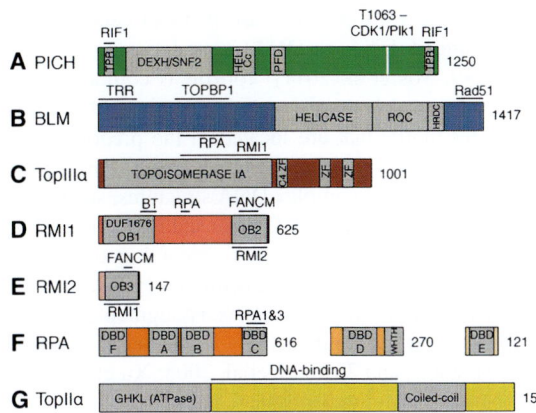

Figure 2. Domain structure of UFB-associated proteins and their interactions. (*A*) PICH: The amino- and carboxy-terminal TPR, SNF2, HeliCc, PICH-family domains are indicated. PICH interacts with RIF1 via the TPRs. The CDK1 phosphorylation site required for interaction with Plk1 is indicated. (*B*) BLM: The RecQ core comprised of the helicase, the RecQ carboxy-terminal, and the HRDC domains, as well as some known interaction sites, are indicated. (*C*) TopIIIα: The conserved topoisomerase IA domain, zinc fingers, and putative interaction region with RMI1 are indicated—the BLM-interaction site is not known. (*D*) RMI1 and (*E*) RMI2: The OB-folds, sites for complex formation, and interaction sites for BLM/TopIIIα, RPA and FANCM are indicated. (*F*) RPA is composed of three subunits. The six DNA-binding sites, the winged-helix-turn-helix domain, and complex-formation sites are indicated. (*G*) TopIIα: The conserved ATPase domain, DNA-binding region, and coiled-coil required for dimerization are indicated. The locations of interacting sites are not known.

et al. 2007; Biebricher et al. 2013). Indeed, this property of PICH may be the main mechanism for how cells normally sense UFBs. Somewhat surprisingly, a PICH mutant lacking ATPase activity does not increase the number of UFBs, although it does prolong their persistence (Nielsen et al. 2015) and also increases the number of chromatin bridges, indicating that UFBs and chromatin bridges have different origins (Kaulich et al. 2012). It should be noted, however, that the ATPase-dead PICH shows altered localization on metaphase chromosomes (Kaulich et al. 2012).

The Bloom Syndrome Protein Complex

BLM is the helicase mutated in Bloom syndrome (BS), a severe autosomal hereditary disorder causing genetic instability and cancer (Ellis et al. 1995; German et al. 2007; Cunniff et al. 2017). BLM belongs to the RecQ family, a group of evolutionary conserved genome caretaking enzymes (Chu and Hickson 2009; Croteau et al. 2014), and comprises a helicase core, flanked by long amino- and carboxy-terminal regions responsible for protein–protein interactions (Fig. 2B; Wu et al. 2000, 2001; Meetei et al. 2003; Doherty et al. 2005; Wang et al. 2013; Blackford et al. 2015). BLM efficiently unwinds various DNA structures such as replication forks (Karow et al. 1997), four-way junctions (Karow et al. 2000), D-loops (Bachrati et al. 2006), or G4 quadruplexes (Sun et al. 1998).

BLM directly interacts with topoisomerase IIIα (TopIIIα) (Wallis et al. 1989; Goulaouic et al. 1999; Wu et al.

2000), a Type 1A topoisomerase that can catalyze only single-stranded DNA (ssDNA) strand passage (Wallis et al. 1989; Vos et al. 2011). TopIIIα is composed of a conserved type 1A topoisomerase domain and multiple zinc-finger motifs that are located in the predominantly disordered carboxyl terminus (Fig. 2C). BLM and TopIIIα together disentangle complex DNA structures, such as the double Holliday junction (dHJ) (Wu and Hickson 2003), a key intermediate in homologous recombination–based DNA repair (Yin et al. 2005; Bizard and Hickson 2014). In higher eukaryotes, the complex is augmented by the RecQ-mediated instability (RMI) 1 (Meetei et al. 2003; Yin et al. 2005) and 2 (Singh et al. 2008; Xu et al. 2008) proteins, forming direct physical interactions with both BLM (Raynard et al. 2006) and TopIIIα (Raynard et al. 2006; Bocquet et al. 2014). RMI1 and RMI2 are both OB-fold-containing proteins (Fig. 2D,E), with no inherent enzymatic activity. However, importantly, RMI1 stimulates the dHJ dissolution by BLM and TopIIIα (Raynard et al. 2006; Wu et al. 2006), whereas RMI2 has a very modest effect on this activity (Singh et al. 2008; Xu et al. 2008). RMI1 and RMI2 form a complex (Hoadley et al. 2010; Wang et al. 2010a) that is required to stabilize TopIIIα. As a result, they form a constitutive heterotrimer (termed the "TRR complex") in vivo.

BLM has been used as a key marker of UFBs in many studies (Chan and Hickson 2009; Chan et al. 2009; Vinciguerra et al. 2010; Ke et al. 2011; Lukas et al. 2011; Barefield and Karlseder 2012; Broderick et al. 2015; Hengeveld et al. 2015). Considering that the BTRR complex has evolved to disentangle complex DNA structures, it is conceivable that this complex is responsible for UFB processing. This is supported by the observation that BS cells, and cells depleted of BLM by short interfering RNAs (siRNAs), display increased levels of all types of UFBs, and that these UFBs often persist into late telophase in these cells (Chan et al. 2007; Barefield and Karlseder 2012).

The recruitment of the BTRR complex to UFBs depends on PICH, and they always seem to coat the same stretch of DNA (Chan et al. 2007). This localization is somewhat curious, considering the fact that PICH binds exclusively to dsDNA (Biebricher et al. 2013), whereas the BTRR prefers ssDNA. The observation that BLM interacts with the carboxyl terminus of PICH suggests that PICH recruits the BTRR complex via direct interactions (Ke et al. 2011).

RPA

Replication protein A is an essential ssDNA binding protein required for most DNA transactions (Wold 1997; Zou et al. 2006). It is composed of three subunits (Fig. 2F) and interacts with the BTRR complex, both functionally and directly via BLM (Brosh et al. 2000; Meetei et al. 2003; Doherty et al. 2005) and RMI1 (Xue et al. 2013). RPA is detectable on a subset of UFBs in anaphase in response to DNA replication stress induced by aphidicolin (Chan and Hickson 2009; Burrell et al. 2013), indicating the presence of ssDNA on some CFS-UFBs (Chan et al. 2009). BLM and RPA show a nonoverlapping pattern of

localization to UFBs, which suggests that the recruitment of BLM to PICH-coated double-stranded UFBs is independent of its interaction with RPA (Porter and Farr 2004; Chan and Hickson 2009). Interestingly, the appearance of RPA-coated UFBs has been shown to be BLM-dependent, implicating BLM in unwinding some structure to create ssDNA (Hengeveld et al. 2015).

Topoisomerase IIα

TopIIα is the enzyme responsible for the majority of decatenation of chromosomes in early mitosis (Porter and Farr 2004). TopIIα is a homodimeric Type IIA topoisomerase (Fig. 2G) that catalyzes the passage of one piece of dsDNA through another in an ATP-dependent manner (Schoeffler and Berger 2008). Even though TopIIα does not seem to directly interact with PICH, it is present at PICH-coated UFBs, and PICH is able to stimulate decatenation by TopIIα in vivo and in vitro (Nielsen et al. 2015). It has been suggested that TopIIα is recruited to a subset of UFBs via direct interaction with TOPBP1 (Broderick et al. 2015).

Other UFB-Associated Proteins

RIF1, TOPBP1, and FANCM were also reported to coat some UFBs (Meetei et al. 2003; Deans and West 2009; Xu et al. 2010; Hoadley et al. 2012; Wang et al. 2013; Blackford et al. 2015). We will not discuss these factors further here, but instead refer readers to relevant publications (German et al. 2007; Vinciguerra et al. 2010; Broderick et al. 2015; Hengeveld et al. 2015; Pedersen et al. 2015).

MODELING UFBs IN A TEST TUBE

Mechanistic insight into the mode of UFB resolution is lacking. To gain a comprehensive understanding of this process, our laboratory is using interdisciplinary approaches to reconstitute mitotic DNA decatenation in vitro. To achieve this, we combine ensemble biochemistry on model DNA substrates, with single-molecule optical tweezers coupled to fluorescence microscopy (Heller et al. 2014).

Modeling UFBs Using Ensemble Biochemistry

CFS and telomeres are both difficult-to-replicate regions, and both CFS-UFBs and T-UFBs are induced by replication stress (Chan and Hickson 2009; Barefield and Karlseder 2012). Therefore, it is thought that these UFBs are composed of underreplicated DNA (Fig. 3A). To study such a UFB in vitro, we created a substrate termed a "late replication intermediate" (LRI), which comprises two interlinked DNA circles mimicking two converging replication forks (A Sarlós, A Biebricher, and AH Bizard, unpubl.). We hypothesized that an LRI would be an ideal substrate for the BTRR complex. Indeed, a similar substrate was shown previously to be processed by the Escherichia coli homologs of the BTRR complex (Suski and Marians 2008).

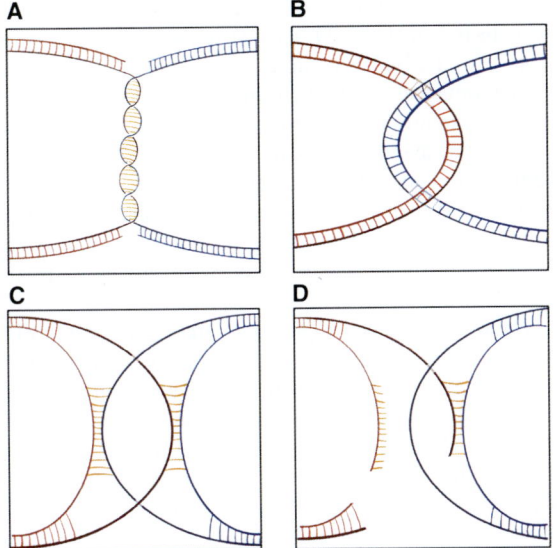

Figure 3. Hypothetical DNA structures that could be present at UFBs. (*A*) A late replication intermediate, (*B*) a complete double-stranded catenane, (*C*) a double Holliday junction, and (*D*) a D-loop.

As discussed above, TopIIα inhibition dramatically increases the number of C-UFBs (Chan et al. 2009). R-UFBs are also induced by ICRF-193 and are thought to arise because of the late condensation of the rDNA locus hindering decatenation (Nielsen and Hickson 2016). Con-

sidering that TopIIα is a dsDNA-specific enzyme, this implies that the majority of C-UFBs and R-UFBs are completely double-stranded catenanes (Fig. 3B). To model this, we used a single-catenane substrate comprised of two interlinked dsDNA circles (Stark et al. 1989; Nielsen et al. 2015). Considering that the BTRR complex prefers ss/dsDNA junctions, it is not clear what the function of BTRR might be on C-UFBs (Chan et al. 2007). Because studies involving the yeast and *E. coli* homologs of BTRR reported some dsDNA catenation activity (Harmon et al. 2003; Cejka et al. 2012), it is conceivable that the BTRR can decatenate ds-UFBs if TopIIα is prohibited.

Most of the enzymes implicated in UFB processing are also involved in DNA repair. Therefore, it cannot be excluded that some UFBs are composed of HR intermediates, such as a dHJ or a D-loop (Fig. 3C,D). Such DNA structures would be expected to be "dissolved" efficiently by the BTRR complex (Wu and Hickson 2003; Bachrati et al. 2006).

Modeling of UFBs in Single-Molecule Experiments

Single-molecule techniques can provide information on enzymatic mechanisms that would be inaccessible by ensemble methods because of the averaging of the activity of thousands of molecules at the same time (Neuman and Nagy 2008). One of the most widely used single-molecule techniques is optical tweezers (Fig. 4A). Using optical tweezers, a single piece of biotinylated DNA can

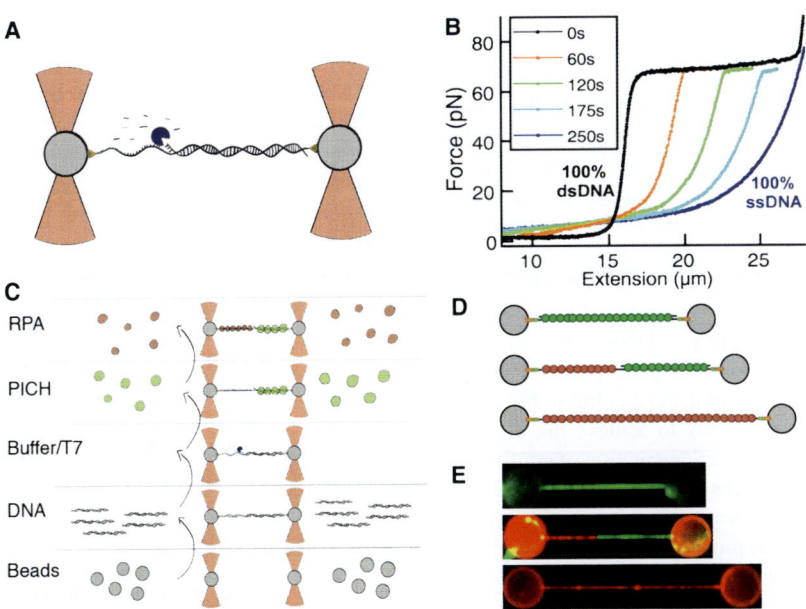

Figure 4. Scheme of optical tweezers experiments. (*A*) A biotin-labeled dsDNA molecule stretched between two streptavidin-coated beads (gray) entrapped by laser beams (orange). The exonuclease activity of T7 DNA polymerase (blue) is shown creating ssDNA. (*B*) A series of DNA force-extension curves, ranging from full-length dsDNA (black) to full-length ssDNA (blue). The orange, green, and cyan curves represent increasing incubation time with T7 DNA polymerase (as shown in the *inset*), which results in increasing lengths of ssDNA tracts. (*C*) Outline of an experimental scheme for use of a flow cell with five channels. First beads are trapped, and then a dsDNA molecule is captured; this can subsequently be converted into an ss/dsDNA hybrid. This is followed by incubation with fluorescently labeled proteins, such as GFP-PICH (green) or RFP-RPA (red). (*D,E*) Schemes (*D*) and fluorescence snapshots (*E*) of a GFP-PICH-coated dsDNA molecule, a RFP-RPA/GFP-PICH-coated ss/dsDNA hybrid molecule, and a RFP-RPA-coated ssDNA stretch. Note that the extension of the DNA molecule is increasing by the introduction of ssDNA, which has a longer contour length than dsDNA.

be tethered between streptavidin-coated polystyrene microspheres and manipulated in the flow cell of a fluorescence microscope (Heller et al. 2014). The effect of tension applied to a model UFB can thus be addressed, and fluorescently tagged proteins can be directly visualized on the stretched DNA.

The simplest UFB model one can imagine is a piece of dsDNA stretched between two beads. Indeed, we have showed using this approach that PICH is a DNA "tension sensor" that binds to stretched dsDNA with high affinity and translocates along the DNA in an ATP-dependent manner (Biebricher et al. 2013). However, as discussed above, presumably not all UFBs are completely double-stranded, and there are also multiple players involved in their processing. It is possible to generate a section of ssDNA with optical tweezers by stretching a dsDNA molecule to forces beyond 65–70 pN, which induces base pair melting. Occasionally, in the presence of a nick (a break in one of the strands), the melted strand dissociates, yielding a permanent stretch of ssDNA (Candelli et al. 2013). A more controlled way to generate a ssDNA/dsDNA hybrid is by using the exonuclease activity of the T7 DNA polymerase induced by putting the DNA under tension (Wuite et al. 2000; Hoekstra et al. 2017). Because reducing the tension can stop the exonuclease activity, a single DNA molecule can be generated containing the desired amount of ssDNA (Fig. 4B). The use of a multichannel flow cell system allows combination of different channels, containing various buffers and proteins, and the same piece of DNA can be freely manipulated between them with minimal contamination (Fig. 4C; Heller et al. 2014). UFBs can be modeled by incubating the DNA with different fluorescently labeled UFB-binding factors. Examples of a full-length dsDNA, a mixed ss/dsDNA, and a full-length ssDNA molecule coated with either GFP-PICH and/or strawberry RPA are shown in Figure 4D,E. By using other fluorescently labeled UFB-factors, such as BLM, the hierarchical recruitment of UFB-processing enzymes to PICH- versus RPA-coated UFBs could in principle be analyzed.

As described above, UFBs are likely to comprise interlinked dsDNA catenanes, and therefore a single piece of DNA, even if it is a mixture of ss/dsDNA, is not suitable for addressing details of the decatenation process. Braiding two DNA molecules together using quadruple-trap optical tweezers would provide a way to achieve this (Brouwer et al. 2017). Furthermore, implementation of confocal microscopy, or recent advances such as super-resolution imaging, allows the monitoring of real-time dynamics of individual molecules on model UFBs even in high fluorescent protein background (Heller et al. 2013).

WHAT HAPPENS IF THE CONVENTIONAL DNA DECATENATION PATHWAYS FAIL?

Cells seem to wait until the last moment to achieve correct segregation and can delay the completion of anaphase/telophase to accomplish this. This is reflected in BS cells, where UFBs persist even in very late telophase,

spanning several microns between the two forming daughter cells (Chan et al. 2007). However, in some cases, when TopIIα or the BTRR complex is overwhelmed or inhibited, endonucleases might ensure that the DNA be cleaved in an apparently less controlled way, but nevertheless in a manner that avoids rupture of the bridge during abscission (Sarbajna et al. 2014; Maciejowski et al. 2015). Unresolved UFBs can also cause cytokinesis delay or the abandonment of cytokinesis, resulting in binucleation (Vinciguerra et al. 2010; Wang et al. 2010b; Germann et al. 2014; Nielsen et al. 2015).

Oncogene-induced replication stress generates UFBs, which is exacerbated in cancer cells (Burrell et al. 2013). Unresolved CFS–UFBs induced by replication stress can lead to the formation of 53BP1 nuclear bodies in the next G_1 phase (Harrigan et al. 2011; Lukas et al. 2011). Interfering with CFS processing mechanisms also induces chromosome missegregation and aneuploidy (Naim and Rosselli 2009; Ying et al. 2013; Minocherhomji et al. 2015). In line with this, depletion of UFB-associated enzymes such as BLM also generates micronuclei (Rosin and German 1985). Micronuclei are a source of chromothripsis, one of the drivers of genomic rearrangements in cancer (Crasta et al. 2012; Zhang et al. 2015).

CONCLUSION

Since their discovery 10 years ago, UFBs have become one of the key markers of genomic instability. It has also become more obvious that cells have to cope with DNA entanglements during mitosis much more frequently than was thought previously. As UFBs arise systematically from specific loci (centromeres, fragile sites, telomeres, and rDNA), specialized recognition and processing machineries have evolved to maintain the stability of these important regions. The significance of understanding fundamental mechanisms ensuring correct chromosome segregation cannot be overestimated, especially in the context of cancer. All essential cellular processes can only be understood properly by combining in vivo observations with in vitro model building studies. In the field of mitosis, comprehensive studies where key aspects of mitosis are reconstituted are still in a very early phase. Clearly, interdisciplinary studies are required to construct a mechanistic model of chromosome segregation. In this review, we have highlighted how the combination of biochemical and single-molecule modeling of DNA structures is a useful tool to study UFBs, impediments of faithful segregation.

ACKNOWLEDGMENTS

We thank all members of the Hickson laboratory for helpful discussions and Hocine W. Mankouri and Anna H. Bizard for valuable comments on the manuscript. Work in the authors' laboratory is supported by the Danish National Research Foundation (DNRF115), the European Research Council, the Nordea Foundation, and a Future and Emerging Technologies grant from the European Union H2020 fund.

REFERENCES

Al Mamun M, Albergante L, Moreno A, Carrington JT, Blow JJ, Newman TJ. 2016. Inevitability and containment of replication errors for eukaryotic genome lengths spanning megabase to gigabase. *Proc Natl Acad Sci* **113**: E5765–E5774.

Arnoult N, Karlseder J. 2015. Complex interactions between the DNA-damage response and mammalian telomeres. *Nat Struct Mol Biol* **22**: 859–866.

Bachrati CZ, Borts RH, Hickson ID. 2006. Mobile D-loops are a preferred substrate for the Bloom's syndrome helicase. *Nucleic Acids Res* **34**: 2269–2279.

Barefield C, Karlseder J. 2012. The BLM helicase contributes to telomere maintenance through processing of late-replicating intermediate structures. *Nucleic Acids Res* **40**: 7358–7367.

Baumann C, Körner R, Hofmann K, Nigg EA. 2007. PICH, a centromere-associated SNF2 family ATPase, is regulated by Plk1 and required for the spindle checkpoint. *Cell* **128**: 101–114.

Baxter J, Aragón L. 2012. A model for chromosome condensation based on the interplay between condensin and topoisomerase II. *Trends Genet* **28**: 110–117.

Biebricher A, Hirano S, Enzlin JH, Wiechens N, Streicher WW, Huttner D, Wang LHC, Nigg EA, Owen-Hughes T, Liu Y, et al. 2013. PICH: A DNA translocase specially adapted for processing anaphase bridge DNA. *Mol Cell* **51**: 691–701.

Bizard AH, Hickson ID. 2014. The dissolution of double Holliday junctions. *CSH Perspect Biol* **6**: a016477.

Blackford AN, Nieminuszczy J, Schwab RA, Galanty Y, Jackson SP, Niedzwiedz W. 2015. TopBP1 interacts with BLM to maintain genome stability but is dispensable for preventing BLM degradation. *Mol Cell* **57**: 1133–1141.

Bocquet N, Bizard AH, Abdulrahman W, Larsen NB, Faty M, Cavadini S, Bunker RD, Kowalczykowski SC, Cejka P, Hickson ID, et al. 2014. Structural and mechanistic insight into Holliday-junction dissolution by Topoisomerase IIIα and RMI1. *Nat Struct Mol Biol* **21**: 261–268.

Broderick R, Nieminuszczy J, Blackford AN, Winczura A, Niedzwiedz W. 2015. TOPBP1 recruits TOP2A to ultra-fine anaphase bridges to aid in their resolution. *Nat Commun* **6**: 6572.

Brosh RM, Li JL, Kenny MK, Karow JK, Cooper MP, Kureekattil RP, Hickson ID, Bohr VA. 2000. Replication protein A physically interacts with the Bloom's syndrome protein and stimulates its helicase activity. *J Biol Chem* **275**: 23500–23508.

Brouwer I, King GA, Heller I, Biebricher AS, Peterman EJG, Wuite GJL. 2017. Probing DNA–DNA interactions with a combination of quadruple-trap optical tweezers and microfluidics. *Methods Mol Biol* **1486**: 275–293.

Burrell RA, McClelland SE, Endesfelder D, Groth P, Weller MC, Shaikh N, Domingo E, Kanu N, Dewhurst SM, Gronroos E, et al. 2013. Replication stress links structural and numerical cancer chromosomal instability. *Nature* **494**: 492–496. Erratum **500**: 490.

Candelli A, Hoekstra TP, Farge G, Gross P, Peterman EJ, Wuite GJ. 2013. A toolbox for generating single-stranded DNA in optical tweezers experiments. *Biopolymers* **99**: 611–620.

Cejka P, Plank JL, Dombrowski CC, Kowalczykowski SC. 2012. Decatenation of DNA by the *S. cerevisiae* Sgs1-Top3-Rmi1 and RPA complex: A mechanism for disentangling chromosomes. *Mol Cell* **47**: 886–896.

Chan KL, Hickson ID. 2009. On the origins of ultra-fine anaphase bridges. *Cell Cycle* **8**: 3065–3066.

Chan KL, North PS, Hickson ID. 2007. BLM is required for faithful chromosome segregation and its localization defines a class of ultrafine anaphase bridges. *EMBO J* **26**: 3397–3409.

Chan KL, Palmai-Pallag T, Ying S, Hickson ID. 2009. Replication stress induces sister-chromatid bridging at fragile site loci in mitosis. *Nat Cell Biol* **11**: 753–760.

Chu WK, Hickson ID. 2009. RecQ helicases: Multifunctional genome caretakers. *Nat Rev Cancer* **9**: 644–654.

Clemente-Blanco A, Mayan-Santos M, Schneider DA, Machin F, Jarmuz A, Tschochner H, Aragon L. 2009. Cdc14 inhibits transcription by RNA polymerase I during anaphase. *Nature* **458**: 219–222.

Crasta K, Ganem NJ, Dagher R, Lantermann AB, Ivanova EV, Pan YF, Nezi L, Protopopov A, Chowdhury D, Pellman D. 2012. DNA breaks and chromosome pulverization from errors in mitosis. *Nature* **482**: 53–58.

Croteau DL, Popuri V, Opresko PL, Bohr VA. 2014. Human RecQ helicases in DNA repair, recombination, and replication. *Annu Rev Biochem* **83**: 519–552.

Cunniff C, Bassetti JA, Ellis NA. 2017. Bloom's syndrome: Clinical spectrum, molecular pathogenesis, and cancer predisposition. *Mol Syndromol* **8**: 4–23.

d'Alcontres MS, Palacios JA, Mejias D, Blasco MA. 2014. TopoIIα prevents telomere fragility and formation of ultra thin DNA bridges during mitosis through TRF1-dependent binding to telomeres. *Cell Cycle* **13**: 1463–1481.

Deans AJ, West SC. 2009. FANCM connects the genome instability disorders Bloom's Syndrome and Fanconi Anemia. *Mol Cell* **36**: 943–953.

Debatisse M, Le Tallec B, Letessier A, Dutrillaux B, Brison O. 2012. Common fragile sites: Mechanisms of instability revisited. *Trends Genet* **28**: 22–32.

Doherty KM, Sommers JA, Gray MD, Lee JW, von Kobbe C, Thoma NH, Kureekattil RP, Kenny MK, Brosh RM. 2005. Physical and functional mapping of the replication protein A interaction domain of the Werner and Bloom syndrome helicases. *J Biol Chem* **280**: 29494–29505.

Doksani Y, de Lange T. 2014. The role of double-strand break repair pathways at functional and dysfunctional telomeres. *Cold Spring Harb Perspect Biol* **6**: a016576.

Doksani Y, Wu JY, de Lange T, Zhuang XW. 2013. Super-resolution fluorescence imaging of telomeres reveals TRF2-dependent T-loop formation. *Cell* **155**: 345–356.

Drosopoulos WC, Kosiyatrakul ST, Schildkraut CL. 2015. BLM helicase facilitates telomere replication during leading strand synthesis of telomeres. *J Cell Biol* **210**: 191–208.

Durkin SG, Glover TW. 2007. Chromosome fragile sites. *Annu Rev Genet* **41**: 169–192.

Ellis NA, Lennon DJ, Proytcheva M, Alhadeff B, Henderson EE, German J. 1995. Somatic intragenic recombination within the mutated locus BLM can correct the high sister-chromatid exchange phenotype of Bloom syndrome cells. *Am J Hum Genet* **57**: 1019–1027.

Gebrane-Younes J, Fomproix N, Hernandez-Verdun D. 1997. When rDNA transcription is arrested during mitosis, UBF is still associated with non-condensed rDNA. *J Cell Sci* **110** (Pt 19): 2429–2440.

German J, Sanz MM, Ciocci S, Ye TZ, Ellis NA. 2007. Syndrome-causing mutations of the *BLM* gene in persons in the Bloom's Syndrome Registry. *Hum Mutat* **28**: 743–753.

Germann SM, Schramke V, Pedersen RT, Gallina I, Eckert-Boulet N, Oestergaard VH, Lisby M. 2014. TopBP1/Dpb11 binds DNA anaphase bridges to prevent genome instability. *J Cell Biol* **204**: 45–59.

Gisselsson D, Pettersson L, Hoglund M, Heidenblad M, Gorunova L, Wiegant J, Mertens F, Dal Cin P, Mitelman F, Mandahl N. 2000. Chromosomal breakage-fusion-bridge events cause genetic intratumor heterogeneity. *Proc Natl Acad Sci* **97**: 5357–5362.

Gisselsson D, Jonson T, Yu C, Martins C, Mandahl N, Wiegant J, Jin Y, Mertens F, Jin C. 2002. Centrosomal abnormalities, multipolar mitoses, and chromosomal instability in head and neck tumours with dysfunctional telomeres. *Br J Cancer* **37**: 202–207.

Glover TW, Berger C, Coyle J, Echo B. 1984. DNA polymerase α inhibition by aphidicolin induces gaps and breaks at common fragile sites in human chromosomes. *Hum Genet* **67**: 136–142.

Goulaouic H, Roulon T, Flamand O, Grondard L, Lavelle F, Riou JF. 1999. Purification and characterization of human DNA topoisomerase IIIα. *Nucleic Acids Res* **27**: 2443–2450.

Griffith JD, Comeau L, Rosenfield S, Stansel RM, Bianchi A, Moss H, de Lange T. 1999. Mammalian telomeres end in a large duplex loop. *Cell* **97**: 503–514.

Harmon FG, Brockman JP, Kowalczykowski SC. 2003. RecQ helicase stimulates both DNA catenation and changes in DNA topology by topoisomerase III. *J Biol Chem* **278:** 42668–42678.

Harrigan JA, Belotserkovskaya R, Coates J, Dimitrova DS, Polo SE, Bradshaw CR, Fraser P, Jackson SP. 2011. Replication stress induces 53BP1-containing OPT domains in G1 cells. *J Cell Biol* **193:** 97–108.

Heller I, Sitters G, Broekmans OD, Farge G, Menges C, Wende W, Hell SW, Peterman EJG, Wuite GJL. 2013. STED nanoscopy combined with optical tweezers reveals protein dynamics on densely covered DNA. *Nature Methods* **10:** 910–916.

Heller I, Hoekstra TP, King GA, Peterman EJ, Wuite GJ. 2014. Optical tweezers analysis of DNA–protein complexes. *Chem Rev* **114:** 3087–3119.

Hengeveld RC, de Boer HR, Schoonen PM, de Vries EG, Lens SM, van Vugt MA. 2015. Rif1 is required for resolution of ultrafine DNA bridges in anaphase to ensure genomic stability. *Dev Cell* **34:** 466–474.

Higa M, Fujita M, Yoshida K. 2017. DNA replication origins and fork progression at mammalian telomeres. *Genes (Basel)* **8:** E112.

Hirano T. 2015. Chromosome dynamics during mitosis. *Cold Spring Harb Perspect Biol* **7:** a015792.

Hoadley KA, Xu DY, Xue YT, Satyshur KA, Wang WD, Keck JL. 2010. Structure and cellular roles of the RMI core complex from the bloom syndrome dissolvasome. *Structure* **18:** 1149–1158.

Hoadley KA, Xue YT, Ling C, Takata M, Wang WD, Keck JL. 2012. Defining the molecular interface that connects the Fanconi anemia protein FANCM to the Bloom syndrome dissolvasome. *Proc Natl Acad Sci* **109:** 4437–4442.

Hoekstra TP, Depken M, Lin SN, Cabanas-Danés J, Gross P, Dame RT, Peterman EJ, Wuite GJ. 2017. Switching between exonucleolysis and replication by T7 DNA polymerase ensures high fidelity. *Biophys J* **112:** 575–583.

Hoffelder DR, Luo L, Burke NA, Watkins SC, Gollin SM, Saunders WS. 2004. Resolution of anaphase bridges in cancer cells. *Chromosoma* **112:** 389–397.

Karow JK, Chakraverty RK, Hickson ID. 1997. The Bloom's syndrome gene product is a 3′-5′ DNA helicase. *J Biol Chem* **272:** 30611–30614.

Karow JK, Constantinou A, Li JL, West SC, Hickson ID. 2000. The Bloom's syndrome gene product promotes branch migration of Holliday junctions. *Proc Natl Acad Sci* **97:** 6504–6508.

Kaulich M, Cubizolles F, Nigg EA. 2012. On the regulation, function, and localization of the DNA-dependent ATPase PICH. *Chromosoma* **121:** 395–408.

Ke Y, Huh JW, Warrington R, Li B, Wu N, Leng M, Zhang J, Ball HL, Li B, Yu H. 2011. PICH and BLM limit histone association with anaphase centromeric DNA threads and promote their resolution. *EMBO J* **30:** 3309–3321.

Kitajima TS, Sakuno T, Ishiguro K, Iemura S, Natsume T, Kawashima SA, Watanabe Y. 2006. Shugoshin collaborates with protein phosphatase 2A to protect cohesin. *Nature* **441:** 46–52.

Kurasawa Y, Yu-Lee LY. 2010. PICH and cotargeted Plk1 coordinately maintain prometaphase chromosome arm architecture. *Mol Biol Cell* **21:** 1188–1199.

Leng M, Besusso D, Jung SY, Wang Y, Qin J. 2008. Targeting Plk1 to chromosome arms and regulating chromosome compaction by the PICH ATPase. *Cell Cycle* **7:** 1480–1489.

Lukas C, Savic V, Bekker-Jensen S, Doil C, Neumann B, Pedersen RS, Grøfte M, Chan KL, Hickson ID, Bartek J, et al. 2011. 53BP1 nuclear bodies form around DNA lesions generated by mitotic transmission of chromosomes under replication stress. *Nature Cell Biology* **13:** 243–253.

Maciejowski J, Li YL, Bosco N, Campbell PJ, de Lange T. 2015. Chromothripsis and kataegis induced by telomere crisis. *Cell* **163:** 1641–1654.

Martinez P, Blasco MA. 2015. Replicating through telomeres: A means to an end. *Trends Biochem Sci* **40:** 504–515.

Martinez P, Thanasoula M, Munoz P, Liao CY, Tejera A, McNees C, Flores JM, Fernandez-Capetillo O, Tarsounas M, Blasco MA. 2009. Increased telomere fragility and fusions resulting from TRF1 deficiency lead to degenerative pathologies and increased cancer in mice. *Genes Dev* **23:** 2060–2075.

McClintock B. 1938. The production of homozygous deficient tissues with mutant characteristics by means of the aberrant mitotic behavior of ring-shaped chromosomes. *Genetics* **23:** 315–376.

McClintock B. 1942. The fusion of broken ends of chromosomes following nuclear fusion. *Proc Natl Acad Sci* **28:** 458–463.

Meetei AR, Sechi S, Wallisch M, Yang DF, Young MK, Joenje H, Hoatlin ME, Wang WD. 2003. A multiprotein nuclear complex connects Fanconi anemia and Bloom syndrome. *Mol Cell Biol* **23:** 3417–3426.

Minocherhomji S, Ying SM, Bjerregaard VA, Bursomanno S, Aleliunaite A, Wu W, Mankouri HW, Shen HH, Liu Y, Hickson ID. 2015. Replication stress activates DNA repair synthesis in mitosis. *Nature* **528:** 286–290.

Moreno A, Carrington JT, Albergante L, Al Mamun M, Haagensen EJ, Komseli ES, Gorgoulis VG, Newman TJ, Blow JJ. 2016. Unreplicated DNA remaining from unperturbed S phases passes through mitosis for resolution in daughter cells. *Proc Natl Acad Sci* **113:** E5757–E5764.

Naim V, Rosselli F. 2009. The FANC pathway and BLM collaborate during mitosis to prevent micro-nucleation and chromosome abnormalities. *Nat Cell Biol* **11:** 761–768.

Nasmyth K. 2011. Cohesin: A catenase with separate entry and exit gates? *Nat Cell Biol* **13:** 1170–1177.

Nera B, Huang HS, Lai T, Xu L. 2015. Elevated levels of TRF2 induce telomeric ultrafine anaphase bridges and rapid telomere deletions. *Nat Commun* **6:** 10132.

Neuman KC, Nagy A. 2008. Single-molecule force spectroscopy: Optical tweezers, magnetic tweezers and atomic force microscopy. *Nat Methods* **5:** 491–505.

Nielsen CF, Hickson ID. 2016. PICH promotes mitotic chromosome segregation: Identification of a novel role in rDNA disjunction. *Cell Cycle* **15:** 2704–2711.

Nielsen CF, Huttner D, Bizard AH, Hirano S, Li TN, Palmai-Pallag T, Bjerregaard VA, Liu Y, Nigg EA, Wang LHC, et al. 2015. PICH promotes sister chromatid disjunction and cooperates with topoisomerase II in mitosis. *Nat Commun* **6:** 8962.

Palm W, de Lange T. 2008. How shelterin protects mammalian telomeres. *Annu Rev Genet* **42:** 301–334.

Pedersen RT, Kruse T, Nilsson J, Oestergaard VH, Lisby M. 2015. TopBP1 is required at mitosis to reduce transmission of DNA damage to G1 daughter cells. *J Cell Biol* **210:** 565–582.

Piskadlo E, Tavares A, Oliveira RA. 2017. Metaphase chromosome structure is dynamically maintained by condensin I-directed DNA (de)catenation. *Elife* **6:** 1328.

Pitchai GP, Kaulich M, Bizard AH, Mesa P, Yao Q, Sarlos K, Streicher W, Nigg E, Montoya G, Hickson I. 2017. A novel TPR-BEN domain interaction mediates PICH-BEND3 association. *Nucleic Acids Res* (in press).

Porter AC, Farr CJ. 2004. Topoisomerase II: Untangling its contribution at the centromere. *Chromosome Res* **12:** 569–583.

Raynard S, Bussen W, Sung P. 2006. A double Holliday junction dissolvasome comprising BLM, topoisomerase IIIα, and BLAP75. *J Biol Chem* **281:** 13861–13864.

Rosin MP, German J. 1985. Evidence for chromosome instability in vivo in Bloom syndrome: Increased numbers of micronuclei in exfoliated cells. *Hum Genet* **71:** 187–191.

Rouzeau S, Cordelieres FP, Buhagiar-Labarchede G, Hurbain I, Onclercq-Delic R, Gemble S, Magnaghi-Jaulin L, Jaulin C, Amor-Gueret M. 2012. Bloom's syndrome and PICH helicases cooperate with topoisomerase IIα in centromere disjunction before anaphase. *PLoS One* **7:** e33905.

Sarbajna S, Davies D, West SC. 2014. Roles of SLX1-SLX4, MUS81-EME1, and GEN1 in avoiding genome instability and mitotic catastrophe. *Genes Dev* **28:** 1124–1136.

Schoeffler AJ, Berger JM. 2008. DNA topoisomerases: Harnessing and constraining energy to govern chromosome topology. *Q Rev Biophys* **41:** 41–101.

Schvartzman JB, Stasiak A. 2004. A topological view of the replicon. *EMBO Rep* **5:** 256–261.

Sfeir A, Kosiyatrakul ST, Hockemeyer D, MacRae SL, Karlseder J, Schildkraut CL, de Lange T. 2009. Mammalian telomeres resemble fragile sites and require TRF1 for efficient replication. *Cell* **138:** 90–103.

Sims AE, Spiteri E, Sims RJ III, Arita AG, Lach FP, Landers T, Wurm M, Freund M, Neveling K, Hanenberg H, et al. 2007. FANCI is a second monoubiquitinated member of the Fanconi anemia pathway. *Nat Struct Mol Biol* **14:** 564–567.

Singh TR, Ali AM, Busygina V, Raynard S, Fan Q, Du CH, Andreassen PR, Sung P, Meetei AR. 2008. BLAP18/RMI2, a novel OB-fold-containing protein, is an essential component of the Bloom helicase-double Holliday junction dissolvasome. *Genes Dev* **22:** 2856–2868.

Singleton MR, Dillingham MS, Wigley DB. 2007. Structure and mechanism of helicases and nucleic acid translocases. *Annu Rev Biochem* **76:** 23–50.

Sirri V, Roussel P, Hernandez-Verdun D. 1999. The mitotically phosphorylated form of the transcription termination factor TTF-1 is associated with the repressed rDNA transcription machinery. *J Cell Sci* **112 (Pt 19):** 3259–3268.

Smogorzewska A, Matsuoka S, Vinciguerra P, McDonald ER, Hurov KE, Luo J, Ballif BA, Gygi SP, Hofmann K, D'Andrea AD, et al. 2007. Identification of the FANCI protein, a monoubiquitinated FANCD2 paralog required for DNA repair. *Cell* **129:** 289–301.

Stark WM, Sherratt DJ, Boocock MR. 1989. Site-specific recombination by Tn3 resolvase: Topological changes in the forward and reverse reactions. *Cell* **58:** 779–790.

Sullivan M, Higuchi T, Katis VL, Uhlmann F. 2004. Cdc14 phosphatase induces rDNA condensation and resolves cohesin-independent cohesion during budding yeast anaphase. *Cell* **117:** 471–482.

Sun H, Karow JK, Hickson ID, Maizels N. 1998. The Bloom's syndrome helicase unwinds G4 DNA. *J Biol Chem* **273:** 27587–27592.

Suski C, Marians KJ. 2008. Resolution of converging replication forks by RecQ and topoisomerase III. *Mol Cell* **30:** 779–789.

Tanaka K, Hao Z, Kai M, Okayama H. 2001. Establishment and maintenance of sister chromatid cohesion in fission yeast by a unique mechanism. *EMBO J* **20:** 5779–5790.

Thompson SL, Compton DA. 2011. Chromosomes and cancer cells. *Chromosome Res* **19:** 433–444.

Uhlmann F, Wernic D, Poupart MA, Koonin EV, Nasmyth K. 2000. Cleavage of cohesin by the CD clan protease separin triggers anaphase in yeast. *Cell* **103:** 375–386.

Vinciguerra P, Godinho SA, Parmar K, Pellman D, D'Andrea AD. 2010. Cytokinesis failure occurs in Fanconi anemia pathway-deficient murine and human bone marrow hematopoietic cells. *J Clin Invest* **120:** 3834–3842.

Voit R, Seiler J, Grummt I. 2015. Cooperative action of Cdk1/cyclin B and SIRT1 is required for mitotic repression of rRNA synthesis. *PLoS Genet* **11:** e1005240.

Vos SM, Tretter EM, Schmidt BH, Berger JM. 2011. All tangled up: How cells direct, manage and exploit topoisomerase function. *Nat Rev Mol Cell Biol* **12:** 827–841.

Wallis JW, Chrebet G, Brodsky G, Rolfe M, Rothstein R. 1989. A hyper-recombination mutation in *S. cerevisiae* identifies a novel eukaryotic topoisomerase. *Cell* **58:** 409–419.

Wang BD, Yong-Gonzalez V, Strunnikov AV. 2004. Cdc14p/FEAR pathway controls segregation of nucleolus in *S. cerevisiae* by facilitating condensin targeting to rDNA chromatin in anaphase. *Cell Cycle* **3:** 960–967.

Wang LHC, Schwarzbraun T, Speicher MR, Nigg EA. 2008. Persistence of DNA threads in human anaphase cells suggests late completion of sister chromatid decatenation. *Chromosoma* **117:** 123–135.

Wang F, Yang YT, Singh TR, Busygina V, Guo R, Wan K, Wang WD, Sung P, Meetei AR, Lei M. 2010a. Crystal structures of RMI1 and RMI2, two OB-fold regulatory subunits of the BLM complex. *Structure* **18:** 1159–1170.

Wang LHC, Mayer B, Stemmann O, Nigg EA. 2010b. Centromere DNA decatenation depends on cohesin removal and is required for mammalian cell division. *J Cell Sci* **123:** 806–813.

Wang JD, Chen JJ, Gong ZH. 2013. TopBP1 controls BLM protein level to maintain genome stability. *Mol Cell* **52:** 667–678.

Wold MS. 1997. Replication protein A: A heterotrimeric, single-stranded DNA-binding protein required for eukaryotic DNA metabolism. *Annu Rev Biochem* **66:** 61–92.

Wu L, Hickson ID. 2003. The Bloom's syndrome helicase suppresses crossing over during homologous recombination. *Nature* **426:** 870–874.

Wu L, Davies SL, North PS, Goulaouic H, Riou JF, Turley H, Gatter KC, Hickson ID. 2000. The Bloom's syndrome gene product interacts with topoisomerase III. *J Biol Chem* **275:** 9636–9644.

Wu L, Davies SL, Levitt NC, Hickson ID. 2001. Potential role for the BLM helicase in recombinational repair via a conserved interaction with RAD51. *J Biol Chem* **276:** 19375–19381.

Wu L, Bachrati CZ, Ou J, Xu C, Yin J, Chang M, Wang W, Li L, Brown GW, Hickson ID. 2006. BLAP75/RMI1 promotes the BLM-dependent dissolution of homologous recombination intermediates. *Proc Natl Acad Sci* **103:** 4068–4073.

Wuite GJL, Smith SB, Young M, Keller D, Bustamante C. 2000. Single-molecule studies of the effect of template tension on T7 DNA polymerase activity. *Nature* **404:** 103–106.

Xu DY, Guo R, Sobeck A, Bachrati CZ, Yang J, Enomoto T, Brown GW, Hoatlin ME, Hickson ID, Wang WD. 2008. RMI, a new OB-fold complex essential for Bloom syndrome protein to maintain genome stability. *Genes Dev* **22:** 2843–2855.

Xu D, Muniandy P, Leo E, Yin J, Thangavel S, Shen X, Ii M, Agama K, Guo R, Fox D III, et al. 2010. Rif1 provides a new DNA-binding interface for the Bloom syndrome complex to maintain normal replication. *EMBO J* **29:** 3140–3155.

Xue XY, Raynard S, Busygina V, Singh AK, Sung P. 2013. Role of replication protein A in double Holliday junction dissolution mediated by the BLM-Topo IIIα-RMI1-RMI2 protein complex. *J Biol Chem* **288:** 14221–14227.

Yin JH, Sobeck A, Xu C, Meetei AR, Hoatlin M, Li L, Wang WD. 2005. BLAP75, an essential component of Bloom's syndrome protein complexes that maintain genome integrity. *EMBO J* **24:** 1465–1476.

Ying SM, Minocherhomji S, Chan KL, Palmai-Pallag T, Chu WK, Wass T, Mankouri HW, Liu Y, Hickson ID. 2013. MUS81 promotes common fragile site expression. *Nat Cell Biol* **15:** 1001–1007.

Zhang CZ, Spektor A, Cornils H, Francis JM, Jackson EK, Liu SW, Meyerson M, Pellman D. 2015. Chromothripsis from DNA damage in micronuclei. *Nature* **522:** 179–184.

Zou Y, Liu YY, Wu XM, Shell SM. 2006. Functions of human replication protein A (RPA): From DNA replication to DNA damage and stress responses. *J Cell Physiol* **208:** 267–273.

Low-Level, Global Transcription during Mitosis and Dynamic Gene Reactivation during Mitotic Exit

Katherine C. Palozola,[1] Hong Liu,[2] Dario Nicetto,[1] and Kenneth S. Zaret[1]

[1]*Department of Cell and Developmental Biology, Perelman School of Medicine, University of Pennsylvania, Philadelphia, Pennsylvania 19104*

[2]*Department of Biochemistry and Molecular Biology, Tulane University School of Medicine, New Orleans, Louisiana 70112*

Correspondence: zaret@upenn.edu

Mitosis is thought to be a period of transcriptional silence due to the compact nature of mitotic chromosomes and the apparent exclusion of RNA Pol II and many transcription factors from mitotic chromatin. Yet accurate reactivation of a cell's specific gene expression program is needed to reestablish functional cell identity after mitosis. The majority of studies on protein regulation and localization during mitosis have relied extensively on antibodies and cross-linking-based approaches that are known to artifactually exclude proteins from mitotic chromatin. Here we show that RNA Pol II localization in mitosis is antibody- and fixation-dependent, and that direct assessment of transcription by pulse-labeling nascent RNA reveals global, low-level mitotic transcription. We also find a hierarchy of gene reactivation as the cells transition from mitosis to their interphase amplitude of gene expression. Resetting of gene transcription during mitotic exit is coincident with enhancer transcription. Our work thus shifts focus from assessing mitotic exit as a binary transcription switch to a more nuanced concert of transcription amplitude and enhancer usage. We suggest that understanding how gene expression patterns are conserved during mitosis rests upon deciphering how transcription is maintained by promoters.

During development, lineage specification and differentiation are performed by the successive expression of key transcription factors. Once a cell's functional identity has been established, it is imperative that its identity be maintained by the inheritance of the cell type–specific gene expression profile through cell division. Mitosis is the period of the cell cycle during which the recently duplicated sister chromatids are divided into two separate nuclei. At the onset of mitosis, chromosomes condense to help ensure separation of the sister chromatids. It has been thought that this condensation excludes transcriptional machinery (Martinez-Balbas et al. 1995; Prasanth et al. 2003) resulting in the termination of transcription (Prescott and Bender 1962; Parsons and Spencer 1997), and others showed a global repression of RNA synthesis by nucleotide incorporation (Prescott and Bender 1962; Konrad 1963; Johnson and Holland 1965; Parsons and Spencer 1997). However, there are caveats to each of these studies surrounding the approaches used and conclusions drawn that may have misinterpreted the transcriptional state of mitotic cells.

The carboxy-terminal domain (CTD) of Rpb1, the largest and enzymatic subunit of RNAP2, is composed of multiple, conserved heptapeptide repeats of Tyr1-Ser2-Pro3-Thr4-Ser5-Pro6-Ser7. Many studies have described the role of CTD posttranslational modifications (PTMs) in the transcription regulatory cycle of recruitment, pausing, initiation, elongation, and termination (Heidemann et al. 2013). In general, phospho-serine 5 RNAP2 (Ser5-P) is found at the transcription start site and diminishes toward the 3′ end of the transcribed region of a gene, whereas phospho-serine 2 RNAP2 (Ser2-P) is low near the transcription start site and increases toward the 3′ end (Komarnitsky et al. 2000; Heidemann et al. 2013). Because of these observations, and Ser2-P's dependence on the elongation factor P-TEFb (Shim et al. 2002), Ser5-P is considered the transcription initiating form of RNAP2, whereas Ser2-P is considered the elongating form of RNAP2. Notwithstanding the implications of these PTMs in transcription, both Ser2-P and Ser5-P were detected in mitotic cell extracts by western blot (WB) (Bregman et al. 1995), although it was unclear whether RNAP2 was transcriptionally engaged.

Notably, Ser2 can be phosphorylated independent of P-TEFb by the atypical kinase, Brd4 (Devaiah et al. 2012), which is associated with mitotic chromatin (Dey et al. 2009). Thus, it is unclear as to whether Ser2-P is indicative of P-TEFb activity. Moreover, these studies have relied on specific CTD antibodies: the H5 monoclonal antibody for Ser2-P and the H14 monoclonal antibody for Ser5-P. However, H5 does not recognize a CTD that is only phosphorylated at serine 2. Rather, the epitope appears to be Ser2-P followed by Ser5-P on the same heptapeptide unit (Chapman et al. 2007). In contrast, H14 recognizes Ser5-P on one heptapeptide unit followed by Ser2-P on the next (Chapman et al. 2007). Thus, given the combinatorial potential for CTD PTMs, the transcriptional status of RNAP2 may not be reduced to a two epitope-based model.

More importantly, RNAP2 and associated proteins have been detected by mass spectrometry of mitotic chromatin (Fig. 1; Ohta et al. 2010). Also, flavopiridol-based tran-

Published by Cold Spring Harbor Laboratory Press; doi: 10.1101/sqb.2017.82.034280

Cold Spring Harbor Symposia on Quantitative Biology, Volume LXXXII

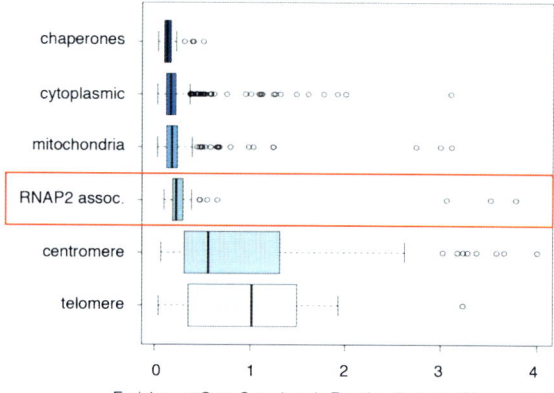

Figure 1. RNA Pol II is a structural constituent of mitotic chromatin. Analysis of mass spectrometry results from HeLa metaphase chromatin generated in Ohta et al. (2010) shows proteins enriched in chromatin compared to cytoplasmic fraction. The red box indicates proteins associated with RNAP2.

scription inhibition studies have shown that RNAP2 is associated with mitotic chromosomes (Liang et al. 2015). Though RNAP2 was not detectable in mitotic HeLa cells by chromatin immunoprecipitation followed by high-throughput sequencing (ChIP-seq), the addition of transcription elongation inhibitor resulted in paused RNAP2 peaks at several promoters in mitotic cells (Liang et al. 2015). These results suggest that low levels of transcribing RNAP2 are present, and that only when elongation is blocked and there is a backup of RNAP2 at the promoter is the signal detected. In addition, recent studies have shown that formaldehyde cross-linking artifactually removes proteins from mitotic chromatin (Lerner et al. 2016; Teves et al. 2016). Even though transcription factors were visibly excluded from mitotic chromatin by immunofluorescence (IF), fluorescently labeled fusions of the same proteins colocalized with mitotic chromatin in live cells.

Here we investigate the role that fixation- and antibody-based methods have played in establishing the traditional paradigm of transcriptional machinery exclusion from mitotic chromatin and the role of pulse-labeling transcripts in uncovering mitotic transcription. We find that there exists a global, low level of mitotic transcription that may be enhancer-independent. We also show that the genes that encode more basic cell functions are the first to increase during mitotic exit, and that enhancer activity apparently coincides with gene reactivation.

RNA POL II DURING MITOSIS

In a previous study, we used the HUH7 human hepatoma cell line to show that the pioneer factor, FoxA1, remains bound to a subset of its interphase binding sites during mitosis (Caravaca et al. 2013). Here, we use this cell line as it is amenable to efficient cell cycle synchronization and enrichment of mitotically arrested cells. Furthermore, upon release from a mitotic block, the cells efficiently reenter the cell cycle. Rather than rely on the

H5 and H14 antibodies that have traditionally been used in the literature, we performed IF for endogenous RNAP2 with an antibody raised against a synthetic RNAP2 Ser2-P peptide. We found RNAP2 Ser2-P colocalized with chromatin at every stage of mitosis (Fig. 2A–L).

Because this observation contradicts previously reported IF studies that show that RNAP2 is evicted from chromatin during mitosis, we sought to confirm the antibody's epitope specificity by preincubating the antibody with excess peptide before IF. Indeed, we found that the RNAP2 Ser2-P signal was lost in both IF (Fig. 2M–P) and WB (Fig. 2Q,R) analyses when the antibody was preincubated with the Ser2-P peptide, and that the level of Ser2-P was similar between mitotic and asynchronous cells when the antibody was not blocked. Given that the antibody we used was raised against a synthetic phosphopeptide, and not endogenous RNAP2, we needed further evidence that the epitope that we were detecting on the mitotic chromatin was in fact RNAP2 and not one of the many other phosphorylated proteins present in mitosis. To this end, we performed reciprocal immunoprecipitation (IP) followed by immunoblot (IB) analyses with the Ser2-P antibody and an antibody against the amino-terminal domain (NTD) of Rbp1, which recognizes RNAP2 independent of the CTD phospho status. In both pulldowns, the antibodies pulled down RNAP2 (Fig. 2S–V). Thus, the epitope detected on mitotic chromatin by IF is indeed a form of RNAP2.

We repeated the mitotic IF with a second Ser2-P antibody raised against the same synthetic Ser2-P peptide. This time, in agreement with previous reports, we found that RNAP2 Ser2-P was largely excluded from the mitotic chromatin (Fig. 2W,X). These results indicate that there is strong antibody dependence on RNAP2 localization in fixed cells, which reinforces the notion that formaldehyde fixation can artifactually exclude epitopes.

To circumvent the issue of fixation, we obtained a functional RNAP2-EGFP fusion construct that was previously shown to rescue temperature-sensitive RNAP2 mutants (Sugaya et al. 2000). Although the construct was lethal to the cells, live-cell imaging showed colocalization of RNAP2-EGFP with mitotic chromatin before cell death (Fig. 2Y,Z). We thus conclude that detection of RNAP2 in mitotic chromatin is both antibody dependent and that cases of exclusion are likely due to fixation artifacts, as previously reported for transcription factors (Lerner et al. 2016; Teves et al. 2016). Importantly, the presence of an active form of RNAP2 in mitosis has been reported at the centromere (Liu et al. 2015), but our observation agrees with various hints throughout the literature that it is also present at low levels throughout mitotic chromatin.

MITOTIC TRANSCRIPTION

Based on our above observation that RNAP2 can be detected on mitotic chromatin, we next sought to directly quantify RNA synthesis during mitosis. Unlike cross-linking-based approaches for detection of RNAP2, which

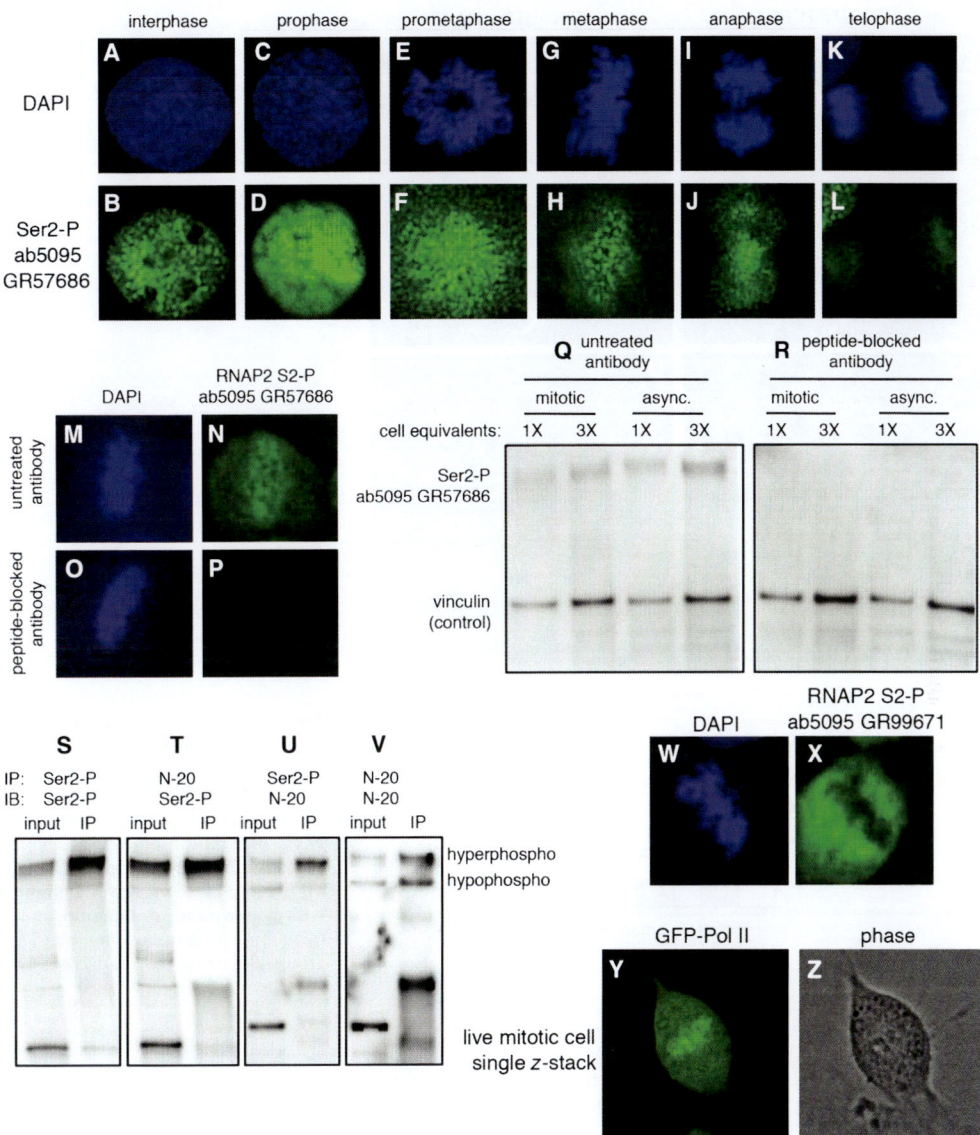

Figure 2. RNA Pol II Ser2-P colocalizes with mitotic chromatin. HUH7 cells stained with DAPI or Ser2-P in (*A,B*) interphase, (*C,D*) prophase, (*E,F*) prometaphase, (*G,H*) metaphase, (*I,J*) anaphase, and (*K,L*) telophase. Metaphase HUH7 cells stained with (*M*) DAPI and (*N*) Ser2-P, or (*O*) DAPI and (*P*) Ser2-P after preincubation with Ser2-P peptide. Mitotic and asynchronous HUH7 whole-cell lysate blotted with (*Q*) Ser2-P or (*R*) Ser2-P after preincubation with Ser2-P peptide; reciprocal immunoprecipitation (IP)-immunoblot (IB) of asynchronous HUH7 cells; (*S*) IP with Ser2-P and blot for Ser2-P; (*T*) IP with amino-terminal domain (NTD) and blot for Ser2-P. (*U*) IP with Ser2-P and blot for NTD. (*V*) IP with NTD and blot for NTD; Ser2-P antibody, ab5095 GR57686; NTD antibody, N-20. Metaphase HUH7 cells stained with (*W*) DAPI and (*X*) Ser2-P; the overlay indicates Ser2-P is enriched around the periphery; Ser2-P antibody, ab5095 GR99671. (*Y*) HUH7 cell transiently overexpressing GFP-RNAP2 with (*Z*) corresponding bright field image.

indirectly assess the transcriptional state of mitotic chromatin in fixed cells, metabolic labeling of nascent transcripts with labeled nucleotide is a direct assessment of RNA synthesis in live, mitotic cells. Previously, HeLa cell metaphase spreads were incubated with FITC-UTP to allow incorporation of the FITC-UTP into any RNA currently synthesized (Liu et al. 2015). The spreads were then fixed and imaged. To aid in the analysis, the original published images were masked such that only the FITC signal that colocalized with centromeres was visible (Liu et al. 2015). Importantly, faint FITC signal was present outside

of centromeres, along metaphase chromosome arms, once the mask was removed (Fig. 3A–D). Furthermore, the detected RNA synthesis was significantly reduced when the cells were treated with the transcriptional inhibitor α-amanitin (Fig. 3E). From these results we conclude that global, low levels of transcription occur throughout mitotic chromosomes, not just at the centromeres.

The FITC-UTP study (Liu et al. 2015) also found that centromeric transcription is dependent on Bub1. The Bub1 kinase phosphorylates H2A T120 at the centromere for the recruitment of Sgo and RNAP2 to H2A T120-P

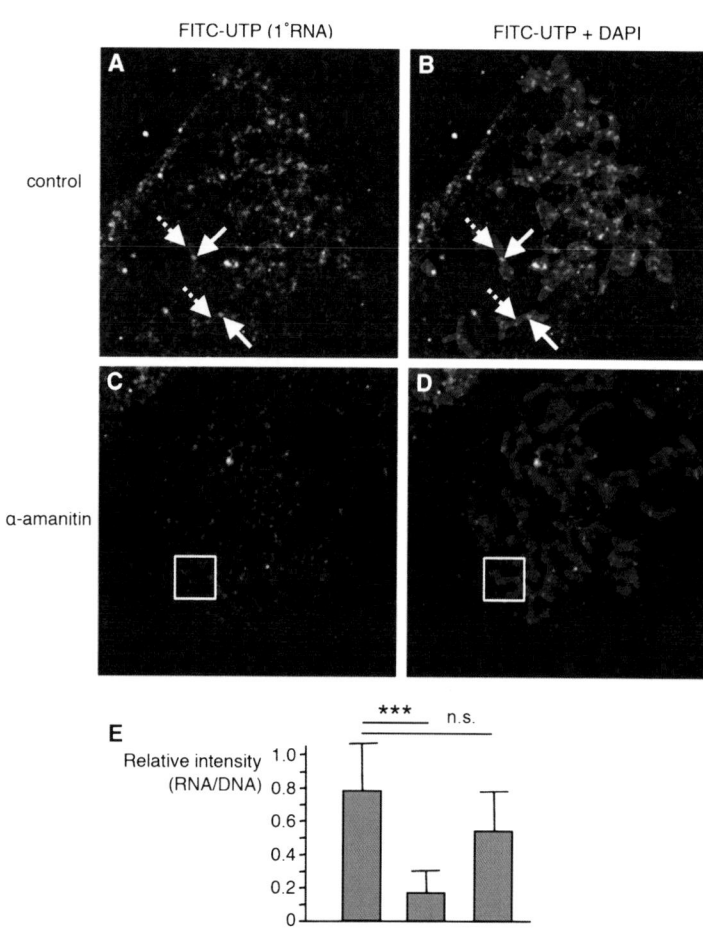

FITC-UTP (1°RNA)   FITC-UTP + DAPI

control

α-amanitin

E
Relative intensity
(RNA/DNA)

***   n.s.

control   α-aman   siBub1
n =14   n =10   n =10

Figure 3. Transcription on metaphase chromosome arms is independent of Bub1. HeLa metaphase spreads labeled with FITC-UTP without (*A,B*) and with (*C,D*) α-amanitin treatment. Solid arrows, FITC at centromere; dashed arrows, FITC on chromosome arms; boxes, no FITC detected above background. (*E*) Quantification of FITC-UTP signal over chromosome arms, with and without α-amanitin or Bub1 siRNA treatment. *n*, chromosome arms per treatment group, *P* < 0.001 (two-tailed *t*-test).

nucleosomes (Liu et al. 2015). When Bub1 activity is diminished by knockdown or enzymatic inhibition, RNAP2 is no longer recruited to the centromere, preventing the transcription of centromeric repeats (Liu et al. 2015). Because the transcription of these repeats is necessary for Sgo recruitment, Sgo is no longer recruited to protect the centromere, causing the cells to fail to segregate properly (Liu et al. 2015). Furthermore, ectopic expression of H2A T120-P was not sufficient to recruit RNAP2 to metaphase chromosome arms (Liu et al. 2015). We find that RNAP2 transcription along metaphase chromosome arms is Bub1-independent (Fig. 3E). Thus, mitotic transcription is occurring distinct from that at the centromere.

To assess the profile of global mitotic transcription, we developed EU-RNA-seq as a way to capture and sequence nascent RNAs in vivo (Yokoyama et al. 2016; Palozola et al. 2017). Mitotic and asynchronous cells were pulse-labeled with the cell permeable uridine analog, 5′-ethynyluridine (EU) for 40 min. The total RNA from each sample was harvested and a click reaction was performed to allow the conjugation of biotin azide to the ethynyl group of EU (Jao and Salic 2008) on any EU-RNAs present in the total RNA harvested. Once the click reaction

was complete, we added custom, biotin-RNA spike-in controls to both mitotic and asynchronous samples (Palozola et al. 2017), and then streptavidin-coated magnetic beads were used to isolate biotinylated RNAs from unlabeled RNAs in each sample. We were then able to generate cDNA libraries directly off the magnetic beads for sequencing.

After global normalization between the asynchronous and mitotic samples, based on the relative spike-in sequence amplification in each replicate, there were approximately 28,000 transcripts expressed in the asynchronous population (Palozola et al. 2017). Of these, 8074 were reproducibly expressed among all three mitotic replicates, accounting for 28% of the genes expressed in an asynchronous population. However, the transcriptome was expressed much lower in mitotic than in asynchronous cells, with a fivefold mean decrement in transcript levels (Palozola et al. 2017). It is important to note that this low-level transcription is not purely the result of the ~3% contaminating interphase cells in the mitotic population. If the entirety of the mitotic signal was due to an interphase population, then we would expect the relative rank of genes within this population to be identical to that of the asynchronous population. However, a pairwise analysis of the Spearman rank correlation coefficient of all mitotic and asynchronous replicates indicated that the two populations were distinct (Palozola et al. 2017). From these studies and other controls (Palozola et al. 2017) we conclude that mitosis is not a period of transcriptional silence, as has long been considered the case.

HIERARCHICAL REACTIVATION DURING MITOTIC EXIT

Previous studies from our laboratory (Caravaca et al. 2013) and others (Kadauke et al. 2012) measured reactivation of individual genes during mitotic exit by real time quantitative polymerase chain reaction (RT-qPCR) with primers targeting adjacent introns and exons, thus assessing primary transcripts. In doing so, it was evident that not all genes are reactivated synchronously and with the same amplitude. However, these studies were limited to a handful of genes. We thus sought to determine the hierarchy with which the transcriptome returned to interphase levels during mitotic exit, on a genome-wide level. We again

used EU-RNA-seq in mitotic cells and as cells exited mitosis (Palozola et al. 2017). Intact HUH7 cells were pulsed-labeled with EU for 40 min during nocodazole-induced mitotic arrest (0′) and at 40′, 80′, 105′, 165′, and 300′ after the nocodazole was washed out and the cells were allowed to resume cycling.

To assess the various dynamics of transcription reactivation, we first removed all genes that are not expressed in asynchronous HUH7 cells. We then performed automated fuzzy c-means clustering (Schwammle and Jensen 2010) for each time point. Clusters were organized from those with genes most highly expressed at 40′ to those with genes most highly expressed in the asynchronous population of cells (Fig. 4). Four main patterns emerged: (1) genes that are highest at 40′ and then continue to decrease (Fig. 4A–D), (2) genes that are highest at 80′ and then continue to decrease (Fig. 4E,G,H), (3) those that increase from 40′ until 105′ or 165′ and then decrease (Fig. 4I–K), and (4) those that continuously increase at each time point after 40′ (Fig. 4M,N). Thus, as indicated by a small number of genes in previous studies (Kadauke et al. 2012; Caravaca et al. 2013), the amplitude of the interphase

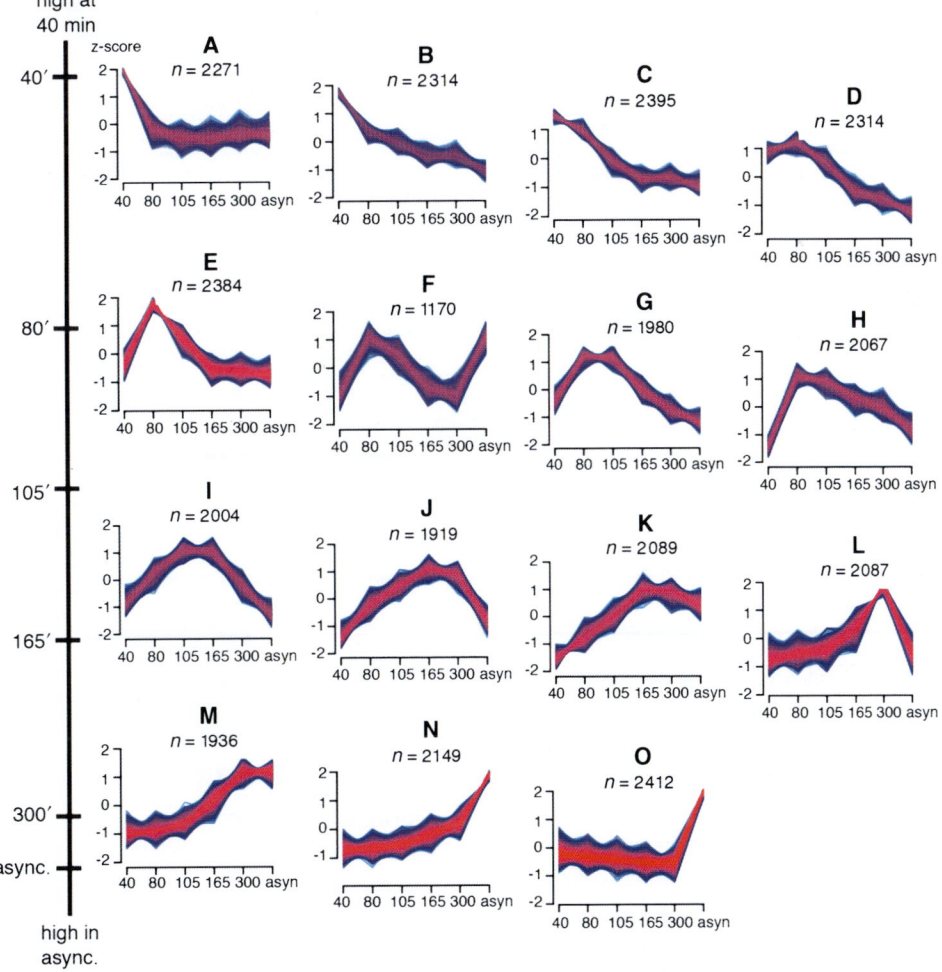

Figure 4. Dynamic patterns of reactivation during mitotic exit. Cluster analysis of the dynamic transition in expression amplitude during mitotic exit. Clusters are ordered from those highly expressed at 40′, to genes highly expressed in asynchronous (asyn).

transcriptome is not achieved synchronously during mitotic exit.

We hypothesized that the first genes to increase expression during mitotic exit in our hepatoma cell line could be enriched for liver specificity. To test this, we performed Gene Ontology (GO) enrichment analysis on early- and late-activated clusters. We were surprised to find that genes highest at 40'–80' (Fig. 5A–E) are those involved in basic cellular functions such as transport and signal transduction. Conversely, genes that are highest at later time points (Fig. 5L–N) are enriched for genes involved in more cell type–specific functions like liver metabolism. However, because EU-RNA-seq measures the rate of tran-

scription by EU incorporation, and shorter transcripts may undergo more rounds of transcription than longer transcripts in a given time point, we sought to determine whether gene length biased the functional gene categories at each time point. To do so, we compared the enriched GO categories of transcripts that first increase ≥1.5-fold over mitosis at 40' or 80' when quantifying the FPKM (fragments per kilobase million) over the entire transcript (Fig. 5P) or the first 10 kb of transcripts that are ≥10 kb in length (Fig. 5Q). Indeed, the enriched GO terms in both scenarios are for basic cell functions. Thus, exiting mitosis and rebuilding daughter cells is prioritized over liver-specific gene expression in the early phases of mitotic exit.

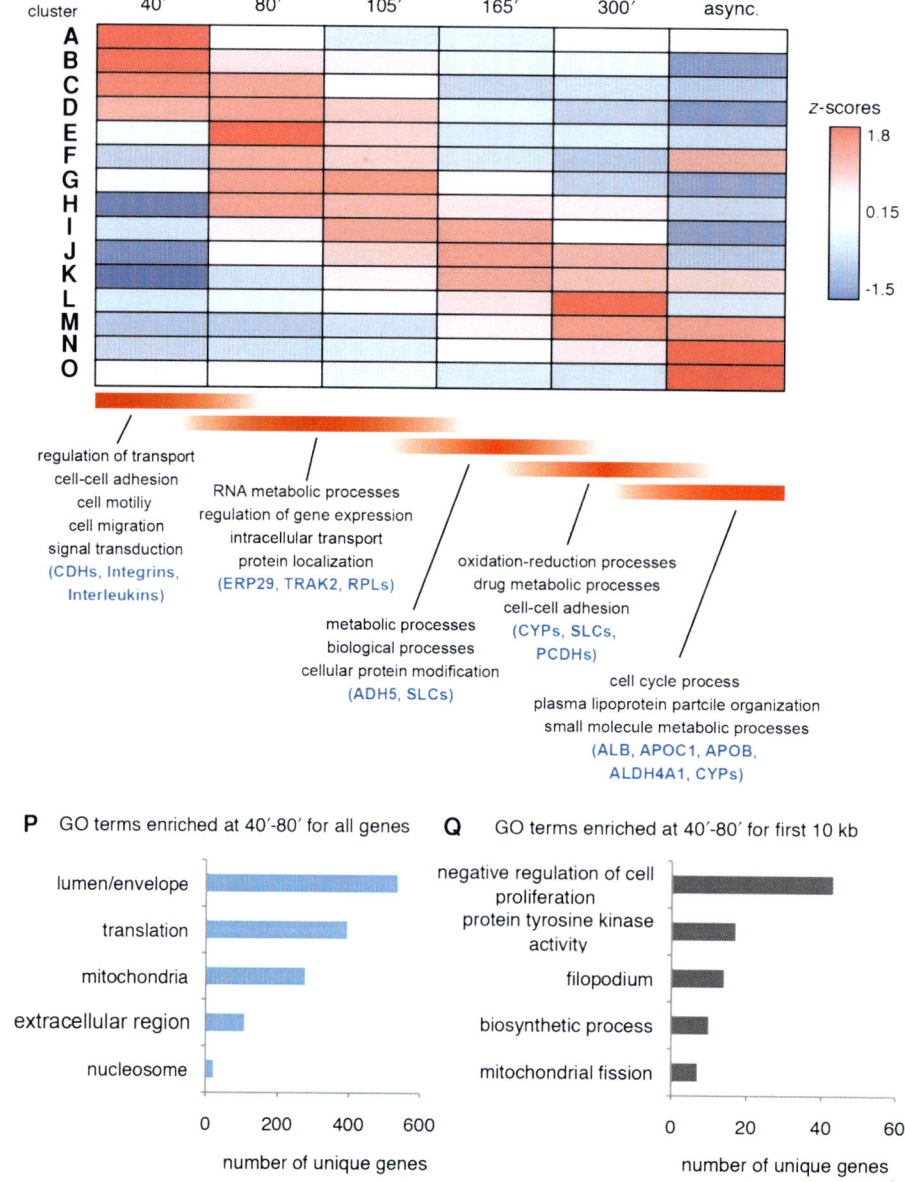

Figure 5. Basic cell functions are prioritized during mitotic exit. (*A–O*) Heatmap of the clusters shown in Figure 4, including representative enriched GO categories. (*P*) Representative enriched GO terms for genes that first increase ≥1.5 fold over mitosis at 40' or 80'. (*Q*) Representative enriched GO terms for genes associated with the 10-kb transcript fragments that first increase ≥1.5 over mitosis at 40' or 80'.

ENHANCER USAGE

Despite the physical condensation of mitotic chromosomes, global ATAC-seq profiles are highly similar between asynchronous and mitotic cells (Teves et al. 2016), indicating an overall retention in accessibility at the molecular level. At a finer level, it has been shown that promoter accessibility is maintained during mitosis (Martinez-Balbas et al. 1995; Hsiung et al. 2015), although enhancer accessibility is lost (Hsiung et al. 2015). Long distance interactions are also lost (Naumova et al. 2013), with enhancer–promoter loops reforming $60' \sim 90'$ after nocodazole washout, concordant with a spike in RNAP2 binding to enhancers and promoters (Hsiung et al. 2016).

Given the sensitivity of EU-RNA-seq, we sought to estimate the timing of enhancer usage during mitotic exit by using eRNA transcription as a surrogate for en-

hancer activity. We curated all previously reported human enhancers (Leung et al. 2015) for those that both are intergenic and have a detectable signal in asynchronous HUH7 cells, and clustered them by their expression at each time point (Fig. 6). We see waves of apparent enhancer reactivation (Fig. 6A), with the largest increase occurring at 80′ (Fig. 6B), akin to the large burst in transcription rate increase at 80′ (Palozola et al. 2017), and the previously reported spike in RNAP2 binding at enhancers and reformation of enhancer–promoter contacts at 90′ (Hsiung et al. 2016). However, the sensitivity of EU-RNA-seq uncovers the dynamics of enhancer usage. Interestingly, it is the eRNAs that are most lowly transcribed in asynchronous cells, which are the first to increase during mitotic exit, with the highest asynchronous eRNAs increasing later (Fig. 6C). Based on these observations, we conclude that enhancers may not be generally used during low-level mitotic transcription and are reactivated as enhancer–pro-

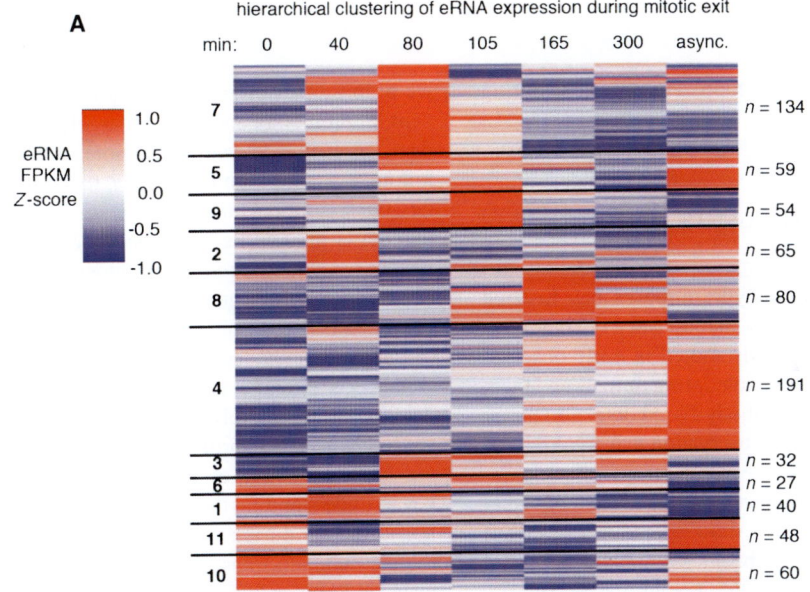

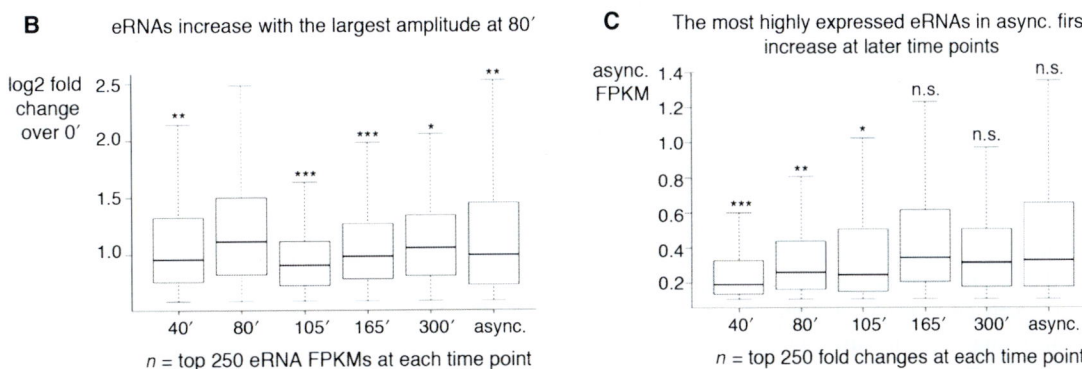

Figure 6. eRNAs are activated in waves during mitotic exit. (*A*) FPKM *Z*-score of intergenic enhancers; hierarchical cluster number on the *left*, number of eRNAs in each cluster on the *right*, $n = 709$. (*B*) $\log_2$ fold increase of each eRNA over mitosis (0′) for eRNAs that first increase at each time point; *n*, top 250 eRNAs most highly expressed at that time point, Wilcoxon rank-sum test between each time point and 80′. (*C*) Asynchronous FPKM of each eRNA that first increases over mitosis (0′) for the first time at each time point; *n*, top 250 eRNAs with the highest fold change at that time point, Wilcoxon rank-sum test between each time point and asynchronous. n.s., not significant; *, $P < 0.05$; **, $P < 0.01$; ***, $P < 0.001$.

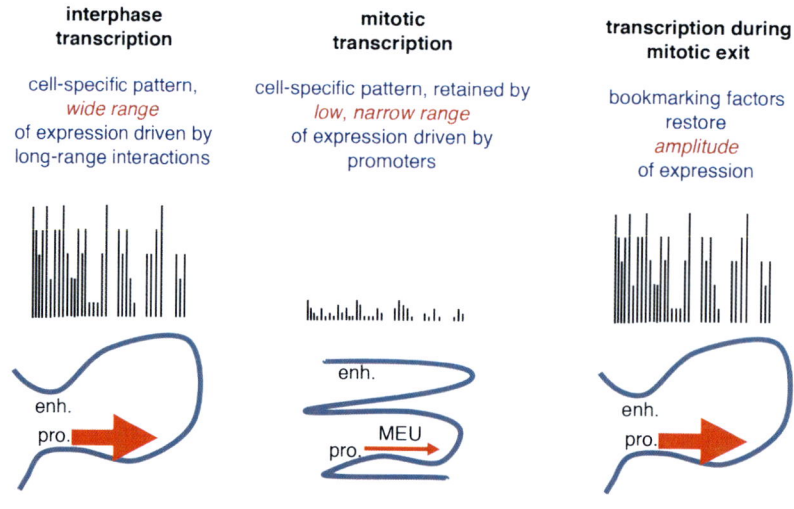

Figure 7. Enhancer usage is diminished in mitosis and is reestablished during mitotic exit. Interphase cells use enhancers to drive the cell type–specific gene expression program, whereas mitotic expression units are enhancerless and allow for continued, low-level gene expression.

moter units as long distance loops are reformed during mitotic exit.

CONCLUSION

Our increasing ability to reprogram cell identity for eventual therapeutic and analytic purposes has fostered interest in the basis of identity maintenance during mitosis. This interest was originally rooted in early observations of reduced RNA production during mitosis and then, with the increasing use of antibody-based detection methods such as IF and ChIP, the observation that RNAP2 and several transcription factors are evicted from mitotic chromatin. These latter studies have dominated the conversation and the original studies showing reduced, but not erased, transcription have been overlooked (Konrad 1963; Johnson and Holland 1965; Gariglio et al. 1974).

Here we have reexamined the mitotic transcriptional state in decades of studies. The localization of RNAP2 is both antibody- and fixation-dependent, and the complete phospho status of the CTD during mitosis remains unclear, although it does carry Ser2-P. Furthermore, global RNAP2-dependent transcription is not entirely erased. Indeed, there exists a low-level pattern of transcription throughout metaphase cells, which is distinct from that at the centromere. We also found that the interphase amplitude of gene expression is not achieved synchronously during the transition from mitosis to interphase, as gene reactivation occurs in waves, with the earliest waves including genes involved in basic cell functions. Gene reactivation appears to be congruent with enhancer usage, as eRNA activity also increases in waves during mitotic exit.

Based on these observations, and on previous reports of loss of long-range interactions (Naumova et al. 2013) and enhancer accessibility during mitosis (Hsiung et al. 2015),

we propose that mitotic transcription is primarily dependent on promoters and may be largely enhancer-independent. In such a scenario, promoters drive enhancerless mitotic expression units (MEUs), much like the rudimentary transcriptional units in yeast, which lack enhancers. In this MEU model, the cell type–specific gene expression pattern is retained throughout mitosis and enhancer usage drives the reestablishment of the interphase amplitude of gene expression during mitotic exit (Fig. 7). Mutagenesis of candidate enhancers will be necessary to confirm an enhancer's putative target, and thus whether enhancer deletion affect interphase, but not mitotic, expression of the target gene. Regardless, our data, together with previous reports (Hsiung et al. 2015, 2016), indicate that the key to understanding the reestablishment of the interphase transcriptome during mitotic exit likely lies in how promoters maintain transcription of many genes during mitosis, and how enhancers are used during mitotic exit.

ACKNOWLEDGMENTS

The Zaret laboratory research described in this chapter was supported by grants from the National Institutes of Health: training grant T32GM00812 to K.C.P. and GM36477 to K.S.Z.

REFERENCES

Bregman DB, Du L, van der Zee S, Warren SL. 1995. Transcription-dependent redistribution of the large subunit of RNA polymerase II to discrete nuclear domains. *J Cell Biol* **129:** 287–298.

Caravaca JM, Donahue G, Becker JS, He X, Vinson C, Zaret KS. 2013. Bookmarking by specific and nonspecific binding of FoxA1 pioneer factor to mitotic chromosomes. *Genes Dev* **27:** 251–260.

Chapman RD, Heidemann M, Albert TK, Mailhammer R, Flatley A, Meisterernst M, Kremmer E, Eick D. 2007. Transcribing

RNA polymerase II is phosphorylated at CTD residue serine-7. *Science* **318:** 1780–1782.

Devaiah BN, Lewis BA, Cherman N, Hewitt MC, Albrecht BK, Robey PG, Ozato K, Sims RJ III, Singer DS. 2012. BRD4 is an atypical kinase that phosphorylates serine2 of the RNA polymerase II carboxy-terminal domain. *Proc Natl Acad Sci* **109:** 6927–6932.

Dey A, Nishiyama A, Karpova T, McNally J, Ozato K. 2009. Brd4 marks select genes on mitotic chromatin and directs postmitotic transcription. *Mol Biol Cell* **20:** 4899–4909.

Gariglio P, Buss J, Green MH. 1974. Sarkosyl activation of RNA polymerase activity in mitotic mouse cells. *FEBS Lett* **44:** 330–333.

Heidemann M, Hintermair C, Voss K, Eick D. 2013. Dynamic phosphorylation patterns of RNA polymerase II CTD during transcription. *Biochim Biophys Acta* **1829:** 55–62.

Hsiung CC, Morrissey CS, Udugama M, Frank CL, Keller CA, Baek S, Giardine B, Crawford GE, Sung MH, Hardison RC, et al. 2015. Genome accessibility is widely preserved and locally modulated during mitosis. *Genome Res* **25:** 213–225.

Hsiung CC, Bartman CR, Huang P, Ginart P, Stonestrom AJ, Keller CA, Face C, Jahn KS, Evans P, Sankaranarayanan L, et al. 2016. A hyperactive transcriptional state marks genome reactivation at the mitosis–G_1 transition. *Genes Dev* **30:** 1423–1439.

Jao CY, Salic A. 2008. Exploring RNA transcription and turnover in vivo by using click chemistry. *Proc Natl Acad Sci* **105:** 15779–15784.

Johnson TC, Holland JJ. 1965. Ribonucleic acid and protein synthesis in mitotic HeLa cells. *J Cell Biol* **27:** 565–574.

Kadauke S, Udugama MI, Pawlicki JM, Achtman JC, Jain DP, Cheng Y, Hardison RC, Blobel GA. 2012. Tissue-specific mitotic bookmarking by hematopoietic transcription factor GATA1. *Cell* **150:** 725–737.

Komarnitsky P, Cho EJ, Buratowski S. 2000. Different phosphorylated forms of RNA polymerase II and associated mRNA processing factors during transcription. *Genes Dev* **14:** 2452–2460.

Konrad CG. 1963. Protein synthesis and RNA synthesis during mitosis in animal cells. *J Cell Biol* **19:** 267–277.

Lerner J, Bagattin A, Verdeguer F, Makinistoglu MP, Garbay S, Felix T, Heidet L, Pontoglio M. 2016. Human mutations affect the epigenetic/bookmarking function of HNF1B. *Nucleic Acids Res* **44:** 8097–8111.

Leung D, Jung I, Rajagopal N, Schmitt A, Selvaraj S, Lee AY, Yen CA, Lin S, Lin Y, Qiu Y, et al. 2015. Integrative analysis of haplotype-resolved epigenomes across human tissues. *Nature* **518:** 350–354.

Liang K, Woodfin AR, Slaughter BD, Unruh JR, Box AC, Rickels RA, Gao X, Haug JS, Jaspersen SL, Shilatifard A. 2015. Mitotic transcriptional activation: Clearance of actively engaged Pol II via transcriptional elongation control in mitosis. *Mol Cell* **60:** 435–445.

Liu H, Qu Q, Warrington R, Rice A, Cheng N, Yu H. 2015. Mitotic transcription installs Sgo1 at centromeres to coordinate chromosome segregation. *Mol Cell* **59:** 426–436.

Martinez-Balbas MA, Dey A, Rabindran SK, Ozato K, Wu C. 1995. Displacement of sequence-specific transcription factors from mitotic chromatin. *Cell* **83:** 29–38.

Naumova N, Imakaev M, Fudenberg G, Zhan Y, Lajoie BR, Mirny LA, Dekker J. 2013. Organization of the mitotic chromosome. *Science* **342:** 948–953.

Ohta S, Bukowski-Wills JC, Sanchez-Pulido L, Alves Fde L, Wood L, Chen ZA, Platani M, Fischer L, Hudson DF, Ponting CP, et al. 2010. The protein composition of mitotic chromosomes determined using multiclassifier combinatorial proteomics. *Cell* **142:** 810–821.

Palozola KC, Donahue G, Liu H, Grant GR, Becker JS, Cote A, Yu H, Raj A, Zaret KS. 2017. Mitotic transcription and waves of gene reactivation during mitotic exit. *Science* **358:** 119–122.

Parsons GG, Spencer CA. 1997. Mitotic repression of RNA polymerase II transcription is accompanied by release of transcription elongation complexes. *Mol Cell Biol* **17:** 5791–5802.

Prasanth KV, Sacco-Bubulya PA, Prasanth SG, Spector DL. 2003. Sequential entry of components of the gene expression machinery into daughter nuclei. *Mol Biol Cell* **14:** 1043–1057.

Prescott DM, Bender MA. 1962. Synthesis of RNA and protein during mitosis in mammalian tissue culture cells. *Exp Cell Res* **26:** 260–268.

Schwammle V, Jensen ON. 2010. A simple and fast method to determine the parameters for fuzzy c-means cluster analysis. *Bioinformatics* **26:** 2841–2848.

Shim EY, Walker AK, Shi Y, Blackwell TK. 2002. CDK-9/cyclin T (P-TEFb) is required in two postinitiation pathways for transcription in the *C. elegans* embryo. *Genes Dev* **16:** 2135–2146.

Sugaya K, Vigneron M, Cook PR. 2000. Mammalian cell lines expressing functional RNA polymerase II tagged with the green fluorescent protein. *J Cell Sci* **113**(Pt 15)**:** 2679–2683.

Teves SS, An L, Hansen AS, Xie L, Darzacq X, Tjian R. 2016. A dynamic mode of mitotic bookmarking by transcription factors. *Elife* **5:** e22280.

Yokoyama Y, Zhu H, Lee JH, Kossenkov AV, Wu SY, Wickramasinghe JM, Yin X, Palozola KC, Gardini A, Showe LC, et al. 2016. BET Inhibitors suppress ALDH activity by targeting ALDH1A1 super-enhancer in ovarian cancer. *Cancer Res* **76:** 6320–6330.

Genome Instability as a Consequence of Defects in the Resolution of Recombination Intermediates

STEPHEN C. WEST AND YING WAI CHAN

The Francis Crick Institute, London NW1 1AT, United Kingdom

Correspondence: stephen.west@crick.ac.uk

The efficient processing of homologous recombination (HR) intermediates, which often contain four-way structures known as Holliday junctions (HJs), is required for proper chromosome segregation at mitosis. Eukaryotic cells possess three distinct pathways of resolution: (i) HJ dissolution mediated by BLM-topoisomerase IIIα-RMI1-RMI2 (BTR) complex, and HJ resolution catalyzed by either (ii) SLX1-SLX4-MUS81-EME1-XPF-ERCC1 (SMX complex) or (iii) GEN1. The BTR pathway acts at all times throughout the cell cycle, whereas the actions of SMX and GEN1 are restrained in S phase and become elevated late in the cell cycle to ensure the resolution of persistent recombination intermediates before mitotic division. By developing a "resolvase-deficient" model system in which the activities of MUS81 and GEN1 are compromised, we have explored the fate of unresolved recombination intermediates. We find that covalently linked sister chromatids promote the formation of a new class of ultrafine bridges at anaphase that we term HR-UFBs. These bridges are broken at cell division, leading to activation of the DNA damage checkpoint and repair by nonhomologous end joining (NHEJ) in the next cell cycle. As a consequence, high levels of gross chromosomal rearrangements and aberrations are observed, together with frequent cell death. These results show that the HJ resolvases provide essential functions for the resolution of recombination intermediates, even in cells that remain proficient for BTR-mediated HJ dissolution.

Homologous recombination provides an essential mechanism for the repair of DNA breaks that arise from the demise of stalled replication forks or are induced by genotoxic agents. Recombination usually takes place between sister chromatids, as one sister provides a template for the error-free repair of the broken DNA. Joint molecule intermediates of homologous recombination often take the form of a four-way junction, also known as a Holliday junction (HJ), that needs to be resolved before chromosome segregation (Holliday 1964; West 2003; Wyatt and West 2014).

In recent years, there have been rapid developments in our understanding of the processes by which recombination intermediates, and in particular HJs, are processed. In contrast to simple organisms such as bacteria, which have a single HJ resolvase known as the RuvC protein (Dunderdale et al. 1991; Iwasaki et al. 1991; West 1997), eukaryotes have three distinct pathways that process junctions. These pathways use two distinct mechanisms: topoisomerase-induced dissolution and nuclease-mediated resolution.

MECHANISMS AND REGULATION OF RECOMBINATION INTERMEDIATE PROCESSING

A four-subunit complex, containing BLM-topoisomerase IIIα-RMI1-RMI2, known as the BTR complex, promotes the dissolution of double HJs. This complex promotes the convergent migration of two HJs to produce a hemicatenane structure that is dissolved by topoisomerase action (Wu and Hickson 2003). BTR-mediated HJ dissolution takes place throughout the cell cycle and gives rise exclusively to noncrossover products (Fig. 1). Cells derived from individuals with Bloom syndrome (BS), which is caused by mutations in BLM, show the diagnostic feature of a high frequency of sister chromatid exchanges and increased genome instability. Consequently, these individuals are predisposed to a broad spectrum of early-onset cancers (Ray and German 1984; Hickson 2003).

Persistent double HJs, and single HJs that cannot serve as a substrate for BLM, are resolved by structure-selective nucleases (resolvases). These enzymes cut HJs by introducing coordinated nicks across the junction and give rise to nicked duplex products that are crossovers (COs) or noncrossovers (NCOs). The elevated frequency of sister chromatid exchanges (i.e., COs between sister chromatids) observed in BLM-deficient cells results from these resolution events (Wechsler et al. 2011).

There are two distinct nucleolytic pathways for resolution (Fig. 1). The first involves a complex that forms in a cell cycle–specific manner by interactions between MUS81-EME1 and a constitutive complex of SLX1-SLX4-XPF-ERCC1 (Wyatt et al. 2017). The resulting complex, which we term SMX, comprises three nuclease activities: SLX1-SLX4, MUS81-EME1, and XPF-ERCC1 (Fig. 2). Direct interactions between SLX4 and MUS81 occur at prometaphase in response to CDK/PLK1-mediated phosphorylation (Svendsen et al. 2009; Castor et al. 2013; Wyatt et al. 2013; Duda et al. 2016). The SMX complex resolves HJs by SLX1-mediated introduction of the first nick and by MUS81-mediated counternicking. Both incisions occur within the lifetime of the

Published by Cold Spring Harbor Laboratory Press; doi: 10.1101/sqb.2017.82.034256

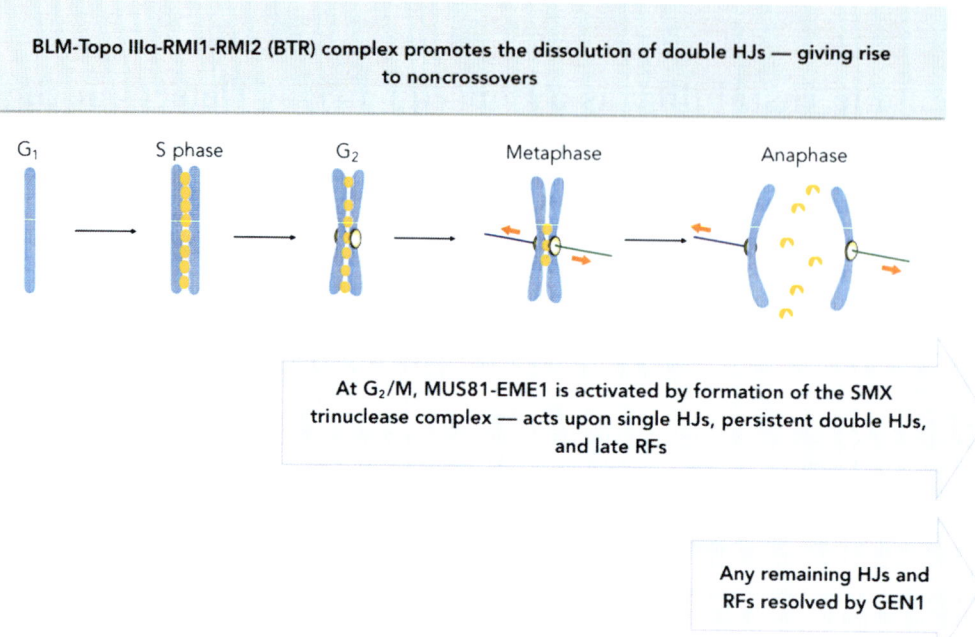

Figure 1. Mechanisms for the processing of recombination intermediates in mitotic human cells. The BTR pathway is active throughout the cell cycle and promotes the dissolution of double Holliday junctions (HJs) to form noncrossover products. Later in the cell cycle, at prometaphase, MUS81-EME1 combines with the SLX1-SLX4-XPF-ERCC1 complex, in response to CDK/PLK1 phosphorylation events, to form the SMX trinuclease complex. SMX acts upon double HJs that have escaped the attention of BTR, single HJs that cannot be dissolved by BTR, and any late replication intermediates such as those present at common fragile sites. Finally, GEN1 protein, which is primarily cytoplasmic, gains access to any remaining recombination/replication intermediates upon breakdown of the nuclear envelope. SMX and GEN1 give rise to crossovers and noncrossovers.

SMX-HJ DNA complex (Wyatt et al. 2013, 2017). The interaction of MUS81-EME1 with SLX4 activates MUS81 for productive cleavage, in reactions that are thought to involve interplay between the amino-terminal HhH self-inhibitory domain of MUS81 and SLX4 (Wyatt et al. 2017). The nuclease activity of XPF is not required for cleavage, although XPF-ERCC1 may play a stimulatory role by contributing to the stability of the complex.

The second pathway of nucleolytic resolution involves the GEN1 HJ resolvase (Fig. 1). GEN1 shows nuclease activity on 5′ flaps, replication fork structures, and HJs (Ip et al. 2008; Rass et al. 2010; Chan and West 2015; Bellendir et al. 2017). The mechanism of HJ cleavage by GEN1 is similar to that showed by the prototypic HJ resolvase RuvC, as symmetrical incisions are introduced across the junction point by the GEN1 homodimer. Like SMX, the actions of GEN1 are restricted to the late stages of the cell cycle, although in this case it is driven by nuclear exclusion (Chan and West 2014). Indeed, GEN1's HJ resolvase activity appears to be restrained until

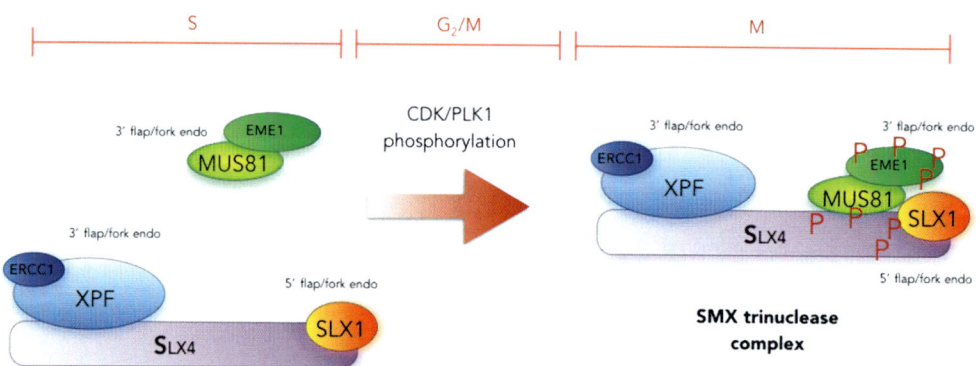

Figure 2. Cell cycle–mediated activation of MUS81-EME1 by formation of the SMX complex. The diagram indicates the interaction of MUS81-EME1 with the SLX1-SLX4-XPF-ERCC1 complex at prometaphase, in response to CDK/PLK1 phosphorylation. The resulting SLX1-SLX4-MUS81-EME1-XPF-ERCC1 (SMX) complex is capable of cleaving a range of replication and recombination intermediates. By association with SLX1-SLX4-XPF-ERCC1, the MUS81-EME1 nuclease is activated.

nuclear membrane breakdown at mitosis. Once the nuclear membrane is dissolved, GEN1 will gain access to, and resolve, any persistent covalent bridges that link sister chromatids and so enable chromosome segregation.

A MODEL SYSTEM FOR RESOLVASE DEFICIENCY

The cellular importance of HJ processing in mammalian cells is apparent from the synthetic lethality observed in cells depleted for BLM (dissolution pathway) and SLX4 (resolution pathway) or SLX4 and GEN1 (both resolution pathways) (Garner et al. 2013; Wyatt et al. 2013; Sarbajna et al. 2014). This mortality is likely to stem from gross chromosomal abnormalities and mitotic defects (Wechsler et al. 2011; Garner et al. 2013; Wyatt et al. 2013; Sarbajna et al. 2014).

To develop a cellular system for the detailed analysis of resolvase deficiency, we recently used short interfering RNA (siRNA) to deplete MUS81 from a $GEN1^{-/-}$ cell line made using CRISPR (clustered regularly interspaced

short palindromic repeat)-Cas9 (Chan and West 2015; Chan et al. 2017). Clonogenic survival assays showed that there was massive synthetic lethality, with <10% survival, and that the cells were highly sensitive to treatment with DNA damaging agents such as cisplatin (Fig. 3A). Moreover, we observed that the metaphase chromosomes were elongated and showed indentations along their length (Fig. 3B) because of the presence of unresolved recombination intermediates (Wechsler et al. 2011; Chan et al. 2017). Similar defects have been observed in BLM-depleted $SLX4$ null cells and in cisplatin-treated cells depleted of SLX4 and GEN1 or MUS81 and GEN1 (Garner et al. 2013; Sarbajna et al. 2014). Segmentation occurred at equivalent positions on the two sister chromatids and was rescued by expression of a bacterial HJ resolvase such as RusA (Wechsler et al. 2011; Garner et al. 2013; Chan et al. 2017). Because the indentations are free of condensins (e.g., SMC2), we previously proposed that the persistent chromatid bridges cause defects in chromosome condensation rather than chromosome breakage (Wechsler et al. 2011).

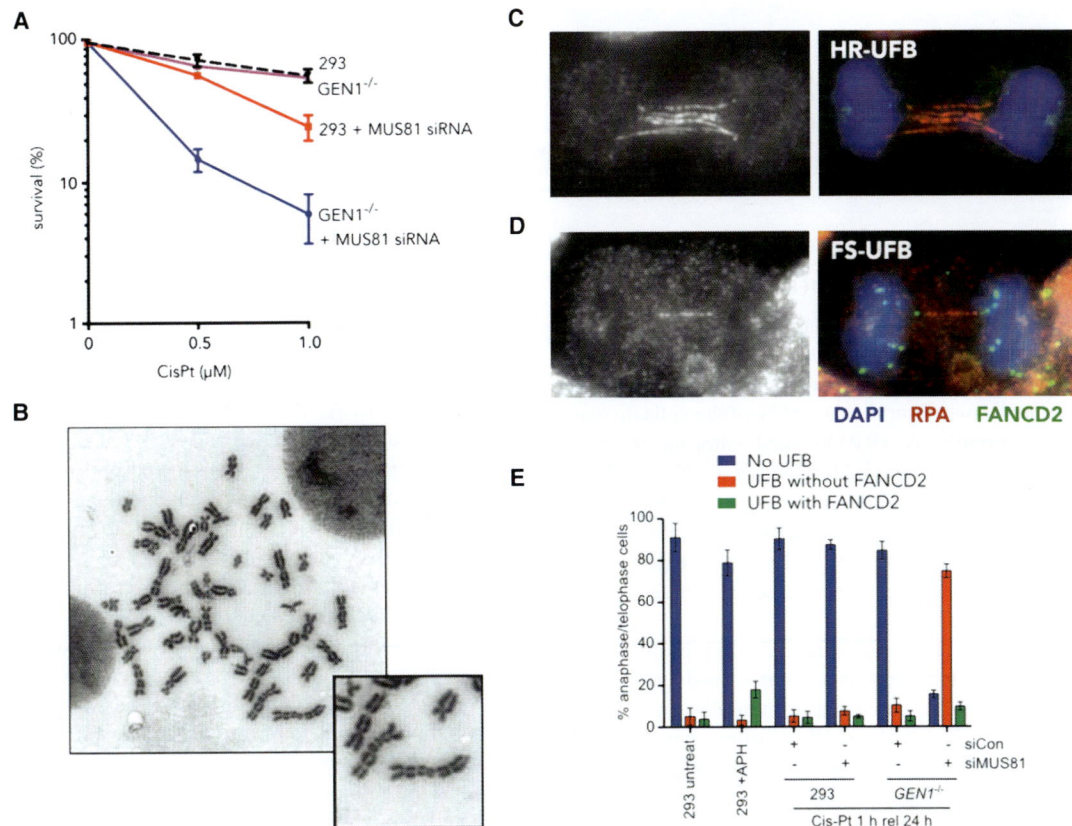

Figure 3. Phenotypic properties of resolvase-deficient cells. (*A*) Sensitivity of 293 and $GEN1^{-/-}$ cells treated with siRNA against MUS81, to cisplatin, measured by clonogenic survival. (*B*) Metaphase spreads from $GEN1^{-/-}$ cells treated with siRNA against MUS81 and a brief cisplatin treatment reveal the presence of segmented chromosomes in which the sister chromatids remain interlinked. (*C*) $GEN1^{-/-}$ cells treated with MUS81 siRNA and cisplatin form ultrafine bridges (UFBs) at anaphase. These are defined as homologous recombination UFBs (HR-UFBs). (*D*) 293 cells treated with aphidicolin, which causes mild replication stress, give rise to fragile site–associated UFBs (FS-UFBs) which are distinguished from HR-UFBs by the presence of FANCD2 foci. RPA2, FANCD2, and DNA were visualized using anti-RPA2 antibody (red), anti-FANCD2 antibody (green), and DAPI (blue). (*E*) Quantification of anaphase/telophase cells with RPA2-positive UFBs, with or without FANCD2 foci, as visualized in *C*. (Adapted from data in Chan et al. 2017.)

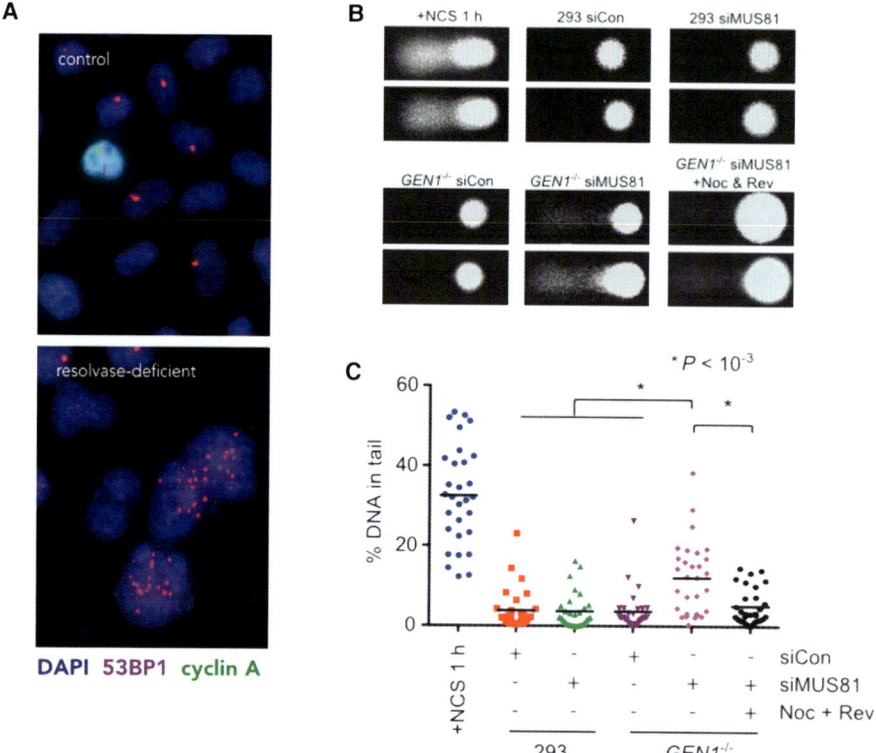

Figure 4. DNA breakage results from mitosis of resolvase-deficient cells. (*A*) 53BP1-positive nuclear bodies (red) in control, and SLX4 + GEN1-depleted, cisplatin-treated G_1 (cyclin A-negative) HeLa cells. (*B,C*) 293 cells and *GEN1*$^{-/-}$ cells were treated with control siRNA or siRNA against MUS81, as well as cisplatin. After further growth, cells were treated with or without nocodazole (Noc) and reversine (Rev). The DNA-PK inhibitor NU7026 was also added to the media. Cells were harvested after 20 h and DNA breaks were analyzed by comet assays. Neocarzinostatin-treated 293 cells were used as a control. (*C*) Quantification of the data shown in *B*. (Adapted from data presented in Sarbajna et al. 2014; Chan et al. 2017.)

HR-UFBs LEAD TO CHROMOSOME ABERRATIONS

Because unresolved recombination intermediates fail to elicit a checkpoint response, the resolvase-deficient cells enter mitosis with their sister chromatid bridges intact. As a consequence, we observe that ~80% of the cells display replication protein A (RPA)-coated ultrafine bridges (UFBs) at anaphase (Fig. 3C,E; Chan et al. 2017). These UFBs, caused by unresolved homologous recombination (HR) intermediates (designated HR-UFBs), are distinct from replication stress–induced UFBs, which arise at common fragile sites (FS-UFBs) and are characterized by the presence of FANCD2 foci (Fig. 3D,E).

The presence of the single-strand binding protein RPA on the HR-UFBs indicates that the unresolved recombination intermediates are processed from duplex into single-stranded DNA. The PICH and BLM helicases were found to be essential for these processing reactions (Chan et al. 2017). Single-stranded UFBs are thought to be fragile enough to be broken by spindle forces at mitosis, and resolvase-deficient cells show high levels of 53BP1 (Sarbajna et al. 2014) or MDC1 (Chan et al. 2017) foci in the following G_1 phase (Fig. 4A). These DNA damage signatures were not observed when cell division was blocked by treatment with nocodazole and reversine, which inhibit spindle assembly and the mitotic checkpoint, respectively. Moreover, direct evidence of DNA breaks was obtained

using alkaline Comet assays (Fig. 4B,C), supporting the proposal that single stranded HR-UFBs are broken at mitotic division (Chan et al. 2017).

Coordinated with the presence of DNA breaks in G_1 of the second cell cycle, we observed high levels of γH2AX and activation of ATM (ataxia telangiectasia mutated) at G_2/M, as measured by CHK2 and KAP1 phosphorylation (Chan et al. 2017). The cells then showed a cell cycle arrest. We did not find evidence of ATR (ataxia telangiectasia and Rad3-related) activation, indicating that the arrest was due to DNA breaks rather than activation of a replication checkpoint. Metaphase spreads prepared from the resolvase-deficient cells revealed a high frequency of end-to-end chromosome fusions and radial chromosomes (Fig. 5A,B; Chan et al. 2017). When the cells were treated with the DNA-PKcs inhibitor NU7026, which inhibits nonhomologous end joining (NHEJ), we observed suppression of the fusion phenotype (Fig. 5C). These results show that the chromosome fusions and rearrangements are generated through the repair of breaks produced at the first mitotic division.

CONCLUSION

In summary, our work with this resolvase-deficient model cell system provides a clear picture of the way that unresolved recombination intermediates lead to cell

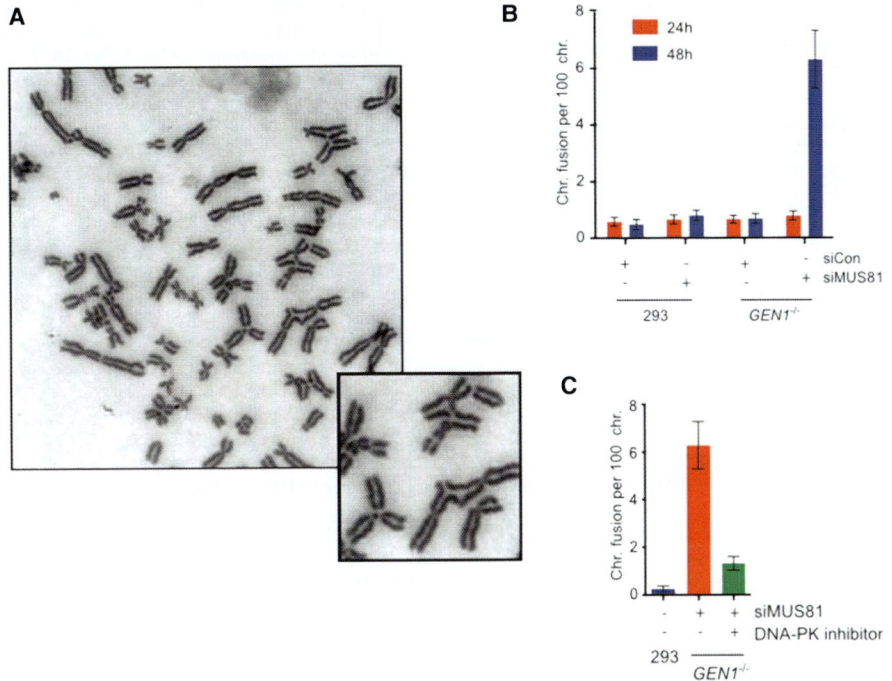

Figure 5. Gross chromosomal abnormalities result from DNA breakage. (*A*) Representative image of a metaphase spread showing end-to-end fusions and chromosome abnormalities in *GEN1*$^{-/-}$ cells treated with MUS81 siRNA and cisplatin. (*B*) Quantification of fusions that arise in resolvase-deficient cells, as in *A*. (*C*) *GEN1*$^{-/-}$ cells were treated as in *A*, except that the DNA-PK inhibitor NU7026 was added 24 h before harvest. Control siRNA-treated 293 cells were used as control. (Adapted from data presented in Chan et al. 2017.)

death. Importantly, the intermediates were found to persist until mitosis where they gave rise to UFBs at anaphase/telophase. Subsequently, the interlinked sister chromatids were acted upon by PICH/BLM helicases, leading to their conversion into single-stranded RPA-coated UFBs. These HR-UFBs were distinct from replication-associated UFBs (FS-UFBs), which are characteristically flanked by the presence of FANCD2 foci. The single-stranded UFBs were then broken at mitotic division, thereby avoiding cytokinesis failure, but resulting in the presence of DNA breaks at G$_1$ in the next cell cycle. These breaks were then repaired by NHEJ, resulting in gross chromosomal fusions and aberrations, leading to cell cycle arrest and cell death.

Although the resolvase-deficient cell system detailed here is somewhat artificial, as complete resolvase deficiency does not occur in nature, it provides a useful model that allows us to study the fate of unresolved recombination intermediates. Such intermediates may persist in highly replicating cancer cells in which the recombination "load" is often elevated compared to normal cells or in cells that have suffered DNA damage that increases the recombination frequency. Similarly, unresolved recombination intermediates may arise in cells that are compromised for their gatekeeper functions (i.e., HR vs. NHEJ) or those that express elevated levels of RAD51 (Xia et al. 1997; Klein 2008). A second important aspect of the work demonstrates that the resolvases do not simply provide a backup for the BTR pathway, because resolvase-deficient cells are effectively inviable. Indeed, they are essential factors that play a critical role in chromosome segregation.

ACKNOWLEDGMENTS

Work in the West laboratory at the Francis Crick Institute is supported by Cancer Research UK, the Medical Research Council, the Wellcome Trust, the European Research Council, and the Louis-Jeannet Foundation.

REFERENCES

Bellendir SP, Rognstad DJ, Morris LP, Zapotoczny G, Walton WG, Redinbo MR, Ramsden DA, Sekelsky J, Erie DA. 2017. Substrate preference of GEN endonucleases highlights the importance of branched structures as DNA damage repair intermediates. *Nucleic Acids Res* **45:** 5333–5348.

Castor D, Nair N, Déclais AC, Lachaud C, Toth R, Macartney TJ, Lilley DMJ, Arthur JS, Rouse J. 2013. Cooperative control of Holliday junction resolution and DNA repair by the SLX1 and MUS81-EME1 nucleases. *Mol Cell* **52:** 221–233.

Chan YW, West SC. 2014. Spatial control of the GEN1 Holliday junction resolvase ensures genome stability. *Nat Commun* **5:** 4844.

Chan YW, West SC. 2015. GEN1 promotes Holliday junction resolution by a coordinated nick and counter-nick mechanism. *Nucleic Acids Res* **43:** 10882–10892.

Chan YW, Fugger K, West SC. 2017. Unresolved recombination intermediates lead to ultra-fine bridges, chromosome breaks and aberrations. *Nat Cell Biol* doi: 10.1058/s41556-017-0011-1.

Duda H, Arter M, Gloggnitzer J, Teloni F, Wild P, Blanco MG, Altmeyer M, Matos J. 2016. A mechanism for controlled breakage of under-replicated chromosomes during mitosis. *Dev Cell* **39:** 740–755.

Dunderdale HJ, Benson FE, Parsons CA, Sharples GJ, Lloyd RG, West SC. 1991. Formation and resolution of recombination intermediates by *E. coli* RecA and RuvC proteins. *Nature* **354:** 506–510.

Garner E, Kim Y, Lach FP, Kottemann MC, Smogorzewska A. 2013. Human GEN1 and the SLX4-associated nucleases MUS81 and SLX1 are essential for the resolution of replication-induced Holliday junctions. *Cell Rep* **5:** 207–215.

Hickson ID. 2003. RecQ helicase: Caretakers of the genome. *Nat Rev Mol Cell Biol* **3:** 169–178.

Holliday R. 1964. A mechanism for gene conversion in fungi. *Genet Res* **89:** 285–307.

Ip SCY, Rass U, Blanco MG, Flynn HR, Skehel JM, West SC. 2008. Identification of Holliday junction resolvases from humans and yeast. *Nature* **456:** 357–361.

Iwasaki H, Takahagi M, Shiba T, Nakata A, Shinagawa H. 1991. *Escherichia coli* RuvC protein is an endonuclease that resolves the Holliday structure. *EMBO J* **10:** 4381–4389.

Klein HL. 2008. The consequences of RAD51 overexpression for normal and tumor cells. *DNA Repair* **7:** 686–693.

Rass U, Compton SA, Matos J, Singleton MR, Ip SCY, Blanco MG, Griffith JD, West SC. 2010. Mechanism of Holliday junction resolution by the human GEN1 protein. *Genes Dev* **24:** 1559–1569.

Ray JH, German J. 1984. Bloom's syndrome and EM9 cells in BrdU-containing medium exhibit similarly elevated frequencies of sister chromatid exchange but dissimilar amounts of cellular proliferation and chromosome disruption. *Chromosoma* **90:** 383–388.

Sarbajna S, Davies D, West SC. 2014. Roles of SLX1-SLX4, MUS81-EME1 and GEN1 in avoiding genome instability and mitotic catastrophe. *Genes Dev* **28:** 1124–1136.

Svendsen JM, Smogorzewska A, Sowa ME, O'Connell BC, Gygi SP, Elledge SJ, Harper JW. 2009. Mammalian BTBD12/SLX4 assembles a Holliday junction resolvase and is required for DNA repair. *Cell* **138:** 63–77.

Wechsler T, Newman S, West SC. 2011. Aberrant chromosome morphology in human cells defective for Holliday junction resolution. *Nature* **471:** 642–646.

West SC. 1997. Processing of recombination intermediates by the RuvABC proteins. *Annu Rev Genet* **31:** 213–244.

West SC. 2003. Molecular views of recombination proteins and their control. *Nat Rev Mol Cell Biol* **4:** 435–445.

Wu L, Hickson ID. 2003. The Bloom's syndrome helicase suppresses crossing over during homologous recombination. *Nature* **426:** 870–874.

Wyatt HDM, West SC. 2014. Holliday junction resolvases. *Cold Spring Harb Perspect Biol* **6:** a023192.

Wyatt HDM, Sarbajna S, Matos J, West SC. 2013. Coordinated actions of SLX1-SLX4 and MUS81-EME1 for Holliday junction resolution in human cells. *Mol Cell* **52:** 234–247.

Wyatt HDM, Laister RC, Martin SR, Arrowsmith CH, West SC. 2017. The SMX DNA repair tri-nuclease. *Mol Cell* **65:** 848–860.

Xia SJJ, Shammas MA, Reis RJS. 1997. Elevated recombination in immortal human cells is mediated by *hs*RAD51 recombinase. *Mol Cell Biol* **17:** 7151–7158.

Reconstitution of Female Germ Cell Fate Determination and Meiotic Initiation in Mammals

So I. Nagaoka[1,2] and Mitinori Saitou[1,2,3,4]

[1]*Department of Anatomy and Cell Biology, Graduate School of Medicine, Kyoto University, Yoshida-Konoe-cho, Sakyo-ku, Kyoto 606-8501, Japan*

[2]*JST, ERATO, Yoshida-Konoe-cho, Sakyo-ku, Kyoto 606-8501, Japan*

[3]*Center for iPS Cell Research and Application, Kyoto University, Shogoin, Sakyo-ku, Kyoto 606-8507, Japan*

[4]*Institute for Integrated Cell-Material Sciences, Kyoto University, Yoshida-Ushinomiya-cho, Sakyo-ku, Kyoto 606-8501, Japan*

Correspondence: saitou@anat2.med.kyoto-u.ac.jp

Meiosis is a fundamental process that underpins sexual reproduction. In mammals, the execution of meiosis is tightly integrated within the complex processes of oogenesis and spermatogenesis, and elucidation of the molecular mechanisms regulating meiotic initiation remains challenging. We have recently developed in vitro culture strategies to induce mouse pluripotent stem cells into germ cells, which successfully contribute to both oogenesis and spermatogenesis and to fertile offspring. The culture strategies faithfully recapitulate transcriptional and epigenetic dynamics as well as signaling principles for germ cell specification, proliferation, and female sex determination/meiotic induction, providing a valuable platform for studies to illuminate the molecular mechanisms underlying such critical processes. Here, we review mammalian gametogenesis with a focus on the implementation of meiosis and, based on our recent studies, discuss new insights into the mechanisms for meiotic initiation and germ cell sex determination in mice.

For most eukaryotic lineages, the creation of haploid gametes through meiosis and fertilization constitutes the foundation for sexual reproduction that assures continual succession of life (Wilkins and Holliday 2009). Through evolution, the regulatory components for meiotic recombination and chromosome segregation have been highly conserved (Marston and Amon 2004; Handel and Schimenti 2010; Watanabe 2012; Baudat et al. 2013); however, the mechanisms controlling meiotic initiation are divergent and involve species-specific and sex-specific developmental contexts (Harigaya and Yamamoto 2007; Lesch and Page 2012). In mammals, the implementation of meiosis is embedded within the complex oogenesis and spermatogenesis processes, and elucidating the molecular mechanisms pertaining to meiotic initiation remains a challenge.

Recently, we devised strategies to successfully reconstitute germ cell specification pathways from pluripotent stem cells (embryonic stem cells [ESCs] and induced pluripotent stem cells [iPSCs]) in mice (Hayashi et al. 2011). The transcriptional and epigenetic constitution of the resultant germ cells closely resembles those of primordial germ cells (PGCs) (Hayashi et al. 2011; Ohta et al. 2017), a group of sexually uncommitted precursor cells that give rise to either oocytes or spermatozoa. Remarkably, the in vitro–derived PGCs (PGC-like cells [PGCLCs]) possess a robust capability to differentiate into functional oocytes and spermatozoa in mice and contribute to fertile offspring (Hayashi et al. 2011, 2012; Hikabe et al. 2016; Ishikura et al. 2016; Ohta et al. 2017). The establishment of robust in vitro germ cell derivation strategies presents unprecedented opportunities for studies to illuminate the mechanisms of epigenetic reprogramming, sex determination, and meiotic initiation during mammalian germ cell development (Kurimoto et al. 2015; Shirane et al. 2016; Miyauchi et al. 2017; Ohta et al. 2017). In this review, we highlight the key components of mammalian germ cell development, the current understanding of the mechanisms of sex determination and meiotic initiation, new findings emerging from the studies using PGCLCs, and outstanding areas of studies that await further investigation.

GERM CELL SPECIFICATION IN MAMMALS

In metazoan lineages, germ cell fate is conferred via two major mechanisms. In model organisms such as *Drosophila melanogaster* and *Caenorhabditis elegans*, the germ cell fate is specified by a mechanism called "preformation," in which the segregation and inheritance of maternal determinants from the egg specifies the future germ cell lineages in developing embryos (Extavour and Akam 2003). On the other hand, in mice and likely in other mammals, germ cell fate is specified by "epigenesis," in which pluripotent embryonic cells are induced to confer the germ cell fate through the actions of inductive and restrictive signals from surrounding tissues and the embryo itself (Extavour and Akam 2003; Saitou and Yamaji 2012). In mice, in which the mechanism has been exten-

Published by Cold Spring Harbor Laboratory Press; doi: 10.1101/sqb.2017.82.033803

Cold Spring Harbor Symposia on Quantitative Biology, Volume LXXXII

sively studied, the germ cell fate is induced in the most proximal posterior epiblast in response to bone morphogenic protein (BMP) signaling emanating from neighboring extraembryonic ectoderm tissues and WNT signaling from the proximal posterior epiblast itself, beginning at around embryonic day 5.5 (E5.5) (Lawson et al. 1999; Saitou et al. 2002; Ohinata et al. 2009). Inhibitory molecules of BMP signaling, such as CER1, are secreted from the anterior visceral endoderm and inhibit germ cell induction in the anterior epiblast, restricting the emergence of the germ cells in the posterior region of the epiblast (Perea-Gomez et al. 2002; Ohinata et al. 2009). Consequently, by E7.25, a cluster of 30–40 PGCs is established within the extraembryonic mesoderm at the base of the allantois, which constitutes the founding population of the germ cell lineages (Fig. 1; Ginsburg et al. 1990; Lawson et al. 1999; Saitou et al. 2002; Ohinata et al. 2009).

SEX DETERMINATION AND MEIOTIC INITIATION IN MAMMALS

After specification, PGCs migrate through the hindgut endoderm toward developing gonads. Concurrently with the completion of PGC migration, the embryonic gonads undergo sex determination and initiate sex-specific developmental programs based on their sex chromosome constitutions. In gonadal somatic cells in XY individuals, the expression of the sex determination gene on the Y chromosome, *Sry*, initiates the formation of testicular structures essential for spermatogenesis (Sinclair et al. 1990; Koopman et al. 1991; Brennan and Capel 2004; Eggers et al. 2014). On the other hand, in the absence of *Sry*, a concerted activation of several ovary-specification genes, such as *Wnt4*, *Rspo1*, and *Foxl2*, drives the development of ovaries (Brennan and Capel 2004; Eggers et al. 2014). Importantly, once the decision is made, networks of activating and repressing signals actively maintain the sexual fate of gonads into adulthood (Uhlenhaut et al. 2009; Matson et al. 2011; Capel 2017).

In contrast to the sex determination mechanisms of the gonadal somatic cells, for germ cells, the commitment to a particular sex is dependent on the surrounding environment regardless of their sex chromosome constitutions (McLaren 1984, 1988; Kimble and Page 2007; Spiller et al. 2017). Intriguingly, in the fetal gonads, sex determination is coupled with the decision to initiate meiotic programs (McLaren 1984, 1988; Kimble and Page 2007; Spiller et al. 2017). In fetal ovaries, after several rounds of mitotic divisions, germ cells (now called oogonia) initiate meiotic prophase. The implementation of chromosomal events of meiotic prophase marks the first visible divergence between female and male germ cells, and it is classically considered as the onset of oogenesis (Hilscher et al. 1974; Speed 1982; McLaren 1984, 1988; Kimble and Page 2007; Spiller et al. 2017). In fetal testes, on the other hand, germ cells (now called prospermatogonia or gonocytes) are actively suppressed from prematurely ini-

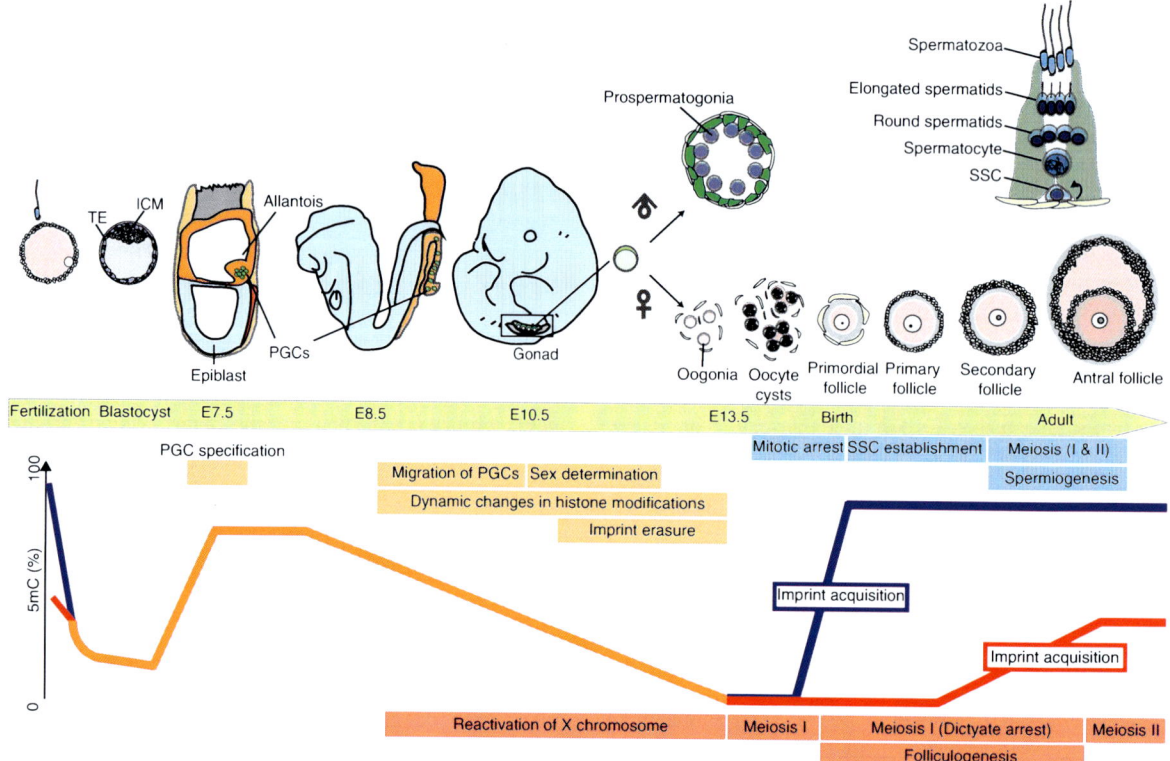

Figure 1. (*Top*) A schematic representation of germ cell development in mice. (*Bottom*) Key events associated with each stage of germ cell development are shown along with the dynamics of the 5mC levels. ICM, inner cell mass; TE, trophectoderm; SSC, spermatogonial stem cell; PGC, primordial germ cell.

tiating meiosis, and this suppression appears critical for the commitment to the male germ cell pathway (Bowles et al. 2006; Koubova et al. 2006; MacLean et al. 2007; Saba et al. 2014b).

For the sexual commitment and meiotic induction, differential availability of retinoic acid (RA) between embryonic ovaries and testes is instrumental in setting up the sexually dimorphic development. RA is an active metabolite of vitamin A that is used in multiple developmental processes, including reproduction (Wilson et al. 1953; Huang and Hembree 1979; Morales and Griswold 1987; Niederreither and Dolle 2008; Li and Clagett-Dame 2009; Griswold et al. 2012; Cunningham and Duester 2015), and the actions of RA are controlled at the level of tissue distribution by the expressions of synthesizing and metabolizing enzymes and also at the molecular level by the actions of RA receptors and a cohort of interacting proteins (Bastien and Rochette-Egly 2004; Niederreither and Dolle 2008; Dolle 2009; Mark et al. 2009; Cunningham and Duester 2015). Developing gonads are capable of synthesizing RA (Bowles et al. 2016), but the predominant sites of RA synthesis appear to be at mesonephroi that develop adjacent to the gonads, with RA likely diffusing into gonads (Bowles et al. 2006; Koubova et al. 2006; Griswold et al. 2012; Feng et al. 2014).

In the germ cells of developing ovaries, the activation of RA signaling leads to the induction of meiotic genes and meiotic prophase (Baltus et al. 2006; Bowles et al. 2006; Koubova et al. 2006; MacLean et al. 2007; Le Bouffant et al. 2010; Childs et al. 2011; Saba et al. 2014b; Soh et al. 2015). Additionally, RA signaling suppresses the activation of several male germ cell specification genes (Bowles et al. 2010; Saba et al. 2014b), thus reinforcing the commencement of the female pathway. In male gonads, on the other hand, SRY and its immediate downstream target, SOX9, up-regulate the expression of CYP26B1, an enzyme that catabolizes RA (Kashimada et al. 2011; Li et al. 2014), thereby suppressing the onset of RA signaling and premature initiation of meiosis in the male germ cells (Bowles et al. 2006; Koubova et al. 2006; MacLean et al. 2007; Saba et al. 2014b). In addition to the suppression of the RA signaling, FGF9, secreted from Sertoli cells, as well as the expression of a translational regulator specific to male germ cells, NANOS2, promote the male germ cell pathway and repress the meiotic program (Suzuki and Saga 2008; Sada et al. 2009; Barrios et al. 2010; Bowles et al. 2010; Saba et al. 2014a; Kato et al. 2016). More recently, activin and nodal signaling have also been shown to reinforce the commitment to the male fate by activating *Nanos2* and other male germ cell genes as well as suppressing the induction of meiosis (Spiller et al. 2012; Wu et al. 2013, 2015).

THE IMPLEMENTATION OF MEIOSIS DURING OOGENESIS AND SPERMATOGENESIS

As discussed above, in females, oogonia initiate meiotic programs and begin meiotic prophase in fetal ovaries, and all the major chromosomal events of meiotic prophase—synapsis between homologous chromosomes,

programmed DNA double-strand break formation, and formation of meiotic recombination—take place in the developing embryo (Gerton and Hawley 2005; Handel and Schimenti 2010; Baudat et al. 2013). Upon completion of the recombination processes, the oocyte enters the dictyate stage and the progression of meiosis becomes suspended (Fig. 1; Handel and Schimenti 2010). Concurrently, oocytes begin to establish intimate associations with surrounding granulosa cells and develop into primordial follicles (McGee and Hsueh 2000; Pepling and Spradling 2001; Matzuk et al. 2002; Li and Albertini 2013). The growth and maturation of the follicle take place postnatally, and intricate bidirectional signaling between the oocyte and granulosa cells is integral for successful folliculogenesis (Matzuk et al. 2002; Albertini 2015). During the growth, oocytes acquire maternal imprinting (Lucifero et al. 2002; Kobayashi et al. 2012) and undergo cytoplasmic maturation to establish the competence for fertilization and embryogenesis (Matzuk et al. 2002; Li and Albertini 2013). As the follicle completes its maturation, the surge of luteinizing hormone induces ovulation and releases the meiotic arrest. During the first meiotic division (also known as the reductional division), crossovers between homologous chromosomes, the product of meiotic recombination during the fetal stage, play a vital role in the segregation of homologous chromosomes (Gerton and Hawley 2005; Handel and Schimenti 2010; Nagaoka et al. 2012). Fertilization triggers the onset of second meiotic division and the production of a haploid egg (Handel and Schimenti 2010; Clift and Schuh 2013). Finally, the fusion between maternal and paternal pronuclei restores diploidy, and the life of a new individual begins. Thus, meiosis encompasses the entire duration of oogenesis in mammals, and intricate control mechanisms at various stages must be in place in order to coordinate the chromosomal events of meiosis and the execution of the developmental programs for generating a competent ovum.

In developing gonads of male embryos, prospermatogonia proliferate through several rounds of mitotic divisions, while actively suppressing the entry into meiosis (Bowles et al. 2006; Koubova et al. 2006; MacLean et al. 2007; Saba et al. 2014b). Subsequently, they enter into a quiescent state of G_0/G_1 mitotic arrest (Manku and Culty 2015; Spiller et al. 2017) and acquire androgenetic epigenome, including paternal imprints (Fig. 1; Davis et al. 2000; Ueda et al. 2000; Kato et al. 2007; Seisenberger et al. 2012; Kobayashi et al. 2013; Kubo et al. 2015). After birth, although many spermatogonia initiate the first wave of spermatogenesis in mice (Yoshida et al. 2006), a small pool establish a spermatogonial stem cell (SSC) population with a lifelong capacity to perform numerous rounds of spermatogenesis (Oatley and Brinster 2008; Yoshida 2012; Kanatsu-Shinohara and Shinohara 2013). At each round of spermatogenesis, SSCs give rise to differentiating spermatogonia, which undergo several rounds of mitotic cell divisions and then initiate meiosis (Griswold 2016). In contrast to meiosis in females, meiotic prophase and the two meiotic divisions in males proceed without a halt and result in the generation of four haploid spermatids (Fig. 1). Subsequently, spermatids undergo morphological transformation, as well as chromatin compaction in the form of histone-to-protamine replacement, and develop into highly motile and fertilization-capable spermatozoa (Toshimori and Eddy 2015). Remarkably, throughout the seminiferous epithelium within the testes, new waves of spermatogenesis are constantly initiated from SSCs, which concurrently and successively mature to haploid spermatozoa, providing a continuous supply for the lifetime of a male (Griswold 2016).

Considering the integration of meiosis within sex-specific gametogenesis steps involving key functions of gonadal somatic cells, the establishment of experimental means that can resolve the intricate interactions between germ cells and somatic cells will be crucial for efforts to understand the mechanism for meiosis.

IN VITRO DERIVATION OF PRIMORDIAL GERM CELLS

The tantalizing capability of the germ cells to create new organisms has been inspiring developmental and stem cell biologists alike to attempt to recreate gametogenesis processes from pluripotent stem cells (Daley 2007; Saitou and Miyauchi 2016). In recent years, the realization of robust in vitro derivation strategies was finally achieved (Hayashi et al. 2011, 2012). This feat was largely attributable to the accumulation of knowledge about germ cell specification (Ginsburg et al. 1990; Lawson et al. 1999; Saitou et al. 2002; Ohinata et al. 2009), refined understanding of different pluripotent states among stem cells propagated in vitro (Hackett and Surani 2014; Martello and Smith 2014), and the development of reproductive technologies that can convincingly test the potency of the derived germ cells (Brinster and Avarbock 1994; Brinster and Zimmermann 1994; Chuma et al. 2005). The successful derivation strategies generate germ cells that closely resemble PGCs in their transcriptional and epigenetic characteristics. Importantly, the derived PGCLCs possess a robust capability to perform successful oogenesis and spermatogenesis that contribute to fertile offspring when transplanted into surrogate animals (for spermatogenesis and oogenesis) (Hayashi et al. 2011, 2012; Ishikura et al. 2016; Ohta et al. 2017) or when cultured in appropriate in vitro growth (IVG) and in vitro maturation (IVM) conditions (for oogenesis) (Fig. 2; Hikabe et al. 2016; Morohaku et al. 2016). Thus, the in vitro germ cell derivation strategy provides a valuable platform for studies to elucidate mechanisms underlying critical processes during mammalian germ cell development.

EXPANSION CULTURE OF PGCs/PGCLCs AND RECONSTITUTION OF THE EPIGENETIC BLANK SLATE OF THE GERM CELLS

Coculturing strategies with embryonic gonadal somatic cells are instructive in promoting the sex determination of PGCLCs and their differentiation toward oocytes or spermatogonia; however, the complexity of germ–soma interaction precludes the elucidation of the precise molecular

mechanisms underlying sex determination. To overcome this obstacle, we have recently developed a culturing system to propagate sexually uncommitted PGCs/PGCLCs without the use of gonadal somatic cells (Ohta et al. 2017). By using the PGCLC system, we conducted systematic chemical screenings to identify compounds that can enhance the proliferative capacity of PGCs/PGCLCs. The screening revealed that cyclic AMP signaling plays a pivotal role in the expansion of PGCs/PGCLCs in culture, as previous studies reported using isolated PGCs (De Felici et al. 1993; Farini et al. 2005). Global transcriptome and cell cycle kinetics of the expanded PGCLCs showed that they maintained characteristics of sexually uncommitted PGCs. Importantly, functional testing upon transplantation into neonatal testes showed their robust capability to contribute to spermatogenesis. The proliferation of PGCLCs was accompanied with a steady decrease in global 5-methylcytosine (5mC) level, presumably driven by progressive dilution of methylation upon DNA replication (Seisenberger et al. 2012; Kagiwada et al. 2013; Kobayashi et al. 2013). At the end of the expansion culture, the global 5mC level of the PGCLCs was comparable to that of gonadal germ cells at E13.5 in mice, a stage at which the genome-wide DNA methylation level reaches its nadir during the germline cycle (Fig. 1; Seisenberger et al. 2012; Kobayashi et al. 2013). It is notable that the demethylation process showed differential kinetics at distinct genomic regions (i.e., promoters of demethylation-resistant "germline genes," ICRs of imprinted genes, and repetitive elements), in a manner reminiscent of PGCs (Seisenberger et al. 2012; Kobayashi et al. 2013). Further, we found that "escapees" that evade DNA demethylation (5mC > 20%) during PGCLC derivation and the expansion culture largely overlapped with those found in E13.5 germ cells (Seisenberger et al. 2012; Ohta et al. 2017). Thus, the induction of PGCLCs and subsequent expansion culture recapitulated the DNA methylation reprogramming events of PGCs. Remarkably, this global hypomethylation state was accompanied by apparent upregulation of histone H3K27me3 at promoters of genes that showed substantial demethylation during the expansion culture, plausibly compensating for the hypomethylation state and preventing aberrant gene expression from such loci. Nonetheless, we observed a partial activation of key gonadal germ cell genes that took place concurrently with the erasure of DNA methylation and reorganization of the histone modification landscape, implying that such epigenetic reorganization leads to a basal activation of some germline genes and might prime the germ cells for the eventual reception of the sex determination signals (Ohta et al. 2017). Taken together, these findings show that expansion culture recapitulates a global epigenetic reorganization process that forms an epigenetic "blank slate" of the germ cells upon which either an androgenetic or a gynogenetic epigenome can be established. The global epigenetic reorganization in the PGCs occurs cell-autonomously, without the guidance from gonadal somatic cells, and thus the reorganization process is an event genetically dissociable from the germ cell sex determination program.

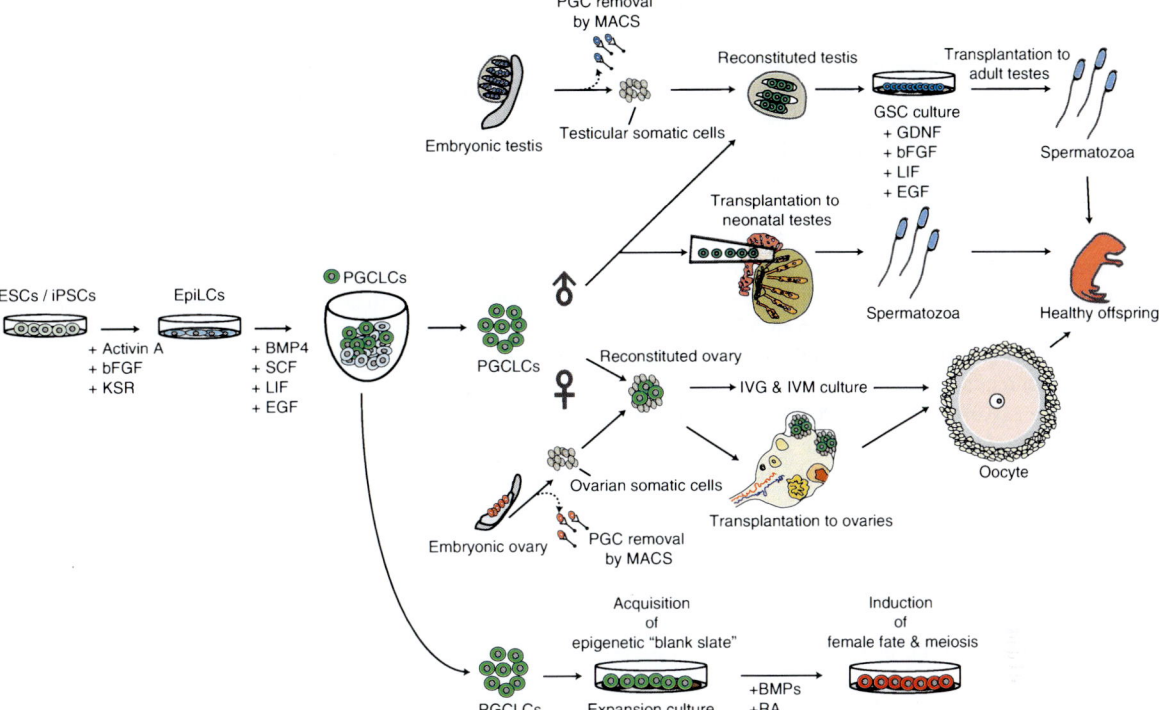

Figure 2. A schematic drawing of the methodologies for mouse germ cell derivation and development from pluripotent stem cells. Embryonic stem cells (ESCs) and induced pluripotent stem cells (iPSCs) cultured under a 2i+LIF condition are induced to epiblast-like cells (EpiLCs). EpiLCs are then induced into primordial germ cell (PGC)-like cells (PGCLCs) in floating aggregates. PGCLCs transplanted into neonatal testes will commit to the male pathway and undergo spermatogenesis successfully (Hayashi et al. 2011). PGCLCs cocultured with embryonic ovarian somatic cells (Reconstituted ovary) commit to the female fate and, upon transplantation to adult ovaries, develop to fertilization-competent oocytes (Hayashi et al. 2012). Alternatively, the reconstituted ovaries can be cultured under appropriate in vitro growth (IVG) and in vitro maturation (IVM) conditions to foster development into competent oocytes completely in vitro (Hikabe et al. 2016). PGCLCs cocultured with embryonic testicular somatic cells (Reconstituted testis) commit to the male fate and differentiate to spermatogonia, which can be subsequently propagated to establish male germline stem cell (GSC) lines that contribute to spermatogenesis upon transplantation to adult testes (Ishikura et al. 2016). PGCLCs propagated in vitro (Expansion culture) maintain a sexually uncommitted state and acquire an epigenetic "blank slate" (Ohta et al. 2017). Subsequently they can be induced to confer the female fate and initiate meiosis upon bone morphogenic protein (BMP) and retinoic acid (RA) treatment (Miyauchi et al. 2017). MACS, magnetic activated cell sorting.

SIGNALING PRINCIPLES FOR FEMALE GERM CELL SEX DETERMINATION AND MEIOTIC INDUCTION

The PGCLC expansion system described above, in which the sexual fate of the germ cells remains uncommitted, provides a valuable platform for studies to reconstitute the processes of germ cell sex determination and meiotic initiation. Using this culture system, we examined the effects of a panel of cytokines that might have an impact on sex determination with or without the treatment of RA (Miyauchi et al. 2017). Among the combinations of molecules we assayed, the combined provision of BMPs and RA resulted in the up-regulation of transcriptional programs for oogenesis and initiation of meiotic prophase. Notably, the transcriptional and cytological progression faithfully recapitulated the events of fetal oocytes, and the initiation of meiosis occurred in a highly synchronous manner within germ cell cysts (Miyauchi et al. 2017), demonstrating a coordinated developmental progress within the oocyte cysts (Fig. 1; Pepling and Spradling 1998). It has previously been shown that developing em-

bryonic ovaries strongly express *Bmp* genes, particularly *Bmp2* under the influence of *Wnt4* (Yao et al. 2004; Ross et al. 2007; Jameson et al. 2012), and germ cell–specific deletion of *Smad4*, a gene coding for an essential signal transducer of BMP/TGFβ signaling, negatively affects the induction of female germ cell fate and meiotic onset (Wu et al. 2016). When considered together with our current results and those of other groups (Farini et al. 2005), these findings clearly show that BMP signaling plays a pivotal role in the induction of female germ cell fate. To our surprise, although RA signaling was proved to be essential for meiotic induction and oogenesis commitment, the addition of RA alone was not sufficient to confer the female fate (Miyauchi et al. 2017). Specifically, treatment with RA alone resulted in the up-regulation of known RA-regulated genes that are involved in meiosis (i.e., *Stra8* and *Rec8*) (Oulad-Abdelghani et al. 1996; Mahony et al. 2011; Koubova et al. 2014; Soh et al. 2015), but the activation of these genes was not sufficient to initiate meiosis. Instead, the RA-only treatment up-regulated genes involved in other developmental programs (e.g., "embryonic organ development"), indicating that the cross talk with BMP signaling is essential not only to activate the transcriptional cascades for the induction of meiosis and oogenic programs but also to repress inappropriate developmental programs elicited by RA. In the absence of *Stra8*, a gene essential for premeiotic DNA replication, combined treatment of BMP and RA failed to fully activate meiotic genes and to repress unnecessary developmental programs (Miyauchi et al. 2017). Notably, however, *Stra8* knockout did not prevent the activation of many fetal oocyte genes, including those involved in oocyte development (e.g., *Figla* and *Sohlh2*), demonstrating that the oocyte developmental program is independent from STRA8 and the chromosomal events of meiotic prophase, as reported recently (Dokshin et al. 2013).

During the expansion culture, PGCLCs globally erase 5mC in a comprehensive manner to a level comparable to E13.5 germ cells in mice (Ohta et al. 2017), a stage immediately preceding meiotic onset in females, and we tested whether such epigenetic states play a permissive role in sex determination and meiotic initiation. The combined BMP and RA treatment failed to induce meiotic genes in PGCLCs immediately after the inception from epiblast-like cells, but the treatment resulted in a robust activation of meiotic genes and commitment to the female fate after PGCLCs completed epigenetic reprogramming, indicating that the global epigenetic changes might make the germ cells more receptive to signals for sex determination and initiation of meiosis (Miyauchi et al. 2017). Interestingly, spermatogonia undergo significant DNA demethylation at promoters of relevant meiotic genes be-

fore meiotic initiation (Kubo et al. 2015; Miyauchi et al. 2017). This demethylation occurs despite the global reacquisition of high 5mC level during male germ cell development (Fig. 1; Kubo et al. 2015), implying that epigenetic requirements for meiotic entry might be common between male and female germ cells (Miyauchi et al. 2017). Such competence acquisition could be partly explained by the erasure of DNA methylation marks at promoters of key germ cell genes (e.g., *Dazl*). An evolutionary conserved RNA binding protein, DAZL, has been shown to be a critical factor for sexual commitment and meiotic initiation in mice (although the influence of DAZL in sex determination and meiotic initiation varies among different genetic backgrounds) (Ruggiu et al. 1997; Lin et al. 2008; Gill et al. 2011). In our work, the reduction of 5mC level at *Dazl* promoter was accompanied by a basal activation of *Dazl* during the expansion culture, which likely contributed to the competence acquisition and thus to the commitment to the female germ cell fate upon activation of RA and BMP signaling (Lin et al. 2008; Gill et al. 2011; Kato et al. 2016; Miyauchi et al. 2017). Collectively, our works show that meiotic initiation requires a delicate coupling between extrinsic inputs and intrinsic molecular circuitry within the germ cell: Germ cell–autonomous epigenetic remodeling is instrumental in initiating the meiotic program and also in inducing female fate upon concerted activation of BMP and RA signaling (Fig. 3). Elucidation of downstream effector(s) of BMP signaling and investigation of the mechanisms by which BMP signaling, in concert with RA signaling, orchestrates

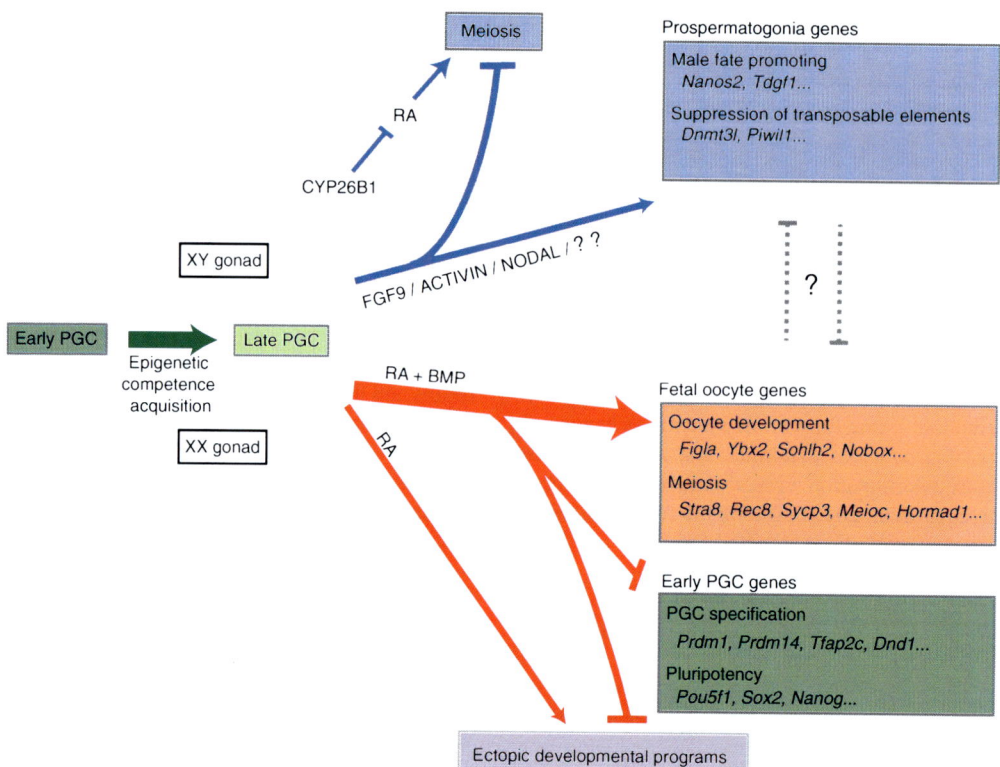

Figure 3. A schematic representation of a potential signaling network controlling the sex determination processes for germ cells.

the onset of female fate and meiotic initiation will be key challenges for the future.

Interestingly, a recent report showed that PGCLCs can initiate meiosis and develop into spermatid-like cells if cocultured with dissociated testicular cells of neonatal mice in the presence of RA, BMP2/4/7, and activin A for 6 days, followed by treatments of follicle stimulating hormone, bovine pituitary extract, and testosterone in the subsequent 8 days (Zhou et al. 2016). Surprisingly, the reported "in vitro spermatogenesis" procedure skips a period of more than 2 weeks of male germ cell development in vivo—that is, the period from PGCs at around E9.5 to spermatogonia before meiotic onset at postnatal day 10 (Bellvé et al. 1977). It is important to note that, during this period, male germ cells undergo epigenetic reprogramming, acquire the androgenetic epigenome, and differentiate into spermatogonia (Davis et al. 2000; Ueda et al. 2000; Kato et al. 2007; Seisenberger et al. 2012; Kobayashi et al. 2013; Kubo et al. 2015; Manku and Culty 2015). Furthermore, the reported procedure was conducted entirely at 37°C, a nonpermissive temperature for spermatogenesis (Steinberger et al. 1964; Sato et al. 2011). Thus, we consider it important to re-examine the validity of this study, including a precise analysis of the induced intermediate cells.

CONCLUSION

In this review, we have provided a brief outline of mammalian germ cell development with a focus on the implementation of meiosis and the latest findings from in vitro germ cell derivation studies. Meiosis is a process that is conserved from unicellular organisms to multicellular organisms, and species-specific gametogenic strategies have evolved for successful execution of meiosis (Harigaya and Yamamoto 2007; Lesch and Page 2012). Recent advances have begun to provide a blueprint of how mammals execute meiosis in the context of epigenetic (Yokobayashi et al. 2013; Miyauchi et al. 2017), transcriptional (Soh et al. 2015; Miyauchi et al. 2017), extracellular signaling (Bowles et al. 2006; Koubova et al. 2006; MacLean et al. 2007; Saba et al. 2014b; Miyauchi et al. 2017), and posttranscriptional regulations (Zheng et al. 2013; Abby et al. 2016; Kato et al. 2016; Hsu et al. 2017; Soh et al. 2017). Moving forward, considering our rudimentary understanding of how sex determination and meiotic onset are coupled in the fetal gonads (Hilscher et al. 1974; McLaren 1984, 1988; Kimble and Page 2007; Spiller et al. 2017), elucidation of the molecular interweaving between these processes will be imperative. Moreover, future studies should aim to address how germ cells commit to the male pathway and establish spermatogonial stem cell molecular circuitry while maintaining the potential for meiosis. Considering the tight integration of meiosis within the intricate oogenesis and spermatogenesis processes, a deeper understanding of the interactions with the surrounding somatic cells will surely hold the key in the coming years. Last, the recent advent of germ cell derivation strategies from human pluripotent stem cells (Irie et al. 2015; Sasaki et al. 2015), elucidation of the cellular

origins of primordial germ cells and molecular signatures defining distinct cell types of peri-implantation primate embryos (Nakamura et al. 2016; Sasaki et al. 2016), and investigations of transcriptome and epigenome characteristics of human germ cells during embryonic development (Gkountela et al. 2015; Guo et al. 2015; Tang et al. 2015; Li et al. 2017) have laid the groundwork for future efforts at the reconstitution of human gametogenesis. If successful, such reconstitution would provide valuable insights into the mechanism of human germ cell development and into the causes of anomalies that hamper human reproduction (Handel and Schimenti 2010; Nagaoka et al. 2012).

ACKNOWLEDGMENTS

We thank H. Ohta, H. Miyauchi, and the members of our laboratory for their discussion and suggestions. The authors were supported in part by a Japan Society for the Promotion of Science (JSPS) Research Fellowship for Young Scientists to S.I.N., by funds from Japan Science and Technology Agency Exploratory Research for Advanced Technology (JST-ERATO) to M.S. (JPMJER1104), and by a Grant-in-Aid for Specially Promoted Research from JSPS to M.S. (17H06098).

REFERENCES

Abby E, Tourpin S, Ribeiro J, Daniel K, Messiaen S, Moison D, Guerquin J, Gaillard JC, Armengaud J, Langa F, et al. 2016. Implementation of meiosis prophase I programme requires a conserved retinoid-independent stabilizer of meiotic transcripts. *Nat Commun* **7:** 10324.

Albertini DF. 2015. The mammalian oocyte. In *Knobil and Neill's physiology of reproduction* (eds. Plant TM, Zeleznik AJ, Albertini DF, et al.), pp. 59–97. Elsevier/Academic, New York.

Baltus AE, Menke DB, Hu YC, Goodheart ML, Carpenter AE, de Rooij DG, Page DC. 2006. In germ cells of mouse embryonic ovaries, the decision to enter meiosis precedes premeiotic DNA replication. *Nat Genet* **38:** 1430–1434.

Barrios F, Filipponi D, Pellegrini M, Paronetto MP, Di Siena S, Geremia R, Rossi P, De Felici M, Jannini EA, Dolci S. 2010. Opposing effects of retinoic acid and FGF9 on Nanos2 expression and meiotic entry of mouse germ cells. *J Cell Sci* **123:** 871–880.

Bastien J, Rochette-Egly C. 2004. Nuclear retinoid receptors and the transcription of retinoid-target genes. *Gene* **328:** 1–16.

Baudat F, Imai Y, de Massy B. 2013. Meiotic recombination in mammals: Localization and regulation. *Nat Rev Genet* **14:** 794–806.

Bellvé AR, Cavicchia JC, Millette CF, O'Brien DA, Bhatnagar YM, Dym M. 1977. Spermatogenic cells of the prepuberal mouse. Isolation and morphological characterization. *J Cell Biol* **74:** 68–85.

Bowles J, Knight D, Smith C, Wilhelm D, Richman J, Mamiya S, Yashiro K, Chawengsaksophak K, Wilson MJ, Rossant J, et al. 2006. Retinoid signaling determines germ cell fate in mice. *Science* **312:** 596–600.

Bowles J, Feng CW, Spiller C, Davidson TL, Jackson A, Koopman P. 2010. FGF9 suppresses meiosis and promotes male germ cell fate in mice. *Dev Cell* **19:** 440–449.

Bowles J, Feng CW, Miles K, Ineson J, Spiller C, Koopman P. 2016. ALDH1A1 provides a source of meiosis-inducing retinoic acid in mouse fetal ovaries. *Nat Commun* **7:** 10845.

Brennan J, Capel B. 2004. One tissue, two fates: Molecular genetic events that underlie testis versus ovary development. *Nat Rev Genet* **5:** 509–521.

Brinster RL, Avarbock MR. 1994. Germline transmission of donor haplotype following spermatogonial transplantation. *Proc Natl Acad Sci* **91:** 11303–11307.

Brinster RL, Zimmermann JW. 1994. Spermatogenesis following male germ-cell transplantation. *Proc Natl Acad Sci* **91:** 11298–11302.

Capel B. 2017. Vertebrate sex determination: Evolutionary plasticity of a fundamental switch. *Nat Rev Genet.* doi: 10.1038/nrg.2017.60.

Childs AJ, Cowan G, Kinnell HL, Anderson RA, Saunders PT. 2011. Retinoic acid signalling and the control of meiotic entry in the human fetal gonad. *PLoS One* **6:** e20249.

Chuma S, Kanatsu-Shinohara M, Inoue K, Ogonuki N, Miki H, Toyokuni S, Hosokawa M, Nakatsuji N, Ogura A, Shinohara T. 2005. Spermatogenesis from epiblast and primordial germ cells following transplantation into postnatal mouse testis. *Development* **132:** 117–122.

Clift D, Schuh M. 2013. Restarting life: Fertilization and the transition from meiosis to mitosis. *Nat Rev Mol Cell Biol* **14:** 549–562.

Cunningham TJ, Duester G. 2015. Mechanisms of retinoic acid signalling and its roles in organ and limb development. *Nat Rev Mol Cell Biol* **16:** 110–123.

Daley GQ. 2007. Gametes from embryonic stem cells: A cup half empty or half full? *Science* **316:** 409–410.

Davis TL, Yang GJ, McCarrey JR, Bartolomei MS. 2000. The H19 methylation imprint is erased and re-established differentially on the parental alleles during male germ cell development. *Hum Mol Genet* **9:** 2885–2894.

De Felici M, Dolci S, Pesce M. 1993. Proliferation of mouse primordial germ cells in vitro: A key role for cAMP. *Dev Biol* **157:** 277–280.

Dokshin GA, Baltus AE, Eppig JJ, Page DC. 2013. Oocyte differentiation is genetically dissociable from meiosis in mice. *Nat Genet* **45:** 877–883.

Dolle P. 2009. Developmental expression of retinoic acid receptors (RARs). *Nucl Recept Signal* **7:** e006.

Eggers S, Ohnesorg T, Sinclair A. 2014. Genetic regulation of mammalian gonad development. *Nat Rev Endocrinol* **10:** 673–683.

Extavour CG, Akam M. 2003. Mechanisms of germ cell specification across the metazoans: Epigenesis and preformation. *Development* **130:** 5869–5884.

Farini D, Scaldaferri ML, Iona S, La Sala G, De Felici M. 2005. Growth factors sustain primordial germ cell survival, proliferation and entering into meiosis in the absence of somatic cells. *Dev Biol* **285:** 49–56.

Feng CW, Bowles J, Koopman P. 2014. Control of mammalian germ cell entry into meiosis. *Mol Cell Endocrinol* **382:** 488–497.

Gerton JL, Hawley RS. 2005. Homologous chromosome interactions in meiosis: Diversity amidst conservation. *Nat Rev Genet* **6:** 477–487.

Gill ME, Hu YC, Lin Y, Page DC. 2011. Licensing of gametogenesis, dependent on RNA binding protein DAZL, as a gateway to sexual differentiation of fetal germ cells. *Proc Natl Acad Sci* **108:** 7443–7448.

Ginsburg M, Snow MH, McLaren A. 1990. Primordial germ cells in the mouse embryo during gastrulation. *Development* **110:** 521–528.

Gkountela S, Zhang KX, Shafiq TA, Liao WW, Hargan-Calvopiña J, Chen PY, Clark AT. 2015. DNA demethylation dynamics in the human prenatal germline. *Cell* **161:** 1425–1436.

Griswold MD. 2016. Spermatogenesis: The commitment to meiosis. *Physiol Rev* **96:** 1–17.

Griswold MD, Hogarth CA, Bowles J, Koopman P. 2012. Initiating meiosis: The case for retinoic acid. *Biol Reprod* **86:** 35.

Guo F, Yan L, Guo H, Li L, Hu B, Zhao Y, Yong J, Hu Y, Wang X, Wei Y et al. 2015. The transcriptome and DNA methylome landscapes of human primordial germ cells. *Cell* **161:** 1437–1452.

Hackett JA, Surani MA. 2014. Regulatory principles of pluripotency: From the ground state up. *Cell Stem Cell* **15:** 416–430.

Handel MA, Schimenti JC. 2010. Genetics of mammalian meiosis: Regulation, dynamics and impact on fertility. *Nat Rev Genet* **11:** 124–136.

Harigaya Y, Yamamoto M. 2007. Molecular mechanisms underlying the mitosis–meiosis decision. *Chromosome Res* **15:** 523–537.

Hayashi K, Ohta H, Kurimoto K, Aramaki S, Saitou M. 2011. Reconstitution of the mouse germ cell specification pathway in culture by pluripotent stem cells. *Cell* **146:** 519–532.

Hayashi K, Ogushi S, Kurimoto K, Shimamoto S, Ohta H, Saitou M. 2012. Offspring from oocytes derived from in vitro primordial germ cell-like cells in mice. *Science* **338:** 971–975.

Hikabe O, Hamazaki N, Nagamatsu G, Obata Y, Hirao Y, Hamada N, Shimamoto S, Imamura T, Nakashima K, Saitou M, et al. 2016. Reconstitution in vitro of the entire cycle of the mouse female germ line. *Nature* **539:** 299–303.

Hilscher B, Hilscher W, Bülthoff-Ohnolz B, Krämer U, Birke A, Pelzer H, Gauss G. 1974. Kinetics of gametogenesis. I. Comparative histological and autoradiographic studies of oocytes and transitional prospermatogonia during oogenesis and prespermatogenesis. *Cell Tissue Res* **154:** 443–470.

Hsu PJ, Zhu Y, Ma H, Guo Y, Shi X, Liu Y, Qi M, Lu Z, Shi H, Wang J, et al. 2017. Ythdc2 is an N6-methyladenosine binding protein that regulates mammalian spermatogenesis. *Cell Res* **27:** 1115–1127.

Huang HF, Hembree WC. 1979. Spermatogenic response to vitamin A in vitamin A deficient rats. *Biol Reprod* **21:** 891–904.

Irie N, Weinberger L, Tang WW, Kobayashi T, Viukov S, Manor YS, Dietmann S, Hanna JH, Surani MA. 2015. SOX17 is a critical specifier of human primordial germ cell fate. *Cell* **160:** 253–268.

Ishikura Y, Yabuta Y, Ohta H, Hayashi K, Nakamura T, Okamoto I, Yamamoto T, Kurimoto K, Shirane K, Sasaki H, et al. 2016. In vitro derivation and propagation of spermatogonial stem cell activity from mouse pluripotent stem cells. *Cell Rep* **17:** 2789–2804.

Jameson SA, Natarajan A, Cool J, DeFalco T, Maatouk DM, Mork L, Munger SC, Capel B. 2012. Temporal transcriptional profiling of somatic and germ cells reveals biased lineage priming of sexual fate in the fetal mouse gonad. *PLoS Genet* **8:** e1002575.

Kagiwada S, Kurimoto K, Hirota T, Yamaji M, Saitou M. 2013. Replication-coupled passive DNA demethylation for the erasure of genome imprints in mice. *EMBO J* **32:** 340–353.

Kanatsu-Shinohara M, Shinohara T. 2013. Spermatogonial stem cell self-renewal and development. *Annu Rev Cell Dev Biol* **29:** 163–187.

Kashimada K, Svingen T, Feng CW, Pelosi E, Bagheri-Fam S, Harley VR, Schlessinger D, Bowles J, Koopman P. 2011. Antagonistic regulation of Cyp26b1 by transcription factors SOX9/SF1 and FOXL2 during gonadal development in mice. *FASEB J* **25:** 3561–3569.

Kato Y, Kaneda M, Hata K, Kumaki K, Hisano M, Kohara Y, Okano M, Li E, Nozaki M, Sasaki H. 2007. Role of the Dnmt3 family in de novo methylation of imprinted and repetitive sequences during male germ cell development in the mouse. *Hum Mol Genet* **16:** 2272–2280.

Kato Y, Katsuki T, Kokubo H, Masuda A, Saga Y. 2016. Dazl is a target RNA suppressed by mammalian NANOS2 in sexually differentiating male germ cells. *Nat Commun* **7:** 11272.

Kimble J, Page DC. 2007. The mysteries of sexual identity. The germ cell's perspective. *Science* **316:** 400–401.

Kobayashi H, Sakurai T, Imai M, Takahashi N, Fukuda A, Yayoi O, Sato S, Nakabayashi K, Hata K, Sotomaru Y, et al. 2012. Contribution of intragenic DNA methylation in mouse gametic DNA methylomes to establish oocyte-specific heritable marks. *PLoS Genet* **8:** e1002440.

Kobayashi H, Sakurai T, Miura F, Imai M, Mochiduki K, Yanagisawa E, Sakashita A, Wakai T, Suzuki Y, Ito T, et al. 2013. High-resolution DNA methylome analysis of primordial germ cells identifies gender-specific reprogramming in mice. *Genome Res* **23:** 616–627.

Koopman P, Gubbay J, Vivian N, Goodfellow P, Lovell-Badge R. 1991. Male development of chromosomally female mice transgenic for Sry. *Nature* **351:** 117–121.

Koubova J, Menke DB, Zhou Q, Capel B, Griswold MD, Page DC. 2006. Retinoic acid regulates sex-specific timing of meiotic initiation in mice. *Proc Natl Acad Sci* **103:** 2474–2479.

Koubova J, Hu YC, Bhattacharyya T, Soh YQ, Gill ME, Goodheart ML, Hogarth CA, Griswold MD, Page DC. 2014. Retinoic acid activates two pathways required for meiosis in mice. *PLoS Genet* **10:** e1004541.

Kubo N, Toh H, Shirane K, Shirakawa T, Kobayashi H, Sato T, Sone H, Sato Y, Tomizawa S, Tsurusaki Y, et al. 2015. DNA methylation and gene expression dynamics during spermatogonial stem cell differentiation in the early postnatal mouse testis. *BMC Genomics* **16:** 624.

Kurimoto K, Yabuta Y, Hayashi K, Ohta H, Kiyonari H, Mitani T, Moritoki Y, Kohri K, Kimura H, Yamamoto T, et al. 2015. Quantitative dynamics of chromatin remodeling during germ cell specification from mouse embryonic stem cells. *Cell Stem Cell* **16:** 517–532.

Lawson KA, Dunn NR, Roelen BA, Zeinstra LM, Davis AM, Wright CV, Korving JP, Hogan BL. 1999. Bmp4 is required for the generation of primordial germ cells in the mouse embryo. *Genes Dev* **13:** 424–436.

Le Bouffant R, Guerquin MJ, Duquenne C, Frydman N, Coffigny H, Rouiller-Fabre V, Frydman R, Habert R, Livera G. 2010. Meiosis initiation in the human ovary requires intrinsic retinoic acid synthesis. *Hum Reprod* **25:** 2579–2590.

Lesch BJ, Page DC. 2012. Genetics of germ cell development. *Nat Rev Genet* **13:** 781–794.

Li R, Albertini DF. 2013. The road to maturation: Somatic cell interaction and self-organization of the mammalian oocyte. *Nat Rev Mol Cell Biol* **14:** 141–152.

Li H, Clagett-Dame M. 2009. Vitamin A deficiency blocks the initiation of meiosis of germ cells in the developing rat ovary in vivo. *Biol Reprod* **81:** 996–1001.

Li Y, Zheng M, Lau YF. 2014. The sex-determining factors SRY and SOX9 regulate similar target genes and promote testis cord formation during testicular differentiation. *Cell Rep* **8:** 723–733.

Li L, Dong J, Yan L, Yong J, Liu X, Hu Y, Fan X, Wu X, Guo H, Wang X, et al. 2017. Single-cell RNA-seq analysis maps development of human germline cells and gonadal niche interactions. *Cell Stem Cell* **20:** 891–892.

Lin Y, Gill ME, Koubova J, Page DC. 2008. Germ cell-intrinsic and -extrinsic factors govern meiotic initiation in mouse embryos. *Science* **322:** 1685–1687.

Lucifero D, Mertineit C, Clarke HJ, Bestor TH, Trasler JM. 2002. Methylation dynamics of imprinted genes in mouse germ cells. *Genomics* **79:** 530–538.

MacLean G, Li H, Metzger D, Chambon P, Petkovich M. 2007. Apoptotic extinction of germ cells in testes of Cyp26b1 knockout mice. *Endocrinology* **148:** 4560–4567.

Mahony S, Mazzoni EO, McCuine S, Young RA, Wichterle H, Gifford DK. 2011. Ligand-dependent dynamics of retinoic acid receptor binding during early neurogenesis. *Genome Biol* **12:** R2.

Manku G, Culty M. 2015. Mammalian gonocyte and spermatogonia differentiation: Recent advances and remaining challenges. *Reproduction* **149:** R139–R157.

Mark M, Ghyselinck NB, Chambon P. 2009. Function of retinoic acid receptors during embryonic development. *Nucl Recept Signal* **7:** e002.

Marston AL, Amon A. 2004. Meiosis: Cell-cycle controls shuffle and deal. *Nat Rev Mol Cell Biol* **5:** 983–997.

Martello G, Smith A. 2014. The nature of embryonic stem cells. *Annu Rev Cell Dev Biol* **30:** 647–675.

Matson CK, Murphy MW, Sarver AL, Griswold MD, Bardwell VJ, Zarkower D. 2011. DMRT1 prevents female reprogramming in the postnatal mammalian testis. *Nature* **476:** 101–104.

Matzuk MM, Burns KH, Viveiros MM, Eppig JJ. 2002. Intercellular communication in the mammalian ovary: Oocytes carry the conversation. *Science* **296:** 2178–2180.

McGee EA, Hsueh AJ. 2000. Initial and cyclic recruitment of ovarian follicles. *Endocr Rev* **21:** 200–214.

McLaren A. 1984. Meiosis and differentiation of mouse germ cells. *Symp Soc Exp Biol* **38:** 7–23.

McLaren A. 1988. Somatic and germ-cell sex in mammals. *Philos Trans R Soc Lond B Biol Sci* **322:** 3–9.

Miyauchi H, Ohta H, Nagaoka S, Nakaki F, Sasaki K, Hayashi K, Yabuta Y, Nakamura T, Yamamoto T, Saitou M. 2017. Bone morphogenetic protein and retinoic acid synergistically specify female germ-cell fate in mice. *EMBO J.* doi: 10.15252/embj.201796875.

Morales C, Griswold MD. 1987. Retinol-induced stage synchronization in seminiferous tubules of the rat. *Endocrinology* **121:** 432–434.

Morohaku K, Tanimoto R, Sasaki K, Kawahara-Miki R, Kono T, Hayashi K, Hirao Y, Obata Y. 2016. Complete in vitro generation of fertile oocytes from mouse primordial germ cells. *Proc Natl Acad Sci* **113:** 9021–9026.

Nagaoka SI, Hassold TJ, Hunt PA. 2012. Human aneuploidy: Mechanisms and new insights into an age-old problem. *Nat Rev Genet* **13:** 493–504.

Nakamura T, Okamoto I, Sasaki K, Yabuta Y, Iwatani C, Tsuchiya H, Seita Y, Nakamura S, Yamamoto T, Saitou M. 2016. A developmental coordinate of pluripotency among mice, monkeys and humans. *Nature* **537:** 57–62.

Niederreither K, Dolle P. 2008. Retinoic acid in development: Towards an integrated view. *Nat Rev Genet* **9:** 541–553.

Oatley JM, Brinster RL. 2008. Regulation of spermatogonial stem cell self-renewal in mammals. *Annu Rev Cell Dev Biol* **24:** 263–286.

Ohinata Y, Ohta H, Shigeta M, Yamanaka K, Wakayama T, Saitou M. 2009. A signaling principle for the specification of the germ cell lineage in mice. *Cell* **137:** 571–584.

Ohta H, Kurimoto K, Okamoto I, Nakamura T, Yabuta Y, Miyauchi H, Yamamoto T, Okuno Y, Hagiwara M, Shirane K, et al. 2017. In vitro expansion of mouse primordial germ cell-like cells recapitulates an epigenetic blank slate. *EMBO J* **36:** 1888–1907.

Oulad-Abdelghani M, Bouillet P, Decimo D, Gansmuller A, Heyberger S, Dolle P, Bronner S, Lutz Y, Chambon P. 1996. Characterization of a premeiotic germ cell-specific cytoplasmic protein encoded by Stra8, a novel retinoic acid-responsive gene. *J Cell Biol* **135:** 469–477.

Pepling ME, Spradling AC. 1998. Female mouse germ cells form synchronously dividing cysts. *Development* **125:** 3323–3328.

Pepling ME, Spradling AC. 2001. Mouse ovarian germ cell cysts undergo programmed breakdown to form primordial follicles. *Dev Biol* **234:** 339–351.

Perea-Gomez A, Vella FD, Shawlot W, Oulad-Abdelghani M, Chazaud C, Meno C, Pfister V, Chen L, Robertson E, Hamada H, et al. 2002. Nodal antagonists in the anterior visceral endoderm prevent the formation of multiple primitive streaks. *Dev Cell* **3:** 745–756.

Ross A, Munger S, Capel B. 2007. Bmp7 regulates germ cell proliferation in mouse fetal gonads. *Sex Dev* **1:** 127–137.

Ruggiu M, Speed R, Taggart M, McKay SJ, Kilanowski F, Saunders P, Dorin J, Cooke HJ. 1997. The mouse Dazla gene encodes a cytoplasmic protein essential for gametogenesis. *Nature* **389:** 73–77.

Saba R, Kato Y, Saga Y. 2014a. NANOS2 promotes male germ cell development independent of meiosis suppression. *Dev Biol* **385:** 32–40.

Saba R, Wu Q, Saga Y. 2014b. CYP26B1 promotes male germ cell differentiation by suppressing STRA8-dependent meiotic and STRA8-independent mitotic pathways. *Dev Biol* **389:** 173–181.

Sada A, Suzuki A, Suzuki H, Saga Y. 2009. The RNA-binding protein NANOS2 is required to maintain murine spermatogonial stem cells. *Science* **325:** 1394–1398.

Saitou M, Miyauchi H. 2016. Gametogenesis from pluripotent stem cells. *Cell Stem Cell* **18**: 721–735.

Saitou M, Yamaji M. 2012. Primordial germ cells in mice. *Cold Spring Harb Perspect Biol* **4**: a008375.

Saitou M, Barton SC, Surani MA. 2002. A molecular programme for the specification of germ cell fate in mice. *Nature* **418**: 293–300.

Sasaki K, Yokobayashi S, Nakamura T, Okamoto I, Yabuta Y, Kurimoto K, Ohta H, Moritoki Y, Iwatani C, Tsuchiya H, et al. 2015. Robust in vitro induction of human germ cell fate from pluripotent stem cells. *Cell Stem Cell* **17**: 178–194.

Sasaki K, Nakamura T, Okamoto I, Yabuta Y, Iwatani C, Tsuchiya H, Seita Y, Nakamura S, Shiraki N, Takakuwa T, et al. 2016. The germ cell fate of cynomolgus monkeys is specified in the nascent amnion. *Dev Cell* **39**: 169–185.

Sato T, Katagiri K, Gohbara A, Inoue K, Ogonuki N, Ogura A, Kubota Y, Ogawa T. 2011. In vitro production of functional sperm in cultured neonatal mouse testes. *Nature* **471**: 504–507.

Seisenberger S, Andrews S, Krueger F, Arand J, Walter J, Santos F, Popp C, Thienpont B, Dean W, Reik W. 2012. The dynamics of genome-wide DNA methylation reprogramming in mouse primordial germ cells. *Mol Cell* **48**: 849–862.

Shirane K, Kurimoto K, Yabuta Y, Yamaji M, Satoh J, Ito S, Watanabe A, Hayashi K, Saitou M, Sasaki H. 2016. Global landscape and regulatory principles of DNA methylation reprogramming for germ cell specification by mouse pluripotent stem cells. *Dev Cell* **39**: 87–103.

Sinclair AH, Berta P, Palmer MS, Hawkins JR, Griffiths BL, Smith MJ, Foster JW, Frischauf AM, Lovell-Badge R, Goodfellow PN. 1990. A gene from the human sex-determining region encodes a protein with homology to a conserved DNA-binding motif. *Nature* **346**: 240–244.

Soh YQ, Junker JP, Gill ME, Mueller JL, van Oudenaarden A, Page DC. 2015. A gene regulatory program for meiotic prophase in the fetal ovary. *PLoS Genet* **11**: e1005531.

Soh YQS, Mikedis MM, Kojima M, Godfrey AK, de Rooij DG, Page DC. 2017. Meioc maintains an extended meiotic prophase I in mice. *PLoS Genet* **13**: e1006704.

Speed RM. 1982. Meiosis in the foetal mouse ovary. I. An analysis at the light microscope level using surface-spreading. *Chromosoma* **85**: 427–437.

Spiller CM, Feng CW, Jackson A, Gillis AJ, Rolland AD, Looijenga LH, Koopman P, Bowles J. 2012. Endogenous Nodal signaling regulates germ cell potency during mammalian testis development. *Development* **139**: 4123–4132.

Spiller C, Koopman P, Bowles J. 2017. Sex determination in the mammalian germline. *Annu Rev Genet.* doi: 10.1146/annurev-genet-120215-035449

Steinberger A, Steinberger E, Perloff WH. 1964. Mammalian testes in organ culture. *Exp Cell Res* **36**: 19–27.

Suzuki A, Saga Y. 2008. Nanos2 suppresses meiosis and promotes male germ cell differentiation. *Genes Dev* **22**: 430–435.

Tang WW, Dietmann S, Irie N, Leitch HG, Floros VI, Bradshaw CR, Hackett JA, Chinnery PF, Surani MA. 2015. A unique gene regulatory network resets the human germline epigenome for development. *Cell* **161**: 1453–1467.

Toshimori K, Eddy EM. 2015. The spermatozoon. In *Knobil and Neill's physiology of reproduction* (eds. Plant TM, Zeleznik AJ, Albertini DF, et al.), pp. 99–148. Elsevier/Academic, New York.

Ueda T, Abe K, Miura A, Yuzuriha M, Zubair M, Noguchi M, Niwa K, Kawase Y, Kono T, Matsuda Y, et al. 2000. The paternal methylation imprint of the mouse H19 locus is acquired in the gonocyte stage during foetal testis development. *Genes Cells* **5**: 649–659.

Uhlenhaut NH, Jakob S, Anlag K, Eisenberger T, Sekido R, Kress J, Treier AC, Klugmann C, Klasen C, Holter NI, et al. 2009. Somatic sex reprogramming of adult ovaries to testes by FOXL2 ablation. *Cell* **139**: 1130–1142.

Watanabe Y. 2012. Geometry and force behind kinetochore orientation: Lessons from meiosis. *Nat Rev Mol Cell Biol* **13**: 370–382.

Wilkins AS, Holliday R. 2009. The evolution of meiosis from mitosis. *Genetics* **181**: 3–12.

Wilson JG, Roth CB, Warkany J. 1953. An analysis of the syndrome of malformations induced by maternal vitamin A deficiency. Effects of restoration of vitamin A at various times during gestation. *Am J Anat* **92**: 189–217.

Wu Q, Kanata K, Saba R, Deng CX, Hamada H, Saga Y. 2013. Nodal/activin signaling promotes male germ cell fate and suppresses female programming in somatic cells. *Development* **140**: 291–300.

Wu Q, Fukuda K, Weinstein M, Graff JM, Saga Y. 2015. SMAD2 and p38 signaling pathways act in concert to determine XY primordial germ cell fate in mice. *Development* **142**: 575–586.

Wu Q, Fukuda K, Kato Y, Zhou Z, Deng CX, Saga Y. 2016. Sexual fate change of XX germ cells caused by the deletion of SMAD4 and STRA8 independent of somatic sex reprogramming. *PLoS Biol* **14**: e1002553.

Yao HH, Matzuk MM, Jorgez CJ, Menke DB, Page DC, Swain A, Capel B. 2004. Follistatin operates downstream of Wnt4 in mammalian ovary organogenesis. *Dev Dyn* **230**: 210–215.

Yokobayashi S, Liang CY, Kohler H, Nestorov P, Liu Z, Vidal M, van Lohuizen M, Roloff TC, Peters AH. 2013. PRC1 coordinates timing of sexual differentiation of female primordial germ cells. *Nature* **495**: 236–240.

Yoshida S. 2012. Elucidating the identity and behavior of spermatogenic stem cells in the mouse testis. *Reproduction* **144**: 293–302.

Yoshida S, Sukeno M, Nakagawa T, Ohbo K, Nagamatsu G, Suda T, Nabeshima Y. 2006. The first round of mouse spermatogenesis is a distinctive program that lacks the self-renewing spermatogonia stage. *Development* **133**: 1495–1505.

Zheng G, Dahl JA, Niu Y, Fedorcsak P, Huang CM, Li CJ, Vagbo CB, Shi Y, Wang WL, Song SH, et al. 2013. ALKBH5 is a mammalian RNA demethylase that impacts RNA metabolism and mouse fertility. *Mol Cell* **49**: 18–29.

Zhou Q, Wang M, Yuan Y, Wang X, Fu R, Wan H, Xie M, Liu M, Guo X, Zheng Y, et al. 2016. Complete meiosis from embryonic stem cell-derived germ cells in vitro. *Cell Stem Cell* **18**: 330–340.

Regulation of Crossover Frequency and Distribution during Meiotic Recombination

TAKAMUNE T. SAITO AND MONICA P. COLAIÁCOVO

Department of Genetics, Harvard Medical School, Boston, Massachusetts 02115

Correspondence: mcolaiacovo@genetics.med.harvard.edu

Crossover recombination is essential for generating genetic diversity and promoting accurate chromosome segregation during meiosis. The process of crossover recombination is tightly regulated and is initiated by the formation of programmed meiotic DNA double-strand breaks (DSBs). The number of DSBs is around 10-fold higher than the number of crossovers in most species, because only a limited number of DSBs are repaired as crossovers during meiosis. Moreover, crossovers are not randomly distributed. Most crossovers are located on chromosomal arm regions and both centromeres and telomeres are usually devoid of crossovers. Either loss or mislocalization of crossovers frequently results in chromosome nondisjunction and subsequent aneuploidy, leading to infertility, miscarriages, and birth defects such as Down syndrome. Here, we will review aspects of crossover regulation observed in most species and then focus on crossover regulation in the nematode *Caenorhabditis elegans* in which both the frequency and distribution of crossovers are tightly controlled. In this system, only a single crossover is formed, usually at an off-centered position, between each pair of homologous chromosomes. We have identified *C. elegans* mutants with deregulated crossover distribution, and we are analyzing crossover control by using an inducible single DSB system with which a single crossover can be produced at specific genomic positions. These combined studies are revealing novel insights into how crossover position is linked to accurate chromosome segregation.

Meiosis is a specialized cell division process that generates haploid gametes from diploid parental germ cells. This reduction in the number of chromosomes is achieved by following a single round of DNA replication with two consecutive cell divisions (meiosis I and II). Homologous chromosomes are separated at meiosis I, and sister chromatids are separated at meiosis II. There are unique chromosomal events that need to take place during prophase to ensure that homologs segregate properly at meiosis I (Fig. 1). Homologous chromosomes need to find each other and pair; these pairing interactions need to be stabilized via the formation of a scaffold known as the synaptonemal complex, which assembles at the interface between paired homologs; and interhomolog recombination needs to take place in order to produce crossovers. Crossover formation is one of the sources of genetic diversity in the population. Moreover, crossovers result in physical attachments (chiasmata) between homologs that, underpinned by cohesion, confer the tension required to properly align the attached homologs (bivalents) at the metaphase plate and then orient them toward opposite poles of the meiosis I spindle.

Errors in crossover formation result in chromosome nondisjunction leading to aneuploidy, which causes infertility, miscarriages, birth defects, and cancers.

Given the impact of crossover formation on human health and reproductive biology, it is therefore not surprising that crossovers are tightly regulated. For example, crossover formation is not frequently observed near centromeres and telomeres, suggesting they may be repressed in these regions. Crossovers at centromere regions lead to aneuploidy in female meiosis and crossovers at telomeres

increase azoospermia (Ottolini et al. 2015; Ren et al. 2016). However, direct testing of how a crossover positioned near centromeres or telomeres might lead to increased errors in chromosome segregation has been challenging in metazoans.

Caenorhabditis elegans is an ideal model organism to study crossover control, because crossover formation is tightly regulated in comparison to other known model organisms. A single off-centered crossover is formed on each of the six pairs of homologous chromosomes in *C. elegans* compared to the one to four crossovers per pair of homologs observed in other species (Barnes et al. 1995; Martinez-Perez and Colaiácovo 2009; Rockman and Kruglyak 2009). Surprisingly, a single DNA double-strand break (DSB) is sufficient to make a crossover in *C. elegans* (Rosu et al. 2011). This property, coupled with the use of a system in which a single DSB can be induced at defined genomic positions, allows us to analyze how crossover position affects meiotic chromosome segregation in *C. elegans*. Here, we review what is known for crossover control from studies in different organisms, our novel findings regarding regulation of crossover position using the single inducible DSB system in *C. elegans*, and the future directions of research using this system aimed at understanding the origin of aneuploidies.

MOLECULAR STEPS IN CROSSOVER FORMATION

Crossover formation starts with the formation of DSBs by a topoisomerase-like protein present from yeast to hu-

Published by Cold Spring Harbor Laboratory Press; doi: 10.1101/sqb.2017.82.034132

Cold Spring Harbor Symposia on Quantitative Biology, Volume LXXXII

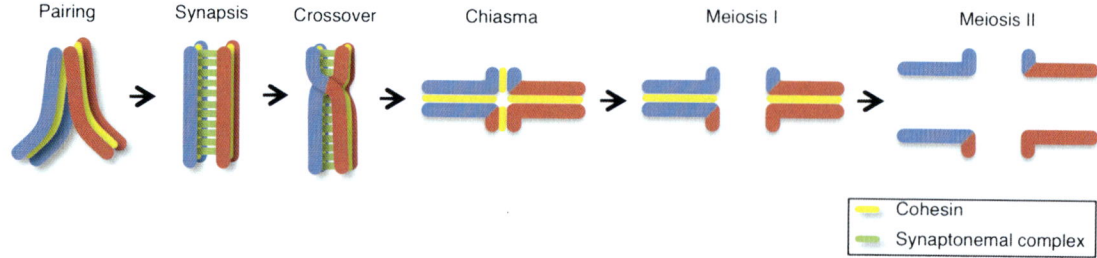

Figure 1. Meiosis and crossover formation. Chromosome dynamics during meiosis. After premeiotic DNA replication, homologous chromosomes find each other (pairing) during the leptotene–zygotene stages. The synaptonemal complex assembles, aligning and holding homologs together throughout their full lengths (synapsis) at the pachytene stage. Repair of DNA double-strand breaks (DSBs) via crossover formation involves the reciprocal exchange of genetic information between homologs. A chiasma is the cytologically visible manifestation of an earlier crossover event underpinned by flanking sister chromatid cohesion and is observed as a cruciform configuration during the diplotene to diakinesis transition. Homologous chromosomes are segregated at the metaphase I to anaphase I transition, and sister chromatids are separated at the metaphase II to anaphase II transition. Paternal chromatids are blue, and maternal chromatids are red. Sister chromatid cohesion is depicted in yellow, and the synaptonemal complex is depicted in green.

mans known as Spo11 (Fig. 2; Keeney et al. 1997). DSBs then undergo 5'-end resection to produce 3' overhangs through the activity of the Mre11/Rad50/Xrs2 exonuclease complex. Rad51 associates with the 3' single-stranded DNA overhangs producing a DNA–protein filament that then engages in a search for homologous DNA sequences. The 3' end invades the homologous template (single-strand invasion resulting in D-loop formation), followed by DNA synthesis. At this point, repair can proceed through different pathways resulting in either the displacement and annealing of the newly synthesized strand to its complementary strand (synthesis-dependent strand annealing [SDSA]) (Fig. 2A) or in second end capture to produce a double Holliday junction (dHJ) intermediate (Fig. 2B). The SDSA pathway results only in noncrossover products. Meanwhile, the asymmetric resolution of dHJs by structure-specific endonucleases (Slx1-Slx4, Mus81-Mms4, and Yen1) produces crossover products (Fig. 2C), whereas their symmetric resolution results in noncrossovers (Fig. 2D). It has been proposed that designated and nondesignated DSBs are converted to dHJs and resolved by different structure-specific endonucleases in yeast to assure and limit the number of crossovers (Zakharyevich et al. 2012). dHJs can also be processed through dissolution mediated by the Sgs1-Top3-Rmi1 complex during which the two Holliday junctions branch migrate toward one another until they form a hemicatenated intermediate that can be decatenated by topoisomerase III, resulting in noncrossovers (Fig. 2E). Finally, all DSBs undergoing intersister repair result in noncrossovers. This detailed blueprint of the molecular requirements for crossover formation allows for assessment of how these may differ during DSB repair depending on the location of the DSB in a metazoan.

CLASS I AND CLASS II CROSSOVERS

Two different classes of crossovers have been identified, namely class I and class II. In yeast, around 90 crossovers are observed in 16 bivalents. Seventy percent (range from 60% to 90%) of crossovers are class I crossovers, which are dependent on the meiosis-specific ZMM pro-

teins (Zip1, Zip2, Zip3, Msh4, Msh5, Mer3, Spo16, and Spo22/Zip4). Zip1 is a structural component of the synaptonemal complex that holds pairs of homologous chromosomes together. Zip2 is a XPF-like helix-hairpin-helix containing protein (Chua and Roeder 1998; Macaisne et al. 2008) and Zip3 is a SUMO E3 ligase, whereas Msh4 and Msh5 are homologs of the *Escherichia coli* mismatch repair protein MutS implicated in stabilizing dHJ intermediates (Snowden et al. 2004). The number of class I crossovers (approximately 60) matches the number of Zip2, Zip3, and Msh4-Msh5 foci observed in yeast pachytene nuclei. Mer3 is a DNA helicase required for Holliday junction branch migration (Nakagawa and Ogawa 1999; Mazina et al. 2004). Finally, Spo16 and Spo22/Zip4 are involved in synaptonemal complex assembly, and they are unique because, in contrast to the other ZMM proteins, they are not essential for crossover interference (Shinohara et al. 2008) (see the subsection Crossover Interference). The remaining ~30% of crossovers fall into class II. Class II crossovers depend on double Holliday junction resolution executed by the structure-specific endonucleases Mus81-Mms4, Slx1-Slx4, and Yen1 in yeast (Fig. 2C,D; Zakharyevich et al. 2012).

MULTIPLE LAYERS OF CROSSOVER REGULATION

Crossover Assurance/Obligate Crossover

Because crossover formation is essential for proper chromosome segregation at meiosis I (Fig. 1), at least one crossover (obligate crossover) has to be formed between each pair of homologous chromosomes. This phenomenon is called "crossover assurance" (Fig. 3A).

A notable exception to this regulation is observed in *Drosophila* in which both male meiosis and female chromosome 4 are devoid of crossover formation (Cooper 1949; Hartmann and Sekelsky 2017) and chromosome segregation occurs randomly at meiosis II. In contrast, recombinant chromatids are preferentially segregated to oocytes and not into the second polar body in human female meiosis (Ottolini et al. 2015). Robust crossover assurance is also observed in *C. elegans*, where only

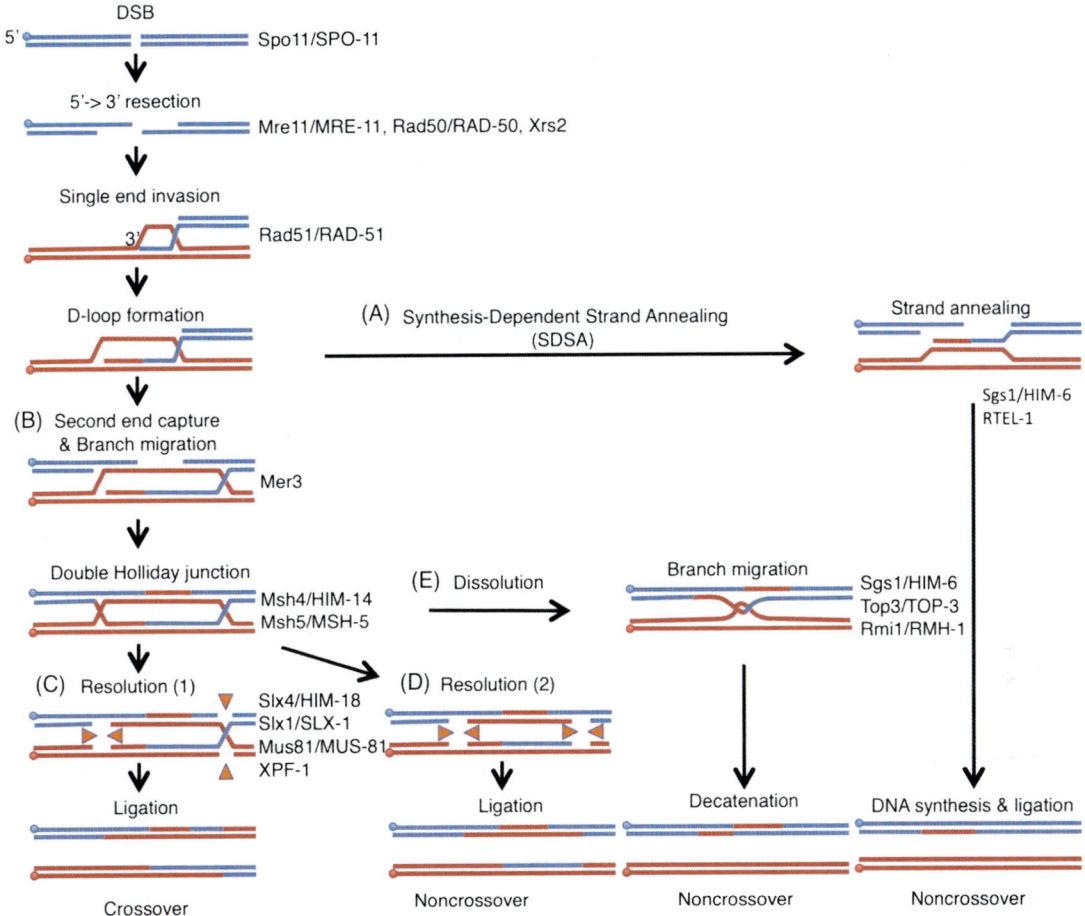

Figure 2. Model of homologous recombination. DNA double strand breaks are generated by the topoisomerase-like protein Spo11. The MRN/X complex (Mre11-Rad50-Nbs1/Xrs2) resects the 5′ ends to expose 3′ overhangs. Single end invasion (SEI) is mediated by Rad51. Homologous recombination can then proceed through the following pathways: (*A*) synthesis-dependent strand annealing resulting in noncrossover products or (*B*) double Holliday junction (dHJ) formation by Mer3 and Msh4-Msh5 resulting in crossover (CO) formation. The DNA helicase ReqQ homologs Sgs1 and RTEL-1 disrupt D-loops to anneal both ends of the DSB. Once double Holliday junctions are formed, they are resolved by the structure-specific endonucleases SLX-1-SLX-4/HIM-18, MUS-81-EME1, and XPF-ERCC1. (*C*) Asymmetric resolution of the dHJ produces crossovers, and (*D*) symmetric resolution results in noncrossovers. (*E*) dHJs can also be processed by the dissolution pathway through the BTR complex (BLM-TOP3-RMI1/2) to make noncrossover products. Paternal DNAs are blue and maternal DNAs are red. Circles indicate the 5′ side of DNA. Orange triangles indicate the direction of catalytic activities of Holliday junction resolvases. Key proteins acting at each step are indicated on the *right*, and both yeast and worm names are indicated.

one DSB per homologous chromosome pair is sufficient to make a crossover (Rosu et al. 2011). Moreover, chromosomes that fail to undergo crossover formation, as a result of either impaired homologous pairing or, in the case of the extra chromosome present in trisomies, lack of synapsis, are preferentially segregated into the polar bodies during both anaphase I and anaphase II of *C. elegans* female meiosis (Cortes et al. 2015; Muscat et al. 2015; Vargas et al. 2017). Therefore, crossover formation is not only important for accurate homolog separation at meiosis I but also acts as a driving force during sister chromatid separation at meiosis II.

Crossover Interference

Crossover interference is a phenomenon in which a crossover at one location reduces the probability of a sec- ond crossover nearby such that when there are two or more crossovers along a bivalent these crossovers are separated away from each other (Fig. 3B). This phenomenon was first described more than 100 years ago (Sturtevant 1915) and is only observed for class I crossover events. A beam-film model has been proposed for crossover interference that simulates establishment and propagation of a mechanical stress along the chromosome axis depicted by an elastic beam plate (metal) covered with a thin brittle film (ceramic) with crossovers being seen as cracks that release the stress locally and thus abrogate crossovers nearby (Kleckner et al. 2004). Topoisomerase II and the meiosis-specific chromosome axis protein Red1 are suggested to be involved in crossover interference through ubiquitination by the histone deacetylase Sir2 and the SUMO-targeted ubiquitin ligases Slx5–Slx8 (Zhang et al. 2014). Synaptonemal complex–dependent crossover interference

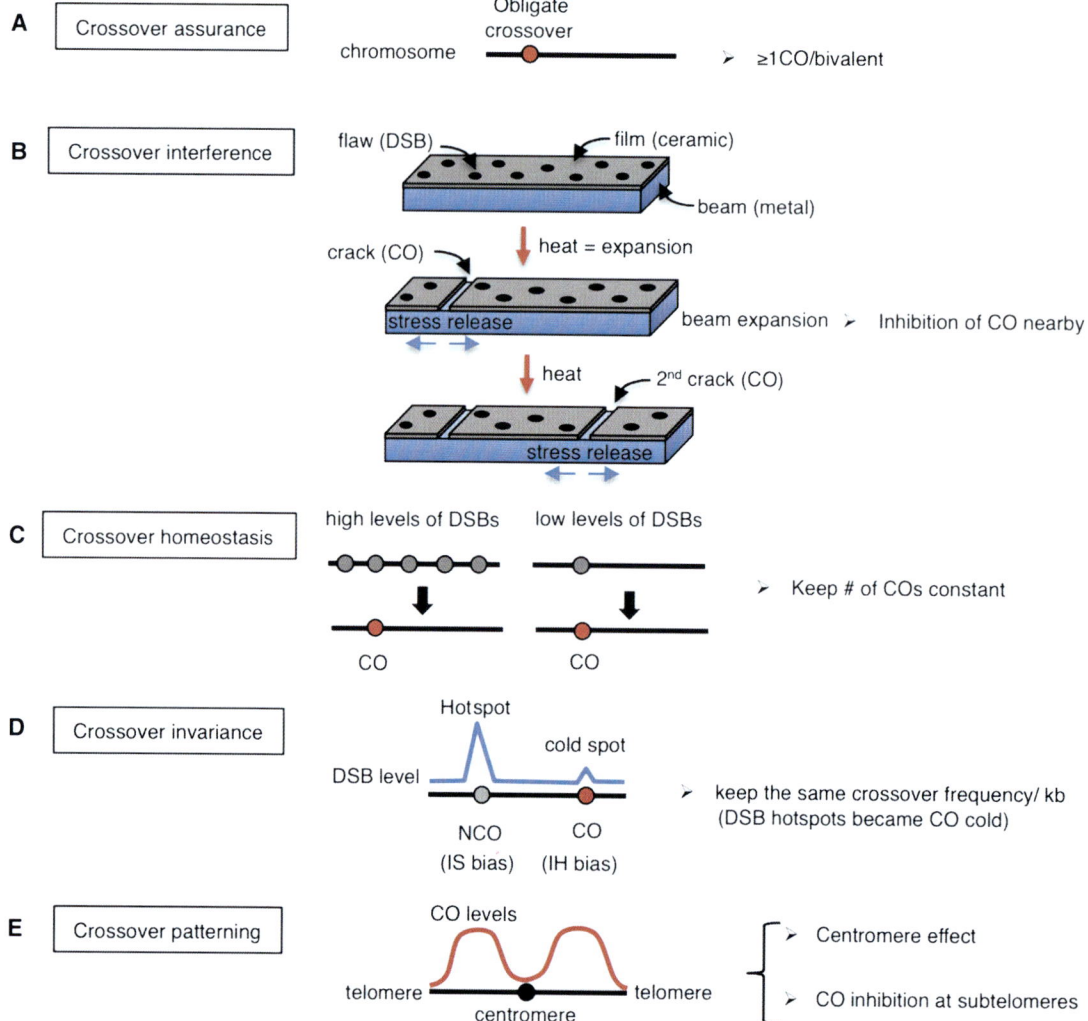

Figure 3. Different types of crossover (CO) control. Five known forms of crossover control are depicted. (*A*) Crossover assurance. At least one crossover per homologous chromosome pair is essential for chiasma formation and proper chromosome segregation at meiosis I. (*B*) Crossover interference. The beam-film model (modified from Kleckner et al. 2004) is represented. Chromosome axes and chromatin loops are likened as metallic beams and ceramic films that are tightly bonded to the beam, respectively. Heating the beam results in a flaw (DSB) being converted into a crack (CO formation), and the release of stress then propagates in both directions. Continued heating generates a second crack away from the first crack resembling interference. (*C*) Crossover homeostasis. Either high or low levels of DSBs per homologous chromosome pair result in the same number of crossovers. Gray circles are DSBs, and orange circles are crossovers. (*D*) Crossover invariance. DSBs at hotspots tend to undergo intersister bias, resulting in a noncrossover outcome, whereas DSBs at cold spots undergo interhomolog bias leading to crossover products in *Schizosaccharomyces pombe*. (*E*) Crossover distribution/centromere effect. Crossovers near centromeres and telomeres are suppressed. Crossovers are also suppressed at the center regions in the holocentric organism *Caenorhabditis elegans*.

is observed in *C. elegans* (Libuda et al. 2013) and in at least one yeast strain (Sym and Roeder 1994; Chen et al. 2008). The biological function of crossover interference is largely unknown, but it may confer a selective advantage because of cosegregation of functionally related linked genes (Wang et al. 2015; Sun et al. 2017).

Crossover Homeostasis

In general, DSBs, which are introduced at the leptotene/zygotene stage of prophase of meiosis I, occur at levels that are 10-fold higher than the number of crossovers detected. The number of crossovers is maintained constant at

the expense of noncrossover events even when DSBs levels are reduced. This phenomenon is termed "crossover homeostasis" and has been observed in yeast (Martini et al. 2006), mice (Cole et al. 2012), worms (Yokoo et al. 2012), and plants (Fig. 3C; Varas et al. 2015).

Crossover Invariance

The choice of repair template is very important to make an interhomolog crossover. In addition, levels of DSBs vary largely across the genome as evidenced by the presence of both hotspots and cold regions of DSBs along chromosomes. In *Schizosaccharomyces pombe*, which lacks crossover interference, a nearly constant level of crossing-over is maintained per unit physical distance across the genome by control of partner choice for DSB repair, a phenomenon referred to as "crossover invariance" (Fig. 3D; Hyppa and Smith 2010; Fowler et al. 2014). At a DSB hotspot, intersister repair is predominant, whereas at a DSB cold region, interhomolog repair is more prevalent. This phenomenon may serve as an alternative mechanism of crossover homeostasis in other organisms to maintain crossover levels constant.

Crossover Patterning/Centromere Effect

Crossover formation is inhibited near centromeres and telomeres in many species including humans (Fig. 3E; Ottolini et al. 2015; Ren et al. 2016). Crossovers at centromeres disrupt cohesion in the pericentric region and affect kinetochore orientation. Interestingly, even gene conversion near centromeres is associated with 60% of segregation errors at meiosis I in *Saccharomyces cerevisiae*, primarily because of premature sister chromatid separation (Sears et al. 1995). The "centromere effect" was first described in *Drosophila* as an inhibition of crossovers at centromeres and pericentromeric euchromatic regions (Beadle 1932). A combination of the centromere effect and crossover suppression by the Blm helicase results in the absence of crossovers on chromosome 4 in *Drosophila* (Hartmann and Sekelsky 2017; Hatkevich and Sekelsky 2017). In plants, cytosines at centromeric regions are highly methylated, thereby suppressing expression of repetitive sequences including transposons. Pollen typing revealed that

crossover frequency is increased in centromeric regions in the DNA methyltransferase *met1* mutants, suggesting that DNA methylation at centromeres is important to suppress crossovers at those regions (Yelina et al. 2012). Yeast Slx4, a regulatory subunit of the structure-specific endonuclease Slx1, is also required for crossover suppression near centromeres in an Slx1-independent manner (Higashide and Shinohara 2016). A similar suppression mechanism was observed in *C. elegans*, where normally crossovers are located on the arms but not at the center of the chromosomes (Fig. 4). However, the crossover suppression observed at the center of the chromosomes in this worm is SLX-1-dependent and SLX4/HIM-18-independent (Saito et al. 2009, 2012, 2013). Taken together, these various layers of crossover control underscore the importance of crossover formation during meiosis and the need for tightly regulating this process throughout species.

CROSSOVER PATTERNING IN *C. ELEGANS*

Although studies of crossover control started a century ago, the molecular mechanisms underlying these processes are still largely unknown, in part because the various layers of crossover control are not independent phenomena. The regulation of obligatory crossover, interference, and homeostasis are thought to be a cooperative process (Wang et al. 2015). To understand these complex phenomena, it is important to use a simple model.

In *C. elegans*, crossovers tend to occur in an off-centered position such that crossover frequencies are 1.3 cM/Mb at the central region of the chromosome, which encompasses

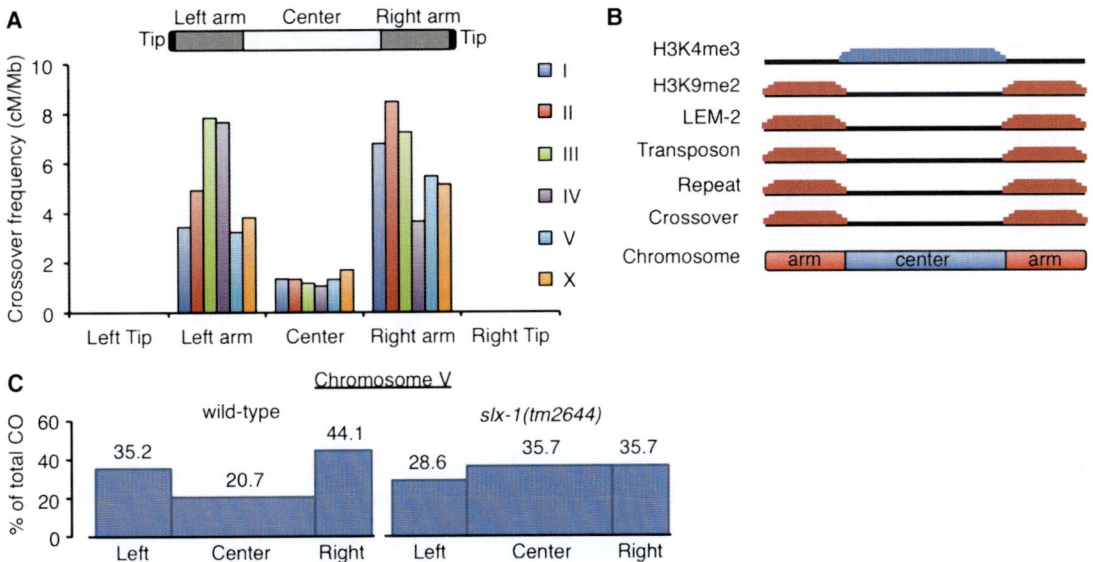

Figure 4. Tight regulation of crossover formation in *Caenorhabditis elegans*. (*A*) Crossovers are enriched at arm regions but suppressed at center regions in both autosomes and the X chromosome in *C. elegans*. No crossovers are observed at subtelomeric regions (average <614 kb from telomeres). Data was adapted from Rockman and Kruglyak (2009). (*B*) Unique features divide chromosome domains in *C. elegans*. Although up to approximately 10 DSBs are distributed in a nonbiased manner along chromosomes (Saito et al. 2012), crossovers occur at the arm regions in which the heterochromatin marker histone H3K9me2, the nuclear membrane protein (LEM-2) binding sequences, transposons, and repeat sequences are enriched. Crossover formation is suppressed at the center region in which the euchromatic marker histone H3K4me3 is enriched. (*C*) Crossover suppression at the center region of autosomes is lost in *slx-1(tm2644)* null mutants. Blue boxes indicate crossover frequencies (Saito et al. 2012). Although the overall crossover frequency is not altered, crossover distribution is altered by increasing at the center region and decreasing at the arms in *slx-1* mutants compared to wild type.

49% of the total length of the autosomes and 36% of the X chromosomes, and 4.7 and 6.1 cM/Mb on the left and right arms, respectively (Fig. 4A; Rockman and Kruglyak 2009). There are some clear differences between the central and arm regions of the chromosomes in this organism. For example, essential genes and the euchromatin marker histone H3K4me3 are enriched at the central region. In contrast, the arm regions are gene-poor, they are enriched for the heterochromatin marker H3K9me2/3, and they are transposon- and repeat-rich and interact with the nuclear membrane via the LEM (LAP2, emerin, MAN1) domain-containing protein LEM-2 (Fig. 4B; *C. elegans* Sequencing Consortium 1998; Gerstein et al. 2010; Ikegami et al. 2010; Liu et al. 2011; Ho et al. 2014). Different mutants have been identified that shift crossover distribution to the center region of the chromosomes in *C. elegans* (Zetka and Rose 1995; Wagner et al. 2010; Meneely et al. 2012; Saito et al. 2012; Chung et al. 2015; Hong et al. 2016; Jagut et al. 2016). However, the molecular mechanism underlying the regulation of crossover position is largely unknown. We found that in mutants for SLX-1, a structure-specific endonuclease that cleaves 5′ flaps, replication forks, and Holliday junctions, crossovers shifted to the center regions, whereas overall crossover frequency was preserved (Fig. 4C; Saito et al. 2012, 2013). This was not due to alterations in DSB distribution because similar frequencies of markers for DSB repair sites were observed cytologically along the arms and at the center of the chromosomes in *slx-1* mutants compared to wild type (Saito et al. 2012). This observation led us to hypothesize that there are mechanisms either inhibiting crossover formation or promoting noncrossover formation after DSB induction at the center region of the chromosomes, such as SDSA, double Holliday junction dissolution, "same sense resolution" of double Holliday junctions, or intersister repair. Based on its Holliday junction resolvase activity, one possibility is that SLX-1 produces noncrossovers at the center region by same sense resolution of double Holliday junctions. An alternative, albeit nonexclusive, possibility is that SLX-1 may act as an epigenetic reader given that it has a PHD/RING-like zinc finger domain implicated in recognizing histone H3K4me3 (Peña et al. 2006; Shi et al. 2006; Matthews et al. 2007; Ramón-Maiques et al. 2007). Given that the euchromatin marker H3K4me3 is enriched at the center region of the autosomes in *C. elegans*, we hypothesize that SLX-1 might be recruited to that region through recognition of H3K4me3 by its PHD/RING finger domain or potentially function as an ubiquitin ligase to activate noncrossover pathways at the center region. It is known that H3K4me2, H3K4me3, and H3K9ac are specifically enriched at the DSB hotspots recognized by the PRDM (PRDI-BF1 [positive regulatory domain I-binding factor 1] and RIZ [retinoblastoma-interacting zinc-finger protein] homology domain)-containing histone H3 methyltransferase Prdm9 in mammals (Hayashi et al. 2005; Buard et al. 2009; Baudat et al. 2010; Myers et al. 2010; Parvanov et al. 2010). After DSBs are introduced, phosphorylation of histone H2AX and hyperacetylation of histone H4 are observed near the hotspots (Mahadevaiah et al. 2001; Buard et al. 2009). H3K4me3 is associated with DSB for-

mation in yeasts and plants (Choi et al. 2013) and the PHD finger of Spp1, a component of the histone H3K4 methyltransferase Set1 complex (also known as COMPASS), reads H3K4me3 to promote DSBs (Acquaviva et al. 2013; Sommermeyer et al. 2013). However, whether H3K4me3 is associated with noncrossover pathways and whether crossover hotspots are marked by H3K9me2/3 remain unknown. One possible explanation for why DSBs are introduced at the arm region is that the DSB machinery could access small euchromatic (H3K4me3) sites dispersed throughout the H3K9me2/3 marked heterochromatin on the arm regions (Ikegami et al. 2010). Combined strategies, described below, allow for direct testing of whether SLX-1 localizes at the center region of chromosomes and the roles of its PHD/RING domain.

INDUCIBLE SINGLE DSB SYSTEM—HOW DOES CROSSOVER POSITION IMPACT CHROMOSOME SEGREGATION IN A METAZOAN?

Although improper crossover distribution has been detected in aneuploid gametes and embryos in humans, the impact of improper crossover distribution has not been tested directly in metazoans. *C. elegans* is a suitable model organism to assess how improper crossover distribution can cause errors in meiotic chromosome segregation given how tightly crossovers are regulated in this organism. We previously showed that, starting in late pachytene, chromosomes remodel around the off-centered crossover, resulting in the characteristic cruciform configuration observed for bivalents in late diakinesis consisting of a long and a short arm intersecting at the chiasma (Fig. 5B; Nabeshima et al. 2005). Aurora B kinase, AIR-2, localizes at the short arms by the end of diakinesis in which it phosphorylates the meiosis-specific cohesin REC-8. This triggers degradation of sister chromatid cohesion along the short arm. LAB-1, a functional shugoshin analog, antagonizes AIR-2 along the long arms, thereby protecting these chromosome subdomains from premature loss of sister chromatid cohesion. Therefore, we hypothesized that achieving proper chromosome remodeling must be critical for an accurate reductional division at meiosis I. To test this directly, we used a single DSB-inducible system in which the *Mos1* transposon from *Drosophila* was introduced into *C. elegans* along with a transposase under a heat shock promoter and a *spo-11* mutation that eliminates endogenous meiotic DSBs (Bessereau et al. 2001). We obtained lines in which a single DSB was produced at either the physical center or at the subtelomeric region in an autosome and compared that to lines in which a single DSB was produced at an off-centered position (Fig. 5A,B). We observed impaired chromosome remodeling based on mislocalization of AIR-2 and LAB-1 as well as subsequent increased nondisjunction when the single DSB was converted into a crossover at the center position. We did not observe a single crossover event at the subtelomeric position although a DSB at such a position did get designated into a crossover, suggesting

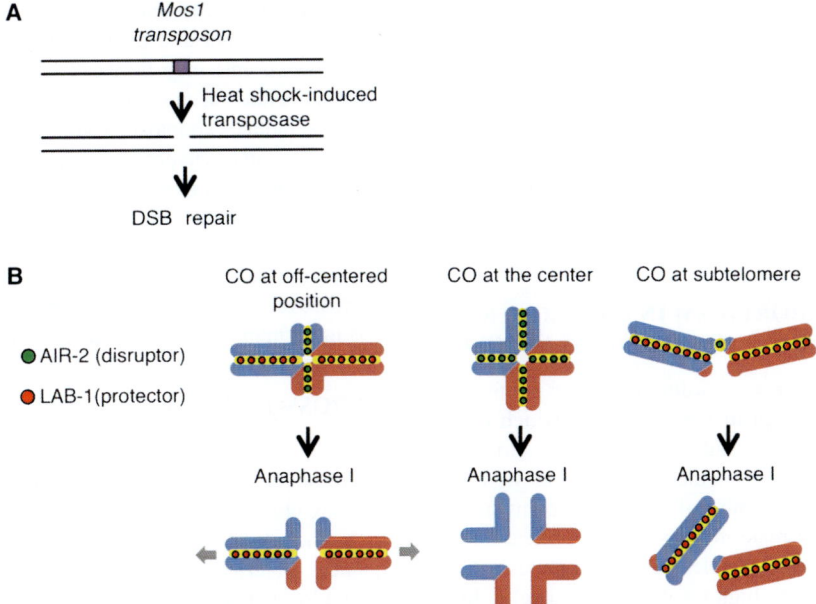

Figure 5. Site-specific analysis of meiotic recombination in *Caenorhabditis elegans*. (*A*) *Mos1*-based single inducible double-strand break (DSB) system. *Mos1* transposons and transposases are integrated into chromosomes in a *spo-11* mutant background. The *Mos1* integrated strain library is available at NemaGENETAG (http://elegans.imbb.forth.gr/nemagenetag/). Heat shock induces expression of the transposase, which excises the *Mos1* transposon resulting in a single DSB at a specific genomic position. (*B*) The single DSB system can be used to investigate the positional effect of crossovers on meiotic chromosome segregation. Blue and pink lines are paternal and maternal chromatids, respectively. The system allows us to analyze the outcome of a single crossover (CO) forming at specific chromosomal sites. A crossover at the very center region disrupts the asymmetric configuration of the bivalent, resulting in premature sister chromatid separation or homolog nondisjunction. Crossovers at subtelomeres result either in potentially fragile connections that are not stably retained at the ends of very short arms or events that fail to mature into crossovers at those positions.

these either cannot be stably maintained or are not effectively processed into a mature crossover at that position (E Altendorfer and MP Colaiácovo, unpubl.). This is supported by the lack of crossovers previously reported at subtelomeric regions by Rockman and Kruglyak (2009). In contrast, a DSB at an off-centered position proceeded as wild type. We are currently assessing whether bivalent formation is increased if a single DSB is induced at the center region of the chromosomes in *slx-1* mutants to confirm our genetic observation that SLX-1 suppresses crossovers at that region. This is being combined with ChIP-seq for epigenetic markers with the goal of unveiling how crossover position impacts chromosome remodeling and proper chromosome segregation and how structure-specific endonucleases are involved in this regulation.

IS THE CLASS II CROSSOVER PATHWAY CONSERVED IN *C. ELEGANS*?

It has been proposed that only class I crossovers exist in *C. elegans*, because almost no crossovers are detected in worm mutants of the *zmm* counterparts (*syp-1, syp-2, syp-3, syp-4, zhp-3, him-14/msh4, msh-5*) (Zalevsky et al. 1999; Kelly et al. 2000; MacQueen et al. 2002; Colaiácovo et al. 2003; Jantsch et al. 2004; Smolikov et al. 2007, 2009) and crossover interference is extremely robust given that only a single crossover occurs per bivalent in normal conditions (Brenner 1974; Barnes et al. 1995). Whereas in budding yeast, the structure-specific endonucleases Slx1–

Slx4 and Mus81–Mms4 are required for class II crossovers (Zakharyevich et al. 2012), in *C. elegans* SLX-1-HIM-18/SLX4 and MUS-81 appear to be involved in 20%–60% of total crossovers together with XPF-1, another structure-specific endonuclease (Saito et al. 2009, 2012, 2013; Agostinho et al. 2013; O'Neil et al. 2013).

Interestingly, class II crossovers are observed in *C. elegans* under conditions in which there is either an excess of DSBs or a mutation in the antirecombinase RTEL-1 helicase (Youds et al. 2010). DSBs induced artificially by γ-irradiation (IR) cause a dose-dependent (10 and 75 Gy) increase in the number of crossovers, whereas the number of crossover-designated sites marked by ZHP-3 is kept at normal levels after IR treatment (Youds et al. 2010). Moreover, these excess crossovers are suppressed back to wild-type levels in *mus-81* mutants (Youds et al. 2010). Therefore, there must be designation-independent and MUS81-dependent class II crossovers in *C. elegans* as well. Furthermore, crossovers in *rtel-1* mutants are randomly distributed, suggesting that RTEL-1-inhibited crossovers show no interference (Youds et al. 2010). Helicase-dependent inhibition of class II crossovers is conserved in budding yeast where it is exerted by Sgs1 (De Muyt et al. 2012; Zakharyevich et al. 2012), in fission yeast by Fml1 (Lorenz et al. 2012), and in plants by FANCM (Crismani et al. 2012; Knoll et al. 2012). These data and the wide conservation of class II crossovers in yeast (de los Santos et al. 2003), plants (Berchowitz et al. 2007; Higgins et al. 2008), and mammals (Holloway et al.

2008; Svetlanov et al. 2008) suggest that a "dormant" class II crossover pathway exists in *C. elegans*.

Mapping of crossover frequency and distribution in either IR-treated combinatorial mutants of structure-specific endonucleases or in mutants for the structure-specific endonucleases combined with an *rtel-1* mutation will reveal whether interference-independent class II crossovers are structure-specific endonuclease-dependent as well in *C. elegans*.

SEXUAL DIMORPHISM IN CROSSOVER REGULATION DURING *C. ELEGANS* MEIOSIS

The biggest difference regarding crossover control between sexes is observed in *Drosophila*, in which crossovers occur in females but not in males. However, differences are observed in other organisms as well. For example, although the number of crossovers seems similar between female and male meiosis in mice, based on the number of MLH1 foci (females = 23.0 and males = 22.7) (Holloway et al. 2008), high-resolution sex-specific linkage maps revealed that crossover frequency on autosomes in females (~71 cM/autosome) is higher than in males (~66 cM/autosome) (Liu et al. 2014). Interestingly, the genomic distribution of crossovers between female and male meiosis is also significantly different; crossover distributions are uniform in females, but subtelomerically enhanced and pericentromerically suppressed in males (Liu et al. 2014). In contrast, opposite results regarding crossover frequency were observed in plants, in which male crossover frequency (2.23 crossovers/bivalent) is 1.7-fold higher than in females (1.33 crossovers/bivalent) (Giraut et al. 2011), even though the distribution of crossovers follows a similar pattern to that observed in mice. Specifically, crossover frequencies at subtelomeric regions are very high in males and very low in females (Giraut et al. 2011). In human females, the number of MLH1 foci is greater than in males, whereas the average crossover frequency is similar between the sexes (Wang et al. 2015). This suggests that crossover maturation after designation may be less efficient in females than in males. In *C. elegans*, crossover frequency is higher in male spermatogenesis than in oogenesis. The ratio of double crossovers is ~4% at chromosomes IV and V in males, whereas almost no double crossovers are detected in oogenesis (Henzel et al. 2011; Gabdank and Fire 2014). Furthermore, *him-8* and *meDf2* mutants, which lack crossovers on the X chromosome because of impaired X chromosome synapsis, show a higher crossover frequency (including double crossovers) on the autosomes compared with wild type (Carlton et al. 2006). These observations suggest that a crossover homeostasis-like regulation that maintains the crossover number per "nucleus" (referred to as an interchromosomal effect [Sturtevant 1919; Lucchesi and Suzuki 1968]) exists in animals that have an XO sex type such as *C. elegans* males as well. How the "additional" crossovers are formed/allowed is an interesting question. The following aspects should be considered: (1) Crossover interference is stronger in hermaphrodite oogenesis than in male spermatogenesis in *C. elegans* (Gab-

dank and Fire 2014); and (2) surveillance mechanisms to check the number and distribution of crossovers (crossover checkpoint) may be weak in XO animals. In fact, although either accumulated DNA damage or aberrant synapsis induces apoptosis resulting in the removal of the affected cell in hermaphrodite gonads, there is no germline apoptosis in *C. elegans* males because of lack of CED-3 activation in the male germline even though a recombination checkpoint is activated in male gonads (Jaramillo-Lambert et al. 2010). Whether there are also differences in crossover control between spermatogenesis and oogenesis in hermaphrodites remains to be investigated.

CROSSOVER REGULATION IN AUTOSOMES AND SEX CHROMOSOMES

Crossover regulation is different between autosomes and sex chromosomes. X and Y chromosomes in mammals, including humans, pair via short homologous sequences referred to as the pseudoautosomal region (PAR) located at their subtelomeres. *C. elegans* provides a useful experimental system to understand the evolution of heteromorphic sex chromosomes (Henzel et al. 2011). The karyotype of the *C. elegans* hermaphrodite consists of five autosomes (5A) and XX, and the male of 5A and XO. An end-to-end fusion of chromosomes X and IV generates a new chromosome named *mnT12*, and males with *mnT12* form a neo-sex body (IV = neo-Y and IV-X fusion = neo-X), in which a portion of chromosome IV mimics the mammalian PAR during meiotic prophase (Sigurdson et al. 1986). Meiotic sex chromosome inactivation (MSCI) occurs during meiotic prophase, and several factors have been identified as required for mammalian MSCI such as γH2AX, ATR kinase, ubiquitin, SUMO, and the BRCA1-A complex (Turner 2007; Lu and Yu 2015). However, the links between MSCI and crossover control remain to be determined. Whether MCSI is conserved in the *C. elegans* neo-sex body also requires further investigation. The unsynapsed X chromosome region of the neo-sex body in *C. elegans* males harbors high levels of the heterochromatic marker H3K9me2, similar to the X–Y bodies in mammals (Henzel et al. 2011). The PAR must receive DSBs to make an obligate crossover like other autosomes. In mice, chromatin axis length at the PAR is long relative to DNA length and in contrast to autosome axes in which DNA content correlates well with axis length (Kauppi et al. 2011). This was proposed to result in shorter chromatin loops on the PAR and contribute to higher DSB levels at this region compared to autosomes. Longer axes are also observed along the paired chromosome IV and *mnT12* in *C. elegans* (Henzel et al. 2011). Because Spo11α-dependent induction of DSBs is observed at the PAR in mice (Kauppi et al. 2011), it would be interesting to investigate whether there are differences between the paired chromosome IV and *mnT12* and other normal chromosome pairs regarding the machinery being engaged for DSB formation.

DSBs are also induced at the nonhomologous regions of sex bodies in mammals and on the X chromosome in males in *C. elegans* that lack a homologous partner. The biological function for DSB repair in these cases is unknown, but interestingly, DSBs on the male X chromosome may be partially inhibited by the structure-specific endonuclease XPF-1 in *C. elegans* (Checchi et al. 2014). Another possible explanation is that XPF-1 functions for repair via single-strand annealing at repeat sequences when RAD-51-dependent homologous recombination is not available on male X chromosomes (Checchi et al. 2014). Further studies are required to understand how DSBs are induced and repaired in XO animals.

In the *C. elegans* hermaphrodite, 0.2% of its progeny are males resulting from X chromosome nondisjunctions during meiosis. Based on the observations that most autosomal aneuploidies are lethal and that nearly 100% of the embryos hatch during hermaphrodite reproduction, X chromosomes in *C. elegans* are more vulnerable to disjoining during meiosis compared with autosomes. We reported that combinatorial mutants of different structure-specific endonucleases caused higher reductions in crossover frequencies on X chromosomes than on autosomes (Saito et al. 2009, 2012, 2013). Understanding the fragility of the X chromosome in the context of crossover control is important. Several known differences were reported between the X and the autosomes. First, the right third of the X chromosome has fewer heterochromatin marks such as H3K9me2/3 and nuclear membrane binding regions compared to the autosomes (Ikegami et al. 2010; Liu et al. 2011). Second, the X chromosome undergoes fewer DSBs than the autosomes (Gao et al. 2015). Further, both the timing of replication and the onset and completion of synapsis are delayed for the X chromosome in *C. elegans* (Jaramillo-Lambert et al. 2007; Mlynarczyk-Evans and Villeneuve 2017). In yeasts and plants, the regulation of DNA replication and induction of meiotic DSBs are well connected (Murakami and Nurse 2001; Higgins et al. 2012; Murakami and Keeney 2014). Whether the delay in replication is related to DSB induction on the X chromosomes in *C. elegans* remains to be analyzed. Now we can trace the fate of specific numbers and locations of DSBs by using an inducible single DSB system in *C. elegans* (Fig. 5A,B). Use of this system, in combination with high-resolution microscopy, single-nucleotide polymorphism (SNP) mapping, and sequence-based genomics approaches, will allow us to address the remaining questions described above.

CONCLUSION

Although crossover control has been a topic of extensive studies, the molecular mechanisms underlying its regulation are largely unknown. Technological advancements now allow for the introduction and analysis of a single (or more) crossover at specific genomic positions in metazoans. In addition to *Mos1* excision, a CRISPR (clustered regularly interspaced short palindromic repeat)-based DSB induction system will be established shortly. These approaches will unveil the positional effects of crossovers, allowing us to understand the origin of aneuploidies. This in turn may have clinical repercussions for treatments of infertility and in finding targets for cancer therapy in humans.

ACKNOWLEDGMENTS

We thank Marina Martinez Garcia for critical reading of this manuscript. This work was supported by the National Institutes of Health grant R01GM105853 to M.P.C. We apologize to authors whose work was not cited because of space constraints.

REFERENCES

Acquaviva L, Székvölgyi L, Dichtl B, Dichtl BS, de La Roche Saint André C, Nicolas A, Géli V. 2013. The COMPASS subunit Spp1 links histone methylation to initiation of meiotic recombination. *Science* **339:** 215–218.

Agostinho A, Meier B, Sonneville R, Jagut M, Woglar A, Blow J, Jantsch V, Gartner A. 2013. Combinatorial regulation of meiotic Holliday junction resolution in *C. elegans* by HIM-6 (BLM) helicase, SLX-4, and the SLX-1, MUS-81 and XPF-1 nucleases. *PLoS Genet* **9:** e1003591.

Barnes TM, Kohara Y, Coulson A, Hekimi S. 1995. Meiotic recombination, noncoding DNA and genomic organization in *Caenorhabditis elegans*. *Genetics* **141:** 159–179.

Baudat F, Buard J, Grey C, Fledel-Alon A, Ober C, Przeworski M, Coop G, de Massy B. 2010. PRDM9 is a major determinant of meiotic recombination hotspots in humans and mice. *Science* **327:** 836–840.

Beadle GW. 1932. A possible influence of the spindle fibre on crossing-over in *Drosophila*. *Proc Natl Acad Sci* **18:** 160–165.

Berchowitz LE, Francis KE, Bey AL, Copenhaver GP. 2007. The role of AtMUS81 in interference-insensitive crossovers in *A. thaliana*. *PLoS Genet* **3:** e132.

Bessereau JL, Wright A, Williams DC, Schuske K, Davis MW, Jorgensen EM. 2001. Mobilization of a *Drosophila* transposon in the *Caenorhabditis elegans* germ line. *Nature* **413:** 70–74.

Brenner S. 1974. The genetics of *Caenorhabditis elegans*. *Genetics* **77:** 71–94.

Buard J, Barthes P, Grey C, de Massy B. 2009. Distinct histone modifications define initiation and repair of meiotic recombination in the mouse. *EMBO J* **28:** 2616–2624.

Carlton PM, Farruggio AP, Dernburg AF. 2006. A link between meiotic prophase progression and crossover control. *PLoS Genet* **2:** e12.

C. elegans Sequencing Consortium. 1998. Genome sequence of the nematode *C. elegans*: A platform for investigating biology. *Science* **282:** 2012–2018.

Checchi PM, Lawrence KS, Van MV, Larson BJ, Engebrecht J. 2014. Pseudosynapsis and decreased stringency of meiotic repair pathway choice on the hemizygous sex chromosome of *Caenorhabditis elegans* males. *Genetics* **197:** 543–560.

Chen SY, Tsubouchi T, Rockmill B, Sandler JS, Richards DR, Vader G, Hochwagen A, Roeder GS, Fung JC. 2008. Global analysis of the meiotic crossover landscape. *Dev Cell* **15:** 401–415.

Choi K, Zhao X, Kelly KA, Venn O, Higgins JD, Yelina NE, Hardcastle TJ, Ziolkowski PA, Copenhaver GP, Franklin FC, et al. 2013. *Arabidopsis* meiotic crossover hot spots overlap with H2A.Z nucleosomes at gene promoters. *Nat Genet* **45:** 1327–1336.

Chua PR, Roeder GS. 1998. Zip2, a meiosis-specific protein required for the initiation of chromosome synapsis. *Cell* **93:** 349–359.

Chung G, Rose AM, Petalcorin MI, Martin JS, Kessler Z, Sanchez-Pulido L, Ponting CP, Yanowitz JL, Boulton SJ. 2015. REC-1 and HIM-5 distribute meiotic crossovers and function redundantly in meiotic double-strand break formation in *Caenorhabditis elegans*. *Genes Dev* **29:** 1969–1979.

Colaiácovo MP, MacQueen AJ, Martinez-Perez E, McDonald K, Adamo A, La Volpe A, Villeneuve AM. 2003. Synaptonemal complex assembly in *C. elegans* is dispensable for loading strand-exchange proteins but critical for proper completion of recombination. *Dev Cell* 5: 463–474.

Cole F, Kauppi L, Lange J, Roig I, Wang R, Keeney S, Jasin M. 2012. Homeostatic control of recombination is implemented progressively in mouse meiosis. *Nat Cell Biol* 14: 424–430.

Cooper KW. 1949. The cytogenetics of meiosis in *Drosophila*; mitotic and meiotic autosomal chiasmata without crossing over in the male. *J Morphol* 84: 81–121.

Cortes DB, McNally KL, Mains PE, McNally FJ. 2015. The asymmetry of female meiosis reduces the frequency of inheritance of unpaired chromosomes. *Elife* 4: e06056.

Crismani W, Girard C, Froger N, Pradillo M, Santos JL, Chelysheva L, Copenhaver GP, Horlow C, Mercier R. 2012. FANCM limits meiotic crossovers. *Science* 336: 1588–1590.

de los Santos T, Hunter N, Lee C, Larkin B, Loidl J, Hollingsworth NM. 2003. The Mus81/Mms4 endonuclease acts independently of double-Holliday junction resolution to promote a distinct subset of crossovers during meiosis in budding yeast. *Genetics* 164: 81–94.

De Muyt A, Jessop L, Kolar E, Sourirajan A, Chen J, Dayani Y, Lichten M. 2012. BLM helicase ortholog Sgs1 is a central regulator of meiotic recombination intermediate metabolism. *Mol Cell* 46: 43–53.

Fowler KR, Sasaki M, Milman N, Keeney S, Smith GR. 2014. Evolutionarily diverse determinants of meiotic DNA break and recombination landscapes across the genome. *Genome Res* 24: 1650–1664.

Gabdank I, Fire AZ. 2014. Gamete-type dependent crossover interference levels in a defined region of *Caenorhabditis elegans* chromosome V. *G3 (Bethesda)* 4: 117–120.

Gao J, Kim HM, Elia AE, Elledge SJ, Colaiácovo MP. 2015. NatB domain-containing CRA-1 antagonizes hydrolase ACER-1 linking acetyl-CoA metabolism to the initiation of recombination during *C. elegans* meiosis. *PLoS Genet* 11: e1005029.

Gerstein MB, Lu ZJ, Van Nostrand EL, Cheng C, Arshinoff BI, Liu T, Yip KY, Robilotto R, Rechtsteiner A, Ikegami K, et al. 2010. Integrative analysis of the *Caenorhabditis elegans* genome by the modENCODE project. *Science* 330: 1775–1787.

Giraut L, Falque M, Drouaud J, Pereira L, Martin OC, Mézard C. 2011. Genome-wide crossover distribution in *Arabidopsis thaliana* meiosis reveals sex-specific patterns along chromosomes. *PLoS Genet* 7: e1002354.

Hartmann MA, Sekelsky J. 2017. The absence of crossovers on chromosome 4 in *Drosophila melanogaster*: Imperfection or interesting exception? *Fly (Austin)* 20: 1–7.

Hatkevich T, Sekelsky J. 2017. Bloom syndrome helicase in meiosis: Pro-crossover functions of an anti-crossover protein. *Bioessays* 39. doi: 10.1002/bies.201700073.

Hayashi K, Yoshida K, Matsui Y. 2005. A histone H3 methyltransferase controls epigenetic events required for meiotic prophase. *Nature* 438: 374–378.

Henzel JV, Nabeshima K, Schvarzstein M, Turner BE, Villeneuve AM, Hillers KJ. 2011. An asymmetric chromosome pair undergoes synaptic adjustment and crossover redistribution during *Caenorhabditis elegans* meiosis: Implications for sex chromosome evolution. *Genetics* 187: 685–699.

Higashide M, Shinohara M. 2016. Budding yeast *SLX4* contributes to the appropriate distribution of crossovers and meiotic double-strand break formation on bivalents during meiosis. *G3 (Bethesda)* 6: 2033–2042.

Higgins JD, Buckling EF, Franklin FC, Jones GH. 2008. Expression and functional analysis of AtMUS81 in *Arabidopsis* meiosis reveals a role in the second pathway of crossing-over. *Plant J* 54: 152–162.

Higgins JD, Perry RM, Barakate A, Ramsay L, Waugh R, Halpin C, Armstrong SJ, Franklin FC. 2012. Spatiotemporal asymmetry of the meiotic program underlies the predominantly distal distribution of meiotic crossovers in barley. *Plant Cell* 24: 4096–4109.

Ho JW, Jung YL, Liu T, Alver BH, Lee S, Ikegami K, Sohn KA, Minoda A, Tolstorukov MY, Appert A, et al. 2014. Comparative analysis of metazoan chromatin organization. *Nature* 512: 449–452.

Holloway JK, Booth J, Edelmann W, McGowan CH, Cohen PE. 2008. MUS81 generates a subset of MLH1-MLH3-independent crossovers in mammalian meiosis. *PLoS Genet* 4: e1000186.

Hong Y, Sonneville R, Agostinho A, Meier B, Wang B, Blow JJ, Gartner A. 2016. The SMC-5/6 complex and the HIM-6 (BLM) helicase synergistically promote meiotic recombination intermediate processing and chromosome maturation during *Caenorhabditis elegans* meiosis. *PLoS Genet* 12: e1005872.

Hyppa RW, Smith GR. 2010. Crossover invariance determined by partner choice for meiotic DNA break repair. *Cell* 142: 243–255.

Ikegami K, Egelhofer TA, Strome S, Lieb JD. 2010. *Caenorhabditis elegans* chromosome arms are anchored to the nuclear membrane via discontinuous association with LEM-2. *Genome Biol* 11: R120.

Jagut M, Hamminger P, Woglar A, Millonigg S, Paulin L, Mikl M, Dello Stritto MR, Tang L, Habacher C, Tam A, et al. 2016. Separable roles for a *Caenorhabditis elegans* RMI1 homolog in promoting and antagonizing meiotic crossovers ensure faithful chromosome inheritance. *PLoS Biol* 14: e1002412.

Jantsch V, Pasierbek P, Mueller MM, Schweizer D, Jantsch M, Loidl J. 2004. Targeted gene knockout reveals a role in meiotic recombination for ZHP-3, a Zip3-related protein in *Caenorhabditis elegans*. *Mol Cell Biol* 24: 7998–8006.

Jaramillo-Lambert A, Ellefson M, Villeneuve AM, Engebrecht J. 2007. Differential timing of S phases, X chromosome replication, and meiotic prophase in the *C. elegans* germ line. *Dev Biol* 308: 206–221.

Jaramillo-Lambert A, Harigaya Y, Vitt J, Villeneuve A, Engebrecht J. 2010. Meiotic errors activate checkpoints that improve gamete quality without triggering apoptosis in male germ cells. *Curr Biol* 20: 2078–2089.

Kauppi L, Barchi M, Baudat F, Romanienko PJ, Keeney S, Jasin M. 2011. Distinct properties of the XY pseudoautosomal region crucial for male meiosis. *Science* 331: 916–920.

Keeney S, Giroux CN, Kleckner N. 1997. Meiosis-specific DNA double-strand breaks are catalyzed by Spo11, a member of a widely conserved protein family. *Cell* 88: 375–384.

Kelly KO, Dernburg AF, Stanfield GM, Villeneuve AM. 2000. *Caenorhabditis elegans msh-5* is required for both normal and radiation-induced meiotic crossing over but not for completion of meiosis. *Genetics* 156: 617–630.

Kleckner N, Zickler D, Jones GH, Dekker J, Padmore R, Henle J, Hutchinson J. 2004. A mechanical basis for chromosome function. *Proc Natl Acad Sci* 101: 12592–12597.

Knoll A, Higgins JD, Seeliger K, Reha SJ, Dangel NJ, Bauknecht M, Schropfer S, Franklin FC, Puchta H. 2012. The Fanconi anemia ortholog FANCM ensures ordered homologous recombination in both somatic and meiotic cells in *Arabidopsis*. *Plant Cell* 24: 1448–1464.

Libuda DE, Uzawa S, Meyer BJ, Villeneuve AM. 2013. Meiotic chromosome structures constrain and respond to designation of crossover sites. *Nature* 502: 703–706.

Liu T, Rechtsteiner A, Egelhofer TA, Vielle A, Latorre I, Cheung MS, Ercan S, Ikegami K, Jensen M, Kolasinska-Zwierz P, et al. 2011. Broad chromosomal domains of histone modification patterns in *C. elegans*. *Genome Res* 21: 227–236.

Liu EY, Morgan AP, Chesler EJ, Wang W, Churchill GA, Pardo-Manuel de Villena F. 2014. High-resolution sex-specific linkage maps of the mouse reveal polarized distribution of crossovers in male germline. *Genetics* 197: 91–106.

Lorenz A, Osman F, Sun W, Nandi S, Steinacher R, Whitby MC. 2012. The fission yeast FANCM ortholog directs non-crossover recombination during meiosis. *Science* 336: 1585–1588.

Lu LY, Yu X. 2015. Double-strand break repair on sex chromosomes: Challenges during male meiotic prophase. *Cell Cycle* 14: 516–525.

Lucchesi J, Suzuki DT. 1968. The interchromosomal control of recombination. *Annu Rev Genet* **2**: 53–86.

Macaisne N, Novatchkova M, Peirera L, Vezon D, Jolivet S, Froger N, Chelysheva L, Grelon M, Mercier R. 2008. SHOC1, an XPF endonuclease-related protein, is essential for the formation of class I meiotic crossovers. *Curr Biol* **18**: 1432–1437.

MacQueen AJ, Colaiácovo MP, McDonald K, Villeneuve AM. 2002. Synapsis-dependent and -independent mechanisms stabilize homolog pairing during meiotic prophase in *C. elegans*. *Genes Dev* **16**: 2428–2442.

Mahadevaiah SK, Turner JM, Baudat F, Rogakou EP, de Boer P, Blanco-Rodríguez J, Jasin M, Keeney S, Bonner WM, Burgoyne PS. 2001. Recombinational DNA double-strand breaks in mice precede synapsis. *Nat Genet* **27**: 271–276.

Martinez-Perez E, Colaiácovo MP. 2009. Distribution of meiotic recombination events: Talking to your neighbors. *Curr Opin Genet Dev* **19**: 105–112.

Martini E, Diaz RL, Hunter N, Keeney S. 2006. Crossover homeostasis in yeast meiosis. *Cell* **126**: 285–295.

Matthews AG, Kuo AJ, Ramón-Maiques S, Han S, Champagne KS, Ivanov D, Gallardo M, Carney D, Cheung P, Ciccone DN, et al. 2007. RAG2 PHD finger couples histone H3 lysine 4 trimethylation with V(D)J recombination. *Nature* **450**: 1106–1110.

Mazina OM, Mazin AV, Nakagawa T, Kolodner RD, Kowalczykowski SC. 2004. *Saccharomyces cerevisiae* Mer3 helicase stimulates 3'-5' heteroduplex extension by Rad51; implications for crossover control in meiotic recombination. *Cell* **117**: 47–56.

Meneely PM, McGovern OL, Heinis FI, Yanowitz JL. 2012. Crossover distribution and frequency are regulated by *him-5* in *Caenorhabditis elegans*. *Genetics* **190**: 1251–1266.

Mlynarczyk-Evans S, Villeneuve AM. 2017. Time-course analysis of early meiotic prophase events informs mechanisms of homolog pairing and synapsis in *Caenorhabditis elegans*. *Genetics* **207**: 103–114.

Murakami H, Keeney S. 2014. DDK links replication and recombination in meiosis. *Cell Cycle* **13**: 3621–3622.

Murakami H, Nurse P. 2001. Regulation of premeiotic S phase and recombination-related double-strand DNA breaks during meiosis in fission yeast. *Nat Genet* **28**: 290–293.

Muscat CC, Torre-Santiago KM, Tran MV, Powers JA, Wignall SM. 2015. Kinetochore-independent chromosome segregation driven by lateral microtubule bundles. *Elife* **4**: e06462.

Myers S, Bowden R, Tumian A, Bontrop RE, Freeman C, MacFie TS, McVean G, Donnelly P. 2010. Drive against hotspot motifs in primates implicates the *PRDM9* gene in meiotic recombination. *Science* **327**: 876–879.

Nabeshima K, Villeneuve AM, Colaiácovo MP. 2005. Crossing over is coupled to late meiotic prophase bivalent differentiation through asymmetric disassembly of the SC. *J Cell Biol* **168**: 683–689.

Nakagawa T, Ogawa H. 1999. The *Saccharomyces cerevisiae* MER3 gene, encoding a novel helicase-like protein, is required for crossover control in meiosis. *EMBO J* **18**: 5714–5723.

O'Neil NJ, Martin JS, Youds JL, Ward JD, Petalcorin MI, Rose AM, Boulton SJ. 2013. Joint molecule resolution requires the redundant activities of MUS-81 and XPF-1 during *Caenorhabditis elegans* meiosis. *PLoS Genet* **9**: e1003582.

Ottolini CS, Newnham L, Capalbo A, Natesan SA, Joshi HA, Cimadomo D, Griffin DK, Sage K, Summers MC, Thornhill AR, et al. 2015. Genome-wide maps of recombination and chromosome segregation in human oocytes and embryos show selection for maternal recombination rates. *Nat Genet* **47**: 727–735.

Parvanov ED, Petkov PM, Paigen K. 2010. Prdm9 controls activation of mammalian recombination hotspots. *Science* **327**: 835.

Peña PV, Davrazou F, Shi X, Walter KL, Verkhusha VV, Gozani O, Zhao R, Kutateladze TG. 2006. Molecular mechanism of histone H3K4me3 recognition by plant homeodomain of ING2. *Nature* **442**: 100–103.

Ramón-Maiques S, Kuo AJ, Carney D, Matthews AG, Oettinger MA, Gozani O, Yang W. 2007. The plant homeodomain finger of RAG2 recognizes histone H3 methylated at both lysine-4 and arginine-2. *Proc Natl Acad Sci* **104**: 18993–18998.

Ren H, Ferguson K, Kirkpatrick G, Vinning T, Chow V, Ma S. 2016. Altered crossover distribution and frequency in spermatocytes of infertile men with azoospermia. *PLoS One* **11**: e0156817.

Rockman MV, Kruglyak L. 2009. Recombinational landscape and population genomics of *Caenorhabditis elegans*. *PLoS Genet* **5**: e1000419.

Rosu S, Libuda DE, Villeneuve AM. 2011. Robust crossover assurance and regulated interhomolog access maintain meiotic crossover number. *Science* **334**: 1286–1289.

Saito TT, Youds JL, Boulton SJ, Colaiácovo MP. 2009. *Caenorhabditis elegans* HIM-18/SLX-4 interacts with SLX-1 and XPF-1 and maintains genomic integrity in the germline by processing recombination intermediates. *PLoS Genet* **5**: e1000735.

Saito TT, Mohideen F, Meyer K, Harper JW, Colaiácovo MP. 2012. SLX-1 is required for maintaining genomic integrity and promoting meiotic noncrossovers in the *Caenorhabditis elegans* germline. *PLoS Genet* **8**: e1002888.

Saito TT, Lui DY, Kim HM, Meyer K, Colaiácovo MP. 2013. Interplay between structure-specific endonucleases for crossover control during *Caenorhabditis elegans* meiosis. *PLoS Genet* **9**: e1003586.

Sears DD, Hegemann JH, Shero JH, Hieter P. 1995. Cis-acting determinants affecting centromere function, sister-chromatid cohesion and reciprocal recombination during meiosis in *Saccharomyces cerevisiae*. *Genetics* **139**: 1159–1173.

Shi X, Hong T, Walter KL, Ewalt M, Michishita E, Hung T, Carney D, Peña P, Lan F, Kaadige MR, et al. 2006. ING2 PHD domain links histone H3 lysine 4 methylation to active gene repression. *Nature* **442**: 96–99.

Shinohara M, Oh SD, Hunter N, Shinohara A. 2008. Crossover assurance and crossover interference are distinctly regulated by the ZMM proteins during yeast meiosis. *Nat Genet* **40**: 299–309.

Sigurdson DC, Herman RK, Horton CA, Kari CK, Pratt SE. 1986. An X-autosome fusion chromosome of *Caenorhabditis elegans*. *Mol Gen Genet* **202**: 212–218.

Smolikov S, Eizinger A, Hurlburt A, Rogers E, Villeneuve AM, Colaiácovo MP. 2007. Synapsis-defective mutants reveal a correlation between chromosome conformation and the mode of double-strand break repair during *Caenorhabditis elegans* meiosis. *Genetics* **176**: 2027–2033.

Smolikov S, Schild-Prüfert K, Colaiácovo MP. 2009. A yeast two-hybrid screen for SYP-3 interactors identifies SYP-4, a component required for synaptonemal complex assembly and chiasma formation in *Caenorhabditis elegans* meiosis. *PLoS Genet* **5**: e1000669.

Snowden T, Acharya S, Butz C, Berardini M, Fishel R. 2004. hMSH4-hMSH5 recognizes Holliday junctions and forms a meiosis-specific sliding clamp that embraces homologous chromosomes. *Mol Cell* **15**: 437–451.

Sommermeyer V, Béneut C, Chaplais E, Serrentino ME, Borde V. 2013. Spp1, a member of the Set1 complex, promotes meiotic DSB formation in promoters by tethering histone H3K4 methylation sites to chromosome axes. *Mol Cell* **49**: 43–54.

Sturtevant AH. 1915. Castle and Wright on crossing over in rats. *Science* **42**: 342.

Sturtevant A. 1919. Contributions to the genetics of *Drosophila melanogaster*. III. Inherited linkage variations in the second chromosome. *Carnegie Inst Wash Pub* **278**: 305–341.

Sun L, Wang J, Sang M, Jiang L, Zhao B, Cheng T, Zhang Q, Wu R. 2017. Landscaping crossover interference across a genome. *Trends Plant Sci* **22**: 894–907.

Svetlanov A, Baudat F, Cohen PE, de Massy B. 2008. Distinct functions of MLH3 at recombination hot spots in the mouse. *Genetics* **178**: 1937–1945.

Sym M, Roeder GS. 1994. Crossover interference is abolished in the absence of a synaptonemal complex protein. *Cell* **79**: 283–292.

Turner JM. 2007. Meiotic sex chromosome inactivation. *Development* **134:** 1823–1831.

Varas J, Sánchez-Morán E, Copenhaver GP, Santos JL, Pradillo M. 2015. Analysis of the relationships between DNA double-strand breaks, synaptonemal complex and crossovers using the Atfas1-4 mutant. *PLoS Genet* **11:** e1005301.

Vargas E, McNally K, Friedman JA, Cortes DB, Wang DY, Korf IF, McNally FJ. 2017. Autosomal trisomy and triploidy are corrected during female meiosis in *Caenorhabditis elegans*. *Genetics* **207:** 911–922.

Wagner CR, Kuervers L, Baillie DL, Yanowitz JL. 2010. xnd-1 regulates the global recombination landscape in *Caenorhabditis elegans*. *Nature* **467:** 839–843.

Wang S, Zickler D, Kleckner N, Zhang L. 2015. Meiotic crossover patterns: Obligatory crossover, interference and homeostasis in a single process. *Cell Cycle* **14:** 305–314.

Yelina NE, Choi K, Chelysheva L, Macaulay M, de Snoo B, Wijnker E, Miller N, Drouaud J, Grelon M, Copenhaver GP, et al. 2012. Epigenetic remodeling of meiotic crossover frequency in *Arabidopsis thaliana* DNA methyltransferase mutants. *PLoS Genet* **8:** e1002844.

Yokoo R, Zawadzki KA, Nabeshima K, Drake M, Arur S, Villeneuve AM. 2012. COSA-1 reveals robust homeostasis and separable licensing and reinforcement steps governing meiotic crossovers. *Cell* **149:** 75–87.

Youds JL, Mets DG, McIlwraith MJ, Martin JS, Ward JD NJ ON, Rose AM, West SC, Meyer BJ, Boulton SJ. 2010. RTEL-1 enforces meiotic crossover interference and homeostasis. *Science* **327:** 1254–1258.

Zakharyevich K, Tang S, Ma Y, Hunter N. 2012. Delineation of joint molecule resolution pathways in meiosis identifies a crossover-specific resolvase. *Cell* **149:** 334–347.

Zalevsky J, MacQueen AJ, Duffy JB, Kemphues KJ, Villeneuve AM. 1999. Crossing over during *Caenorhabditis elegans* meiosis requires a conserved MutS-based pathway that is partially dispensable in budding yeast. *Genetics* **153:** 1271–1283.

Zetka MC, Rose AM. 1995. Mutant *rec-1* eliminates the meiotic pattern of crossing over in *Caenorhabditis elegans*. *Genetics* **141:** 1339–1349.

Zhang L, Wang S, Yin S, Hong S, Kim KP, Kleckner N. 2014. Topoisomerase II mediates meiotic crossover interference. *Nature* **511:** 551–556.

Oocyte Quality Control: Causes, Mechanisms, and Consequences

Neil Hunter[1,2,3,4]

[1]Howard Hughes Medical Institute, University of California, Davis, Davis, California 95616

[2]Department of Microbiology and Molecular Genetics, University of California, Davis, Davis, California 95616

[3]Department of Molecular and Cellular Biology, University of California, Davis, Davis, California 95616

[4]Department of Cell Biology and Human Anatomy, University of California, Davis, Davis, California 95616

Correspondence: nhunter@ucdavis.edu

Oocyte quality and number are key determinants of reproductive life span and success. These variables are shaped in part by the elimination of oocytes that experience problems during the early stages of meiosis. Meiotic prophase-I marks an extended period of genome vulnerability in which epigenetic reprogramming unleashes retroelements and hundreds of DNA double-strand breaks (DSBs) are inflicted to initiate the programmed recombination required for accurate chromosome segregation at the first meiotic division. Expression of LINE-1 retroelements perturbs several aspects of meiotic prophase and is associated with oocyte death during the early stages of meiotic prophase I. Defects in chromosome synapsis and recombination also trigger oocyte loss, but typically at a later stage, as cells transition into quiescence and form primordial follicles. Interrelated pathways that signal defects in DSB repair and chromosome synapsis mediate this late oocyte attrition. Here, I review our current understanding of early and late oocyte attrition based on studies in mouse and describe how these processes appear to be both distinct and overlapping and how they help balance the quality and size of oocyte reserves to maximize fecundity.

The common fate of a mammalian oocyte is an early death, with ~80% of human oocytes being lost before or shortly after birth (Fig. 1; Baker 1963; Kurilo 1981). The size and quality of the surviving pool of primordial follicles are important determinants of female fecundity and reproductive life span (Broekmans et al. 2007). Oogenesis begins during fetal development following the establishment of primordial germ cells in the undifferentiated gonads. The ensuing oogonia expand by mitosis such that very large numbers of primary oocytes enter meiosis, approximately six to seven million in humans. However, by birth, oocyte numbers have already crashed down to approximately one to two million and at the onset of puberty only approximately 200,000 to 300,000 remain (Fig. 1; Block 1953; Baker 1963; Forabosco et al. 1991). This finite ovarian reserve (Gleicher et al. 2011) comprises nongrowing primordial follicles arrested in the dictyate stage of meiosis, before the first meiotic division. Ovarian reserves are continually depleted through ongoing recruitment of primordial follicles to the growing follicle pool, and cyclical follicle-stimulating hormone (FSH)-dependent activation of cohorts of antral follicles to reenter meiosis (McGee and Hsueh 2000). For both recruitment and activation phases, the default outcome is again cell death through an apoptotic process termed atresia (Kaipia and Hsueh 1997), with only the dominant ovulatory follicle(s) completing the meiosis I division. Thus, only approximately 350 oocytes will escape cell death and be ovulated during the human reproductive life span, corresponding to <0.006% of the 6–7 million potential eggs initially formed during fetal development. Follicle depletion is associated with reduced production of the hormones estrogen and inhibin by the ovary, disrupting the hypothalamic–pituitary–gonadal (HPG) hormonal axis and eventually leading to menopause (Honour 2018). Thus, follicle depletion serves as a timer for the major landmarks of female reproduction (Broekmans et al. 2007). Moreover, genetic and environmental factors that influence the size of initial ovarian reserves and rates of follicle recruitment can significantly alter reproductive life span (Gleicher et al. 2011; Tilly and Sinclair 2013; Aiken et al. 2015; Findlay et al. 2015; Grive and Freiman 2015; Laven 2015, 2016; Laven et al. 2016).

The causes, mechanisms, and roles of the massive oocyte culling that occurs during fetal and early postnatal life have been the subject of much study and debate (Tilly 2001; Hartshorne et al. 2009). However, cumulative evidence from studies in mouse indicates that much of the oocyte death during this period is the result of quality control processes that eliminate potentially defective cells and nurture the cells that will survive (Di Giacomo et al. 2005; Lei and Spradling 2013, 2016; Malki et al. 2014). Two major stages of oocyte loss can be inferred in mouse (Fig. 2). Early oocyte attrition (EOA) occurs between embryonic days E15.5 and E8.5 and causes the loss of ~50% of all oocytes (Malki et al. 2014). Late oocyte attrition (LOA) follows in the early postnatal period as cells transition into quiescence to establish the pool of

Published by Cold Spring Harbor Laboratory Press; doi: 10.1101/sqb.2017.82.035394

Cold Spring Harbor Symposia on Quantitative Biology, Volume LXXXII

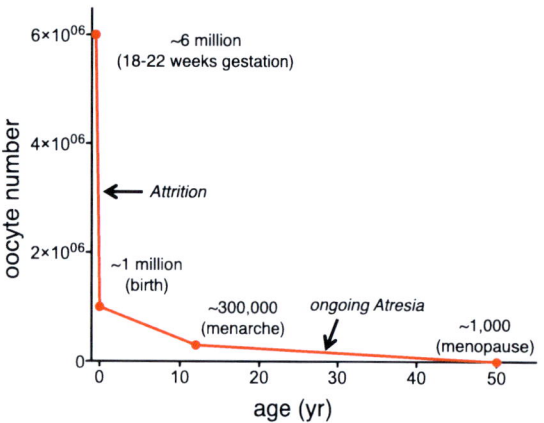

Figure 1. Oocyte decline in human fetal ovaries. Approximately 80% of the oocytes that enter meiosis are culled by birth and less than half of those remaining survive through puberty. Through ongoing atresia, following recruitment and hormonal activation, oocyte numbers continually decline until the hypothalamic–pituitary–gonadal (HPG) hormonal axis can no longer be supported and menopause ensues.

related with the activation of LINE-1 transposons (Malki et al. 2014), whereas LOA appears to be a response to errors in meiotic prophase I (Fig. 2; Di Giacomo et al. 2005; Bolcun-Filas et al. 2014; Rinaldi et al. 2017a). However, oocyte attrition is also generally associated with developmental processes in the nascent ovary (Pepling and Spradling 1998; Lei and Spradling 2013, 2016). These processes begin with the formation of clonal cysts comprising approximately 30 primordial germ cells interconnected in a syncytium. Coincident with meiotic prophase I, cysts then break down and reorganize into nonclonal nests. Concurrently, interconnected cells are differentiating into oocytes targeted for survival and nurse-like cells that nurture the developing oocytes by donating organelles and cytoplasm (Lei and Spradling 2016). Analogous to the well-characterized nurse cells of the *Drosophila* ovary (Jenkins et al. 2013), this self-sacrifice of mouse nurse-like cells concludes in programmed cell death prior to follicle formation.

DEREPRESSION OF LINE-1 ELEMENTS AND EARLY OOCYTE ATTRITION

Long interspersed element 1, LINE-1 (L1), is the only autonomous retroelement that remains active in the human genome (Cordaux and Batzer 2009). It belongs to the non–

resting follicles (Di Giacomo et al. 2005; Klinger et al. 2015; Qiao et al. 2018). Not only are these two stages of oocyte death temporally distinct, but they also appear to have different underlying causes. In mouse, EOA is cor-

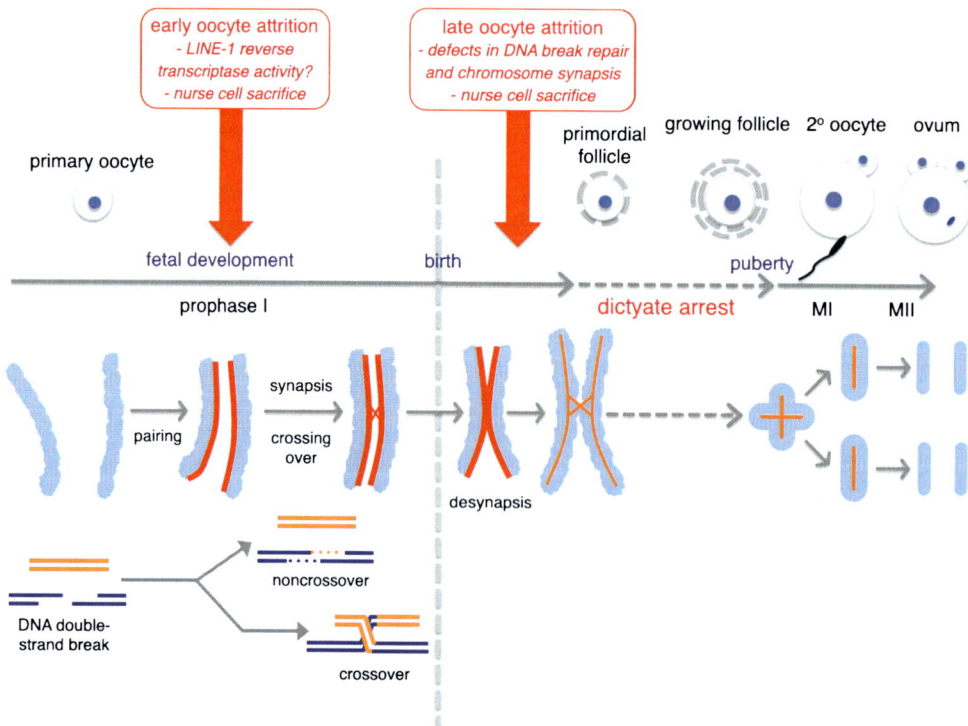

Figure 2. Time line of oocyte meiosis in mouse highlighting nuclear (*top*), chromosomal (*middle*), and recombination (*bottom*) landmarks. Pairing, synapsis, and crossing-over between homologous chromosomes are facilitated by programmed homologous recombination initiated by hundreds of DNA double-strand breaks. Early oocyte attrition (EOA) occurs during early meiotic prophase-I and is correlated with LINE-1 activity. Late oocyte attrition (LOA) occurs after homologs desynapse and oocytes begin to transition into dictyate arrest. LOA can be triggered by unrepaired DNA breaks and transcriptional silencing induced at sites where synapsis has failed. Meiosis only resumes in oocytes that have been recruited for growth and activated by follicle-stimulating hormone (FSH). Only the dominant ovulatory follicle will complete the meiosis-I division before arresting in metaphase II until fertilization.

long terminal repeat retrotransposon class of transposable elements that also includes nonautonomous *Alu* and SVA elements and makes up around one-third of the human genome. L1 alone accounts for ~17% of the human genome—more than 500,000 copies—but less than 100 copies are thought to remain functional (Brouha et al. 2003). These full-length L1 elements are ~6 kb in length and comprise a 5′ UTR containing an RNA Pol II promoter, two open reading frames (ORFs), and a 3′ UTR that terminates with a poly(A) tail and contains a polyadenylation signal (Babushok and Kazazian 2007). The ORFs encode activities required for transposition: ORF1 is a nucleic acid chaperone, and ORF2 has endonuclease and reverse transcriptase activities.

L1 transcripts are exported from the nucleus and translated. The ORF1 and ORF2 proteins bind with a strong *cis* bias to the mRNA that encoded them, assembling into ribonucleoprotein (RNP) complexes that reenter the nucleus to mediate transposition. In a process termed target-site-primed reverse transcription, the ORF2 endonuclease nicks the target-site DNA at the consensus sequence (5′-TTTT/AA-3′). The resulting 5′-TTTT-3′-OH strand anneals to the poly(A)-tail of the associated L1 mRNA to prime the ORF2 reverse transcriptase (Cost et al. 2002). Nicking of the second DNA strand of the target DNA is thought to provide a primer for second strand synthesis to complete L1 integration. Although the ORF1 and ORF2 proteins tend to bind to their cognate L1 mRNA, trans-activation of abundant *Alu* and SVA nonautonomous retroelements and other cellular mRNAs can occur (Esnault et al. 2000; Dewannieux et al. 2003; Ostertag et al. 2003; Beck et al. 2011).

Consequences of LINE-1 Activity

The activities of L1 and other transposable elements have had profound effects on cellular function, genome stability, and evolution. *Cis* and *trans* activity of L1 is a major source of insertional mutagenesis, both somatic and germline, and is implicated in numerous diseases including cancer (Cordaux and Batzer 2009). Over the course of human evolution, insertion of more than 8000 human processed pseudogenes and more than a million nonautonomous retroelements are attributed to L1 activity (Zhang et al. 2003; Vinckenbosch et al. 2006).

ORF2 endonuclease activity can serve as a source of genome instability by generating DNA double-strand breaks (DSBs) uncoupled from successful transposition (Belgnaoui et al. 2006; Gasior et al. 2006). L1 insertion can also bypass the need for the ORF2 endonuclease by hijacking endogenous DSBs (Morrish et al. 2002). Although this endonuclease-independent insertion may facilitate DSB repair, it also has the potential to interfere with endogenous repair pathways as well as causing potentially deleterious insertions. Both endonuclease-dependent and -independent insertion has been associated with local DNA rearrangements including deletions, duplications, inversions, and translocations (Belancio et al. 2008). L1 itself is a source of unstable microsatellite sequences and insertion generates additional poly(A) microsatellites de novo (Grandi and An 2013).

L1 insertion also influences transcription, both positively and negatively. An antisense promoter in the 5′ UTR can drive transcription of flanking cellular genes and appears to be broadly used in tissue-specific gene regulation (Nigumann et al. 2002). Oppositely, as targets of heterochromatin formation, L1 elements can locally silence gene expression (Slotkin and Martienssen 2007). Intragenic L1 elements can impede transcription (Han et al. 2004) and serve as sense and antisense promoters (Faulkner et al. 2009), alternative splice junctions (Belancio et al. 2006, 2008), and termination sites (Lee et al. 2008). More generally, L1 expression is associated with cellular responses such as activation of the innate-immune response (Crow 2010; Goodier et al. 2015) and the DNA-damage response (Belgnaoui et al. 2006) and their downstream effects such as cell cycle arrest, senescence, and apoptosis (Belgnaoui et al. 2006; Gasior et al. 2006; Wallace et al. 2008).

Multiple Layers of LINE-1 Repression

Given the potentially catastrophic effects of L1 activity, cells have evolved defense mechanisms to interfere with each step of its life cycle (Goodier 2016; Pizarro and Cristofari 2016; Liu et al. 2018). At the level of initiation, the L1 promoter is suppressed by CpG DNA methylation (Hata and Sakaki 1997; Bourc'his and Bestor 2004) and the 5′ UTR has come under the control of host transcription factors (Tchenio et al. 2000; Yang et al. 2003; Athanikar et al. 2004). Consistently, L1 transcriptional repression also involves a variety of histone modifications (Castañeda et al. 2011) including de novo H3 Lys9 trimethylation by the "human silencing hub" complex (Liu et al. 2018). Involvement of the SMC complex condensin in restricting retrotransposons, including L1, also indicates a repressive role for higher-order chromatin structure (Schuster et al. 2013; Ward et al. 2017). piwi-interacting small RNA (piRNA) biogenesis is a particularly important mechanism for silencing transposable elements, including L1, in the germline and is also required for their de novo remethylation (Aravin et al. 2008; Kuramochi-Miyagawa et al. 2008; Castañeda et al. 2011). Posttranscriptionally, RNA interference, degradation, premature polyadenylation, and editing can attenuate LINE-1 activity (Perepelitsa-Belancio and Deininger 2003; Yang and Kazazian 2006; Schumann 2007; Zhang et al. 2014; Hamdorf et al. 2015; Orecchini et al. 2018).

At the posttranslational level, the ubiquitin-proteasome system targets the L1ORF1 protein via TEX19.1 and the E3-ligase UBR2 (MacLennan et al. 2017). A DNA exonuclease, TREX1, is thought to destroy the reverse transcribed cDNA strand of L1 (Stetson et al. 2008). TREX1 also has an exonuclease-independent function that reduces L1 ORF1 protein levels (Li et al. 2017). SAMHD1, a dNTP triphosphohydrolase, is inferred to impede nuclear import of L1 RNPs by enhancing their sequestration in stress granules (Hu et al. 2015), cytoplasmic structures that accumulate untranslated mRNAs when cells are stressed (Sheinberger and Shav-Tal 2017). TREX1, SAMHD1, and several other repressors of L1 activity are downstream components of the interferon (IFN) response pathway (Goodier et al. 2015). Involvement of these innate restriction factors implies that L1 induces an innate immune response, perhaps through detection of L1 RNA: DNA hybrids or L1 encoded proteins (Crow 2010; Rigby et al. 2014). Host DNA repair factors have also been implicated in suppression of late steps of L1 integration (Pizarro and Cristofari 2016; Servant et al. 2017; Liu et al. 2018). Notably, the ERCC1-XPF endonuclease has been proposed to remove the branched intermediates of L1ORF2 reverse transcription (Gasior et al. 2008).

Epigenetic Reprogramming Is Coincident with Meiosis in Females

As primordial germ cells migrate, DNA methylation (5 mC) is removed throughout the genome as a primary step in the global epigenetic reprogramming required for gametogenesis and ensuing embryogenesis (Seisenberger et al. 2012; Messerschmidt et al. 2014). In oocytes, the demethylated state is sustained throughout fetal development and only restored after birth. As such, meiotic prophase-I in females occurs in the context of a globally hypomethylated genome. In contrast, males restore DNA methylation at the prospermatogonial stage during the perinatal period, long before meiosis initiates. Hypomethylation causes a burst of L1 expression. In mouse, although LINE-1 RNA is detectable in primordial germ cells, it may not be translated until oocytes are in meiotic prophase (Trelogan and Martin 1995; Seisenberger et al. 2012). Also, intriguingly, although L1 expression peaks during meiosis, retrotransposition appears to occur mainly during embryogenesis and may be mediated by L1 RNA that was transmitted via the gametes (Kano et al. 2009).

Along with DNA demethylation, the global changes in chromatin organization and transcriptional reprogramming that characterize meiotic prophase may render primary oocytes uniquely disposed to high levels of L1 expression. Early meiotic prophase is characterized by interactions between pairs of homologous chromosomes (homologs) and their physical connection by crossovers (Fig. 2; Hunter 2015). Following S phase, meiotic chromosomes organize into stereotypic structures comprising linear arrays of chromatin loops, the bases of which organize into cores or axes that defines the interfaces for homologous interactions (Zickler and Kleckner 1999, 2015). Importantly, this global, semicompact loop-axis organization supersedes the higher-order structures (chromatin loops, topologically associating domains, and epigenetic compartments) that mediate transcriptional regulation in somatic cells (Dekker and Mirny 2016; Hansen et al. 2018). Other chromatin changes such as nucleosome composition and histone modifications, including those associated with meiotic recombination, could also enhance L1 activity during meiosis (Saitou et al. 2012; Ng et al. 2013; Székvölgyi et al. 2015; Izquierdo-Bouldstridge et al. 2017).

LINE-1 Derepression during Oogenesis Is Associated with Multiple Defects

The major events of meiotic prophase—homolog pairing, synapsis, and crossing-over—are mediated by programmed homologous recombination initiated by DNA breakage (Fig. 2; Lam and Keeney 2014; Hunter 2015). During the leptotene stage, SPO11 protein catalyzes DSB formation, inflicting on the order of 200–300 DNA DSBs per nucleus in mouse and human (Cole et al. 2012; Gruhn et al. 2013). Ensuing interhomolog DNA pairing and strand exchange bring chromosomes into close juxtaposition, enabling formation of synaptonemal complexes (SCs), densely packed transverse filaments that connect homologous chromosomes along their lengths (Zickler and Kleckner 1999). SCs form during zygotene and crossovers form in the context of fully synapsed SCs during the pachytene stage. Crossover formation is tightly regulated such that each pair of chromosomes becomes connected by at least one exchange, as required for accurate segregation at the meiosis-I division. Homologs then desynapse and cells enter diplotene. At this stage, oocytes then transition onto the protracted dictyate stage and become surrounded by a single layer of supporting granulosa cells to establish the reserve of primordial follicles (Fig. 2). After puberty, meiosis resumes in FSH-activated follicles and meiosis-I ensues in the dominant ovulatory cell.

L1ORF1 protein is readily detected in the nuclei of mouse fetal oocytes during early prophase stages, and striking cell-to-cell variation is observed implying stochastic variation in L1 activity (Malki et al. 2014). However, cytoplasmic transfer from syncytial nurse-like cells could help lower L1 activity in differentiating oocytes and contribute to the observed variation in L1ORF1 levels (Lei and Spradling 2016). Several lines of evidence point to a causal role for L1 in EOA and imply that oocytes with excessive L1 expression are targeted for killing (Malki et al. 2014): (i) EOA is enhanced when the piRNA pathway is defective; (ii) ORF1 protein levels correlate with oocyte survival; (iii) expression of an L1 transgene enhances EOA; and (iv) the nucleoside analog azidothymidine AZT, which presumptively inhibits of L1ORF2 reverse transcriptase activity (Jones et al. 2008; Dai et al. 2011), prevents EOA at normal times (between embryonic day E13.5 and E18.5), suggesting that a reverse transcriptase intermediate triggers EOA during this period (Fig. 3). However, oocyte death is not permanently rescued by AZT, and by 2 days postpartum (dpp), numbers drop to levels seen in untreated controls, suggesting that reverse transcriptase–independent activities of L1 can contribute to LOA (see below).

L1 expression during meiosis is associated with the formation of SPO11-independent DSBs (Soper et al. 2008; Carofiglio et al. 2013) and provokes a variety of prophase errors including persistent DNA damage, synapsis defects, reduced crossing-over, and elevated chromosome missegregation (Fig. 3; Malki et al. 2014). The DNA repair and synapsis defects are not relieved by AZT treatment consistent with the possibility that they result from

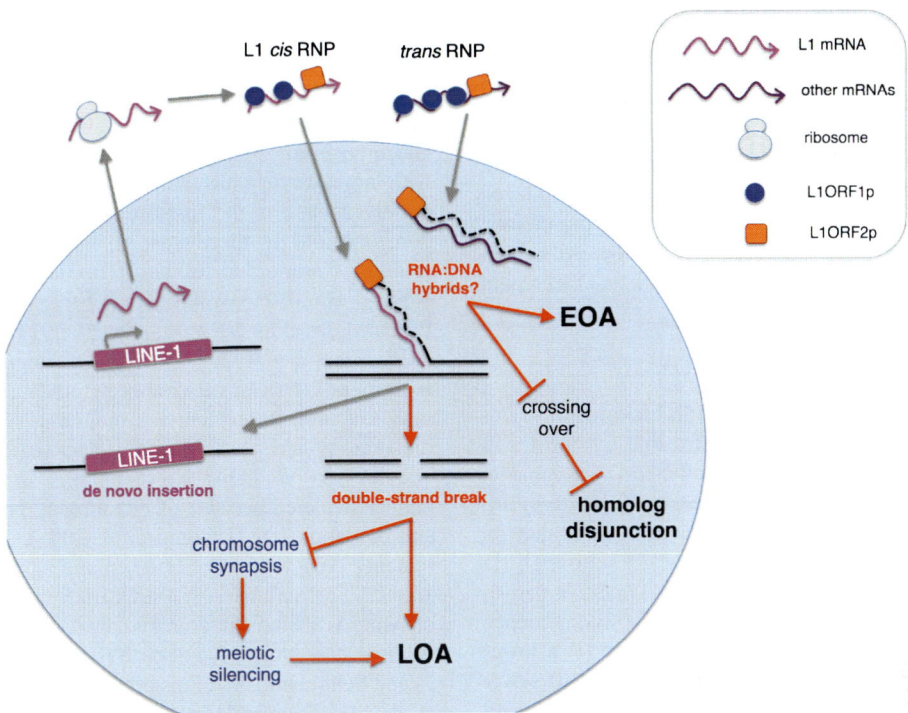

Figure 3. LINE-1 activities during early prophase-I perturb meiosis and may trigger early oocyte attrition (EOA) and late oocyte attrition (LOA). DNA demethylation leads to derepression of full-length LINE-1 (L1) elements during oocyte meiosis. L1ORF1 and ORF2 proteins are translated and assemble into RNP complexes, showing a string *cis* bias for the encoding L1 mRNA, but transactivation of non-L1 mRNAs can also occur. Target site nicking by the L1ORF2 enodonuclease primes the L1ORF2 reverse transcriptase to synthesize an RNA:DNA hybrid intermediate. Nicking of the other strand and second-strand synthesis leads to L1 integration. RNA:DNA hybrids trigger EOA and interfere with crossing-over, elevating the risk of chromosome missegregation. Aberrant L1 activity may also lead to double-strand break formation and defective synapsis leading to LOA.

DNA damage inflicted by the L1ORF2 endonuclease. Intriguingly, AZT does suppress the crossover deficit associated with excessive L1 expression. In fact, wild-type oocytes make ~10% more crossovers when treated with AZT, indicating that L1ORF2 reverse transcriptase activity somehow interferes with the process of meiotic crossing-over (Malki et al. 2014).

Notably, meiosis in human females shows a high degree of heterogeneity relative to males, with frequent synaptic errors and highly variable crossover numbers (Hassold and Hunt 2001; Lenzi et al. 2005; Tease et al. 2006; Hassold et al. 2007; Hunt and Hassold 2008; Gruhn et al. 2013). Moreover, crossover patterning in females is suboptimal, producing outcomes that are at risk for chromosome missegregation, such as nonexchange chromosomes and chromosomes with a single crossover located close to a centromere or telomere (Cheng et al. 2009; Nagaoka et al. 2012; Herbert et al. 2015). A key factor underlying this defect is inefficient maturation of crossovers, which is inferred to occur with only ~75% efficiency in females (Wang et al. 2017). The association between L1 activity, increased synaptic errors, and decreased crossover formation in mouse (Malki et al. 2014) raises the possibility that L1 and/or other transposable elements contribute to the inefficient crossover maturation and elevated aneuploidy seen in human females.

In sum, analysis in mouse suggests that L1 activity induces EOA via a reverse transcriptase–dependent inter-

mediate, although direct evidence for this inference remains an important goal (Fig. 3; Malki et al. 2014). One possibility is that L1-catalyzed RNA:DNA hybrids trigger an innate immune response. Alternatively, L1-mediated RNA:DNA hybrids may interfere with protein translation as observed in human platelets (Schwertz et al. 2018). Also, whether EOA occurs via canonical apoptotic pathways and its relationship to nurse cell sacrifice remain unclear.

By inflicting additional DNA damage on top of the hundreds of programmed DSBs, L1 activity could push damage signaling above a threshold not normally reached during early prophase and sufficient to induce cell death. However, the AZT sensitivity of EOA suggests that L1-induced DNA damage is not the cause of early oocyte death. In fact, the early prophase period of oocyte meiosis appears to be relatively insensitive to DNA damage–induced apoptosis. For example, recombination mutants that fail to repair DSBs, even those with elevated DSB levels such as $Atm^{-/-}$, induce oocyte culling at the pre-follicle stage (Di Giacomo et al. 2005), not during early prophase as seen for L1-associated EOA. Also, zygotene/pachytene-stage oocytes are relatively resistant to irradiation-induced apoptosis compared to those in diplotene (Hanoux et al. 2007; Kim and Suh 2014; Hunter N, unpubl. data). It makes sense that oocytes should suppress DNA damage–induced apoptosis pathways during these stages while hundreds of programmed meiotic

DSBs are processed. Consistently, oocyte culling in AZT-treated fetal ovaries now occurs in the early postnatal period (Fig. 3; Malki et al. 2014), during the diplotene-to-dictyate transition when programmed DSBs have mostly been repaired and cells become sensitive to DNA damage–induced apoptosis (Hanoux et al. 2007). This observation raises the possibility that L1-induced DNA lesions are not efficiently repaired by the meiotic recombination machinery, or that some other AZT-independent activity of L1 interferes with meiotic DSB repair.

Advantages of EOA

In early meiotic prophase, the high background of ongoing DSB repair may make it unfeasible for oocytes to monitor L1 activity based on L1-inflicted DNA damage. The ability of oocytes to detect excessive L1 activity independently of DNA damage overcomes this challenge, enabling selective early culling of oocytes with high L1 activity. Despite the associated payoff of massive oocyte loss, follicles selected for low L1 activity likely experience a several-fold advantage because they have (i) a lower probability to experience de novo insertion; (ii) fewer SPO11-independent DSBs that could lead to LOA and other defects (see below) (Carofiglio et al. 2013; Rinaldi et al. 2017a); (iii) fewer synapsis defects and associated meiotic silencing, which can also cause oocyte death (see below) (Royo et al. 2013); (iv) higher crossover levels and thus a lower chance of homolog missegregation at meiosis I and aneuploidy in the resulting zygote; (v) a lower probability to transmit L1 RNA to the zygote and inflict insertion during embryogenesis (Kano et al. 2009); and (vi) possibly, a higher capacity to silence L1 postnatally and in the next generation. It should be noted that EOA occurs around the time of cytoplasmic transfer between interconnected sister oocytes and the breakdown of cysts (Lei and Spradling 2013, 2016). The relationship between the level of L1 activity and the differentiation of oocytes that will survive and the nurse cells that will nurture them and then die is unknown and an important question for the future.

Potential Advantages of L1 Derepression in the Germline

It is also speculated that DNA demethylation and the resulting derepression of L1 elements play positive roles for germ cell differentiation, meiosis, and genotype diversity during development (Kano et al. 2009; van der Heijden and Bortvin 2009; Castañeda et al. 2011; Chuma 2014; Malki et al. 2014). For example, transcriptional regulation mediated by insertionally inactive L1 elements could be important for the meiotic program. The sheer abundance of L1 elements and their enriched association with chromosome axes and SCs (Pearlman et al. 1992; Hernández-Hernández et al. 2008) raises the possibility they help organize meiotic chromosome structure and facilitate homolog recognition and pairing (van der Heijden and Bortvin 2009).

DNA DAMAGE AND SYNAPSIS DEFECTS TRIGGER LOA

LOA was originally delineated in meiotic mutants defective for DSB repair and/or homolog synapsis (Di Giacomo et al. 2005). Such mutants are typically born with large oocyte pools (at least 50% of wild type), which rapidly decline in the first few days after birth, as cells transition into the dictyate stage, arrest, and establish the reserve of primordial follicles (Di Giacomo et al. 2005; Kogo et al. 2012a; Wojtasz et al. 2012; Kerr et al. 2013; Bolcun-Filas et al. 2014; Malki et al. 2014; Cloutier et al. 2015; Rinaldi et al. 2017a; Hunter N, unpubl. data). The severity of LOA can vary between mutants. For example, in Spo11 mutants, which lack programmed DSBs, chromosome synapsis is severely defective, but ~15% of oocytes survive LOA. However, this small reserve is rapidly depleted because of recruitment, ensuing atresia, and ovulation (Di Giacomo et al. 2005). In contrast, in mutants such as Dmc1, Msh4, Msh5, Atm, Trip13, and Mcmdc2, recombination is initiated, but DSB repair and/or synapsis are defective, and wholesale culling of oocytes is observed within 2–5 d of birth (Di Giacomo et al. 2005; Li and Schimenti 2007; Finsterbusch et al. 2016; McNairn et al. 2017; Hunter N, unpubl. data). Spo11 mutation is epistatic to DSB repair mutants with respect to the severity of oocyte loss, indicating that defective repair of SPO11-dependent DSBs is a potent trigger of LOA.

A Single DNA Damage Response Pathway May Trigger LOA Caused by Meiotic Defects

The observations described above led to the concept of two distinct checkpoint processes that can lead to LOA, one that monitors DSB repair and another that responds to defective synapsis (Di Giacomo et al. 2005; Wojtasz et al. 2012; Bolcun-Filas et al. 2014; Cloutier et al. 2015). However, more recent studies have revealed the existence of a significant population of SPO11-independent DSBs that can also trigger oocyte death via the DNA damage response (Carofiglio et al. 2013; Rinaldi et al. 2017a). This class of DSBs is highly variable in number, mirroring the variable expression of the L1ORF1 protein and consistent with the idea that a possible source of SPO11-independent DSBs is DNA cleavage initiated by the L1ORF2 endonuclease (Malki et al. 2014). Thus, a single pathway that signals persistent DNA breaks, regardless of their cause, may account for the majority of LOA (Rinaldi et al. 2017a). This meiotic DNA damage response pathway is mediated by a chromosome axis-based, kinase-signaling cascade whose key components include the proximal kinase ATR and the effector kinase CHK2 (Fig. 4; Bolcun-Filas et al. 2014; Subramanian and Hochwagen 2014; Rinaldi et al. 2017a). CHK2 activates the apoptotic regulators p53 and an isoform of p63 (a p53 paralog) called TAp63 (the Trans-Activation isoform), which appear to trigger apoptosis through canonical pathways (Morita et al. 1999; Kerr et al. 2012; Bolcun-Filas et al. 2014; Klinger et al. 2015; Omari et al. 2015). TAp63 is first expressed in late pachytene and diplotene oocytes and re-

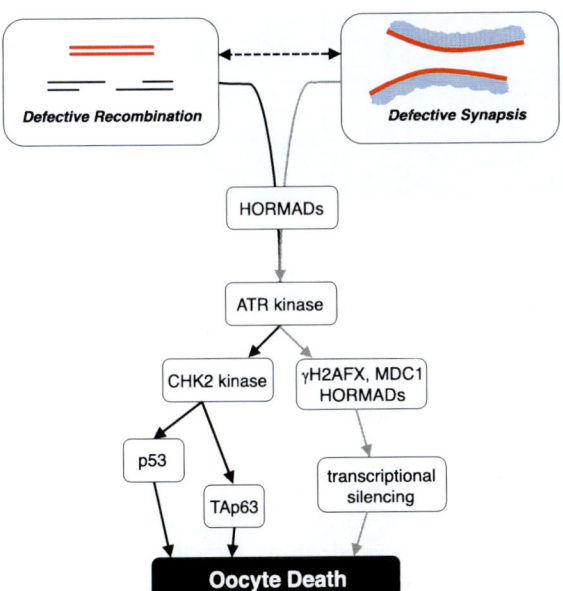

Figure 4. Meiotic DNA damage response and transcriptional silencing pathways. Double-strand breaks and defective synapsis are signaled via the HORMADs and the ATR kinase. ATR activates the effector kinase CHK2 to induce apoptosis via p53 and TAp63. ATR also mediates transcriptional silencing by nucleating and amplifying H2AFX phosphorylation in conjunction with MCD1 and the HORMAD proteins. The dashed line indicates the interdependence between defective recombination and defective chromosome synapsis.

mains constitutively expressed in resting follicles conferring exquisite sensitivity to DNA damage (Suh et al. 2006; Livera et al. 2008; Bolcun-Filas et al. 2014; Kim and Suh 2014). p53 may be the primary mediator of apoptosis during the diplotene-to-dictyate transition, with TAp63 largely superseding this function in quiescent oocytes. However, a more complex regulatory relationship between p53 and TAp63 is also suggested (Bolcun-Filas et al. 2014).

A key observation implicating the meiotic DNA damage response in LOA is the striking rescue of oocyte death and infertility in *Trip13* mutant females by *Chk2* mutation (Bolcun-Filas et al. 2014). However, rescue is incomplete, suggesting that other kinases may partially substitute for CHK2 function, especially when residual DSB levels are high; candidates include CHK1, DNA-PK, or direct signaling by ATR or the related PI3K-like kinase, ATM. Notably, the efficiency of oocyte rescue by *Chk2* mutation negatively correlates with the number of residual DSBs present in different mutant backgrounds and/or as a result of ionizing irradiation (Rinaldi et al. 2017a). These observations point to a damage-threshold model for oocyte elimination in which oocytes with at least 10 residual DSBs are culled.

Meiosis-Specific HORMA-Domain Proteins Facilitate LOA

Other key players in LOA include HORMAD1 and HORMAD2, two members of a conserved family of meiosis-specific HORMA (Hop1, Rev7, and Mad2)-domain proteins (HORMADs) (Daniel et al. 2011; Kogo

et al. 2012b; Shin et al. 2013). HORMADs play central roles in regulating the major events of meiotic prophase-I, including meiotic DSB formation, homolog pairing, and synapsis, checkpoint signaling, transcriptional silencing (described below), and biasing meiotic recombination to occur between homologs by impeding intersister DSB repair (Carballo et al. 2008; Shin et al. 2010; Royo et al. 2013; Vader and Musacchio 2014; Stanzione et al. 2016; Rinaldi et al. 2017a). As discussed below, the latter two functions appear to be central to the role of the HORMADs in promoting LOA (Rinaldi et al. 2017a; Hunter N, unpubl. results).

HORMADs initially associate with unsynapsed chromosome axes during leptotene but are locally depleted at regions of synapsis during zygotene and excluded from fully synapsed chromosomes during pachytene (Fig. 5; Wojtasz et al. 2009; Fukuda et al. 2010). By analogy to the orthologous budding yeast Hop1 protein (Niu et al. 2005; Carballo et al. 2008; Goldfarb and Lichten 2010; Lao and Hunter 2010), and supported by indirect evidence, mammalian HORMADs are inferred to function in early meiotic prophase to impede DSB repair between sister chromatids and thereby promote interhomolog interactions (Fig. 6; Daniel et al. 2011; Kogo et al. 2012b; Shin et al. 2013; Rinaldi et al. 2017a). HORMAD depletion is coupled to synapsis via the AAA+ ATPase, TRIP13 (a.k.a. PCH2 in nonmammalian species) (Li and Schimenti 2007; Wojtasz et al. 2009; Roig et al. 2010), which likely disrupts protein–protein interactions at the homolog axis, including the ability of HORMADs to oligomerize (Ye et al. 2017; West et al. 2018). In asynaptic mutants and when TRIP13 function is compromised, HORMADs remain associated with the homolog axes and impede DSB repair (Fukuda et al. 2010; Shin et al. 2010; Daniel et al. 2011; Shin et al. 2013; Rinaldi et al. 2017a). Thus, oocyte culling in mutants defective for homolog synapsis and DSB repair can be suppressed to varying degrees by mutation of *Hormad1* or *Hormad2* (Daniel et al. 2011; Kogo et al. 2012b; Shin et al. 2013; Rinaldi et al. 2017a). However, the mode of suppression by *Hormad1/2* mutation appears to be distinct from that of *Chk2* and *p53/TAp63* mutations.

The high levels of DNA damage that persist in dictyate-stage oocytes from *Chk2 Trip13* double mutants are consistent with CHK2 acting as a bona fide checkpoint protein, allowing cells to progress but not affecting the efficiency of DSB repair (Bolcun-Filas et al. 2014). However, this damage is subsequently repaired, oocytes remain viable, and *Chk2 Trip13* mutants are fertile. This contrasts the effects of *Hormad1/2* mutation, which decrease DSB levels and accelerate DSB repair (Fukuda et al. 2010; Shin et al. 2010; Daniel et al. 2011; Shin et al. 2013; Rinaldi et al. 2017a). HORMAD1 has a twofold effect on DSB levels, facilitating the formation of SPO11-dependent DSB and impeding their repair until homologs have synapsed. HORMAD2 shares only the latter function and its localization to unsynapsed chromosome axes requires HORMAD1. Thus, absence of HORMADs is inferred to reduce DSB load (*Hormad1* mutation) and enhance DSB-repair capacity (*Hormad1* and *Hormad2* mutations) by allowing recombination between sister chromatids such that damage signaling is diminished

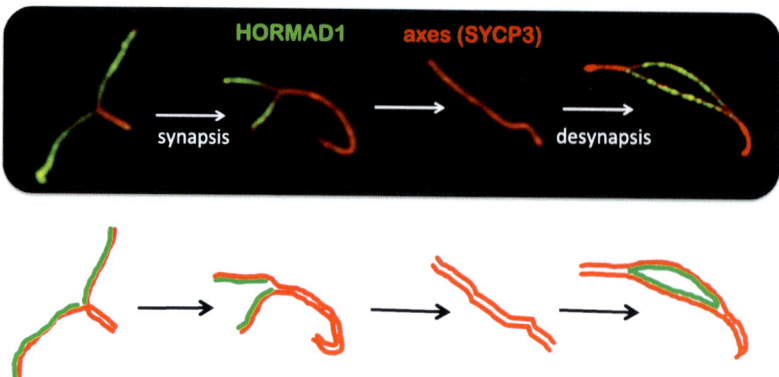

Figure 5. Dynamic localization of HORMADs during meiotic prophase I. Images show mouse oocyte chromosomes immunostained for the axes marker SYCP3 and HORMAD1.

(Daniel et al. 2011; Kogo et al. 2012b; Shin et al. 2013; Rinaldi et al. 2017a). Put another way, by blocking the repair of DSBs between sister chromatids, HORMADs help maintain damage signaling at levels required to trigger LOA. These observations indicate that CHK2 and the HORMAD proteins collaborate to signal defective interhomolog interactions and trigger LOA (Fig. 4).

Meiotic Silencing Is Mediated by HORMADs and Components of the DNA Damage Response

Chromosome asynapsis in *Spo11* mutants, and in a variety of other contexts that cause partial asynapsis, also triggers a distinct response termed meiotic silencing that results in transcriptional inactivation and is inferred to contribute to oocyte death by silencing genes required for survival (Mahadevaiah et al. 2008; Burgoyne et al. 2009; Garcia-Cruz et al. 2009; Kouznetsova et al. 2009; Blanco-Rodríguez 2012; Royo et al. 2013; Cloutier et al. 2015; Turner 2015). Sensing of asynapsis involves homolog axis components, the HORMAD proteins, and BRCA1. ATR and its activators ATRIP and TOPBP1 are then recruited to establish a reversible asynapsis signaling step that involves phosphorylation of the HORMADs (Fukuda et al. 2012; Royo et al. 2013). Silencing initiates via the ATR-catalyzed phosphorylation the histone variant H2AFX. This γH2AFX mark spreads throughout the unsynapsed region via a signal-amplification step mediated by ATR, HORMADs, and the γH2AFX binding factor

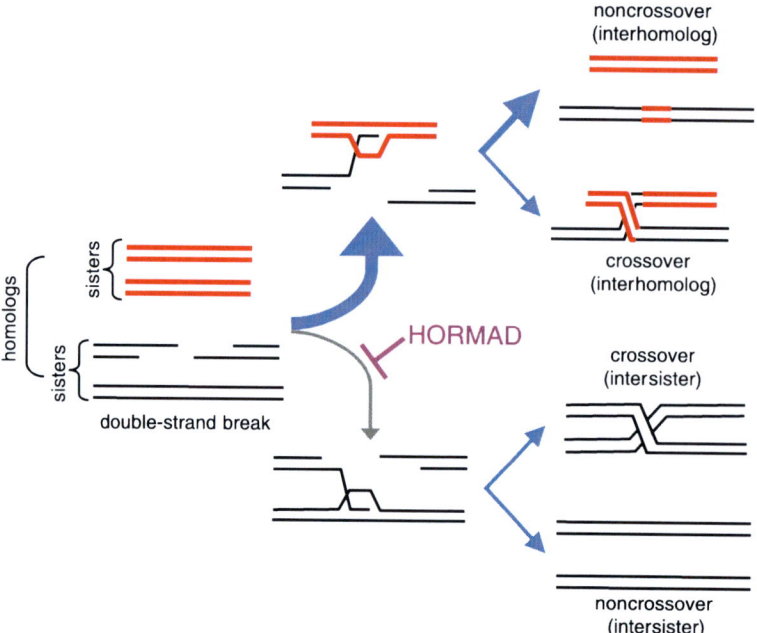

Figure 6. HORMADs bias recombination to occur between homologs rather than sister chromatids. Meiotic recombination must occur between homologs to achieve pairing, synapsis, and crossing-over. HORMADs promote interhomolog interactions through unknown mechanisms. One model posits that HORMADs create a barrier to intersister recombination (Hollingsworth 2010). Alternatively, HORMADs may prevent recombination from progressing beyond the initial nascent DNA strand-exchange step and couple progression to homolog synapsis (Goldfarb and Lichten 2010; Lao and Hunter 2010).

MDC1 (Ichijima et al. 2011; Kogo et al. 2012a). Ensuing heterochromatinization results in exclusion of RNA Pol II. Persistent DSBs within unsynapsed regions, both SPO11-dependent and independent, are thought to serve as initiation sites for meiotic silencing (Carofiglio et al. 2013; Ellnati et al. 2017). However, the enrichment of DSB markers observed within silenced regions is also likely to reflect the inhibitory role of HORMADs on DSB repair.

Physiological LOA and the Role of HORMADs

LOA was defined and is typically studied in a variety of pathological conditions that cause severe defects in DSB repair and/or synapsis and wholesale oocyte death. In these settings, initial association of HORMADs persists and promotes oocyte death by impeding DSB repair between sister chromatids (Rinaldi et al. 2017a). In contrast, in a physiological setting, DSBs engage in interhomolog interactions, synapsis generally occurs efficiently, and HORMADs dissociate, presumably enabling recombination to progress. HORMADs then reassociate with desynapsing homolog axes during diplonema when DSBs have been repaired and crossing-over has occurred (Fig. 5; Fukuda et al. 2010; Niu et al. 2005; Wojtasz et al. 2009). The role of HORMADs at this stage is unknown, but a role in signaling recombination defects can be envisioned. Residual DSBs present in diplotene may be the result of failed interhomolog recombination or de novo DSB formation (e.g., via L1 activity or aberrant processing of recombination intermediates by nucleases). Reassociation of HORMADs at this stage could block intersister repair to robustly signal such defects and enable a quality control decision to be made. Nascent meiotic silencing events could also be reinforced by HORMAD reloading. Thus, the reassociation of HORMADs is potentially a key event that primes oocytes for LOA specifically during the diplotene-to-dictyate transition and may explain why oocytes become sensitive to irradiation-induced apoptosis at this time (Hanoux et al. 2007; Hunter N, unpubl. data). Other events that may potentiate LOA during this transition include the onset of expression and ability to activate apoptotic factors, including TAp63 and caspase 2 (Suh et al. 2006; Hanoux et al. 2007; Livera et al. 2008; Kim and Suh 2014). Consequently, oocytes that have experienced defects in synapsis and/or recombination are eliminated before they become established in the ovarian reserve as primordial follicles, thereby minimizing nonproductive ovulation and the risk of meiotic errors.

CONCLUSION

Oocyte attrition occurs in response to L1 activity, meiotic errors, and self-sacrifice of nurse-like cells. These processes are estimated to cull up to 80% of all oocytes in mouse (Malki et al. 2014; Hunter N, unpubl. data), mirroring the dramatic reduction in oocyte numbers seen in human females between ~20 wk gestation and birth (Findlay et al. 2015). Thus, oocyte attrition in human females is presumed to occur via equivalent quality control

processes. However, significant differences likely exist. For example, numbers of full-length, potentially mobile L1 elements differ between mouse and human, with an estimated 11,000 in mouse compared to just 100 in human (Goodier and Kazazian 2008). As such, the L1 load may be lower in humans. An important goal for the future is to attain a clearer understanding of the causes and mechanisms of oocyte culling in humans or nonhuman primates.

Studies of EOA and LOA described here raise many burning questions. What are the relationships between L1 expression, meiotic errors, and oocyte/nurse cell differentiation? Is EOA triggered by L1-mediated RNA: DNA hybrids and, if so, what is the mechanism? Does EOA occur via apoptosis or an alternative cell death pathway? Why is L1 expression so variable between oocytes and what factors influence this ostensibly cell-autonomous phenomenon? What is the contribution of the various suppressive mechanisms to L1 inhibition during meiosis? Does the L1ORF2 reverse-transcriptase activity interfere with crossing-over, and, if so, does this activity contribute to the crossover maturation defect seen in human females? Is L1ORF2 endonuclease activity responsible for SPO11-independent DSBs? What are the contributions of L1-dependent and independent prophase defects to LOA and to meiotic errors in oocytes that escape EOA? Can genetic and environmental factors be identified that specifically influence EOA and LOA? Is L1-regulated gene expression important for mammalian meiosis? Does treatment with reverse-transcriptase inhibitors during pregnancy impact EOA in humans? Do kinases other than CHK2 contribute to physiological LOA? What is the contribution of meiotic silencing to physiological LOA? How do HORMAD proteins impede DSB repair? How does synapsis trigger HORMAD dissociation and how are they reloaded onto diplotene chromosomes? Finally, can ovarian reserves be enlarged to extend reproductive life span, or can they be rescued from death induced by cancer therapeutic agents without sacrificing oocyte quality, fertility, and ovary function (Livera et al. 2008; Kerr et al. 2012; Bolcun-Filas et al. 2014; Rinaldi et al. 2017b)?

Despite these oocyte quality control processes, errors in meiosis are the leading cause of pregnancy miscarriage and congenital disease in humans (Hassold et al. 2007; Nagaoka et al. 2012; Hunter 2015). By interfering with the normal processes of meiotic prophase, the activities of L1 and other retroelements represent one source of meiotic errors. L1 insertion also poses a tangible threat to genomic integrity: The rate of insertion is estimated at greater than one in eight births in laboratory mice (Richardson et al. 2017), and at approximately one in 20 births in the humans (Kazazian 1999). Thus, it remains important to better understand the causes of meiotic errors, the processes that work to minimize them, and the quality-control processes that selectively eliminate defective gametes and nurture and protect those that survive.

ACKNOWLEDGMENTS

We thank Alex Bortvin, Ewelina Bolcun-Filas, Vera Rinaldi, John Schimenti, James Turner, and members of my laboratory for enlightening discussions. Research in my laboratory is supported by the Howard Hughes Medical Institute and the National Institutes of Health National Institute of General Medical Sciences (NIH NIGMS) under award GM074223.

REFERENCES

Aiken CE, Tarry-Adkins JL, Ozanne SE. 2015. Transgenerational developmental programming of ovarian reserve. *Sci Rep* **5:** 16175.

Aravin AA, Sachidanandam R, Bourc'his D, Schaefer C, Pezic D, Toth KF, Bestor T, Hannon GJ. 2008. A piRNA pathway primed by individual transposons is linked to de novo DNA methylation in mice. *Mol Cell* **31:** 785–799.

Athanikar JN, Badge RM, Moran JV. 2004. A YY1-binding site is required for accurate human LINE-1 transcription initiation. *Nucleic Acids Res* **32:** 3846–3855.

Babushok DV, Kazazian HH Jr. 2007. Progress in understanding the biology of the human mutagen LINE-1. *Hum Mutat* **28:** 527–539.

Baker TG. 1963. A quantitative and cytological study of germ cells in human ovaries. *Proc R Soc Lond B Biol Sci* **158:** 417–433.

Beck CR, Garcia-Perez JL, Badge RM, Moran JV. 2011. LINE-1 elements in structural variation and disease. *Annu Rev Genomics Hum Genet* **12:** 187–215.

Belancio VP, Hedges DJ, Deininger P. 2006. LINE-1 RNA splicing and influences on mammalian gene expression. *Nucleic Acids Res* **34:** 1512–1521.

Belancio VP, Roy-Engel AM, Deininger P. 2008. The impact of multiple splice sites in human L1 elements. *Gene* **411:** 38–45.

Belgnaoui SM, Gosden RG, Semmes OJ, Haoudi A. 2006. Human LINE-1 retrotransposon induces DNA damage and apoptosis in cancer cells. *Cancer Cell Int* **6:** 13.

Blanco-Rodríguez J. 2012. Programmed phosphorylation of histone H2AX precedes a phase of DNA double-strand break-independent synapsis in mouse meiosis. *Reproduction* **144:** 699–712.

Block E. 1953. A quantitative morphological investigation of the follicular system in newborn female infants. *Acta Anat (Basel)* **17:** 201–206.

Bolcun-Filas E, Rinaldi VD, White ME, Schimenti JC. 2014. Reversal of female infertility by Chk2 ablation reveals the oocyte DNA damage checkpoint pathway. *Science* **343:** 533–536.

Bourc'his D, Bestor TH. 2004. Meiotic catastrophe and retrotransposon reactivation in male germ cells lacking Dnmt3L. *Nature* **431:** 96–99.

Broekmans FJ, Knauff EA, te Velde ER, Macklon NS, Fauser BC. 2007. Female reproductive ageing: Current knowledge and future trends. *Trends Endocrinol Metab* **18:** 58–65.

Brouha B, Schustak J, Badge RM, Lutz-Prigge S, Farley AH, Moran JV, Kazazian HH Jr. 2003. Hot L1s account for the bulk of retrotransposition in the human population. *Proc Natl Acad Sci* **100:** 5280–5285.

Burgoyne PS, Mahadevaiah SK, Turner JM. 2009. The consequences of asynapsis for mammalian meiosis. *Nat Rev Genet* **10:** 207–216.

Carballo JA, Johnson AL, Sedgwick SG, Cha RS. 2008. Phosphorylation of the axial element protein Hop1 by Mec1/Tel1 ensures meiotic interhomolog recombination. *Cell* **132:** 758–770.

Carofiglio F, Inagaki A, de Vries S, Wassenaar E, Schoenmakers S, Vermeulen C, van Cappellen WA, Sleddens-Linkels E, Grootegoed JA, Te Riele HP, et al. 2013. SPO11-independent DNA repair foci and their role in meiotic silencing. *PLoS Genet* **9:** e1003538.

Castañeda J, Genzor P, Bortvin A. 2011. piRNAs, transposon silencing, and germline genome integrity. *Mutat Res* **714:** 95–104.

Cheng EY, Hunt PA, Naluai-Cecchini TA, Fligner CL, Fujimoto VY, Pasternack TL, Schwartz JM, Steinauer JE, Woodruff TJ,

Cherry SM, et al. 2009. Meiotic recombination in human oocytes. *PLoS Genet* **5:** e1000661.

Chuma S. 2014. LINE-1 of evidence for fetal oocyte attrition by retrotransposon. *Dev Cell* **29:** 501–502.

Cloutier JM, Mahadevaiah SK, ELInati E, Nussenzweig A, Tóth A, Turner JM. 2015. Histone H2AFX links meiotic chromosome asynapsis to prophase I oocyte loss in mammals. *PLoS Genet* **11:** e1005462.

Cole F, Kauppi L, Lange J, Roig I, Wang R, Keeney S, Jasin M. 2012. Homeostatic control of recombination is implemented progressively in mouse meiosis. *Nat Cell Biol* **14:** 424–430.

Cordaux R, Batzer MA. 2009. The impact of retrotransposons on human genome evolution. *Nat Rev Genet* **10:** 691–703.

Cost GJ, Feng Q, Jacquier A, Boeke JD. 2002. Human L1 element target-primed reverse transcription in vitro. *EMBO J* **21:** 5899–5910.

Crow MK. 2010. Long interspersed nuclear elements (LINE-1): Potential triggers of systemic autoimmune disease. *Autoimmunity* **43:** 7–16.

Dai L, Huang Q, Boeke JD. 2011. Effect of reverse transcriptase inhibitors on LINE-1 and Ty1 reverse transcriptase activities and on LINE-1 retrotransposition. *BMC Biochem* **12:** 18.

Daniel K, Lange J, Hached K, Fu J, Anastassiadis K, Roig I, Cooke HJ, Stewart AF, Wassmann K, Jasin M, et al. 2011. Meiotic homologue alignment and its quality surveillance are controlled by mouse HORMAD1. *Nat Cell Biol* **13:** 599–610.

Dekker J, Mirny L. 2016. The 3D genome as moderator of chromosomal communication. *Cell* **164:** 1110–1121.

Dewannieux M, Esnault C, Heidmann T. 2003. LINE-mediated retrotransposition of marked Alu sequences. *Nat Genet* **35:** 41–48.

Di Giacomo M, Barchi M, Baudat F, Edelmann W, Keeney S, Jasin M. 2005. Distinct DNA-damage-dependent and -independent responses drive the loss of oocytes in recombination-defective mouse mutants. *Proc Natl Acad Sci* **102:** 737–742.

ElInati E, Russell HR, Ojarikre OA, Sangrithi M, Hirota T, de Rooij DG, McKinnon PJ, Turner JMA. 2017. DNA damage response protein TOPBP1 regulates X chromosome silencing in the mammalian germ line. *Proc Natl Acad Sci* **114:** 12536–12541.

Esnault C, Maestre J, Heidmann T. 2000. Human LINE retrotransposons generate processed pseudogenes. *Nat Genet* **24:** 363–367.

Faulkner GJ, Kimura Y, Daub CO, Wani S, Plessy C, Irvine KM, Schroder K, Cloonan N, Steptoe AL, Lassmann T, et al. 2009. The regulated retrotransposon transcriptome of mammalian cells. *Nat Genet* **41:** 563–571.

Findlay JK, Hutt KJ, Hickey M, Anderson RA. 2015. How is the number of primordial follicles in the ovarian reserve established? *Biol Reprod* **93:** 111.

Finsterbusch F, Ravindranathan R, Dereli I, Stanzione M, Tränkner D, Tóth A. 2016. Alignment of homologous chromosomes and effective repair of programmed DNA double-strand breaks during mouse meiosis require the minichromosome maintenance domain containing 2 (MCMDC2) protein. *PLoS Genet* **12:** e1006393.

Forabosco A, Sforza C, De Pol A, Vizzotto L, Marzona L, Ferrario VF. 1991. Morphometric study of the human neonatal ovary. *Anat Rec* **231:** 201–208.

Fukuda T, Daniel K, Wojtasz L, Toth A, Höög C. 2010. A novel mammalian HORMA domain-containing protein, HORMAD1, preferentially associates with unsynapsed meiotic chromosomes. *Exp Cell Res* **316:** 158–171.

Fukuda T, Pratto F, Schimenti JC, Turner JM, Camerini-Otero RD, Höög C. 2012. Phosphorylation of chromosome core components may serve as axis marks for the status of chromosomal events during mammalian meiosis. *PLoS Genet* **8:** e1002485.

Garcia-Cruz R, Roig I, Robles P, Scherthan H, Garcia Caldés M. 2009. ATR, BRCA1 and gammaH2AX localize to unsynapsed chromosomes at the pachytene stage in human oocytes. *Reprod Biomed Online* **18:** 37–44.

Gasior SL, Wakeman TP, Xu B, Deininger PL. 2006. The human LINE-1 retrotransposon creates DNA double-strand breaks. *J Mol Biol* **357**: 1383–1393.

Gasior SL, Roy-Engel AM, Deininger PL. 2008. ERCC1/XPF limits L1 retrotransposition. *DNA Repair (Amst)* **7**: 983–989.

Gleicher N, Weghofer A, Barad DH. 2011. Defining ovarian reserve to better understand ovarian aging. *Reprod Biol Endocrinol* **9**: 23.

Goldfarb T, Lichten M. 2010. Frequent and efficient use of the sister chromatid for DNA double-strand break repair during budding yeast meiosis. *PLoS Biol* **8**: e1000520.

Goodier JL. 2016. Restricting retrotransposons: A review. *Mob DNA* **7**: 16.

Goodier JL, Kazazian HH Jr. 2008. Retrotransposons revisited: The restraint and rehabilitation of parasites. *Cell* **135**: 23–35.

Goodier JL, Pereira GC, Cheung LE, Rose RJ, Kazazian HH Jr. 2015. The broad-spectrum antiviral protein ZAP restricts human retrotransposition. *PLoS Genet* **11**: e1005252.

Grandi FC, An W. 2013. Non-LTR retrotransposons and microsatellites: Partners in genomic variation. *Mob Genet Elements* **3**: e25674.

Grive KJ, Freiman RN. 2015. The developmental origins of the mammalian ovarian reserve. *Development* **142**: 2554–2563.

Gruhn JR, Rubio C, Broman KW, Hunt PA, Hassold T. 2013. Cytological studies of human meiosis: Sex-specific differences in recombination originate at, or prior to, establishment of double-strand breaks. *PLoS One* **8**: e85075.

Hamdorf M, Idica A, Zisoulis DG, Gamelin L, Martin C, Sanders KJ, Pedersen IM. 2015. miR-128 represses L1 retrotransposition by binding directly to L1 RNA. *Nat Struct Mol Biol* **22**: 824–831.

Han JS, Szak ST, Boeke JD. 2004. Transcriptional disruption by the L1 retrotransposon and implications for mammalian transcriptomes. *Nature* **429**: 268–274.

Hanoux V, Pairault C, Bakalska M, Habert R, Livera G. 2007. Caspase-2 involvement during ionizing radiation-induced oocyte death in the mouse ovary. *Cell Death Differ* **14**: 671–681.

Hansen AS, Cattoglio C, Darzacq X, Tjian R. 2018. Recent evidence that TADs and chromatin loops are dynamic structures. *Nucleus* **9**: 20–32.

Hartshorne GM, Lyrakou S, Hamoda H, Oloto E, Ghafari F. 2009. Oogenesis and cell death in human prenatal ovaries: What are the criteria for oocyte selection? *Mol Hum Reprod* **15**: 805–819.

Hassold T, Hunt P. 2001. To err (meiotically) is human: The genesis of human aneuploidy. *Nat Rev Genet* **2**: 280–291.

Hassold T, Hall H, Hunt P. 2007. The origin of human aneuploidy: Where we have been, where we are going. *Hum Mol Genet* **16**: R203–R208.

Hata K, Sakaki Y. 1997. Identification of critical CpG sites for repression of L1 transcription by DNA methylation. *Gene* **189**: 227–234.

Herbert M, Kalleas D, Cooney D, Lamb M, Lister L. 2015. Meiosis and maternal aging: Insights from aneuploid oocytes and trisomy births. *Cold Spring Harb Perspect Biol* **7**: a017970.

Hernández-Hernández A, Rincón-Arano H, Recillas-Targa F, Ortiz R, Valdes-Quezada C, Echeverría OM, Benavente R, Vázquez-Nin GH. 2008. Differential distribution and association of repeat DNA sequences in the lateral element of the synaptonemal complex in rat spermatocytes. *Chromosoma* **117**: 77–87.

Hollingsworth NM. 2010. Phosphorylation and the creation of interhomolog bias during meiosis in yeast. *Cell Cycle* **9**: 436–437.

Honour JW. 2018. Biochemistry of the menopause. *Ann Clin Biochem* **55**: 18–33.

Hu S, Li J, Xu F, Mei S, Le Duff Y, Yin L, Pang X, Cen S, Jin Q, Liang C, et al. 2015. SAMHD1 Inhibits LINE-1 Retrotransposition by Promoting Stress Granule Formation. *PLoS Genet* **11**: e1005367.

Hunt PA, Hassold TJ. 2008. Human female meiosis: What makes a good egg go bad? *Trends Genet* **24**: 86–93.

Hunter N. 2015. Meiotic recombination: The essence of heredity. *Cold Spring Harb Perspect Biol* **7**: a016618.

Ichijima Y, Ichijima M, Lou Z, Nussenzweig A, Camerini-Otero RD, Chen J, Andreassen PR, Namekawa SH. 2011. MDC1 directs chromosome-wide silencing of the sex chromosomes in male germ cells. *Genes Dev* **25**: 959–971.

Izquierdo-Bouldstridge A, Bustillos A, Bonet-Costa C, Aribau-Miralbés P, García-Gomis D, Dabad M, Esteve-Codina A, Pascual-Reguant L, Peiró S, Esteller M, et al. 2017. Histone H1 depletion triggers an interferon response in cancer cells via activation of heterochromatic repeats. *Nucleic Acids Res* **45**: 11622–11642.

Jenkins VK, Timmons AK, McCall K. 2013. Diversity of cell death pathways: Insight from the fly ovary. *Trends Cell Biol* **23**: 567–574.

Jones RB, Garrison KE, Wong JC, Duan EH, Nixon DF, Ostrowski MA. 2008. Nucleoside analogue reverse transcriptase inhibitors differentially inhibit human LINE-1 retrotransposition. *PLoS One* **3**: e1547.

Kaipia A, Hsueh AJ. 1997. Regulation of ovarian follicle atresia. *Annu Rev Physiol* **59**: 349–363.

Kano H, Godoy I, Courtney C, Vetter MR, Gerton GL, Ostertag EM, Kazazian HH Jr. 2009. L1 retrotransposition occurs mainly in embryogenesis and creates somatic mosaicism. *Genes Dev* **23**: 1303–1312.

Kazazian HH Jr. 1999. An estimated frequency of endogenous insertional mutations in humans. *Nat Genet* **22**: 130.

Kerr JB, Hutt KJ, Michalak EM, Cook M, Vandenberg CJ, Liew SH, Bouillet P, Mills A, Scott CL, Findlay JK, et al. 2012. DNA damage-induced primordial follicle oocyte apoptosis and loss of fertility require TAp63-mediated induction of Puma and Noxa. *Mol Cell* **48**: 343–352.

Kerr JB, Myers M, Anderson RA. 2013. The dynamics of the primordial follicle reserve. *Reproduction* **146**: R205–R215.

Kim DA, Suh EK. 2014. Defying DNA double-strand break-induced death during prophase I meiosis by temporal TAp63α phosphorylation regulation in developing mouse oocytes. *Mol Cell Biol* **34**: 1460–1473.

Klinger FG, Rossi V, De Felici M. 2015. Multifaceted programmed cell death in the mammalian fetal ovary. *Int J Dev Biol* **59**: 51–54.

Kogo H, Tsutsumi M, Inagaki H, Ohye T, Kiyonari H, Kurahashi H. 2012a. HORMAD2 is essential for synapsis surveillance during meiotic prophase via the recruitment of ATR activity. *Genes Cells* **17**: 897–912.

Kogo H, Tsutsumi M, Ohye T, Inagaki H, Abe T, Kurahashi H. 2012b. HORMAD1-dependent checkpoint/surveillance mechanism eliminates asynaptic oocytes. *Genes Cells* **17**: 439–454.

Kouznetsova A, Wang H, Bellani M, Camerini-Otero RD, Jessberger R, Höög C. 2009. BRCA1-mediated chromatin silencing is limited to oocytes with a small number of asynapsed chromosomes. *J Cell Sci* **122**: 2446–2452.

Kuramochi-Miyagawa S, Watanabe T, Gotoh K, Totoki Y, Toyoda A, Ikawa M, Asada N, Kojima K, Yamaguchi Y, Ijiri TW, et al. 2008. DNA methylation of retrotransposon genes is regulated by Piwi family members MILI and MIWI2 in murine fetal testes. *Genes Dev* **22**: 908–917.

Kurilo LF. 1981. Oogenesis in antenatal development in man. *Hum Genet* **57**: 86–92.

Lam I, Keeney S. 2014. Mechanism and regulation of meiotic recombination initiation. *Cold Spring Harb Perspect Biol* **7**: a016634.

Lao JP, Hunter N. 2010. Trying to avoid your sister. *PLoS Biol* **8**: e1000519.

Laven JS. 2015. Genetics of early and normal menopause. *Semin Reprod Med* **33**: 377–383.

Laven JS. 2016. Primary ovarian insufficiency. *Semin Reprod Med* **34**: 230–234.

Laven JSE, Visser JA, Uitterlinden AG, Vermeij WP, Hoeijmakers JHJ. 2016. Menopause: Genome stability as new paradigm. *Maturitas* **92**: 15–23.

Lee JY, Ji Z, Tian B. 2008. Phylogenetic analysis of mRNA polyadenylation sites reveals a role of transposable elements in evolution of the 3′-end of genes. *Nucleic Acids Res* **36**: 5581–5590.

Lei L, Spradling AC. 2013. Mouse primordial germ cells produce cysts that partially fragment prior to meiosis. *Development* **140**: 2075–2081.

Lei L, Spradling AC. 2016. Mouse oocytes differentiate through organelle enrichment from sister cyst germ cells. *Science* **352**: 95–99.

Lenzi ML, Smith J, Snowden T, Kim M, Fishel R, Poulos BK, Cohen PE. 2005. Extreme heterogeneity in the molecular events leading to the establishment of chiasmata during meiosis i in human oocytes. *Am J Hum Genet* **76**: 112–127.

Li XC, Schimenti JC. 2007. Mouse pachytene checkpoint 2 (trip13) is required for completing meiotic recombination but not synapsis. *PLoS Genet* **3**: e130.

Li P, Du J, Goodier JL, Hou J, Kang J, Kazazian HH Jr, Zhao K, Yu XF. 2017. Aicardi-Goutieres syndrome protein TREX1 suppresses L1 and maintains genome integrity through exonuclease-independent ORF1p depletion. *Nucleic Acids Res* **45**: 4619–4631.

Liu N, Lee CH, Swigut T, Grow E, Gu B, Bassik MC, Wysocka J. 2018. Selective silencing of euchromatic L1s revealed by genome-wide screens for L1 regulators. *Nature* **553**: 228–232.

Livera G, Petre-Lazar B, Guerquin MJ, Trautmann E, Coffigny H, Habert R. 2008. p63 null mutation protects mouse oocytes from radio-induced apoptosis. *Reproduction* **135**: 3–12.

MacLennan M, García-Cañadas M, Reichmann J, Khazina E, Wagner G, Playfoot CJ, Salvador-Palomeque C, Mann AR, Peressini P, Sanchez L, et al. 2017. Mobilization of LINE-1 retrotransposons is restricted by Tex19.1 in mouse embryonic stem cells. *eLife* **6**: e26152.

Mahadevaiah SK, Bourc'his D, de Rooij DG, Bestor TH, Turner JM, Burgoyne PS. 2008. Extensive meiotic asynapsis in mice antagonises meiotic silencing of unsynapsed chromatin and consequently disrupts meiotic sex chromosome inactivation. *J Cell Biol* **182**: 263–276.

Malki S, van der Heijden GW, O'Donnell KA, Martin SL, Bortvin A. 2014. A role for retrotransposon LINE-1 in fetal oocyte attrition in mice. *Dev Cell* **29**: 521–533.

McGee EA, Hsueh AJ. 2000. Initial and cyclic recruitment of ovarian follicles. *Endocr Rev* **21**: 200–214.

McNairn AJ, Rinaldi VD, Schimenti JC. 2017. Repair of meiotic DNA breaks and homolog pairing in mouse meiosis requires a minichromosome maintenance (MCM) paralog. *Genetics* **205**: 529–537.

Messerschmidt DM, Knowles BB, Solter D. 2014. DNA methylation dynamics during epigenetic reprogramming in the germline and preimplantation embryos. *Genes Dev* **28**: 812–828.

Morita Y, Perez GI, Maravei DV, Tilly KI, Tilly JL. 1999. Targeted expression of Bcl-2 in mouse oocytes inhibits ovarian follicle atresia and prevents spontaneous and chemotherapy-induced oocyte apoptosis in vitro. *Mol Endocrinol* **13**: 841–850.

Morrish TA, Gilbert N, Myers JS, Vincent BJ, Stamato TD, Taccioli GE, Batzer MA, Moran JV. 2002. DNA repair mediated by endonuclease-independent LINE-1 retrotransposition. *Nat Genet* **31**: 159–165.

Nagaoka SI, Hassold TJ, Hunt PA. 2012. Human aneuploidy: Mechanisms and new insights into an age-old problem. *Nat Rev Genet* **13**: 493–504.

Ng JH, Kumar V, Muratani M, Kraus P, Yeo JC, Yaw LP, Xue K, Lufkin T, Prabhakar S, Ng HH. 2013. In vivo epigenomic profiling of germ cells reveals germ cell molecular signatures. *Dev Cell* **24**: 324–333.

Nigumann P, Redik K, Matlik K, Speek M. 2002. Many human genes are transcribed from the antisense promoter of L1 retrotransposon. *Genomics* **79**: 628–634.

Niu H, Wan L, Baumgartner B, Schaefer D, Loidl J, Hollingsworth NM. 2005. Partner choice during meiosis is regulated by Hop1-promoted dimerization of Mek1. *Mol Biol Cell* **16**: 5804–5818.

Omari S, Waters M, Naranian T, Kim K, Perumalsamy AL, Chi M, Greenblatt E, Moley KH, Opferman JT, Jurisicova A. 2015. Mcl-1 is a key regulator of the ovarian reserve. *Cell Death Dis* **6**: e1755.

Orecchini E, Frassinelli L, Galardi S, Ciafrè SA, Michienzi A. 2018. Post-transcriptional regulation of LINE-1 retrotransposition by AID/APOBEC and ADAR deaminases. *Chromosome Res* **26**: 45–59.

Ostertag EM, Goodier JL, Zhang Y, Kazazian HH Jr. 2003. SVA elements are nonautonomous retrotransposons that cause disease in humans. *Am J Hum Genet* **73**: 1444–1451.

Pearlman RE, Tsao N, Moens PB. 1992. Synaptonemal complexes from DNase-treated rat pachytene chromosomes contain $(GT)_n$ and LINE/SINE sequences. *Genetics* **130**: 865–872.

Pepling ME, Spradling AC. 1998. Female mouse germ cells form synchronously dividing cysts. *Development* **125**: 3323–3328.

Perepelitsa-Belancio V, Deininger P. 2003. RNA truncation by premature polyadenylation attenuates human mobile element activity. *Nat Genet* **35**: 363–366.

Pizarro JG, Cristofari G. 2016. Post-transcriptional control of LINE-1 retrotransposition by cellular host factors in somatic cells. *Front Cell Dev Biol* **4**: 14.

Qiao H, Rao HBDP, Yun Y, Sandhu S, Fong JH, Sapre M, Nguyen M, Tham A, Van BW, Chng TYH, et al. 2018. Impeding DNA break repair enables oocyte quality control. *bioRxiv* doi: 10.1101/277913.

Richardson SR, Gerdes P, Gerhardt DJ, Sanchez-Luque FJ, Bodea GO, Muñoz-Lopez M, Jesuadian JS, Kempen MHC, Carreira PE, Jeddeloh JA, et al. 2017. Heritable L1 retrotransposition in the mouse primordial germline and early embryo. *Genome Res* **27**: 1395–1405.

Rigby RE, Webb LM, Mackenzie KJ, Li Y, Leitch A, Reijns MA, Lundie RJ, Revuelta A, Davidson DJ, Diebold S, et al. 2014. RNA:DNA hybrids are a novel molecular pattern sensed by TLR9. *EMBO J* **33**: 542–558.

Rinaldi VD, Bolcun-Filas E, Kogo H, Kurahashi H, Schimenti JC. 2017a. The DNA damage checkpoint eliminates mouse oocytes with chromosome synapsis failure. *Mol Cell* **67**: 1026–1036 e1022.

Rinaldi VD, Hsieh K, Munroe R, Bolcun-Filas E, Schimenti JC. 2017b. Pharmacological inhibition of the DNA damage checkpoint prevents radiation-induced oocyte death. *Genetics* **206**: 1823–1828.

Roig I, Dowdle JA, Toth A, de Rooij DG, Jasin M, Keeney S. 2010. Mouse TRIP13/PCH2 is required for recombination and normal higher-order chromosome structure during meiosis. *PLoS Genet* **6**: e1001062.

Royo H, Prosser H, Ruzankina Y, Mahadevaiah SK, Cloutier JM, Baumann M, Fukuda T, Höög C, Toth A, de Rooij DG, et al. 2013. ATR acts stage specifically to regulate multiple aspects of mammalian meiotic silencing. *Genes Dev* **27**: 1484–1494.

Saitou M, Kagiwada S, Kurimoto K. 2012. Epigenetic reprogramming in mouse pre-implantation development and primordial germ cells. *Development* **139**: 15–31.

Schumann GG. 2007. APOBEC3 proteins: Major players in intracellular defence against LINE-1-mediated retrotransposition. *Biochem Soc Trans* **35**: 637–642.

Schuster AT, Sarvepalli K, Murphy EA, Longworth MS. 2013. Condensin II subunit dCAP-D3 restricts retrotransposon mobilization in *Drosophila* somatic cells. *PLoS Genet* **9**: e1003879.

Schwertz H, Rowley JW, Schumann GG, Thorack U, Campbell RA, Manne BK, Zimmerman GA, Weyrich AS, Rondina MT. 2018. Endogenous LINE-1 (Long Interspersed Nuclear Element-1) reverse transcriptase activity in platelets controls translational events through RNA–DNA hybrids. *Arterioscler Thromb Vasc Biol* **38**: 801–815.

Seisenberger S, Andrews S, Krueger F, Arand J, Walter J, Santos F, Popp C, Thienpont B, Dean W, Reik W. 2012. The dynamics of genome-wide DNA methylation reprogramming in mouse primordial germ cells. *Mol Cell* **48**: 849–862.

Servant G, Streva VA, Derbes RS, Wijetunge MI, Neeland M, White TB, Belancio VP, Roy-Engel AM, Deininger PL. 2017. The nucleotide excision repair pathway limits L1 retrotransposition. *Genetics* **205**: 139–153.

Sheinberger J, Shav-Tal Y. 2017. mRNPs meet stress granules. *FEBS Lett* **591:** 2534–2542.

Shin YH, Choi Y, Erdin SU, Yatsenko SA, Kloc M, Yang F, Wang PJ, Meistrich ML, Rajkovic A. 2010. Hormad1 mutation disrupts synaptonemal complex formation, recombination, and chromosome segregation in mammalian meiosis. *PLoS Genet* **6:** e1001190.

Shin YH, McGuire MM, Rajkovic A. 2013. Mouse HORMAD1 is a meiosis i checkpoint protein that modulates DNA double-strand break repair during female meiosis. *Biol Reprod* **89:** 29.

Slotkin RK, Martienssen R. 2007. Transposable elements and the epigenetic regulation of the genome. *Nat Rev Genet* **8:** 272–285.

Soper SF, van der Heijden GW, Hardiman TC, Goodheart M, Martin SL, de Boer P, Bortvin A. 2008. Mouse maelstrom, a component of nuage, is essential for spermatogenesis and transposon repression in meiosis. *Dev Cell* **15:** 285–297.

Stanzione M, Baumann M, Papanikos F, Dereli I, Lange J, Ramlal A, Trankner D, Shibuya H, de Massy B, Watanabe Y, et al. 2016. Meiotic DNA break formation requires the unsynapsed chromosome axis-binding protein IHO1 (CCDC36) in mice. *Nat Cell Biol* **18:** 1208–1220.

Stetson DB, Ko JS, Heidmann T, Medzhitov R. 2008. Trex1 prevents cell-intrinsic initiation of autoimmunity. *Cell* **134:** 587–598.

Subramanian VV, Hochwagen A. 2014. The meiotic checkpoint network: Step-by-step through meiotic prophase. *Cold Spring Harb Perspect Biol* **6:** a016675.

Suh EK, Yang A, Kettenbach A, Bamberger C, Michaelis AH, Zhu Z, Elvin JA, Bronson RT, Crum CP, McKeon F. 2006. p63 protects the female germ line during meiotic arrest. *Nature* **444:** 624–628.

Székvölgyi L, Ohta K, Nicolas A. 2015. Initiation of meiotic homologous recombination: Flexibility, impact of histone modifications, and chromatin remodeling. *Cold Spring Harb Perspect Biol* **7:** a016527.

Tchenio T, Casella JF, Heidmann T. 2000. Members of the SRY family regulate the human LINE retrotransposons. *Nucleic Acids Res* **28:** 411–415.

Tease C, Hartshorne G, Hulten M. 2006. Altered patterns of meiotic recombination in human fetal oocytes with asynapsis and/or synaptonemal complex fragmentation at pachytene. *Reprod Biomed Online* **13:** 88–95.

Tilly JL. 2001. Commuting the death sentence: How oocytes strive to survive. *Nat Rev Mol Cell Biol* **2:** 838–848.

Tilly JL, Sinclair DA. 2013. Germline energetics, aging, and female infertility. *Cell Metab* **17:** 838–850.

Trelogan SA, Martin SL. 1995. Tightly regulated, developmentally specific expression of the first open reading frame from LINE-1 during mouse embryogenesis. *Proc Natl Acad Sci* **92:** 1520–1524.

Turner JM. 2015. Meiotic silencing in mammals. *Annu Rev Genet* **49:** 395–412.

Vader G, Musacchio A. 2014. HORMA domains at the heart of meiotic chromosome dynamics. *Dev Cell* **31:** 389–391.

van der Heijden GW, Bortvin A. 2009. Transient relaxation of transposon silencing at the onset of mammalian meiosis. *Epigenetics* **4:** 76–79.

Vinckenbosch N, Dupanloup I, Kaessmann H. 2006. Evolutionary fate of retroposed gene copies in the human genome. *Proc Natl Acad Sci* **103:** 3220–3225.

Wallace NA, Belancio VP, Deininger PL. 2008. L1 mobile element expression causes multiple types of toxicity. *Gene* **419:** 75–81.

Wang S, Hassold T, Hunt P, White MA, Zickler D, Kleckner N, Zhang L. 2017. Inefficient crossover maturation underlies elevated aneuploidy in human female meiosis. *Cell* **168:** 977–989 e917.

Ward JR, Vasu K, Deutschman E, Halawani D, Larson PA, Zhang D, Willard B, Fox PL, Moran JV, Longworth MS. 2017. Condensin II and GAIT complexes cooperate to restrict LINE-1 retrotransposition in epithelial cells. *PLoS Genet* **13:** e1007051.

West AMV, Komives EA, Corbett KD. 2018. Conformational dynamics of the Hop1 HORMA domain reveal a common mechanism with the spindle checkpoint protein Mad2. *Nucleic Acids Res* **46:** 279–292.

Wojtasz L, Daniel K, Roig I, Bolcun-Filas E, Xu H, Boonsanay V, Eckmann CR, Cooke HJ, Jasin M, Keeney S, et al. 2009. Mouse HORMAD1 and HORMAD2, two conserved meiotic chromosomal proteins, are depleted from synapsed chromosome axes with the help of TRIP13 AAA-ATPase. *PLoS Genet* **5:** e1000702.

Wojtasz L, Cloutier JM, Baumann M, Daniel K, Varga J, Fu J, Anastassiadis K, Stewart AF, Remenyi A, Turner JM, et al. 2012. Meiotic DNA double-strand breaks and chromosome asynapsis in mice are monitored by distinct HORMAD2-independent and -dependent mechanisms. *Genes Dev* **26:** 958–973.

Yang N, Kazazian HH Jr. 2006. L1 retrotransposition is suppressed by endogenously encoded small interfering RNAs in human cultured cells. *Nat Struct Mol Biol* **13:** 763–771.

Yang N, Zhang L, Zhang Y, Kazazian HH Jr., 2003. An important role for RUNX3 in human L1 transcription and retrotransposition. *Nucleic Acids Res* **31:** 4929–4940.

Ye Q, Kim DH, Dereli I, Rosenberg SC, Hagemann G, Herzog F, Tóth A, Cleveland DW, Corbett KD. 2017. The AAA+ ATPase TRIP13 remodels HORMA domains through N-terminal engagement and unfolding. *EMBO J* **36:** 2419–2434.

Zhang Z, Harrison PM, Liu Y, Gerstein M. 2003. Millions of years of evolution preserved: A comprehensive catalog of the processed pseudogenes in the human genome. *Genome Res* **13:** 2541–2558.

Zhang A, Dong B, Doucet AJ, Moldovan JB, Moran JV, Silverman RH. 2014. RNase L restricts the mobility of engineered retrotransposons in cultured human cells. *Nucleic Acids Res* **42:** 3803–3820.

Zickler D, Kleckner N. 1999. Meiotic chromosomes: Integrating structure and function. *Annu Rev Genet* **33:** 603–754.

Zickler D, Kleckner N. 2015. Recombination, pairing, and synapsis of homologs during meiosis. *Cold Spring Harb Perspect Biol* **7:** a016626.

Cellular and Molecular Mechanisms of Centromere Drive

Michael A. Lampson[1] and Ben E. Black[2]

[1]Department of Biology, University of Pennsylvania, Philadelphia, Pennsylvania 19104
[2]Department of Biochemistry and Biophysics, Perelman School of Medicine, University of Pennsylvania, Philadelphia, Pennsylvania 19104-6059

Correspondence: lampson@sas.upenn.edu; blackbe@pennmedicine.upenn.edu

The asymmetric outcome of female meiosis I, whereby an entire set of chromosomes are discarded into a polar body, presents an opportunity for selfish genetic elements to cheat the process and disproportionately segregate to the egg. Centromeres, the chromosomal loci that connect to spindle microtubules, could potentially act as selfish elements and "drive" in meiosis. We review the current understanding of the genetic and epigenetic contributions to centromere identity and describe recent progress in a powerful model system to study centromere drive in mice. The progress includes mechanistic findings regarding two main requirements for a centromere to exploit the asymmetric outcome of female meiosis. The first is an asymmetry between centromeres of homologous chromosomes, and we found this is accomplished through massive changes in the abundance of the repetitive DNA underlying centromeric chromatin. The second requirement is an asymmetry in the meiotic spindle, which is achieved through signaling from the oocyte cortex that leads to asymmetry in a posttranslational modification of tubulin, tyrosination. Together, these two asymmetries culminate in the biased segregation of expanded centromeres to the egg, and we describe a mechanistic framework to understand this process.

Sexual reproduction in eukaryotes depends on a haploid–diploid life cycle. Meiosis, the process by which haploids are generated, provides an opportunity for genetic elements to compete for transmission to the offspring because each gamete carries only one of the two alleles of a gene. According to Mendel's Law of Segregation (First Law), alleles are transmitted with equal probability, but it is increasingly clear that this law can be violated, and segregation can be manipulated by selfish genetic elements through meiotic drive. The impact of meiotic drive on many aspects of evolution and genetics is now recognized, with examples widespread across eukaryotes (Werren 2011; Rice 2013; Helleu et al. 2015; Lindholm et al. 2016), but the underlying cell biological mechanisms are largely unknown. Selfish elements can drive by eliminating competing gametes (e.g., sperm killing or spore killing) or by increasing their transmission to the egg in female meiosis, which is the focus of this review.

Because of its inherent asymmetry, female meiosis provides a clear opportunity for selfish elements to cheat: Only chromosomes that segregate to the egg can be transmitted to offspring, whereas the rest are degraded in polar bodies (Sandler and Novitski 1957; Pardo-Manuel de Villena and Sapienza 2001). The centromere drive hypothesis (Fig. 1A–C; Henikoff et al. 2001) proposes that a centromere, as the locus that directs chromosome segregation, can act as a selfish element by increasing its own transmission through female meiosis at the expense of the homologous chromosome. The hypothesis was formulated to explain the paradox that although centromere function is essential for eukaryotic cell division and highly conserved, both repetitive centromere DNA and centromere-binding proteins have evolved rapidly. Because the basic mechanisms of kinetochore assembly at centromeres and interactions with spindle microtubules (MTs) are similar across many eukaryotes, the expectation is that purifying selection would minimize amino acid changes. Key centromere proteins such as CENP-A and CENP-C, however, show strong signatures of positive selection based on the ratio of synonymous to nonsynonymous substitutions in their coding sequences (Malik and Henikoff 2001; Talbert et al. 2004; Schueler et al. 2010; Zedek and Bureš 2016).

The centromere drive hypothesis has two parts. First, evolution of centromere DNA is driven by competition to orient toward the spindle pole that will remain in the egg, and expansion of repetitive sequences (or other changes to these sequences) at a centromere somehow leads to preferential orientation. The second part explains the evolution of centromere proteins through conflict between individual centromeres, which expand to gain an advantage in female meiosis, and the rest of the genome. Differences between centromeres of homologous chromosomes, which lead to biased segregation in female meiosis, may also impose a fitness cost such as reduced male fertility. This cost would provide selective pressure favoring alleles of centromere-binding proteins that equalize centromeres and suppress drive by binding independent of sequence.

This review focuses on cell biological and molecular mechanisms for centromere drive in female meiosis, based on recent work in a mouse model system. Conceptually, drive depends on three conditions (Fig. 1D). The first is asymmetry in female meiotic cell division and cell fate, which is well established and a universal feature of sexual reproduction in animals (Gorelick et al. 2016). The second

Published by Cold Spring Harbor Laboratory Press; doi: 10.1101/sqb.2017.82.034298

Cold Spring Harbor Symposia on Quantitative Biology, Volume LXXXII

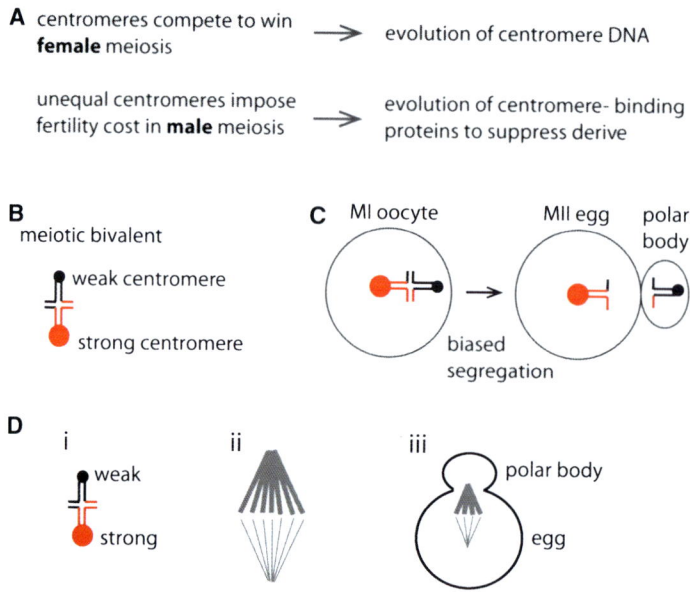

Figure 1. The centromere drive hypothesis. (*A*) Hypothesis for the evolution of centromere DNA and proteins. (*B*) A meiotic bivalent with unequal centromeres (red and black circles). (*C*) Biased segregation with the stronger centromere remaining in the egg. (*D*) A meiotic bivalent with weaker and stronger centromeres (i) preferentially orients on an asymmetric spindle (ii), which orients relative to the cortex where the polar body forms (iii).

is a difference between the centromeres of homologous chromosomes that influences their segregation, discussed in Part 1 below. The third is asymmetry in the meiotic spindle that can be exploited by selfish centromeres if they preferentially attach to the egg side, discussed in Part 2.

PART 1: CENTROMERES

A classic example of selfish chromosome behavior involves the heterochromatic knob locus in maize (Rhoades 1942; Yu et al. 1997). In other situations, centromeres can represent the selfish element (Fishman and Saunders 2008; Iwata-Otsubo et al. 2017). For both noncentromeric and centromeric selfish elements in meiotic chromosome drive, the unifying concept is that they direct advantageous molecular interactions with the meiotic spindle that positively bias the transmission of the chromosome in which they reside. Knobs generate a new structure that interacts directly with spindle MTs (Yu et al. 1997) using a special minus-end-directed kinesin motor protein (K Dawe, pers commun), bypassing the typical connection that occurs through a centromere-localized kinetochore. Centromeric selfish elements could, in principal, function by biasing connections to the egg-oriented side of an asymmetric spindle.

Centromeres have long been considered the "black box" of the chromosome because in animals and most eukaryotes the underlying DNA is highly repetitive. Thus, the balance of genetic and epigenetic contributions to centromere identity and function remains difficult to nail down (Fig. 2). For instance, does the DNA sequence at centromeres even matter? As mentioned above, centromeric DNA is very rapidly evolving (Henikoff et al. 2001; Du-

mont and Fachinetti 2017). In addition, it has long been appreciated that the repetitive DNA at human centromeres is neither necessary nor sufficient for centromere function (Earnshaw and Migeon 1985; Depinet et al. 1997; du Sart et al. 1997; Warburton et al. 1997; Eichler 1999). On the other hand, human artificial chromosomes (HACs) have been reported to require specific forms of higher-order structures of human centromeric repeats (Harrington et al. 1997; Schueler et al. 2001) or artificial amplification of non-HAC forming higher-order centromere repeats (Hayden et al. 2013). Further, there is a requirement in HAC formation for the only known DNA sequence specific centromere-binding protein in metazoans, CENP-B (Ohzeki et al. 2002; Okada et al. 2007), arguing for an important genetic contribution. CENP-B also has direct physical connections to CENP-A and CENP-C at the centromere that contribute to centromere function (Fachinetti et al. 2015; Hoffmann et al. 2016). Strong data support a major epigenetic contribution to centromere identity, where the histone H3 variant, CENP-A, forms nucleosomes that are fundamental to specifying centromere location (for review, see Black and Cleveland 2011). To this point, it is possible to seed a new functional centromere capable of epigenetic propagation by initially directing a local high density of CENP-A nucleosome assembly (Barnhart et al. 2011; Mendiburo et al. 2011; Ohzeki et al. 2012; Hori et al. 2013; Chen et al. 2014; Logsdon et al. 2015; Tachiwana et al. 2015). It is attractive to think that the balance of genetic and epigenetic forces at the centromere could be at the center of a molecular "tug-of-war" that drives rapid centromere evolution (Henikoff et al. 2001). At a bare minimum, the strong evidence for both genetic and epigenetic contributions to centromere function requires that both should be considered when

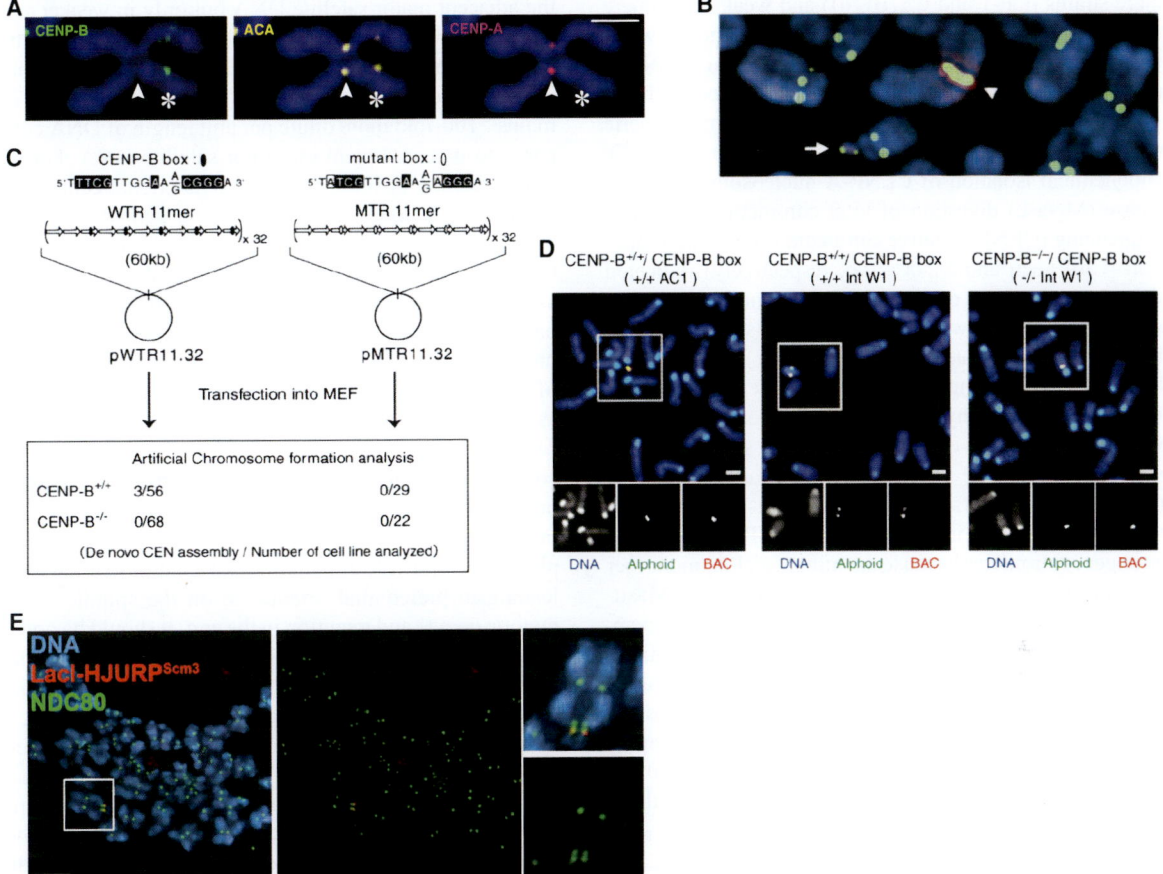

Figure 2. Foundations for our understanding of a balance between genetic and epigenetic contributions to human centromere identity. (*A*) Example of centromere silencing and de novo formation on the same chromosome. Anti-centromere antisera (ACA) recognize both CENP-A at the neocentromere (arrowhead) and CENP-B at the silenced centromere at the original location (asterisk). Scale bar, 2 μm. (*B*) Example of a HAC assay for functional centromeric DNA. The small arrow denotes the HAC formed by functional X chromosome centromere DNA, whereas the arrowhead indicates the centromere from a natural copy of the X chromosome. X chromosome centromere DNA FISH is in red. CENP-E immunofluorescence is in green. (*C*) HAC formation results indicating a requirement for both CENP-B and the CENP-B box within α-satellite DNA. WTR, wild type repeat; MTR, mutant repeat. (*D*) Example of HAC formation (artificial chromosome, AC) in mouse cells (*left*), and instances where a HAC failed to form, integrating (Int) into a natural chromosome instead (*center* and *right*). Scale bar, 2 μm. (*E*) Seeding centromeric chromatin that can form a functional kinetochore. Tethering a Lac repressor (LacI) fusion with the CENP-A binding domain (HJURP^Scm3) of HJURP to a Lac repressor (LacO) array assembles CENP-A nucleosomes, leading to formation of functional centromeric chromatin that can recruit kinetochore components such as the microtubule-binding protein, Ndc80. Scale bar, 5 μm. (*A*, Adapted, with permission, from Bassett et al. 2010; *B*, adapted from Schueler et al. 2001, with permission from The American Association for the Advancement of Science; *C*,*D*, adapted from Okada et al. 2007, with permission from Elsevier; *E*, adapted, with permission, from Barnhart et al. 2011.)

trying to decipher the molecular mechanisms of centromere drive in any particular branch of the eukaryotic evolutionary tree.

The mouse model has emerged as an exciting system to reveal the balance of epigenetic and genetic influences in the molecular arms race at centromeres. It has already provided a strong system to interrogate the cell biological basis for meiotic drive. A key early finding by one of our laboratories (Lampson's), was that different natural and laboratory strains of mice have "stronger" and "weaker" centromeres. Two key specific findings were that (1) stronger centromeres accumulated more of the kinetochore protein, Hec1/Ndc80, and (2) the imbalanced centromeres between homologous chromosomes caused aberrant alignment between the poles of the metaphase spindle of meiosis I (Chmátal et al. 2014). A hypothesis to emerge

from these studies was that the larger kinetochore of the stronger centromere strains was built on centromeric chromatin that was somehow more attractive to recruiting kinetochore components than their weaker centromere strain counterparts.

With the CENP-A nucleosome as the candidate to serve this centromeric chromatin role, we initially considered three possible ways in which this chromatin could be altered: one implicating the CENP-A protein and two implicating the DNA that wraps the CENP-A-containing histone octamer. For the CENP-A protein, because it is so extremely long-lived that it has no measurable turnover at mouse oocyte centromeres (Smoak et al. 2016), we considered that it could have substitutions in its primary sequence that might affect centromere strength through the germline of a hybrid animal. However, the strong centromere strains (CF-1 and C57BL/6J) and weak centromere strain (CHPO) all have identical protein sequences (Iwata-Otsubo et al. 2017). Thus, we focused on the CENP-A nucleosomal DNA (Fig. 3), and we considered both the sequence and the abundance of the repeating monomeric unit of mouse centromere DNA (termed "minor satellite"). Biochemical isolation of CENP-A nucleosomes and nuclease (MNase) digestion of total chromatin coupled to sequencing (CENP-A native chromatin immunoprecipitation [ChIP]-seq and MNase-seq, respectively) indicated that the sequences of minor satellite monomers were very homogeneous within a strain and also very similar between strains (Iwata-Otsubo et al. 2017). On the other hand, the sequencing experiments and complementary fluorescence in situ hybridization (FISH) experiments revealed that stronger centromeres contain six- to 10-fold more minor satellite DNA than do weaker centromeres (Fig. 3A,B; Iwata-Otsubo et al. 2017). This increase leads to a similarly large increase of the CENP-B protein on stronger centromeres because the minor satellite monomer contains its recognition element, the CENP-B box (Masumoto et al. 1989). CENP-A abundance at a centromere, and downstream centromere components (e.g., its direct binding partner, CENP-C), is limited by the expression levels of itself and proteins, such as its chromatin assembly factor, HJURP, in the epigenetic pathway for centromeric chromatin assembly (Zasadzińska and Foltz 2017). Nonetheless, CENP-A and CENP-C are increased on stronger centromeres (Iwata-Otsubo et al. 2017) to a similar extent, as is Hec1/Ndc80 (Chmátal et al. 2014). The massive differential in CENP-B levels facilitated tracking stronger and weaker centromeres of bivalent chromosomes in meiosis I. Remarkably, we observed biased orientation of the stronger centromeres toward the egg pole, with the weaker centromeres oriented toward the cortex and destined to be discarded into the polar body (Fig. 3E; Iwata-Otsubo et al. 2017). Thus, our findings explain the molecular basis for strengthening centromeres through expansion of the assembly site for CENP-A nucleosomes.

Two other important findings emerged from our genomic analysis of centromeric chromatin in the stronger and weaker centromere strains. First, nucleosomes containing CENP-A, but not those containing its canonical counterpart, histone H3, are faithful to a single nucleosome assembly site within the monomer minor satellite sequence (Fig. 3C; Iwata-Otsubo et al. 2017). Outside of the strictly genetically defined centromere of budding yeast (Clarke and Carbon 1980; Furuyama and Biggins 2007), this is the first example of which we are aware to indicate such a strong and specific positioning of CENP-A nucleosomes. Further, the center of the nucleosomal DNA (the so-called nucleosomal dyad position) falls within the CENP-B box (Fig. 3C). It is now important to test the possibility that the CENP-B protein plays an important role in positioning CENP-A nucleosomes and in mediating centromere drive. Such tests are made feasible by the fact that CENP-B is nonessential in mice (Hudson et al. 1998), unlike other centromere proteins (e.g., CENP-A and CENP-C) (Kalitsis et al. 1998; Howman et al. 2000). Second, we found that a substantial fraction of CENP-A nucleosomes shift to the adjacent major satellite DNA but only in weaker centromeres (Iwata-Otsubo et al. 2017). It is not clear if this weak centromere strain-specific shift has any functional consequence in major satellite DNA, because there it remains >100-fold more dilute per unit length of DNA compared to its enrichment on minor satellite DNA. For the minor satellite positions where the functional centromere resides, our cytological (Fig. 3D) and genomic findings that weaker centromere strains have a far higher density of CENP-A nucleosomes (Iwata-Otsubo et al. 2017) (as high as half of all nucleosomes on minor satellite DNA), compared to stronger centromere strains where CENP-A is only a very minor chromatin component (found in <8% of nucleosomes), suggest that minor satellite nucleosome assembly sites account for centromere strength.

Very broadly, our findings add an important genetic contribution on top of epigenetic contributions known to be important for centromere identity and strength. Our favored model is that the abundance of minor satellite plays a central role in determining centromere strength, leading to preferential orientation on the spindle of the meiotic oocyte and retention in the egg. It should be noted, however, that evidence in flies supports the notion that simply altering the levels of CENP-A in the male germline in one generation can influence the amount of CENP-A in the next generation (Raychaudhuri et al. 2012). A pressing issue to resolve in the future is the extent to which such epigenetic forces shape the strength of centromeric chromatin in the female germline in flies, in our mouse model for drive, and in other eukaryotes. When thinking broadly about eukaryotic evolution, it is important to note that such a fundamental chromosomal process as chromosome inheritance through the germline could be biased through multiple pathways. Sorting out the molecular basis (genetic and/or epigenetic) of centromere selfishness between closely related strains/species will likely require direct experimental interrogation in every case. Thinking back to the classic example of meiotic chromosome drive in maize where noncentromeric heterochromatic knobs direct connections to the spindle (Rhoades 1942; Yu et al. 1997), one must also always consider potential mechanisms of biased chromosome inheritance that bypass the centromere altogether.

PART 2: MEIOTIC SPINDLE OF THE OOCYTE

The biased orientation observed with hybrid bivalents in the CHPO × CF-1 model system (Fig. 3E) implies some asymmetry within the spindle that is exploited by stronger centromeres to increase their transmission to the egg. Spindle asymmetry has been reported in grasshopper (Hewitt 1976), and there are examples in other organisms that were not analyzed in depth (Crowder et al. 2015). In mouse oocytes we showed that the MI spindle is asymmetric for a specific posttranslational modification of tubulin (Janke 2014), tyrosinated α-tubulin, with more tyrosinated MTs oriented toward the cortex (Fig. 4A; Akera et al. 2017). The cortex near the spindle is enriched for CDC42 and RAC GTPases, and this cortical polarization is established by RAN^{GTP} generated around the chro-

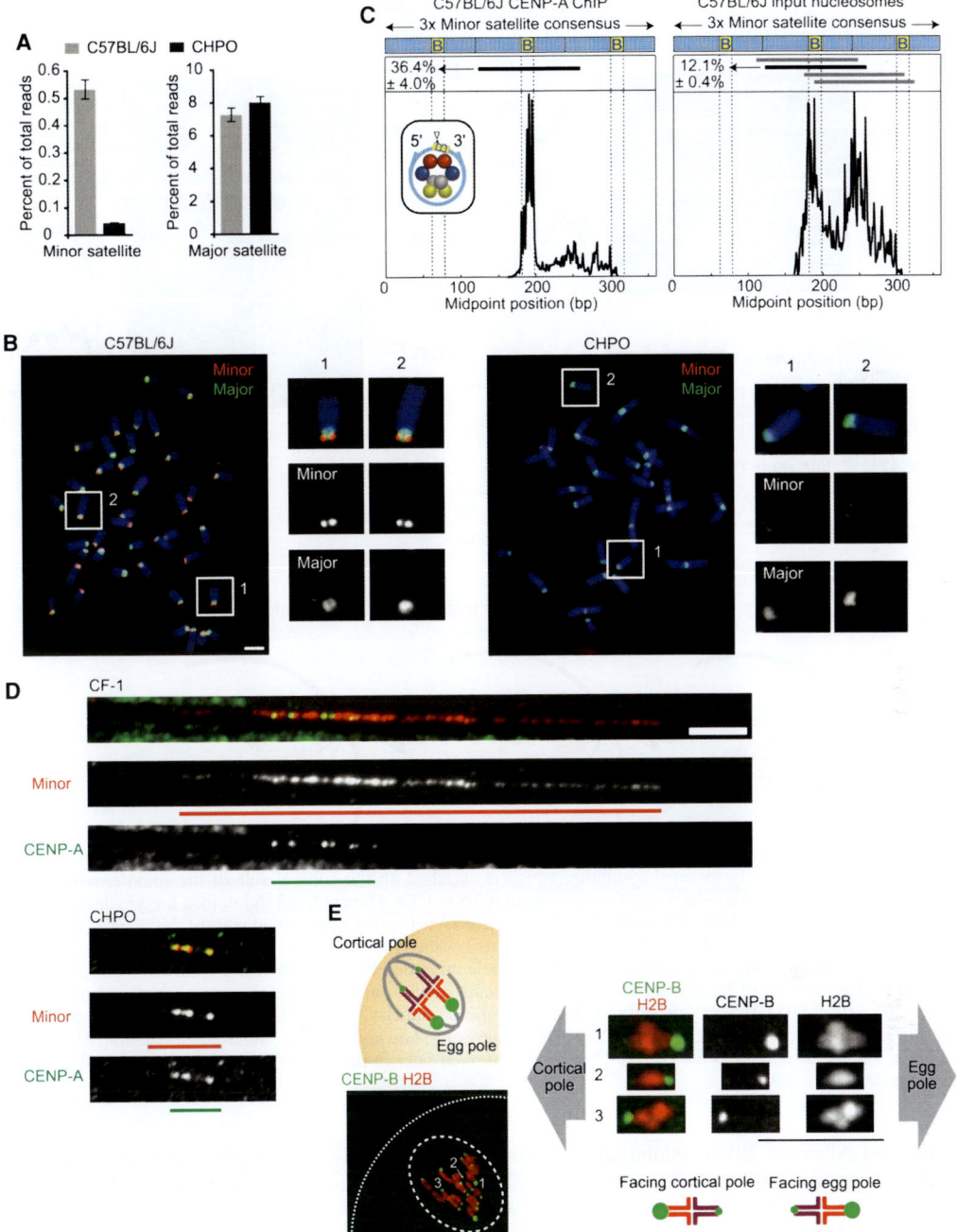

Figure 3. Amplified mouse minor satellite repeats act as selfish elements in female meiosis. (*A*) Quantitation of the MNase-seq reads from a stronger centromere strain (C57BL/6J) and weaker centromere strain (CHPO). Weaker centromeres have a much smaller amount of minor satellite DNA while maintaining a similar level of major satellite DNA relative to stronger centromeres. (*B*) FISH analysis also shows that weaker centromeres have only very low levels of minor satellite DNA. Scale bar, 5 µm. (*C*) CENP-A nucleosomes are specifically phased with one primary position within the minor satellite repeat monomer unit. CENP-A ChIP-seq analysis of midpoints of CENP-A nucleosomes shows a striking enrichment for a single primary assembly site within the monomer unit of mouse minor satellite (three tandem monomers are shown in the diagram at *top*; the horizontal black line indicates the primary CENP-A nucleosome assembly site). Notably, the midpoint (dyad axis of symmetry of the nucleosome, marked by a triangle in the *inset* nucleosome schematic) is within the CENP-B box. Canonical nucleosomes are the major form of nucleosome on minor satellite DNA in stronger centromere mouse strains, and the input "bulk" nucleosomes are not nearly as well-phased compared to CENP-A nucleosomes. The CENP-A specific phasing suggests a specific connection between the genetic and epigenetic factors involved in specifying mouse centromeres. (*D*) CENP-A nucleosomes fill the minor satellite region at weaker centromeres. Images of CENP-A and minor satellite localized by immunofluorescence and FISH, respectively, on extended chromatin fibers from stronger (CF1, *top* images) or weaker (CHPO, *bottom* images) centromere strains. Green and red bars show the length of CENP-A and minor satellite signals, respectively. Scale bar, 5 µm. CENP-A nucleosomes fill the minor satellite region at weaker centromeres (*bottom* panels) but not stronger centromeres (*top* panels). (*E*) Stronger centromeres orient preferentially to the egg in meiosis I. Schematic shows bivalents in CF-1 × CHPO oocytes, with CF-1 centromeres facing the egg. Image shows a CF-1 × CHPO oocyte expressing CENP-B-EGFP and H2B-mCherry, shortly before anaphase onset; dashed white lines show cortex and spindle outline. The orientation of each bivalent was determined using CENP-B-EGFP intensity to distinguish CF-1 (brighter) and CHPO (dimmer) centromeres. Asterisk indicates that it is significantly different from 50%. Scale bar, 10 µm. (Adapted, with permission, from Iwata-Otsubo et al. 2017.)

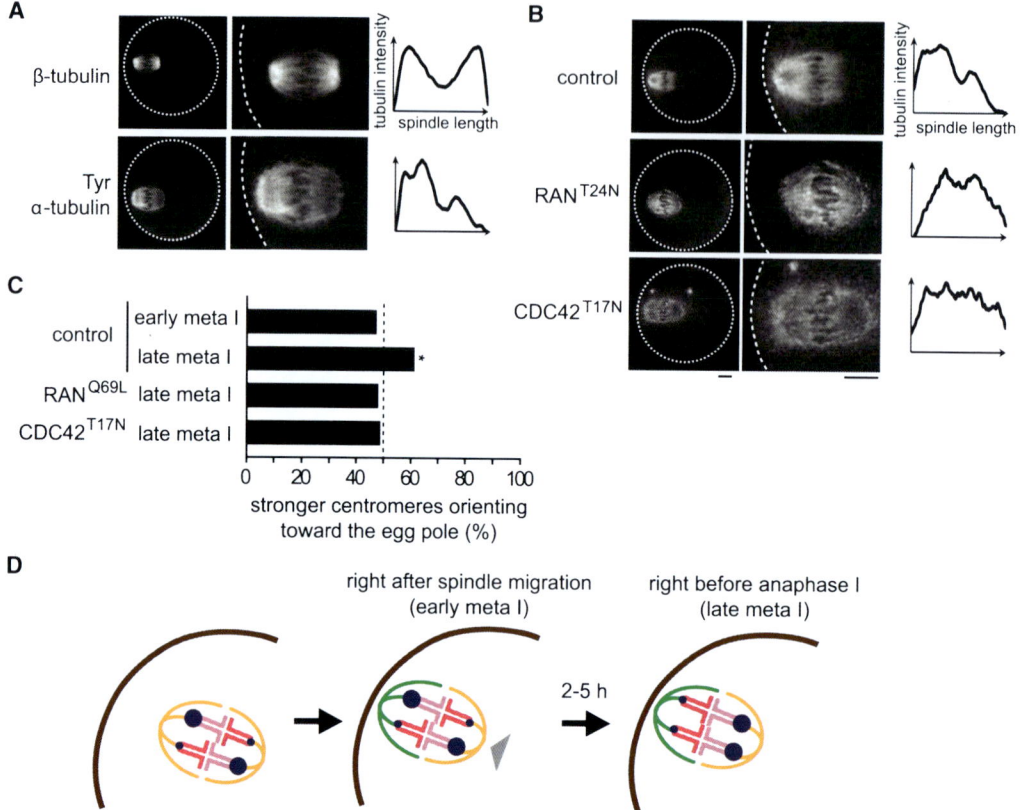

Figure 4. Spindle asymmetry and biased orientation of selfish centromeres. (*A*) Oocytes fixed and stained at metaphase I show asymmetry within the spindle for tyrosinated α-tubulin, which is enriched on the cortical side of the spindle, whereas β-tubulin is symmetric. (*B*) Oocytes expressing dominant negative mutants of RAN or CDC42 were fixed and stained for tyrosinated α-tubulin. Both mutants prevent spindle asymmetry. Images *A* and *B* show the whole oocyte (*left*) or a magnified view of the spindle (*right*); dashed line, cortex; scale bars, 10 μm. Graphs are line scans of tubulin intensity across the spindle. (*C*) Bivalent orientation was measured in CHPO × CF-1 hybrid oocytes as in Fig. 3, either shortly after spindle migration to the cortex (early meta I), or shortly before anaphase onset (late meta I). The fraction of bivalents with the stronger centromere oriented toward the egg is shown (*indicates significant deviation from 50%, *P* < 0.005). Biased orientation is lost if spindle asymmetry is prevented by expression of a constitutively active RAN mutant or a dominant negative CDC42 mutant. (*D*) Schematic showing initially unbiased orientation of hybrid bivalents immediately after spindle migration to the cortex. The bias is observed later in metaphase I. (Portions reproduced from Akera et al. 2017, with permission from AAAS.)

mosomes (Li and Albertini 2013). Inhibition of either RAN or CDC42 prevents both spindle asymmetry and biased orientation of hybrid bivalents in our model system (Fig. 4B,C), which has two important implications (Akera et al. 2017). First, spindle asymmetry is established by localized CDC42 activity at the cortex (Fig. 5A). Second, centromeres can interact with the asymmetric spindle to bias their orientation. Because the asymmetry is generated by a signal from the cortex, it has a consistent orientation with the tyrosinated side toward the cortex, and therefore provides spatial cues that can be exploited by selfish centromeres. One important outstanding question arising from this work is how CDC42 regulates tubulin tyrosination, but here we will focus on a different question: how bivalents with weaker and stronger centromeres interact with the asymmetric spindle to bias their transmission to the egg.

One possible model is that MTs from the egg side of the spindle preferentially capture stronger centromeres, and these attachments are maintained until anaphase. If MTs on the egg side were more dynamic, for example, they

might initially interact with centromeres that present a larger kinetochore target with more MT-binding proteins (Chmátal et al. 2017). Weaker centromeres would then capture MTs from the cortical side, leading to biased orientation. A second model is that bivalents sample both configurations, and the one with stronger centromeres toward the more tyrosinated MTs on the cortical side is labile and will tend to reorient, whereas the opposite configuration is more stable (Fig. 5B). This trial-and-error mechanism, in which the preferred configuration is selectively stabilized, is analogous to the long-standing model for how correct, bi-oriented attachments are stabilized by tension (Nicklas 1997; Lampson and Grishchuk 2017).

The timing of meiotic events in our model system supports the second model. The MI spindle forms around the chromosomes, which are initially positioned in the center of the oocyte, and initial kinetochore-MT attachments are established at this time (Kitajima et al. 2011). The chromosomes and spindle then migrate together to the cortex to allow the highly asymmetric cell division that preserves most of the cytoplasm in the egg while extruding a rela-

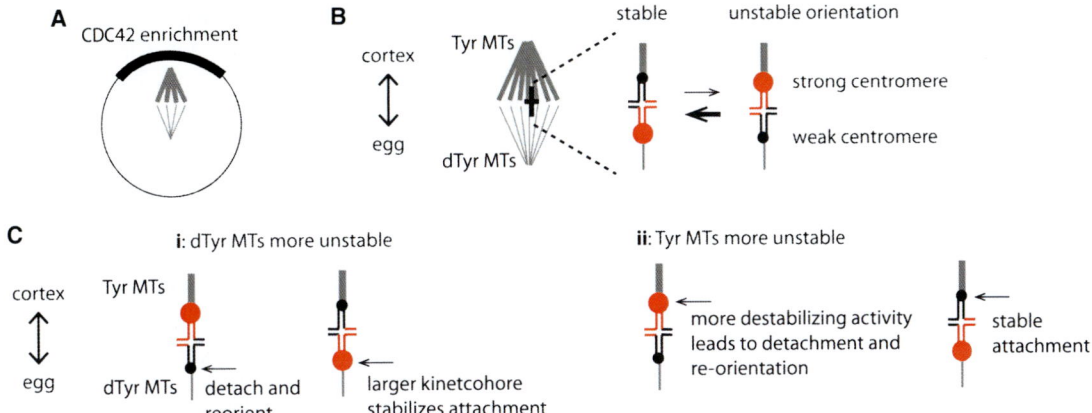

Figure 5. Model for biased chromosome orientation. (*A*) Spindle asymmetry and cortical CDC42 enrichment. The cortical side of the spindle has higher levels of tyrosinated α-tubulin compared to the egg side. (*B*) The orientation with the stronger centromere binding to more tyrosinated (Tyr) MTs and the weaker centromere to more detyrosinated (dTyr) MTs is labile and will tend to reorient. The opposite configuration is more stable and persists. (*C*) Models for how one bivalent configuration is selectively stabilized relative to the other. If dTyr MTs are more unstable (i), weaker centromeres may detach to initiate reorientation on the egg side. Conversely, stronger centromeres stabilize interactions on that side of the spindle to prevent reorientation. Alternatively, if Tyr MTs are more unstable (ii), stronger centromeres may initiate reorientation by destabilizing interactions with the cortical side of the spindle.

tively small polar body. We find that spindles are asymmetric only when positioned near the cortex. Furthermore, orientation of hybrid bivalents is unbiased immediately after migration to the cortex, and the bias is established later (Fig. 4C,D; Akera et al. 2017). These findings indicate that bivalents preferentially reorient on the asymmetric spindle to bias stronger centromeres toward the egg side.

Preferential reorientation indicates both differences between stronger and weaker centromeres in how they interact with spindle MTS and a difference between the two sides of the spindle, such that one configuration is selectively stabilized relative to the other. One model is that detyrosinated MTs on the egg side of the spindle tend to make more unstable kinetochore attachments, but stronger centromeres stabilize interactions with the egg side by building kinetochores with more MT-binding activity (Fig. 5Ci). The other configuration, with weaker centromeres connected to these unstable MTs, would be labile and likely to reorient. In this model, stronger centromeres win by preventing reorientation after reaching their preferred configuration. Supporting this model, stronger centromeres do build kinetochores with increased levels of kinetochore proteins such as CENP-A and CENP-C and the major MT-binding protein, HEC1 (Chmátal et al. 2014; Iwata-Otsubo et al. 2017). Conversely, an alternative model is that tyrosinated MTs on the cortical side make more unstable kinetochore attachments (Fig. 5Cii). In this case stronger centromeres can win by preferentially destabilizing the configuration where they face the cortical side, triggering reorientation toward the egg side. The configuration with weaker centromeres toward the cortical side would persist if these centromeres have less destabilizing activity.

Our findings are consistent with the second model, in which stronger centromeres destabilize their interactions with the cortical side of the spindle rather than stabilizing interactions with the egg side. Briefly, we showed that kinetochore MTs attached to stronger centromeres are

more unstable than those of weaker centromeres, based on sensitivity to low temperature, and the attachments are most unstable when facing the cortical side. Furthermore, manipulating tubulin tyrosination levels showed that more tyrosinated MTs are more unstable (Akera et al. 2017). These results suggest that increased MT destabilizing activity at stronger centromeres drives preferential reorientation when facing the cortical side, which is more tyrosinated. Predictions of this model, for example that reorientation events are initiated by the stronger centromere, still need to be tested directly.

Overall, our results provide a cell biological framework for centromere drive. Chromosome positioning near the cell cortex is crucial for the highly asymmetric division in female meiosis (Li and Albertini 2013). The chromosomes direct cortical polarization by producing RAN^{GTP}, and the resulting enrichment of CDC42 on the polarized cortex generates asymmetry in α-tubulin tyrosination in the spindle. Spindle asymmetry is thus inherent to female meiosis, at least in this system, and can be exploited by centromeres or other selfish elements to bias their transmission to the egg. Observations in our model system indicate that centromeres can bias their orientation to the egg side of the spindle by preferentially destabilizing their interactions with tyrosinated MTs. Whether this strategy represents a broader theme in female drive systems is currently unknown because of the paucity of cell biological studies. One prominent example is knobs that drive in maize meiosis, which are not centromere linked and have adopted a different strategy based on recruiting a motor that positions them favorably on the spindle (Dawe 2009).

A natural question that follows from our model is the molecular identity of the MT destabilizing activity. Candidates include Aurora B kinase, which localizes to centromere chromatin and phosphorylates kinetochore substrates that bind MTs to destabilize these interactions, and kinesin-13 family members such as MCAK (mitotic centromere-associated kinesin), which catalyze MT depolymerization. Intriguingly, MCAK prefers tyrosinated MTs as a substrate (Peris et al. 2009; Sirajuddin et al. 2014). If stronger centromeres are enriched for MCAK activity, they would preferentially destabilize interactions with the cortical side of the spindle, which is more tyrosinated, driving reorientation. Future work will address this molecular interpretation of the destabilization model.

Progress in our model system for centromere drive raises additional pressing questions for future research. Our findings suggest that natural selection favors centromeres that increase destabilizing activity, specifically acting on tyrosinated spindle MTs. What are the underlying mechanisms that determine this activity, and what is the molecular link to increased levels of centromere chromatin at stronger centromeres, as indicated by CENP-A and CENP-C levels? Our molecular understanding of centromere assembly and function has advanced rapidly in recent years, and integrating this knowledge into models of centromere drive and selective pressures shaping centromere evolution promises to be an exciting research direction.

CONCLUSION

The centromere drive hypothesis originated as an explanation for why signatures of positive selection are detected in centromere proteins. To briefly restate the hypothesis, natural selection favors changes in centromere DNA that increase transmission to the egg in female meiosis, but there is a fitness cost associated with unbalanced centromeres, so selection also favors changes in centromere proteins that equalize centromeres and suppress drive. Work in our mouse model system so far has focused on the first part: understanding how stronger centromeres win in female meiosis. If there is a fitness cost associated with such drive, it has not yet been revealed, and how centromere proteins may have evolved to minimize this cost is unknown. Thus, the primary question underlying the drive hypothesis—whether suppression of drive is the selective pressure responsible for rapid evolution of centromere proteins—remains unanswered. Based on our results, such suppression might reduce the influence of DNA sequence on centromere function (e.g., by minimizing the contribution of CENP-B, which is the only protein known to bind directly to minor satellite sequences). Negating any sequence dependence of interactions between DNA and either CENP-A or CENP-C would also suppress drive, although such dependence has yet to be demonstrated. Alternatively, if centromeres win by increasing MT destabilizing activity, drive could be suppressed by weakening the link between centromere expansion and recruitment of those activities. Testing these models is an important challenge for future investigations of centromere drive.

ACKNOWLEDGMENTS

We thank K. Dawe for sharing unpublished data, H. Willard, H. Masumoto, and D. Foltz for permission to reproduce data in Fig. 2, T. Akera and J. Dawicki-McKenna for comments on the manuscript, and members of the Black and Lampson labs for helpful discussions. Work in our labs is supported by the National Institutes of Health (NIH) (GM082989 to B.E.B. and GM122475 to M.A.L.).

REFERENCES

Akera T, Chmátal L, Trimm E, Yang K, Aonbangkhen C, Chenoweth DM, Janke C, Schultz RM, Lampson MA. 2017. Spindle asymmetry drives non-Mendelian chromosome segregation. *Science* **358:** 668–672.

Barnhart MC, Kuich PH, Stellfox ME, Ward JA, Bassett EA, Black BE, Foltz DR. 2011. HJURP is a CENP-A chromatin assembly factor sufficient to form a functional de novo kinetochore. *J Cell Biol* **194:** 229–243.

Bassett EA, Wood S, Salimian KJ, Ajith S, Foltz DR, Black BE. 2010. Epigenetic centromere specification directs Aurora B accumulation but is insufficient to efficiently correct mitotic errors. *J Cell Biol* **190:** 177–185.

Black BE, Cleveland DW. 2011. Epigenetic centromere propagation and the nature of CENP-A nucleosomes. *Cell* **144:** 471–479.

Chen CC, Dechassa ML, Bettini E, Ledoux MB, Belisario C, Heun P, Luger K, Mellone BG. 2014. CAL1 is the *Drosophila* CENP-A assembly factor. *J Cell Biol* **204:** 313–329.

Chmátal L, Gabriel SI, Mitsainas GP, Martínez-Vargas J, Ventura J, Searle JB, Schultz RM, Lampson MA. 2014. Centromere strength provides the cell biological basis for meiotic drive and karyotype evolution in mice. *Curr Biol* **24:** 2295–2300.

Chmátal L, Schultz RM, Black BE, Lampson MA. 2017. Cell biology of cheating—Transmission of centromeres and other selfish elements through asymmetric meiosis. *Prog Mol Subcell Biol* **56:** 377–396.

Clarke L, Carbon J. 1980. Isolation of a yeast centromere and construction of functional small circular chromosomes. *Nature* **287:** 504–509.

Crowder ME, Strzelecka M, Wilbur JD, Good MC, von Dassow G, Heald R. 2015. A comparative analysis of spindle morphometrics across metazoans. *Curr Biol* **25:** 1542–1550.

Dawe RK. 2009. Maize centromeres and knobs (neocentromeres). In *Handbook of maize genetics and genomics* (ed. Bennetzen JL, Hake S), pp. 239–250. Springer, New York.

Depinet TW, Zackowski JL, Earnshaw WC, Kaffe S, Sekhon GS, Stallard R, Sullivan BA, Vance GH, Van Dyke DL, Willard HF, et al. 1997. Characterization of neo-centromeres in marker chromosomes lacking detectable alpha-satellite DNA. *Hum Mol Genet* **6:** 1195–1204.

du Sart D, Cancilla MR, Earle E, Mao JI, Saffery R, Tainton KM, Kalitsis P, Martyn J, Barry AE, Choo KH. 1997. A functional neo-centromere formed through activation of a latent human centromere and consisting of non-α-satellite DNA. *Nat Genet* **16:** 144–153.

Dumont M, Fachinetti D. 2017. DNA sequences in centromere formation and function. *Prog Mol Subcell Biol* **56:** 305–336.

Earnshaw WC, Migeon BR. 1985. Three related centromere proteins are absent from the inactive centromere of a stable isodicentric chromosome. *Chromosoma* **92:** 290–296.

Eichler EE. 1999. Repetitive conundrums of centromere structure and function. *Hum Mol Genet* **8:** 151–155.

Fachinetti D, Han JS, McMahon MA, Ly P, Abdullah A, Wong AJ, Cleveland DW. 2015. DNA sequence-specific binding of CENP-B enhances the fidelity of human centromere function. *Dev Cell* **33:** 314–327.

Fishman L, Saunders A. 2008. Centromere-associated female meiotic drive entails male fitness costs in monkeyflowers. *Science* **322:** 1559–1562.

Furuyama S, Biggins S. 2007. Centromere identity is specified by a single centromeric nucleosome in budding yeast. *Proc Natl Acad Sci* **104:** 14706–14711.

Gorelick R, Carpinone J, Derraugh LJ. 2016. No universal differences between female and male eukaryotes: Anisogamy and asymmetrical female meiosis. *Biol J Linn Soc* **120**: 1–21.

Harrington JJ, Van Bokkelen G, Mays RW, Gustashaw K, Willard HF. 1997. Formation of de novo centromeres and construction of first-generation human artificial microchromosomes. *Nat Genet* **15**: 345–355.

Hayden KE, Strome ED, Merrett SL, Lee HR, Rudd MK, Willard HF. 2013. Sequences associated with centromere competency in the human genome. *Mol Cell Biol* **33**: 763–772.

Helleu Q, Gérard PR, Montchamp-Moreau C. 2015. Sex chromosome drive. *Cold Spring Harb Perspect Biol* **7**: a017616.

Henikoff S, Ahmad K, Malik HS. 2001. The centromere paradox: Stable inheritance with rapidly evolving DNA. *Science* **293**: 1098–1102.

Hewitt GM. 1976. Meiotic drive for B-chromosomes in the primary oocytes of *Myrmeleotettix maculatus* (Orthopera: Acrididae). *Chromosoma* **56**: 381–391.

Hoffmann S, Dumont M, Barra V, Ly P, Nechemia-Arbely Y, McMahon MA, Hervé S, Cleveland DW, Fachinetti D. 2016. CENP-A is dispensable for mitotic centromere function after initial centromere/kinetochore assembly. *Cell Rep* **17**: 2394–2404.

Hori T, Shang W-H, Takeuchi K, Fukagawa T. 2013. The CCAN recruits CENP-A to the centromere and forms the structural core for kinetochore assembly. *J Cell Biol* **200**: 45–60.

Howman EV, Fowler KJ, Newson AJ, Redward S, MacDonald AC, Kalitsis P, Choo KH. 2000. Early disruption of centromeric chromatin organization in centromere protein A (Cenpa) null mice. *Proc Natl Acad Sci* **97**: 1148–1153.

Hudson DF, Fowler KJ, Earle E, Saffery R, Kalitsis P, Trowell H, Hill J, Wreford NG, de Kretser DM, Cancilla MR, et al. 1998. Centromere protein B null mice are mitotically and meiotically normal but have lower body and testis weights. *J Cell Biol* **141**: 309–319.

Iwata-Otsubo A, Dawicki-McKenna JM, Akera T, Falk SJ, Chmátal L, Yang K, Sullivan BA, Schultz RM, Lampson MA, Black BE. 2017. Expanded satellite repeats amplify a discrete CENP-A nucleosome assembly site on chromosomes that drive in female meiosis. *Curr Biol* **27**: 2365–2373.e8.

Janke C. 2014. The tubulin code: Molecular components, readout mechanisms, and functions. *J Cell Biol* **206**: 461–472.

Kalitsis P, Fowler KJ, Earle E, Hill J, Choo KH. 1998. Targeted disruption of mouse centromere protein C gene leads to mitotic disarray and early embryo death. *Proc Natl Acad Sci* **95**: 1136–1141.

Kitajima TS, Ohsugi M, Ellenberg J. 2011. Complete kinetochore tracking reveals error-prone homologous chromosome biorientation in mammalian oocytes. *Cell* **146**: 568–581.

Lampson MA, Grishchuk EL. 2017. Mechanisms to avoid and correct erroneous kinetochore-microtubule attachments. *Biology (Basel)* **6**: 1.

Li R, Albertini DF. 2013. The road to maturation: Somatic cell interaction and self-organization of the mammalian oocyte. *Nat Rev Mol Cell Biol* **14**: 141–152.

Lindholm AK, Dyer KA, Firman RC, Fishman L, Forstmeier W, Holman L, Johannsesson H, Knief U, Kokko H, Larracuente AM, et al. 2016. The ecology and evolutionary dynamics of meiotic drive. *Trends Ecol Evol* **31**: 315–326.

Logsdon GA, Barrey EJ, Bassett EA, DeNizio JE, Guo LY, Panchenko T, Dawicki-McKenna JM, Heun P, Black BE. 2015. Both tails and the centromere targeting domain of CENP-A are required for centromere establishment. *J Cell Biol* **208**: 521–531.

Malik HS, Henikoff S. 2001. Adaptive evolution of Cid, a centromere-specific histone in *Drosophila*. *Genetics* **157**: 1293–1298.

Masumoto H, Masukata H, Muro Y, Nozaki N, Okazaki T. 1989. A human centromere antigen (CENP-B) interacts with a short specific sequence in alphoid DNA, a human centromeric satellite. *J Cell Biol* **109**: 1963–1973.

Mendiburo MJ, Padeken J, Fülöp S, Schepers A, Heun P. 2011. *Drosophila* CENH3 is sufficient for centromere formation. *Science* **334**: 686–690.

Nicklas RB. 1997. How cells get the right chromosomes. *Science* **275**: 632–637.

Ohzeki J, Nakano M, Okada T, Masumoto H. 2002. CENP-B box is required for de novo centromere chromatin assembly on human alphoid DNA. *J Cell Biol* **159**: 765–775.

Ohzeki J, Bergmann JH, Kouprina N, Noskov VN, Nakano M, Kimura H, Earnshaw WC, Larionov V, Masumoto H. 2012. Breaking the HAC barrier: Histone H3K9 acetyl/methyl balance regulates CENP-A assembly. *EMBO J* **31**: 2391–2402.

Okada T, Ohzeki J, Nakano M, Yoda K, Brinkley WR, Larionov V, Masumoto H. 2007. CENP-B controls centromere formation depending on the chromatin context. *Cell* **131**: 1287–1300.

Pardo-Manuel de Villena F, Sapienza C. 2001. Nonrandom segregation during meiosis: The unfairness of females. *Mamm Genome* **12**: 331–339.

Peris L, Wagenbach M, Lafanechère L, Brocard J, Moore AT, Kozielski F, Job D, Wordeman L, Andrieux A. 2009. Motor-dependent microtubule disassembly driven by tubulin tyrosination. *J Cell Biol* **185**: 1159–1166.

Raychaudhuri N, Dubruille R, Orsi GA, Bagheri HC, Loppin B, Lehner CF. 2012. Transgenerational propagation and quantitative maintenance of paternal centromeres depends on Cid/Cenp-A presence in *Drosophila* sperm. *PLoS Biol* **10**: e1001434.

Rhoades MM. 1942. Preferential segregation in maize. *Genetics* **27**: 395–407.

Rice WR. 2013. Nothing in genetics makes sense except in light of genomic conflict. *Annu Rev Ecol Evol Syst* **44**: 217–237.

Sandler L, Novitski E. 1957. Meiotic drive as an evolutionary force. *Am Nat* **91**: 105–110.

Schueler MG, Higgins AW, Rudd MK, Gustashaw K, Willard HF. 2001. Genomic and genetic definition of a functional human centromere. *Science* **294**: 109–115.

Schueler MG, Swanson W, Thomas PJ; NISC Comparative Sequencing Program, Green ED. 2010. Adaptive evolution of foundation kinetochore proteins in primates. *Mol Biol Evol* **27**: 1585–1597.

Sirajuddin M, Rice LM, Vale RD. 2014. Regulation of microtubule motors by tubulin isotypes and post-translational modifications. *Nat Cell Biol* **16**: 335–344.

Smoak EM, Stein P, Schultz RM, Lampson MA, Black BE. 2016. Long-term retention of CENP-A nucleosomes in mammalian oocytes underpins transgenerational inheritance of centromere identity. *Curr Biol* **26**: 1110–1116.

Tachiwana H, Müller S, Blümer J, Klare K, Musacchio A, Almouzni G. 2015. HJURP involvement in de novo CenH3 (CENP-A) and CENP-C recruitment. *Cell Rep* **11**: 22–32.

Talbert PB, Bryson TD, Henikoff S. 2004. Adaptive evolution of centromere proteins in plants and animals. *J Biol* **3**: 18.

Warburton PE, Cooke CA, Bourassa S, Vafa O, Sullivan BA, Stetten G, Gimelli G, Warburton D, Tyler-Smith C, Sullivan KF, et al. 1997. Immunolocalization of CENP-A suggests a distinct nucleosome structure at the inner kinetochore plate of active centromeres. *Curr Biol* **7**: 901–904.

Werren JH. 2011. Selfish genetic elements, genetic conflict, and evolutionary innovation. *Proc Natl Acad Sci* **108**(Suppl): 10863–10870.

Yu H-G, Hiatt EN, Chan A, Sweeney M, Dawe RK. 1997. Neocentromere-mediated chromosome movement in maize. *J Cell Biol* **139**: 831–840.

Zasadzińska E, Foltz DR. 2017. Orchestrating the specific assembly of centromeric nucleosomes. *Prog Mol Subcell Biol* **56**: 165–192.

Zedek F, Bureš P. 2016. CenH3 evolution reflects meiotic symmetry as predicted by the centromere drive model. *Sci Rep* **6**: 33308.

Hierarchical Regulation of Centromeric Cohesion Protection by Meikin and Shugoshin during Meiosis I

Seira Miyazaki,[1,2] Jihye Kim,[3] Takeshi Sakuno,[1,2] and Yoshinori Watanabe[1,2]

[1]*Graduate Program in Biophysics and Biochemistry, Graduate School of Science, University of Tokyo, 1-1-1Yayoi, Tokyo 113-0032, Japan*

[2]*Laboratory of Chromosome Dynamics, Institute of Molecular and Cellular Biosciences, University of Tokyo, 1-1-1Yayoi, Tokyo 113-0032, Japan*

[3]*Research Institute, National Cancer Center, Goyang, Gyeonggi 410-769, Republic of Korea*

Correspondence: ywatanab@iam.u-tokyo.ac.jp

The kinetochore is the key apparatus regulating chromosome segregation. Particularly in meiosis, unlike in mitosis, sister kinetochores are captured by microtubules emanating from the same spindle pole (mono-orientation), and sister chromatid cohesion mediated by cohesin is protected at centromeres in the following anaphase. Shugoshin, which localizes to centromeres depending on the phosphorylation of histone H2A by Bub1 kinase, plays a central role in protecting meiotic cohesin Rec8 from separase cleavage. Another key meiotic kinetochore factor, Moa1 (meikin), which was initially characterized as a mono-orientation factor in fission yeast, also regulates cohesion protection. Moa1, which associates stably with CENP-C during meiosis I, recruits Plo1 (polo-like kinase) to the kinetochores and phosphorylates Spc7 (KNL1), inducing the persistent accumulation of Bub1 at kinetochores. The meiotic Bub1 pool ensures robust Sgo1 (shugoshin) localization and cohesion protection at centromeres by cooperating with heterochromatin protein Swi6, which binds and stabilizes Sgo1. Further, molecular genetic analyses reveal a hierarchical regulation of centromeric cohesion protection by meikin and shugoshin during meiosis I.

For the proper transmission of the genetic information, faithful chromosome segregation is essential in all organisms. During the cell cycle, sister chromatid cohesion is established in S phase dependent on the cohesin complex, and is maintained until metaphase. Sister chromatid cohesion at centromeres is essential to establish chromosome biorientation, in which sister kinetochores are captured by spindle microtubules from opposite poles. In the transition from metaphase to anaphase, the anaphase-promoting complex (APC) triggers the degradation of securin, an inhibitory chaperone for separase that cleaves cohesin and removes cohesin along the entire chromosome (Uhlmann et al. 1999; Onn et al. 2008; Peters et al. 2008; Nasmyth and Haering 2009). During meiosis, however, one round of DNA replication is followed by two consecutive nuclear divisions, resulting in the production of four haploid nuclei or gametes. In meiosis I, sister chromatids are captured from the same pole (mono-orientation), whereas homologous chromosomes (homologs) connected by chiasmata are captured by spindle microtubules emanating from the opposite poles (Fig. 1). At the onset of anaphase I, cohesin is cleaved by separase along the arm regions but protected at centromeres until metaphase II (Buonomo et al. 2000; Kitajima et al. 2003a; Tachibana-Konwalski et al. 2010). Thus, mono-orientation and the protection of centromeric cohesion are two hallmarks of the regulation of meiotic chromosome segregation (Moore and Orr-Weaver 1998; Petronczki et al. 2003; Brar and Amon 2008; Watanabe 2012; Duro and Marston 2015).

COHESION PROTECTION

Cohesin complexes are modified in meiosis. Especially, the Rad21(Scc1) subunit is largely replaced by its meiotic counterpart, Rec8 (Klein et al. 1999; Watanabe and Nurse 1999). During anaphase of meiosis I, Rec8 is cleaved only along the chromosome arms by separase, whereas centromeric Rec8 is preserved until meiosis II. If Rec8 is replaced by Rad21 during meiosis, sister chromatid cohesion, but not protection at the centromeres, is restored, leading to the separation of sister chromatids at meiosis I. Therefore, an intrinsic property of the Rec8 subunit absent from Rad21 contributes to centromeric protection at meiosis I (Toth et al. 2000; Yokobayashi et al. 2003). Pericentric heterochromatin plays a crucial role in enriching cohesin complexes and, thereby, strengthens centromeric cohesion in mitosis (Bernard et al. 2001b; Nonaka et al. 2002; Fukagawa et al. 2004). This might be applicable in meiosis, because the localization of the Rec8 complex is reduced from the pericentric regions in heterochromatin mutants (Kitajima et al. 2003b).

A functional screening of fission yeast identified the Rec8 protector as a gene that causes the disjunction of chromosomes, and thus it is toxic during mitotic growth only when co-expressed with Rec8 but not with Rad21 (Kitajima et al. 2004). This gene encodes a meiosis-specific protein named shugoshin (Sgo1), which means "guardian spirit" in Japanese. Sgo1 localizes exclusively at pericentric heterochromatin regions, the site at which

Published by Cold Spring Harbor Laboratory Press; doi: 10.1101/sqb.2017.82.033811

Cold Spring Harbor Symposia on Quantitative Biology, Volume LXXXII

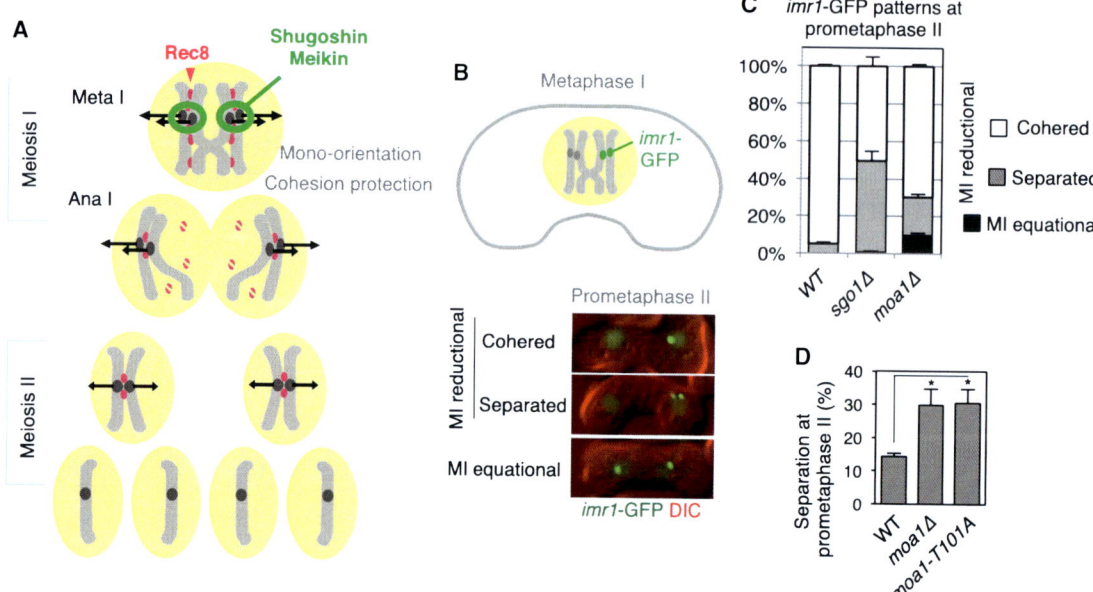

Figure 1. Moa1–Plo1 promotes cohesion protection at centromeres. (*A*) Schematic image of chromosome segregation during meiosis. (*B*) The green dot represents an *imr1*-GFP-labeled centromere. In normal meiosis I (MI), sisters are captured by spindles from the same pole, and heterozygous *imr1*-GFP signals move to one side (reductional segregation). The majority of *imr1*-GFP signals in prometaphase II zygote are observed as one dot (cohered) because centromeric cohesion is protected, whereas some cells exhibit two dots (separated) because of the failure of cohesion protection. If mono-orientation is compromised at meiosis I, some sisters would segregate to opposite sides (equational segregation). (*C*) The chromosome segregation pattern during meiosis I and the splitting of centromeres were counted by observing heterozygous *imr1*-GFP signals in prophase II arrest (by the *mes1-B44* mutation) in the indicated cells. Error bars, SD, *n* > 150 cells, three independent experiments. (*D*) The frequency of the splitting of centromeres was measured by observing heterozygous *imr1*-GFP signals in prometaphase II (prometa II) arrest (*mes1*) in the indicated cells. WT, wild type. (Modified, with permission, from Miyazaki et al. 2017, © John Wiley & Sons, Inc.)

Rec8 was predicted to play a role in the centromeric protection at meiosis I (Kitajima et al. 2003b). Independent knockout screening in fission yeast and budding yeast also identified the *sgo1/SGO1* gene (Marston et al. 2004; Rabitsch et al. 2004) as a Rec8 protector in meiosis. Remarkably, it turns out that shugoshin shares a hitherto unperceived limited similarity to MEI-S332, a *Drosophila* protein that was previously shown to be required for the persistence of centromeric cohesion during meiosis I (Davis 1971; Kerrebrock et al. 1995; Lee and Orr-Weaver 2001). Analyses in several eukaryotic organisms indicate that meiotic cohesin protection at centromeres is mediated by the centromeric protein shugoshin (Sgo1 in fission yeast). Shugoshin forms a complex with protein phosphatase 2A (PP2A) at the centromeres (Kitajima et al. 2006; Riedel et al. 2006; Tang et al. 2006; Lee et al. 2008; Llano et al. 2008) and antagonizes Rec8 phosphorylation, a prerequisite for cleavage by separase in meiosis I (Ishiguro et al. 2010; Katis et al. 2010).

Bub1, a well-conserved spindle checkpoint kinase, is required to preserve centromeric protection during meiosis I in fission yeast (Bernard et al. 2001a). Indeed, Sgo1 fails to localize at centromeres in *bub1* mutants (Kitajima et al. 2004). A biochemical approach in fission yeast identified serine 121 of histone 2A (H2A) as a Bub1 substrate (Kawashima et al. 2010). Crucially, Sgo1 binds nucleosomes including phosphorylated H2A-S121, accounting for the mechanism of Sgo1 localization at centromeres (Kawashima et al. 2010). Sgo1 localization is also pro-

moted by pericentric heterochromatin protein Swi6 (HP1), which binds directly to Sgo1 (Yamagishi et al. 2008). In mitosis, Bub1 as a complex with Bub3 localizes to kinetochores through its interaction with Spc7 (KNL1) only when Spc7 is phosphorylated by Mph1 (MPS1). Mph1 is a conserved protein kinase required for the spindle assembly checkpoint (SAC) locating at unattached or misaligned kinetochores (Yamagishi et al. 2012; London and Biggins 2014; Musacchio 2015; Sacristan and Kops 2015). Therefore, Bub1 is largely released from kinetochores when chromosomes are aligned during metaphase in mitotic cells. Recent reports reveal that KNL1 is also phosphorylated by PLK1 in humans and *Caenorhabditis elegans* (Espeut et al. 2015; von Schubert et al. 2015), suggesting the existence of a versatile regulation of Bub1 localization.

MONO-ORIENTATION VERSUS BIORIENTATION

The geometric aspect of kinetochores has been long recognized in vertebrates (Östergren 1951). The staining of human interphase nuclei with anticentromere antibodies revealed that the centromere is duplicated and resolved by the end of interphase (Brenner et al. 1981). This physical separation or resolution of sister centromeres would be important for the back-to-back assembly of sister kinetochores, thus facilitating bipolar attachment to microtubules in mitosis. In contrast to mitosis, cytological analyses of several animal germ cells have shown that sister kinetochores orient side by side and fuse in meiosis I (Goldstein 1981; Moore and Orr-Weaver 1998; Lee et al. 2000; Parra et al. 2004). The molecular mechanism underlying the regulation of kinetochore geometry has been studied recently (Watanabe 2012).

Spo13-MONOPOLIN IN BUDDING YEAST

In budding yeast a set of proteins called monopolin, which are required for mono-orientation, have been identified and extensively studied. Monopolin includes Csm1 (chromosome segregation in meiosis protein 1), Lrs4 (loss of rDNA silencing protein 4), Mam1 (monopolar microtubule attachment during meiosis I protein 1), and CK1, and localizes to centromeres specifically in meiosis I (Toth et al. 2000; Rabitsch et al. 2003). The structural analysis of Csm1–Lrs4 suggests that this complex forms a V shape with two pairs of kinetochore-binding domains that indeed bind kinetochore component Dns1. These results suggest that Csm1–Lrs4 may bring kinetochores together, which favors the clamp model (Corbett et al. 2010; Corbett and Harrison 2012). Evidence suggests that the enrichment of CK1 activity at kinetochores, which depends on the presence of Csm1–Lrs4, might be an ultimate requirement for the establishment of mono-orientation in budding yeast (Petronczki et al. 2006). However, which CK1 substrates are required for mono-orientation and the mechanism by which it occurs remain elusive. Budding yeast Spo13 (sporulation-specific protein 13), another factor that is required for mono-orientation as well as cohesion protection, associates with the Polo-like kinase Cdc5 and acts to recruit or stabilize the monopolin complex at centromeres (Clyne et al. 2003; Katis et al. 2004, 2010; Lee et al. 2004; Monje-Casas et al. 2007). In budding yeast, paired sister centromeres assemble a single kinetochore and bind only one microtubule (Sarangapani et al. 2014). Therefore, monopolin would conjoin two microtubule attachment sites and thereby make them into a one "point centromere."

Rec8 IS REQUIRED FOR MONO-ORIENTATION IN FISSION YEAST

In fission yeast, homologs of Csm1 and Lrs4 (named Pcs1 and Mde4, respectively) are dispensable for mono-orientation in meiosis I. However, in mitosis, *psc1* or *mde4* mutant cells show merotelic attachment, in which a single kinetochore is attached by microtubules emanating from both spindle poles (Gregan et al. 2007). Indeed, Pcs1–Mde4 recruits condensin, which may act to clamp together adjacent microtubule attachment sites, although the meiosis I–specific mono-orientation function is not conserved (Tada et al. 2011).

The finding that a *rec8* mutation in fission yeast causes equational, rather than reductional, division at meiosis I raises the possibility that cohesin complexes regulate kinetochore orientation. Similarly, plant and worm Rec8 have been shown to play an essential role in establishing

monopolar attachment at meiosis I (Yu and Dawe 2000; Chelysheva et al. 2005; Severson et al. 2009). In fission yeast, a mitotic cohesin complex that includes Rad21 accumulates preferentially at the pericentromeric heterochromatin, whereas the meiotic Rec8–cohesin complex accumulates additionally at the core centromere, the region where the kinetochore assembles. When Rec8 is removed and replaced by Rad21 in meiosis I, the Rad21–cohesin complex accumulates at the pericentromeric region, but much less at the central core region, causing equational rather than reductional division at meiosis I. When Rec8 is inactivated specifically only at the core centromere, but its other functions are preserved, kinetochores become bioriented at meiosis I, proving the essential role of Rec8 at the central core region for mono-orientation (Yokobayashi and Watanabe 2005).

Based on the above evidence, it has been proposed that physical attachment of sister chromatids or cohesion at the centromeric core conjoins the two kinetochore domains at meiosis I, whereas the core regions open to opposite sides when not establishing this cohesion at mitosis and meiosis II. A direct observation of cohesion at the core centromere in fission yeast was enabled by popping out this DNA region from the neighboring chromosomal domains during prophase I, which is before the attachment of kinetochores to spindle microtubules (Sakuno et al. 2009). This analysis revealed that cohesion at the core centromere is indeed established and maintained, particularly during meiosis I, whereas this cohesion is lost in *rec8Δ* or *moa1Δ* (see below), which is defective in mono-orientation. Importantly, cohesion at the central core region is not detected during mitosis or meiosis II in wild-type cells, whereas cohesion at the pericentromeric region is intact. Finally, when a proteinous artificial tether is introduced at the core centromere, monopolar attachment is restored in meiotic *rec8Δ* cells and even in normal mitotic cells. These results imply that mono-orientation of kinetochores is promoted ultimately by conjoining DNA duplexes underlying the kinetochores rather than the action of a kinetochore protein itself (Sakuno et al. 2009).

CONSERVED MEIOTIC KINETOCHORE FACTOR MEIKIN (Moa1, Spo13, MEIKIN)

Genetic screening to search for factors that regulate mono-orientation has identified a meiosis-specific kinetochore protein, Moa1 (monopolar attachment) in fission yeast (Yokobayashi and Watanabe 2005). Moa1 interacts with the conserved kinetochore protein Cnp1 (CENP-C homolog) and localizes exclusively at the central core of the centromere from prophase I to metaphase I but disappears in anaphase I (Tanaka et al. 2009). Moa1 also interacts with Rec8 and plays a role in establishing cohesion at the core centromere regions (mono-orientation) and some cohesion protection in pericentric regions. Taking advantage of the knowledge that Moa1 binds to conserved kinetochore protein CENP-C (Cnp3), two-hybrid screening using CENP-C as bait identified a meiosis-specific kinetochore protein MEIKIN in mice (Kim et al. 2015). Although there is no significant sequence homology between MEIKIN and Moa1, significant biochemical and functional similarities were identified between these two factors. Both MEIKIN and Moa1 recruit polo-like kinase (PLK) to kinetochores and the kinase activity of PLK is crucial for mono-orientation and cohesion protection (also see below). It turned out that these meiotic functions are reminiscent of those of budding yeast Spo13. Thus, the conserved meiosis-specific kinetochore regulator, meikin (MEIKIN in vertebrates; Moa1 in fission yeast; Spo13 in budding yeast) and its associated PLK play a crucial role in promoting mono-orientation and, at least partly, cohesion protection.

FISSION YEAST MEIKIN AFFECTS COHESION PROTECTION

Although fission yeast Moa1 was identified as a mono-orientation factor, a role in cohesion protection was also implicated (Yokobayashi and Watanabe 2005). Indeed, in *moa1Δ* cells, although a small population of cells undergo equational segregation at meiosis I (because of defects in mono-orientation), the majority undergo reductional segregation because of the presence of chiasmata and tension exerted across homologs (Fig. 1B). Strikingly, 22% of these "reductional" *moa1Δ* cells showed the separation of the GFP-marked centromere (*imr1*-GFP) in prometaphase II (Fig. 1B). This separation value is significantly higher than in wild-type cells (<5%), although lower than in *sgo1Δ* cells (50%), in which cohesion protection is completely abolished (Fig. 1B). These results suggest that *moa1Δ* cells show partial defects in cohesion protection during reductional division at meiosis I.

Moa1 associates with Plo1 (PLK1 homolog), and the centromeric Plo1 is required for reductional segregation at meiosis I (Kim et al. 2015). Here we examined whether Plo1 is responsible for defects in cohesion protection at prometaphase II, as is observed in *moa1Δ* cells. For this purpose, we first analyzed the *moa1-T101A* mutant, in which Moa1 localizes at kinetochores but fails to recruit Plo1 (Kim et al. 2015). Indeed, the separation of *imr1*-GFP in prometaphase II was observed in *moa1-T101A* cells to the same extent as in *moa1Δ* cells (Fig. 1C). These results indicate that Plo1 recruited to kinetochores by Moa1 is responsible for the protection of centromeric cohesion during anaphase I.

Bub1 REGULATES COHESION PROTECTION

Bub1, a well-conserved spindle checkpoint kinase, is required to preserve centromeric protection during meiosis I (Bernard et al. 2001a). Indeed, Sgo1 fails to localize at centromeres in *bub1* mutants (Kitajima et al. 2004; Fernius and Hardwick 2007). A biochemical approach in fission yeast identified serine 121 of histone 2A (H2A) as a Bub1 substrate (Kawashima et al. 2010). Crucially, Sgo1 binds nucleosomes, including phosphorylated H2A-S121, accounting for the mechanism of Sgo1 localization at centromeres (Kawashima et al. 2010). Sgo1 localization is also promoted by the pericentric heterochromatin protein Swi6

(HP1), which binds directly to Sgo1 (Yamagishi et al. 2008). In mitosis, Bub1 as a complex with Bub3 localizes to kinetochores through its interaction with Spc7 (KNL1) only when Spc7 is phosphorylated by Mph1 (MPS1) at unattached or misaligned kinetochores. Therefore, Bub1 is largely released from kinetochores when chromosomes are aligned during metaphase in mitotic cells.

Moa1–Plo1 REGULATES MEIOTIC Bub1 LOCALIZATION

KNL1 is also phosphorylated by PLK1 in humans and *C. elegans* (Espeut et al. 2015; von Schubert et al. 2015), suggesting the existence of a versatile regulation of Bub1 localization. It is reasonable to speculate that Moa1–Plo1, which is required for cohesion protection, may play a role in Bub1 enrichment during meiosis I. In fission yeast mitosis, Bub1 is enriched at kinetochores only when Mph1 (MPS1 homolog) accumulates at unattached kinetochores and phosphorylates the MELT repeats of the kinetochore protein Spc7 (KNL1) (Fig 2A; Yamagishi et al. 2012). Curiously, however, meiotic Bub1 signals are retained at kinetochores throughout metaphase I until anaphase I in wild-type cells, and the signals persist even in *mph1Δ* cells (Fig. 2B). Instead, centromeric Bub1 signals decline around late metaphase I in *moa1Δ* cells and largely disappear in *mph1Δ moa1Δ* cells (Fig. 2B). Thus, Moa1-associated Plo1 may play a key role in the accumulation of Bub1 at kinetochores in a redundant capacity with Mph1. The time-lapse live cell imaging of Bub1-GFP indicates that the duration of metaphase I is shortened in *moa1Δ*

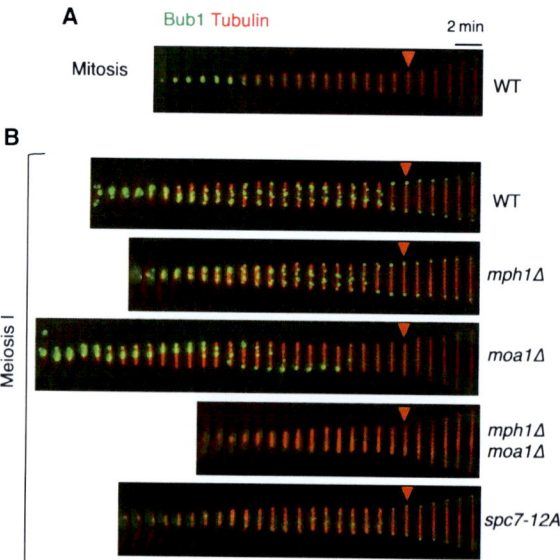

Figure 2. Bub1 is dispersed in *mph1Δ moa1Δ* cells as in *spc7-12A* cells. (*A,B*) The indicated cells expressing Bub1-GFP and mCherry-Atb2 (tubulin) were observed by time-lapse imaging during mitosis (*A*) or meiosis I (*B*) at 1-min intervals. A representative example of indicated cell is shown. The red arrowhead denotes the onset of spindle elongation at anaphase. WT, wild type. (Modified, with permission, from Miyazaki et al. 2017, © John Wiley & Sons, Inc.)

$mph1\Delta$ cells (Fig. 2B), suggesting that Moa1–Plo1, like Mph1, has a function in SAC activation in meiotic cells.

LIKE Mph1, Moa1–Plo1 PHOSPHORYLATES Spc7

Is Moa1–Plo1 indeed responsible for the phosphorylation of MELT repeats in Spc7? Bacterially purified Plo1, like Mph1, phosphorylates the amino-terminal domain of Spc7 (Spc7-N), which contains MELT repeats, but not other domains that lack MELT repeats (Fig. 3A). Consistently, centromeric Bub1 signals are largely dispersed in $spc7\text{-}12A$ cells as is seen in $mph1\Delta \ moa1\Delta$ cells during meiosis I (Fig. 2B). It is also shown that centromere-tethering Plo1 enables Bub1 recruitment to kinetochores even in mitotic interphase, and this is also the case in $mph1\Delta$ cells (Miyazaki et al. 2017). These results strongly support the notion that meiosis-specific Bub1 recruitment to kinetochores relies on the phosphorylation of the MELT repeats on Spc7, a process mediated by Moa1–Plo1 and Mph1.

KINETOCHORE-BOUND Bub1 AND CENTROMERIC HETEROCHROMATIN REDUNDANTLY SUSTAIN Sgo1 LOCALIZATION

Notably, centromeric Sgo1 signals show an ~70% reduction in $spc7\text{-}12A$ cells, in which Bub1 is largely dispersed from kinetochores, whereas few defects in cohesion protection are observed in $spc7\text{-}12A$ cells (Fig. 4A). Moreover, the expression of the Bub1 kinase catalytic domain, which cannot bind kinetochores, allows Sgo1 localization at centromeres depending on the heterochromatin protein Swi6 (Kawashima et al. 2010). Therefore, the residual Sgo1 localization (~30%; yet functional) in $spc7\text{-}12A$ cells might be produced by centromeric Swi6 together with dispersed Bub1 kinase activity, which would mildly phosphorylate histone H2A along the whole chromosomes including centromeres (Kawashima et al. 2010). Indeed, the cohesion protection defect in $spc7\text{-}12A$ cells is much enhanced by introducing the $sgo1\text{-}VE$ mutation, which renders Sgo1 unable to interact with Swi6 (Fig. 4A; Yamagishi et al. 2008). Thus, Bub1 enrichment mediated by Mph1 and Moa1–Plo1 (Spc7 phosphorylation) plays a crucial role in Sgo1 localization and cohesion protection in a redundant capacity with centromeric heterochromatin (Fig. 4B).

Moa1–Plo1 MAY ENHANCE Sgo1 FUNCTION IN ADDITION TO ITS LOCALIZATION

A remarkable observation is that protection defects are prominent in $moa1\Delta$ cells but not in $spc7\text{-}12A$ cells (Fig. 4C), whereas the opposite is true for Bub1 reduction (Fig. 2B). Therefore, in addition to the enhancement of Bub1 localization, Moa1 may play another key role in cohesion protection. This role of Moa1 might differ from the heterochromatin pathway because $moa1\Delta$ and $sgo1\text{-}VE$ show additive defects in cohesion protection (Fig. 4C). Given that $moa1\Delta$ shows no additive defects with $sgo1\Delta$ (Fig, 4C), it is reasonable to speculate that Moa1 may somehow enhance Sgo1 function directly as well as by its localization (Fig. 4B).

CONCLUSION

The mono-orientation of sister kinetochores and protection of centromeric cohesion are two hallmarks of the regulation of meiotic chromosome segregation that are widely conserved among eukaryotic organisms. Meikin is a recently emerging protein family that may regulate both mono-orientation and cohesion protection. Although fission yeast Moa1 was initially identified as a mono-orientation factor, it regulates cohesion protection similarly to mouse MEIKIN and budding yeast Spo13. Especially, in fission yeast, Moa1–Plo1 together with Mph1 play a crucial role in the enrichment of Bub1 at kinetochores throughout meiosis I. This Bub1 pool ensures robust Sgo1 localization and cohesion protection at centromeres by cooperating with heterochromatin protein Swi6, which binds and stabilizes Sgo1. Although the meiosis-specific Bub1-Sgo1 enrichment mechanism is conserved in mouse (Miyazaki et al. 2017), the contribution of heterochromatin to Sgo1 stabilization is not yet proven in mammalian meiosis. In summary, accumulating evidence indicates that not only shugoshin but also meikin contribute to the meiosis I–specific cohesion protection mechanism. Previous studies suggest that the ectopic or physiological localization of shugoshin at centromeres in meiosis II cannot protect cohesin from separase cleavage at the onset of

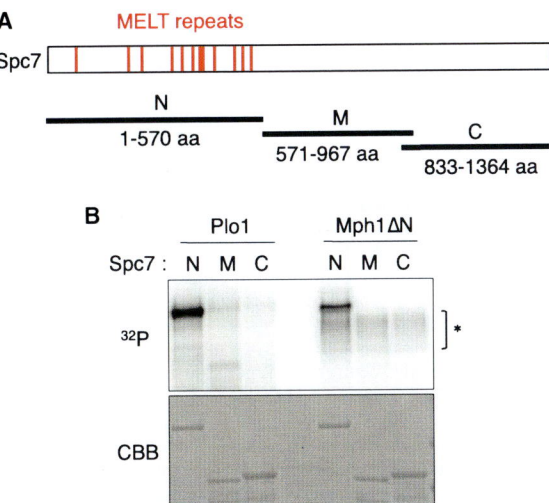

Figure 3. Spc7 phosphorylation at MELT sequence by Plo1 and Mph1 promotes Spc7 and Bub1/Bub3 binding. (*A*) Schematic depiction of Spc7 fragments used in the in vitro phosphorylation assay. (*B*) Recombinant GST-fused Spc7 fragments were incubated with recombinant GST-Plo1 (*left*) or GST-Mph1ΔN (kinase domain only) (*right*) in the presence of [γ-^{32}P] ATP. The incorporation of radioactive phosphate groups was visualized by autoradiography (^{32}P) and compared with protein levels (Coomassie brilliant blue [CBB]). The asterisk indicates Mph1ΔN autophosphorylation. (Modified, with permission, from Miyazaki et al. 2017, © John Wiley & Sons, Inc.)

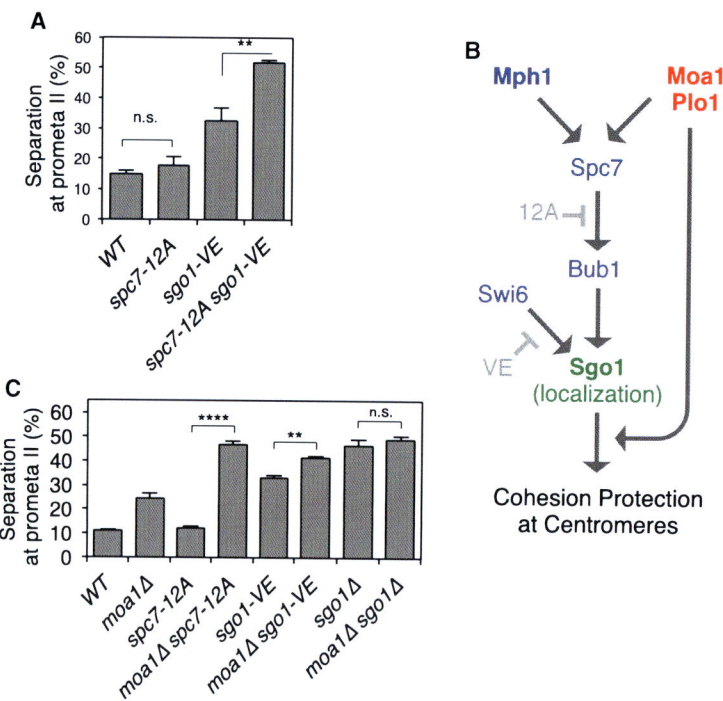

Figure 4. Moa1–Plo1 enhances Sgo1 function in addition to its localization. (*A,C*) The frequency of centromere splitting was measured by observing heterozygous *imr1*-GFP signals in prometaphase II (prometa II) arrest (*mes1*) in the indicated cells. Error bars, SD from three independent experiments. Total cells: *n* > 140. *n.s.*; not significant; **, *P* < 0.01; ***, *P* < 0.005; ****, *P* < 0.001; one-way ANOVA with Bonferroni's multiple comparisons test. (*B*) Schematic depiction of Moa1–Plo1 function in the regulation of cohesion protection at centromeres during meiosis I. Moa1–Plo1 facilitates Sgo1 localization at centromeres through Spc7 phosphorylation in a redundant capacity with Mph1. Swi6 stabilizes Sgo1 localization. Moa1–Plo1 also regulates another pathway required for cohesion protection in meiosis I. (Modified, with permission, from Miyazaki et al. 2017, © John Wiley & Sons, Inc.)

anaphase II (Moore et al. 1998; Rabitsch et al. 2004). This could be accounted for at least in part, by the fact that meikin is absent from kinetochores in meiosis II.

ACKNOWLEDGMENTS

We thank all past members of the Watanabe laboratory for their work done in the laboratory and valuable discussion. This work was supported by MEXT KAKENHI grant number 25000014 (Y.W.).

REFERENCES

Bernard P, Maure JF, Javerzat JP. 2001a. Fission yeast Bub1 is essential in setting up the meiotic pattern of chromosome segregation. *Nat Cell Biol* **3**: 522–526.

Bernard P, Maure JF, Partridge JF, Genier S, Javerzat JP, Allshire RC. 2001b. Requirement of heterochromatin for cohesion at centromeres. *Science* **294**: 2539–2542.

Brar GA, Amon A. 2008. Emerging roles for centromeres in meiosis I chromosome segregation. *Nat Rev Genet* **9**: 899–910.

Brenner S, Pepper D, Berns MW, Tan E, Brinkley BR. 1981. Kinetochore structure, duplication, and distribution in mammalian cells: Analysis by human autoantibodies from scleroderma patients. *J Cell Biol* **91**: 95–102.

Buonomo SB, Clyne RK, Fuchs J, Loidl J, Uhlmann F, Nasmyth K. 2000. Disjunction of homologous chromosomes in meiosis I depends on proteolytic cleavage of the meiotic cohesin Rec8 by separin. *Cell* **103**: 387–398.

Chelysheva L, Daiallo S, Vezon D, Gendrot G, Vrielynck N, Belcram K, Rocques N, Marquez-Lema A, Bhatt AM, Horlow C, et al. 2005. AtREC8 and AtSCC3 are essential to the monopolar orientation of the kinetochores during meiosis. *J Cell Sci* **118**: 4621–4632.

Clyne RK, Katis VL, Jessop L, Benjamin KR, Herskowitz I, Lichten M, Nasmyth K. 2003. Polo-like kinase Cdc5 promotes chiasmata formation and cosegregation of sister centromeres at meiosis I. *Nature Cell Biol* **5**: 480–485.

Corbett KD, Harrison SC. 2012. Molecular architecture of the yeast monopolin complex. *Cell Rep* **1**: 583–589.

Corbett KD, Yip CK, Ee LS, Walz T, Amon A, Harrison SC. 2010. The monopolin complex crosslinks kinetochore components to regulate chromosome-microtubule attachments. *Cell* **142**: 556–567.

Davis BK. 1971. Genetic analysis of a meiotic mutant resulting in precocious sister-centromere separation in *Drosophila melanogaster*. *Mol Gen Genet* **113**: 251–272.

Duro E, Marston AL. 2015. From equator to pole: Splitting chromosomes in mitosis and meiosis. *Gene Dev* **29**: 109–122.

Espeut J, Lara-Gonzalez P, Sassine M, Shiau AK, Desai A, Abrieu A. 2015. Natural loss of Mps1 kinase in nematodes uncovers a role for polo-like kinase 1 in spindle checkpoint initiation. *Cell Rep* **12**: 58–65.

Fernius J, Hardwick KG. 2007. Bub1 kinase targets Sgo1 to ensure efficient chromosome biorientation in budding yeast mitosis. *PLoS Genet* **3**: e213.

Fukagawa T, Nogami M, Yoshikawa M, Ikeno M, Okazaki T, Takami Y, Nakayama T, Oshimura M. 2004. Dicer is essential for formation of the heterochromatin structure in vertebrate cells. *Nature Cell Biol* **6**: 784–791.

Goldstein LS. 1981. Kinetochore structure and its role in chromosome orientation during the first meiotic division in male *D. melanogaster*. *Cell* **25**: 591–602.

Gregan J, Riedel CG, Pidoux AL, Katou Y, Rumpf C, Schleiffer A, Kearsey SE, Shirahige K, Allshire RC, Nasmyth K. 2007. The kinetochore proteins Pcs1 and Mde4 and heterochromatin are required to prevent merotelic orientation. *Curr Biol* **17:** 1190–1200.

Ishiguro T, Tanaka K, Sakuno T, Watanabe Y. 2010. Shugoshin-PP2A counteracts casein-kinase-1-dependent cleavage of Rec8 by separase. *Nat Cell Biol* **12:** 500–506.

Katis VL, Matos J, Mori S, Shirahige K, Zachariae W, Nasmyth K. 2004. Spo13 facilitates monopolin recruitment to kinetochores and regulates maintenance of centromeric cohesion during yeast meiosis. *Curr Biol* **14:** 2183–2196.

Katis VL, Lipp JJ, Imre R, Bogdanova A, Okaz E, Habermann B, Mechtler K, Nasmyth K, Zachariae W. 2010. Rec8 phosphorylation by casein kinase 1 and Cdc7-Dbf4 kinase regulates cohesin cleavage by separase during meiosis. *Dev Cell* **18:** 397–409.

Kawashima SA, Yamagishi Y, Honda T, Ishiguro K, Watanabe Y. 2010. Phosphorylation of H2A by Bub1 prevents chromosomal instability through localizing shugoshin. *Science* **327:** 172–177.

Kerrebrock AW, Moore DP, Wu JS, Orr-Weaver TL. 1995. MEI-S332, a *Drosophila* protein required for sister-chromatid cohesion, can localize to meiotic centromere regions. *Cell* **83:** 247–256.

Kim J, Ishiguro K, Nambu A, Akiyoshi B, Yokobayashi S, Kagami A, Ishiguro T, Pendas AM, Takeda N, Sakakibara Y, et al. 2015. Meikin is a conserved regulator of meiosis-I-specific kinetochore function. *Nature* **517:** 466–471.

Kitajima TS, Miyazaki Y, Yamamoto M, Watanabe Y. 2003a. Rec8 cleavage by separase is required for meiotic nuclear divisions in fission yeast. *EMBO J* **22:** 5643–5653.

Kitajima TS, Yokobayashi S, Yamamoto M, Watanabe Y. 2003b. Distinct cohesin complexes organize meiotic chromosome domains. *Science* **300:** 1152–1155.

Kitajima TS, Kawashima SA, Watanabe Y. 2004. The conserved kinetochore protein shugoshin protects centromeric cohesion during meiosis. *Nature* **427:** 510–517.

Kitajima TS, Sakuno T, Ishiguro K, Iemura S, Natsume T, Kawashima SA, Watanabe Y. 2006. Shugoshin collaborates with protein phosphatase 2A to protect cohesin. *Nature* **441:** 46–52.

Klein F, Mahr P, Galova M, Buonomo SBC, Michaelis C, Nairz K, Nasmyth K. 1999. A central role for cohesins in sister chromatid cohesion, formation of axial elements, and recombination during yeast meiosis. *Cell* **98:** 91–103.

Lee BH, Kiburz BM, Amon A. 2004. Spo13 maintains centromeric cohesion and kinetochore coorientation during meiosis I. *Curr Biol* **14:** 2168–2182.

Lee JY, Orr-Weaver TL. 2001. The molecular basis of sister-chromatid cohesion. *Annu Rev Cell Dev Biol* **17:** 753–777.

Lee J, Miyano T, Dai Y, Wooding P, Yen TJ, Moor RM. 2000. Specific regulation of CENP-E and kinetochores during meiosis I/meiosis II transition in pig oocytes. *Mol Reprod Dev* **56:** 51–62.

Lee J, Kitajima TS, Tanno Y, Yoshida K, Morita T, Miyano T, Miyake M, Watanabe Y. 2008. Unified mode of centromeric protection by shugoshin in mammalian oocytes and somatic cells. *Nature Cell Biol* **10:** 42–52.

Llano E, Gomez R, Gutierrez-Caballero C, Herran Y, Sanchez-Martin M, Vazquez-Quinones L, Hernandez T, de Alava E, Cuadrado A, Barbero JL, et al. 2008. Shugoshin-2 is essential for the completion of meiosis but not for mitotic cell division in mice. *Genes Dev* **22:** 2400–2413.

London N, Biggins S. 2014. Signalling dynamics in the spindle checkpoint response. *Nat Rev Mol Cell Biol* **15:** 736–748.

Marston AL, Tham WH, Shah H, Amon A. 2004. A genome-wide screen identifies genes required for centromeric cohesion. *Science* **303:** 1367–1370.

Miyazaki S, Kim J, Yamagishi Y, Ishiguro T, Okada Y, Tanno Y, Sakuno T, Watanabe Y. 2017. Meikin-associated polo-like kinase specifies Bub1 distribution in meiosis I. *Genes Cells* **22:** 552–567.

Monje-Casas F, Prabhu VR, Lee BH, Boselli M, Amon A. 2007. Kinetochore orientation during meiosis is controlled by Aurora B and the monopolin complex. *Cell* **128:** 477–490.

Moore DP, Orr-Weaver TL. 1998. Chromosome segregation during meiosis: Building an unambivalent bivalent. *Curr Top Dev Biol* **37:** 263–299.

Moore DP, Page AW, Tang TT, Kerrebrock AW, Orr-Weaver TL. 1998. The cohesion protein MEI-S332 localizes to condensed meiotic and mitotic centromeres until sister chromatids separate. *J Cell Biol* **140:** 1003–1012.

Musacchio A. 2015. The molecular biology of spindle assembly checkpoint signaling dynamics. *Curr Biol* **25:** R1002–R1018.

Nasmyth K, Haering CH. 2009. Cohesin: Its roles and mechanisms. *Annu Rev Genet* **43:** 525–558.

Nonaka N, Kitajima T, Yokobayashi S, Xiao G, Yamamoto M, Grewal SI, Watanabe Y. 2002. Recruitment of cohesin to heterochromatic regions by Swi6/HP1 in fission yeast. *Nat Cell Biol* **4:** 89–93.

Onn I, Heidinger-Pauli JM, Guacci V, Unal E, Koshland DE. 2008. Sister chromatid cohesion: A simple concept with a complex reality. *Annu Rev Cell Dev Bi* **24:** 105–129.

Östergren G. 1951. The mechanism of co-orientation in bivalents and multivalents. *Hereditas* **37:** 85–156.

Parra MT, Viera A, Gomez R, Page J, Benavente R, Santos JL, Rufas JS, Suja JA. 2004. Involvement of the cohesin Rad21 and SCP3 in monopolar attachment of sister kinetochores during mouse meiosis I. *J Cell Sci* **117:** 1221–1234.

Peters JM, Tedeschi A, Schmitz J. 2008. The cohesin complex and its roles in chromosome biology. *Genes Dev* **22:** 3089–3114.

Petronczki M, Siomos MF, Nasmyth K. 2003. Un menage a quatre: The molecular biology of chromosome segregation in meiosis. *Cell* **112:** 423–440.

Petronczki M, Matos J, Mori S, Gregan J, Bogdanova A, Schwickart M, Mechtler K, Shirahige K, Zachariae W, Nasmyth K. 2006. Monopolar attachment of sister kinetochores at meiosis I requires casein kinase 1. *Cell* **126:** 1049–1064.

Rabitsch KP, Petronczki M, Javerzat JP, Genier S, Chwalla B, Schleiffer A, Tanaka TU, Nasmyth K. 2003. Kinetochore recruitment of two nucleolar proteins is required for homolog segregation in meiosis I. *Dev Cell* **4:** 535–548.

Rabitsch KP, Gregan J, Schleiffer A, Javerzat JP, Eisenhaber F, Nasmyth K. 2004. Two fission yeast homologs of *Drosophila* Mei-S332 are required for chromosome segregation during meiosis I and II. *Curr Biol* **14:** 287–301.

Riedel CG, Katis VL, Katou Y, Mori S, Itoh T, Helmhart W, Galova M, Petronczki M, Gregan J, Cetin B, et al. 2006. Protein phosphatase 2A protects centromeric sister chromatid cohesion during meiosis I. *Nature* **441:** 53–61.

Sacristan C, Kops GJ. 2015. Joined at the hip: Kinetochores, microtubules, and spindle assembly checkpoint signaling. *Trends Cell Biol* **25:** 21–28.

Sakuno T, Tada K, Watanabe Y. 2009. Kinetochore geometry defined by cohesion within the centromere. *Nature* **458:** 852–858.

Sarangapani KK, Duro E, Deng Y, Alves FD, Ye Q, Opoku KN, Ceto S, Rappsilber J, Corbett KD, Biggins S, et al. 2014. Sister kinetochores are mechanically fused during meiosis I in yeast. *Science* **346:** 248–251.

Severson AF, Ling L, van Zuylen V, Meyer BJ. 2009. The axial element protein HTP-3 promotes cohesin loading and meiotic axis assembly in *C. elegans* to implement the meiotic program of chromosome segregation. *Genes Dev* **23:** 1763–1778.

Tachibana-Konwalski K, Godwin J, van der Weyden L, Champion L, Kudo NR, Adams DJ, Nasmyth K. 2010. Rec8-containing cohesin maintains bivalents without turnover during the growing phase of mouse oocytes. *Genes Dev* **24:** 2505–2516.

Tada K, Susumu H, Sakuno T, Watanabe Y. 2011. Condensin association with histone H2A shapes mitotic chromosomes. *Nature* **474:** 477–483.

Tanaka K, Chang HL, Kagami A, Watanabe Y. 2009. CENP-C functions as a scaffold for effectors with essential kinetochore functions in mitosis and meiosis. *Dev Cell* **17:** 334–343.

Tang Z, Shu H, Qi W, Mahmood NA, Mumby MC, Yu H. 2006. PP2A is required for centromeric localization of Sgo1 and proper chromosome segregation. *Dev Cell* **10:** 575–585.

Toth A, Rabitsch KP, Galova M, Schleiffer A, Buonomo SB, Nasmyth K. 2000. Functional genomics identifies monopolin: A kinetochore protein required for segregation of homologs during meiosis I. *Cell* **103:** 1155–1168.

Uhlmann F, Lottspeich F, Nasmyth K. 1999. Sister-chromatid separation at anaphase onset is promoted by cleavage of the cohesin subunit Scc1. *Nature* **400:** 37–42.

von Schubert C, Cubizolles F, Bracher JM, Sliedrecht T, Kops GJPL, Nigg EA. 2015. Plk1 and Mps1 cooperatively regulate the spindle assembly checkpoint in human cells. *Cell Rep* **12:** 66–78.

Watanabe Y. 2012. Geometry and force behind kinetochore orientation: Lessons from meiosis. *Nat Rev Mol Cell Biol* **13:** 370–382.

Watanabe Y, Nurse P. 1999. Cohesin Rec8 is required for reductional chromosome segregation at meiosis. *Nature* **400:** 461–464.

Yamagishi Y, Sakuno T, Shimura M, Watanabe Y. 2008. Heterochromatin links to centromeric protection by recruiting shugoshin. *Nature* **455:** 251–255.

Yamagishi Y, Yang CH, Tanno Y, Watanabe Y. 2012. MPS1/Mph1 phosphorylates the kinetochore protein KNL1/Spc7 to recruit SAC components. *Nat Cell Biol* **14:** 746–752.

Yokobayashi S, Watanabe Y. 2005. The kinetochore protein Moa1 enables cohesion-mediated monopolar attachment at meiosis I. *Cell* **123:** 803–817.

Yokobayashi S, Yamamoto M, Watanabe Y. 2003. Cohesins determine the attachment manner of kinetochores to spindle microtubules at meiosis I in fission yeast. *Mol Cell Biol* **23:** 3965–3973.

Yu H-G, Dawe RK. 2000. Functional redundancy in the maize meiotic kinetochore. *J Cell Biol* **151:** 131–141.

Topologically Associating Domains in Chromosome Architecture and Gene Regulatory Landscapes during Development, Disease, and Evolution

RAFAEL GALUPA[1] AND EDITH HEARD[1,2]

[1]*Institut Curie, PSL Research University, CNRS, INSERM, Genetics and Developmental Biology Unit, Mammalian Developmental Epigenetics Group, 75248 Paris Cedex 05, France*

[2]*Collège de France, 75231 Paris Cedex 05, France*

Correspondence: edith.heard@curie.fr; rafael.galupa@curie.fr

The packaging of genetic material into chromatin and chromosomes has been recognized for more than a century, thanks to microscopy and biochemical approaches. This was followed by the progressive realization that chromatin organization is critical for genome functions such as transcription and DNA replication and repair. The recent discovery that chromosomes are partitioned at the submegabase scale into topologically associating domains (TADs) has implications for our understanding of gene regulation during developmental processes such as X-chromosome inactivation, as well as for evolution and for the search for disease-associated loci. Here we discuss our current knowledge about this recently recognized level of mammalian chromosome organization, with a special emphasis on the potential role of TADs as a structural basis for the function and evolution of mammalian regulatory landscapes.

The spatial organization of the eukaryotic genome is tightly linked to its functions, including the regulation of gene activity. Microscopy studies have been key for our understanding of how chromosome folding in the nucleus may be related to transcriptional regulation of different parts of the genome but have been limited by their relatively low throughput and resolution. The recent advent of chromosome conformation capture (3C) technologies (Cullen et al. 1993; Dekker et al. 2002; Denker and de Laat 2016) and next-generation sequencing has allowed a degree of molecular resolution that enables us to assess how DNA elements physically interact with each other, opening up exciting new perspectives for our understanding of long-range regulation of gene expression. The first 3C-based analysis at the whole-genome scale (termed Hi-C), using a human cell line, revealed the genome-wide existence of spatially segregated compartments of at least two different types, associated with open or closed chromatin, and spanning very large chromosomal regions (Lieberman-Aiden et al. 2009). Subsequent higher-resolution "C" studies, genome-wide and at the *X-inactivation center* locus, discovered that at the submegabase level, mammalian chromosomes are partitioned into domains of high frequency interactions, which were named topological domains (Dixon et al. 2012) or topologically associating domains (TADs) (Nora et al. 2012). TADs and their boundaries were found to be very well conserved across syntenic regions of mammalian chromosomes, first shown in human and mouse (Dixon et al. 2012) and later in macaque, dog, and rabbit as well (Vietri Rudan et al. 2015). Similar genomic domains were also

identified in the fruit fly—called "physical domains" (Sexton et al. 2012)—and later in different organisms, too; however, these are not necessarily equivalent, both structurally and/or functionally, to mammalian TADs (see Dekker and Heard 2015 for a review on topological domains across different organisms). TADs, as initially defined in mammalian cells, span an average of 900 kb (Dixon et al. 2012) and are often partitioned into smaller sub-TADs of a few hundred kilobases (Phillips-Cremins et al. 2013). The highest resolution "C" maps to date in mammals show the existence of even smaller domains, ~10–100-kb-long "contact domains" (Rao et al. 2014; Bonev et al. 2017). Among these different domains of chromosome folding, the relative stability of TADs across cell types and the conserved locations of TAD boundaries between man and mouse has led to their being proposed as "a structural basis for regulatory landscapes" (Nora et al. 2013). Indeed, in genome-wide association studies (GWASs), TADs can now be taken into account as a means of identifying disease-risk loci, based on potential regulatory variants in candidate single-nucleotide polymorphism (SNP) sets (Way et al. 2017). The concept of TADs, although still debated, has received increasing support from recent functional studies and will be discussed here. We will cover some specific examples, including the *X-inactivation center*, one of the loci where the implication of TAD organization in developmental gene expression was first put forward (Nora et al. 2012), and other complex regulatory loci, including *Shh* and *Hox* loci, where more recent insights have emerged.

Published by Cold Spring Harbor Laboratory Press; doi: 10.1101/sqb.2017.82.035030

Cold Spring Harbor Symposia on Quantitative Biology, Volume LXXXII

TADs: A FUNCTIONALLY PRIVILEGED SCALE IN THE CHROMOSOME FOLDING HIERARCHY?

The discovery of TADs was based on "C" techniques and supported by microscopy approaches: DNA fluorescence in situ hybridization (FISH) confirmed that sequences in the same TAD showed significantly higher overlap than sequences on either side of a TAD boundary (Dixon et al. 2012; Nora et al. 2012; Sexton et al. 2012). TADs are commonly defined as "domains within which sequences interact preferentially with each other, compared to their interactions with sequences outside." However, this could be applied to almost all domains identified using C-technologies. Compartments, TADs, subTADs, and contact domains represent preferentially interacting sequences that seem to be insulated from adjacent interacting sequences. So are there any specific characteristics of TADs that would support their definition as distinct entities? Several functional properties have been attributed to TADs, including enrichment of active histone marks (Dixon et al. 2012), CTCF clustering at boundaries (Dixon et al. 2012; Berlivet et al. 2013; Fraser et al. 2015), transcriptional co-regulation (Le Dily et al. 2014; Nora et al. 2012), and enhancer–promoter communication (Shen et al. 2012; Nora et al. 2013). Do TADs correlate with these properties more than other types of domains?

Defining TADs: Function beyond Structure

In most studies so far, the algorithms used to identify TADs take into account changes in the direction, or sum, of the interaction frequencies (as a measure of insulation), but they also impose limitations such as domain length (see Forcato et al. 2017 for a comparison study of different algorithms). It has therefore remained unclear whether TADs are simply an arbitrary scale within a continuum of insulation levels of a nested hierarchy of domains, or whether they represent an intrinsic level of chromosome organization. Recently, this question was addressed by developing an algorithm that identifies and stratifies topological domains from genome-wide interaction frequency maps (Zhan et al. 2017), using a single parameter, reciprocal insulation between adjacent domains, to generate multiple sets of domains from a given map. With this algorithm, compartments, TADs, subTADs, and contact domains emerged at different values of reciprocal insulation, in a continuous spectrum of nested self-interacting domains (Zhan et al. 2017). TADs are therefore not a structurally privileged level within the spectrum of insulated domains. However, the authors found that functional properties previously attributed to TADs but sometimes also enriched at other domains (e.g., transcriptional co-regulation, active histone marks, CTCF clustering at boundaries, enhancer–promoter communication) were maximized at the TAD scale of reciprocal insulation (Zhan et al. 2017).

In summary, TADs cannot be defined solely based on their structural features, as they represent an arbitrary level within a continuum of increasingly/decreasingly insulated domains. Instead a definition of TADs should take into account their functional significance as well. We thus propose that *TADs represent submegabase domains in which genomic elements interact preferentially with each other, maximizing specific functional properties of the genome.* Importantly the reference TAD "atlas" (Dixon et al. 2012) is actually a good approximation of the scale at which functional features are maximized (Zhan et al. 2017). For the sake of simplicity this is the working definition we use when referring to TADs in this review.

TADs Represent the Combined Interaction Frequencies from an Average Population

Most 3C-based techniques involve analyses of millions of pooled cells. What do TADs represent at the single-cell level? Polymer modeling recapitulating interaction maps within the *X-inactivation center* (*Xic*) region, has predicted highly variable, but nonrandom, conformations within TADs (Giorgetti et al. 2014), suggesting that TADs represent an averaged ensemble of multiple conformations across the cell population. DNA FISH confirmed this and highlighted the cell-to-cell variability regarding the shape, compaction, and spatial separation of Xic TADs (Nora et al. 2012; Giorgetti et al. 2014). Recently, single-cell Hi-C experiments confirmed variable cell-to-cell chromosome structures suggested by microscopy studies, with individual contacts at the megabase scale rarely surpassing TAD boundaries (Stevens et al. 2017; Nagano et al. 2013, 2017). Such cell-to-cell variability in TAD organization has important implications for the role of TADs in transcriptional regulation (Nagano et al. 2013; Giorgetti et al. 2014)—if TADs were stable entities in every cell within a population, *cis*-regulatory elements would be confined within a static chromatin configuration and the regulatory input in each cell would be equivalent. If instead TADs reflect an average of the interactions at the single-cell level, enhancer–promoter contacts emerge as probabilistic events in a fluctuating environment (Fudenberg and Mirny 2012; Nora et al. 2013; Giorgetti et al. 2014), providing variable regulatory input across the cell population. This could also explain cell-to-cell transcriptional heterogeneity (Amano et al. 2009; Vera et al. 2016). The relationship between the structural dynamics of TADs and the dynamics of enhancer–promoter interactions within them is still poorly explored and will have to be addressed with live imaging, which remains challenging in mammalian systems (see Fukaya et al. 2016 for a recent example in *Drosophila* embryos). The idea that such interactions are probabilistic as opposed to stable and directed is still debated and is discussed below in the context of recent models for domain and loop formation.

MECHANISMS AND DYNAMICS OF TAD ORGANIZATION

Scanning the topological map of a mammalian chromosome, TADs are as noticeable as the transitions between them. These are generally referred to as "boundaries," but they do not always demarcate sharp transitions and can sometimes correspond to rather large regions, with long transitions between TADs (Rocha et al. 2015). It should be noted as well that boundaries correspond to regions across which interactions are markedly reduced but not completely absent. Some TADs seem more insulated than others, which might reflect the strength of their boundaries. One study found that the level of enrichment of architectural proteins at TAD boundaries is correlated with their level of insulation (Van Bortle et al. 2014). It remains unclear, however, what forms a TAD boundary, and whether TAD boundaries represent insulatory elements per se, with specific characteristics such as binding of certain factors (see more below), or whether they arise as a result of adjacent, self-interactions between intra-TAD sequences. Probably both scenarios are possible and may even occur together. Structural elements within TADs have been proposed to help defining boundaries between TADs, by organizing the internal TAD structure and thereby preventing interactions with neighboring TADs (Giorgetti et al. 2014).

CTCF: A Role in Insulation and Loop Formation?

TAD boundaries are enriched in binding of CTCF (Dixon et al. 2012), a zinc-finger nucleic acid–binding protein, which nevertheless binds intra-TAD elements as well. Originally described as a transcription factor and insulator, CTCF was quickly recognized as a candidate for mediating TAD organization and loop formation (for review, see Merkenschlager and Nora 2016). Interestingly, conserved CTCF sites are mostly located at TAD boundaries, whereas species-specific CTCF sites, sometimes derived from retrotransposon expansion (Schmidt et al. 2012), are more often found inside TADs (Gómez-Marín et al. 2015; Vietri Rudan et al. 2015). It is unknown why some CTCF sites seem to participate in the formation of boundaries whereas others do not. Functional studies trying to address CTCF contribution to the topological organization of chromosomes suffer from CTCF being essential for development and cell proliferation, rendering knockout approaches difficult to interpret, whereas knockdown experiments are not efficient enough to completely deplete CTCF (Zuin et al. 2014). A recent study managed to overcome this by using an inducible degron system, which acutely depletes CTCF and can be reversible (Nora et al. 2017). Nora et al. (2017) found that CTCF is absolutely required for insulation of most TADs and loops between CTCF target sites in mouse embryonic stem cells and derived differentiated cells. Interestingly, loss of TADs was not accompanied by loss of active and inactive genomic compartments (Nora et al. 2017), suggesting that genomic organization in compartments does not depend on its folding in TADs, and that different mechanisms underlie their establishment and maintenance.

Molecular Models for the Establishment of TAD Organization

Models of "loop extrusion" (Riggs 1990; Blackwood and Kadonaga 1998; Kimura et al. 1999; Nasmyth 2001;

Alipour and Marko 2012) have been proposed to explain how TADs and chromatin loops arise at the molecular level (Sanborn et al. 2015; Fudenberg et al. 2016; Goloborodko et al. 2016)—an "extruding factor," able to engulf two DNA chains and move along them, would extrude DNA until it reaches "stalling factors," that block its progression; a DNA loop would thus be formed and stabilized. CTCF has been suggested as a "stalling factor," whereas cohesin is proposed as the "extruding factor" (for review, see Merkenschlager and Nora 2016). Cohesin is a protein complex that forms a "ring," involved in holding sister chromatids together after DNA replication and regulating their separation during cell division. Cohesin-bound sites very often overlap with CTCF-bound sites and several examples show that cohesin also participates in long-range cis-interactions (Hadjur et al. 2009; Degner et al. 2011; Guo et al. 2012). Acute depletion of either CTCF or cohesin lead to loss of TADs and loops (Nora et al. 2017; Rao et al. 2017) and mutations in factors involved in the loading or release of cohesin from DNA lead to differences in the length of the loops formed (Busslinger et al. 2017; Haarhuis et al. 2017; Schwarzer et al. 2017; Wutz et al. 2017).

Factors other than CTCF and cohesin might also contribute to the formation of boundaries in mammalian genomes. A proportion of TAD boundaries (<20%) remained unaffected upon acute depletion of CTCF (Nora et al. 2017), indicating that CTCF-independent mechanisms to establish and/or maintain TADs exist. This also implies that there is some heterogeneity among the domains globally classified as TADs. The highest-resolution Hi-C maps available to date show three classes of TAD boundaries: (i) CTCF-bound, (ii) no CTCF and proximity to active promoters, and (iii) no CTCF and no active marks, corresponding to repeat regions (Bonev et al. 2017), consistent with previous studies (Dixon et al. 2012). Cohesin is present in the CTCF-bound boundaries, as well as in those associated with active promoters (Bonev et al. 2017). Transcription seems therefore highly correlated with local chromatin insulation, and cell type–specific TAD boundaries are often associated with the activity of cell type–specific genes (Bonev et al. 2017). Transcription itself can influence the deposition of cohesin along the genome (Busslinger et al. 2017) or could possibly create torsional constraints leading to local insulation (Remeseiro et al. 2016). However, transcriptional activation per se was not sufficient to induce chromatin insulation and create a TAD boundary (Bonev et al. 2017). Considering that loss of boundary elements might lead to dramatic and severe phenotypic consequences (discussed later), the fact that TAD boundaries might be composed of different elements and/or involve different mechanisms suggests that this might represent an evolutionary strategy to buffer the potential effects of mutations at single elements (Lupiáñez et al. 2016).

TAD Dynamics during the Cell Cycle

Exploring how TADs are established (and maintained) at the mechanistic level is particularly relevant in the context of cell cycle, given that TADs have to be reestablished after each cell division, as revealed by the absence of a compartmentalized organization in mitotic (metaphasic) chromosomes (Naumova et al. 2013). Furthermore, single-cell Hi-C revealed that TADs (as well as compartments, contact insulation, and long-range loops) are dynamic during the cell cycle (Nagano et al. 2017). Specific distributions of short- and long-range contracts characterize each phase of the cell cycle (Nagano et al. 2017), suggesting that the genome is not stably folded at any particular stage. This probably explains at least partially the high cell-to-cell variability in chromosome conformation observed by single-cell Hi-C (Nagano et al. 2013) or polymer modeling and DNA FISH (Giorgetti et al. 2014). Is there a topological memory from the previous cell cycle, or does loop extrusion (or other mechanisms) act completely de novo after each cell cycle? The cell cycle itself seems to be important to establish TAD organization. Acute depletion of CTCF in nondividing cells revealed that TAD structure collapsed to the same extent as in dividing cells (Nora et al. 2017), indicating that topological organization can be lost independently of cell division. However, when CTCF was restored, and contrary to dividing cells, structure was not fully recovered in nondividing cells (Nora et al. 2017), implying that the establishment of TADs depends on passage through cell cycle.

TAD Dynamics during Preimplantation Development

Another fundamental question concerns when and how TAD organization emerges during development, given that just after fertilization, the embryonic genome undergoes intensive remodeling in terms of its chromatin, transcription, and organization. Is TAD organization inherited from the gametes or established de novo in the developing embryo? The murine sperm genome, wrapped mostly with protamins, is bound by CTCF and cohesin and shows a topological organization similar to that of other cell types, such as mESCs and fibroblasts, but with a higher proportion of long-distance interactions (Battulin et al. 2015; Du et al. 2017; Jung et al. 2017; Ke et al. 2017). However, mature oocytes (arrested in metaphase of meiosis II) lack high-order chromosome structures (such as TADs) (Du et al. 2017; Ke et al. 2017) and their uniform folding configuration resembles that of the mitotic chromosomes (Naumova et al. 2013). Based on single-cell or low-input Hi-C protocols, the topological organization of the early embryo has been recently explored (Du et al. 2017; Flyamer et al. 2017; Ke et al. 2017). Although a full picture is still missing, certain trends emerge from these studies: (1) the organization of the zygotic genome and two-cell embryo is rather unique, with a "diffuse" structure; (2) the organization found in somatic cells (including TADs) seems to be gradually acquired and consolidated from one stage to the other during preimplantation development; and (3) the establishment of this organization is independent of transcription/zygotic genome activation but requires DNA replication. Which TADs appear first and are they correlated with specific transcriptional programs and/or chromatin states? When a TAD is established at a specific stage, is it maintained in subsequent stages? Are there stage-specific TAD boundaries? Interestingly, cohesin and Wapl (a cohesin release factor) are involved in shaping the topological organization of the zygotic genome, compatible with the "loop extrusion" model (Gassler et al. 2017). Further refinement and functional studies will be necessary to complete our understanding of how TAD organization is set up during development.

TADs: A STRUCTURAL BASIS FOR REGULATORY LANDSCAPES?

The term "regulatory landscape" in the context of gene regulation was first used by the Duboule lab (Monge et al. 2003; Spitz et al. 2003) to refer to large genomic regions containing clusters of enhancers and the promoters within their reach. A reporter gene inserted at hundreds of genomic locations in mice revealed that the activity range of *cis*-regulatory elements extends over large domains, which strongly correlate with TADs (Symmons et al. 2014). Complete regulatory landscapes of many genes are indeed in the range of hundreds kilobases or more, as is the case for the *X-inactivation center*, the full extent of which, however, is still unknown—for an extensive review on the *Xic*, see Augui et al. (2011). Indeed, even the largest *Xic* transgenes tested so far (~460 kb) were found to be unable to function as single-copy ectopic *Xic*s to induce X inactivation, and they do not fully recapitulate normal *Xist* expression patterns (Heard et al. 1999). Thus, the discovery that the *Xic* is partitioned into at least two TADs, spanning a total of ~800 kb, suggests that this might be the minimal *Xic* interval (Nora et al. 2012). Despite the largely invariant positioning of TADs across cell types (Dixon et al. 2012; Nora et al. 2012; Smith et al. 2016), variations in the internal conformation of TADs are frequently and reproducibly observed across different cell types (Nora et al. 2012; Dixon et al. 2015; Bonev et al. 2017), suggesting that TADs might represent a structural scaffold within which cell type–specific interactions can occur. In this section, we will discuss the accumulating evidence in support of the role of TADs in shaping and/or reflecting regulatory landscapes, as proposed when they were first discovered (Nora et al. 2013).

Physical and Functional Communication between Enhancers and Promoters

Many known pairs of enhancers and target promoters seem to be found within the same TAD (Shen et al. 2012; Nora et al. 2013), although several exceptions exist (Lower et al. 2009), and developmental genes often have enhancers spread across two neighboring TADs (see later). Several models have been proposed to explain the modes of action of enhancers (Kolovos et al. 2012), the most commonly cited being the "looping" model, whereby physical proximity in the nucleus between enhancer and promoter is required for their function. Numerous studies, especially using 3C-based technologies support looping by showing that enhancers and promoters establish spatial interactions, with intervening chromatin looping out (Tolhuis et al. 2002; Sanyal et al. 2012; Shen et al. 2012), and at the β-globin locus, forcing looping between the β-major promoter and the locus control region (LCR) induces transcription (Deng et al. 2012). However, specific looping at the time of promoter up-regulation has rarely been shown, and despite the focus on interactions between promoters and enhancers, the most prominently detected interactions in mammalian genomes are those between CTCF bound sites. Whether this is because of the more dynamic or labile nature of the former or because of detection biases with 3C-based technologies that privilege the latter is still not clear. A recent digestion- and ligation-free method for capturing chromatin contacts (GAM, genome architecture mapping) found a particular enrichment for pairwise interactions between enhancer elements and active genes (Beagrie et al. 2017), suggesting that 3C-derived might be less efficient in capturing this type of loops. Interactions between CTCF sites have also been shown to contribute to promoter–enhancer communication in some cases (for review, see Merkenschlager and Odom 2013 and Merkenschlager and Nora 2016). This function of CTCF might be more directly related to its role in shaping TADs than in directly linking promoters and enhancers, as only a few enhancers bind CTCF and many reside far away from CTCF sites (Cuadrado et al. 2015).

Contacts between promoters and their *cis*-regulatory elements may be constitutive, present in all cell types with no association to transcriptional activation (Amano et al. 2009; Montavon et al. 2011; Jin et al. 2013; Ghavi-Helm et al. 2014), or established de novo in a cell type–specific manner, accompanying cell type–specific transcriptional activation (Tolhuis et al. 2002; Simonis et al. 2006; Bonev et al. 2017). The latter—the "instructive model" (de Laat and Duboule 2013)—implies that looping events depend on factors present only in specific cell types (Spilianakis and Flavell 2004; Vakoc et al. 2005; Deng et al. 2012; Bonev et al. 2017) and/or that cell type–specific epigenetic modifications influence whether or not a DNA sequence can be bound by TFs or architectural proteins. Constitutive, preformed interactions—the "permissive model" (de Laat and Duboule 2013)—can nevertheless be of functionally relevance, as they can be associated with paused RNA polymerase (Ghavi-Helm et al. 2014), probably in a transcriptional poised state because of the lack of a specific set of transcription factors. This could be a rapid way to render cells permissive for transcriptional activation at specific stages or in specific tissues once those transcription factors become present (Amano et al. 2009). Another possible role for these preformed contacts between promoters and enhancers might be to prevent them from establishing interactions with other elements, as proposed in (Lonfat and Duboule 2015)—indeed, ectopic interactions might lead to dramatic phenotypic consequences, as discussed in the last section. The instructive and permissive models are not mutually exclusive: in some contexts, de novo contacts can accompany transcriptional activation within a preformed interacting domain

(Montavon et al. 2011). Here, the preformed contacts might contribute to the stabilization of the overall structure—the TAD structural scaffold—to allow the new interactions to be properly established and/or maintained.

In summary, based on current findings, it is likely that correct spatiotemporal gene expression is most often achieved via a combination of insulation from ectopic interactions and permissiveness for appropriate interactions, which usually but not always occurs within the limits of a TAD. A major part of what drives specificity of promoter-enhancer interactions within a TAD is likely to be factors binding to specific DNA sequences; however, cooperativity between transcription factors, presence of RNAs, and preexisting or induced chromatin states may all play a role in enabling the finely tuned usage of *cis*-regulatory elements for appropriate gene expression.

TADs Facilitate Enhancer–Promoter Communication

TADs seem to help the action of remote enhancers by reducing the effects of genomic distances: a recent study investigating a series of chromosomal rearrangements within the ∼1-Mb-size TAD that contains the *Shh* gene (Symmons et al. 2016), revealed that reduced or increased intra-TAD distances had no impact on *Shh* expression nor on correct *Shh*-dependent limb development. However, if the TAD was disrupted as a consequence of genomic inversions that place a TAD boundary between *Shh* promoter and its limb-specific enhancer, the contact frequencies between enhancer and promoter—and their transcriptional output—became distance-dependent, leading to a spectrum of phenotypical alterations (Symmons et al. 2016). In this case it would appear that TAD organization indeed promotes distance-independent interactions between distant elements, which would otherwise interact only very sporadically, failing to trigger appropriate gene expression (Symmons et al. 2016). Consistent with a model whereby stochastic fluctuations within a TAD bring regulatory elements and target promoters into closer proximity, favoring transcriptional activation, we have reported fluctuations in TAD conformation coupled to fluctuations in transcription at the *X-inactivation center* locus (Giorgetti et al. 2014). Using sequential RNA–DNA FISH, we found that a more compact TAD correlated with higher expression levels of one of its genes, suggesting a scenario in which enhancers and promoter are in closer proximity (Giorgetti et al. 2014).

TADs and Transcriptional Co-Regulation

The description of TADs at the *Xic* was accompanied by the observation that, during differentiation of mouse embryonic stem cells, expression of genes within the same TAD showed coordinated dynamics (up- or down-regulation) (Nora et al. 2012). This correlation (median correlation coefficient [mcc] of 0.40) was significantly higher than for genes in different TADs (mcc of 0.03) or randomly selected (mcc of 0.09). This was later confirmed beyond the *Xic*, genome-wide (Zhan et al. 2017), and also reported upon hormone stimulation, with up to 20% of the TADs showing coordinated up-regulation or down-regulation of the majority of the genes therein (Le Dily et al. 2014). TADs might be able to constrain diffusion of factors required for transcriptional activity and therefore may respond to transcriptional stimuli as a whole, as proposed in Nora et al. (2013) and Remeseiro et al. (2016). Physical clustering within TADs could therefore be used to coordinate gene expression programs during development. As mentioned previously, in the hierarchical folding of chromosomes, it is at the scale of TADs that the likelihood of genes within a domain being co-regulated during differentiation is maximized (Zhan et al. 2017).

Another curious example of an intimate link between TADs and transcription is that of the inactive X chromosome in mammals. This almost silent chromosome is mostly devoid of TAD structures (Minajigi et al. 2015; Giorgetti et al. 2016), except at the limited number of loci that retain transcriptional activity (Giorgetti et al. 2016). This raises the question of whether transcription at these loci drives their topological architecture, or whether these loci are transcribed because they retain their three-dimensional organization. Whichever may be cause or consequence, addressing such questions will provide additional insights into understanding the tight association between TADs and transcription.

TADs and Chromatin States

Differentially marked chromatin domains demarcate active and silent regions of the genome and the boundaries of such domains are often demarcated by TAD boundaries (Nora et al. 2012). Do such chromatin domains underlie TAD formation? In mouse embryonic stem cells (mESCs) deleted for the modifiers of H3K9me2 or H3K27me3, TAD organization at the *Xic*—which harbors a large domain of H3K9me2/H3K27me3—was unaffected (Nora et al. 2012), suggesting that TADs form independently of chromatin domains. Could TADs instead define such chromatin domains, by serving as modular units for the action of chromatin modifiers and limiting their spread beyond TAD boundaries (Nora et al. 2013; Ciabrelli and Cavalli 2015)? A recent study using a degron system to deplete the CTCF protein, which abolished most TADs in ES cells, revealed that H3K27me3 domains remained largely unchanged in this context (Nora et al. 2017). Furthermore, deleting a boundary CTCF element in the mouse *HoxA* locus did not lead to spread of H3K27me3, despite a shift in the interaction border (Narendra et al. 2015). However, the authors observed spreading of the active mark, H3K4me3, concomitant with the aberrant activation of previously repressed genes, affected by the boundary shift (Narendra et al. 2015). H3K4me3 spread could simply be due to ectopic gene expression, or it might reflect mechanisms regulating local spread of this chromatin mark (Narendra et al. 2015). Whether transitions between chromatin states can be directly dictated by TADs remains to be

disentangled from indirect effects of rewiring transcriptional activity, and the role of TADs and chromatin states in each other's formation or maintenance needs to be further explored.

TADs AND REGULATORY LANDSCAPES DURING DEVELOPMENT AND EVOLUTION

TAD organization may represent an evolutionary strategy to regulate developmental genes (for review, see Lonfat and Duboule 2015). Mammalian developmental genes often have multiple functions that are cell type– or stage-specific. It is therefore not surprising that they are frequently accompanied by complex regulatory landscapes. Examples of developmentally regulated loci are discussed below, although complex regulatory landscapes can also be found for broadly transcribed genes, including *Myc*, which has a different set of enhancers in ES versus B cells (Ruf et al. 2011; Kieffer-Kwon et al. 2013).

Developmental Genes Often Lie in Regions with Bipartite TAD Organization

We note that many developmentally regulated loci share a remarkably similar topological organization, with the gene promoter(s) lying close to or at the boundary between two TADs that harbor important *cis*-regulatory elements for their regulation (Fig. 1). Such locus architecture is found at the *Hox* clusters (Andrey et al. 2013; Lonfat et al. 2014), the *Xic* (Nora et al. 2012; Giorgetti et al. 2014), the *Six* genes (Gómez-Marín et al. 2015), and the *Tfap2c/Bmp7* locus (Tsujimura et al. 2015). The bipartite structure found at these developmental loci seems to be conserved across evolution, at least in certain animal lineages. The organization of the *Six* cluster is very similar in mouse, zebrafish, and sea urchin (Gómez-Marín et al. 2015), whereas *Hox* clusters are also found partitioned into two domains from mammals to fish (Woltering et al. 2014) and the *Xic/XIC* shows a boundary at the *Xist/Tsix* unit in both human and mouse (Dixon et al. 2012; Nora et al. 2012). This suggests that there are evolutionary constraints to maintain this highly conserved and particular organization in two adjacent TADs. For the *Hox* clusters, these constraints seem easier to understand—located at the boundary between two TADs, the locus needs to switch interactions at specific developmental stages to ensure proper segmentation of the developing limb (see Fig. 1 legend for more details). In the other cases (the *Six* clusters; *Tfap2c* and *Bmp7*; the *Xic*), however, the bipartite organization apparently serves to segregate distinct regulatory elements in two different TADs, either oppositely regulated or with different tissue specificities. Why then keep these domains adjacent to each other across millions of years of evolution if they represent different regulatory landscapes? To us, this suggests that cross-talk regulatory mechanisms probably exist between the two TADs at those loci, imposing evolutionary constraints and favoring the conservation of two adjacent TADs. Spitz and colleagues have reported that at the *Tfap2c/Bmp7* locus

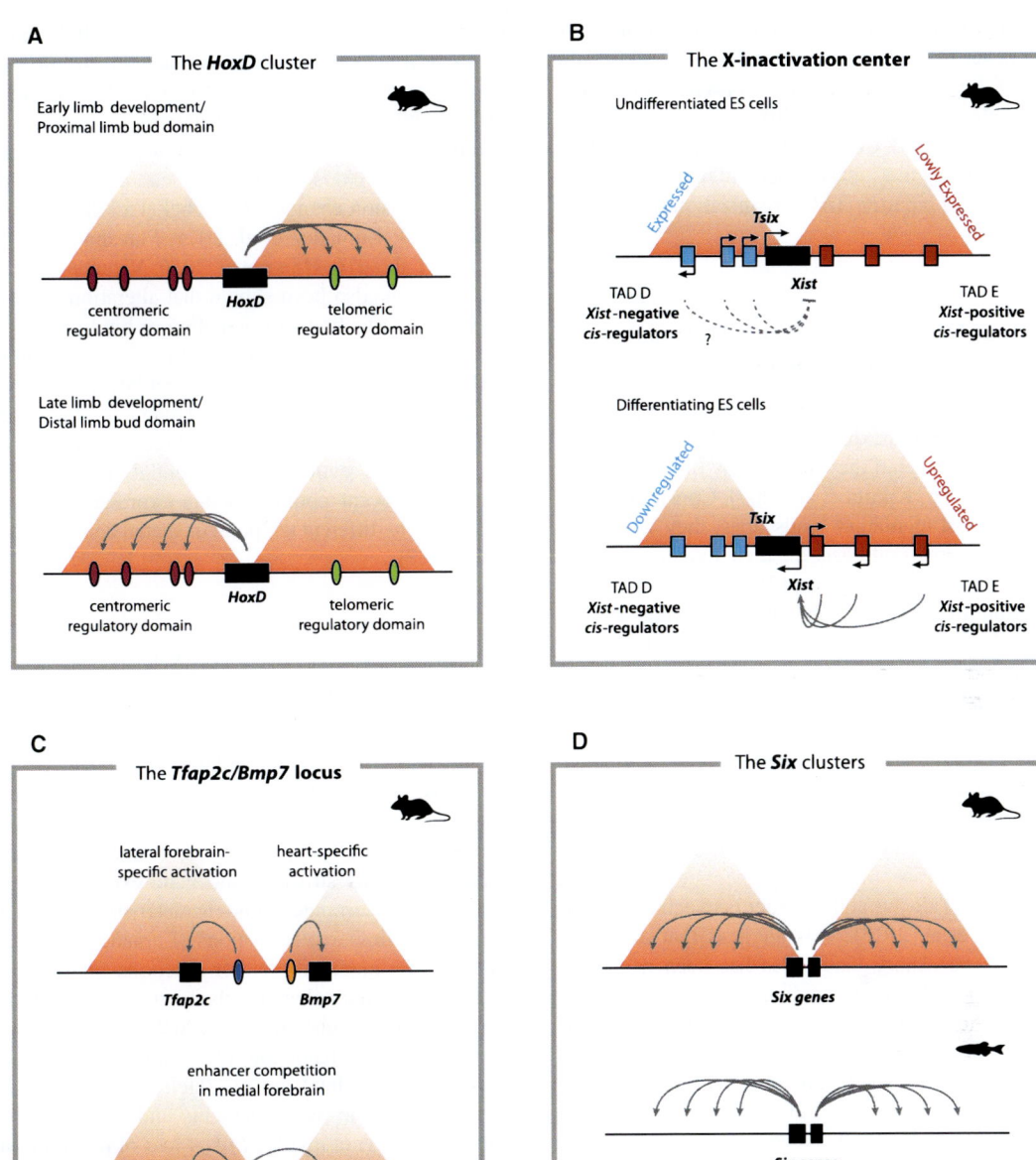

Figure 1. Bipartite TAD organization of developmental genes. (*A*) Regulation of limb development by the *HoxD* cluster occurs in two phases, which depend on the usage of two different regulatory landscapes, one on each side of the cluster. Accordingly, the locus shows a bipartite TAD structure and lies precisely on the boundary between them (Andrey et al. 2013; Lonfat and Duboule 2015). Although genes in either extremity of the cluster interact preferentially with the TAD they are closer to, central *Hox* genes undergo a topological switch from one TAD to the other at specific developmental stages in specific regions of the developing limb (Andrey et al. 2013). This switch is accompanied by specific gene expression patterns, critical for the patterning of vertebrate limbs (Andrey et al. 2013; Woltering et al. 2014). This type of regulation seems to be present at the *HoxA* cluster as well, which shows similar expression patterns during limb development and a similar TAD organization (Lonfat et al. 2014; Woltering et al. 2014). (*B*) The *Xic*, the master regulatory locus of X-chromosome inactivation, is also partitioned into two TADs in the mouse (Nora et al. 2012), with an antisense transcription unit at the boundary, composed of the noncoding *Xist* locus and its negative *cis*-regulator *Tsix*. Although the *Xist* promoter is within a TAD with some of its known positive *cis*-regulators, the *Tsix* promoter seems to lie under the influence of the adjacent TAD. Like for the *HoxA* and *HoxD* clusters, each TAD seems important at a specific stage: Whereas genes within the Tsix-TAD are coordinately down-regulated during embryonic stem cell differentiation, the genes within the Xist-TAD are up-regulated (Nora et al. 2012). (*C*) *Tfap2c* and *Bmp7* lie close to each other and are active in multiple tissues during embryogenesis, independently regulated by a distinct set of enhancers (Tsujimura et al. 2015). Each enhancer set lies in a different TAD, controlling different tissue-specificities, with a "transition zone" (corresponding to a TAD boundary) in between the *Tfap2c* and *Bmp7* genes, which allows for enhancer competition in specific tissues (Tsujimura et al. 2015). (*D*) The *Six* homeobox gene clusters are also organized in two TADs, lying close or at the border between them, and the expression patterns of genes on each side of the boundary are markedly different (Gómez-Marín et al. 2015). This organization seems to be conserved in fish and in sea urchin (Gómez-Marín et al. 2015).

the insulation between the two TADs is not absolute—it probably never is between any given TADs—and that the position of *Bmp7* influences the expression of *Tfap2c* in *cis*, despite being located in different, adjacent TADs (Tsujimura et al. 2015). Cross-TAD communication might thus represent the evolutionary constraint that explains this highly conserved bipartite TAD organization of developmental genes.

TADs as Modular Units during Mammalian Genome Evolution

The presence of *cis*-regulatory elements can impose evolutionary constraints, disfavoring the break of synteny with their targets, even if they lie far away from each other (Ahituv et al. 2005; Kikuta et al. 2007). Considering that TADs might host regulatory landscapes, they would not only provide a structural basis for their function but also for their evolution (Nora et al. 2013; Acemel et al. 2017). The position of TADs is robustly conserved, at least in mammals (Dixon et al. 2012; Vietri Rudan et al. 2015), and current available data from evolutionary distant species suggests that synteny breaks found within TADs are rather uncommon (Ahituv et al. 2005; Dixon et al. 2012; Nora et al. 2013; Vietri Rudan et al. 2015; Acemel et al. 2017). Hadjur and colleagues uncovered a number of complex rearrangements between the genomes of mouse and dog, involving duplications, insertions, and inversions, and in each case, the rearrangement never occurred within a TAD but always at the border between two TADs (Vietri Rudan et al. 2015). This suggests that disrupting TADs and *cis*-regulatory landscapes has been negatively selected during evolution. Accordingly, most—if not all—examples in the literature that report disruptions of TADs are associated with deleterious effects (Groschel et al. 2014; Northcott et al. 2014; Lupiáñez et al. 2015; Flavahan et al. 2016; Franke et al. 2016; Hnisz et al. 2016; Vicente-García et al. 2017). Nevertheless, evolutionary changes can also happen within TADs, provided that gain or loss of regulatory elements do not strongly affect TAD structural organization (Acemel et al. 2017)—in support of this, interaction domains in mouse and zebrafish seem to have conserved boundaries but quite different TAD lengths (Woltering et al. 2014). The emerging scenario is that TADs can indeed be reorganized during evolution as intact modules, as proposed in Nora et al. (2013), and also that TAD boundaries can constitute important hotspots for genomic rearrangements during evolution, as noted elsewhere (Acemel et al. 2017).

Genome organization in TADs might also provide new possible mechanisms for the evolution of gene regulation. A recent study has shown that large-scale duplications can lead to the formation of new TADs, bringing together previously insulated regions and leading to aberrant gene expression and limb malformations (Franke et al. 2016). From an evolutionary perspective, as suggested by Gómez-Skarmeta and colleagues, this indicates that processes such as gene duplication and neofunctionalization (the process by which a gene acquires a new function

upon a gene duplication event), classically thought to occur in a stepwise manner, can actually occur simultaneously with the formation of neo-TADs (Acemel et al. 2017). Other chromosome mutations or rearrangements (such as deletions, inversion, and translocations) that split, fuse, or alter TADs in some way can also easily lead to gene expression changes (Groschel et al. 2014; Northcott et al. 2014; Lupiáñez et al. 2015; Flavahan et al. 2016; Franke et al. 2016; Hnisz et al. 2016; Vicente-García et al. 2017). It has also been shown that alterations in a single CTCF site, affecting its orientation or binding of the protein, might be enough to reshape the organization of the TAD and the loops between *cis*-regulatory elements and their targets (de Wit et al. 2015; Guo et al. 2015; Sanborn et al. 2015; Tang et al. 2015). TAD reorganization has thus a considerable evolutionary potential, if even a single mutational event could generate expression patterns in substantially different temporal or spatial contexts (Acemel et al. 2017).

TADs AND REGULATORY LANDSCAPES DURING DISEASE

Chromosomal rearrangements involving disruption or displacement of TAD boundaries, or fusion or fission of TADs, can result in gene expression alterations that underlie specific pathologies. A recent study illustrated this very elegantly by exploring in mouse models the changes in TAD structure and gene expression induced by genomic rearrangements characteristic of human patients with limb malformations (Lupiáñez et al. 2015). These rearrangements—either deletions or inversions—included boundary elements flanking an ~2-Mb TAD harboring a single gene, *Epha4*, which is expressed in the developing limb. Knockout of *Epha4*, however, does not lead to limb skeletal defects (Helmbacher et al. 2000), implying that the rearrangements underlying the human pathologies do not involve *Epha4* loss of function. Mundlos and colleagues showed that these rearrangements allowed a set of enhancers within the same TAD as *Epha4*, probably responsible for the limb-specific expression pattern of *Epha4*, to establish new contacts with genes in the neighboring TADs. These genes (*Wnt6*, *Pax3*, *Ihh*) are normally not expressed in the developing limb at the same stage as *Epha4*, but the new ectopic interactions were accompanied by limb-specific activation of these genes, which likely underlies the limb malformations observed in mutant mice at birth. This study highlights the importance of TADs and their boundaries to restrict the range of action of *cis*-regulatory elements, but also how structural variants affecting TAD organization can lead to aberrant gene expression and morphological alterations in vivo. Other examples also illustrate this (Montavon et al. 2012; Franke et al. 2016).

Evidence for disruption of TADs in cancer contexts has also been reported, as a result of either chromosomal rearrangements or compromised CTCF binding. An excess of somatic mutations at CTCF sites is found in essentially all types of cancers, especially at sites involved in higher-order chromatin structures, such as TAD boundaries (Katainen et al. 2015; Hnisz et al. 2016; Kaiser et al. 2016). CTCF sites can also be affected by DNA hypermethylation, as observed in a subset of gliomas, where this is associated with impaired CTCF binding, loss of insulation between TADs and aberrant oncogene activation (Flavahan et al. 2016). Chromosomal rearrangements, a hallmark of cancers, can also lead to activation of oncogenes by exposing them to the activity of new enhancer elements, a phenomenon known as "enhancer adoption" or "enhancer hijacking" (Groschel et al. 2014; Northcott et al. 2014) and a likely result of TAD reorganization. In addition, chromosomal rearrangements can result in loss of interactions, leading to functional haploinsufficiency (Groschel et al. 2014), a known cause of some cancer syndromes. Recently, microdeletions that eliminate TAD boundaries were recurrently found in T-cell acute lymphoblastic leukemia genomes (Hnisz et al. 2016). Young and colleagues have further showed that in nonmalignant cells, perturbing boundaries of TADs containing protooncogenes was sufficient to activate them (Hnisz et al. 2016). Together, these observations suggest that, at least in some types of cancer, disruption of TADs might actually represent a driver phenomenon of tumorigenesis (Kaiser and Semple 2017).

CONCLUSION

The discovery that mammalian genomes are partitioned in TADs (Dixon et al. 2012; Nora et al. 2012) has influenced our understanding of long-range gene regulation, shedding further light on the mechanisms that govern the communication between enhancers and promoters, as well as on the evolution of the regulatory landscapes they underlie. Several open questions remain, however. Are regulatory landscapes restricted by TADs or can they span several TAD boundaries? How deterministic are fluctuations in TAD structure in regulating enhancer–promoter communication? How dynamic is the establishment of TAD structure during preimplantation development in mammals? What is the functional significance of topological domains beyond mammals, how equivalent are they, and which mechanisms underlie the different domains? Further investigations will allow the disentanglement of context-specific mechanisms and pave the way to establish general rules orchestrating the dynamic establishment and maintenance of TADs during development, disease, and evolution.

ACKNOWLEDGMENTS

We would like to apologize to all the authors whose original work we could not cite here because of space constraints. We would like to thank Katia Ancelin for critical reading of the manuscript; Katia Ancelin, Elphège Nora, and Luca Giorgetti for discussions about TADs and their meaning; the Duboule, Gómez-Skarmeta, and Spitz laboratories for inspirational reviews. R.G. was supported by grants from Région Ile-de-France (DIM Bio-thérapies) and Fondation pour la Recherche Médicale (FDT20160435295). E.H. is supported by ERC Advanced Investigator award (ERC-2014-AdG no. 671027), Labelisation La Ligue, ANR DoseX 2017, Labex DEEP (ANR-11-LBX-0044), part of the IDEX Idex PSL (ANR-10-IDEX-0001-02 PSL), and ABS4NGS (ANR-11-BINF-0001).

REFERENCES

Acemel RD, Maeso I, Gómez-Skarmeta JL. 2017. Topologically associated domains: A successful scaffold for the evolution of gene regulation in animals. *Wiley Interdiscip Rev Dev Biol* **6**: e265.

Ahituv N, Prabhakar S, Poulin F, Rubin EM, Couronne O. 2005. Mapping cis-regulatory domains in the human genome using multi-species conservation of synteny. *Hum Mol Genet* **14**: 3057–3063.

Alipour E, Marko JF. 2012. Self-organization of domain structures by DNA-loop-extruding enzymes. *Nucleic Acids Res* **40**: 11202–11212.

Amano T, Sagai T, Tanabe H, Mizushina Y, Nakazawa H, Shiroishi T. 2009. Chromosomal dynamics at the Shh locus: Limb bud-specific differential regulation of competence and active transcription. *Dev Cell* **16**: 47–57.

Andrey G, Montavon T, Mascrez B, Gonzalez F, Noordermeer D, Leleu M, Trono D, Spitz F, Duboule D. 2013. A switch between topological domains underlies HoxD genes collinearity in mouse limbs. *Science* **340**: 1234167.

Augui S, Nora EP, Heard E. 2011. Regulation of X-chromosome inactivation by the X-inactivation centre. *Nat Rev Genet* **12**: 429–442.

Battulin N, Fishman VS, Mazur AM, Pomaznoy M, Khabarova AA, Afonnikov DA, Prokhortchouk EB, Serov OL. 2015. Comparison of the three-dimensional organization of sperm and fibroblast genomes using the Hi-C approach. *Genome Biol* **16**: 77.

Beagrie RA, Scialdone A, Schueler M, Kraemer DCA, Chotalia M, Xie SQ, Barbieri M, de Santiago I, Lavitas L-M, Branco MR, et al. 2017. Complex multi-enhancer contacts captured by genome architecture mapping. *Nature* **543**: 519–524.

Berlivet S, Paquette D, Dumouchel A, Langlais D, Dostie J, Kmita M. 2013. Clustering of tissue-specific sub-TADs accompanies the regulation of HoxA genes in developing limbs. *PLoS Genet* **9**: e1004018.

Blackwood EM, Kadonaga JT. 1998. Going the distance: A current view of enhancer action. *Science* **281**: 60–63.

Bonev B, Mendelson Cohen N, Szabo Q, Fritsch L, Papadopoulos GL, Lubling Y, Xu X, Lv X, Hugnot J-P, Tanay A, et al. 2017. Multiscale 3D genome rewiring during mouse neural development. *Cell* **171**: 557–572.e24.

Busslinger GA, Stocsits RR, van der Lelij P, Axelsson E, Tedeschi A, Galjart N, Peters J-M. 2017. Cohesin is positioned in mammalian genomes by transcription, CTCF and Wapl. *Nature* **544**: 503–507.

Ciabrelli F, Cavalli G. 2015. Chromatin-driven behavior of topologically associating domains. *J Mol Biol* **427**: 608–625.

Cuadrado A, Remeseiro S, Grana O, Pisano DG, Losada A. 2015. The contribution of cohesin-SA1 to gene expression and chromatin architecture in two murine tissues. *Nucleic Acids Res* **43**: 3056–3067.

Cullen KE, Kladde MP, Seyfred MA. 1993. Interaction between transcription regulatory regions of prolactin chromatin. *Science* **261**: 203–206.

Degner SC, Verma-Gaur J, Wong TP, Bossen C, Iverson GM, Torkamani A, Vettermann C, Lin YC, Ju Z, Schulz D, et al. 2011. CCCTC-binding factor (CTCF) and cohesin influence the genomic architecture of the Igh locus and antisense transcription in pro-B cells. *Proc Natl Acad Sci* **108**: 9566–9571.

Dekker J, Heard E. 2015. Structural and functional diversity of topologically associating domains. *FEBS Lett* **589**: 2877–2884.

Dekker J, Rippe K, Dekker M, Kleckner N. 2002. Capturing chromosome conformation. *Science* **295:** 1306–1311.

de Laat W, Duboule D. 2013. Topology of mammalian developmental enhancers and their regulatory landscapes. *Nature* **502:** 499–506.

Deng W, Lee J, Wang H, Miller J, Reik A, Gregory P, Dean A, Blobel G. 2012. Controlling long-range genomic interactions at a native locus by targeted tethering of a looping factor. *Cell* **149:** 1233–1244.

Denker A, de Laat W. 2016. The second decade of 3C technologies: Detailed insights into nuclear organization. *Genes Dev* **30:** 1357–1382.

de Wit E, Vos ESM, Holwerda SJB, Valdes-Quezada C, Verstegen MJAM, Teunissen H, Splinter E, Wijchers PJ, Krijger PHL, de Laat W. 2015. CTCF binding polarity determines chromatin looping. *Mol Cell* **60:** 676–684.

Dixon JR, Selvaraj S, Yue F, Kim A, Li Y, Shen Y, Hu M, Liu JS, Ren B. 2012. Topological domains in mammalian genomes identified by analysis of chromatin interactions. *Nature* **485:** 376–380.

Dixon JR, Jung I, Selvaraj S, Shen Y, Antosiewicz-Bourget JE, Lee AY, Ye Z, Kim A, Rajagopal N, Xie W, et al. 2015. Chromatin architecture reorganization during stem cell differentiation. *Nature* **518:** 331–336.

Du Z, Zheng H, Huang B, Ma R, Wu J, Zhang X, He J, Xiang Y, Wang Q, Li Y, et al. 2017. Allelic reprogramming of 3D chromatin architecture during early mammalian development. *Nature* **547:** 232–235.

Flavahan WA, Drier Y, Liau BB, Gillespie SM, Venteicher AS, Stemmer-Rachamimov AO, Suvà ML, Bernstein BE. 2016. Insulator dysfunction and oncogene activation in IDH mutant gliomas. *Nature* **529:** 110–114.

Flyamer IM, Gassler J, Imakaev M, Brandão HB, Ulianov SV, Abdennur N, Razin SV, Mirny LA, Tachibana-Konwalski K. 2017. Single-nucleus Hi-C reveals unique chromatin reorganization at oocyte-to-zygote transition. *Nature* **544:** 110–114.

Forcato M, Nicoletti C, Pal K, Livi CM, Ferrari F, Bicciato S. 2017. Comparison of computational methods for Hi-C data analysis. *Nat Methods* **14:** 679–685.

Franke M, Ibrahim DM, Andrey G, Schwarzer W, Heinrich V, Schöpflin R, Kraft K, Kempfer R, Jerković I, Chan WL, et al. 2016. Formation of new chromatin domains determines pathogenicity of genomic duplications. *Nature* **538:** 265–269.

Fraser J, Ferrai C, Chiariello AM, Schueler M, Rito T, Laudanno G, Barbieri M, Moore BL, Kraemer DC, Aitken S, et al. 2015. Hierarchical folding and reorganization of chromosomes are linked to transcriptional changes in cellular differentiation. *Mol Syst Biol* **11:** 852.

Fudenberg G, Mirny LA. 2012. Higher-order chromatin structure: Bridging physics and biology. *Curr Opin Genet Dev* **22:** 115–124.

Fudenberg G, Imakaev M, Lu C, Goloborodko A, Abdennur N, Mirny L. 2016. Formation of chromosomal domains by loop extrusion. *Cell Rep* **15:** 2038–2049.

Fukaya T, Lim B, Levine M. 2016. Enhancer control of transcriptional bursting. *Cell* **166:** 358–368.

Gassler J, Brandão HB, Imakaev M, Flyamer IM, Ladstätter S, Bickmore WA, Peters J-M, Mirny LA, Tachibana K. 2017. A mechanism of cohesin-dependent loop extrusion organizes zygotic genome architecture. *EMBO J* **36:** 3600–3618.

Ghavi-Helm Y, Klein FA, Pakozdi T, Ciglar L, Noordermeer D, Huber W, Furlong EEM. 2014. Enhancer loops appear stable during development and are associated with paused polymerase. *Nature* **512:** 96–100.

Giorgetti L, Galupa R, Nora EP, Piolot T, Lam F, Dekker J, Tiana G, Heard E. 2014. Predictive polymer modeling reveals coupled fluctuations in chromosome conformation and transcription. *Cell* **157:** 950–963.

Giorgetti L, Lajoie BR, Carter AC, Attia M, Zhan Y, Xu J, Chen CJ, Kaplan N, Chang HY, Heard E, et al. 2016. Structural organization of the inactive X chromosome in the mouse. *Nature* **535:** 575–579.

Goloborodko A, Marko JF, Mirny LA. 2016. Chromosome compaction by active loop extrusion. *Biophys J* **110:** 2162–2168.

Gómez-Marín C, Tena JJ, Acemel RD, López-Mayorga M, Naranjo S, de la Calle-Mustienes E, Maeso I, Beccari L, Aneas I, Vielmas E, et al. 2015. Evolutionary comparison reveals that diverging CTCF sites are signatures of ancestral topological associating domains borders. *Proc Natl Acad Sci* **112:** 7542–7547.

Groschel S, Sanders MA, Hoogenboezem R, De Wit E, Bouwman BAM, Erpelinck C, Van der Velden VHJ, Havermans M, Avellino R, Van Lom K, et al. 2014. A single oncogenic enhancer rearrangement causes concomitant EVI1 and GATA2 deregulation in leukemia. *Cell* **157:** 369–381.

Guo Y, Monahan K, Wu H, Gertz J, Varley KE, Li W, Myers RM, Maniatis T, Wu Q. 2012. CTCF/cohesin-mediated DNA looping is required for protocadherin α promoter choice. *Proc Natl Acad Sci* **109:** 21081–21086.

Guo Y, Xu Q, Canzio D, Shou J, Li J, Gorkin D, Jung I, Wu H, Zhai Y, Tang Y, et al. 2015. CRISPR inversion of CTCF sites alters genome topology and enhancer/promoter function. *Cell* **162:** 900–910.

Haarhuis JHI, van der Weide RH, Blomen VA, Yáñez-Cuna JO, Amendola M, van Ruiten MS, Krijger PHL, Teunissen H, Medema RH, van Steensel B, et al. 2017. The cohesin release factor WAPL restricts chromatin loop extension. *Cell* **169:** 693–707.e14.

Hadjur S, Williams LM, Ryan NK, Cobb BS, Sexton T, Fraser P, Fisher AG, Merkenschlager M. 2009. Cohesins form chromosomal cis-interactions at the developmentally regulated IFNG locus. *Nature* **460:** 410–413.

Heard E, Mongelard F, Arnaud D, Avner P. 1999. Xist yeast artificial chromosome transgenes function as X-inactivation centers only in multicopy arrays and not as single copies. *Mol Cell Biol* **19:** 3156–3166.

Helmbacher F, Schneider-Maunoury S, Topilko P, Tiret L, Charnay P. 2000. Targeting of the EphA4 tyrosine kinase receptor affects dorsal/ventral pathfinding of limb motor axons. *Development* **127:** 3313–3324.

Hnisz D, Weintraub AS, Day DS, Valton A-L, Bak RO, Li CH, Goldmann J, Lajoie BR, Fan ZP, Sigova AA, et al. 2016. Activation of proto-oncogenes by disruption of chromosome neighborhoods. *Science* **351:** 1454–1458.

Jin F, Li Y, Dixon JR, Selvaraj S, Ye Z, Lee AY, Yen C-A, Schmitt AD, Espinoza CA, Ren B. 2013. A high-resolution map of the three-dimensional chromatin interactome in human cells. *Nature* **503:** 290–294.

Jung YH, Sauria MEG, Lyu X, Cheema MS, Ausio J, Taylor J, Corces VG. 2017. Chromatin states in mouse sperm correlate with embryonic and adult regulatory landscapes. *Cell Rep* **18:** 1366–1382.

Kaiser VB, Semple CA. 2017. When TADs go bad: Chromatin structure and nuclear organisation in human disease. *F1000Res* **6:** 314.

Kaiser VB, Taylor MS, Semple CA. 2016. Mutational biases drive elevated rates of substitution at regulatory sites across cancer types. *PLoS Genet* **12:** e1006207.

Katainen R, Dave K, Pitkänen E, Palin K, Kivioja T, Välimäki N, Gylfe AE, Ristolainen H, Hänninen UA, Cajuso T, et al. 2015. CTCF/cohesin-binding sites are frequently mutated in cancer. *Nat Genet* **47:** 818–821.

Ke Y, Xu Y, Chen X, Feng S, Liu Z, Sun Y, Yao X, Li F, Zhu W, Gao L, et al. 2017. 3D chromatin structures of mature gametes and structural reprogramming during mammalian embryogenesis. *Cell* **170:** 367–381.e20.

Kieffer-Kwon K-R, Tang Z, Mathe E, Qian J, Sung M-H, Li G, Resch W, Baek S, Pruett N, Grøntved L, et al. 2013. Interactome maps of mouse gene regulatory domains reveal basic principles of transcriptional regulation. *Cell* **155:** 1507–1520.

Kikuta H, Laplante M, Navratilova P, Komisarczuk AZ, Engstrom PG, Fredman D, Akalin A, Caccamo M, Sealy I, Howe K, et al. 2007. Genomic regulatory blocks encompass multiple neighboring genes and maintain conserved synteny in vertebrates. *Genome Res* **17:** 545–555.

Kimura K, Rybenkov VV, Crisona NJ, Hirano T, Cozzarelli NR. 1999. 13S condensin actively reconfigures DNA by introducing global positive writhe: Implications for chromosome condensation. *Cell* **98**: 239–248.

Kolovos P, Knoch TA, Grosveld FG, Cook PR, Papantonis A. 2012. Enhancers and silencers: An integrated and simple model for their function. *Epigenetics Chromatin* **5**: 1.

Le Dily F, Baù D, Pohl A, Vicent GP, Serra F, Soronellas D, Castellano G, Wright RHG, Ballare C, Filion G, et al. 2014. Distinct structural transitions of chromatin topological domains correlate with coordinated hormone-induced gene regulation. *Genes Dev* **28**: 2151–2162.

Lieberman-Aiden E, van Berkum NL, Williams L, Imakaev M, Ragoczy T, Telling A, Amit I, Lajoie BR, Sabo PJ, Dorschner MO, et al. 2009. Comprehensive mapping of long-range interactions reveals folding principles of the human genome. *Science* **326**: 289–293.

Lonfat N, Duboule D. 2015. Structure, function and evolution of topologically associating domains (TADs) at *HOX* loci. *FEBS Lett* **589**: 2869–2876.

Lonfat N, Montavon T, Darbellay F, Gitto S, Duboule D. 2014. Convergent evolution of complex regulatory landscapes and pleiotropy at Hox loci. *Science* **346**: 1004–1006.

Lower KM, Hughes JR, De Gobbi M, Henderson S, Viprakasit V, Fisher C, Goriely A, Ayyub H, Sloane-Stanley J, Vernimmen D, et al. 2009. Adventitious changes in long-range gene expression caused by polymorphic structural variation and promoter competition. *Proc Natl Acad Sci* **106**: 21771–21776.

Lupiáñez DG, Kraft K, Heinrich V, Krawitz P, Brancati F, Klopocki E, Horn D, Kayserili H, Opitz JM, Laxova R, et al. 2015. Disruptions of topological chromatin domains cause pathogenic rewiring of gene-enhancer interactions. *Cell* **161**: 1012–1025.

Lupiáñez DG, Spielmann M, Mundlos S. 2016. Breaking TADs: How alterations of chromatin domains result in disease. *Trends Genet* **32**: 225–237.

Merkenschlager M, Nora EP. 2016. CTCF and cohesin in genome folding and transcriptional gene regulation. *Annu Rev Genomics Hum Genet* **17**: 17–43.

Merkenschlager M, Odom DT. 2013. CTCF and cohesin: Linking gene regulatory elements with their targets. *Cell* **152**: 1285–1297.

Minajigi A, Froberg JE, Wei C, Sunwoo H, Kesner B, Colognori D, Lessing D, Payer B, Boukhali M, Haas W, et al. 2015. Chromosomes. A comprehensive Xist interactome reveals cohesin repulsion and an RNA-directed chromosome conformation. *Science* **349**. doi: 10.1126/science.aab2276 aab2276.

Monge I, Kondo T, Duboule D. 2003. An enhancer-titration effect induces digit-specific regulatory alleles of the HoxD cluster. *Dev Biol* **256**: 212–220.

Montavon T, Soshnikova N, Mascrez B, Joye E, Thevenet L, Splinter E, de Laat W, Spitz F, Duboule D. 2011. A regulatory archipelago controls Hox genes transcription in digits. *Cell* **147**: 1132–1145.

Montavon T, Thevenet L, Duboule D. 2012. Impact of copy number variations (CNVs) on long-range gene regulation at the HoxD locus. *Proc Natl Acad Sci* **109**: 20204–20211.

Nagano T, Lubling Y, Stevens TJ, Schoenfelder S, Yaffe E, Dean W, Laue ED, Tanay A, Fraser P. 2013. Single-cell Hi-C reveals cell-to-cell variability in chromosome structure. *Nature* **502**: 59–64.

Nagano T, Lubling Y, Várnai C, Dudley C, Leung W, Baran Y, Mendelson Cohen N, Wingett S, Fraser P, Tanay A. 2017. Cell-cycle dynamics of chromosomal organization at single-cell resolution. *Nature* **547**: 61–67.

Narendra V, Rocha PP, An D, Raviram R, Skok JA, Mazzoni EO, Reinberg D. 2015. CTCF establishes discrete functional chromatin domains at the Hox clusters during differentiation. *Science* **347**: 1017–1021.

Nasmyth K. 2001. Disseminating the genome: Joining, resolving, and separating sister chromatids during mitosis and meiosis. *Annu Rev Genet* **35**: 673–745.

Naumova N, Imakaev M, Fudenberg G, Zhan Y, Lajoie BR, Mirny LA, Dekker J. 2013. Organization of the mitotic chromosome. *Science* **342**: 948–953.

Nora EP, Lajoie BR, Schulz EG, Giorgetti L, Okamoto I, Servant N, Piolot T, van Berkum NL, Meisig J, Sedat J, et al. 2012. Spatial partitioning of the regulatory landscape of the X-inactivation centre. *Nature* **485**: 381–385.

Nora EP, Dekker J, Heard E. 2013. Segmental folding of chromosomes: A basis for structural and regulatory chromosomal neighborhoods? *Bioessays* **35**: 818–828.

Nora EP, Goloborodko A, Valton A-L, Gibcus JH, Uebersohn A, Abdennur N, Dekker J, Mirny LA, Bruneau BG. 2017. Targeted degradation of CTCF decouples local insulation of chromosome domains from genomic compartmentalization. *Cell* **169**: 930–944.e22.

Northcott PA, Lee C, Zichner T, Stutz AM, Erkek S, Kawauchi D, Shih DJH, Hovestadt V, Zapatka M, Sturm D, et al. 2014. Enhancer hijacking activates GFI1 family oncogenes in medulloblastoma. *Nature* **511**: 428–434.

Phillips-Cremins JE, Sauria MEG, Sanyal A, Gerasimova TI, Lajoie BR, Bell JSK, Ong C-T, Hookway TA, Guo C, Sun Y, et al. 2013. Architectural protein subclasses shape 3D organization of genomes during lineage commitment. *Cell* **153**: 1281–1295.

Rao SSP, Huntley MH, Durand NC, Stamenova EK, Bochkov ID, Robinson JT, Sanborn AL, Machol I, Omer AD, Lander ES, et al. 2014. A 3D map of the human genome at kilobase resolution reveals principles of chromatin looping. *Cell* **159**: 1665–1680.

Rao SSP, Huang S-C, Glenn St Hilaire B, Engreitz JM, Perez EM, Kieffer-Kwon K-R, Sanborn AL, Johnstone SE, Bascom GD, Bochkov ID, et al. 2017. Cohesin loss eliminates all loop domains. *Cell* **171**: 305–320.e24.

Remeseiro S, Hornblad A, Spitz F. 2016. Gene regulation during development in the light of topologically associating domains. *Wiley Interdiscip Rev Dev Biol* **5**: 169–185.

Riggs AD. 1990. DNA methylation and late replication probably aid cell memory, and type 1 DNA reeling could aid chromosome folding and enhancer function. *Philos Trans R Soc Lond B Biol Sci* **326**: 285–297.

Rocha PP, Raviram R, Bonneau R, Skok JA. 2015. Breaking TADs: Insights into hierarchical genome organization. *Epigenomics* **7**: 523–526.

Ruf S, Symmons O, Uslu VV, Dolle D, Hot C, Ettwiller L, Spitz F. 2011. Large-scale analysis of the regulatory architecture of the mouse genome with a transposon-associated sensor. *Nat Genet* **43**: 379–386.

Sanborn AL, Rao SSP, Huang S-C, Durand NC, Huntley MH, Jewett AI, Bochkov ID, Chinnappan D, Cutkosky A, Li J, et al. 2015. Chromatin extrusion explains key features of loop and domain formation in wild-type and engineered genomes. *Proc Natl Acad Sci* **112**: E6456–E6465.

Sanyal A, Lajoie BR, Jain G, Dekker J. 2012. The long-range interaction landscape of gene promoters. *Nature* **489**: 109–113.

Schmidt D, Schwalie PC, Wilson MD, Ballester B, Gonçalves Â, Kutter C, Brown GD, Marshall A, Flicek P, Odom DT. 2012. Waves of retrotransposon expansion remodel genome organization and CTCF binding in multiple mammalian lineages. *Cell* **148**: 335–348.

Schwarzer W, Abdennur N, Goloborodko A, Pekowska A, Fudenberg G, Loe-Mie Y, Fonseca NA, Huber W, H Haering C, Mirny L, et al. 2017. Two independent modes of chromatin organization revealed by cohesin removal. *Nature* **551**: 51–56.

Sexton T, Yaffe E, Kenigsberg E, Bantignies F, Leblanc B, Hoichman M, Parrinello H, Tanay A, Cavalli G. 2012. Three-dimensional folding and functional organization principles of the *Drosophila* genome. *Cell* **148**: 458–472.

Shen Y, Yue F, McCleary DF, Ye Z, Edsall L, Kuan S, Wagner U, Dixon J, Lee L, Lobanenkov VV, et al. 2012. A map of the cis-regulatory sequences in the mouse genome. *Nature* **488**: 116–120.

Simonis M, Klous P, Splinter E, Moshkin Y, Willemsen R, de Wit E, van Steensel B, de Laat W. 2006. Nuclear organization of active and inactive chromatin domains uncovered by chromosome conformation capture–on-chip (4C). *Nat Genet* **38:** 1348–1354.

Smith EM, Lajoie BR, Jain G, Dekker J. 2016. Invariant TAD boundaries constrain cell-type-specific looping interactions between promoters and distal elements around the CFTR locus. *Am J Hum Genet* **98:** 185–201.

Spilianakis CG, Flavell RA. 2004. Long-range intrachromosomal interactions in the T helper type 2 cytokine locus. *Nat Immunol* **5:** 1017–1027.

Spitz F, Gonzalez F, Duboule D. 2003. A global control region defines a chromosomal regulatory landscape containing the HoxD cluster. *Cell* **113:** 405–417.

Stevens TJ, Lando D, Basu S, Atkinson LP, Cao Y, Lee SF, Leeb M, Wohlfahrt KJ, Boucher W, O'Shaughnessy-Kirwan A, et al. 2017. 3D structures of individual mammalian genomes studied by single-cell Hi-C. *Nature* **544:** 59–64.

Symmons O, Uslu VV, Tsujimura T, Ruf S, Nassari S, Schwarzer W, Ettwiller L, Spitz F. 2014. Functional and topological characteristics of mammalian regulatory domains. *Genome Res* **24:** 390–400.

Symmons O, Pan L, Remeseiro S, Aktas T, Klein F, Huber W, Spitz F. 2016. The Shh topological domain facilitates the action of remote enhancers by reducing the effects of genomic distances. *Dev Cell* **39:** 529–543.

Tang Z, Luo OJ, Li X, Zheng M, Zhu JJ, Szalaj P, Trzaskoma P, Magalska A, Wlodarczyk J, Ruszczycki B, et al. 2015. CTCF-mediated human 3D genome architecture reveals chromatin topology for transcription. *Cell* **163:** 1611–1627.

Tolhuis B, Palstra R-J, Splinter E, Grosveld F, de Laat W. 2002. Looping and interaction between hypersensitive sites in the active α-globin locus. *Mol Cell* **10:** 1453–1465.

Tsujimura T, Klein FA, Langenfeld K, Glaser J, Huber W, Spitz F. 2015. A discrete transition zone organizes the topological and regulatory autonomy of the adjacent Tfap2c and Bmp7 genes. *PLoS Genet* **11:** e1004897.

Vakoc CR, Letting DL, Gheldof N, Sawado T, Bender MA, Groudine M, Weiss MJ, Dekker J, Blobel GA. 2005. Proximity among distant regulatory elements at the β-globin locus requires GATA-1 and FOG-1. *Mol Cell* **17:** 453–462.

Van Bortle K, Nichols MH, Li L, Ong C-T, Takenaka N, Qin ZS, Corces VG. 2014. Insulator function and topological domain border strength scale with architectural protein occupancy. *Genome Biol* **15:** R82.

Vera M, Biswas J, Senecal A, Singer RH, Park HY. 2016. Single-cell and single-molecule analysis of gene expression regulation. *Annu Rev Genet* **50:** 267–291.

Vicente-García C, Villarejo-Balcells B, Irastorza-Azcárate I, Naranjo S, Acemel RD, Tena JJ, Rigby PWJ, Devos DP, Gómez-Skarmeta JL, Carvajal JJ. 2017. Regulatory landscape fusion in rhabdomyosarcoma through interactions between the PAX3 promoter and FOXO1 regulatory elements. *Genome Biol* **18:** 106.

Vietri Rudan M, Barrington C, Henderson S, Ernst C, Odom DT, Tanay A, Hadjur S. 2015. Comparative Hi-C reveals that CTCF underlies evolution of chromosomal domain architecture. *Cell Rep* **10:** 1297–1309.

Way GP, Youngstrom DW, Hankenson KD, Greene CS, Grant SF. 2017. Implicating candidate genes at GWAS signals by leveraging topologically associating domains. *Eur J Hum Genet* **25:** 1286–1289.

Woltering JM, Noordermeer D, Leleu M, Duboule D. 2014. Conservation and divergence of regulatory strategies at Hox loci and the origin of tetrapod digits. *PLoS Biol* **12:** e1001773.

Wutz G, Várnai C, Nagasaka K, Cisneros DA, Stocsits RR, Tang W, Schoenfelder S, Jessberger G, Muhar M, Hossain MJ, et al. 2017. Topologically associating domains and chromatin loops depend on cohesin and are regulated by CTCF, WAPL, and PDS5 proteins. *EMBO J* **36:** 3573–3599.

Zhan Y, Mariani L, Barozzi I, Schulz EG, Blüthgen N, Stadler M, Tiana L, Giorgetti L. 2017. Reciprocal insulation analysis of Hi-C data shows that TADs represent a functionally but not structurally privileged scale in the hierarchical folding of chromosomes. *Genome Res* **27:** 479–490.

Zuin J, Dixon JR, van der Reijden MIJA, Ye Z, Kolovos P, Brouwer RWW, van de Corput MPC, van de Werken HJG, Knoch TA, van IJcken WFJ, et al. 2014. Cohesin and CTCF differentially affect chromatin architecture and gene expression in human cells. *Proc Natl Acad Sci* **111:** 996–1001.

Dynamic Control of Chromosome Topology and Gene Expression by a Chromatin Modification

Qian Bian,[1] Erika C. Anderson, Katjuša Brejc, and Barbara J. Meyer

Howard Hughes Medical Institute and Department of Molecular and Cell Biology, University of California at Berkeley, Berkeley, California 94720-3204

Correspondence: bjmeyer@berkeley.edu

The function of chromatin modification in establishing higher-order chromosome structure during gene regulation has been elusive. We dissected the machinery and mechanism underlying the enrichment of histone modification H4K20me1 on hermaphrodite X chromosomes during *Caenorhabditis elegans* dosage compensation and discovered a key role for H4K20me1 in regulating X-chromosome topology and chromosome-wide gene expression. Structural and functional analysis of the dosage compensation complex (DCC) subunit DPY-21 revealed a novel Jumonji C demethylase subfamily that converts H4K20me2 to H4K20me1 in worms and mammals. Inactivation of demethylase activity in vivo by genome editing eliminated H4K20me1 enrichment on X chromosomes of somatic cells, increased X-linked gene expression, reduced X-chromosome compaction, and disrupted X-chromosome conformation by diminishing the formation of topologically associated domains. H4K20me1 is also enriched on the inactive X of female mice, making our studies directly relevant to mammalian development. Unexpectedly, DPY-21 also associates specifically with autosomes of nematode germ cells in a DCC-independent manner to enrich H4K20me1 and trigger chromosome compaction. Thus, DPY-21 is an adaptable chromatin regulator. Its H4K20me2 demethylase activity can be harnessed during development for distinct biological functions by targeting it to diverse genomic locations through different mechanisms. In both somatic cells and germ cells, H4K20me1 enrichment modulates three-dimensional chromosome architecture, demonstrating the direct link between chromatin modification and higher-order chromosome structure.

Chromatin modification and three-dimensional (3D) chromosome structure play central roles in regulating gene expression (Vieux-Rochas et al. 2015; Dekker and Mirny 2016; Soshnev et al. 2016; Nora et al. 2017; Schuettengruber et al. 2017). However, the role of histone modifications in establishing higher-order chromosome structure during gene regulation has been elusive.

Histone modifications are known to alter the interactions between nucleosomes that govern the compaction state of a chromatin fiber (Francis et al. 2004; Lu et al. 2008; Kalashnikova et al. 2013). However, the effect of histone modifications on higher-order chromosome organization beyond chromatin-fiber compaction is not well understood. Only recently has super-resolution imaging shown that chromatin domains enriched in H3K27me3 and Polycomb proteins adopt unique folded states (Boettiger et al. 2016). Once the Polycomb repressive complex 1 is recruited to H3K27me3-enriched chromatin, it generates chromatin domains 20–140 kb in size (Kundu et al. 2017). These domains are distinct from topologically associating domains (TADs), which are self-interacting genomic regions up to 1 Mb in size that permit loci within a domain to interact with each other but insulate these loci from interactions with loci in other TADs (Nora et al. 2013).

We analyzed X-chromosome dosage compensation in the nematode *Caenorhabditis elegans* to explore the relationship between chromatin modification and higher-order chromosome structure during chromosome-wide gene regulation. Dosage compensation is achieved by a 10-subunit dosage compensation complex (DCC) that binds to both hermaphrodite X chromosomes via sequence-dependent recruitment elements on X (*rex* sites) to reduce transcription by half (Fig. 1A,C; Csankovszki et al. 2009; Jans et al. 2009; Mets and Meyer 2009; Meyer 2010). This chromosome-wide modulation of gene expression ensures that males (XO) and hermaphrodites (XX) express equivalent levels of X-chromosome products despite their unequal dose of X chromosomes. Five DCC subunits (Fig. 1B) are homologous to subunits of condensin, a complex conserved from yeast to man that compacts and resolves chromosomes in preparation for chromosome segregation during mitosis and meiosis (Csankovszki et al. 2009; Mets and Meyer 2009; Meyer 2010; Hirano 2016). The similarity between the DCC and condensin suggested that the DCC might remodel the structure of X to achieve gene repression.

Indeed, the DCC imposes a distinct higher-order structure onto both hermaphrodite X chromosomes while repressing gene expression. It remodels the topology of X into a sex-specific spatial conformation distinct from that of autosomes or male X chromosomes by forming TADs of ~1 Mb (Crane et al. 2015). The highest-affinity *rex* sites are located at TAD boundaries and several observations support a model in which DCC-dependent looping interactions between these high-affinity *rex* sites direct TAD formation (Crane et al. 2015). (1) Interactions between *rex* sites are

---

[1]Present address: Shanghai Institute of Precision Medicine, Ninth People's Hospital, Shanghai Jiao Tong University School of Medicine, Shanghai, China

Published by Cold Spring Harbor Laboratory Press; doi: 10.1101/sqb.2017.82.034439

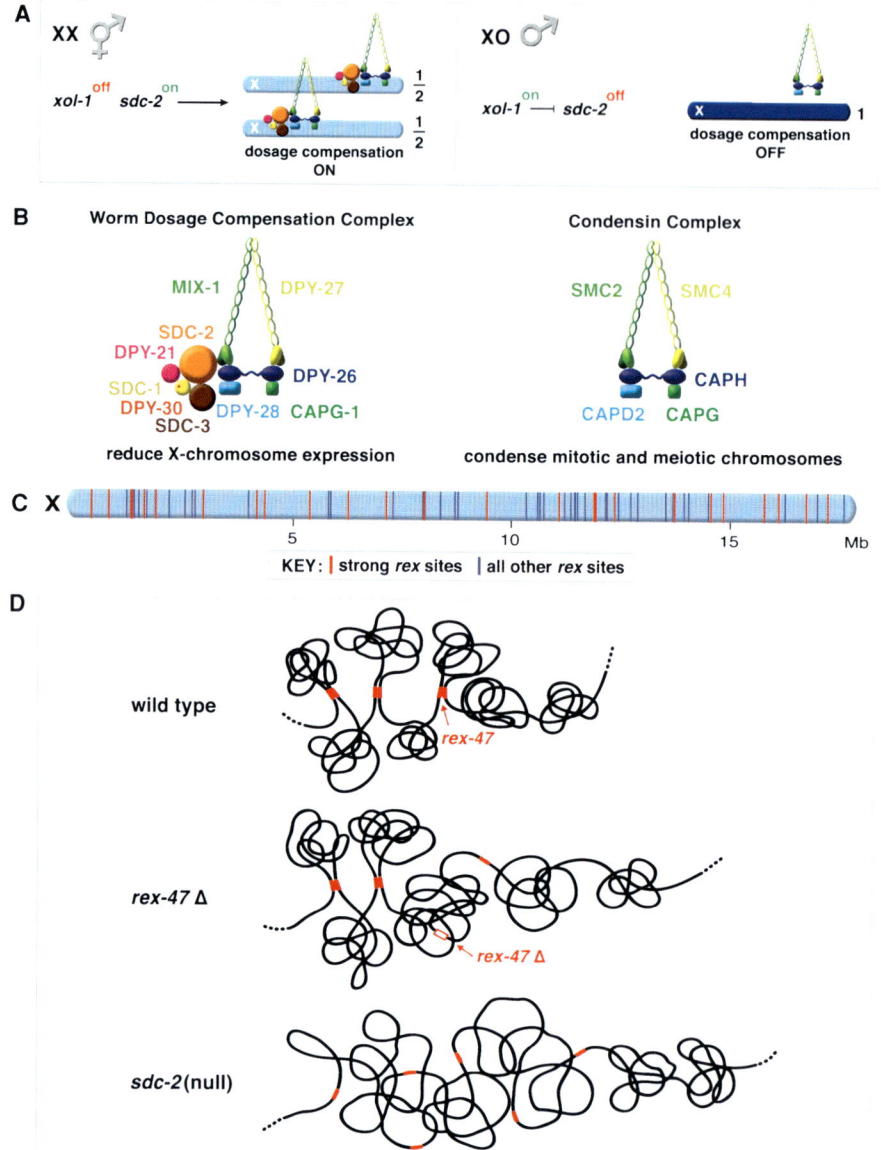

Figure 1. Overview of dosage compensation in *Caenorhabditis elegans*. (*A*) In XX hermaphrodites, a dosage compensation complex (DCC) binds to both X chromosomes to reduce gene expression by half, thereby equalizing expression with that from the single male X. The product of the XX-specific gene *sdc-2* triggers assembly of the DCC onto X. In XO males, *xol-1*, the male-specific regulator of sex determination and dosage compensation, represses *sdc-2*, thereby preventing the DCC from binding to the male X. (*B*) The DCC compared with condensin I of other eukaryotes. The DCC condensin subunits (MIX-1, DPY-27, DPY-26, DPY-28, and CAPG-1) are color matched to their condensin I homologs (Csankovszki et al. 2009; Mets and Meyer 2009; Meyer 2010). All DCC condensin subunits except DPY-27 also function in other condensins that act in *C. elegans* mitosis and meiosis. The DPY-27 paralog SMC-4 (not shown) replaces DPY-27 in mitotic and meiotic condensins. The DCC likely arose by duplicating the gene encoding SMC-4 and modifying it to create DPY-27 for a specific role in gene expression (Hagstrom et al. 2002). In addition to condensin subunits, the DCC also includes a novel XX-specific protein with a large coiled-coil domain (SDC-2) (Dawes et al. 1999) that triggers assembly of the DCC onto X chromosomes. Two DCC subunits aid SDC-2 in recruiting the complex to X, SDC-3 (a zinc finger protein) and DPY-30 (a subunit of the MLL/COMPASS H3K4me3 methyltransferase complex) (Klein and Meyer 1993; Hsu et al. 1995; Davis and Meyer 1997; Pferdehirt et al. 2011). Two subunits, SDC-1 (a zinc finger protein) and DPY-21, are required for DCC activity but not assembly (Nonet and Meyer 1991; Yonker and Meyer 2003). DPY-21 is a Jumonji C H4K20me2 demethylase described here and in (Brejc et al. 2017). (*C*) DCC recruitment sites across X chromosomes. The DCC recruitment elements on X (*rex*) were discovered by the combination of genome-wide approaches (ChIP-chip and ChIP-seq) to identify DCC-binding sites without regard to autonomous recruitment ability and a functional approach in vivo to assess DCC binding to sites detached from X (Jans et al. 2009; Crane et al. 2015). *rex* sites are distributed across X and confer X-chromosome specificity to dosage compensation. DCC binding to *rex* sites facilitates DCC spreading across X to sites that cannot bind the complex if detached from X (Pferdehirt et al. 2011). Several of the strongest *rex* sites (red) are essential for formation of topologically associated domains (TADs). (*D*) Cartoon model of TAD formation on a segment of X. (*Top*) The DCC remodels the topology of X into a hermaphrodite-specific conformation by forming TADs. DCC-dependent looping interactions between high-affinity *rex* sites located at TAD boundaries direct TAD formation (Crane et al. 2015). (*Middle*) Deletion of the high-affinity site *rex-47* located at a DCC-dependent TAD boundary eliminates boundary formation (Crane et al. 2015). (*Bottom*) Severe disruption of DCC binding by an *sdc-2* mutation eliminates formation of all DCC-dependent TADs on X (Crane et al. 2015).

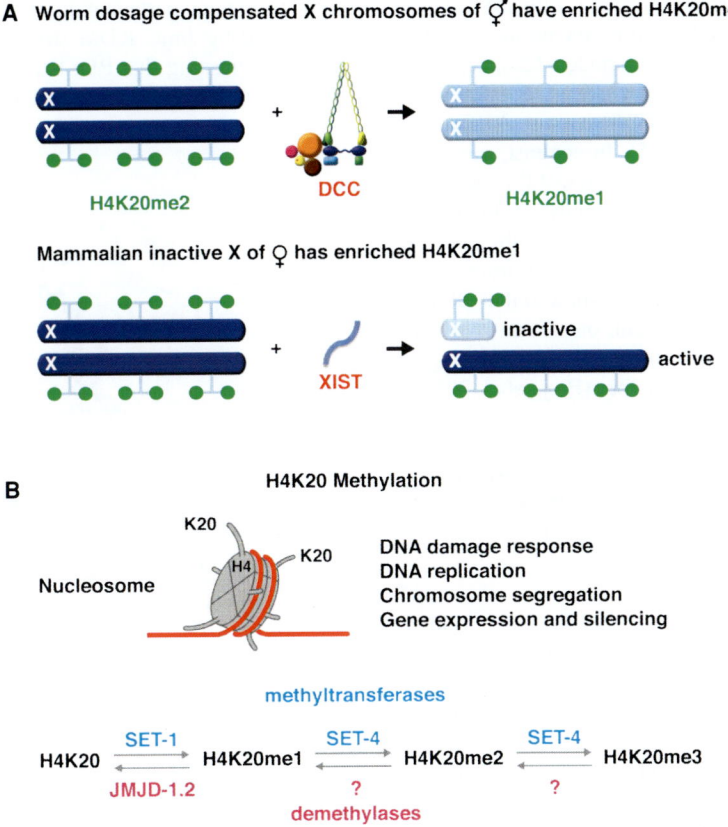

Figure 2. H4K20me1 enrichment on the repressed X chromosomes is a shared feature of diverse dosage compensation strategies. (*A*) Dosage-compensated X chromosomes of *C. elegans* hermaphrodites have dosage compensation complex (DCC)-dependent H4K20me1 enrichment. H4K20me1 enrichment on the inactive X chromosome of female mammals requires the long noncoding RNA XIST that triggers X inactivation. For neither strategy had the mechanism of H4K20me1 enrichment been determined. (*B*) H4K20 methylation controls myriad nuclear functions, but the mechanisms that regulate different H4K20 methylation states are not well understood. H4K20me2/me3 demethylases had not been identified.

among the most prominent long-range interactions along the X chromosome, and disruption of the DCC abolishes the interactions and TAD structure. (2) The stronger the *rex* sites, the more frequent are the DCC-dependent interactions among them. (3) Deletion of a single high-affinity *rex* site at a TAD boundary eliminates formation of the TAD boundary. Thus, not only does a condensin complex play a key role in compacting and resolving mitotic and meiotic chromosomes, it plays a central role in shaping the 3D landscape of interphase chromosomes.

During this chromosome remodeling process, the chromatin modification H4K20me1 becomes selectively enriched on both hermaphrodite X chromosomes in a DCC-dependent manner (Fig. 2A; Liu et al. 2011; Vielle et al. 2012; Wells et al. 2012; Kramer et al. 2015). H4K20me1 is also selectively enriched on the inactive X chromosome of female mammals, highlighting a shared feature of diverse dosage compensation strategies (Fig. 2A; Kohlmaier et al. 2004). In neither case had the mechanism of H4K20me1 enrichment been discovered. Furthermore, the effect of the H4K20me1 modification on gene regulation and chromosome structure had not been determined.

In general, the role of H4K20me1 in gene regulation has remained a mystery because of its contribution to both

gene activation and gene repression in different contexts (Beck et al. 2012). Although H4K20 methylation has been implicated in many nuclear functions beyond gene regulation, such as DNA replication and repair, mitotic chromosome condensation, and cell cycle control, the mechanisms that regulate different H4K20 methylation states and transduce them into correctly executed nuclear functions are not understood (Fig. 2B; Beck et al. 2012; Jorgensen et al. 2013; van Nuland and Gozani 2016). We discovered the machinery and mechanism that catalyze H4K20me1 enrichment on nematode X chromosomes and the impact of H4K20me1 on higher-order chromosome structure and gene regulation (Brejc et al. 2017).

In principle, H4K20me1 enrichment on nematode X chromosomes could occur by activating the methyltransferase that converts H4K20 to H4K20me1 (SET-1), by blocking the methyltransferase that converts H4K20me1 to H4K20me2/me3 (SET-4), by blocking the demethylase that converts H4K20me1 to H4K20 (JMJD-1.1/1.2), or by activating an unknown demethylase that converts H4K20me2 to H4K20me1 (Fig. 2B). Although H4K20me2 is the most abundant form of H4K20 in eukaryotic cells (Pesavento et al. 2008), only a neuron-specific H4K20me2 demethylase had been reported

(Wang et al. 2015). No H4K20me2 demethylase had been identified that could function during the dosage compensation process. Indeed, published models featured the inhibition of SET-4 as the likely mechanism for H4K20me1 enrichment on X in *C. elegans* (Vielle et al. 2012; Wells et al. 2012; Kramer et al. 2015). In contrast, our X-ray crystallography and biochemical assays of the DCC subunit DPY-21 revealed a new subfamily of Jumonji C (JmjC) histone demethylases that converts H4K20me2 to H4K20me1 in vitro and is widely conserved from worms to mammals (Brejc et al. 2017). We showed that DPY-21 catalyzes H4K20me1 enrichment on X in vivo, and H4K20me1 enrichment, in turn, helps remodel the higher order of structure of X chromosomes (Brejc et al. 2017).

DCC SUBUNIT DPY-21 AND MOUSE ROSBIN PROTEIN ARE THE FIRST JUMONJI C (JmjC) DEMETHYLASES THAT CONVERT H4K20me2 TO H4K20me1

Although amino acid sequence analysis failed to identify a demethylase domain in any of the 10 DCC subunits, structure prediction programs suggested homology between the carboxy-terminal domain of DPY-21 and JmjC domain–containing lysine demethylases (KDMs), despite low (15%) sequence identity. JmjC KDMs are Fe^{2+} and α-ketoglutarate (α-KG)-dependent dioxygenases that demethylate lysines in histone and nonhistone proteins (Markolovic et al. 2016). This potential connection was investigated by determining a 1.8 Å crystal structure of the *C. elegans* DPY-21$^{1210-1617}$ fragment that encompasses the putative JmjC domain (Brejc et al. 2017). Comparisons to known structures (Holm and Rosenstrom 2010) revealed DPY-21 to be a JmjC domain–containing protein most similar to the JmjC KDMs (Fig. 3A,B).

Like other JmjC KDMs, the DPY-21$^{1210-1617}$ structure includes a JmjC domain that is folded into a double-stranded β-helix (DSBH) and surrounded by a Jumonji N (JmjN) domain, a β-hairpin motif, and a mixed domain (Fig. 3A; Chen et al. 2006). The DPY-21 DSBH core bears facial triad residues (H1452, D1454, and H1593) that chelate Fe^{2+} and form an active site (Fig. 3B). An α-KG molecule coordinates Fe^{2+} in a bidentate manner, and a water molecule completes the octahedral coordination of Fe^{2+} (Fig. 3B). The α-KG is further stabilized by hydrogen bonding to side chains of T1449 and Y1585 and also by hydrogen bonding to side chains of W1410, K1526, and S1603 via water molecules. In addition, α-KG forms van der Waals contacts with A1499, L1587, and F1595.

The DPY-21 carboxy-terminal domain is similar (33% identity) to carboxy-terminal domains of metazoan proteins named round spermatid basic proteins 1 (ROSBIN) (Fig. 3C; Yonker and Meyer 2003). Mouse ROSBIN, an essential protein with limited biochemical characterization, is expressed in embryos and male gonads (Takahashi et al. 2004; Koscielny et al. 2014). Structure-guided sequence comparison of ROSBIN proteins with DPY-21 showed that all residues that interact with Fe^{2+} and α-KG are conserved. Moreover, sequence conservation ex-

tends to regions participating in histone peptide binding, as defined by JmjC KDM structures (Markolovic et al. 2016), suggesting that ROSBINs may also be JmjC demethylases that share substrate specificity.

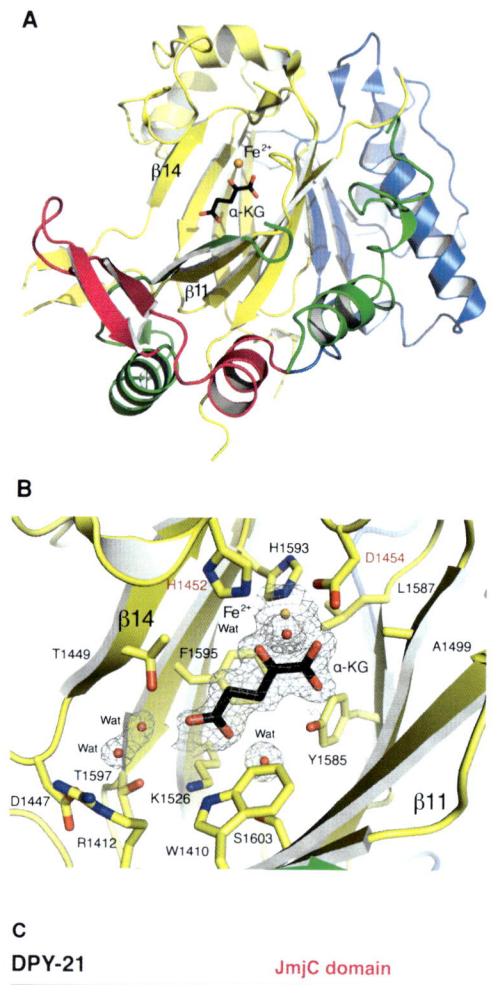

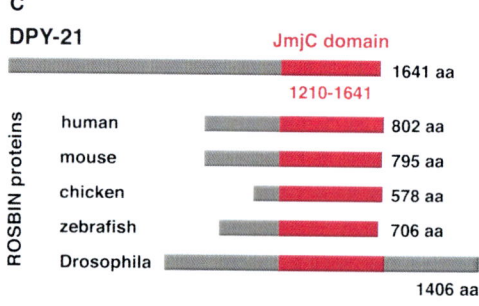

Figure 3. 1.8 Å structure of the DPY-21 JmjC demethylase domain. (*A*) DPY-21$^{1210-1617}$ structure in complex with α-KG (black) and Fe^{2+} (orange) showing JmjC domain (yellow), JmjN (blue), β-hairpin (magenta), and mixed domain (green). (Adapted from Brejc et al. 2017.) (*B*) Active site of DPY-21$^{1210-1617}$ showing JmjC domain residues (yellow) in complex with Fe^{2+} (orange), α-KG (black), and water molecules (red). Facial triad residues H1452 and D1454 (red letters) were changed to alanines for in vitro and in vivo studies. The electron density, $2F_o$–F_c (mesh), contoured at 1.0 σ above the mean is shown for Fe^{2+}, α-KG, and water molecules. (Adapted from Brejc et al. 2017.) (*C*) Evolutionary conservation of DPY-21 JmjC domain (magenta) in ROSBIN proteins across species.

Assays performed in vitro with modified histones showed that both DPY-21 and mROSBIN convert H4K20me2 to H4K20me1, demonstrating the evolutionary conservation of a novel H4K20me2 JmjC demethyase activity (Brejc et al. 2017). mRosbin also demethylated an H4K20me2 peptide in vitro (Brejc et al. 2017). Substitution of facial triad residues (H1452A and D1454A) in DPY-21 abolished enzymatic activity, and both proteins were inactive without their Fe^{2+} and α-KG cofactors (Brejc et al. 2017). Furthermore, both DPY-21 and mROS-BIN have a strong preference for H4K20me2 as a substrate, because neither protein was active in vitro with H4K20me1, H4K20me3, or histone H3 methyl substrates (Brejc et al. 2017).

The single previously identified H4K20me2 demethylase (nLSD1) is a neuronal-specific isoform of LSD1 that uses different chemistry from JmjC domains (Wang et al. 2015). nLSD1 is a flavin-dependent monoamine oxidase that demethylates both H4K20me2 and H4K20me1. The only JmjC H4K20 demethylases identified before DPY-21 and ROSBIN have specificity for H4K20me3 (PHF2) and H4K20me1 (PHF8) (Liu et al. 2010; Qi et al. 2010; Stender et al. 2012). Both have an amino-terminal PHD domain required for demethylase activity in vitro, which occurs only in the context of nucleosomes. In contrast, DPY-21 and mROSBIN lack PHD and other chromatin-interacting domains and only require the JmjC domain for substrate recognition, allowing them to demethylate H4K20me2 in vitro on histone and histone peptides. These structural and biochemical findings merit classifying DPY-21 and ROSBIN proteins as a new subfamily of JmjC demethylases, the KDM9 subfamily.

THE DPY-21 JmjC DEMETHYLASE ACTS IN A CELL-CYCLE-DEPENDENT MANNER TO ENRICH H4K20me1 ON X CHROMOSOMES OF SOMATIC CELLS

H4K20me1 enrichment on hermaphrodite X chromosomes occurs in a cell cycle–dependent manner, during interphase but not mitosis, where H4K20me1 levels are uniformly elevated on all chromosomes (Fig. 4A,B; Brejc et al. 2017). H4K20me1 enrichment is not evident on interphase X chromosomes before the 200-cell stage of embryogenesis and only becomes reliably detectable on X in most interphase cells around the 300–350-cell stage, long after initial recruitment of SDC-2 and other DCC subunits to X (30–40-cell stage) (Fig. 4A; Brejc et al. 2017). In contrast to other DCC subunits, DPY-21's association with X is precisely coincident with the timing of H4K20me1 enrichment on X. Moreover, DPY-21 is not bound to mitotic chromosomes, unlike other DCC subunits (Fig. 4B). These results suggest that DPY-21 demethylates H4K20me2 on X to enrich H4K20me1.

In three different genome-engineered strains expressing DPY-21 variants with substitutions in JmjC facial triad amino acids (H1452A, D1454A, or H1452A/D1454A), sex-specific enrichment of H4K20me1 was absent from interphase X chromosomes (Fig. 4A). In contrast, HK20me1 levels were unaffected on mitotic chromosomes (Fig. 4A,B; Brejc et al. 2017).

In chromatin immunoprecipitation sequencing (ChIP-seq) experiments measuring the genome-wide distribution of H4K20me1 in wild-type and dpy-21(JmjC)-mutant embryos using a spike-in control to normalize read count, H4K20me1 was enriched on X compared to autosomes in wild-type but not in demethylase-mutant embryos (Fig. 4C; Brejc et al. 2017). Loss of H4K20me1 enrichment on X in dpy-21(JmjC) mutants coupled with the timing of DPY-21's association with X during interphase and DPY-21's absence from mitotic chromosomes indicate that JmjC demethylase activity is responsible for X enrichment of H4K20me1.

LOSS OF H4K20me2 DEMETHYLASE ACTIVITY DISRUPTS DOSAGE COMPENSATION AND ELEVATES X-LINKED GENE EXPRESSION

Prior studies aimed at understanding the role of H4K20me1 enrichment on X during dosage compensation used RNAi-mediated knockdown of the methyltransferase gene set-1 to reduce H4K20me1 levels genome-wide or mutation of set-4 to block the progression of H4K20me1 to H4K20me2/me3, thereby increasing H4K20me1 levels across all chromosomes (Vielle et al. 2012; Wells et al. 2012; Kramer et al. 2015). The resulting genome-wide changes in gene expression made it difficult to assess the specific importance of H4K20me1 levels on X during dosage compensation. Furthermore, interpretation of genetic assays evaluating dosage compensation defects (Vielle et al. 2012; Wells et al. 2012) was confounded, in retrospect, by the finding that set-1 knockdown causes synergistic lethality in both sexes from defective mitosis (Brejc et al. 2017).

In contrast, the highly specific dpy-21(JmjC) mutations enabled us to eliminate H4K20me1 enrichment selectively from X and show through genetic assays and direct RNA measurements that loss of demethylase activity disrupts dosage compensation. The role of H4K20me1 was assessed by first asking whether dpy-21(JmjC) mutations suppress the XO-specific lethality caused by xol-1 mutations, which inappropriately activate the DCC in males. xol-1, the master regulator of sex-determination and dosage compensation, acts in XO embryos to turn off the hermaphrodite pathway of sexual differentiation and to prevent DCC binding to the single male X by repressing sdc-2, the XX-specific trigger of DCC binding to X (Miller et al. 1988; Rhind et al. 1995; Dawes et al. 1999). xol-1 mutant XO animals die from reduced X-linked gene expression. If the JmjC catalytic activity is important for reducing X expression during dosage compensation, mutations that inactivate this demethylase should suppress the xol-1 XO-specific lethality. Indeed, JmjC mutations prevented the death of xol-1 XO males, indicating that dpy-21(JmjC) mutations disrupt dosage compensation, and H4K20me1 enrichment is important for the dosage compensation mechanism (Brejc et al. 2017).

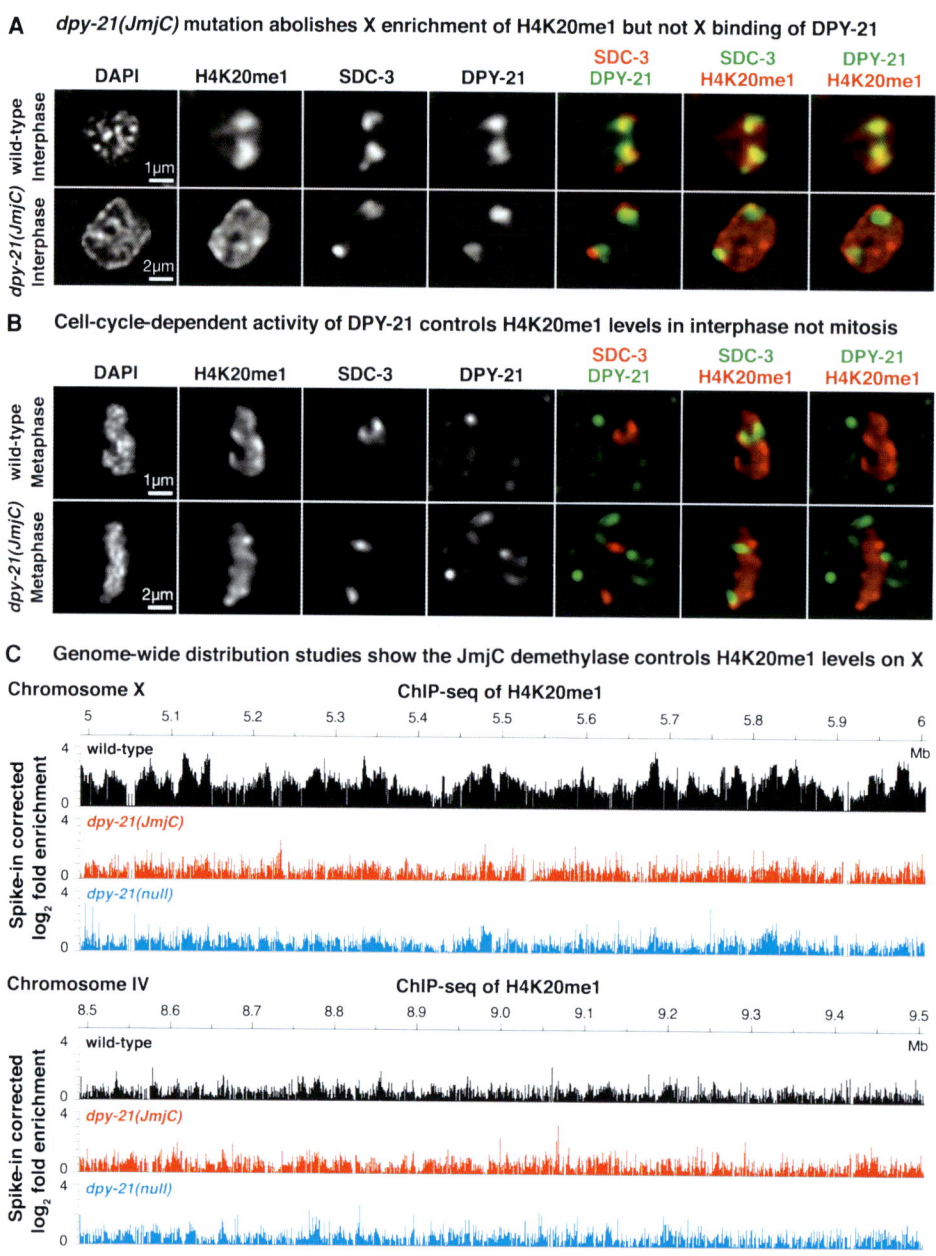

Figure 4. DPY-21 JmjC H4K20me2 demethylase enriches H4K20me1 on X in vivo. (*A*) Confocal images of an interphase nucleus from a 376-cell wild-type embryo (*top*) and an interphase nucleus from a 335-cell *dpy-21(JmjC)* mutant embryo (*bottom*) stained with DAPI and antibodies to DPY-21, dosage compensation complex (DCC) subunit SDC-3, and H4K20me1. The JmjC mutation does not affect binding of DPY-21 to X, but it does disrupt the H4K20me1 enrichment on X. (*B*) Metaphase nucleus from a 376-cell wild-type embryo (*top*) and metaphase nucleus from a 335-cell *dpy-21(JmjC)* mutant embryo (*bottom*) stained as in *A*. During mitosis, DPY-21 dissociates from X, but SDC-3 remains bound. The JmjC mutation does not affect the H4K20me1 level on mitotic chromosomes. (*C*) ChIP-seq profiles show spike-in-corrected H4K20me1 enrichment in representative regions of chromosome X and chromosome IV in wild-type, *dpy-21(JmjC)*, and *dpy-21(null)* mutant embryos. X enrichment of H4K20me1 is lost in *dpy-21* mutants, but H4K20me1 levels are unchanged on autosomes. (Based on data from Brejc et al. 2017.)

The importance of demethylase activity in repressing gene expression predicts that blocking H4K20me2/me3 production via a *set-4* mutation should prevent the rescue of *xol-1* XO males by *dpy-21(JmjC)* mutations. This prediction held true, providing strong genetic evidence that production of H4K20me2/me3 on X is an intermediate step in the enrichment of H4K20me1 on X (Brejc et al. 2017).

The effect of *dpy-21(JmjC)* mutations on X-chromosome gene expression was shown directly by RNA-seq experiments. Cumulative plots comparing the distribution of fold changes in gene expression on X chromosomes and autosomes between wild-type and *dpy-21* mutant embryos revealed that X-chromosome gene expression was significantly elevated relative to that of each individual autosome in *dpy-21(JmjC)* mutants (Fig. 5). Similar re-

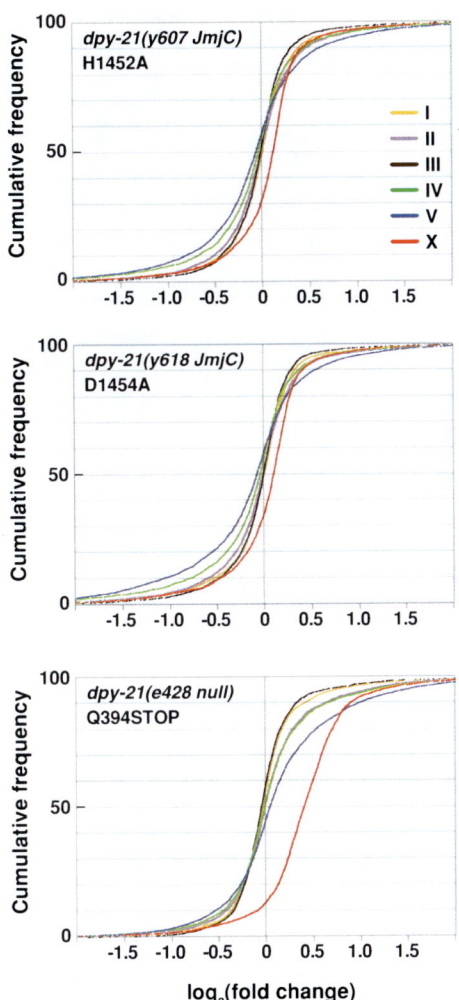

Figure 5. X-chromosome gene expression is elevated relative to autosomes in *dpy-21(JmjC)* mutants. Cumulative plots show the distribution of expression changes for genes on X and for each individual autosome in *dpy-21(JmjC)* mutant versus wild-type embryos as assayed by RNA-seq. The *x*-axis represents the log$_2$ fold change in expression. X-chromosome gene expression is elevated compared to that of each autosome in *dpy-21(y607 JmjC)*, *dpy-21(y618 JmjC)*, *dpy-21(e428 null)* mutants ($P < 2.2 \times 10^{-16}$, one-sided Wilcoxon rank-sum test). (Based on data from Brejc et al. 2017.)

sults were obtained for *dpy-21(null)* mutants, but the relative elevation in X expression was somewhat greater (Fig. 5), consistent with the more severe mutant phenotypes for null mutants (Brejc et al. 2017). These findings show that DPY-21's demethylase activity, and by extension the enrichment of H4K20me1 on X, functions in vivo to help repress X-chromosome gene expression.

Chromatin with posttranslationally modified histones can recruit specialized proteins ("readers") to control the structure of nucleosome arrays and regulate gene expression (Soshnev et al. 2016). In particular, chromatin readers of the malignant brain tumor (MBT) repeat protein family associate with nucleosomes enriched in H4K20me1 or H4K20me2, help compact the chromatin fiber, and repress gene expression (Trojer et al. 2007; Blanchard et al. 2014).

However, knockout of the two *C. elegans* MBT repeat proteins, *mbtr-1* and *lin-61*, failed to suppress *xol-1* XO lethality, unlike *dpy-21(JmjC)* mutations, indicating a lack of dosage compensation defects (Brejc et al. 2017). This result implies that if H4K20me1-binding proteins modulate X-chromosome conformation and gene repression, they must belong to an unidentified family of H4K20me1 readers. As alternatives, H4K20me1 might antagonize other proteins with chromatin-modifying activities, or less likely, it might control chromatin folding directly.

H4K20me1-DRIVEN REMODELING OF X-CHROMOSOME TOPOLOGY

X chromosomes undergo changes in conformation during dosage compensation. Cytological studies measuring the volumes of chromosome territories showed that DCC binding increases the compaction state of X chromosomes (Lau et al. 2014). Genome-wide chromosome conformation capture (Hi-C) studies showed the DCC remodels X chromosomes into a unique, sex-specific spatial conformation, distinct from that of autosomes by inducing TADs using its highest-affinity binding sites (*rex* sites) to mediate long-range chromatin interactions (Crane et al. 2015). We found that the H4K20me2 demethylase activity of DPY-21 contributes to both the compaction of X and the DCC-driven remodeling of X topology.

Chromosome volume measurements performed in gut nuclei of wild-type and mutant adults showed that the fraction of total chromosome volume occupied by X increased by 30% in all three *dpy-21(JmjC)* mutants ($P < 10^{-6}$, two-sided Wilcoxon rank-sum test) (Brejc et al. 2017). In contrast, X volume was not significantly changed ($P = 0.2$) in a dosage compensation mutant (*sdc-1(null)*) that has normal H4K20me1 enrichment on X and comparably modest dosage compensation defects as in *dpy-21(JmjC)* mutants. Thus, the increase in X-chromosome volume is not a byproduct of changes in gene expression, and H4K20me1 contributes directly to X-chromosome compaction.

The effect of H4K20me1 enrichment on the remodeling of X-chromosome topology was determined by investigating the timing of TAD boundary formation relative to the timing of DPY-21 binding to X and by genome-wide conformation capture experiments (Hi-C) to determine the position and degree of TAD boundary formation in *dpy-21(JmjC)* mutants.

Fluorescence in situ hybridization (FISH) experiments revealed that the DCC-dependent TAD boundary at *rex-14* started to form by the 80–120-cell stage and was well formed by the 180–250-cell stage, before DPY-21 was reliably detected on X chromosomes by immunofluorescence (Brejc et al. 2017). Thus, H4K20me2 demethylase activity is not essential to initiate the formation of DCC-dependent TADs. However, Hi-C experiments comparing TADs in wild-type and *dpy-21(JmjC)* mutants revealed that DCC-dependent TAD boundaries were diminished in the mutant (Fig. 6A–C). Insulation profiles showed that TAD boundaries were formed at the same 17 locations on X chromosomes of wild-type and mutant animals, but

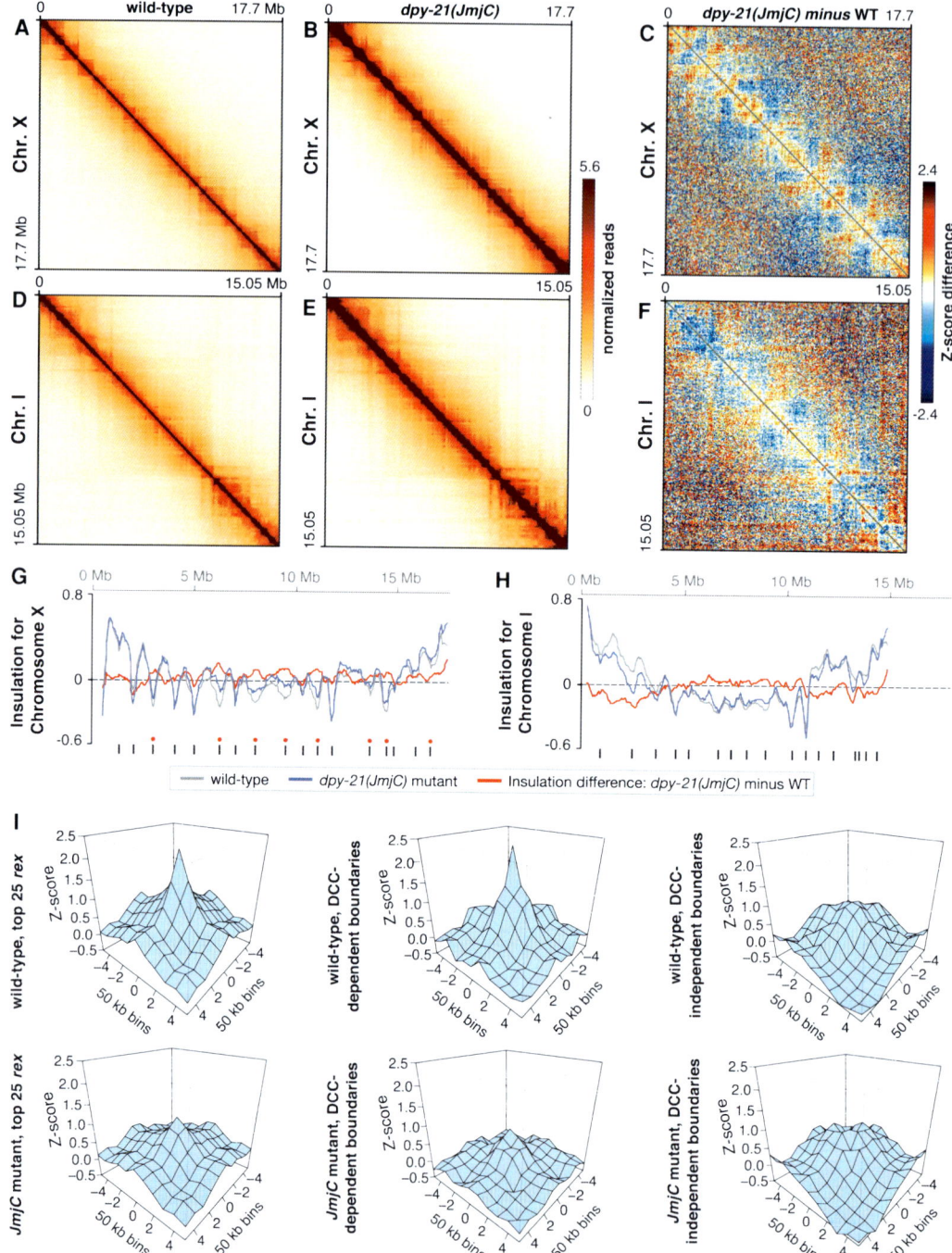

Figure 6. DPY-21 demethylase activity modulates X-chromosome topology. (*A,B,D,E*) Heat maps of Hi-C data binned at 50-kb resolution show chromatin interaction frequencies on chromosomes X and I in wild-type and *dpy-21(JmjC)* mutant embryos. These heat maps show results from one of two replicates. Heat maps combining both replicates are presented in Brejc et al. (2017), with the same conclusions. (*C,F*) Z-score difference heat maps of Hi-C data in *A,B* and *D,E*, respectively, binned at 50-kb resolution show chromatin interactions that increase (orange-red) and decrease (blue) on X (C) and I (*F*) in *dpy-2(JmjC)* mutant versus wild-type embryos. (*G,H*) Insulation plots for chromosomes X or I of wild-type (gray) or *dpy-21(JmjC)* mutant (blue) embryos and insulation difference plots (red) from data in *A–E*. Black bars, location of topologically associated domain (TAD) boundaries in wild-type embryos. Red dots, dosage compensation complex (DCC)-dependent boundaries greatly diminished or eliminated upon DCC depletion (Crane et al. 2015). Each has a high-affinity *rex* site. An insulation score reflects the cumulative interactions occurring across each interval. Minima denote areas of high insulation classified as TAD boundaries. The difference between insulation profiles of wild-type and *dpy-21(JmjC)* mutants reflects the change in boundary strength. An increase in insulation score at a TAD boundary means less insulation in the mutant, indicating a weakening of the boundary. DCC-dependent TAD boundaries on X were reduced in *dpy-21(JmjC)* mutants, but DCC-independent boundaries on X and autosomes were not significantly changed. (*I*) Three-dimensional profiles of average Hi-C interaction frequencies (Z-scores) in 50-kb bins around pairs of top 25 *rex* sites, DCC-dependent boundaries on X, and DCC-independent boundaries on X in wild-type or *dpy-21(JmjC)* mutant embryos. Profiles are centered at 0. Interactions between *rex* sites and DCC-dependent boundaries are reduced in demethylase mutants. (Adapted from Brejc et al. 2017.)

the difference in insulation profiles showed that the strength of all eight DCC-dependent TAD boundaries on X was significantly reduced in *JmjC* mutants (Fig. 6G; Brejc et al. 2017). In contrast, the nine DCC-independent TAD boundaries on X (Fig. 6G) and the TAD boundaries on autosomes (Fig. 6D–F,H) were not significantly weakened by the *dpy-21(JmjC)* mutation (Brejc et al. 2017).

Analysis of Hi-C interaction frequencies (Z-scores) revealed an underlying cause for the defect in TAD boundary formation. In *dpy-21(JmjC)* mutants, interactions between sites within 1 Mb were increased significantly, whereas interactions between sites >1 Mb apart were decreased significantly (Fig. 6C; Brejc et al. 2017). Among the reduced long-range interactions (>1 Mb) were those between the DCC-dependent TAD boundaries and between the highest-affinity *rex* sites that mediate TAD boundary formation. Reduction in these interactions contributed to the loss in TAD boundary strength (Fig. 6I). In contrast, interactions between DCC-independent boundaries on X were unchanged (Fig. 6I), and interactions on autosomes showed moderate changes in the opposite direction (Fig. 6F).

These results support a two-tiered model for TAD formation during dosage compensation. In the first stage, the condensin DCC initiates TAD formation prior to DPY-21 recruitment and hence via a demethylase-independent mechanism that promotes long-range interactions between the highest-affinity DCC-binding sites. In the second stage, DPY-21 binds to X and catalyzes enrichment of H4K20me1, thereby enhancing long-range DNA interactions across X. X-chromosome compaction is generally increased, and the increase in long-range interactions among *rex* sites strengthens TAD boundaries.

IN GERM CELLS, H4K20me2 DEMETHYLASE ENRICHES H4K20me1 ON AUTOSOMES BY A DCC-INDEPENDENT MECHANISM TO PROMOTE CHROMOSOME COMPACTION

As in somatic cells, H4K20me1 shows a dynamic pattern of enrichment in germ cells. H4K20me1 levels are high on all mitotic chromosomes in the gonad and become reduced as nuclei enter meiosis (Fig. 7A,B). By early pachytene, H4K20me1 is enriched only on autosomes of both sexes and absent from X chromosomes (Fig. 7B), raising the question of whether DPY-21 also controls H4K20me1 levels in the germline, where the DCC does not assemble or function. In gonads of *dpy-21(JmjC)* mutants, H4K20me1 levels remained high on mitotic chromosomes, but H4K20me1 was absent from all pachytene chromosomes, indicating that the JmjC demethylase is essential in germ cells for autosomal enrichment of H4K20me1 during meiosis but not mitosis (Fig.7A,B). The localization of DPY-21 met all expectations for a demethylase that regulates autosomal levels of H4K20me1 during meiosis but not mitosis: DPY-21 bound to autosomes but not X chromosomes of pachytene nuclei and failed to bind mitotic chromosomes (Fig. 7A,C).

These findings prompted the question of whether DPY-21 and H4K20me1 regulate chromosome compaction during germ cell development. A change in meiotic chromosome compaction should be reflected in a change in length of the chromosome axis. Indeed, measurements of axis lengths in 3D for chromosomes X and I in wild-type and *dpy-21(JmjC)* mutants revealed that X axis length was not different in the mutants, as predicted by the absence of H4K20me1 and DPY-21 (Fig. 7D). However, chromosome I axis length was extended by 20% in mutants (Fig. 7D), indicating that selective binding of DPY-21, and hence H4K20me1 enrichment, regulates higher-order chromosome structure in germ cells. Because the key DCC subunits that recruit DPY-21 to X chromosomes in XX somatic cells are not expressed in the germline, the DPY-21 demethylase activity must operate via a DCC-independent mechanism using factors that either recruit DPY-21 to autosomes or repel it from X. Thus, DPY-21 is an adaptable chromatin regulator that is harnessed during development for distinct biological functions through different binding partners (Fig. 8). In both somatic cells and germ cells, H4K20me1 modulates 3D chromosome architecture, demonstrating the direct link between chromatin modification and higher-order chromosome structure.

CONCLUSION

We showed the function of chromatin modification in establishing higher-order chromosome structure during gene regulation. Moreover, discovery of ubiquitous H4K20me2 demethylases that act in worms and mammals to enhance H4K20me1 levels further expands our opportunity to dissect the dynamic regulation of H4K20 methylation essential for DNA replication, DNA damage response, chromosome segregation, gene regulation, and development. Prior studies suggested that H4K20 methylation was likely to be controlled by demethylases as well as previously characterized methyltransferases, but the H4K20me2/me3 demethylases responsible for such regulation had not been identified (van Nuland and Gozani 2016).

The finding that the DPY-21 and mROSBIN demethylases lack obvious DNA and chromatin-binding domains to confer target specificity provides the flexibility for these enzymes to be recruited to different genomic locations by diverse mechanisms to control essential nuclear processes. In general, though, only a limited number of DNA-binding factors are known to target histone demethylases to chromatin (Dimitrova et al. 2015), making the recruitment of an H4K20me2 demethylase to X chromosomes by a specialized condensin complex a highly tractable model. DCC binding and spreading distributes demethylase activity chromosome-wide. Thus, the DCC acts as an "eraser" of a prominent histone posttranslational modification, increasing H4K20me1 levels along X by activating an H4K20me2 demethylase rather than regulating an H4K20 methyltransferase.

DPY-21's role in chromatin modification further illustrates how dosage compensation evolved by co-opting conserved machinery used in other biological processes for the new task of fine-tuning X-chromosome expression. Our observations suggest that other condensin and structural maintenance of chromosomes (SMC) complexes

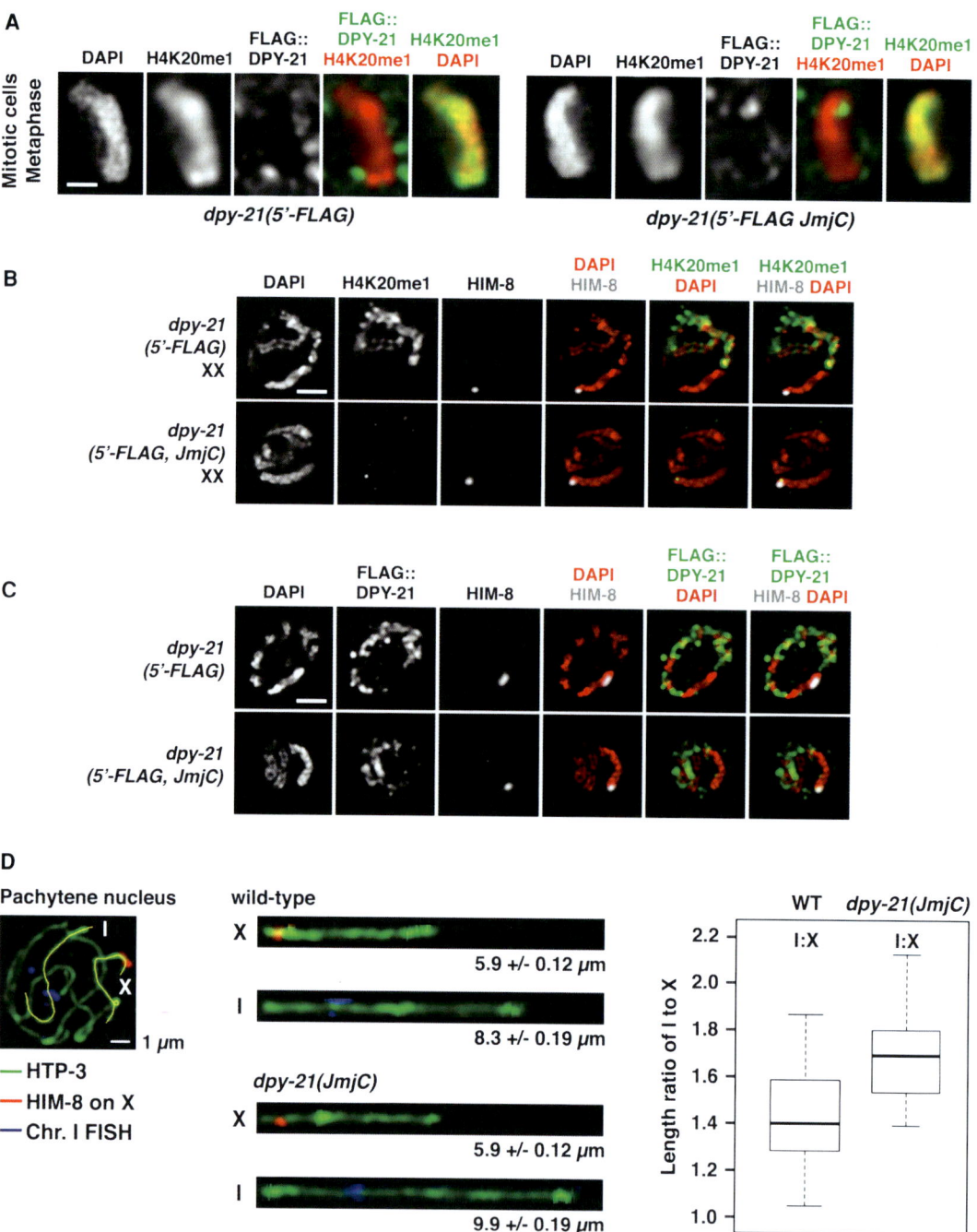

Figure 7. DPY-21 JmjC demethyase acts in germ cells to enrich H4K20me1 and compact autosomes. (*A*) Confocal images of metaphase chromosomes from the mitotic zones of *dpy-21(5'-FLAG)* and *dpy-21(5'-FLAG, JmjC)* gonads costained with H4K20me1 and FLAG antibodies. H4K20me1 is highly enriched on all mitotic chromosomes, but DPY-21 does not localize to mitotic chromosomes. Inactivation of the DPY-21 demethylase does not reduce H4K20me1 levels on mitotic chromosomes. Scale bar, 1 μm. (*B*) Pachytene nuclei from *dpy-21(5'-FLAG)* and *dpy-21(5'-FLAG, JmjC)* XX gonads stained with antibodies against H4K20me1 and the X-chromosome-specific marker HIM-8. H4K20me1 is selectively enriched on autosomes in pachytene nuclei in the *dpy-21(5'-Flag)* strain but absent from X (*top*). H4K20me1 is absent from all chromosomes in pachytene nuclei of *dpy-21(JmjC)* mutant XX gonads (*bottom*). (*C*) Pachytene nuclei from *dpy-21(5'-Flag)* and *dpy-21(5'-Flag, JmjC)* XX gonads stained with antibodies against FLAG-tagged DPY-21 and HIM-8. DPY-21 specifically localizes to autosomes but not to X, consistent with its demethylation of only autosomes. Scale bars (*B*, *C*), 2 μm. (*D*) DPY-21 demethylase is required for full compaction of autosomes in germ cells. (*Left*) High-resolution image of a pachytene nucleus from a wild-type gonad stained with a FISH probe to chromosome I (blue) and antibodies to axis protein HTP-3 (green) and X-specific protein HIM-8 (red). The 3D traces of X and I (yellow) were used to straighten each chromosome. (*Middle*) Computationally straightened X and I from wild-type and *dpy-21(JmjC)* gonads are displayed horizontally. Chromosome tracing was performed on nuclei in the last quarter of pachytene. Average total axis length and standard error of the mean (SEM) are shown below each axis. (*Right*) Box plots show the ratios of I to X axis lengths in wild-type and *dpy-21(JmjC)* pachytene nuclei. The I to X ratios are significantly lower in wild-type versus *dpy-21(JmjC)* gonads ($P = 1.7 \times 10^{-8}$, two-sided Wilcoxon rank-sum test). (Adapted from Brejc et al. 2017.)

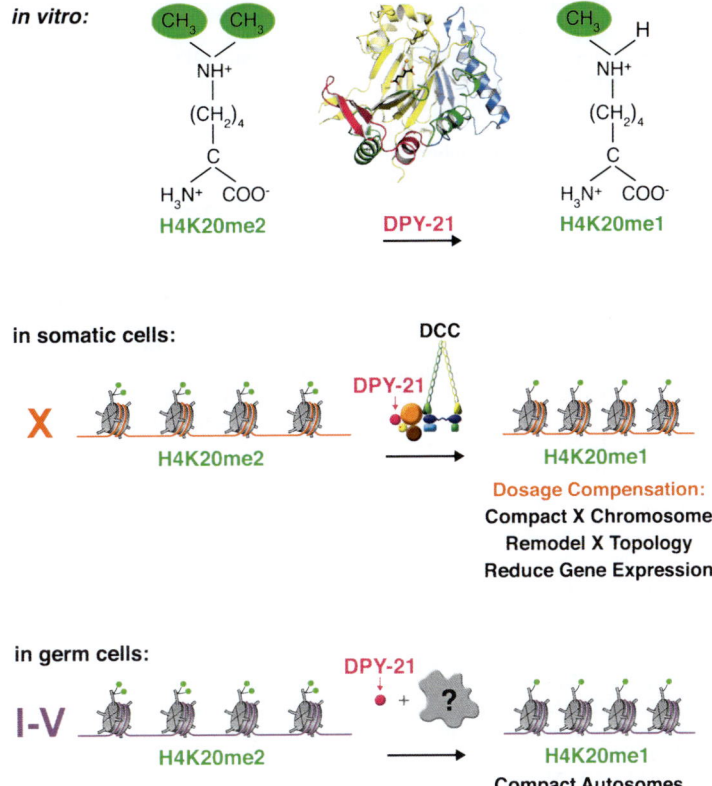

Figure 8. An H4K20me2 histone demethylase regulates 3D chromosome structure and gene expression by modulating the dynamic enrichment of H4K20me1. The 1.8 Å crystal structure and biochemical activity of DPY-21 revealed a new, highly conserved H4K20me2 JmjC demethylase subfamily that converts H4K20me2 to H4K20me1 in vitro. In somatic cells, DPY-21 binds to X chromosomes via the dosage compensation complex (DCC) and enriches H4K20me1 to repress gene expression. The H4K20me1 enrichment controls the higher-order structure of X chromosomes by facilitating compaction and topologically associated domain (TAD) formation. In germ cells, DPY-21 enriches HK20me1 on autosomes in a DCC-independent manner to promote chromosome compaction. (Adapted from Brejc et al. 2017.)

could act as scaffolds to recruit proteins with chromosome-modifying activities.

The enrichment of H4K20me1 on the inactive X chromosome of female mice (Kohlmaier et al. 2004) makes our discoveries directly relevant for mammalian development. The long noncoding RNA *XIST*, the trigger of mammalian X inactivation, induces accumulation of H4K20me1 on X, but X inactivation per se is not required for H4K20me1 enrichment, suggesting that H4K20me1 might contribute to establishing inactivation. Consistent with this interpretation, knockout of the mouse H4K20me1 methyltransferase causes decondensation of X (Oda et al. 2009). Thus, our observations offer new directions for unraveling the regulation and function of H4K20me1 in X-chromosome inactivation and other mechanisms of gene control that act over long distances. Our findings also provide new avenues for understanding the mechanisms by which chromatin modifications help remodel higher-order chromosome structure during development.

ACKNOWLEDGMENTS

We thank Deborah Stalford for drafting the figures. This work was supported by National Institutes of Health grant R01 GM030702 to B.J.M. and by the Howard Hughes Medical Institute.

REFERENCES

Beck DB, Oda H, Shen SS, Reinberg D. 2012. PR-Set7 and H4K20me1: At the crossroads of genome integrity, cell cycle, chromosome condensation, and transcription. *Genes Dev* **26:** 325–337.

Blanchard DP, Georlette D, Antoszewski L, Botchan MR. 2014. Chromatin reader L(3)mbt requires the Myb-MuvB/DREAM transcriptional regulatory complex for chromosomal recruitment. *Proc Natl Acad Sci* **111:** E4234–E4243.

Boettiger AN, Bintu B, Moffitt JR, Wang S, Beliveau BJ, Fudenberg G, Imakaev M, Mirny LA, Wu C-T, Zhuang X. 2016. Super-resolution imaging reveals distinct chromatin folding for different epigenetic states. *Nature* **529:** 418–422.

Brejc K, Bian Q, Uzawa S, Wheeler BS, Anderson EC, King DS, Kranzusch PJ, Preston CG, Meyer BJ. 2017. Dynamic control of X chromosome conformation and repression by a histone H4K20 demethylase. *Cell* **171:** 85–102 e123.

Chen Z, Zang J, Whetstine J, Hong X, Davrazou F, Kutateladze TG, Simpson M, Mao Q, Pan CH, Dai S, et al. 2006. Structural insights into histone demethylation by JMJD2 family members. *Cell* **125:** 691–702.

Crane E, Bian Q, McCord RP, Lajoie BR, Wheeler BS, Ralston EJ, Uzawa S, Dekker J, Meyer BJ. 2015. Condensin-driven remodelling of X chromosome topology during dosage compensation. *Nature* **523:** 240–244.

Csankovszki G, Collette K, Spahl K, Carey J, Snyder M, Petty E, Patel U, Tabuchi T, Liu H, McLeod I, et al. 2009. Three distinct condensin complexes control *C. elegans* chromosome dynamics. *Curr Biol* **19:** 9–19.

Davis TL, Meyer BJ. 1997. SDC-3 coordinates the assembly of a dosage compensation complex on the nematode X chromosome. *Development* **124:** 1019–1031.

Dawes HE, Berlin DS, Lapidus DM, Nusbaum C, Davis TL, Meyer BJ. 1999. Dosage compensation proteins targeted to X chromosomes by a determinant of hermaphrodite fate. *Science* **284:** 1800–1804.

Dekker J, Mirny L. 2016. The 3D genome as moderator of chromosomal communication. *Cell* **164:** 1110–1121.

Dimitrova E, Turberfield AH, Klose RJ. 2015. Histone demethylases in chromatin biology and beyond. *EMBO Rep* **16:** 1620–1639.

Francis NJ, Kingston RE, Woodcock CL. 2004. Chromatin compaction by a Polycomb group protein complex. *Science* **306:** 1574–1577.

Hagstrom KA, Holmes VF, Cozzarelli NR, Meyer BJ. 2002. *C. elegans* condensin promotes mitotic chromosome architecture, centromere organization, and sister chromatid segregation during mitosis and meiosis. *Genes Dev* **16:** 729–742.

Hirano T. 2016. Condensin-based chromosome organization from bacteria to vertebrates. *Cell* **164:** 847–857.

Holm L, Rosenstrom P. 2010. Dali server: Conservation mapping in 3D. *Nucleic Acids Res* **38:** W545–W549.

Hsu DR, Chuang PT, Meyer BJ. 1995. DPY-30, a nuclear protein essential early in embryogenesis for *Caenorhabditis elegans* dosage compensation. *Development* **121:** 3323–3334.

Jans J, Gladden JM, Ralston EJ, Pickle CS, Michel AH, Pferdehirt RR, Eisen MB, Meyer BJ. 2009. A condensin-like dosage compensation complex acts at a distance to control expression throughout the genome. *Genes Dev* **23:** 602–618.

Jorgensen S, Schotta G, Sorensen CS. 2013. Histone H4 lysine 20 methylation: key player in epigenetic regulation of genomic integrity. *Nucleic Acids Res* **41:** 2797–2806.

Kalashnikova AA, Porter-Goff ME, Muthurajan UM, Luger K, Hansen JC. 2013. The role of the nucleosome acidic patch in modulating higher order chromatin structure. *J R Soc Interface* **10:** 20121022.

Klein RD, Meyer BJ. 1993. Independent domains of the SDC-3 protein control sex determination and dosage compensation in *C. elegans*. *Cell* **72:** 349–364.

Kohlmaier A, Savarese F, Lachner M, Martens J, Jenuwein T, Wutz A. 2004. A chromosomal memory triggered by Xist regulates histone methylation in X inactivation. *PLoS Biol* **2:** E171.

Koscielny G, Yaikhom G, Iyer V, Meehan TF, Morgan H, Atienza-Herrero J, Blake A, Chen CK, Easty R, Di Fenza A, et al. 2014. The international mouse phenotyping consortium web portal, a unified point of access for knockout mice and related phenotyping data. *Nucleic Acids Res* **42:** D802–D809.

Kramer M, Kranz AL, Su A, Winterkorn LH, Albritton SE, Ercan S. 2015. Developmental dynamics of X-chromosome dosage compensation by the DCC and H4K20me1 in *C. elegans*. *PLoS Genet* **11:** e1005698.

Kundu S, Ji F, Sunwoo H, Jain G, Lee JT, Sadreyev RI, Dekker J, Kingston RE. 2017. Polycomb repressive complex 1 generates discrete compacted domains that change during differentiation. *Mol Cell* **65:** 432–446 e435.

Lau AC, Nabeshima K, Csankovszki G. 2014. The *C. elegans* dosage compensation complex mediates interphase X chromosome compaction. *Epigenetics Chromatin* **7:** 31.

Liu T, Rechtsteiner A, Egelhofer TA, Vielle A, Latorre I, Cheung MS, Ercan S, Ikegami K, Jensen M, Kolasinska-Zwierz P, et al. 2011. Broad chromosomal domains of histone modification patterns in *C. elegans*. *Genome Res* **21:** 227–236.

Liu W, Tanasa B, Tyurina OV, Zhou TY, Gassmann R, Liu WT, Ohgi KA, Benner C, Garcia-Bassets I, Aggarwal AK, et al. 2010. PHF8 mediates histone H4 lysine 20 demethylation

events involved in cell cycle progression. *Nature* **466:** 508–512.

Lu X, Simon MD, Chodaparambil JV, Hansen JC, Shokat KM, Luger K. 2008. The effect of H3K79 dimethylation and H4K20 trimethylation on nucleosome and chromatin structure. *Nat Struct Mol Biol* **15:** 1122–1124.

Markolovic S, Leissing TM, Chowdhury R, Wilkins SE, Lu X, Schofield CJ. 2016. Structure–function relationships of human JmjC oxygenases-demethylases versus hydroxylases. *Curr Opin Struct Biol* **41:** 62–72.

Mets DG, Meyer BJ. 2009. Condensins regulate meiotic DNA break distribution, thus crossover frequency, by controlling chromosome structure. *Cell* **139:** 73–86.

Meyer BJ. 2010. Targeting X chromosomes for repression. *Curr Opin Genet Dev* **20:** 179–189.

Miller LM, Plenefisch JD, Casson LP, Meyer BJ. 1988. *xol-1*: A gene that controls the male modes of both sex determination and X chromosome dosage compensation in *C. elegans*. *Cell* **55:** 167–183.

Nonet ML, Meyer BJ. 1991. Early aspects of *Caenorhabditis elegans* sex determination and dosage compensation are regulated by a zinc-finger protein. *Nature* **351:** 65–68.

Nora EP, Dekker J, Heard E. 2013. Segmental folding of chromosomes: A basis for structural and regulatory chromosomal neighborhoods? *Bioessays* **35:** 818–828.

Nora EP, Goloborodko A, Valton AL, Gibcus JH, Uebersohn A, Abdennur N, Dekker J, Mirny LA, Bruneau BG. 2017. Targeted degradation of CTCF decouples local insulation of chromosome domains from genomic compartmentalization. *Cell* **169:** 930–944 e922.

Oda H, Okamoto I, Murphy N, Chu J, Price SM, Shen MM, Torres-Padilla ME, Heard E, Reinberg D. 2009. Monomethylation of histone H4-lysine 20 is involved in chromosome structure and stability and is essential for mouse development. *Mol Cell Biol* **29:** 2278–2295.

Pesavento JJ, Yang H, Kelleher NL, Mizzen CA. 2008. Certain and progressive methylation of histone H4 at lysine 20 during the cell cycle. *Mol Cell Biol* **28:** 468–486.

Pferdehirt RR, Kruesi WS, Meyer BJ. 2011. An MLL/COMPASS subunit functions in the *C. elegans* dosage compensation complex to target X chromosomes for transcriptional regulation of gene expression. *Genes Dev* **25:** 499–515.

Qi HH, Sarkissian M, Hu GQ, Wang Z, Bhattacharjee A, Gordon DB, Gonzales M, Lan F, Ongusaha PP, Huarte M, et al. 2010. Histone H4K20/H3K9 demethylase PHF8 regulates zebrafish brain and craniofacial development. *Nature* **466:** 503–507.

Rhind NR, Miller LM, Kopczynski JB, Meyer BJ. 1995. *xol-1* acts as an early switch in the *C. elegans* male/hermaphrodite decision. *Cell* **80:** 71–82.

Schuettengruber B, Bourbon HM, Di Croce L, Cavalli G. 2017. Genome regulation by Polycomb and trithorax: 70 years and counting. *Cell* **171:** 34–57.

Soshnev AA, Josefowicz SZ, Allis CD. 2016. Greater than the sum of parts: Complexity of the dynamic epigenome. *Mol Cell* **62:** 681–694.

Stender JD, Pascual G, Liu W, Kaikkonen MU, Do K, Spann NJ, Boutros M, Perrimon N, Rosenfeld MG, Glass CK. 2012. Control of proinflammatory gene programs by regulated trimethylation and demethylation of histone H4K20. *Mol Cell* **48:** 28–38.

Takahashi T, Tanaka H, Iguchi N, Kitamura K, Chen Y, Maekawa M, Nishimura H, Ohta H, Miyagawa Y, Matsumiya K, et al. 2004. Rosbin: a novel homeobox-like protein gene expressed exclusively in round spermatids. *Biol Reprod* **70:** 1485–1492.

Trojer P, Li G, Sims RJ 3rd, Vaquero A, Kalakonda N, Boccuni P, Lee D, Erdjument-Bromage H, Tempst P, Nimer SD, et al. 2007. L3MBTL1, a histone-methylation-dependent chromatin lock. *Cell* **129:** 915–928.

van Nuland R, Gozani O. 2016. Histone H4 Lysine 20 (H4K20) methylation, expanding the signaling potential of the proteome one methyl moiety at a time. *Mol Cell Proteomics* **15:** 755–764.

Vielle A, Lang J, Dong Y, Ercan S, Kotwaliwale C, Rechtsteiner A, Appert A, Chen QB, Dose A, Egelhofer T, et al. 2012.

H4K20me1 contributes to downregulation of X-linked genes for *C. elegans* dosage compensation. *PLoS Genet* **8:** e1002933.

Vieux-Rochas M, Fabre PJ, Leleu M, Duboule D, Noordermeer D. 2015. Clustering of mammalian Hox genes with other H3K27me3 targets within an active nuclear domain. *Proc Natl Acad Sci* **112:** 4672–4677.

Wang J, Telese F, Tan Y, Li W, Jin C, He X, Basnet H, Ma Q, Merkurjev D, Zhu X, et al. 2015. LSD1n is an H4K20 demeth-

ylase regulating memory formation via transcriptional elongation control. *Nat Neurosci* **18:** 1256–1264.

Wells MB, Snyder MJ, Custer LM, Csankovszki G. 2012. *Caenorhabditis elegans* dosage compensation regulates histone H4 chromatin state on X chromosomes. *Mol Cell Biol* **32:** 1710–1719.

Yonker SA, Meyer BJ. 2003. Recruitment of *C. elegans* dosage compensation proteins for gene-specific versus chromosome-wide repression. *Development* **130:** 6519–6532.

Polytene Chromosome Structure and Somatic Genome Instability

ALLAN C. SPRADLING

Department of Embryology, Howard Hughes Medical Institute, Carnegie Institution for Science, Baltimore, Maryland 21218

Correspondence: spradling@carnegiescience.edu

Polytene chromosomes have for 80 years provided the highest resolution view of interphase genome structure in an animal cell nucleus. These chromosomes represent the normal genomic state of nearly all *Drosophila* larval and many adult cells, and a better understanding of their striking banded structure has been sought for decades. A more recently appreciated characteristic of *Drosophila* polytene cells is somatic genome instability caused by unfinished replication (UR). Repair of stalled forks generates enough deletions in polytene salivary gland cells to alter 10%–90% of the DNA strands within more than 100 UR regions comprising 20% of the euchromatic genome. We accurately map UR regions and show that most approximate large polytene bands, indicating that replication forks frequently stall near band boundaries in late S phase. Chromosome conformation capture has recently identified dense topologically associated domains (TADs) in many genomes and most UR bands are similar or slightly smaller than a cognate *Drosophila* TAD. We argue that bands serve the evolutionarily ancient function of coordinating genome replication with local gene activity. We also discuss the relatively recent evolution of polyteny and somatic instability in Diptera and propose that these processes helped propel the amazing success of two-winged flies in becoming the most ecologically diverse insect group, with 200 times the number of species as mammals.

Polyploid cells are produced during normal development when progenitors switch to a cell cycle without cytokinesis, a process that occurs at some level in most or all species. Diploid cells also may become polyploid to repair tissue damage (Losick et al. 2013, 2016). By endocycling (cycling but not dividing), polyploid cells grow in balance with genome copy number, a strategy that scarcely perturbs cellular physiology or gene regulation and explains why polyploid versions of diverse common cell types can be found across the phylogenetic spectrum (Nagl 1978; Edgar et al. 2014; Neiman et al. 2017). In contrast, polyploid cells that can return to the mitotic cycle, cells often generated by cytokinesis failure or from certain programmed endocycles (Fox et al. 2010), are susceptible to increased genomic instability and prone to oncogenesis (Fujiwara et al. 2005; Duncan et al. 2010; Schoenfelder et al. 2014). Why polyploid cells are so common has remained a matter of debate, but large cells may be mechanically advantageous and stress-resistant (Orr-Weaver 2015; Schoenfelder and Fox 2015; Neiman et al. 2017).

Dipteran polyploid cell chromosomes are termed "polytene" because they maintain replicated sister strands and homologs in exceptionally close association compared with other polyploid cells. Enhanced alignment is associated with resetting the endocycle near the end of S phase, rather than in G_2 or M phase as in most polyploid cells. Their origin from a cell cycle ending in S phase suggests that polytene chromosomes keep multiple sister chromosomes tightly aligned by stabilizing normal S phase pairing involving Cohesins (Fig. 1A,B). Polytene chromosomes have been uniquely valuable for analyzing higher-

order chromatin organization (Bridges 1935; Ashburner 1970, Lefevre 1976; Zhimulev 1996; Stormo and Fox 2017). The normal functioning of diverse cell types possessing such chromosomes, which fold in a characteristic manner but lack consistent contact points between different chromosome arms (Mathog et al. 1984), sounds a cautionary note for theories postulating intricate three-dimensional interactions between distant genomic regions.

Consistent with S phase resets, polytene cells in Dipterans generally show "underreplication" of satellite-rich regions of centromeric heterochromatin, whereas *Drosophila* and some other higher Dipteran species also underreplicate-specific euchromatic regions that normally duplicate late in S phase (Gall et al. 1971; Hammond and Laird 1985; Karpen and Spradling 1990; Spradling 1993; Moshkin et al. 2001; Belyakin et al. 2005; Nordman et al. 2011; Yarosh and Spradling 2014). *Drosophila* euchromatic unfinished replication (UR) regions were originally characterized in the larval salivary gland, but most of the same major regions undergo UR in other polytene tissues examined, including larval midgut, larval fat body, and adult ovary (Nordman et al. 2011; Yarosh and Spradling 2014). UR regions in several tissues contain few expressed genes, are enriched for repressive chromatin marks, and are depleted for candidate replication origins (Sher et al. 2012). The suppressor-of-underreplication gene product (SUUR) is essential for euchromatic UR (Belyaeva et al. 1998); it binds the replication fork protein PCNA and may slow elongation (Nordman et al. 2014). Copy-number changes generated by underreplication have been proposed to result from persistent nets of unfinished

Published by Cold Spring Harbor Laboratory Press; doi: 10.1101/sqb.2017.82.033670

Cold Spring Harbor Symposia on Quantitative Biology, Volume LXXXII

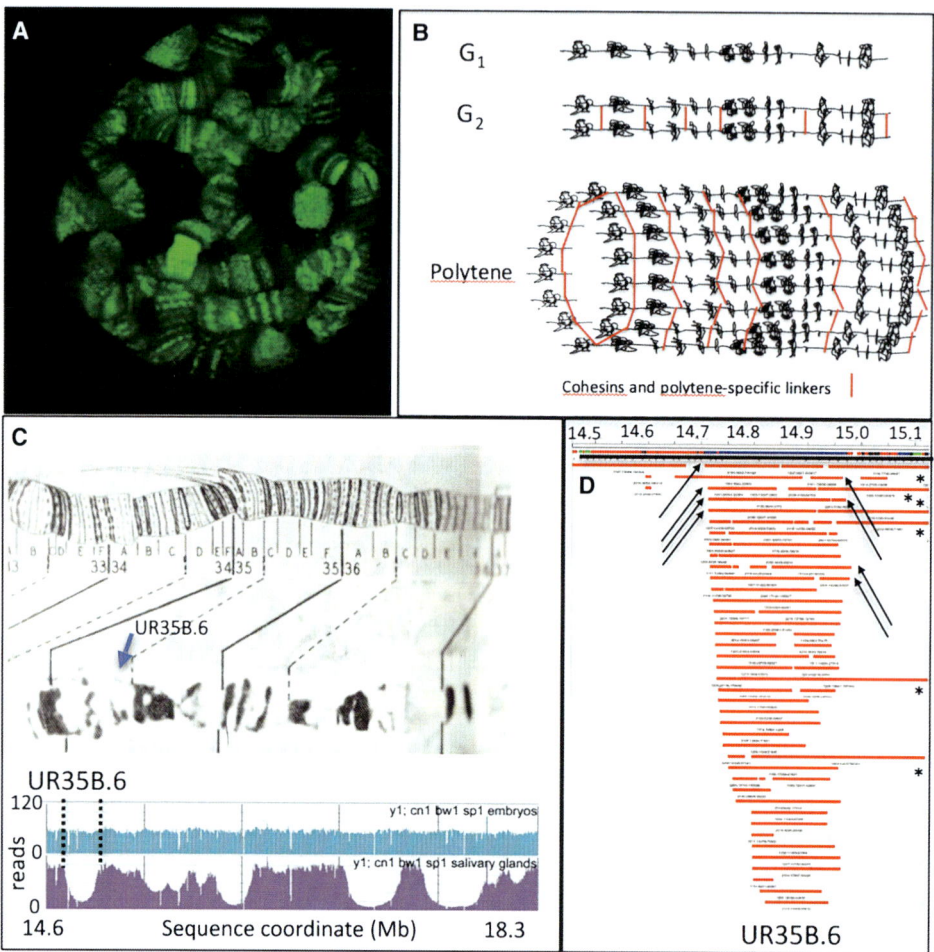

Figure 1. Polytene chromosome structure and somatic instability. (*A*) Polytene chromosome within an intact larval salivary gland nucleus, as revealed by green fluorescent protein (GFP) fluorescence of the protein trap line CC00258 (Buszczak et al. 2007). (*B*) G₁: Model of a G₁ diploid chromosome—a unit chromatin fiber containing highly folded territories separated by more extended regions. G₂: Model of a G₂ diploid chromosome—two unit fibers held together by cohesions (red). Polytene: Model of a polytene chromosome—multiple unit fibers arrayed to form a hollow cylinder and held together by cohesins and novel polytene pairing factors (red). (*C*) Bridges map (Bridges 1935) and Lefevre photograph (Lefevre 1976) of region 34–36 on chromosome 2L, the most somatically unstable region in euchromatin. Sequence read profiles from this region are shown below for diploid (*upper*) and salivary gland (*lower*) DNA. Under-replicated UR regions are revealed as smooth domains of reduced copy number; UR35B.6 is indicated. (*D*) Genome region around UR35B.6 showing chromatin domains (bars at *top*) and deletions (red bars) that are the underlying cause of sequence underrepresentation in UR regions, identified from polytene DNA sequence reads (from Yarosh and Spradling 2014). Arrows show deletion boundaries used to calculate UR boundaries (Table 1), whereas asterisks show six deletions form the region that continued to the next adjacent UR region.

replication forks (Laird 1980; Nordman and Orr-Weaver 2012), but efforts to detect such structures were unsuccessful (Spierer and Spierer 1984; Glaser et al. 1992). Recently, high-throughput sequencing showed that during each endocycle, fork breakage and ligation are responsible for UR by generating deletions that covalently alter each UR region, leaving as few as 10% of their DNA strands intact (Yarosh and Spradling 2014).

The organization of metazoan genomes into discrete territories is probably an ancient and conserved property that was first revealed in the bands of giant polytene chromosomes. Only a handful of small differences in band patterns have been identified between different tissues (Hochstrasser 1987; Heino 1989; Richards 1980), suggesting that banding corresponds to a general aspect of

genome organization minimally related to tissue differentiation. How bands and interbands correspond to functional genomic features has been studied in a few favorable chromosome regions (Vatolina et al. 2011; Zhimulev et al. 2014; Zielke et al. 2016). Putative interband regions are enriched in specific chromatin proteins, active histone marks, transcribed genes, replication origins, and P element insertion sites. Comparison of features such as chromatin marks and gene activity suggests that the domain structure of the *Drosophila* genome is highly similar between polytene and diploid cells (Vatolina et al. 2011; Zielke et al. 2016). Insulator proteins were localized at the junction of bands and interbands and were proposed to organize chromatin domains (Pai et al. 2004; Gerasimova et al. 2007).

Table 1. Relation of unfinished replication (UR) regions to bands using P insertions

UR or P name	N	L avg	SD	R avg	SD	Band	Interval
1.2.05341		595				21C7-D1	34
21D.1	12	629	10	764	39	**21D1-2**	134
1.2.01855		702				21D1-2	
1.2.04723				827		21D3-4	63
1.2.k05428		870				21D4-E1	51
21E.1 s	6	921	1	1010	14	**21E1-2**	89
1.2.k06921				1063		21E2-3	53
1.2.k00619		1159				21F1-2	70
22A.1	15	1229	24	1447	29	**22A1-2**	219
1.2.k11704				1614		22A3-4	167
1.2.k09624		1737				22B1-2	21
22B.2 s	24	1758	5	1817	2	**22B2**	59
1.2.k09932				2046		22C1-2	229
1.2.10638		2455				22F-A1-2	30
23A.1 t	14	2485	5	2704	10	**23A1-2**	219
1.2.k05909				2808		23B1-2	104
1.2.05965		3825				24C8	56
24D.1	38	3881	6	4030	16	**24D1-2**	149
1.2.k01102				4031		24D3-4	1
1.2.k08903		4390				24F1-2	165
25A.3	36	4555	6	4775	12	**25A1-2**	220
1.2.k10004				4853		25B1-2	78
1.2.03771		5327				25D4-6	66
25E.1	8	5393	13	5493	17	**25E1-2**	99
1.2.k11511				5542		25E5-6	49
1.2.k11511		5542				25E5-6	26
25F.1	15	5568	65	5713	4	**251-2**	145
1.2.k06502				5725		25F3-4	12
1.2.10642		6083				26B8-9	56
26C.1	25	6139	4	6304	8	**26C1-2**	164
1.2.k13720				6324		26C2-3	20
1.2.k06704		7040				27D1-2	248
27F.1	7	7288	3	7364	44	**27F1-2**	77
1.2.02657		7037				27F1-2	
1.2.k10113				7424		27F4-6	60
1.2.k10113		7424				27F4-6	192
28C.1 st	5	7616	23	7702	46	**28C1-2**	86
1.2.rL220				7810		28C4-6	108
1.2.03424		8528				29D4-5	6
29E.1	7	8534		8660		**29E1-2**	126
1.2.k13702		8544				29E1-2	
1.2.k04003				8687		29E3-4	27
1.2.s2978		8989				29F8-A1	24
30A.1	18	9013	2	9108	12	**30A1-2**	95
1.2.k05809				9176		30A3-6	68
1.2.k10307		10,517				31F4-5	37
32A.1	28	10,554	2	10,707	7	**32A1-3**	153
1.2.k13206				10,767		32A4-5	60
1.2.k02807		11,221				32E1-2	100
32F.1	19	11,321	17	11,445	29	**32F1**	124
1.2.03602				11,446		32F1-2	1
1.2.03602		11,446				32F1-2	96
32F.3	58	11,542	6	11,778	3	**32F3-4**	235
1.2.04418				11,805		33A1-2	27
1.2.06470		11,808				33A2-3	40
33B.1 s	3	11,848	27	11,908	60	**33B1-4**	61
1.2.01810				12,028		33B8-12	120

Continued

Table 1. *Continued*

UR or P name	N	L avg	SD	R avg	SD	Band	Interval
1.2.08323		12,108				33D1-2	100
33E.1 s	40	12,208	12	12,318	18	**33E1**	110
1.2.k06909				12,435		33E5-7	117
1.2.k00612		12,545				33F1-2	12
33F.1	8	12,557	19	12,662	37	**33F1-2**	105
1.2.k05448				12,704		33F1-2	42
1.2.k05448		12,704				33F1-2	52
34A.1	31	12,756	6	12,948	14	**34A1-2**	192
1.2.k09035		12,822				34A1-2	
1.2.01510				12,975		34A1-2	27
1.2.06646		13,878				34E1-2	70
34F.1 s	21	13,948	16	14,071	53	**34F1**	123
1.2.k11509				14,233		34F3-4	162
1.2.k13218		14,689				35B3-5	23
35B.6	73	14,712	3	14,972	11	**35B6**	260
1.2.k08808				15,008		35B6-10	36
1.2.05441		15,111				35C1-2	36
35C.3	8	15,147	8	15,267	111	**35C3-4**	120
1.2.06430				15,271		35D1-4	4
1.2.06430		15,271				35D1-4	-4
35D.1	44	15,267	21	15,472	98	**35D1-2**	204
13 inserts		15,333		15,338		35D1-2	
1.2.k05305				15,496		35D3-4	24
1.2.k05305		15,496				35D3-4	19
35D.3 s	84	15,515	2	15,659	7	**35D3-4**	144
1.2.05206				15,746		35D3-4	87
1.2.k09033		15,763				35D6-7	17
35E.1	52	15,780	4	15,914	52	**35E1-2**	134
1.2.k07829				16,287		35F1-2	373
1.2.k09033		15,763				35D6-7	171
35E.2	43	15,934	14	16,194	44	**35E3-4**	260
1.2.k07829				16,287		35F1-2	93
1.2.k04216		16,352				35F11-12	39
36A.1	9	16,391	11	16,467	16	**36A1**	76
1.2.k07510				16,493		36A2-3	26
1.2.k16215		16,526				36A4-5	26
36A.6	6	16,552	32	16,684	101	**36A6-7**	131
1.2.k15102				16,691		36A10-11	7
1.2.k03902		16,825				36B1-2	100
36C.1	69	16,925	3	17,349	23	**36C1-2**	425
1.2.k10816				17,450		36D1-3	101
1.2.k10816		17,450				36D1-3	61
36E.1	94	17,511	2	17,953	11	**36E1**	441
1.2.k09927				18,320		36E3-4	367
1.2.k10816		17,450				36D1-3	514
36E.2	20	17,964	4	18,173	37	**36E2**	209
1.2.k09927				18,320		36E3-4	147
1.2.k10816		17,450				36D1-3	715
36E.3 st	15	18,165	2	18,269	4	**36E3**	103
1.2.k09927				18,320		36E3-4	51
1.2.k06028		19,166				37C6-7	46
37D.1 s	10	19,212	3	19,292	22	**37D1-2**	80
1.2.01068				19,575		37F1-2	283
1.2.03552		20,085				38B4-6	19
38C.1	23	20,104	28	20,255	30	**38C1-2**	151
1.2.01820				20,243		38C1-2	
1.2.k07219				20,382		38C5-6	127

Continued

Table 1. *Continued*

UR or P name	N	L avg	SD	R avg	SD	Band	Interval
l.2.k07219		20,382				38C5-6	103
38D.1 s	15	20,485	5	20,608	15	**38D1-2**	123
l.2.k02501				20,639		38D1-2	31
l.2.02074		21,659				39F1-2	62
40D.1	55	21,721	78	22,105	2	**40A1-4**	383
l.2.k16406		21,828				40A1-4	
l.2.04319		21,829				40A1-4	

The table shows the mapped UR regions by name (bold), the number of associated salivary gland deletions in reads analyzed here (*N*), the average left (L) and right (R) boundaries and standard deviations (SDs) in R6 coordinates (kb) usually determined from four deletion end points, and the associated band (bold) from this study (band). An "s" following the name indicates a UR that may be smaller than its associated band, whereas a "t" indicates that less than half of Hi-C studies predicted a similar topologically associated domain. Also shown are relevant P element insertions (P name), their R6 insertion site in L or R, and their cytogenetic position assigned by Laverty (band). Insertions in bold are internal to the UR. The size of the UR (Interval) or the distance (Interval) between the L and R flanking insertions and the start or end of the UR are shown.

This emerging picture of how metazoan genomes are arranged into specific territories has received strong support from studies using chromosome conformation capture methods including Hi-C (see Ghirlando and Felsenfeld 2016; Rowley and Corces 2016). Regional units with a greater probability of interaction known as topologically associated domains (TADs) represent a common feature of animal cell chromatin, including *Drosophila* (Hou et al. 2012; Sexton et al. 2012; Eagen et al. 2015; Ullanov et al. 2016). TADs were nearly identical in polytene and diploid DNA, confirming that polytene chromosomes are good models of diploid genome organization (Eagen et al. 2015) and suggesting that TADs are bands (Eagen et al. 2015; Ullanov et al. 2016). Subsequent studies at higher resolution have revealed up to several thousand TAD boundaries (Eagen et al. 2017; Hug et al. 2017; Ramirez et al. 2017; Stadler et al. 2017). The idea that TADs correspond generally to bands remains attractive, but the boundaries identified by different groups often differ, and some recent studies may be resolving structures smaller than bands.

MAPPING UR REGIONS AND POLYTENE BANDS USING HIGH-RESOLUTION IN SITU HYBRIDIZATION

A major limitation of understanding bands has been the difficulty of mapping them precisely onto the genome sequence (see Zielke et al. 2016). Most in situ hybridization mapping as summarized in FlyBase (Marygold et al. 2016) has been performed at a resolution far lower than that of individual bands and with sometimes-contradictory results due to the difficulty of high-resolution polytene mapping. To circumvent this problem, we investigated the cytogenetic location of UR domains, bands, and TADs using a collection of P element insertions mapped by in situ hybridization with exceptional accuracy and consistency. The ~1100 insertions in the collection were all localized in situ by Todd Laverty using polytene chromosomes stretched to provide single-band resolution and photographically recorded as part of the Drosophila Genome Project (Laverty and Rubin 2000) before their insertion points on the genome sequence or relevance to this project were known.

UR DOMAINS CORRESPOND TO MAJOR BANDS

Before mapping, we reexamined the number and exact location of UR regions determined previously using copy-number data (Yarosh and Spradling 2014). UR region boundaries are imperfectly defined using copy-number profiles, because boundaries represent the point where the UR signal goes to zero (Fig. 1C). We reasoned that UR boundaries could be measured more accurately using the deletions that give rise to UR (Fig. 1D). Although deletion end points are not confined to the edges of UR regions, frequently a series of similar breakpoints are located near UR boundaries (Fig. 1D, arrows). Consequently, using a collection of 4459 deletions identified from polytene larval salivary gland sequence reads essentially as described (Yarosh and Spradling 2014), we averaged the end coordinates of boundary-associated deletions for each UR region's beginning and end point and determined their average values. We found that using clusters of polytene-specific deletions as the criterion for identifying a UR region improved sensitivity and increased the total number of identified URs from 115 (Yarosh and Spradling 2014) to 203. We then compared these end points with the most closely flanking and any internal P element insertions from the Laverty collection.

The results for chromosome 2L (Table 1) show that 26 of 37 UR regions mappable by in situ hybridization correspond closely to individual, generally large, dense bands (usually denoted as doublets by Bridges). For example (Fig. 2A), the 164-kb-long UR26C.1 is flanked 56 kb on the left by insertion 1(2)10642 at 26B8-9. Fifty-six kilobases is a reasonable amount of DNA to encompass the remaining sequences in region 26B, suggesting that the UR corresponds to the dark band 26C1-2. Just 20 kb beyond the UR lies insertion 1(2)k13720 which was mapped just past 26C1-2, at "26C2-3." Although the precise start and end points of the UR cannot be determined from two flanking sites, the great majority of the 164-kb

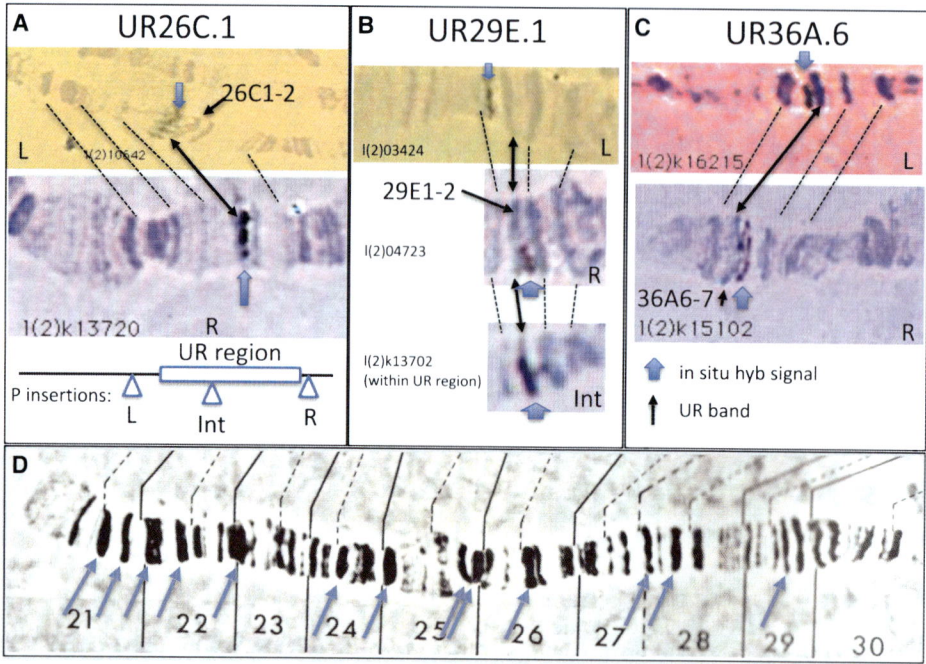

Figure 2. UR regions correspond to single dense bands. (*A*) Mapping UR26C.1 to band 26C1-2 using in situ hybridizations of the indicated flanking P element insertions (as diagrammed below) lying closest to the left (L) and (R) boundaries (Table 1). Dashed arrows indicate equivalent bands in the two panels, whereas the heavy double arrow points to band 25C1-2 that is flanked by in situ hybridization signals (blue arrows). (*B*) Mapping of UR29E.1 to band 29E1-2 as in *A*. This UR containing an internal (int) P insertion, l(2)k13702, whose localization to band 29E1-2 is shown. (*C*) Mapping of UR36A.6 to band 36A6-7. (*D*) A summary photograph of the distal half of chromosome 2L with arrows showing the strong bands mapped here to UR regions. In situ hybridizations are from the Laverty collection (Laverty and Rubin 2000).

UR must reside in the dark 26C1-2 band. It is plausible to assume that replication forks proceeding from outside (Sher et al. 2012) usually stall at or just inside 26C1-2 boundaries to generate the observed deletions and UR profile. Likewise, UR29E.1 (Fig. 2B) is flanked by even closer insertions that link it to the band 29E1-2, and UR36A.6 (Fig. 2C) maps to band 36A6-7. When a UR, such as 29E.1 (Fig. 2B), also has an internal insertion (l(2)01855), it invariably maps to the expected large band, whereas the flanking insertions map to either side of that band (Table 1).

The other 11 URs show a similar pattern with insertions flanking a relatively large band positioned to contain the UR. However, either the UR size or its location relative to flanking insertions suggests that the UR comprises only part of the large band. For example, the 192-kb UR34A.1 corresponds to band 34A1-2 but insertion l(2)01510, located 27 kb past the end of the UR, still localized to 34A1-2. Across the genome as a whole there are also a few atypical URs that span several bands, such as those comprising the two major homeotic gene clusters on chromosome 3, ANT-C and BX-C.

As can be seen from the summary for distal chromosome arm 2L (Fig. 2D), ~33% of the large dense bands on the chromosome comprise UR regions. A substantial number of euchromatic regions defined by preferential breakage, late replication, etc., have been termed "intercalary heterochromatin" (Kaufmann 1939; Zhimulev and Belyaeva 2003) and ~25% of these correspond to UR

regions (Belyakin et al. 2005). Consistent with this, the mapping presented here shows that a region's chromatin state is not a good predictor of whether it will underreplicate. The great majority of genomic Polycomb domains are not URs, and only ~12% of URs correspond to Polycomb domains, most quite weak in their effects. Black chromatin is widespread, and many dense bands with black chromatin are not URs, but the majority of UR bands contain mostly "black" chromatin. Domains enriched in H3K9me3 outside the centric regions would be expected to act like intercalary heterochromatin, and these relatively rare zones include some of the strongest URs such as 36C.1 and 36E.1. However, most URs are not enriched in H3K9me3, and some H3K9me3-rich zones, despite perhaps replicating fairly late in S, are not URs. Thus, the only consistent feature that defines a region as a UR is failing to complete replication during a significant number of endocycles.

The observation (Yarosh and Spradling 2014) that UR regions contain many large genes was further investigated by calculating the abundance of genes of various sizes in URs versus in the genome as a whole. Genes (protein-coding or all annotated genes) are progressively enriched up to fourfold as a function of size in UR regions relative to their frequency in the genome as a whole (Table 2). URs were also reported to be enriched in genes encoding IgG superfamily and other cell surface proteins in both *Drosophila* and mammals (Hannibal et al. 2014; Yarosh and Spradling 2014). The nature of the 1939 UR genes as a

Table 2. Unfinished replication (UR) enrichment versus gene size

Gene size	Total	Euch	UR genes	UR euch	UR/total	UR euch/euch
Protein-coding genes						
Any size	13,762	13,563	1404	1328	0.10	0.10
>10 kb	1948	1874	239	212	0.12	0.11
>50 k	305	277	83	73	0.27	0.26
>100 kb	86	68	30	24	0.35	0.35
>150 kb	25	17	9	7	0.36	0.41
Protein and RNA genes						
Any size	16,106	15,837	1940	1826	0.12	0.12
>10 kb	1994	1917	259	229	0.13	0.12
>50 kb	317	287	91	79	0.29	0.28
>100 kb	91	72	36	28	0.40	0.39
>150 kb	28	19	12	9	0.43	0.47

The table shows the number of protein coding (*upper*) or protein and RNA (*lower*) genes based on R6 genome annotation within the indicated size class, either throughout the genome (total), within euchromatin (euch), within UR regions as defined by Yarosh and Spradling (2014) and updated here (UR genes), or within euchromatic UR regions (UR euch). The ratios of genes in the indicated size classes within UR regions to total genes (UR/total), and the ratios of genes in euchromatic UR regions to total euchromatic genes (UR euch/euch) are shown. Enrichment of genes of a given size class in UR regions is estimated by comparing the ratios of that size class to "Any size."

group was further investigated using gene ontology (Huang et al. 2009). The results (Table 3) confirm the large enrichment in IgG superfamily genes, many of which have functions in neural development and pathfinding. Additionally, UR regions are significantly overrepresented with genes encoding membrane proteins and with genes that sense odorants and function in olfaction.

BANDS SHOWING UR USUALLY CORRESPOND TO TADs

The relationship between *Drosophila* TADs and bands was similarly analyzed using the Laverty collection and the TAD boundary values reported in recent publications. URs defined by deletion end points (Table 1) were compared with TADs reported in six recent studies (Fig. 3). In general, there were many differences between the TAD boundaries identified in these studies, which were performed over a span of 5 years and differed substantially in resolution. However, among the subclass of large bands showing UR, there was usually close correspondence to a TAD "consensus." For example, band 26C1, which comprises much or all of UR26C.1, matches approximately to a single large TAD in five of the six studies. Four studies agree that the left boundary of this TAD is near 6100 kb, 39 kb from the measured left UR boundary but still 17 kb to the right of the flanking l(2)10642 P element at 26B8-9. The right boundaries of four TAD measurements are all about 10–15 kb to the right of the UR, whereas two are past the flanking element l(2)k13720 at 26C2-3. Thus, a TAD similar to band 26C1-2 and UR26C.1 probably exists, but the UR may be slightly smaller and the TAD may comprise or be slightly larger than this band. It is easy to

Table 3. Gene ontology (GO) analysis of UR region genes

Category	Term	Count	P_{Value}	Benjamini
Annotation Cluster 1	Enrichment Score: 10.69			
INTERPRO	IPR007110:Immunoglobulin-like domain	50	5.90E−19	5.85E−16
INTERPRO	IPR013783:Immunoglobulin-like fold	52	8.30E−15	2.75E−12
Annotation Cluster 2	Enrichment Score: 6.69			
GOTERM_CC_DIRECT	GO:0005956~protein kinase CK2 complex	14	1.52E−10	4.79E−08
GOTERM_BP_DIRECT	GO:0080163~regulation of protein serine/threonine phosphatase activity	13	4.81E−10	6.36E−07
Annotation Cluster 3	Enrichment Score: 4.8			
GOTERM_BP_DIRECT	GO:0007606~sensory perception of chemical stimulus	26	1.52E−08	6.71E−06
GOTERM_MF_DIRECT	GO:0005549~odorant binding	27	4.39E−07	1.17E−04
Annotation Cluster 4	Enrichment Score: 4.1			
INTERPRO	IPR003591:Leucine-rich repeat, typical subtype	21	9.10E−06	7.51E−04
Annotation Cluster 5	Enrichment Score: 3.90			
GOTERM_CC_DIRECT	GO:0016021~integral component of membrane	305	4.66E−08	4.88E−06
Annotation Cluster 7	Enrichment Score: 2.59			
GOTERM_MF_DIRECT	GO:0005549~odorant binding	27	4.39E−07	1.17E−04
GOTERM_MF_DIRECT	GO:0004984~olfactory receptor activity	25	1.63E−05	2.16E−02
Annotation Cluster 8	Enrichment Score: 2.48			
UP_SEQ_FEATURE	DNA-binding region:Homeobox	16	1.24E−06	7.48E−04
Annotation Cluster 10	Enrichment Score: 2.30			
GOTERM_BP_DIRECT	GO:0050896~response to stimulus	8	3.45E−04	5.54E−02
INTERPRO	IPR006170:Pheromone/odorant binding protein	12	1.44E−03	5.35E−02

The 1404 protein-coding genes located with UR regions as defined by Yarosh and Spradling (2014) and updated here were subjected to GO analysis using the National Institutes of Health (NIH) DAVID website (Huang et al. 2009) and the release 6 annotation of the *Drosophila* genome. For brevity, some redundant or less relevant matches are not shown.

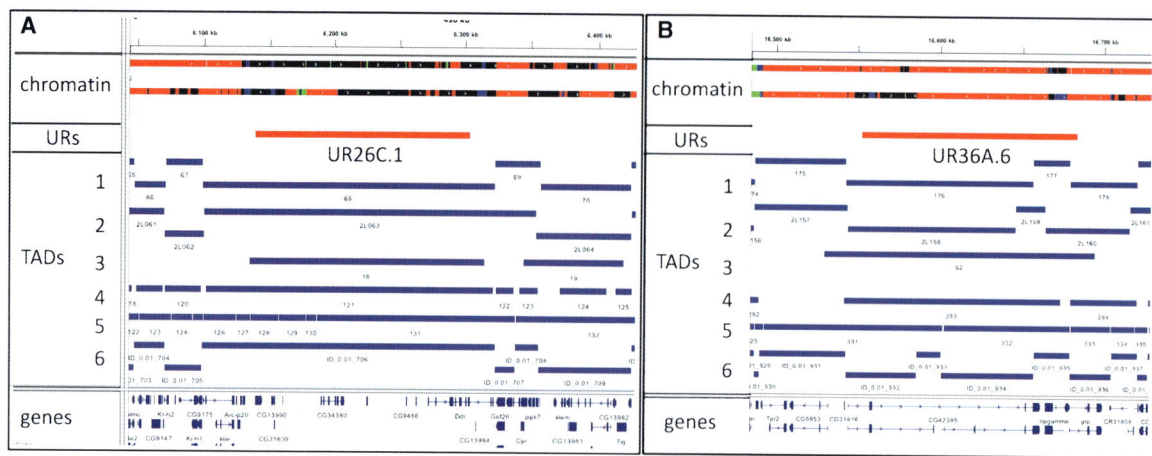

Figure 3. Correspondence between UR bands and topologically associated domains (TADs). (*A*) The 2L chromosome region surrounding UR26C.1, followed by two tracks showing chromatin types from S2 and BG3 tissue culture cells colored as in Filion et al. (2010). Below, the position of UR regions ("URs") mapped in Table 1 are shown in red. TADs mapped to this region by the indicated publications (1–6) are mapped below (blue). (1) Sexton et al. (2012); (2) Hou et al. (2012); (3) Eagen et al. (2015); (4) Eagen et al. (2017); (5) Stadler et al. (2017); (6) Ramirez et al. (2017). Genes in the region are also plotted. In five of six studies, UR26C.1 approximates one TAD. (*B*) The same tracks as in *A* are plotted for the genomic region surrounding UR36A.6. In this case, the UR is reported to contain from one to four TADs.

understand why a UR would be slightly smaller than a band, because replication forks might not stall instantly after encountering the band edge, whereas the slightly larger size of a TAD measurement might be due to insufficient Hi-C resolution. In general, the agreement between the UR, band, and TAD is quite striking.

Similar approximate agreement with a TAD was observed in at least 21 of the other 38 UR regions in Table 1. In many of these, the UR was consistently smaller than the TAD, especially in the case of URs that were judged to comprise only part of a large band. Nonetheless, it was also common to have URs split into two or more TADs in at least some of the reported data. For example, UR36A.6 (Fig. 3B) comprises band 36A6-7 and is unusual in being weak (six deletions) and in corresponding to a largely red ("active") chromatin domain. Although two studies identified TADs in this region with similar dimensions, the others split UR36A.6 into two to four TADs. Thus, it appears that UR regions associated with single strong bands frequently do correspond to TADs, but that high-resolution Hi-C may be required to precisely map their end points. However, Hi-C data sometimes go farther and break a UR and its corresponding band into subregions. The biological meaning of such subdivision is currently unclear.

TISSUE COMPARISON OF URs AND TADs

We compared the relationship of UR regions and bands between tissues by analyzing DNA from the *Drosophila* midgut, which contains 8C polytene enterocytes as a major cell type. After sequencing to a depth of 272 million reads, the midgut read depth profile showed clear underreplication (Fig. 4A). Moreover, the size and location of URs appeared to be about the same in salivary gland and midgut. However, midgut UR regions were less underreplicated in general than salivary gland URs, and individ-

ual URs varied substantially in their relative levels of UR between the two tissues. For example, UR34A.1, UR34F.1, UR35D.1, and UR35D.3 were much more fully replicated in midgut, suggesting that origin usage or timing differs in these regions between the tissues (Fig. 4A). We identified 1729 unique reads that define deletions between 10 and 500 kb from the midgut data, observed that they preferentially mapped to UR regions, and used the deletion end points to calculate the precise boundaries of 14 midgut UR regions on chromosome 2L (Fig. 4B). The values obtained in the case of all 14 midgut UR zones were the same within measurement error as the corresponding coordinates in salivary gland DNA. For example, the right boundary of UR35B.6 of 14970 ± 11.43 kb in salivary gland compared closely with 14920 ± 19.79 kb in midgut. Thus, consistent with the early appearance of a fixed pattern of TADs during development (Hug et al. 2017) and the invariant pattern of bands and TADs between tissues, the dimensions of URs are the same in midgut and salivary gland, reflecting a fundamental aspect of *Drosophila* genome structure. However, the level of underreplication does vary in particular UR zones, most likely because of differences in origin usage and timing between these tissues.

CONCLUSION

Bands Are Fundamental Genomic Units That May Coordinate Replication and Transcription

Our results help clarify the nature of the genomic domains that are visualized in polytene cells as chromosomal bands. We showed that many of the strongest bands correspond almost exactly to UR regions, implicating bands in replication control, not just in gene regulation. These bands also largely match major TADs determined by Hi-C

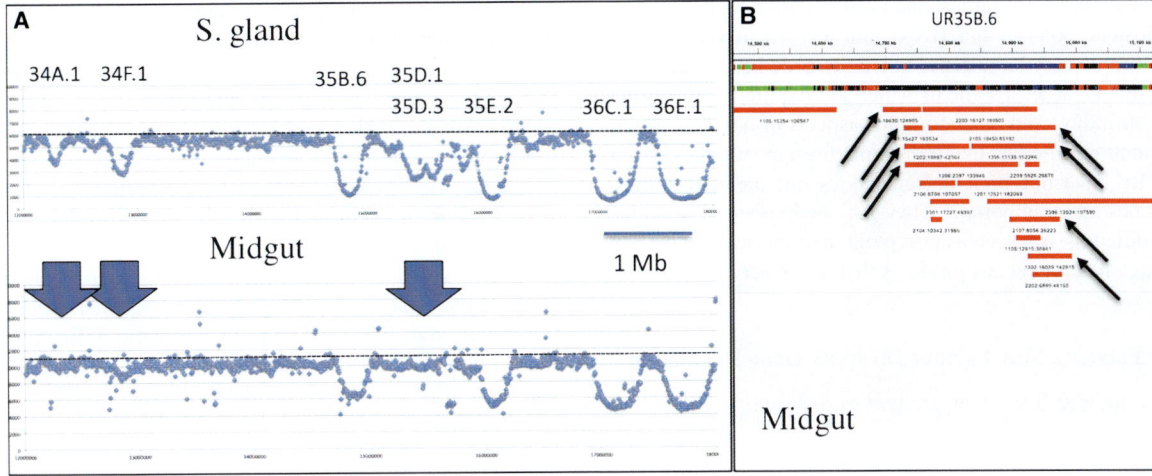

Figure 4. Precise correspondence in coordinates but not in depth of UR regions in larval salivary gland (S. gland) and adult midgut DNA. (*A*) Plot of read depth in 5-kb bins from region 34–36 of chromosome 2L from larval salivary gland (*upper*) and adult midgut (*lower*) DNA. Major UR regions show decreased read depth and are labeled. Blue arrows indicate four UR regions that are greatly diminished in midgut compared with salivary gland. (*B*) Plot of region surrounding UR35B.6 (see Fig. 1D), showing chromatin from BG3 and S2 cells (colored according to Filion et al. 2010, but H3K27me3-rich is green and H3K9me3-rich is blue), and deletions determined from midgut sequence reads. Arrows indicate the deletions used to calculate the *left* and *right* end coordinates of the midgut UR35B.6, which proved indistinguishable from salivary gland UR35B.6 end points (see text).

studies. Neither band structure nor TAD organization varies significantly between different tissues, and we showed that UR domains are also precisely the same between different cell types. Bands and TADs likely represent a sequence-encoded aspect of metazoan genome structure.

We propose that these fixed genomic domains serve to organize the temporal program of replication (Gilbert et al. 2010; Pope et al. 2014) in a manner that promotes appropriate gene expression. It is well-known from studies of unicellular organisms that DNA replication and transcription have the potential to significantly interfere with each other if not subject to regulation (see Merrikh et al. 2012). Highly transcribed genes including rRNA genes in bacteria are usually positioned so that replication forks will travel in the same direction as transcription to minimize interference, rather than in the opposite direction. In metazoans there is also a strong tendency for highly expressed genes (i.e., genes whose products are needed in large amounts) to be replicated early in S phase. Early replication is expected to increase product production by generating transcripts from a second template for as long as possible. The organization of the genome into domains, coupled with regulated origin activation, may help ensure these outcomes.

Bands May Promote Favorable Replication Timing for Genes Transcribed at All Levels

Whether replication timing is important for genes transcribed at a moderate or low rate has remained unclear. However, there are several ways in which replication timing might assist such genes independent of enhancing gene dosage. Interactions between enhancers and promoters within a TAD that involve the formation of looped contacts would likely be susceptible to disruption by replication fork passage. Afterward, gene expression would remain off for however long was required to restore these interactions and complete RNA polymerase passage. Genomic domains controlling replication timing might ensure that their component genes shut down at a time in the cell cycle that minimizes the impact on developmental and physiological events in which they are participating.

There may even be a currently unrecognized class of "slow-activating" genes that would benefit from a program of late replication. These might include genes with extremely complex regulation involving large transcription units and multiple enhancers, such that product production can only begin after a long delay while appropriate regulatory structures are established and the length of the gene is traversed by slowly moving polymerases. If slow-activating genes were to replicate early in S phase, this intricate regulatory organization along with nascent transcripts would be disrupted by fork passage quite soon after product production had finally begun. Only after a second lag of equal length would such genes be able to resume transcription, perhaps leaving insufficient time in the cell cycle for adequate production. We propose that such slow-activating genes would benefit from replicating late in the cell cycle, as this would ensure that they could produce product with only one major pause per cell cycle for regulatory assembly and initial transcription. These considerations might explain why the genome would evolve tissue-invariant late-replicating bands. Furthermore, these domains might contain some of the largest, most complex, and highly regulated genes in the genome.

Our studies suggest that UR domains house genes that could in this way benefit from late replication. URs are enriched for large genes and encode genes with apparently complex regulation in embryonic and neural development.

For example, the largest and most complex homeotic gene clusters, ANT-C and BX-C, are located in UR regions. Multiple very large genes in URs encode IgG superfamily proteins, at least some of which are required for neuronal pathfinding and other complex aspects of nervous system function. The fact that transcripts from most genes within URs have not been detected does not mean that these genes are unimportant. They may be expressed in understudied tissues such as polyploid neurons and glia or produce heterogeneous products that differ between cells.

Polyteny May Further Diversify Gene Regulation

The idea that bands are part of a system to time DNA replication and that late replicating bands house a subclass of genes with complex regulation provides a speculative but interesting rationale for the evolution of polyteny and underreplication after the origin of Diptera in the early Mesozoic Era. Extant cytological surveys indicate that polyploidy, but not polyteny, is widespread throughout the phylogenetic tree and in particular within many insect orders. Non-Dipteran polyploid cells, to the extent known, reset the endocycle near or within M phase and fully replicate their satellite DNA (Gage 1974). In contrast, polytene chromosomes are essentially confined among insects to the Diptera, one of the higher insect orders that arose in the Triassic (Grimaldi and Engel 2005), long after insect body plans and general physiology had been established. Polytene cells in most Dipterans probably underreplicate centromeric heterochromatin, based on their chromosome morphology (White 1973). Thus, the evolution of flies was associated with alterations in the endocycle that produced polyteny and centromeric UR. These changes were not needed to make a generic insect but are likely to have been part of the evolutionary innovation that put two-winged flies on the path to success across the globe.

Compartmentalized genome territories are widespread in metazoans and were presumably in place long before Dipteran evolution. We propose that the advent of polyteny and centromeric UR provided new flexibility in gene expression not available in other insect groups. By bringing multiple aligned copies of the genome close together in a polytene chromosome, opportunities for interstrand enhancer–promoter interactions via looping would be greatly increased. Novel regulatory interactions might be further multiplied by bringing homologs into proximity, which might also explain the origin of somatic pairing in both polytene and diploid cells, another characteristic of Diptera. Pairing-dependent interactions have been widely documented in *Drosophila* genetics and can frequently be explained by cross-strand enhancer action (Lewis 1954; Gelbart and Wu 1982; Bingham and Zachar 1985; Lee and Wu 2006; Mellert and Truman 2012).

Somatic Genome Instability May Have Contributed to Dipteran Evolution

Whether somatic genome instability within euchromatic UR regions evolved at the same time as polyteny within

early Dipteran groups or had a more recent origin during the expansion of the higher Diptera after the Cretaceous (Wiegmann et al. 2011) is currently unknown. Very few species have been tested for euchromatic UR, and no reliable correlate of this process in polytene chromosome morphology is currently available. A late origin is suggested by the limited phylogenetic distribution of clear SUUR homologs only within the genus *Drosophila* and a few other higher flies.

Despite specialized mouthparts limiting them to liquid food, Diptera have adapted with unprecedented success (Grimaldi and Engel 2005). Euchromatic UR might have contributed to their adaptability in concert with polyteny by further enhancing regulatory flexibility. UR-generated chromosomal deletions may alter gene regulation and protein structure by juxtaposing novel genomic regions. In addition, UR deletions are expected to perturb the local alignment of sister strands in a manner that would further expand the range of possible enhancer–promoter interactions. Somatic instability within euchromatic UR regions may increase the ability of flies to adapt to novel environments because these regions preferentially contain large genes with complex regulation that affect neural function and behavior, as indicated by GO analysis (Table 3). The diversification of UR region genes involved in "sensory perception of chemical stimulus," "odorant binding," and "olfactory receptor activity" may have helped flies to identify and exploit rare and transient food sources. It may have allowed them to adapt their sensory perception and behavior rapidly, contributing to their success in dominating diverse ecological niches. SUUR mutants might survive in laboratory conditions, but have difficulty finding dispersed food resources in wild environments. Thus, a deeper understanding of bands, polyteny, and UR may help us learn how flies have been able to make an outsized impact on Earth and on human health.

Polytene Chromosomes Remain Valuable for Understanding Chromosome and Nuclear Biology

In conclusion, polytene chromosomes have contributed significantly to our understanding of chromosome organization and genome function for the last 80 years. Today, however, the opinion has grown that polytene cells differ greatly from mammalian cells, and that giant chromosomes have been surpassed as tools by chromosome conformation capture and other genomic techniques. This study has emphasized that polytene chromosomes arose primarily as a cell cycle change, not a developmental change, and that strong experimental evidence shows they differ little in structure and function from diploid chromosomes. These natural quasicrystals of unit chromatin strands provide a high resolution view of structural and functional processes that function in nuclei throughout the animal kingdom. The ability to directly visualize the fundamental mechanisms of cell nuclei, in conjunction with molecular and genomic techniques, will continue to reward those who choose to use polytene cells to investigate the many outstanding questions in chromosome biology.

ACKNOWLEDGMENTS

The author is especially grateful to Will Yarosh and Don Fox, former members of the Spradling laboratory, for DNA sequencing of *Drosophila* tissues and to Shelly Paterno and Mike Buszczak for the protein trap photo in Figure 1A. He thanks current laboratory members Chenhui Wang, Steve Deluca, and Bob Levis for comments on the manuscript. He is particularly grateful to Todd Laverty and Gerry Rubin for sharing data from the Laverty P element in situ hybridization collection. Allison Pinder provided expert assistance with DNA sequencing and Fred Tan gave valuable advice on sequence analysis.

REFERENCES

Ashburner M. 1970. Function and structure of polytene chromosomes during insect development. *Adv Insect Physiol* **7:** 1–95.

Belyaeva ES, Zhimulev IF, Volkova EI, Alekseyenko AA, Moshkin YM, Koryakov DE. 1998. Su(UR)ES: A gene suppressing DNA underreplication in intercalary and pericentric heterochromatin of *Drosophila melanogaster* polytene chromosomes. *Proc Natl Acad Sci* **95:** 7532–7537.

Belyakin SN, Christophides GK, Alekseyenko AA, Kriventseva EV, Belyaeva ES, Nanayev RA, Makunin IV, Kafatos FC, Zhimulev IF. 2005. Genomic analysis of *Drosophila* chromosome underreplication reveals a link between replication control and transcriptional territories. *Proc Natl Acad Sci* **102:** 8269–8274.

Bingham PM, Zachar Z. 1985. Evidence that two mutations, wDZL and z1, affecting synapsis-dependent genetic behavior of white are transcriptional regulatory mutations. *Cell* **40:** 819–825.

Bridges CB. 1935. Salivary chromosome maps: With a key to the banding of the chromosomes of *Drosophila melanogaster*. *J Hered* **26:** 60–64.

Buszczak M, Paterno S, Lighthouse D, Bachman J, Plank J, Owen S, Skora A, Nystul T, Ohlstein B, Allen A, et al. 2007. The Carnegie protein trap library: A versatile tool for *Drosophila* developmental studies. *Genetics* **175:** 1505–1531.

Duncan AW, Taylor MH, Hickey RD, Hanlon-Newell AE, Lenzi ML, Olson SB, Finegold MJ, Grompe M. 2010. The ploidy conveyor of mature hepatocytes as a source of genetic variation. *Nature* **467:** 707–710.

Eagen KP, Hartl TA, Kornberg RD. 2015. Stable chromosome condensation revealed by chromosome conformation capture. *Cell* **163:** 934–946.

Eagen KP, Aiden EL, Kornberg RD. 2017. Polycomb-mediated chromatin loops revealed by a subkilobase-resolution chromatin interaction map. *Proc Natl Acad Sci* **114:** 8764–8769.

Edgar BA, Zielke N, Gutierrez C. 2014. Endocycles: A recurrent evolutionary innovation for post-mitotic cell growth. *Nat Rev Mol Cell Biol* **15:** 197–210.

Filion GJ, van Bemmel JG, Braunschweig U, Talhout W, Kind J, Ward LD, Brugman W, de Castro IJ, Kerkhoven RM, Bussemaker HJ, et al. 2010. Systematic protein location mapping reveals five principal chromatin types in *Drosophila* cells. *Cell* **143:** 212–224.

Fox D, Gall JG, Spradling AC. 2010. Error-prone polyploid mitosis during normal *Drosophila* development. *Genes Dev* **24:** 2294–2302.

Fujiwara T, Bandi M, Nitta M, Ivanova EV, Bronson RT, Pellman D. 2005. Cytokinesis failure generating tetraploids promotes tumorigenesis in p53-null cells. *Nature* **437:** 1043–1047.

Gage LP. 1974. Polyploidization of the silk gland of *Bombyx mori*. *J Mol Biol* **86:** 97–108.

Gall JG, Cohen EH, Polan ML. 1971. Repetitive DNA sequences in *Drosophila*. *Chromosoma* **33:** 319–344.

Gelbart WM, Wu CT. 1982. Interactions of zeste mutations with loci exhibiting transvection effects in *Drosophila melanogaster*. *Genetics* **102:** 179–189.

Gerasimova TI, Lei EP, Bushey AM, Corces VG. 2007. Coordinated control of dCTCF and gypsy chromatin insulators in *Drosophila*. *Mol Cell* **28:** 761–772.

Ghirlando R, Felsenfeld G. 2016. CTCF: Making the right connections. *Genes Dev* **30:** 881–891.

Gilbert DM, Takebayashi SI, Ryba T, Lu J, Pope BD, Wilson KA, Hiratani I. 2010. Space and time in the nucleus: Developmental control of replication timing and chromosome architecture. *Cold Spring Harb Symp Quant Biol* **75:** 143–153.

Glaser RL, Karpen GH, Spradling AC. 1992. Replication forks are not found in a *Drosophila* mini-chromosome demonstrating a gradient of polytenization. *Chromosoma* **102:** 15–19.

Grimaldi D, Engel MS. 2005. *Evolution of the insects*. Cambridge University Press, Cambridge.

Hammond MP, Laird CD. 1985. Chromosome structure and DNA replication in nurse and follicle cells of *Drosophila melanogaster*. *Chromosoma* **91:** 267–278.

Hannibal RL, Chuong EB, Rivera-Mulia JC, Gilbert DM, Valouev A, Baker JC. 2014. Copy number variation is a fundamental aspect of the placental genome. *PLoS Genet* **10:** e1004290.

Heino TI. 1989. Polytene chromosomes from ovarian pseudonurse cells of the *Drosophila melanogaster otu* mutant. *Chromosoma* **97:** 363–373.

Hochstrasser M. 1987. Chromosome structure in four wild-type polytene tissues of *Drosophila melanogaster*. The 87A and 87C heat shock loci are induced unequally in the midgut in a manner dependent on growth temperature. *Chromosoma* **95:** 197–208.

Hou C, Li L, Qin ZI, Corces VG. 2012. Gene density, transcription and insulators contribute to the partition of the *Drosophila* genome into physical domains. *Mol Cell* **48:** 471–484.

Huang DW, Sherman BT, Lempicki RA. 2009. Systematic and integrative analysis of large gene lists using DAVID bioinformatics resources. *Nat Protoc* **4:** 44–57.

Hug CB, Grimaldi AG, Kruse K, Vaquerizas JM. 2017. Chromatin architecture emerges during zygotic genome activation independent of transcription. *Cell* **169:** 216–228.e19.

Karpen GH, Spradling AC. 1990. Reduced DNA polytenization of a minichromosome region undergoing position-effect variegation in *Drosophila*. *Cell* **63:** 97–107.

Kaufmann BP. 1939. Distribution of induced breaks along the X-chromosome of *Drosophila melanogaster*. *Proc Natl Acad Sci* **25:** 571–577.

Laird CD. 1980. Structural paradox of polytene chromosomes. *Cell* **22:** 869–874.

Laverty T, Rubin GM. 2000. Encyclopedia of *Drosophila*, vol. 2. CD distributed by the Berkeley *Drosophila* Genome Project, http://www.fruitfly.org/index.html.

Lee AM, Wu CT. 2006. Enhancer-promoter communication at the *yellow* gene of *Drosophila melanogaster*: Diverse promoters participate in and regulate trans interactions. *Genetics* **174:** 1867–1880.

Lefevre G Jr. 1976. A photographic representation and interpretation of the polytene chromosomes of *Drosophila melanogaster* salivary glands. In *The genetics and biology of* Drosophila, Vol. 1a (ed. Ashburner M, Novitski E), pp. 32–64. Academic, New York.

Lewis EB. 1954. The theory and application of a new method of detecting chromosomal rearrangements in *Drosophila melanogaster*. *Am Nat* **88:** 225–239.

Losick VP, Fox DT, Spradling AC. 2013. Polyploidization and cell fusion contribute to wound healing in the adult *Drosophila* epithelium. *Curr Biol* **23:** 2224–2232.

Losick VP, Jun AS, Spradling AC. 2016. Wound-induced polyploidization: Regulation by hippo and JNK signaling and conservation in mammals. *PLoS ONE* **11:** e0151251.

Marygold SJ, Crosby MA, Goodman JL; FlyBase Consortium. 2016. Using FlyBase, a database of *Drosophila* genes and genomes. *Methods Mol Biol* **1478:** 1–31.

Mathog E, Hochstrasser M, Gruenbaum Y, Saumweber H, Sedat J. 1984. Characteristic folding pattern of polytene chromosomes in *Drosophila* salivary gland nuclei. *Nature* **308:** 414–421.

Mellert DJ, Truman JW. 2012. Transvection is common throughout the *Drosophila* genome. *Genetics* **191:** 1129–1141.

Merrikh H, Zhang Y, Grossman AD, Wang J. 2012. Replication-transcription conflicts in bacteria. *Nat Rev Microbiol* **10:** 449–458.

Moshkin YM, Alekseyenko AA, Semeshin VF, Spierer A, Spierer P, Makarevich GF, Balyaeva ES, Zhimulev IF. 2001. The bithorax complex of *Drosophila melanogaster*: Underreplication and morphology in polytene chromosomes. *Proc Natl Acad Sci* **98:** 570–574.

Nagl W. 1978. *Endopolyploidy and polyteny in differentiation and evolution*. Elsevier Science, Amsterdam.

Neiman M, Beaton MJ, Hessen DO, Jeyasingh PD, Weider LJ. 2017. Endopolyploidy as a potential driver of animal ecology and evolution. *Biol Rev* **92:** 234–247.

Nordman J, Orr-Weaver TL. 2012. Regulation of DNA replication during development. *Development* **139:** 455–464.

Nordman J, Li S, Eng T, Macalpine D, Orr-Weaver TL. 2011. Developmental control of the DNA replication and transcription programs. *Genome Res* **21:** 175–181.

Nordman JT, Kozhevnikova EN, Verrijzer CP, Pindyurin AV, Andreyeva EN, Shloma VV, Zhimulev IF, Orr-Weaver TL. 2014. DNA copy-number control through inhibition of replication fork progression. *Cell Rep* **9:** 841–849.

Orr-Weaver TL. 2015. When bigger is better: The role of polyploidy in organogenesis. *Trends Genet* **31:** 307–315.

Pai CY, Lei EP, Ghosh D, Corces VG. 2004. The centrosomal protein CP190 is a component of the gypsy chromatin insulator. *Mol Cell* **16:** 737–748.

Pope BD, Ryba T, Dileep V, Yue F, Wu W, Denas O, Vera DL, Wang Y, Hansen R, Canfield TK, et al. 2014. Topologically associating domains are stable units of replication-timing regulation. *Nature* **515:** 402–405.

Ramirez F, Bhardwaj V, Villaveces J, Arrigoni L, Grüning BA, Lam KC, Habermann B, Akhtar A, Manke T. 2017. High-resolution TADs reveal DNA sequences underlying genome organization in flies. *bioRxiv* doi: http://dx.doi.org/10.1101/115063; see also http://chorogeome.ie-freiburg.mpg.de.

Richards G. 1980. The polytene chromosomes in the fat body nuclei of *Drosophila melanogaster*. *Chromosoma* **79:** 241–250.

Rowley MJ, Corces VG. 2016. The three-dimensional genome: Principles and roles of long distance interactions. *Cell* **162:** 703–705.

Schoenfelder KP, Fox DT. 2015. The expanding implications of polyploidy. *J Cell Biol* **209:** 485–491.

Schoenfelder KP, Montague RA, Paramore SV, Lennox AL, Mahowald AP, Fox DT. 2014. Indispensable pre-mitotic endocycles promote aneuploidy in the *Drosophila* rectum. *Development* **141:** 3551–3560.

Sexton T, Yaffe E, Kenigsberg E, Bantignies F, Leblanc B, Hoichman M, Parrinello H, Tanay A, Cavalli G. 2012. Three-dimensional folding and functional organization principles of the *Drosophila* genome. *Cell* **148:** 458–472.

Sher N, Bell GW, Li S, Nordman J, Eng T, Eaton ML, MacAlpine DM, Orr-Weaver TL. 2012. Developmental control of gene copy number by repression of replication initiation and fork progression. *Genome Res* **22:** 64–75.

Spierer A, Spierer P. 1984. Similar level of polyteny in bands and interbands of *Drosophila* giant chromosomes. *Nature* **307:** 176–178.

Spradling AC. 1993. Position effect variegation and genomic instability. *Cold Spring Harbor Symp Quant Biol* **58:** 585–596.

Stadler MR, Jaines JE, Eisen MB. 2017. Convergence of topological domain boundaries, insulators and polytene interbands revealed by high-resolution mapping of chromatin contacts in the early *Drosophila melanogaster* embryo. *bioRxiv* doi.org/10.1101/149344.

Stormo BM, Fox DT. 2017. Polyteny: Still a giant player in chromosome research. *Chromosome Res* doi: 10.1007/s10577-017-9562-z.

Ullanov SV, Khrameeva EE, Gavirlov AA, Flyamer IM, Kos P, Mikhaleva EA, Penin AA, Logacheva MD, Imakaev MV, Chertovich A, et al. 2016. Active chromatin and transcription play a key role in chromosome partitioning into topologically associating domains. *Genome Res* **26:** 70–84.

Vatolina TY, Boldyreva LV, Demakova OV, Demakov SA, Kokoza EB, Semeshin VF, Babenko VN, Goncharov FP, Balyaeva ES, Zhimulev IF. 2011. Identical functional organization of nonpolytene and polytene chromosomes in *Drosophila melanogaster*. *PLoS ONE* **6:** e25960.

White MJD. 1973. *Animal cytology and evolution*. Cambridge University Press, Cambridge.

Wiegmann BM, Trautwein MD, Winkler IS, Barr NB, Kim JW, Lambkin C, Bertone MA, Cassel BK, Bayless KM, Heimberg AM, et al. 2011. Episodic radiations in the fly tree of life. *Proc Natl Acad Sci* **108:** 5690–5695.

Yarosh W, Spradling A. 2014. Incomplete replication generates somatic DNA alterations within *Drosophila* polytene salivary gland cells. *Genes Dev* **28:** 1840–1855.

Zhimulev IF. 1996. Morphology and structure of polytene chromosomes. *Adv Genet* **34:** 1–490.

Zhimulev IF, Belyaeva ES. 2003. Intercalary heterochromatin and genetic silencing. *BioEssays* **25:** 1040–1051.

Zhimulev IF, Zykova TY, Goncharov FP, Khoroshkoa VA, Demakov OV, Semeshin VF, Pokholkova GV, Boldyreva LV, Demidova DS, Babenko VN, et al. 2014. Genetic organization of interphase chromosome bands and interbands in *Drosophila melanogaster*. *PLoS ONE* **9:** e101631.

Zielke T, Glotov A, Saumweber H. 2016. High-resolution in situ hybridization analysis on the chromosomal interval 61C7-61C8 of *Drosophila melanogaster* reveals interbands as open chromatin domains. *Chromosoma* **125:** 423–435.

Symmetry from Asymmetry or Asymmetry from Symmetry?

Elizabeth W. Kahney,[1,2] Rajesh Ranjan,[1,2] Ryan J. Gleason,[1,2] and Xin Chen[1]

[1]Department of Biology, The Johns Hopkins University, Baltimore, Maryland 21218-2685

Correspondence: xchen32@jhu.edu

The processes of DNA replication and mitosis allow the genetic information of a cell to be copied and transferred reliably to its daughter cells. However, if DNA replication and cell division were always performed in a symmetric manner, the result would be a cluster of tumor cells instead of a multicellular organism. Therefore, gaining a complete understanding of any complex living organism depends on learning how cells become different while faithfully maintaining the same genetic material. It is well recognized that the distinct epigenetic information contained in each cell type defines its unique gene expression program. Nevertheless, how epigenetic information contained in the parental cell is either maintained or changed in the daughter cells remains largely unknown. During the asymmetric cell division (ACD) of *Drosophila* male germline stem cells, our previous work revealed that preexisting histones are selectively retained in the renewed stem cell daughter, whereas newly synthesized histones are enriched in the differentiating daughter cell. We also found that randomized inheritance of preexisting histones versus newly synthesized histones results in both stem cell loss and progenitor germ cell tumor phenotypes, suggesting that programmed histone inheritance is a key epigenetic player for cells to either remember or reset cell fates. Here, we will discuss these findings in the context of current knowledge on DNA replication, polarized mitotic machinery, and ACD for both animal development and tissue homeostasis. We will also speculate on some potential mechanisms underlying asymmetric histone inheritance, which may be used in other biological events to achieve the asymmetric cell fates.

Asymmetric inheritance of cell fate determinants in developing organisms is known to play a major role in cellular differentiation, and it is a fundamental process in generating cellular diversity. Our current understandings of the mechanisms that orchestrate asymmetric cell division (ACD) have been gathered from a wide variety of developmental model organisms, including yeast, flies, worms, and mice, among others. As early as 1905, cell lineage analysis of the ascidian *Styela partita* identified cytoplasmic determinants derived from the egg that segregate to distinct cell lineages responsible for generating five specialized tissue types (Conklin 1905). Despite examples of intrinsic segregation of cell fate determinants, it was not until 1994 that the first determinant, Numb, was molecularly characterized (Rhyu et al. 1994). To date, key determinants of cell fate found to be distributed unequally in ACDs include cell surface receptors, transcription factors, mRNA, DNA, histones, and organelles such as endosomes, centrosomes, and mitochondria (Carmena 2008; Knoblich 2008; Tran et al. 2013; Katajisto et al. 2015). During development, this asymmetry is critical for generating divergent cell fates and progenitor cell self-renewal. Failure of these mechanisms can lead to severe defects in cell proliferation, which manifest as tissue degeneration or tumorigenesis.

The asymmetric inheritance of DNA molecules as a cell fate determinant during ACD has been considered previously. In 1975, John Cairns proposed the "immortal strand" hypothesis, suggesting that the stem cell continually inherits the old DNA strands to minimize accumulation of random DNA replication errors that could change cell fate (Cairns 1975). However, the immortal strand hypothesis has not been widely accepted because of the lack of solid supporting in vivo evidence. Two similar (and more accepted) models, named the "strand-specific imprinting and selective chromatid segregation" (Klar 1994, 2007) and "silent sister chromatid" (Lansdorp 2007) hypotheses suggest epigenetic differences between sister chromatids are required to direct the asymmetric outcomes during ACD.

In this review, we will discuss how the processes of DNA replication, chromosomal segregation, and cell division lead to asymmetric outcomes and how organisms are able to develop, maintain homeostasis, and adapt to a changing environment through these asymmetric processes. We argue that the symmetric outcome of making exact copies of DNA and daughter cells is necessary but not sufficient for the propagation and diversification of life. We then hypothesize that the development and homeostasis of multicellular organisms depend on modified molecular and cellular processes to generate asymmetry from the mechanisms that control the otherwise equal distribution of cellular components into the two daughter cells. We will discuss studies that have reported on asymmetric inheritance of cell fate determinants in diverse organisms with a focus on epigenetic differences between sister chromatids, and we will give examples of nonrandom segregation of sister chromatids.

---

[2]These authors contributed equally to this work.

Published by Cold Spring Harbor Laboratory Press; doi: 10.1101/sqb.2017.82.034272

Cold Spring Harbor Symposia on Quantitative Biology, Volume LXXXII

DNA REPLICATION IS AN ASYMMETRIC PROCESS THAT CAN BE BIASED

The asymmetric outcomes of DNA replication and cell division rely heavily on modifications that lead to heritable changes in gene expression and, hence, cell fate. Such modifications occur without altering the primary sequence of the DNA and are collectively referred to as epigenetics (Jacobs and van Lohuizen 2002; Turner 2002; Ringrose and Paro 2004; Probst et al. 2009). It is possible that DNA replication has a heretofore underappreciated role in establishing distinct epigenomes between sister chromatids that will be inherited by each daughter cell upon cell division. DNA consists of two antiparallel strands containing a deoxyribose sugar–phosphate backbone that supports varying sequences of four bases that pair in a complementary way. Through elegant studies, we know that DNA is synthesized in a semiconservative manner, meaning that each daughter DNA will inherit one template strand and one newly synthesized strand as double-stranded DNA (dsDNA) (Meselson and Stahl 1958).

The components of DNA replication machinery bind to DNA in pairs and initiate DNA replication in a bidirectional manner. Because DNA can only be synthesized in the 5′→3′ direction, the DNA polymerase responsible for creating the new strand is required to read the single-stranded (ss) template in the 3′→5′ direction, beginning from an existing 3′-OH overhang. Interestingly, this creates an inherent asymmetry as to how the new strands are synthesized. One strand, the leading strand, begins with a single RNA primer and can be synthesized continuously as the advancing replication fork exposes more ss template and the template is read in the 3′→5′ direction (Bessman et al. 1956, 1958; Lehman et al. 1958; Meselson and Stahl 1958; Kornberg et al. 1989). However, the other template strand, termed the lagging strand, runs antiparallel to the leading strand and cannot be read by the polymerase in the same direction as the advancing replication fork. Thus, the lagging strand is synthesized in short segments, called Okazaki fragments, in the direction opposite to the advancing replication fork. Each Okazaki fragment begins with an RNA primer and is, in fact, synthesized by DNA polymerases different from those of the leading strand (Sakabe and Okazaki 1966; Okazaki et al. 1968; Balakrishnan and Bambara 2013). Furthermore, the lagging strand must undergo additional processing to remove ss "flaps" left behind. To explain, DNA polymerase δ displaces nucleotides from the previously synthesized DNA polymerase α fragment, and nicks left between fragments must be sealed by DNA ligase (Table 1, Fig. 1; Bambara et al. 1997; Rossi et al. 2008; Cerritelli and Crouch 2009).

Although much is known about replication fork licensing and elongation, it is interesting to consider that the origins of replication have yet to be well defined within most eukaryotes, and that transcription can have a direct effect on the localization of prereplication complexes (Vashee et al. 2003; Cayrou et al. 2011). For each cell type, the transcriptional machinery may affect the density and location of replication origins and the length of the replicons in between them and may bias the replication

Table 1. Leading- versus lagging-strand enriched molecules with their function in brief

Replication component	Strand enrichment	Function
DNA polymerase ε	Leading	Synthesizes the leading strand
MCM2–7 helicase	Leading	Unwinds DNA for replication
Cdc45	Leading	Interacts with Mcm proteins; converts the prereplicative complex to the initiation complex
GINS	Leading	Essential for the interaction of Mcm proteins and Cdc45 during initiation and elongation
MCM10	Leading	Activates the Cdc45–MCM–GINS helicase at DNA replication origins
DNA polymerase α	Lagging	Begins replication by synthesizing an RNA primer and adding approximately 20 DNA nucleotides
DNA polymerase δ	Lagging	Synthesizes the lagging strand
PCNA	Lagging	Ring-shaped clamp that stabilizes DNA polymerases onto DNA
RFC	Lagging	Loads PCNA onto the DNA
RPA	Lagging	Binds ssDNA to prevent secondary structure formation
RNase H	Lagging	Removes any remaining RNA nucleotides
DNA2 and FEN1	Lagging	Remove "flaps" of DNA created by DNA Pol δ advancing into and lifting the previous Okazaki fragment
Ligase	Lagging	Seals nicks in the DNA backbone between segments of newly synthesized DNA

Data from Langston et al. 2014; Yu et al. 2014.

of certain genes to either the leading or the lagging strand. Known fork-blocking proteins could also serve to bias the length and direction of replication forks. Paradoxically, although transcription may affect DNA replication, DNA replication is also likely to affect transcription. That is, transcription machinery is displaced as DNA is unwound and must rebind following fork passage. Now, however, it can only bind one of the two copies of DNA present after replication. Studies have shown that rebinding of the transcription machinery can be biased to either the leading or the lagging strand, depending on the rate of fork progression and the inherent maturation of the two strands after fork passage (Alabert and Groth 2012; Vasseur et al. 2016). It has also been shown that this biased rebinding event can lead to heritable changes in gene expression where one daughter cell "remembers" its transcriptional state and the other daughter cell lags behind, with the need to reestablish its transcriptional state (Ferraro et al. 2016).

HISTONE RECYCLING AFTER DNA REPLICATION COULD BIAS CELL FATE

In addition to the inherent asymmetries between the leading and lagging strands of DNA replication, one

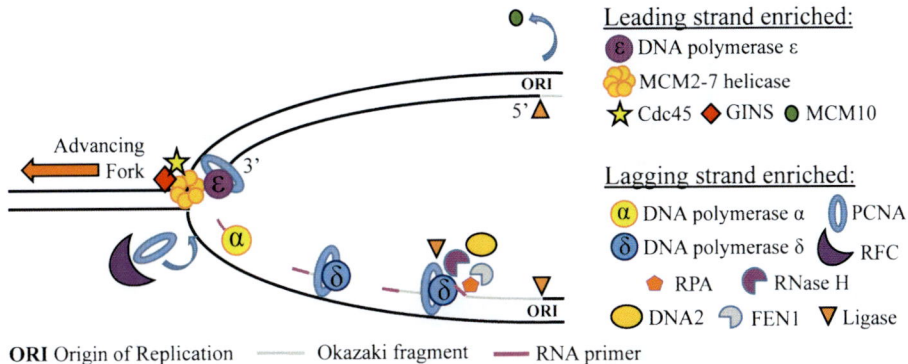

ORI Origin of Replication ——— Okazaki fragment ——— RNA primer

Figure 1. A schematic cartoon of DNA replication fork. DNA replication, although intended to create two equal copies of a double-stranded DNA template, does possess inherent asymmetries relative to strand synthesis. Indeed, several experiments have found functional outcomes that arise from asymmetric leading- versus lagging-strand synthesis. For example, the discontinuous synthesis of the lagging strand has been postulated as a more error-prone process. This could potentially lead to an increased mutation rate that may be evolutionarily beneficial without compromising the genomic stability of the continuously synthesized leading strand (Furusawa 2014). Additionally, it has been shown that molecular lesions created during lagging-strand synthesis contribute to mating type switching in both the budding yeast *Saccharomyces cerevisiae* and the fission yeast *Schizosaccharomyces pombe* (Hanson and Wolfe 2017).

must also consider asymmetries in the epigenetic modification of the DNA itself, as well as the nucleosome, the basic packaging unit of DNA. Methylation of DNA has been well studied and is generally associated with transcriptional repression. Just as with the transcriptional machinery, recovery of DNA methylation appears more slowly on the lagging strand than it does on the leading strand, perhaps allowing time for the two sisters to be differentially recognized or for the methylome on the lagging strand to be rewritten (Stancheva et al. 1999; Tajbakhsh and Gonzalez 2009).

Another major epigenetic information carrier for cell fate is the nucleosome structure, which is comprised of eight histone proteins (two H2A–H2B dimers and one H3–H4 tetramer). Posttranslational modifications of histone proteins have profound effects on cell fate and transcriptional activity (Peterson and Laniel 2004). Of note, nucleosomes must be disassembled ahead of the replication fork and reassembled onto one of the two new dsDNA templates that now exist in the wake of the fork (McKnight and Miller 1977; Sogo et al. 1986). Although the process of new histone deposition onto the DNA has been well studied, how preexisting histones are recycled during DNA replication is less clear (Burgess and Zhang 2013). Elucidating this mechanism is essential to understanding how DNA replication may impact epigenetic information partitioning.

To date, three possible models of histone recycling after fork progression have been proposed. First, the semiconservative model suggests that the H3–H4 tetramer is split into two dimers such that the four dimers of the nucleosome (two H2A–H2B and two H3–H4) are evenly distributed between the two new dsDNA strands. This mechanism was thought to be an elegant solution to evenly distributing epigenetic information such that both daughter strands would inherit equal posttranslational histone modifications, predominantly carried by the H3 and H4 tails (Zhu and Reinberg 2011). However, several lines of evidence have surfaced against the semiconservative

model of histone recycling. For example, it has been found that the H3–H4 tetramer rarely, if ever, splits into two dimers once the tetramer has been assembled (Xu et al. 2010). Furthermore, the tails of H3 and H4 within each tetramer are not symmetrically modified (Chen et al. 2011; van Rossum et al. 2012; Voigt et al. 2012). Thus, even if the tetramer does split, with each new dsDNA inheriting one H3–H4 dimer, then the epigenetic information of the previously unreplicated region would not be preserved equally between the two daughter strands (Fig. 2A). Second, the dispersive model of histone recycling proposes that the H3–H4 tetramer remains intact, but the tetramers and dimers disassembled ahead of the fork are still randomly distributed between the leading and lagging strands behind the fork (Jackson and Chalkley 1981, 1985; Alabert and Groth 2012; Herz et al. 2014; Alabert et al. 2015; Hammond et al. 2017). Additionally, histone modifying enzymes use the posttranslational modifications present on these recycled tetramers to appropriately modify the new H3–H4 tetramers that become incorporated nearby (Fig. 2B; Ayyanathan et al. 2003; Hansen et al. 2008; Margueron et al. 2009; Alabert and Groth 2012; Alabert et al. 2015; Audergon et al. 2015; Ragunathan et al. 2015). Third, the conservative model of histone recycling suggests that preexisting H3–H4 tetramers can be biased to incorporate nonrandomly into either the leading or the lagging strand (Seale 1976; Weintraub 1976; Leffak et al. 1977; Riley and Weintraub 1979; Seidman et al. 1979; Roufa and Marchionni 1982). This mechanism could provide one daughter strand with the same epigenetic information as that in the mother cell, whereas the other daughter strand could predominantly incorporate new, unmarked histones devoid of such epigenetic information (Fig. 2C).

Strong evidence supports both the dispersive model and the conservative model of histone recycling during DNA replication. Of note, these studies have been done in various organisms and cell types, as well as in vitro. It is important to consider that histone recycling may be dif-

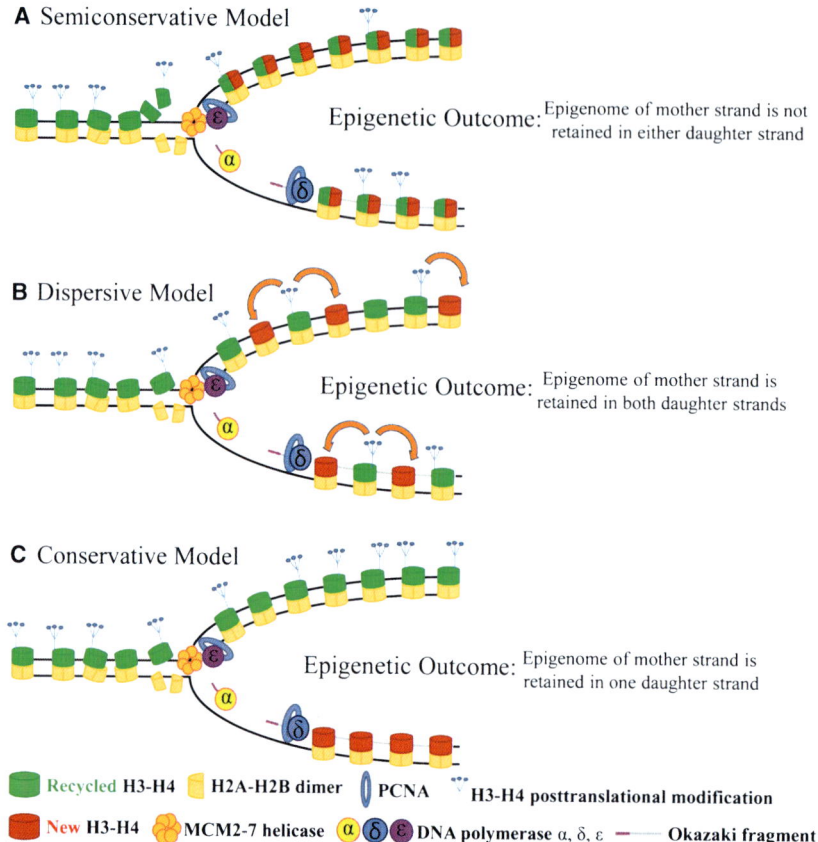

Figure 2. Models of histone recycling and the epigenetic effect of each. (*A*) Semiconservative model. The semiconservative model suggests that the H3–H4 tetramer is split into two dimers that are evenly distributed between the two new double-stranded DNA daughters. Although it is a potentially elegant solution to evenly distributing relevant epigenetic information between the two new daughter strands, it has been shown that the H3–H4 tetramer rarely, if ever, splits and the tails of H3 and H4 within each tetramer are not symmetrically modified (Xu et al. 2010; Chen et al. 2011; van Rossum et al. 2012; Voigt et al. 2012). Thus, the epigenetic information of that previously unreplicated region would not be preserved equally onto the two daughter strands. (*B*) Dispersive model. The dispersive model of histone recycling proposes that the H3–H4 tetramer remains intact, but the tetramers and dimers disassembled ahead of the fork are still randomly distributed onto the leading and lagging strands behind the fork (Jackson and Chalkley 1981, 1985; Alabert and Groth 2012; Herz et al. 2014; Alabert et al. 2015; Hammond et al. 2017). Additionally, histone modifying enzymes use the posttranslational modifications present on these recycled tetramers to appropriately modify the new H3–H4 tetramers that become incorporated nearby and the epigenome of the mother strand is retained in both daughter strands (Ayyanathan et al. 2003; Hansen et al. 2008; Margueron et al. 2009; Alabert and Groth 2012; Alabert et al. 2015; Audergon et al. 2015; Ragunathan et al. 2015). (*C*) Conservative model. The conservative model of histone recycling suggests that preexisting H3–H4 tetramers can be biased to incorporate nonrandomly into either the leading or the lagging strand (Seale 1976; Weintraub 1976; Leffak et al. 1977; Riley and Weintraub 1979; Seidman et al. 1979; Roufa and Marchionni 1982). Such a mechanism could provide one daughter strand with the same epigenetic information as that in the mother cell, whereas the other daughter strand predominantly incorporates new, unmarked histones devoid of such epigenetic information.

ferent during DNA replication depending on the biological context. For example, how stem cells maintain their stemness through many rounds of mitosis has been a long-standing question in the epigenetics field. Our finding that preexisting histones are selectively retained in the renewed stem cell daughter, whereas newly synthesized histones are enriched in the differentiating daughter cell in *Drosophila* male germline stem cells (GSCs) suggests that the predominant mechanism of histone recycling may be the conservative model (Tran et al. 2012; Tran et al. 2013; Xie et al. 2015; Snedeker et al. 2017; Xie et al. 2017). Our finding also indicates that the asymmetric epigenome established during DNA replication needs to be recognized and properly segregated by potentially polar-

ized mitotic machinery. Next, we will discuss how chromatin-bound *cis*-factors and non-chromatin-bound *trans*-regulators coordinate to ensure nonrandom sister chromatid segregation.

THE CENTROMERE: AN EPIGENETIC BASIS TO DISTINGUISH ASYMMETRIC SISTER CHROMATIDS

Centromeres direct chromosome segregation during mitosis, which is mediated by the recruitment of the kinetochore as well as microtubules. Centromeres are epigenetically defined in most eukaryotes by a centromere-

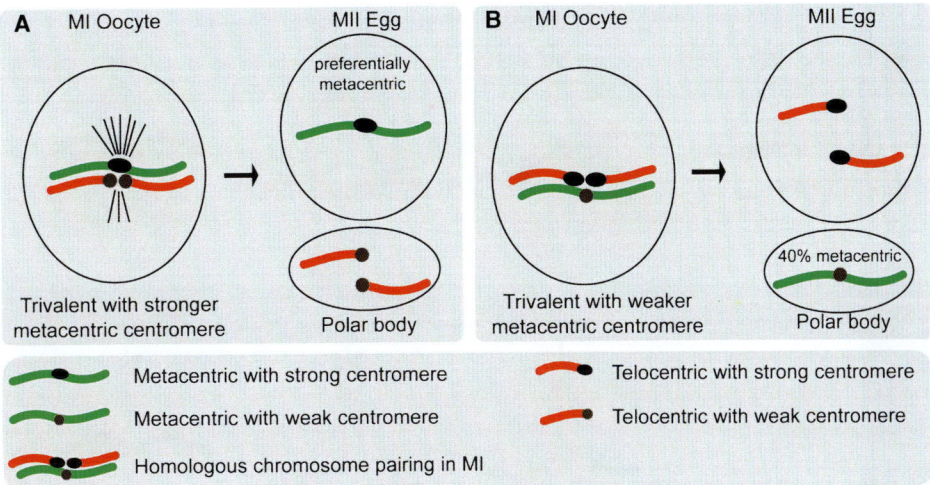

Figure 3. The "centromere drive" hypothesis holds that centromere strength directs chromosome segregation in meiosis I (MI). When two telocentric chromosomes fuse in a natural population to create one metacentric chromosome, (*A*) if telocentric chromosome fusion creates a metacentric chromosome with a stronger centromere, then the metacentric chromosomes preferentially segregate to the egg in MI, or (*B*) if telocentric chromosome fusion creates a metacentric chromosome with a weaker centromere, then 40% of the metacentric chromosomes segregate to the polar body in MI (Chmátal et al. 2014).

specific histone H3 variant known as the centromere identifier in flies and CENP-A in mammals (Palmer et al. 1987; Allshire and Karpen 2008). The centromeric histones have undergone rapid evolution and the length of DNA defined as the centromeric region has greatly increased through a positive selection process termed "centromere drive" (Henikoff et al. 2001; Henikoff and Malik 2002; Malik 2009). An expansion of the centromeric DNA by recombination could create a centromere that has increased microtubule binding ability, which could, in turn, lead to preferential chromosome transmission, such as that found during female meiosis. For example, the mouse karyotype typically consists of $2n = 40$ telocentric chromosomes, but numerous natural populations show dramatically reduced chromosome numbers in which $2n = 22$ chromosomes, a phenomenon attributed to Robertsonian (Rb) fusion, a chromosomal rearrangement that joins two telocentric chromosomes to create one metacentric chromosome (White et al. 2010). Retention of a metacentric chromosome in offspring depends on the direction of chromosome segregation during meiosis I (MI). The direction of chromosome segregation depends on centromere strength; stronger centromeres have more CENP-A protein and outer kinetochore components and, hence, higher microtubule binding ability. Therefore, the stronger centromeres are preferentially retained in the egg, whereas the weaker centromeres are preferentially segregated in the polar body during meiosis (Fig. 3; Chmátal et al. 2014). A similar event has also been reported in the budding yeast *Saccharomyces cerevisiae*, in which the inner and outer kinetochore components show asymmetric segregation in a lineage-specific manner during meiosis (Thorpe et al. 2009).

However, this phenomenon has not been reported during mitosis. Therefore, it would be worth testing a hypothesis similar to "centromere drive" during ACD. Because two distinct daughter cells arise from ACD, it is plausible that epigenetic asymmetry on sister chromatids would in-

clude sister centromeres. This would allow the mitotic machinery to distinguish the sister chromatids.

CHROMATIN ORGANIZATION: A MECHANISM FOR *TRANS*-NUCLEAR MEMBRANE COMMUNICATION

Chromosomes contain euchromatic and heterochromatic domains, which have distinct nuclear functions and organization throughout development (Misteli 2007; Mekhail and Moazed 2010; Rajapakse and Groudine 2011). The most striking examples of chromatin organization are the centromere cluster (Rabl configuration), heterochromatin cluster (chromocenter), and telomere cluster (bouquet configuration) at the nuclear periphery (Funabiki et al. 1993; Jin et al. 1998; Zickler and Kleckner 1998; Scherthan and Schönborn 2001; Fransz et al. 2002; Guenatri et al. 2004; Fang and Spector 2005; Zickler 2006). The centromere cluster and the pericentromeric heterochromatin region could provide a location where specific factors are concentrated to facilitate communication between chromosomes and microtubules. For example, kinetochore proteins and heterochromatin factors, such as HP1 and H3K9me2/3, could concentrate at the centromere or pericentromeric regions (Bernard et al. 2001; Kawashima et al. 2007). Interestingly, mutations of kinetochore components Mis6 and Nuf2 (NDC80 complex) result in centromere declustering (Appelgren et al. 2003; Asakawa et al. 2005). Nevertheless, how kinetochore components are linked to the nuclear envelope and mediate centromere dynamics remains elusive.

Mitotic hallmarks, such as phosphorylation of key histone residues including H3S10P, H3T3/T6P, H3.1/2S28P, and H1.4S26P, are shown to be predominantly associated with old histones at early mitosis in cultured human cell lines (Lin et al. 2016). We wanted to define the mechanism

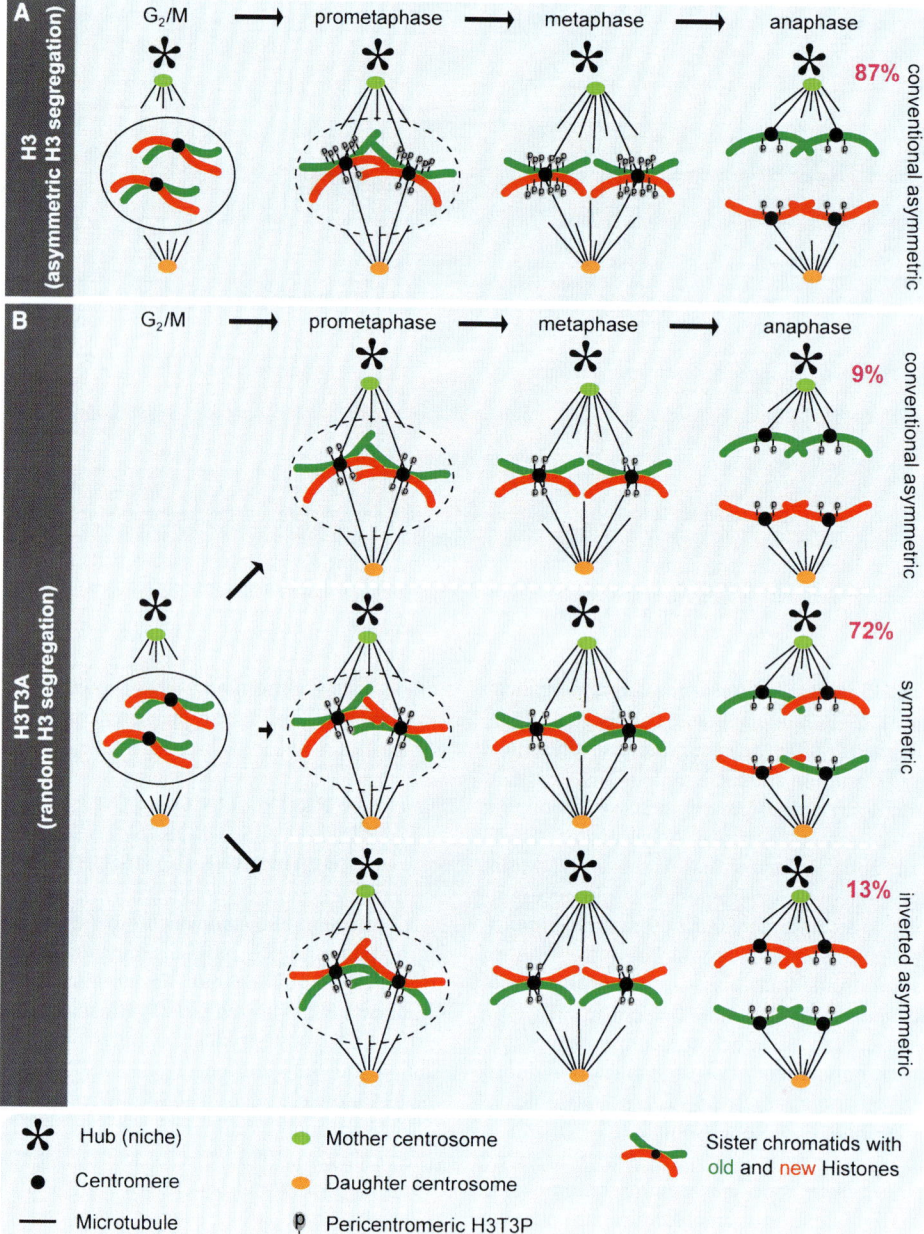

Figure 4. Pericentromeric heterochromatin modifications could regulate centromere and microtubule interactions, as well as nonrandom sister chromatid segregation. (*A*) Sequential phosphorylation of T3 on old histone H3 before new H3 at the pericentromeric region during mitosis in *Drosophila* male GSCs ensures nonrandom sister chromatid segregation. (*B*) Expression of H3T3A, where T3 of histone H3 is mutated to alanine (A), randomizes the segregation pattern of sister chromatids (Xie et al. 2015, 2017).

(s) by which sister chromatids might be recognized and segregated in an asymmetric manner. To begin to address this question, we recently reported that the H3T3P mark at pericentromeric regions distinguishes old and new histones in *Drosophila* male GSCs (Fig. 4A; Xie et al. 2015). Furthermore, misregulation of this phosphorylation leads to randomized inheritance of old and new H3, as well as both GSC loss and progenitor germ cell tumor phenotypes. This suggests that asymmetric phosphorylation of H3T3 at the pericentromeric regions may be one mechanism by which the mitotic machinery can recognize and faithfully segregate asymmetric sister chromatids (Fig. 4B).

CENTROSOMES AND MICROTUBULES: MECHANICAL TOOLS FOR NONRANDOM SISTER CHROMATID SEGREGATION

The centrosome is a complex molecular structure that functions as the major microtubule-organizing center in the cell. Recent studies have revealed intriguing asymmetry between mother and daughter centrosomes, as well as the involvement of such asymmetry in a number of critical cellular processes. Two centrosomes are distinct from each other, partly resulting from their microtubule nucleation activity and their age differences. Interestingly, the older of

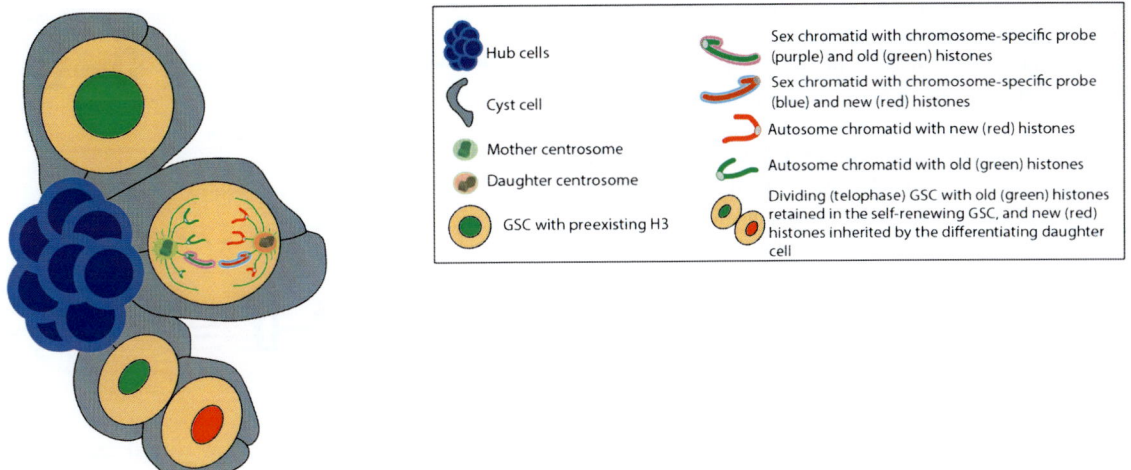

Figure 5. Nonrandom segregation of sister chromatids, asymmetric centrosome inheritance, and asymmetric histone inheritance during *Drosophila* male germline stem cell (GSC) asymmetric cell division (ACD). Asymmetric GSC divisions give rise to two daughter cells: a self-renewed GSC that remains in proximity to the niche (blue hub cells), and a differentiating daughter cell that migrates to the distal side of the cell, resulting from a perpendicular spindle orientation relative to the niche. During this division, mother (green) and daughter (orange) centrosomes are asymmetrically inherited by the self-renewed GSC and the differentiating daughter cell, respectively. Nonrandom sister chromatid segregation of the X/Y sex chromosomes was shown in male GSCs using CO-FISH (chromosome-oriented fluorescence in situ hybridization) in combination with strand-specific probes to distinguish sister chromatids. Sex chromosomes (purple and blue outlined chromatids) show an ~85:15 bias segregation during male GSC cell division (Yadlapalli and Yamashita 2013). A dual-color labeling strategy to distinguish preexisting (green) versus newly synthesized (red) canonical histone H3 revealed that old histone H3 is selectively retained in the self-renewed GSC (green nuclei), whereas newly synthesized H3 is enriched in the differentiating daughter cell (red nuclei) (Tran et al. 2012).

the two centrosomes has been shown to nucleate microtubules considerably earlier, which is correlated with a differential response to signaling molecules (Fig. 5; Rebollo et al. 2007; Rusan and Peifer 2007; Anderson and Stearns 2009; Pelletier and Yamashita 2012). The observation that centrosome age could be associated with differential response to various signaling cues has raised the possibility that the inherent asymmetry of centrosomes could contribute to the determination of distinct cell fates during ACD.

In the fission yeast *Schizosaccharomyces pombe* during meiosis, LINC (linker of nucleoskeleton and cytoskeleton) connects the centrosome with telomeres (Tomita and Cooper 2007), whereas during mitosis LINC connects the centrosome with centromere, rather than telomeres (Fernández-Álvarez et al. 2016). Loss of such contacts during meiosis or mitosis abolishes normal spindle formation (Tomita et al. 2013; Fennell et al. 2015), suggesting that the *trans*-nuclear envelope contacts through LINC play an important role in mediating cross talk between centrosomes and chromosomes, residing in the cytoplasm and the nucleus, respectively.

Studies have revealed that the *Drosophila* male GSCs and mouse neural glial progenitor cells inherit the mother centrosome (Yamashita et al. 2007; Wang et al. 2009), whereas *Drosophila* neuroblasts and female GSCs inherit the daughter centrosome (Conduit and Raff 2010; Januschke et al. 2011). This difference in centrosome inheritance patterns during ACD provokes the speculation that the developmentally programmed centrosome located at the stem cell side, whether mother or daughter, might bear fate determinants or other characteristics that contribute to stem cell fate. Therefore, it is conceivable that early

microtubule nucleation at one of the two centrosomes, either age-dependent or in differential response to signaling molecules at the stem cell side, could engage in cross talk with centromeric chromatin through LINC to ensure preferential sister chromatid attachment.

In summary, a cohort of *trans* factors, such as centrosomes, microtubules, nuclear membrane, and kinetochore complex, and *cis* factors, such as centromeres and epigenetic modifications on chromatin, act together in differential recognition and nonrandom segregation of sister chromatids. In the future, more studies are needed to understand how this axis of asymmetry, centrosome–microtubule–nuclear membrane kinetochore–centromere–chromatid, is regulated and whether disruption of this axis leads to any cellular defects.

ASYMMETRIC INHERITANCE OF CELL FATE DETERMINANTS IN UNICELLULAR ORGANISMS

Over the past three decades, the mating type switching behavior in two eukaryotic yeasts, the budding yeast *S. cerevisiae* and the fission yeast *S. pombe,* has served as an exceptional model to dissect the mechanisms of asymmetric cell fate specification (Klar 2007). Mating type switching is accomplished by two functionally similar but molecularly distinct processes in *S. cerevisiae* and *S. pombe.* The genomes of these species encode a three-cassette gene structure containing one active and two silent copies of the mating type locus. These cells can alter their mating type through a programmed DNA rearrangement process and execute it through the cleavage of the active locus, whereas a copy of a silent locus serves as a donor for synthesis-dependent strand annealing (Dalgaard and Klar 2001). In *S. pombe*, the mating type switching pattern results from the inheritance of a specific parental DNA strand, which is dependent on a strand-specific epigenetic imprint that occurs during lagging-strand DNA synthesis (Dalgaard and Klar 2001). Generation of this imprinting phenomenon is dependent on the orientation of DNA replication at the active mating type locus, *mat1* (Dalgaard and Klar 1999). Specifically, it is thought that one or two ribonucleotides form the imprint and that these RNA residues may have been originally used to prime DNA synthesis on the lagging strand, which may be ligated and not removed by the DNA repair machinery during the first S phase (Vengrova and Dalgaard 2006). This imprint is then maintained until the next S phase, when the leading-strand replication complex is stalled at the imprint locus (Fig. 6; Vengrova and Dalgaard 2004). This stalled fork induces a recombination event between *mat1* and one of the two silent donor cassettes, *mat2P* or *mat2M*, leading to mating type switching. As a result of this mechanism, one of the two daughters of a newly switched cell inherits a switch in mating locus, and one of the four granddaughters has the switched mating type. Together, these findings show that asymmetries inherent in DNA replication can be developmentally regulated to ensure distinct cell fate determination after cell division. Considering that *S. pombe* is a haploid organism, it does not require selective segregation of sister chromatids; instead, the daughter cell inherits the cell fate epigenetic determinant from the chromosome randomly from the parental cell. For such mechanism to function reliably in a diploid organism, selective recognition of the distinct chromatid would be necessary.

Mating type gene switching in *S. cerevisiae* is mediated by an enzyme not found in *S. pombe*, the HO endonuclease, which is responsible for the programmed creation of a site-specific double-strand break at the active mating type locus, *MAT* (Haber 2012). Mating type DNA rearrangements occur exclusively in mother cells, not in the daughter cells or spores. The discovery of the exclusive expression of HO in the mother cells led to another breakthrough in asymmetric inheritance of cell fate determinants (Bobola et al. 1996; Sil and Herskowitz 1996; Long et al. 1997). *S. cerevisiae* achieves mitotic proliferation through a highly polarized ACD, giving rise to a smaller daughter (bud) cell. After each ACD, the cell-fate determinant *asymmetric synthesis of HO* (*Ash1*) mRNA, encoding a HO-specific transcriptional repressor, is asymmetrically inherited by the bud cell (Long et al. 1997). This asymmetric localization is mediated by a ribonucleoprotein complex, which is transported across the actin cytoskeleton to the distal tip where translation of *Ash1* mRNA occurs (Cosma 2004). To date, more than 20 mRNAs have been found to be asymmetrically inherited during *S. cerevisiae* cell division (Shepard et al. 2003; Jambhekar et al. 2005).

Cell polarity and spindle orientation are coordinated before mitosis and mediated by three polarized cytoskeletal systems, including actin, septins, and microtubules (Bi and Park 2012). Orientation of the yeast spindle pole

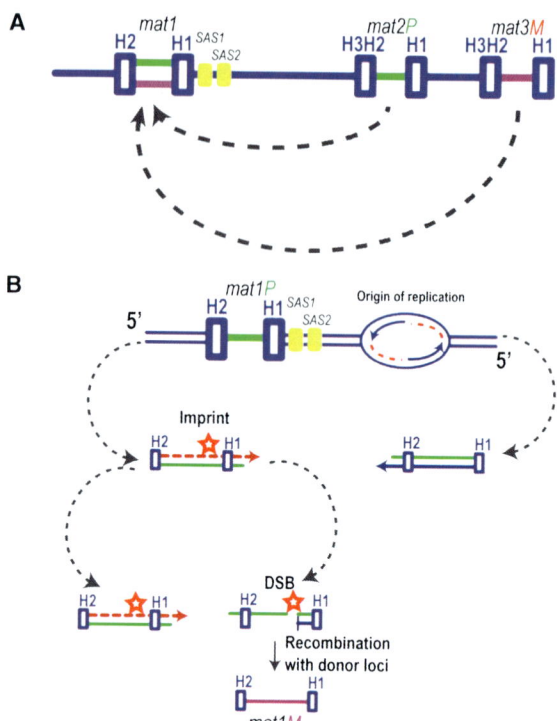

Figure 6. Mating type switching in *Schizosaccharomyces pombe*. (*A*) Arrangement of the mating type region, displaying switchable *mat1* and silent donor loci *mat2P* and *mat3M*. The *mat1* locus contains either plus (green coding) or negative (purple coding) regions derived from the donor cassettes, *mat2P* (plus information) or *mat3M* (minus information). Homology domains H1, H2, and H3 are annotated. Arrows indicate recombination events leading to mating type switching. SAS1 and SAS2 are *cis*-acting sequences involved in imprinting. (*B*) Generation of the imprint is regulated by the direction of *mat1* DNA replication. The imprint (shown as a red star) is installed if the specific strand is the lagging strand (designated by the dashed red line). During the next S phase, when the leading-strand (designated by the blue arrow) replication complex encounters the imprint, a double-strand break (DSB) occurs. The broken strand then uses one of the two the donor loci, either *mat2P* or *mat3M*, as a template to repair the *mat1* locus.

body (SPB), the equivalent of the centrosome, is linked to a stereotypic pattern of SPB inheritance (Pereira and Yamashita 2011). Spindle formation starts in the mother cell body with the older centrosome oriented toward the bud, which establishes spindle polarity, directing orientation of the mitotic spindle along the mother–bud axis and the inheritance of the old SPB by the daughter cell. A similar phenomenon is observed in the ACDs of the *Drosophila* male GSCs, as well as mouse neural glial progenitor cells, where the mother centrosome is preferentially retained near the hub–GSC interface and by the radial glial progenitors that remain in the ventricular zone, respectively (Yamashita et al. 2007; Wang et al. 2009). The molecular mechanisms that govern the establishment of this cell polarity and spindle orientation have been highly conserved throughout evolution (Pereira and Yamashita 2011). This role in establishing polarity, as well as preferential asymmetric inheritance, raises the intriguing possibility that distinct centrosomes may be associated with recognizing and segregating cell fate determinants, such as individual chromatids, which has been previously suggested for adult stem cells, including *Drosophila* male GSCs and mouse skeletal muscle satellite cells (Shinin et al. 2006; Xie et al. 2015).

tones in dividing GSCs, suggest that epigenetic differences that distinguish sister chromatids might coordinate selective chromatid segregation and direct cell fate after mitosis. Together, these findings reveal the presence of asymmetric epigenetic inheritance during cell division, which may maintain GSC cell identity, while concomitantly resetting the chromatin structure in the differentiating daughter cell to ensure proper cell fate specification.

ACD IN *DROSOPHILA*—A MODEL OF NONRANDOM SEGREGATION OF SISTER CHROMATIDS AND ASYMMETRIC EPIGENETIC INHERITANCE

Nonrandom segregation of sister chromatids occurring in the germline was uncovered by using chromosome-oriented fluorescence in situ hybridization (CO-FISH) to resolve individual chromatid inheritance (Yadlapalli and Yamashita 2013). The CO-FISH method revealed that sex chromosomes (X and Y) show an ∼85:15 strand bias during male GSC ACD (e.g., 85% of GSCs inherit the Watson strand [W] and 15% of GSCs inherit the Crick strand [C] at each GSC division). Autosomes including both the second and the third chromosomes, however, display a random segregation pattern (50:50), but show a consistent co-segregation mode (i.e., WW:CC instead of WC:CW) (Fig. 5). Of note, earlier studies using BrdU to test global DNA segregation showed that male GSCs do not follow the immortal strand model (Yadlapalli and Yamashita 2013). These results, combined with our work showing that H3T3P distinguishes old versus new his-

NONRANDOM SEGREGATION OF DNA STRANDS IN MAMMALIAN CELLS AND DISEASE

Adult skeletal muscle in mammals has an extraordinary regenerative capacity after injury. The regenerative precursor cells originate from the satellite muscle cells, a tissue-specific adult stem cell population (Collins et al. 2005). Mononucleated satellite cells are mitotically quiescent and reside in a niche under the basal lamina, or basement membrane, juxtaposed to the muscle fiber. As skeletal muscle stem cells, satellite cells divide asymmetrically to maintain the stem cell population and differentiate, leading to new myofiber formation (Fig. 7; Kuang et al. 2007). Interestingly, the orientation of the cell division within the satellite cell niche determines cell fate (Kuang et al. 2007). Sister cells arising from a satellite cell division are found in either a planar orientation where both cells remain in direct contact with the basal lamina and myofiber or in an apical–basal orientation where one

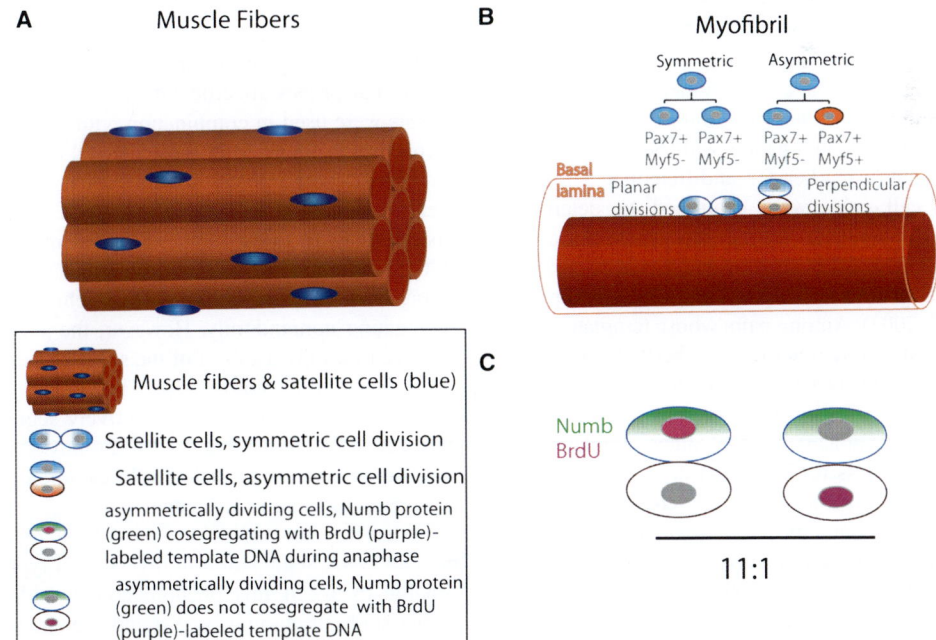

Figure 7. Orientation of muscle satellite cell division determines cell fate. (*A*) Satellite cells (blue) reside under the basal lamina adjacent to the myofibril (red). (*B*) Satellite cells enter either a symmetric cell division or an ACD. ACD (i.e., when the mitotic spindle is oriented perpendicular to the muscle fiber) generates a self-renewing (expressing Pax7+/Myf5−, blue cell) and a differentiating daughter cell (Pax7+/Myf5+, red cell). Symmetric divisions (i.e., when the mitotic spindle is oriented parallel to the muscle fiber) generate two self-renewing cells that are both Pax7+/Myf5−. (*C*) Co-segregation of template DNA strands labeled with BrdU (purple) and asymmetric distribution of Numb (green) to one daughter cell were observed in asymmetrically dividing myoblasts derived from satellite cells during anaphase in an 11:1 bias.

daughter cell is pushed toward the basal lamina and the other is oriented apically toward the myofiber, occurring with 92% and 8% frequency, respectively. Further studies revealed that apical–basal orientation of cell divisions resulted in asymmetric cell fates. The daughter cell attached to the basal lamina remained a stem cell, whereas the daughter cell that loses contact with the basal lamina becomes a committed myogenic cell. In contrast, stem cell divisions in a planar orientation were symmetric and generated identical daughter cells.

Satellite cells express the transcription factor Pax7, but not Myf5 (Pax7+/Myf5−), whereas the asymmetric differentiating daughter cell expresses both Pax7 and Myf5 (Pax7+/Myf5+) (Kuang et al. 2007). Established experimental approaches to track segregation of old versus nascent DNA involve single or consecutive rounds of halogenated nucleotide analog labels. The thymidine analog, 5-bromo-2′deoxyuridine (BrdU), can be used to label newly synthesized DNA strands in freshly isolated satellite cells from single fibers. Three days after labeling, single parental cells generated two daughter cells, one BrdU+ and another BrdU−, indicating that template DNA strands can be co-segregated in adult muscle stem cells at a 7% frequency (Shinin et al. 2006). Furthermore, template DNA strands were found to co-segregate with the asymmetric cell fate determinant Numb (Fig. 7C). The low frequency observed may be an underestimate owing to the ex vivo culture conditions, but it could be higher if tested in an in vivo tissue context or if investigated to specifically determine DNA inheritance in asymmetrically dividing Pax7+/Myf5− cells. Indeed, later experiments observed an increased frequency (38%) of asymmetric template DNA inheritance in Pax7+/Myf5− cells (Kuang et al. 2007).

During myogenic lineage commitment, satellite cells differentiate and express Desmin, a muscle-specific intermediate filament. In combination with Sca-1 (stem cell antigen-1 protein, a marker for undifferentiated muscle progenitors), cell pairs were examined to determine whether specific templates segregated with specific cell fates. Strikingly, 79% of pairs showed Desmin expression only in the daughter inheriting nascent BrdU+ templates (Conboy et al. 2007). Among pairs whose templates were labeled by BrdU as symmetrically inherited, nearly all were symmetric for Desmin expression. Furthermore, 84% of asymmetric Desmin-positive cells showed asymmetry of Sca-1, demonstrating that older templates co-segregate with the less differentiated cells. An independent study using CO-FISH with single-chromatid resolution showed that asymmetric DNA segregation includes all chromosomes. Based on relative Pax7 levels, a population of high Pax7-expressing satellite cells was characterized to perform template strand co-segregation at a higher frequency (Rocheteau et al. 2012). Together, these experiments provide the evidence of template strand co-segregation based on template age, demonstrating that asymmetric co-segregation is associated with cell fate determination. It remains to be elucidated (1) how template strand age is monitored and recognized during cell lineage progression and (2) whether this co-segregation of tem-

plate DNA is linked to gene regulation or silencing of specific loci in the satellite cells.

Cardiac resident stem cells in neonatal and adult mammalian hearts have been identified by distinct membrane markers and transcription factors, including c-kit and Nkx2.5, respectively (Beltrami et al. 2003). These c-kit-positive endogenous cardiac stem cells (eCSC) are self-renewing, are multipotent, and can divide through ACD (Beltrami et al. 2003; Urbanek et al. 2006). Furthermore, these eCSCs have been shown to be necessary and sufficient for myocyte regeneration, leading to anatomical and functional myocardial recovery following myocardial damage (Ellison et al. 2013). The c-kit-positive CSCs were isolated and tested for asymmetric chromatid segregation using the thymidine analogs BrdU and IdU in combination with different pulse-chase time points to detect old versus nascent DNA strands (Kajstura et al. 2012; Sundararaman et al. 2012). From 4% to 7% of c-kit-positive CSCs isolated from myocardial samples displayed asymmetric inheritance of nascent DNA detected during anaphase and telophase in two independent studies (Kajstura et al. 2012; Sundararaman et al. 2012). This range significantly exceeds the probability that a random segregation of chromatids would yield an asymmetrical distribution of labeled nucleotides. Therefore, further characterization is necessary to determine whether a subpopulation of c-kit-positive CSCs exists and, similar to muscle satellite cells, shows increased ACD and nonrandom chromatid segregation. CO-FISH experiments using chromosome-specific probes could address individual chromosome inheritance upon ACD.

Recent examples have also shown chromatid-biased DNA segregation in colon crypt cells (Falconer et al. 2010). To identify sister chromatids, CO-FISH with unidirectional probes specific for centromere and telomere repeats were used in combination with BrdU to label nascent chromosomes. Mice were injected with BrdU hourly for 12 h to label actively dividing cells; colon tissue was then fixed, sectioned, and subjected to CO-FISH probes. Sister nuclei showing reciprocal, asymmetric CO-FISH fluorescence were found throughout the colon crypt, indicating that sister chromatids of most chromosomes were segregating nonrandomly. However, the asymmetry was observed for only a subset of the sister chromatids in any cell pair within the colon crypt. This reflects a possibility that a subset of colon cells selectively segregates sister chromatids from most, but not all, chromosomes. Whether specific chromatids are selectively captured within these cells remains to be investigated.

To date, several studies have provided evidence for nonrandom DNA segregation in diverse cell types. It is reasonable to ask if this phenomenon is widespread. An earlier study of chromosome strand segregation based on site-specific recombination markers in mouse embryonic stem cells revealed a nonrandom distribution of chromosome 7 (Armakolas and Klar 2006). However, two recent studies using CO-FISH indicate that this is not the case and that chromosomes are randomly segregated (Falconer et al. 2012; Sauer et al. 2013). Studies using BrdU to follow the ancestral DNA during the first ACD of the

Caenorhabditis elegans embryo also failed to detect asymmetric segregation of DNA (Ito and McGhee 1987). Furthermore, examples of adult stem cells that do not asymmetrically segregate chromosomes include hair follicle stem cells (Sotiropoulou et al. 2008) and hematopoietic stem cells (Kiel et al. 2007). Together, studies so far indicate that asymmetric segregation of DNA strands occurs in some, but not all, stem cell types.

Many types of adult stem cells undergo ACD to balance self-renewal and differentiation for normal tissue homeostasis. Misregulation of any of the molecular mechanisms that control the asymmetric segregation of cell fate determinants during stem cell divisions may result in hyperproliferation of the stem cell compartment, leading to tumorigenesis, or a loss of the stem cell population, resulting in tissue dystrophy (Knoblich 2010). Previous studies suggest that tumors contain rare cell populations that have stem cell properties, and when injected into immunocompromised mice, they are able to self-renew and generate heterogeneous tumors (Cho and Clarke 2008; Vermeulen et al. 2008; Charafe-Jauffret et al. 2009). These studies indicate that a subpopulation of tumor cells can self-renew and repopulate the heterogeneous tumor, suggesting tumor cell repopulation may occur via ACD within subpopulations of tumor cells. Recent studies in both primary lung cancer cells and cell lines indicate a subset of cells that divide asymmetrically, segregating their template DNA strands exclusively to one daughter cell (Pine et al. 2010). Specifically, double-label experiments using IdU and CldU (chlorodeoxyuridine) in combination with real-time imaging in non–small cell lung cancer (NSCLC) cells showed that old template strand DNA segregated asymmetrically in anaphase and telophase cells. Of the seven NSCLC cell lines examined, asymmetric segregation of template DNA ranging from 0.5% to 6.8% was observed. Furthermore, primary NSCLC tumors displayed an enriched population of cells that asymmetrically segregated template DNA, ranging from 12.5% to 18%, which could reflect an increased concentration of asymmetrically dividing cells within the primary tumor or upon ex vivo expansion. Segregation of the template DNA strands correlated significantly with distinct cell fate markers, including co-segregation with cell fate marker CD113, labeling a tumor subpopulation that could repopulate the entire cell population of lung tumor cells in vitro and in vivo (Eramo et al. 2008; Bertolini et al. 2009). Although these studies have uncovered a significant population of lung cancer primary cells and cell lines able to coordinate asymmetric segregation of template DNA, our understanding of the cell fate choices influenced by this asymmetry is limited and awaits further investigation.

CONCLUSION

One of the greatest discoveries in the 20th century was the double helix structure and semiconservative duplicating process of DNA, providing an elegant and fundamental principle of life. However, as discussed here and reviewed in Snedeker et al. (2017), the inherent asymmetry of DNA replication and the increasing knowledge about the polarity in mitosis raise some questions. Further research will explore whether symmetric outcomes arise from tightly regulated asymmetric molecular and cellular processes, or whether symmetry is the default pathway and is then broken by asymmetric processes.

In reality, both symmetric and asymmetric outcomes are required to build up a multicellular organism originating from a single cell, a fertilized egg, to produce an individual human being made up of hundreds of cell types. Even though most cells in our bodies carry identical DNA sequences, only a subset of these sequences turn on expression at the proper time, in the right place, and with the precise level during development and homeostasis. It is well recognized that the distinct epigenetic information contained in each cell type defines its unique gene expression program. However, how the epigenetic information contained in the parental cell can be maintained, or changed, in the daughter cells remains largely unknown. This question is extremely difficult to address because the epigenome is composed of numerous components that dynamically change their composition. Nonetheless, this question is central to our understanding of the fundamental principles of biology and our ability to develop new treatments against human diseases including birth defects, neurodegenerative disease, tissue dystrophy, infertility, and cancers. Asymmetric histone inheritance could represent the mechanism that maintains equilibrium between the rigidity of genetic information and the plasticity of epigenetic information. We anticipate that future work will address whether this mechanism is used at specific gene loci for differential gene expression upon ACD and whether this mechanism is also applicable to other cell types or in other organisms.

ACKNOWLEDGMENTS

This work was supported by the National Institutes of Health (NIH) grants F32GM119347-02 (R.J.G.), F31 GM122339 (E.W.K.), RO1GM112008, and R21HD084959; the David and Lucile Packard Foundation; Faculty Scholar from Howard Hughes Medical Institute, Bill & Melinda Gates Foundation, and the Simons Foundation and Johns Hopkins University start-up (X.C.).

REFERENCES

Alabert C, Groth A. 2012. Chromatin replication and epigenome maintenance. *Nat Rev Mol Cell Biol* **13:** 153–167.

Alabert C, Barth TK, Reverón-Gómez N, Sidoli S, Schmidt A, Jensen ON, Imhof A, Groth A. 2015. Two distinct modes for propagation of histone PTMs across the cell cycle. *Genes Dev* **29:** 585–590.

Allshire RC, Karpen GH. 2008. Epigenetic regulation of centromeric chromatin: Old dogs, new tricks? *Nat Rev Genet* **9:** 923–937.

Anderson CT, Stearns T. 2009. Centriole age underlies asynchronous primary cilium growth in mammalian cells. *Curr Biol* **19:** 1498–1502.

Appelgren H, Kniola B, Ekwall K. 2003. Distinct centromere domain structures with separate functions demonstrated in live fission yeast cells. *J Cell Sci* **116:** 4035–4042.

Armakolas A, Klar AJ. 2006. Cell type regulates selective segregation of mouse chromosome 7 DNA strands in mitosis. *Science* **311:** 1146–1149.

Asakawa H, Hayashi A, Haraguchi T, Hiraoka Y. 2005. Dissociation of the Nuf2-Ndc80 complex releases centromeres from the spindle-pole body during meiotic prophase in fission yeast. *Mol Biol Cell* **16:** 2325–2338.

Audergon PN, Catania S, Kagansky A, Tong P, Shukla M, Pidoux AL, Allshire RC. 2015. Epigenetics. Restricted epigenetic inheritance of H3K9 methylation. *Science* **348:** 132–135.

Ayyanathan K, Lechner MS, Bell P, Maul GG, Schultz DC, Yamada Y, Tanaka K, Torigoe K, Rauscher FJ. 2003. Regulated recruitment of HP1 to a euchromatic gene induces mitotically heritable, epigenetic gene silencing: A mammalian cell culture model of gene variegation. *Genes Dev* **17:** 1855–1869.

Balakrishnan L, Bambara RA. 2013. Okazaki fragment metabolism. *Cold Spring Harb Perspect Biol* **5:** a010173.

Bambara RA, Murante RS, Henricksen LA. 1997. Enzymes and reactions at the eukaryotic DNA replication fork. *J Biol Chem* **272:** 4647–4650.

Beltrami AP, Barlucchi L, Torella D, Baker M, Limana F, Chimenti S, Kasahara H, Rota M, Musso E, Urbanek K, et al. 2003. Adult cardiac stem cells are multipotent and support myocardial regeneration. *Cell* **114:** 763–776.

Bernard P, Maure JF, Partridge JF, Genier S, Javerzat JP, Allshire RC. 2001. Requirement of heterochromatin for cohesion at centromeres. *Science* **294:** 2539–2542.

Bertolini G, Roz L, Perego P, Tortoreto M, Fontanella E, Gatti L, Pratesi G, Fabbri A, Andriani F, Tinelli S, et al. 2009. Highly tumorigenic lung cancer CD133⁺ cells display stem-like features and are spared by cisplatin treatment. *Proc Natl Acad Sci* **106:** 16281–16286.

Bessman MJ, Kornberg A, Lehman IR, Simms ES. 1956. Enzymic synthesis of deoxyribonucleic acid. *Biochim Biophys Acta* **21:** 197–198.

Bessman MJ, Lehman IR, Simms ES, Kornberg A. 1958. Enzymatic synthesis of deoxyribonucleic acid. II. General properties of the reaction. *J Biol Chem* **233:** 171–177.

Bi E, Park HO. 2012. Cell polarization and cytokinesis in budding yeast. *Genetics* **191:** 347–387.

Bobola N, Jansen RP, Shin TH, Nasmyth K. 1996. Asymmetric accumulation of Ash1p in postanaphase nuclei depends on a myosin and restricts yeast mating-type switching to mother cells. *Cell* **84:** 699–709.

Burgess RJ, Zhang Z. 2013. Histone chaperones in nucleosome assembly and human disease. *Nat Struct Mol Biol* **20:** 14–22.

Cairns J. 1975. Mutation selection and the natural history of cancer. *Nature* **255:** 197–200.

Carmena A. 2008. Signaling networks during development: The case of asymmetric cell division in the *Drosophila* nervous system. *Dev Biol* **321:** 1–17.

Cayrou C, Coulombe P, Vigneron A, Stanojcic S, Ganier O, Peiffer I, Rivals E, Puy A, Laurent-Chabalier S, Desprat R, et al. 2011. Genome-scale analysis of metazoan replication origins reveals their organization in specific but flexible sites defined by conserved features. *Genome Res* **21:** 1438–1449.

Cerritelli SM, Crouch RJ. 2009. Ribonuclease H: The enzymes in eukaryotes. *FEBS J* **276:** 1494–1505.

Charafe-Jauffret E, Ginestier C, Iovino F, Wicinski J, Cervera N, Finetti P, Hur MH, Diebel ME, Monville F, Dutcher J, et al. 2009. Breast cancer cell lines contain functional cancer stem cells with metastatic capacity and a distinct molecular signature. *Cancer Res* **69:** 1302–1313.

Chen X, Xiong J, Xu M, Chen S, Zhu B. 2011. Symmetrical modification within a nucleosome is not required globally for histone lysine methylation. *EMBO Rep* **12:** 244–251.

Chmátal L, Gabriel SI, Mitsainas GP, Martinez-Vargas J, Ventura J, Searle JB, Schultz RM, Lampson MA. 2014. Centromere strength provides the cell biological basis for meiotic drive and karyotype evolution in mice. *Curr Biol* **24:** 2295–2300.

Cho RW, Clarke MF. 2008. Recent advances in cancer stem cells. *Curr Opin Genet Dev* **18:** 48–53.

Collins CA, Olsen I, Zammit PS, Heslop L, Petrie A, Partridge TA, Morgan JE. 2005. Stem cell function, self-renewal, and behavioral heterogeneity of cells from the adult muscle satellite cell niche. *Cell* **122:** 289–301.

Conboy MJ, Karasov AO, Rando TA. 2007. High incidence of non-random template strand segregation and asymmetric fate determination in dividing stem cells and their progeny. *PLoS Biol* **5:** e102.

Conduit PT, Raff JW. 2010. Cnn dynamics drive centrosome size asymmetry to ensure daughter centriole retention in *Drosophila* neuroblasts. *Curr Biol* **20:** 2187–2192.

Conklin E. 1905. The organization and cell lineage of the ascidian egg. *J Acad Nat Sci Phila* **13:** 1–119.

Cosma MP. 2004. Daughter-specific repression of *Saccharomyces cerevisiae* HO: Ash1 is the commander. *EMBO Rep* **5:** 953–957.

Dalgaard JZ, Klar AJ. 1999. Orientation of DNA replication establishes mating-type switching pattern in *S. pombe*. *Nature* **400:** 181–184.

Dalgaard JZ, Klar AJ. 2001. A DNA replication-arrest site *RTS1* regulates imprinting by determining the direction of replication at *mat1* in *S. pombe*. *Genes Dev* **15:** 2060–2068.

Ellison GM, Vicinanza C, Smith AJ, Aquila I, Leone A, Waring CD, Henning BJ, Stirparo GG, Papait R, Scarfo M, et al. 2013. Adult c-kit(pos) cardiac stem cells are necessary and sufficient for functional cardiac regeneration and repair. *Cell* **154:** 827–842.

Eramo A, Lotti F, Sette G, Pilozzi E, Biffoni M, Di Virgilio A, Conticello C, Ruco L, Peschle C, De Maria R. 2008. Identification and expansion of the tumorigenic lung cancer stem cell population. *Cell Death Differ* **15:** 504–514.

Falconer E, Chavez EA, Henderson A, Poon SS, McKinney S, Brown L, Huntsman DG, Lansdorp PM. 2010. Identification of sister chromatids by DNA template strand sequences. *Nature* **463:** 93–97.

Falconer E, Hills M, Naumann U, Poon SS, Chavez EA, Sanders AD, Zhao Y, Hirst M, Lansdorp PM. 2012. DNA template strand sequencing of single-cells maps genomic rearrangements at high resolution. *Nat Methods* **9:** 1107–1112.

Fang Y, Spector DL. 2005. Centromere positioning and dynamics in living *Arabidopsis* plants. *Mol Biol Cell* **16:** 5710–5718.

Fennell A, Fernández-Álvarez A, Tomita K, Cooper JP. 2015. Telomeres and centromeres have interchangeable roles in promoting meiotic spindle formation. *J Cell Biol* **208:** 415–428.

Fernández-Álvarez A, Bez C, O'Toole ET, Morphew M, Cooper JP. 2016. Mitotic nuclear envelope breakdown and spindle nucleation are controlled by interphase contacts between centromeres and the nuclear envelope. *Dev Cell* **39:** 544–559.

Ferraro T, Esposito E, Mancini L, Ng S, Lucas T, Coppey M, Dostatni N, Walczak AM, Levine M, Lagha M. 2016. Transcriptional memory in the *Drosophila* embryo. *Curr Biol* **26:** 212–218.

Fransz P, De Jong JH, Lysak M, Castiglione MR, Schubert I. 2002. Interphase chromosomes in Arabidopsis are organized as well defined chromocenters from which euchromatin loops emanate. *Proc Natl Acad Sci* **99:** 14584–14589.

Funabiki H, Hagan I, Uzawa S, Yanagida M. 1993. Cell cycle-dependent specific positioning and clustering of centromeres and telomeres in fission yeast. *J Cell Biol* **121:** 961–976.

Furusawa M. 2014. The disparity mutagenesis model predicts rescue of living things from catastrophic errors. *Front Genet* **5:** 421.

Guenatri M, Bailly D, Maison C, Almouzni G. 2004. Mouse centric and pericentric satellite repeats form distinct functional heterochromatin. *J Cell Biol* **166:** 493–505.

Haber JE. 2012. Mating-type genes and MAT switching in *Saccharomyces cerevisiae*. *Genetics* **191:** 33–64.

Hammond CM, Strømme CB, Huang H, Patel DJ, Groth A. 2017. Histone chaperone networks shaping chromatin function. *Nat Rev Mol Cell Biol* **18:** 141–158.

Hansen KH, Bracken AP, Pasini D, Dietrich N, Gehani SS, Monrad A, Rappsilber J, Lerdrup M, Helin K. 2008. A model for transmission of the H3K27me3 epigenetic mark. *Nat Cell Biol* **10:** 1291–1300.

Hanson SJ, Wolfe KH. 2017. An evolutionary perspective on yeast mating-type switching. *Genetics* **206:** 9–32.

Henikoff S, Malik HS. 2002. Centromeres: Selfish drivers. *Nature* **417:** 227.

Henikoff S, Ahmad K, Malik HS. 2001. The centromere paradox: Stable inheritance with rapidly evolving DNA. *Science* **293:** 1098–1102.

Herz HM, Morgan M, Gao X, Jackson J, Rickels R, Swanson SK, Florens L, Washburn MP, Eissenberg JC, Shilatifard A. 2014. Histone H3 lysine-to-methionine mutants as a paradigm to study chromatin signaling. *Science* **345:** 1065–1070.

Ito K, McGhee JD. 1987. Parental DNA strands segregate randomly during embryonic development of *Caenorhabditis elegans*. *Cell* **49:** 329–336.

Jackson V, Chalkley R. 1981. A new method for the isolation of replicative chromatin: Selective deposition of histone on both new and old DNA. *Cell* **23:** 121–134.

Jackson V, Chalkley R. 1985. Histone segregation on replicating chromatin. *Biochemistry* **24:** 6930–6938.

Jacobs JJ, van Lohuizen M. 2002. Polycomb repression: From cellular memory to cellular proliferation and cancer. *Biochim Biophys Acta* **1602:** 151–161.

Jambhekar A, McDermott K, Sorber K, Shepard KA, Vale RD, Takizawa PA, DeRisi JL. 2005. Unbiased selection of localization elements reveals *cis*-acting determinants of mRNA bud localization in *Saccharomyces cerevisiae*. *Proc Natl Acad Sci* **102:** 18005–18010.

Januschke J, Llamazares S, Reina J, Gonzalez C. 2011. *Drosophila* neuroblasts retain the daughter centrosome. *Nat Commun* **2:** 243.

Jin Q, Trelles-Sticken E, Scherthan H, Loidl J. 1998. Yeast nuclei display prominent centromere clustering that is reduced in nondividing cells and in meiotic prophase. *J Cell Biol* **141:** 21–29.

Kajstura J, Bai Y, Cappetta D, Kim J, Arranto C, Sanada F, D'Amario D, Matsuda A, Bardelli S, Ferreira-Martins J, et al. 2012. Tracking chromatid segregation to identify human cardiac stem cells that regenerate extensively the infarcted myocardium. *Circ Res* **111:** 894–906.

Katajisto P, Dohla J, Chaffer CL, Pentinmikko N, Marjanovic N, Iqbal S, Zoncu R, Chen W, Weinberg RA, Sabatini DM. 2015. Stem cells. Asymmetric apportioning of aged mitochondria between daughter cells is required for stemness. *Science* **348:** 340–343.

Kawashima SA, Tsukahara T, Langegger M, Hauf S, Kitajima TS, Watanabe Y. 2007. Shugoshin enables tension-generating attachment of kinetochores by loading Aurora to centromeres. *Genes Dev* **21:** 420–435.

Kiel MJ, He S, Ashkenazi R, Gentry SN, Teta M, Kushner JA, Jackson TL, Morrison SJ. 2007. Haematopoietic stem cells do not asymmetrically segregate chromosomes or retain BrdU. *Nature* **449:** 238–242.

Klar AJ. 1994. A model for specification of the left–right axis in vertebrates. *Trends Genet* **10:** 392–396.

Klar AJ. 2007. Lessons learned from studies of fission yeast mating-type switching and silencing. *Annu Rev Genet* **41:** 213–236.

Knoblich JA. 2008. Mechanisms of asymmetric stem cell division. *Cell* **132:** 583–597.

Knoblich JA. 2010. Asymmetric cell division: Recent developments and their implications for tumour biology. *Nat Rev Mol Cell Biol* **11:** 849–860.

Kornberg A, Lehman IR, Bessman MJ, Simms ES. 1989. Enzymic synthesis of deoxyribonucleic acid. 1956. *Biochim Biophys Acta* **1000:** 57–58.

Kuang S, Kuroda K, Le Grand F, Rudnicki MA. 2007. Asymmetric self-renewal and commitment of satellite stem cells in muscle. *Cell* **129:** 999–1010.

Langston LD, Zhang D, Yurieva O, Georgescu RE, Finkelstein J, Yao NY, Indiani C, O'Donnell ME. 2014. CMG helicase and DNA polymerase ε form a functional 15-subunit holoenzyme for eukaryotic leading-strand DNA replication. *Proc Natl Acad Sci* **111:** 15390–15395.

Lansdorp PM. 2007. Immortal strands? Give me a break. *Cell* **129:** 1244–1247.

Leffak IM, Grainger R, Weintraub H. 1977. Conservative assembly and segregation of nucleosomal histones. *Cell* **12:** 837–845.

Lehman IR, Bessman MJ, Simms ES, Kornberg A. 1958. Enzymatic synthesis of deoxyribonucleic acid. I. Preparation of substrates and partial purification of an enzyme from *Escherichia coli*. *J Biol Chem* **233:** 163–170.

Lin S, Yuan ZF, Han Y, Marchione DM, Garcia BA. 2016. Preferential phosphorylation on old histones during early mitosis in human cells. *J Biol Chem* **291:** 15342–15357.

Long RM, Singer RH, Meng X, Gonzalez I, Nasmyth K, Jansen RP. 1997. Mating type switching in yeast controlled by asymmetric localization of ASH1 mRNA. *Science* **277:** 383–387.

Malik HS. 2009. The centromere-drive hypothesis: A simple basis for centromere complexity. *Prog Mol Subcell Biol* **48:** 33–52.

Margueron R, Justin N, Ohno K, Sharpe ML, Son J, Drury WJ III, Voigt P, Martin SR, Taylor WR, De Marco V, et al. 2009. Role of the polycomb protein EED in the propagation of repressive histone marks. *Nature* **461:** 762–767.

McKnight SL, Miller OL Jr. 1977. Electron microscopic analysis of chromatin replication in the cellular blastoderm *Drosophila melanogaster* embryo. *Cell* **12:** 795–804.

Mekhail K, Moazed D. 2010. The nuclear envelope in genome organization, expression and stability. *Nat Rev Mol Cell Biol* **11:** 317–328.

Meselson M, Stahl FW. 1958. The replication of DNA in *Escherichia coli*. *Proc Natl Acad Sci* **44:** 671–682.

Misteli T. 2007. Beyond the sequence: Cellular organization of genome function. *Cell* **128:** 787–800.

Okazaki R, Okazaki T, Sakabe K, Sugimoto K, Sugino A. 1968. Mechanism of DNA chain growth. I. Possible discontinuity and unusual secondary structure of newly synthesized chains. *Proc Natl Acad Sci* **59:** 598–605.

Palmer DK, O'Day K, Wener MH, Andrews BS, Margolis RL. 1987. A 17-kD centromere protein (CENP-A) copurifies with nucleosome core particles and with histones. *J Cell Biol* **104:** 805–815.

Pelletier L, Yamashita YM. 2012. Centrosome asymmetry and inheritance during animal development. *Curr Opin Cell Biol* **24:** 541–546.

Pereira G, Yamashita YM. 2011. Fly meets yeast: Checking the correct orientation of cell division. *Trends Cell Biol* **21:** 526–533.

Peterson CL, Laniel MA. 2004. Histones and histone modifications. *Curr Biol* **14:** R546–R551.

Pine SR, Ryan BM, Varticovski L, Robles AI, Harris CC. 2010. Microenvironmental modulation of asymmetric cell division in human lung cancer cells. *Proc Natl Acad Sci* **107:** 2195–2200.

Probst AV, Dunleavy E, Almouzni G. 2009. Epigenetic inheritance during the cell cycle. *Nat Rev Mol Cell Biol* **10:** 192–206.

Ragunathan K, Jih G, Moazed D. 2015. Epigenetics. Epigenetic inheritance uncoupled from sequence-specific recruitment. *Science* **348:** 1258699.

Rajapakse I, Groudine M. 2011. On emerging nuclear order. *J Cell Biol* **192:** 711–721.

Rebollo E, Sampaio P, Januschke J, Llamazares S, Varmark H, Gonzalez C. 2007. Functionally unequal centrosomes drive spindle orientation in asymmetrically dividing *Drosophila* neural stem cells. *Dev Cell* **12:** 467–474.

Rhyu MS, Jan LY, Jan YN. 1994. Asymmetric distribution of numb protein during division of the sensory organ precursor cell confers distinct fates to daughter cells. *Cell* **76:** 477–491.

Riley D, Weintraub H. 1979. Conservative segregation of parental histones during replication in the presence of cycloheximide. *Proc Natl Acad Sci* **76:** 328–332.

Ringrose L, Paro R. 2004. Epigenetic regulation of cellular memory by the Polycomb and Trithorax group proteins. *Annu Rev Genet* **38:** 413–443.

Rocheteau P, Gayraud-Morel B, Siegl-Cachedenier I, Blasco MA, Tajbakhsh S. 2012. A subpopulation of adult skeletal muscle stem cells retains all template DNA strands after cell division. *Cell* **148:** 112–125.

Rossi ML, Pike JE, Wang W, Burgers PM, Campbell JL, Bambara RA. 2008. Pif1 helicase directs eukaryotic Okazaki fragments toward the two-nuclease cleavage pathway for primer removal. *J Biol Chem* **283:** 27483–27493.

Roufa DJ, Marchionni MA. 1982. Nucleosome segregation at a defined mammalian chromosomal site. *Proc Natl Acad Sci* **79:** 1810–1814.

Rusan NM, Peifer M. 2007. A role for a novel centrosome cycle in asymmetric cell division. *J Cell Biol* **177:** 13–20.

Sakabe K, Okazaki R. 1966. A unique property of the replicating region of chromosomal DNA. *Biochim Biophys Acta* **129:** 651–654.

Sauer S, Burkett SS, Lewandoski M, Klar AJ. 2013. A CO-FISH assay to assess sister chromatid segregation patterns in mitosis of mouse embryonic stem cells. *Chromosome Res* **21:** 311–328.

Scherthan H, Schönborn I. 2001. Asynchronous chromosome pairing in male meiosis of the rat (*Rattus norvegicus*). *Chromosome Res* **9:** 273–282.

Seale RL. 1976. Studies on the mode of segregation of histone nu bodies during replication in HeLa cells. *Cell* **9:** 423–429.

Seidman MM, Levine AJ, Weintraub H. 1979. The asymmetric segregation of parental nucleosomes during chromosome replication. *Cell* **18:** 439–449.

Shepard KA, Gerber AP, Jambhekar A, Takizawa PA, Brown PO, Herschlag D, DeRisi JL, Vale RD. 2003. Widespread cytoplasmic mRNA transport in yeast: Identification of 22 bud-localized transcripts using DNA microarray analysis. *Proc Natl Acad Sci* **100:** 11429–11434.

Shinin V, Gayraud-Morel B, Gomès D, Tajbakhsh S. 2006. Asymmetric division and cosegregation of template DNA strands in adult muscle satellite cells. *Nat Cell Biol* **8:** 677–687.

Sil A, Herskowitz I. 1996. Identification of asymmetrically localized determinant, Ash1p, required for lineage-specific transcription of the yeast HO gene. *Cell* **84:** 711–722.

Snedeker J, Wooten M, Chen X. 2017. The inherent asymmetry of DNA replication. *Annu Rev Cell Dev Biol* **33:** 291–318.

Sogo JM, Stahl H, Koller T, Knippers R. 1986. Structure of replicating simian virus 40 minichromosomes. The replication fork, core histone segregation and terminal structures. *J Mol Biol* **189:** 189–204.

Sotiropoulou PA, Candi A, Blanpain C. 2008. The majority of multipotent epidermal stem cells do not protect their genome by asymmetrical chromosome segregation. *Stem Cells* **26:** 2964–2973.

Stancheva I, Koller T, Sogo JM. 1999. Asymmetry of Dam remethylation on the leading and lagging arms of plasmid replicative intermediates. *EMBO J* **18:** 6542–6551.

Sundararaman B, Avitabile D, Konstandin MH, Cottage CT, Gude N, Sussman MA. 2012. Asymmetric chromatid segregation in cardiac progenitor cells is enhanced by Pim-1 kinase. *Circ Res* **110:** 1169–1173.

Tajbakhsh S, Gonzalez C. 2009. Biased segregation of DNA and centrosomes: Moving together or drifting apart? *Nat Rev Mol Cell Biol* **10:** 804–810.

Thorpe PH, Bruno J, Rothstein R. 2009. Kinetochore asymmetry defines a single yeast lineage. *Proc Natl Acad Sci* **106:** 6673–6678.

Tomita K, Cooper JP. 2007. The telomere bouquet controls the meiotic spindle. *Cell* **130:** 113–126.

Tomita K, Bez C, Fennell A, Cooper JP. 2013. A single internal telomere tract ensures meiotic spindle formation. *EMBO Rep* **14:** 252–260.

Tran V, Lim C, Xie J, Chen X. 2012. Asymmetric division of *Drosophila* male germline stem cell shows asymmetric histone distribution. *Science* **338:** 679–682.

Tran V, Feng L, Chen X. 2013. Asymmetric distribution of histones during *Drosophila* male germline stem cell asymmetric divisions. *Chromosome Res* **21:** 255–269.

Turner BM. 2002. Cellular memory and the histone code. *Cell* **111:** 285–291.

Urbanek K, Cesselli D, Rota M, Nascimbene A, De Angelis A, Hosoda T, Bearzi C, Boni A, Bolli R, Kajstura J, et al. 2006. Stem cell niches in the adult mouse heart. *Proc Natl Acad Sci* **103:** 9226–9231.

van Rossum B, Fischle W, Selenko P. 2012. Asymmetrically modified nucleosomes expand the histone code. *Nat Struct Mol Biol* **19:** 1064–1066.

Vashee S, Cvetic C, Lu W, Simancek P, Kelly TJ, Walter JC. 2003. Sequence-independent DNA binding and replication initiation by the human origin recognition complex. *Genes Dev* **17:** 1894–1908.

Vasseur P, Tonazzini S, Ziane R, Camasses A, Rando OJ, Radman-Livaja M. 2016. Dynamics of nucleosome positioning maturation following genomic replication. *Cell Rep* **16:** 2651–2665.

Vengrova S, Dalgaard JZ. 2004. RNase-sensitive DNA modification(s) initiates *S. pombe* mating-type switching. *Genes Dev* **18:** 794–804.

Vengrova S, Dalgaard JZ. 2006. The wild-type *Schizosaccharomyces pombe* mat1 imprint consists of two ribonucleotides. *EMBO Rep* **7:** 59–65.

Vermeulen L, Sprick MR, Kemper K, Stassi G, Medema JP. 2008. Cancer stem cells—Old concepts, new insights. *Cell Death Differ* **15:** 947–958.

Voigt P, LeRoy G, Drury WJ III, Zee BM, Son J, Beck DB, Young NL, Garcia BA, Reinberg D. 2012. Asymmetrically modified nucleosomes. *Cell* **151:** 181–193.

Wang X, Tsai J-W, Imai JH, Lian W-N, Vallee RB, Shi S-H. 2009. Asymmetric centrosome inheritance maintains neural progenitors in the neocortex. *Nature* **461:** 947–955.

Weintraub H. 1976. Cooperative alignment of nu bodies during chromosome replication in the presence of cycloheximide. *Cell* **9:** 419–422.

White TA, Bordewich M, Searle JB. 2010. A network approach to study karyotypic evolution: The chromosomal races of the common shrew (*Sorex araneus*) and house mouse (*Mus musculus*) as model systems. *Syst Biol* **59:** 262–276.

Xie J, Wooten M, Tran V, Chen B-C, Pozmanter C, Simbolon C, Betzig E, Chen X. 2015. Histone H3 threonine phosphorylation regulates asymmetric histone inheritance in the *Drosophila* male germline. *Cell* **163:** 920–933.

Xie J, Wooten M, Tran V, Chen X. 2017. Breaking symmetry— Asymmetric histone inheritance in stem cells. *Trends Cell Biol* **27:** 527–540.

Xu M, Long C, Chen X, Huang C, Chen S, Zhu B. 2010. Partitioning of histone H3-H4 tetramers during DNA replication-dependent chromatin assembly. *Science* **328:** 94–98.

Yadlapalli S, Yamashita YM. 2013. Chromosome-specific nonrandom sister chromatid segregation during stem-cell division. *Nature* **498:** 251–254.

Yamashita YM, Mahowald AP, Perlin JR, Fuller MT. 2007. Asymmetric inheritance of mother versus daughter centrosome in stem cell division. *Science* **315:** 518–521.

Yu C, Gan H, Han J, Zhou ZX, Jia S, Chabes A, Farrugia G, Ordog T, Zhang Z. 2014. Strand-specific analysis shows protein binding at replication forks and PCNA unloading from lagging strands when forks stall. *Mol Cell* **56:** 551–563.

Zhu B, Reinberg D. 2011. Epigenetic inheritance: Uncontested? *Cell Res* **21:** 435–441.

Zickler D. 2006. From early homologue recognition to synaptonemal complex formation. *Chromosoma* **115:** 158–174.

Zickler D, Kleckner N. 1998. The leptotene-zygotene transition of meiosis. *Annu Rev Genet* **32:** 619–697.

Function of Junk: Pericentromeric Satellite DNA in Chromosome Maintenance

Madhav Jagannathan[1] and Yukiko M. Yamashita[1,2,3]

[1]Life Sciences Institute, University of Michigan, Ann Arbor, Michigan 48109
[2]Department of Cell and Developmental Biology, University of Michigan, Ann Arbor, Michigan 48109
[3]Howard Hughes Medical Institute, University of Michigan, Ann Arbor, Michigan 48109
Correspondence: yukikomy@umich.edu

Satellite DNAs are simple tandem repeats that exist at centromeric and pericentromeric regions on eukaryotic chromosomes. Unlike the centromeric satellite DNA that comprises the vast majority of natural centromeres, function(s) for the much more abundant pericentromeric satellite repeats are poorly understood. In fact, the lack of coding potential allied with rapid divergence of repeat sequences across eukaryotes has led to their dismissal as "junk DNA" or "selfish parasites." Although implicated in various biological processes, a conserved function for pericentromeric satellite DNA remains unidentified. We have addressed the role of satellite DNA through studying chromocenters, a cytological aggregation of pericentromeric satellite DNA from multiple chromosomes into DNA-dense nuclear foci. We have shown that multivalent satellite DNA-binding proteins cross-link pericentromeric satellite DNA on chromosomes into chromocenters. Disruption of chromocenters results in the formation of micronuclei, which arise by budding off the nucleus during interphase. We propose a model that satellite DNAs are critical chromosome elements that are recognized by satellite DNA-binding proteins and incorporated into chromocenters. We suggest that chromocenters function to preserve the entire chromosomal complement in a single nucleus, a fundamental and unquestioned feature of eukaryotic genomes. We speculate that the rapid divergence of satellite DNA sequences between closely related species results in discordant chromocenter function and may underlie speciation and hybrid incompatibility.

HISTORY OF SATELLITE DNA RESEARCH

The term "heterochromatin" was first introduced by Emil Heitz (Heitz 1928) to describe regions of the genome that remained condensed even during interphase. A major constituent of heterochromatin is "satellite DNA," AT-rich tandem repeats, which are found in centromeric and pericentromeric regions of most of eukaryotic chromosomes. They were identified in early cesium chloride density centrifugation experiments as "satellite" bands that sedimented at different densities compared to the rest of genomic DNA because of skewed AT/GC contents (Kit 1961; Sueoka 1961; Sybalski 1968).

Despite the fact that the majority of eukaryotic centromeres are comprised of satellite DNA (Wong and Rattner 1988; Joseph et al. 1989; Willard 1990; Sun et al. 1997, 2003; Schueler et al. 2001), the lack of conservation in centromere repeat sequences, the "centromere paradox" (Henikoff et al. 2001), allied with the identification of neocentromeres lacking repeat sequences (Voullaire et al. 1993; du Sart et al. 1997; Barry et al. 1999) has led to the prevailing model that centromeres are defined in an epigenetic manner (Karpen and Allshire 1997; Allshire and Karpen 2008; Fukagawa and Earnshaw 2014). Nonetheless, satellite repeats underlying centromeres are speculated to have certain functionality to support centromeric function (Rosin and Mellone 2017).

Pericentromeric satellite DNA far surpasses centromeric satellite DNA in abundance, constituting up to 50% of genomes in certain cases (Garrido-Ramos 2017). Despite its abundance, the function of pericentromeric satellite DNA remains poorly understood. Unlike centromeric satellite DNA, whose role in chromosome segregation is indicated in kinetochore function, pericentromeric satellite DNA has been often dismissed as "junk DNA" (Ohno 1972), "selfish parasitic DNA" (Orgel and Crick 1980), or "fossils of centromeric evolution" (Malik 2009). However, many species contain substantial amounts of pericentromeric satellite DNA in their genome, whose maintenance poses a significant burden on the cell's resources. This has led researchers to speculate that pericentromeric satellite DNA may serve critical, yet unidentified roles that justify their large burden on the cell.

Cytologically, pericentromeric satellite DNA is organized into chromocenters within eukaryotic nuclei (Mayer et al. 2005). Initially, chromocenters were identified in plant cells as strongly stained foci when treated with nucleic acid dyes (Baccarini 1908). The composition of these foci remained unknown until the invention of in situ hybridization, which showed that chromocenters are dense aggregations of pericentromeric satellite DNA from multiple heterologous chromosomes (Jones 1970; Pardue and Gall 1970). Chromocenters show DNA methylation and histone methylation on H3K9/H4K20, epigenetic modifications associated with chromatin compaction and tran-

Published by Cold Spring Harbor Laboratory Press; doi: 10.1101/sqb.2017.82.034504
Cold Spring Harbor Symposia on Quantitative Biology, Volume LXXXII

scriptional repression (Saksouk et al. 2015; Nishibuchi and Déjardin 2017). Much like pericentromeric satellite DNA, the role of chromocenters as a cytological structure has also remained obscure.

PERICENTROMERIC HETEROCHROMATIN/ SATELLITE DNA: EXIST TO BE REPRESSED?

Both centromeric and pericentromeric satellite DNA are heterochromatinized via similar mechanisms involving epigenetic modifications, RNAi machinery, and heterochromatin-associated proteins (Allshire and Madhani 2017; Nishibuchi and Déjardin 2017 and references therein). In brief, heterochromatin is characterized by DNA methylation and histone modifications associated with transcriptional silencing such as histone H3K9 methylation. H3K9 trimethylation, which is catalyzed by a histone methyl transferase (Su(var)3-9 in *Drosophila*, Suv39 and SETDB1 in mammals), is recognized by heterochromatin protein 1 (HP1), a hallmark of constitutive heterochromatin. Recruitment of heterochromatin proteins to the underlying satellite DNA requires transcription via a few mechanisms (Hall et al. 2012; Biscotti et al. 2015; Saksouk et al. 2015). First, transcription of satellite DNA leads to Dicer-dependent siRNA production, which is required for the establishment and maintenance of heterochromatin state (Bühler and Moazed 2007; Moazed 2011). Second, recent studies showed that noncoding satellite DNA transcripts recruit Suv39 enzymes to centromeric and pericentromeric sequences, leading to heterochromatinization (Johnson et al. 2017; Shirai et al. 2017; Velazquez Camacho et al. 2017). Interestingly, satellite DNA transcription has also been implicated in the formation of chromocenters in early mouse embryos (Probst et al. 2010).

Whereas heterochromatinization of centromeres has clear functional implications in chromosome segregation (Allshire and Karpen 2008; Fukagawa and Earnshaw 2014), it remains unclear why pericentromeric satellite DNA must be silenced. Its derepression has been shown to cause a plethora of problems, such as DNA damage through the accumulation of RNA:DNA hybrids (Zeller et al. 2016), chromosome missegregation, and meiotic hyper recombination (Peters et al. 2001; Bouzinba-Segard et al. 2006; Hahn et al. 2013; Tasselli et al. 2016), although it is often unclear whether these defects are caused by perturbation of centromeric or pericentromeric heterochromatin. Transcriptional derepression of satellite DNA is also associated with human pathologies such as cancer (Eymery et al. 2009; Ting et al. 2011; Zhu et al. 2011; Bersani et al. 2015), aging (Shumaker et al. 2006; De Cecco et al. 2013a, 2013b), senescence (Enukashvily et al. 2007), and cardiomyopathy (Gaubatz and Cutler 1990; Haider et al. 2012). Accordingly, most studies thus far have investigated the mechanisms of heterochromatin transcriptional silencing, based on the reasoning that its derepression is problematic. However, if pericentromeric heterochromatin serves no function, the easiest solution would be the removal of these sequences, rather than silencing through heterochromatinization. Thus, why satellite DNA/constitutive heterochromatin exists in the first place, and whether pericentromeric satellite DNA serves any fundamental function, is poorly understood.

PROPOSED ROLES OF PERICENTROMERIC SATELLITE DNA

Based on the reasoning that such a large component of the genome cannot be nonfunctional, many researchers have speculated roles of satellite DNAs—for example, playing structural roles or mediating meiotic homologous pairing (Yunis and Yasmineh 1971; Walker 1971; John and Miklos 1979; Kuhn et al. 2011).

One prevailing model of satellite DNA function is their involvement in meiotic homologous chromosome pairing. Cytologically, satellite DNA show strong pairing in meiotic prophase in a broad range of species (Yunis and Yasmineh 1971; Hawley et al. 1992; Dernburg et al. 1996), leading to a popular idea that satellite DNA may regulate meiotic chromosome pairing. However, earlier studies using various chromosomal deficiencies that delete most of satellite DNA (for review, see John and Miklos 1979; Miklos and John 1979) concluded that satellite DNA/heterochromatin is not required for chromosome pairing. It is possible that satellite DNA-mediated pairing may function redundantly in parallel with other mechanisms that facilitate chromosomal pairing.

Another example of putative satellite DNA function is found in the *Drosophila* Y chromosome. Although *Drosophila* Y chromosomes are highly heterochromatic, they contain several embedded genes that are required for male fertility. These genes contain megabase-sized introns consisting of satellite DNA that are transcribed in spermatocytes, forming a lamp brush–like structure termed Y-loops (John and Miklos 1979; Bonaccorsi et al. 1988, 1990; Pisano et al. 1993). It has been proposed that these large intronic transcripts may fulfill a protein-binding function and sequester proteins required for later meiotic stages (Bonaccorsi et al. 1988, 1990; Pisano et al. 1993). Alternatively, intronic heterochromatin may serve to repress expression of these fertility factors in nonspermatocyte cells. However, it should be noted that it has not been possible to delete intronic satellite DNA to determine effects on the expression of fertility genes. Interestingly, a recent study suggested that intronic satellite DNA is a phenomenon specific to genes on the Y chromosome. By comparing the Y chromosomes of *Drosophila melanogaster* and *Drosophila pseudoobscura*, it was found that orthologous genes translocated from the *D. melanogaster* Y to a *D. pseudoobscura* autosome have reduced intron sizes (Chang and Larracuente 2017). Yet *D. pseudoobscura* spermatocytes show cytologically recognizable Y-loops, indicating that a distinct set of genes on the *D. pseudoobscura* Y chromosome have acquired mega intron(s). This implies that satellite DNA may play a role in gene expression specifically for Y chromosome genes, which are always maintained in a heterochromatic environment, except for once in their lifetime during spermatocyte development.

Additional studies have indicated a role for satellite DNA transcription in response to cellular stress, most prominently during the heat shock response (Jolly et al. 2004; Rizzi et al. 2004; Valgardsdottir et al. 2008; Pezer and Ugarkovic 2012). In human cells, heat shock induces the expression of the HSF1 protein, which accumulates on nuclear stress bodies and up-regulates the transcription of the pericentromeric SatIII satellite DNA (Jolly et al. 2004; Rizzi et al. 2004). It has been proposed that the massive activation of SatIII transcription might sequester many factors required for global transcription and splicing, thus leading to down-regulation of transcription under stress (Biamonti and Vourc'h 2010). Likewise, human SatII, whose transcription is up-regulated in cancer cells, has been shown to sequester the MeCP2 (methyl CpG binding) protein, demonstrating the ability of satellite DNA transcripts to regulate the epigenetic state (Hall et al. 2017). In yet another example, satellite DNA transcripts from the *Drosophila* X chromosome are processed into small RNAs, promoting X chromosome recognition for dosage compensation in somatic cells (Menon et al. 2014; Joshi and Meller 2017).

These studies have highlighted the importance of satellite DNA in various biological processes. However, most of these examples of satellite DNA function pertain to species-specific phenomenon, and a unifying theme underlying satellite DNA function across eukaryotes is still lacking.

CHROMOCENTER BUNDLING PROTEINS

Studying the function of satellite DNA is challenging for obvious reasons. Foremost is the inability to remove these repetitive sequences, which can span megabases on chromosomes. Therefore, we initiated our study on the function of satellite DNA by focusing on satellite DNA-binding proteins, specifically, D1 in *Drosophila* and HMGA1 in mouse. Both proteins contain multiple AT-hook motifs, which bind the minor groove of AT-rich DNA (Reeves and Nissen 1990; Huth et al. 1997). D1 was identified as a nonhistone chromosomal protein (Rodriguez Alfageme et al. 1980), which was subsequently shown to have affinity for the 1.672 g/cm³ satellite DNA ($\{AATAT\}_n$) from *Drosophila* (Rodriguez Alfageme et al. 1980; Levinger and Varshavsky 1982a, 1982b). The $\{AATAT\}_n$ satellite comprises a substantial fraction of *Drosophila* genome at nearly ~8% of the genome (Lohe et al. 1993). A role for D1 in constitutive heterochromatin has been suggested based on its effects on position effect variegation (Aulner et al. 2002).

HMGA1 (HMG-1/Y) was also identified as a mammalian nonhistone chromosomal protein (Goodwin et al. 1973; Lund et al. 1983). Further studies revealed its ability to bind satellite DNAs (Strauss and Varshavsky 1984), primarily the abundant pericentromeric satellite DNA called "major satellite" in mouse (Vissel and Choo 1989; Radic et al. 1992), which comprises nearly 6% of the genome (Lyon and Searle 1989). Consistent with the notion of pericentromeric heterochromatin forming chromocenter (Jones 1970; Pardue and Gall 1970; Guenatri

et al. 2004), HMGA1 localizes to mouse chromocenters (Brocher et al. 2010).

We have found that both D1 and HMGA1 are required for clustering of chromocenter, suggesting a role of these satellite DNA-binding proteins in chromocenter formation (Fig. 1). In vitro experiments showed that HMGA1 could cross-link DNA molecules through its multiple AT-hooks (Vogel et al. 2011), suggesting HMGA1's biochemical capacity to cross-link multiple DNA strands. In an in vivo experiment, it was shown that exogenous expression of D1 in *Drosophila* salivary glands resulted in the ectopic fusion of heterologous chromosomes, likely at sites of intercalary heterochromatin (Smith and Weiler 2010). These published results and the fact that D1/HMGA1 are required for chromocenter formation led us to propose a model that D1 and HMGA1 are multivalent DNA-binding proteins capable of bundling multiple DNA strands, allowing clustering of pericentromeric satellite DNA from multiple chromosomes into chromocenters (Fig. 1). We found that exogenous expression of D1 protein, which can bind to mouse major satellite DNA, was sufficient to enhance the clustering of chromocenter in mouse cells, demonstrating the ability of D1 to facilitate the association/cross-linking of satellite DNA.

More direct evidence for D1 cross-linking its target DNA came from a protein fusion of D1 and LacI, which was sufficient to recruit euchromatic LacO repeat sequences into chromocenters. Finally, we observed chromatin threads, which are positive for D1/HMGA1 proteins, as well as their target satellite DNA sequences, connecting multiple chromosomes in early prophase cells, when individual chromosomes start condensing in preparation for mitosis. We speculate that these threads connecting prophase chromosomes are remnants of interchromosomal connections during interphase, which bundles chromosomes together into chromocenters. These threads are reminiscent of earlier observations of threads connecting mitotic chromosomes (Takayama 1975; Burdick 1976) that have later been shown to contain satellite DNA (Kuznetsova 2007).

Taken together, we propose that these satellite DNA-binding proteins (and perhaps other uncharacterized proteins that are known to bind to satellite DNA) participate in chromocenter formation by cross-linking their target DNA on multiple chromosomes.

FUNCTION OF CHROMOCENTER IN ENCAPSULATING THE FULL COMPLEMENT OF GENOME INTO A SINGLE NUCLEUS

The disruption of chromocenter formation in D1 mutant flies and HMGA1 knockdown cells provided a unique opportunity to interrogate the function of chromocenter/satellite DNA. We found that disruption of chromocenter upon D1/HMGA1 depletion resulted in a dramatic increase in micronuclei formation. Time-lapse live imaging revealed that the micronuclei formed during interphase, where micronuclei bud off from the rest of the nucleus. These results indicate that cross-linking of pericentro-

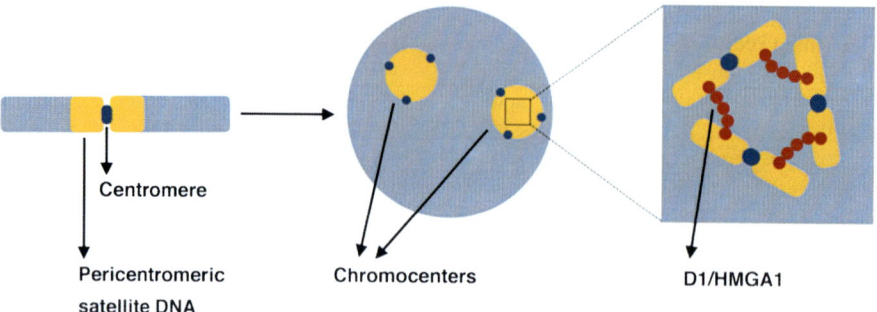

Figure 1. Model of chromocenter formation. Pericentromeric satellite DNA (yellow) is bound by satellite DNA-binding proteins (D1/HMGA1), which cross-link satellite DNA from multiple chromosomes, leading to chromocenter formation.

meric heterochromatin is important to form a physical network of chromosomes such that individual chromosomes do not float away from each other, leading to micronuclei formation (Fig. 2).

Packaging of the full complement of the genome into a single nucleus is a fundamental requirement for eukaryotic cells, whose genetic material is split into multiple chromosomes. Packaging of the genome in a single compartment must be a prerequisite for concerted gene function, for example, to allow a transcription factor to find its target genes. The importance of maintaining a single nucleus (i.e., prevention of micronuclei formation) has been shown by series of studies that demonstrated that (1) micronuclei exhibit defective DNA replication/DNA repair and increased DNA damage (Crasta et al. 2012), (2) micronuclei exhibit defective nuclear envelope integrity (Hatch et al. 2013), and (3) chromosomes within micronuclei are susceptible to chromosome shattering and rearrangements, termed chromothripsis (Hatch and Hetzer 2015; Zhang et al. 2015), which has been suggested to be a driver of oncogenesis. Similar to these micronuclei-associated defects, D1 mutation/HMGA1 knockdown resulted in accumulation of DNA damage and chromosomal breaks, leading to cell death. These results indicate that chromocenter formation, mediated by D1 and HMGA1, plays a critical role in encapsulating the full genome into a single nucleus, a failure of which leads to micronuclei formation and the resulting cellular defects.

Interestingly, even major nuclei showed defective integrity of the nuclear envelope in D1 mutant/HMGA1 knockdown cells. This suggested that chromocenter formation might be required not only for encapsulating all chromosomes into single nucleus, but also for maintaining nucle-

ar envelope integrity. Early cytological studies showed that constitutive heterochromatin was juxtaposed to the nuclear envelope and around nucleoli (Rae and Franke 1972). Because of the enrichment of heterochromatin at the nuclear envelope, it is thought to function as a gene repressive compartment. Global analysis of gene expression at the nuclear envelope has shown that genes within lamin-associated domains (LADs) generally show decreased expression compared to genes outside of lamin-associated domains (Guelen et al. 2008). How do changes in constitutive heterochromatin clustering affect the integrity of the nuclear envelope? Interestingly, constitutively localized LADs are enriched for H3K9 methylation, are AT-rich and have been suggested to function as a structural backbone for the organization of interphase chromosomes (Meuleman et al. 2013). In addition, recent studies have shown that impairing heterochromatin using either RNAi of Prdm3/16 (Pinheiro et al. 2012) or chemical inhibitors of histone methyltransferases (Stephens et al. 2018) results in defects of the nuclear envelope. Because the majority of chromocenters are localized to the nuclear periphery (Mayer et al. 2005), chromocenter-forming proteins such as D1/HMGA1 may perhaps function by linking constitutive heterochromatin to nuclear envelope, or by facilitating nuclear envelope formation around the constitutive heterochromatin.

IS CHROMOCENTER FUNCTION UNIVERSAL?

We have presented the hypothesis that pericentromeric satellite DNA mediates chromocenter formation, which

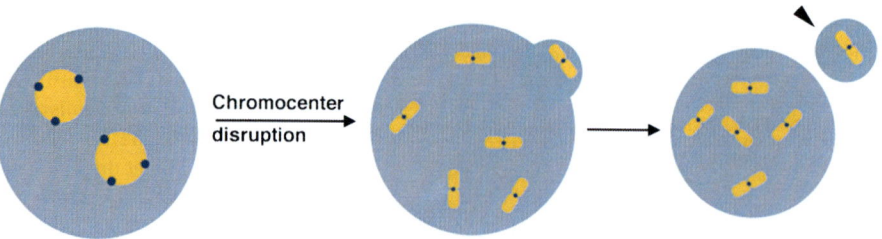

Figure 2. Chromocenter disruption leads to micronuclei formation. Loss of satellite DNA-binding proteins, D1/HMGA1, leads to micronuclei formation. Based on this result, we propose that bundling of multiple chromosomes via formation of chromocenter is critical to maintain the full complement of the genome into a single nucleus.

functions to encapsulate the full set of chromosomes into a single nucleus. Given the ubiquity of satellite DNA, we postulate that chromocenter formation and function may be a universal mechanism conserved across many species.

This hypothesis leads to a few speculations that may provide further insights into the biology of chromocenter. First, in certain cases, eukaryotic chromosomes appear to be completely lacking satellite DNA (e.g., horse chromosome 11, orangutan chromosome 12) (Wade et al. 2009; Piras et al. 2010; Locke et al. 2011) raising the question of if/how these chromosomes may be incorporated into chromocenters. Interestingly, a recent report has shown that transposable element (TE)-derived tandem repeats are a component of mouse chromocenters (Kuznetsova et al. 2016). TE-derived sequences are abundant, heterochromatinized and typically interspersed throughout eukaryotic chromosomes (Saksouk et al. 2015; Nishibuchi and Déjardin 2017), including pericentromeric heterochromatin, where they are present in complex islands within large satellite DNA tracts (Sun et al. 1997, 2003). In addition, satellite DNA is derived from transposable elements in certain cases (Heikkinen et al. 1995; Kapitonov et al. 1998; Kidwell 2002). Recognition of TE-derived repeats or their heterochromatic nature may help chromocenter formation, possibly mediating incorporation of satellite DNA-free chromosomes into chromocenters. Thus, chromosomes lacking typical satellite DNA might still participate in chromocenter formation via their resident TEs. Also, if TE-derived sequences indeed function to mediate chromocenter formation, it may force us to reconsider the biological status of TEs: instead of pure parasites, they may be symbionts of eukaryotic genomes.

The hypothesis that pericentromeric satellite DNA may be a critical component of chromosomes to ensure its maintenance leads to another interesting consideration. Generally, repetitive sequences are thought to be inherently unstable because of sporadic loss of copy number caused by intrachromatid recombination (Charlesworth et al. 1994; Stephan and Cho 1994). If satellite DNA were a critical structural component of the chromosome, it would have to be actively maintained. This is consistent with the fact that satellite DNA shows little within-species variation in the genomes of *Drosophila* species despite its potential instability (Bosco et al. 2007). Strikingly, it has been shown that satellite DNA can expand in cancer cells (Bersani et al. 2015). Although it was regarded as an "abnormality" of cancer cells, it might reflect the acquired immortality of cancer cells. Similar to the telomere maintenance mechanism, which is confined to immortal cells (germ cells, some somatic stem cells, and cancer cells), maintenance (expansion) of satellite DNA may be a privileged process that can only occur in immortal cells.

SPECULATION: SATELLITE DNA AND SPECIATION

It is widely appreciated that satellite DNA sequences are highly divergent even among closely related species (Ugarković and Plohl 2002). Rapid changes in the spectrum of satellite DNAs (dramatic expansion and shrinkage of a particular repeat sequence, leading to a vastly distinct landscape of satellite DNA composition in closely related species) has been attributed to processes such as intrachromatid recombination, unequal sister chromatid exchange, replication slippage, rolling circle amplification and reinsertion, and gene conversion (Charlesworth et al. 1994). The "library" hypothesis complements this idea, by postulating that each species does not entirely lose or acquire certain satellite repeat sequences but instead the related species share a common library of satellite sequences that merely fluctuate in copy number (Salser et al. 1976; Fry and Salser 1977).

Irrespective of the mechanisms that explain birth and death of satellite DNA repeat sequences, the highly divergent sequences of satellite DNA reinforced the idea that they are junk (Ohno 1972; Orgel and Crick 1980). On the other hand, the very same fact led to a speculation that rapid changes in satellite DNA sequences may underlie speciation, making two species incompatible with each other (Walker 1971; Yunis and Yasmineh 1971; John and Miklos 1979). Our hypothesis that satellite DNA is required for chromocenter formation can potentially offer an answer to this paradox. If satellite DNA functions as a platform for proteins that have the capacity to cross-link multiple DNA strands, the repeat sequence per se might not matter as much as its ability to be bound by cross-linking proteins (i.e., chromocenter proteins). If this is the case, evolutionary pressure will select for recognition of satellite DNA by binding proteins with less stringency for the exact sequence, which may rapidly change because of the inherent instability of repetitive DNA.

Now we are presented with a striking possibility: The rapid change in satellite DNA sequences over the course of evolution might quickly separate two populations within a species, because rapid changes in satellite DNA sequences must be matched by rapid changes in satellite DNA-binding proteins. Two populations may therefore be reproductively isolated because their satellite DNA and chromocenter-forming proteins become incompatible. If this is the case, hybrid incompatibility may arise from the incompatibility of chromocenter-forming factors (satellite DNA-binding proteins and their cognate repeat sequences) (Fig. 3). This is certainly a testable prediction, and encouragingly, the majority of speciation genes identified thus far bind constitutive heterochromatin (Sawamura 2012). For example, two hybrid incompatibility genes in *Drosophila*, *Hmr* and *Lhr* (Watanabe 1979; Hutter and Ashburner 1987; Barbash et al. 2003; Brideau et al. 2006), are known to localize to the chromocenters of polytene chromosomes (Satyaki et al. 2014) and pericentromeric heterochromatin of diploid nuclei (Blum et al. 2017), indicating that they may participate in chromocenter formation or its regulation. It would be of particular interest in the future to examine cell biological aspects of hybrid incompatibility to see whether they might present phenotypes that are consistent with these ideas (e.g., chromocenter disruption). It will also be of interest to examine the relationship between Hmr/ Lhr and D1.

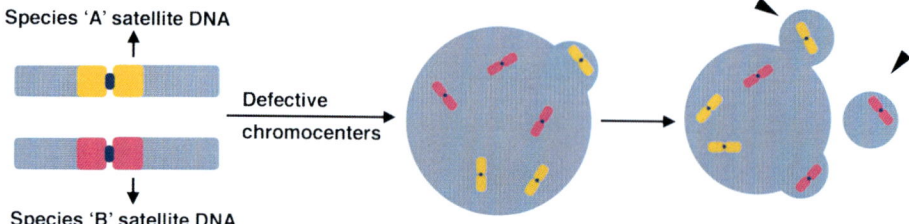

Figure 3. Hypothetical model of incompatibility between chromosomes from distinct species in forming chromocenter. The well-known fact that satellite DNA sequences are highly divergent between closely related species, combined with our finding that satellite DNA is critical for chromocenter formation and thus chromosome maintenance, leads to the question of whether chromosomes from different species can form chromocenters. We speculate that the incompatibility in chromocenter formation may underlie hybrid incompatibility.

CONCLUSION

As described above, here we present our hypothesis that chromocenter formation mediated by bundling of pericentromeric satellite DNA is a universal mechanism of eukaryotic cells to encapsulate the full complement of the genome into a single nucleus. Although satellite DNA function has remained enigmatic for a long time, our model offers explanations to many aspects of satellite DNA: lack of DNA sequence conservation, yet persistent presence in a broad range of eukaryotes, and cytological behavior of satellite DNA (chromocenter formation). Our model also provides several testable predictions, which need to be addressed in future studies. In summary, we propose that satellite DNA plays a fundamental role in the maintenance of the eukaryotic genome.

ACKNOWLEDGMENTS

The research in the Yamashita laboratory is supported by the Howard Hughes Medical Institute and the National Institute of General Medical Sciences (R01GM118308). M.J. is supported by a postdoctoral fellowship from the American Heart Association. We thank the Yamashita laboratory members for discussions.

REFERENCES

Allshire RC, Karpen GH. 2008. Epigenetic regulation of centromeric chromatin: Old dogs, new tricks? *Nat Rev Genet* **9:** 923–937.

Allshire RC, Madhani HD. 2017. Ten principles of heterochromatin formation and function. *Nat Rev Mol Cell Biol* doi: 10.1038/nrm.2017.119.

Aulner N, Monod C, Mandicourt G, Jullien D, Cuvier O, Sall A, Janssen S, Laemmli UK, Käs E. 2002. The AT-hook protein D1 is essential for *Drosophila melanogaster* development and is implicated in position-effect variegation. *Mol Cell Biol* **22:** 1218–1232.

Baccarini P. 1908. Sulle cinesi vegetative del "Cynomorium coccineum L." *N Giorn Bot Ital N Ser* **15:** 189–203.

Barbash DA, Siino DF, Tarone AM, Roote J. 2003. A rapidly evolving MYB-related protein causes species isolation in *Drosophila*. *Proc Natl Acad Sci* **100:** 5302–5307.

Barry AE, Howman EV, Cancilla MR, Saffery R, Choo KH. 1999. Sequence analysis of an 80 kb human neocentromere. *Hum Mol Genet* **8:** 217–227.

Bersani F, Lee E, Kharchenko PV, Xu AW, Liu M, Xega K, MacKenzie OC, Brannigan BW, Wittner BS, Jung H, et al. 2015. Pericentromeric satellite repeat expansions through RNA-derived DNA intermediates in cancer. *Proc Natl Acad Sci* **112:** 15148–15153.

Biamonti G, Vourc'h C. 2010. Nuclear stress bodies. *Cold Spring Harb Perspect Biol* **2:** a000695.

Biscotti MA, Canapa A, Forconi M, Olmo E, Barucca M. 2015. Transcription of tandemly repetitive DNA: Functional roles. *Chromosome Res* **23:** 463–477.

Blum JA, Bonaccorsi S, Marzullo M, Palumbo V, Yamashita YM, Barbash DA, Gatti M. 2017. The hybrid incompatibility genes *Lhr* and *Hmr* are required for sister chromatid detachment during anaphase but not for centromere function. *Genetics* **207:** 1457–1472.

Bonaccorsi S, Pisano C, Puoti F, Gatti M. 1988. Y chromosome loops in *Drosophila melanogaster*. *Genetics* **120:** 1015–1034.

Bonaccorsi S, Gatti M, Pisano C, Lohe A. 1990. Transcription of a satellite DNA on two Y chromosome loops of *Drosophila melanogaster*. *Chromosoma* **99:** 260–266.

Bosco G, Campbell P, Leiva-Neto JT, Markow TA. 2007. Analysis of *Drosophila* species genome size and satellite DNA content reveals significant differences among strains as well as between species. *Genetics* **177:** 1277–1290.

Bouzinba-Segard H, Guais A, Francastel C. 2006. Accumulation of small murine minor satellite transcripts leads to impaired centromeric architecture and function. *Proc Natl Acad Sci* **103:** 8709–8714.

Brideau NJ, Flores HA, Wang J, Maheshwari S, Wang X, Barbash DA. 2006. Two Dobzhansky-Muller genes interact to cause hybrid lethality in *Drosophila*. *Science* **314:** 1292–1295.

Brocher J, Vogel B, Hock R. 2010. HMGA1 down-regulation is crucial for chromatin composition and a gene expression profile permitting myogenic differentiation. *BMC Cell Biol* **11:** 64.

Bühler M, Moazed D. 2007. Transcription and RNAi in heterochromatic gene silencing. *Nat Struct Mol Biol* **14:** 1041–1048.

Burdick AB. 1976. Somatic cell chromosome interconnections in trypan preparations of Chinese hamster testicular cells. *Exp Cell Res* **99:** 425–428.

Chang C-H, Larracuente AM. 2017. Genomic changes following the reversal of a Y chromosome to an autosome in *Drosophila pseudoobscura*. *Evolution* **71:** 1285–1296.

Charlesworth B, Sniegowski P, Stephan W. 1994. The evolutionary dynamics of repetitive DNA in eukaryotes. *Nature* **371:** 215–220.

Crasta K, Ganem NJ, Dagher R, Lantermann AB, Ivanova EV, Pan Y, Nezi L, Protopopov A, Chowdhury D, Pellman D. 2012. DNA breaks and chromosome pulverization from errors in mitosis. *Nature* **482:** 53–58.

De Cecco M, Criscione S, Peterson A, Neretti N, Sedivy J, Kreiling J. 2013a. Transposable elements become active and mobile in the genomes of aging mammalian somatic tissues. *Aging* **5:** 867–883.

De Cecco M, Criscione SW, Peckham EJ, Hillenmeyer S, Hamm EA, Manivannan J, Peterson AL, Kreiling JA, Neretti N, Sedivy JM. 2013b. Genomes of replicatively senescent cells undergo global epigenetic changes leading to gene silencing and activation of transposable elements. *Aging Cell* **12:** 247–256.

Dernburg AF, Sedat JW, Hawley RS. 1996. Direct evidence of a role for heterochromatin in meiotic chromosome segregation. *Cell* **86:** 135–146.

du Sart D, Cancilla MR, Earle E, Mao JI, Saffery R, Tainton KM, Kalitsis P, Martyn J, Barry AE, Choo KH. 1997. A functional neo-centromere formed through activation of a latent human centromere and consisting of non-α-satellite DNA. *Nat Genet* **16:** 144–153.

Enukashvily NI, Donev R, Waisertreiger IS, Podgornaya OI. 2007. Human chromosome 1 satellite 3 DNA is decondensed, demethylated and transcribed in senescent cells and in A431 epithelial carcinoma cells. *Cytogenet Genome Res* **118:** 42–54.

Eymery A, Horard B, El Atifi-Borel M, Fourel G, Berger F, Vitte AL, Van den Broeck A, Brambilla E, Fournier A, Callanan M, et al. 2009. A transcriptomic analysis of human centromeric and pericentric sequences in normal and tumor cells. *Nucleic Acids Res* **37:** 6340–6354.

Fry K, Salser W. 1977. Nucleotide sequences of HS-α satellite DNA from kangaroo rat *Dipodomys ordii* and characterization of similar sequences in other rodents. *Cell* **12:** 1069–1084.

Fukagawa T, Earnshaw WC. 2014. The centromere: Chromatin foundation for the kinetochore machinery. *Dev Cell* **30:** 496–508.

Garrido-Ramos MA. 2017. Satellite DNA: An evolving topic. *Genes (Basel)* **8:** E230 doi: 10.3390/genes8090230.

Gaubatz JW, Cutler RG. 1990. Mouse satellite DNA is transcribed in senescent cardiac muscle. *J Biol Chem* **265:** 17753–17758.

Goodwin GH, Sanders C, Johns EW. 1973. A new group of chromatin-associated proteins with a high content of acidic and basic amino acids. *Eur J Biochem* **38:** 14–19.

Guelen L, Pagie L, Brasset E, Meuleman W, Faza MB, Talhout W, Eussen BH, de Klein A, Wessels L, de Laat W, et al. 2008. Domain organization of human chromosomes revealed by mapping of nuclear lamina interactions. *Nature* **453:** 948–951.

Guenatri M, Bailly D, Maison C, Almouzni G. 2004. Mouse centric and pericentric satellite repeats form distinct functional heterochromatin. *J Cell Biol* **166:** 493–505.

Hahn M, Dambacher S, Dulev S, Kuznetsova AY, Eck S, Wörz S, Sadic D, Schulte M, Mallm JP, Maiser A, et al. 2013. Suv4-20h2 mediates chromatin compaction and is important for cohesin recruitment to heterochromatin. *Genes Dev* **27:** 859–872.

Haider S, Cordeddu L, Robinson E, Movassagh M, Siggens L, Vujic A, Choy M-K, Goddard M, Lio P, Foo R. 2012. The landscape of DNA repeat elements in human heart failure. *Genome Biol* **13:** R90.

Hall LE, Mitchell SE, O'Neill RJ. 2012. Pericentric and centromeric transcription: A perfect balance required. *Chromosome Res* **20:** 535–546.

Hall LL, Byron M, Carone DM, Whitfield TW, Pouliot GP, Fischer A, Jones P, Lawrence JB. 2017. Demethylated HSATII DNA and HSATII RNA foci sequester PRC1 and MeCP2 into cancer-specific nuclear bodies. *Cell Rep* **18:** 2943–2956.

Hatch EM, Hetzer MW. 2015. Chromothripsis. *Curr Biol* **25:** R397–R399.

Hatch EM, Fischer AH, Deerinck TJ, Hetzer MW. 2013. Catastrophic nuclear envelope collapse in cancer cell micronuclei. *Cell* **154:** 47–60.

Hawley RS, Irick H, Zitron AE, Haddox DA, Lohe A, New C, Whitley MD, Arbel T, Jang J, McKim K, Childs G. 1992. There are two mechanisms of achiasmate segregation in *Drosophila* females, one of which requires heterochromatic homology. *Dev Genet* **13:** 440–467.

Heikkinen E, Launonen V, Müller E, Bachmann L. 1995. The pvB370 BamHI satellite DNA family of the *Drosophila* virilis group and its evolutionary relation to mobile dispersed genetic pDv elements. *J Mol Evol* **41:** 604–614.

Heitz E. 1928. Das Heterochromatin der Moose.I. *Jahrb Wiss Bot* **69:** 762–818.

Henikoff S, Ahmad K, Malik H. 2001. The centromere paradox: Stable inheritance with rapidly evolving DNA. *Science* **293:** 1098–1102.

Huth JR, Bewley CA, Nissen MS, Evans JN, Reeves R, Gronenborn AM, Clore GM. 1997. The solution structure of an HMG-I(Y)-DNA complex defines a new architectural minor groove binding motif. *Nat Struct Biol* **4:** 657–665.

Hutter P, Ashburner M. 1987. Genetic rescue of inviable hybrids between *Drosophila melanogaster* and its sibling species. *Nature* **327:** 331–333.

John B, Miklos GL. 1979. Functional aspects of satellite DNA and heterochromatin. *Int Rev Cytol* **58:** 1–114.

Johnson WL, Yewdell WT, Bell JC, McNulty SM, Duda Z, O'Neill RJ, Sullivan BA, Straight AF. 2017. RNA-dependent stabilization of SUV39H1 at constitutive heterochromatin. *Elife* **6:** e25299.

Jolly C, Metz A, Govin J, Vigneron M, Turner BM, Khochbin S, Vourc'h C. 2004. Stress-induced transcription of satellite III repeats. *J Cell Biol* **164:** 25–33.

Jones KW. 1970. Chromosomal and nuclear location of mouse satellite DNA in individual cells. *Nature* **225:** 912–915.

Joseph A, Mitchell AR, Miller OJ. 1989. The organization of the mouse satellite DNA at centromeres. *Exp Cell Res* **183:** 494–500.

Joshi SS, Meller VH. 2017. Satellite repeats identify X chromatin for dosage compensation in *Drosophila melanogaster* males. *Curr Biol* **27:** 1393–1402.e2.

Kapitonov VV, Holmquist GP, Jurka J. 1998. L1 repeat is a basic unit of heterochromatin satellites in cetaceans. *Mol Biol Evol* **15:** 611–612.

Karpen GH, Allshire RC. 1997. The case for epigenetic effects on centromere identity and function. *Trends Genet* **13:** 489–496.

Kidwell MG. 2002. Transposable elements and the evolution of genome size in eukaryotes. *Genetica* **115:** 49–63.

Kit S. 1961. Equilibrium sedimentation in density gradients of DNA preparations from animal tissues. *J Mol Biol* **3:** 711–716.

Kuhn G, Küttler H, Moreira-Filho O. 2011. The 1.688 repetitive DNA of *Drosophila*: Concerted evolution at different genomic scales and association with genes. *Mol Biol Evol* **29:** 7–11.

Kuznetsova IS, Enukashvili NI, Noniashvili EM, Shatrova AN, Aksenov ND, Zenin VV, Dyban AP, Podgornaya OI. 2007. Evidence for the existence of satellite DNA-containing connection between metaphase chromosomes. *J Cell Biochem* **101:** 1046–1061.

Kuznetsova IS, Ostromyshenskii DI, Komissarov AS, Prusov AN, Waisertreiger IS, Gorbunova AV, Trifonov VA, Ferguson-Smith MA, Podgornaya OI. 2016. LINE-related component of mouse heterochromatin and complex chromocenters' composition. *Chromosome Res* **24:** 309–323.

Levinger L, Varshavsky A. 1982a. Protein D1 preferentially binds A + T-rich DNA in vitro and is a component of *Drosophila melanogaster* nucleosomes containing A + T-rich satellite DNA. *Proc Natl Acad Sci* **79:** 7152–7156.

Levinger L, Varshavsky A. 1982b. Selective arrangement of ubiquitinated and D1 protein-containing nucleosomes within the *Drosophila* genome. *Cell* **28:** 375–385.

Locke DP, Hillier LW, Warren WC, Worley KC, Nazareth LV, Muzny DM, Yang S-P, Wang Z, Chinwalla AT, Minx P, et al. 2011. Comparative and demographic analysis of orangutan genomes. *Nature* **469:** 529–533.

Lohe AR, Hilliker AJ, Roberts PA. 1993. Mapping simple repeated DNA sequences in heterochromatin of *Drosophila melanogaster*. *Genetics* **134:** 1149–1174.

Lund T, Holtlund J, Fredriksen M, Laland SG. 1983. On the presence of two new high mobility group-like proteins in HeLa S3 cells. *FEBS Lett* **152:** 163–167.

Lyon MF, Searle AG. 1989. *Genetic variants and strains of the laboratory mouse.* Oxford University Press, Oxford.

Malik HS. 2009. The centromere-drive hypothesis: A simple basis for centromere complexity. *Prog Mol Subcell Biol* **48:** 33–52.

Mayer R, Brero A, von Hase J, Schroeder T, Cremer T, Dietzel S. 2005. Common themes and cell type specific variations of higher order chromatin arrangements in the mouse. *BMC Cell Biol* **6:** 44.

Menon DU, Coarfa C, Xiao W, Gunaratne PH, Meller VH. 2014. siRNAs from an X-linked satellite repeat promote X-chromosome recognition in *Drosophila melanogaster*. *Proc Natl Acad Sci* **111:** 16460–16465.

Meuleman W, Peric-Hupkes D, Kind J, Beaudry JBB, Pagie L, Kellis M, Reinders M, Wessels L, van Steensel B. 2013. Constitutive nuclear lamina-genome interactions are highly conserved and associated with A/T-rich sequence. *Genome Res* **23:** 270–280.

Miklos GL, John B. 1979. Heterochromatin and satellite DNA in man: Properties and prospects. *Am J Hum Genet* **31:** 264–280.

Moazed D. 2011. Mechanisms for the inheritance of chromatin states. *Cell* **146:** 510–518.

Nishibuchi G, Déjardin J. 2017. The molecular basis of the organization of repetitive DNA-containing constitutive heterochromatin in mammals. *Chromosome Res* **25:** 77–87.

Ohno S. 1972. So much "junk" DNA in our genome. *Brookhaven Symp Biol* **23:** 366–370.

Orgel LE, Crick FHC. 1980. Selfish DNA: The ultimate parasite. *Nature* **284:** 604–607.

Pardue ML, Gall JG. 1970. Chromosomal localization of mouse satellite DNA. *Science* **168:** 1356–1358.

Peters AH, O'Carroll D, Scherthan H, Mechtler K, Sauer S, Schöfer C, Weipoltshammer K, Pagani M, Lachner M, Kohlmaier A, et al. 2001. Loss of the Suv39h histone methyltransferases impairs mammalian heterochromatin and genome stability. *Cell* **107:** 323–337.

Pezer Z, Ugarkovic D. 2012. Satellite DNA-associated siRNAs as mediators of heat shock response in insects. *RNA Biol* **9:** 587–595.

Pinheiro I, Margueron R, Shukeir N, Eisold M, Fritzsch C, Richter FM, Mittler G, Genoud C, Goyama S, Kurokawa M, et al. 2012. Prdm3 and Prdm16 are H3K9me1 methyltransferases required for mammalian heterochromatin integrity. *Cell* **150:** 948–960.

Piras FM, Nergadze SG, Magnani E, Bertoni L, Attolini C, Khoriauli L, Raimondi E, Giulotto E. 2010. Uncoupling of satellite DNA and centromeric function in the genus *Equus*. *PLoS Genet* **6:** e1000845.

Pisano C, Bonaccorsi S, Gatti M. 1993. The kl-3 loop of the Y chromosome of *Drosophila melanogaster* binds a tektin-like protein. *Genetics* **133:** 569–579.

Probst AV, Okamoto I, Casanova M, El Marjou F, Le Baccon P, Almouzni G. 2010. A strand-specific burst in transcription of pericentric satellites is required for chromocenter formation and early mouse development. *Dev Cell* **19:** 625–638.

Radic MZ, Saghbini M, Elton TS, Reeves R, Hamkalo BA. 1992. Hoechst 33258, distamycin A, and high mobility group protein I (HMG-I) compete for binding to mouse satellite DNA. *Chromosoma* **101:** 602–608.

Rae MM, Franke WW. 1972. The interphase distribution of satellite DNA-containing heterochromatin in mouse nuclei. *Chromosoma* **39:** 443–456.

Reeves R, Nissen MS. 1990. The A.T-DNA-binding domain of mammalian high mobility group I chromosomal proteins. A novel peptide motif for recognizing DNA structure. *J Biol Chem* **265:** 8573–8582.

Rizzi N, Denegri M, Chiodi I, Corioni M, Valgardsdottir R, Cobianchi F, Riva S, Biamonti G. 2004. Transcriptional activation of a constitutive heterochromatic domain of the human genome in response to heat shock. *Mol Biol Cell* **15:** 543–551.

Rodriguez Alfageme C, Rudkin GT, Cohen LH. 1980. Isolation, properties and cellular distribution of D1, a chromosomal protein of *Drosophila*. *Chromosoma* **78:** 1–31.

Rosin LF, Mellone BG. 2017. Centromeres drive a hard bargain. *Trends Genet* **33:** 101–117.

Saksouk N, Simboeck E, Déjardin J. 2015. Constitutive heterochromatin formation and transcription in mammals. *Epigenetics Chromatin* **8:** 3.

Salser W, Bowen S, Browne D, el-Adli F, Fedoroff N, Fry K, Heindell H, Paddock G, Poon R, Wallace B, et al. 1976. In-

vestigation of the organization of mammalian chromosomes at the DNA sequence level. *Fed Proc* **35:** 23–35.

Satyaki PR, Cuykendall TN, Wei KH, Brideau NJ, Kwak H, Aruna S, Ferree PM, Ji S, Barbash DA. 2014. The *Hmr* and *Lhr* hybrid incompatibility genes suppress a broad range of heterochromatic repeats. *PLoS Genet* **10:** e1004240.

Sawamura K. 2012. Chromatin evolution and molecular drive in speciation. *Int J Evol Biol* **2012:** 301894.

Schueler MG, Higgins AW, Rudd MK, Gustashaw K, Willard HF. 2001. Genomic and genetic definition of a functional human centromere. *Science* **294:** 109–115.

Shirai A, Kawaguchi T, Shimojo H, Muramatsu D, Ishida-Yonetani M, Nishimura Y, Kimura H, Nakayama JI, Shinkai Y. 2017. Impact of nucleic acid and methylated H3K9 binding activities of Suv39h1 on its heterochromatin assembly. *Elife* **6:** e25317.

Shumaker DK, Dechat T, Kohlmaier A, Adam SA, Bozovsky MR, Erdos MR, Eriksson M, Goldman AE, Khuon S, Collins FS, et al. 2006. Mutant nuclear lamin A leads to progressive alterations of epigenetic control in premature aging. *Proc Natl Acad Sci* **103:** 8703–8708.

Smith MB, Weiler KS. 2010. *Drosophila* D1 overexpression induces ectopic pairing of polytene chromosomes and is deleterious to development. *Chromosoma* **119:** 287–309.

Stephan W, Cho S. 1994. Possible role of natural selection in the formation of tandem-repetitive noncoding DNA. *Genetics* **136:** 333–341.

Stephens AD, Liu PZ, Banigan EJ, Almassalha LM, Backman V, Adam SA, Goldman RD, Marko JF. 2018. Chromatin histone modifications and rigidity affect nuclear morphology independent of lamins. *Mol Biol Cell.* **29:** 220–233.

Strauss F, Varshavsky A. 1984. A protein binds to a satellite DNA repeat at three specific sites that would be brought into mutual proximity by DNA folding in the nucleosome. *Cell* **37:** 889–901.

Sueoka N. 1961. Variation and heterogeneity of base composition of deoxyribonucleic acids: A compilation of old and new data. *J Mol Biol* **3:** 31–40.

Sun X, Wahlstrom J, Karpen G. 1997. Molecular structure of a functional *Drosophila* centromere. *Cell* **91:** 1007–1019.

Sun X, Le HD, Wahlstrom JM, Karpen GH. 2003. Sequence analysis of a functional *Drosophila* centromere. *Genome Res* **13:** 182–194.

Sybalski W. 1968. Use of cesium sulfate for equilibrium density gradient centrifugation. *Methods Enzymol* **12**(Pt B): 330–360.

Takayama S. 1975. Interchromosomal connectives in squash preparations of L cells. *Exp Cell Res* **91:** 408–412.

Tasselli L, Xi Y, Zheng W, Tennen RI, Odrowaz Z, Simeoni F, Li W, Chua KF. 2016. SIRT6 deacetylates H3K18ac at pericentric chromatin to prevent mitotic errors and cellular senescence. *Nat Struct Mol Biol* **23:** 434–440.

Ting DT, Lipson D, Paul S, Brannigan BW, Akhavanfard S, Coffman EJ, Contino G, Deshpande V, Iafrate AJ, Letovsky S, et al. 2011. Aberrant overexpression of satellite repeats in pancreatic and other epithelial cancers. *Science* **331:** 593–596.

Ugarković D, Plohl M. 2002. Variation in satellite DNA profiles —Causes and effects. *EMBO J* **21:** 5955–5959.

Valgardsdottir R, Chiodi I, Giordano M, Rossi A, Bazzini S, Ghigna C, Riva S, Biamonti G. 2008. Transcription of satellite III non-coding RNAs is a general stress response in human cells. *Nucleic Acids Res* **36:** 423–434.

Velazquez Camacho O, Galan C, Swist-Rosowska K, Ching R, Gamalinda M, Karabiber F, De La Rosa-Velazquez I, Engist B, Koschorz B, Shukeir N, et al. 2017. Major satellite repeat RNA stabilize heterochromatin retention of Suv39h enzymes by RNA-nucleosome association and RNA:DNA hybrid formation. *Elife* **6:** e25293.

Vissel B, Choo KH. 1989. Mouse major (γ) satellite DNA is highly conserved and organized into extremely long tandem arrays: Implications for recombination between nonhomologous chromosomes. *Genomics* **5:** 407–414.

Vogel B, Löschberger A, Sauer M, Hock R. 2011. Cross-linking of DNA through HMGA1 suggests a DNA scaffold. *Nucleic Acids Res* **39:** 7124–7133.

Voullaire LE, Slater HR, Petrovic V, Choo KH. 1993. A functional marker centromere with no detectable α-satellite, satellite III, or CENP-B protein: Activation of a latent centromere? *Am J Hum Genet* **52:** 1153–1163.

Wade CM, Giulotto E, Sigurdsson S, Zoli M, Gnerre S, Imsland F, Lear TL, Adelson DL, Bailey E, Bellone RR, et al. 2009. Genome sequence, comparative analysis, and population genetics of the domestic horse. *Science* **326:** 865–867.

Walker PM. 1971. Origin of satellite DNA. *Nature* **229:** 306–308.

Watanabe TK. 1979. A gene that rescues the lethal hybrids between *Drosophila melanogaster* and *Drosophila simulans*. *Jpn J Genet* **54:** 325–331.

Willard HF. 1990. Centromeres of mammalian chromosomes. *Trends Genet* **6:** 410–416.

Wong A, Rattner J. 1988. Sequence organization and cytological localization of the minor satellite of mouse. *Nucleic Acids Res* **16:** 11645–11661.

Yunis JJ, Yasmineh WG. 1971. Heterochromatin, satellite DNA, and cell function. Structural DNA of eucaryotes may support and protect genes and aid in speciation. *Science* **174:** 1200–1209.

Zeller P, Padeken J, van Schendel R, Kalck V, Tijsterman M, Gasser SM. 2016. Histone H3K9 methylation is dispensable for *Caenorhabditis elegans* development but suppresses RNA:DNA hybrid-associated repeat instability. *Nat Genet* **48:** 1385–1395.

Zhang C-Z, Spektor A, Cornils H, Francis JM, Jackson EK, Liu S, Meyerson M, Pellman D. 2015. Chromothripsis from DNA damage in micronuclei. *Nature* **522:** 179–184.

Zhu Q, Pao GM, Huynh AM, Suh H, Tonnu N, Nederlof PM, Gage FH, Verma IM. 2011. BRCA1 tumour suppression occurs via heterochromatin-mediated silencing. *Nature* **477:** 179–184.

Homologous Recombination and Replication Fork Protection: BRCA2 and More!

WEIRAN FENG[1,2] AND MARIA JASIN[1,2]

[1]Developmental Biology Program, Memorial Sloan Kettering Cancer Center, New York, New York 10065
[2]Louis V. Gerstner, Jr. Graduate School of Biomedical Sciences, Memorial Sloan Kettering
Cancer Center, New York, New York 10065

Correspondence: m-jasin@ski.mskcc.org

BRCA2 is a breast and ovarian tumor suppressor that guards against genome instability, a hallmark of cancer. Significant progress has been made in improving our understanding of BRCA2 function from biochemical, cellular, and mouse studies. The knowledge gained has been actively exploited to develop therapeutic strategies, including PARP inhibition, which has shown promising clinical outcomes. Recently, tremendous excitement has been generated by the findings of the roles of BRCA2 and other proteins in suppressing replication stress through homologous recombination and in the protection of stalled replication forks. Processes such as mitotic DNA synthesis and fork reversal have taken center stage in these studies. Here, we discuss our recent findings in the context of these advances.

BRCA2 is a well-known tumor suppressor that was identified more than two decades ago (Wooster et al. 1995) yet interest in this protein continues to grow. Mono-allelic inheritance of a deleterious *BRCA2* mutation confers up to a 70% risk for breast cancer and a 40% risk for ovarian cancer before age 70 (Antoniou et al. 2003), and a lower risk for other tumor types. Although retained in some cases (Maxwell et al. 2017), the wild-type allele is typically lost in tumor cells from *BRCA2* mutation carriers. Therefore, tumor formation is typically associated with a severe disruption in BRCA2 function. Somatic *BRCA2* mutations have also been identified more recently in several tumor types, including ovary (Pennington et al. 2014) and prostate (Robinson et al. 2015). Biallelic *BRCA2* germline mutations predispose to Fanconi anemia, a syndrome characterized by developmental defects and tumor susceptibility (D'Andrea 2010). In these cases, at least one of the *BRCA2* alleles is expected to be hypomorphic, since complete loss of BRCA2 function causes embryonic lethality in mice (Evers and Jonkers 2006).

It is widely accepted that BRCA2 suppresses tumor formation by preventing genome instability, a hallmark of cancer (Hanahan and Weinberg 2011). Nevertheless, it is not fully understood how BRCA2 loss, or genome instability in general, promotes tumor formation. There is a paradox in that BRCA2 deficiency leads to cell lethality in mouse models (Patel et al. 1998; Evers and Jonkers 2006; Kuznetsov et al. 2008; Badie et al. 2010), rather than unrestrained proliferation as might be expected by loss of a tumor suppressor gene. BRCA2 has a well-established function in maintaining the integrity of the genome through its role in homologous recombination (HR). More recently, a related but separable function, replication fork protection (FP), has been discovered, and the

relative contributions of these two pathways to genome integrity maintenance and cell viability are under active investigation. Here, we summarize the recent findings from several laboratories, including our own, on the function of BRCA2 with implications for cancer development and treatment.

BRCA2, GENOME INTEGRITY MAINTENANCE, AND CANCER

The role of BRCA2 in HR has been a subject of active investigation for many years (Prakash et al. 2015). HR repairs DNA lesions including DNA double-strand breaks (DSBs) using a homologous DNA sequence, typically the sister chromatid in mitotic cells. One critical step of HR is strand invasion, which primes subsequent DNA synthesis using the homologous sequence as a template, thereby ensuring an error-free repair outcome (Fig. 1). BRCA2 plays an essential role in this process by loading RAD51 recombinase onto single-stranded DNA formed at DSBs; the RAD51 nucleoprotein filaments then catalyze the subsequent strand invasion reaction (Jensen et al. 2010; Liu et al. 2010; Thorslund et al. 2010). Due to the HR deficiency, cells with impaired BRCA2 function are hypersensitive to cross-linking agents, such as cisplatin, and to poly(ADP-ribose) polymerase (PARP) inhibitors, which are being extensively explored as cancer therapeutics (Lord and Ashworth 2016).

FP is an additional BRCA2-mediated process that helps safeguard genomic integrity (Schlacher et al. 2011). Under replication stress, nascent DNA strands at stalled forks are susceptible to degradation by nucleases such as MRE11. BRCA2 prevents such nascent strand degrada-

Published by Cold Spring Harbor Laboratory Press; doi: 10.1101/sqb.2017.82.035006

Cold Spring Harbor Symposia on Quantitative Biology, Volume LXXXII

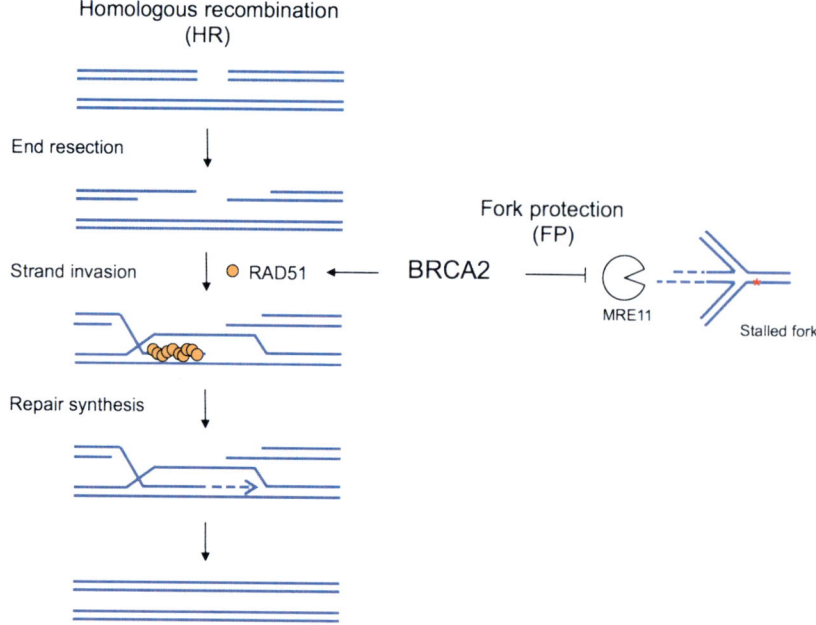

Figure 1. HR and FP functions of BRCA2 to protect genome integrity. (*Left*) HR repair of a DSB is initiated by resection of the break ends to generate 3′ single-stranded DNA overhangs. Subsequent strand invasion into a homologous DNA is critical for repair DNA synthesis, which ultimately promotes an error-free repair outcome. Although RAD51 is crucial for strand invasion, BRCA2 plays an essential role by recruiting RAD51 onto the resected DNA. (*Right*) In the FP process, BRCA2 prevents the nascent strands of a stalled replication fork from being degraded by nucleases such as MRE11.

tion thereby protecting stalled forks (Fig. 1). In addition to BRCA2, other HR proteins, such as RAD51 itself, the breast and ovarian cancer suppressor BRCA1, and Fanconi anemia proteins also play important roles in the FP pathway (Schlacher et al. 2011, 2012). However, FP and HR are functionally separable, as evidenced by multiple approaches that specifically manipulate one process without affecting the other (Schlacher et al. 2011; Ding et al. 2016; Ray Chaudhuri et al. 2016; Dungrawala et al. 2017; Feng and Jasin 2017; Taglialatela et al. 2017). Of note, the role of RAD51 in FP seems to be more complicated than BRCA2: disruption of its function compromises FP in some studies (Schlacher et al. 2011; Kolinjivadi et al. 2017), but does not affect FP activity under other circumstances (Thangavel et al. 2015; Feng and Jasin 2017; Lemacon et al. 2017; Mijic et al. 2017). The seeming discrepancy can be attributed to an additional function of RAD51, that of fork reversal (Zellweger et al. 2015), which seems to be required for nascent strand degradation, a subject that is currently a focus of the field (discussed in detail below).

Since the initial identification of BRCA1 and BRCA2, a series of other HR factors, in particular PALB2, which bridges BRCA1 and BRCA2, and RAD51 paralogs RAD51C and RAD51D have been identified as tumor suppressors (Erkko et al. 2007; Rahman et al. 2007; Tischkowitz et al. 2007; Meindl et al. 2010; Loveday et al. 2011). Where tested, these proteins also have roles in FP (Fig. 3; Somyajit et al. 2015). Similar to BRCA1 and BRCA2 deficiency, disruption of these proteins also causes sensitivity to cross-linking agents and PARP inhibitors (Prakash et al. 2015). *BRCA1/2* mutant cancers exhibit a

particular pattern of base substitutions and genome rearrangements—that is, "mutational signatures" (Nik-Zainal et al. 2016; Polak et al. 2017). A computational model based on these mutational signatures has been used to predict additional patient tumors with HR-deficiency beyond those containing mutations in known HR genes, thus expanding the pool of cancers that may respond to platinum or PARP inhibitor therapy (Davies et al. 2017).

Despite encouraging success in the clinic, resistance to platinum drugs and PARP inhibitor therapy can eventually be acquired by tumors. Secondary mutations in the mutated HR genes are frequently observed that reestablish the reading frame and, when checked, restore protein function (Edwards et al. 2008; Sakai et al. 2008; Norquist et al. 2011; Kondrashova et al. 2017; Chen et al. 2018). Remarkably, in a recent study of circulating tumor DNA from prostate cancer patients, 34 secondary mutations were identified in a single patient that restored the BRCA2 reading frame to confer therapy resistance (Quigley et al. 2017). HR restoration is usually considered to be the underlying mechanism, although FP is likely to be restored as well in most instances. Interestingly, FP restoration has been associated with chemoresistance in experimental models, even without restoration of HR (Ray Chaudhuri et al. 2016). Validating the causal relationship between FP and therapy resistance in patients will be critical to understand the clinical relevance of FP.

CONSEQUENCES OF BRCA2 DEFICIENCY

One puzzle in the field is that, while predisposing to cancer, BRCA2 deficiency paradoxically leads to inviability in mice, both in embryos and in cells (Patel et al. 1998; Evers and Jonkers 2006; Kuznetsov et al. 2008; Badie et al. 2010). Therefore, a gap in our understanding needs to be bridged between the immediate consequence of cell inviability and the long-term tumor susceptibility from BRCA2 deficiency. In addition, while BRCA2 loss is expected to impair HR in all tissues (Kass et al. 2016), it predominantly predisposes to cancer in the breast and ovary. Resolution of these paradoxes requires a cellular model from a disease-relevant tissue, such as human mammary epithelial cells.

We recently set out to approach these questions by generating BRCA2 conditional models in a nontransformed human mammary epithelial cell line with a relatively stable genome (MCF10A; [Soule et al. 1990; Cowell et al. 2005]). This study reveals that BRCA2-deficiency-triggered cell lethality is conserved in these relatively normal human mammary cells (Feng and Jasin 2017). BRCA2 deficiency leads to cell cycle arrest in G_1, a surprising result considering that BRCA2 functions in genome integrity maintenance pathways that are active in S and G_2. To reconcile these seemingly counterintuitive results, we traced the source of DNA lesions that occur upon BRCA2 loss. We found that BRCA2 inactivation leads to DNA under replication, which in turn causes abnormalities during mitosis and 53BP1 nuclear body formation in the subsequent G_1 phase associated with a p53-dependent cell cycle arrest (Fig. 2; Feng and Jasin 2017). Independent results from other groups are in agreement with this model (Lai et al. 2017; Schoonen et al. 2017). Notably, while mitotic abnormalities have previously been associated with BRCA2 deficiency (Tutt et al. 1999; Daniels et al. 2004; Laulier et al. 2011; Choi et al. 2012; Mondal et al. 2012), we establish that it is the premitotic stresses from the S and G_2 phases that cause the subsequent aberrations (chromosome missegregation, 53BP1 nuclear body formation), given that delaying mitotic entry abrogates these abnormalities (Feng and Jasin 2017).

What is the cause of DNA under replication? BRCA2-deficient cells accumulate single-stranded DNA lesions in G_2 (Feng and Jasin 2017). Fork reversal, hyperresection at DNA breaks (Feng and Jasin 2017), and single-stranded DNA formed at or behind the forks (Kolinjivadi et al. 2017) can all contribute to these lesions. We propose that these persistent, unrepaired DNA lesions prevent timely completion of DNA replication. Thus, the sequelae of BRCA2 deficiency can be traced from S/G_2 DNA lesions as the source of the G_1 arrest to cell inviability as the consequence (Fig. 2; Feng and Jasin 2017).

INVOLVEMENT OF p53 AFTER BRCA2 LOSS

The above model predicts that the lethal phenotypes of BRCA2 disruption may be mitigated by inactivation of cell cycle arrest processes. The p53 pathway is induced upon BRCA2 loss and is responsible for the subsequent G_1 arrest. Indeed, p53 loss partially restores cell proliferation to BRCA2-deficient cells (Feng and Jasin 2017). These observations provide insight into the frequent asso-

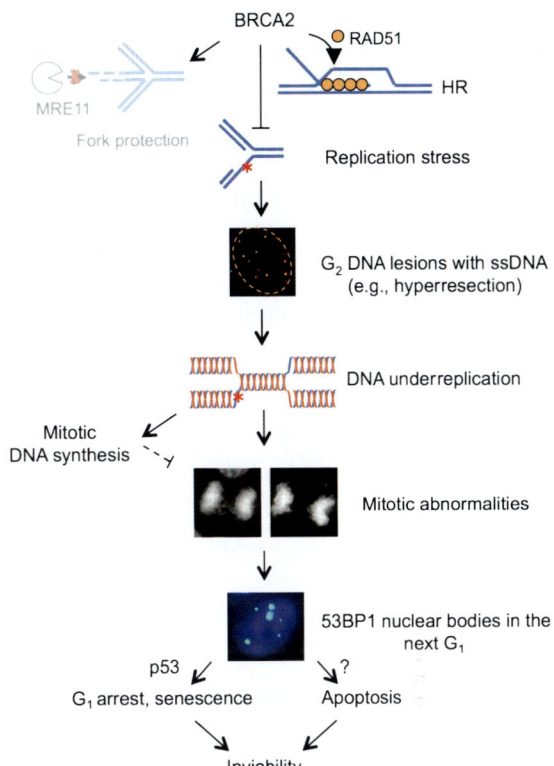

Figure 2. Consequences of BRCA2 deficiency in multiple cell cycle phases. As proposed in our recent study using MCF10A cells (Feng and Jasin 2017), even in unperturbed situations, BRCA2 deficiency causes replication stress that compromises the timely completion of DNA replication. The resulting underreplicated DNA leads to G_2 DNA lesions and single-stranded DNA (ssDNA) formation, which in turn leads to abnormalities in mitosis and 53BP1 nuclear body formation in the subsequent G_1 phase. Such G_1 lesions trigger p53-dependent G_1 arrest and cellular senescence as well as p53-independent apoptosis, resulting in cell inviability. At the functional level, suppression of replication stress is primarily mediated by the HR, rather than the FP, activity of BRCA2. (Reprinted from Feng and Jasin 2017 under a Creative Commons license http://creativecommons.org/licenses/by/4.0/.)

ciation of *TP53* mutations with *BRCA2*-mutated cancers (Ramus et al. 1999; Greenblatt et al. 2001; Roy et al. 2011), which parallels results from mouse systems (Jonkers et al. 2001; Bouwman et al. 2010). However, p53 inactivation only partially rescues MCF10A cells (Feng and Jasin 2017). While senescence and G_1 arrest are substantially abrogated by p53 loss, apoptosis is not (Feng and Jasin 2017), indicating that a p53-independent apoptotic pathway prevents full rescue of viability in the absence of BRCA2. One possibility is the involvement of other p53 family members (i.e., p63 and p73), which play similar roles to p53 in multiple processes, including apoptosis, and have been shown to act redundantly in some contexts (Dotsch et al. 2010; Wang et al. 2017). Therefore, multiple pathways, both p53-dependent and -independent, work together to ultimately result in inviability of BRCA2-deficient cells (Figs. 2, 5A).

p53 pathway activation in G_1 could occur through several nonmutually exclusive ways. First, 53BP1 directly interacts with p53 and regulates p53-dependent G_1 checkpoint arrest (Iwabuchi et al. 1994; Cuella-Martin et al. 2016). Therefore, p53 can be activated as a direct response to 53BP1 nuclear bodies. However, this interaction may not be the sole explanation for p53 activation, given that 53BP1 inactivation neither restores viability nor diminishes p53 induction in BRCA2-deficient mouse cells (Bouwman et al. 2010). Second, 53BP1 nuclear bodies mark DNA breaks due to improper resolution of underreplicated DNA during mitosis (Lukas et al. 2011; Naim et al. 2013; Ying et al. 2013); p53 activation can thus be triggered as a downstream event of DNA damage signaling. Consistent with this notion, a number of DNA damage response proteins reside in 53BP1 nuclear bodies (Lukas et al. 2011). In addition to 53BP1-related mechanisms, p53-dependent G_1 arrest may also be activated by mitotic errors independently of DNA damage (Kuffer et al. 2013; Ganem et al. 2014; Pedersen et al. 2016). Collectively, we propose that the replication stress in BRCA2-deficient cells leads to abnormalities in the subsequent M and G_1 phases, which

in turn trigger apoptosis together with p53-dependent G_1 arrest and cellular senescence (Fig. 2).

FORK REVERSAL AND NASCENT STRAND DEGRADATION

Unlike other canonical HR factors such as BRCA2, RAD51 plays a more complex role during FP. Manipulating RAD51 filament stability itself clearly impacts FP: Disrupting RAD51 filaments in wild-type cells, by overexpressing a BRC repeat from BRCA2, impairs FP, while stabilizing the filament, by expressing RAD51 K133R, can restore FP in deficient cells (Schlacher et al. 2011, 2012). Paradoxically, depletion of RAD51 itself, however, does not affect FP (Thangavel et al. 2015; Feng and Jasin 2017). Emerging evidence suggests that the resolution of this conundrum lies in the process of fork reversal that precedes nascent strand degradation (Fig. 3A). Fork rever-

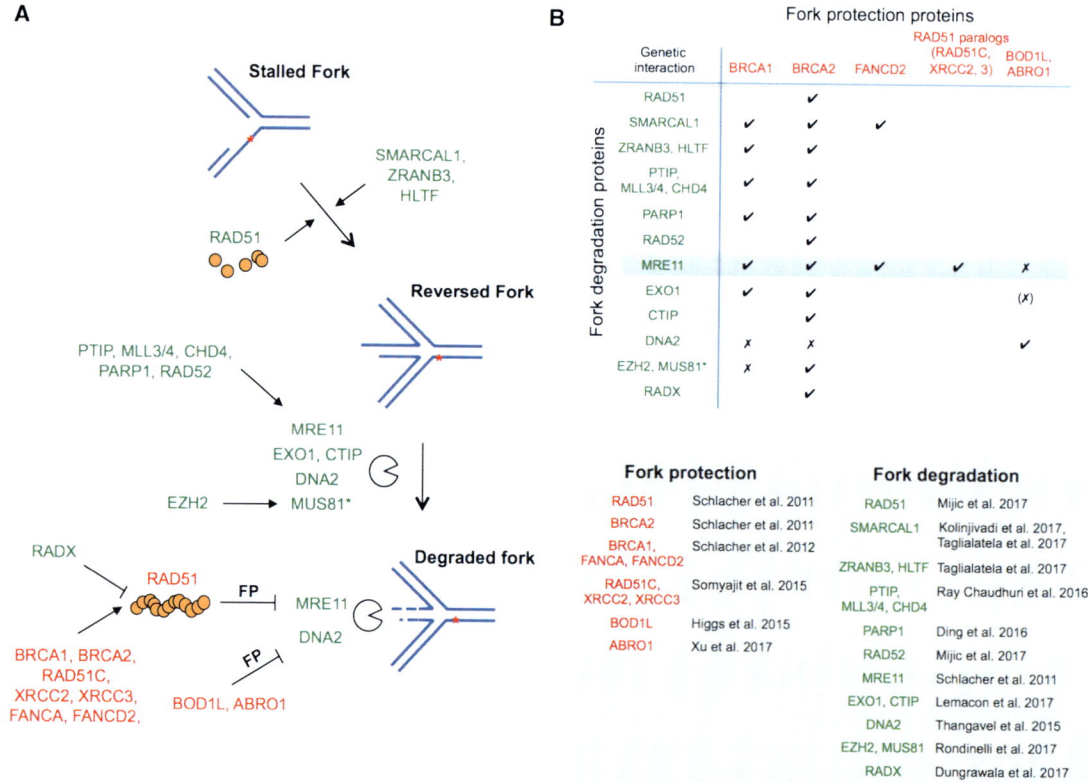

Figure 3. Mechanisms of fork degradation and protection. (*A*) Fork degradation occurs on a reversed fork. Reversal of stalled replication forks is promoted by RAD51 recombinase and DNA translocases (SMARCAL1, ZRANB3, HLTF). In the absence of FP factors like BRCA2, MRE11 and other nucleases can lead to fork degradation. Different pathways regulate MRE11 and MUS81 recruitment to stalled forks. MRE11 recruitment is promoted by a number of proteins, including PARP1 and the PTIP-MLL3/4 axis, whereas MUS81 recruitment relies on chromatin modifier EZH2. Prevention of MRE11-mediated fork degradation requires RAD51, which is facilitated by HR-Fanconi anemia proteins but antagonized by RADX. Thus, RAD51 both promotes and prevents fork degradation, in particular, at the steps of fork reversal and through the formation of stabilized filaments, respectively. Prevention of DNA2-mediated fork degradation involves BOD1L and ABRO1. Proteins that contribute to FP and degradation are labeled with a red and green color, respectively. Asterisk, MUS81's role in fork degradation is not always observed (discussed in the text). (*B*) Summary of the reported genetic interactions between FP proteins and proteins that directly or indirectly promote fork degradation, simplified here as "fork degradation proteins." A checkmark indicates that the absence of a FP protein leads to nascent strand degradation involving the corresponding fork degradation protein. An *x* indicates that evidence exists that a given fork degradation protein is not responsible for fork degradation in the absence of the corresponding FP protein. (*x*) indicates that EXO1 was not tested for BOD1L. Citations of proteins involved in FP and degradation are listed.

sal refers to the remodeling of replication forks into a four-way junction structure, a process that ensures proper resumption of replication observed in a variety of species/cell types including mammalian cells (Neelsen and Lopes 2015; Berti and Vindigni 2016; Quinet et al. 2017). Surprisingly, RAD51 is critical for reversing forks, but BRCA2 does not seem to play a role in this process (Zellweger et al. 2015; Mijic et al. 2017). Therefore, the fork reversal activity of RAD51 is distinguishable from its strand invasion function during the canonical HR process.

Reversed forks have emerged as the entry point for subsequent nascent strand degradation (Kolinjivadi et al. 2017; Lemacon et al. 2017; Mijic et al. 2017; Taglialatela et al. 2017). Indeed, RAD51 depletion in BRCA2-deficient cells precludes nascent strand degradation (Lemacon et al. 2017; Mijic et al. 2017), which we also observe in our system (Feng and Jasin 2017). Moreover, disruption of DNA translocases that have established fork reversal activity, such as SMARCAL1 (Bétous et al. 2012, 2013), ZRANB3 (Ciccia et al. 2012), and HLTF (Kile et al. 2015), also restores FP to BRCA2-deficient cells or BRCA2-depleted *Xenopus* egg extracts (Kolinjivadi et al. 2017; Lemacon et al. 2017; Mijic et al. 2017; Taglialatela et al. 2017). Together, these recent studies converge on a two-step model in which stalled replication forks are first reversed by RAD51 and DNA translocases, which are then vulnerable to nascent strand degradation, a step antagonized by the FP process mediated by BRCA2-dependent RAD51 filament formation/stabilization (Fig. 3A).

The fork reversal and protection activities of RAD51 prove to be functionally separable, as evidenced by the RAD51 T131P mutant expressed in a patient-derived cell line (Wang et al. 2015), which is proficient at fork reversal, but impaired in FP (Mijic et al. 2017).

The FP defect in these cells is presumably due to a failure to form stable RAD51 filaments (Wang et al. 2015), consistent with our previous findings using a BRCA2 separation of function mutant (Schlacher et al. 2011). Whether RAD51 has intrinsic fork reversal activity or promotes the activity of a translocase is not clear. Taken together, the ability of RAD51 to act at distinct biochemical steps may explain the diverse outcomes when RAD51 is perturbed in different ways.

NUCLEASES INVOLVED IN NASCENT STRAND DEGRADATION

Along with requiring a reversed fork as a substrate, nascent strand degradation involves multiple nucleases and epigenetic control. MRE11 was the first nuclease characterized to be involved in the resection of stalled forks in BRCA/FA-deficient backgrounds and remains the gold standard for analyzing FP pathways (Schlacher et al. 2011, 2012). MRE11 recruitment to stalled replication forks is mediated by histone H3 lysine 4 methyltransferases MLL3 and MLL4 (MLL3/4) and interacting protein PTIP (Ray Chaudhuri et al. 2016) and also relies on a continuously expanding list of proteins, such as PARP1 (Ding et al. 2016), CHD4 (Ray Chaudhuri et al.

2016), and RAD52 (Fig. 3A; Mijic et al. 2017). Either directly inhibiting MRE11 activity with small molecules or disrupting its recruitment by depleting any of these proteins restores FP to BRCA2- (and where tested BRCA1-) deficient cells (Fig. 3B), underscoring the critical role of MRE11 in nascent strand degradation.

In addition to MRE11, resection enzymes EXO1 and CTIP also independently mediate nascent strand degradation in BRCA1- and BRCA2-deficient cells (Fig. 3; Lemacon et al. 2017). In contrast, another resection nuclease DNA2 does not play a major role in these mutants (Thangavel et al. 2015; Ray Chaudhuri et al. 2016), although it has been shown to play a role in other contexts (i.e., with deficiencies of the FP factors BOD1L and ABRO1 [Fig. 3]). ABRO1 protects stalled forks independent of RAD51 (Xu et al. 2017a); while BOD1L promotes RAD51 chromatin loading (Higgs et al. 2015), whether stabilized RAD51 filaments are required in the absence of BOD1L remains to be tested. DNA2 also degrades stalled forks even in wild-type U2OS cells under prolonged replication stress resulting from hydroxyurea treatment, while MRE11, EXO1, and CTIP are not involved in this process (Thangavel et al. 2015). Therefore, multiple nucleolytic pathways are differentially activated in response to varying stresses and genetic perturbations.

The structure-specific nuclease MUS81 has a more complex involvement in the FP pathway. Parallel to the MLL3/4-PTIP recruitment path for MRE11 is the recently described EZH2-MUS81 axis (Fig. 3A), which provides one context in considering the role of MUS81. EZH2 is recruited to stalled forks and mediates methylation of histone H3 at lysine 27, and MUS81 is able to interact with this histone modification.

Perturbation of either EZH2 or MUS81 restores FP to BRCA2-defective cells (Rondinelli et al. 2017). This discovery further extends the scope of nucleases and epigenetic alterations at the fork that promote nascent strand degradation, although it remains to be determined whether MUS81 nuclease activity is directly involved. Interestingly, the EZH2-MUS81 axis only operates in BRCA2-, but not BRCA1-, deficient backgrounds (Fig. 3B; Rondinelli et al. 2017), indicating that the FP functions of BRCA1 and BRCA2 are separable. Accordingly, MUS81 chromatin recruitment is induced upon loss of BRCA2 (Bhowmick et al. 2016; Lai et al. 2017; Lemacon et al. 2017; Rondinelli et al. 2017), but not BRCA1 (Lemacon et al. 2017; Xu et al. 2017b). It is well appreciated that BRCA1 acts at a distinct step of HR, upstream of BRCA2 (Chen et al. 2018), but thus far, the biology behind their different roles in FP remains unclear and warrants future investigation.

While providing a risk to genome integrity, one potentially positive role for nascent strand degradation is to promote fork restart. BRCA2-deficient cells undergo extensive fork degradation but maintain normal fork restart activity, at least in some contexts (Schlacher et al. 2011). Two recent studies reported that MUS81 nuclease supports fork restart in the absence of BRCA2 (Lemacon et al. 2017; Rondinelli et al. 2017). MUS81 acts by cleaving degraded forks to mediate fork restart, a conclusion that is supported by the observation that MUS81 depletion in BRCA2-deficient cells leads to fewer DSBs and a concomitant increase of reversed forks, especially those with a single-stranded arm, indicative of extensive fork resection (Lemacon et al. 2017). Consistent with this, MUS81 depletion did not rescue FP in the BRCA2-deficient U2OS cells and in fact the cells are sensitized to hydroxyurea (Lemacon et al. 2017), in line with a synthetic lethal interaction between the two genes observed in another study (Lai et al. 2017).

However, Rondinelli et al., using a panel of other cell lines, has argued for a synthetic viable, rather than synthetic lethal interaction: MUS81 disruption restores FP upon BRCA2 deficiency to confer resistance to PARP inhibition (Rondinelli et al. 2017), although this was not observed by Lemacon et al. (2017). The discrepant roles of MUS81 in fork degradation could be related to its ability to process diverse types of substrates (Dehe and Gaillard 2017).

CONTRIBUTIONS OF HR AND FP TO CELL FITNESS

One key mechanistic question related to BRCA2 function is how it supports cell proliferation. HR factors like BRCA1 and BRCA2 are essential for embryonic cell survival, which has implicated HR as an essential process in mammalian cells (Moynahan and Jasin 2010). More recently, however, FP in the absence of HR has been reported to sustain viability of BRCA2-deficient cells, both mouse embryonic stem (ES) cells and tumor cells subjected to chemotherapy, arguing that HR may not be essential in these contexts (Ding et al. 2016; Ray Chaudhuri et al. 2016). Subsequent studies have also observed improved cell fitness (viability, chemoresistance) when nascent strand degradation is prevented, by suppressing either fork reversal (Taglialatela et al. 2017) or the RAD51 antagonist RADX (Dungrawala et al. 2017).

However, the correlation between FP and cell fitness does not always hold true. For example, cells with a mutation at the BRCA2 S3291 residue are severely disrupted in FP, but remain proficient in both unperturbed survival and preventing genome instability under PARP inhibitor/platinum treatment, presumably because HR is intact (Schlacher et al. 2011; Feng and Jasin 2017). In fact, fully restoring FP to BRCA2-deficient cells by perturbing different proteins (ZRANB3, MUS81, RADX) (Fig. 3) varies dramatically in the extent to which it leads to acquired chemoresistance, ranging from no resistance (Lemacon et al. 2017; Mijic et al. 2017) to partial or substantial resistance (Dungrawala et al. 2017; Rondinelli et al. 2017). Different cellular contexts could contribute to the varying experimental outcomes. Therefore, it is critical to determine the functional relationship between HR and FP for viability of normal human cells, particularly in BRCA1/2-relevant mammary cells.

To approach this question, we generated three independent separation-of-function systems in MCF10A cells (Fig. 4). FP, but not HR, is selectively disrupted through expression of a mutant BRCA2 peptide (BRCA2 S3291E)

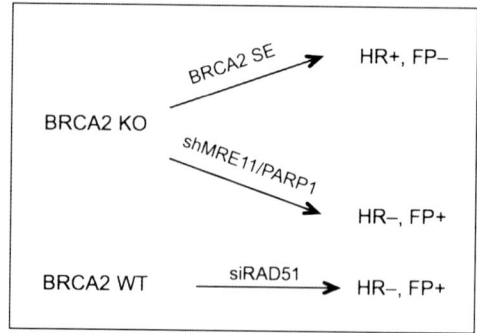

Figure 4. Separation-of-function systems to study HR and FP functions. Three separation-of-function systems to dissect the individual contributions of HR and FP pathways have been described (Feng and Jasin 2017).

or restored by MRE11 or PARP1 ablation in BRCA2-deficent cells (Feng and Jasin 2017). In the third system, HR, but not FP, is selectively impaired by RAD51 depletion in BRCA2 wild-type cells (Feng and Jasin 2017). All three systems unambiguously show that HR, but not FP is critical to suppress replication stress and support cell viability in this nontransformed human mammary cell line (Fig. 2; Feng and Jasin 2017).

Thus, these two pathways, HR and FP, are differentially required for cell viability and genome integrity in different cell lines. How easy it is to meet the survival threshold, and thereby dictate the outcome of pathway restoration, may depend on the cellular context (Fig. 5). For example, the p53 pathway is compromised in mouse ES cells (Aladjem et al. 1998), such that these cells may have a higher tolerance to DNA damage; accordingly, preventing nascent strand degradation alone may be sufficient for their survival. Similarly, having gone through the stresses of BRCA2 loss, tumor cells have evolved to tolerate HR loss, such that FP restoration alone may suffice to reduce genome instability and thereby confer chemoresistance (Fig. 5). In agreement with a context-dependent pathway requirement, SMARCAL1 inactivation, preventing fork degradation, alleviates sensitivity to PARP inhibitors and cisplatin of BRCA1-depleted breast cancer MDA-MB-231 cells, but not MCF10A cells (Taglialatela et al. 2017). Future studies are needed to understand the relationship between these pathways and various cellular milieus.

IMPLICATIONS FOR THERAPY

More than a decade ago, seminal work on PARP inhibition opened a new avenue of exploiting synthetic lethality to target cancers with BRCA1/2 deficiency (Bryant et al. 2005; Farmer et al. 2005). PARP inhibitors have clearly shown promise with FDA approval for ovarian and, more recently, breast cancer treatment. Underlying mechanisms for the hypersensitivity of *BRCA1/2*-mutant tumor cells to PARP inhibition include accumulation of single-strand breaks and especially trapping of PARP protein on DNA to impede replication, both of which lead to lesions that require HR to be repaired (Lord and Ashworth

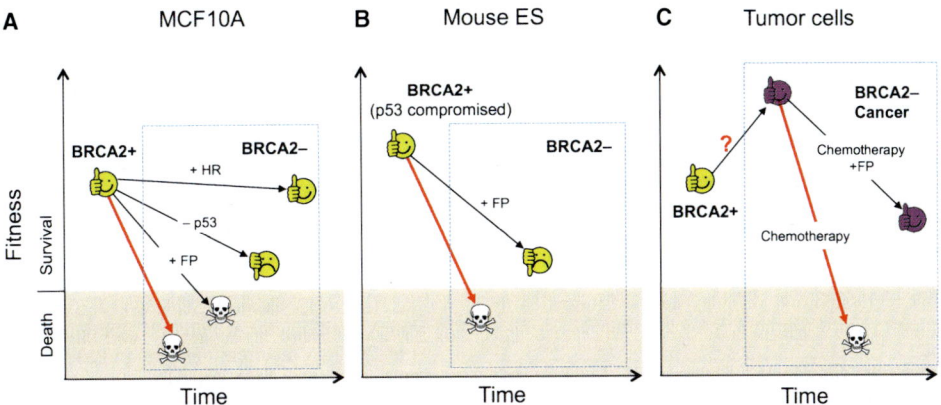

Figure 5. Context-dependent requirement for HR versus FP for cell survival. Differential requirements for HR and FP are observed to reach the threshold of cell viability in different contexts of BRCA2 deficiency, as highlighted in the dashed blue boxes. BRCA2 loss alone (untreated or with chemotherapy) (red arrows) is shown as a baseline for comparison to BRCA2 loss with the indicated additional genetic alterations (black arrows). (*A*) Nontransformed mammary epithelial cells MCF10A do not survive BRCA2 loss. Restoration of HR, but not FP, re-establishes cell viability. p53 loss fails to suppress apoptosis and only partially rescues cell proliferation. (*B*) In mouse ES cells, FP restoration is sufficient to rescue cell survival after BRCA2 loss. These cells may have a higher tolerance to DNA damage than MCF10A cells due to a compromised p53 response. (*C*) In BRCA2-deficient tumor cells, FP is also sufficient to overcome the viability threshold to confer chemoresistance. Presumably, these cells have evolved to survive BRCA2 loss and aberrantly proliferate (purple face) and thus may have a higher tolerance to DNA damage.

2017). Impaired mitotic progression may also promote PARP inhibition-associated cytotoxicity (Schoonen et al. 2017).

Despite the promise, PARP inhibitor therapies are challenged by tumor relapse with acquired resistance (Lord and Ashworth 2017). Thus, novel independent strategies to target *BRCA1/2* cancers may complement current therapies. Exploiting an increasing dependence of HR-deficient cells on minor DSB repair pathways has led to the discovery of novel targets for synthetic lethality: POLθ, which plays a role in microhomology-mediated end joining (Ceccaldi et al. 2015; Mateos-Gomez et al. 2015), and RAD52, a nonessential HR factor which becomes essential for the residual HR in BRCA1/2-deficient cells (Feng et al. 2011; Lok et al. 2013). More recently, another strategy to target has been suggested that involves the phenomenon of mitotic DNA synthesis. Even under unperturbed conditions, BRCA2 prevents DNA under replication (Feng and Jasin 2017; Lai et al. 2017), such that in the absence of BRCA2 mitotic DNA synthesis is activated (Feng and Jasin 2017; Lai et al. 2017), presumably as a last resort to complete DNA replication and prevent mitotic catastrophe (Minocherhomji et al. 2015; Bhowmick et al. 2016). A potential target in this pathway is RAD52, which is proposed to help prime mitotic DNA synthesis (Bhowmick et al. 2016). Other mitotic DNA synthesis components include MUS81 or SLX4, (Minocherhomji et al. 2015), loss of which impairs survival of BRCA2-deficient cancer cells (Lai et al. 2017). Interestingly, in contrast to the case of otherwise unperturbed cell proliferation, upon PARP inhibition MUS81 depletion paradoxically improves the viability of BRCA2-deficient cells, as discussed above (Rondinelli et al. 2017). Given that the critical pathways for cell survival may vary under different stresses, the importance of the accurate design of therapy strategies cannot be overstated.

CONCLUSION

Significant progress has been made in advancing our understanding of BRCA2 functions, in particular, the interplay between replication and repair, therapy design, and resistance mechanisms. Despite the recent progress, we are only starting to learn the functional relationship between HR and FP, and processes like fork reversal and mitotic DNA synthesis. Also remaining largely unexplored are the complexity of the differential contributions of these pathways in diverse contexts and its possible implications in different biological processes (i.e., development, tumor formation, therapy response, and resistance acquisition). Furthermore, the additional tumor suppressive barriers that impede cell survival after *BRCA2* loss are not fully understood. Future studies are needed to provide insight into these aspects to improve the picture of BRCA2 biology as well as strengthen cancer therapy.

ACKNOWLEDGMENTS

We thank members of the Jasin laboratory for discussions and suggestions. W.F. was supported by an Olayan Fellowship. The Jasin laboratory is supported by MSK Cancer Center Support Grant/Core Grant P30 CA008748, a Geoffrey Beene Cancer Research Center Grant, and National Institutes of Health (NIH) grant R01 CA185660.

REFERENCES

Aladjem MI, Spike BT, Rodewald LW, Hope TJ, Klemm M, Jaenisch R, Wahl GM. 1998. ES cells do not activate p53-dependent stress responses and undergo p53- independent apoptosis in response to DNA damage. *Curr Biol* **8:** 145–155.
Antoniou A, Pharoah PDP, Narod S, Risch HA, Eyfjord JE, Hopper JL, Loman N, Olsson H, Johannsson O, Borg Å,

et al. 2003. Average risks of breast and ovarian cancer associated with BRCA1 or BRCA2 mutations detected in case series unselected for family history: A combined analysis of 22 studies. *Am J Hum Genet* **72**: 1117–1130.

Badie S, Escandell JM, Bouwman P, Carlos AR, Thanasoula M, Gallardo MM, Suram A, Jaco I, Benitez J, Herbig U, et al. 2010. BRCA2 acts as a RAD51 loader to facilitate telomere replication and capping. *Nat Struct Mol Biol* **17**: 1461–1469.

Berti M, Vindigni A. 2016. Replication stress: Getting back on track. *Nat Struct Mol Biol* **23**: 103–109.

Bétous R, Mason AC, Rambo RP, Bansbach CE, Badu-Nkansah A, Sirbu BM, Eichman BF, Cortez D. 2012. SMARCAL1 catalyzes fork regression and Holliday junction migration to maintain genome stability during DNA replication. *Genes Dev* **26**: 151–162.

Bétous R, Couch FB, Mason AC, Eichman BF, Manosas M, Cortez D. 2013. Substrate-selective repair and restart of replication forks by DNA translocases. *Cell Rep* **3**: 1958–1969.

Bhowmick R, Minocherhomji S, Hickson ID. 2016. RAD52 facilitates mitotic DNA synthesis following replication stress. *Mol Cell* **64**: 1117–1126.

Bouwman P, Aly A, Escandell JM, Pieterse M, Bartkova J, van der Gulden H, Hiddingh S, Thanasoula M, Kulkarni A, Yang Q, et al. 2010. 53BP1 loss rescues BRCA1 deficiency and is associated with triple-negative and BRCA-mutated breast cancers. *Nat Struct Mol Biol* **17**: 688–695.

Bryant HE, Schultz N, Thomas HD, Parker KM, Flower D, Lopez E, Kyle S, Meuth M, Curtin NJ, Helleday T. 2005. Specific killing of BRCA2-deficient tumours with inhibitors of poly (ADP-ribose) polymerase. *Nature* **434**: 913–917.

Ceccaldi R, Liu JC, Amunugama R, Hajdu I, Primack B, Petalcorin MI, O'Connor KW, Konstantinopoulos PA, Elledge SJ, Boulton SJ, et al. 2015. Homologous- recombination-deficient tumours are dependent on Polθ-mediated repair. *Nature* **518**: 258–262.

Chen CC, Feng W, Lim PX, Kass EM, Jasin M. 2018. Homology-directed repair and the role of BRCA1, BRCA2, and related proteins in genome integrity and cancer. *Annu Rev Cancer Biol* **2**: 313–336.

Choi E, Park PG, Lee HO, Lee YK, Kang GH, Lee JW, Han W, Lee HC, Noh DY, Lekomtsev S, et al. 2012. BRCA2 fine-tunes the spindle assembly checkpoint through reinforcement of BubR1 acetylation. *Dev Cell* **22**: 295–308.

Ciccia A, Nimonkar AV, Hu Y, Hajdu I, Achar YJ, Izhar L, Petit SA, Adamson B, Yoon JC, Kowalczykowski SC, et al. 2012. Polyubiquitinated PCNA recruits the ZRANB3 translocase to maintain genomic integrity after replication stress. *Mol Cell* **47**: 396–409.

Cowell JK, LaDuca J, Rossi MR, Burkhardt T, Nowak NJ, Matsui S. 2005. Molecular characterization of the t(3;9) associated with immortalization in the MCF10A cell line. *Cancer Genet Cytogenet* **163**: 23–29.

Cuella-Martin R, Oliveira C, Lockstone HE, Snellenberg S, Grolmusova N, Chapman JR. 2016. 53BP1 integrates DNA repair and p53-dependent cell fate decisions via distinct mechanisms. *Mol Cell* **64**: 51–64.

D'Andrea AD. 2010. Susceptibility pathways in Fanconi's anemia and breast cancer. *N Engl J Med* **362**: 1909–1919.

Daniels MJ, Wang Y, Lee M, Venkitaraman AR. 2004. Abnormal cytokinesis in cells deficient in the breast cancer susceptibility protein BRCA2. *Science* **306**: 876–879.

Davies H, Glodzik D, Morganella S, Yates LR, Staaf J, Zou X, Ramakrishna M, Martin S, Boyault S, Sieuwerts AM, et al. 2017. HRDetect is a predictor of BRCA1 and BRCA2 deficiency based on mutational signatures. *Nat Med* **23**: 517–525.

Dehe PM, Gaillard PH. 2017. Control of structure-specific endonucleases to maintain genome stability. *Nat Rev Mol Cell Biol* **18**: 315–330.

Ding X, Ray Chaudhuri A, Callen E, Pang Y, Biswas K, Klarmann KD, Martin BK, Burkett S, Cleveland L, Stauffer S, et al. 2016. Synthetic viability by BRCA2 and PARP1/ARTD1 deficiencies. *Nat Commun* **7**: 12425.

Dotsch V, Bernassola F, Coutandin D, Candi E, Melino G. 2010. p63 and p73, the ancestors of p53. *Cold Spring Harb Perspect Biol* **2**: a004887.

Dungrawala H, Bhat KP, Le Meur R, Chazin WJ, Ding X, Sharan SK, Wessel SR, Sathe AA, Zhao R, Cortez D. 2017. RADX promotes genome stability and modulates chemosensitivity by regulating RAD51 at replication forks. *Mol Cell* **67**: 374–386 e375.

Edwards SL, Brough R, Lord CJ, Natrajan R, Vatcheva R, Levine DA, Boyd J, Reis-Filho JS, Ashworth A. 2008. Resistance to therapy caused by intragenic deletion in BRCA2. *Nature* **451**: 1111–1115.

Erkko H, Xia B, Nikkila J, Schleutker J, Syrjakoski K, Mannermaa A, Kallioniemi A, Pylkas K, Karppinen SM, Rapakko K, et al. 2007. A recurrent mutation in PALB2 in Finnish cancer families. *Nature* **446**: 316–319.

Evers B, Jonkers J. 2006. Mouse models of BRCA1 and BRCA2 deficiency: Past lessons, current understanding and future prospects. *Oncogene* **25**: 5885–5897.

Farmer H, McCabe N, Lord CJ, Tutt AN, Johnson DA, Richardson TB, Santarosa M, Dillon KJ, Hickson I, Knights C, et al. 2005. Targeting the DNA repair defect in BRCA mutant cells as a therapeutic strategy. *Nature* **434**: 917–921.

Feng W, Jasin M. 2017. BRCA2 suppresses replication stress-induced mitotic and G_1 abnormalities through homologous recombination. *Nat Commun* **8**: 525.

Feng Z, Scott SP, Bussen W, Sharma GG, Guo G, Pandita TK, Powell SN. 2011. Rad52 inactivation is synthetically lethal with BRCA2 deficiency. *Proc Natl Acad Sci* **108**: 686–691.

Ganem NJ, Cornils H, Chiu SY, O'Rourke KP, Arnaud J, Yimlamai D, Thery M, Camargo FD, Pellman D. 2014. Cytokinesis failure triggers hippo tumor suppressor pathway activation. *Cell* **158**: 833–848.

Greenblatt MS, Chappuis PO, Bond JP, Hamel N, Foulkes WD. 2001. TP53 mutations in breast cancer associated with BRCA1 or BRCA2 germ-line mutations: Distinctive spectrum and structural distribution. *Cancer Res* **61**: 4092–4097.

Hanahan D, Weinberg RA. 2011. Hallmarks of cancer: The next generation. *Cell* **144**: 646–674.

Higgs MR, Reynolds JJ, Winczura A, Blackford AN, Borel V, Miller ES, Zlatanou A, Nieminuszczy J, Ryan EL, Davies NJ, et al. 2015. BOD1L is required to suppress deleterious resection of stressed replication forks. *Mol Cell* **59**: 462–477.

Iwabuchi K, Bartel PL, Li B, Marraccino R, Fields S. 1994. Two cellular proteins that bind to wild-type but not mutant p53. *Proc Natl Acad Sci* **91**: 6098–6102.

Jensen RB, Carreira A, Kowalczykowski SC. 2010. Purified human BRCA2 stimulates RAD51-mediated recombination. *Nature* **467**: 678–683.

Jonkers J, Meuwissen R, van der Gulden H, Peterse H, van der Valk M, Berns A. 2001. Synergistic tumor suppressor activity of BRCA2 and p53 in a conditional mouse model for breast cancer. *Nat Genet* **29**: 418–425.

Kass EM, Lim PX, Helgadottir HR, Moynahan ME, Jasin M. 2016. Robust homology- directed repair within mouse mammary tissue is not specifically affected by Brca2 mutation. *Nat Commun* **7**: 13241.

Kile AC, Chavez DA, Bacal J, Eldirany S, Korzhnev DM, Bezsonova I, Eichman BF, Cimprich KA. 2015. HLTF's ancient HIRAN domain binds 3′ DNA ends to drive replication fork reversal. *Mol Cell* **58**: 1090–1100.

Kolinjivadi AM, Sannino V, De Antoni A, Zadorozhny K, Kilkenny M, Techer H, Baldi G, Shen R, Ciccia A, Pellegrini L, et al. 2017. Smarcal1-mediated fork reversal triggers Mre11-dependent degradation of nascent DNA in the absence of Brca2 and stable Rad51 nucleofilaments. *Mol Cell* **67**: 867–881.e7.

Kondrashova O, Nguyen M, Shield-Artin K, Tinker AV, Teng NNH, Harrell MI, Kuiper MJ, Ho GY, Barker H, Jasin M, et al. 2017. Secondary somatic mutations restoring RAD51C and RAD51D associated with acquired resistance to the PARP inhibitor rucaparib in high-grade ovarian carcinoma. *Cancer Discov* **7**: 984–998.

Kuffer C, Kuznetsova AY, Storchova Z. 2013. Abnormal mitosis triggers p53- dependent cell cycle arrest in human tetraploid cells. *Chromosoma* **122:** 305–318.

Kuznetsov SG, Liu P, Sharan SK. 2008. Mouse embryonic stem cell-based functional assay to evaluate mutations in BRCA2. *Nat Med* **14:** 875–881.

Lai X, Broderick R, Bergoglio V, Zimmer J, Badie S, Niedzwiedz W, Hoffmann JS, Tarsounas M. 2017. MUS81 nuclease activity is essential for replication stress tolerance and chromosome segregation in BRCA2-deficient cells. *Nat Commun* **8:** 15983.

Laulier C, Cheng A, Stark JM. 2011. The relative efficiency of homology-directed repair has distinct effects on proper anaphase chromosome separation. *Nucleic Acids Res* **39:** 5935–5944.

Lemacon D, Jackson J, Quinet A, Brickner JR, Li S, Yazinski S, You Z, Ira G, Zou L, Mosammaparast N, et al. 2017. MRE11 and EXO1 nucleases degrade reversed forks and elicit MUS81-dependent fork rescue in BRCA2-deficient cells. *Nat Commun* **8:** 860.

Liu J, Doty T, Gibson B, Heyer WD. 2010. Human BRCA2 protein promotes RAD51 filament formation on RPA-covered single-stranded DNA. *Nat Struct Mol Biol* **17:** 1260–1262.

Lok BH, Carley AC, Tchang B, Powell SN. 2013. RAD52 inactivation is synthetically lethal with deficiencies in BRCA1 and PALB2 in addition to BRCA2 through RAD51-mediated homologous recombination. *Oncogene* **32:** 3552–3558.

Lord CJ, Ashworth A. 2016. BRCAness revisited. *Nat Rev Cancer* **16:** 110–120.

Lord CJ, Ashworth A. 2017. PARP inhibitors: Synthetic lethality in the clinic. *Science* **355:** 1152–1158.

Loveday C, Turnbull C, Ramsay E, Hughes D, Ruark E, Frankum JR, Bowden G, Kalmyrzaev B, Warren-Perry M, Snape K, et al. 2011. Germline mutations in RAD51D confer susceptibility to ovarian cancer. *Nat Genet* **43:** 879–882.

Lukas C, Savic V, Bekker-Jensen S, Doil C, Neumann B, Pedersen RS, Grofte M, Chan KL, Hickson ID, Bartek J, et al. 2011. 53BP1 nuclear bodies form around DNA lesions generated by mitotic transmission of chromosomes under replication stress. *Nat Cell Biol* **13:** 243–253.

Mateos-Gomez PA, Gong F, Nair N, Miller KM, Lazzerini-Denchi E, Sfeir A. 2015. Mammalian polymerase θ promotes alternative NHEJ and suppresses recombination. *Nature* **518:** 254–257.

Maxwell KN, Wubbenhorst B, Wenz BM, De Sloover D, Pluta J, Emery L, Barrett A, Kraya AA, Anastopoulos IN, Yu S, et al. 2017. BRCA locus-specific loss of heterozygosity in germline BRCA1 and BRCA2 carriers. *Nat Commun* **8:** 319.

Meindl A, Hellebrand H, Wiek C, Erven V, Wappenschmidt B, Niederacher D, Freund M, Lichtner P, Hartmann L, Schaal H, et al. 2010. Germline mutations in breast and ovarian cancer pedigrees establish RAD51C as a human cancer susceptibility gene. *Nat Genet* **42:** 410–414.

Mijic S, Zellweger R, Chappidi N, Berti M, Jacobs K, Mutreja K, Ursich S, Ray Chaudhuri A, Nussenzweig A, Janscak P, et al. 2017. Replication fork reversal triggers fork degradation in BRCA2-defective cells. *Nat Commun* **8:** 859.

Minocherhomji S, Ying S, Bjerregaard VA, Bursomanno S, Aleliunaite A, Wu W, Mankouri HW, Shen H, Liu Y, Hickson ID. 2015. Replication stress activates DNA repair synthesis in mitosis. *Nature* **528:** 286–290.

Mondal G, Rowley M, Guidugli L, Wu J, Pankratz VS, Couch FJ. 2012. BRCA2 localization to the midbody by filamin A regulates cep55 signaling and completion of cytokinesis. *Dev Cell* **23:** 137–152.

Moynahan ME, Jasin M. 2010. Mitotic homologous recombination maintains genomic stability and suppresses tumorigenesis. *Nat Rev Mol Cell Biol* **11:** 196–207.

Naim V, Wilhelm T, Debatisse M, Rosselli F. 2013. ERCC1 and MUS81-EME1 promote sister chromatid separation by processing late replication intermediates at common fragile sites during mitosis. *Nat Cell Biol* **15:** 1008–1015.

Neelsen KJ, Lopes M. 2015. Replication fork reversal in eukaryotes: From dead end to dynamic response. *Nat Rev Mol Cell Biol* **16:** 207–220.

Nik-Zainal S, Davies H, Staaf J, Ramakrishna M, Glodzik D, Zou X, Martincorena I, Alexandrov LB, Martin S, Wedge DC, et al. 2016. Landscape of somatic mutations in 560 breast cancer whole-genome sequences. *Nature* **534:** 47–54.

Norquist B, Wurz KA, Pennil CC, Garcia R, Gross J, Sakai W, Karlan BY, Taniguchi T, Swisher EM. 2011. Secondary somatic mutations restoring BRCA1/2 predict chemotherapy resistance in hereditary ovarian carcinomas. *J Clin Oncol* **29:** 3008–3015.

Patel KJ, Yu VP, Lee H, Corcoran A, Thistlethwaite FC, Evans MJ, Colledge WH, Friedman LS, Ponder BA, Venkitaraman AR. 1998. Involvement of Brca2 in DNA repair. *Mol Cell* **1:** 347–357.

Pedersen RS, Karemore G, Gudjonsson T, Rask MB, Neumann B, Heriche JK, Pepperkok R, Ellenberg J, Gerlich DW, Lukas J, et al. 2016. Profiling DNA damage response following mitotic perturbations. *Nat Commun* **7:** 13887.

Pennington KP, Walsh T, Harrell MI, Lee MK, Pennil CC, Rendi MH, Thornton A, Norquist BM, Casadei S, Nord AS, et al. 2014. Germline and somatic mutations in homologous recombination genes predict platinum response and survival in ovarian, fallopian tube, and peritoneal carcinomas. *Clin Cancer Res* **20:** 764–775.

Polak P, Kim J, Braunstein LZ, Karlic R, Haradhavala NJ, Tiao G, Rosebrock D, Livitz D, Kubler K, Mouw KW, et al. 2017. A mutational signature reveals alterations underlying deficient homologous recombination repair in breast cancer. *Nat Genet* **49:** 1476–1486.

Prakash R, Zhang Y, Feng W, Jasin M. 2015. Homologous recombination and human health: The roles of BRCA1, BRCA2, and associated proteins. *Cold Spring Harb Perspect Biol* **7:** a016600.

Quigley D, Alumkal JJ, Wyatt AW, Kothari V, Foye A, Lloyd P, Aggarwal R, Kim W, Lu E, Schwartzman J, et al. 2017. Analysis of circulating cell-free DNA identifies multiclonal heterogeneity of BRCA2 reversion mutations associated with resistance to PARP inhibitors. *Cancer Discov* **7:** 999–1005.

Quinet A, Lemacon D, Vindigni A. 2017. Replication fork reversal: Players and guardians. *Mol Cell* **68:** 830–833.

Rahman N, Seal S, Thompson D, Kelly P, Renwick A, Elliott A, Reid S, Spanova K, Barfoot R, Chagtai T, et al. 2007. PALB2, which encodes a BRCA2-interacting protein, is a breast cancer susceptibility gene. *Nat Genet* **39:** 165–167.

Ramus SJ, Bobrow LG, Pharoah PD, Finnigan DS, Fishman A, Altaras M, Harrington PA, Gayther SA, Ponder BA, Friedman LS. 1999. Increased frequency of TP53 mutations in BRCA1 and BRCA2 ovarian tumours. *Genes Chromosomes Cancer* **25:** 91–96.

Ray Chaudhuri A, Callen E, Ding X, Gogola E, Duarte AA, Lee JE, Wong N, Lafarga V, Calvo JA, Panzarino NJ, et al. 2016. Replication fork stability confers chemoresistance in BRCA-deficient cells. *Nature* **535:** 382–387.

Robinson D, Van Allen EM, Wu YM, Schultz N, Lonigro RJ, Mosquera JM, Montgomery B, Taplin ME, Pritchard CC, Attard G, et al. 2015. Integrative clinical genomics of advanced prostate cancer. *Cell* **161:** 1215–1228.

Rondinelli B, Gogola E, Yucel H, Duarte AA, van de Ven M, van der Sluijs R, Konstantinopoulos PA, Jonkers J, Ceccaldi R, Rottenberg S, et al. 2017. EZH2 promotes degradation of stalled replication forks by recruiting MUS81 through histone H3 trimethylation. *Nat Cell Biol* **19:** 1371–1378.

Roy R, Chun J, Powell SN. 2011. BRCA1 and BRCA2: Different roles in a common pathway of genome protection. *Nat Rev Cancer* **12:** 68–78.

Sakai W, Swisher EM, Karlan BY, Agarwal MK, Higgins J, Friedman C, Villegas E, Jacquemont C, Farrugia DJ, Couch FJ, et al. 2008. Secondary mutations as a mechanism of cisplatin resistance in BRCA2-mutated cancers. *Nature* **451:** 1116–1120.

Schlacher K, Christ N, Siaud N, Egashira A, Wu H, Jasin M. 2011. Double-strand break repair-independent role for BRCA2 in blocking stalled replication fork degradation by MRE11. *Cell* **145:** 529–542.

Schlacher K, Wu H, Jasin M. 2012. A distinct replication fork protection pathway connects Fanconi anemia tumor suppressors to RAD51-BRCA1/2. *Cancer Cell* **22:** 106–116.

Schoonen PM, Talens F, Stok C, Gogola E, Heijink AM, Bouwman P, Foijer F, Tarsounas M, Blatter S, Jonkers J, et al. 2017. Progression through mitosis promotes PARP inhibitor-induced cytotoxicity in homologous recombination-deficient cancer cells. *Nat Commun* **8:** 15981.

Somyajit K, Saxena S, Babu S, Mishra A, Nagaraju G. 2015. Mammalian RAD51 paralogs protect nascent DNA at stalled forks and mediate replication restart. *Nucleic Acids Res* **43:** 9835–9855.

Soule HD, Maloney TM, Wolman SR, Peterson WD Jr, Brenz R, McGrath CM, Russo J, Pauley RJ, Jones RF, Brooks SC. 1990. Isolation and characterization of a spontaneously immortalized human breast epithelial cell line, MCF-10. *Cancer Res* **50:** 6075–6086.

Taglialatela A, Alvarez S, Leuzzi G, Sannino V, Ranjha L, Huang JW, Madubata C, Anand R, Levy B, Rabadan R, et al. 2017. Restoration of replication fork stability in BRCA1- and BRCA2-deficient cells by inactivation of SNF2-family fork remodelers. *Mol Cell* **68:** 414–430 e418.

Thangavel S, Berti M, Levikova M, Pinto C, Gomathinayagam S, Vujanovic M, Zellweger R, Moore H, Lee EH, Hendrickson EA, et al. 2015. DNA2 drives processing and restart of reversed replication forks in human cells. *J Cell Biol* **208:** 545–562.

Thorslund T, McIlwraith MJ, Compton SA, Lekomtsev S, Petronczki M, Griffith JD, West SC. 2010. The breast cancer tumor suppressor BRCA2 promotes the specific targeting of RAD51 to single-stranded DNA. *Nat Struct Mol Biol* **17:** 1263–1265.

Tischkowitz M, Xia B, Sabbaghian N, Reis-Filho JS, Hamel N, Li G, van Beers EH, Li L, Khalil T, Quenneville LA, et al. 2007. Analysis of PALB2/FANCN-associated breast cancer families. *Proc Natl Acad Sci* **104:** 6788–6793.

Tutt A, Gabriel A, Bertwistle D, Connor F, Paterson H, Peacock J, Ross G, Ashworth A. 1999. Absence of Brca2 causes genome instability by chromosome breakage and loss associated with centrosome amplification. *Curr Biol* **9:** 1107–1110.

Wang AT, Kim T, Wagner JE, Conti BA, Lach FP, Huang AL, Molina H, Sanborn EM, Zierhut H, Cornes BK, et al. 2015. A dominant mutation in human RAD51 reveals its function in DNA interstrand crosslink repair independent of homologous recombination. *Mol Cell* **59:** 478–490.

Wang Q, Zou Y, Nowotschin S, Kim SY, Li QV, Soh CL, Su J, Zhang C, Shu W, Xi Q, et al. 2017. The p53 family coordinates Wnt and nodal inputs in mesendodermal differentiation of embryonic stem cells. *Cell Stem Cell* **20:** 70–86.

Wooster R, Bignell G, Lancaster J, Swift S, Seal S, Mangion J, Collins N, Gregory S, Gumbs C, Micklem G. 1995. Identification of the breast cancer susceptibility gene BRCA2. *Nature* **378:** 789–792.

Xu S, Wu X, Wu L, Castillo A, Liu J, Atkinson E, Paul A, Su D, Schlacher K, Komatsu Y, et al. 2017a. Abro1 maintains genome stability and limits replication stress by protecting replication fork stability. *Genes Dev* **31:** 1469–1482.

Xu Y, Ning S, Wei Z, Xu R, Xu X, Xing M, Guo R, Xu D. 2017b. 53BP1 and BRCA1 control pathway choice for stalled replication restart. *Elife* **6:** e30523.

Ying S, Minocherhomji S, Chan KL, Palmai-Pallag T, Chu WK, Wass T, Mankouri HW, Liu Y, Hickson ID. 2013. MUS81 promotes common fragile site expression. *Nat Cell Biol* **15:** 1001–1007.

Zellweger R, Dalcher D, Mutreja K, Berti M, Schmid JA, Herrador R, Vindigni A, Lopes M. 2015. Rad51-mediated replication fork reversal is a global response to genotoxic treatments in human cells. *J Cell Biol* **208:** 563–579.

Fork Protection and Therapy Resistance in Hereditary Breast Cancer

SHARON B. CANTOR AND JENNIFER A. CALVO

Department of Molecular, Cell, and Cancer Biology, University of Massachusetts Medical School,
UMASS Memorial Cancer Center, Worcester, Massachusetts 01605

Correspondence: sharon.cantor@umassmed.edu

The BRCA-Fanconi anemia (FA) pathway preserves the genome and suppresses cancer and is a main determinant of chemotherapeutic efficacy. The hereditary breast cancer genes *BRCA1* and *BRCA2* function in DNA double-strand break repair mediating distinct steps of homologous recombination (HR). More recently, independent of DNA repair, functions in the replication stress response have come to light, providing insight as to how the BRCA-FA pathway also balances genome preservation with proliferation. The BRCA-FA proteins associate with the replisome and contribute to the efficiency and recovery of replication following perturbations that slow or arrest DNA replication. Although the full repertoire of functions in the replication stress response remains to be elucidated, the function of *BRCA1* and *BRCA2* in protecting stalled replication forks contributes along with HR to the sensitivity of BRCA-associated tumors to chemotherapy. Moreover, chemoresistance evolves from restoration of either HR and/or fork protection. Although mechanisms underlying the restoration of HR have been characterized, it remains less clear how restoration of fork protection is achieved. Here, we outline mechanisms of "rewired" fork protection and chemotherapy resistance in BRCA cancer. We propose that mechanisms are linked to permissive replication that limits fork remodeling and therefore opportunities for fork degradation. Combating this chemoresistance mechanism will require drugs that inactivate replication bypass mechanisms.

BRCA-FA PATHWAY FUNCTIONS BEYOND DNA REPAIR

Deficiency in the BRCA-Fancomi anemia (FA) pathway has widespread physiological consequences. Germline mutations in the hereditary breast cancer genes such as *BRCA1* and *BRCA2* are highly penetrant and predispose 20%–80% of carriers to breast and ovarian cancer (Apostolou and Fostira 2013). In addition, biallelic inactivation of BRCA genes causes FA, a rare multigenic disease in which loss of any of the 21 distinct complementation groups, FANC-A, FANC-B, FANC-C, ..., FANC-T, confers disease (for reviews, see Bhattacharjee and Nandi 2017; Cheung and Taniguchi 2017). Patients are characterized by progressive bone marrow failure, developmental defects, and cancer predisposition. FA patient cells are also exquisitely sensitive to agents that induce DNA interstrand cross-links (ICLs). Consequently, research has focused on understanding the function of the BRCA-FA pathway in ICL repair. In particular, given the role of BRCA proteins in fixing double-strand DNA breaks by homologous recombination (HR), this function was thought to extend to the repair of ICL-induced breaks. However, essential functions in preserving replication have emerged that are independent of HR. This role was not readily apparent because BRCA-FA-deficient cells are not universally sensitive to fork slowing or stalling drugs. Recent advances in studying the biology of DNA replication and associated machinery have clarified roles in replication and the replication stress response and provide a new perspective for understanding BRCA-FA disease and the function of BRCA-FA proteins in cell survival.

BRCA-FA PROTEINS PROTECT STALLED REPLICATION FORKS FROM DEGRADATION

A key initial finding linking the BRCA-FA proteins to the replication stress response was the observation that BRCA2 protects stalled replication forks from break formation following treatment with hydroxyurea (HU) that decreases the generation of deoxyribonucelotides through inhibition of ribonucleotide reductase (Lomonosov et al. 2003). More recently, BRCA1, BRCA2, FANCD2, and FANCA were shown to protect nascent DNA strands of replication forks stalled by HU. Mechanistically, BRCA1 and BRCA2 protect nascent DNA at stalled forks from MRE11-mediated degradation by loading RAD51 on ssDNA exposed at stalled forks (Schlacher et al. 2011; Ying et al. 2012; Chaudhuri et al. 2016). Correspondingly, in the absence of BRCA1 or BRCA2 nascent DNA strands of stalled replication forks undergo extensive nucleolytic degradation leading to long stretches of single-stranded DNA (ssDNA) (Schlacher et al. 2011, 2012; Kolinjivadi et al. 2017b). Consistent with this mode of action, a degradation phenotype also underlies a RAD51 mutation in patients with an FA-like phenotype (Ameziane et al. 2015; Wang et al. 2015; Zadorozhny et al. 2017). Thus, RAD51 is a key mediator of BRCA function in HR at DNA breaks and in protecting nascent DNA at stalled forks. In both settings, RAD51 loading on ssDNA could restrict MRE11

Published by Cold Spring Harbor Laboratory Press; doi: 10.1101/sqb.2017.82.034413

Cold Spring Harbor Symposia on Quantitative Biology, Volume LXXXII

resection to initiate recombination-based mechanisms to promote repair at breaks and restart at stalled forks. The mechanistic overlap of HR proteins operating in DNA repair and DNA replication were recently reviewed (Kolinjivadi et al. 2017a).

Synchrony in fork processing is established in part by the fact that factors regulating DNA end resection also contribute to fork slowing and subsequent restart mechanisms. Consequently, loss of several BRCA-FA factors causes not only fork degradation phenotypes but also slowing and restart defects following replication stress. In particular, RAD51, FANCD2, FANCM, FAN1, and the FANCD2–FAN1 interaction ensure that replication elongation is slowed in response to HU (Luke-Glaser et al. 2010; Lossaint et al. 2013; Zellweger et al. 2015; Lachaud et al. 2016). BRCA-FA-deficient cells also display defects in slowing replication in response to ICLs that were proposed to underlie the pronounced G_2/M arrest and sensitivity (Sala-Trepat et al. 2000). Even during unperturbed replication, RAD51 loss leads to the accumulation of ssDNA gaps that are visible by electron microscopy (EM) indicating its role in normal replication (Lopes et al. 2006; Hashimoto et al. 2010). Gaps could result from discontinuous replication if DNA synthesis skips over barriers and reinitiates downstream by employing the primase, PrimPol (Guilliam and Doherty 2017). Whatever the source of gaps, they could lead to extensive resection that underlies fork degradation phenotypes that characterize loss of BRCA-FA proteins. Gaps could be initiating sites for remodelers that mediate aberrant processing in BRCA-FA cells (Kolinjivadi et al. 2017a). Future work will ideally reveal how fork slowing, reversal, protection, and recombination are ultimately coordinated and/or linked with other BRCA-FA pathway functions including replication restart (Petermann et al. 2010; Schwab et al. 2013; Raghunandan et al. 2015; Lemacon et al. 2017; Rondinelli et al. 2017), regulation of new origin firing (Thompson et al. 2017), and traversing the replisome past replication blocking lesions (Huang et al. 2013).

DNA REPAIR AND FORK PROTECTION UNDERLIES *BRCA1/2* GENOME STABILITY AND TUMOR SUPPRESSION FUNCTIONS

These recently identified roles for BRCA-FA pathway in replication fork stability raise the possibility that maintenance of replication fork stability may contribute to its genome preservation and tumor suppression functions. The idea that replication dysfunction independent of HR drives cancer formation is in part supported by analysis of mammary epithelial cells from *BRCA1* mutation carriers that have not yet developed cancer. These premalignant cells were found to show functional HR and DNA damage checkpoint signaling but defects in the protection of stalled replication forks (Pathania et al. 2014). Moreover, cells carrying *BRCA2* heterozygous truncating mutations showed extensive replication stress–induced fork degrada-

tion (Tan et al. 2017). Thus, loss of DNA repair may not be the only crucial factor in the etiology of BRCA cancer. More likely, it could be a combined loss of and/or coordination between DNA repair and DNA replication that undermines the maintenance of genome integrity and tumor suppression.

Until recently, restoration of the HR pathway was the only described mechanism by which *BRCA1/2*-mutant cancers survive genotoxins, such as cisplatin or inhibitors of poly(ADP-ribose) polymerase (PARPi) (Edwards et al. 2008; Sakai et al. 2008, 2009). In *BRCA1*-mutant cells, restoration of HR is achieved by several mechanisms including reversion mutations or loss of the nonhomologous end joining (NHEJ) factor, 53BP1 (Bunting et al. 2010, 2012). Conversely, in *BRCA2*-mutant cells, restoration of HR is solely achieved by reversion mutations (Edwards et al. 2008; Sakai et al. 2008, 2009). The finding that approximately half of *BRCA2*-mutant cancers develop chemoresistance in the absence of restored HR suggested that reversion-independent resistance mechanisms awaited identification (Norquist et al. 2011). To define genetic determinants of cisplatin resistance, we performed a genome-wide short-hairpin RNA (shRNA) screen and found that loss of the chromatin remodeler, CHD4, mediates resistance to cisplatin and PARPi in *BRCA2*-mutated cancer (Guillemette et al. 2015). CHD4 depletion did not restore HR but reduced the levels of chromosomal aberrations in *BRCA2*-mutant cells exposed to cisplatin (Guillemette et al. 2015). The elevated resistance and genomic stability correlated with reduced MRE11 chromatin association and increased fork protection (Chaudhuri et al. 2016). A series of recent publications concur that rewired fork protection confers chemoresistance independent of HR. This is achievable in *BRCA1/2*-deficient cells by several mechanisms, including loss of the chromatin modifier complex MLL3–4/PTIP/MRE11, the fork remodelers, SMARCAL1, HLTF, or ZRANB3, PARP1, the methyltransferase EZH2, and the negative regulator of RAD51, RADX (Chaudhuri et al. 2016; Ding et al. 2016; Dungrawala et al. 2017; Kolinjivadi et al. 2017b; Taglialatela et al. 2017; Vujanovic et al. 2017), as well as through up-regulation of FANCD2 (Kais et al. 2016; Michl et al. 2016). The clinical significance of this mechanism is confirmed by the poor patient response and survival outcomes observed upon restored fork protection (Guillemette et al. 2015; Chaudhuri et al. 2016; Rondinelli et al. 2017). Thus, these factors may prove useful as potential biomarkers of *BRCA1/2*-deficient tumor response to chemotherapy.

DEFINING DEGRADATION FACTORS AND FORK DYNAMICS IN *BRCA1/2*-DEFICIENT CELLS

A common feature of rewired fork protection is that nuclease activity is tempered by various mechanisms. In particular, in *BRCA1/2*-mutant cells, suppressing MRE11-nucleolytic degradation restores fork protection. This can be done with the MRE11 inhibitor mirin or by reestablishing Rad51 filament formation at stalled forks through expression of a RAD51 mutant lacking ATPase activity or by depletion of the anti-RAD51 factor, RADX (Schlacher et al. 2011; Dungrawala et al. 2017). In addition, loss of factors that facilitate the recruitment of MRE11 to stalled forks such as PARP1, PTIP, or CHD4 restores fork protection in *BRCA1/2*-deficient cells (Table 1; Chaudhuri et al. 2016; Ding et al. 2016). Given that MRE11 has limited nucleolytic activity (Cannavo and Cejka 2014), other nucleases likely contribute to fork degradation. In particular, CtIP initiates the MRE11-dependent degradation that is extended by EXO1 (Lemacon et al. 2017). Although it remains debated whether DNA2 contributes to degradation in *BRCA1/2*-deficient cells (Chaudhuri et al. 2016; Spies et al. 2016; Kolinjivadi et al. 2017b; Lemacon et al. 2017), DNA2 degrades forks in cells deficient in Abro1, a paralog of the *BRCA1*-interacting protein, Abraxas (Xu et al. 2017). An MRE11-independent mechanism has also been described in which the nuclease MUS81 contributes to fork degradation in a pathway with EZH2 (Table 1; Rondinelli et al. 2017).

Replication fork structures also play a prominent role in nucleolytic degradation. In particular, a reversed replication fork, a so-called "chicken foot," is the target of MRE11 digestion in *BRCA1/2*-deficient cells (Mijic et al. 2017). Accordingly, fork reversal has been shown to be a prerequisite for fork degradation in *BRCA1/2*-deficient cells (Kolinjivadi et al. 2017b). In addition to the remodelers HLTF, SMARCAL1, and ZRANB3 (Kolinjivadi et al. 2017b; Taglialatela et al. 2017), RAD52, PARP1 and RAD51 drive fork reversal–dependent degradation in *BRCA1/2*-deficient cells (Table 1; Lemacon et al. 2017; Mijic et al. 2017; Ray Chaudhuri et al. 2012). There are numerous proteins known to generate reversed forks such as DNA helicases FBH1, BLM, WRN, RECQL5 and DNA translocases FANCM and RAD54 (Blastyák et al. 2010; Bétous et al. 2012; Ciccia et al. 2012; Fugger et al. 2015; Kile et al. 2015; Neelsen and Lopes 2015). The relevance of these factors to fork degradation in BRCA cancer remains to be investigated. How loss of each remodeler can individually restore fork protection in *BRCA1/2*-deficient cancer also awaits further analysis. Fork remodelers may act in a common pathway; however, HLTF, ZRANB3, and SMARCAL1 do not form a complex (Taglialatela et al. 2017). It also remains to be determined how fork remodelers interact with the other pathways of chemoresistance such as EZH2 or MLL3–4/PTIP/MRE11, both of which act independently of each other (Rondinelli et al. 2017). Given that the reversed fork structure has exposed DNA ends that mimic DNA double-stranded break ends, it will also be important to determine if in addition to MRE11, these ends are an entry point for NHEJ factors. If so, the reported elevated NHEJ activity that causes genomic instability in BRCA-FA cells (Adamo et al. 2010; Bunting et al. 2012; Pace et al. 2010) may be suppressed by depletion of fork remodelers. However, as described below, preventing fork reversal and subsequent degradation or NHEJ reactions is not always sufficient for chemoresistance in *BRCA1/2*-deficient tumors.

COMPLEXITY OF REWIRED FORK PROTECTION MECHANISMS

The emerging literature indicates that rewired fork protection mechanisms and their relationship to cell viability, genome stability, and chemoresistance are complex. Rewired fork protection in *BRCA1/2*-deficient cells is achieved by loss of either of three remodelers (HLTF, SMARCAL1, and ZRANB3), PTIP, or PARP1, but loss of CHD4 or EZH2 is restricted to *BRCA2*-mutant cells (Guillemette et al. 2015; Ding et al. 2016; Ray Chaudhuri et al. 2016; Rondinelli et al. 2017; Taglialatela et al. 2017). These findings suggest that stalled fork structures or compensating pathways in *BRCA1* and *BRCA2* cells may be distinct, but the overall strategy of rewired fork protection remains consistent; forks that escape nucleases are protected from degradation. However, not all routes that escape degradation will confer chemotherapy resistance. RAD51, MRE11, or RAD52 are required for viability of *BRCA1/2*-deficient cells, so their loss will not result in chemoresistance but rather synthetic lethality (Fig. 1; Feng et al. 2011; Ying et al. 2012; Lok et al. 2013). The balance between lethality and survival/chemoresistance also appears dependent on the sequence by which a fork degradation factor is lost, before or after BRCA deficiency. Depletion of PARP1 in *BRCA2*-deficient cells leads to synthetic lethality (Feng and Jasin 2017), whereas deletion of PARP1 before *BRCA2* depletion provides some protective effects (Ding et al. 2016).

An additional level of complexity derives from the fact that some proteins possess multiple functions in the replication stress response. For example, RAD51 participates in both the formation and protection of reversed forks (Hashimoto et al. 2010; Zellweger et al. 2015). Therefore, depending on which RAD51 function is disrupted, fork protection or degradation could be altered. Further findings show the importance of cellular context, specifically mutational status, for therapy resistance. For example, CHD4 depletion conferred cisplatin resistance in *BRCA2*-mutant cell lines harboring truncated *BRCA2*, but did not improve cisplatin resistance in *BRCA2*-depleted cells, indicating resistance requires the maintenance of a residual *BRCA2*-mutant species (SB Cantor, unpubl.). As different cell types show differing levels of reversed forks (Ahuja et al. 2016), this may explain the findings that SMARCAL1 depletion resulted in chemoresistance in the MCF10A cancer cell line but not in nonmalignant mammary epithelial cells, despite both cell lines being *BRCA1*-deficient (Taglialatela et al. 2017). Likewise, *Ptip* depletion in *Brca2*-null mouse B cells improves genome stability and resistance to HU, but in *BRCA2*-null primary human cells fails to restore viability (Chaudhuri et al. 2016; Feng and Jasin 2017). Ideally future work will unravel these distinctions so that expectations for therapy are better understood.

Finally, it is important to realize that mechanisms of fork protection may vary for different genotoxic agents; mechanisms revealed following treatment of cells with HU may not be informative for mechanism following treatment of cells with cisplatin or PARPi, as these agents

Table 1. Simplified overview of factors whose loss restores fork protection and/or chemoresistance in BRCA1/2-deficient cells

Factor	Function	Fork Protection	Chromosome stability	Chemoresistance	Clinical Relevance	Publication(s)
MRE11	Nuclease that degrades nascent DNA in stalled replication forks	Inhibition restores fork protection in BRCA2-depleted (V-C8), BRCA1-depleted U2OS and BRCA1-mutant (UWB1) cells.				Schlacher et al. 2011 and 2012; Ying et al. 2012; Lemacon et al. 2017; Mijic et al. 2017
CHD4	Recruits MRE11 to stalled fork	Depletion restores fork protection in BRCA2-mutant (PEO1) cells.	Depletion protects against chromosomal aberrations in BRCA2-mutant (PEO1) cells (cisplatin).	Depletion increased resistance in BRCA2-mutant (PEO1, CapanI, FA-D1) cells (cisplatin/PARPi).	Low CHD4 correlated with reduced progression-free survival (PFS) in BRCA2-mutant ovarian cancers	Guillemette et al. 2015; Chaudhuri et al. 2016
PARP1	Recruits MRE11 to stalled fork, fork reversal activity	Deletion restores fork protection in $Brca1^{-/-}$ or $Brca2^{-/-}$ B cells and $Brca2^{-/-}$ mESC.	Deletion protects against chromosomal aberrations in $Brca2^{-/-}$ mESC and $Brca1^{-/-}$ B cells (cisplatin, camptothecin).		PARP1 deletion reduces tumor-free survival in $Brca2^{-/-}$ mouse model.	Chaudhuri et al. 2016; Ding et al. 2016
PTIP	Recruits MRE11 to stalled fork	Depletion restores fork protection in $Brca1^{-/-}$ or $Brca2^{-/-}$ B cells.	Deletion protects against chromosomal aberrations in $Brca1^{-/-}$ or $Brca2^{-/-}$ B cells (HU, cisplatin, camptothecin).	Deletion increased resistance in $BRCA1^{-/-}$ B cells (HU).	Low PTIP correlated with reduced PFS in BRCA2-mutant ovarian cancer	Chaudhuri et al. 2016
HLTF	Fork reversal activity, E3 ubiquitin ligase for PCNA	Depletion restores fork protection in BRCA1- or BRCA2-depleted MCF10A cells.				Taglialatela et al. 2017
SMARCAL1	Fork reversal activity	Depletion restores fork protection in BRCA1- or BRCA2-depleted MCF10A cells or in BRCA2-depleted *Xenopus* extracts.	Depletion protects against chromosomal aberrations in BRCA1- and BRCA2-depleted MCF10A cells (camptothecin).	Depletion increased resistance in BRCA1-mutant (MDA-MB-436) cells (cisplatin/PARPi).	Low SMARCAL1 correlated with reduced overall survival (OS) in BRCA1-mutant breast cancer	Kolinjivadi et al. 2017b; Taglialatela et al. 2017
ZRANB3	Fork reversal activity	Depletion restores fork protection in BRCA1- or BRCA2-depleted MCF10A cells.	Depletion protects against chromosomal aberrations in BRCA1- and BRCA2-depleted MCF10A cells (camptothecin), Deletion causes increased chromatid breaks/gaps in BRCA2-depleted U2OS cells (HU).			Vujanovic et al. 2017; Taglialatela et al. 2017; Mijic et al. 2017

Continued

Table 1. *Continued*

Factor	Function	Fork Protection	Chromosome stability	Chemoresistance	Clinical Relevance	Publication(s)
RAD52	Fork reversal activity, loads Rad51 onto ssDNA, recruits MRE11 to stalled fork	Depletion restores fork protection in BRCA2-depleted U2OS cells	Depletion protects against chromosomal breakage in BRCA2-depleted U2OS cells (HU).			Mijic et al. 2017
RAD51	Fork reversal activity, binds and protects ssDNA	Depletion restores fork protection in BRCA2-depleted RPE-1 cells				Mijic et al. 2017
EXO1	Nuclease that extends MRE11 fork degradation	Depletion restores fork protection in BRCA2-mutant (PEO1) or BRCA1-mutant (UWB1) cells, and BRCA1- or BRCA2-depleted U2OS cells.	Depletion protects against chromosomal aberrations in BRCA2-depleted U2OS (HU).			Lemacon et al. 2017
MLL4	Induces histone H3 methylation at Lysine 4 (H3K4) to recruit MRE11 to stalled fork	Deletion restores fork protection in $Brca1^{-/-}$ or $Brca2^{-/-}$ B cells.	Deletion protects against chromosomal aberrations in $Brca2^{-/-}$ B cells (PARPi, cisplatin).			Chaudhuri et al. 2016
RADX	Inhibits the accumulation of Rad51 at forks	Depletion restores fork protection in BRCA2-depleted U2OS cells, BRCA2 mutant Capan1 cells		Depletion increased resistance in BRCA2-depleted U2OS cells (PARPi).		Dungrawala et al. 2017
EZH2	Induces trimethylation of histone H3 at Lysine 27 (H3K27me3) to recruit Mus81 to stalled fork	Depletion restores fork protection in BRCA2-mutant (VU423) cells.	Inhibition protects against chromosomal aberrations in BRCA2-depleted Hela or BRCA2-mutant (VU423) cells (mitomycin C).	Depletion increases resistance in BRCA2-depleted HeLa cells (PARPi, cisplatin).	Low EZH2 correlated with reduced PFS in BRCA2-mutant ovarian cancer EZH2 inhibition promotes relapse in $Brca2^{-/-}$ breast tumor mouse model (PARPi).	Rondinelli et al. 2017

have distinct modes of disrupting replication. For example, EZH2 and MUS81 are required for the restart of replication in *BRCA2*-deficient cells following their release from HU (Lemacon et al. 2017; Rondinelli et al. 2017). HU treatment promotes replication fork reversal (Zellweger et al. 2015), and reversed forks are extensively resected by nucleases in *BRCA2*-deficient cells. MUS81 cleaves the over-resected reversed DNA forks in *BRCA2*-deficient cells and promotes POLD3-dependent fork rescue, explaining their dependence on MUS81 for replication fork progression and resistance to HU (Lai et al. 2017; Lemacon et al. 2017). Conversely, following treatment with PARPi, it is less clear how MUS81 depletion

achieves PARPi resistance in *BRCA2*-deficient cells (Rondinelli et al. 2017). In particular, MUS81 is probably not required for fork restart upon treatment with PARPi because PARPi suppresses fork slowing and reversal depriving cells of a MUS81 substrate (Sugimura et al. 2008; Ray Chaudhuri et al. 2012). Likewise, it remains to be determined if similar to depletion of SMARCAL1, ZRANB3 depletion confers cisplatin/PARPi resistance. ZRANB3 depletion restores fork protection in distinct cell systems; however, genomic instability is increased or decreased in *BRCA1/2*-deficient cells depending on the source of replication stress (Mijic et al. 2017; Taglialatela et al. 2017).

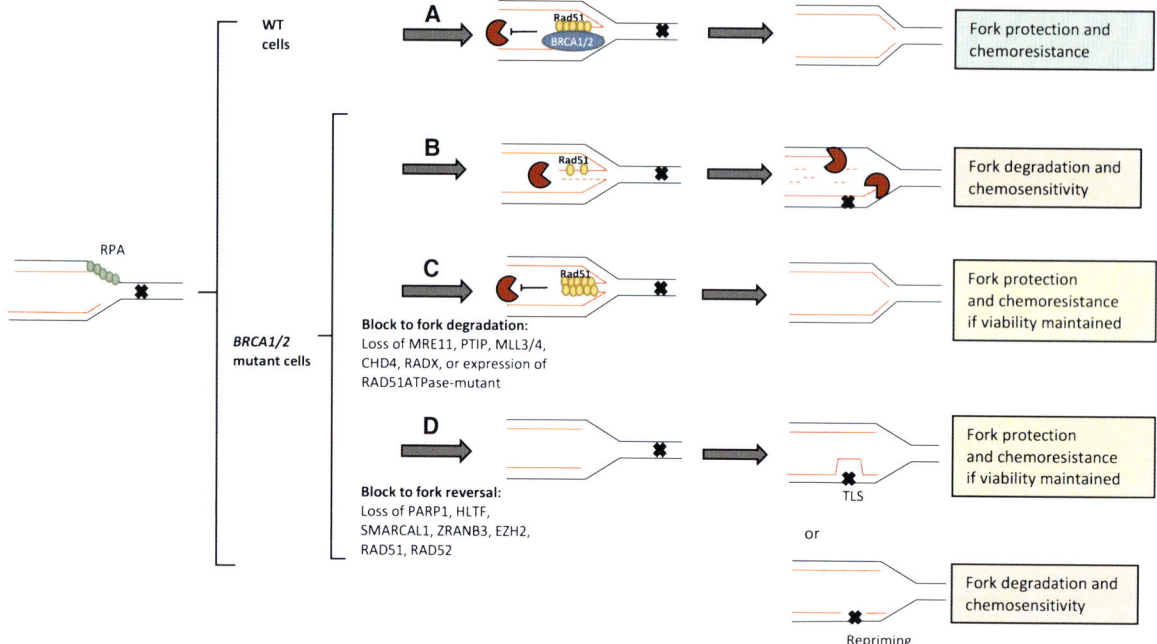

Figure 1. Model of the mechanisms and potential consequences of fork protection in *BRCA1/2*-deficient cells. (*A*) In wild-type (WT cells), when replication stress (represented by X) is encountered in the course of cancer therapy, there is a reversal of replication forks, and the protection of nascent DNA in a BRCA1-, BRCA2-, and Rad51-dependent manner, that limits resection by nucleases such as MRE11, CtIP, and EXO1. Following TS or recombination-based restart, the replication fork is restarted, thereby conferring chemoresistance. In absence of BRCA1/2, (*B*) Rad51 is no longer stabilized on reversed forks, allowing access of nucleases to nascent DNA, resulting in extensive degradation and chemosensitivity. (*C*) Extensive nascent cell degradation can be avoided by either loss of fork degradation factors or gain in stabilization factors that will allow forks to restart via template switch (TS) and confer chemotherapy resistance if viability is not also compromised. (*D*) Loss of fork reversal will limit fork degradation if translesion synthesis (TLS) is active at the fork and gaps generated by repriming reactions are avoided.

REWIRED FORK PROTECTION AND REPLICATION PROGRESSION WITHOUT FORK REVERSAL

When rewired fork protection confers therapy resistance in *BRCA1/2*-mutant cells, the mechanism of replication should be quite different than as found in *BRCA1/2*-proficient cells. In this regard, it is important to note that upon replication stress, replication forks typically slow and reverse (Zellweger et al. 2015). As described above, rewired fork protection achieved through suppression of HLTF, SMARCAL1, ZRANB3, PARP1, and RAD51 is expected to suppress fork reversal. Even though reversed forks are dangerous species that present DNA ends for resection, ssDNA generation, and fork degradation, they also serve to liberate DNA lesions from replisome entanglements, allowing more efficient DNA repair processing. DNA ends in reversed forks also provide DNA substrates for recombination. In addition, the reversed fork is thought to facilitate other strategies for repair or bypass of DNA lesions, such as template switch (TS). TS is an error-free DNA damage tolerance pathway that uses the newly synthesized daughter strand DNA for homology-directed repair to bypass lesions (for review, see Lovett 2017). Access to recombination, TS, and other repair mechanisms through fork remodeling activities is an important genome stabilizing response to stress that limits fork breakage (Ray Chaudhuri et al. 2012; Neelsen and Lopes

2015). Thus, rewired fork protection that limits replication fork slowing and reversal should compromise genomic stability. In BRCA-deficient cancer cells, however, fork reversal could have dire consequences given that its loss provides genomic stability and chemoresistance.

Clues to replication progression mechanisms that confer therapy resistance come from how other activities are altered in *BRCA1/2*-deficient cells that have rewired fork protection. In particular, depletion of HLTF not only results in loss of fork reversal, but TS is also disrupted. This stems from the role of HLTF as an E3 ubiquitin ligase responsible for the addition of ubiquitination chains on PCNA, an essential step in TS (Branzei 2011; Lin et al. 2011). Rewired fork protection established through ZRANB3 depletion will also alter the replication stress response. Polyubiquitinated PCNA recruits ZRANB3 to sites of replication stress, both of which are critical for fork reversal (Ciccia et al. 2012; Vujanovic et al. 2017) and proposed to mediate TS. In addition to catalyzing the regression of stalled replication forks, ZRANB3 also prevents inappropriate recombination (Ciccia et al. 2012). Therefore, loss of ZRANB3 could elevate aberrant recombination as well as interfere with a polyUb-PCNA-dependent axis required for engagement of TS. Furthermore, loss of ZRANB3-PCNA interaction may allow access of de-ubiquitination enzymes thereby decreasing polyUb-PCNA and in turn increase monoUb-PCNA. If so, loss of ZRANB3, similar to loss of HLTF, could enhance the overall monoUb-PCNA. This result could have great significance as monoUb-PCNA serves as a molecular switch for the error-prone mechanism of translesion synthesis (TLS), a DNA tolerance mechanism that allows the timely bypass of DNA lesions using error-prone TLS polymerases (Choe and Moldovan 2017).

REWIRED FORK PROTECTION VIA TLS AT THE FORK

Could TLS operate at the fork to limit gap formation, maintain replication, and promote chemoresistance in *BRCA1/2*-deficient cells (Fig. 1)? TLS is best described for filling in gaps postreplication and operates effectively when restricted from S phase (Karras and Jentsch 2010). However, TLS may operate at the fork, as the replisome is passing through the site of DNA damage or other barrier to replication (Sale et al. 2012). Indeed, TLS polymerases enable replication to continue despite DNA damage or other barriers induced by chemotherapy because they have low-fidelity, which allows the insertion of nucleotides opposite to bulky DNA lesions that block high-fidelity replicative DNA polymerases. Moreover, when nucleotide pools are reduced as in HU-treated cells, TLS could also tolerate insertion of mismatched nucleotide (Edmunds et al. 2008; Quinet et al. 2014). Consistent with the role of TLS reactions occurring at the elongating fork, replication fork stalling is observed in cells depleted of TLS polymerases (Quinet et al. 2014, 2016). TLS activity at the fork may also be restricted to a subset of polymerases. Whereas Rev3L operates in postreplicative gap filling, Rev1 and Polη are required for TLS at stalled forks (Quinet et al. 2016).

In considering the role of TLS in rewired fork protection, it is important to note that TLS may be compromised in BRCA-FA cells. Indeed, in some respects, loss of the BRCA-FA pathway phenocopies loss of TLS and vice versa. Sensitivity to ICL-inducing agents is a phenotype attributed to loss of either pathway. Moreover, FA-like phenotypes in the hematopoietic stem cells result from loss of Ub-PCNA (Pilzecker et al. 2017). Pathway interconnections between BRCA-FA and TLS are established by interactions and the employment of common ubiquitin-modifying enzymes (for reviews, see Kim and D'Andrea 2012; Kim et al. 2012; Fu et al. 2013; Tian et al. 2013; Boisvert and Howlett 2014; Choe and Moldovan 2017). Notably the pathways are also genetically linked. Biallelic inactivation of Rev7(FANCV), the regulatory subunit of the TLS polymerase Polζ, underlies the genetic defect in the FA-V complementation group 1 (Bluteau et al. 2016). One could speculate that TLS defects in *BRCA-FA*-deficient cells selects for compensatory mechanisms that activate TLS downstream from the BRCA-FA pathway. This could underlie how BRCA-FA cells overcome proliferation defects due to endogenous lesions such as aldehyde-induced damage and R-loops, and/or how bone marrow failure transforms into leukemia in FA patients (García-Rubio et al. 2015). Moreover, a gain in TLS could eliminate mitotic DNA synthesis that compensates for under-

replication (Bhowmick et al. 2016) especially in *BRCA2*-deficient cells (Lai et al. 2017). Indeed, an investigation of the mutational signature in cells deficient in *BRCA1* suggests a compensatory up-regulation of TLS (Zamborszky et al. 2017).

Although it may be easier to tip the balance toward TLS in BRCA-FA cancer cells, the rewired replisome may take this a step further. In particular, TLS could suffice to replace replicative polymerases during replication if gaps are very short, which could be possible if nuclease-dependent resection at the fork is blocked or limited. In addition, the simultaneous loss of RAD51 and associated Pol α (Kolinjivadi et al. 2017b) could enable other gap-filling pathways to compensate as the replisome stalls. Indeed, monoUb-PCNA and RAD51 have nonredundant functions in gap filling (Hashimoto et al. 2010). A switch to TLS through a series of modifications on PCNA (Kannouche and Lehmann 2004) and increased chromatin access to stalled forks may underlie the mechanism by which CHD4 loss confers therapy resistance. Indeed, CHD4 loss in *BRCA2*-mutant cells reduced chromatin bound MRE11 and RAD51, while also elevating focal accumulation of the E3 ubiquitin ligase RAD18 that is necessary for PCNA ubiquitination (Guillemette et al. 2015; Chaudhuri et al. 2016). Notably, chromatin-bound RAD51 in *BRCA2*-mutant cells is already very low, but CHD4 depletion reduces this further (Guillemette et al. 2015). Thus, by disrupting the residual RAD51 that operates independent of BRCA2 to promote fork reversal (Kolinjivadi et al. 2017b) and/or that captures ssDNA for gap-filling reactions postreplication (González-Prieto et al. 2013), CHD4 loss could block fork reversal and degradation as well as liberate ssDNA for TLS reactions at the fork. Other routes to elevated TLS through changes in chromatin access or RAD18 complexes have been described (Kim et al. 2014; Yamada et al. 2014). We also found that the FANCJ DNA helicase has the ability to aberrantly promote TLS and suppress HR when its interaction with BRCA1 or its carboxy-terminal acetylation are disrupted (Cantor and Nayak 2016; Xie et al. 2010a, 2012). Conceivably, when not properly regulated by BRCA1, FANCJ disrupts fork-remodeling pathways and/or improves TLS efficiency by unfolding DNA secondary structures that interfere with replication. Interestingly, the mismatch repair protein MSH2, which binds secondary structures formed at stalled replication forks, blocks TLS pathways in FANCJ-deficient cells (Peng et al. 2014).

Further characterization will be required to decipher the contribution of TLS to chemoresistance and rewired fork protection in BRCA cancers. Nevertheless, it is an important therapeutic target (Yamanaka et al. 2017) given the growing body of evidence illustrating a causative role for TLS in the development of chemoresistance (Doles et al. 2010; Xie et al. 2010b; Srivastava et al. 2015). Importantly, significant progress has been made in the development of TLS inhibitors (Actis et al. 2013; Korzhnev and Hadden 2016; Sail et al. 2017; Sanders et al. 2017). TLS inhibitors may be efficacious as a first-line combination chemotherapy or prevent chemoresistance via restored fork protection when used in combination with cisplatin or PARPi. It will also be important to determine whether alternative strategies to block HR and/or fork protection mechanisms of chemoresistance such as inhibitors of ATR (Yazinski et al. 2017) disrupt permissive replication mediated by TLS or other pathways at the replication fork.

CONCLUSION

The BRCA-FA pathway displays indispensable roles in maintaining genome stability, suppressing tumors, and mediating chemoresistance. It does this, not only through integral roles in DNA repair, but growing evidence indicates the BRCA/FA pathway contributes to the maintenance of replication forks during times of replication stress. In the absence of replication fork maintenance, *BRCA1/2*-mutant cancers are capable of "rewiring" the replication fork to allow replication to proceed. We propose that the TLS bypass pathway plays an important part in the mechanism by which replication forks proceed through replication stress. The role of the FA-BRCA pathway in maintaining replication forks and the newly established role for rewired replication forks in chemoresistance provide exciting new possibilities for the development of new chemotherapeutic interventions.

ACKNOWLEDGMENTS

We thank the members of the Cantor laboratory for helpful discussions. Special thanks to Dr. George-Lucien Moldovan for a critical reading of the review. This work was supported by National Institutes of Health (NIH) grant R01 CA176166-01A as well as charitable contributions from the Lipp Family Foundation.

REFERENCES

Actis M, Inoue A, Evison B, Perry S, Punchihewa C, Fujii N. 2013. Small molecule inhibitors of PCNA/PIP-box interaction suppress translesion DNA synthesis. *Bioorg Med Chem* **21:** 1972–1977.

Adamo A, Collis SJ, Adelman CA, Silva N, Horejsi Z, Ward JD, Martinez-Perez E, Boulton SJ, La Volpe A. 2010. Preventing nonhomologous end joining suppresses DNA repair defects of Fanconi anemia. *Mol Cell* **39:** 25–35.

Ahuja AK, Jodkowska K, Teloni F, Bizard AH, Zellweger R, Herrador R, Ortega S, Hickson ID, Altmeyer M, Mendez J, et al. 2016. A short G_1 phase imposes constitutive replication stress and fork remodelling in mouse embryonic stem cells. *Nat Commun* **7:** 10660.

Ameziane N, May P, Haitjema A, van de Vrugt HJ, van Rossum-Fikkert SE, Ristic D, Williams GJ, Balk J, Rockx D, Li H, et al. 2015. A novel Fanconi anaemia subtype associated with a dominant-negative mutation in *RAD51*. *Nat Commun* **6:** 8829.

Apostolou P, Fostira F. 2013. Hereditary breast cancer: The era of new susceptibility genes. *Biomed Res Int* **2013:** 747318.

Bétous R, Mason AC, Rambo RP, Bansbach CE, Badu-Nkansah A, Sirbu BM, Eichman BF, Cortez D. 2012. SMARCAL1 catalyzes fork regression and Holliday junction migration to maintain genome stability during DNA replication. *Genes Dev* **26:** 151–162.

Bhattacharjee S, Nandi S. 2017. DNA damage response and cancer therapeutics through the lens of the Fanconi Anemia DNA repair pathway. *Cell Commun Signal* **15:** 41.

Bhowmick R, Minocherhomji S, Hickson ID. 2016. RAD52 facilitates mitotic DNA synthesis following replication stress. *Mol Cell* **64:** 1117–1126.

Blastyák A, Hajdú I, Unk I, Haracska L. 2010. Role of double-stranded DNA translocase activity of human HLTF in replication of damaged DNA. *Mol Cell Biol* **30:** 684–693.

Bluteau D, Masliah-Planchon J, Clairmont C, Rousseau A, Ceccaldi R, Dubois d'Enghien C, Bluteau O, Cuccuini W, Gachet S, Peffault de Latour R, et al. 2016. Biallelic inactivation of REV7 is associated with Fanconi anemia. *J Clin Invest* **126:** 3580–3584.

Boisvert RA, Howlett NG. 2014. The Fanconi anemia ID2 complex: Dueling saxes at the crossroads. *Cell Cycle* **13:** 2999–3015.

Branzei D. 2011. Ubiquitin family modifications and template switching. *FEBS Lett* **585:** 2810–2817.

Bunting SF, Callén E, Wong N, Chen HT, Polato F, Gunn A, Bothmer A, Feldhahn N, Fernandez-Capetillo O, Cao L, et al. 2010. 53BP1 inhibits homologous recombination in Brca1-deficient cells by blocking resection of DNA breaks. *Cell* **141:** 243–254.

Bunting SF, Callén E, Kozak ML, Kim JM, Wong N, López-Contreras AJ, Ludwig T, Baer R, Faryabi RB, Malhowski A, et al. 2012. BRCA1 functions independently of homologous recombination in DNA interstrand crosslink repair. *Mol Cell* **46:** 125–135.

Cannavo E, Cejka P. 2014. Sae2 promotes dsDNA endonuclease activity within Mre11-Rad50-Xrs2 to resect DNA breaks. *Nature* **514:** 122–125.

Cantor SB, Nayak S. 2016. FANCJ at the FORK. *Mutat Res* **788:** 7–11.

Chaudhuri AR, Callen E, Ding X, Gogola E, Duarte AA, Lee JE, Wong N, Lafarga V, Calvo JA, Panzarino NJ, et al. 2016. Replication fork stability confers chemoresistance in BRCA-deficient cells. *Nature* **535:** 382–387.

Cheung RS, Taniguchi T. 2017. Recent insights into the molecular basis of Fanconi anemia: Genes, modifiers, and drivers. *Int J Hematol* **106:** 335–344.

Choe KN, Moldovan GL. 2017. Forging ahead through darkness: PCNA, still the principal conductor at the replication fork. *Mol Cell* **65:** 380–392.

Ciccia A, Nimonkar AV, Hu Y, Hajdu I, Achar YJ, Izhar L, Petit SA, Adamson B, Yoon JC, Kowalczykowski SC, et al. 2012. Polyubiquitinated PCNA recruits the ZRANB3 translocase to maintain genomic integrity after replication stress. *Mol Cell* **47:** 396–409.

Ding X, Ray Chaudhuri A, Callen E, Pang Y, Biswas K, Klarmann KD, Martin BK, Burkett S, Cleveland L, Stauffer S, et al. 2016. Synthetic viability by BRCA2 and PARP1/ARTD1 deficiencies. *Nat Commun* **7:** 12425.

Doles J, Oliver TG, Cameron ER, Hsu G, Jacks T, Walker GC, Hemann MT. 2010. Suppression of Rev3, the catalytic subunit of Polζ, sensitizes drug-resistant lung tumors to chemotherapy. *Proc Natl Acad Sci* **107:** 20786–20791.

Dungrawala H, Bhat KP, Le Meur R, Chazin WJ, Ding X, Sharan SK, Wessel SR, Sathe AA, Zhao R, Cortez D. 2017. RADX promotes genome stability and modulates chemosensitivity by regulating RAD51 at replication forks. *Mol Cell* **67:** 374–386 e375.

Edmunds CE, Simpson LJ, Sale JE. 2008. PCNA ubiquitination and REV1 define temporally distinct mechanisms for controlling translesion synthesis in the avian cell line DT40. *Mol Cell* **30:** 519–529.

Edwards SL, Brough R, Lord CJ, Natrajan R, Vatcheva R, Levine DA, Boyd J, Reis-Filho JS, Ashworth A. 2008. Resistance to therapy caused by intragenic deletion in BRCA2. *Nature* **451:** 1111–1115.

Feng W, Jasin M. 2017. BRCA2 suppresses replication stress-induced mitotic and G_1 abnormalities through homologous recombination. *Nat Commun* **8:** 525.

Feng Z, Scott SP, Bussen W, Sharma GG, Guo G, Pandita TK, Powell SN. 2011. Rad52 inactivation is synthetically lethal with BRCA2 deficiency. *Proc Natl Acad Sci* **108:** 686–691.

Fu D, Dudimah FD, Zhang J, Pickering A, Paneerselvam J, Palrasu M, Wang H, Fei P. 2013. Recruitment of DNA polymerase η by FANCD2 in the early response to DNA damage. *Cell Cycle* **12:** 803–809.

Fugger K, Mistrik M, Neelsen KJ, Yao Q, Zellweger R, Kousholt AN, Haahr P, Chu WK, Bartek J, Lopes M, et al. 2015. FBH1 catalyzes regression of stalled replication forks. *Cell Rep.* doi: 10.1016/j.celrep.2015.02.028

García-Rubio ML, Pérez-Calero C, Barroso SI, Tumini E, Herrera-Moyano E, Rosado IV, Aguilera A. 2015. The Fanconi anemia pathway protects genome integrity from R-loops. *PLoS Genet* **11:** e1005674.

González-Prieto R, Muñoz-Cabello AM, Cabello-Lobato MJ, Prado F. 2013. Rad51 replication fork recruitment is required for DNA damage tolerance. *EMBO J* **32:** 1307–1321.

Guillemette S, Serra RW, Peng M, Hayes JA, Konstantinopoulos PA, Green MR, Cantor SB. 2015. Resistance to therapy in *BRCA2* mutant cells due to loss of the nucleosome remodeling factor CHD4. *Genes Dev* **29:** 489–494.

Guilliam TA, Doherty AJ. 2017. PrimPol-prime time to reprime. *Genes (Basel)* **8:** 20.

Hashimoto Y, Ray Chaudhuri A, Lopes M, Costanzo V. 2010. Rad51 protects nascent DNA from Mre11-dependent degradation and promotes continuous DNA synthesis. *Nat Struct Mol Biol* **17:** 1305–1311.

Huang J, Liu S, Bellani MA, Thazhathveetil AK, Ling C, de Winter JP, Wang Y, Wang W, Seidman MM. 2013. The DNA translocase FANCM/MHF promotes replication traverse of DNA interstrand crosslinks. *Mol Cell* **52:** 434–446.

Kais Z, Rondinelli B, Holmes A, O'Leary C, Kozono D, D'Andrea AD, Ceccaldi R. 2016. FANCD2 maintains fork stability in *BRCA1/2*-deficient tumors and promotes alternative end-joining DNA repair. *Cell Rep* **15:** 2488–2499.

Kannouche PL, Lehmann AR. 2004. Ubiquitination of PCNA and the polymerase switch in human cells. *Cell Cycle* **3:** 1011–1013.

Karras GI, Jentsch S. 2010. The RAD6 DNA damage tolerance pathway operates uncoupled from the replication fork and is functional beyond S phase. *Cell* **141:** 255–267.

Kile AC, Chavez DA, Bacal J, Eldirany S, Korzhnev DM, Bezsonova I, Eichman BF, Cimprich KA. 2015. HLTF's ancient HIRAN domain binds 3′ DNA ends to drive replication fork reversal. *Mol Cell* **58:** 1090–1100.

Kim H, D'Andrea AD. 2012. Regulation of DNA cross-link repair by the Fanconi anemia/BRCA pathway. *Gene Dev* **26:** 1393–1408.

Kim H, Yang K, Dejsuphong D, D'Andrea AD. 2012. Regulation of Rev1 by the Fanconi anemia core complex. *Nat Struct Mol Biol* **19:** 164–170.

Kim H, Dejsuphong D, Adelmant G, Ceccaldi R, Yang K, Marto JA, D'Andrea AD. 2014. Transcriptional repressor ZBTB1 promotes chromatin remodeling and translesion DNA synthesis. *Mol Cell* **54:** 107–118.

Kolinjivadi AM, Sannino V, de Antoni A, Técher H, Baldi G, Costanzo V. 2017a. Moonlighting at replication forks—A new life for homologous recombination proteins BRCA1, BRCA2 and RAD51. *FEBS Lett* **591:** 1083–1100.

Kolinjivadi AM, Sannino V, De Antoni A, Zadorozhny K, Kilkenny M, Técher H, Baldi G, Shen R, Ciccia A, Pellegrini L, et al. 2017b. Smarcal1-mediated fork reversal triggers Mre11-dependent degradation of nascent DNA in the absence of Brca2 and stable Rad51 nucleofilaments. *Mol Cell* **67:** 867–881 e867.

Korzhnev DM, Hadden MK. 2016. Targeting the translesion synthesis pathway for the development of anti-cancer chemotherapeutics. *J Med Chem* **59:** 9321–9336.

Lachaud C, Moreno A, Marchesi F, Toth R, Blow JJ, Rouse J. 2016. Ubiquitinated Fancd2 recruits Fan1 to stalled replication forks to prevent genome instability. *Science* **351:** 846–849.

Lai X, Broderick R, Bergoglio V, Zimmer J, Badie S, Niedzwiedz W, Hoffmann JS, Tarsounas M. 2017. MUS81 nuclease activity is essential for replication stress tolerance and chromosome segregation in *BRCA2*-deficient cells. *Nat Commun* **8:** 15983.

Lemacon D, Jackson J, Quinet A, Brickner JR, Li S, Yazinski S, You Z, Ira G, Zou L, Mosammaparast N, et al. 2017. MRE11 and EXO1 nucleases degrade reversed forks and elicit MUS81-dependent fork rescue in *BRCA2*-deficient cells. *Nat Commun* **8:** 860.

Lin JR, Zeman MK, Chen JY, Yee MC, Cimprich KA. 2011. SHPRH and HLTF act in a damage-specific manner to coordinate different forms of postreplication repair and prevent mutagenesis. *Mol Cell* **42:** 237–249.

Lok BH, Carley AC, Tchang B, Powell SN. 2013. RAD52 inactivation is synthetically lethal with deficiencies in BRCA1 and PALB2 in addition to BRCA2 through RAD51-mediated homologous recombination. *Oncogene* **32:** 3552–3558.

Lomonosov M, Anand S, Sangrithi M, Davies R, Venkitaraman AR. 2003. Stabilization of stalled DNA replication forks by the BRCA2 breast cancer susceptibility protein. *Genes Dev* **17:** 3017–3022.

Lopes M, Foiani M, Sogo JM. 2006. Multiple mechanisms control chromosome integrity after replication fork uncoupling and restart at irreparable UV lesions. *Mol Cell* **21:** 15–27.

Lossaint G, Larroque M, Ribeyre C, Bec N, Larroque C, Décaillet C, Gari K, Constantinou A. 2013. FANCD2 binds MCM proteins and controls replisome function upon activation of s phase checkpoint signaling. *Mol Cell* **51:** 678–690.

Lovett ST. 2017. Template-switching during replication fork repair in bacteria. *DNA Repair (Amst)* **56:** 118–128.

Luke-Glaser S, Luke B, Grossi S, Constantinou A. 2010. FANCM regulates DNA chain elongation and is stabilized by S-phase checkpoint signalling. *EMBO J* **29:** 795–805.

Michl J, Zimmer J, Buffa FM, McDermott U, Tarsounas M. 2016. FANCD2 limits replication stress and genome instability in cells lacking BRCA2. *Nat Struct Mol Biol* **23:** 755–757.

Mijic S, Zellweger R, Chappidi N, Berti M, Jacobs K, Mutreja K, Ursich S, Ray Chaudhuri A, Nussenzweig A, Janscak P, et al. 2017. Replication fork reversal triggers fork degradation in *BRCA2*-defective cells. *Nat Commun* **8:** 859.

Neelsen KJ, Lopes M. 2015. Replication fork reversal in eukaryotes: From dead end to dynamic response. *Nat Rev Mol Cell Biol* **16:** 207–220.

Norquist B, Wurz KA, Pennil CC, Garcia R, Gross J, Sakai W, Karlan BY, Taniguchi T, Swisher EM. 2011. Secondary somatic mutations restoring BRCA1/2 predict chemotherapy resistance in hereditary ovarian carcinomas. *J Clin Oncol* **29:** 3008–3015.

Pace P, Mosedale G, Hodskinson MR, Rosado IV, Sivasubramaniam M, Patel KJ. 2010. Ku70 corrupts DNA repair in the absence of the Fanconi anemia pathway. *Science* **329:** 219–223.

Pathania S, Bade S, Le Guillou M, Burke K, Reed R, Bowman-Colin C, Su Y, Ting DT, Polyak K, Richardson AL, et al. 2014. BRCA1 haploinsufficiency for replication stress suppression in primary cells. *Nat Commun* **5:** 5496.

Peng M, Xie J, Ucher A, Stavnezer J, Cantor SB. 2014. Crosstalk between BRCA-Fanconi anemia and mismatch repair pathways prevents MSH2-dependent aberrant DNA damage responses. *EMBO J* **33:** 1698–1712.

Petermann E, Orta ML, Issaeva N, Schultz N, Helleday T. 2010. Hydroxyurea-stalled replication forks become progressively inactivated and require two different RAD51-mediated pathways for restart and repair. *Mol Cell* **37:** 492–502.

Pilzecker B, Buoninfante OA, van den Berk P, Lancini C, Song JY, Citterio E, Jacobs H. 2017. DNA damage tolerance in hematopoietic stem and progenitor cells in mice. *Proc Natl Acad Sci* **114:** E6875–E6883.

Quinet A, Vessoni AT, Rocha CR, Gottifredi V, Biard D, Sarasin A, Menck CF, Stary A. 2014. Gap-filling and bypass at the replication fork are both active mechanisms for tolerance of low-dose ultraviolet-induced DNA damage in the human genome. *DNA Repair (Amst)* **14:** 27–38.

Quinet A, Martins DJ, Vessoni AT, Biard D, Sarasin A, Stary A, Menck CF. 2016. Translesion synthesis mechanisms depend on the nature of DNA damage in UV-irradiated human cells. *Nucleic Acids Res* **44:** 5717–5731.

Raghunandan M, Chaudhury I, Kelich SL, Hanenberg H, Sobeck A. 2015. FANCD2, FANCJ and BRCA2 cooperate to promote replication fork recovery independently of the Fanconi anemia core complex. *Cell Cycle* **14:** 342–353.

Ray Chaudhuri A, Hashimoto Y, Herrador R, Neelsen KJ, Fachinetti D, Bermejo R, Cocito A, Costanzo V, Lopes M. 2012. Topoisomerase I poisoning results in PARP-mediated replication fork reversal. *Nat Struct Mol Biol* **19:** 417–423.

Ray Chaudhuri A, Callen E, Ding X, Gogola E, Duarte AA, Lee JE, Wong N, Lafarga V, Calvo JA, Panzarino NJ, et al. 2016. Replication fork stability confers chemoresistance in *BRCA*-deficient cells. *Nature* **535:** 382–387.

Rondinelli B, Gogola E, Yucel H, Duarte AA, van de Ven M, van der Sluijs R, Konstantinopoulos PA, Jonkers J, Ceccaldi R, Rottenberg S, et al. 2017. EZH2 promotes degradation of stalled replication forks by recruiting MUS81 through histone H3 trimethylation. *Nat Cell Biol* **19:** 1371–1378.

Sail V, Rizzo AA, Chatterjee N, Dash RC, Ozen Z, Walker GC, Korzhnev DM, Hadden MK. 2017. Identification of small molecule translesion synthesis inhibitors that target the Rev1-CT/RIR protein–protein interaction. *ACS Chem Biol* **12:** 1903–1912.

Sakai W, Swisher EM, Karlan BY, Agarwal MK, Higgins J, Friedman C, Villegas E, Jacquemont C, Farrugia DJ, Couch FJ, et al. 2008. Secondary mutations as a mechanism of cisplatin resistance in *BRCA2*-mutated cancers. *Nature* **451:** 1116–1120.

Sakai W, Swisher EM, Jacquemont C, Chandramohan KV, Couch FJ, Langdon SP, Wurz K, Higgins J, Villegas E, Taniguchi T. 2009. Functional restoration of BRCA2 protein by secondary *BRCA2* mutations in *BRCA2*-mutated ovarian carcinoma. *Cancer Res* **69:** 6381–6386.

Sala-Trepat M, Rouillard D, Escarceller M, Laquerbe A, Moustacchi E, Papadopoulo D. 2000. Arrest of S-phase progression is impaired in Fanconi anemia cells. *Exp Cell Res* **260:** 208–215.

Sale JE, Lehmann AR, Woodgate R. 2012. Y-family DNA polymerases and their role in tolerance of cellular DNA damage. *Nat Rev Mol Cell Biol* **13:** 141–152.

Sanders MA, Haynes B, Nangia-Makker P, Polin LA, Shekhar MP. 2017. Pharmacological targeting of RAD6 enzyme-mediated translesion synthesis overcomes resistance to platinum-based drugs. *J Biol Chem* **292:** 10347–10363.

Schlacher K, Christ N, Siaud N, Egashira A, Wu H, Jasin M. 2011. Double-strand break repair-independent role for BRCA2 in blocking stalled replication fork degradation by MRE11. *Cell* **145:** 529–542.

Schlacher K, Wu H, Jasin M. 2012. A distinct replication fork protection pathway connects Fanconi anemia tumor suppressors to RAD51-BRCA1/2. *Cancer Cell* **22:** 106–116.

Schwab RA, Nieminuszczy J, Shin-Ya K, Niedzwiedz W. 2013. FANCJ couples replication past natural fork barriers with maintenance of chromatin structure. *J Cell Biol* **201:** 33–48.

Spies J, Waizenegger A, Barton O, Sürder M, Wright WD, Heyer WD, Löbrich M. 2016. Nek1 regulates Rad54 to orchestrate homologous recombination and replication fork stability. *Mol Cell* **62:** 903–917.

Srivastava AK, Han C, Zhao R, Cui T, Dai Y, Mao C, Zhao W, Zhang X, Yu J, Wang QE. 2015. Enhanced expression of DNA polymerase η contributes to cisplatin resistance of ovarian cancer stem cells. *Proc Natl Acad Sci* **112:** 4411–4416.

Sugimura K, Takebayashi S, Taguchi H, Takeda S, Okumura K. 2008. PARP-1 ensures regulation of replication fork progression by homologous recombination on damaged DNA. *J Cell Biol* **183:** 1203–1212.

Taglialatela A, Alvarez S, Leuzzi G, Sannino V, Ranjha L, Huang JW, Madubata C, Anand R, Levy B, Rabadan R, et al. 2017.

Restoration of replication fork stability in *BRCA1*- and *BRCA2*-deficient cells by inactivation of SNF2-family fork remodelers. *Mol Cell* **68:** 414–430 e418.

Tan SLW, Chadha S, Liu Y, Gabasova E, Perera D, Ahmed K, Constantinou S, Renaudin X, Lee M, Aebersold R, et al. 2017. A class of environmental and endogenous toxins induces BRCA2 haploinsufficiency and genome instability. *Cell* **169:** 1105–1118 e1115.

Thompson EL, Yeo JE, Lee EA, Kan Y, Raghunandan M, Wiek C, Hanenberg H, Scharer OD, Hendrickson EA, Sobeck A. 2017. FANCI and FANCD2 have common as well as independent functions during the cellular replication stress response. *Nucleic Acids Res* **45:** 11837–11857.

Tian F, Sharma S, Zou J, Lin SY, Wang B, Rezvani K, Wang H, Parvin JD, Ludwig T, Canman CE, et al. 2013. BRCA1 promotes the ubiquitination of PCNA and recruitment of translesion polymerases in response to replication blockade. *Proc Natl Acad Sci* **110:** 13558–13563.

Vujanovic M, Krietsch J, Raso MC, Terraneo N, Zellweger R, Schmid JA, Taglialatela A, Huang JW, Holland CL, Zwicky K, et al. 2017. Replication fork slowing and reversal upon DNA damage require PCNA polyubiquitination and ZRANB3 DNA translocase activity. *Mol Cell* **67:** 882–890.

Wang AT, Kim T, Wagner JE, Conti BA, Lach FP, Huang AL, Molina H, Sanborn EM, Zierhut H, Cornes BK, et al. 2015. A dominant mutation in human RAD51 reveals its function in DNA interstrand crosslink repair independent of homologous recombination. *Mol Cell* **59:** 478–490.

Xie J, Litman R, Wang S, Peng M, Guillemette S, Rooney T, Cantor SB. 2010a. Targeting the FANCJ-BRCA1 interaction promotes a switch from recombination to polη-dependent bypass. *Oncogene* **29:** 2499–2508.

Xie K, Doles J, Hemann MT, Walker GC. 2010b. Error-prone translesion synthesis mediates acquired chemoresistance. *Proc Natl Acad Sci* **107:** 20792–20797.

Xie J, Peng M, Guillemette S, Quan S, Maniatis S, Wu Y, Venkatesh A, Shaffer SA, Brosh RM Jr, Cantor SB. 2012. FANCJ/BACH1 acetylation at lysine 1249 regulates the DNA damage response. *PLoS Genet* **8:** e1002786.

Xu S, Wu X, Wu L, Castillo A, Liu J, Atkinson E, Paul A, Su D, Schlacher K, Komatsu Y, et al. 2017. Abro1 maintains genome stability and limits replication stress by protecting replication fork stability. *Genes Dev* **31:** 1469–1482.

Yamada M, Masai H, Bartek J. 2014. Regulation and roles of Cdc7 kinase under replication stress. *Cell Cycle* **13:** 1859–1866.

Yamanaka K, Chatterjee N, Hemann MT, Walker GC. 2017. Inhibition of mutagenic translesion synthesis: A possible strategy for improving chemotherapy? *PLoS Genet* **13:** e1006842.

Yazinski SA, Comaills V, Buisson R, Genois MM, Nguyen HD, Ho CK, Todorova Kwan T, Morris R, Lauffer S, Nussenzweig A, et al. 2017. ATR inhibition disrupts rewired homologous recombination and fork protection pathways in PARP inhibitor-resistant BRCA-deficient cancer cells. *Genes Dev* **31:** 318–332.

Ying S, Hamdy FC, Helleday T. 2012. Mre11-dependent degradation of stalled DNA replication forks is prevented by BRCA2 and PARP1. *Cancer Res* **72:** 2814–2821.

Zadorozhny K, Sannino V, Belan O, Mlcouskova J, Spirek M, Costanzo V, Krejci L. 2017. Fanconi-anemia-associated mutations destabilize RAD51 filaments and impair replication fork protection. *Cell Rep* **21:** 333–340.

Zamborszky J, Szikriszt B, Gervai JZ, Pipek O, Poti A, Krzystanek M, Ribli D, Szalai-Gindl JM, Csabai I, Szallasi Z, et al. 2017. Loss of BRCA1 or BRCA2 markedly increases the rate of base substitution mutagenesis and has distinct effects on genomic deletions. *Oncogene* **36:** 746–755.

Zellweger R, Dalcher D, Mutreja K, Berti M, Schmid JA, Herrador R, Vindigni A, Lopes M. 2015. Rad51-mediated replication fork reversal is a global response to genotoxic treatments in human cells. *J Cell Biol* **208:** 563–579.

How the Genome Folds, Divides, Lives, and Dies

Whitney L. Johnson,[1,2] Kathleen T. Xie,[1,2] Mijung Kwon,[1,2,5] Shiwei Liu,[1,2,5]
and David Pellman[1,2,3,4]

[1]*Department of Pediatric Oncology, Dana-Farber Cancer Institute, Boston, Massachusetts 02215*
[2]*Department of Cell Biology, Harvard Medical School, Boston, Massachusetts 02215*
[3]*Broad Institute of MIT and Harvard, Cambridge, Massachusetts 02142*
[4]*Howard Hughes Medical Institute, Chevy Chase, Maryland 20815*
Correspondence: david_pellman@dfci.harvard.edu

The 82nd Cold Spring Harbor Symposium was the first to focus on Chromosome Segregation and Structure. In one sense, this was something new. There has been significant recent progress on chromosome segregation and chromosome structure, and it was a propitious time to bring these fields together. But in another sense, the topics discussed at the meeting were classic ones that have been central to many prior Cold Spring Harbor Symposia, going back decades. In looking back over prior meetings, it is stunning to see the progress that has been made in understanding genes, chromatin, cell division, cancer, and chromosome biology in general. Remarkably, about 80 years ago, it was still valid to ask whether chromosomes are even worth studying. Summing up the 1941 meeting, H.J. Muller laconically noted, "The time has come when a few … of the physicists and biochemists realize that genes and chromosomes do form valid subjects of inquiry, and that what the biologists have been saying about their mysterious properties calls for some more looking into on their part." Fortunately, long before the current symposium, Muller's view prevailed, and the debate was settled on whether chromosomes are indeed a valid subject for inquiry. Even so, for many years, chromosome biology research was limited by the available tools. In summarizing the 1973 meeting, Hewson Swift said, with some frustration, that eukaryotic chromosomes under the electron microscope looked, "… even at their best, something like a bad day at a macaroni factory." (Pasta has long been a favored metaphor for chromosomes.) Nowadays, we can make a bit more sense from our observations of genomic noodles. For example, at the current meeting, we learned that the "vermicelli"-like appearance of WAPL-depleted chromatin can provide deep insights into the organizing principles of chromosomes. New technologies have been applied to classic problems and have pushed the field forward in exciting directions. So, by 2017, we had a better day at the "macaroni factory."

The 82nd Symposium was a concentrated five-day burst of information. In addition to highlighting a cross section of the most exciting cutting-edge work, the meeting was broad and provided a crash course on relevant topics for scientists looking to bridge fields. The organizers—Terri Grodzicker, David Stewart, and Bruce Stillman—put together an impressive program with an exciting list of speakers. An interesting side note: Cold Spring Harbor Symposia used to be more of a way of life than a focused event. In the 1941 meeting summary, it was noted that Milislav Demerec made a major organizational shift by shortening the meeting from the customary five weeks to two! Wistfulness over a bygone era of leisurely science aside, the 82nd Symposium illustrated what can be accomplished in five days.

Why have a meeting focused on chromosome segregation and structure now? Many important long-standing questions are being addressed—from the epigenetic specification of centromeres to the requirement of mitotic chromosome compaction for proper segregation. However, one new theme stood out: Taking what we have learned about self-assembly from the spindle and applying it to chromosome structure. Many aspects of spindle assembly are driven by molecular motor proteins that can sort and orient microtubules, measure their lengths, and create intermediate structures such as asters or antiparallel microtubule arrays. Motors can also control signaling within the spindle by concentrating signaling molecules and generating signal gradients. New methods are elucidating these self-assembly design principles, including new live cell imaging methods, structural studies, biochemical reconstitution, and single-molecule imaging. The 82nd Cold Spring Harbor Symposium occurred at an inflection point for the chromosome structure field where, for the first time, a similar understanding of chromosome self-assembly seems within grasp. Here again, motor proteins organize structure, folding the chromatin polymers into intermediate structures using simple biophysical principles and basic biochemical activities. Below, we summarize self-assembly and other major themes that emerged during the meeting, while hoping to still convey the beauty of many individual stories.

---

[5]These authors contributed equally to this work.

Published by Cold Spring Harbor Laboratory Press; doi: 10.1101/sqb.2017.82.035527

SELF-ASSEMBLY AND REGULATION OF THE CHROMOSOME SEGREGATION MACHINERY

The symposium had numerous talks highlighting progress on understanding the mitotic spindle and its coordination with chromosomes, nuclear envelopes, and cytokinesis.

How to Build a Spindle and Know if You Got It Right

Tubulin molecules assemble into dynamic microtubule polymers, which must organize into higher-order spindle structures—a classic nonequilibrium self-assembly process. **Tarun Kapoor** reviewed work from his laboratory demonstrating that a microtubule motor (kinesin-4) and a nonmotor microtubule-associated protein (PRC1) can serve as measuring devices, by accumulating at microtubule ends in a cluster whose size is proportional to the length of the microtubule (Subramanian et al. 2013). Using light sheet microscopy, his group is studying how these tagged microtubules self-assemble into a functioning spindle midzone capable of driving anaphase B spindle elongation. He also described the viscoelastic material properties of the spindle. For fast-acting forces, the spindle behaves as an elastic solid that adapts to applied force. For slow-acting forces comparable to the movement of chromosomes, the spindle is more fluid, allowing chromosomes to pass through while maintaining spindle integrity.

Microtubules assemble into spindles, but they must also interact with chromosomes to facilitate chromosome segregation. **Stephen Harrison** addressed the design principles of the kinetochore, the site of attachment of the chromosome to the mitotic spindle (Jenni et al., this volume). He presented structures of several kinetochore subcomplexes. The subcomplexes are often assembled into higher-order structures via "peptide-in-groove" contacts, where an unstructured peptide of one subcomplex component docks into a structured binding site of another. This type of connection has many benefits: It is amenable to regulation because the peptide is easily accessed by kinases and phosphatases, and the contact also provides mechanical compliance and structural adaptability.

In addition to linking the chromosome and spindle, the kinetochore is also regulatory. Kinetochores that have not yet properly attached to microtubules send a "wait anaphase" signal via the spindle assembly checkpoint (SAC), providing time to correct segregation errors. The SAC prevents mitotic exit by inhibiting the APC/C E3 ligase that degrades mitotic cyclins. Through biochemical and structural studies, there has been tremendous recent progress in understanding the SAC. However, talks at the meeting illustrated that there are also still surprises. **Arshad Desai** showed that not only can the kinetochore inhibit the APC/C and put the brakes on anaphase entry, but that it can also activate the APC/C and accelerate anaphase progression (Lara-Gonzalez et al., this volume). Both of these functions are mediated by Cdc20, a key regulator of the APC/C. At unattached kinetochores,

Cdc20 is loaded onto the mitotic checkpoint complex, which inhibits the APC/C. However, Cdc20 dynamically associates with the kinetochore even when microtubules are properly attached. Desai showed that while fluxing through the attached kinetochore, Cdc20 is dephosphorylated and activated, which accelerates mitotic exit. Therefore, it is kinetochore-microtubule attachment that flips the kinetochore from an APC/C "brake" to an APC/C "activator." **Jennifer DeLuca** also discussed how kinetochores modulate the strength of kinetochore–microtubule attachments during mitosis. Early in mitosis, dynamic attachments are beneficial, so that errors can be readily corrected. Later, stable attachments are beneficial, so that correctly attached chromosomes do not detach from the spindle. It is well known that the mitotic kinase Aurora B facilitates the early lability of attachment through inhibitory phosphorylation of kinetochore proteins. DeLuca presented surprising findings that its cousin Aurora A is also important in this process, although Aurora A is famous for its functions at the centrosome, not the kinetochore (DeLuca, this volume). Thus, although the localization of the majority population of a protein can provide clues about function, minor populations can be important, and should not be overlooked. At anaphase onset, the cohesin remaining between sister chromatids is released in an APC/C-dependent manner. **Yoshinori Watanabe** found that the conserved meiotic kinetochore protein Moa1/Meikin, although initially identified as a factor that facilitates monoorientation in meiosis I, also protects pericentric cohesin from removal until anaphase II (Miyazaki et al., this volume). Moa1-mediates cohesin protection by increasing both the localization and the activity of Shugoshin (Sgo1), a known cohesin protector. This work uncovers a new protector of meiotic cohesin and shows how one protein can regulate multiple meiotic-specific functions.

Three Ways of Looking at a Centrosome

There were new insights into the structure and assembly of the centrosome. In most animal cells, centrosomes organize the poles of the mitotic spindle. Centrosomes are typically composed of microtubule-containing centrioles surrounded by an amorphous protein mass called the pericentriolar material (PCM). Using recombinant *C. elegans* proteins, **Tony Hyman**'s group showed that even in the absence of microtubule-containing centrioles, PCM can spontaneously self-assemble and nucleate microtubules. The scaffold protein SPD-5 undergoes a phase transition, forming liquid-like droplets, which can sequentially recruit other PCM proteins and concentrate enough unpolymerized tubulin to nucleate microtubule asters. **Tim Stearns** showed that centriole triplet microtubules are required to maintain the structural integrity of centrioles (Wang and Stearns, this volume). Human cells that lack δ-tubulin or ε-tubulin form defective centrioles without normal triplet microtubules, fail to undergo centriole maturation, and disintegrate during mitotic exit by an interesting but still mysterious process. **Mónica Bettencourt-Dias** addressed how centrioles are eliminated in female meiosis so that the zygote has a normal centriole number after fertilization. Her work demonstrates that centrosome reduction starts with transcriptional repression of Polo kinase during oogenesis (Nabais et al., this volume). Loss of Polo leads to the loss of PCM, which in turn is required to maintain centrioles. The reputation of the PCM has suffered somewhat by association with the adjective "amorphous." The talks described above provided a rehabilitation of sorts, highlighting PCM's central role in centrosome integrity and in development.

The Spindle Is the Center of the Universe?

Spindle function has to be coordinated with many other cellular processes, which was the topic of several talks. **Tim Mitchison** described how, in large frog eggs, microtubules asters from each spindle pole grow all the way to the cell cortex, and their region of overlap defines the site of cytokinetic furrow formation (Mitchison and Field, this volume). Interestingly, in polyspermic eggs, the cleavage furrow only forms between asters from the same spindle, with chromatin in between them. Here, again, motors are key. Motors transport the chromosome passenger complex (CPC) from chromosomes to the aster overlap zone, and only overlapping asters with the CPC define the site of the cleavage furrow. Thus, chromosomes can direct cytokinetic furrow positioning, even at large distances in huge egg cells. As meeting participant Bill Earnshaw has previously noted (Earnshaw and Bernat 1991), these experiments illustrate that Daniel Mazia, although witty, got it wrong when he said, "Indeed, the role in mitosis of the chromosome arms, which carry most of the genetic material, may be compared with that of a corpse at a funeral: they provide the reason for the proceedings but do not take an active part in them" (Mazia 1961).

At the end of chromosome segregation in metazoan cells, the nuclear envelope (NE) reforms around the daughter chromosomes masses, and all of the chromosomes must be encapsulated to form a single nucleus. **Daniel Gerlich** used an imaging-based RNAi screen to identify mutants that form multiple nuclei. The screen revealed that loss of BAF (barrier-to-autointegration factor) resulted in the formation of micronuclei, which are miniature additional nuclei that contain one or a few chromosomes and are commonly observed in cancer. A satisfying combination of imaging and biophysical experiments showed that BAF forms a stiff shell around decondensing chromosomes, thus keeping them together as one mass. Interestingly, the mechanical properties of BAF can be regulated by phosphorylation, which ensures that the stiff shell persists for only a short period, until the NE reforms. Our laboratory (**David Pellman**'s laboratory) addressed the coordination between chromosome segregation and NE assembly, which is relevant to understanding the properties of micronuclei. We had previously shown that chromosomes encapsulated in micronuclei are prone to DNA damage and chromothripsis, a type of large-scale genome rearrangement. Spontaneous disruption of the micronu-

clear envelope triggers DNA damage (Hatch et al. 2013). We found that this fragility of micronuclei can be explained by a defect in NE assembly. Chromosomes that separate from the main chromosome mass lag within the spindle, where spindle microtubules occlude key NE components. The data suggest that rather than the tight, checkpoint-mediated coordination between chromosome segregation and NE assembly that was previously proposed, there is only loose coordination from the normal timing of spindle disassembly. The absence of precise regulatory controls may explain why errors during mitotic exit are frequent and a major source of catastrophic genome rearrangement.

Recently, it has been shown that NE disruption (e.g., in micronuclei) can activate an innate proinflammatory response. This response occurs because cytoplasmic DNA, which the cell interprets to be foreign DNA, activates the cytoplasmic DNA sensor cGAS (cyclic GMP-AMP synthase). However, in mitosis, genomic DNA is also exposed to the cytoplasm and yet does not trigger innate immune signaling. Using both *Xenopus* extracts and cultured mammalian cells, **Hiro Funabiki** showed that while mitotic chromatin can bind cGAS, the presence of nucleosomes inhibits cGAS activation, providing a crude first discrimination between self (chromatinized) DNA and foreign (naked) DNA (Funabiki et al., this volume). However, in the presence of drugs that prolong mitosis, cGAS does eventually become activated, promoting apoptosis.

Life after HeLa Cells

We have learned a tremendous amount about cell division and chromosome biology from tissue culture cells and single cell model organisms. However, participants at the meeting also sought to understand the diversity of cell functions within normal tissue and development. Multiple speakers addressed how cell extrinsic factors—such as in vivo tissue architecture and cell–cell contacts—impact chromosome segregation and cell division. **Angelika Amon** showed that complex three-dimensional (3D) tissue architecture promotes accurate chromosome segregation and genome stability. She reported that the frequency of chromosome missegregation is higher in 2D tissue cultures than in 3D organoid cultures or in tissues. Cell nonautonomous mechanisms, such as cell–cell contacts and cell–matrix adhesion, are known to influence the fidelity of chromosome segregation. Dissecting other features of the normal 3D tissue microenvironment that enhance segregation fidelity remains a fascinating problem. **Eric Nigg** reported new effects of the centrosome on tissue architecture (Nigg et al., this volume). He found that inducing centrosome structural aberrations (overexpression of NLP) disrupted 3D spheroid structure. This disruption occurs because cells with centrosome aberrations are stiff and squeeze dividing cells out of the epithelium, affecting even those cells without NLP overexpression. This study provides an interesting physical example of a cell nonautonomous effect. Finally, **Jan van Deursen** addressed the 100-year-old hypothesis from Theodor Boveri that centrosome abnormalities can cause aneuploidy and promote tumorigenesis. He reviewed the many mouse models that have been generated to test the hypothesis and highlighted that we still do not understand why some models generate tumors and others do not. He proposed that defects in the separation of centrosomes, which cause abnormal spindle geometry, could be an important feature of the more tumorigenic models.

SELF-ASSEMBLY AND REGULATION OF CHROMOSOME STRUCTURE

The 82nd Cold Spring Harbor Symposium marked and celebrated a major step forward in the field of chromosome structure. The causes and consequences of chromosome structure have been a central mystery in biology since the discovery of the mitotic chromosome cycle by Walther Flemming and the subsequent observation that the interphase nucleus is partitioned into euchromatin and heterochromatin by Emil Heitz. Now, a quantitative and testable model—termed the loop extrusion model—appears to explain key aspects of chromosome structure as a self-organization process driven by motor proteins. It is proposed that the dramatic difference in the appearance of mitotic and interphase chromosomes may be something of an illusion. Instead of differing by mechanistically distinct organizing principles, mitotic and interphase chromosomes may instead represent a continuum in the degree of looped chromosome organization. A host of new technologies have driven these breakthroughs.

The Basics of Loop Extrusion

Leonid Mirny described the loop extrusion model and summarized its history, starting with earlier versions by Arthur Riggs, John Marko, and meeting participant Kim Nasmyth (Riggs 1990; Nasmyth 2001; Alipour and Marko 2012). The core idea of the model is that chromosomes become organized into loops, by motor proteins ("loop extruders") that pump DNA processively to generate loops (Fudenberg et al., this volume). These loop extruders can accumulate in certain regions of the genome if they encounter a barrier, either by colliding into each other, or the transcription regulator CTCF. Because CTCF binding is sequence-dependent, this particular barrier creates a pattern of loops that is similar between individual cells and is therefore detectable by population-average experiments. These semistable loops form insulated regions termed topologically associated domains (TADs) that have been detected in chromosome conformation capture experiments (mostly Hi-C).

The function of the molecules driving loop extrusion are rapidly being defined. The major, if not only, loop extruders are the SMC complexes (structural maintenance of chromosomes), including cohesin and condensin. These complexes share an overall similar ring-shaped structure, with two proteins (each forming long antiparallel coiled coils) being joined at a "hinge" on one end, and two ATPase "head" domains at the other end. The heads can dimerize in the presence of ATP and are additionally linked by kleisin proteins, forming a double gate. As discussed by Mirny, the current loop extrusion model makes quantitative and testable predictions about the consequences of eliminating loop extruders, increasing their density or processivity, and eliminating extrusion barriers. Indeed, other talks at the symposium and work published after the meeting validate these predictions. SMC-dependent loop extrusion has now been observed by live-cell imaging in *Bacillus subtilis* (Wang et al. 2017), and single molecule imaging experiments show that SMC proteins are processive DNA motors capable of loop extrusion in vitro (Terakawa et al. 2017; Ganji et al. 2018).

Jan-Michael Peters also presented work validating the loop extrusion model, showing that CTCF is required for TAD patterning in interphase. Additionally, his group observed that if cohesin cannot be removed from chromosomes (by depletion of WAPL, a protein required for cohesin removal), chromosomes condense into structures that visually resemble prophase chromosomes; these prophase-like structures are termed "vermicelli" (Tedeschi et al. 2013). Hi-C analysis revealed that WAPL-depleted chromosomes show extensive looping which can even extend beyond CTCF boundaries (presumably due to the increased lifetime of cohesin on chromatin after its removal factor has been depleted) (Haarhuis et al. 2017; Wutz et al. 2017). Interestingly, Hi-C analysis in mitotic cells also revealed extensive looping with boundaries not determined by CTCF at all (boundaries may instead primarily be demarcated by the accumulation of loop extruders colliding with each other) (Gibcus et al. 2018). WAPL-depleted interphase chromosomes thus resemble mitotic chromosomes structurally. These findings suggest the simplifying view that interphase and mitotic chromosome organization exist on a continuum, differing mainly through the activity and concentration of loop extruders as well as the integrity of the extrusion barriers. Why cohesins are the main loop extruders in interphase whereas condensins are the primary extruders in mitosis remains to be determined.

What are the minimal requirements to make a mitotic chromosome? **Tatsuya Hirano** used biochemical reconstitution to show that condensins (I and II) are sufficient to organize DNA into a mitotic chromatid-like structures, even in the absence of histones (Shintomi and Hirano, this volume). Although these nucleosome-depleted structures were sparser and more fragile, linear compaction of mitotic DNA occurs efficiently, presumably via DNA looping. Hirano reminded the participants that condensins serving as the fundamental architects of mitotic chromosome should not be that surprising, given that condensins are conserved in bacteria whereas histones are not.

Kerry Bloom described the improvement of a statistical mechanics model of DNA structure by including loop extruding motors into his simulations of chromatin (Lawrimore et al., this volume). The entropy-driven tendency of floppy chromatin fibers to form loops suggests that loop extruders may be able to make local jumps, which might explain the large step sizes recently observed during loop extrusion (Ganji et al. 2018). Moreover, Bloom noted that local tethers on chromatin can cause an accumulation of loop extruders. Because local tethers occur at centromeres due to microtubule attachment, this model may explain the observed pattern of condensin accumulation around centromeres. Thus, Bloom's model suggests ways in which the spindle microtubules can pattern the organization of chromatin.

Put a Ring on It

SMC proteins can be loaded onto DNA in several ways. If the SMC ring is opened, DNA can enter the ring, and after the ring closes, the SMC encloses the DNA (a topological interaction). Alternatively, DNA can bind the ring directly or enter the ring as a loop, without the SMC ring opening (a nontopological interaction). Talks from **Kim Nasmyth** and **Frank Uhlmann** addressed the mechanisms that enable cohesins to toggle between the two different modes of interaction with DNA. Topological interactions are important for sister chromatin cohesion, whereas nontopological interactions likely enable loop extrusion. Supporting this idea, Nasmyth described separation-of-function mutations of cohesin that enable nontopological interactions but abolish topological interactions and sister chromatid cohesion. These results provide definitive evidence for cohesin loading onto DNA in two ways.

For dynamic topological interactions to occur, DNA must be able to enter and exit the cohesin ring. All groups working on the problem agree on the cohesin release mechanism, which either involves kleisin cleavage at anaphase or dissociation of the kleisin-SMC3 interface. However, there is no consensus on the DNA entry mechanism. Protein cross-linking and protein fusion experiments from Nasmyth's laboratory suggest that cohesin is a one-way street, with DNA entry through the hinge and exit from SMC head-kleisin gate. In contrast, Frank Uhlmann's work suggests that the SMC head-kleisin gate is reversible, used for both entry and exit. It is possible that both junctions can allow entry, but which is the dominant path remains to be determined. To obtain functional cohesion, after the DNA from one chromatid loads, the sister chromatid must also be captured. A previous model for sister chromatid capture involved "replication-through-a-ring," in which the replisome passes through the ring, and thus both daughter strands of DNA are encircled without necessarily requiring ring opening. However, Uhlmann presented appealing biochemical data for an alternative model, where the sister chromatids may be incorporated into the ring in two steps. First, SMC opens to capture double-stranded DNA from the leading strand during replication. Then, SMC opens again to capture single-stranded DNA from the lagging strand.

Organization and Regulation at All Length Scales

Ideally one would want to understand the principles of chromatin self-organization at all scales from the nucleosome all the way up to a single chromosome territory. **Guohong Li** described his work on cryo-EM structures of small arrays of nucleosomes that form 30 nm filaments (Chen and Li, this volume). The new high-resolution structures discriminate between competing models in the field, establishing that 12 nucleosome arrays form a zigzag two-start nucleosome stack. Recalling Stephen Harrison's talk, knowing the binding junctions that define nucleosome compaction enables biophysical experiments to better define sites of regulation—for example, how histone variants influence compaction. It makes sense that these fundamental biochemical properties of nucleosome arrays in vitro will be relevant in vivo, but this still remains something of a leap of faith, as recent studies have not universally detected 30-nm filaments in vivo (Lieberman-Aiden et al. 2009; Fussner et al. 2012; Nishino et al. 2012). It is possible that short segments of 30 nm fibers do form in vivo but that current methods fail to detect them. **Ollie Rando** described Micro-C, a method that enables small scale analysis of the structure of nucleosome arrays in vivo. Micro-C data could be consistent with the formation of short segments of 30-nm chromatin fibers in yeast. Additionally, although yeast do not have typical TADs, with Micro-C Rando's group was able to detect the organization of small numbers of genes into TAD-like "gene crumples."

The functional impact of chromosome organization on gene expression is still a complex and open question. Several talks addressed the functional consequences of chromosome organization and revealed new wrinkles on the interplay between chromatin organization and the regulation of gene expression. **Edith Heard** showed that during X-chromosome inactivation, the inactive X chromosome (Xi) is largely devoid of TADs (Galupa and Heard, in this volume). The Xi is instead organized into two large self-interacting mega-domains, separated by a region containing the DXZ4 macrosatellite sequence. Like TADs, the formation of such mega-domains is proposed to facilitate the expression of certain genes on Xi that can escape silencing, as loss of megadomains by DXZ4 deletion prevents facultative gene expression escape on the mouse Xi (Giorgetti et al. 2016). The resulting large-scale restructuring of the chromosome therefore generates changes in gene expression, albeit on a modest scale.

TADs and mega-domains affect gene expression, but **Gerd Blobel** suggested that a transcriptional regulator can also help form TAD boundaries (Hsu and Blobel, this volume). He showed that a chromatin reader protein of the BET family, BRD2, colocalizes with CTCF genome wide. BRD2 helps form chromatin boundaries, both by limiting the spread of enhancer activity, and by maintaining the structural integrity of domain boundaries. Clinically relevant BET inhibitors may therefore perturb gene expression through disruption of TAD boundaries, not just by direct transcriptional effects. **Allan Spradling** described the linkage between TAD boundaries and DNA replication timing domains in Dipteran polytene chromosomes (Spradling, this volume). Here, some TAD boundaries are thought to cause replication fork stalling and breakage, leading to the formation of underreplicated regions and a heterogeneous array of deletions within these sequences. Under-replication and the loss of late replicating heterochromatic sequences could provide a genetic basis for gene silencing.

EPIGENETIC REGULATION
OF THE GENOME

In addition to chromatin loops and TADs, other mechanisms contribute to genome regulation. Specialized chromatin states are established by DNA modifications, histone modifications, or histone variants. Chromatin states can propagate epigenetically, independent of DNA sequence. Talks at the Symposium highlighted important progress in this area, in several cases defining interesting connections to chromosome segregation.

Centromeres and Chromatin States Required
for Chromosome Segregation

A classic example of epigenetic regulation is the centromere, a site on each chromosome where the microtubule-binding kinetochore is built during mitosis. Centromeres in most organisms are thought to be defined by a histone H3 variant, CENP-A, which not only recruits kinetochore proteins but also facilitates its own assembly into chromatin, thus propagating epigenetically. Centromeres are so crucial to proper chromosome segregation that one might expect their composition and organization to be highly conserved. However, **Steven Henikoff** reminded us of the remarkable diversity of centromere organization across organisms (Henikoff et al., this volume). For example, centromeres in worms and some insects are holocentric (distributed across the entire chromosome), whereas centromeres in other species are monocentric (localized at a single primary constriction). Centromere DNA sequences are also one of the fastest evolving regions of the genome. To explain the rapid evolution of centromeric DNA sequences and CENP-A, Henikoff proposed the "centromere drive hypothesis," which he developed with Harmit Malik (Henikoff et al. 2001; Henikoff and Malik 2002). The centromere drive hypothesis posits that centromeric sequence variants act as selfish DNA elements, competing to be retained in the egg rather than discarded into a polar body during female meiosis. This favors the emergence of "stronger" and sometimes larger centromeres. What prevents these centromeres from taking over the chromosome is a fitness cost during male meiosis, which selects for new CENP-A variants that neutralize the effects of the expanding centromere.

The centromere drive hypothesis is not only a fascinating genetic phenomenon but also a fundamental cell biology problem. The model requires that stronger centromeres exploit spindle asymmetry to enable their preferential segregation to the egg. **Ben Black** and **Michael Lampson** reported exciting progress in elucidating the cell biological basis of centromere drive in mouse meiosis I (Lampson and Black, this volume). Previous work showed that RAN gradients around chromosomes establish a concentration of the CDC42 GTPase at the cortex near the spindle. Black and Lampson showed that localized activity of CDC42 establishes asymmetric tubulin modifications within the spindle, and these modified microtubules form less stable attachments to stronger (larger) centromeres/kinetochores. This preferentially ori-

ents stronger centromeres toward the egg pole. Although many questions remain, this work initiates a mechanistic understanding of how selfish DNA elements can drive centromere evolution in mice. As discussed by Henikoff, there are likely to be many variations on this theme. He suggested that holocentric chromosome architecture, as is found in nematodes, may have resulted from pressure to escape centromere drive. Moreover, epigenetically defined centromeres may themselves have originated to fight off selfish DNA elements that would otherwise promote their own segregation.

Two other talks also discussed cell division asymmetry and how it may be influenced by centromeres and histone modifications. **Xin Chen** found that *Drosophila* male germline stem cells show a biased distribution of histones during their asymmetric cell division (Kahney et al., this volume). After mitosis, newly translated histones end up in the differentiating daughter cell, whereas older histones partition into the daughter stem cell. Chen showed that this asymmetric distribution of histones is possible because of the asymmetry of two distinct processes; first, older histones are preferentially loaded onto the same chromatid during DNA replication, and second, the sister chromatids with older histones are preferentially segregated into the daughter stem cell during mitosis. Importantly, mutation of a specific residue of histone H3 led to loss of biased chromosome segregation, causing germline stem cell loss and progenitor stem cell tumor phenotypes, and indicating that asymmetric histone inheritance is an important mechanism that determines cell fate. In another talk, **Yves Barral** provided evidence that the centromere has a global impact on chromosome structure and segregation in the inherently asymmetric cell divisions of budding yeast. He found that without centromeres, chromosomes are unable to recruit Aurora B during mitosis and are therefore unable to phosphorylate histone 3 serine 10 (H3S10) and properly condense chromosomes. Interestingly, artificially tethering Aurora B to acentric chromosomes rescues condensation, and chromosomes become more likely than before to segregate into budding daughter cells.

In addition to recruiting the kinetochore and facilitating chromosome segregation, CENP-A serves as an epigenetic mark by promoting its own local assembly into chromatin. Several talks addressed the determinants for positioning CENP-A on the chromosome. Precise mapping of centromere locations on genomic DNA has been historically difficult due to the highly repetitive nature of centromeric DNA sequences. **Steve Henikoff**, **Don Cleveland**, and **Ben Black** all discussed data on CENP-A positioning in mammalian cells, using de novo clustering of ChIP-seq reads and newly assembled models of centromeric repeats. They found that CENP-A is positioned on centromeric repeats in a precise and stereotypical manner. CENP-A is phased between CENP-B boxes, which are binding sites for CENP-B, the only known sequence-specific centromeric DNA binding protein in humans. Surprisingly, however, Don Cleveland showed that the phasing of CENP-A nucleosomes between CENP-B boxes does not depend on the presence of CENP-B protein. This observation could be explained either by CENP-B being a vestigial mechanism for CENP-A positioning that is no longer in use, or by CENP-B impacting CENP-A positioning only over long timescales, after many cell divisions. Steve Henikoff also speculated that CENP-B may impact the local conformation of chromatin to promote CENP-A assembly.

Although CENP-A position appears precise on the chromosome, initial CENP-A loading is not precise. **Don Cleveland** showed that during CENP-A loading in G_1, some CENP-A is incorporated incorrectly onto chromosome arms. However, ectopic centromere formation is prevented, because incorrectly placed CENP-A on chromosomes arms is cleared out during S phase, while correctly placed CENP-A on α-satellite repeats remains precisely positioned. Interestingly, **Genevieve Almouzni** discussed how cancer cells become overly reliant on the system for targeting CENP-A, thus creating a potential therapeutic vulnerability (Sitbon et al., this volume). Both CENP-A and HJURP, a CENP-A assembly chaperone, are overexpressed in cancer cells. If CENP-A is overexpressed, cancer cells become acutely sensitive to HJURP depletion, presumably because HJURP prevents deleterious CENP-A deposition on chromosome arms. Therefore, cancer cells can exhibit an addition to nononcogene, epigenetic factors.

Epigenetic Regulation of Genes and "Junk"

Epigenetic mechanisms not only ensure maintenance of centromeres but also control transcription. Transcriptional states need to propagate through many cell divisions to maintain cell identity, dosage compensation, and imprinting patterns. **Marisa Bartolomei** discussed how allele-specific gene expression is disrupted in mouse models of imprinting disorders. She showed that although mutations in imprinting control regions (ICRs) cause the loss of imprinting in primary MEFs, this effect is not homogenous; distinct cell subpopulations show either normal monoallelic or abnormal biallelic expression. These subpopulations show different methylation patterns at their ICRs and never interconvert. This result implies a mechanism where ICR mutant cells make stochastic decisions about ICR methylation status that are stably propagated epigenetically. This mechanism may explain mosaicism seen in human imprinting disorders. Another classic epigenetic phenomenon was discussed by **Barbara Meyer**, who reported the discovery of a general class of epigenetic regulators from studies of *C. elegans* X chromosome dosage compensation (Bian et al., this volume). In the worm, the dosage compensation complex (DCC) reduces gene expression by half from the X chromosome in XX hermaphrodites. DCC contains condensin-like subunits that facilitate dosage compensation by forming large (~1-Mb) TADs. In addition, the silenced X is enriched for a specific histone mark, H4K20me1. From its crystal structure, Meyer discovered that the DCC subunit DPY-21 belongs to a new Jumonji C demethylase subfamily. DPY-21 converts H4K20me2 to H4K20me1 and is required for proper dosage compensation. These demethylases are conserved, providing new

tools for studying other functions of H4K20 methylation, such as in DNA replication. This study highlighted the awesome power of structural biology to inform genetics.

Because transcription has been thought to be repressed during mitosis, how are transcriptional programs reestablished to propagate cell fate? **Ken Zaret** proposed that a low level of transcription is maintained in mitosis, which they detected by ethynyl uridine (EU) labeling (Palozola et al., this volume). Although mitosis disrupts enhancer-promoter interactions, promoters maintain low amplitude transcription. Zaret proposes that this low-level transcription is the long speculated "bookmark" in mitosis that carries the template for the cell's transcriptional program into the next cell cycle. Consistent with earlier talks on loop extrusion and chromosome structure, this work also emphasizes the emerging concept that the organization of mitotic chromosomes is not completely unique, but rather lies on a functional continuum with interphase chromosomes.

In addition to controlling expression of protein-coding genes, epigenetic mechanisms are required to constitutively silence the gene-poor, repetitive regions of genomes packaged into heterochromatin. Although often referred to as "junk DNA," repeats silenced by constitutive heterochromatin have been shown to show both helpful and harmful functions. **Antoine Peters** and colleagues investigated how Polycomb chromatin modifiers and the DAXX/ATRX chaperone load histone H3.3 and establish constitutive heterochromatin at regions surrounding centromeres during mouse development, a function important for proper chromosome segregation and embryonic development. **Yukiko Yamashita** explored the function of repetitive satellite DNA in flies, finding that the AATAT-binding protein D1 is required for clustering pericentric heterochromatin regions on different chromosomes together into chromocenters (Jagannathan and Yamashita, this volume). D1 mutant cells show increased numbers of micronuclei, leading to the proposal that D1-mediated heterochromatin clustering is necessary to prevent nuclear budding and maintain genome integrity. **Aaron Straight** discussed how, similar to mechanisms observed in fission yeast, proper constitutive heterochromatin silencing in human cells actually requires some noncoding transcription from these loci. He described that pericentric alpha-satellite DNA repeats, although packaged into heterochromatin, transcribe low levels of RNA. This RNA remains bound to chromatin even during mitosis and directly binds to the H3K9 methyltransferase SUV39H1. This interaction stabilizes SUV39H1's localization at heterochromatin and is necessary to maintain histone methylation and heterochromatic states. Finally, **Susan Gasser** talked about using *C. elegans* to study the role of H3K9 methylation in preventing DNA damage (Zeller and Gasser, this volume). Surprisingly, worms with complete loss of H3K9 methylation are viable but are temperature-sensitive sterile due to damage-induced germline apoptosis. They find that increased transcription of repetitive elements forms RNA:DNA hybrids, which when unresolved cause replication stress and DNA damage. Future work in this area will likely focus on how repetitive, gene-poor genome regions must strike a balance between too much and too little transcription, and the roles of constitutive heterochromatin in preventing DNA damage in a variety of contexts.

CONTROL OF GENOME INTEGRITY BY DNA BREAKS

We have discussed how chromosomes interface with the chromosome segregation machinery, form higher-order structures, and are regulated epigenetically by histones. These processes depend on the chromosome being continuous. However, DNA can break. DNA breaks can be beneficial, by introducing variation into sexually reproducing populations. These breaks can also be harmful, by causing genome instability or cell death. The fate of DNA breaks depends on the context in which they occur. DNA breaks are not just a local issue for genome stability but also have a global impact on chromosome segregation during meiosis and mitosis.

"Good Cop" Breaks in Meiosis and Germ Cells

Although DNA breaks can lead to genome damage and mutations, DNA breaks are also essential in meiosis. They facilitate homologous chromosome pairing and crossovers, ensuring both proper chromosome segregation and the shuffling of genetic variation.

Several talks addressed how the location of meiotic DNA breaks is specified. One key protein that controls the locations of meiotic recombination hotspots is PRDM9, a histone methyltransferase with an array of zinc-finger domains. It is one of the fastest evolving proteins in the genome and drives speciation, as different alleles of PRDM9 can be incompatible and cause hybrid sterility in mice. **Bernard de Massy** described transgenic mice carrying a catalytically inactive PRDM9 variant, which showed that the methyltransferase activity of PRDM9 is essential for H3K4me3 and H3K36me3 deposition and specifying the locations of double-strand breaks (DSBs) in vivo, which in turn supports meiosis progression. How these epigenetic modifications interact with the nucleases that actually introduce DNA breaks is a key issue for future research. Although DSBs are required for meiotic crossovers, far more DSBs occur than crossovers, and how crossovers are controlled is also still a mystery. In *C. elegans*, only one crossover occurs per homolog pair. **Monica Colaiácovo** reported that these crossovers primarily occur off-center, away from the ends and middle of chromosomes (Saito and Colaiácovo, this volume). The off-center positioning of crossovers results in asymmetric anaphase I bivalent chromosomes. Interestingly, the short and long arms of these bivalents recruit different synaptonemal complex proteins, an asymmetry that appears to be important for the normal timing of cohesion loss between sister chromatids. To test the effect of crossover position, she described an inducible transposon system to target a crossover to specific genomic locations, revealing that a centered crossover leads to premature loss of cohesion. Therefore, asymmetric

crossovers and differential protein recruitment to the long and short arms of the resulting bivalents control important aspects of *C. elegans* meiosis. These findings raise fascinating questions about how chromosome asymmetry is established and sensed.

Additional studies of meiotic crossover control were reported using both in vitro and in vivo approaches. **Scott Keeney** described progress in recombinant protein purification of the yeast endonuclease Spo11 in complex with its obligate partners Rec102, Rec104, and Ski8, to enable biochemical and structural studies. Structural information will be key to understanding how the Spo11 complex breaks DNA and facilitates homologous recombination (HR). **Neil Hunter** addressed the mechanisms that trigger the culling of oocytes that fail to successfully complete the resolution of programmed meiotic breaks (Hunter, this volume). Surprisingly, if breaks are not repaired within the normal time frame, a SUMO and HORMA protein signaling system prevents the repair of meiotic DSBs. These unrepaired breaks then trigger apoptosis. Culling of defective oocytes therefore occurs by throwing an ingenious judo-like move; persistent DNA breaks are used against the oocytes that generated them, serving as a memory to survey germline health.

In addition to cell intrinsic properties, germ cell development also requires cues from the tissue microenvironment. Although primordial germ cell (PGC) fate had been thought to be restricted to gametes (i.e., PGCs are thought to be unipotent) and PGC commitment to sexual specialization is thought to be cell autonomous, **David Page** described work showing that PGCs are actually pluripotent and that sexual specialization is dependent on signals from the genital ridge. A mouse conditional knockout of GATA4—which obliterates the genital ridge—showed that without the genital ridge, PGCs still migrate but stay pluripotent and do not commit to oogenesis or spermatogenesis. The signaling from the genital ridge and the response in the PGCs is the subject of ongoing studies. The complexity of these germ-soma interactions is interesting, but also makes the precise mechanistic study of sex determination and germ cell differentiation difficult. Towards this end, **Mitinori Saitou** described using pluripotent stem cells to generate progenitor germ cell-like cells (PGCLCs), which mimic PGCs but do not require co-culturing with gonadal somatic cells (Nagaoka and Saitou, this volume). Using systematic chemical screening, they determined that cyclic AMP signaling plays a pivotal role in PGCLC proliferation. They also find that cell cycle kinetics, transcription profiles, and DNA methylation profiles mimic those of PGCs, making PGCLCs a powerful tool for future study of germ cell fates.

"Bad Cop" Breaks in Mitosis, Somatic Cells, and Cancer

Whereas the adverse effects of DNA damage are balanced with the essential roles of DNA breaks in meiosis, DNA damage is generally deleterious in mitosis. Several talks described important progress in understanding the mechanisms of how DNA damage leads to chromosome missegregation, cell death, or cancer.

DNA damage checkpoints do not catch all errors during DNA replication/S phase, and these uncaught errors then need to be resolved in mitosis or they will cause defects during chromosome segregation. Because of forces from the chromosome segregation machinery, incompletely replicated chromosomal segments, incompletely decatenated DNA or unresolved recombination intermediates in anaphase become DNA structures called ultrafine bridges (UFBs). UFBs are difficult to detect by standard DNA dyes and recruit a specific set of proteins involved in their metabolism. **Stephen West** described the consequences of persistent UFBs on genome instability (West and Chang, this volume). His group generated a cell line system for studying unresolved Holliday junctions during mitosis (by depleting the nuclease Mus81 and the resolvase GEN1). Importantly, these unresolved Holliday junctions do not appear to trigger the DNA damage response; they persist into mitosis and become UFBs. The UFBs then become single-stranded, a process dependent on the DNA translocase PICH and the helicase BLM. During mitosis, the UFBs break, which normally occurs enzymatically (via Mus81 and GEN1). However, West provided evidence that they can also be broken by mitotic spindle forces. It is estimated that although spindle forces may not be strong enough to break double-stranded DNA, they may be strong enough to break single-stranded DNA. This processing of UFBs is therefore the cell's last chance to resolve joined sister chromatids. However, the process is error-prone, with a high probability of chromosome deletions or chromosome arm-level copy number alterations. A challenge for studying these processes in cells is the short time window for mitosis and the complication that proteins involved in UFB metabolism have functions outside of mitosis. To address this challenge, **Ian Hickson** is pioneering much needed in vitro study of UFBs (Sarlós et al., this volume). He described single molecule modeling of UFBs using DNA substrates held by multiple optical tweezers. This setup allows visualization of both the identity and the order of binding of UFB proteins, and the optical tweezers allow control of the physical and mechanical properties of DNA. Optical tweezer manipulations revealed that the translocase PICH has a high affinity for stretched dsDNA, which may be a crucial signal for its recruitment to UFBs.

Elucidating the basic mechanisms of DNA repair is not only of fundamental importance for understanding how cells work, but also has significant implications for cancer therapy. Although DNA metabolism defects lead to genomic instability and promote tumorigenesis, these same defects are also therapeutic vulnerabilities. The poster child for this version of nononcogene addiction is in breast and ovarian cancer, where tumors lacking functional BRCA genes are sensitive to DNA cross-linking agents (such as cisplatin) or to inhibitors of poly ADP-ribose polymerase (PARP). Although PARP inhibitors have been fast-tracked into the clinic, drug resistance is a major barrier to efficacy. Future clinical progress now requires a mechanistic understanding of drug resistance.

Several talks addressed the roles of HR proteins in chemosensitivity and drug resistance. Although BRCA2 is most famous for its established role in HR, recent studies revealed that BRCA2 and its binding partner RAD51 are also important in protecting stressed replication forks from resection and degradation by MRE11. **Sharon Cantor** described shRNA screens for chemo-resistance in BRCA-deficient tumors (Cantor and Calvo, this volume). She identified chromatin regulators that appear to modulate the replication fork protection function of BRCA2, supporting the idea that restoration of fork protection can be a chemo-resistance mechanism. However, **Maria Jasin** cautioned that the essential function of BRCA2 in HR should not be forgotten. In contrast to studies with mouse embryonic stem cells or human tumor cells (Guillemette et al. 2015; Ray Chaudhuri et al. 2016), Jasin showed that it is the HR functions and not fork protection functions that are essential for the viability of nontransformed mammary epithelial cells (Feng and Jasin, this volume). She elucidated a mechanism explaining the lethality of BRCA2 loss involving incomplete DNA replication, UFBs, and cell cycle arrest after mitotic UFB processing. She further showed the important role of BRCA2's HR function in chemoresistance. This work highlights cell type differences in the wiring of essential processes. **Daniel Durocher** used a comprehensive screening approach to find new genetic vulnerabilities to PARP inhibition and new synthetic lethal interactions with BRCA loss. Surprisingly, cells treated with PARP inhibitors show striking sensitivity to RNaseH2 loss. RNaseH2 corrects the misincorporation of ribonucleotides into DNA; without RNaseH2, a backup process involving topoisomerase I corrects these lesions instead, and topoisomerase I can leave single-stranded DNA nicks. Inhibited PARP become trapped on this nicked DNA, which appears more cytotoxic than nicks alone. Because the RNase H2 gene is in a locus frequently deleted in certain cancers, PARP inhibition may prove a good therapeutic treatment for those types of cancers.

In summary, organisms maintain a balance between beneficial and detrimental DNA breaks. In germline cells, the evolutionary benefits appear to outweigh the risks, because successful meiosis requires the generation of DNA breaks and crossovers. Nevertheless, extensive damage still requires surveillance, possibly to prevent overexpenditure of energy on inviable offspring. In somatic cells, unintended DNA damage needs to be resolved, and unresolved structures can lead to cell death or the propagation of mutations that may lead to cancer.

NEW METHODS DRIVE NEW DISCOVERY

Important discoveries can come from the clever application of existing technologies. However, major new progress often originates from new technical approaches—many of which were featured at the meeting. **Tom Misteli** presented a host of new strategies for high-content deep imaging, including SpotLearn to analyze high-throughput FISH signals and enable large-scale interrogation of genome organization at the single-cell level (Gudla et al., this volume). Although chromosome conformation capture methods have provided important information about 3D genome structure, Misteli's experiments show that they miss the true dynamics of chromosome behavior. **Martin Hetzer** is addressing difficult problems about organelle and tissue aging. He discussed the identification of long-lived proteins and their subcellular localization by combining isotope labeling strategies and serial block-face scanning electron microscopy (SBEM-SEM) in a new technique called nanoSIMS (nano secondary ion mass spec). **Iain Cheeseman** discussed large-scale genome editing strategies to define what is "essential" about essential genes, and how their functions differ between cell types. **Steve Henikoff** described CUT&RUN (Cleavage Under Targets and Release Using Nuclease), a sensitive and powerful alternative to ChIP-based methods for mapping chromatin-associated complexes (Henikoff et al., this volume). CUT&RUN is already widely adopted. Finally, **Aaron Straight** described ChAR-seq (chromatin-associated RNA sequencing), a new method for mapping the genomic interactions of all chromatin-associated RNAs genome-wide. These techniques and others will undoubtedly enable discovery of previously unexplored biological phenomena and open doors to new fields of research.

CONCLUSION

The 82nd Cold Spring Harbor Symposium gave us a vantage point from which to view the future of chromosome biology research. Prior related symposia have swung back and forth in their emphasis on either the description of new phenomena or the elucidation of mechanism, with technological advances usually triggering each swing of the pendulum. The 82nd symposium caught the pendulum about halfway. A complete description of chromosome structure is coming into focus through genomic approaches, advanced imaging, and mathematical modeling. These techniques have provided a level of information about chromosome organization that, until only recently, was unimaginable. There is certainly more to be learned about chromosome structure genome-wide, but a major task going forward will be to figure out in molecular and mechanistic detail how chromosome substructure forms. The 82nd symposium highlighted that this effort is well underway with huge progress from biochemical and biophysical studies. By the next symposium, we predict that we will have the same satisfying mechanistic level understanding of motors that generate chromosome loops and link sister chromatids that we have for the motors that organize the mitotic apparatus.

The current state of the chromosome segregation field provides a perspective on the future of the chromosome structure field. First, in part through genetic screens, we know most of the players that enable proper chromosome segregation. The field is now focusing on the assembly principles and how intermediate level structures—for example, spindle midzones and the PCM—are assembled

from these parts. Similar assembly principles are now also being used for chromosome structure to describe how loop extrusion motors form intermediate structures like chromosome loops. Second, mathematical modeling of spindle microtubule dynamics and interactions has enabled the formulation of quantitative models with testable predictions. The spectacular success of the quantitative iteration of the loop extrusion model shows that this era has already arrived for chromosome structure. Third, as work on the phase transition of pericentriolar proteins illustrated, studying the material properties of biopolymers can identify new functions that emerge from protein ensembles. This kind of work is in its infancy for chromosome biology, but similar work on the synaptonemal complex, for example, might lead to insight about how the numbers and position of crossovers are regulated in meiosis. Finally, how asymmetry is generated during cell division has long been a focus in the cell polarity and chromosome segregation fields. By the next symposia, we hope to have this level of understanding for chromosome structure problems such as the asymmetric distribution of histones and the asymmetric organization of *C. elegans* bivalents reported at the 82nd Symposium.

Although the current symposium highlighted great progress in defining chromosome structure, much remains to be understood about its functional consequences. For example, it is clear that cohesin proteins are required to form TADs, and that genome editing of TAD boundaries has established effects on transcription. Therefore, the role of mutated cohesin subunits in tumorigenesis and cancer has been attributed to misregulated transcription. However, transcriptional profiling of cells after cohesin inactivation shows surprisingly subtle effects, with only small-scale changes affecting many genes (Rao et al. 2017). Either tumorigenesis results from alterations in a small number of exquisitely dosage sensitive genes or, probably more likely, it emerges from the aggregate alteration of many genes. Understanding the combinatorial effects of many small changes is a challenging general problem in biology. Tackling this in the future will require new methods; perhaps analytic approaches used to study polygenic traits will be relevant (Boyle et al. 2017). Also, in addition to transcription, perhaps by the next meeting it will be apparent that many other long-range chromosome interaction phenomena, such as VDJ recombination, require loop extrusion.

The chromosome segregation field has long focused on microtubules and their assembly into spindles, but the field is now bringing the nucleus back into focus. The coordination between chromosome segregation and the nuclear envelope is still being elucidated in metazoan cells. We have known for 100 years that mitotic defects can lead to aneuploidy; however, we have only recently fully appreciated the impact mitotic errors have on the nuclear envelope and the structural integrity of chromosomes. For example, mitotic errors can lead to chromosome breakage at common fragile sites, as well as the mutational process termed chromothripsis. Mitotic errors can therefore drive rapid genome changes and evolution. This has obvious relevance to cancer. By the next symposium, we will have a much better, perhaps comprehensive, understanding of abnormal nuclear architecture in cancer and the mutational processes that drive cancer genome evolution. These insights will hopefully lead to new ideas about therapy; for example, recent work has revealed unexpected links between nuclear envelope integrity and innate immune signaling that may, in some contexts, serve as a unique cancer vulnerability (Harding et al. 2017; Mackenzie et al. 2017; Bakhoum et al. 2018).

Although much has been learned from model systems like HeLa cells, it is now clear that there is a host of important unsolved chromosome segregation problems that must be studied in their natural habitat. As mentioned by Iain Cheeseman, cell type–specific regulation of chromosome segregation is clearly important, but relatively understudied. We also need to understand how the 3D architecture of tissues impacts chromosome segregation—hopefully in quantitative terms. Fully understanding how cell biological processes impact organism-level processes such as development or disease will require the combination of in vivo models with new advances in cellular imaging.

Summing up a meeting as rich and diverse as the 82nd Cold Spring Harbor Symposium is a dangerous business. The temptation is to shoehorn the presentations into a few predefined themes. We have tried to highlight important themes in this summary, but we also do not want to ignore the pleasing scientific diversity. As Isaiah Berlin pointed out in his famous essay *The Hedgehog and the Fox*, many writers and thinkers want to be "hedgehogs" with one big idea (e.g., Plato, Hegel, Dostoevsky). However, it is the "foxes" (e.g., Shakespeare, Goethe, Balzac) that appreciate the plurality of human types and experiences, and who may in fact get closer to the truth. Looking over past meetings, one is struck by the frequency with which unexpected findings, often in obscure model systems, led to creative work that changed the direction of the whole field. In his Dorcas Cummings Lecture, David Page not only took us on an entertaining tour of the checkered evolutionary history of the human Y chromosome but also threw a pitch from way out in left field. He proposed that by ignoring the possibility of sex-specific cell autonomous differences, we may be missing fundamental information about cells and chromosomes. Although we cannot predict with certainty which "out-there" ideas from the 82nd symposium will become a dominant theme in the next chromosome segregation and structure symposium, we will all be back to look for them.

ACKNOWLEDGMENTS

We would like to thank the National Institutes of Health (5R01 CA213404-20) and the Howard Hughes Medical Institute for support.

REFERENCES

Alipour E, Marko JF. 2012. Self-organization of domain structures by DNA-loop-extruding enzymes. *Nucleic Acids Res* **40**: 11202–11212.

Bakhoum SF, Ngo B, Laughney AM, Cavallo J-A, Murphy CJ, Ly P, Shah P, Sriram RK, Watkins TBK, Taunk NK, et al. 2018. Chromosomal instability drives metastasis through a cytosolic DNA response. *Nature* **553**: 467–472.

Boyle EA, Li YI, Pritchard JK. 2017. An expanded view of complex traits: From polygenic to omnigenic. *Cell* **169**: 1177–1186.

Earnshaw WC, Bernat RL. 1991. Chromosomal passengers: Toward an integrated view of mitosis. *Chromosoma* **100**: 139–146.

Fussner E, Strauss M, Djuric U, Li R, Ahmed K, Hart M, Ellis J, Bazett-Jones DP. 2012. Open and closed domains in the mouse genome are configured as 10-nm chromatin fibres. *EMBO Rep* **13**: 992–996.

Ganji M, Shaltiel IA, Bisht S, Kim E, Kalichava A, Haering CH, Dekker C. 2018. Real-time imaging of DNA loop extrusion by condensin. *Science* **360**: 102–105.

Gibcus JH, Samejima K, Goloborodko A, Samejima I, Naumova N, Nuebler J, Kanemaki MT, Xie L, Paulson JR, Earnshaw WC, et al. 2018. A pathway for mitotic chromosome formation. *Science* **359**: eaao6135.

Giorgetti L, Lajoie BR, Carter AC, Attia M, Zhan Y, Xu J, Chen CJ, Kaplan N, Chang HY, Heard E, et al. 2016. Structural organization of the inactive X chromosome in the mouse. *Nature* **535**: 575–579.

Guillemette S, Serra RW, Peng M, Hayes JA, Konstantinopoulos PA, Green MR, Cantor SB. 2015. Resistance to therapy in BRCA2 mutant cells due to loss of the nucleosome remodeling factor CHD4. *Genes Dev* **29**: 489–494.

Haarhuis JHI, van der Weide RH, Blomen VA, Yáñez-Cuna JO, Amendola M, van Ruiten MS, Krijger PHL, Teunissen H, Medema RH, Van Steensel B, et al. 2017. The cohesin release factor WAPL restricts chromatin loop extension. *Cell* **169**: 693–700.e14.

Harding SM, Benci JL, Irianto J, Discher DE, Minn AJ, Greenberg RA. 2017. Mitotic progression following DNA damage enables pattern recognition within micronuclei. *Nature* **548**: 466–470.

Hatch EM, Fischer AH, Deerinck TJ, Hetzer MW. 2013. Catastrophic nuclear envelope collapse in cancer cell micronuclei. *Cell* **154**: 47–60.

Henikoff S, Malik HS. 2002. Centromeres: Selfish drivers. *Nature* **417**: 227.

Henikoff S, Ahmad K, Malik HS. 2001. The centromere paradox: Stable inheritance with rapidly evolving DNA. *Science* **293**: 1098–1102.

Lieberman-Aiden E, van Berkum NL, Williams L, Imakaev M, Ragoczy T, Telling A, Amit I, Lajoie BR, Sabo PJ, Dorschner MO, et al. 2009. Comprehensive mapping of long-range interactions reveals folding principles of the human genome. *Science* **326**: 289–293.

Mackenzie KJ, Carroll P, Martin C-A, Murina O, Fluteau A, Simpson DJ, Olova N, Sutcliffe H, Rainger JK, Leitch A, et al. 2017. cGAS surveillance of micronuclei links genome instability to innate immunity. *Nature* **548**: 461–465.

Mazia D. 1961. *Mitosis and the physiology of cell division.* Elsevier, New York.

Nasmyth K. 2001. Disseminating the genome: Joining, resolving, and separating sister chromatids during mitosis and meiosis. *Annu Rev Genet* **35**: 673–745.

Nishino Y, Eltsov M, Joti Y, Ito K, Takata H, Takahashi Y, Hihara S, Frangakis AS, Imamoto N, Ishikawa T, et al. 2012. Human mitotic chromosomes consist predominantly of irregularly folded nucleosome fibres without a 30-nm chromatin structure. *EMBO J* **31**: 1644–1653.

Rao SSP, Huang S-C, Hilaire BGS, Engreitz JM, Perez EM, Kieffer-Kwon K-R, Sanborn AL, Johnstone SE, Bascom GD, Bochkov ID, et al. 2017. Cohesin loss eliminates all loop domains. *Cell* **171**: 305–309.e24.

Ray Chaudhuri A, Callen E, Ding X, Gogola E, Duarte AA, Lee J-E, Wong N, Lafarga V, Calvo JA, Panzarino NJ, et al. 2016. Replication fork stability confers chemoresistance in BRCA-deficient cells. *Nature* **535**: 382–387.

Riggs AD. 1990. DNA methylation and late replication probably aid cell memory, and type I DNA reeling could aid chromosome folding and enhancer function. *Philos Trans R Soc Lond B Biol Sci* **326:** 285–297.

Subramanian R, Ti S-C, Tan L, Darst SA, Kapoor TM. 2013. Marking and measuring single microtubules by PRC1 and kinesin-4. *Cell* **154:** 377–390.

Tedeschi A, Wutz G, Huet S, Jaritz M, Wuensche A, Schirghuber E, Davidson IF, Tang W, Cisneros DA, Bhaskara V, et al. 2013. Wapl is an essential regulator of chromatin structure and chromosome segregation. *Nature* **501:** 564–568.

Terakawa T, Bisht S, Eeftens JM, Dekker C, Haering CH, Greene EC. 2017. The condensin complex is a mechanochemical motor that translocates along DNA. *Science* **358:** 672–676.

Wang X, Brandão HB, Le TBK, Laub MT, Rudner DZ. 2017. *Bacillus subtilis* SMC complexes juxtapose chromosome arms as they travel from origin to terminus. *Science* **355:** 524–527.

Wutz G, Várnai C, Nagasaka K, Cisneros DA, Stocsits RR, Tang W, Schoenfelder S, Jessberger G, Muhar M, Hossain MJ, et al. 2017. Topologically associating domains and chromatin loops depend on cohesin and are regulated by CTCF, WAPL, and PDS5 proteins. *EMBO J* **36:** 3573–3599.

DORCAS CUMMINGS LECTURE

Video recording of the Dorcas Cummings Lecture is available on Cold Spring Harbor Laboratory's Leading Strand channel on YouTube (playlist 82nd Symposium Interview Series Chromosome Segregation and Structure).

Dorcas Cummings Lecture

David Page

Dr. David Page presented the Dorcas Cummings lecture entitled "Sex and Disease: Do Males and Females Read Their Genomes Differently" to friends and neighbors of Cold Spring Harbor Laboratory and Symposium participants on Saturday, June 3, 2017. Dr. Page is a Professor of Biology at the Massachusetts Institute of Technology, an Investigator of the Howard Hughes Medical Institute, and the Director of the Whitehead Institute for Biomedical Research.

Thank you very much for the invitation to speak and for your generous introduction. I would like to thank all of you for your presence here this evening. It is an honor and a privilege to speak at Cold Spring Harbor Laboratory on any occasion and especially in memory of Dorcas Cummings.

I invite you now to join me on a journey from the past to the future with my favorite chromosome. Actually, with my favorite pair of chromosomes. On the left, the X chromosome: proud, statuesque, respectable. On the right, with its head down, the Y chromosome: diminutive, demure, downtrodden. Truth be told, I have spent my entire career defending the honor of the Y chromosome in the face of innumerable insults to its character and its future prospects. I ask you, men and women of the Cold Spring Harbor community, how could the Y chromosome get such a bad rap?

To understand the tragic past of the Y chromosome we've got to go back more than a hundred years, to 1904, when Charles Benedict Davenport became director of this laboratory. Just a few years later, he would establish Cold Spring Harbor as a leader in human genetics, then framed as eugenics, when he founded the Eugenics Records Office here. During those opening years of the 20th century, the principles of inheritance deduced in the 1860s by Mendel in his garden of peas were rediscovered and rose to prominence. In quick succession, three great modes of inheritance were reported in our species: autosomal recessive, autosomal dominant, and X-linked recessive.

It turns out that in a paper from 1907 several investigators claimed a fourth mode of inheritance: Y-linked inheritance. This report [Tomassi, *Arch Psichaitr Neuropat Antropol Crim Med Leg* **28**: 60 (1907)], which I'm sure that Charles Davenport read with considerable interest, was published in the *Archives of Psychiatry, Neuropathology, Anthropology, Criminology, Medicine, and Law*. This was one of the early interdisciplinary journals (you thought *Nature* was broad). The trait under consideration was "hairy ears," big tufts of hair growing from the earlobe, and the argument for its Y-linked inheritance looked pretty

decent, with father-to-son transmission across the family tree—maybe a few guys in the last generation shaved their ears—but otherwise, it looked quite promising. Over the ensuing 50 years, a number of other traits were also claimed to show Y-linked inheritance. The first half of the 20th century was a heady time for the Y chromosome.

But the good times for the Y chromosome came to a crashing halt in 1957 in Ann Arbor, Michigan, at the annual meeting of the American Society of Human Genetics. There, the society's president, Curt Stern—who actually was a *Drosophila* geneticist from the University of California at Berkeley—delivered a colorful presidential address that was entitled "On Porcupine Skin and Hairy Ears or, The Alleged Sins of the Y Chromosome," although the editor of the society's journal cleaned up the title prior to publication to the more pedestrian "The Problem of Complete Y-Linkage in Man" [Stern, *Am J Hum Genet* **9**: 147 (1957)].

In his presidential address, in front of all the human geneticists of North America, Stern cataloged and debunked "all seventeen presumably or possibly Y-linked traits," including porcupine skin and hairy ears, showing all of them to be flimsy claims that were based on shoddy pedigree analysis. By the end of Curt Stern's presidential address, no genes were left standing on the Y chromosome. The best that Stern could do to cheer up the chromosome was to suggest that since it exists, it must have a function, concluding, "That the Y chromosome has a function of its own is attested by its very existence. What it is still must be discovered." Actually, Stern's scholarly debunking was absolutely right; none of the previous claims of Y-linked genes withstood scrutiny. It was not Stern's intention, but his defrocking of these spurious claims led others to a new understanding of the Y chromosome: It must be a genetic wasteland.

This was not the low point for the Y chromosome. It would get much worse, and our lab was partly to blame. In the 1990s, one of my graduate students, Bruce Lahn—now a professor at the University of Chicago—showed that our X and Y chromosomes had evolved from an

Published by Cold Spring Harbor Laboratory Press; doi: 10.1101/sqb.2017.82.035550

ordinary pair of chromosomes (autosomes), which had been identical in males and females of our reptilian ancestors. Subsequent work in my lab confirmed what was feared: Over evolutionary time the X chromosome had done a superb job of nurturing and preserving the genes of the ancestral autosome, while the Y chromosome had callously and carelessly frittered them away.

It turns out that 300 million years ago, when we were reptiles, we had no sex chromosomes. We had only ordinary chromosomes, and they came in pairs. One such pair of ordinary chromosomes did not know it, but they would evolve to become our X and Y chromosomes. They were a happy pair: They engaged in free trade, they swapped information, all sorts of stuff. Then something happened: One member of the pair sustained a mutation, giving rise to the sex-determining gene on what would become the Y chromosome. The two chromosomes had formerly been in close communication with each other, but the nascent Y chromosome then changed its behavior. It said, "Enough of this. I am going to adopt isolationist strategies." The Y chromosome decided to go its own way. These isolationist strategies of the Y chromosome led, not surprisingly, to the decline of its economy. The Y chromosome lost many of its genes, becoming a shadow of its former self. Simultaneously, the X chromosome expanded enormously, at the expense of the Y chromosome. So we ended up with a much smaller Y chromosome containing the sex-determining gene, overshadowed by the giant X chromosome.

In 2002, two colleagues in the field saw an opportunity to deal a truly fatal blow to the Y chromosome and published a punishing editorial in a weekly journal of some repute. In a *Nature* editorial grandly titled "The Future of Sex" [*Nature* **415**: 963 (2002)], John [Aitken] and Jenny [A. Marshall Graves], my good friends, concluded that "… the Y chromosome is particularly vulnerable … because it is not a matching partner for the X chromosome, so it cannot retrieve lost genetic information…." After recounting the tale of the chromosome's diminishment that I have told you today, they delivered a devastating punchline: "At the present rate of decay, the Y chromosome will self-destruct in around 10 million years."

I had been planning to make a career out of the Y chromosome.

I was not the first in my lab to read this editorial. It was one of my graduate students, who came running into my office with tears streaming down his face. We held an emergency lab meeting, and we resolved to pick up the pace of our research.

We could not move quickly enough. A comic book series called *Y: The Last Man* burst onto the scene. The series consistently made subtle use of "Y" symbolism and inspired the production of a decidedly bad movie (*The Last Man on Planet Earth* [1999]), whose premise was that, "… feeling they were better off without males, the women of Earth decided to outlaw men because they were too violent. They developed a weapon called the Y-bomb, which resulted in the deaths of 97% of men." I tell you, no other chromosome has had to put up with such attacks. "Twenty years later, a scientist …"—and what else could she be called except Hope Chayse?—"… conducts a clon-

ing experiment to produce a new male whom she names Adam. When Adam reaches maturity, he finds himself on the run, hiding out with rebel bands of the last remaining men." Actually, the last remaining men end up hiding in an abandoned NFL football stadium. I highly recommend this movie to you, if you can find a copy of it. Believe me, it is not available on any of the streaming services.

Anyway, it got worse. The Internet became littered with models of the Y chromosome like this one, with "genes" like the "channel surfing gene" (*FLP*), which sometimes is up here and then it flips down to here; the "balls, two" gene (*BLZ-2*), which confers self-confidence unlinked to ability; the *DC10* gene, which confers the ability to identify aircraft in the sky; and the *MOM-4U* gene, which drives young sons to present spiders and snakes to their mothers. Also included in this model of the Y chromosome is the well-known *P2E* ("ptui") gene (codes for spitting); and then one that my wife is convinced is closely linked to the inability to remember anniversaries and birthdates, the *HUH?* gene for selective hearing loss.

This is what I have had to deal with. Something had to be done to stop this public humiliation of the Y chromosome, so our lab responded with help from our sister species. Here I would like to tell you a tale of three primates, or at least their Y chromosomes. We turned to the rhesus monkey, a chimp named Clint, and, last but not least, a human. I thought this would be an appropriate time and place to identify the man whose Y chromosome we sequenced: none other than [Symposium organizer] Bruce [Stillman].

I don't want to drag you through the details of the DNA sequence analysis, but I had an opportunity to discuss our results on *The Colbert Report* [3/26/12; http://www.cc .com/video-clips/rc1xqe/the-colbert-report-david-page], where I summarized what we learned by comparing in detail the Y chromosomes of these three species. We found that the human Y chromosome and the rhesus Y chromosome carry essentially the same genes. This suggests that nothing much has happened to the Y chromosome in the last 25 million years. The Y chromosome was in a steep nosedive, losing genes at a furious pace, but then it leveled out and has been flying at a low but steady altitude since, so men are going to be okay.

Thus ends Part One of this lecture. Now, having rescued the Y chromosome from a century of misunderstanding, let me suggest that the Y chromosome, together with its partner the X chromosome, may play a critical role in the future of medicine.

What do we know today about the role of the Y chromosome in medicine? The Y chromosome is known to carry a single gene that causes a human embryo to develop testes rather than ovaries, and deletions of the Y chromosome's sperm production genes are the most common known genetic cause of male infertility in our species. However, what I want to tell you about today extends far beyond the reproductive tract and to diseases that occur in females as well as males. My topic is sex and disease (but not what you think). I would like to share with you what I think is the really important link between sex and disease, a connection that is not talked about enough.

Let me get the ball rolling by sharing with you three observations that might surprise you. First, I am going to suggest that our concept of the human genome is off the mark. Second, that males and females are not equal. Third, that the study of disease is flawed in significant ways.

With respect to my first point, there are times in history when scientists, as brilliant as they sometimes are, have gotten things wrong. For centuries many smart people thought the world was flat. We also thought the Sun revolved around the Earth. In this age of the genomic revolution, I am going to suggest that we are missing something vitally important.

Let's begin where we all began: a fertilized egg. All the cells in your body—your heart cells, your brain cells, even your skin cells—all derive from this one cell, the fertilized egg. This special cell divides to become two, four, eight, and so on until we reach the roughly 10 trillion cells that make up our body. What's amazing is that within the nucleus of each of your 10 trillion cells, you carry the same 23 pairs of chromosomes, which contain all of your DNA, all the instructions your body needs to function. There are 22 pairs that are the same in males and females, and then comes the 23rd pair, which in females is a nicely matched pair of X chromosomes. In males, that 23rd pair is a mismatched X and Y.

As I have told you, the Y chromosome has always been underestimated. Even today, most scientists and physicians think that the Y chromosome is important only within the cells of our reproductive tract. This erroneous assumption has led them to believe that, apart from the reproductive tract, the genomes of males and females are functionally equivalent. In fact, the Human Genome Project and the resultant recent initiatives in precision medicine are based on our being 99.9% the same at the genomic level. This idea has gained traction for many reasons. It sounds great politically to say that we are all 99.9% the same. In fact, Bill Clinton actually used this idea to bring the country together in his 2000 State of the Union speech: "This fall at the White House, we had this very distinguished scientist there, who is an expert in this whole work in the human genome." I will not name any names. "He said that we are all, regardless of race, genetically 99.9% the same."

This sounds great, and it is even true if the two individuals you are comparing are both males. It is also true if the two individuals you are comparing are both females. If you make a mistake, and you compare a male and a female, they are only 98.5% identical. Let's flip this around: Instead of talking about degree of identity, let's talk about the degree of difference. In other words, between two males, it is a 0.1% difference; between two females, a 0.1% difference. It is a 15 times greater difference in the genomes of male and female: 1.5%. What is that difference? Of course, it is XX versus XY.

What is called "precision medicine" today is really the study of the 0.1% genetic differences between two men or between two women. And we are now committing, quite appropriately, at the national level, hundreds of millions of dollars to this study of precision medicine. But by comparison, the 1.5% genetic difference between males and females has no name or federal program devoted to it, no banner or slogan. Let's call it "sex differences." The area of sex differences has barely begun to receive funding or the focused attention of dedicated researchers.

But how biologically or clinically significant are these sex differences—these genetic differences between males and females? It turns out that a male human is as closely related to a female human as he is to a male chimp: there is a 1.5% difference between male and female humans, just as there is a 1.5% difference between male and female chimpanzees. The human genetics revolution has missed this important fact. Our field has instead created a unisex model, when in fact males and females are not equal— they are not equal in their genomes, and they are not equal in the face of disease.

What do I mean by this and why does it matter? Let me give you a handful of examples. Take rheumatoid arthritis. For every man who has rheumatoid arthritis, there are two or three women with the disease. Is rheumatoid arthritis a disease of the reproductive tract? No. Is it anatomically obvious why women should suffer from this disease two to three times as frequently as men? No. There is no simple explanation to be found in our anatomy.

Let's flip it around: Autism spectrum disorder. The latest statistics suggest that for every girl that has an autism spectrum disorder, four boys are affected. Why is that the case? Let's flip it around again. Lupus. For every man who suffers with lupus, there are six women who suffer with the disease. And there are many other disorders that, like lupus or autism, are more common in females or in males. For other diseases where the incidence is similar in males and females, the severity or consequences of the disease may be greater in one sex than the other.

Let's examine the example of dilated cardiomyopathy to illustrate why sex differences matter in medicine. A specific genetic defect causes a thinning of the wall of the heart and a dangerous ballooning. If you look at the survival curves—the "death" curves, if you will—for women and men with this disease and the same underlying genetics (autosomal dominant genetics, for the scientists in the audience), men die at a much younger age, about 10 years earlier [Herman et al., *N Engl J Med* **366**: 619 (2012)]. Nobody knows why. When I query medical specialists, academics, and researchers about this disorder, or any of the others I have shown or any of dozens of others that I could mention, when I ask, "Why is it that one sex is more commonly or more severely affected than the other?" I almost always get the same response: "I don't have a clue."

This is in an age of precision medicine.

If I press harder, the answer that many physicians and scientists come up with is, "Maybe it's sex hormones." It turns out that the human genetics revolution has provided us researchers with powerful tools to ask why one man is at higher or lower risk than another man for a given disease, or why one woman is at greater or lesser risk than another woman. However, incredible as it may sound, we do not yet have a toolkit to ask why males as a group are at higher or lower risk than females as a group. This is a big, big question, but no one has a clue about the answer. But maybe the answer has been staring us in the face all along:

That is, the individuals who tend to get diseases such as autism and dilated cardiomyopathy are XY, and the individuals who tend to get diseases such as lupus and rheumatoid arthritis are XX. This is a fundamental difference, right? It is present in all our cells, but we in the scientific community have been operating for about 60 years on a faulty assumption: that the Y chromosome is functionally important only in the reproductive tract.

I am going to give you a one-slide crash course in how sex differentiation is taught in every medical school in the world. Instructors teach that being XX or XY is of direct biological consequence only in the nether regions, in the reproductive organs. According to this longstanding view, all nonreproductive differences between males and females, including differences in disease susceptibility, should be attributed to the sex hormones: the androgens and estrogens that are produced by the reproductive organs and circulate throughout the body.

In recent years, however, my research group at Whitehead Institute has discovered that the Y chromosome is actually operating throughout the entire body, as is the X chromosome. The cells of your heart, your pancreas, your brain, your skin—they know whether they contain XX or XY chromosomes. I want to hybridize the old sex differences model with a new one, which acknowledges how the X and Y chromosomes have roles throughout the body. Accepting this reality will lead to a far better way to study disease.

I go to my colleagues performing laboratory research at medical schools, universities, drug companies, and even at Whitehead Institute, and I ask scientists who are working with human cells, "Are you working with XX cells or XY cells?" The answer I most frequently get is, "I don't know." How could you figure things out if you do not know or have not thought to ask whether you are working with XX or XY cells? This means that, in many cases, the research that is being done to discover the underlying causes of diseases, or new treatments for them, is not taking into account this most fundamental of differences between males and females. This is why I suggested rather provocatively at the beginning of this talk that the study of disease is fundamentally flawed.

What can we do about this? How can we rethink the relationship between sex and disease? First, I believe that XX and XY cells may do their molecular business a bit differently from each other, throughout the body. Scientists around the world need to incorporate this distinction into their research for treatments and cures. At my lab at Whitehead Institute, we are already doing this, and we have preliminary evidence that the way proteins are made may be slightly different in XX and XY cells.

We need a better toolkit for scientists and drug developers to use, one that recognizes and includes this fundamental difference between male XY and female XX cells, tissues, organs, and bodies. If we take these steps—and I believe we can—we will arrive at an entirely new paradigm for treating disease. It will really matter whether a patient is a female or a male—and not just to physicians with deep understandings of the reproductive tract, not just to gynecologists and urologists; it will matter to cardiologists, to endocrinologists, to dermatologists. I anticipate that a full appreciation of the roles of the X and Y chromosomes will fundamentally change the way that you, your children, and your grandchildren will experience healthcare in the future.

This work is going to require the efforts of many scientists in many laboratories in many countries. It is just getting started. Let me introduce you to a few of the early adopters in my own lab who have joined the cause: Winston [Bellott], Jen [Hughes], and Helen [Skaletsky], who, with our colleagues at Washington University [Richard Wilson, Wes Warren, Tina Graves, Robert Fulton] and Baylor [Richard Gibbs, Donna Muzny, Shannon Dugan], pioneered the comparisons among the human, primate, and other mammalian Y chromosomes that brought many of these questions to the fore. Lukáš Chmátal, a postdoctoral fellow whose previous work on centromere strength with Mike Lampson and Richard Schultz at Penn has been described by several speakers at this meeting. Lukáš is now examining sexual dimorphism in the human heart. Emily Jackson is a recently arrived grad student who, in her pre–grad-school life, trained with David Pellman, whose summary will close this scientific meeting.

Let me offer a glimpse of the directions in which my lab is taking this work. We are examining the roles of microRNAs in these processes; Sahin Naqvi, a graduate student, has found that conserved microRNA targeting reveals preexisting heterogeneities in gene dosage sensitivity that shaped sex chromosome evolution in mammals and birds [Naqvi et al., *Genome Res* **28**: 47 (2018)]. Working with 12 different tissues from five species, Sahin is beginning to scope out sex differences in gene expression across the body. I mentioned Lukáš' work on sex differences in the human heart [with Jon and Christine Seidman and Rick Mitchell at Brigham and Women's Hospital; Steve Gygi at Harvard Medical School]. We're examining the brains of mice and humans (looking at microglia in particular) [with Richard Ransohoff at Biogen; Chris Glass at the University of California–San Diego]; postdoctoral fellow Laura Blanton is exploring sex differences in immune cells in both mice and humans [with Dan Kastner at NIH; Andrew Lane at the Dana–Farber Cancer Institute]; and postdoctoral fellow Adrianna San Roman is studying those not-so-rare individuals who carry not two sex chromosomes, but one (XO), three (XXY, XYY, XXX), four, or even five (XXXXY, XYYYY) sex chromosomes, and their effects on global gene expression [with Max Muenke at NIH; Carole Samango–Sprouse at Focus Foundation].

In closing, whether you are a clinician or a lab researcher or a supporter of biomedical investigation, I want to challenge you to thoughtfully consider the approach that I have presented today, and how the knowledge of genetically based sex differences can transform our understanding of human health and disease. Thank you very much.

CONVERSATIONS AT THE SYMPOSIUM

Video recording of the Symposium Conversations is available on Cold Spring Harbor Laboratory's Leading Strand channel on YouTube (playlist 82nd Symposium Interview Series Chromosome Segregation and Structure).

A Conversation with Marisa Bartolomei

INTERVIEWER: BETH MOOREFIELD

Senior Editor, Nature Structural and Molecular Biology

Marisa Bartolomei is a Professor in the Department of Cell and Developmental Biology
and Co-Director of the Epigenetics Institute, University of Pennsylvania Perelman
School of Medicine, Philadelphia, Pennsylvania.

Beth Moorefield: You study a unique process of gene
expression in mammalian cells known as genomic imprinting, which directs expression specifically from either
maternal or paternal alleles. I thought we could speak
about the function of the imprinted genes and the mechanisms that govern their regulation.

Dr. Bartolomei: Imprinting is a mammalian phenomenon, and it affects ~100–200 genes. It's nicely conserved
in mammals, which gives us the opportunity to use mouse
as a good model to study imprinting in humans. These
genes have very important processes in growth, but they
also have functions in postnatal energy homeostasis, in
behavior, and in other processes. So, when these genes
are missing or defective, you end up with very broad
changes—broad sorts of imprinting disorders—if there
are defects in humans. That would include Beckwith–
Wiedemann or Silver–Russell syndromes—those are
growth imprinting disorders—or Angelman and Prader–
Willi syndromes, and those are neurobehavioral disorders.
So they have a broad range of functions, and absence of
these genes causes these disorders, which is why we want
to really understand their regulation.

Beth Moorefield: Are the maternal or paternal alleles
imprinted with equal frequency, or is there a bias toward
one or the other?

Dr. Bartolomei: These imprinted genes are regulated by
regions that we call "imprinting control regions." These
imprinting control regions are discrete elements in the
genome that experience epigenetic modifications, so
they have DNA methylation that's put on either during
male or female gametogenesis. If you look at these regions, there are many more of these methylated imprint-
ing-control regions that come from the female germline,
and just a few in the male germline, so a lot of the action
seems to be happening in the female germline. That said,
these imprinted genes are found in large clusters through
the genome and they seem to be pretty well equally represented as maternally expressed imprinted genes and paternally expressed imprinted genes. There doesn't seem to
be a bias in that sense.

Beth Moorefield: How would they distinguish either of
these alleles? How are they identified as maternal or paternal so that they're differentially recognized by the transcriptional machinery?

Dr. Bartolomei: That's the question. We know that these
differential epigenetic modifications that are put on in the
germline are what helps us to say, "If this comes from the
maternal allele, either express or repress off the maternal
allele," or modifications that are put on in the paternal
germline help us to recognize something as being paternal
and either expressed or repressed. These germline modifications are actually key to allowing the somatic cells to
say, "Maternal, express; paternal, repress," or vice versa.
These modifications that are put on in the germline are
really the key to the imprinting. When something is perturbed, that's when you see dysregulation of imprinted
genes, and that can occur in the germline, or it also can
occur postfertilization.

Beth Moorefield: Do each of the alleles share common
regulatory elements themselves?

Dr. Bartolomei: When imprinted genes were first identified ~25 years ago, there actually were three imprinted
genes that were published in the same year: IGF-2 (Insulin-like Growth Factor 2), Insulin-like Growth Factor 2
Receptor, and H19, which encodes a noncoding RNA. It
was thought that these imprinted genes were going to
contain this primary sequence that was identified either
in the male or the female germline and that was what was
going to be the key to imprinting. In fact, we thought there
was going to be a "silver bullet": "We're going to identify
the imprinting box!"… like the TATA box, there was an
"imprinting box." It was going to be a simple sequence.
But there is no such simple sequence, so maybe there's
something more complex that we just haven't figured out
with secondary structure or some other kind of sequence.
Each imprinted region, each imprinted gene, seems to
have a different type of control sequence that does different things and we really don't know what that specific
signal is that says, "This is going to be an imprinted
gene and this is not going to be an imprinted gene." We

Published by Cold Spring Harbor Laboratory Press; doi: 10.1101/sqb.2017.82.034447

Cold Spring Harbor Symposia on Quantitative Biology, Volume LXXXII

have little bits and pieces, but the absolute mechanism hasn't been worked out completely.

Beth Moorefield: Are there actually affected areas? The ones that are differentially modified, are they always in the same position within these alleles or within the different genes themselves?

Dr. Bartolomei: That gets a little bit to the mechanism. The real central piece of imprinting, these imprinting control regions or ICRs, those are the ones that have the methylation put on either in the male or female germline and it's maintained as long as imprinting is maintained. There're two regions that people where these ICRs can be found. One is intergenic—so, in between genes—and the other is actually in promoters. Often, it's in the promoters of long noncoding RNAs. One of the first uses that was found for long noncoding RNAs was that their transcription drove imprinting genes in *cis*. The methylation is in the promoter and if it's highly methylated, the long noncoding RNA is not made. If it's unmethylated, it is made. That's an example of how these imprinting control regions are not only giving you maternal- or paternal-like differentiation, but also functioning to do something.

The intergenic imprinting control regions are a little different. For the ones that I study, H19 and IGF-2, the imprinting control region is actually an insulator region and it binds CTCF. When it's unmethylated, CTCF binds and it helps to confer the imprinting of the region. When it's methylated, CTCF can't bind, so there's no insulator formed. These ICRs have these two sorts of broad functions. There are other mechanisms involved, but these are the two main functions.

Beth Moorefield: If you have the differential pattern, how does it actually get propagated to the daughter cells upon division?

Dr. Bartolomei: When the methylation is put on in the germline, it's through the de novo DNA methyltransferases. It's likely that differential chromatin modifications are involved too, but the propagation is through the maintenance methyltransferase, DNMT1. It's actually one of the first experiments done by the Jaenisch laboratory showing that in the absence of DNMT1, there was failure to maintain imprinting. As the cells divide, DNMT1 migrates with the replication fork and puts the methylation on the newly replicated strands so that the imprints are maintained. The key for imprinting is that it's epigenetically maintained. It's maintained throughout the life of the organism, in most cases.

Beth Moorefield: Not only maintenance but also the modifications have to be reversed to permit germline development: How is that accomplished?

Dr. Bartolomei: The mammalian embryo is very interesting in that there are two major times in development when there is this very large-scale reprogramming that occurs. The first time is after fertilization. The gametes come in with their own methylation patterns, and there is reprogramming as these cells are going to become pluripotent.

At that time, imprints are maintained. There is something very special about them that enables that differential methylation to be maintained.

With mammals, as the embryo develops, the germline is set aside from the somatic cells. At early postimplantation, some cells that are recruited from the somatic cells, based on their location, are going to go on to form the germline. What happens there is then there is a second time of reprogramming. In that case, everything that will be reprogrammed is reprogrammed, including imprinting control regions. That's because if you're taking the germline from somatic cells that have maternal imprints and paternal imprints and you're going to become a germ cell in a female, you want to erase the maternal and paternal imprints and put on the maternal imprints; if you're in a male, you erase those and put on paternal imprints. That reprogramming occurs as the primordial germ cells are being specified and are replicating and migrating to enter the genital ridge.

Because a lot of methylation is lost at that time—DNMT1 is down-regulated and moved outside the nucleus for the most part—it was originally thought that what would happen is that the cells would replicate and there wouldn't be maintenance methylation. You would have a passive loss of methylation: Things would be diluted out, they enter the genital ridge, and they're remethylated. We know now that sometimes the demethylation is faster. In the last 10 years, a whole new mechanism of demethylation has been described through the TET enzymes. It's a family of three enzymes that will oxidize methylcytosines—TET1, -2, and -3—TET1 and -2 are expressed at the time when the methylation imprints are being erased. What we, and others, have shown now is that in the absence of TET1, some germ cells fail to erase all of the DNA methylation imprints; others do it fine. It seems that active and passive methylation are working together to reprogram, to erase imprints. In some pools of germ cells, we see a lot of remaining methylation; some, we see hardly any. We don't really understand what's going on in that case. The more you dig, the more mysteries you find in this field.

It's really an incredibly exciting field, a lot to be learned. That's still the basis of on-going research. Why is it? Why is it that we need two mechanisms, an active and a passive? Are they redundant? Are they helping each other? Are there some regions of the genome that are just so densely methylated, have such high affinity for DNMTs, that you need this extra mechanism in place to demethylate? We don't know yet, and we'd like to really learn that.

Beth Moorefield: It's not clear yet whether or not the residual methylation is actually precluding the de novo methylation events or what the functional interactions might be there?

Dr. Bartolomei: We do know that if we look at TET mutants that come from female and we look at maternally methylated regions that we would expect to be methylated in oocytes, they're methylated fine, but if we look at paternally methylated regions that should be *un*methylated in oocytes, they are not. Sometimes they're still methylated; sometimes they're not. It doesn't look like residual methylation—whatever might be happening—is preventing de novo methylation from occurring. That's only looking at early in development and later in development. We don't know what's happening in between. If we could live-image cells… There's always this "on-paper experiment" you can think about where you could follow a cell and see demethylation occurring at every different stage. We can't do anything like that yet.

Beth Moorefield: Is there any evidence that environmental factors can influence any of these activities and actually alter patterns?

Dr. Bartolomei: That's another area of active investigation in my lab and lots of other labs. There's a whole field called Developmental Origins of Health and Disease. The idea there is that insults in utero to an embryo can be maintained or remembered and expressed later on in development. We're very interested in this hypothesis. A lot of it was originally defined by David Barker, and it was shown epidemiologically during the Dutch Hunger Winter when there was minimal caloric availability to people at that time. Women who had first-trimester babies during those times of minimal calories, these children were born and 50 years later had very high levels of heart disease and metabolic diseases. A lot of that was epidemiologically shown, but it has now been moved to animal models.

We've been very interested in this idea that very early gestational exposures may be the most dramatic. That's because that's when there's a lot of reprogramming going on. We've looked at endocrine disrupters and some assisted reproductive technologies where there're exposures in the early embryos. We use imprinted genes because for us they're our canary in the coal mine. We've shown with some of these exposures there are defects in the ability to maintain the methylation imprints or maintain other kinds of methylation, and there're defects in other kinds of reprogramming. Environmental perturbations during that early time in development can have a lasting influence on offspring. We're trying to look at that in more detail. It's another area of very active investigation as people have the capacity now to profile what's happening in the early embryo. Before we couldn't do that but now we can take small numbers of cells, look at DNA methylation, look at chromatin modifications, compare normal to environmental exposure of some kind. Science is such that we can do a lot of things that we hoped we could do 20 years ago.

A Conversation with Mónica Bettencourt-Dias

Interviewer: Jan Witkowski

Cold Spring Harbor Laboratory

Mónica Bettencourt-Dias is a Principal Investigator at the Instituto Gulbenkian de Ciência, Oeiras, Portugal.

Jan Witkowski: You spoke on the first evening about centrioles. Is that what you consider your field? Or do you consider your field more broadly, cell biology?

Dr. Bettencourt-Dias: My field is, broadly, cell biology and cytoskeleton. I'm very interested in centrosomes in cilia. Centrosomes are actually the major microtubule organizing center in animal cells, and they are very tiny: over a hundred times smaller than the diameter of a human hair. But they are very important for the cytoskeleton of the cell. Within the centrosomes, there are structures called the centrioles that also have a different functionality in the cells. They can migrate to the membrane where they form cilia and flagella, which are very important for cell movement, and also to move particles, like expelling particles from our trachea. Cilia also serve as antennas in many of our cells: They are very important to sense light in your eyes, to sense smell, to sense whether you've eaten enough or not in the last meal. They do tons of things in our body, and participate at many times in our life.

Jan Witkowski: In the context of this meeting, they're important for organizing the spindle.

Dr. Bettencourt-Dias: Exactly. With centrosomes, I mostly focus on how cells count the structures, because normally to have accurate chromosome segregation, you would have two centrosomes, one at each spindle pole. What happens in several diseases is that you have a deregulated number. Normally, animal cells control very well the number of these structures—much as they control the amount of DNA that they have—so that when you have cell division, each daughter cell inherits exactly the same amount. Each daughter cell should also inherit exactly the same number of centrosomes: one centrosome per daughter cell. It's very important that cells duplicate these structures in a coordinated fashion with the DNA replication, so that when you have chromosome segregation, you also have centrosome segregation, and you inherit the right amount of DNA *and* the right amount of centrosomes.

Jan Witkowski: Presumably, when the numbers of centrosomes get mixed up, you get inaccurate segregation of chromosomes.

Dr. Bettencourt-Dias: Yes. More than a hundred years ago, Theodore Boveri proposed that a deregulated centrosome number could lead to transformation. You could get a new ploidy, inaccurate chromosome segregation, and then transformation into tumors. Now we know that that's the case. It has been shown very recently that if you have a deregulated centrosome number, you can have inaccurate chromosome segregation and tumors. But we also know that having more centrosomes—not just during cell division, but also in interphase when the cell is not dividing—can be deleterious for our bodies, although maybe not for the cell, because it could have advantages. For example, it can break the adhesions between cells and they will become more invasive, so that's one of the properties that they may get. Also, the nucleation of more microtubules is important for processes like inflammation and many other phenomena.

Jan Witkowski: You do your research on the control of centriole numbers and why there are never two per cell. Centrioles are complicated things, so how do they replicate?

Dr. Bettencourt-Dias: These are beautiful structures. Their structure was elucidated in the '50s, when we started to have electron microscopy. They have this beautiful ninefold symmetry, which is actually conserved throughout the eukaryotic Tree of Life. When you think about DNA replication, you have one strand, and then the other strand forms complementary to the first one, so it's very obvious how it replicates. But centrioles are slightly different. You have one centriole, which is a barrel, and the new centriole forms close by in an orthogonal fashion. There's been a huge discussion whether they actually form in a "templated" fashion, whether one helps the other to form. The self-replication of centrioles and perpetuation of central centrosomes was actually proposed by Boveri and van Beneden. But at the same time, other people saw that these structures can form de novo, so they can appear, for example, in sea urchins.

It's not very clear why the centriole leads to the formation of another one close by, given that they can form without any centriole present. The idea that we have now-

Published by Cold Spring Harbor Laboratory Press; doi: 10.1101/sqb.2017.82.034520

adays is that centrioles can self-assemble: if you just have them on their own, at least part of the molecules that constitute centrioles can start forming the beginning of the structure. We think that the parental centriole, the one that already exists, recruits the components that form centrioles, so that it catalyzes the formation of new centrioles close to itself.

Jan Witkowski: But "catalyzes" only in the sense of attracting the necessary components?

Dr. Bettencourt-Dias: Exactly. And then they will do their business. There're a lot of positive feedback loops that enforce that things will happen there and not elsewhere in the cytoplasm. This is beautiful, because something that already exists dictates the place where the new ones will form, and because you have regulatory molecules localizing there, it will also dictate the time when these things happen.

Jan Witkowski: Does each centrosome have two centrioles?

Dr. Bettencourt-Dias: Exactly. During replication they come slightly apart, and this is also important to regulate the process. Each one of them will form a new "partner" close by, so that when they migrate to opposite poles during mitosis, you'll have one centrosome, each with two centrioles, at opposite ends of the cell.

Jan Witkowski: What's the mechanism by which the proteins that form centrioles are attracted to a preexisting one?

Dr. Bettencourt-Dias: About 15 years ago we started to know the molecules that play a role in centriole formation. Before, it was almost impossible to address this problem; now we know what the molecules are. We know that they are recruited, so the ones that exist at the already-existing centriole will recruit the other molecules that are needed to form the new one, and they will perpetuate the structure where it already exists.

Jan Witkowski: But what's the mechanism by which those proteins that are needed are attracted?

Dr. Bettencourt-Dias: I think it's just a question of maintenance. If you have molecules that are already there that have affinity for the other ones, they'll be recruited. I think some of them are not even recruited by microtubules. It's just that they have more affinity for what is there, so they'll be retained at the old structure and form a new structure. Of course, a big question is why do they form at a certain time, which is when the DNA's also replicating, and this is what we are also studying. We want to make the link between these molecules that we have identified that play a role in forming a new centriole, and we want to know whether they are regulated by the cell cycle machinery that also regulates DNA replication, so that the two things are coordinated. What we have identified is that the major cell cycle players—the cyclin-dependent kinases that promote advancing the cell cycle—also regulate these molecules. They actually prevent this structure from being formed in mitosis; it only starts being formed at the beginning of the next cell cycle.

Jan Witkowski: In mutants that have abnormalities of the cell cycle, do you get abnormalities in centrosome formation?

Dr. Bettencourt-Dias: You can. For example, a major cyclin-dependent kinase, CDK-1, actually prevents centrioles from being formed in mitosis, and in fly mutants for CDK-1, you get many more centrosomes. So, yes.

Jan Witkowski: Are there well-recognized "centrosome-opathies" or "centriole-opathies"?

Dr. Bettencourt-Dias: Definitely. Since the beginning of 2000, there're a variety of diseases like microcephaly and primordial dwarfism where the mutated genes that are causing the diseases were identified as genes that localize to the centrosome and are important for centrosome biogenesis. We now know that microcephaly is strongly associated: It's a "centrosome-opathy" because it's associated with problems with the centrosome and in the division of the stem cells that give rise to the brain. You have more death of cells in the brain that are supposed to populate the brain, therefore there's a smaller brain. Also, there can be precocious differentiation of stem cells in the brain. Therefore you exhaust the stem cell pool, and you have fewer cells in the brain, and therefore you have a smaller brain. Recently, two groups also linked Zika virus infection to effects in the centrosome that linked also to microcephaly. Definitely, there are centrosome-opathies. Cancer is similar. There were some recent studies pointing that if you have transient deregulation of centrosome number, you can actually induce cancer in mice.

Jan Witkowski: Are centrioles present in all multicellular organisms?

Dr. Bettencourt-Dias: No. Centrioles are very interesting, because if you look throughout the eukaryotic Tree of Life, you see that in all the different branches, you have centrioles. But in many of the different branches, there are species that have lost them. For example, higher plants, some amoebas, some fungi have lost them. But because you have it in all the different branches, it's most likely that the last common ancestor of eukaryotes already had these structures. Actually, the molecules that are needed to form centrioles are present in all of these different branches, so it's really very plausible that the last common ancestor already had them.

Jan Witkowski: Even in organisms that no longer have centrioles?

Dr. Bettencourt-Dias: We don't know exactly how the loss of centrioles occurs. Centrosomes participate in cell division, but actually, the ancestral form of cell division did not rely on centrosomes. Species initially likely had centrioles because they needed to make cilia for movement, because this was critical. Then, because the centrioles were needed for the cilia and the centrioles needed to be symmetrically inherited by the daughter cells, they became associated with the poles of the spindle so that they could be inherited symmetrically. Once there, they might have been co-opted to participate in cell division in certain organisms. For example, they are not required for somatic cell divisions in the fruit fly, but they are required in embryonic divisions, and also in male meiosis.

Jan Witkowski: Presumably, human cells can't divide without centrioles?

Dr. Bettencourt-Dias: They can divide, even without centrioles. There're many different studies—the first ones using laser ablation of the centrosomes, and more recently knocking down the centriole components because we know the machinery—and they can divide normally. Again, I think it's because there's this ancestral mode of cell division. Basically, you have microtubule nucleation from factors that exist around the chromatin. Once you have microtubules, more microtubules are nucleated from them, forming microtubules that organize the spindle, so that you have normal chromosome segregation.

However, what happens in human cells is that once you remove the centrioles, they'll actually divide, but then they will arrest in the cell cycle once they don't have centrioles. Human cells have ways to sense whether they have these structures or not, and they will arrest. That's not the case in fly cells. They are fine without centrosomes, and their somatic cells keep dividing.

Jan Witkowski: I seem to recall that there was talk that centrioles had their own DNA.

Dr. Bettencourt-Dias: There's a long history in the centriole field of people looking for DNA and claiming there was DNA. So far, centriole duplication doesn't seem to rely on its own DNA, but there are recent reports claiming that there is RNA for at least some of the centrosomal proteins on the centriole. I think that's part of the future. Even though the question was raised long ago, I think it still needs to be answered.

A Conversation with Kerry Bloom

INTERVIEWER: JAN WITKOWSKI

Cold Spring Harbor Laboratory

Kerry Bloom is the Thad L. Beyle Distinguished Professor of Biology at the University of North Carolina at Chapel Hill, Chapel Hill, North Carolina.

Jan Witkowski: Could you explain the centromere spring?

Dr. Bloom: The chromosome, in the public domain right now, is "the sequence." Of course, there's much more in the chromosome than the DNA sequence. That's going to be the challenge for us moving forward: How do you store information? How do you propagate information beyond just the sequence of nucleic acids?

Scientists have watched chromosomes move for hundreds of years, and it's a ballet. They do this beautiful movement back and forth, literally dancing around each other until they finally all line up—I use that word loosely—and then go to what will be daughter cells. The accuracy that they must achieve to segregate all 46 chromosomes roughly 10 trillion times for all the cells in our body is beyond anything that's ever been man-made. To get to your explicit question, they achieve that accuracy by building a spring between the two microtubule attachment sites.

However, they don't count chromosomes. You could have imagined a mechanism where they're counting: "Is 1 lined up? Is 2 lined up? ... Is 45 lined up? When 46 is lined up, let's go." They don't do that. They build a little spring between the two microtubules from opposite spindle poles, and when that spring is under some tension, it quenches a checkpoint that is responsible for delaying the next phase of the cell cycle, when chromosomes segregate. If even one chromosome is still left behind (i.e., the spring is not under tension), that'll suffice to delay the cell cycle. We're interested in how that chromosome spring works. That region of the chromosome is called the centromere. There's 6 feet of DNA in one cell. How do you take this very floppy molecule and build a molecular spring? That, basically, was our challenge.

Jan Witkowski: What are the physical components in the cell of the spring? What proteins are involved? Are they all known?

Dr. Bloom: Many of them are known. There're just a few key proteins called SMCs for "structural maintenance of chromosome" proteins. They comprise the cohesin proteins that hold sister chromatids together and condensin proteins. These are all ring-like proteins. It's really as simple as protein rings and the DNA. It's the DNA that's the major physical component itself and how we organize that DNA into loops that makes the spring.

Jan Witkowski: Your approach is a biophysical one. Don't you need to have a lot of knowledge about the molecules involved and their properties before you can start modeling?

Dr. Bloom: You would think so. If one is thinking of molecular dynamics and modeling protein structure and proteins docking or, in the pharmaceutical world of drugs docking into protein binding sites, you have to have a lot of knowledge of the atomic structure of a protein.

But let's go through the chromosome very quickly: DNA? 1953. The nucleosome? 1976, something like that. Beyond the nucleosome, there has been remarkably little progress in understanding higher-order chromosome organization. I think that reflects the fact that despite our molecular understanding of the proteins, we haven't really incorporated the physical properties of the DNA as a polymer.

A Nobel laureate, Pierre[-Gilles] de Gennes, pioneered this around the '50s or '60s. If you look at a PubMed of his work, it sort of dribbles, dribbles, dribbles along and then it's rocketed, because now the biologists have realized we can learn a lot with remarkably little incorporation of the proteins into the kinds of polymer models that de Gennes pioneered. These models ... there's no atoms, there's no molecules in these models. These are very simple bead-spring models because we're trying to model—in the human—3×10^9 bp. If we start putting atoms in there, there's no computer in the world that'll come close.

Jan Witkowski: Tell me about your approach. What did you start with? When you were doing your modeling, how did you set about building a theoretical spring?

Dr. Bloom: The workflow was in a room with computer scientists, applied mathematicians, and physicists. We would meet once a week and the meetings would go for 2 or 3 hours. I was shocked that all these people would sit there for 2 or 3 hours and do this. It started on a chalkboard where I would start drawing the position of the DNA.

Published by Cold Spring Harbor Laboratory Press; doi: 10.1101/sqb.2017.82.034488
Cold Spring Harbor Symposia on Quantitative Biology, Volume LXXXII

I'm an experimental biologist, so it started with being able to visualize the centromere DNA in live cells. Basically, what that entails is looking at fluorescent spots relative to the spindle microtubules and watching them jiggle around in real time. Then we started literally drawing how we imagined the DNA might exist within this structure that would be compatible with the motion that we see in the microscope. From those drawings, we started writing down equations.

The key in this is it's a very reductionist perspective. There's way more complexity than we're building in our models, but we're trying to capture the gestalt of the system in as few parameters as possible. It started with the math people and the physicists and myself drawing structures and trying to understand: Does the DNA go like this? Does the DNA make loops? What do I know from the biology that's accurate? What can the physicist tell me about whether this makes physical sense or not?

Jan Witkowski: You've distinguished between drawing models, as opposed to ones that you built from molecular dynamics. I still don't quite get how you go from this very reductionist, few-parameters-as-possible view to a very detailed model.

Dr. Bloom: It's called a bead-spring model. There's these bead-springs, and there's thousands and thousands of these bead-springs. The model has physical parameters. For instance in yeast, the nucleus has a certain size. We model it as a sphere. When I talk about DNA, in the model it's actually these bead-spring chains. The landscape that has changed in the past decade is the computer graphics that now go with the models. Now we can add into the models: "A bead looks like this," and then the computer graphics allow us to basically seamlessly go from the mathematical model into a computer-based model that mimics the behavior of these beads.

Jan Witkowski: So the mathematical model can be turned into a diagram, but then you add the components the cell biologist knows about.

Dr. Bloom: Exactly.

Jan Witkowski: In the pictures based on molecular dynamics, where you try to account for what the molecules are up to … Can that be dangerous, in the sense that the first type of model can be taken as a true representation of what's going on, when it really shouldn't be?

Dr. Bloom: There's not a single geometric solution in our model. We do computer graphics of this model. We then take that model—and again, this is because of the rapid rise in computer science—and I can put fluorophores in that model and simulate what those fluorophores look like in the microscope. Now I have a direct comparison of what that model predicts, relative to what I see in the microscope.

We have "condensin" in the model … "cohesin" in the model … They're not the molecules of condensin and cohesin. What they are is bead-spring circles that have the *dimensions* of cohesin and condensin, respectively. That's what they are. I let them wiggle around due to

Brownian dynamics. Then I ask, after they reach some equilibrium, what does the model predict they would look like? Then we tweak the model until the model matches the experiment.

Jan Witkowski: That leads to my next point. You're tweaking your model so that it behaves in the way that fits what you see experimentally. Doesn't that mean that all you've done is given another description of the experimental observations? Does your model tell you things that you would not have thought of otherwise?

Dr. Bloom: We started with a long, floppy DNA chain and "How do you build a centromere spring?" It turns out that the centromere spring is a bottle brush. You know the bottle brushes that we clean test tubes with? That's the centromere spring. You have an axis and there's a bunch of little hairbrushes. What the spring is, is there's a central axis. Now I have a bunch of hairs. These are all bead-springs. If I take a piece of DNA—again, a bead-spring—and let it wiggle around, it's going to wiggle, wiggle, wiggle, and make a random chain. It's going to have some radius defined by some mathematical equation. If I take that same bead-spring and now put these hairs on it that are also bead-springs, when it bends up, the hairs interfere with each other; when it bends down, the hairs interfere with each other. Of course, this is in 3D. That bottle brush builds stiffness. That's the centromere spring. It's a bottle brush spring that's stiff, throwing the two centromeres—the regions that attach the microtubules—that are predisposed to be on the opposite sides of each other before you even get to a microtubule. That intuition we would not have considered.

Jan Witkowski: In the cell, what are these loop pieces? Is it the DNA?

Dr. Bloom: They're DNA loops, *a la* Joe Gall lamp brush chromosomes or Oscar Hertwig 1890s lampbrush chromosomes. What's exciting to me is that the reason why we have not made considerable progress past the nucleosome is that there *is* no ordered structure. I think these springs exist all throughout the chromosome. The springs are a consequence of the density of these loops. You can stiffen the chromatin fiber, or you can weaken the chromatin fiber. We used to talk about heterochromatin and euchromatin—literally, dark and light chromatin. I think the conversation will turn to regions of high loop density, regions of low loop density, stiff regions of the chromosome, weak, floppy region of the chromosome, that allow DNA mechanics to occur.

Jan Witkowski: You mentioned starting this project in a room with mathematicians and physicists. How did they regard you, and how did you find their understanding and willingness to appreciate biology? What was it like dealing with mathematicians and physicists with a very sloppy science like biology?

Dr. Bloom: I shielded them from the complexity of all the genes that contribute to a process, because that would short-circuit the conversation. It turns out that this is why physicists don't become biologists. They're very uncomfortable with the complexity and the uncertainty that we are very comfortable with. We both just had to get over it. We left our egos at the door and we got over our inhibitions of what we don't like. That's how they dealt with it. They would come up all the time with "The strands *must* do this! Why don't they collide with each other?" Well, there's topoisomerase, but let's not talk about that now, okay?

Jan Witkowski: When Jim [Watson] and Francis [Crick] came up with the double helix, Max Delbrück, a physicist, was very worried about how the replicated strands could separate. How could they make concatamers? Francis said, "Well, the cell does it." There's something that does it. We won't worry about that problem.

Dr. Bloom: The strategy's the same.

Jan Witkowski: The other side is about biologists thinking in these mathematical, biophysical terms about their work. I've always felt the biologists almost envy mathematicians and physicists. As soon as a biologist sees a nice equation in a paper, they think "Oh, this must be really good." How can we get biologists out of that mind-set?

Dr. Bloom: You get them out of the mind-set by showing them that there's a lot to be learned. Cold Spring Harbor was prescient in quantitative biology for, now, the 82nd year. Obviously, all the colleges across the nation are now "quantitative biology" this, "quantitative biology" that. They're finally figuring out that we need to get there.

Jan Witkowski: I think it's going to be difficult, though. Biology has always been quantitative, but "quantitative" is not simply doing terabytes of sequencing. Too often at the moment, that's what quantitative biology is thought to be: just produce tons of data. It's not quantitative in the sense that you're using it.

Dr. Bloom: Computer graphics are what's going to span that gap. Now we don't just have to look at the equations. The equations can actually be under the rug and the biologist can look at the outcome graphically. In my mind, pictures are what bridge the gap.

The other thing that was very inhibitory was the language problem. "Stress" and "strain" have very different mathematical definitions. In conversation, "I'm under strain," "I'm under stress"—they're almost interchangeable. It was the language barrier that was huge. This is what I mean by leaving egos at the door. We really had to dive into "What do you mean by this word?" or "What do you mean by this equation?," and we were patient with each other. The problem is it took years to get past that, until we could really talk with each other. These disciplines are as complex as biology. I'm a biologist, yet I could walk out and not recognize anything with a green leaf from any other thing with a green leaf, right? It's the same thing with physicists. It's not "Physics" in general. I'm working with polymer physicists. Before this, I didn't even know that was a field.

A Conversation with Iain Cheeseman

INTERVIEWER: LARA SZEWCZAK

Scientific Editor, Cell

Iain Cheeseman is an Associate Professor of Biology at the Massachusetts Institute of Technology, and a member of the Whitehead Institute for Biomedical Research, Cambridge, Massachusetts.

Lara Szewczak: You've worked a lot on the kinetochore, which is this massive, dynamic assembly of proteins that helps anchor chromosomes to microtubules, to allow chromosome segregation. What got you interested in studying that complex?

Iain Cheeseman: When I was a grad student, I'd always found the cell cycle and cell division to be interesting. I joined Georjana Barnes and Dave Drubin's lab in 1997. At that point, I don't think I'd even heard of what the kinetochore was as a thing, and I had a project where I was going to work on this microtubule-associated protein in budding yeast. About a year or two into that, I started to realize that this protein was probably doing something else. This was a protein called Dam1 and I realized that in addition to holding on to a microtubule, it was at this structure called the kinetochore.

At that point, I kind of just fell in love with that structure. It was everything I wanted to try to understand about a cell. It's a very physical structure. There're a lot of proteins there. It's a very mechanical structure; it has to do stuff. There's a business end of what it has to achieve, and it's a highly controlled structure. You need to coordinate that with the cell and the cell cycle and ensure that you're not generating errors. I really loved that set of challenges. You have this machine: You have to build it, and have it physically do stuff, and you have to control it. When I started in '97, there were maybe 12 proteins that we knew about that were kinetochore components, so this was quite early days. It was just beautiful to see this expanding list of molecular complexity and to really understand how this process works. I fell in love with this Dam1 complex and the entire kinetochore and have never really stopped.

Lara Szewczak: You said there were 12 proteins when you started. How many are there now?

Iain Cheeseman: In human cells, I would say ~110, but it varies in different organisms. The Dam1 complex is a protein that is this beautiful ring-like machine that holds onto the microtubule and carries a chromosome with it. I loved studying that, but one of my advisors, Dave Drubin,

pretty much every week would walk into the lab and say, "Okay, so have you found the human homolog yet?" I felt like I was really bad at BLAST searches for many years, but it turns out it actually doesn't exist, so there certainly are differences between organisms. Some things are true in fungi, but slightly different in human cells. Then I transitioned to Arshad Desai's lab, where I focused on *C. elegans* kinetochores, which are holocentric; they occupy the entire chromosome. They're very similar to yeast and very similar to human cells, yet also different in other ways, in terms of that machine. But probably, the largest number of proteins we know is in human cells. I think you can argue a bit about, "What is a kinetochore component?" or "What is a protein that is at that site?" but I think, as I would define it, ~110.

Lara Szewczak: What makes something a kinetochore component as opposed to "just in the vicinity"?

Iain Cheeseman: The way I would define that is the connectivity between the chromosomal DNA and a microtubule polymer. You have to hold on to the nucleic acid, the chromosome. You have to be able to help mediate its segregation. You have to hold on to the microtubule. The things that are localized to the site that are playing a role in those interactions and holding that entire structure together, I would put in that category. But it's a really complex part of the DNA as well, so there's going to be heterochromatin there, and other specific things that participate in that way. I probably wouldn't define those as kinetochore components.

Lara Szewczak: Can give a bit of an overview of your present work?

Iain Cheeseman: I love what this molecular machine is, and the complexity of how you build this, and assemble it, and drive things. But I also loved when I was working in budding yeasts, where you could do clean and clear genetic experiments. So I transitioned to *C. elegans*, and then about halfway through in Arshad's lab, I started working on human cells. About 95% of our lab's work now is in human cell culture.

Published by Cold Spring Harbor Laboratory Press; doi: 10.1101/sqb.2017.82.034454

Cold Spring Harbor Symposia on Quantitative Biology, Volume LXXXII

Human cells are gorgeous. They're beautiful. They're big. You can really nicely see the spindle and the kinetochore structure. But it was really hard to do a yeast-like experiment there, and I've always kind of been jealous about that: the clarity and clearness by which you can create a replacement, or really ask what a protein in doing. I think the CRISPR-based approaches are really revolutionizing how we can think about doing that sort of experiment. We still cannot backcross human cells and do some of the beautiful kind of genetics that you'd want, but I think a lot of things are accessible there, in a way that wasn't even a year or two ago. We've been exploiting that approach, and it's caused us to reimagine a lot of our expectations.

For example, take those 110 proteins. The kinetochore is critical for what it means to be a cell. Every time you make two cells, you have to distribute your chromosomes. You have to do that process correctly. You have to make sure that you not only physically achieve that, but you do that with high fidelity. That should be an essential process, right? For a cell to continue to survive and duplicate, the kinetochore should be essential. Yet in budding yeast, out of, say, 65 yeast kinetochore proteins, there's probably about 25 or 30 that you can eliminate, and that yeast cell will apparently grow fine.

We know of a lot of human kinetochore proteins, but it would be very hard to say which of those are actually essential. They're important, but could you get rid of them and have the cell still survive? RNAi was a wonderful way to be able to do functional experiments, but that really simple thing—"Is this thing essential or not?"—is not something that we could access. That's really been transformed by the ability to go in and clearly and carefully knock stuff out with CRISPR. We've been exploiting that strategy to go through things and a lot of those things have conformed to our expectations, but a lot of those things have been a surprise. That simple metric—"Is this essential?"—really changes the way you see a process.

There's a signaling pathway called the spindle assembly checkpoint. This pathway senses when there is a problem in chromosome attachment. This holds up the cell until you get things right, and then you can progress, and segregate, and distribute your chromosomes. In yeast, that's a nonessential pathway. Those proteins were first identified in budding yeast—beautiful work from Andrew Murray's and Andy Hoyt's labs. They could make null mutants in those and the yeast were fine, but they behaved like they could no longer hold up mitosis when there was a problem. In most papers that are out there, the simplistic statement would be, "Well, these proteins are nonessential in yeast but they're essential in human cells. They're essential in vertebrate cells, so the checkpoint is an essential pathway in human cells."

I don't think that's true. It's true that many of those proteins, if you eliminate them, are essential, but you can create a cell that is genetically checkpoint-defective, and it will grow just fine in culture. That makes you think very differently about what that pathway is, and what its role is. It meets, much more, that classical definition of what a checkpoint is, where it's a pathway that gets activated when there's a problem but that you don't necessarily need it under all circumstances. So many of those

proteins are critical for additional roles, but I think the checkpoint itself is not essential.

Lara Szewczak: What was the protein that was the biggest surprise for you? That you thought, "This has to be essential, and it's not."

Iain Cheeseman: Probably, these checkpoint proteins. The ones that we weren't able to eliminate completely—and when I say "we," the person who did this work was Kara McKinley, a graduate student in our lab. Kara made a collection of knockouts where she targeted >220 different genes that play roles in cell division. These were things that we would assume would be critical. Maybe 40% of them didn't show a strong and potent phenotype in cell culture, although we're quite confident that we're getting rid of these things. This complex that we *could* completely eliminate, which I really thought was going to be central and important, was this Rod–Zw10–Zwilch complex. But you can get rid of those proteins, and the cell is fine.

There've been a lot of other things like that, because it's a question of whether it's a surprise. At the time, it's, "Okay, something's wrong. There should be something happening here. These cell should be dying." But then you start to think about it: "What was the data that made us think that it *would* be important or essential?" I think that's not only been true with knockouts, but also with specific mutations. I think that there're a lot of things that can compromise a process without killing the cell. The cells are actually pretty robust, and they can tolerate things in cell culture, but some of these low rates of chromosome desegregation, in an organism, would be problematic.

Lara Szewczak: Do you see a pattern? Are things that are more structurally integral more "essential"—whatever that is—than things that are either more peripheral, or that have catalytic activity? Can you categorize your hits that way?

Iain Cheeseman: I continue to be enamored with the kinetochore, but I think one of the features of it that is fascinating is how flexible and plastic it is as a molecular machine.

Lara Szewczak: Do you mean that literally, in terms of deformability, or do you mean in terms of what can be accommodated in a functional unit?

Iain Cheeseman: Literally, it is a structure that can be highly deformed under force and tension, and is very flexible in that way, to be able to resist and handle that force. But mostly, figuratively, in the sense that it is dramatically rewired during evolution. During the cell cycle, it's a dynamic structure: it's assembling and disassembling. I think those answers vary between one organism and the next. I think it's important to define in human cells and in human cell culture what is essential for cell division, what is essential for chromosome segregation, but I think the answer's going to be very different from what that is in budding yeast, or what that is even in mouse cells.

We have a protein that we've worked on where everything that we do to it in human cells seems really central and critical. Then we realized that there was a mouse knockout of that, but the mice seemed to grow fine. That really surprised us. I don't think it's so much about a critical catalytic domain or structural role. It's about the way that the human kinetochore has chosen to do that. There are structural proteins that are nonessential in human cells, but those same proteins are structural proteins that are essential in budding yeast. I love that rewiring. It really makes you think. You take something like the ribosome or the proteasome, and these things are largely invariant across eukaryotes. Ribosomes are even pretty similar when you think about the way bacteria do that. And here's this structure that, even over a short evolutionary distance, is changing and adjusting in different ways.

Lara Szewczak: It sounds like you've done this essentiality screen that's opening up a lot. What do you feel is beyond your reach? What's the thing you want to know but you can't quite get there yet?

Iain Cheeseman: That's what keeps you going, right? When I began, we didn't even have the molecular players. There was a historic phase of the field from Fleming through the '60s that was very visual: understanding what the cell division was as a structure, and being able to see the kinetochore early.

The modern molecular age of the kinetochore and cell division began about 1987 when Bill Earnshaw used the CREST autoimmune serum to find some of the first human kinetochore proteins, and there was comparable work in other model organisms. In '97 we still only had a dozen. The number exploded over the next decade. We're largely at that point where we've done the molecular cataloguing, where we know what those parts are. That opens this new door, which is, "Okay, we have all these players. What are they doing?" There's all these activities you need to achieve. How do they work individually and in combination to make that happen? For a lot of the work that we care about, that really is the focus: bringing these together, not thinking about them in isolation, not thinking, "I found this new protein." You have a machine that is an integrated molecular machine. How do you achieve that? I'm really enjoying the stuff that we're playing with for that.

I would also really love to think about how you modulate what this structure is, and how it works in an organismal context. I've been careful about saying "in cell culture." I love human cell culture, and I love our HeLa cells, and, they're a very powerful place to explore things, but what's even more beautiful is to think about the 30 trillion cells in our body: the diversity of tissues and organs, and circumstances and developmental situations, and how you take these core cell biological processes and think about them in an organism. I'm really excited to be able to tackle that. Some of that is starting to be accessible, but there still need to be some better tools.

Lara Szewczak: If there's one tool that you could have to let you ask some new questions, what would it be? What would it let you do?

Iain Cheeseman: For mice to have a 24-hour doubling time, or something like that? Maybe ease our ability to access tissues and control things genetically in mice. There

are people who are excellent at that. Our lab needs to learn how to do those. Organoid systems have opened those doors. The tools are there, and they've been developed by diverse labs. There's something that we as a field need to do a better job of taking advantage of, and harnessing. What is the missing reagent? I don't know that there is one. I think that it's possible to make those. Now it's just time to do that.

Lara Szewczak: If you had all your resources at hand and were starting your lab today, brand new, would you ask the same questions, or is there something different that you would go after?

Iain Cheeseman: I teach undergrads cell biology at MIT, and I also love new people coming by our lab. It's really fun for me to talk about these processes and these ideas. I'm surrounded by this unbelievably smart group of people, but most of the questions they ask, my answer is, "I don't know." I love that. As much as we had been able to visually see this structure for a hundred years, the basic things of what it means to be a kinetochore, and what it means to segregate chromosomes and divide a cell (which is such a central biological process): These are things that we simply don't know. I'm so enamored by those ideas and questions that it's like this addiction where, of course, I'd choose to do these again.

I love the physical nature of biology. I love structures. I also love regulatory circuits, and the cell cycle's a nice example of that. I particularly like the interplay between those two. I think that's why I continue to love the kinetochore: because it's a complex and beautiful example of that. You have to achieve this as a very physical thing, but it needs to be intricately controlled and regulated. If it were to be all solved tomorrow, which it's not going to be, I think that there would be other cell biological structures that I would also find fascinating.

Lara Szewczak: Given the stages that the kinetochore has gone through in terms of understanding it, are there lessons there for other large macromolecular dynamic complexes?

Iain Cheeseman: Yeah. In some ways, we're behind some of these. I really love reading about the proteasome and the ribosome. There's excellent work to reconstitute DNA replication in vitro. But we've known about DNA replication for a long time, and many of those factors were identified earlier than kinetochore components and there are still basic questions about them. I think that it did take reconstituting those processes to really understand that. I look at those fields with envy but also to try to learn how we can apply those lessons to whatever this structure is.

I have two excellent DNA replication colleagues in my department: Steve Bell, who has reconstituted many of these things, and Terry Orr-Weaver, who looks at these in a developmental context and trying to understand how different cells and tissues alter the way they're replicating things to create polyploidy or other kinds of amplifications. It's nice to look at those other processes and try to learn from them. I think that there are a lot of parallels for the cilia and the centriole. And there's the nuclear pore, and many other beautiful machines in the cell.

A Conversation with Monica Colaiácovo

INTERVIEWER: LARA SZEWCZAK

Scientific Editor, Cell

Monica Colaiácovo is a Professor in the Department of Genetics at Harvard Medical
School, Boston, Massachusetts.

Lara Szewczak: You work on meiosis, in particular on the
process of sorting out where crossovers happen. Maybe
we can just start with a brief explanation of what that
actually means.

Dr. Colaiácovo: Crossovers are extremely important
because they provide a way to exchange genetic information between the chromosomes during meiosis. That
means that it provides genetic diversity. But also, in the
context of meiosis it creates a locked-in interaction between the pairs of homologous chromosomes so that, as
they progress and finally align at the metaphase plate and
microtubules attach, you'll have enough tension there to
ensure that the chromosomes will segregate away from
each other appropriately to opposite ends of the spindle.
It is important to get that crossover. It has to be tightly
regulated. It is incredibly important for successful and
accurate segregation of the chromosomes during meiosis.

It is very clear that crossovers are tightly regulated.
They're not randomly distributed, and that's universal.
We see that in mammals, we see it in *C. elegans*—the
model organism we use in my lab—and budding yeast,
and plants, and flies. There is always an obligate crossover, which means that there's a push in the system so that
you will always get at least one crossover per homolog pair
to ensure that there'll be that attachment.

There is also the phenomenon of crossover interference.
Although chromosomes are undergoing multiple programmed meiotic double-strand breaks (which is a unique
feature of meiosis; you actually want to make those breaks
so that you will engage into a crossover outcome), only a
subset of those breaks will then get designated or destined
to become a crossover. Once one is selected to become a
crossover, other breaks nearby are not repaired as a crossover event; that single crossover interferes with the probability of getting crossovers nearby.

All these different layers of regulation told us this is a
system that's very robust. It has to be tightly regulated, but
why is it that the position of the crossover might matter?
We were intrigued by that question because there's a lot of
information from human studies showing an association
with aneuploidy in trisomy 21, as well as the majority of
examples of trisomy 16, with the position of the crossover.

If it's very terminally located, close to the end of a chromosome, it doesn't sustain normal or accurate chromosome segregation. Why is that?

We decided to look at that in a metazoan, because that
had not been directly tested in a metazoan system before.
To do that, we co-opted the power of transposons. It turns
out that there is a Mariner transposon that was imported
into *C. elegans* from *Drosophila*; this was beautiful work
done by Erik Jorgensen and [Jean-Louis] Bessereau's lab.
Basically, they generated these lines where you would
have a single transposon insertion at a discrete place in
the genome. We screened many of these lines to find specific ones where we knew the transposon was excised at a
very high frequency in an efficient manner. The way these
are excised is because you have a transposase under a heat
shock promoter. To make sure that the only source of the
break was going to be the system that we were going to be
using, the inducible system, we crossed that into a *spo-11*
mutant background where you no longer have the programmed formation of meiotic double-strand breaks.

Now we could look at a break generated at a specific
location and ask the question: "When that break gets resolved as a crossover, what is going on?" Is it really going
to succeed in generating a functional attachment between
the homologs that will facilitate the accurate segregation of
the homologs away from each other? What happens if we
make a break right at the physical center of a chromosome,
or right at the end, which, anecdotally, should be problematic...

Lara Szewczak: So, telomeric?

Dr. Colaiácovo: Subtelomeric, and then what would happen if it happened in an off-centered position, which is
normally where crossovers happen in *C. elegans*, because
that's an important control. It would tell us whether the
Mos system, this transposon system, was working as a
normal SPO-11 break might. What we found was that
the control of an off-centered event is productive. It results
in a functional chiasma and all the chromosome remodeling that has to happen after the crossover proceeds in a
normal fashion. But if the break that got selected to become a crossover was a break right in the physical center

Published by Cold Spring Harbor Laboratory Press; doi: 10.1101/sqb.2017.82.034462

Cold Spring Harbor Symposia on Quantitative Biology, Volume LXXXII

of a chromosome, that did not succeed in sustaining an accurate attachment between the chromosomes, although we would still see the ability to form a cruciform structure.

What that means is: you've got the break. It becomes a crossover, and what happens next is you're going to have massive remodeling of these chromosomes. You're going to lose some proteins and gain other novel proteins at very specific chromosomal subdomains. Ultimately, there's also increased compaction of the chromosomes. What you acquire is a very characteristic configuration. It actually looks like a cruciform with the short and long axes, and there's a reason for that. Because of how chromosomes have to be aligned at the metaphase plate, the short arms are aligned equatorially; the long arms face the opposite spindles. What you want is to remove cohesin at that interface between the short arms so that now the homologs can segregate away from each other.

We examined if that process of remodeling occurs normally. We found that when that break is in the middle and a crossover is the outcome of that, you still formed a cruciform, but it was no longer an asymmetric cruciform. It was now a symmetric bivalent, and some of the proteins that should have been retained at a specific subdomain were no longer there. These proteins… in this case, LAB-1, which is a functional ortholog of Shugoshin and plays a very important role in insulating a specific domain (the long arms) from premature loss of sister chromatid cohesion, that domain was no longer protected. Now you had a mislocalization of Aurora B kinase to both the short and the long arms of the bivalent.

So we took it a step further, and monitored the metaphase-to-anaphase transition to see whether that mislocalization indeed resulted in premature loss of sister chromatid cohesion, and it does. We concluded that maybe part of the reason you don't normally use breaks at the center region of chromosomes to become a crossover is that you won't be able to sustain normal remodeling. You'll result in a bivalent, but that attachment is no longer productive, in the sense that it can't promote accurate chromosome segregation. That was very interesting, because it gave us a more logical idea of the basic mechanism as to why there's a disconnect between [the fact that there is the formation of] all these breaks, but you tend to select certain ones and not others [as crossovers].

Then we asked the question, "Well, what happens if this is now located at a subtelomeric region?" We've looked at a lot of events, a lot of oocytes, and we cannot find a bivalent at the end of diakinesis. Remember, this is a pipeline. We can see that we succeeded in making a break at the subtelomeric end; we've got markers for that. We know that a subset of those is getting selected to undergo a crossover path because we do see procrossover markers at that region, but we can't see an actual chiasma at the end of diakinesis. This tells us that these subtelomeric events are incredibly unable to actually retain that attachment. Something is sliding off; something is not working properly there. It is very analogous to what we think goes on in mammalian systems where, if you've got these events that are designated to become a crossover, it doesn't really result in an effective, functional, accurate segregation event.

Lara Szewczak: Is one way to think about it is that if you're subtelomeric, there's just not enough DNA to work with?

Dr. Colaiácovo: That's one possibility.

Lara Szewczak: How does the measuring work? How do you know where you are, and does that have to do with crossover interference?

Dr. Colaiácovo: We hypothesize that there has to be a way for you to be measuring distances. We did a screen to try to understand how that might work where we took advantage of Aurora B kinase fused to GFP [green fluorescence protein]. We looked at genes that, when depleted by RNAi, would interfere with the ability of Aurora B kinase to localize only on one but not the other arm. We found a lot of interesting candidates where either Aurora B kinase cannot localize at all on these bivalents, or completely mislocalizes and is now in both axes. We focused on the subset with Aurora B kinase on both axes and we noticed that a related set of proteins was coming up. These genes encode for a specific set of proteins, all of which tell us that there has to be a way for you to communicate with telomere ends: these are telomere-associated factors. That's one end of the equation. That could be one way you're communicating this. We're currently testing that hypothesis.

The other question we had was, "Are there any other features?" There could be DNA topology that's playing a role here. There could be issues about the chromatin landscape and epigenetic readouts: Are you reading differences between one direction versus the other direction? We've already been testing this using the inducible system. We found that if the break that gets selected to become a crossover is right at the middle of a chromosome (versus an off-center choice), before the formation of that break and then after the formation of that break, the marks that we see in both directions—the histone methyl marks and acetyl marks—toward the chromosome ends are not the same. When you have an event that happens at the physical middle, the changes throughout the full length of these chromosomes are very different from the changes we see when it's an off-centered event. Something isn't working properly when you've got that selection right at the middle.

Lara Szewczak: When you have an off-center event, are the marks on either side different, and when you make it in the middle, do the two sides look like one or the other of the off-center?

Dr. Colaiácovo: There's a difference when it's an off-centered event, which makes sense, because that could be part of how you're reading that distance. That difference is gone if it's in the middle. We're trying to figure out if there is a problem with who gets recruited that might be working as specific methyltransferases and so on that regulate this process. Is that part of why you get miscommunication, an inability to signal or read distances toward one end? How that gets superimposed onto these telomere-associated proteins, we don't know yet. We're trying to figure out how these two might be interfacing.

Lara Szewczak: I'm just trying to understand this one point. When you have an event in the middle and you get different marks, does it look like one side of the off-center? Is it that it can no longer distinguish one side from the other, or is it completely different from the two sides?

Dr. Colaiácovo: It can no longer distinguish one side from the other. It just looks the same in both directions, as opposed to one kind of mark acquisition in one direction versus not having that on the other.

Lara Szewczak: You've done this in worms and want to extrapolate to mammalian cells, our cells. What are the differences between the two systems that you're going to have to account for?

Dr. Colaiácovo: In *C. elegans* you're dealing with holocentric chromosomes. We don't have a discretely positioned centromere. In *C. elegans* chromosomes, you're incredibly dependent on this remodeling, because it then allows you to bypass the fact that you don't have a localized centromere to promote that proper positioning, attachment, tension, and so on. You have to be really precise with the remodeling in *C. elegans*, because that will dictate a lot of how precisely you'll be able to segregate in the metaphase-to-anaphase transition.

Having said that, the chromosome remodeling observation is not unique to worms. It's been demonstrated in flies, in yeast, and in mammals. We know that when the synaptonemal complex [SC] disassembles—it's one of the other features associated with this remodeling process—it does so in an asymmetric manner. It does so in the same way in these other systems, except that there you retain those SC proteins on the centromere. It makes sense that, in our case, they get retained on the short arm, particularly if we think about what allows you to properly align at the metaphase plate and where there's a lot of microtubule polymerization that then allows you to push these chromosomes apart from each other; it behaves a lot like a centromere. Work from Arshad Desai's lab supports that; he thinks there's good assembly and enrichment for certain factors that promote segregation in terms of kinetochore activity and so on. Although *C. elegans* may be more reliant on this because it is key for it to achieve accurate segregation, I don't think this is going to be restricted to worms, because the remodeling is pretty universal.

The beauty of this is that, if you think about what we already know about aneuploidy in humans, this directly correlates with what we see now in a testable setup in the context of what we can do in the worm.

Lara Szewczak: You talked about figuring out this measuring mechanism and unpacking potential for communication to the telomere. Thinking beyond that, what's the next question you want to ask?

Dr. Colaiácovo: We are trying to understand how that scaffold, the zipper-like synaptonemal complex, actually deals with all of these changes. How do you regulate this so that it knows where to go in an asymmetric manner, and what is the triggering point for this? That's one of the main directions we're headed: trying to focus more on the synaptonemal complex as another key element here.

Lara Szewczak: Are there particular technologies that are going to help you do that?

Dr. Colaiácovo: We started looking a lot at chromosome dynamics, so we've done a lot of FRAP [fluorescence recovery after photobleaching] experiments and measuring the ability to repopulate certain subdomains when they're depleted of a certain protein. That gives us important information about how dynamic or not these arms are before or after certain key events during meiosis. In addition, we are now doing a lot of superresolution imaging. In these cruciform structures in particular, we're going to be able to see a lot of information that—even with high-resolution microscopy—we are missing. That will hopefully allow us to tease apart more of the details of what's going on with the remodeling.

Lara Szewczak: What's the one thing that you're taking away from this meeting that is new information or a new way of thinking about your own work that's going to influence what you do?

Dr. Colaiácovo: The concept of how we can integrate a lot of what we're seeing with really beautiful 3D modeling. The concept of whether something is a gel, or a liquid. I think all of that is fascinating and is not normally the way we think about things or the way we present our information, and that's definitely something that I look forward to incorporating into what we do.

A Conversation with Bernard de Massy

INTERVIEWER: LARA SZEWCZAK

Scientific Editor, Cell

Bernard de Massy is a Researcher at the Institute of Human Genetics of the National Center for Scientific Research in Montpellier, France.

Lara Szewczak: Your lab focuses on meiosis and the process of recombination. What got you into the field in the first place?

Dr. de Massy: My first interest was evolution, and then I got trained into DNA metabolism: DNA replication, DNA recombination. I realized that there was one area where I could really try to develop an understanding of the molecular mechanism, but with evolutionary implications: the process of recombination during meiosis. It's really a molecular machinery that has very long-term consequences on the transmission of genetic information, on genome diversity. That was really what made me choose this field.

Lara Szewczak: What was the question that you started with when you started working on meiosis?

Dr. de Massy: The first question…that was in budding yeast…was to characterize some specific sites of meiotic recombination in the yeast genome that were just discovered a few years before. I've been really continuing this trajectory from yeast, but now it's already 20 years. I shifted to mouse and ask similar question in mammals, in mouse meiosis.

Lara Szewczak: You've really focused on how the breaks form. Tell us a little about that.

Dr. de Massy: Seven years ago, we discovered a factor—PRDM9 [PRD and RIZ homology DoMain containing protein 9]—and I think it's been a dream for me. In characterizing this protein and the molecular mechanism of breaks that are induced in meiosis, it unexpectedly turned out that it has enormous evolutionary implications. This protein works by specifying the sites in the genome where recombination takes place. It's amazing, because it really breaks some established concepts about the genetic map: the genetic linkages that were thought to be fixed and follow some rules according to, I don't know, gene functions, for instance. The way this gene works in mammals shows that the genetic map is actually incredibly plastic.

Lara Szewczak: How does it work?

Dr. de Massy: This protein in particular, it binds to DNA. It has some specific DNA binding recognition, but we've evaluated that it binds to maybe 10 to 15 thousand potential sites in the genome. Then it modifies the histones that are on the nucleosomes adjacent to its binding sites. So, it modifies the nucleosome, the histones, and then…we don't know very well, actually. We assume that it interacts with other proteins. There are some postulated interactions we are currently trying to validate to attract the protein that catalyzes the formation of double strand breaks and the catalytic activity. We know which protein encodes for the catalytic activity—it's SPO11—but we don't know how these proteins get together. That's the part we don't know how it works.

Lara Szewczak: There doesn't seem to be any redundancy. When you don't have PRDM9, you aren't positioning the breaks. That seems really dangerous, when you think about the process of meiosis and the fact that you do want to be able to have this process happen and diversity get introduced. Was that surprising to you?

Dr. de Massy: Let me put it slightly differently. There is still a big surprise and still an unsolved paradox with this protein, but for a slightly different reason. This protein is found in many vertebrates. We are not exactly sure of the phylogeny, but it probably arose with the emergence of vertebrates or maybe earlier. But it has been lost, independently, in many vertebrate lineages. It has been lost in birds, in some mammals, in some fish, and those species without this PRDM9 protein do perfectly well. Recombination takes place at sites that are apparently accessible chromatin and maybe they have some modification in particular, but the pattern of recombination is completely different in mammals like humans, versus another mammal, like dogs—Canidae—where they don't have PRDM9. There are two alternative pathways; that's the way it looks. And PRDM9 is a very specific and unusual pathway because its activity destroys its own binding sites across generations. It's very puzzling: Why is PRDM9 doing this, whereas other species can do perfectly well without it? Is there an advantage for this specific function? What PRDM9 does is that it makes the position of recombination events dynamic…I mean dynamic during evolution, meaning that they change over time.

© 2017 de Massy. This article is distributed under the terms of the Creative Commons Attribution-NonCommercial License, which permits reuse and redistribution, except for commercial purposes, provided that the original author and source are credited.

Published by Cold Spring Harbor Laboratory Press; doi: 10.1101/sqb.2017.82.034538
Cold Spring Harbor Symposia on Quantitative Biology, Volume LXXXII

384

Lara Szewczak: So because it destroys its own sites, you can't then rebind.

Dr. de Massy: Exactly, and then there are new variants that probably come in that are selected and design for recombination to take place at different sites, whereas species that don't have PRDM9 have fixed and stable sites of recombination. That's very exciting for me. To really understand this molecule, we need to take into account its long-term evolutionary implications. I'm now constantly discussing and interacting with evolutionary biologists to try to appreciate this. That's a challenge.

Lara Szewczak: On the one hand, you're interacting with evolutionary biologists, on the other hand, you're trying to understand the molecular interactions that let you bridge from PRDM9 in specific binding sites to SPO11 that's actually doing the cleavage. What approaches are you taking to look at that?

Dr. de Massy: Biochemistry. It's not easy because of the cells that we are working on—mouse oocytes or mouse spermatocytes—but it's a completely different approach with a different experimental procedure. More and more, it's a biochemical approach.

Lara Szewczak: Are you bringing in other model organisms, or are you sticking with mouse, because of the concerns of differences of mechanism?

Dr. de Massy: That's an important point: whether the system we are using is the best one to answer the question. In this specific context, if we want to understand where and how PRDM9 works, I don't know which alternative I could take. There are other organisms where the molecular understanding of recombination is very well developed, like yeast, but yeast doesn't have PRDM9, so that's not an option. So, right now that's my choice and we have to continue and use this system if we want to progress.

Lara Szewczak: If those are the questions that you are thinking about now, what's just beyond reach? What questions do you want to be answering in 3 years?

Dr. de Massy: I'm interested in the molecular mechanism of meiosis, but not in all the detail. I don't think I will go into the deepest details of a particular molecular reaction. That's not my interest. I'm rather interested in how this

event at the DNA level takes place in the context of the nucleus. Meiosis, and meiotic prophase in particular, is really fantastic. There are plenty of completely unsolved questions. Somehow, when the breaks are formed, they have to be repaired and interact with homologous sequences, which are present on the homologous chromosomes. There is, I think, a relatively black box about how broken DNA finds the homologous sequence. Nobody really knows how it works. Chromosomes have to find each other. The chromosomes have to move, and there already is some data on that movement. To answer your question, the idea would be to link this molecular event with chromosome movement and dynamics, and ideally to follow the movement live in the nucleus.

Lara Szewczak: To get to the point where you can answers those questions, what are you gearing up to do in the lab?

Dr. de Massy: For that, we need new tools. We have to develop new technologies or use technologies that have maybe been used for different purposes and design them for this particular system, which is mouse meiosis.

Lara Szewczak: What question would you ask first if you could pluck a new technology out of the air?

Dr. de Massy: I want to follow a specific DNA sequence as the meiosis is progressing, and to follow two sequences from two homologous chromosomes in the nucleus during the process of meiotic prophase. That would be a challenge.

Lara Szewczak: If you couldn't work on meiosis anymore, what would you work on? If somebody handed you a pile of money and said "Go do something new," what would you do?

Dr. de Massy: I would still get back to some evolutionary aspects. I've been thinking about two things. First, would be to reconstitute a molecular ancestry for the proteins we are studying now. That's one aspect. The other is similar, but in opposite direction. I would like to take cells and have them go for fifty thousand generations, to have a system for that. I know some people do that with bacteria because with the regeneration times of 20 minutes, you can do those. But I would like to have an evolution experiment to read the future.

A Conversation with Arshad Desai

INTERVIEWER: LARA SZEWCZAK

Scientific Editor, Cell

Arshad Desai is a Professor of Cellular and Molecular Medicine at the University of California, San Diego and the Leader of the Laboratory of Chromosome Biology at the Ludwig Institute for Cancer Research.

Lara Szewczak: Part of your lab focuses on how kinetochores tie in to the cell cycle. What got you interested in that?

Dr. Desai: One of the most striking things when you watch a movie of mitosis is this incredibly synchronous segregation of the chromosomes where somehow they seem to know exactly when to go. I've always been fascinated by how you coordinate such a complex cellular-scale process such that it occurs with remarkable synchrony. Over the years we've appreciated that the kinetochore is the key orchestrator of this process and we now know that one of its mechanisms is by controlling the activity of the main enzyme that drives cells out of mitosis. We recently found that kinetochores are also promoting the activity that drives cells out of mitosis and we believe this may allow a real fine tuning or local regulation of how we control the decision to make sure all the chromosomes are properly lined up, and then you go out of mitosis.

Lara Szewczak: The main enzyme being the APC [anaphase-promoting complex] cyclosome: How do we think about that now?

Dr. Desai: APC/cyclosome [APC/C] was this forbiddingly large complex—16 subunits or so—and we've been struggling to understand how it works. The recent revolution in structural biology has given us a precise picture of its architecture, but also the reasons for its complexity because it has to both drive this reaction of catalysis of specific substrates in a very rapid manner and also be extremely sensitively regulated by the chromosomes so that it doesn't inappropriately get turned on. That's the reason why this is such a complex enzyme. Fundamentally, its main activity is to do ubiquitin ligation on two key substrates, cyclin and securin, and it's the degradation of those substrates that drives cells out of mitosis.

Lara Szewczak: How do the chromosomes talk to the APC/C?

Dr. Desai: We've known for over 20 years that one way that chromosomes talk to the APC/C is that the kinetochores, when they're unattached, are generating a "wait anaphase" signal and the point of the signal is to really prevent the APC/C from degrading its substrates. There's a lot of beautiful mechanistic work on that. In particular, the kinetochore is catalyzing a conformational conversion of a key protein called Mad2 [mitotic arrest deficient 2] and this then becomes part of a complex that goes on to inhibit the APC/C.

Our recent work suggests there's another role of the kinetochore in controlling the APC/C. What the kinetochore is doing is that it's altering the main enzyme activator of the APC/C known as CDC20 [cell division cycle protein 20] and it's activating CDC20 to turn on the APC/C. And so there's a balance between positive and negative regulation that's actually anchored on a specific binding site on the kinetochore.

Lara Szewczak: What are the contexts where you'd want the two different functional roles working?

Dr. Desai: If a kinetochore is unattached then the chromosome has not yet found the spindle, and you do not want the cell to go out of mitosis. In that situation, you want to make sure that the kinetochore is stopping the APC/C. If the kinetochore is attached, you also want to make sure the APC/C turns on relatively quickly so the cells don't spend an inappropriately long time in mitosis. One way to achieve that is to convert the negative signal into a positive signal. This may also enable local control of APC/C activation, which is something we'd like to explore in the future.

Lara Szewczak: And when you say "local," you mean spatial?

Dr. Desai: Spatially localized control. Yes.

Lara Szewczak: So that you're making this decision for one chromosome microtubule attachment at a time…?

Dr. Desai: Right. You don't want to fully relieve cohesion between paired sisters immediately after attachment, but possibly you may alter locally the cohesin dynamics. This is just speculation at this stage but we know that there is positive and negative regulation going on, so one of our goals is to understand exactly what this is doing in cells.

Published by Cold Spring Harbor Laboratory Press; doi: 10.1101/sqb.2017.82.034702
Cold Spring Harbor Symposia on Quantitative Biology, Volume LXXXII

Lara Szewczak: What toggles between the activating and the inhibiting?

Dr. Desai: We believe the toggle is through microtubule attachment to kinetochores. It's been appreciated for a long time that when microtubules attach in a certain configuration the checkpoint signaling reaction is turned off. This would immediately tip the balance toward the activation reaction. A key part of the activation reaction is actually dephosphorylation, so we also suspect that microtubule attachment is delivering the phosphatase to the kinetochore. The mechanism of that is unknown but that would also alter the balance based on attachment.

Lara Szewczak: And that's dephosphorylation of CDC20?

Dr. Desai: Dephosphorylation of CDC20 to go on and bind to the APC/C and then drive cells out of mitosis.

Lara Szewczak: So it's a different way of thinking about CDC20.

Dr. Desai: Actually, our work was inspired by classic biochemical work, which had shown that CDC20 could be negatively regulated by phosphorylation. The in vivo follow-up on that has not really been there, but I think what we've contributed is to show that it's really controlling the timing of APC/C activation, and that the kinetochore is influencing that reaction.

Lara Szewczak: Taking a look beyond figuring out this local mechanism, what do you think are the next big questions for understanding this interplay between getting things attached right and moving the cell on through the cell cycle?

Dr. Desai: There's this very localized interplay between mechanical events where the kinetochore is the major machine that's responsible for coupling to the microtubules and driving the movement of the chromosomes, and these signaling events that are deciding what the cell should do. I think the really big questions are to understand precisely how the mechanics influence the signaling pathways and how they're talking to each other. That's a little challenging because we are not good at integrating forces into studies of signaling mechanisms. That's an area where we still understand relatively little about how the process is working.

I work in the Ludwig Institute for Cancer Research and we know that most solid cancers have aneuploidy or incorrect chromosome numbers and they also have chromosomal rearrangements and the ultimate origin of those is problems with segregation in mitosis. The signaling reactions that are occurring in this very confined space are also relevant to understanding the genesis of that phenotype, which is still quite mysterious in the sense that, unlike DNA repair, we don't have clear links with mutations and the phenotype. There are rare mutations in chromosome segregation machinery in cancer but not at the prevalence that would explain the phenotype that we see in solid cancers. So, we also are very interested in linking this sort of detailed mechanistic understanding to potential dysregulation in cancer.

Lara Szewczak: Have the rare mutations pointed toward any likely suspects?

Dr. Desai: There's mutations in the APC/C that were just recently described. There are also mutations in proteins that are involved in the spindle checkpoint, like Bub1, but their prevalence is extremely low; it's not a high prevalence mutation. Obviously, dysregulation by changes in expression or amplification in this kind of very tightly coordinated pathway could be equally deleterious. That's an area we're interested in because you do see misregulation of a lot of these components in different cancers, but we need to model that in a clean way so we can interpret it and see if there is a link to the genesis of aneuploidy.

Lara Szewczak: Misregulation at the level of gene expression, or at the level of protein stability, or…?

Dr. Desai: Most of the analysis is from genomics so there're both amplifications that are found, and also gene expressions changes. A lot of this is coming out of things like the TCGA [The Cancer Genome Atlas] analysis. We're also curious whether there are germline changes that could explain susceptibility or explain how tumors are different. That's something that we're also exploring.

Lara Szewczak: In a solid tumor, forces can be different because of the stiffness of the cellular matrix. You said we're not good at integrating force with these kinds of studies. What kind of technology is that going to take?

Dr. Desai: We can do force analysis in purified systems with tools like optical traps but it's very challenging to integrate analysis of signaling in there. One exciting area is to use force biosensors in a cellular context. While that's always a bit difficult to interpret, I think that's the best tool we have at this moment to assess potential forces and how they influence reactions.

The other big challenge is that a lot of these reactions involve things with rapid half-lives and posttranslational modifications, so I think that's another challenge as to how we can understand those. It's a golden era for high-resolution cell biology. We have beautiful tools and systems to really try to understand how these complex pathways work but at the same time we are always in the need for new technologies and so we're looking for clever people to come up with things we don't anticipate to try and tackle these questions.

Lara Szewczak: The work we've discussed was largely in worms, but you don't just work on worms. What kinds of questions are you going to take into different systems?

Dr. Desai: We also work with human cells quite a bit, but there our work is much more directed at understanding how the centrosome that organizes the spindle is involved in formation of the spindle and also how it contributes to the process of accurate division. That involves collaborative work where we've developed very clean small molecule inhibitors that allow us to remove centrosomes from cells and then see what happens to division, or the accuracy of chromosome segregation, and also potential therapeutic approaches that derive from that. We're not analyzing detailed kinetochore biology in human cells; our colleagues are doing fantastic work on that. We focus much more on trying to get at this question of genesis of cancer cell aneuploidy and cell division-targeted therapies. It's more exploratory and more targeted on centrosomes and spindles.

Lara Szewczak: For the aneuploidy field, what do you think are the most exciting things going on right now?

Dr. Desai: I think the realization that a single lagging chromosome in mitosis can trigger these vast genomic sorts of rearrangements has been a major change in our thinking. The other thing that's really exciting is getting a sense of what aneuploidy is doing to cells. That's a very complex field. One of the struggles we have is that we don't have a good measure of aneuploidy. We can count something with one extra chromosome as aneuploidy or something with 40 extra chromosomes as aneuploidy and we haven't really figured out how to address that challenge and compare that in a clean way to phenotypes. There's been beautiful work done by modeling aneuploidy with single extra chromosomes, but in terms of relating back to the spectrum of changes we see in cancers, that's still been challenging to tackle.

The biggest question I still have personally is, where does it come from? What's the genesis of this? It's clear that once you have these events that then you can actually trigger more aneuploidy and more problems. Is this just chance, or is this actually reflecting some vulnerability that we don't understand yet? This is a big debate in the cancer field, and that's something we would love to contribute to in the future.

Lara Szewczak: So, "chance" versus something that may be "predetermined"... Where are you starting from?

Dr. Desai: We're trying to take cells that are most "wildtype" for us: For human cells, that would be ES [embryonic stem] cells; that's the closest. We are very influenced by a "genetic model" sort of thinking: trying to alter specific classic cancer-related pathways and assess whether there's any obvious induction of aneuploidy phenotypes. The bigger challenge we have is we want to do a quantitative assay that gives us a quick readout. That's something we're trying to work on quite intensively right now. I think that will open up a lot of work. With the advent of single-cell sequencing we can do analyses of genome imbalances but this is still not a very high-throughput approach. We need to be able to score tens of thousands, if not millions, of events to look at the relevant frequencies here. From model organisms, we know the frequency of missegregation is extremely low but we still don't have a very good handle on that in mammals. That's a big question for the aneuploidy field: to get a rigorous baseline measurement.

Lara Szewczak: I want to switch gears completely. Imagine that you're just starting your lab. You have funding and resources, but you can't work on kinetochores. Given the scope of what's going on in science, what would you work on?

Dr. Desai: I do really like the work on microbial communities. I really enjoy that and learning how they function as a unit more than themselves and also how they interact and influence, say, hosts, etc. I'm talking more on the functional side, not so much the genomics side. I think that's a fascinating problem that is getting its due recognition because they are everywhere. When people talk about biomass, it's all microbes, really. There's so much unexplored biology there in terms of the kinds of reactions and the kinds of components and pathways. It's just open terrain in terms of exploration. That area excites me a lot.

One other area that I've actually been stimulated to go into partly by our observations where we found that a lot of the chromosome segregation machinery is doing really unexpected stuff in development, stuff we didn't anticipate. That's also made me think a lot about how things work inside an embryo where you have a lot of cells and a lot of neighbors and you have all these complex morphogenetic events occurring, and how do you achieve that? In *C. elegans*, we can watch it go from one cell that we just leave in a microscope overnight and there's a larvae the next day, so that's truly remarkable. Trying to take this sort of mechanistic worldview into those systems... that's something I would probably do more.

One other area I think is really exciting is the detailed biophysics of macromolecular machines. This is an area that I have always admired. I have a lot of friends who work in that area and I think you can derive really fundamental principles from those kinds of approaches. That's another area I find really exciting.

A Conversation with Daniel Durocher

INTERVIEWER: JAN WITKOWSKI

Cold Spring Harbor Laboratory

Daniel Durocher is a Senior Investigator at the Lunenfeld-Tanenbaum Research Institute,
Mount Sinai Hospital and the Faculty of Medicine, University of Toronto.

Jan Witkowski: Perhaps you could tell us a little bit about your work.

Dr. Durocher: It really is a continuation of our long-standing interest on how cells maintain the stability of the genome. Over the years, we've explored many aspects of it, but with the advent of using CRISPR as a genetic tool to really map out pathways, we've recently gone back and asked the questions: Do we have all the parts of DNA damage response proteins, and how can we find new ones? We've applied the CRISPR-based technology to do this, and it's been extremely successful.

Jan Witkowski: CRISPR is a very powerful technique. How are you using it? Are you knocking out genes at random and seeing the effect on repair? What strategy do you use?

Dr. Durocher: We're using guide RNA libraries. The guide RNA is the little piece of RNA that guides Cas9 to do cuts. We use these libraries in pools in lentiviruses, so that allows us to introduce them in cells very efficiently. Essentially, we do a population of cells. Each of the cells has a different guide RNA, and each of these guide RNAs has a different knockout for the gene of interest. Then, we subject this population to DNA damaging agents and read out which guide RNAs caused selective sensitivity to the DNA damage in question. That gives us a very unbiased view of the response. What we're seeing—and this was anticipated by the body of work—is that this is very profound. It impacts many aspects of cell biology from DNA repair itself, to chromatin biology, to the cell cycle. For the first time, we're having a genetic view of the response to these DNA damaging agents in human cells.

Jan Witkowski: The agents used are both chemical and radiation?

Dr. Durocher: Exactly. DNA is very susceptible to attack, both from the physical environment, from chemical mutagens, but also from the biology. In fact, we have enzymes that attack DNA or use DNA as well, so we're using that whole space for our study. We've done radiation, we've done chemicals, and we've also done genetic perturbation that leads to DNA damage as a tool.

Jan Witkowski: Presumably, you have some sort of positive control in the sense that you're picking up genes and proteins that you would expect to.

Dr. Durocher: Exactly. What was so exciting when we started doing these studies is that we essentially in one fell swoop could rediscover what was known. In addition, we were able to pick up new things, and also pick up things that were really surprising as well. The power of the technique is such that you can really believe these results, especially because you've got all the known components as well.

Jan Witkowski: Do you pick up different things for different chemical agents depending on the precise damage to the DNA that's done?

Dr. Durocher: Absolutely. It's really remarkable. As you would expect, agents that have similar modes of action will pick up similar genes. We can even build relationships between agents based on the profiles they're generating. In many ways, we could use an agent that is an unknown, put it on cells, and by the profile, we could say what type of DNA damage it would cause.

Jan Witkowski: Ever since the Ames test, there's been a long-standing desire to have better screens for chemical mutagens.

Dr. Durocher: That's one of the objectives that we have in doing these types of experiments. We can probably develop signatures that will really tell us what types of lesions are generated. In some cases it's mixed, and we see this and we can start deconvolving that as well.

Jan Witkowski: Partly with an eye toward screening for environmental hazards, is this a particularly sensitive way of detecting damaging agents?

Dr. Durocher: Absolutely. Also, we can modify the system to make it even more sensitive. We can also partly disable certain genome stability mechanisms, to make the cells even more sensitive. This is something we've not yet completely explored, but the potential is certainly there to really build very sensitive biosensors through this approach.

Published by Cold Spring Harbor Laboratory Press; doi: 10.1101/sqb.2017.82.034546
Cold Spring Harbor Symposia on Quantitative Biology, Volume LXXXII

Jan Witkowski: There's been a tremendous argument about "how much" makes an environmental hazard actually dangerous. If you can't have zero chemicals as the only safe level, as you develop ever more sensitive tests, this argument becomes even more contentious.

Dr. Durocher: That's a very good point. From my side, we're interested in the genetic architecture of these responses. Doing this as a readout is an interesting idea, but the interpretation of what that readout is something we think about more.

Jan Witkowski: Let's get back to the biology, then. You said you've been picking up things that were either not known or you would not have been expected to be involved. Can you give a couple of examples?

Dr. Durocher: In one particular DNA damaging condition that effects DNA replication, we find ribosome quality control genes coming up. This suggests there's cross talk between translation and some aspects of the quality control pathways that govern DNA replication. This is something we're actively working on. We don't really know exactly what is the nature of this cross talk, but this is the type of new biology we're hoping to find.

We're also seeing a number of RNA-binding proteins involved in multiple aspects of gene expression. That goes in line with this theme that we're seeing: this meeting of chromosome organization, gene expression, and the functional genome as well. That's maybe a little more expected, but this is something that we'll dig into also.

Jan Witkowski: These are proteins that are involved in the quality control of ribosome structure or function?

Dr. Durocher: These are involved in the rescue of stalled ribosomes when they counter either structures or problems during translation. Essentially, it was the pathway that leads to its rescue, or its degradation, or its removal from the nascent transcript. Probably, what we're having is aberrant proteins being produced, and then these somehow interfere with DNA replication, in this case. What these proteins are is an active area of study in the lab.

Jan Witkowski: How does DNA repair play into the overall structural organization of the chromosome? Or is it the other way around?

Dr. Durocher: I think there's a very profound interaction between these: the quality control of the DNA molecule itself, and both the overall chromosome organization and chromosome segregation. As a quick example, homologous recombination is a DNA repair process that uses homology on other chromosomes to repair DNA damage. In the process of homologous recombination, you create these joint molecules. If these are not resolved in time, they lead to problems in chromosome segregation, because the chromosomes are physically linked together. There are a number of instances where you have this type of feedback.

The other way, during, for example, homologous recombination, the damaged DNA needs to find a homologous sequence. How does it search for this homologous sequence in the nuclear environment? Now you're going from the whole chromosome, to helping with the repair. Chromosome organization, in that sense, is also very useful.

Another thing we've been working on the past few years is that a number of DNA repair pathways are actually suppressed during chromosome segregation. Why? Maybe it's to avoid these contacts between chromosomes. These are the type of questions where there's a really nice interplay between chromosome organization and structure and DNA repair.

Jan Witkowski: Are there processes involved in DNA repair that are integral to chromosome dynamics?

Dr. Durocher: Yes. For example, the cohesin complex, which is really important for shaping the structure of the chromosomes and is also involved in keeping the newly replicated sister chromatids together. They play an important role in the DNA repair process as well. Almost at every stage of organization, from the unit of the nucleosome to the unit of the chromosome and how it's folded in the nucleus, there's a really intimate association with the DNA repair processes.

A Conversation with Steve Henikoff

INTERVIEWER: RICHARD SEVER

Assistant Director, Cold Spring Harbor Laboratory Press

Steve Henikoff is a Member, Basic Sciences Division of the Fred Hutchinson Cancer
Center and an Investigator at the Howard Hughes Medical Institute.

Richard Sever: You work with a veritable alphabet soup of centromere-associated proteins. Could you tell us a bit about the history of centromeres?

Dr. Henikoff: Walter Fleming introduced the term "chromatin" in 1882, so chromatin actually precedes the rediscovery of Mendel's laws, precedes the beginning of genetics. What he called chromatin was in these drawings of chromosome segregation in, I think it was newt: beautiful drawings of chromosomes that were being pulled to the poles at mitosis. What was particularly striking was that you could see everything. You saw the centrosome; you saw the spindle fibers that were holding onto the chromosomes. He was very careful to show a little darkening for each of the connections between the spindle fiber and the chromosome. He actually saw centromeres, or heterochromatin as we call it now, pericentric regions, whatever. He saw them and he colored them in. So actually, centromeres were discovered before the chromosome theory of heredity. Centromeres were really the first genetic loci, after Mendel's work.

The beginning of the molecular study of centromeres was this paper from Louise Clark and John Carbon where they showed that if you take a little piece of DNA that comes from the genetically mapped centromere and put it on a plasmid, you make a chromosome that will segregate normally during both mitosis and meiosis. That proved that that piece of DNA was the centromere. I remember when that paper came out, it was quite revolutionary. I thought, "Oh! They're all gonna be that way." We were studying heterochromatin in *Drosophila* at the time, and I thought it would really be very simple because in there among all those repeated sequences that we can't make any sense out of, there's going to be some little magic sequence. But that was not to be.

A lot of progress was made studying the budding yeast centromere, and even fission yeast turned out to be different. It turned out that nearly all multicellular organisms have centromeres that are highly repetitive tandem sequences. There was a lot of skepticism at the time that centromeres were just going to be homogeneous alpha satellite repeats, for example—which turns out to be what our centromeres are composed of. Other people thought it might be other satellites, until there was the

experiment by Hunt Willard and his colleagues showing that you can actually make artificial centromeres. It's the equivalent of the yeast experiment, but you can do it with human centromeres. But you needed long arrays of these alpha satellite sequences, so that was a problem.

Then there was the fact that in 2000 or 2001, the draft human genome was published. It's still a gap at all of our centromeres. We haven't assembled it yet, because it's so homogeneous you can't do much with it. It's been difficult to study. But a lot of progress has been made in understanding centromeres, particularly in yeast. That's our best example.

Richard Sever: How long is that dedicated sequence in budding yeast?

Dr. Henikoff: It's one hundred twenty base pairs. It varies a little bit between chromosomes; there are sixteen chromosomes. They have canonical sequences that comprise the first element, which is about eight base pairs long, and that's a binding site for a transcription factor. On the other end, there is the binding site for a kinetochore-specific complex that recruits the centromeric nucleosome that sits in the middle, but there are only about eighty base pairs for that. People have been wondering, how do you get a nucleosome at eighty base pairs? This has been a bit of a controversy. It's pretty well established that it's the smaller particles. They're only four histones in there, probably, because that's all that you can really wrap eighty base pairs around.

I don't think that's been so much of a problem. Where it's really turned out to be a problem are the satellite centromeres: the ones in humans, plants, and most higher eukaryotes.

Richard Sever: These "satellite centromeres" are the alpha satellite sequences? What exactly are those?

Dr. Henikoff: The alpha satellite is a one hundred seventy or so base pair unit that gets repeated multiple times. It's almost homogeneous, but importantly, there are these higher order repeats. In our genome, for example, the majority of the alpha satellites for most of our chromosomes is a fundamental dimeric unit. These units might only be about 60% or so identical, one to the next, so

Published by Cold Spring Harbor Laboratory Press; doi: 10.1101/sqb.2017.82.034785

Cold Spring Harbor Symposia on Quantitative Biology, Volume LXXXII

they're really very different from each other. But then you go to the next one, where that dimer gets repeated, et cetera, in the same orientation. It's very homogeneous, but then sometimes there's enough divergence that you'll see that there are higher orders above that: there's dimers, there's tetramers, et cetera. Usually, it's even-numbered ones for most of the chromosomes, but three of our chromosomes have an odd number. Five is the basic number. So it gets very complicated from chromosome to chromosome. That's been one of the problems.

Mouse centromeres are much simpler. They only have two different satellite repeats. One of them, the minor, has a one hundred twenty base pair repeat unit, and the major has a two hundred thirty-four base repeat unit, and they don't see higher order structures. The higher order structures are something that you find in primates but you don't find it in some others. We've been looking a lot into *Arabidopsis*; plants have the same kind of structure. They have a 178 base pair repeat unit, but they don't have higher order structures.

I can go through all these examples, but the one thing that's in common is that there are these highly repetitive sequences that are there, and that makes for an interesting challenge to study them. But it also asks the question: Why is it like that?

Richard Sever: Given this is such a fundamental thing that's shared by anything that needs to segregate its sister chromatids—which is everything—why is it all so different?

Dr. Henikoff: We have a hypothesis that we refer to as centromere meiotic drive. It only occurs in female meiosis. Female meiosis is asymmetric. Of the four products of meiosis, only one of them will get chosen to go to the next generation. There's the egg pole, and then there's the pole that will give you the polar bodies. In other words, only one of the four products will move to the egg pole. It turns out that there is a reorientation process, because there's a competition between centromeres from the maternal side, and centromeres from the paternal side.

Richard Sever: So we're back to selfish genes.

Dr. Henikoff: It's absolutely selfish. We'd actually proposed that centromere drive is a process in which there's a reorientation. There's a competition between the maternal and the paternal side and because of that competition centromeres are competing to make it into the egg pole. They're the most selfish elements that can be. They're just repeat sequences, and so the bigger centromeres in some organisms make it to the egg pole. That was shown very nicely in work from Ben Black and from Mike Lampson, who showed that this actually occurs in mouse, and they worked out a lot of the details of the process. So, we think that the "why" is that centromeres are competing, and they're rapidly evolving because of it. It's a Darwinian process of cheating in female meiosis.

Richard Sever: The next level, which is all of this alphabet soup of proteins coming in, is probably more conserved.

Dr. Henikoff: No, actually. It turns out it's an arms race. Harmit Malik, when he was a postdoc in the lab, discovered that CENP-A—centromere protein A—which is the centromeric histone, which is absolutely essential to everything...

Richard Sever: This is the thing that replaces the regular histone H...

Dr. Henikoff: Right. The H3 variant—we were working on in *Drosophila*, but it turns out to be more general—itself was showing rapid evolution. What Malik discovered was that you see a large excess of replacement, over synonymous, changes. We know about arms races where this can occur between a pathogen and the host immune system. A virus' coat protein will mutate to evade immune surveillance, and then the host immune system will clobber it, so you get this arms race going, and it builds up changes on the surface of the viral coat part. We see the same kind of process must have been occurring for the centromeric histone. But how could that be? That's going to be deleterious, and you'll be dead, because every cell division's going to die.

We found that for CENP-A, for CENP-C, basic centromeric proteins are undergoing an arms race, and female meiosis is the only time when you can actually have an arms race. We call that the centromere paradox: Why is it that centromeric sequences are evolving so rapidly, both in the size of the centromeres—your Y chromosome and mine might differ by an order of magnitude in size; they differ tremendously—yet they're doing the most basic process we know of in genetics, which is chromosome segregation, which is very stable. How could that be? We think it's this arms race of female meiosis. That was fifteen or so years ago, and I think it's holding up pretty well.

Richard Sever: You mentioned CENP-A, and you've noted there are others: CENP-B, -C, -TWSX, -Z. This is a giant protein complex on top of the DNA that links to the kinetochore or attaches to microtubules, correct?

Dr. Henikoff: That's right. It's on the DNA. The DNA wraps around the nucleosome part; it also wraps around CENP-TWSX. There's a complex of histone fold proteins that the DNA wraps around. What's important about that is there're two connections to the outer kinetochore. There's been some question as to where those connections come from, at the chromatin level. One connection is CENP-T, and one is CENP-C. That's where we get into alphabet soup. CENP-T has partners: W, S, and X. CENP-A has partners: H4, H2A, and H2B. If you think about it like that, you've got two parts that connect. What we show is that they're all actually part of one coherent complex. We showed this in a couple different ways, including a new method that we call "CUT&RUN" [Cleavage Under Targets and Release Using Nuclease].

Richard Sever: Everybody was familiar with ChIP-seq, where you bind your antibody onto the protein and you sonicate away all DNA. This seems like a much cleaner way to get a better signal.

Dr. Henikoff: This is a method that was introduced in 2004 by Ulrich Laemmli, however it really hadn't been used since until we started working on this and improving the protocol. We're starting out with intact cells, as opposed to breaking everything up like you do for ChIP. Everything stays intact. We then permeabilize them in such a way that we can get antibodies to go in and find their targets in the chromatin, say, for a transcription factor or a centromere protein. Then we add a fusion between protein A and micrococcal nuclease. Protein A will bind to the immunoglobulin G of the antibody, so therefore you're tethering this micrococcal nuclease, which is our standard tool for looking at chromatin.

Once it's tethered, it has some very nice properties. We can put it at low temperature so it'll just stay in place during the reaction. Micrococcal nuclease requires calcium, so it's not going to do anything until you add the calcium. You add the calcium, let it go for a few seconds or an hour or something, it doesn't seem to matter. It will cut out on both sides the particle, and then it'll float out through the nuclear pores and we just take the supernatants and extract the DNA, and it's very simple. Our current protocol takes about half a day. It could even take just a couple hours.

Being clean is important when you're looking at something like a centromere where there's only one per chromosome, or transcription factors where you have one here that has a footprint of about twenty base pairs and another here of twenty base pairs, and you've got ten kilobases in between. Basically, when you chew up the whole DNA like you do with ChIP, you're going to have a certain probability of every little piece of DNA getting sequenced, and that's going to give you a lot of background. It might be a low background, but you'll have to do a lot of sequencing in order to see those peaks. With CUT&RUN, because we leave the vast majority of the DNA behind on the beads and only sequence what gets released and floats out, the background really goes way down, such that we only have to sequence about a tenth as deeply. And because the antibody is not looking at the whole cell contents that get ground up when you do ChIP but only seeing the intact cells and the nucleus, it's only seeing the surfaces. That means that the antibody binding's usually very efficient. For a histone modification, we can get down to a hundred cells, so it should be good for low cell numbers. It's all new, but a lot of people are trying it, and I've been getting good reports. I think it's going to catch on.

Also, the ChIP method is destructive because at some point, before you add the antibody, you solubilize everything and you can lose some parts. That's what we're finding: No matter how you solubilize it, whether you chew everything up with micrococcal nuclease or grind it up, it's going to cause some damage. With CUT&RUN, the antibody's added when everything is intact. In fact, it's added at the time that you permeabilize the cell. They were live, and we get them as quickly as possible, and everything is intact. Because of that, I think we're getting a more accurate picture. What we see is quite compatible with what we've seen with ChIP, but I think it shows us that what we'd been looking at were, sort of, eroded particles. Actually, the intact particles are much larger than what we thought typically: over one hundred eighty base pairs over this dimeric repeat unit of three hundred and forty. Because they're larger, they can accommodate a lot of the alphabet soup of proteins that we see there, including the ones I've just talked about. We can see that they're all there.

A Conversation with Martin Hetzer

Chief Editor, Nature Reviews Molecular Cell Biology

Martin Hetzer is the Jesse and Caryl Philips Foundation Chair, a Professor in the Molecular and Cell Biology Laboratory, and Vice President and Chief Science Officer of the Salk Institute for Biological Studies in La Jolla, California.

Kim Baumann: One focus of your lab is long-lived proteins in the nucleus. Could you tell us how you discovered them and also what interests you in terms of the biology of these proteins?

Dr. Hetzer: It dates back to a study a postdoc performed in the lab in 2009 in nematodes—*C. elegans*—and he had a very basic question related to a subset of nuclear proteins called nucleoporins. They form the only gateway/transport channel that mediates all molecular communication between the nucleus and the cytoplasm. He was interested in the question, "What happens to the expression of those genes that encode for proteins for the nuclear pore when cells leave the cell cycle and when they finally differentiate?"

Since the adult worm is entirely made of postmitotic or nondividing cells, he thought it was a great model system and he made a very surprising discovery. What he found is that a subset of nuclear pore proteins are present in the adult animal in every cell, very highly abundant, but the mRNA that encodes for those proteins is no longer detectable. He was very intrigued by that, because typically if a protein is present in a cell, then the mRNA is also present. Through further studies, he could show that all these nuclear pore complex proteins, which are large protein assemblies, are actually built during embryogenesis only in dividing cells in the embryo but no longer in the adult animals. The consequence of that was that those proteins have to last for the entire life span of the worm.

Kim Baumann: The nuclear pore is embedded in the nuclear envelope: that's the core of the pore, and then there are more peripheral structures. Is it true for all of the proteins that make the nuclear pores, or only for a subset?

Dr. Hetzer: Only for a subset of them. Every nuclear pore is made up of thirty different polypeptides and its only about half of them that form the scaffold structure of the nuclear pore that are no longer made. The other ones are still produced, so that is another level of complexity. Some are apparently very long-lived, while others are replaced and turn over quite rapidly.

The striking finding was that we had what we thought was the first example of a large protein assembly that lasts

through the entire life span of the animal. Then, we thought, maybe that's specific for *C. elegans*, because the life span of a nematode is about two to three weeks. We wondered if this could be true for an animal that has a much longer life span. Could this also be true for humans? It's an interesting question, because many nondividing cells in adult organisms don't turn over. For instance, our neurons in our cortex and cerebellum are as old as we are. Could it be possible that this protein complex could actually last for years?

Kim Baumann: Are there are other long-lived proteins in tissues that have very low turnover, or is it specific to the nuclear pore?

Dr. Hetzer: We wondered the same thing. There are only a handful of proteins known to be extremely long-lived: Collagen or elastin—those are extracellular proteins— and β-crystallin, which forms the lenses of our eyes. Those are either outside of cells, or in very specialized environments.

We decided to do a pulse-chase labeling experiment in rodents. We fed the animals with a nonradioactive isotope, ^{15}N, chased with ^{14}N for 2 years, and then performed quantitative mass spectrometry on isolated organs. We compared liver and brain initially. We did find about 50 very long-lived proteins. We've found, gratifyingly, the nuclear pore proteins again, very similar to what we saw in *C. elegans*. But, we also found proteins that are involved in gene regulation, such as histones and very specific histone variants. The most surprising part is we even found enzymes: We found histone deacyltransferase to be long-lived.

Kim Baumann: I would think that these kinds of proteins would be structural proteins, rather than involved in any regulation or regulatory processes. Is there any advantage to having them? Are there any common features?

Dr. Hetzer: There are no common features. There are no specific sequences. They can all be degraded in dividing cells, for instance. They are not necessarily structural proteins. We even first thought—for the nuclear pore, which is sort of a scaffold structure—that this would be an im-

---

© 2017 Hetzer. This article is distributed under the terms of the Creative Commons Attribution-NonCommercial License, which permits reuse and redistribution, except for commercial purposes, provided that the original author and source are credited.

Published by Cold Spring Harbor Laboratory Press; doi: 10.1101/sqb.2017.82.034793
Cold Spring Harbor Symposia on Quantitative Biology, Volume LXXXII

portant structural feature. It turns out that nuclear pores can actually form without those very long-lived components, so there must be another reason.

The best idea we have so far is for long-lived histones. We developed a system where you can distinguish between young and old protein in a differentiated cell in vitro. This trick allowed us to perform chromatin immunoprecipitation and genome sequencing. We found that long-lived histones are associated with repressed genes. One hypothesis that we are currently pursuing is that maybe the reason why very long-lived cells use long-lived histones is to reduce transcriptional noise. Whenever a nucleosome assembles, the cell risks that transcription factors might bind to the site and trigger transcription. It is well described across many species that, during aging, transcription noise increases. But a neuron in our brain has to stay in the same state. It always wants to be a neuron and perform its genetic program over decades. Maybe those cells use very, very long-lived proteins to very robustly repress genes.

Kim Baumann: You're saying that it's a way of stabilizing gene expression or regulating cell identity so that there is more stability.

Dr. Hetzer: Of course, we have no tests for this. One way to test this would be to force the turnover of those proteins and then we would expect transcription noise to increase during aging. That's one hypothesis. The other hypothesis that we are pursuing is linked to a phenomenon that was actually developed by French entomologists. It's something that I only recently learned and it's a phenomenon called "stigmergy." Most people I talk to have never heard of it. I'd never heard of it.

The principle is very interesting. There was a French entomologist who studied self-assembly in nature. He studied ants and termites and realized that insects, despite the fact that they have no intelligence, can produce a highly organized assembly. Termites can build buildings, but they also can self-assemble around a food source and different paths. He found that ants leave a pheromone trail to a food source. The next ant comes and leaves a stronger trail. They use this "stigmatus." So it's a "stigmergy" for self-assembly. One hypothesis that we are pursing is maybe the longevity of these proteins is part of that stigma: a guiding principle for the large protein assemblies. Even within a tissue, it might provide a principle—"the Stigmergy Principle"—of tissue assembly. You might have a subset of very long-lived proteins, or subcellular structures, or cells, to organize complex biological systems.

Kim Baumann: Is this a way of stabilizing structures?

Dr. Hetzer: It's more than stabilizing. It's really organizing. One outcome of our studies is that vasculature—capillary endothelial cells—are very, very old, even in organs such as the liver where hepatocytes turn over. So, maybe these long-lived cells are used to assemble and coordinate tissue separation within a tissue at the molecular level. Maybe these very stable nuclear pore proteins in the center are actually the organizing principle for the other components. That's possible, because the peripheral nuclear pore proteins actually do turn over and their residence time at the nuclear pore can be measured in seconds or minutes. You have six orders of magnitude difference in the time for which components are present in the same complex. That's unheard of. So, maybe you need that as sort of an organizing principle.

We have other ideas that are very speculative. For instance, in collagen and crystallin, people have identified chemical processes such as deamidation. Those are chemical reactions that can occur without any external factors and they cause changes in the amino acids such that even protein structure can change. Both occur naturally in long-lived proteins. There are enzymes that can reverse that process, so perhaps there are processing places that allow proteins to change function over time. Perhaps even protein repair mechanisms are in place to allow those proteins to function over long periods of time. We're working on this, but we don't have a definitive answer.

Kim Baumann: Are these proteins subject to less damage so that they can be kept in a functional state longer? Is there any correlation between how much they could be exposed to damage and therefore they would be long-lived?

Dr. Hetzer: That's a very good question. The extracellular proteins and the β-crystallin are really protected from cellular components, but these other long-lived proteins are fully exposed to potentially harmful metabolites. We do see signs of oxidative damage of those proteins. The problem is that we see the same level of damage in young animals and old animals, so we don't think that the functional decline is associated with damage. A still unexplained phenomenon is that we do see the specific loss of some of those long-lived proteins during aging and we don't know why.

Kim Baumann: Loss of the whole protein?

Dr. Hetzer: In the nuclear pore, for instance, it's specifically these very long-lived subcomplexes that are lost both in old neurons, and also in old long-lived pancreas cells. Maybe they're damaged and thus they are lost, or there is an age-dependent decline and a very, very slow replacement process. That's because they are maintained in a way that each individual component is slowly removed. We're talking about replacement rates for some of those, like, once in a lifetime for a mouse: very, very slow replacement. Since we know that protein homeostasis generally declines during aging, perhaps that's a failure to replace those components, but the jury's still out.

Kim Baumann: During aging, the machinery that ensures that there is a healthy protein declines. There is misfolding, or there is accumulation, or aggregation of proteins. Are any of these proteins involved in anything like that, given that they are possibly accumulating damage over time? Do they aggregate? Are they removed in the same ways that other proteins are removed?

Dr. Hetzer: We don't know the fate of those. We know that they disappear. Whether they aggregate somewhere in a different cellular location, that's possible. We do see, as have many other people before us, an increase in aggregate formation in old cells, but whether they are accumulating in various aggregates or they are degraded once they are removed from the complex is an open question.

There is one link that is very interesting. One of the main functions of the nuclear pores is to form a barrier between the nucleoplasm and the cytoplasm, to only allow proteins and other molecules to pass that have a specific signal. In old cells, we found that other components can actually leak into the nucleus in an uncontrolled manner. One of those proteins is tubulin B3, a cytoplasmic protein; it should only be in the cytoplasm. We do see it inside the nuclei in old cells. In many cases, it forms very large aggregates, so they're not microtubules. They are large paracrystalline structures. We're currently analyzing the composition of those. Whether some of those long-lived proteins might actually be participating or enabling some of those aggregates is an open question.

Kim Baumann: What do you think is the next big question regarding these proteins? Are they still being identi-

fied? Do we know how many of them there are, and in which places?

Dr. Hetzer: We are looking in different organs. We have comprehensively only analyzed the liver and the brain, but we are now analyzing the heart and we have identified a few. The really big question is: Are they potentially driving the aging process? Are they really important, or are they just like many other things in the cell that also start to fail. To test that, we face a very difficult problem. You can't use RNAi or any other method to deplete those proteins because, obviously, they are long-lived. We have to develop—which we are doing now—methods to extract those proteins from the complexes and trigger their turnover. When we can do this, we can go back into an animal and force the turnover and then ask if there is any consequence, either on longevity or survival of neurons. That is the biggest challenge. We're really tied into a causation potential between this longevity and age-dependent decline. That's the next big work. Another area that might be tied into this is protein repair mechanisms. These are very poorly understood. There might even be new biology involving proteins that might repair those long-lived proteins.

A Conversation with Ian Hickson

INTERVIEWER: BETH MOOREFIELD

Senior Editor, Nature Structural and Molecular Biology

Ian Hickson is a Professor in the Department of Cellular and Molecular Medicine and the Director of the Center for Chromosome Stability at the University of Copenhagen.

Beth Moorefield: Your group works on a mammalian translocase called PICH [Plk1-interacting checkpoint helicase], which is required to separate sister chromatids at mitotic anaphase, and the physical and functional properties of PICH that makes it uniquely suited to this purpose. I thought we could talk about what DNA structures PICH targets.

Dr. Hickson: PICH recognizes an unusual structure in mitosis called the "ultrafine anaphase bridge." When the separating sister masses of DNA come apart, they sometimes leave behind a thread of DNA that has no histones on it. It also doesn't stain with DNA dyes. The only way you can see it is by the proteins that are bound to the structure. That's why they're called "ultrafine bridges," because you can't see them without staining for the proteins. The first protein that was identified that recognizes these structures is PICH. It, to some extent, defines the structure.

Beth Moorefield: But it's not the only protein that localizes to it.

Dr. Hickson: No, it isn't. There're a lot of proteins, mainly DNA repair proteins. The main complex that binds is called the "Bloom syndrome complex," named after proteins defective in a cancer predisposition disorder called Bloom syndrome. That's the BLM protein. It has two partner proteins. One is a topoisomerase, called Topo III, and the other is a protein called RIF1 [Replication Timing Regulatory Factor 1]. They bind to these ultrafine bridges. However, the binding of BLM and Topo III and RIF1 to these bridges is entirely dependent upon PICH. PICH controls the recruitment of all the other factors to the bridge DNA.

Beth Moorefield: So, it's not actually actively disentangling the DNA?

Dr. Hickson: We don't think so. We think the topoisomerase III is actually the key enzyme for disentangling the DNA. Topoisomerases are enzymes that can unknot and disentangle catenated or entangled DNA.

Beth Moorefield: What are the physical properties of the DNA that PICH recognizes?

Dr. Hickson: We can only speculate about that. We think the DNA is conventional double-stranded DNA but at its center is entangled or catenated DNA, such that the DNA cannot come apart, and what you have to do is you have to disentangle that to allow the DNA masses to come apart. We don't think it recognizes the entanglement. It decorates the bridge all the way along its length. The question is, how could it possibly do that? How does it know where to be on the bridge?

We think there's a reason for that. We collaborate with a group in the Netherlands headed by Gijs Wuite and Erwin Peterman. They use a technique called "optical tweezers," which is a method by which you can manipulate DNA molecules using laser light in a flow cell and you can apply force to the DNA. They showed that if you stretch DNA the PICH binds to it much more efficiently than if you don't stretch the DNA. In mitosis, the DNA masses are being pulled apart and the bridge is in between. We believe the bridge is under tension; the bridge is being stretched. PICH actually recognizes the physical stretching of the DNA. We see it as a form of tension sensor. That's a little exaggerated, but that's how we see it.

The PICH has the fundamental ability to find a piece of stretched DNA in mitosis, and only in mitosis. It's an enzyme that's actually in the cytoplasm in interphase, so it only gets access to the DNA when the nuclear envelope breaks down in mitosis. It's a purely mitotic factor that's sitting outside the nucleus waiting to rush in onto the DNA.

Beth Moorefield: So, by actually using these stretched molecules, you can mimic the cellular system?

Dr. Hickson: We're doing that now. We replace the sister masses of DNA with polystyrene beads, and then we stretch a piece of DNA in between them.

Beth Moorefield: Once PICH is actually bound to the DNA, what is it actually doing to it?

Dr. Hickson: It's a DNA translocase, which means that it runs along the DNA duplex without separating the two strands. It's not a DNA helicase, which separates the two strands. It runs along the DNA, which can change the

---

© 2017 Hickson. This article is distributed under the terms of the Creative Commons Attribution-NonCommercial License, which permits reuse and redistribution, except for commercial purposes, provided that the original author and source are credited.

Published by Cold Spring Harbor Laboratory Press; doi: 10.1101/sqb.2017.82.034777
Cold Spring Harbor Symposia on Quantitative Biology, Volume LXXXII

397

structure of the DNA. We think that its main job is to recruit all the other factors, including the Bloom's complex, that then do the real business. Without protein–protein interactions, they don't get recruited.

We've reproduced that using the optical tweezer setup, where we've asked how does the Bloom protein recognize the DNA, and how do Topo III and RIF1 recognize the DNA. The thing is, they sit on a piece of DNA that they cannot recognize themselves. If you give them double-stranded DNA in vitro, they cannot bind to that. PICH binds to double-stranded DNA; but they bind to single-stranded DNA. If you coat the DNA with PICH, they now bind to the PICH-coated DNA. PICH allows the complex to bind to a piece of DNA it would never bind to normally. We don't think it manipulates the DNA structure, because if you make a catalytically dead PICH, it still recruits the factors to these bridges. It's the physical presence of PICH that then attracts all the other proteins onto there, and then they find the abnormality, the entanglement, and they disentangle it. That's our working model.

Beth Moorefield: So, it's not actually activating those factors, but rather it's working as an adapter to recruit them to the sites where they're necessary?

Dr. Hickson: We think so, although it is an enzyme, so it must be doing something to the DNA. We haven't actually worked out what that something is. It is possible, and we do have some evidence for it, that it changes DNA supercoiling. It has the ability to alter DNA supercoiling in combination with the topoisomerase. It's fairly preliminary at this stage, but we think it might be manipulating the DNA structure through tracking along the helix and imposing supercoiling on the DNA. We don't understand why that would help the situation, but a speculation would be that if the DNA is supercoiled instead of being relaxed, it actually drives decatenation more efficiently. That's been shown for other topoisomerases.

Beth Moorefield: Can you demonstrate this by using decatenation assays?

Dr. Hickson: We are doing that currently.

Beth Moorefield: Can these individual factors form complexes off the DNA?

Dr. Hickson: Yes, they can. We demonstrated that they form protein interactions. PICH bind directly to both BLM and to topoisomerase III, and it does so through conserved domains. We're fairly confident it has the ability to recognize the two proteins in solution, but they never see each other outside of mitosis because PICH is in cytoplasm and the others are in the nucleus.

Beth Moorefield: Are there specific sites of recruitment for PICH on the ultrafine bridges? Do you know how it's spreading to cover the whole bridge?

Dr. Hickson: It seems as if it gets recruited almost instantaneously to the entire length of the bridge. As soon as the bridge forms you can see PICH along the length, and however long the bridge gets, it just coats the entire thing.

We are trying to test that in vitro with these optical tweezers, but it doesn't seem like it nucleates and grows. It seems like when the tension reaches a point that's suitable, it then just coats the DNA very quickly. One possibility is that it forms a filament on the DNA and so you end up with a cooperative loading of a large amount of protein in a very short period of time forming some kind of ultrastructure on the DNA. That will require us to do some more detailed structural analysis.

Beth Moorefield: In that regard, how does it compare with other DNA binding proteins involved in the DNA recombination?

Dr. Hickson: Completely differently. They generally recognize in almost equimolar amounts a particular type of substrate, whereas there'll be tens of thousands of molecules of PICH on a long ultrafine bridge. The question is, why do you need tens of thousands? And you probably don't, but for some reason it may be that that's the only way that you could recognize the DNA quickly in anaphase, which happens very rapidly. You have to find the DNA; you have to decatenate it. The only way to do that is to basically cover the DNA and then get the Bloom and Topo III proteins covering the DNA as well, and then wherever the problem is you can deal with it. It can't be done by some slow tracking process, because anaphase would have finished by then; it'd be disastrous for the cell. We see it as a mechanism by which you waste a lot of protein on the DNA in order to get to the place you need to be as fast as you possibly can. That may be naïve, but that's how I see it.

Beth Moorefield: And it doesn't have to contend with nucleosomes; it would have to displace them?

Dr. Hickson: There was a hypothesis that PICH actively removes the nucleosomes, but we don't think that this is the case. The catalytically dead protein recognizes naked DNA with no nucleosomes, so it can't have removed the nucleosomes from that. The DNA has no nucleosomes on it, and that's probably because it's stretched. If you pull on nucleosomal DNA, the nucleosomes pop off the DNA simply because of pulling forces.

Beth Moorefield: What other properties of DNA have you been able to elucidate using the optical tweezer approach?

Dr. Hickson: You can measure a whole series of parameters to do with DNA structure, but it's not something that we've pursued in detail. That's a major interest of the lab we collaborate with, but we don't really analyze issues to do with how much you can stretch the DNA or what happens to it.

For example, perhaps counter-intuitively, if you stretch a piece of DNA too much, it actually starts to denature. The strands come apart. You can actually make bubbles in the DNA by pulling very hard on it. Of course, if there are any nicks in the DNA then the DNA strand will unravel, and the DNA will become single-stranded. Eventually, you can basically denature a piece of double-stranded

DNA by pulling hard on it and eventually one strand will unravel from the other. That's what happens with extreme forces.

One of the problems that we have is that we don't really know how strong the spindle forces are in a cell. They've been estimated, but the estimates vary enormously. We don't know whether we're in the right ballpark with the optical tweezers. They are limited by how hard you could pull on a piece of DNA. So above about sixty piconewtons, it's really difficult to do anything. You can go down to about 0.5 piconewtons in the setup. That's the range that we have. If the mitotic spindle isn't in that range then we're not necessary studying the right phenomenon. How one would test accurately the strength of the mitotic spindle forces, I don't know. There've been estimates that vary by a hundredfold as to how strong it is.

Beth Moorefield: How is PICH regulated?

Dr. Hickson: PICH is regulated very strongly by a kinase called PLK1, polo-like kinase. It's phosphorylated by polo kinase, and polo kinase is required for its correct localization. It also localizes to the centromeres before anaphase.

Actually, PICH seems to regulate the localization of polo-like kinase as well, which is curious, given how important polo is for mitosis. You can inactivate PICH in a human cell and the cells are still alive, but polo-like kinase seems not to be regulated properly and not to localize quite properly, so it's remarkable the cells are still alive. But that seems to be the key player in its regulation: PLK1.

Beth Moorefield: It's not subject to additional regulation by association with any of the other factors? It only works with ultrafine bridges?

Dr. Hickson: Not that we're aware of. The other factors on the bridge tend to be enzymes. They tend not to be modifying proteins. The only one that's a candidate for doing something interesting would be this protein called RIF1, because RIF1 is a targeting subunit for protein phosphatase 1. Actually, RIF1 might be part of the mechanism by which dephosphorylation events occur during mitosis around the bridge. We're studying that in some detail now. That's the only real candidate for a regulator. We've no evidence that RIF1 regulates PICH itself, but PICH does recruit it to ultrafine bridges.

A Conversation with Maria Jasin

Interviewer: Lara Szewczak

Scientific Editor, Cell

Maria Jasin is a Member of the Developmental Biology Program and Gerstner Sloan Kettering Graduate School of Biomedical Sciences at Memorial Sloan Kettering Cancer Center in New York and Professor at the Weill Cornell Graduate School of Medical Sciences.

Lara Szewczak: Let's start off with the main focus of your lab.

Dr. Jasin: We study homologous recombination, more for its genetic requirement rather than its biochemistry. Homologous recombination is a key DNA repair pathway in mammalian cells. We study it in cell culture and in mice, but also in germ cell development, because it's essential for meiosis. Homologous recombination is critical in a broad range of systems: DNA repair, cancer suppression, mouse development, and meiosis. Each area is very exciting!

In the cell culture work that we've been doing, critical proteins in the recombination process are tumor suppressors: in fact, the breast cancer tumor suppressors—primarily BRCA1 and BRCA2. Both proteins are also ovarian cancer suppressors, but mutations in these genes have been found more recently to be important in other cancers, somatic mutations as well as germ line mutations, which can affect the development of a tumor and/or its treatment.

We want to know what's so important about BRCA2 in cellular processes, especially in tumor suppression. Our model is a relatively normal breast cell line: a nontransformed cell line called MCF10A. Luckily, we now have the tools to make conditional knockouts fairly easily in mammalian cells, especially in human cells that were more difficult to work with in the past. We made conditional alleles and showed that BRCA2 is absolutely essential for viability. Now we want to find out why, and how tumor cells get around the fact that this is an essential protein.

Lara Szewczak: There's actually a fair bit known about BRCA2. How is what you're finding different from how we conventionally think about that protein?

Dr. Jasin: There's been a lot of work in the area of homologous recombination and the biochemical roles of BRCA2 and other proteins. BRCA2 is very critical for loading the RAD51 protein, the main strand exchange protein in the cell, at damage sites. A lot of people study it, if not biochemically, then in cells by looking at standard repair assays, some of which we developed to look at double-strand break repair. We also discovered that it has an important role in protecting stalled replication forks. A critical thought for the whole field is that homologous recombination is critical for every replication, every S phase, to repair spontaneous DNA damage that comes up, so we wanted to really understand as best we could what's being disrupted.

We had focused on these two pathways for a while, considering homologous recombination per se as quite a critical pathway. But we also discovered another pathway of replication fork protection that, in fact, does not require the repair role of BRCA2. Moreover, this fork protection pathway recently was proposed to be the reason why BRCA2 is essential in mouse cells. Yet, when we started looking at separation-of-function alleles in these human breast cancer cells, we really do find that homologous recombination, and not protection of nascent strands, is really critical for the survival of those cells. We don't think it's necessarily for repair of the kind of standard two-ended DNA break, but from replication structures that in the end will need homologous recombination for repair.

Lara Szewczak: Why do you think there's a difference between human and mouse?

Dr. Jasin: The mouse system was an ES cell system, and mouse ES cells—embryonic stem cells—have different responses to DNA damage. They're primarily in S phase, and have very short gap phases. An interesting thing for us was to find when we knocked out BRCA2, the cells did not arrest in S phase, but they actually continued to the next G_1 and accumulated in G_1. That solves a major issue, which is that often p53 mutations are associated with BRCA1 and BRCA2 tumors: If the cells are accumulating in G_1, then an easy way to get past that is to have a p53 mutation. It's very surprising that S phase damage can continue on through G_2, through M phase, and even into the following G_1 and create problems for the cells.

Lara Szewczak: If there's S phase damage that persists, do you see aberrancies in mitosis?

Published by Cold Spring Harbor Laboratory Press; doi: 10.1101/sqb.2017.82.035345

Cold Spring Harbor Symposia on Quantitative Biology, Volume LXXXII

Dr. Jasin: Yes. A lot of the very typical mitotic aberrations: bridges, lagging chromosomes, micronuclei. All those classic things, and then interestingly, an attempt to complete DNA replication in mitosis. We see mitotic DNA synthesis in the BRCA2 mutant cells. We think that may be an attempt to deal with aberrant structures that make it into mitosis, but that it's not sufficient. It may help alleviate some of the damage, but it can't deal with all of the remaining damage left over from S phase.

In G_2, we see markers of DNA damage, specifically single-stranded DNA. We see metaphase chromosomes with replication foci, and also FANCD2 foci, which are associated with replication damage, and then the mitotic structures lead to 53BP1 bodies in the next G_1: 53BP1 bodies are huge conglomerations of repair proteins, which is different from smaller 53BP1 foci. Again, this could be an attempt to repair this persistent damage, but it's insufficient, maybe because the cells can't do recombination at G_1. They can't do recombination in the previous S phase, and they can't do it in the subsequent S phase, so things just fall apart.

Lara Szewczak: Now that you're thinking about two different ways that BRCA2 can operate, what's the next step for you?

Dr. Jasin: We're still interested in the particular issues. We don't think the issues are necessarily at common fragile sites, or centromeres. Are there particular sequences at risk with loss of BRCA2? Do those have any common features, and why is recombination particularly necessary for those? That's a particular interest.

p53 loss will actually rescue *BRCA2* mutant cells, but the colonies are tiny. What else do cells need for a full rescue? In a breast, for example, when you have mutant cells that've lost heterozygosity—and so they have now two *BRCA2* mutant alleles—what are those first steps that may occur that allow those cells to survive and go on to form a tumor?

Lara Szewczak: What are your early indications on that front? You said you do work not just in tissue culture, but also with specific human samples.

Dr. Jasin: One of the projects we'd like to develop involves determining the actual rates of loss of heterozygosity (loss of the wild-type *BRCA2* allele), and does a breast differ in that compared to other tissues? We've developed an in vivo tissue recombination assay, and at least in our initial assay the mammary gland in mice doesn't seem to have any preferential use of BRCA1 or BRCA2 in recombination, but the mammary gland seems to use recombination at a higher level. Is that because there is generally more damage, so mammary tissue has amplified recombination? Are these bursts of replication? Does that cause a different dynamic in those cells compared to other cells?

Lara Szewczak: How does that assay work?

Dr. Jasin: We introduce a break into a GFP reporter gene and if the cells undergo recombination they become GFP$^+$. We can very nicely see basal cells and luminal cells in the mammary tissue architecture becoming GFP positive, and we can quantify that. It's a nice in vivo assay. It showed us how robust recombination can be even in the background of a proficient, nonhomologous end-joining pathway. We think that will be useful for parsing out some of the questions.

Lara Szewczak: Looking at recombination in mammalian cells, was it obvious that it was going to be as abundant and vital?

Dr. Jasin: Actually, when I set up my lab I told everybody I was going to work on meiotic recombination and that was because, coming from a background of doing gene targeting, we didn't think that homologous recombination was very important in mammalian somatic cells for DNA repair. However, soon after starting my lab, we introduced a break into the genome of cultured cells and asked how it's repaired: the basis now for CRISPR–Cas9. We saw nonhomologous end-joining events, which were mutagenic, but then we also saw, quite robustly, homologous recombination events. That actually altered my focus at the very beginning of my career to studying homologous recombination in cell culture, at a time when this field had no protein players known to be involved in the process. Very little was known overall. People only would study very rare spontaneous events, but when we introduced damage in the form of a break in the chromosome we had a very robust assay. Luckily, a few years later there were connections being made to the recently discovered breast cancer suppressors, and genomic analyses led to the identification of mammalian homologs of yeast recombination genes. The field has been incredibly interesting and stimulating ever since. I feel lucky that even though I started as a young PI in a field that was so small and considered obscure to some, and the field has really blossomed: lots of colleagues coming from different angles, and lots of implications for human health, and interesting mechanisms.

Lara Szewczak: When you have trainees leaving your lab, do you advise them to try and find a similar situation?

Dr. Jasin: Ideally, yes, but post-doctoral fellows invest a lot of time in their own projects, and typically they try to find a niche from that to focus on in their lab. In a way, it was quite risky when I started my career to try an area that was so undeveloped. NIH funding was very poor at that time, with no preference for new investigators, and application for HHMI awards was restricted, unlike today. But, luckily, Sloan Kettering had flexible funding at that time to allow me to develop risky projects. But I do think it's a little scarier these days, even more so because of the sustained funding limitations at NIH, to try to develop something completely different. Nonetheless, I think it's particularly rewarding to do so and it certainly can lead to a very impactful scientific career.

Lara Szewczak: Let me switch gears a bit. CRISPR is changing a lot of areas of science, but I would think for this field, in particular, it would have a big impact.

Dr. Jasin: People have not used CRISPR in a huge way in terms of actually asking questions about DNA repair, but I think that's changing. We started out in the '90 s using the I-SceI endonuclease, which is a single-chain, 200-some amino acid nuclease. I-SceI is very specific, so that's been the workhorse, and we can use it or Cas9 interchangeably in our GFP assays to introduce breaks. We've also used Cas9 to generate defined chromosomal rearrangements to study repair mechanisms.

The main power of CRISPR, compared with TALENs, is the genome-wide applications, and attempts are being made to do genome-wide DNA repair assays to understand repair more globally. I think that kind of thing is going to develop quickly. Of course, everybody in the DNA repair field is using it for all of the standard approaches, in particular to knockout repair proteins or make conditional alleles, just like everybody else in every other field.

Lara Szewczak: If you were thinking three to five years down the road, what question do you want to be able to answer, regardless of whether or not the technology exists now?

Dr. Jasin: If in 3 to 5 years I had an answer for why some tissue types are so prone to tumorigenesis when homologous recombination is disrupted, I'd be thrilled for finally understanding that conundrum. That's a major question for me.

I also love meiotic recombination, which we work on in collaboration with my colleague Scott Keeney. During meiosis hundreds of double-strand breaks are introduced in the genome, a really huge amount of genome-wide damage. That's another aspect that I'm hoping in the next three to five years we'll have some new insights.

Lara Szewczak: The work that you do in meiosis, and the massive number of breaks: What's the thing in the lab that you're most excited about right now?

Dr. Jasin: We have nucleotide resolution maps of where double-strand breaks occur, and we're learning a lot about the proteins that are involved with that. We have done assays at specific sites of break formation to understand how they're repaired, and the importance of crossovers so the homologs can stay together, but also the importance of

noncrossovers so the homologs can get together to begin with. There are a lot of proteins that are involved in the process that are problematic in patients. ATM is a major one that leads to infertility in males and females.

Lara Szewczak: I didn't know that.

Dr. Jasin: Patients have neurological issues well before their reproductive age, so the germ cell issues are often overlooked. ATM is an incredibly interesting protein in meiosis because of its unique role in regulating the number of double-strand breaks. There's about tenfold more double-strand breaks in an ATM mutant compared to normal. Where do all those breaks go? What is happening? Are there issues even in heterozygous individuals? That's something that we like to know.

It's also not just the number of breaks; it's their spatial distribution. We think ATM is probably in part acting like it normally does in mitosis: responding to breaks and signaling. But in this case, it's to provide feedback to regulate the number of breaks, but also probably the spatial positioning. Is it really the case that once you get a break on one chromatid it's going to suppress breaks on the other chromatids and nearby on the same chromosome? Is there some redistribution of breaks that is going to cause genome rearrangements? I'd really like to be able to explore these fine maps and our knowledge about break formation to understand what happens in some of these pathological situations.

Lara Szewczak: If you were starting over brand new: You don't have a lab, you can ask any question you want. Somebody is handing you money to do it, but you can't work on recombination. What do you do?

Dr. Jasin: At heart, there's the biologist in me, and one of the things I really love about CRISPR is that you can so readily go to other organisms—nonmodel organisms—and make them models, to understand their specific biology. That would be great fun for me. I would love to take a year off and explore all the options that are out there. It doesn't quite fit the cancer biology emphasis of Sloan Kettering, but I think that would be a fun way to develop a new aspect in my career.

A Conversation with Leonid Mirny

INTERVIEWER: BETH MOOREFIELD

Senior Editor, Nature Structural and Molecular Biology

Leonid Mirny is a Professor at the Institute for Medical Engineering and Science and the Department of Physics at the Massachusetts Institute of Technology.

Beth Moorefield: Your group has been elucidating the 3D organization of the genome using a combination of biophysical approaches to determine the dynamic properties of chromatin fibers that govern their folding and interactions, as well as Hi-C approaches to probe their structures and organization within the cell. Maybe we could talk about how the polymer models are actually helping to inform genome structure.

Dr. Mirny: My thinking about modeling in biology in general, and the role of modeling in chromosome biology, begins with realizing that models can test mechanistic hypothesis. If an experimental paper ends with a "Figure 4" and a hypothesis of how things work, a modeling paper starts with this hypothesis in Figure 1. The hypothesis of how it may happen is the starting point for our modeling; and experimental data are used to select and to validate the models. In our works on Chromosome Biology, we're not really building models of chromosomes based on Hi-C, instead we're using Hi-C and other experimental data to test and select models that can best reproduce these experimental data.

The starting point is usually a hypothesis—maybe something that has been published, or maybe our own hypothesis of how we're thinking things may work—and then we test this hypothesis by putting suggested mechanism in the simulation box. For example, polymers should be of this length and have this type of interaction: Can this actually give rise to something that's observed experimentally? That's the overall framework of testing many possible models and finding those that can reproduce experimental observations. This follows my favorite Sherlock Holmes saying: "Once you eliminate the impossible, whatever remains, no matter how improbable, must be the truth."

Beth Moorefield: So basically, you're defining the parameters that then limit or constrict their movements or interactions?

Dr. Mirny: For example, parameters that control how different elements of the genome interact with each other or with DNA-bound proteins (e.g., cohesin or CTCF) or parameters of the chromatin fiber: diameter of the fiber, its flexibility, density of different chromatin fibers in the volume. If we cannot simulate a whole nucleus, we simulate a small volume where a few chromosomes or chromosomal regions interact. These are physical parameters. Plus, we add some biological interactions. We can say, "What if euchromatin is attracted to other euchromatic regions? Or what if heterochromatin is attracted to other heterochromatic regions. What's going to happen?" We run simulations forward and we see what happens.

For example, if we want to compare simulations with Hi-C data, we compute frequencies of interactions between different regions. This gives us a simulated Hi-C map that we can then compare to the real Hi-C map. We usually sweep a broad range of parameters of the model and find a range of parameters where it agrees with Hi-C data. If we see that it does not agree, we'll say, "Okay, we need to go back and revisit the assumptions of the model. Maybe it's actually heterochromatin that needs to interact with heterochromatin," or "We need to add some other interactions, for example interaction of heterochromatin with the nuclear lamina." That's our general approach.

Beth Moorefield: That approach actually has been revealing different levels of organization, but then also this reciprocal influence of chromatin fiber activity and structure.

Dr. Mirny: One of our main hypotheses now is that beyond pairwise interactions, active processes (i.e., ATP-consuming motors) are shaping genome structure. For example, loop extrusion is a process where the action of a molecular motor is shaping chromatin. Loop extrusion leads to formation of chromosomal domains, it can also facilitate and moderate enhancer–promoter interactions. But this requires a motor that can extrude loops, a motor that nobody has actually observed. That's why it's still a hypothesis. We believe that cohesin and other SMC [structural maintenance of chromosomes] proteins perform this particular mechanical function of being molecular motors that can extrude loops. We were lucky to establish collaborations with several experimental labs that started testing this model and the general biological roles of different components of the loop extrusion machinery, as well as other proteins that play a role in shaping chromatin architecture.

Beth Moorefield: Are these two parameters actually universal features of the organization of genomes, regardless of cell type or species?

Published by Cold Spring Harbor Laboratory Press; doi: 10.1101/sqb.2017.82.034850

Dr. Mirny: While specific parameters may differ, the process of loop extrusion can be rather universal. We see signatures of loop extrusion in eukaryotes as well as in bacteria. The most direct evidence of the loop extrusion mechanism actually comes from studies of bacteria: the work of David Rudner's group in *Bacillus*, and also works of Mike Laub's group in *Caulobacter*, who demonstrated that loop extrusion by SMC is essential for juxtaposition of chromosomal arms in bacteria and measured the speed of extrusion. We're focusing on mammalian systems and looking for evidence of similar mechanisms in other organisms. We and our collaborators have found that cohesin (another SMC protein) is needed for formation of domains in mouse cells, requiring one loop-extruding enzyme every few hundreds of kilobases of DNA, while, in *S. cerevisiae*, we and Jon Baxter's group found very low levels of loop extrusion. So, that's as far as we can go. As theoreticians we can say what's needed and what's not needed to reproduce the data. Occam's Razor logic would tell us that if you don't need it, it doesn't exist. But it's biology, so mechanisms can be very complicated... (shrugs)

Beth Moorefield: The size of the genome is going to affect how it's going to be packaged and then accessed. Do you find that there is strong relationship with size?

Dr. Mirny: The sizes of the genome and the nuclus certainly play major roles in shaping chromosomal organization. We and others have seen that in *S. cerevisiae* you can explain imaging and Hi-C data by just assuming that chromosomal arms are freely jointed chains. They are attached to the spindle pole body, but otherwise they're just free polymers, random walk polymers, and are sufficiently short and yeast nuclear volume is sufficiently large to accommodate uncompacted chromosomes.

At the same time, in mammal genomes, chromosomes are incredibly long and nuclear volumes are not enough to allow them to be open polymers, so chromosomes need to be compacted even during interphase, and this is certainly reflected in Hi-C data and microscopy. However, we noticed that in cells with larger nuclei, for example, mouse oocytes, chromosomes are much more open, almost like chromosomes in *S. cerevisiae*. We made this observation in our collaboration with the group of Kikuë Tachibana-Konwalski in Vienna that did truly amazing single nuclei Hi-C. One of the interesting insights from the physics point of view is that when nuclei become much larger, chromosomes change their shape and organization. But they still have loop extrusion as evident from the presence of topological-associated domains there. So, mouse chromosomes even in very large nucleis are not really like yeast chromosomes.

Beth Moorefield: To always test the predictions of the physical models requires a lot of matching to the Hi-C data itself, which requires a lot of communication. What has been something that's actually helped to advance that kind of communication between the theoretical and the practical sides of these investigations?

Dr. Mirny: What really helped us is that we're not only developing models, but we're actually analyzing Hi-C data

from point zero, essentially from FASTQ files that come out of the sequencing machine. We decided to invest a lot of efforts in developing tools for processing and analysis of Hi-C data, for correcting various biases in Hi-C data— and this way we actually get first-hand experience with what's in the data—and also make sure that we're analyzing real signal rather than various artifacts that may have crawled into the data during processing. That also helps our communication with experimental colleagues, helps to develop a common language, common understanding. We constantly exchange ideas, brainstorm together. Again, our approach starts with a hypothesis. That's something where we certainly need lots of input from our colleagues on the biology side, who are closer to a specific biological system. From these interactions with our colleagues we also learned not to overgeneralize our findings from biophysical modeling; we try to avoid saying: "All chromosomes are like this, or all chromosomes are like that," because we understand that in different biological systems, the rules of the game can be very different. Nevertheless, we aim at finding sufficiently general physical principles and mechanisms that govern chromosome organization. And if we are lucky such mechanisms can be sufficiently universal; operate in different cell types, different organisms, and have multiple biological functions. We believe that loop extrusion is one of them. I'm very grateful to all the people who worked with us, particularly to Job Dekker, as well as Kikue Tachibana, Francois Spitz, Irina Solovei, and Bill Earnshaw, and the many others who were patient with us, helping and guiding us in developing models, analyzing and processing data, helping us develop intuition about chromosome organization in different biological systems.

Beth Moorefield: What would you say are some of the biggest advances to driving this field forward, in terms of working on the 3D genome organization?

Dr. Mirny: It may surprise you, but the biggest advance in this field, in my opinion, has become the progress in using archives for sharing papers prior to publication. *bioRxiv* and other archives have really accelerated the pace at which knowledge is disseminated. And now, practically everything appears on *bioRxiv*. I know from our own experience that it takes us a year or more to publish a paper. It's really unfortunate when you go to a conference and somebody tells you about some exciting findings, but it takes a year till you can read the paper and examine the figures when the paper is published. This has changed dramatically in the last year. Now, that practically everything goes into *bioRxiv* prior to publication we, as a field, are a year—if not more than a year—ahead of other fields that do not use *bioRxiv*. Other fields of biology may be behind, while physics and mathematics, have been using archives for decades.

I think it also changes the level of interactions between people. The field might become less competitive, in the negative connotation of competition. People go, "Okay, we're going to post this on *bioRxiv*, everybody's going to see it, we're going to get some feedback." I think it also improves the quality of publications because we get additional feedback from colleagues who can suggest something, who can be critical, or praise your work. You really get a sense of where your work stands early on in the publication process. In the long run, I think it will accelerate other fields of biology because they'll have to catch up. If they depend on chromosome biology, they wouldn't want to be a year behind, so they will also start posting on *bioRxiv*.

Beth Moorefield: It gives another voice whose expertise is invaluable to critically evaluate the data, really.

Dr. Mirny: Exactly! But it's a bit more complicated when it comes to sharing data. At this point, *bioRxiv* only con-

tains publications; people usually keep their data private until the date of the publication. In principle, there's no reason why everybody shouldn't release their data right away when they post a *bioRxiv* paper, but people don't want to be the first to do this. That's what happened with *bioRxiv* in the past: Initially, people were reluctant to post preprints, because they worried that this will give unfair advantage to people who don't post on *bioRxiv*. But if almost everybody's posting, you don't want to hold on to your paper and *not* post it; you want to catch up with everybody else. There is a critical transition, and I hope with data release the same will happen within the next year or so.

A Conversation with Timothy Mitchison

INTERVIEWER: RICHARD SEVER

Assistant Director, Cold Spring Harbor Laboratory Press

Timothy Mitchison is the Hasib Sabbagh Professor of Systems Biology, co-Director of the Initiative in Systems Pharmacology, and Deputy Chair for the Department of Systems Biology at Harvard Medical School.

Richard Sever: Can you just remind everyone the arrangement of microtubules in the cell? In a typical animal cell, you have a radial array in interphase, and it changes during mitosis. What's the arrangement?

Dr. Mitchison: Microtubules are these long, thin protein polymers. We use the word "cytoskeleton" and they're sort of like the bones of the cell, although they're very dynamic. They grow and shrink, and they're transport tracks for material. Let's say a white blood cell is crawling around in your body: the microtubules radiate out from a central point called the centrosome and transport material and help the cell stay organized as it moves.

When a cell divides, the DNA compacts into individual chromosomes and the microtubules reorganize to build a structure called a mitotic spindle, which sort of looks spindle-shaped. The chromosomes attach to microtubules, and the microtubules actually pull the separated chromosomes into the two daughter cells. It's this beautiful dance of chromosomes, where microtubules are actually doing the work of moving. That was always my interest in cell division. It was less the DNA chromosome part, and more this beautiful dynamic organization of microtubules that orchestrates the movement.

Richard Sever: You mentioned that they're dynamic and we should think of them very differently from the image that the notion of a skeleton immediately conjures up. What controls the dynamics of how they grow and shrink at one end?

Dr. Mitchison: That was actually my big break in science. As a Ph.D. student, I got to be the first person to discover that microtubules grow and shrink continuously, powered by hydrolysis of a high-energy supply, GTP. We call that "dynamic instability." It's out of sync, so different microtubules are growing and shrinking. That allows them to rearrange very quickly. GTP hydrolysis also is the energy that the microtubules use to pull the chromosomes apart in mitosis.

Richard Sever: What's regulating those?

Dr. Mitchison: With a single microtubule growing and shrinking in the cell, we believe it's spontaneous, actually.

It's sort of probabilistic, although often when a microtubule hits the edge of the cell, it will mechanically trigger it to shrink. It's either spontaneous, or it hits a mechanical barrier. For the microtubules that touch a chromosome, they actually make a link through a structure on the DNA and there's a complicated protein assembly that holds on to the microtubules. They attract it, sort of like fly paper, and the microtubules join the DNA and then stick. So you could say that these proteins attached to the DNA are 'capturing' the microtubules.

Richard Sever: In both instances—during interphase before cell division, and at cell division—what's at the other end of the tubule? For that to work, it has to be anchored.

Dr. Mitchison: The canonical answer to that question is a structure called the centrosome, which is a small sphere in the center of the cell with a beautiful thing called a centriole in the middle, and the microtubules are nucleated there. They're generated there, and they're held on there. A lot of them are anchored to the centrosome, but it's more complicated than that. In fact, there isn't a centrosome in the meiotic spindles of the unfertilized frog egg. The minus ends of them—we call the ends that attach to the chromosomes the plus end and the other end, conventionally, the minus end—all gather together at the poles of the spindle and they're held in place probably by the motor protein dynein, that "walks down" and gathers them together. I'm not sure we know the details at that end.

Richard Sever: At some point, something must happen in the cell cycle to go from having one anchorage point to two to create the spindle.

Dr. Mitchison: Again, the conventional view of them—and it's very true in a simple system like a budding or a fission yeast, where a lot of the genetics of the cell cycle were first worked out—is that the microtubule organizing center, which is the equivalent of the centrosome, actually splits in two and separates, so you have two points that grow out.

Richard Sever: So you have a replication event first.

Dr. Mitchison: You could call it a replication, but it's not a nucleic acid; it's a protein assembly. It's more like a

Published by Cold Spring Harbor Laboratory Press; doi: 10.1101/sqb.2017.82.035378

growth and splitting. It's a little less precise than a replication. That's one way of doing it. But in the frog egg system we use, it seems to actually be more a case of self-organization. You have a jumble of microtubules, and they magically sort themselves out. In that case, the drive to be "bipolar"—having plus ends in the middle and minus ends at the extremes—a lot of that actually comes from a motor protein called kinesin-5 which, if two microtubules cross each other, the motor protein in the middle will push them apart and they self-organize.

Richard Sever: When most people think of motor proteins, they think of things running up and down these tubules transporting things. This is basically the same, but it's connecting two together?

Dr. Mitchison: Exactly. It transports one microtubule on another. The bipolar organization of the frog egg spindle is really created by one set of motors pushing the microtubules apart in the middle, and another set gathering the minus ends. It's a motor that transports one microtubule on another, and there are two flavors that go in opposite directions. That's a broad outline. How that works in detail, to build a mathematical model of it… I think we're not quite there yet.

Richard Sever: You mentioned looking at it in frog eggs. One feature of them is that they're extremely big compared with regular cells: about a millimeter. What was your reason for looking at frog eggs? Was it because of this size?

Dr. Mitchison: Not in the first place. I don't know how many people have seen it with their eyes, but a frog egg dividing is beautiful. I remember seeing it in biology class at school. I view a frog egg dividing as a kind of icon in biology. But for me, I was trained as a biochemist, and I like to use biochemical techniques. An appealing thing about a frog egg is you can get a whole flask full of them and grind them up, and have a lot of the proteins that make cells divide. You can have a whole bushel of those proteins. It's quantity. Also, for studying cell division, a frog egg is set up to divide. It does nothing except divide for the first twelve or thirteen divisions. Then, things get more complicated and the cells start changing into different cell types. In a frog egg, like other kinds of egg cells, the complement of protein inside them is very much specialized to allow them to divide. It's a rich source of cell division proteins. From a biochemical perspective, it's a good place to start. My friend, Bob Palazzo, instead of using frog eggs liked sea urchin eggs and clam eggs, and for the same reason. Except instead of getting a cupful like you can from frog eggs, you could get a bucketful. But frog eggs are a little closer to man. They're vertebrates. They're amphibians. So frog egg's a nice choice. It's somewhat close to human biology we would like to understand, and you can get it in biochemical quantities. That was really the reason for starting in frog eggs.

In the 1990s, we very much thought of frog eggs as a model for human cells dividing. We thought we were discovering conserved processes that are really the same in frog eggs and a dividing human embryo cell or a human cancer cell. In my head, I tried to minimize the differences. Now, as we know more about these different kinds of cells, the differences start to get interesting. One of the striking things about a frog egg is just how big it is. A typical dividing human cell might be 10 microns, so that's a 1/100th of a millimeter. It's a hundred times smaller linearly. By volume, you have to cube that number.

Richard Sever: So, a million times smaller.

Dr. Mitchison: By volume, right. One of the overarching problems in cell biology is how do molecules, which are tiny, organize cells, which are large. If the cell is really large, that problem is even more so. The frog egg divides every half hour, roughly, and in that half hour it has to assemble a physical structure. For example, the first cleavage in the frog egg cuts exactly down the middle, and the second cleavage cuts exactly at right angles. It's a system of molecules positioning themselves in space—microtubules, specifically—that positions that cleavage furrow, and how it can do it on this huge spatial scale is kind of mind-bending.

Richard Sever: What's the specific challenge for the organization of the microtubule structure in a large cell? Is there a structural challenge? And also, when does it stop?

Dr. Mitchison: There's a structural challenge and also a temporal challenge. One question is how long can a microtubule get? Can a microtubule grow right from the centrosome in the middle to the outside, which is roughly half a millimeter? We studied that process pretty carefully. We sometimes use eggs, but they're difficult to study because they're opaque and we have to fix them, and stain them, and clear them. But we often make an extract. We pack a bunch of eggs in a tube, crush them, and get this layer of cytoplasm and do a lot of the same biology. That's how most of our experiments go.

The egg is organized by radial arrays of microtubules called asters. They look like Fourth of July fireworks, the kind that shoots a whole bunch of lines out from a central point. Going in, we assumed that the microtubules would go all the way from the middle to the outside. But when we look carefully using proteins that marked the tips of the microtubules, we realized that actually the asters are made up of a whole lot of little microtubules that are nested. So they grow out from the sides of each other.

Richard Sever: So, it's more like a tree?

Dr. Mitchison: Exactly. There are bushes where every shoot grows from the roots right up, but if you want to build a big tree, you make branches. I think the reason you have to do that just may be mechanical, actually. Just like with a tree, it probably wouldn't be physically possible to build it with every shoot going down to the root.

Richard Sever: Does that mean you need different things in frogs to create the branches more than you would in a human?

Dr. Mitchison: It's a good question. We haven't nailed the molecular basis of creating a microtubule. Nucleating it, we call it. You seed it. We think there's a kind of molecular seed, and we have not solved the molecular mechanism of this growth from the side of microtubules. I would like to do that. We don't have, for example, genetics in frog eggs, so while it's a really good system for microscopy, for studying the physiology of proteins you already know about, it's not quite as good for discovering new proteins for the first time. You can, but other systems tend to be better for that.

Richard Sever: Returning to what you were saying earlier, you have this distance problem, but then a geometry problem as well.

Dr. Mitchison: The geometry problem really fascinates me. How do you find the middle, for example? From the outside, when a frog egg divides you see these cleavage furrows that cut down. They cut the egg into two, and then in four. It turns out that the cleavage furrows, where they're positioned, is when two of these microtubule asters touch each other, where the plus ends overlap. Imagine two domes touching each other, and in the middle, that defines a plane. There're several proteins we've been working on that are recruited at that plane. As the asters grow, that plane grows out to the surface. When it touches the surface, it stimulates the cortex to contract. Understanding this plane that's formed where the two asters touch has been one of the many things that we do.

Richard Sever: And the fact that it's perpendicular to the axis is determined by the microtubules?

Dr. Mitchison: Yes. You have to know the positions of the centers of the asters, and the positions of the microtubules. One part is that when the microtubules grow into something that is a boundary, they stop growing. When the two asters grow into each other, that's another boundary. They stop growing in the middle. We've been working on that.

A second factor is how the nucleating sites are positioned. We've made less progress on that. We know another motor protein, dynein, is involved in that. For example, one version of the problem is that you've cleaved the egg in two. Now, the next furrow's going to be at ninety degrees. Where does that angle come from? It comes from finding the long axis of the cell, and we think that the geometry of the microtubules controls the net forces that are acting on them. That's what positions the next furrow. There's actually a lot we still don't know about that. I don't want to pretend we know more than we do.

One of the fun things is there are people who have been looking at frog eggs and wondering how they make those cleavage furrows for more than a hundred years, so there are lovely experiments we've referred to from the 1890s and 1910s. We actually repeated those experiments. We got the same results, but we're able to take it further now that we have some of the molecules. That stepping in the tracks of the people who came before us is part of the fun of it.

A Conversation with Tim Stearns

INTERVIEWER: JAN WITKOWSKI

Cold Spring Harbor Laboratory

Tim Stearns is the Frank Lee and Carol Hall Professor and a Professor of Genetics at Stanford University.

Jan Witkowski: What are some of the key points about centriole structure and regulation that are of research interest at the moment?

Dr. Stearns: The centrosomes are segregated in mitosis and that is one of the interesting things about them. If you look through history, the realization that they are important organelles in cells happened more than a hundred years ago, but a modern understanding of what they do and how you ensure that each cell gets one at the beginning of the cell cycle has only occurred more recently. They, like chromosomes, are segregated on the mitotic spindle. So, by analogy to chromosomes, that means they have to be duplicated exactly once per cell cycle. The duplication and the segregation of centrosomes and the centrioles within them is really something that you can think about in parallel with thinking about chromosomes.

Jan Witkowski: A centriole looks almost as though it buds off from… I was going to say its parent, but that's not right. The original?

Dr. Stearns: We do use the word "mother," that's true.

Jan Witkowski: How does that process happen? It comes off at right angles.

Dr. Stearns: It's really quite remarkable. For DNA, we've known for a long time—and the original notion of this came from the Watson and Crick double-strand helix—that you replicate DNA by pulling it apart and then use those strands as templates to create new strands. For the centriole, it's been a mystery as to how it duplicates because you don't have that obvious mechanism that comes from the structure. What you do have is the centriole—which is a proteinaceous structure, not nucleic acid—and it has this very characteristic structure, the centriole, which has this beautiful ninefold symmetry and other proteins attach to that. That duplicates just like chromosomes, once every cell cycle.

As you say, it looks somewhat like a new centriole grows as a bud on the side of an existing one. What does happen is that when you initiate centriole formation in a cell in the G_1 phase of the cell cycle, each centriole grows exactly one from its side. The new one grows at a right angle from the side of the existing one. The mother centriole would be the existing one and a new daughter grows from the side of the mother. That's true in many organisms, so it's really a quite well conserved aspect of how duplication works. And yet, why it works that way, why you only form one new centriole on the side of a mother is still not clear. Even though we know a lot about the molecules involved in duplication, the reason it's constrained to occur exactly there and the way you constrain it to occurring exactly once, are still things that are active areas of work.

Jan Witkowski: What are the molecules that make up a centriole?

Dr. Stearns: The structure of the centriole itself, when you look at the classical image that you can find in textbooks or on Google Images, is this beautiful ninefold symmetric microtubule-based structure, where the microtubules are in triplets. This is unusual. You only find these triplets where there's an A tubule and then a B and a C tubule that grow on the side of the A tubule. The centriole is the only place you find that. Both the ninefold symmetry and having these triplets are things that really define the centriole.

The ninefold symmetry comes from a protein called SAS-6. There's been a lot of really quite beautiful work on this that shows that there's intrinsic behavior of the protein that results in that ninefold symmetry. The SAS-6 protein forms ninefold symmetric oligomers both in vitro and, presumably, in vivo and that's the basis for this ninefold symmetry of the centriole, which is absolutely conserved from all eukaryotes that have centrioles, and presumably in the last common ancestor of all eukaryotes. Those microtubules that this cartwheel that SAS-6 forms are attached to are formed from α- and β-tubulin just like all microtubules in the cell, except that, again, you never find triplet microtubules in any other setting.

For a long time it's been thought that there must be other proteins: δ-tubulin and ε-tubulin are members of the tubulin family but are not the canonical members of the ones that make microtubules, and the specialized tubulin, γ-tubulin, that nucleates microtubules, or caps them at least. These are different.

© 2017 Stearns. This article is distributed under the terms of the Creative Commons Attribution-NonCommercial License, which permits reuse and redistribution, except for commercial purposes, provided that the original author and source are credited.

Published by Cold Spring Harbor Laboratory Press; doi: 10.1101/sqb.2017.82.035360
Cold Spring Harbor Symposia on Quantitative Biology, Volume LXXXII

409

Jan Witkowski: The A, B, and C tubules use the same tubulin molecules?

Dr. Stearns: As regular microtubules, yes. The bulk of those A, B, and the C tubules are made up of α- and β-tubulin, which make this heterodimer that polymerizes to make microtubules. It seems that those centriolar microtubules in this triplet form are made largely of α- and β-tubulin. I say "largely" just to leave open the possibility that there's something else that might make the triplet microtubules special.

Jan Witkowski: Why don't they go on to make a quadruplet?

Dr. Stearns: That is a great question. Why is it limited to being a triplet? That is not clear. It's not clear why they don't ever go beyond being triplets.

Jan Witkowski: All the things that link the nine, the bundles to keep them in a cylindrical form: Are there things between those triplets?

Dr. Stearns: Yes. Part of what keeps the structure intact is this cartwheel that you see when you do electron microscopy; that's really the only way you can see it in its full glory in terms of its structure. That's when you see the ninefold symmetry of this cartwheel at the center of the centriole. Again, there are spokes that radiate out from that and they contact these microtubules that are arranged in ninefold symmetry. Other proteins that are part of that have been seen in the electron microscope but we don't know the identity of them. There's something called the A–C linker, which links adjoining triplet microtubules, but no component has ever been identified for those yet.

Jan Witkowski: You're using genetic tools to dissect structure.

Dr. Stearns: Right. It's one of the revolutions of cell biology: the ability, using CRISPR–Cas9, to manipulate the genomes of mammalian cells. This ability really started with zinc fingers, which were hard to use and then TALENs, which were easier. But now CRISPR–Cas9 makes it tremendously easy to manipulate the genome of mammalian cells. Not as easy as many people think. If you're interested in turning mammalian cells into the equivalent of a yeast cell, that's not really feasible, but this ability to make mutations and then assess the phenotype is key to modern cell biology.

Jan Witkowski: How have you been using CRISPR–Cas9 in your work?

Dr. Stearns: These proteins, δ-tubulin and ε-tubulin, which are variant members of the tubulin family are not in all eukaryotes, but they're in most that have centrioles and cilia. We have been working on these proteins for a long time. We identified them—and others did too—in the late '90s. Susan Dutcher first identified δ-tubulin and then we identified ε-tubulin and more recently we, and others, identified ζ-tubulin: the last member of the tubulin superfamily. I can say that conclusively because we have a lot of sequenced genomes, so it's clear that there are no other

members of the tubulin superfamily. Because of the conservation of them, we know that the last common ancestor of eukaryotes had all six of these. Some have been lost in various branches of the eukaryotic tree, but if you look at human cells, we have α, β, γ—which all eukaryotes have—and then we have δ- and ε-tubulin.

We've known this for a long time. The limitation has been, how do you determine the function of proteins in mammalian cells without the ability to manipulate the genome? RNAi, of course, was one of the tools that came along and made some forms of manipulation of gene expression possible. We tried that with δ- and ε-tubulin but it never worked very well. Our thought about that is that there's really a very small number of molecules of these proteins, such that depleting the RNA, even to the levels you typically get with good RNAi, was never sufficient to yield a phenotype. Now, we can make real deletions using Cas9 that allow one to have a defined genotype with a predicted null mutation homozygous form and then assess the phenotype of the cells.

Jan Witkowski: When you knock out these, what happens to the cells? First, what happens to the centriole? Then what's the effect on the cell?

Dr. Stearns: Using Cas9, we tried hard to knock out both δ-tubulin and ε-tubulin and could never identify cell lines where they were clearly knocked out. It turns out that the reason for that is because cells that have a problem with their centrioles, if they are p53-plus, p53 senses a problem and the cells arrest in G_1 of the cell cycle. We didn't realize at the beginning of our experiment that that was true, but we did soon afterward from the work of others. Once we started doing the experiments in p53-minus cell lines, then we could easily get null mutations. Right away, that hinted that there was going to be something wrong with centrioles that was being sensed by p53. Sure enough, when we looked closely at the phenotype of these mutant cell lines—and both the δ-tubulin and ε-tubulin null mutants have very similar phenotypes—the phenotype at the level of the centrioles is that the structure of the centrioles that do exist is different. They only have singlet microtubules instead of triplet microtubules. Remember, those triplets are one of the key features of what makes a centriole special. These lack triplet microtubules and only have singlets.

There's another aspect to it that caught our eye immediately, which was that if you look at a population of cells that are null mutants for these genes, there's a very interesting distribution of centrioles in these cells. A normal cell has two centrioles in the G_1 phase of the cell cycle. They duplicate prior to mitosis; you have four. Then they segregate on the spindle and you get a daughter cell with two centrioles again. In this case, about half of the cells had zero centrioles: none detectable whatsoever. The other half of the cells had more than four centrioles: five, six, seven, eight. Right away again, that suggested that there was something very odd about centriole duplication in these cells.

We determined that what was going on, is that that half having centrioles and half having no centrioles is what you see when you look at an asynchronous population, with cells at all stages of the cell cycle. If you look at specific stages of the cell cycle, you see that in G_1 there are zero centrioles. They form in S Phase. They persist through mitosis and then they disintegrate at some point between mitosis and the next interphase. Actually, probably the start of the next interphase.

It's one of the interesting things about centrioles. Just as we talked about the seeming budding that occurs when you grow a new centriole from the side of an existing one, because that's such a common feature of centriole duplication people have long assumed that you had to have an existing centriole to grow a new one. We know that that is not true because it's been shown many different ways that for many cells from animals and other organisms too, you can create new centrioles de novo with no preexisting centriole.

Jan Witkowski: So, you've got these two populations in asynchronous cultures, but if you followed a single cell, you'd find that it at one stage had none and at another stage had several?

Dr. Stearns: Exactly. If you look at a single cell as it progressed through the cell cycle, you start with zero, and you would create centrioles de novo with no existing centrioles. Those would persist until mitosis but then they would... I use the word "disintegrate," although actually seeing that happen is not something that we've been able to do, but that's our interpretation of the results.

Jan Witkowski: Are there cells that never have any centrioles? Can cells survive and grow and multiply without centrioles?

Dr. Stearns: It is true that centrioles duplicate exactly once per cell cycle and they participate in the mitotic spindle. Because of their close association with the spindle, people had always assumed that they were required to make the spindle. That is clearly not true, because you can force mammalian cells to lose their centrioles, either genetically or now with a small molecule inhibitor of a required kinase, and in both of those cases the cells will be able to grow without centrioles. So, no centrioles, no centrosome, but they can do mitosis. They have to be p53-minus though, because if you don't have centrosomes, there is a problem—probably with mitosis—that p53 senses. If you get rid of p53, the cells can divide just fine with no centrosomes.

Jan Witkowski: p53 having that sort of oversight absolutely makes a lot of sense in terms of trying to avoid chromosomal abnormalities and such.

Dr. Stearns: Right.

Jan Witkowski: You've shown by knocking out these tubulins that they are associated or involved in important aspects of centrioles. Do you know where they are in the centriole?

Dr. Stearns: We've been unable to really get convincing localization of these proteins. It's not for lack of trying. We, and others, have tried to see where in the cell these proteins are. For cell biologists, that's one of the key clues to what a protein does. It's been difficult either with antibodies or with tagged proteins of various sorts. I think it's because there're very few molecules of these proteins in cells. That's an assessment based not only on failure to localize the proteins but also on this small amount of RNA and the very low level of protein as assessed with antibodies on western blots. If you look at the centriole structure, where these proteins might fit into that where they could help to make or to stabilize the triplet microtubules, there's several ways you could imagine very low-abundance proteins playing critical roles in that structure.

Jan Witkowski: You mentioned that you've not really been able to follow the disintegration or destruction of these centrioles that end. What do you think is going on? Have you tried messing around with the ubiquitination system to see if you can force them through?

Dr. Stearns: We've gotten close to seeing the disintegration. We certainly can see it by light microscopy, but centrioles are very small structures and light microscopy doesn't really get you to the level you need to be to tell what's going on. By electron microscopy, which we'd done in collaboration with Jadranka Loncarek, we can see that as cells enter mitosis the centrioles elongate, even these aberrant centrioles. They don't have triplet microtubules; they only have singlet microtubules so they're narrower, but they do elongate to the full length. In the electron microscope, it's clear that there's something wrong with them. Having singlets only, it seems that they don't recruit some proteins that are parts of what a maturing centriole would have. They have gaps in the structure. We believe that what is resulting in disintegration is the depolymerization of these singlet microtubules. Thus, the triplet microtubules are really required to be stable. That's what provides the long-term stability of these triplet microtubules, which persist for a very long time. These microtubules don't turn over, unlike most microtubules in the cell.

Jan Witkowski: Centrioles are also a part of cilia?

Dr. Stearns: Yes.

Jan Witkowski: Cilia also get reformed after division?

Dr. Stearns: They do. Most cells in our body that are in G_0 phase—so they're doing their thing as a differentiated cell—will have a primary cilium projecting from the surface from the cell. The microtubule component of that cilium, which is the core structure, is contiguous with the centriole, so it grows from the end of the centriole. When one says that the centriole—or it's often called the basal body in this context—is part of the cilium, it literally is. It is part of the microtubule structure of the cilium. That grows every cell cycle after mitosis in a cycling cell and again, is present in most differentiated cells.

Jan Witkowski: Your "funny" ones don't get involved in the cilium?

Dr. Stearns: We've never seen one with a cilium. That, in part, is because we've also seen that they never have these appendages, which are structures that are added onto the centriole to functionalize it. In the case of making a cilium, they allow it to interact with the plasma membrane. That interaction is critical to creating the ciliary compartment and you have to do that to be able to traffic molecules specifically into the ciliary compartment to make the cilium. These centrioles never have that ability to attach to the membrane and we never see them associated with a cilium, so that's another defect that these centrioles have.

Jan Witkowski: So those cells never develop cilia?

Dr. Stearns: Right. These null cells never develop cilia, which would clearly be a bad thing for the organism because the primary cilium, at least in vertebrates or mammals, is absolutely required for development.

Jan Witkowski: If I could grant you one wish, what would you really like as your next bit of knowledge?

Dr. Stearns: I think the most useful thing would be to know exactly where those proteins are because when you're working at the level of trying to relate protein function to structure as we are, you really need to be able to say where these proteins—which could be structural components of this complex thing, the centriole—are in the structure. That will inform models for how they actually work. Speculatively, without any evidence yet, this linker between the triplet microtubules—the "A–C linker" it's called, because it links the A tubule to the C tubule of the next triplet—could be a structure that these δ- and ε-tubulins are either a component of, or required to make. The absence of that structure might explain the instability of the centriole. That's just a hypothesis and we need to have data about that, but the localization of the proteins would be a great start for that.

A Conversation with Aaron Straight

INTERVIEWER: KIM BAUMANN

Chief Editor, Nature Reviews Molecular Cell Biology

Aaron Straight is an Associate Professor of Biochemistry at Stanford University.

Kim Baumann: Your laboratory works on different aspects of chromosome biology that are related to understanding how the segregation of chromosomes occurs in a correct way. This includes work on the centromere and how it drives assembly of kinetochores, as well as on RNA biology and how RNAs regulate the structure and organization of chromosomes. Could you start by telling us about the connections between these two fields?

Dr. Straight: Those can seem like somewhat disparate fields and not necessarily connected. I started out trying to understand chromosome segregation because I was interested in the mechanics of chromosome movement and how and what controls the dynamic movement of chromosomes during anaphase and spindle elongation. Those problems are still interesting problems, but in some ways, I feel like we have gotten away from that because something else caught my attention.

A really fascinating problem in cell biology that will take decades to play out is that both of those examples that you gave—which are near and dear to my heart—are not so much about the specific mechanics of moving chromosomes and controlling the regulation of when they separate, but more about how a cell sets up specific functional domains of chromosomes, and in the face of all the other cellular processes—whether that is cellular mitosis or differentiation—how it decides how to specify a particular function and then attempt to reinforce that function over time, whether it's on a short timescale like a cell cycle, or the lifetime of an organism. For the most part, what I work on is chromosome-based functionalization. The centromere is a great example of that because it's made of DNA and histones and chromosome proteins, but it has a very unique function that's distinct from the content of the other parts of the chromosome. As far as we know that function is, at least in humans, primarily encoded not at the DNA sequence level but by the proteins that associate there.

Kim Baumann: And it's a function that needs to be maintained spatially in the chromosome. It's a region of the chromosome.

Dr. Straight: That gets right to the heart of it. If it's within some part of the chromosome but it's not the sequence defining it, then what gives it boundaries and what gives it identity? As the cells go through division, if the cell is duplicating and diluting it every time it replicates its DNA and chromatin, how is it maintained or reassessed? We spent several years trying to understand that problem. Really, the core of that problem for the centromeres is that there's a specific histone that's distinct from the rest of the histones, so there must be some cellular mechanism that can recognize that and say, "This is where we put more of that identifying histone and we don't put it elsewhere and we don't put other things here that shouldn't be here." Even though that sounds trivial, it turns out to be quite a complicated problem to solve. We are starting to touch on the machinery that does that sort of recognition of what a chromosome does, but I think we are far from understanding how you couple that recognition to the reassembly and the protection of that part of the chromosome from the rest of the chromosome.

Kim Baumann: How does RNA fit into this?

Dr. Straight: We're very new to that world. We've really been working on that problem seriously for maybe four or five years. We fell into that almost because of an accidental study of the centromere region. We were interested in a general concept, which is that there are certain RNAs that are well established that bind chromosomes and change their function. The best known is female dosage compensation: One X chromosome gets coated by RNA and silenced and the other doesn't. So we asked the generic question: Are there other RNAs that might act in that kind of mechanism to control the chromosome fate or chromosome structure?

Kim Baumann: Do you mean specifically heterochromatic regions, or more than that?

Dr. Straight: I wish we'd been that well-informed when we started, but I don't think we had thought through the problem very well at that point, so we just took an unbiased look: What RNAs stick to chromosomes? But we did that in mitosis when transcription is, for the most part, repressed. This is in human cells. We jump between a lot of different organisms in the lab, but this RNA stuff has been in human cells, although we are now moving in

Published by Cold Spring Harbor Laboratory Press; doi: 10.1101/sqb.2017.82.035022

Cold Spring Harbor Symposia on Quantitative Biology, Volume LXXXII

to other systems. The idea was: Can we take an unbiased look? Can we just essentially take a census or a visual assay to ask where do RNAs bind on human chromosomes? We chose mitotic chromosomes because transcription's shut down for the most part. We developed a labeling system to look at it, and we saw something pretty extraordinary, which was that the RNAs stayed associated with mitotic chromosomes. In particular, they concentrated around centromeres. We didn't anticipate that would be the case, but we knew how to work with that part of the chromosome so that helped us get more interested in the problem. We've spent a few years now working on that.

Kim Baumann: These RNAs are transcribed from those regions and then they remain there? How does that work?

Dr. Straight: That is an extremely good question, because they are exactly as you just described it. They are RNAs that are transcribed from those regions, but maintain their associations with those regions. We've recently discovered that those RNAs are involved in bringing a series of histone-modifying enzymes to that locus to actually modify the chromatin and help repress it. That's how it comes back to the centromere, because these are, in some sense, nonsequence-dependent events. We don't think the RNA sequence is important.

Kim Baumann: Is there any specific structure to the RNAs? Are there characteristics of these RNAs that make them work the way they work?

Dr. Straight: We've tried very hard to understand that. We've even done some specific structure mapping on the RNAs. We really have no evidence that there is a specific structure involved. Instead, we are thinking that the RNAs are acting as a platform just to increase the local concentration of the enzymes that bind them.

Kim Baumann: So it's mostly a physical function?

Dr. Straight: We think so.

Kim Baumann: Do they also regulate the activity of the enzymes, or do they just bring them?

Dr. Straight: We have tested that, and our data so far is that there's not specific regulation of the enzymes. Rather than changing the activity, they essentially localize and concentrate the activity. So, it's a way of redistributing the activity to a particular site—as opposed to modifying an activity—so it acts in one place and not another.

Coming back to that larger problem: How these are related? I think both of those problems are problems in how you maintain an epigenetically defined locus. I would say the RNA project that we've started to work on is really still in its infancy. I think that's a general problem for people studying noncoding RNAs in any organism, in that now there are tens of thousands identified, and we know the functions of a handful.

Kim Baumann: Do you know anything about how they are activated? Are they constitutively active because these regions are defined and they remain there? This is a region that is maintained. Are they being constantly produced?

Dr. Straight: We've looked at that, and we know that they're being synthesized in one particular part of the cell cycle, in G_2, before the cell goes into mitosis. Maybe that's just to make sure that the transcript is there so that, when the chromosomes condense, it's at the right place at the right time. We don't know what the regulation of their synthesis is. We just haven't done those experiments. That's an interesting problem: whether there is regulated behavior. For instance, if you go to an early embryo or pluripotent cell, they lose a lot of that heterochromatin. There has to be some regulation of that domain, whether it's through RNA or whether it's through something else.

Kim Baumann: Do they function also in other parts of the chromosome, or are they specific to the centromere?

Dr. Straight: They're not specific to the centromere.

Kim Baumann: So it's maintaining, or perhaps defining, different regions?

Dr. Straight: We think for this particular class of these modifying enzymes that it is really involved in repressing all kinds of different repeat sequences in the human genome, whether those are centromeric, transposable elements, or endogenous retroviruses. Whereas, we don't think these RNAs affect the euchromatic genome very much. We think this process is quite specific to making heterochromatin and keeping it.

Kim Baumann: What is the next step in studying these?

Dr. Straight: We realized we were studying one RNA. We just had this one thing we're studying, and that's essentially how the field has been going. You'll find an RNA that's interesting and you'll bang away on it for a while and learn something about the biology about that one RNA. But what we wanted to do is ask, can we develop a method that allows us to look at all the RNAs? Can we globally find all the RNAs that bind to human chromosomes, and then understand where they are acting and maybe try to understand how they function? So just over the past two years, we've developed a method called chromosome-associated RNA sequencing, that allows us to essentially link an RNA to the site where it's bound to the DNA through a direct linkage. It's a physical link between the RNA and DNA. That's advantageous because it doesn't matter if it's coming from the site where it is transcribed or if it's going to some other part of the chromosome and acting, as long as we can capture that interaction. We've just recently done that and shown that we can, using high-throughput sequencing, identify all the other RNAs and all of the sites where they are bound. By doing that you immediately begin to generate hypotheses about how an RNA might bind.

For instance, we developed this technique in *Drosophila* because the genome is a bit smaller so it is easier to do, and the first thing that jumped out at us is that the dosage compensation RNAs, which control the expression of the X and Y balance, just beautifully coat the entire X chromosome. So we know that we can detect *cis* interactions for those RNAs. But we also see those RNAs going to other parts of the genome. And we see a lot of other RNAs with interesting behaviors. I feel like this is just beginning to scratch the surface.

Kim Baumann: It will be interesting to see how they are targeted against certain regions. It's a chicken-and-egg thing, because they target factors to the chromatin and vice versa, but if they are not just where they are transcribed, they need to reach their destination.

Dr. Straight: That is one the most important questions in the field right now. For noncoding RNAs, most of them we do not understand how they're targeted to where they act, and we don't understand how they are maintained there. We know that RNAs can bind to different complexes and RNA-binding proteins, but what gives them selectivity for a place in the genome is really a black box. I don't think that's a problem that is going to be solved by one lab. That's a problem for the whole field to try and understand.

Kim Baumann: There can be mechanisms that involve sequence-specific and nonsequence-specific events, right? With many factors, perhaps?

Dr. Straight: There are good examples of sequence-specific and nonsequence-specific factors that are involved in the process, but I think we still don't understand the specificity problem. If it is sequence-specific, we do understand it to some extent. For the ones that aren't, that's a much tougher problem, but it's fun to think about.

Kim Baumann: Going back to the RNAs that function in the centromeric area, how are they kept there? You say that there is no sequence-specificity, so what does it tell you about how that region functions? If you were to disrupt any of those mechanisms, what happens?

Dr. Straight: The poster child to really understand these problems is studies in fission yeast. That, arguably, is where mechanisms for RNAs controlling chromatin have been best worked out in these regions around centromeres. In that case though, it's known that there are sequence-dependent interactions that help control shutting down the chromatin, and if you destroy that system in fission yeast it causes chromosome desegregation.

Kim Baumann: The whole machinery is disrupted?

Dr. Straight: Exactly. We've just recently been able to test that in humans, and because now you can do things like modify genes with CRISPR and actually make mutants in the genome like you would in yeast, and you can do those experiments now. We find that when we compromise that RNA-dependent system, we start to derepress all of these heterochromatic regions that depend on our repeat sequence. That's part of why we think one of the key features of these systems is to ensure that you maintain the repression of heterochromatic regions, which is a bit chicken-and-egg because it requires transcription in order to repress the transcription that's occurring there. Understanding how that balance plays out, whether that's a way for the cell to be able to regulate the relative expression from some of these regions, is also an interesting problem to think about.

Kim Baumann: Is this going to be a major focus for your lab for now?

Dr. Straight: We will keep pushing on this for a while. We haven't abandoned the centromere yet. I love that part of the chromosome and we'll keep working there. I think this represents maybe a broader class of problems, and I'm hopeful that we'll get some interesting insight.

A Conversation with Yukiko Yamashita

INTERVIEWER: KIM BAUMANN

Chief Editor, Nature Reviews Molecular Cell Biology

Yukiko Yamashita is the James Playfair McMurrich Collegiate Professor of the Life Sciences and a Professor in the Department of Cell and Developmental Biology at the University of Michigan Medical School and the University of Michigan Life Sciences Institute.

Kim Baumann: The main focus of your laboratory is studying the mechanisms that regulate asymmetric cell division, but you also work on chromocenters. I know they have been described for about a century from a cytological point of view, but could you tell us what they are and how you became interested in them?

Dr. Yamashita: The last question first. We don't get to dictate what to study. As is so often the case, our data lead us in a direction. We pretty much bumped into this. We had been studying asymmetric stem cell division for a very long time. Then four years back, we published one paper about asymmetric stem cell division. For that study, we needed a chromosome-specific repetitive sequence as a probe. When we looked into the probe sequence as a tool, inevitably we noticed these satellite DNAs had very interesting distributions across all those chromosomes. That really led us to look into this satellite DNA, which is a major constituent of pericentromeric heterochromatin. They're repetitive DNA, like very short telomere repeats. That's why it skews ATGC contents. As a result, when you do the centrifugation on the cesium chloride, those DNA settle in a different way compared to the major euchromatic DNA. That's the original reason they were called satellite DNA. These repeats exist mainly in the pericentromeric heterochromatin. This pericentromeric heterochromatin has been known to form chromocenters for, as you said, about a century. Of course, the function of the centromere is very well known and established, but the pericentromeric heterochromatin that excludes the centromeric sequence, as well as the chromocenter that it forms, neither of their functions have been shown.

We decided to look into its potential function and we studied a protein that binds to satellite DNA. Nowadays, you can do pretty much anything with CRISPR-Cas9, but one thing you probably can't do is to "CRISPR out" the entire pericentromeric heterochromatin because that spans ten megabases or more sometimes. What you can do, is you can potentially remove the protein that binds to the chromocenter. We thought that'd be a really good starting point. Luckily for us, once we removed those chromocen-ter-binding proteins, what happens is that the chromocenter becomes dispersed.

Kim Baumann: Are these proteins that bind to satellite DNAs specific to the pericentromeric area, or is it where there are other satellite DNAs?

Dr. Yamashita: The reason we got into this protein was that, probably a few decades back, people had identified it biochemically, starting from a simple repeat sequence to see what proteins might bind to it. Through that, this protein's biochemical function has been known for a really long time. This protein binds to simple AATAT repeat satellites, but nobody really understood the actual cell biological function of that protein. So that's why this time we decided to go into a little bit of detail to look at its phenotype and to try to tie it back to actual cell biological function.

Kim Baumann: What happens if you remove this protein?

Dr. Yamashita: Normally, the chromocenter is the association of pericentromeric heterochromatin from many chromosomes. For example, if you have ten chromosomes in your cell, that could cluster into three. But if you remove this protein, instead of three chromocenters, there can be six or seven. The number increases, suggesting that pericentromeric heterochromatin association is somehow perturbed.

The next thing that happens after you remove the chromocenter protein is that individual chromosomes start floating out of the nucleus to form micronuclei, which was quite a surprise to us.

Kim Baumann: Is this during cell division?

Dr. Yamashita: In interphase. If you take a look at the wild-type nucleus, it is very stable: always round, and it stays that shape. If you perturb the chromocenter formation, all of a sudden this round nucleus starts budding off to spit out a few chromosomes. This really told us that the chromocenter formation is a way to somehow link all those chromosomes into a sort of network.

Published by Cold Spring Harbor Laboratory Press; doi: 10.1101/sqb.2017.82.034983

Kim Baumann: You think that all cell types have some sort of chromocenter?

Dr. Yamashita: Yes. That is actually a very conserved feature. The chromocenter has been observed in essentially almost all cell types. Plant cells, mammalian cells, *Drosophila* cells, essentially any.

Kim Baumann: Can you tell us a bit more about micronuclei? My understanding is that they can form at the end of cell division if something was wrong. How do you think that links to your observations?

Dr. Yamashita: We also looked at all the bad things that can happen when micronuclei form. For example, when nuclear envelope integrity is disrupted, DNA damage happens inside. There is similarity, but also some differences, between those lagging-chromosome–induced micronuclei and chromocenter-disruption–induced micronuclei. When you disrupt the chromocenter, it's not just micronuclei; even the major macronuclei have all sorts of problems with nuclear envelope integrity as well as protection from DNA damage.

Kim Baumann: So, there is increased DNA damage in those cases?

Dr. Yamashita: Two pathways lead to this final consequence. One: you have the lagging chromosome that can lead to micronuclei formation and then lead to a nuclear integrity problem. The other way, the nuclear integrity can be heavily disrupted if you don't have a chromocenter, and that can lead to micronuclei formation. That said, there are very interesting parallels: similarities, but a little bit of difference to it.

Kim Baumann: In which cell types have you studied this? Is it across more than one species?

Dr. Yamashita: First, we looked into germ cells in *Drosophila* testis, simply because that's the system in which we always work. Of course, we wanted to know if there are weird things about our model system, so we expanded this to mouse cell culture. Once you make a probe for a certain pericentromeric satellite, it's easier to go across many different mouse cells instead of going to other species. We took a look at fibroblasts, immune cells, many different kinds of cell types in the mouse, and the phenotype is always very consistent.

Kim Baumann: Do you think this is going to become a major interest for your life? What are the next steps?

Dr. Yamashita: For a while, because not much has been known about the chromocenter or pericentromeric heterochromatin, so there is a lot we can do. I'm looking forward to it. I think we're going to work on many aspects of this pericentromeric heterochromatin in the coming years. But at the same time, I think what we try in the laboratory is that everybody really takes on a different project so that when they become independent, not only are they independent of me but independent of each other. That said, if we extend from the pericentromeric heterochromatin, I'm looking into the way that it's going to be very discrete from what we discover here, now.

Kim Baumann: In terms of satellite DNAs, do you have any other lines of research that are related to these repetitive sequences across the genome?

Dr. Yamashita: We are finally convinced that pericentromeric heterochromatin has an important function. It was believed for a very long time to be junk. Not many people really cared how pericentromeric heterochromatin was taken care of by the host cells or actual cells, because they were believed to be junk or parasites: selfish DNA elements.

Kim Baumann: But these areas are very conserved, right?

Dr. Yamashita: It's conserved in terms of the characteristics, but if you look at the sequence, it's very different across species. I believe that was one reason that, for a long time, people didn't really appreciate that pericentromeric heterochromatin has such a fundamental biological function. Now that we know it has a function, I'm very interested how this vast array of the pericentromeric heterochromatin is maintained. This is probably a similar question to telomere maintenance. That has been studied back when Liz Blackburn and Carol Greider started to look into how this is really maintained, what kind of enzyme was doing it.

Kim Baumann: You mean maintain in terms of length?

Dr. Yamashita: Yes. The actual biochemical problem would be different, because pericentromeric satellite DNA is in the middle of the chromosome, so you don't have to worry about the end-replication problem, but how to maintain a repeat: that's conceptually similar. I believe there has to be some mechanism to maintain it. That's one thing I'm very keen on doing in the near future.

Kim Baumann: Now that you've discovered that it plays such an important function, do you think there will be more studies on the chromocenter?

Dr. Yamashita: You mean, from other people? That, I very much look forward to. Of course, that may get into competition. But if you're afraid of competition that means you'll be making discoveries all by yourself. That will be slow. I would actually look forward to more people joining this endeavor.

Author Index

Subject Index